Frontiers in Antimicrobial Resistance

A Tribute to Stuart B. Levy

Frontiers in

Antimicrobial Resistance

A Tribute to Stuart B. Levy

Editors

David G. White
Division of Animal and Food Microbiology
Office of Research
Center for Veterinary Medicine
U.S. Food and Drug Administration
Laurel, Maryland

Michael N. Alekshun
Paratek Pharmaceuticals, Inc.
Boston, Massachusetts

Patrick F. McDermott
Division of Animal and Food Microbiology
Office of Research
Center for Veterinary Medicine
U.S. Food and Drug Administration
Laurel, Maryland

Washington, D.C.

Address editorial correspondence to ASM Press, 1752 N St. NW, Washington, DC 20036-2904, USA

Send orders to ASM Press, P.O. Box 605, Herndon, VA 20172, USA
Phone: (800) 546-2416 or (703) 661-1593
Fax: (703) 661-1501
E-mail: books@asmusa.org
Online: estore.asm.org

Library of Congress Cataloging-in-Publication Data

Frontiers in antimicrobial resistance: a tribute to Stuart B. Levy / editors, David G. White, Michael N. Alekshun, Patrick F. McDermott.
p. ; cm.
Includes index.
ISBN 1-55581-329-1
1. Drug resistance in microorganisms. 2. Drug resistance in cancer cells.
[DNLM: 1. Drug Resistance, Microbial. QW 45 F935 2005] I. White, David G., Ph.D. II. Alekshun, Michael N. III. McDermott, Patrick F. IV. Levy, Stuart B.

QR177.F76 2005
616.9'041—dc22

2005015030

10 9 8 7 6 5 4 3 2 1

Printed in the United States of America

Cover photo: (Background) Two-dimensional crystals of the tetracycline efflux protein. (Inset, left to right) Three-dimensional structure of *Escherichia coli* MarR; tetracycline efflux protein; vancomycin-intermediate *Staphylococcus aureus* (courtesy of Janice Carr, CDC); disk diffusion antimicrobial susceptibility test of a mixed culture of *Escherichia coli* and a *Proteus* sp. at pH 7.2 (courtesy of CDC).

CONTENTS

Contributors • ix
Foreword • xiii
Preface • xv
Introduction • xvii
Photographs • xix

I. Tetracyclines and Resistance
Introduction: *Michael N. Alekshun*

1. Tetracycline Resistance: Efflux, Mutation, and Other Mechanisms • 3
Frederic M. Sapunaric, Mila Aldema-Ramos, and Laura M. McMurry

2. Tetracycline Resistance Due to Ribosomal Protection Proteins • 19
Marilyn C. Roberts

3. Discovery and Industrialization of Therapeutically Important Tetracyclines • 29
Mark L. Nelson and Steven J. Projan

II. Single Antimicrobial Resistance Mechanisms
Introduction: *Patrick F. McDermott*

4. Fluoroquinolone Resistance • 41
Jordi Vila

5. β-Lactam Resistance in the 21st Century • 53
George Jacoby and Karen Bush

6. Macrolide, Lincosamide, Streptogramin, Ketolide, and Oxazolidinone Resistance • 66
Marilyn C. Roberts and Joyce Sutcliffe

7. Aminoglycoside Resistance Mechanisms • 85
David D. Boehr, Ian F. Moore, and Gerard D. Wright

8. Glycopeptide Resistance in Enterococci • 101
Florence Depardieu and Patrice Courvalin

9. Phenicol Resistance • 124
Stefan Schwarz and David G. White

III. Multiple Antimicrobial Resistance Mechanisms
Introduction: *Patrick F. McDermott*

10. Multiple Antimicrobial Resistance • 151
Anthony M. George

11. Metal Resistance Loci of Bacterial Pathogens • 165
Anne O. Summers

12. Biocides and Resistance • 174
Bonnie M. Marshall and Laura M. McMurry
Appendix: *Merri C. Moken*

13. The Nexus of Oxidative Stress Responses and Antibiotic Resistance Mechanisms in *Escherichia coli* and *Salmonella* • 191
Bruce Demple

14. The *mar* Locus • 198
Thamarai Schneiders, Herbert Haechler, and William Yan

15. The *mar* Regulon • 209
Teresa M. Barbosa and Pablo J. Pomposiello

16. Identification of Mar Mutants among Clinical Bacterial Isolates • 224
JoAnn Dzink-Fox and Margret Oethinger

17. Structure and Function of MarA and Its Homologs • 235
Robert G. Martin and Judah L. Rosner

18. Function and Structure of MarR Family Members • 247
Michael N. Alekshun and James F. Head

19. Role, Structure, and Function of Multidrug Efflux Pumps in Gram-Negative Bacteria • 261
Hiroshi Nikaido

20. Role of Multidrug Efflux Pumps in Gram-Positive Bacteria • 275
Glenn W. Kaatz

IV. Antimicrobial Resistance in Select Pathogens
Introduction: *David G. White*

21. Advances in Vancomycin Resistance: Research in *Staphylococcus aureus* • 289
Keiichi Hiramatsu, Maria Kapi, Yutaka Tajima, Longzhu Cui, Suwanna Trakulsomboon, and Teruyo Ito

22. *Enterococcus* • 299
Roland Leclercq and Patrice Courvalin

23. *Streptococcus pneumoniae* • 314
Joyce Sutcliffe and Marilyn C. Roberts

24. *Helicobacter* and *Campylobacter* • 330
Patrick F. McDermott and Diane E. Taylor

25. Anaerobes • 340
Isabelle Podglajen, Jacques Breuil, and Ekkehard Collatz

26. *Pseudomonas aeruginosa* • 355
Keith Poole

27. Antibiotic Resistance in *Salmonella* and *Shigella* • 367
E. J. Threlfall

28. Antibiotic Resistance in *Escherichia coli* • 374
Mark A. Webber and Laura J. V. Piddock

29. Epidemiology and Treatment Options for Select Community-Acquired and Nosocomial Antibiotic-Resistant Pathogens • 387
John E. Gustafson and John D. Goldman

30. Drug-Resistant Falciparum Malaria: Mechanisms, Consequences, and Challenges • 401
Karen Hayton, Rick M. Fairhurst, Bronwen Naudé, Xin-zhuan Su, and Thomas E. Wellems

31. Antimicrobial Susceptibility Testing of Bacterial Agents of Bioterrorism: Strategies and Considerations • 414
Fred C. Tenover

V. Ecology and Fitness of Drug Resistance
Introduction: *David G. White*

32. Fitness Traits in Soil Bacteria • 425
Fabrice N. Gravelat, Steven R. Strain, and Mark W. Silby

33. Ecology of Antibiotic Resistance Genes • 436
Abigail A. Salyers, Hera Vlamakis, and Nadja B. Shoemaker

34. Resistance in the Food Chain and in Bacteria from Animals: Relevance to Human Infections • 446
Vincent Perreten

35. Antimicrobial Use in Plant Agriculture • 465
Anne K. Vidaver

VI. Drug Resistance in Cancer Cells
Introduction: *Michael N. Alekshun*

36. Mechanisms of Resistance to Anticancer Agents • 473
Michael P. Draper, Graham K. Jones, Christopher J. Gould, and David E. Modrak

37. Development of Resistance to Anticancer Agents • 500
David E. Modrak, Graham K. Jones, and Michael P. Draper

VII. Policy, Education, and Exploration
Introduction: *Michael N. Alekshun*

Commentary. The Birth of the Alliance for the Prudent Use of Antibiotics (APUA) • 517
Ellen Koenig

38. Alliance for the Prudent Use of Antibiotics: Scientific Vision and Public Health Mission • 519
Kathleen T. Young and Thomas F. O'Brien

39. Antimicrobial Use in Animals in the United States: Developments in Policy and Practice • 528
Stephen J. DeVincent and Christina Viola

40. From Perplexing Proteins to Paratek Pharmaceuticals: One Scientific Front Forged by Stuart B. Levy • 537
Mark L. Nelson and S. Ken Tanaka

Afterword. Learning and Teaching: a Personal Reflection • 551
Julian Davies

Concluding Remarks • 555
Jay A. Levy

Index • 559

CONTRIBUTORS

Mila Aldema-Ramos
Department of Molecular Biology and Microbiology and Center for Adaptation Genetics and Drug Resistance Research, Tufts University School of Medicine, Boston, MA 02111

Michael N. Alekshun
Paratek Pharmaceuticals, Inc., Boston, MA 02111

Teresa M. Barbosa
Instituto de Tecnologia Química e Biológica, Universidade Nova de Lisboa, Apartado 127, 2781-901 Oeiras Codex, Portugal

David D. Boehr
Antimicrobial Research Centre, Department of Biochemistry, McMaster University, West Hamilton, Ontario, Canada L8N 3Z5

Jacques Breuil
Centre Hospitalier Intercommunal, 94190 Villeneuve Saint Georges, France

Karen Bush
Johnson & Johnson Pharmaceutical Research & Development, L.L.C., Raritan, NJ 08869

Ekkehard Collatz
INSERM U655–Laboratoire de Recherche sur les Antibiotiques, University Paris VI, 75270 Paris Cedex 06, France

Patrice Courvalin
Unite des Agents Antibacteriens, Institut Pasteur, 75724 Paris Cedex 15, France

Longzhu Cui
Department of Bacteriology, Juntendo University, 2-1-1 Bunkyo-Ku, Tokyo, Japan 113-8421

Julian Davies
Department of Microbiology and Immunology, University of British Columbia, Vancouver, British Columbia, Canada

Bruce Demple
Department of Genetics and Complex Diseases, Harvard School of Public Health, Boston, MA 02115

Florence Depardieu
Unite des Agents Antibacteriens, Institut Pasteur, 75724 Paris Cedex 15, France

Stephen J. DeVincent
26 Montgomery Street, Boston, MA 02116

Michael P. Draper
Paratek Pharmaceuticals, Boston, MA 02111

JoAnn Dzink-Fox
Novartis Institute for BioMedical Research, Inc., Cambridge, MA 02139

Rick M. Fairhurst
Laboratory of Malaria and Vector Research, National Institute of Allergy and Infectious Diseases, National Institutes of Health, Bethesda, MD 20892

Anthony M. George
Department of Cell and Molecular Biology, University of Technology Sydney, Broadway, N.S.W., Australia 2007

John Goldman
Department of Internal Medicine, Harrisburg, PA 17104

Christopher J. Gould
Brandeis University, Department of Biochemistry, Waltham, MA 02454-9110

Fabrice N. Gravelat
Center for Adaptation Genetics and Drug Resistance, Department of Molecular Biology and Microbiology, Tufts University School of Medicine, Boston, MA 02111

John E. Gustafson
Department of Biology, New Mexico State University, Las Cruces, NM 88003-8001

Herbert Haechler
Institute of Veterinary Bacteriology, University of Berne, 3012 Berne, Switzerland

Karen Hayton
Laboratory of Malaria and Vector Research, National Institute of Allergy and Infectious Diseases, National Institutes of Health, Bethesda, MD 20892

James F. Head
Department of Physiology and Biophysics, Boston University School of Medicine, Boston, MA 02118

Keiichi Hiramatsu
Department of Bacteriology, Juntendo University, 2-1-1 Bunkyo-Ku, Tokyo, Japan 113-8421

Teruyo Ito
Department of Bacteriology, Juntendo University, 2-1-1 Bunkyo-Ku, Tokyo, Japan 113-8421

George Jacoby
Lahey Clinic, Burlington, MA 01805

Graham K. Jones
Paratek Pharmaceuticals, Boston, MA 02111

Glenn W. Kaatz
John D. Dingell VA Medical Center and Wayne State University School of Medicine, Detroit, MI 48201

Maria Kapi
Department of Bacteriology, Juntendo University, 2-1-1 Bunkyo-Ku, Tokyo, Japan 113-8421

Ellen Koenig
Dominican Institute of Virological Studies (IDEV), Zona Universitario, Santo Domingo, Dominican Republic

Roland Leclercq
Service de Microbiologie, CHU Côte de Nacre, Caen, France

Bonnie Marshall
Department of Molecular Microbiology and Microbiology, Center for Adaptation Genetics and Drug Resistance, Tufts University, School of Medicine, Boston MA 02111

Robert G. Martin
Laboratory of Molecular Biology, NIDDK, National Institutes of Health, Bethesda, MD 20892

Patrick F. McDermott
Office of Research, Center for Veterinary Medicine, U. S. Food and Drug Administration, Laurel, MD 20708

Laura M. McMurry
Department of Molecular Biology and Microbiology and Center for Adaptation Genetics and Drug Resistance Research, Tufts University School of Medicine, Boston, MA 02111

David E. Modrak
Garden State Cancer Center, Belleville, NJ 07109

Merri C. Moken
78 Bay State Road #4, Boston, MA 02215

Ian Moore
Antimicrobial Research Centre, Department of Biochemistry, McMaster University, West Hamilton, Ontario, Canada L8N 3Z5

Bronwen Naudé
Laboratory of Malaria and Vector Research, National Institute of Allergy and Infectious Diseases, National Institutes of Health, Bethesda, MD 20892

Mark L. Nelson
Paratek Pharmaceuticals, Boston, MA 02111

Hiroshi Nikaido
Department of Molecular and Cell Biology, University of California, Berkeley, CA 94720-3202

Thomas F. O'Brien
Microbiology Laboratory, Brigham & Women's Hospital, Boston, MA 02115-6110

Margret Oethinger
Yale University School of Medicine, Yale-New Haven Hospital, Department of Laboratory Medicine, New Haven, CT 06504

Vincent Perreten
Institute of Veterinary Bacteriology, Faculty of Veterinary Medicine, University of Berne, CH-3001 Berne, Switzerland

Laura. J. V. Piddock
Antimicrobial Agents Research Group, Division of Immunity and Infection, The University of Birmingham, Edgbaston, Birmingham, United Kingdom

Isabelle Podglajen
INSERM U655–Laboratoire de Recherche sur les Antibiotiques, University Paris VI, 75270 Paris Cedex 06, and Hôpital Européen Georges Pompidou, 75015 Paris, France

Pablo J. Pomposiello
Department of Microbiology, University of Massachusetts, Amherst, MA 01003

Keith Poole
Department of Microbiology and Immunology, Queen's University, Kingston, Ontario, Canada K7L 3N6

Steven J. Projan
Wyeth Research,Cambridge, MA 02140

Marilyn C. Roberts
Department of Pathobiology, University of Washington, Seattle, WA 98195

Judah L. Rosner
Laboratory of Molecular Biology, NIDDK, National Institutes of Health, Bethesda, MD 20892

Abigail A. Salyers
Department of Microbiology, University of Illinois, Urbana, IL 61801

Frederic M. Sapunaric
Department of Molecular Biology and Microbiology and Center for Adaptation Genetics and Drug Resistance Research, Tufts University School of Medicine, Boston, MA 02111

Thamarai Schneiders
Department of Molecular Microbiology and Microbiology, Center for Adaptation Genetics and Drug Resistance, Tufts University, School of Medicine, Boston MA 02111

Stefan Schwarz
Institut für Tierzucht, Bundesforschungsanstalt für Landwirtschaft (FAL), 31535 Neustadt-Mariensee, Germany

Nadja B. Shoemaker
Department of Microbiology, University of Illinois, Urbana, IL 61801

Mark W. Silby
Center for Adaptation Genetics and Drug Resistance, Department of Molecular Biology and Microbiology, Tufts University School of Medicine, Boston, MA 02111

Steven R. Strain
Department of Biology, Slippery Rock University of Pennsylvania, Slippery Rock, PA 16057

Xin-zhuan Su
Laboratory of Malaria and Vector Research, National Institute of Allergy and Infectious Diseases, National Institutes of Health, Bethesda, MD 20892

Anne O. Summers
Department of Microbiology, The University of Georgia, Athens, GA 30602-2605

Joyce Sutcliffe
Rib-X Pharmaceuticals, Inc., New Haven, CT 06511

Yutaka Tajima
Department of Bacteriology, Juntendo University, 2-1-1 Bunkyo-Ku, Tokyo, Japan 113-8421

S. Ken Tanaka
Paratek Pharmaceuticals, Boston, MA, 02111

Diane E. Taylor
Department of Medical Microbiology and Immunology, University of Alberta, Edmonton, T6G 2 H7 Alberta, Canada

Fred C. Tenover
Division of Healthcare Quality Promotion, Centers for Disease Control and Prevention, Atlanta, GA 30333

E. J. Threlfall
Health Protection Agency, Laboratory of Enteric Pathogens, Centre for Infections, London NW9 5HT, United Kingdom

Suwanna Trakulsomboon
Department of Medicine, Faculty of Medicine Siriraj Hospital, Mahidol University, Bangkoknoi, Bangkok 10700, Thailand

Anne K. Vidaver
Department of Plant Pathology, University of Nebraska, Lincoln, NE 68583-0722

Jordi Vila
Department of Microbiology, Hospital Clinic, University of Barcelona, School of Medicine, 08036 Barcelona, Spain

Christina Viola
22 Highland Road, Somerville, MA 02144

Hera Vlamakis
Department of Microbiology, University of Illinois, Urbana, IL 61801

Mark. A. Webber
Antimicrobial Agents Research Group, Division of Immunity and Infection, The University of Birmingham, Edgbaston, Birmingham, United Kingdom

*Thomas E. Wellems**
Laboratory of Malaria and Vector Research, National Institute of Allergy and Infectious Diseases, National Institutes of Health, Bethesda, MD 20892

David G. White
Division of Animal and Food Microbiology, Office of Research, Center for Veterinary Medicine, U. S. Food & Drug Administration, Laurel, MD 20708

Gerard D. Wright
Antimicrobial Research Centre, Department of Biochemistry, McMaster University, West Hamilton, Ontario, Canada L8N 3Z5

William Yan
Food Directorate, Health Canada, Sir Frederick Banting Research Centre, Tunney's Pasture, Ottawa, Ontario K1A 0L2, Canada

Kathleen T. Young
Alliance for the Prudent Use of Antibiotics, Boston, MA 02111

FOREWORD

Microbial resistance to growth inhibitors has been observed since the earliest days of chemotherapy, e.g., C.H. Browning's work on trypanosomes in Paul Ehrlich's laboratory. The biological basis of "drug-fastness" was controversial for many years. Some believed that the drug induced adaptive changes in the (then still obscure) genetic structures of the treated microbes; others proposed the random occurrence of resistant mutants which were merely selected by the drug. A variety of experiments in the decade 1943 to 1952, ranging from the fluctuation test introduced by Luria and Delbruck (1943) to indirect selection using replica plating (Lederbergs, 1952), provided abundant evidence for the selectionist theory. It was no surprise to the genetically informed that resistance became a growing problem not long after the introduction of sulfonamides and then penicillin into clinical practice.

Random mutation was by no means the whole story. Landmark studies by Tsutomu Watanabe and coworkers in the late 1950s uncovered resistance transfer factors (RTFs) in *Shigella* spp. that conferred resistance to multiple antibiotics, including streptomycin, chloramphenicol, tetracycline, and sulfonamide. In clinical practice, the borrowing of RTFs from other pre-existing resistant strains was often more important than the selection of de novo mutants in the treated host, especially for multiple drug resistance. The RTFs fit into an overall framework of infective heredity, founded on the prevalence of plasmids (extranuclear autonomous fragments of DNA that allow for lateral transfer of genetic information across strains, species, and even phyla and that may encode any part of the genome, including tumorigenesis and conjugational competence genes).

The discovery of RTFs gave Dr. Stuart B. Levy the impetus to work in Watanabe's laboratory and to embark on a career devoted to the study of antibiotic resistance and its public health impact. While this public health problem was originally confined to Japan, it was realized early on that it would be only a matter of time before it become a worldwide threat. Stuart's contributions to the field of antibiotic resistance over the past four decades have included many important discoveries regarding various resistance mechanisms. His pioneering efforts in the arena of tetracycline resistance led to the seminal discovery of active drug efflux as a mechanism of resistance. Stuart is also credited with the discovery of the intrinsic (chromosomal) and inducible multiple antibiotic resistance (*mar*) locus in *Escherichia coli.* This system regulates susceptibility to multiple antibiotics, oxidative stress compounds, detergents, organic solvents, and biocides through a network of genes termed the Mar regulon (but primarily involving mechanisms of decreased drug uptake and increased drug efflux). These findings have helped to create new avenues in the field of intrinsic drug resistance and have led to important new understandings of the bacterium's natural defense mechanisms to noxious stimuli. The parallels between bacterial tetracycline-specific efflux mechanisms and multiple antibiotic efflux mechanisms led Stuart to investigate whether similar mechanisms were operative in eukaryotic cells during growth in the presence of cancer chemotherapeutics.

Early studies on selective antimicrobial pressure in the environment, performed in the Levy laboratory at Tufts University School of Medicine, are considered groundbreaking experiments on the subject. They have fostered a large body of ongoing research on agricultural antimicrobial use, which continues to be pursued today. Over 30 years ago, Stuart led a prototypic prospective study on the effects of antibiotic-laced feed on the carriage of antibiotic-resistant bacteria in farm animals and staff that was published in the *New England Journal of Medicine.* Subtherapeutic amounts of oxytetracycline, often used as a growth promoter for chickens, led to their colonization by tetracycline-resistant and then multiply resistant *E. coli*. Within 3 to 5 months, farm staff were excreting high numbers of tetracycline- and multidrug-resistant *E. coli* as well. Thus, feed for chickens impacts the flora carried by neighboring humans and, surely, the overall environment. One recent estimate put the tonnage of antibiotics used in animals and animal feed in the millions (M. Mellon, C. Benbrook, and K. L. Benbrook, *Hogging It: Estimates of Antimicrobial Abuse in Livestock,* UCS Publications, Cambridge, United Kingdom, 2001), matching the amount used for human therapeutics (S. A. McEwen and P. J. Fedorka-Cray, *Clin. Infect. Dis.* **34**:S93-S106, 2002).

Stuart is recognized as one of the leading researchers in the tetracycline class of agents. He was among the first to show genetic heterogeneity among drug resistance determinants by defining different tetracycline resistance genes. Moreover, in some of the earliest studies in molecular epidemiology, he showed the distribution of different tetracycline resistance determinants among bacteria. His team led the effort to produce a nomenclature system for the different tetracycline resistance determinants, which is regularly revised with new discoveries. Stuart's more recent laboratory forays have led him into studies designed to evaluate resistance to household disinfectants and biocides. Pine oil, a common component of detergents and other household cleaning fluids, was used initially by Stuart and colleagues as a selective agent to recover resistance to multiple antibiotics in vitro. Biocides, including agents such as triclosan, were previously thought to kill bacteria by physical disruption. Well-crafted laboratory experiments convincingly demonstrated that triclosan does indeed have a specific site of action within the cell at a protein (FabI) that is involved in fatty acid biosynthesis. These studies imply that the overuse of antibacterial-containing household products could lead to the unintended selection of antibiotic-resistant bacteria.

The optimism of the early period of antimicrobial discovery has been tempered by the emergence of bacterial strains displaying resistance to multiple antimicrobial agents, an unfortunate legacy of past decades of antimicrobial use and abuse. As a practicing physician and scientist, Stuart has been an ardent spokesperson for the recognition of this problem over the past 30 years. Drug resistance presents an ever-increasing global public health threat that involves all major microbial pathogens and antimicrobial drugs. The scarcity of new antimicrobials on the horizon and the increasing prevalence of multidrug resistance mean that we must redouble our efforts to judiciously preserve the agents at hand, while intensifying the search for new therapeutics. For many years, Stuart was a lone voice speaking out for prudent use of antibiotics while receiving little responsive action. In the present day, he continues to serve as a passionate advocate for alternatives to growth-promoting antimicrobials in food animal production, as well as for prudent antibiotic use in all settings. The public health consequences of a misguided inclination for "sterilizing" our environment with biocides and other broad-spectrum household antimicrobials, in terms of potential adverse effects on microbial ecology and drug resistance, have become a contemporary point of concern for Stuart. He speaks and writes on the need for greater awareness of how our short-sighted overuse and sometimes misuse of these therapeutics generate multidrug-resistant bacteria that today confront clinicians and patients worldwide.

His message, directed initially to professional groups, has moved to the public in the form of lectures and his widely cited book *The Antibiotic Paradox: How Miracle Drugs Are Destroying the Miracle*. First published in 1992, this book has since been updated and translated into French, Korean, and Chinese. Stuart's dedicated leadership in the policy and education arenas is also exemplified by the Alliance for the Prudent Use of Antibiotics (APUA) which he founded in 1981. APUA is a growing international organization with a presence in 100 countries and networks of chapters and individuals working closely on communication and research to foster prudent antibiotic use and to curtail antibiotic resistance. They recently published a report entitled "Shadow Epidemic," which quite clearly illustrates the high prevalence of resistance to front-line agents in common microbial pathogens. Stuart's message reaches wide audiences through print and visual media and appearances before national and international audiences. Now through this publication, Stuart's colleagues, collaborators, and students have the opportunity to honor his past and present achievements and, perhaps of greatest import to him, to carry his message still further.

Joshua Lederberg, Ph.D.
Raymond and Beverly Sackler Foundation Scholar
The Rockefeller University
New York, NY
May 2005

PREFACE

Antimicrobial resistance in pathogenic microorganisms is recognized as one of the chief threats to human health worldwide. While antimicrobial resistance has long been recognized as a biological phenomenon, early observations were generally considered interesting laboratory events with little, if any, clinical relevance. Even when resistance became more common, the advent of new anti-infective agents was able to outpace incipient resistance phenotypes, which typically appeared at low frequency and increased in usually small increments. Today, the situation is much different. Beginning in the late 1980s, a large number of resistance phenotypes began to appear, and there is now widespread resistance to many antimicrobial agents, with multiple antimicrobial resistance phenotypes being the rule rather than the exception for many pathogens. More recently, multidrug resistant pathogens indigenous to the hospital environment have moved into the community (e.g., methicillin-resistant *Staphylococcus aureus*). The CDC estimates that 70% of bacterial infections are caused by organisms resistant to at least one antimicrobial, and the Institute of Medicine estimates the total annual cost of antimicrobial resistance in the U.S. at nearly $5 billion.

Some researchers recognized the signs and the potential pitfalls of indiscriminate antibiotic use early on and persisted for years in explaining the looming problem to a largely silent audience. One such pioneer was Dr. Stuart B. Levy. In addition to providing a thorough review of key pathogens and drug resistance in various classes of antimicrobials, this book is also intended as a tribute to Stuart's unique contributions to the field. Through his own diligence, imagination, and charisma, he has combined basic microbiology research with public education in an ongoing effort to limit antimicrobial resistance, to preserve the power of these precious natural resources, and to offer new therapies to future generations. The enthusiastic replies we received from the contributors when first proposing this project are a testimony to Stuart's influence on the science of resistance and the scientific community.

The sections in this book were chosen with the goal of providing a comprehensive overview of antibacterial and anticancer drug resistance, with special emphasis on those areas where Stuart stimulated scientific advancements through his own research and that of his students, research fellows, and colleagues. Thus, the authors of the respective chapters have either spent time in his laboratory or collaborated with him on research, policy, or educational initiatives.

This book is divided into 7 sections and 40 chapters that focus on the current knowledge of mechanisms of antimicrobial and anticancer drug resistance, the ecology of resistant bacteria, the factors influencing antimicrobial resistance, and efforts in developing and implementing relevant policy and education. While the book is unified in theme, each chapter is meant to stand alone as the author's own perspectives on the resistance problem. We are confident that the reader will find these chapters to be an excellent review of the subject.

In addition to the contributing authors and a dedicated ASM publication staff, the editors would like to thank Neil Woodford, Marcus Zervos, Christopher Ohl, Barry Hurlburt, Judah L. Rosner, David Hecht, Tom Shryock, Douglas Hutchinson, Kenneth Thompson, Qijing Zhang, and Steven Projan for critical review of portions of this volume.

David G. White
Michael N. Alekshun
Patrick F. McDermott
April 2005

INTRODUCTION

I am honored and somewhat awed by this book. To see so many of my colleagues, associates, and students contributing to this compendium of knowledge and insight into the drug resistance field is more than any research scientist can ever expect. What an impressive integration of findings, concepts, and new ideas provided by so many distinguished investigators in this broad and multifaceted field!

My decision to pursue a research career in microbiology was largely influenced by my introduction to the phenomenon of antibiotic resistance in 1963. I was spending a year away from medical school as a predoctoral student with Professor Raymond Latarjet at the Laboratoire Pasteur of the Institut du Radium in Paris. In a journal club, I was fascinated to learn about R factors, curious means of drug resistance transfer among bacteria of different genera. I decided to understand firsthand the findings in this field. I wrote to Tsutomu Watanabe, an early pioneer in drug resistance research to ask if I could come work with him during the summer of my third year of medical school. I was delighted to be accepted and to spend several months in 1964 in his laboratory at Keio University in Tokyo, Japan.

Side-by-side with him at the bench or singing with him and his group after work at a karaoke restaurant, we established a strong friendship and collegiality up until his untimely death in 1973. He remains an influence on me. I am passionately taken by the mysteries of drug resistance and am not alone, as evidenced by this book. The problem has engaged individuals in multiple disciplines, including genetics, molecular biology, biochemistry, pharmacology, epidemiology, ecology, and evolution. There is also, of course, the economic cost as well as the public and private health impact of resistance in the treatment of infectious diseases. More and more, we understand the globality of resistance—how easily resistance genes can move among genera and species and how resistant bacteria spread from country to country.

Another individual, Joshua Lederberg, also influenced my entry into this field. His discovery of gene transfer among bacteria formed the foundation for understanding the transfer of drug resistance on R factors. I had a chance meeting with him in Stanford, Calif., when I was looking for residency programs. Josh has remained a friend, a valued supporter of prudent antibiotic use, and a steadfast active participant in the infectious disease area. Among other early pioneers with whom I have had the pleasure to interact are Richard Novick, Julian Davies, and Stanley Falkow. Their studies and outreach activities stimulated and provided a firm foundation for many of us in the field.

I thank Michael Alekshun, Patrick McDermott, and David White for their organizing and steerage of this book and their continued dedication to the drug resistance problem. Each of these researchers has achieved recognition for independent accomplishments in the field. I am proud and pleased to have hosted them in my laboratory. For many years, I have argued for more attention to drug resistance, not just from the vantage point of public health, but also from my concern that the antibiotic resistance field was losing investigators and failing to attract new ones, even as the problem magnified. Mike, Pat, and David are fine examples of the next generation.

On a larger scale, it is gratifying to see that a new Study Section on Drug Discovery and Mechanisms of Antimicrobial Resistance has been formed at the National Institutes of Health. This action alone in response to urging by many of us will encourage new investigators to enter the field and support those already in it.

I am also indebted to my long-standing colleague Laura McMurry, who for three decades has worked closely with me in our studies, particularly on the Tet protein and the *mar* regulatory locus. My second "right hand" is Bonnie Marshall, whose expertise in clinical microbiology has been complemented by her striking ability to skillfully learn genetics and molecular biology. She has been there for whatever ecology study we have undertaken, whether it be spread of bacteria and plasmids on a farm, release of bacteria from autoclaves, biocide impact on household microbiology, or contamination of wind instruments. Finally, my third long-standing valuable support is Mark Nelson, whose willingness to work, often alone, in identifying and creating new tetracyclines provided the means to move forward in the tetracycline "renaissance." This effort eventually led to the founding of Paratek Pharmaceuticals, among whose aims are to identify new tetracyclines not subject to current resistance mechanisms and to uncover novel approaches to prevent and treat infectious diseases.

Our knowledge of antibiotic resistance has come a long way. The genetics of resistance and its causative transfer elements were just being discovered in the 1960s when I was first introduced to the problem. Over these past four decades, we have learned about new genetic elements, new transfer mechanisms, new kinds of resistances, new organisms, and new insights into the origins of some resistance genes, particularly among soil organisms. More importantly, we have reached a period when antimicrobial resistance is finally receiving the close attention that it merits from government and nongovernment organizations and individuals. The problem has climbed to the priority list of most public health agendas. Several of us place it at the pinnacle, since it impacts so many areas of infectious diseases and health. As the *Alliance for the Prudent Use of Antibiotics* states, drug resistance is a disease in itself—a "shadow epidemic"—which casts a disheartening gloom over the successful treatment of all kinds of infectious diseases throughout the world.

Nonetheless, I believe we are at the beginning of an exciting new era, hopefully replete with new molecular findings in drug resistance, new drugs for the physicians' armamentarium, and a more expansive appreciation of the power of microbes. We need to approach infectious diseases with a new paradigm that seeks control, not elimination, of microbes and disease prevention, not sterilization of our environment. The contributors to this volume are among the leaders facing these issues and providing the knowledge needed to find a solution to this important health challenge.

Stuart B. Levy
March 2005

Samuel Beckett and Stuart, Paris, 1963

Stuart and Raymond Latarjet, Laboratoire Pasteur, Institut du Radium, 1973

Stuart and Tsutomu Watanabe; Keio University, Tokyo, Japan, 1964

Stuart and Yves Chabbert, Pasteur Institut, 1976

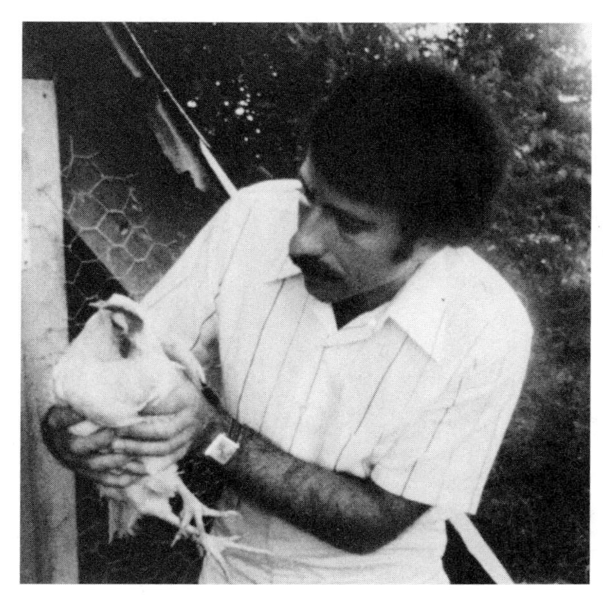

Stuart at the chicken farm, Sherborn, Mass., 1975

Stuart at the end of the chicken study, Sherborn, Mass., 1975

Naomi Datta and Stuart, Kavouri, Greece, 1976

Francois Gros and Stuart, Institut Pasteur, Paris, 1976

Clinical study unit, Tufts—New England Medical Center, Boston, Mass., 1979

George Jacoby, Stuart, and Wally Gilbert at a press conference after the release of *Antibiotic Misuse Statement*, Boston, Mass., 1981

Karen Ippen-Ihler and Stuart singing "Genetic Capers," Gordon Research Conference, 1980

Stuart and Lady Amelia Fleming, Athens, Greece, 1983

Wolfgang Hillen and Stuart, Germany, 1985

Stuart with The Fifth Dimension, ICAAC, Atlanta, Ga., 1990

Release of *The Antibiotic Paradox*, 1992

Mark Richmond and Stuart, Madrid, Spain, 1994

Opening of the Center for Adaptation Genetics and Drug Resistance, 1992. Back row: Murat Akova, Petri Viljanen, David White, and Eamonn Nulty. Middle row: Stuart Levy, Asun Seone, Amy Hegg, and Audrey Bernstein. Front row: Mark Nelson, Claire Sherman, Anne Happel, Marianne Williams, Laura McMurry, and Tom Clancy.

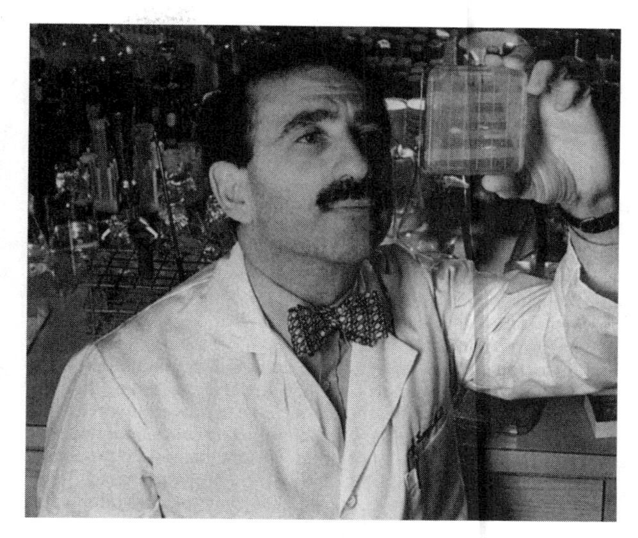

From *Newsweek,* 1994

Stuart receiving the Hoechst-Roussel award from Michael Scheld, 1995

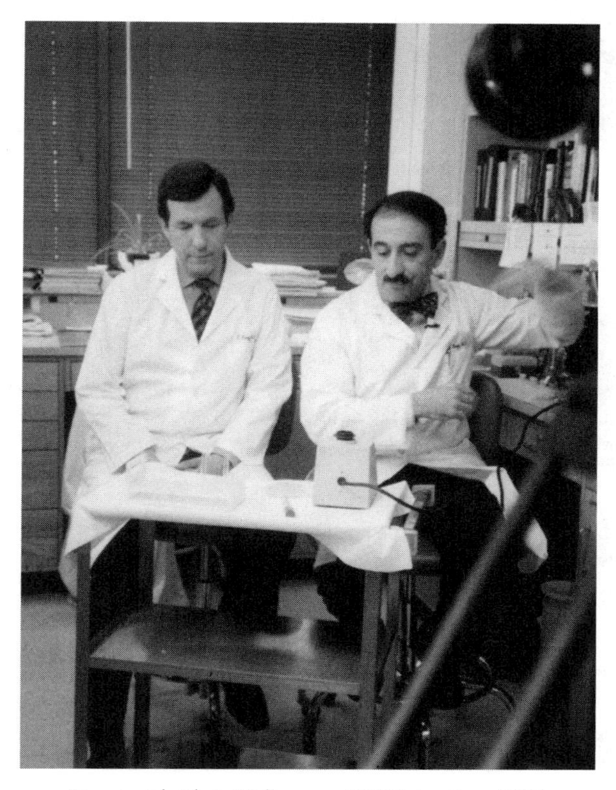

Stuart with Chris Wallace on *ABC Primetime*, 1994

Patrice Courvalin, Claude James Soussy, and Stuart, ICAAC, 1995

Akihito Yamaguchi and Stuart, Boston, Mass., 1991

Stuart and Steve Lerner, ICAAC, 1995

Hiroshi Nikaido and Stuart, 1996

Rita Colwell and Stuart, 1996

Stuart with his sister, Ellen Levy Koenig, ICAAC, 1995

IACMAC/ASM Conference, Moscow, Russia, 2000

Honorary degree, Wesleyan University (with guest speaker Oprah Winfrey, center), 1998

Giuseppe Cornaglia and Suart, Italy, 1999

Jay and Stuart Levy, Centennial ASM Celebration, Chicago, Ill., 1999

APUA. Back row: Michèle Kulick, Phakdey Chea, Michael Feldgarden, Amelie Peryea, Anibal Sosa, Smita Desai, Stefanie Valovic, and Flora Traub. Front row: Bonnie Marshall, Thomas O'Brien, Stuart Levy, Kathleen Young, and Ron Lanoue.

Members of the Center for Adaptation Genetics and Drug Resistance, Tufts University, Boston, Mass. 2003. From left to right: Mila Aldema-Ramos, Tamarai Dorai-Schneiders, Frederic Sapunaric, Fabrice Gravelat, Rupa Udani, Bonnie Marshall, Mark Silby, Stuart Levy, Christopher Gould, Jason Smith, Laura McMurry, Herve Nicoloff.

APUA Awards Reception, ICAAC, 2003. From left to right: David Bell, Patrick McDermott, Stuart, Karen Travers, and Richard Besser.

Paratek Pharmaceuticals, Boston, Mass., 2000. Row 1: Mark Nelson, Tom Bigger, Stuart, and George Hillman; row 2: JoAnn Gifford, Karen Stapleton, Manching Ku, and Mohamed Ismail; row 3: Ted Andrews, Delora Schneider, Laura (McIntyre) Honeyman, and Susan Weir; row 4: Michael Alekshun, Graham Jones, Steve Bryant, and Chris Esposito; row 5: David Messersmith and Paul Hawkins.

The Department of Molecular Biology and Microbiology, Tufts University School of Medicine, 2002. Row 1: Carol Kumamoto, Cathy Squires, Joan Mecsas; row 2: Dean Dawson, Stuart Levy, Linc Sonenshein, Claire Moore; row 3: Edward Goldberg, Ralph Isberg, Andrew Camilli, John Coffin; row 4: Matt Waldor, Mike Malamy, David Lazinski, Ted Park, Andrew Wright.

Still together . . . Laura McMurray (since 1971), Stuart, and Bonnie Marshall (since 1977)

I. TETRACYCLINES AND RESISTANCE

SECTION EDITOR: Michael N. Alekshun

Antibiotics within the tetracycline family have historically possessed a broad spectrum of activity and are currently used in the management of infections caused by gram-positive, gram-negative, and parasitic organisms. Success, however, is not without a cost; extensive use of these agents has resulted in the widespread selection of tetracycline-resistant pathogens.

The contributions within this section provide an overview of the mechanisms that bacteria use to evade the antimicrobial activity of the tetracycline family. Included are descriptions of efflux and ribosome protection mechanisms, which are the most important from the clinical perspective. The seminal contributions of Stuart Levy's laboratory in characterizing drug efflux as a mechanism of tetracycline (antibiotic) resistance continue to influence the perception of antibiotic discovery, activity, and resistance. Also discussed are mechanisms of drug degradation, chromosomal mutations, and other unknown systems that microorganisms use to thwart tetracycline action. The section concludes with a concise history of tetracycline discovery and perspectives on the modern-day tetracyclines (minocycline and doxycycline). The newest member of the tetracycline class, tigecycline, which is in advance clinical development, is also presented.

Frontiers in Antimicrobial Resistance: a Tribute to Stuart B. Levy
Edited by D. G. White, M. N. Alekshun, and P. F. McDermott

Chapter 1

Tetracycline Resistance: Efflux, Mutation, and Other Mechanisms

Frederic M. Sapunaric, Mila Aldema-Ramos, and Laura M. McMurry

TETRACYCLINE ACTION AND MODES OF RESISTANCE

The structure of tetracycline is shown in chapter 3. When complexed with a divalent cation such as Mg^{2+} in the cytoplasm, the drug binds reversibly to the 16S ribosomal RNA (rRNA) of prokaryotes near the ribosomal acceptor A site (6, 76), preventing binding of aminoacyl-tRNA to this site. The elongation step of protein synthesis is thereby curtailed and microbial growth is inhibited (see also references 56, 57, 91). Four mechanisms have evolved to counteract tetracycline. They are active efflux (keeping tetracycline out of the cytoplasm), inactivation of the tetracycline molecule, rRNA mutations (preventing tetracycline from binding to the ribosome), and ribosomal protection (preventing tetracycline from binding to the ribosome). The first three mechanisms are covered here, while the last one is discussed in chapter 2. The structure and function of tetracycline and its analogs are described in chapter 3.

ACTIVE EFFLUX OF TETRACYCLINE

Overview

Most of the genes that exclusively encode tetracycline efflux are positioned on transferable plasmids and/or transposons. To date, about 21 classes of proteins (≤80% amino acid identity as the dividing line between classes) that pump tetracycline out of the cells have been identified in aerobic and anaerobic gram-negative or gram-positive bacteria, demonstrating their wide distribution among the bacterial kingdom (48, 56, 57, 78, 80). The classes fall into six groups (25, 56, 57). Most of our discussion of efflux concerns members of groups 1 and 2.

Historical Perspective

When the Levy group began its work, tetracycline resistance was known to be inducible by subinhibitory concentrations of tetracycline and to be associated with a lower uptake of the drug. At that time the minicell system was used to study proteins made by naturally occurring resistance plasmids called R-factors. Minicells are formed by abberant cell division in certain *Escherichia coli* mutants, contain only plasmid DNA, and synthesize only (radiolabelled) plasmid-encoded proteins. In 1973 we discovered a protein that was synthesized only by those R-factors which conferred tetracycline resistance and only in response to induction by tetracycline. This protein, which we creatively named TET, was in the cytoplasmic membrane (45).

Members of our group then sorted the known tetracycline resistance determinants into four different group 1 classes A-D by DNA hybridization (63). In each of these classes, "TetA" now designates the efflux pump and "TetR" the transcriptional repressor, while the class label is appended in parentheses, e.g., TetA(B) is the pump from class B (48). Subsequently we found class E (50) and helped to characterize class H (26). Since then many other tetracycline resistance determinants, including those not involving efflux, have been found by various laboratories (see chapter 2) (48, 56, 57) (http://faculty.washington.edu/marilynr).

Because the uptake of tetracycline in resistant cells was decreased, it seemed logical to focus on this as the cause of resistance. Resistant cells had an altered energy dependence for accumulation of tetracycline (46). At the time most people, including us, thought of a reduction in uptake of a drug as springing from a decrease in permeability of the cell wall or membrane. Hence we imagined tetracycline being kept out of the cells by the Tet protein. Using right-side-out membrane vesicles

Frederic M. Sapunaric, Mila Aldema-Ramos, and Laura M. McMurry • Department of Molecular Biology and Microbiology and Center for Adaptation Genetics and Drug Resistance, Tufts University School of Medicine, Boston, MA 02111.

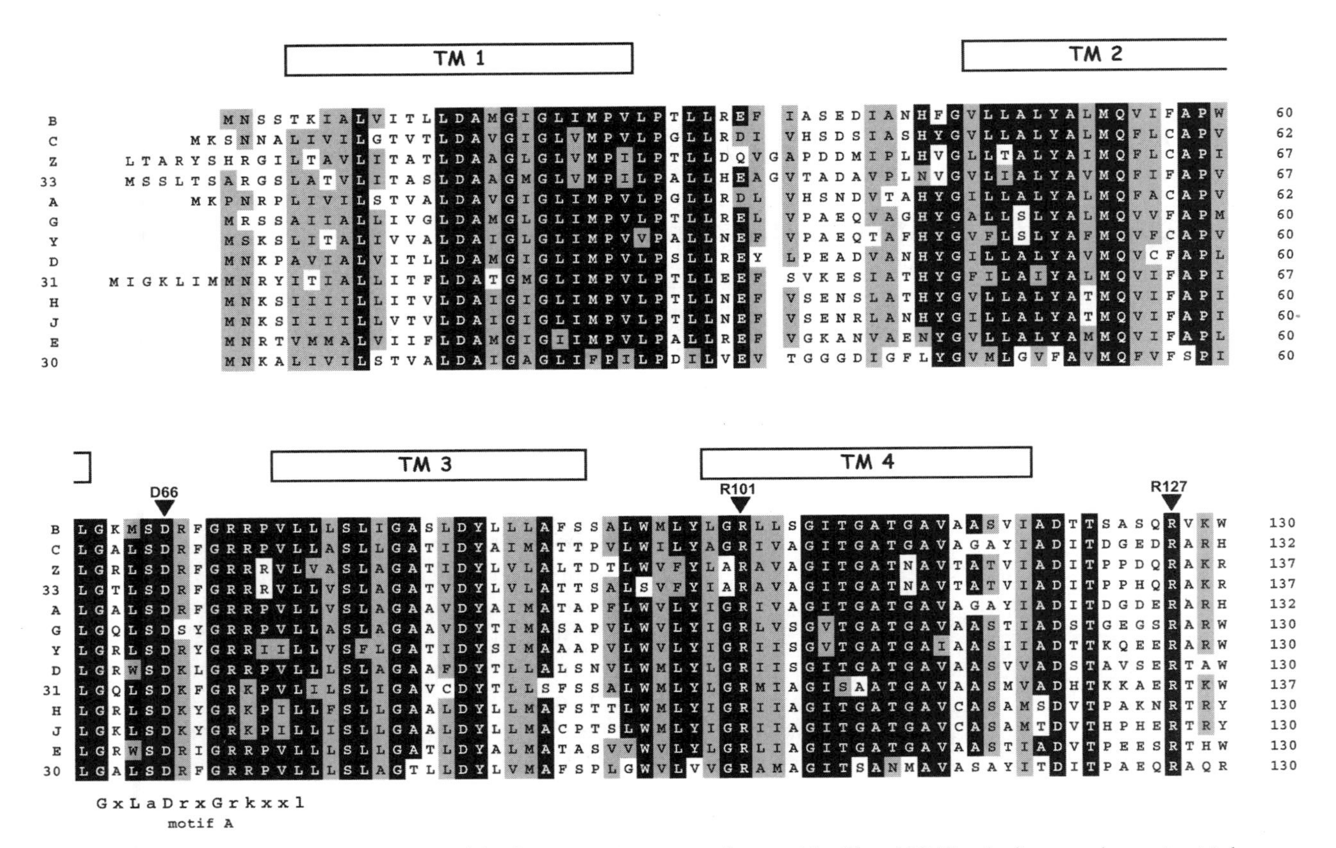

Figure 1. Multiple sequence alignment of the thirteen TetA proteins of group 1 by Clustal W. The shading was done using Multiple Align Show (http://www.cbio.psu.edu/sms/multi_align.html). Residues in vertical columns in which ≥70% of the residues are identical have black backgrounds; ≥70% similar residues have gray backgrounds. Similarity groups are (Gly), (Asp, Glu), (Arg, Lys), (Ala, Phe, Ile, Leu, Met, Pro, Val, Trp), and (Cys, His, Asn, Gln, Ser, Thr, Tyr). Transmembrane helices (TM) are shown as rectangles above the alignment. Black arrowheads show critical residues in TetA of class B, and a white arrow shows the position of Fenton cleavage in classes B and D. The start codon predicted in GenBank for class Z corresponds to a Val within TM1, so we have extended the translated sequence upstream 13 amino acids to be in step with its closest relative, class 33. For convenience we have moved the better-characterized class B and class C proteins from their true positions, just above classes D and G, respectively, to the top of the alignment. Gaps assigned by Clustal W (28) in regions before TM1 and after TM12 have been eliminated. The GenBank accession numbers are B, P02980 (Bertrand sequence [70]); C, J01749; Z, AF121000; 33, AJ420072; A, X00006; G, S52437; Y, AF070999; D, X65876; 31, AJ250203; H, U00792; J, AF038993; E, L06940; 30, AF090987.

prepared from cells by a standard penicillin lysis method of Ron Kaback, we looked for energy-dependent "keep out" in vitro. However, the energized vesicles from resistant cells and sensitive cells accumulated the same amount of drug. Then, one day, we raised the pH from the standard pH 6.6 up to pH 8.0. Now the amount of uptake of tetracycline by resistant vesicles was much higher—not lower—than that for sensitive vesicles! A small idea flickered: were some of our vesicles actually inside-out, representing *efflux* in the intact cell, and were these the ones transporting at pH 8.0? Purposefully making inside-out membrane vesicles by the French pressure cell method used (at pH 8.0) by Barry Rosen in studies of Ca^{2+} and Na^{+} efflux (81), we again indeed saw much more uptake by the vesicles from tetracycline-resistant cells than by those from sensitive cells. The same result happened for all four of the resistance classes (59). Clearly many of the vesicles in our original right-side-out preparations had actually been inside-out. In this way we became the first to discover that a drug resistance could be caused by active efflux.

The difference in tetracycline accumulation between resistant and sensitive cells was still notably smaller than the difference in resistance. At similar degrees of inhibition, resistant cells accumulated much more drug than sensitive cells (47). After looking without success for an additional resistance mechanism, such as drug sequestration, degradation, or resistant ribosomes, we concluded that much of the energy-dependent tetracycline "accumulation" in cells may be in the membrane or periplasm. A considerable basal resistance persisted even when the proton motive force was drastically reduced (53).

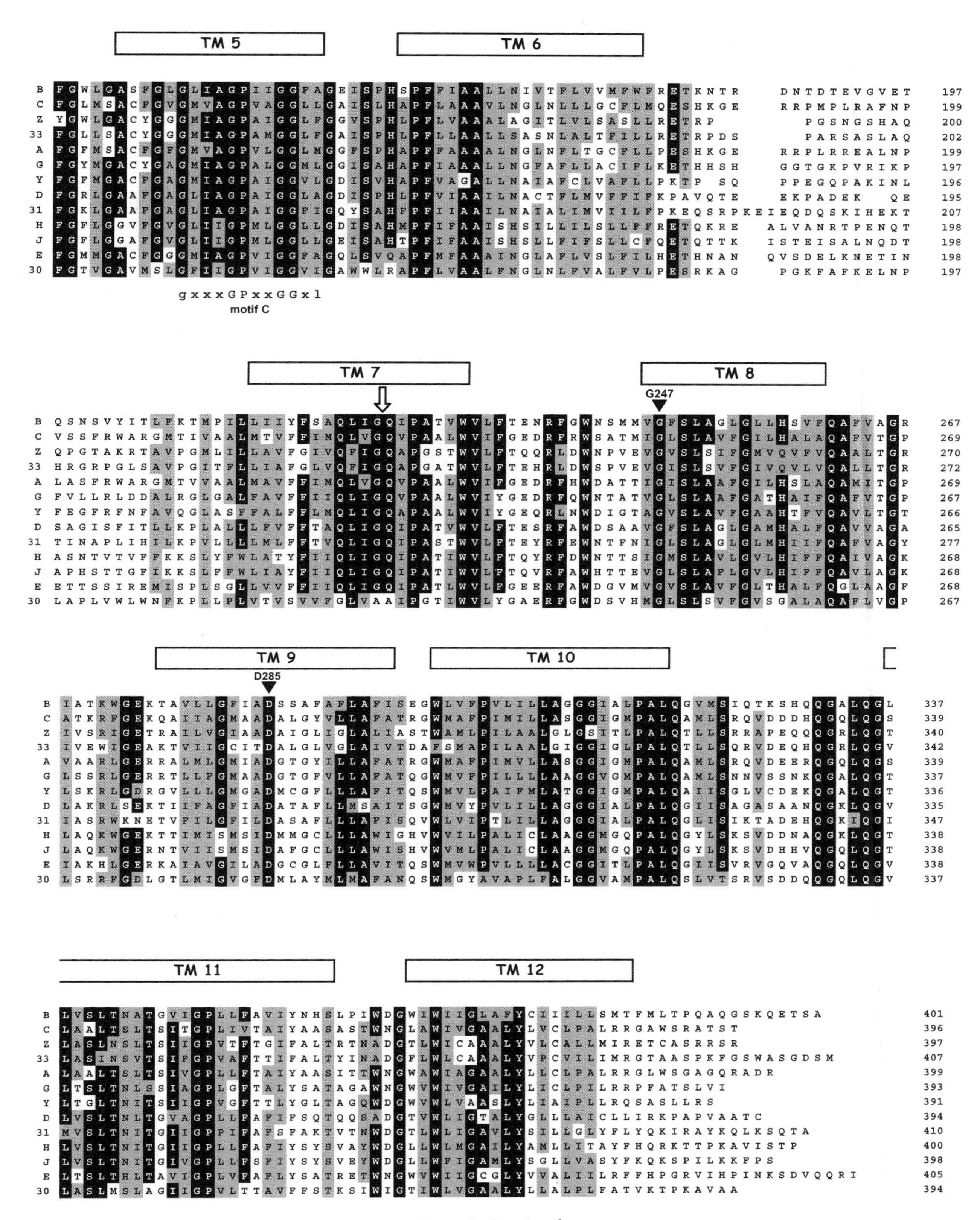

Figure 1. *Continued*

Early work also showed that extracts of resistant cells contained an inhibitor of synthesis of the TetA protein as assayed by in vitro transcription-translation; inhibition could be reversed by low concentrations of tetracycline (122). This factor was later characterized as the repressor TetR.

A single frame shift error in the earliest sequence of the *tetA*(C) region of pBR322 had led to the presumption of two Tet polypeptides. This led to our genetic complementation studies in class B, where mutations which reduced resistance fell into two separate and contiguous genetic complementation groups (10), consistent with two adjacent genes for two proteins. We indeed "found the second Tet protein" in minicells—but it turned out to be simply the repressor, which we had thought would be too sparse to detect so easily. The class C sequence was finally corrected, while that for class B was reported by both the Hillen and Bertrand groups (with four significant amino acid differences; the latter sequence [GenBank P02980] seems to have prevailed and is what we have subsequently obtained). TetA was only one polypeptide after all. The complementation was then attributed to multimerization of TetA(B), as discussed more fully below.

Our group showed that the first and second halves of TetA are related by both topology and sequence, suggesting an evolutionary gene duplication and fusion (85).

Efflux Mechanism, Group 1

There are now 13 known classes of tetracycline efflux pumps in group 1. Their alignment is shown in Fig. 1. All such TetA proteins have 12 transmembrane putative α helix(es) (TM) (see Fig. 2). The best characterized of these proteins are those from class B (Tn*10*) and class C (pBR322), and we shall focus on these.

We showed that the efflux of tetracycline requires a proton motive force, has a K_m for tetracycline of around 6 μM, and requires Mg^{2+} (59). That a tetracycline-divalent cation complex is the actual substrate was shown using radiolabeled Co^{2+} (121). Transport is electroneutral, that is, exit from the cell of one tetracycline-divalent cation complex (net charge +1) is energized by the downhill entry of one proton (113). The class C TetA protein is able to also transport K^+ into the cell (see reference 21). This is not true for

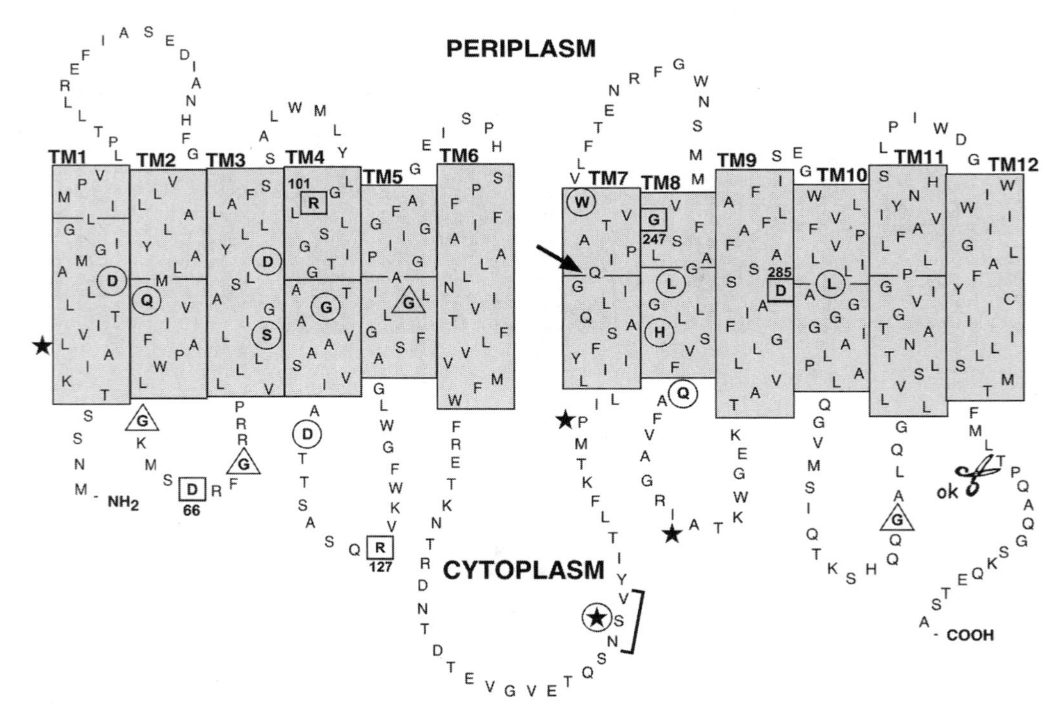

Figure 2. The TetA(B) tetracycline/H^+ antiporter. Transmembrane α-helices (TM) are indicated by gray rectangles numbered TM1-TM12. Circles indicate residues of class B possibly involved in substrate binding. The bracket indicates interdomain loop region corresponding to site in class A possibly involved in substrate binding. Squares indicate amino acids critical for tetracycline efflux. Triangles indicate amino acids possibly involved in local conformation change. The arrow indicates Fenton cleavage site. Scissors indicate the position where a truncation permits good activity. Horizontal lines indicate the permeability barrier for hydrophilic molecules. The circled star indicates the location in class C of its homologous Ser202. The other stars are placed at sites corresponding to those of class C and designate secondary suppressor mutations of the inactivating Ser202Phe substitution of class C.

class B and has not been described for any of the other group 1 proteins, possibly for lack of study, although it is true for group 2 proteins (see below). Strikingly, the first two to three TM of the class C protein are sufficient for K^+ transport (21). This function has not been well characterized.

TetA protein topology and functional relationships

The 12-TM model predicted by the Kyte-Doolittle algorithm for detecting transmembrane domains was confirmed by initial topology studies of TetA of class B (13) and class C (3, 65).

More recent studies with class B using cysteine scanning mutagenesis, cross-linking, and suppressors of inactivating mutations have provided additional information about the precise ends of the TM, their locations with respect to each other, and their amphipathic "sidedness." Specifically, in a mammoth undertaking by the Yamaguchi group over the years, each of the 401 residues of TetA(B) (except for the first Met) was singly mutated to Cys and studied in numerous ways. The reactivity of the Cys residues with sulfhydral reagents, which requires a water molecule, yielded exquisite detail about the structure and topology of the protein, identifying the boundaries of the transmembrane helices and exposure of each residue to aqueous or hydrophobic regions (100). Four TM (3, 6, 9, and 12) had no reactive residues and hence were possibly embedded completely in a nonpolar environment and were presumed to sit at four "outer corners" of the protein. Four other TM (2, 5, 8, and 11) had one vertical water-exposed "stripe" along the entire length of the TM; these probably packed internally to provide an aqueous channel that could span the membrane bilayer. The last four internal TM crossed the membrane as two sets of scissor-like pairs: TM1 and TM7 had an aqueous stripe only for the half of the TM that was nearest the periplasm, while TM4 and TM10 had an aqueous side only in the half nearest the cytoplasm. Addition of a membrane-impermeable sulfhydral reagent on the cytoplasmic or periplasmic side of the membrane defined a "permeability barrier" in the transporter about halfway across the membrane, involving the eight TM having aqueous sides (100) (Fig. 2).

Creation of a set of double Cys mutants, in which each mutant had a Cys in two different periplasmic loops or two different cytoplasmic loops, allowed interhelix Cys-Cys crosslinking studies, defining neighboring regions (44, 101). The location of secondary "suppressor" mutations (which reverse the inactivating effect of primary mutations) sometimes also indicated adjacent regions. For example, suppressors of a mutation in TM8 were found only in TM5, 8, 11, or adjacent cytoplasmic loops (described in reference 89, according to updated topology of reference 100).

A model of TetA(B) resulted from these studies (100), in which the gross tertiary structure of TetA has some similarity to that recently determined by X-ray crystallography for the glycerol phosphate transporter GlpT (31) and the lactose permease LacY (1), although differences have been noted (106).

TetA protein purification

We constructed a variety of TetA(B) fusions for purification, including TetA-col-LacZ where col was a collagenase cleavage site, and fusions to maltose binding protein, glutathione-S-transferase, and chitin binding domain with an intervening self-cleaving intein motif. All were unsatisfactory due to breakdown products, low yield, and/or binding of TetA cleavage or breakdown products to the full length protein. A Tet-6H fusion finally proved satisfactory. The protein was solubilized from membranes with dodecylmaltoside and purified by nickel chelation chromatography (2). The purified protein had the circular dichroism spectrum expected from its predicted 68% α helical content, suggesting it has a native form (2). Although we reported a substrate-dependent shift in Tet-6H fluorescence indicating substrate binding (2), this later proved to have been an artifact caused by the interaction of the antibiotic with detergent. However, that the Fenton reaction (see below) cleaved the purified protein somewhat suggested that it binds substrate.

Structure of TetA(B) based on the 17 Å resolution of two-dimensional (2D) crystals of Tet-6H

2D crystallization was done with the Tet-6H fusion protein, which was overexpressed in *E. coli* and purified as described (2), with a subsequent strong anion exchange chromatography step. Tet-6H was then reconstituted into a lipid bilayer to form protein arrays in two dimensions, which were negatively stained and examined by electron microscopy. Two different crystal forms (I and II) were generated (124). Type I was vesicular and tubular and smaller (~2 μm). Type II was observed more frequently and gave large sheet-like crystals (~10 μm). A projection map calculated with all Fourier terms to 17 Å resolution (after correction for lattice distortions [27]) from type II crystals is shown in Fig. 3.

The crystal has p321 symmetry: one unit cell encloses two protein complexes of threefold symmetry. This packing arrangement strongly suggests that Tet-6H forms trimers when reconstituted in a phospholipid environment. The two complexes per unit cell are also related by an in-plane twofold axis, meaning that one trimeric complex in the unit cell is "upside-down" in the bilayer with respect to the other trimer, which is typical for 2D crystals grown from

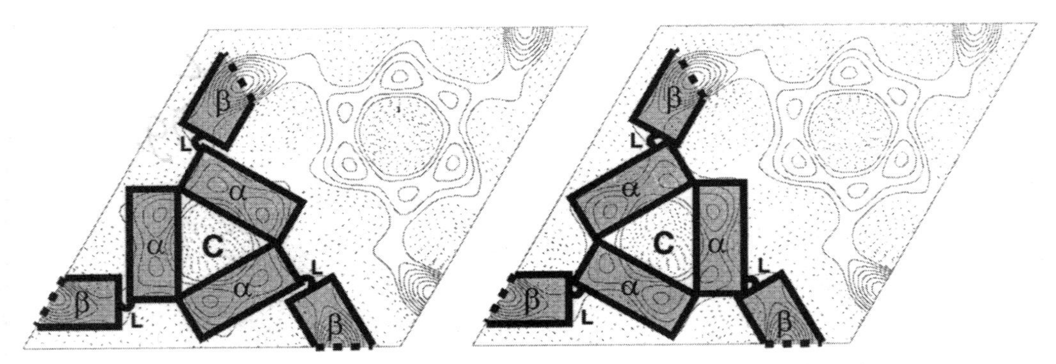

Figure 3. Projection map of 2D crystals of Tet-6H. The possible locations of the α and β halves of Tet-6H are superposed in two different trimer arrangements, each constrained to have α-α interactions. The page represents the plane of the membrane. Reprinted from Molecular Microbiology (124) with permission of Blackwell Publishing.

isotropic solution (27). The outlined envelope of one threefold complex has a triangular shape with a perimeter of approximately 105 Å and a triangular center that formed a stain-filled depression of ~35 Å diameter (Fig. 3) (124). These dimensions are consistent with those measured for single trimeric particles found outside of crystals. The overlay in Fig. 3 is based on our findings that TetA(B)-TetA(B) oligomerization interactions occurred between α domains only (54). Recalling that the N-terminal half of TetA(B) has been termed "α" and the C-terminal half "β," the central ring of density thus may represent α domains of protein involved in forming intermolecular contacts, while the arms stretching out from the central ring may represent the β domains (124). Our attempts to produce Tet-6H crystals for 3D X-ray diffraction analysis have as of this writing yielded crystals, but with poor internal order.

The two halves of the TetA protein work specifically together

The intracistronic complementation results (9) suggested that the N-terminal and C-terminal halves of the protein might function somewhat independently. To test this possibility, chimeric genes were constructed to express hybrid Tet products whose α and β halves came from different classes (A, B, or C). Although resistance was retained by hybrid Tet(A/C), where the halves came from closely related determinants, it was lost in hybrids Tet(B/C) and Tet(C/B), derived from more distant relatives (83). This requirement for specific interactions between the two halves of the protein suggests that the N- and C-terminal domains have coevolved in each member of the Tet family. The N-terminal half of TetA(B) by itself appeared to be much more stable than the C-terminal half, while the C/B hybrid was more stable than the B/C hybrid (54).

The nonconserved cytoplasmic interdomain loop of TetA has a role in resistance

The N- and C-terminal halves of all the TetA proteins in group 1 are linked by a large cytoplasmic region, designated the interdomain loop, from amino acid 179 to 213 for TetA(B) (Fig. 2). This is the longest of all the loops that connect helices. Even though it is not conserved among the different classes (Fig. 1), it has recently been shown to have a role in protein function. Substitutions of Cys within the interdomain loop of TetA(B) reduce tetracycline resistance 8-32-fold at Asp190, Glu192, Ser201, and Met210 (100). Certain substitutions within this loop also alter substrate specificity (87a). An alteration in a block of three adjacent amino acids in the interdomain loop of the class A protein changes its substrate specificity, increasing the efflux of minocycline and glycylcyclines (105) (Fig. 2). The insertion of four residues into the interdomain loop of TetA(C) eliminates protein function (33), and a Ser202Phe substitution (i.e., Ser 202 replaced by Phe) in this loop leads to a 12-fold decrease in tetracycline resistance (88). Second site suppressor mutations of Ser202Phe of TetA(C) were found at Leu11Phe (TM1), Ala213Thr (interdomain loop), and Ala270Val (cytoplasmic loop 8-9) (Fig. 2) and may indicate a direct physical interaction between Ser202 and the suppressing sites (87). Altogether, these results indicate that the interdomain loop is more than a simple linker between the two halves.

Multimerization of TetA

There are a number of results suggesting that TetA forms a multimer. In 1982 we reported that mutations in the first half of the TetA(B) protein complement (weakly) those in the second half (10), as described above. Subsequently we showed that a wild-type half protein could complement a point mutant located in the corresponding half of an otherwise wild-type full-length protein (9). We had also found that C/B and B/C hybrid proteins, each inactive, complement for resistance when coexpressed in the same cell (83). The unstable B/C hybrid protein is stabilized by the C/B protein (54). A dimer of TetA(B) is often seen in SDS-PAGE gels. Such findings led to the idea that TetA might function as a dimer (or higher multimer). Two individual monomers could be cross-linked into dimers in vivo (43), although similar cross-linking occurs for LacY (14) even though this protein is almost certainly a monomer.

On the other hand, more recent efforts in the Levy laboratory have failed to show other features expected if TetA(B) functioned as a multimer. Mutant proteins do not reduce the resistance caused by wild-type protein in vivo, as would be expected by "negative dominance" from faulty mixed dimers, nor does TetA act as a dimerization domain when fused to the DNA-binding region of lambda repressor (54). Biochemical studies using membrane extracts show that, although α-α interactions do occur between different monomers of TetA, the expected intermolecular α-β interactions do not (54). The 2D structure of TetA(B) reveals trimers (124), but this may reflect packing requirements rather than biology since trimers are also seen in 2D crystals of LacY (125), which is believed to be a monomer. AcrAB crystallizes in 3D as a trimer (67), but it represents a completely different family of proteins. Therefore, it is possible that TetA functions normally as a monomer but also exists in low amounts as a multimer, which could account for the low levels of complementation usually seen. Alternatively, the low levels of complementation could result from interactions between small amounts of N-terminal and C-terminal wild-type half-proteins formed by degradation, since we know that TetA activity can be obtained by coexpression of both half wild-type protein domains (84, 119). Other 12-TM members of the "major facilitator superfamily" (MFS) of transporters seem to be monomers, such as LacY (1, 8) and GlpT (31).

Mutational analysis to study TetA function: overview

From other transporters, we suspect that TetA undergoes a conformational change that is dependent on binding of tetracycline and/or a proton and that results in pick-up of tetracycline from the inside of the cell followed by release to the outside in exchange for an incoming proton. TetA is a member of the MFS (see above) and, like most members of this superfamily (75), has a motif A in cytoplasmic loop 2-3 (Fig. 1). Antiporters also share motif C in TM 5 (75) (Fig. 1). The 13 TetA proteins of class 1 have additional regions of high identity, which might be specific for tetracycline transport (Fig. 1). This information has been used together with mutagenesis to find which regions of the TetA protein might be involved in substrate binding, proton binding, and conformational changes.

Cysteine scanning mutagenesis at every position of TetA(B) by the Yamaguchi group (see "TetA protein topology and functional relationships" above) showed that there are only 17 residues in TetA of class B where replacement by Cys severely reduces activity without altering protein stability: Pro24, Asp66, Arg101, Gly105, Gly111, Asp120, Arg127, Gly139, Gly141, Gly150, Gly247, Gly254, Asp285, Gly336, Gly346, Pro350, and Tyr357 (100). Of these, no other substitutions are allowed at Asp66 (cytoplasmic loop 2-3) (112, 115, 120), Arg101 (TM4) (40), or Gly247 (TM8) (88, 89, 100), indicating a primary importance of these positions. Arg127 (cytoplasmic loop 4-5) can be replaced only by Lys (32, 40), and Asp285 (TM9) can be replaced only by Glu (42, 111), showing the need for conservation of charge at Arg127 (+) and Asp285 (−). At the remaining 12 sites, other residues are permitted or no detailed studies have been done. Other site-directed mutageneses at specific (usually conserved) amino acids revealed other important residues, as did random chemical mutagenesis (61) and site-directed mutagenesis (62) of TetA(C). For convenience, the following discussion on the effect of specific mutations on function has been divided into three categories, substrate binding, conformational change, and proton coupling. Presumably these could overlap.

Residues involved in substrate binding by TetA

Candidate residues for substrate binding are those where a substitution causes a change in substrate affinity or specificity. Such sites are located in both halves of the protein (Fig. 2). In TetA(B) they include Asp15 (TM 1) (60), Gln54 (TM2) (110), Ser77 (TM3) (116), Asp84 (TM3) (111), Gly111 (TM4) (105), Asp120 (cytoplasmic loop 4-5) (93), Trp231 (TM7) (22), Leu253 (TM8) (105), His257 (TM8) (109), and Gln261 (TM8) (110). Also included are Leu308 (TM10) (22) and the Ala220Glu/Asp285Asn pair (TM 7/9) (114). The interdomain loop also has residues involved in specificity, as mentioned above in the section on that loop.

Substrate-dependent cleavage of TetA(B) has also given information about the binding site. Fe^{2+} (and Co^{2+}) ions under certain conditions can break covalent bonds in what is known as the Fenton reaction. Both of these divalent cations are chelated by tetracycline, and as such are able to specifically cut the backbone of the class B and closely related class D Tet-6H fusion proteins at one major site, Gln225, which is therefore presumably near the tetracycline binding site (52). Gln225 is near the center of TM7 close to the thin permeability barrier defined by Yamaguchi et al. (100) and in the second half of the protein (Fig. 2). No cleavage is found for the less-related protein of class C, possibly due to protection of the backbone target by a side chain (52). The C/B hybrid protein (see above), which contains the α domain from class C and the β domain from class B, is cut as though it were the class B protein, not the class C. There is a second minor site of cleavage, possibly in TM10 in all three proteins (52). In spite of the cut site in TM7, and maybe in TM10, neither TM has residues critical for resistance (100), although mutations at Trp231 (at the periplasmic end of TM7) and Leu308 (TM10) do affect substrate specificity (22) and so may have a role in modulating substrate binding. Since Fenton cleavage can occur up to 12 Å from the divalent iron cation chelated by tetracycline (52), the "true" binding region may involve a TM or loop adjacent to TM7 (TM10).

Protection by tetracycline of (mutant) Cys residues from attack by N-ethylmaleimide (NEM) is another way of identifying the substrate binding site. Only eight residues in the entire protein were protected by tetracycline (100). Of these residues, six were in the first half of the protein, most being near the permeability barrier crossing the middle of the protein more or less in the plane of the membrane, where a substrate might reside in the absence of energy. It was therefore suggested that the substrate binding site might be largely in the first half of the protein (100). Alternatively, the protection of the six Cys residues in the first half of the protein by tetracycline may have been caused by their occlusion via a conformational change resulting from binding in the second half. If TetA exists as a multimer, binding to one monomer might even cause cleavage in its partner(s). All in all, the conclusion from these studies is that we are not sure where the substrate binds to TetA.

We note that the recent structures of LacY and GlpT suggest that their substrate binding sites lie at the interface between the N- and C-terminal domains of the transporters and involve residues of both halves (1, 31). Whether TetA might follow this paradigm is unclear, since a model of TetA based upon the crystal structures of LacY and GlpT differs somewhat from that predicted by the Yamaguchi group (106).

Alterations of the tetracycline molecule which prevent it from binding to TetA occur almost exclusively in the positions 10, 11, and 12, which change the "bottom" of the molecule so that Mg^{2+} cannot be chelated (69). This is consistent with the fact that it is the chelated form of tetracycline that is transported by TetA (see above). Many modifications of side groups along the "top" either have no effect on binding or may even enhance it (69). Binding of tetracycline to the Tet repressor suggests possibilities for binding to the transporter TetA, since the binding site of the TetA likely has a hydrophilic component. The repressor uses His and Glu to bind to the Mg^{2+} region (bottom) of the tetracycline-Mg^{2+} cation complex, while various hydrophobic groups plus Arg bind to the aromatic D ring (29). Residues of TetA(B) within TM and similar to His or Glu include Asp15, Asp84, His257, and Asp285. However, Asp15 can be replaced by other residues without complete inactivation (60, 111), as can Asp84 (see 100). His257 is not present in all of the TetA proteins (Fig. 1); moreover, residual activity occurred in some mutants (117). Asp285 is the most critical negatively charged residue within a TM. Its replacement led to no activity unless the replacement was Glu (111). In TetA(C), an inactivating replacement of this residue by Asn was obtained by random mutagenesis (61). In TetA(B), Asp285Asn (middle of TM9) can be partly suppressed by Ala220Glu (middle of TM7) (94, 114), suggesting the possible importance of a carboxylic acid in this region, perhaps projecting locally into the putative aqueous passageway of which TM9 otherwise plays little part. In the quite different group 2 proteins, Tet(K) (17) and Tet(L) (36), a negative charge is required within TM1, TM5, and TM13 (see below), where TM13 probably corresponds to TM10 of class B.

Residues of TetA involved in conformational changes

These residues we define as those where primary mutations are suppressed by secondary ones at positions not likely to be in direct contact with the primary site, so that a conformational change is involved. They also include residues where Gly appears to allow a turn. For example, mutations in cytoplasmic loop 2-3 in or near motif A of TetA(B) can be suppressed by ones in the middle of TM6 (95) or in periplasmic loop 1-2, on the other side of the membrane (41). The relative activities of mutant TetA(B) with different substitutions at Gly62 and Gly69 in loop 2-3 suggested that these two Gly residues facilitate turns (120), and possibly a turn also occurs at Gly332 (cytoplasmic loop 10-11) (37). In TetA(C), a mutation in cytoplasmic loop 2-3 was suppressed by only two mutations, both in periplasmic loop 9-10, on the other side of the membrane (62). Mutations in TetA(C) of Gly in the conserved GlyPro dipeptide of the "antiporter" motif C cause loss of activity; modeling indicated that GlyPro is possibly involved in helix bending (107). Therefore, motifs A and C may be involved in conformational changes common to many transporters.

Residues of TetA involved in proton transport

Amino acids which act to relay protons presumably are able to be protonated or to form hydrogen bonds, possible with water. Mutations of His257 in TM8 of TetA(B) permit downhill tetracycline transport in vesicles but no proton antiport, suggesting a role for this residue in proton exchange (109, 117). However, the importance of His257 becomes unclear, considering that some substitutions at His257 confer significant tetracycline resistance and three group 1 proteins do not have this histidine (Fig. 1). The His257-substitution proteins may have become inactivated during preparation of membrane vesicles, as has been occasionally seen for other mutant TetA(B) proteins. In TetA(C), those mutant proteins which functioned better with a larger transmembrane pH gradient, and which therefore might be defective in proton transport, had substitutions scattered throughout the protein (61). Recently a study on LmrP, a multidrug transporter which is also in the major facilitator superfamily, has shown that the equivalent of Asp66 in motif A is involved in proton coupling but not substrate binding (51).

Other interesting residues in TetA

Truncation of the protein at amino acid 388 (terminal cytoplasmic tail) still permits activity, but truncation at amino acid 382 (TM12) does not (90). An inactivating loss of positive charge at Arg70 (cytoplasmic loop 2-3) is compensated for by a loss of negative charge at Asp120 (cytoplasmic loop 4-5), suggesting that a bridge occurs there which could be either ionic or nonionic; these two cytoplasmic loops can be crosslinked via Cys mutations at residues 70 and 120 (93).

Regulation of *tet* transcription in group 1

Tetracycline resistance is inducible upon addition of a subinhibitory amount of tetracycline. Expression of the *tetA* gene and the divergently transcribed *tetR* (repressor) gene are both repressed by the TetR repressor, which binds to a palindromic operator in each of the promoters and blocks initiation of transcription. The structures for the repressor bound to tetracycline or to DNA have been determined (29, 73). The promoters for *tetA* and *tetR* are quite strong, and *tetA* is repressed at several-fold lower concentrations of repressor than is *tetR* (5). At about 1 nM tetracycline, a tetracycline-divalent cation complex interacts with TetR, releasing this dimeric protein from the DNA and allowing transcription of both *tetA* and *tetR* (73). The system is evidently poised so that the concentration of tetracycline in the cell remains below the 1 μM estimated as the amount needed to inhibit protein synthesis, while the amount of TetA is kept in check to avoid overproduction toxicity. One might then expect that the TetA efflux system should function best below 1 μM substrate, yet in vitro measurements using inside-out membrane vesicles show the K_m for tetracycline is about 20 μM; the explanation is not apparent.

Efflux Mechanism, Group 2

Tet(K) and Tet(L) comprise the principal members of group 2 tetracycline efflux proteins and share 59% identity. The location of their 14 TM was elucidated by computational and genetic analyses (20). The Tet K and Tet L determinants are predominately found in gram-positive species, typically on conjugative plasmids and/or on the chromosome (80). A newly annotated 14-TM protein Tet(38) also belongs in this group because of a 25% identity with both Tet(K) and Tet(L) proteins (104a). It was suggested that, like the 12-TM pumps, the 14-TM tetracycline pumps arose from a duplication of an ancestral 6-TM-encoding gene; however, for the 14-TM pumps, an additional two TM (7 and 8) from another source were later inserted between the two 6-TM domains (75).

Like the group 1 proteins, both Tet(K) and Tet(L) proteins catalyze the exchange of extracellular H^+ for a cytoplasmic complex of tetracycline-divalent metal ion (24, 58, 118). However, this exchange is electrogenic, which means that more than one proton enters per exiting tetracycline-metal complex (23, 24).

Interestingly, Tet(K) and Tet(L) proteins are also capable of two other transport activities. First, they can catalyze electrogenic exchange of a single Na^+ and K^+ for more than one proton. In *Bacillus subtilis,* Tet(L) plays a major role in Na^+ resistance and in pH homeostasis (7, 24). Second, each of the "cytoplasmic substrates," Na^+, K^+, and the tetracycline-metal complex, can be exchanged in an apparently electrogenic manner for external K^+ instead of H^+ (23, 34). The use of K^+ as an external coupling ion may contribute not only to the organism's K^+ uptake capacity but also to its ability to exclude Na^+ and tetracycline at elevated pH values (108).

In both proteins, site-directed mutagenesis at three conserved transmembrane glutamate residues (Glu30-TM1, Glu152-TM5, and Glu397-TM13) has demonstrated the necessity of a negative charge at these positions for tetracycline-metal/H^+ activity (15, 36). In Tet(L), studies focusing on Glu397 showed that its negative charge is also required for function of the Na^+, K^+ catalytic modes (36). A negative charge is needed at Asp318 (cytoplasmic loop10-11) for tetracycline-metal/H^+ transport by Tet(K) (16) but not Tet(L) (36). In addition, transport of tetracycline-metal/H^+ by Tet(K) activity is greatly reduced or abolished by substitutions at Pro156 and Gly159. In Tet(L), Gly70 and Asp74, both in cytoplasmic loop 2-3, and Lys366 in TM12 are important for tetracycline efflux (36). In contrast, Asp200Cys and Arg110Cys lower only the Tet(L) tetracycline resistance but not the Na^+ resistance or K^+ uptake (15). The deletion of both TM7 and TM8 from Tet(L) inactives tetracycline efflux but not the monovalent cation transports (35).

In conclusion, these data suggest that a common set of amino acids is required for all three transport activities of Tet(K) and Tet(L), while additional residues are uniquely critical for tetracycline-metal/H^+ antiport.

The Tet(L) protein associates in vivo as a stable dimer which persists in vitro in a number of detergents and over a wide pH range (86). However, unlike the case for TetA(B), coexpression in the same cell of two different Tet(L) mutant proteins, each presumably mutated in a different half, does not restore tetracycline efflux, that is, there is no complementation suggestive of functional dimerization (86). Therefore it was speculated that Tet(L) can transport as a monomer, but that a dimer may permit cooperativity among the several Tet(L) substrates (86).

The *tet*(K) and *tet*(L) genes are each inducible by tetracycline and are probably regulated by translation attenuation and translation reinitiation, respectively. For a review of translation attenuation, see reference 49. For *tet*(K) on plasmid pT181, the transcriptional start site is 196 bp upstream of the translational start codon. In the 5′ part of the mRNA are two ribosome-binding sites (RBS) (39); see also reference 66. The region of the mRNA containing these RBS can theoretically form two mutually exclusive stem-loop structures. One contains RBS2 and so occludes it, preventing translational initiation. According to the translation attenuation model, in the absence of tetracycline, the ribosome would bind to RBS1, synthesizing a leader peptide until reaching the stem-loop structure containing RBS2, where translation would terminate ("attenuate"). When tetracycline is present, the ribosomes reading from RBS1 would stall and thereby allow the formation of an alternate, more stable stem-loop structure that does not mask RBS2 and permits the translation of the efflux pump from RBS2.

The chromosomal Tet L determinant is thought to be regulated by a translation reinitiation mechanism (99). In this case, the presence of tetracycline is proposed to induce stalling of ribosomes during the early translation from RBS1 of a leader peptide 20 amino acids in length, permitting the formation of a critical stem-loop structure (−11.9 kcal/mol). This stem-loop structure is proposed to then guide the ribosome to RBS2, thereby reinitiating translation, now of the mRNA for the pump itself (99).

Efflux Mechanism, Other Groups

Various other proteins unrelated to those of groups 1 or 2 or to each other also exclusively efflux tetracycline (57). The class P determinant from *Clostridium perfringens* encodes both an efflux protein [TetA(P)] and a ribosomal protection protein [TetB(P)] on overlapping genes (92). TetA(P) has little homology to other Tet proteins, and characterization of TetA(P) by mutagenesis has been difficult because of the low intrinsic activity and the instability of many mutant proteins. Nevertheless, it was found that a Glu or Asp within putative TM2 and TM4, and a Glu within putative TM3, are essential for efflux (4, 38). The efflux protein has a modified motif A in cytoplasmic loop 2-3 where Pro61 and Arg71 are essential (4).

We resequenced *otrB* from *Streptomyces rimosus* and found it to have 14 TM, not 7.5 TM (55). Tet(V) from *Mycobacterium smegmatis* has at least 10 putative TM, and active efflux was demonstrated (11). The protein encoded by the Tet 35 determinant from *Vibrio harveyi* has no homology to any tetracycline efflux protein but clearly causes active efflux (103). It has nine putative TM, gives only low level resistance, and has low homology to Na^+/H^+ antiporters. OtrC from *S. rimosus* (unpublished; GenBank AY509111) is a two-protein putative ABC transporter, with the putative ATP-binding protein and putative 6-TM domain on separate polypeptides. The only antibiotic to which OtrC is known to confer resistance is tetracycline (79). TetAB from *Corynebacterium striatum* is a two-part putative ABC transporter where each protein has both a 5-6 putative TM and an ATP-binding cassette and both proteins are required for resistance (102). This determinant also confers oxacillin resistance, so TetAB is actually a multidrug transporter. Tetracycline efflux is also mediated by a variety of other multidrug transporters such as AcrAB and MexAB (71) and LmrP (51).

DEGRADATIVE INACTIVATION OF TETRACYCLINE

Degradation is not a commonly found mechanism for tetracycline resistance. Breakdown of tetracyclines by the ascomycete *Xylaria digitata* was reported in 1962 (64). More recently, three degrading proteins from bacteria have been described, TetX (74, 97, 98), Tet37 (12), and an unnamed determinant

(68), from *Bacteroides fragilis,* total oral flora, and *Pseudomonas aeruginosa,* respectively. The first two are not related by sequence, but both were reported to be oxidoreductases that require NAD(P) to alter and inactivate tetracycline (12, 96). The TetX protein cannot confer resistance in the anaerobic *Bacteroides* organism but can in well-aerated *E. coli.* More recently, TetX has been shown to be a FAD-containing monooxygenase that causes hydroxylation at position 11a of tetracyclines with attendant breakdown of the drug; TetX is related to uncharacterized open reading frames in aerobic soil microorganisms (123). The determinant from *P. aeruginosa* has not yet been sequenced but has a different substrate specificity from TetX (68).

RIBOSOMAL 16S RNA MUTATIONS CAUSING TETRACYCLINE RESISTANCE

In 1998 a new resistance mechanism was discovered in clinical isolates of *Propionibacterium acnes.* It involved a single base mutation in the 16S rRNA at a position equivalent to the *E. coli* 16S rRNA base 1058 (82). Since genetic manipulation in these cutaneous bacteria was not possible, the mutation was recreated in the cloned ribosomal *rrnB* operon of *E. coli* and expressed on a high copy number plasmid. The *E. coli* strain with the plasmid showed increased tetracycline resistance, albeit much less than that in *P. acnes,* possibly related to the presence of the seven chromosomal copies of the *E. coli* rRNA gene (82). Nucleotide 1058 is located in helix 34 of the 16S rRNA adjacent to the primary tetracycline binding site that leads to inhibition of protein synthesis (6, 76). This suggested that a disturbance in the base pairing of helix 34 interfered with the binding of tetracycline.

Recently, independent clinical isolates of highly tetracycline-resistant strains of *Helicobacter pylori* were found to have a triple base pair substitution AGA to TTC at bases 965-967 (*E. coli* coordinates) on both copies of the 16s rRNA gene (19, 104). These mutations are in helix 31 of the 16S rRNA, which also overlaps the primary tetracycline binding site (76). Site-directed mutagenesis generating all combinations of single and double substitutions in isogenic *H. pylori* strains confirmed that the effect on resistance was proportional to the severity of the change (18).

H. pylori strains engineered to have a single 16S rRNA were passaged several times on tetracycline in the laboratory to obtain resistant mutants in which the resistance was attributed to a deletion of base G942 of the 16S rRNA gene (104). No mutations in helix 31 or helix 34 were identified by this method.

TETRACYCLINE RESISTANCE BY UNKNOWN MECHANISMS

Tet 34 is a unique determinant specifiying resistance to oxytetracycline and was isolated from *Vibrio* from cultivated fish (72). The sequence indicates a relationship to xanthine-guanine phosphoribosyltransferase, a Mg^{2+}-requiring protein which converts salvaged purines into XMP and GMP. Tetracycline resistance by Tet 34 also requires Mg^{2+}, but the mechanism is unknown (72). Tet U, from *Enterococcus,* (77), is another unusual determinant and has no apparent sequence homology to any other proteins.

IN APPRECIATION

F.S.: I joined the Levy group on 1 February 2002 as a postdoctoral fellow just after finishing my thesis at the Center for Protein Engineering, Liege, Belgium. Since the first phone call, which was in fact my interview for the job, until now, nearly 3 years later, all my expectations about Stuart and the tetracycline project have been reached. Stuart is very enthusiastic about the negative and positive results that I bring to him. After each Tet group meeting or seminar, I walk out revived and full of energy, ready to solve the unsolved mystery of the Tet proteins. This book is my opportunity to tell him, Merci Stuart pour tous tes encouragements. J'aimerai rajouter que pour une première expérience de postdoc je ne pouvais pas rêver mieux. Encore MERCI.

M. A.-R.: I joined the TetA protein group of the Levy lab as a postdoc in July 1992. I thought I was lucky to join the group at the time when it had just moved into a much nicer, more spacious place with a better view of the business district of Boston. This was the seventh floor of the South Cove building, better known as the "seventh heaven." A year later, the laboratory was inaugurated as "The Center for Adaptation Genetics and Drug Resistance," with Stuart as director. That event had a little fanfare where a hired musician was playing classical music in the hallway and there was plenty of food for all the guests.

When I first came, I had just finished graduate school in the Department of Chemistry at the University of Connecticut, and I had high hopes that I could finish the task of crystallizing the TetA protein in no time. I thought that if I just applied all the science and techniques I had learned when I successfully crystallized two membrane proteins for my dissertation, then it would be no sweat. But months slowly have turned into years, and high-quality X-ray diffracting 3D crystals are still unattainable. Sometimes it can get so frustrating, but Stuart, the optimistic person with happy disposition that he is, can always make us feel that we are still on the right track and that we can do this. On Stuart's birthday, I always want to make myself inconspicuous because he always includes me with his birthday cake because it's my birthday too (not necessarily the same year though!). But that's just how he is, always making you feel that you belong to the group. Thank you Stuart for everything.

L. M.: I have worked with Stuart since 1972, shortly after he set up his laboratory at Tufts. The group totaled three—counting Stuart. Now there are about 15. Before he married Cecile, he often used to be in his office until 9 pm. The office oddly had a tiny lab bench. This is where he did what I think was his last benchwork at Tufts, as an offshoot of the tetracycline resistance project. He selected spontaneous (chromosomal) mutants of *Escherichia coli* on plates containing tetracycline. Like cells containing certain R-factors, these mutants turned out to be resistant to other antibiotics (though at lower levels). They led to Tony George's identification of the mar locus and eventually to a whole new area of research (see section III).

Stuart's manner of directing is to discuss and decide on the general plan for a project, and then to let you go work on it. He doesn't hover. On the other hand, when I want advice, I just stop by his office for what is almost always a fun, fruitful discussion of results and ideas. Unless of course he is away, which can happen—he might be anywhere in the world, giving a scientific presentation or cruising with his brother Jay on a ship full of nothing but sets of twins. Stuart likes adventure and challenges. He worked in laboratories in Japan, France, and Italy and can speak several languages, he started an organization to lessen overuse of antibiotics and also a pharmaceutical company, and he is the father of three children—to mention a nano-percent of his adventures. Stuart has been a stimulating and generous colleague. My association with him and working in his group has been very satisfying.

Acknowledgments. Work in the S.B.L. laboratory was supported in part by grants from the National Institutes of Health. F.S. was supported by a Fellowship of the Belgian American Educational Foundation.

REFERENCES

1. **Abramson, J., I. Smirnova, V. Kasho, G. Verner, H. R. Kaback, and S. Iwata.** 2003. Structure and mechanism of the lactose permease of *Escherichia coli*. *Science* **301:**610–615.
2. **Aldema, M. L., L. M. McMurry, A. R. Walmsley, and S. B. Levy.** 1996. Purification of the Tn*10*-specified tetracycline efflux antiporter TetA in a native state as a polyhistidine fusion protein. *Mol. Microbiol.* **19:**187–195.
3. **Allard, J. D., and K. P. Bertrand.** 1992. Membrane topology of the pBR322 tetracycline resistance protein: TetA-PhoA gene fusions and implications for the mechanism of TetA membrane insertion. *J. Biol. Chem.* **267:**17809–17819.
4. **Bannam, T. L., P. A. Johanesen, C. L. Salvado, S. J. Pidot, K. A. Farrow, and J. I. Rood.** 2004. The *Clostridium perfringens* TetA(P) efflux protein contains a functional variant of the Motif A region found in major facilitator superfamily transport proteins. *Microbiology* **150:**127–134.
5. **Bertrand, K. P., K. Postle, L. V. Wray, Jr., and W. S. Reznikoff.** 1984. Construction of a single-copy promoter vector and its use in analysis of regulation of the transposon Tn*10* tetracycline resistance determinant. *J. Bacteriol.* **158:**910–919.
6. **Brodersen, D. E., W. M. Clemons, Jr., A. P. Carter, R. J. Morgan-Warren, B. T. Wimberly, and V. Ramakrishnan.** 2000. The structural basis for the action of the antibiotics tetracycline, pactamycin, and hygromycin B on the 30S ribosomal subunit. *Cell* **103:**1143–1154.
7. **Cheng, J., A. A. Guffanti, and T. A. Krulwich.** 1994. The chromosomal tetracycline resistance locus of *Bacillus subtilis* encodes a Na^+/H^+ antiporter that is physiologically important at elevated pH. *J. Biol. Chem.* **269:**27365–27371.
8. **Costello, M. J., J. Escaig, K. Matsushita, P. V. Viitanen, D. R. Menick, and H. R. Kaback.** 1987. Purified *lac* permease and cytochrome *o* oxidase are functional as monomers. *J. Biol. Chem.* **262:**17072–17082.
9. **Curiale, M., L. M. McMurry, and S. B. Levy.** 1984. Intracistronic complementation of the tetracycline resistance membrane protein specified by Tn*10*. *J. Bacteriol.* **157:**211–217.
10. **Curiale, M. S., and S. B. Levy.** 1982. Two complementation groups mediate tetracycline resistance determined by Tn*10*. *J. Bacteriol.* **151:**209–215.
11. **De Rossi, E., M. C. Blokpoel, R. Cantoni, M. Branzoni, G. Riccardi, D. B. Young, K. A. De Smet, and O. Ciferri.** 1998. Molecular cloning and functional analysis of a novel tetracycline resistance determinant, tet(V), from *Mycobacterium smegmatis*. *Antimicrob. Agents Chemother.* **42:**1931–1937.
12. **Diaz-Torres, M. L., R. McNab, D. A. Spratt, A. Villedieu, N. Hunt, M. Wilson, and P. Mullany.** 2003. Novel tetracycline resistance determinant from the oral metagenome. *Antimicrob. Agents Chemother.* **47:**1430–1432.
13. **Eckert, B., and C. F. Beck.** 1989. Topology of the transposon Tn*10*-encoded tetracycline resistance protein within the inner membrane of *Escherichia coli*. *J. Biol. Chem.* **264:**11663–11670.
14. **Ermolova, N., L. Guan, and H. R. Kaback.** 2003. Intermolecular thiol cross-linking via loops in the lactose permease of *Escherichia coli*. *Proc. Natl. Acad. Sci. USA* **100:**10187–10192.
15. **Fujihira, E., T. Kimura, Y. Shiina, and A. Yamaguchi.** 1996. Transmembrane glutamic acid residues play essential roles in the metal-tetracycline/H^+ antiporter of *Staphylococcus aureus*. *FEBS Lett.* **391:**243–246.
16. **Fujihira, E., T. Kimura, and A. Yamaguchi.** 1997. Roles of acidic residues in the hydrophilic loop regions of metal-tetracycline/H^+ antiporter Tet(K) of *Staphylococcus aureus*. *FEBS Lett.* **419:**211–214.
17. **Fujihira, E., N. Tamura, and A. Yamaguchi.** 2002. Membrane topology of a multidrug efflux transporter, AcrB, in *Escherichia coli*. *J. Biochem.* (*Tokyo*) **131:**145–151.
18. **Gerrits, M. M., M. Berning, A. H. Van Vliet, E. J. Kuipers, and J. G. Kusters.** 2003. Effects of 16S rRNA gene mutations on

tetracycline resistance in *Helicobacter pylori. Antimicrob. Agents Chemother.* **47**:2984–2986.

19. **Gerrits, M. M., M. R. de Zoete, N. L. Arents, E. J. Kuipers, and J. G. Kusters.** 2002. 16S rRNA mutation-mediated tetracycline resistance in *Helicobacter pylori. Antimicrob. Agents Chemother.* **46**:2996–3000.
20. **Ginn, S. L., M. H. Brown, and R. A. Skurray.** 1997. Membrane topology of the metal-tetracycline/H^+ antiporter TetA (K) from *Staphylococcus aureus. J. Bacteriol.* **179**:3786–3789.
21. **Griffith, J. K., T. Kogoma, D. L. Corvo, W. L. Anderson, and A. L. Kazim.** 1988. An N-terminal domain of the tetracycline resistance protein increases susceptibility to aminoglycosides and complements potassium uptake defects in *Escherichia coli. J. Bacteriol.* **170**:598–604.
22. **Guay, G. G., M. Tuckman, and D. M. Rothstein.** 1994. Mutations in the *tetA*(B) gene that cause a change in substrate specificity of the tetracycline efflux pump. *Antimicrob. Agents Chemother.* **38**:857–860.
23. **Guffanti, A. A., J. Cheng, and T. A. Krulwich.** 1998. Electrogenic antiport activities of the Gram-positive Tet proteins include a $Na^+(K^+)/K^+$ mode that mediates net K^+ uptake. *J. Biol. Chem.* **273**:26447–26454.
24. **Guffanti, A. A., and T. A. Krulwich.** 1995. Tetracycline/H^+ antiport and Na^+/H^+ antiport catalyzed by the *Bacillus subtilis* TetA(L) transporter expressed in *Escherichia coli. J. Bacteriol.* **177**:4557–4561.
25. **Guillaume, G., V. Ledent, W. Moens, and J. M. Collard.** 2004. Phylogeny of efflux-mediated tetracycline resistance genes and related proteins revisited. *Microb. Drug Resist.* **10**:11–26.
26. **Hansen, L. M., L. M. McMurry, S. B. Levy, and D. C. Hirsh.** 1993. A new tetracycline resistance determinant, Tet H, from *Pasteurella multocida* specifying active efflux of tetracycline. *Antimicrob. Agents Chemother.* **37**:2699–2705.
27. **Henderson, R., J. M. Baldwin, K. H. Downing, J. Lepault, and F. Zemlin.** 1986. Structure of purple membrane from *Halobacterium halobium:* recording, measurement and evaluation of electron micrographs at 3.5 angstroms resolution. *Ultramicroscopy* **19**:147–178.
28. **Higgins, D., J. Thompson, T. Gibson, J. D. Thompson, D. G. Higgins, and T. J. Gibson.** 1994. CLUSTAL W: improving the sensitivity of progressive multiple sequence alignment through sequence weighting, position-specific gap penalties and weight matrix choice. *Nucleic Acids Res.* **22**:4673–4680.
29. **Hinrichs, W., C. Kisker, M. Duevel, A. Mueller, K. Tovar, W. Hillen, and W. Saenger.** 1994. Structure of the Tet repressor-tetracycline complex and regulation of antibiotic resistance. *Science* **264**:418–420.
30. Reference deleted.
31. **Huang, Y., M. J. Lemieux, J. Song, M. Auer, and D. N. Wang.** 2003. Structure and mechanism of the glycerol-3-phosphate transporter from *Escherichia coli. Science* **301**:616–620.
32. **Iwaki, S., N. Tamura, T. Kimura-Someya, S. Nada, and A. Yamaguchi.** 2000. Cysteine-scanning mutagenesis of transmembrane segments 4 and 5 of the Tn*10*-encoded metal-tetracycline/H^+ antiporter reveals a permeability barrier in the middle of a transmembrane water-filled channel. *J. Biol. Chem.* **275**: 22704–22712.
33. **Jewell, J. E., J. Orwick, J. Liu, and K. W. Miller.** 1999. Functional importance and local environments of the cysteines in the tetracycline resistance protein encoded by plasmid pBR322. *J. Bacteriol.* **181**:1689–1693.
34. **Jin, J., A. A. Guffanti, D. H. Bechhofer, and T. A. Krulwich.** 2002. Tet(L) and Tet(K) tetracycline-divalent metal/H^+ antiporters: characterization of multiple catalytic modes and a mutagenesis approach to differences in their efflux substrate and coupling ion preferences. *J. Bacteriol.* **184**:4722–4732.
35. **Jin, J., A. A. Guffanti, C. Beck, and T. A. Krulwich.** 2001. Twelve-transmembrane-segment (TMS) version (DeltaTMS VII-VIII) of the 14-TMS Tet(L) antibiotic resistance protein retains monovalent cation transport modes but lacks tetracycline efflux capacity. *J. Bacteriol.* **183**:2667–2671.
36. **Jin, J., and T. A. Krulwich.** 2002. Site-directed mutagenesis studies of selected motif and charged residues and of cysteines of the multifunctional tetracycline efflux protein Tet(L). *J. Bacteriol.* **184**:1796–1800.
37. **Kawabe, T., and A. Yamaguchi.** 1999. Transmembrane remote conformational suppression of the Gly-332 mutation of the Tn*10*-encoded metal-tetracycline/H^+ antiporter. *FEBS Lett.* **457**:169–173.
38. **Kennan, R. M., L. M. McMurry, S. B. Levy, and J. I. Rood.** 1997. Glutamate residues located within putative transmembrane helices are essential for TetA(P)-mediated tetracycline efflux. *J. Bacteriol.* **179**:7011–7015.
39. **Khan, S. A., and R. P. Novick.** 1983. Complete nucleotide sequence of pT181, a tetracycline-resistance plasmid from *Staphylococcus aureus. Plasmid.* **10**:251–259.
40. **Kimura, T., M. Nakatani, T. Kawabe, and A. Yamaguchi.** 1998. Roles of conserved arginine residues in the metal-tetracycline/H^+ antiporter of *Escherichia coli. Biochemistry* **37**: 5475–5480.
41. **Kimura, T., T. Sawai, and A. Yamaguchi.** 1997. Remote conformational effects of the Gly-62 → Leu mutation of the Tn*10*-encoded metal-tetracycline/H^+ antiporter of *Escherichia coli* and its second-site suppressor mutation. *Biochemistry* **36**: 6941–6946.
42. **Kimura, T., M. Suzuki, T. Sawai, and A. Yamaguchi.** 1996. Determination of a transmembrane segment using cysteine-scanning mutants of transposon Tn*10*-encoded metal-tetracycline/H^+ antiporter. *Biochemistry* **35**:15896–15899.
43. **Kimura-Someya, T., S. Iwaki, S. Konishi, N. Tamura, Y. Kubo, and A. Yamaguchi.** 2000. Cysteine-scanning mutagenesis around transmembrane segments 1 and 11 and their flanking loop regions of Tn*10*-encoded metal-tetracycline/H^+ antiporter. *J. Biol. Chem.* **275**:18692–18697.
44. **Kubo, Y., S. Konishi, T. Kawabe, S. Nada, and A. Yamaguchi.** 2000. Proximity of periplasmic loops in the metal-tetracycline/H^+ antiporter of *Escherichia coli* observed on site-directed chemical cross-linking. *J. Biol. Chem.* **275**:5270–5274.
45. **Levy, S. B., and L. McMurry.** 1974. Detection of an inducible membrane protein associated with R-factor-mediated tetracycline resistance. *Biochem. Biophys. Res. Comm.* **56**:1060–1068.
46. **Levy, S. B., and L. McMurry.** 1978. Plasmid-determined tetracycline resistance involves new transport systems for tetracycline. *Nature* **276**:90–92.
47. **Levy, S. B., and L. McMurry.** 1978. Probing the expression of plasmid-mediated tetracycline resistance in *Escherichia coli,* p. 177–180. *In* D. Schlessinger (ed.), *Microbiology—1978.* American Society for Microbiology, Washington, D.C.
48. **Levy, S. B., L. M. McMurry, T. M. Barbosa, V. Burdett, P. Courvalin, W. Hillen, M. C. Roberts, J. I. Rood, and D. E. Taylor.** 1999. Nomenclature for new tetracycline resistance determinants. *Antimicrob. Agents Chemother.* **43**:1523–1524.
49. **Lovett, P. S., and E. J. Rogers.** 1996. Ribosome regulation by the nascent peptide. *Microbiol. Rev.* **60**:366–385.

50. **Marshall, B., S. Morrissey, P. Flynn, and S. B. Levy.** 1986. A new tetracycline-resistance determinant, class E, isolated from Enterobacteriaceae. *Gene* **50:**111–117.
51. **Mazurkiewicz, P., G. J. Poelarends, A. J. Driessen, and W. N. Konings.** 2004. Facilitated drug influx by an energy-uncoupled secondary multidrug transporter. *J. Biol. Chem.* **279:**103–108.
52. **McMurry, L. M., M. L. Aldema-Ramos, and S. B. Levy.** 2002. Fe^{2+}-tetracycline-mediated cleavage of the Tn*10* tetracycline efflux protein TetA reveals a substrate binding site near glutamine 225 in transmembrane helix 7. *J. Bacteriol.* **184:**5113–5120.
53. **McMurry, L. M., M. Hendricks, and S. B. Levy.** 1986. Effects of toluene permeabilization and cell deenergization on tetracycline resistance in *Escherichia coli*. *Antimicrob. Agents Chemother.* **29:**681–686.
54. **McMurry, L. M., and S. B. Levy.** 1995. The NH_2-terminal half of the tetracycline efflux protein from Tn*10* contains a functional dimerization domain. *J. Biol. Chem.* **270:**22752–22757.
55. **McMurry, L. M., and S. B. Levy.** 1998. Revised sequence of OtrB (Tet347) tetracycline efflux protein from *Streptomyces rimosus*. *Antimicrob. Agents Chemother.* **42:**3050.
56. **McMurry, L. M., and S. B. Levy.** Tetracycline resistance determinants in gram-positive bacteria. *In* V. A. Fischetti, R. P. Novick, J. J. Ferretti, D. A. Portnoy, and J. I. Rood (ed.), *Gram-Positive Pathogens,* 2nd ed., in press. ASM Press, Washington, D.C.
57. **McMurry, L. M., and S. B. Levy.** 2000. Tetracycline resistance in Gram-positive bacteria, p. 660–677. *In* V. A. Fischetti, R. P. Novick, J. J. Ferretti, D. A. Portnoy, and J. I. Rood (ed.), *Gram-Positive Pathogens*. ASM Press, Washington, D.C.
58. **McMurry, L. M., B. H. Park, V. Burdett, and S. B. Levy.** 1987. Energy-dependent efflux mediated by class L (*tet*L) tetracycline resistance determinant from streptococci. *Antimicrob. Agents Chemother.* **31:**1648–1650.
59. **McMurry, L. M., R. E. Petrucci, Jr., and S. B. Levy.** 1980. Active efflux of tetracycline encoded by four genetically different tetracycline resistance determinants in *Escherichia coli*. *Proc. Natl. Acad. Sci. USA* **77:**3974–3977.
60. **McMurry, L. M., M. Stephan, and S. B. Levy.** 1992. Decreased function of the class B tetracycline efflux protein Tet with mutations at aspartate 15, a putative intramembrane residue. *J. Bacteriol.* **174:**6294–6297.
61. **McNicholas, P., I. Chopra, and D. M. Rothstein.** 1992. Genetic analysis of the *tetA*(C) gene on plasmid pBR322. *J. Bacteriol.* **174:**7926–7933.
62. **McNicholas, P., M. McGlynn, G. G. Guay, and D. M. Rothstein.** 1995. Genetic analysis suggests functional interactions between the N- and C-terminal domains of the TetA(C) efflux pump encoded by pBR322. *J. Bacteriol.* **177:**5355–5357.
63. **Mendez, B., C. Tachibana, and S. B. Levy.** 1980. Heterogeneity of tetracycline resistance determinants. *Plasmid* **3:**99–108.
64. **Meyers, E., and D. A. Smith.** 1962. Microbiological degradation of tetracyclines. *J. Bacteriol.* **84:**797–802.
65. **Miller, K. W., P. L. Konen, J. Olsen, and K. M. Ratanavanich.** 1993. Membrane protein topology determination by proteolysis of maltose binding protein fusions. *Anal. Biochem.* **215:**118–128.
66. **Mojumdar, M., and S. A. Khan.** 1988. Characterization of the tetracycline resistance gene of plasmid pT181 of *Staphylococcus aureus*. *J. Bacteriol.* **170:**5522–5528.
67. **Murakami, S., R. Nakashima, E. Yamashita, and A. Yamaguchi.** 2002. Crystal structure of bacterial multidrug efflux transporter AcrB. *Nature* **419:**587–593.
68. **Nakamura, A., M. Nakagawa, H. Yoshikoshi, S. Shoutou, K. O'Hara, and T. Sawai.** 2002. Novel enzymatically determined minocycline-resistance in a *Pseudomonas aeruginosa* clinical isolate, abstr. C1-1603. *In Interscience Conference on Antimicrobial Agents and Chemotherapy*. American Society for Microbiology.
69. **Nelson, M. L., B. H. Park, and S. B. Levy.** 1994. Molecular requirements for the inhibition of the tetracycline antiport protein and the effect of potent inhibitors on the growth of tetracycline-resistant bacteria. *J. Med. Chem.* **37:**1355–1361.
70. **Nguyen, T. T., K. Postle, and K. P. Bertrand.** 1983. Sequence homology between the tetracycline-resistance determinants of Tn*10* and pBR322. *Gene* **25:**83–92.
71. **Nikaido, H.** 1998. Multiple antibiotic resistance and efflux. *Curr. Opin. Microbiol.* **1:**516–523.
72. **Nonaka, L., and S. Suzuki.** 2002. New Mg^{2+}-dependent oxytetracycline resistance determinant tet 34 in *Vibrio* isolates from marine fish intestinal contents. *Antimicrob. Agents Chemother.* **46:**1550–1552.
73. **Orth, P., D. Schnappinger, W. Hillen, W. Saenger, and W. Hinrichs.** 2000. Structural basis of gene regulation by the tetracycline inducible Tet repressor-operator system. *Nat. Struct. Biol.* **7:**215–219.
74. **Park, B. H., and S. B. Levy.** 1988. The cryptic tetracycline resistance determinant on Tn*4400* mediates tetracycline degradation as well as tetracycline efflux. *Antimicrob. Agents Chemother.* **32:**1797–1800.
75. **Paulsen, I. T., M. H. Brown, and R. A. Skurray.** 1996. Proton-dependent multidrug efflux systems. *Microbiol. Rev.* **60:**575–608.
76. **Pioletti, M., F. Schlunzen, J. Harms, R. Zarivach, M. Gluhmann, H. Avila, A. Bashan, H. Bartels, T. Auerbach, C. Jacobi, T. Hartsch, A. Yonath, and F. Franceschi.** 2001. Crystal structures of complexes of the small ribosomal subunit with tetracycline, edeine and IF3. *EMBO J.* **20:**1829–1839.
77. **Ridenhour, M. B., H. M. Fletcher, J. E. Mortensen, and L. Daneo-Moore.** 1996. A novel tetracycline-resistant determinant, tet(U), is encoded on the plasmid pKQ10 in *Enterococcus faecium*. *Plasmid* **35:**71–80.
78. **Roberts, M. C.** 1997. Genetic mobility and distribution of tetracycline resistance determinants, p. 206–218. *In* D. J. Chadwick (ed.), *Antibiotic Resistance: Origins, Evolution, Selection, and Spread*. John Wiley and Sons, Chichester, U.K.
79. **Roberts, M. C.** 2004. Personal communication.
80. **Roberts, M. C.** 1996. Tetracycline resistance determinants: mechanisms of action, regulation of expression, genetic mobility, and distribution. *FEMS Microbiol. Rev.* **19:**1–24.
81. **Rosen, B. P., and T. Tsuchiya.** 1979. Preparation of everted membrane vesicles from *Escherichia coli* for the measurement of calcium transport. *Methods Enzymol.* **56:**233–241.
82. **Ross, J. I., E. A. Eady, J. H. Cove, and W. J. Cunliffe.** 1998. 16S rRNA mutation associated with tetracycline resistance in a gram-positive bacterium. *Antimicrob. Agents Chemother.* **42:**1702–1705.
83. **Rubin, R. A., and S. B. Levy.** 1990. Interdomain hybrid tetracycline proteins confer tetracycline resistance only when they are derived from closely related members of the *tet* gene family. *J. Bacteriol.* **172:**2303–2312.
84. **Rubin, R. A., and S. B. Levy.** 1991. Tet protein domains interact productively to mediate tetracycline resistance when present on separate polypeptides. *J. Bacteriol.* **173:**4503–4509.

85. **Rubin, R. A., S. B. Levy, R. L. Heinrikson, and F. J. Kézdy.** 1990. Gene duplication in the evolution of the two complementing domains of Gram-negative bacterial tetracycline efflux proteins. *Gene* **87:**7–13.
86. **Safferling, M., H. Griffith, J. Jin, J. Sharp, M. De Jesus, C. Ng, T. A. Krulwich, and D. N. Wang.** 2003. TetL tetracycline efflux protein from *Bacillus subtilis* is a dimer in the membrane and in detergent solution. *Biochemistry* **42:**13969–13976.
87. **Sapunaric, F. M., and S. B. Levy.** 2003. Second-site suppressor mutations for the serine 202 to phenylalanine substitution within the interdomain loop of the tetracycline efflux protein Tet(C). *J. Biol. Chem.* **278:**28588–28592.
87a. **Sapunaric, F. M., and S. B. Levy.** Substitutions in the interdomain loop of the Tn*10* TetA efflux transporter alter tetracycline resistance and substrate specificity. *Microbiology,* in press.
88. **Saraceni-Richards, C. A., and S. B. Levy.** 2000. Evidence for interactions between helices 5 and 8 and a role for the interdomain loop in tetracycline resistance mediated by hybrid Tet proteins. *J. Biol. Chem.* **275:**6101–6106.
89. **Saraceni-Richards, C. A., and S. B. Levy.** 2000. Second-site suppressor mutations of inactivating substitutions at Gly247 of the tetracycline efflux protein, Tet(B). *J. Bacteriol.* **182:** 6514–6516.
90. **Sato, K., M. H. Sato, A. Yamaguchi, and M. Yoshida.** 1994. Tetracycline/H^+ antiporter was degraded rapidly in *Escherichia coli* cells when truncated at last transmembrane helix and this degradation was protected by overproduced GroEL/ES. *Biochem. Biophys. Res. Commun.* **202:**258–264.
91. **Schnappinger, D., and W. Hillen.** 1996. Tetracyclines: antibiotic action, uptake, and resistance mechanisms. *Arch. Microbiol.* **165:**359–369.
92. **Sloan, J., L. M. McMurry, D. Lyras, S. B. Levy, and J. I. Rood.** 1994. The *Clostridium perfringens* Tet P determinant comprises two overlapping genes: *tetA*(P), which mediates active tetracycline efflux, and *tetB*(P), which is related to the ribosomal protection family of tetracycline-resistance determinants. *Mol. Microbiol.* **11:**403–415.
93. **Someya, Y., T. Kimura-Someya, and A. Yamaguchi.** 2000. Role of the charge interaction between Arg(70) and Asp(120) in the Tn*10*-encoded metal-tetracycline/H^+ antiporter of *Escherichia coli. J. Biol. Chem.* **275:**210–214.
94. **Someya, Y., A. Niwa, T. Sawai, and A. Yamaguchi.** 1995. Site-specificity of the second-site suppressor mutation of the Asp-285 → Asn mutant of metal-tetracycline/H^+ antiporter of *Escherichia coli* and the effects of amino acid substitutions at the first and second sites. *Biochemistry* **34:**7–12.
95. **Someya, Y., and A. Yamaguchi.** 1997. Second-site suppressor mutations for the Arg70 substitution mutants of the Tn*10*-encoded metal-tetracycline/H^+ antiporter of *Escherichia coli. Biochim. Biophys. Acta* **1322:**230–236.
96. **Speer, B. S., L. Bedzyk, and A. A. Salyers.** 1991. Evidence that a novel tetracycline resistance gene found on two *Bacteroides* transposons encodes an NADP-requiring oxidoreductase. *J. Bacteriol.* **173:**176–183.
97. **Speer, B. S., and A. A. Salyers.** 1988. Characterization of a novel tetracycline resistance that functions only in aerobically grown *Escherichia coli. J. Bacteriol.* **170:**1423–1429.
98. **Speer, B. S., and A. A. Salyers.** 1989. Novel aerobic tetracycline resistance gene that chemically modifies tetracycline. *J. Bacteriol.* **171:**148–153.
99. **Stasinopoulos, S. J., G. A. Farr, and D. H. Bechhofer.** 1998. *Bacillus subtilis* tetA(L) gene expression: evidence for regulation by translational reinitiation. *Mol. Microbiol.* **30:**923–932.
100. **Tamura, N., S. Konishi, S. Iwaki, T. Kimura-Someya, S. Nada, and A. Yamaguchi.** 2001. Complete cysteine-scanning mutagenesis and site-directed chemical modification of the Tn*10*-encoded metal-tetracycline/H^+ antiporter. *J. Biol. Chem.* **276:**20330–20339.
101. **Tamura, N., S. Konishi, and A. Yamaguchi.** 2003. Mechanisms of drug/H^+ antiport: complete cysteine-scanning mutagenesis and the protein engineering approach. *Curr. Opin. Chem. Biol.* **7:**570–579.
102. **Tauch, A., S. Krieft, A. Puhler, and J. Kalinowski.** 1999. The *tetAB* genes of the *Corynebacterium striatum* R-plasmid pTP10 encode an ABC transporter and confer tetracycline, oxytetracycline and oxacillin resistance in *Corynebacterium glutamicum. FEMS Microbiol. Lett.* **173:**203–209.
103. **Teo, J. W., T. M. Tan, and C. L. Poh.** 2002. Genetic determinants of tetracycline resistance in *Vibrio harveyi. Antimicrob. Agents Chemother.* **46:**1038–1045.
104. **Trieber, C. A., and D. E. Taylor.** 2002. Mutations in the 16S rRNA genes of *Helicobacter pylori* mediate resistance to tetracycline. *J. Bacteriol.* **184:**2131–2140.
104a. **Truong-Bolduc, Q. C., P. M. Dunman, J. Strahilevitz, S. J. Projan, and D. C. Hooper.** 2005. MgrA is a multiple regulator of two new efflux pumps in *Staphylococcus aureus. J. Bacteriol.* **187:**2395–2405.
105. **Tuckman, M., P. J. Petersen, and S. J. Projan.** 2000. Mutations in the interdomain loop region of the *tetA*(A) tetracycline resistance gene increase efflux of minocycline and glycylcyclines. *Microb. Drug Resist.* **6:**277–282.
106. **Vardy, E., I. T. Arkin, K. E. Gottschalk, H. R. Kaback, and S. Schuldiner.** 2004. Structural conservation in the major facilitator superfamily as revealed by comparative modeling. *Protein Sci.* **13:**1832–1840.
107. **Varela, M. F., C. E. Sansom, and J. K. Griffith.** 1995. Mutational analysis and molecular modelling of an amino acid sequence motif conserved in antiporters but not symporters in a transporter superfamily. *Mol. Memb. Biol.* **12:**313–319.
108. **Wang, W., A. A. Guffanti, Y. Wei, M. Ito, and T. A. Krulwich.** 2000. Two types of *Bacillus subtilis tetA*(L) deletion strains reveal the physiological importance of TetA(L) in K^+ acquisition as well as in Na^+, alkali, and tetracycline resistance. *J. Bacteriol.* **182:**2088–2095.
109. **Yamaguchi, A., K. Adachi, T. Akasaka, N. Ono, and T. Sawai.** 1991. Metal-tetracycline/H^+ antiporter of *Escherichia coli* encoded by a transposon Tn*10*: histidine 257 plays an essential role in H^+ translocation. *J. Biol. Chem.* **266:**6045–6051.
110. **Yamaguchi, A., T. Akasaka, T. Kimura, T. Sakai, Y. Adachi, and T. Sawai.** 1993. Role of the conserved quartets of residues located in the N- and C-terminal halves of the transposon Tn*10*-encoded metal-tetracycline/H^+ antiporter of *Escherichia coli. Biochemistry* **32:**5698–5704.
111. **Yamaguchi, A., T. Akasaka, N. Ono, Y. Someya, M. Nakatani, and T. Sawai.** 1992. Metal-tetracycline/H^+ antiporter of *Escherichia coli* encoded by transposon Tn*10*. Roles of the aspartyl residues located in the putative transmembrane helices. *J. Biol. Chem.* **267:**7490–7498.
112. **Yamaguchi, A., Y. Inagaki, and T. Sawai.** 1995. Second-site suppressor mutations for the Asp-66 → Cys mutant of the transposon Tn*10*-encoded metal-tetracycline/H^+ antiporter of *Escherichia coli. Biochemistry* **34:**11800–6.
113. **Yamaguchi, A., Y. Iwasaki-Ohba, N. Ono, M. Kaneko-Ohdera, and T. Sawai.** 1991. Stoichiometry of metal-tetracycline/H^+ antiport mediated by transposon Tn*10*-encoded tetracycline resistance protein in *Escherichia coli. FEBS Lett.* **282:**415–418.

114. **Yamaguchi, A., R. O'yauchi, Y. Someya, T. Akasaka, and T. Sawai.** 1993. Second-site mutation of Ala-220 to Glu or Asp suppresses the mutation of Asp-285 to Asn in the transposon Tn*10*-encoded metal-tetracycline/H^+ antiporter of *Escherichia coli*. *J. Biol. Chem.* **268:**26990–26995.
115. **Yamaguchi, A., N. Ono, T. Akasaka, T. Noumi, and T. Sawai.** 1990. Metal-tetracycline/H^+ antiporter of *Escherichia coli* encoded by a transposon, Tn*10*: the role of the conserved dipeptide, Ser^{65}-Asp^{66} in tetracycline transport. *J. Biol. Chem.* **265:**15525–15530.
116. **Yamaguchi, A., N. Ono, T. Akasaka, and T. Sawai.** 1992. Serine residues responsible for tetracycline transport are on a vertical stripe including Asp-84 on one side of transmembrane helix 3 in transposon Tn*10*-encoded tetracycline/H^+ antiporter of *Escherichia coli*. *FEBS. Lett.* **307:**229–232.
117. **Yamaguchi, A., T. Samejima, T. Kimura, and T. Sawai.** 1996. His257 is a uniquely important histidine residue for tetracycline/H^+ antiport function but not mandatory for full activity of the transposon Tn*10*-encoded metal-tetracycline/H^+ antiporter. *Biochemistry* **35:**4359–4364.
118. **Yamaguchi, A., Y. Shiina, E. Fujihira, T. Sawai, N. Noguchi, and M. Sasatsu.** 1995. The tetracycline efflux protein encoded by the *tet*(K) gene from *Staphylococcus aureus* is a metal-tetracycline/H^+ antiporter. *FEBS Lett.* **365:**193–197.
119. **Yamaguchi, A., Y. Someya, and T. Sawai.** 1993. The *in vivo* assembly and function of the N- and C-terminal halves of the Tn*10*-encoded TetA protein in *Escherichia coli*. *FEBS Lett.* **324:**131–135.
120. **Yamaguchi, A., Y. Someya, and T. Sawai.** 1992. Metal-tetracycline/H^+ antiporter of *Escherichia coli* encoded by transposon Tn*10*. The role of a conserved sequence motif, GXXXXRXGRR, in a putative cytoplasmic loop between helices 2 and 3. *J. Biol. Chem.* **267:**19155–19162.
121. **Yamaguchi, A., T. Udagawa, and T. Sawai.** 1990. Transport of divalent cations with tetracycline as mediated by the transposon Tn*10*-encoded tetracycline resistance protein. *J. Biol. Chem.* **265:**4809–4813.
122. **Yang, H. L., G. Zubay, and S. B. Levy.** 1976. Synthesis of an R plasmid protein associated with tetracycline resistance is negatively regulated. *Proc. Natl. Acad. Sci. USA* **73:**1509–1512.
123. **Yang, W., I. F. Moore, K. P. Koteva, D. C. Bareich, D. W. Hughes, and G. D. Wright.** 2004. TetX is a flavin-dependent monooxygenase conferring resistance to tetracycline antibiotics. *J. Biol. Chem.* **279:**52346–52352.
124. **Yin, C. C., M. L. Aldema-Ramos, M. I. Borges-Walmsley, R. W. Taylor, A. R. Walmsley, S. B. Levy, and P. A. Bullough.** 2000. The quarternary molecular architecture of TetA, a secondary tetracycline transporter from *Escherichia coli*. *Mol. Microbiol.* **38:**482–492.
125. **Zhuang, J., G. G. Prive, G. E. Werner, P. Ringler, H. R. Kaback, and A. Engel.** 1999. Two-dimensional crystallization of *Escherichia coli* lactose permease. *J. Struct. Biol.* **125:**63–75.

Frontiers in Antimicrobial Resistance: a Tribute to Stuart B. Levy
Edited by D. G. White, M. N. Alekshun, and P. F. McDermott

Chapter 2

Tetracycline Resistance Due to Ribosomal Protection Proteins

Marilyn C. Roberts

There have been 11 ribosomal protection genes identified which code for cytoplasmic proteins that protect the ribosomes from the action of tetracycline in vitro and in vivo and confer resistance to tetracycline, doxycycline, and minocycline to the host (19). They have sequence similarity to the ribosomal elongation factors, EF-G and EF-Tu, and are grouped into the translation factor superfamily of GTPases (42). Eight of the genes have G+C% ranging between 32 and 40%, which is consistent with a gram-positive origin and included the *tet*(Q) gene, which was first described in gram-negative *Bacteroides* and more recently the *tet*(36) gene, which has only been found in *Bacteroides* (Tables 1 and 2). The *Streptomycetes otr*(A) and *tet* genes have a G+C% of 72 to 76%, which is similar to the *Streptomyces* genome (Table 1). The one exception is the *tet*(W) gene with a 50 to 55% G+C%. The ancestral source of this gene is not clear. However, in the last few years the *tet*(W) gene has been identified in a variety of genera including gram-positive, gram-negative, and *Streptomycetes* (Table 2). The protein produced by this gene is most closely related to the Tet(M) group of proteins, which includes the Tet(M), Tet(O), Tet(S), and Tet(32) proteins. These five proteins share between 68 and 78% amino acid identity (5, 52, 80). Interestingly, recombination events between a *tet*(O) and a *tet*(W) gene have been identified in tetracycline resistant (Tc^r) *Megasphaera elsdenii* (85). Two different mosaic genes were identified and sequenced. Both their 3′, between base pairs 225 and 243, and their 5′, between base pairs 286 and 610, were homologous (99 to 100% identity) to the *tet*(O) gene, while the middle section, comprising 57 or 73% of the DNA sequence, were 98% identical to the *tet*(W) gene. For both mosaic genes, the overall G+C% was >50% and each had >80% homology with other *tet*(W) genes and are considered *tet*(W) genes by the current nomenclature system (44). It is not known whether the recombination between the two genes occurred within *M. elsdenii* or if these mosaic genes are found in other genera. To determine this, more *tet*(O) and *tet*(W) genes will need to be sequenced completely.

The Tet(T) protein shares 49% amino acids identity with both the Tet(Q) and Tet(36) proteins, while Tet(Q) and Tet(36) proteins share 60% amino acid identity with each other (22, 94) and all three have G+C% of 32 to 39% (Table 1). The TetB(P) protein is most related to the *Streptomycetes* Otr(A) protein though its gene has a G+C% of 32%, which is similar to the G+C% of the *Clostridium* chromosome, while the *Streptomycetes* genes have G+C% >70%. The ancestral source of the *tetB*(P) is not known, though it has been postulated to have come from the antibiotic producers (84).

MECHANISM OF RESISTANCE

How these proteins protect the host bacteria has been investigated by a number of laboratories. However, most studies on the mechanism of resistance in this group of soluble proteins have used either the Tet(M) and/or Tet(O) proteins, with the assumption that all 11 proteins should have the same mechanism of resistance (19, 24). The first clue to how these proteins worked came from Sanchez-Pescador et al. (76) when they identified amino acid homology between the Tet(M) protein from *Ureaplasma urealyticum,* a cell-wall-free microbe, and translational elongation factors. Their studies suggested that the ribosomal protection proteins might function as tetracycline resistant elongation factors. This turned out to be too simplistic of a model. Burdett was able to purify and characterized the Tet(M) protein and found that it could not substitute for the elongation factors in vivo or in vitro (12–14). Connell et al. (24, 25) suggest that the ribosomal protection proteins may be evolutionarily derived from the elongation factors because

Marilyn C. Roberts • Department of Pathobiology, Box 357238, University of Washington, Seattle, WA 98195.

Table 1. G+C content of ribosomal protection genes

Gene	G+C%[a]
tet(M)	35–36
tet(O)	39–40
tetB(P)	32
tet(Q)	39
tet(S)	33–34
tet(T)	32
tet(W)	50–55
tet(32)	40
tet(36)	37
otr(A)	72
tet	76

[a] Percentages are rounded off to the next whole number.

they do promote the release of tetracycline from the ribosome as well as bind and hydrolyze GTP in a ribosome-dependent manner. One characteristic of these proteins is their ability to protect the ribosomes both in vitro and in whole cells. This differs from bacterial tetracycline resistance due to the tetracycline efflux proteins, since they require intact membranes to function and are unable to protect the ribosome in vitro (19, chapter 1). These proteins also work in a wide range of bacteria which differ in G+C%, susceptibility to ambient air, and ecology (Table 2).

To understand how these proteins work requires the understanding of how tetracycline inhibits protein synthesis. This has been investigated by looking at the interaction between tetracycline and the 30S ribosomal subunit in a 30S subunit-tetracycline crystal. Two structures have been analyzed with one showing tetracycline bound to two sites in the 30S subunit (9) and a second structure showing tetracycline bound to six sites (Tet-1 through Tet-6) on the subunit (58). The Tet-1 and primary site in the two structures are located in the ribosomal A site, and these plus the secondary and Tet-5 binding sites are the most likely candidates for the tetracycline binding site (25). A model describing the Tet(O)-mediated tetracycline resistance using the biochemical and structural data from both Tet(O) and Tet(M) proteins has been proposed (25, 77), though neither the Tet(M) nor the Tet(O) protein has been crystallized. In this model, the ribosomes without tetracycline function normally to produce proteins. When tetracycline is added to the bacteria, it binds to the ribosomes, altering the ribosome's conformational state which in turn interrupts the elongation cycle and protein synthesis stops. The ribosomal protection proteins are thought to interact with the base of h34 protein, within the ribosome, causing an allosteric disruption of the primary tetracycline binding site(s). This releases the tetracycline molecules from the ribosome and allows the ribosome to return to its normal post-translocational conformational state, which was altered by the binding of tetracycline. Once the normal conformation returns, protein synthesis can proceed. Whether the ribosomal proteins actively prevent tetracycline from rebinding to the ribosomes after they have been released is not known, nor is it known if once the tetracycline is released it can rebind to the same or a different ribosome and inhibit protein syn-

Table 2. Distribution of ribosomal protection genes

Gene	No. of genera	Genera
tet(M)	37	*Abiotrophia, Acinetobacter, Actinomyces, Aerococcus, Afipia, Bacteroides, Bacillus, Bacterionema, Bifidobacterium, Clostridium, Corynebacterium, Eikenella, Enterococcus, Erysipelothrix, Eubacterium, Fusobacterium, Gardnerella, Gemella, Haemophilus, Kingella, Lactobacillus, Listeria, Microbacterium, Mitsuokella, Mobiluncus, Mycoplasma, Mycobacterium, Neisseria, Pasteurella, Peptostreptococcus, Prevotella, Staphylococcus, Streptococcus, Streptomyces, Selenomonas, Ureaplasma, Veillonella*
tet(O)	10	*Aerococcus, Butyrivibrio, Campylobacter, Enterococcus, Lactobacillus, Mobiluncus, Neisseria, Peptostreptococcus, Staphylococcus, Streptococcus*
tet(S)	5	*Enterococcus, Lactococcus, Lactobacillus, Listeria, Veillonella*
tetB(P)	1	*Clostridium*
tet(Q)	15	*Bacteroides, Capnocytophaga, Clostridium, Eubacterium, Gardnerella, Lactobacillus, Mitsuokella, Mobiluncus, Neisseria, Peptostreptococcus, Prevotella, Porphyromonas, Selenomonas, Streptococcus, Veillonella*
tet(T)	1	*Streptococcus*
tet(W)	18	*Actinomyces, Arcanobacterium, Bacillus, Bifidobacterium, Butyrivibrio, Clostridium, Fusobacterium, Lactobacillus, Mitsuokella, Megasphaera, Neisseria, Prevotella, Porphyromonas, Selenomonas, Staphylococcus, Streptococcus, Streptomyces, Veillonella*
tet(32)	1	*Clostridium*
tet(36)	1	*Bacteroides*
tet	1	*Streptomyces*
otr(A)	2	*Mycobacterium, Streptomyces*

[a] Data from references 1, 3, 5, 8, 11, 15–23, 26–30, 32, 34, 35–37, 38, 40, 41, 43, 46, 48, 49, 52, 53, 55–57, 61, 63–72, 74, 75, 78–82, 84–86, 90, 91, 93–95.

theses again. A detailed review of the various experiments done to elucidate the mechanism of these proteins can be found in Connell et al. (25).

Very recently the interaction between tigecycline and glycylcycline, a new derivative of minocycline which is not affected by ribosomal protection proteins, has been examined (6). Glycylcyclines compete with tetracycline for ribosomal binding sites because they have either identical or overlapping sites on the ribosome and have the same mode of action as tetracycline. However, the glycylcyclines have a higher binding affinity for ribosomes than does tetracycline (19). The most recent derivative, tigecycline, is able to overcome Tc^r in bacteria due to the tetracycline ribosomal protection proteins and/or tetracycline efflux proteins because of steric hindrance of the tigecycline due to the large substitution at position 9 of the molecule. If tigecycline becomes available for therapy, it will be of interest to see if natural mutations in the genes for the ribosomal protection proteins will appear that are able to compensate and allow for tigecycline resistance. This is certainly possible, since Tc^r *Salmonella* and *Shigella* isolates, carrying tetracycline efflux genes, that are resistant to some of the glycylcycline derivatives have already been identified (19).

DISTRIBUTION OF THE RIBOSOMAL PROTECTION GENES

The ribosomal protection genes dominate mechanisms of tetracycline resistance in the oral bacterial flora, while they have been less common in classical gram-negative enteric genera (19, 20–23, 30, 34, 35, 40, 41, 46, 63, 65, 72, 91, 95) (Table 2). These genes are also commonly found in anaerobic gram-negative bacteria in both the oral and urogenital tract (3, 19–21, 26, 32, 41, 43, 45, 55, 56, 63, 67, 71, 91). Some species also seem to preferentially carry these types of tetracycline resistance genes. For example, of 105 animal *Staphylococcus intermedius* isolates, 99 (94%) carried the *tet*(M) gene, while of the 123 other staphylococci, only 6 (5%) carried one of the tetracycline resistance ribosomal protection genes (79). This type of species preference has not been well studied nor is it understood. As a group, the ribosomal protection genes have host ranges which vary from narrow with one genus carrying the *tetB*(P), *tet*(T), *tet*(32), *tet*(36), and *tet* genes to very broad such as the *tet*(M) gene which is found in 37 genera (Table 2). The differences found in host range may be due to lack of study and/or biologic differences. For example, the *otr*(A) gene, originally identified in the antibiotic producing *Streptomyces* (28), has been identified in rapidly growing *Mycobacterium* obtained from patients (57) but has not been used for screening in other studies. The five remaining genes have been identified in 5 to 37 different genera and are found in gram-positive, gram-negative, and anaerobic genera and will be the focus of this section (Table 2).

The differences in host range may be partially due to the type of element that each gene is associated with. This may be relevant because conjugative transposons, in general, have less host specificity than do plasmids and are able to be transferred to unrelated species and genera. Most natural plasmids, however, can not pass between unrelated genera or from a gram-positive to a gram-negative bacteria, or vice versa (23, 27, 59, 63–72). The ribosomal protection genes have not yet been found in gene cassettes, the newest class of mobile elements primarily found in gram-negative species and *Staphylococcus* spp. (60). Of the five ribosomal protection genes, three—*tet*(M), *tet*(Q), and *tet*(W)—are often associated with conjugative transposons, have the largest host range (15–37 genera), have been identified in a variety of ecosystems, and have been conjugally transferred into other genera (Table 2) (1–3, 5, 7–8, 11, 18, 20–21, 23, 26, 32, 35–37, 39–41, 43, 45, 51, 53–56, 59, 61, 63–75, 79–85, 91, 94, 95). In contrast, the *tet*(O) and *tet*(S) genes, found in 10 and 5 genera, respectively, have not been previously found on conjugative transposons and have primarily been associated with gram-positive genera (8a, 15, 16, 17, 33, 46, 88–90) (Table 2). However in a recent report, we identified Tc^r clinical *Streptococcus pyogenes* isolates that have *tet*(O) genes which are associated with a conjugative transposon which also carried the *mef*(A) and *msr*(D) genes, both coding for efflux proteins which confer macrolide resistance (34). The new location of the *tet*(O) gene has made it easier for us to move it by conjugation in the laboratory, and one wonders if this configuration will make it easier for the *tet*(O) gene to extend its natural host range as well.

The first ribosomal protection gene described, the *tet*(M) gene, was identified in *Streptococcus* in a 1980s publication (37). However, the *tet*(M) gene has more recently been identified in a study of clinical *Enterococcus* spp. isolated between 1954 and 1955 (4). This suggests that the *tet*(M) gene has been present in at least in one gram-positive genera of bacteria for over 50 years. The *tet*(M) gene has now been identified in 37 genera from a variety of ecosystems, including humans, animals, and the soil. This gene has been transferred into other species not naturally found to carry the *tet*(M) gene (7, 23, 39). Though the *tet*(M) gene is thought to be of gram-positive origin, it has also been identified in gram-negative, cell-wall free, aerobic, and anaerobic genera (Table 2) (1, 4, 11, 18, 19, 23, 26, 29, 30, 35–37, 40–41, 51, 53–54, 56, 59, 61, 66–72, 78–79, 83, 91). The *tet*(M) gene has the widest host range of any of the acquired tetracycline resistance genes, including the tetracycline resistance efflux genes (67a; http://faculty.washington.edu/marilynr/). In addition, the *tet*(M) gene has been conjugally transferred in the laboratory to species and genera which have not been found naturally to carry the gene, suggesting the potential for an even wider host range in nature of this particular gene (7, 23).

Low levels of tetracycline added to mating experiments have increased the frequency of transfer of the *tet*(M) gene from undetectable levels ($<10^{-10}$/recipient) in *Listeria* spp. donors to detectable levels of 10^{-5}/recipient, also using *Listeria* recipients, and 10^{-7}/recipient with an *Enterococcus faecalis* recipient (29). Similar increases in transferability have been demonstrated with the *tet*(Q) gene (see below).

A number of studies have shown that the *tet*(M) gene is commonly found in the oral and urogenital bacterial flora (19, 55, 56, 63, 65), including a recent study by Villedieu et al. (91), which found that the *tet*(M) gene was carried in 79% of the 105 Tc^r oral isolates examined from 20 healthy adults. They even found the *tet*(M) gene in *Streptomyces* isolate for the first time. The wide host range of the *tet*(M) gene may be due, in part, to its association with a family of promiscuous conjugative transposons (23).

The *tet*(O) gene was first described in gram-positive isolates but has also been found on plasmids from *Campylobacter* spp. (19, 24, 25, 46, 50, 77, 88–90, 92, 95). Originally this gene was associated with plasmids but not conjugative transposons, though it still has been identified in 10 genera including aerobes, anaerobes, and gram-positive and gram-negative bacteria (Table 2). What is not evident from the Table is that the *tet*(O) gene host range has changed with the identification in the 1990s of *Streptococcus pneumoniae* from both Washington state and South Africa, which carried the *tet*(O) gene rather than the *tet*(M) gene, which had previously been the only acquired tetracycline resistant gene found in the species (46, 95). Why the switch in recent years is not clear, though a similar switch from the carriage of the *erm*(B) rRNA methylase, to the carriage of the efflux *mef*(A) gene, has also been demonstrated in a number of geographical areas for *S. pneumoniae* (46). Perhaps more significant is the recent discovery of the *tet*(O) gene within a conjugative transposon that also carries a macrolide resistant efflux *mef*(A) and *msr*(D) genes in *S. pyogenes* (34). This has allowed us to move the *tet*(O) gene between unrelated genera.

The *tet*(Q) gene was first identified in the anaerobic gram-negative *Bacteroides* spp. and has a G+C% that is similar to that of the *Bacteroides* chromosome of 40% (73, 74). A recent paper by Shoemaker et al. (82) identified the *tet*(Q) gene in *Bacteroides* spp. isolated prior to 1970. They also found that the carriage of the *tet*(Q) gene increased from 28% in the pre-1970s to 83% in the 1990s isolates. Currently, the *tet*(Q) gene is found in 15 genera including gram-positive and gram-negative bacteria, aerobes, and anaerobes (Table 2). Eleven of the genera are anaerobic with most being gram negative and include both the oral and urogenital tract species.

The *tet*(W) gene was first reported in 2000 (81). Since then it has been identified in 18 different genera from a variety of ecosystems (81, 85, 91), with a predominance of anaerobic genera (Table 2). The first *tet*(W) genes sequenced were from the anaerobic rumen and intestinal tract bacteria. These genes were very similar to each other and had <1% sequence differences. The authors suggested that the low level of sequence variation was due to recent transfer of the *tet*(W) genes into these bacteria (5, 81). The *tet*(W) genes are also common in oral bacteria as illustrated by a recent study (91) where this gene was identified in 21% of the 105 oral isolates taken from 20 healthy adults and was the second most commonly found tetracycline resistance gene, after the *tet*(M) gene. The *tet*(W) genes are interesting because they have a much higher G+C% (≥50%) compared to the other non-*Streptomycetes* ribosomal protection genes, yet these genes have been identified in low G+C% gram-positive streptococci, in gram-negative *Neisseria* sp. with 50% G+C%, and *Streptomyces* sp. with >70% G+C% (91) (Table 2). The ancestral source for the *tet*(W) gene is not clear.

In early reports, the *tet*(S) gene was found in gram-positive cocci, *Enterococcus, Lactococcus,* and gram-positive rods *Lactobacillus* and *Listeria* (16, 17, 19, 33). It has only been in a very recent study that this gene was found in one gram-negative anaerobe *Veillonella* sp. (91). Few previous studies have actually looked for the presence of this gene in gram-negative isolates; therefore, it is not clear if the *tet*(S) gene is currently present in other aerobic, facultative species, or even anaerobic isolates. The *tet*(S) gene has been associated with both conjugative and nonconjugative plasmids but has not been found in conjugative transposons until recently (41a). However, with the recent identification of the *tet*(O) gene associated with a conjugative element (34), this eventual possibility can not be ruled out Clearly, more work on the distribution and what sort of elements it is associated with is needed for the *tet*(S) gene.

MOBILE ELEMENTS AND GENE LINKAGES

The ribosomal protection genes are associated with plasmids, transposons, and conjugative transposons. Interestingly they have not yet been found in integrons, which function as a general gene-capture system and allow multiple antibiotic resistance genes to be linked. Integrons have been identified in gram-negative genera and staphylococci (60). Limited information is available for the *Streptomyces* genes, *otr*(A) and *tet*, though they are linked to genes that produce tetracycline type compounds and are found in the chromosome (28). While the *otr*(A) gene has been found in *Mycobacterium* spp. taken from patients (57), it is unclear if the *otr*(A) and/or the *tet* genes are associated with a mobile element(s). Similarly, we know little about the association of the *S. pyogenes tet*(T) gene with mobile element(s), though it was identified seven years ago (22). The *tet*(S) genes are associated with conjugative and nonconjugative plasmids and also found in the chromosome, where they did not appear to be associated with conjugative transposons until recently (15–17, 19, 41a). No linkages have been identified currently, though there is always the potential this will change, as has recently occurred with the *tet*(O) gene (34).

The Tet P determinant is unique among the ribosomal protection genes because it generally consists of a *tetA*(P) gene, which codes for a functional inducible efflux protein, and a *tetB*(P) gene, which codes for a ribosomal protection gene. The two genes are transcribed from a single promoter, P3, which is located 529 bp upstream of the *tetA*(P) start codon (38). Deletion of P3 or alteration of the spacing between the −35 and −10 regions reduced the level of transcription, while deletion and/or mutation of the T, a factor-independent terminator, increased read-through transcription in a reporter construct. Both tetracycline resistance genes have been associated with conjugative and nonconjugative plasmids in *Clostridium* spp. (84), while neither has been identified in other anaerobic or aerobic gram-positive bacteria (38, 48, 49). Whether this operon is associated with anaerobic and/or aerobic gram-negative bacteria has not been examined. Lyras and Rood (48) have postulated that the progenitor *tetA*(P) gene evolved by the replacement of the 3′ end of the gene with the *tetB*(P) coding region. The *tetA*(P) gene has been found alone in some Tc^r *Clostridium* spp., and it is responsible for the host's tetracycline resistance. Similarly, when the *tetA*(P) gene was cloned away from the *tetB*(P) gene, it was able to confer tetracycline resistance. Like the majority of other tetracycline resistance efflux proteins, this protein conferred tetracycline but not minocycline resistance (19, 84). In contrast, when the *tetB*(P) gene was cloned separately from the *tetA*(P) gene and introduced into *Clostridium perfringens* or *Escherichia coli* recipients, only low-level resistance to tetracycline and minocycline was observed (84). Doxycycline was not examined though it is likely that the *tetB*(P) gene would also confer low-level resistance to this antibiotic as do other ribosomal protection genes (19). The *tetB*(P) gene has not been found naturally without the *tetA*(P) gene. It is not clear if the *tetB*(P) gene represents a single event which has been maintained even though it may not contribute to the host's antibiotic resistance.

The *tet*(M) gene is often associated with conjugative transposons from the Tn*916*-Tn*1545* transposon family, and both Tn*916* and Tn*1545* have been extensively studied (23, 87). These elements carry a gene which codes for putative antirestriction functions, and few restriction enzyme cleavage sites have been identified (11, 23, 31, 87). Unlike many plasmids which can prevent related plasmids from entering the bacteria where it resides, the Tn*916*-Tn*1545* transposon family does not possess this property. Thus, the presence of a resident element of this family does not prevent entry and establishment of a second related element. The smallest element characterized is Tn*916*, which is 18 kb and contains 24 *orfs* (31). This element carries a recombinase (Int) that belongs to the lambda integrase family. The protein makes staggered cleavages at the ends of the element and is adjacent to the *xis* gene, which codes for a protein that is a similar to the lambda Xis protein. Both of these genes are on the left end of the transposon located upstream from the structural *tet*(M) gene. The transcription of the *tet*(M) gene is moderately induced by tetracycline through an antitermination mechanism upstream of the structural gene (87). This may explain why directly upstream from the structural gene is a region which is highly conserved with 96 to 100% identity found among seven different *tet*(M) sequences taken from a diverse group of bacteria (64). In contrast, less sequence homology is found directly downstream of the structural gene. Though the gene is officially considered to have moderate induction potential (83, 87), we saw significantly higher levels of gene transfer, with transfer rates going from $<10^{-10}$/recipient to 10^{-5} to 10^{-7}/recipient depending on the recipient host used, when the *Listeria* sp. donor was pretreated with tetracycline (29). These levels are similar to the induction levels found in *Bacteroides* spp. donors carrying the *tet*(Q) gene (see below). Downstream of the *tet*(M) gene are *orf13* and *orf23*, which are thought to be involved with conjugal transfer (31). Transcription of the *tet*(M) gene is from the *tet* promoter (31, 87).

Conjugative transfer appears to be influenced by the flanking sequences which correlate with the location of the Tn*916*-Tn*1545* family inserted into the chromosome (83, 87, 93). Some bacterial species appear to have preferred sites of insertion while others do not. Conjugal transfer is via a circular intermediate and occurs by excision from the donor, conjugal DNA transfer to the recipient and integration into the recipient's chromosome. The sequences flanking the *tet*(M) element in the *Listeria* sp. donor described above were not examined, though it is unlikely they would explain the large change in the transfer frequency seen with the tetracycline treatment. The *tet*(M) conjugative transposon has been shown to cotransfer a plasmid from *Bacillus subtilis* into *Bacillus thuringiensis* (54), which is a similar plasmid cotransfer with the *tet*(Q) elements in *Bacteroides* (see below). Significant work has been done examining the process of conjugal transfer from one cell to another, and a nice recent review of the process can be found in reference 93.

The Tn*916*-Tn*1545* family of transposons is able to form composite chromosomal elements where one transposon is integrated into another. With this arrangement, either the complete composite element may be transferred or just the inserted element can transfer. These composite elements have primarily been studied in gram-positive cocci (23, 93). This group of transposons often carry other antibiotic resistance genes including the *erm*(B) gene, which is a rRNA methylase that confers resistance to macrolides, lincosamides, and streptogramin B (MLS_B); a chloramphenicol acetyltransferase coding for chloramphenicol resistance [(*cat*(*pC221*))]; and/or aminoglycoside phosphotransferase, *aphA-3*, coding kanamycin resistance (23, 37, 93). The presence of the *cat* and *aphA-3* genes within these promiscuous transposons may explain why both chloramphenicol and kanamycin resistant *S. pneumoniae* strains continue to be isolated in areas where the use of these antibiotics has been stopped (46). Recently, Tn*2009* (GenBank AF376746) was sequenced from a *S. pneumoniae* isolate. This sequence is interesting because for the first time a linkage between the macrolide resistance efflux genes, *mef*(A) and *msr*(D), occurred upstream of the *tet*(M) gene. It is not clear if the Tn*2009*-like element was created within *S. pneumoniae,* perhaps in response to the increase carriage of the *mef*(A) and *msr*(D) genes (46), or if it was created in another species. It is also unclear if this element is unique to *S. pneumoniae* or can be found in other gram-positive or gram-negative genera. Regardless, the presence of Tn*2009* demonstrates that the Tn*916*-Tn*1545* family of transposons continues to evolve and acquire new antibiotic resistance genes.

The *tet*(O) gene was first associated with plasmids rather than with conjugative transposons. A 400 bp region directly upstream of the *tet*(O) structural gene is required for full expression of tetracycline resistance (92). Recently, we found the *tet*(O) gene integrated into conjugative transposons carrying the macrolide resistance efflux genes *mef*(A) and *msr*(D), the same genes found in Tn*2009* (see above). All three of the antibiotic resistance genes can be conjugally transferred in the laboratory to *S. pyogenes* and *E. faecalis* recipients (34). We have just finished sequencing a 11, 945 bp region of this element including upstream of the *tet*(O) structural gene through the *orf8* gene of the *mef* element (8a) (47). The *tet*(O) gene from this element shared >99% homology with the *tet*(O) genes from *S. pneumoniae* (GenBank Y07780) and *Streptococcus mutans* (GenBank M20905). The sequence upstream of the *tet*(O) gene, between base pair 4-342, was 99% identical to the sequence upstream of the *tet*(O) structural gene from *Campylobacter jejuni* (GenBank M74450) and, from bp 219-342, 100% identical with the DNA sequences upstream of the *tet*(O) structural gene in *S. pneumoniae* (GenBank Y07780). Whether the transposon was created in *S. pyogenes* is not clear, and it will be of interest to determine if this element can be identified in other streptococci, other gram-positive bacteria, or even in gram-negative species. Once again this new element shows that the bacterial conjugative elements carrying tetracycline resistance ribosomal protection genes do continue to evolve and demonstrates the role these elements play in linking and moving antibiotic resistance genes in the bacterial population.

The *tet*(Q) gene is often associated with large (65 to >150 kb) conjugative transposons and is found on a family of conjugative Tc^r *Bacteroides* elements as well as transposons (73, 74). This gene is often linked to an rRNA methylase, *erm*(F) gene, has a low G+C%, and is found in a number of gram-positive and gram-negative genera (20–21, 43, 45, 55, 56). The *tet*(Q) gene with a G+C% of 39% could be of gram-positive ancestry, though we can not rule out that it originated in *Bacteroides* with a 40% G+C% (Tables 1 and 2). The *tet*(Q) containing conjugative transposons are able to mobilize coresident plasmids and mediate excision and circularization of discrete nonadjacent segments of chromosomal DNA in *Bacteroides* (73, 74). Introduction of low levels of tetracycline induce an increase in conjugal transfer of these elements from donor to recipient (73, 74).

The *tet*(W) gene has been identified in the chromosome of *Butyrivibrio fibrisolvens* and *Arcanobacterium pyogenes* isolates (5, 8, 80, 81). No additional sequence information is available for the *B. fibrisolvens* gene, while a 12 kb fragment from *A. pyogenes* has been sequenced (8, 80). In *A. pyogenes,* the *tet*(W) genes have been associated with *mob* genes, which encode mobilization proteins, and an origin of transfer, the *oriT.* Two *tet*(W) genes, from *A. pyogenes,* had G+C% of 52.2%, which was significantly lower than the *A. pyogenes* chromosomal housekeeping genes, which averaged 62.5%. This suggests that these *tet*(W) genes were not innate to the species (8). The *tet*(W) *mob+ A. pyogenes* strains were able to conjugally transfer the *tet*(W) gene to tetracycline susceptible *A. pyogenes*. The resulting transconjugants had 10-fold higher frequency of transfer than the original donors when used in mating experiments (8). This suggested that where the *tet*(W) gene inserts and its flanking sequences might influence the frequency of transfer similar to what has been described above for the *tet*(M) elements. Upstream of the structural gene, the DNA sequences showed similarity to the DNA sequences that control the regulation of induction in the *tet*(M) gene, but no induction was observed with the *A. pyogenes* strains. The *tet*(W) gene from *B. fibrisolvens* was also able to conjugally transfer to susceptible recipients, suggesting that this gene is associated with conjugative elements in multiple genera (80). It is unknown whether the *tet*(W) elements are able to cotransfer resident plasmids.

CONCLUSIONS

Eleven tetracycline ribosomal protection genes have been identified, compared to 22 tetracycline resistant efflux genes, yet these genes are some of the most widely distributed tetracycline resistance genes in the bacterial world (19; http://faculty.washington.edu/marilynr/) (Table 2). They are found almost everywhere Tc^r bacteria live. Many of the ribosomal protection genes have low G+C% ($\leq$40%) and are thought to be of gram-positive ancestry, with analogous genes in the antibiotic producers. These genes have also been in the bacterial population for at least 50 years for the *tet*(M) gene in *Enterococcus* sp. (4) and for over 30 years for the *tet*(Q) gene in *Bacteroides* spp. (82). It is certainly possible that these genes could be found in even older isolates, if such strains were available for examination. Many of these genes are associated with conjugative transposons which allow for horizontal gene transfer to unrelated genera, access to bacteria from many different ecosystems, and linkages with new antibiotic resistance genes. It is interesting that many of these conjugative elements, including the two recently described ones, tetracycline resistance ribosomal protection genes and macrolide resistance genes, either code for rRNA methylase or macrolide efflux genes (see chapter 6). These conjugative elements may allow antibiotic resistance genes to be maintained in bacterial populations without selective pressure (46), while allowing flexibility and the ability to form new linkages between genes in the host bacteria. This is important since the data suggest that the type of element the resistance genes are associated with may impact their ability to spread in the bacterial population and their potential bacterial host range. In some cases, low doses of tetracycline have been shown to increase the frequency of conjugal transfer, which suggests the continual use of tetracyclines, whether for treatment of infectious or noninfectious diseases or as growth promoters in the food industry, could directly impact the spread of these genes through bacterial communities.

In the last few years, three new ribosomal protection genes have been identified, *tet*(W), *tet*(32), and *tet*(36), and an effort has been made to look at these genes in other ecological niches (2). Therefore, it is likely that additional new genes from this family's tetracycline resistance proteins may be characterized in the future. We have seen these genes spread to new species and genera over time, and it is likely that the ribosomal protection genes will continue to spread to new species and genera in the coming years, especially if the overall use of tetracycline does not change. New linkages between this group of genes and other antibiotic and nonantibiotic resistant genes will likely continue to evolve. To keep up on changes, visit http://faculty.washington.edu/marilynr/, which is updated as new information becomes available for all acquired tetracycline resistance genes.

REFERENCES

1. **Agerso, Y., L. B. Jensen, M. Givskov, and M. C. Roberts.** 2002. The identification of a tetracycline resistance gene *tet*(M), on a Tn*916*-like transposon, in the *Bacillus cereus* group. *FEMS Micobiol. Lett.* **214:**251–256.
2. **Aminov, R. I., N. Garrigues-Jeanjean, and R. I. Mackie.** 2001. Molecular ecology of tetracycline resistance: development and validation of primers of primers for detection of tetracycline resistance genes encoding ribosomal protection proteins. *Appl. Environ. Microbiol.* **67:**22–32.
3. **Andres, M. T., W. O. Chung, M. C. Roberts, and J. F. Fierro.** 1998. Antimicrobial susceptibilities of *Porphyromonas gingivalis, Prevotella intermedia* and *Prevotella nigrescens* isolated in Spain. *Antimicrob. Agents Chemother.* **42:**3022–3023.
4. **Atkinson, B. A., A. Abu-Al-Jaibat, and D. J. LeBlanc.** 1997. Antibiotic resistance among enterococci isolated from clinical specimens between 1953 and 1954. *Antimicrob. Agents Chemother.* **41:**1598–1600.
5. **Barbosa, T. M., K. P. Scott, and H. J. Flint.** 1999. Evidence for recent intergeneric transfer of a new tetracycline resistance gene, *tet*(W), isolated from *Butyrivibrio fibrisolens,* and the occurrence of *tet*(O) in ruminal bacteria. *Environ. Microbiol.* **1:**53–64.
6. **Bauer, G., C. Berens, S. J. Projan, and W. Hillen.** 2004. Comparison of tetracycline and tigecycline binding to ribosomes mapped by dimethylsulphate and drug-directed Fe^{2+} cleavage of 16S rRNA. *J. Antimicrob. Chemother.* **53:**592–599.
7. **Bertram, J., M. Stratz, and P. Durre.** 1991. Natural transfer of conjugative transposon Tn*916* between Gram-positive and Gram-negative bacteria. *J. Bacteriol.* **173:**443–448.
8. **Billington, S. J., J. G. Songer, and B. H. Jost.** 2002. Widespread distribution of a Tet W determinant among tetracycline-resitant isolates of the animal pathogen *Arcanobacterium pyogenes. Antimicrob. Agents Chemother.* **46:**1281–1287.

8a. **Brenciani, A., K. K. Ojo, A. Monachetti, S. Menzo, M. C. Roberts, P. E. Varaldo, and E. Giovanetti.** 2004. Distribution and molecular analysis of *mef*(A)-containing elements in tetracycline-susceptible and -resistant *Streptococcus pyogenes* clinical isolates with efflux-mediated erythromycin resistance. *J. Antimicrob. Chemother.* **54:**991–998.

9. **Brodersen, D. E., W. M. Clemons, A. P. Carter, R. J. Morgan-Warren, B. T. Wimberly, and V. T. Ramakrishnan.** 2000. The structural basis for the action of the antibiotic tetracycline, pactamycin, and hygromycin B on the 30S ribosomal subunit. *Cell* **103:**1143–1154.

10. **Brown, B. A., R. J. Wallace, and G. Onyi.** 1996. Activities of the glycylcyclines *N, N*-dimethylglycylamido-minocycline and *N, N*-dimethylglycylamido-6-demethyl-6-deoxytetracycline against *Nocardia* spp. and tetracycline-resistant isolates of rapidly growing mycobacteria. *Antimicrob. Agents Chemother.* **40:**874–878.

11. **Brown, J. T., and M. C. Roberts.** 1987. Cloning and characterization of *tetM* from a *Ureaplasma urealyticum* strain. *Antimicrob. Agents Chemother.* **31:**1852–1854.

12. **Burdett, V.** 1991. Purification and characterization of Tet(M), a protein that renders ribosomes resistant to tetracycline. *J. Biol. Chem.* **266:**2872–2877.

13. **Burdett, V.** 1993. tRNA modification activity is necessary for Tet(M)-mediated tetracycline resistance. *J. Bacteriol.* **175:**7209–7215.

14. **Burdett, V.** 1996. Tet(M)-promoted release of tetracycline from ribosomes is GTP dependent. *J. Bacteriol.* **178:**3246–3251.

15. **Charpentier, E., G. Gerbaud, and P. Courvalin.** 1994. Presence of the *Listeria* tetracycline resistance gene *tet*(S) in *Enterococcus faecalis. Antimicrob. Agents Chemother.* **38:**2330–2335.

16. **Charpentier, E., and P. Courvalin.** 1994. Presence of the *Listeria* tetracycline resistance gene *tet*(S) in *Enterococcus faecalis. Antimicrob. Agents Chemother.* **38:**2330–2335.

17. **Charpentier, E., and P. Courvalin.** 1999. Antibiotic resistance in *Listeria* spp. *Antimicrob. Agents Chemother.* **43:**2103–2108.

18. **Chaslus-Dancla, E., M.-C. Lesage-Descauses, S. Leroy-Setrin, J.-L. Marel, and J.-P. Lafont.** 1995. Tetracycline resistance determinants, Tet B and Tet M detected in *Pasteurella haemolytica* and *Pasteurella multocida* from bovine herds. *J. Antimicrob. Chemother.* **36:**815–819.

19. **Chopra, I., and M. C. Roberts.** 2001. Tetracycline antibiotics: Mode of action, applications, molecular biology and epidemiology of bacterial resistance. *Microbiol. Mol. Biol. Rev.* **65:**232–260.

20. **Chung, W. O., K. Young, Z. Leng, and M. C. Roberts.** 1999. Mobile elements carrying *ermF* and *tetQ* genes in Gram-positive and Gram-negative bacteria. *J. Antimicrob. Chemother.* **44:**329–335.

21. **Chung, W. O., J. Gabany, G. R. Persson, and M. C. Roberts.** 2002. Distribution of *erm*(F) and *tet*(Q) genes in four oral bacterial species and genotypic variation between resistant and susceptible isolates. *J. Clin. Periodontol.* **29:**152–158.

22. **Clermont, D., O. Chesneau, G. De Cespedes, and T. Horaud.** 1997. New tetracycline resistance determinants coding for ribosomal protection in streptococci and nucleotide sequence of *tet*(T) isolated from *Streptococcus pyogens* A498. *Antimicrob. Agents Chemother.* **41:**112–116.

23. **Clewell, D. B., S. E. Flannagan, and D. Jaworski.** 1995. Unconstrained bacterial promiscuity: the Tn*916*-Tn*1545* family of conjugative transposons. *Trends Microbiol.* **3:**229–236.

24. **Connell, S. R., D. M. Tracz, K. H. Nierhaus, and D. E. Taylor.** 2003. Ribosomal protection proteins and their mechanism of tetracycline resistance. *Antimicrob. Agents Chemother.* **47:**3675–3681.

25. **Connell, S. R., C. A. Trieber, E. Einfeldt, D. M. Tracz, D. E. Taylor, and K. Nierhaus.** 2003. Mechanism of Tet(O), perturbs the conformation of the ribosomal decoding center. *Mol. Microbiol.* **45:**1463–1472.

26. **de Barbeyrac, B., B. Dutilh, C. Quentin, H. Renaudin, and C. Bebear.** 1991. Susceptibility of *Bacteroides ureolyticus* to antimicrobial agents and identification of a tetracycline resistance determinant related to *tetM. J. Antimicrob. Chemother.* **27:**721–731.

27. **DePaola, A., and M. C. Roberts.** 1995. Class D and E tetracycline resistance determinants in gram-negative catfish pond bacteria. *Mol. Cell. Probes* **9:**311–313.

28. **Dittrich, W., and H. Schrempf.** 1992. The unstable tetracycline resistance gene of *Streptomyces lividans* 1326 encodes a putative protein with similarities to translational elongation factors and Tet (M) and Tet (O) proteins. *Antimicrob. Agents Chemother.* **36:**1119–1124.

29. **Facinelli, B., M. C. Roberts, E. Giovanetti, C. Casolari, U. Fabio, and P. E. Varaldo.** 1993. Genetic basis of tetracycline resistance in food borne isolates of *Listeria innocua. Appl. Environ. Microbiol.* **59:**614–616.

30. **Fitzgerald, G. F., and D. B. Clewell.** 1985. A conjugative transposon (Tn*919*) in *Streptococcus sanguis. Infect. Immun.* **47:**415–420.

31. **Flannagan, S. E., L. A. Zitzow, Y. A. Su, and D. B. Clewell.** 1994. Nucleotide sequence of the 18-kb conjugative transposon Tn*916* from *Enterococcus faecalis. Plasmid* **32:**350–354.

32. **Fletcher, H. M., and F. L. Macrina.** 1991. Molecular survey of clindamycin and tetracycline resistance determinants in *Bacteroides* species. *Antimicrob. Agents Chemother.* **35:**2415–2418.

33. **Francois, B., M. Charles, and P. Courvalin.** 1997. Conjugative transfer of *tet*(S) between strains of *Enterococcus faecalis* is associated with the exchange of large fragments of chromosomal DNA. *Microbiology* **143:**2145–2154.

34. **Giovanetti, E., A. Brenciani, R. Lupidi, M. C. Roberts, and P. E. Varaldo.** 2003. The presence of the *tet*(O) gene in erythromycin and tetracycline-resistant strains of *Streptococcus pyogenes. Antimicrob. Agents Chemother.* **47:**2844–2849.

35. **Hartley, D. L., K. R. Hones, J. A. Tobian, D. J. LeBlanc, and F. L. Macrina.** 1984. Disseminated tetracycline resistance in oral streptococci: Implication of a conjugative transposon. *Infect. Immun.* **45:**13–17.

36. **Huang, R., D. M. Gascoyne-Binzi, P. M. Hawkey, M. Yu, J. Heritage, and A. Eley.** 1997. Molecular evolution of the *tet*(M) gene in *Gardnerella vaginalis. J. Antimicrob. Chemother.* **40:**561–565.

37. **Inamine, J. M., and V. Burdett.** 1985. Structural organization of a 67-kilobase streptococcal conjugative element mediating multiple antibiotic resistance. *J. Bacteriol.* **161:**620–626.

38. **Johanesen, P. A., D. Lyras, T. L. Bannam, and J. I. Rood.** 2001. Transcriptional analysis of the *tet*(P) operon from *Clostridium perfringens. J. Bacteriol.* **183:**7110–7119.

39. **Kauc, L., and S. H. Goodgal.** 1989. Introduction of transposon Tn*916* DNA into *Haemophilus influenzae* and *Haemophilus parainfluenzae. J. Bacteriol.* **171:**6625–6628.

40. **Knapp, J. S., S. R. Johnson, J. M. Zenilman, M. C. Roberts, and S. A. Morse.** 1988. High-level tetracycline resistance resulting from TetM in strains of *Neisseria* species, *Kingella denitrificans,* and *Eikenella corrodens. Antimicrob. Agents Chemother.* **32:**765–767.

41. **Lacroix, J.-M., and C. B. Walker.** 1993. Detection and incidence of the tetracycline resistance determinant *tet*(M) in the microflora associated with adult periodontitis. *J. Periodontol.* **66:**102–108.

41a. **Lancaster, H., A. P. Roberts, R. Bedi, M. Wilson, and P. Mullany.** 2004. Characterization of Tn*916S*, a Tn*916*-like element containing the tetracycline resistance determinant *tet*(S). *J. Bacteriol.* **186:**4395–4398.
42. **Leipe, D., Y. I. Wolf, E. V. Koonin, and L. Aravind.** 2002. Classification of and evolution of P-loop GTPases and related ATPases. *J. Mol. Biol.* **317:**41–72.
43. **Leng, Z., D. E. Riley, R. E. Berger, J. N. Krieger, and M. C. Roberts.** 1997. Distribution and mobility of the tetracycline resistant determinant Tet Q. *J. Antimicrob. Chemother.* **40:**551–559.
44. **Levy, S. B., L. M. McMurry, T. M. Barbosa, V. Burdett, P. Courvalin, W. Hillen, M. C. Roberts, J. I. Rood, and D. E. Taylor.** 1999. Nomenclature for new tetracycline resistance determinants. *Antimicrob. Agents Chemother.* **43:**1523–1524.
45. **Li, L.-Y., N. B. Shoemaker, and A. A. Salyers.** 1995. Location and characteristics of the transfer region of *Bacteriodes* conjugative transposon and regulation of transfer genes. *J. Bacteriol.* **177:**4992–4999.
46. **Luna, V. A., and M. C. Roberts.** 1998. The presence of the *tetO* gene in a variety of tetracycline resistant *Streptococcus pneumoniae* serotypes from Washington State. *J. Antimicrob. Chemother.* **42:**613–619.
47. **Luna, V. A., P. Coates, E. A. Eady, J. Cove, T. T. H. Nguyen, and M. C. Roberts.** 1999. A variety of Gram-positive bacteria carry mobile *mef* genes. *J. Antimicrob. Chemother.* **44:**19–25.
48. **Lyras, D., and J. I. Rood.** 1996. Genetic organization and distribution of tetracycline resistance determinants in *Clostridium perfringes. Antimicrob. Agents Chemother.* **40:**2500–2504.
49. **Lyras, D., and J. I. Rood.** 2000. Transposition of Tn*4451* and Tn*4453* involves a circular intermediate that forms a promoter for the large resolvase, TnpX. *Mol. Microbiol.* **38:**588–601.
50. **Manavathu, E. K., C. L. Fernandez, B. S. Cooperman, and D. E. Taylor.** 1990. Molecular studies on the mechanism of tetracycline resistance mediated by Tet(O). *Antimicrob. Agents Chemother.* **34:**71–77.
51. **Manganelli, R., L. Romano, S. Ricci, M. Zazzi, and G. Pozzi.** 1995. Dosage of Tn*916* circular intermediates in *Enterococcus faecalis. Plasmid* **34:**48–57.
52. **Melville, C. M., K. P. Scott, D. K. Mercer, and H. J. Fling.** 2001. Novel tetracycline resistance gene, *tet*(32), in the *Clostridium*-related human colonic anaerobe K10 and its transmission in vitro to the rumen anaerobe *Butyrivibrio fibrisolvens. Antimicrob. Agents Chemother.* **45:**3246–3249.
53. **Morse, S. A., S. J. Johnson, J. W. Biddle, and M. C. Roberts.** 1986. High-level tetracycline resistance in *Neisseria gonorrhoeae* due to the acquisition of the *tetM* determinant. *Antimicrob. Agents Chemother.* **30:**664–670.
54. **Naglich, J. G., and R. E. Andrews, Jr.** 1988. Tn*916*-dependent conjugal transfer of pC194 and pUB110 from *Bacillus subtillis* into *Bacillus thuringiensis* subsp. *israelensis. Plasmid* **20:**113–126.
55. **Olsvik, B., I. Olsen, and F. C. Tenover.** 1994. The *tet(Q)* gene in bacteria isolated from patients with refractory periodontal disease. *Oral Microbiol. Immun.* **9:**251–255.
56. **Olsvik, B., I. Olsen, and F. C. Tenover.** 1995. Detection of *tet*(M) and *tet*(O) using the polymerase chain reaction in bacteria isolated from patients with periodontal disease. *Oral Microbiol. Immun.* **10:**87–92.
57. **Pang, Y., B. A. Brown, V. A. Steingrube, R. J. Wallace Jr., and M. C. Roberts.** 1994 Tetracycline resistance determinants in *Mycobacterium* and *Streptomyces* species. *Antimicrob. Agents Chemother.* **38:**1408–1412.
58. **Pioletti, M., F. Schlunzen, J. Harms, R. Zarivach, M. Gluhmann, H. Avila, A. Bashan, H. Bartels, T. Auerbach, C. Jacobi, T. Hartsch, A. Yonath, and F. Franceschi.** 2001. Crystal structures of complexes of the small ribosomal subunit with tetracycline, edeine and IF3. *EMBO J.* **20:**1829–1839.
59. **Poyart, C., J. Celli, and P. Trieu-Cuot.** 1995. Conjugative transposition of Tn*916*-related elements from *Enterococcus faecalis* to *Escherichia coli* and *Pseudomonas fluorescens. Antimicrob. Agents Chemother.* **39:**500–506.
60. **Recchia, G. D., and R. M. Hall.** 1995. Gene cassettes: a new class of mobile element. *Microbiology* **141:**3015–27.
61. **Ribera, A., J. Ruiz, and J. Vila.** 2003. Presence of the Tet M determinant in a clinical isolate of *Acinetobacter baumannii. Antimicrob. Agents Chemother.* **47:**2310–2312.
62. **Riley, D. E., M. C. Roberts, T. Takayama, and J. N. Krieger.** 1992. Development of polymerase chain reaction-based (PCR) diagnosis of *Trichomonas vaginalis* using cloned, genomic sequences. *J. Clin. Microbiol.* **30:**465–472.
63. **Roberts, M. C.** 1989. Gene transfer in the urogenital and respiratory tract, p. 347–375. *In* S. B. Levy and R. V. Miller (ed.), *Gene Transfer in the Environment.* McGraw-Hill Publishing Co., New York, N.Y.
64. **Roberts, M. C.** 1996. Tetracycline resistant determinants: mechanisms of action, regulation of expression, genetic mobility and distribution. *FEMS Microbiol. Rev.* **19:**1–24.
65. **Roberts, M. C.** 1998. Antibiotic resistance in oral/respiratory bacteria. *Crit. Rev. Oral Biol. Med.* **9:**522–540.
66. **Roberts, M. C.** 1989. Plasmid-mediated Tet M in *Haemophilus ducreyi. Antimicrob. Agents Chemother.* **33:**1611–1613.
67. **Roberts, M. C.** 1991. Tetracycline resistance in *Peptostreptococcus* species. *Antimicrob. Agents Chemother.* **35:**1682–1684.
67a. **Roberts, M. C.** 2005. Minireview: update on acquired tetracycline resistance genes. *FEMS Microbiol. Lett.* **245:**195–203.
68. **Roberts, M. C., and J. S. Knapp.** 1988. Host range of the conjugative 25.2 Mdal tetracycline resistance plasmid from *Neisseria gonorrhoeae. Antimicrob. Agents Chemother.* **32:**488–491.
69. **Roberts, M. C., and J. S. Knapp.** 1988. Transfer of β-lactamase plasmids from *Neisseria gonorrhoeae* to *Neisseria meningitidis* and commensal *Neisseria* species by the 25.2-megadalton conjugative plasmid. *Antimicrob. Agents Chemother.* **32:**1430–1432.
70. **Roberts, M. C., and J. S. Knapp.** 1989. Transfer frequency of various 25.2 Mdal TetM-containing plasmids in *Neisseria gonorrhoeae. Sex. Trans. Dis.* **16:**91–94.
71. **Roberts, M. C., and J. Lansciardi.** 1990. Transferable Tet M in *Fusobacterium nucleatum. Antimicrob. Agents Chemother.* **34:**1836–1838.
72. **Roberts, M. C., and B. J. Moncla.** 1988. Tetracycline resistance and TetM in oral anaerobic bacteria and *Neisseria perflava-N. sicca. Antimicrob. Agents Chemother.* **32:**1271–1273.
73. **Salyers, A. A., N. B. Shoemaker, A. M. Stevens, and L.-Y. Li.** 1995. Conjugative transposons: an unusual and diverse set of integrated gene transfer elements. *Microbiol. Rev.* **59:**579–590.
74. **Salyers, A. A., and N. B. Shoemaker.** 1992. Chromosomal gene transfer elements of the *Bacteroides* group. *Eur. J. Clin. Microbiol. Infect. Dis.* **11:**1032–1038.
75. **Sanchez-Pescador, R., J. T. Brown, M. Roberts, and M. S. Urdea.** 1988. The nucleotide sequence of the tetracycline resistance determinant tetM from *Ureaplasma urealyticum. Nucleic Acids Res.* **16:**1216–1217.
76. **Sanchez-Pescador, R., J. T. Brown, M. Roberts, and M. S. Urdea.** 1988. Homology of the TetM with translational elongation factors: implications for potential modes of tetM conferred tetracycline resistance. *Nucleic Acids Res.* **16:**1218.
77. **Spahn, C. M. T., G. Blaha, R. K. Agrawal, P. Penczek, R. A. Grassucci, C. A. Trieber, S. R. Connell, D. E. Taylor, K. H. Nierhaus, and J. Frank.** 2001. Localization of the ribosomal

protection protein Tet(O) on the ribosome and the mechanism of tetracycline resistance. *Mol. Cell* **7:**1037–1045.

78. **Sarafian, S. K., C. A. Genco, M. C. Roberts, and J. S. Knapp.** 1990. Acquisition of β-lactamase and Tet M-containing conjugative plasmids by phenotypically different strains of *Neisseria gonorrhoeae*. *Sex. Trans. Dis.* **17:**67–71.
79. **Schwarz, S., M. C. Roberts, C. Werckenthin, Y. Pang, and C. Lange.** 1998. Tetracycline resistance in *Staphylococcus* spp. from domestic and pet animals. *Vet. Microbiol.* **63:**217–228.
80. **Scott, K. P., T. M. Barbosa, K. J. Forbes, and H. J. Flint.** 1997. High-frequency transfer of a naturally occurring chromosomal tetracycline resistance element in the ruminal anaerobe *Butyrivibrio fibrisolvens*. *Appl. Environ. Microbiol.* **63:**3405–3411.
81. **Scott, K. P., C. M. Melville, T. M. Barbosa, and H. J. Flint.** 2000. Occurrence of the new tetracycline resistance gene *tet*(W) in bacteria from the human gut. *Antimicrob. Agents Chemother.* **44:**775–777.
82. **Shoemaker, N. B., H. Vlamakis, K. Hayes, and A. A. Salyers.** 2001. Evidence for extensive resistance gene transfer among *Bacteroides* spp. and among *Bacteroides* and other genera in the human colon. *Appl. Environ. Microbiol.* **67:**561–568.
83. **Showsh, S. A., and R. E. Andrews.** 1992. Tetracycline enhances Tn*916*-mendiated conjugal transfer. *Plasmid* **28:**213–224.
84. **Sloan, J., L. M. McMurry, D. Lyras, S. B. Levy, and J. I. Rood.** 1994. The *Clostridium perfringens* Tet P determinant comprises two overlapping genes: *tetA*(P), which mediates active tetracycline efflux, and *tetB*(P), which is related to the ribosomal protection family of tetracycline-resistant determinants. *Mol. Microbiol.* **11:**403–415.
85. **Stanton, T. B., and S. B. Humphrey.** 2003. Isolation of tetracycline-resistant *Megasphaera elsdenii* strains with novel mosaic gene combinations of *tet*(O) and *tet*(W) from swine. *Appl. Environ. Microbiol.* **69:**3874–3882.
86. **Stevens, A. M., N. B. Shoemaker, L.-Y. Li, and A. A. Salyers.** 1993. Tetracycline regulation of genes on *Bacteroides* conjugative transposons. *J. Bacteriol.* **175:**6134–6141.
87. **Su, Y. A., P. He, and D. B. Clewell.** 1992. Characterization of the *tet(M)* determinant of Tn*916:* evidence for regulation by transcription attenuation. *Antimicrob. Agents Chemother.* **36:** 769–778.
88. **Taylor, D. E.** 1986. Plasmid-mediated tetracycline resistance in *Campylobacter jejuni:* Expression in *Escherichia coli* and identification of homology with streptococcal class M determinant. *J. Bacteriol.* **165:**1037–1039.
89. **Taylor, D. E., and A. Chau.** 1996. Tetracycline resistance mediated by ribosomal protection. *Antimicrob. Agents Chemother.* **40:**1–5.
90. **Trieber, C. A., N. Burkhardt, K. H. Nierhaus, and D. E. Taylor.** 1998. Ribosomal protection from tetracycline mediated by Tet(O): Tet(O) interaction with ribosomes is GTP-dependent. *Biol. Chem.* **379:**847–855.
91. **Villedieu, A., M. L. Diaz-Torres, N. Hunt, R. McNab, D. A. Spratt, M. Wilson, and P. Mullany.** 2003. Prevalence of tetracycline resistance genes in oral bacteria. *Antimicrob. Agents Chemother.* **47:**878–882.
92. **Wang, Y., and D. E. Taylor.** 1991. A DNA sequence upstream of the *tet*(O) gene is required for full expression of tetracycline resistance. *Antimicrob. Agents Chemother.* **35:**2020–2025.
93. **Weaver, K. E., L. B. Rice, and G. Churchward.** 2002. Plasmids and transposons, p. 219–263. *In* M. S. Gilmore (ed.), *The Enterococci: Pathogenesis, Molecular Biology, and Antibiotic Resistance.* ASM Press, Washington, D.C.
94. **Whittle, G., T. R. Whitehead, N. Hamburger, N. B. Shoemaker, M. A. Cotta, and A. A. Salyers.** 2003. Identification of a new ribosomal protection type of tetracycline resistance gene, *tet*(36), from swine manure pits. *Appl. Environ. Microbiol.* **69:**4151–4158.
95. **Widdowson, C. A., K. P. Klugman, and D. Hanslo.** 1996. Identification of the tetracycline resitance gene, *tetO*, in *Streptococcus pneumoniae. Antimicrob. Agents Chemother.* **40:**2891–2893.

Frontiers in Antimicrobial Resistance: a Tribute to Stuart B. Levy
Edited by D. G. White, M. N. Alekshun, and P. F. McDermott

Chapter 3

Discovery and Industrialization of Therapeutically Important Tetracyclines

MARK L. NELSON AND STEVEN J. PROJAN

The discovery and clinical use of the tetracycline family of antibiotics emerged from efforts in research and development that were a leap of faith for the times in the 1930s. The heavy industrial chemical business American Cyanamid purchased Lederle Laboratories and began a high-risk effort to discover and study new drugs of commercial and therapeutic value. Until then, Cyanamid had made substantial profits licensing and manufacturing calcium cyanamide, a synthetic fertilizer, and sought to diversify its markets by entering into the pharmaceutical field with the acquisition of the Lederle Antitoxin Laboratories, originally founded in 1906 to raise immune antisera in horses as anti-infective therapeutics. In 1938, William Bell, American Cyanamid president, announced his decision to expand further into pharmaceutical research with directives to his executives, "You may come up with nothing, but you may discover a single drug that may conquer even one major disease, then the public will be well served and our company will prosper" (32). With that statement, American Cyanamid and the Lederle laboratories division further increased research efforts in Pearl River, New York, and began the intensive and systematic search for commercially valuable chemicals, particularly antibiotics. In 1944 the general manager of Lederle, Wilbur Malcolm, requested that the head of research, Chandra Bose SubbaRow, begin a search for an antibiotic better than streptomycin, which had been identified and isolated by Selman Waksman and Albert Schatz at Cook College (Rutgers University) in New Jersey some 50 miles south of Pearl River, in 1943 (37).

DISCOVERY OF THE TETRACYCLINE ANTIBIOTICS

The search for antibiotic-producing microorganisms began with the discovery of penicillin, and in an effort to study therapeutic substances from soil microorganisms, SubbaRow and Lederle a year earlier, in 1943, hired 71-year-old Benjamin Minge Duggar (Fig. 1), a retired professor of plant physiology and economic botany from the University of Wisconsin, to head their soil screening department. An energetic and scholarly scientist, Duggar was world renowned for his study of soil fungi, their classification and economic importance, and in particular, the cultivation and mass production of edible mushrooms using methods that are still in practice today (8). He had built an extensive network of colleagues throughout the years and asked them, in their different locations, to send him soil samples from sites that had not been recently disturbed or exposed to the elements. The samples were routinely screened for their activity via a culture and broth dilution assay in which the microorganisms were isolated and the colonies checked for antibiotic activity against a panel of gram-negative and gram-positive bacteria.

While many antibiotic-producing organisms and their substances were known at the time, most were too toxic for medical use and were quickly discarded. However, from a soil sample labeled A377, dug from a dormant timothy hay field outside of Columbia, Missouri, and collected by William Albrecht, a soils microbiologist and friend of Duggar, came an unusual bronze-colored colony, an actinomycete Duggar named *Streptomyces aureofaciens* that produced a compound with desirable properties compared to the usual chemical toxins isolated. In vitro the unknown compound had unprecedented antibacterial activity, inhibiting the growth of both the gram-negative and gram-positive bacteria alike and, even more surprising, was potent against rickettsial diseases such as Rocky Mountain spotted fever, both in vitro and in vivo (32). The researchers were enthusiastic about this newly found compound and its increased spectrum against a wide panel of bacteria compared to other marketed antibiotics, not to mention its oral bioavailability, and quickly began the

Mark L. Nelson • Paratek Pharmaceuticals, 75 Kneeland Street, Boston, MA 02111. **Steven J. Projan** • Wyeth Research, 200 Cambridge Park Drive 5001C, Cambridge, MA 02140.

Figure 1. Professor Benjamin Minge Duggar (1872–1956), discoverer of chlortetracycline.

process of large-scale fermentation and chemical isolation of the active substance by one of the fermentation innovators and pioneers of the time, J. R. D. McCormick (25, 28). The compound was soon mass produced and subjected to further animal testing and, by 1947, was being used experimentally in patients who had contracted Rocky Mountain spotted fever, a tick-borne infection for which there was no cure and a high mortality rate. The therapeutic activity and success of this compound gave it status as a "wonder drug" and it was launched by Lederle with the approval of the U.S. Food and Drug Administration in late 1948 under the trade name Aureomycin (Fig. 2, panel I, chlortetracycline), a name chosen in reference to the bronze-colored soil bacterium from which it was produced. Duggar reported the process for fermentation and the isolation and preliminary chemical and spectral characteristics of the substance first in 1948 in the *Annals of the New York Academy of Sciences* (9), followed by its first and generative patent in 1949 titled "Aureomycin and Preparation of the Same" (U.S. patent 2,482,055). Even before its chemical structure was fully elucidated, this first tetracycline was brought to market and met with instant success in the clinical treatment of numerous previously unmanageable infectious diseases such as typhoid fever and typhus and infections caused by invasive *Streptococcus pneumoniae* and β-hemolytic streptococci.

During this same time period, other pharmaceutical companies were announcing their own discoveries of new, bio-prospected antibiotics. Research teams at Parke-Davis, together with academic collaborators at Yale, isolated chlormycetin, an antibiotic with extraordinary activity against typhus caused by rickettsial infections (36), and began its marketing, while Charles Pfizer Pharmaceuticals, Inc., a chemical company also involved in fermentation technology, primarily producing penicillin and feedstock citric acid for the food and beverage industry, likewise learned quickly the lessons being taught by American Cyanamid's success with Aureomycin and began panning for wonder drugs on its own.

By 1950 and through extensive screening of soil samples from all over the world, Pfizer isolated another antibiotic soil bacterium which produced a compound not too different chemically than Cyanamid's Aureomycin. The organism *Streptomyces rimosus* was isolated from a sample collected in Terre Haute, Indiana, and produced an antibiotic similar in chemical structure to *Aureomycin* which they named Terramycin (Fig. 2, panel II, oxytetracycline), but it was slightly more active with the same oral bioavailability and broad spectrum of activity (11). It also possessed fewer side effects, which would give it a clinical advantage in the treatment of infectious diseases.

Pfizer's Terramycin, named in reference to *terra*, Latin for *Earth* (and perhaps its origin, Terre Haute,

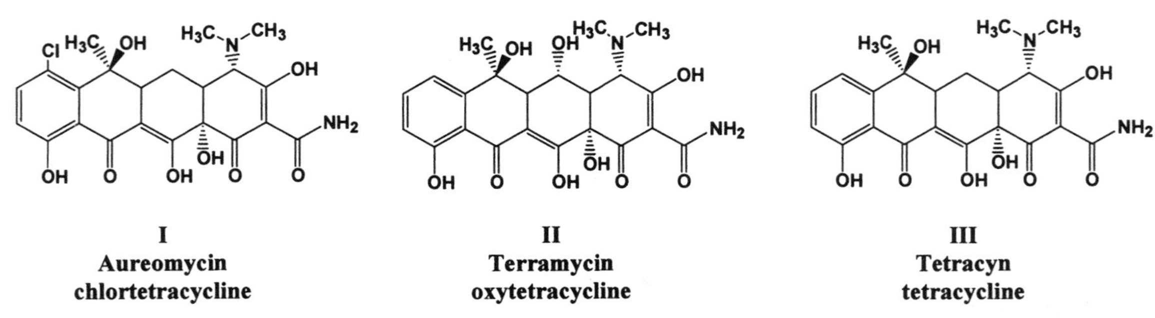

Figure 2. The chemical structures, trade name, and common name of chlortetracycline (I), oxytetracycline (II), and tetracycline (III).

Indiana), was launched with much fanfare in 1950. Within months it became a serious competitor with Aureomycin, accounting for and cutting into a large proportion of the U.S. antibiotic market (32).

Chemical elucidation of the chemical structure of Terramycin by the Pfizer chemists indicated the presence of a linearly arranged naphthacene ring system, and they coined the term "tetracycline," describing its basic chemical and structural makeup (14). However, the complete chemical structures of both Aureomycin and Terramycin remained to be fully chemically elucidated, and a competition arose between the leading organic chemists of the time, including those at Lederle, led by Joseph Boothe, and those at Pfizer, led by Lloyd Conover in collaboration with one of the all-time pioneers in chemistry, Nobel laureate Robert Burns Woodward of Harvard University. Both companies had teams of chemists racing to prove the chemical structures of the drugs, and by the end of 1953 the chemical structure was beginning to unfold, with the Pfizer chemists postulating a tetracyclic naphthacene core structure for both Terramycin and Aureomycin (14, 15, 38). They also noted that Terramycin possessed an additional hydroxyl group and was devoid of an aromatic chlorine atom, while Aureomycin possessed a chlorine atom and was devoid of a hydroxyl group. As part of the structural proof they also chemically modified Aureomycin, in Conover's laboratory, to the even more effective antibiotic, tetracycline (Fig. 2, panel III), a reaction and technological breakthrough in organic chemistry and semisynthesis that became the subject of patent litigation with Lederle as well as recognition of the Pfizer group and Woodward for the discovery of the most described molecule of its class, tetracycline. The first semisynthetic derivative, tetracycline (Fig. 2, panel III, tradenamed Tetracyn by Pfizer) gained the distinction as an even more potent and less toxic antibiotic than its progenitors and is the generic structure for which the entire collective family is named. Interestingly, through strain selection and process improvement, tetracycline today is produced solely by fermentation, obviating the need for semisynthesis and making it available at far lower cost than the totally synthetic route developed by Boothe and Andrew Kende in 1959 (3).

The Pfizer team with Woodward described the complete chemical structures of the tetracycline antibiotics in a landmark paper published in the *Journal of the American Chemical Society* titled "The Structure of Aureomycin" (39). In it they were able to correctly describe the chemical structures of Aureomycin (Fig. 2, panel I) and Terramycin (Fig. 2, panel II) by "albeit small" differences in their UV spectrum, compared to Aureomycin that was chemically modified by catalytic hydrogenation to produce Tetracyn (Fig. 2, panel III).

In parallel, the biosynthetic pathways of the three compounds were studied and elucidated by Lederle and Pfizer where McCormick examined primarily *S. aureofaciens* (25, 28), optimizing fermentation conditions and the yields of Aureomycin, while Pfizer studied *S. rimosus* for its production of Terramycin. By the mid-1950s these three new antibiotics, Aureomycin, Terramycin, and Tetracyn (Fig. 2, panels I, II, and III, respectively), were prescribed widely, selling tens of millions of dollars worth of drugs every year in the Unites States alone and increasing profits and research expansion of both companies into both antibiotics and other therapeutic areas.

BIOSYNTHESIS OF THE TETRACYCLINES

The biochemistry of the production of the natural tetracyclines was studied by McCormick at Lederle in the 1960s (27) and Hostelak (16) and others (17, 19), who manipulated the polyketide skeleton by adding different precursors during fermentation, in order to increase yields and to produce other, more novel tetracycline molecules. Radiolabel incorporation experiments with malonate and acetate subunits indicated that the naphthacene ring system is formed through addition of one malonyl-CoA unit and eight acetyl-CoA units (Fig. 3, panel I), followed by cyclization through a concerted series of enzyme-mediated foldings, ring closures, and stereochemical transformations forming the basic tetracycline scaffold (Fig. 3, panels II and III). Exocyclic position 6 methylation was found to produce 6-methylpretetramide (Fig. 3, panel III), a stable precursor of the tetracyclines, which was further modified to the intermediate 4-hydroxy-6-methylpretetramide (Fig. 3, panel IV). This pivotal intermediate is a substrate for hydroxylase and ketoreductase enzymes that ultimately leads to a structurally changed intermediate, 4-oxo-anydrotetracycline (Fig. 3, panel V). The critical enzymatic introduction of the 12a-hydroxyl group at this stage changes the structurally planar pretetramide molecule to a structure typical of bioactive tetracyclines, where the A ring is now angular compared to the BCD rings and the three-dimensional shape of antibacterial tetracyclines is created (Fig. 3, panel VI). Bioamination of position 4 forms 4-aminotetracycline, which is then methylated forming 4-dimethylaminoanhydrotetracycline (Fig. 3, panel VII). Hydroxylation at position 6 and the creation of an unsaturated center between positions 5a and 11a produces a pivotal intermediate, 6-hydroxy-5a(11a)-dehydrotetracycline (Fig. 3, panel VIII). Using *S. rimosus,* it was found that the unsaturated substrate is

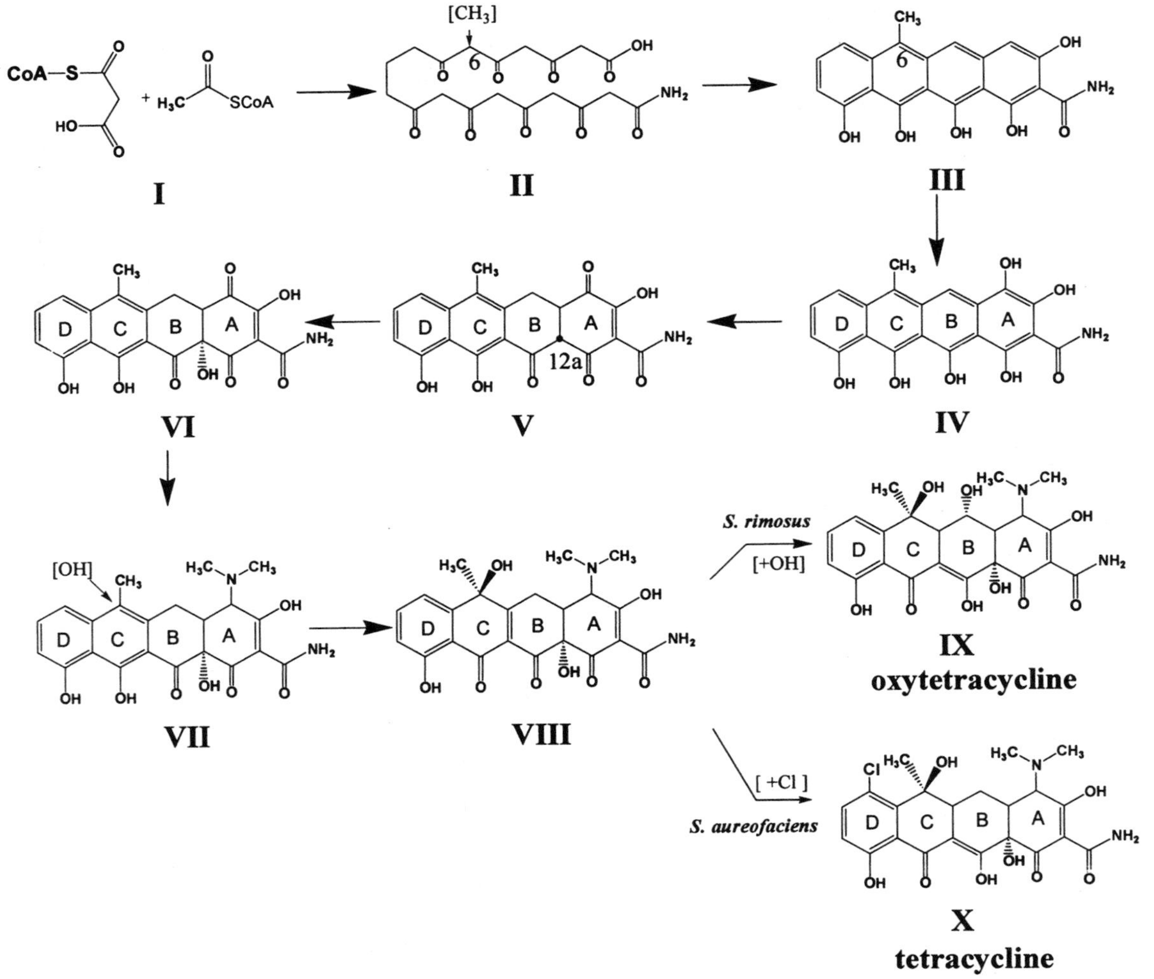

Figure 3. The biosynthetic pathway of the tetracyclines producing oxytetracycline (IX) and tetracycline (X) by *Streptomyces* species.

hydroxylated at position 5, producing oxytetracycline (Fig. 3, panel IX), while using *S. aureofaciens* hydroxylation does not occur, forming tetracycline, which upon halogenation via haloperoxidase enzymes produces 7-chlortetracycline (Fig. 3, panel X).

SEMISYNTHETIC MODIFICATIONS AND PRODUCTION OF THE CLINICALLY USED TETRACYCLINES

As soon as the tetracyclines were discovered, both the Lederle and Pfizer chemists began modifying their chemical structures in an effort to increase their potency, decrease their side effects, improve upon their pharmacokinetics, and create a molecule with sufficient novelty and patentability to gain a greater share of the antibiotic market. Both companies chose different tetracycline starting materials towards their goal and used different semisynthetic pathways to produce more active drugs, while neither chose total synthesis of the tetracycline scaffold as an efficient route to new compounds. Natural chemical products can be semisynthetically modified in one of two ways. By reacting tetracyclines with different reagents, one can chemically modify its functional groups and hopefully derive an increase in bioactivity or derive other suitable drug-like properties; or one can introduce into a tetracycline a chemical group modification that can be further used as a linker subunit upon which to build further chemical diversity (Fig. 4).

Both methods were used in the search for new semisynthetic tetracycline derivatives with Pfizer choosing the first method, modifying chemical functional groups along the core scaffold to produce methacycline (Fig. 4, panel I) and doxycycline (Fig. 4, panel II), while Lederle chose both methods, first chemically modifying key functional groups and then chemically

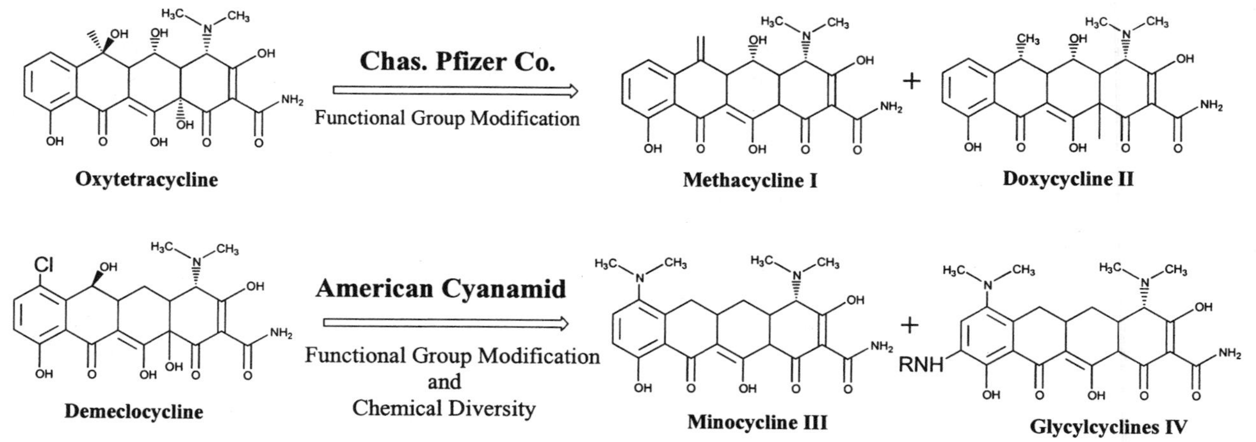

Chas. Pfizer Co.

Functional Group Modification

Oxytetracycline

Methacycline I

Doxycycline II

American Cyanamid

Functional Group Modification and Chemical Diversity

Demeclocycline

Minocycline III

Glycylcyclines IV

Figure 4. The semisynthetic pathway chosen by the Chas. Pfizer Co. for the production of methacycline (I) and doxycycline (II) and the pathway chosen by American Cyanamid for the production of minocycline (III) and the glycylcyclines (IV).

adding a reactive linker moiety upon which to build further chemical diversity, producing minocycline (Fig. 4, panel III). Both pathways led to commercially valuable and successful tetracyclines for both companies until their patent expirations. Recently, one semisynthetic method used by American Cyanamid (now named Wyeth) has produced a new class of tetracyclines, the glycylcyclines (Fig. 4, panel IV), which will be discussed.

EFFORTS OF THE CHARLES PFIZER CO.: THE SEMISYNTHESIS OF METHACYCLINE AND DOXYCYCLINE

Pfizer used its *S. rimosus* product oxytetracycline (Fig. 5, panel I) as starting material for further semisynthetic chemistry. Position 6-hydroxytetracyclines, such as oxytetracycline, are chemically unstable in acids and bases and result in the formation of C-ring

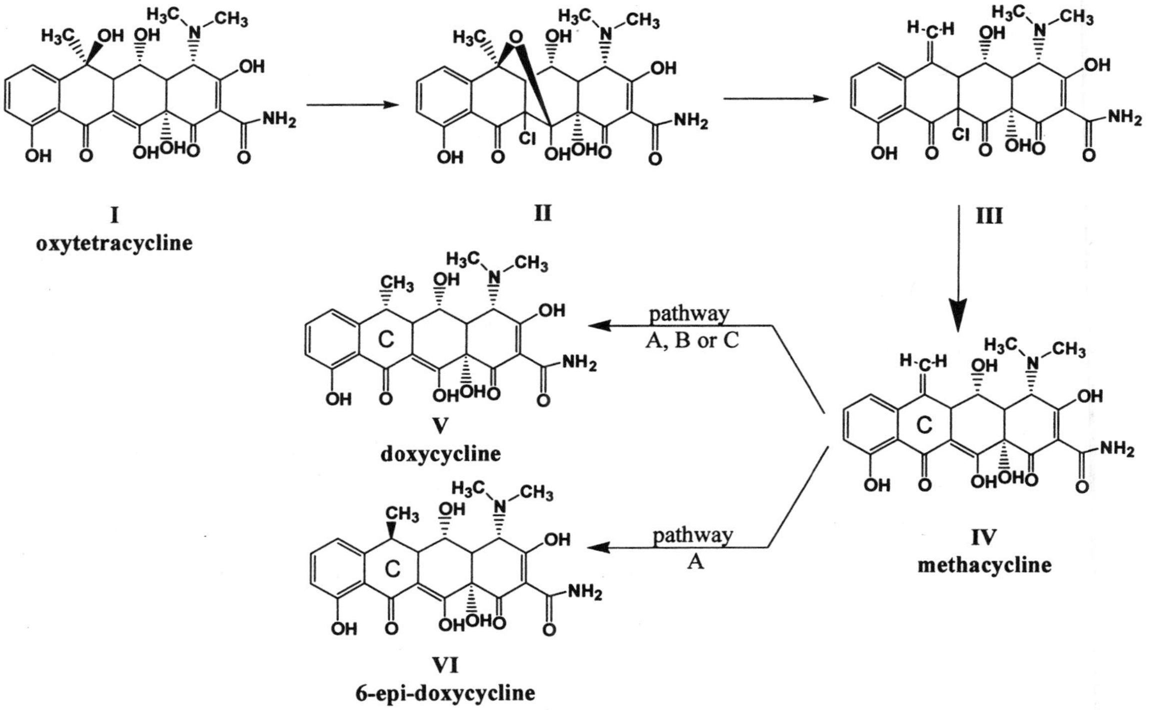

I oxytetracycline

II

III

pathway A, B or C

pathway A

V doxycycline

VI 6-epi-doxycycline

IV methacycline

Figure 5. Semisynthetic pathway of methacycline (IV), doxycycline (V), and 6-epi doxycycline (VI) achieved by the Chas. Pfizer Co.

modified anhydrotetracyclines (44) in acid and B-ring modified isotetracyclines in inorganic bases (18), diminishing their antibacterial activity and producing compounds of no commercial value. Pfizer chemists removed the possibility of chemical changes and degradation due to extremes of pH by stabilizing the C-ring via chlorination of the 11a position, forming the intermediate 11a-Cl-6, 12-hemiketal derivative of oxytetracycline (Fig. 5, panel II) (1). Further reaction in anhydrous HF formed an exocyclic bond between positions C6 and C13, producing 11a-Cl-6-methylenetetracycline (Fig. 5, panel III). Deblocking the 11a-Cl position under reducing conditions results in the antibiotic 6-methylenetetracycline (Fig. 5, panel IV, methacycline), a clinically used antibiotic possessing a broad spectrum of activity with a more favorable stability and pharmacokinetic profile than any of the previous marketed tetracyclines. Today methacycline is produced on an industrial scale worldwide by many different companies as a generic agent since its patent expiration in 1978 and is sold under many trade names, including Rondomycin. Methacycline is effective in treating a variety of infectious diseases in human health (21), veterinary medicine (47), and agriculture (6) and as a synthetic precursor to another commercially and therapeutically valuable compound, doxycycline (Fig. 5, panel V).

Methacycline was further used as a starting material for the semisynthesis of other derivatives of tetracycline at Pfizer. The chemical reduction of the exocyclic double bond of methacycline to a methyl group results in the production of a mixture of C6 positional isomers of 6-α-methyl (Fig. 5, panel V) and 6-α-methyl-6-deoxytetracyclines (Fig. 5, panel VI, reaction pathway A) (40). Chemical separation and testing of both isomers revealed that the 6-α-methyl derivative was much more potent against all strains of bacteria examined, but the chemical process lacked the ability for routine scale-up and manufacturing. The Pfizer chemists changed the reaction pathway and created a commercially valuable product through the reaction of methacycline with benzylmercaptan via an anti-Markovnikov free radical addition to yield the 13-α-benzylthiol derivative of methacycline (Fig. 5, reaction pathway B) (2). Catalytic reduction to the desirable 6-α-methyl group led to stereospecific production of 6α-deoxytetracycline, which was given the generic name doxycycline (Fig. 5, panel V) and whose synthesis pathway proved robust enough even on an industrial scale. Today the catalytic reduction of methacycline to doxycycline is possible using the Pfizer pathway (Fig. 5, pathway B) or by using more modern catalysts to stereospecifically reduce methacycline directly to doxycycline (Fig. 5, pathway C), methods that have been exploited by other pharmaceutical companies and chemical manufacturers to produce doxycycline on an industrial scale (10, 34). By chemical group modification of oxytetracycline, the Pfizer chemists showed it was possible to improve the stability and activity of a natural product and created value in a derivative of therapeutic importance.

Today doxycycline is still used widely for a variety of community acquired bacterial infections but has gained special notoriety for use versus *Bacillus anthracis* infections (anthrax) as one of only two drugs approved for such use (ciprofloxacin being the other) (5). More commonly, doxycycline is used for the treatment of Lyme disease (20), an increasingly common bacterial infection caused by the spirochete *Borrelia burgdorferi* and transmitted by tick bites. In fact, it has been shown that a single dose of doxycycline very soon after a tick bite can almost uniformly prevent the emergence of a *Borrelia* infection (29).

Methacycline and doxycycline were successful as second-generation tetracyclines in the world antibiotic market, again generating significant revenue for the Chas. Pfizer Co. until their patent expirations in the late 1980s.

In 2001, doxycycline was approved by the FDA for use in periodontal disease under the trade name Periostat, which is a low-dose formulation (20 mg/day) with activity as a gingival matrix-metalloproteinase inhibitor (12) marketed by Collagenex Pharmaceuticals, of Newtown, Pennsylvania. The low dose formulation was deemed a non-antibiotic formulation (43) and has been extensively marketed as the first tetracycline used against a disease that does not have an infectious etiology, despite the fact that minocycline has been widely used for a decade in the treatment of rheumatoid arthritis (22). There have been many observations concerning the non-antibiotic uses of doxycycline, which has shown a wide range of unexpected activities against a range of diseases related to inflammation and tissue degenerative processes, and clinical trials are ongoing to test doxycycline or minocycline for activity versus certain cancers, Lou Gehrig's disease (amyotrophic lateral sclerosis), and Huntington's disease (see http://www.clin.gov).

EFFORTS OF THE AMERICAN CYANAMIDE CO. AND LEDERLE LABORATORIES: THE SEMISYNTHESIS OF MINOCYCLINE AND TIGECYCLINE, THE FIRST THIRD-GENERATION TETRACYLINE IN CLINICAL TRIALS BY WYETH

McCormick and Niles Sjolander at Lederle experimented with blocked mutant strains of *Streptomyces* and those that were exposed to mutagenic sources in an effort to produce novel tetracyclines and to study their biosynthetic pathways. In 1956, demeclocycline (Fig. 6, panel I) was discovered (24) as part of that screening program from strains of *Streptomyces* species and was initially marketed as Declomycin by 1959, a compound representing only a marginal improvement over tetracycline. This tetracycline was restricted in use because of its increased phototoxicity, but it became quite important as a starting material for semisynthetic manipulations of the tetracycline molecule, as well as the center of an incident of corporate espionage, where the producing strain was sold to European industrialists by a disgruntled Lederle chemist (32).

Demeclocycline was further chemically modified via catalytic reduction, producing 6α-deoxy-6-demethyl tetracycline, thus removing the labile 6-hydroxy group while forming a tetracycline with the most basic chemical features necessary for antibiotic activity, known today as sancycline (Fig. 6, panel II) (26). With the 6-deoxy group removed, the chemists had also made the molecule stable for harsh acidic reagents, especially those capable of modifying the aromatic D-ring by electrophilic aromatic substitution. Reaction of the D-ring with nitration reagents in strong acid was now possible, producing nitro group derivatives at the C7 and C9 positions (Fig. 6, panels III and IV, respectively), which upon hydrogenation to the amino group affords C7 (Fig. 6, panel V) and C9 (Fig. 6, panel VI) amino derivatives of sancycline in approximately equal yields (4). Chemical separation of the isomers allowed the Cyanamid chemists to further explore substituent space at both positions C7 and C9, based upon the reactivity of the amino group with typical amine-modifying reagents. The C7-amino derivative of sancycline (Fig. 6, panel V), under reductive alkylation conditions of formaldehyde in the presence of catalysts, produced one of the most potent antibacterial tetracyclines described to date, minocycline (Fig. 6, panel VII) (23). The semisynthetic process for the manufacture of minocycline had limitations, however, and significant advances in the conversion of the 9-aminosancycline regioisomer to 7-aminosancycline resulted in a cost-effective and scalable process that is still applicable today (7). Other methods of manufacturing minocycline rely on the inherent regiospecific chemical reactivity of sancycline to produce 7-amino sancycline, processes which appear in both the scientific and the patent literature (13, 46).

Minocycline has advantages as an antibiotic; it is effective against tetracycline susceptible as well as some strains of tetracycline resistant bacteria (30, 31). It is also used for the treatment of severe and chronic acne, where it is believed to also act by an anti-inflammatory mechanism in addition to its antibacterial activity (45).

Wyeth also took advantage of the production of C7 and C9 amino position isomers to produce an even more active antibiotic currently in phase III clinical

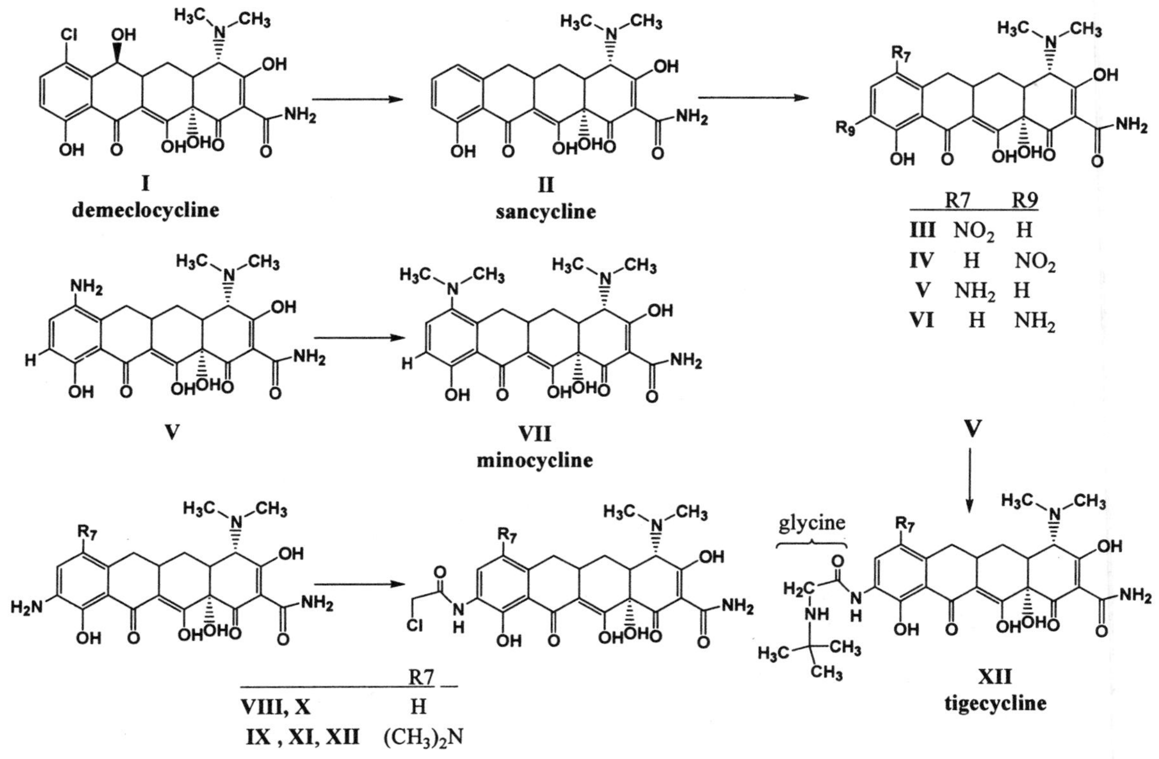

I
demeclocycline

II
sancycline

	R7	R9
III	NO_2	H
IV	H	NO_2
V	NH_2	H
VI	H	NH_2

V

VII
minocycline

	R7
VIII, X	H
IX, XI, XII	$(CH_3)_2N$

glycine

XII
tigecycline

Figure 6. Semisynthetic pathway of sancycline (II), minocycline (VII), and tigecycline (VII) achieved by the American Cyanamid Co. (Wyeth).

Figure 7. P. E. Sum led the chemistry team at Wyeth responsible for the glycylcyclines and tigecycline.

trials in the United States and in centers throughout the world. In a program initiated by Frank Tally and Wyeth chemists led by P. E. Sum (Fig. 7), 9-aminosancycline and 9-aminominocycline (Fig. 6, panels VIII and IX, respectively) were expeditiously reacted with the linking reagent chloroacetylchloride, producing an amide bond with the C9 amino group producing 9-chloroglycinylamino sancycline and minocycline (Fig. 6, panels V and VI, respectively) (41). The reactive halogen group was then further derivatized with amines to expand the chemical diversity at this position, resulting in two series of compounds based upon sancycline and minocycline and forming amine derivatives with a glycine subunit, the glycylcyclines. A glycine derivative of minocycline with an attached t-butyl amine group was found to be one of the most potent antibacterial compounds synthesized of an extensively studied series (Fig. 6, panel XII), and was found to be effective in preclinical studies. The compound, named tigecycline, currently is the focus of phase III human clinical trials by Wyeth.

ACTIVITY OF THE GLYCYLCYCLINES AND TIGECYCLINE

The glycylcyclines as a class have exhibited excellent activity versus both tetracycline-susceptible and -resistant bacteria (33); it therefore promises to be useful for infections caused by multidrug-resistant bacteria including gram-positive, gram-negative, and atypical bacteria. Tigecycline is one of the most potent members of this group and is characterized by a long half-life in patients and by broad tissue distribution (35). Tigecycline shares the same antibacterial properties of its antecedent tetracyclines; however, unlike the previous generations of tetracyclines, oral bioavailability has been poor thereby restricting tigecycline to intravenous use. To date, tigecycline has been demonstrated to be efficacious in phase II studies in complicated skin and skin structure infections as well as intra-abdominal infections.

In defining the structure-activity relationships of the glycylcyclines, it was initially observed that relatively small substituents at the C9 position showed increased potency versus bacterial strains carrying tetracycline efflux proteins (42). This key observation made by Wyeth scientists directed the chemistry of Sum and her colleagues, and hundreds of derivatives were synthesized and tested against a panel of strains carrying every major tetracycline resistance gene. In that manner the ideal chain length and substituent at the C9 position was determined, leading to the ultimate selection of tigecycline as a clinical candidate. It is anticipated that tigecycline will be available for the treatment of specific serious infections as a new hospital-based antibiotic in the year 2006, the first tetracycline to be approved for medical use in over 30 years.

CONCLUSION

The tetracyclines were some of the first antibiotics discovered and mass marketed throughout the world for the treatment of a broad spectrum of infectious disease states and represent a chronological progression of the discovery of natural products as drugs to semisynthetic derivatives of better potency and properties. American Cyanamid and Lederle Laboratories led the discovery, while the Chas. Pfizer Co. also expanded the tetracycline family of antibiotics. Both companies were responsible for bringing newer generations of tetracyclines to the clinic and the marketplace. The efforts of both companies in bringing newer tetracyclines to clinical practice has been a valiant one, with Lederle standing above the rest and persevering towards newer tetracycline antibiotics. However, the last 30 years have seen a sharp decline in the discovery of other, newer and more potent tetracyclines, and only two companies have had active chemistry efforts to discover new tetracyclines, Wyeth and Paratek Pharmaceuticals of Boston, Massachusetts. It is hoped that a novel semisynthetic tetracycline antibiotic will be available for use against bacterial pathogens in the early 21st century. In fitting with this theme it is appropriate that the discoverers of the first tetracycline used in the clinic and the newest tetracycline in clinical development are both from American Cyanamid (Wyeth), and both Duggar and Sum enjoy this distinction and their efforts are enthusiastically applauded.

REFERENCES

1. **Blackwood, R. K., J. J. Beereboom, H. H. Rennhard, M. Schach von Wittenau, and C. R. Stephens.** 1961. 6–Methylenetetracyclines. A new class of tetracycline antibiotics. *J. Am. Chem. Soc.* **83:**2773–2775.
2. **Blackwood, R. K., J. J. Beereboom, H. H. Rennhard, M. Schach von Wittenau, and C. R. Stephens.** 1963. 6-Methylenetetracyclines. III. Preparation and properties. *J. Am. Chem. Soc.* **85:** 3943–3953.
3. **Boothe, J. H., A. S. Kende, T. L. Fields, and R. G. Wilkinson.** 1959. Total synthesis of tetracyclines. I. (+/−)-Dedimethylamino-12a-deoxy-6-demethylanhydrochlortetracycline. *J. Am. Chem. Soc.* **81:**1006–1007.
4. **Boothe, J. H., J. J. Hlavka, J. P. Petisi, and J. L. Spencer.** 1960. 6-Deoxytetracyclines. I. Chemical modification by electrophilic substitution. *J. Am. Chem. Soc.* **82:**1253–1254.
5. **Centers for Disease Control and Prevention.** 2001. Update: Investigation of bioterrorism-related anthrax and interim guidelines for exposure management and antimicrobial therapy. *Centers for Disease Control and Prevention Morb. Mortal. Wkly. Rep.* **50:**909–919.
6. **Chiyowski, L. N.** 1973. Effectiveness of antibiotics applied as postinoculation sprays against clover phyllody and aster yellows. *J. Plant Sci.* **53:**87–91.
7. **Church, R. F. R., R. E. Schaub, and M. J. Weiss.** 1971. Synthesis of 7-dimethylamino-6-demethyl-6-deoxytetracycline (Minocycline) via 9-nitro-6-demethyl-6-deoxytetracycline. *J. Org. Chem.* **36:**723–725.
8. **Duggar, B. M.** 1905. The principles of mushroom growing and mushroom spawn making. *USDA Bureau Plant Industry Bull.* **35:**1–60.
9. **Duggar, B. M.** 1948. Aureomycin: a product of the continuing search for new antibiotics. *Ann. New York Acad. Sci. USA* **51:** 171–181.
10. **Felekidis, A., M. Goblet-Stachow, J. F. Liegeois, B. Pirotte, J. Delarge, A. Demonceau, M. Fontaine, A. F. Noels, I. T. Chizhevsky, T. V. Zinevich, V. I. Bregadze, F. M. Dolgushin, A. I. Yanovsky, and T. Y. Struchkov.** 1997. Ligand effects in the hydrogenation of methacycline to doxycycline and epi-doxycycline catalyzed by rhodium complexes. Molecular structure of the key catalyst [closo-3,3-(η2,3-$C_7H_7CH_2$)-3,1,2-$RhC_2B_9H_{11}$]. *J. Organometallic Chem.* **536/537:**405–412.
11. **Findlay, A. C., G. L. Hobby, S. Y. Pan, J. B. Regna, D. B. Routien, D. B. Seeley, G. M. Shull, B. A. Sobin, I. A. Solomens, J. W. Vinson, and J. H. Kane.** 1950. Terramycin, a new antibiotic. *Science* **111:**85.
12. **Golub, L. M., T. Sorsa, H. M. Lee, S. Ciancio, D. Sorbi, N. S. Ramamurthy, B. Gruber, T. Salo, and Y. T. Konttinen.** 1995. Doxycycline inhibits neutrophil (PMN)-type matrix metalloproteinases in human adult periodontitis gingiva. *J. Clin. Periodontol.* **22:**100–109.
13. **Hlavka, J., A. Schneller, H. Krazinski, and J. H. Boothe.** 1962. The 6–deoxytetracyclines. III. Electrophilic and nucleophilic substitution. *J. Am. Chem. Soc.* **84:**1426–1430.
14. **Hochstein, F. A., C. R. Stephens, L. H. Conover, P. P. Regna, R. Pasternack, K. J. Brunings, and R. B. Woodward.** 1952. Terramycin. VIII. The structure of Terramycin. *J. Am. Chem. Soc.* **74:**3708–3709.
15. **Hochstein, F. A., C. R. Stephens, L. H. Conover, P. P. Regna, R. Pasternack, P. N. Gordon, F. J. Pilgrim, K. J. Brunings, and R. B. Woodward.** 1953. The structure of Terramycin. *J. Am. Chem. Soc.* **75:**5455–5475.
16. **Hostelak, Z., M. Blumauerova, and Z. Vanek.** 1979. Tetracycline antibiotics, p. 293–353. *In* A. H. Rose (ed.), *Economic Microbiology,* vol. 3. *Secondary Products of Metabolism.* Academic Press, London, United Kingdom.
17. **Hunter, I. S., and R. A. Hill.** 1997. Tetracyclines, p. 659–682. *In* W. R. Strohl (ed.), *Biotechnology of Antibiotics,* 2nd Edition, vol. 82, *Drugs and Pharmaceutical Sciences,* Marcel Dekker, Inc. New York, N.Y.
18. **Hutchings, B. L., C. W. Waller, R. W. Broschard, C. F. Wolf, P. W. Fryth, and J. H. Williams.** 1952. Degradation of Aureomycin. V. Aureomycinic acid. *J. Am. Chem. Soc.* **74:**4980.
19. **Hutchinson, C. R.** 1981. The biosynthesis of tetracycline and anthracycline antibiotics, p. 1–11. *In* J. W. Corcoran, (ed.), *Antibiotics,* vol. IV, *Biosynthesis.* Springer, Berlin Heidelberg, New York.
20. **Johnson, S. E., G. C. Klein, G. P. Schmid, and J. C. Feeley.** 1984. Susceptibility of the Lyme disease spirochete to seven antimicrobial agents. *Yale J. Biol. Med.* **57:**549–553.
21. **Kirby, W. M. M., C. E. Roberts, Jr., and R. E. Burdick.** 1961. Comparison of two new tetracyclines with tetracycline and demethylchlortetracycline. *Antimicrob. Agents Chemother.* 286–292.
22. **Kloppenburg, M., B. A. C. Dijkmans, C. L. Verweij, and F. C. Breedveld.** 1996. Inflammatory and immunological parameters of disease activity in rheumatoid arthritis patients treated with minocycline. *Immunopharmacology* **31:**163–169.
23. **Martell, M. J., and J. H. Boothe.** 1967. The 6–deoxytetracyclines. VII. Alkylated aminotetracyclines possessing unique antibacterial activity. *J. Med. Chem.* **10:**44–46.
24. **McCormick, J. R. D., N. O. Sjolander, U. Hirsch, E. R. Jensen, and A. P. Doerschuk.** 1957. A new family of antibiotics: the demethyltetracyclines. *J. Am. Chem. Soc.* **79:**4561–4563.
25. **McCormick, J. R. D., N. O. Sjolander, S. Johnson, and A. P. Doershuk.** 1959. Biosynthesis of tetracyclines. II. Simple defined media for growth of *Streptomyces aureofaciens* and elaboration. *J. Bacteriol.* **77:**475–477.
26. **McCormick, J. R. D., E. R. Jensen, P. A. Miller, and A. P. Doerschuk.** 1960. The 6-deoxytetracyclines. Further studies on the relationship between structure and antibacterial activity in the tetracycline series. *J. Am. Chem. Soc.* **82:**3381–3386.
27. **McCormick, J. R. D.** 1966. Biosynthesis of the tetracyclines: an integrated biosynthetic scheme (Part I and II), p. 556–574. *In* M. Herold and Z. Gabriel (ed.), *Antibiotics. Advances in Research, Production and Clinical Use.* Butterworths, London, United Kingdom.
28. **Miller, P. A., J. R. D. McCormick, and A. P. Doershuk.** 1956. Studies of chlorotetracycline biosynthesis and the preparation of chlorotetracycline-C14. *Science* **123:**1030–1031.
29. **Nadelman, R. B., J. Nowakowski, D. Fish, R. C. Falco, K. Freeman, D. McKenna, P. Welch, R. Marcus, M. E. Aguero-Rosenfeld, D. T. Dennis, and G. P. Wormser.** 2001. Tick Bite Study Group. Prophylaxis with single-dose doxycycline for the prevention of Lyme disease after an Ixodes scapularis tick bite. *N. Engl. J. Med.* **345:**79–84.
30. **Nakazawa, S., H. Ono, T. Nishino, S. Kuwahara, and S. Goto.** 1970. In vitro and in vivo laboratory evaluation of minocycline—a new tetracycline derivative, p. 353–360. *In* Progr. Antimicrob. Anticancer Chemother., Proceedings of the International Congress on Chemotherapy, 6th Meeting Date 1969, University Park Press, Baltimore, Md.
31. **Noble, J. F.** 1972. Minocycline. Laboratory and clinical studies, p. 4–15. *In* E. Lauschner (ed.), *Minocyclin-Symposium.* Georg Thieme, Stuttgart, Germany.
32. **Pearson, M.** 1969. *The Million Dollar Bugs.* Putnam, New York, N.Y.
33. **Petersen, P. J., N. V. Jacobus, W. J. Weiss, P. E. Sum, and R. T. Testa.** 1999. In vitro and in vivo antibacterial activities of a

novel glycylcycline, the 9-t-butylglycylamido derivative of minocycline (GAR-936). *Antimicrob. Agents Chemother.* **43:** 738–744.

34. **Pirotte, B., A. Felekidis, M. Fontaine, A. Demonceau, A. F. Noels, J. Delarge, L. T. Chizhevsky, T. V. Zinevich, I. V. Pisareva, and V. I. Bregadze.** 1993. Stereoselective hydrogenation of methacycline to doxycycline catalyzed by rhodium-carborane complexes. *Tetrahedron Lett.* **34:**1471–1474.
35. **Postier, R. G., S. L. Green, S. R. Klein, E. J. Ellis-Grosse, and E. Loh.** Tigecycline 200 Study Group. Results of a multicenter, randomized, open-label efficacy and safety study of two doses of tigecycline for complicated skin and skin-structure infections in hospitalized patients. *Clin. Ther.* **26:**704–714.
36. **Rebstock, M. C., H. M. Crooks, J. Controulis, Bartz, and Q. R. Bartz.** 1949. Chloramphenicol (chloromycetin). IV. Chemical studies. *J. Am. Chem. Soc.* **71:**2458–2462.
37. **Schatz, A., and S. A. Waksman.** 1944. Effect of streptomycin and other antibiotic substances upon *Mycobacterium tuberculosis* and related organisms. *Proc. Soc. Exp. Biol. Med.* **57:**244–248.
38. **Stephens, C. R., L. H. Conover, F. A. Hochstein, P. P. Regna, F. J. Pilgrim, K. J. Brunings, and R. B. Woodward.** 1952. Terramycin. VIII. Structure of Aureomycin and Terramycin. *J. Am. Chem. Soc.* **74:**4976–4977.
39. **Stephens, C. R., L. H. Conover, R. Pasternak, F. A. Hochstein, W. T. Moreland, P. P. Regna, F. J. Pilgrim, K. J. Brunings, and R. B. Woodward.** 1954. The structure of Aureomycin. *J. Am. Chem. Soc.* **76:**3568–3575.
40. **Stephens, C. R., J. J. Beereboom, H. H. Rennhard, P. N. Gordon, K. Murai, R. K. Blackwood, and M. Schach von Wittenau.** 1963. 6-deoxytetracyclines. IV. Preparation, C-6 stereochemistry, and reactions. *J. Am. Chem. Soc.* **85:**2643–2652.
41. **Sum, P. E., V. J. Lee, R. T. Testa, J. J. Hlavka, G. A. Ellestad, J. D. Bloom, Y. Gluzman, and F. P. Tally.** 1994. Glycylcyclines. 1. A new generation of potent antibacterial agents through modification of 9-aminotetracyclines. *J. Med. Chem.* **37:**184–188.
42. **Sum, P. E., and P. Petersen.** 1999. Synthesis and structure-activity relationship of novel glycylcycline derivatives leading to the discovery of GAR-936. *Bioorg. Med. Chem. Lett.* **9:**1459–1462.
43. **Thomas, J. G., R. J. Metheny, J. M. Karakiozis, J. M. Wetzel, and R. J. Krout.** 1998. Long-term sub-antimicrobial doxycycline (Periostat) as adjunctive management in adult periodontitis: Effects on subgingival bacterial population dynamics. *Adv. Den. Res.* **12:**32–39.
44. **Waller, C. W., B. L. Hutchings, R. W. Broschard, A. A. Goldman, W. J. Stein, C. F. Wolf, and J. H. Williams.** 1952. Degradation of Aureomycin. VII. Aureomycin and anhydroaureomycin. *J. Am. Chem. Soc.* **74:**4981–4982.
45. **Webster, G. F., J. J. Leyden, K. J. McGinley, and W. P. McArthur.** 1982. Suppression of polymorphonuclear leukocyte chemotactic factor production in Propionibacterium acnes by subminimal inhibitory concentrations of tetracycline, ampicillin, minocycline and erythromycin. *Antimicrob. Agents Chemother.* **21:**770–772.
46. **Zambrano, R. T., and K. Butler.** 1962. Reductive alkylation process. U.S. Patent 3,483,251.
47. **Ziv, G., and F. G. Sulman.** 1971. Analysis of pharmacokinetic properties of nine tetracycline analogs in dairy cows and ewes. *Am. J. Vet. Res.* **35:**1197–1201.

II. SINGLE ANTIMICROBIAL RESISTANCE MECHANISMS

SECTION EDITOR: Patrick F. McDermott

THE PREDOMINANT MECHANISMS EMPLOYED BY BACTERIA to resist the inhibitory effects of antimicrobials can be grouped by function into a fairly small number of categories: lowered permeability, target site mutation, enzymatic inactivation, and efflux. Each of these mechanisms, alone or in concert, plays a role in resistance. This section provides an overview of resistance to the major classes of antimicrobials.

Frontiers in Antimicrobial Resistance: a Tribute to Stuart B. Levy
Edited by D. G. White, M. N. Alekshun, and P. F. McDermott

Chapter 4

Fluoroquinolone Resistance

JORDI VILA

STRUCTURE-ACTIVITY RELATIONSHIPS

Fluoroquinolones are an important class of wide-spectrum antibacterial agents. Their chemical structure is based on the 1, 4-dihydro-4-oxo-pyridine molecule, which has a carboxylic acid substituent at position 3. This substituent with the carbonyl group at position 4 appears to be essential for the activity of quinolones. In fact, several models have implicated these two substituents in the interaction of the quinolone with DNA. Currently, quinolones can be classified into three generations (Table 1 and Fig. 1). First generation quinolones, such as nalidixic acid related compounds, lack the fluorine atom at position 6, except for flumequine, which some authors consider as belonging to the second generation. These quinolones reach high concentrations in the urinary tract and hence were used for treatment of urinary tract infections caused by gram-negative bacilli, except for *Pseudomonas aeruginosa*. They lack activity against gram-positive cocci and anaerobes. The second-generation quinolones incorporated a cyclic diamine at position 7 and a fluorine atom at position 6 in the quinolone nucleus. This group is characterized by excellent activity gram-negative bacteria. However, in spite of presenting enhanced antibacterial activity in comparison with the first generation group, even against gram-positive cocci, they show a moderate activity against *Staphylococcus aureus* and *Streptococcus pneumoniae* and no activity against anaerobes. The third generation group has greater activity against gram-positive cocci, and some are also active against anaerobes (12). The main differences in the structure of the third generation with respect to the second generation quinolones are found in the substituents located at positions 1, 7, and 8 of the quinolone nucleus (Fig. 1).

Mechanism of Action

At present, DNA gyrase and topoisomerase IV are considered the protein targets of the quinolones. The DNA gyrase is a tetrameric enzyme with two A subunits and two B subunits encoded by the *gyrA* and *gyrB* genes, respectively. The main activity of this enzyme is to catalyze the negative supercoiling of DNA, with the hydrolysis of ATP being required to obtain the energy necessary to perform this process. This hydrolysis is catalyzed by the B subunit. The DNA gyrase binds to a short segment of DNA and cleaves the double-stranded DNA, which creates a DNA gate. DNA contiguous with the DNA gate is wrapped around the A subunits of the DNA gyrase in a positive superhelical sense and presents a segment of DNA (called the T segment) to the open terminal of the B subunits (called the ATP-operated clamp). Upon the ATP binding, the ATP-operated clamp closes, capturing the T segment. The DNA gate opens, pulling the broken ends of the DNA gate apart facilitating the passage of the T segment through the gap. To complete the strand-passage cycle, the DNA gate is religated upon closure, the T segment is expelled via opening of the exit gate (C-terminal of the A subunit), and the ATP-operated clamp reopens upon ATP hydrolysis. These actions by the DNA gyrase change the superhelical state of the DNA (51). The DNA gyrase is known to play an important role in both the transcription and replication of DNA (Fig. 2). Topoisomerase IV has also been found to be a protein target for quinolones. This enzyme also has a tetrameric structure with two A subunits and two B subunits encoded by the *parC* and *parE* genes, respectively. The major role of this enzyme seems to be in decatenating daughter replicons following DNA replication (63) (Fig. 2). However, the DNA gyrase

Jordi Vila • Department of Microbiology, Hospital Clinic, University of Barcelona, School of Medicine, Villarroel, 170, 08036 Barcelona, Spain.

Table 1. Classification of main commercialized quinolones

First generation	Second generation	Third generation
Nalidixic acid Pipemidic acid	Norfloxacin Ciprofloxacin Ofloxacin Perfloxacin	Levofloxacin[a] Moxifloxacin Gatifloxacin

[a] Levofloxacin has been included in the third generation because it is more active than ofloxacin against gram-positive cocci and also because the serum peak allows a good therapeutic index against these microorganisms.

also shows some decatenating activity (63). In fact, some microorganims, such as *Mycobacterium tuberculosis, Campylobacter* spp., *Corynebacterium* spp., and *Helicobacter pylori,* do not possess topoisomerase IV, and it has recently been shown that the DNA gyrase of *Mycobacterium smegmatis* presents an enhanced decatenating activity, and hence, it likely assumes the role of topoisomerase IV in these microorganisms (32).

Although the mechanism of action of quinolones has been studied in detail, some aspects of this phenomenon remain controversial. It is accepted that quinolones bind to DNA gyrase in the presence of DNA and probably Mg^{2+}, forming a complex. However, they do not bind each element independently. The amino acid Tyr 122 of the A subunit of the DNA gyrase forms a transient phosphotyrosine linkage with a broken DNA strand. It is now generally accepted that the quinolone has two binding domains, one interacting with DNA and the other with DNA gyrase. Therefore, the inhibition of DNA replication is thought to be due to the formation of this DNA gyrase-quinolone-DNA complex, which acts as a lesion, blocking fork movement. This phenomenon has been called the "poison hypothesis" (26). Different models have been proposed to explain the formation of the complex between a quinolone, DNA, and DNA gyrase. The model proposed by Shen (57) suggested three functional domains on the quinolone molecule: the DNA binding domain, through which the quinolone binds DNA by hydrogen bridges; the drug self-association domain; and the drug-enzyme interaction domain. The C-7 substituent in this latter domain would interact with

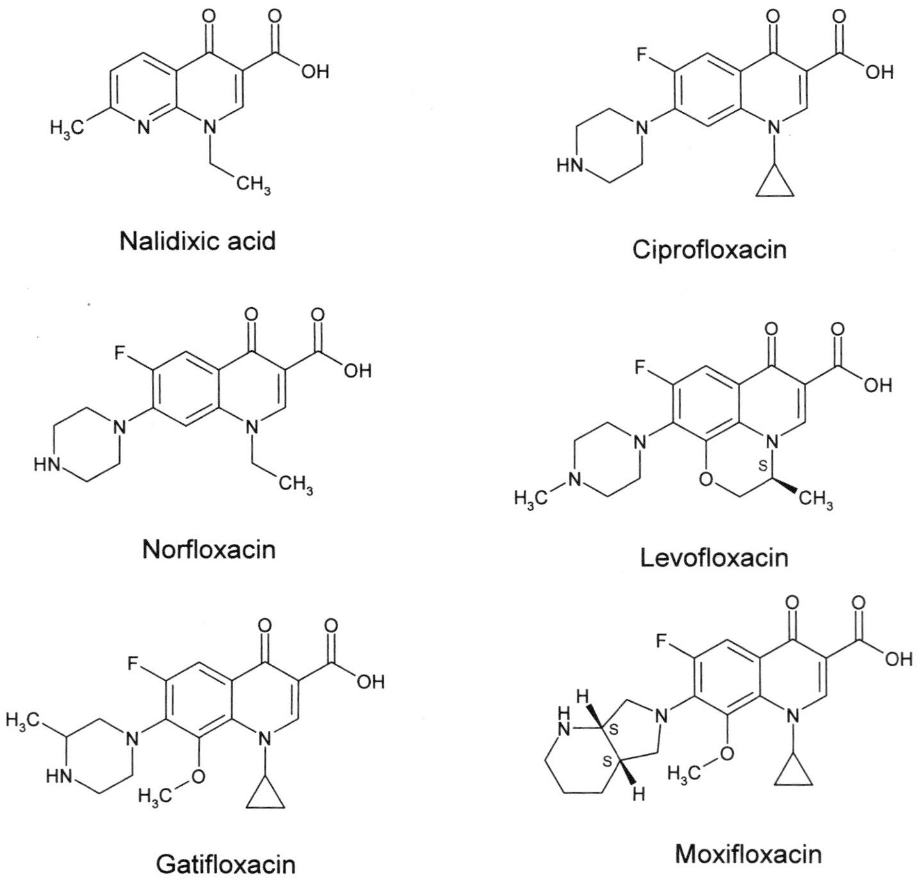

Figure 1. Chemical structure of the main commercialized quinolones.

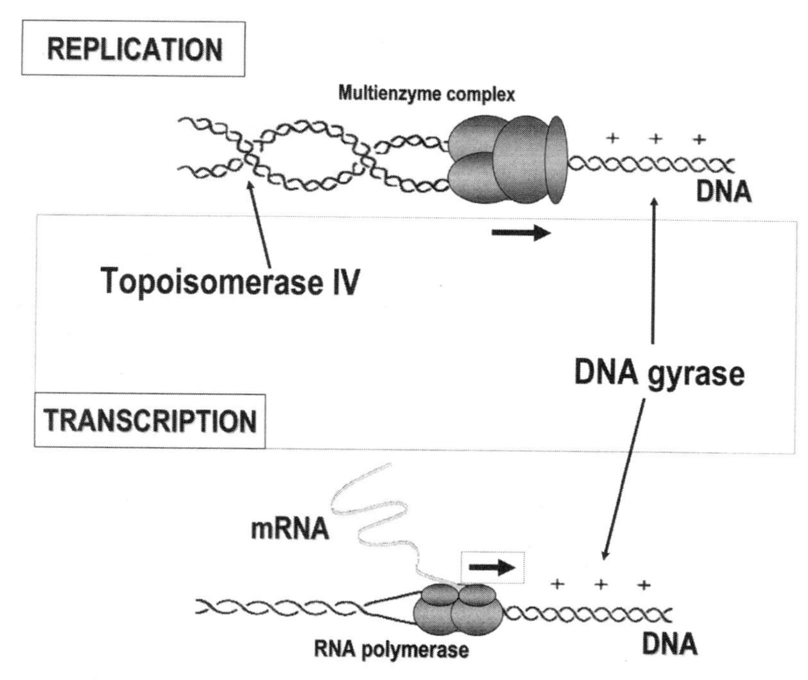

Figure 2. Process of DNA replication and DNA transcription involving DNA gyrase and topoisomerase IV.

the B subunit of the DNA gyrase (Fig. 3). The Palumbo model (45) proposes that the interaction between the quinolone and DNA would take place through the formation of a complex between substituents at 3 and 4, Mg^{2+} ions, and phosphate groups of the DNA. Moreover, the planar structure of the double ring of the quinolone would interact with the same planar structure of the DNA bases. This model also suggests that substituents at positions 1 and 7 of the quinolone could interact with the DNA gyrase. In agreement with the last proposal, with molecular modeling we have calculated the distance between the substituent at C-7 and the substituent at N-1 of ciprofloxacin (ca. 9 Å [0.9 nm]) and the distance between Ser-83 and Asp-87 of the GyrA protein. This latter calculation was performed for different protein conformations ranging from an α-helix to complete extended form (3.5 and 14 Å [0.35 and 1.4 nm] for α-helix and extended form, respectively) (Fig. 4). Therefore, the distance between both substituents of the quinolone fits well with the range calculated for the distance between the Ser-83 and Asp-87 of GyrA.

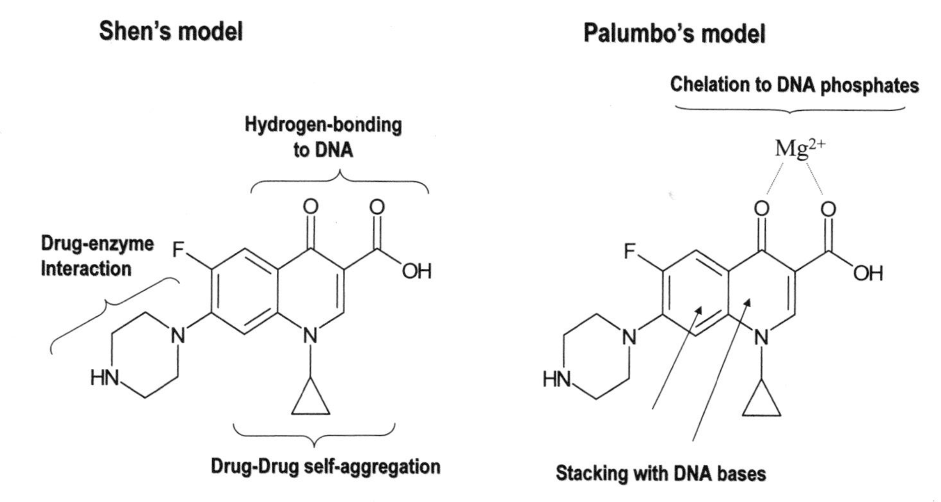

Figure 3. Functional domains of the quinolone molecule according to Palumbo's and Shen's models.

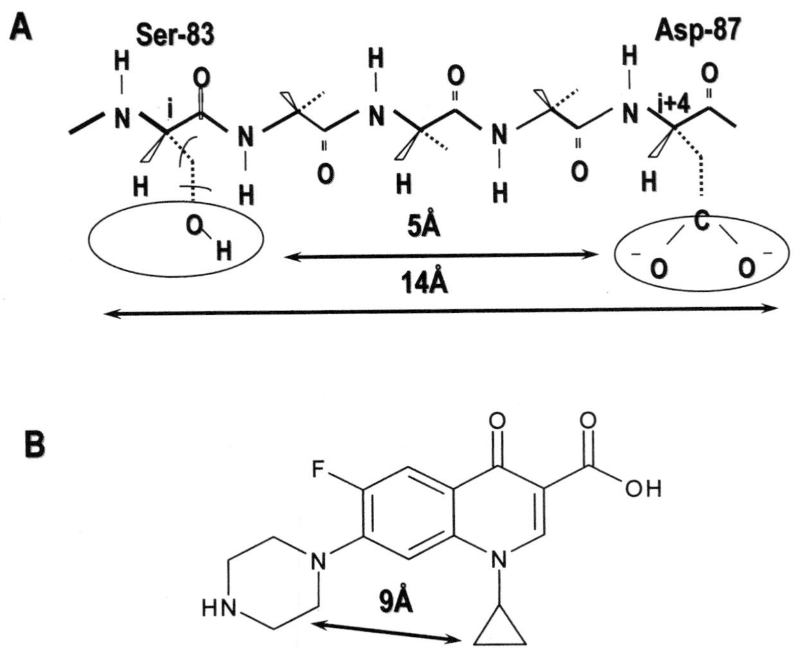

Figure 4. (A) Distance between Ser-83 and Asp-87 of the GyrA protein. (B) Distance between substituents at positions 1 and 7 of the quinolone molecule.

MECHANISMS OF RESISTANCE

The acquisition of quinolone resistance is mainly due to chromosomal mutations, although a plasmid mediating quinolone resistance has recently been described (21). The chromosomal mutations can be distributed into two groups: (i) mutations in topoisomerases genes (*gyrA, gyrB, parC, and parE*), and (ii) mutations causing reduced drug accumulation, either by a decreased uptake or by increased efflux (Table 2).

Topoisomerase Mutations

Escherichia coli and *P. aeruginosa* are the two microorganisms in which the mechanisms of resistance to quinolones have been investigated in depth. In *E. coli,* mutations related to quinolone resistance acquisition have been located in a region of the *gyr*A gene known as the quinolone resistance determining region (QRDR). This region spans Ala-67 to Gln-106 and is situated near the Tyr-122, where the DNA gyrase binds to DNA. Changes involving Gly-81, Ser-83, Ala-84, and Asp-87 have been found in resistant mutants obtained in laboratory and clinical isolates, whereas mutations at amino acid codons Ala-67, Asp-82, and Gln-106 have only been described in laboratory-obtained resistant mutants (68). Substitution outside this region, such as the amino acid change Ala-51 to Val, has also been described in an in vitro resistant mutant (16). However, among these changes, those at Ser-83 and Asp-87 are the most frequently found in clinical isolates, and they have been related to a moderate level of resistance to quinolones (64). Mutations in the *parC* QRDR have also been shown in clinical isolates of *E. coli* (66). These changes involve amino acid substitutions of Ser-80 to Arg or Ile and Glu-84 to Val or Lys. Mutations in the *gyrB* and *parE* genes are not important in the acquisition of quinolone resistance in *E. coli* clinical isolates. Overall, one change in GyrA is related to a decreased susceptibility (ciprofloxacin MIC of c. 0.25 mg/liter), while two changes, one in GyrA and a second in ParC, are associated with a moderate level of resistance to ciprofloxacin (MIC, 0.5 to 1 mg/liter). Three amino acid substitutions (two in GyrA and one in ParC or one in GyrA and two in ParC) are associated with a high level of resistance to ciprofloxacin (MIC, 4 mg/liter), and four changes (two in GyrA and two in ParC) are associated with the highest level of resistance (MIC 16-32 mg/liter).

Membrane Permeability

Variable MICs of different quinolones in two strains showing the same substitutions in GyrA and/or ParC can be attributed to differences in drug accumulation, either by a decrease in permeability of the outer membrane or an increase of efflux of the drug out of the cell. The penetration of quinolones through the outer membrane seems to take place in two ways, one

Table 2. Mechanisms of resistance to fluoroquinolones in different microorganisms

Microorganism	MIC (mg/liter) CIP	Amino acid substitutions[a]		Efflux pumps[b]
		GyrA	ParC	
Escherichia coli[c]	0.03			AcrAB-TolC, AcrEF-TolC, MdfA, YdhE
	0.25	Ser83-Leu		
	1	Ser83-Leu	Ser80-Ile	
	8	Ser83-Leu // Asp87-Asn	Ser80-Ile	
	64	Ser83-Leu // Asp87-Asn	Ser80-Ile // Glu84-Lys	
Pseudomonas aeruginosa	0.1			MexAB-OprM, MexCD-OprJ, MexEF-OprN, MexXY-OprM, MexVW-OprM
	2	Thr83-Ile		
	16	Thr83-Ile	Ser80-Leu	
	64	Thr83-Ile // Asp87-Asn	Ser80-Leu	
Campylobacter spp.	0.06		NP[d]	CmeABC
	>32	Thr86-Ile	NP	
	>32	Thr86-Ile // Asp90-Asn		
Streptococcus pneumoniae	0.5			PmrA
	2		Ser79-Phe	
	16	Ser81-Phe	Ser79-Phe	
Staphylococcus aureus	0.06			NorA
	1		Ser80-Phe	
	4	Ser84-Leu	Ser80-Phe	
	32	Ser84-Leu	Ser80-Phe // Glu84-Gly	

[a] These are the most frequently found substitutions.
[b] The main efflux pumps for each microorganism are mentioned but are not related to the MIC in the table.
[c] Can be extrapolated to the remaining *Enterobacteriaceae*.
[d] This microorganism does not have topoisomerase IV.

dependent and the other independent of porins. It has been suggested that the more hydrophilic quinolones cross the outer membrane mainly through the porins (40), whereas the more hydrophobic quinolones penetrate through porin and nonporin pathways (6). However, a decrease in membrane permeability in *Enterobacteriaceae* is mainly related to a minor or complete loss of expression of porins (41). Nevertheless, this contribution of the outer membrane barrier is not seen clearly unless its effect is multiplicatively amplified by additional resistance mechanisms, such as drug efflux (39).

Efflux

Multidrug efflux systems often pump out a wide variety of drugs with different chemical structures (48). In *E. coli,* the operon *acrAB* encodes a stress-induced efflux system. AcrB is an efflux transport spanning the inner membrane, and AcrA is a membrane fusion protein. AcrAB functions with an outer membrane protein (TolC) encoded at a separate location of the chromosome. Clinical isolates of *E. coli* showing high-level fluoroquinolone resistance contain mutations in the *gyrA* and *parC* genes and overproduce AcrAB (35, 71). The inhibition of AcrAB or the loss of expression of the *acrAB* operon can decrease the MIC of quinolones below the breakpoint, despite the presence of mutation(s) in the *gyrA* gene (42, 55), implying a certain basal expression of this efflux pump in intrinsic resistance. To date, AcrAB-TolC is the best characterized drug efflux pump in *E. coli* clinical isolates. Other efflux pumps in this microorganism have been shown to extrude quinolones; however, their role as a mechanism of quinolone resistance in *E. coli* clinical isolates has not been deeply investigated. Several transcriptional factors such as MarA, SoxS, and Rob have been found in *E. coli* as well as in other *Enterobacteriaceae* (8, 9). MarA, SoxS, and Rob activate the transcription of the *acrAB* operon. However, Rob appears to be synthesized constitutively, unlike MarA and SoxS. Concomitantly, overexpression of MarA and SoxS decreases the synthesis of the porin OmpF of *E. coli* by induction of *micF,* whose antisense RNA products interacts with *ompF* mRNA to prevent OmpF translation.

Plasmid-Mediated Resistance

Recently, a *Klebsiella pneumoniae* clinical isolate carrying a plasmid which confers low level quinolone resistance has been reported (34). This phenotype was subsequently associated with the presence of the *qnr* gene, which was located within an integron-like environment, and encoded a protein that protects DNA gyrase from quinolones (62). Although it was first thought that the prevalence of this gene among *K. pneumoniae* and *E. coli* isolates was very low (23), plasmid-mediated quinolone resistance associated with *qnr* has recently been found in 8% of the quinolone-resistant clinical strains of *E. coli* from Shanghai, China (72) and 11.1% of the quinolone-resistant clinical strains of *K. pneumoniae* from the United States (73).

FLUOROQUINOLONE RESISTANCE IN THE CLINICAL SETTING

The consumption of quinolones in the last decade has been spectacular. The level of quinolone resistance has been associated with the use of these antibacterial agents both in humans and animals (14). Thus, in industrialized countries, where the use of quinolones is a common practice, the most frequently consumed antibacterial agent is currently ciprofloxacin (27), although the consumption of new fluoroquinolones such as levofloxacin and moxifloxacin has increased in the last years (27).

Traveler's Diarrhea

In our hospital, in 1988 all *E. coli* strains isolated were susceptible to fluoroquinolones; however, the quinolone resistance has steadily risen, reaching a level of 26% in 2003. The most spectacular example of the increase in fluoroquinolone resistance has been observed in *Campylobacter jejuni,* with 76% of the isolates in our hospital now being resistant to quinolones (52), whereas the antimicrobial resistance in *C. jejuni* causing traveler's diarrhea was only 12.5% (70). The high level of fluoroquinolone resistance observed in some countries is probably due to two factors. First, this microorganism lacks topoisomerase IV; therefore a mutation in the *gyrA* gene is sufficient to increase the MIC of ciprofloxacin and levofloxacin above 32 mg/liter, whereas two mutations are necessary in the *gyrA* gene to generate a high level (>32 mg/liter) of resistance to moxifloxacin (54). Second, the constitutive expression of the CmeABC pump contributes to the intrinsic resistance (31). A similar phenomenon is observed in other microorganisms, such as *Corynebacterium* spp., lacking topoisomerase IV (60).

The high rate of quinolone resistance in *C. jejuni* has been associated with the introduction of enrofloxacin in veterinary usage (14), and obviously antibiotics are not used in animals in most developing countries. Overall, in the developing countries, the levels of quinolone resistance among gram-negative bacteria are very low (70), probably due to the low use of these antimicrobial agents, associated with their high cost. However, on analyzing enterotoxigenic *E. coli* as a cause of traveler's diarrhea, we found that 17% of enterotoxigenic *E. coli* isolated from travellers to India showed nalidixic acid resistance (70). This is likely due to the increase in the use of quinolones in this country in the last few years. It is important to point out that in several countries, such as Spain, an increase has been observed in the resistance to nalidixic acid in some enteropathogens, such as *Y. enterocolitica, Salmonella* spp., and *Aeromonas* spp. However, this is not accompanied by a high level of fluoroquinolone resistance in these microorganisms (5, 44, 56, 69).

Nonfermentative Gram-Negative Bacilli

P. aeruginosa, Acinetobacter baumannii, and *Stenotrophomonas maltophilia* are the three most relevant nonfermentative gram-negative bacteria. These microorganisms usually present multiresistance. The quinolone-susceptible strains of these microorganisms show higher baseline MICs for quinolones and other antimicrobial agents in comparison with *Enterobacteriaceae.* The presence of a few small-sized porins in the outer membrane of these microorganisms can explain, in part, this intrinsic resistance; however, the decreased permeability presents a synergistic interplay with constitutively efflux pump expression. Five different efflux pumps have been shown in *P. aeruginosa* (MexAB-OprM, MexCD-OprJ, MexEF-OprN, MexXY-OprM, MexVW-OprM), all of which affect fluoroquinolones (48). Based on the genome sequence, more efflux pumps are encoded in *P. aeruginosa,* but their potential roles in antibacterial resistance have not been investigated yet. Among the quinolone-resistant associated efflux pumps, the MexAB-OprM pump can be constitutively expressed, thereby playing an important role in intrinsic resistance. In *A. baumannii,* the AdeABC pump has been shown to be related to quinolone resistance; however, its constitutive expression in wild-type strains has not been shown. The role of mutations in the *gyrA* and *parC* genes as quinolone resistant determinants in these microorganisms is similar to that described in *Enterobacteriaceae.* Thus, the first point mutation is in amino acid codon 83, which consists of a serine residue in *A. baumannii* and threonine instead of serine in *P. aeruginosa.* Other mutations in the *gyrA* gene (amino acid codon Asp-87) and mutations in the *parC* gene (amino acid codons Ser-80 and Glu-84) in these species contribute to increasing the level of resistance to quinolones (65, 67). Similar to what has been described in *Enterobacteriaceae,* the overexpression of one or more of the efflux pumps in *P. aeruginosa* and *A. baumannii* can contribute, in collaboration with mutations in the *gyrA* and *parC* genes, to modulate the final fluoroquinolone MIC (38).

Quinolone resistance in *S. maltophilia* is slightly different from that described above for *P. aeruginosa* and *A. baumannii.* In this species, glutamine instead of serine or threonine is found in position 83 in both quinolone-susceptible and -resistant strains (50). This may explain, in part, the decreased susceptibility to quinolones in this microorganism, since serine or threonine seem to be the optimal amino acid for the formation of the complex quinolone-DNA-gyrase. To date, no mutations in the *parC* gene have been found to be associated with quinolone resistance in *S. maltophilia.* Recent evidence indicates that quinolone efflux may contribute to the intrinsic and acquired resistance of this microorganism (50). Two different efflux pumps have been described so far in *S. maltophilia,* SmeDEF (2), and SmeABD (28). SmeDEF seems to play an important role in the intrinsic resistance of these microorganisms (76). The overexpression of SmeDEF affects the resistance to quinolones, tetracyclines, and chloramphenicol (1, 76), whereas the overexpression of SmeC (28) produces a resistance phenotype characterized by resistance to aminoglycosides, quinolones, and some β-lactam antibiotics.

Streptococcus pneumoniae

The emergence of multiresistant *S. pneumoniae* clinical isolates may make the new fluoroquinolones, such as levofloxacin and moxifloxacin, good alternatives for the treatment of infections caused by this microorganism in geographical areas with a high rate of multiresistance (4). In the last few years a slight decrease has been observed in the susceptibility to ciprofloxacin (29). In a study carried out in Canada, the prevalence of *S. pneumoniae* with reduced susceptibility to fluoroquinolones (defined as MIC of ciprofloxacin ≥4 mg/liter) increased from 0% in 1993 to 1.7% in 1997–98 (7). Recently, another study also performed in Canada with *S. pneumoniae* clinical isolates collected in 2000 showed that 20 (62%) out of 32 ciprofloxacin-resistant strains were also resistant to levofloxacin (30). In the United States, the resistance of *S. pneumoniae* to fluoroquinolones remains below 1%. However, the number of strains with a mutation in the *parC* gene has risen from 0.4% in 1992 to 1996 to 4.5 in 1999 to 2000 (74). These strains, which already show a mutation in the *parC* gene, can easily acquire a second mutation in the *gyrA* gene, increasing the MIC of fluoroquinolones above the breakpoint (11). In a study developed in Hong Kong, 12.1% of *S. pneumoniae* clinical isolates were resistant to fluoroquinolones; however, it was further observed that this was due to a dissemination of a multiresistant clone (19, 20).

The resistance to fluoroquinolones in this microorganism is associated with mutations in the *parC* and *gyrA* genes and overexpression of some efflux pump(s). Although strains showing a mutation in the *gyrA* gene without mutations in the *parC* genes have been isolated, most clinical isolates first present a mutation in the *parC* gene. The overexpression of an efflux pump, such as that encoded in the *pmrA* gene (18, 47) and probably others not yet characterized, may play a supplementary role in the final MIC. In in vitro selected quinolone-resistant mutants, efflux may play a primary role. These efflux pump(s) do not seem to affect resistance to different fluoroquinolones to the same degree (46).

Staphylococcus aureus

At the time of their introduction, fluoroquinolones showed good activity against methicillin-susceptible and -resistant *S. aureus* (17). However, the sporadic emergence of resistance to the new fluoroquinolones has been reported (22) in *S. aureus* isolated from the skin flora of patients during ciprofloxacin therapy, as well as in the skin flora during the treatment of methicillin-resistant *S. aureus* carriers (37). Overall, the resistance is higher in methicillin-susceptible than in methicillin-resistant *S. aureus* isolates (61). Although it has been reported that the primary target for quinolones in *S. aureus* is topoisomerase IV, due to the fact that the first mutation associated with quinolone resistance is found in the *parC* gene (*grlA* gene in *S. aureus*), some recent studies have suggested that the primary target may be different depending on the structure of quinolones (53). Single ParC amino acid change seems to be sufficient to cause clinical resistance to ciprofloxacin but not to the newest fluoroquinolones such as sparfloxacin, levofloxacin, and trovafloxacin (58). Efflux encoded by the *norA* gene plays a role in *S. aureus* fluoroquinolone resistance (24, 75). NorA is one of the efflux systems related to fluoroquinolone resistance, and it seems that the increase in the level of resistance provided by NorA is due to the overexpression of the gene associated with a mutation in the promoter region (25). Therefore, as in other microorganisms the biochemical mechanisms of quinolone resistance are substitutions in certain amino acids of the A subunits of DNA gyrase (Ser-84 and Glu-88) and topoisomerase IV (Ser-80 and Glu-84) and fluoroquinolone efflux.

FACTORS FAVORING THE EMERGENCE OF QUINOLONE RESISTANCE

Resistance can emerge not only in the target pathogen, but also in the normal flora of the gut, skin, and throat. Several factors influence the emergence of quinolone resistance in both ecological niches (Table 3). These can be categorized as (i) factors dependent on the drug, (ii) factors dependent on the bacterium, and (iii) factors dependent on the host and others. Among the factors dependent on the drug, the first to be considered is the concentration of drug achieved in the infection site or in the epithelia, which, to be optimum, should be at least 10-fold the MIC of the drug for the microorganism causing the infection (3). Another factor is the time of exposure, considered as the time during which the concentration of the drug is above the MIC. Moreover, long courses of antimicrobial agents are more likely to encourage the emergence of resistance than shorter courses, and finally, the mutagenicity of quinolones can promote the emergence of resistance. We have recently shown that levofloxacin and moxifloxacin can select in vitro *S. aureus* and *S. pneumoniae* resistant mutants in a low rate in comparison with other fluoroquinolones such as ciprofloxacin, trovafloxacin, and clinafloxacin, and this event may be associated with the fact that moxifloxacin and levofloxacin are the least mutagenic quinolones (59).

An important factor dependent on the bacterium is the inoculum. The higher the inoculum, the greater the probability of spontaneous mutation associated with quinolone resistance. For some microorganisms, fluoroquinolone resistance occurs at high frequencies in vitro. For example, mutations in *P. aeruginosa* fluoroquinolone resistance occurs at a frequency of 10^{-7} to 10^{-8} cells per generation when exposed to a concentration of fourfold the MIC. A high bacterial inoculum is often present in infections such as abscesses, empyemas, and chronic lung infections in patients

Table 3. Factors favoring emergence of quinolone resistance

Factors dependent on the quinolone
- Relationship between the concentration of drug in the infection or colonization site and the MIC of the quinolone for the bacteria causing infection or colonization
- Ratio between time of drug exposure and rapidity of bactericidal activity
- Duration of the treatment
- Mutagenicity of the quinolone

Factors dependent on the microorganism
- Density of bacterial population
- Stationary-phase growth, "quorum sensing"[a]
- Hypermutability
- Capability and possibility to produce biofilm[a]
- Facility to acquire quinolone resistance

Factors dependent on the host
- Pharmacokinetics of the drug at the infection or colonization site
- Immune status
- Environment at the infection site (e.g., pH and low oxygen tension)

[a] These factors generate a transitory resistance.

with cystic fibrosis. In fact, *P. aeruginosa* clinical isolates causing lung infections in patients with cystic fibrosis have been shown to often display high mutation rates (hypermutable) (43). These strains are probably generated by alterations in DNA repair and error-avoidance genes (36) and may explain, in part, the high antibiotic resistance in *P. aeruginosa* isolates from cystic fibrosis patients. In addition, many cells in a large inoculum are slow growing stationary-phase cells. During the late log and early stationary phases of growth, synthesis of the MexAB-OprM efflux pump in *P. aeruginosa* has been shown to increase up to sevenfold (15), thereby decreasing quinolone accumulation. The resulting low level of resistance enables some cells to survive the initial killing effect of quinolones, favoring the development of higher resistance by mutations in the *gyrA* and/or *parC* genes.

Quorum sensing allows bacteria to detect the density of their own species. *P. aeruginosa* have at least three quorum sensing systems that interact with each other. These systems are LasR/LasI, LuxR/LuxI, and 2-heptyl-3-hydroxy-4-quinolone (13). Some of these proteins act as inducers, and when large numbers of bacteria are present the concentration of these inducers reaches a critical threshold and various genes become up- and down-regulated. Therefore, quorum sensing controls the expression of virulence factors, the formation of biofilms, the entry into a stationary phase, the conjugal transfer of plasmid DNA, spore formation, and transformation competence (13). SdiA is an *E. coli* protein homologous to LuxR that regulates cell division in a cell density-dependent, or quorum sensing manner. It has recently been shown that overproduction of SdiA increases levels of AcrAB and, hence, confers multidrug resistance, including decreased fluoroquinolone susceptibility (49).

Biofilm can be defined as aggregates of cells, the bottom of the biofilm being attached to a surface and embedded in a protective polysaccharide. An important feature of all biofilms is their remarkable resistance to antibiotic eradication, due to mechanisms that remain unclear (10). When cells exist in biofilm they can become 10 to 100 times more resistant to the effects of antimicrobial agents than planktonic cells. Interestingly, despite the probable in vivo importance of the biofilm mode of growth, known multidrug efflux systems do not play a significant role in the antimicrobial resistance in biofilms of *E. coli* or *P. aeruginosa* (10). However, this situation does not rule out the possibility that additional, not yet characterized efflux systems may be necessary to assess the role of efflux pumps in biofilm resistance to antimicrobial agents. Differences in bacterial density throughout the biofilm determine the gradients of nutrient and oxygen availability within the biofilm structure. Bacteria located in the biofilm periphery have better access to nutrients and oxygen than bacteria located deeper within the biofilm community. Transition from exponential to slow or no growth is generally accompanied by an increase in resistance to antimicrobial agents. This metabolic heterogeneity can lead to differences in susceptibility to antimicrobial agents. In fact, it has recently been described that antimicrobial agents are able to kill only the bacteria located in zones with increased metabolic activity and oxygen concentration, suggesting that these two situations are more relevant in biofilm formation (33).

SUMMARY

Fluoroquinolone resistance has dramatically increased in the last decade, particularly in some microorganisms. The molecular bases of fluoroquinolone resistance are similar in all microorganisms mainly due to mutations in the *gyrA* and *parC* genes, encoding the A subunits of the DNA gyrase and topoisomerase IV, the protein targets for quinolones. However, mutations in genes associated with a decrease in drug accumulation or an increased efflux of the drug can also contribute to the final level of quinolone resistance. Recently, a plasmid mediating quinolone resistance has been described in *E. coli* and *K. pneumoniae*. Several models of microorganisms can be considered according to the effect of mutations on resistance to fluoroquinolones. Among these we can emphasize (i) *Enterobacteriaceae*, which are among this group of microorganisms in which multiple mutations in the *gyrA* and *parC* genes are required to generate clinically important resistance; (ii) microorganisms such as *P. aeruginosa* and *A. baumannii*, in which a single mutation is sufficient to cause clinically important levels of resistance to fluoroquinolones since they already show an intrinsic resistance to these antibacterial agents; (iii) *C. jejuni* and other microorganisms lacking topoisomerase IV, in which a single mutation in the *gyrA* gene produces a high level of resistance; and (iv) *S. pneumoniae* and *S. aureus*, in which the mutations in the *parC* gene seem to be a priority over mutations in the *gyrA* gene.

REFERENCES

1. **Alonso, A., and J. L. Martínez.** 1997. Multiple antibiotic resistance in *Stenotrophomonas maltophilia*. *Antimicrob. Agents Chemother.* **41:**1140–1142.
2. **Alonso, A., and J. L. Martinez.** 2000. Cloning and characterization of SmeDEF, a novel multidrug efflux pump from *Stenotrophomonas maltophilia*. *Antimicrob. Agents Chemother.* **44:** 3079–3086.
3. **Andes, D. R., and W. Craig.** 1998. Pharmacodynamics of fluoroquinolone in experimental models of endocarditis. *Clin. Infect. Dis.* **27:**47–50.

4. **Bartlett, J. G., R. F. Breiman, L. A. Mandell, and T. M. File, Jr.** 1998. Community-acquired pneumonia in adults: guidelines for management. *Clin. Infect. Dis.* **26:**811–838.
5. **Capilla, S., J. Ruiz, P. Goñi, J. Castillo, M. C. Rubio, M. T. Jiménez de Anta, R. Gómez-Lus, and J. Vila.** 2004. Characterization of the molecular mechanisms of quinolone resistance in *Yersinia enterocolitica* O:3 clinical isolates. *J. Antimicrob. Chemother.*, in press.
6. **Chapman, J. S., and N. H. Geogopapadokou.** 1988. Routes of quinolone permeation in *Escherichia coli. Antimicrob. Agents Chemother.* **32:**438–442.
7. **Chen, D. K., A. McGeer, J. C. de Azavedo, D. E. Low, and the Canadian Bacterial Surveillance Network.** 1999. Decreased susceptibility of *Streptococcus pneumoniae* to fluoroquinolones in Canada. *N. Engl. J. Med.* **341:**233–239.
8. **Cohen, S. P., L. M. McMurry, D. C. Hooper, J. S. Wolfson, and S. B. Levy.** 1989. Cross-resistance to fluoroquinolones in multiple-antibiotic-resistant (Mar) *Escherichia coli* selected by tetracycline or chloramphenicol: decreased drug accumulation associated with membrane changes in addition to OmpF reduction. *Antimicrob. Agents Chemother.* **33:**1318–1325.
9. **Cohen, S. P., H. Hächler, and S. B. Levy.** 1993. Genetic and functional analysis of the multiple antibiotic resistance (mar) locus in *Escherichia coli. J. Bacteriol.* **175:**1484–1492.
10. **Davies, D.** 2003. Understanding biofilm resistance to antibacterial agents. *Nature Rev.* **2:**114–122.
11. **Davies, T. A., A. Evangelista, S. Pfleger, K. Bush, D. F. Sahm, and R. Goldschmidt.** 2002. Prevalence of single mutations in topoisomerase type II genes among levofloxacin-susceptible clinical strains of *Streptococcus pneumoniae* isolated in the United States in 1992 to 1996 and 1999 to 2000. *Antimicrob. Agents Chemother.* **46:**119–124.
12. **Drlica, K., and X. Zhao.** 1997. DNA gyrase, topoisomerase IV, and the 4-quinolones. *Microbiol. Mol. Biol. Rev.* **61:**377–392.
13. **Duny, G. M., and S. C. Winans** (ed). 1999. *Cell-Cell Signalling in Bacteria.* American Society for Microbiology, Washington, D.C.
14. **Endtz, H. P., G. J. Ruijs, B. Van Kingleren, W. H. Jansen, T. Van der Reyden, and R. P. Mouton.** 1991. Quinolone resistance in campylobacter isolated from man and poultry following the introduction of fluoroquinolones in veterinary practice. *J. Antimicrob. Chemother.* **27:**199–208.
15. **Evans, K., and K. Poole.** 1999. The MexA-MexB-OprM multidrug efflux system of *Pseudomonas aeruginosa* is growth phase regulated. *FEMS Microbiol. Let.* **173:**35–39.
16. **Friedman, S. M, T. Lu, and K. Drlica.** 2001. Mutation in the DNA gyrase A gene of *Escherichia coli* that expands the quinolone resistance-determining region. *Antimicrob. Agents Chemother.* **45:**2378–2380.
17. **Gahin-Hausen, B., P. Joogard, and M. Arpi.** 1987. In vitro activity of ciprofloxacin against methicillin-sensitive and methicillin-resistant staphylocci. *Eur. J. Clin. Microbiol.* **6:**581–584.
18. **Gill, M. J., N. P. Brenwald, and R. Wise.** 1999. Identification of an efflux pump gene, pmrA, associated with fluoroquinolone resistance in *Streptococcus pneumoniae. Antimicrob. Agents Chemother.* **43:**187–189.
19. **Ho, P. L., T. L. Que, D. N. Tsang, T. K. Ng, K. H. Chow, and W. H. Seto.** 1999. Emergence of fluoroquinolone resistance among multiple resistant strains of *Streptococcus pneumoniae* in Hong Kong. *Antimicrob. Agents Chemother.* **43:**1310–1313.
20. **Ho, P. L., W. C. Yam, T. L. Que, D. N. Tsang, T. K. Ng, K. H. Chow, and W. H. Seto.** 2001. Target site modifications and efflux phenotype in clinical isolates of *Streptococcus pneumoniae* from Hong Kong with reduced susceptibility to fluoroquinolones. *J. Antimicrob. Chemother.* **48:**731–734.
21. **Hooper, D.** 2001. Emerging mechanisms of fluoroquinolone resistance. *Emerg. Infect. Dis.* **7:**337–341.
22. **Humphryes, H., and E. Mulvihill.** 1985. Ciprofloxacin-resistant *Staphylococcus aureus. Lancet* **ii:**383.
23. **Jacoby, G. A., N. Chow, and K. B. Waites.** 2003. Prevalence of plasmid-mediated quinolone resistance. *Antimicrob. Agents Chemother.* **47:**559–562.
24. **Kaatz, G. W., S. M. Seo, and C. A. Ruble.** 1993. Efflux-mediated fluoroquinolone resistance in *Staphylococcus aureus. Antimicrob. Agents Chemother.* **37:**1086–1094.
25. **Kaatz, G. W., and S. M. Seo.** 1997. Mechanisms of fluoroquinolone resistance in genetically related strains of *Staphylococcus aureus. Antimicrob. Agents Chemother.* **41:**2733–2737.
26. **Kreuzer, K. N., and N. R. Cozzarelli.** 1979. *E. coli* mutants thermosensitive for deoxyribonucleic acid gyrase subunit A: Effects on deoxyribonucleic acid replication, transcription and bacteriophage growth. *J. Bacteriol.* **140:**424–435.
27. **Lázaro, E., J. Oteo, G. Baquero, F. J. de Abajo, and J. Campos.** 2004. Evolución del consumo y resistencia a quinolonas en la comunidad en España. *In XI Congreso de la SEIMC.* Bilbao, Spain.
28. **Li, X. Z., L. Zhang, and K. Poole.** 2002. SmeC, an outer membrane multidrug efflux protein of *Stenotrophomonas maltophilia. Antimicrob. Agents Chemother.* **46:**333–343.
29. **Liñares, J., A. G. de la Campa, and R. Pallares.** 1999. Fluoroquinolone resistance in *Streptococcus pneumoniae. N. Engl. J. Med.* **341:**1546–1548.
30. **Low, D. E, J. de Azavedo, K. Weiss, T. Mazzulli, M. Kuhn, D. Church, K. Forward, G. Zhanel, A. Simor, and A. McGeer.** 2002. Antibiotic resistance among clinical isolates of *Streptococcus pneumoniae* in Canada during 2000. *Antimicrob. Agents Chemother.* **46:**1295–1301.
31. **Luo, N., O. Sahin, J. Lin, L. O. Michel, and Q. Zhang.** 2003. In vivo selection of *Campylobacter* isolates with high levels of fluoroquinolone resistance associated with *gyrA* mutations and the function of the CmeABC efflux pump. *Antimicrob. Agents Chemother.* **47:**390–394.
32. **Manjunatha, U. H., M. Dalal, M. Chatterji, D. R. Radha, S. S. Wiweswariah, and V. Nagaraja.** 2002. Functional characterization of mycobacterial DNA gyrase: an efficient decatenase. *Nucleic Acids Res.* **30:**2144–2153.
33. **Marshall, C., I. Walters, F. Roe, A. Bugnicourt, M. J. Franklin and P. S. Steward.** 2003. Contribution of antibiotic penetration, oxygen limitation and low metabolic activity to tolerance of *Pseudomonas aeruginosa* biofilm to ciprofloxacin and tobramycin. *Antimicrob. Agents Chemother.* **47:**317–321.
34. **Martínez-Martínez, L., A. Pascual, and G. A. Jacoby.** 1998. Quinolone resistance from a transferable plasmid. *Lancet* **351:**797–799.
35. **Mazzariol, A., Y. Tokue, T. M. Kanegawa, G. Cornaglia, and H. Nikaido.** 2000. High-level fluoroquinolone-resistant clinical isolates of *Escherichia coli* overproduce multidrug efflux AcrA. *Antimicrob. Agents Chemother.* **44:**3441–3443.
36. **Miller, J. H.** 1996. Spontaneous mutators in bacteria: insights into pathways of mutagenesis and repair. *Annu. Rev. Microbiol.* **50:**625–643.
37. **Mulligan, M. E., P. J. Ruane, and L. Johnson.** 1987. Ciprofloxacin for eradication of methicillin-resistant *Staphylococcus aureus* colonization. *Am. J. Med.* **82(suppl. 4):**580–589.
38. **Nakano, M., T. Deguchi, T. Kawamura, M. Yasuda, M. Kimura, Y. Okano, and Y. Kawada.** 1997. Mutations in the *gyrA* and *parC* genes in fluoroquinolone-resistant clinical isolates of *Pseudomonas aeruginosa. Antimicrob. Agents Chemother.* **41:**2289–2291.

39. **Nikaido, H.** 1996. Multidrug efflux pumps of gram-negative bacteria. *J. Bacteriol.* **178:**5853–5859.
40. **Nikaido, H.** 2001. Preventing drug access to targets: cell surface permeability barriers and active efflux in bacteria. *Semin. Cell Dev. Biol.* **12:**215–223.
41. **Nikaido, H.** 2003. Molecular basis of bacterial outer membrane permeability revisited. *Microbiol. Mol. Biol. Rev.* **67:**593–656.
42. **Oethinger, M., W. V. Kern, A. S. Jellen-Ritter, L. McMurry, and S. B. Levy.** 2000. Ineffectiveness of topoisomerase mutations in mediating clinically significant fluoroquinolone resistance in *Escherichia coli* in the absence of the AcrAB efflux pump. *Antimicrob. Agents Chemother.* **44:**10–13.
43. **Oliver, A., R. Cantón, P. Campo, F. Baquero, and J. Blázquez.** 2000. High frequency of hypermutable *Pseudomonas aeruginosa* in cystic fibrosis lung infection. *Science.* **288:**1251–1253.
44. **Olivera, S., F. J. Castillo, M. T. Llorente, A. Clavel, M. Varea, C. Seral, and M. C. Rubio.** 2002. Antimicrobial resistanse of clinical strains of *Salmonella enterica* isolated in Zaragoza. *Rev. Esp. Quimioter.* **15:**152–155.
45. **Palumbo, M., B. Gatto, G. Zagotto, and G. Palú.** 1993. On the mechanisms of action of quinolone drugs. *Trends Microbiol.* **1:**232–234.
46. **Pestova, E., J. J. Millichap, F. Siddiqui, G. A. Noskin, L. R. Peterson.** 2002. Non-pmrA-mediated multidrug resistance in *Streptococcus pneumoniae. J. Antimicrob. Chemother.* **49:**553–556.
47. **Piddock, L. J. V., M. M. Johnson, S. Simjee, and L. Pumbwe.** 2000. Expression of effluxs gene pmrA in fluoroquinolone-resistant and -susceptible clinical isolates of *Streptococcus pneumoniae. Antimicrob. Agents Chemother.* **46:**808–812.
48. **Poole, K.** 2000. Efflux-mediated resistance to fluoroquinolones in gram-negative bacteria. *Antimicrob. Agents Chemother.* **44:**2233–2241.
49. **Rahmati, S., S. Yang, A. L. Davidson, and E. L. Zechiedrich.** 2002. Control of the AcrAB multidrug efflux pump by quorum-sensing regulator SdiA. *Mol. Microbiol.* **43:**677–685.
50. **Ribera, A., A. Doménech-Sánchez, J. Ruiz, V. J. Benedi, M. T. Jiménez de Anta, and J. Vila.** 2002. Mutations in *gyrA* and *parC* QRDRs are not relevant for quinolone resistance in epidemiological unrelated *Stenotrophomonas maltophilia* clinical isolates. *Microb. Drug Resist.* **8:**245–251.
51. **Roca, J.** 1995. The mechanisms of DNA topoisomerases. *Trends Biochem. Sci.* **20:**156–160.
52. **Ruiz, J., P. Goñi, F. Marco, F. Gallardo, B. Mirelis, M. T. Jiménez de Anta, and J. Vila.** 1998. Increased resístanse in *Campylobacter jejuni.* A genetic analysis of the gyrA gene mutation in ciprofloxacin resistant clinical isolates. *Microbiol. Immunol.* **42:**223–226.
53. **Ruiz, J., J. M. Sierra, M. T. Jimenez de Anta, and J. Vila.** 2001. Characterization of in vitro obtained sparfloxacin-resistant mutants of *Staphylococcus aureus. Int. J. Antimicrob. Chemother.* **18:**107–112.
54. **Ruiz, J., A. Moreno, M. T. Jiménez de Anta, and J. Vila.** A double mutation in the *gyr*A gene is necessary to produce a high level of resistance to moxifloxacin in *Campylobacter* spp. clinical isolates. *Int. J. Antimicrob. Agents,* in press.
55. **Saenz, Y., J. Ruiz, M. Zarazaga, M. Teixidó, C. Torres, and J. Vila.** 2004. Effect of the efflux pump inhibitor Phe-Arg-β–naphthylamide on the MIC values of quinolones, tetracycline and chloramphenicol, in *Escherichia coli* isolates of different origin. *J. Antimicrob. Chemother.* **53:**544–545.
56. **Sánchez-Cespedes, J., M. M. Navia, R. Martínez, B. Orden, B. Millán, J. Ruiz, and J. Vila.** 2003. Clonal dissemination of *Yersinia enterocolitica* strains with various susceptibilities to nalidixic acid. *J. Clin. Microbiol.* **41:**1769–1771.
57. **Shen, L. L., L. A. Mitscher, P. N. Sharma, T. J. O'Donnell, D. W. T. Chu, C. S. Cooper, T. Rosen, and A. G. Pernet.** 1989. Mechanism of inhibition of DNA gyrase by quinolone antibacterials: a cooperative drug-DNA binding model. *Biochemistry* **28:**3886–3894.
58. **Sierra, J. M., F. Marco, J. Ruiz, M. T. Jiménez de Anta, and J. Vila.** 2002. Correlation between the activity of different fluoroquinolones and the presence of mechanisms of quinolone resistance in epidemiologically related and unrelated strains of methicillin-susceptible and -resistant *Staphylococcus aureus. Clin. Microbiol. Infect.* **8:**781–790.
59. **Sierra, J. M., J. G. Cabeza, M. Ruiz Chaler, T. Montero, J. Hernández, J. Mensa, M. Llagostera, and J. Vila.** The selection of resistanse to and the mutagenicity of different fluoroquinolones in *Staphylococcus aureus* and *Streptococcus pneumoniae. Clin. Microbiol. Infect.,* in press.
60. **Sierra, J. M., L. Martínez-Martínez, F. Vázquez, E. Giralt, and J. Vila.** 2005. Relationship between mutations in the *gyrA* gene and quinolone resistance in clinical isolates of *Corynebacterium striatum* and *Corynebacterium amycolatum. Antimicrob. Agents Chemother.* **49:**1714–1719.
61. **Thornsberry, C.** 1994. Susceptibility of clinical isolates to ciprofloxacin in the United States. *Infection* **22(suppl. 4):**215–219.
62. **Tran, J. H., and G. A. Jacoby.** 2002. Mechanism of plasmid-mediated quinolone resistance. *Proc. Natl. Acad. Sci. USA* **99:**5638–5642.
63. **Ullsperger, C., and N. R. Cozzarelli.** 1996. Contrasting enzymatic activities of topoisomerase IV and DNA gyrase from *Escherichia coli. J. Biol. Chem.* **271:**31549–31555.
64. **Vila, J., J. Ruiz, F. Marco, A. Barceló, P. Goñi, E. Giralt, and M. T. Jiménez de Anta.** 1994. Association between double mutation in the *gyrA* gene of ciprofloxacin-resistant clinical isolates of *Escherichia coli* and minimal inhibitory concentration. *Antimicrob. Agents Chemother.* **38:**2477–2479.
65. **Vila, J., J. Ruiz, P. Goñi, M. A. Marcos, and M. T. Jimenez de Anta.** 1995. Mutations in the *gyrA* gene of quinolone-resistant clinical isolates of *Acinetobacter baumannii. Antimicrob. Agents Chemother.* **39:**1201–1203.
66. **Vila, J., J. Ruiz, P. Goñi, and M. T. Jiménez de Anta.** 1996. Detection of mutations in *parC* in quinolone-resistant clinical isolates of *Escherichia coli. Antimicrob. Agents Chemother.* **40:** 491–493.
67. **Vila, J., J. Ruiz, P. Goñi, and M. T. Jiménez de Anta.** 1997. Quinolone resistance in the topoisomerase IV parC gene in *Acinetobacter baumannii. J. Antimicrob. Chemother.* **39:**757–762.
68. **Vila, J., J. Ruiz, and M. M. Navia.** 1999. Molecular bases of quinolone resistance acquisition in gram-negative bacteria. *Recent Res. Devel. Antimicrob. Agents Chemother.* **3:**323–344.
69. **Vila, J., F. Marco, L. Soler, M. Chacón, and M. J. Figueras.** 2002. *In vitro* antimicrobial susceptibility of clinical isolates of *Aeromonas caviae, Aeromonas hydrophila* and *Aeromonas veronii* biotype sobria. *J. Antimicrob. Chemother.* **49:**697–702.
70. **Vila, J., and S. B. Levy.** 2003. Antimicrobial resistance, p. 58–75. *In* C. D. Ericsson, H. L. Dupont, and R. Steffen (ed.), *Travelers' Diarrhea.* BC. Decker Inc., Hamilton, Ontario, Canada.
71. **Wang, H., J. L. DzinkFox, M. J. Chen, and S. B. Levy.** 2001. Genetic characterization of highly fluoroquinolone-resistant clinical isolates of *Escherichia coli* strains from China: role of *acrR* mutations. *Antimicrob. Agents Chemother.* **45:**1515–1521.
72. **Wang, M., J. H. Tran, G. A. Jacoby, Y. Zhang, F. Wang, and D. C. Hooper.** 2003. Plasmid-mediated quinolone resistance in clinical isolates of *Escherichia coli* from Shangai, China. *Antimicrob. Agents Chemother.* **47:**2242–2248.

73. **Wang, M., D. F. Sahm, G. A. Jacoby, and D. C. Hooper.** 2004. Emerging plasmid-mediated quinolone resistance associated with the *qnr* gene in *Klebsiella pneumoniae* clinical isolates in the United States. *Antimicrob. Agents Chemother.* **48:**1295–1299.

74. **Whitney, C. G., M. M. Farley, J. Hadler, L. H. Harrison, C. Lexau, A. Reingold, L. Lefkowitz, P. R. Cieslak, M. Cetron, E. R. Zell, J. H. Jorgensen, A. Schuchat, and the Active Bacterial Core Surveillance Program of the Emerging Infections Program Network.** 2000. Increasing prevalence of multidrug-resistant *Streptococcus pneumoniae* in the United States. *N. Engl. J. Med.* **343:**1917–1924.

75. **Yoshida, H., M. Bogaki, S. Nakamura, K. Ubukata, and M. Cono.** 1990. Nucleotide sequence and characterization of the *Staphylococcus aureus norA* gene, which confers resistance to quinolones. *J. Bacteriol.* **172:**6942–6949.

76. **Zhang, L., X. Z. Li, and K. Poole.** 2001. SmeDEF multidrug efflux pump contributes to intrinsic multidrug resistance in *Stenotrophomonas maltophilia. Antimicrob. Agents Chemother.* **45:**3497–3503.

Frontiers in Antimicrobial Resistance: a Tribute to Stuart B. Levy
Edited by D. G. White, M. N. Alekshun, and P. F. McDermott

Chapter 5

β-Lactam Resistance in the 21st Century

George Jacoby and Karen Bush

β-Lactamases are the major defense against β-lactam antibiotics, especially in gram-negative bacteria. In gram-positive organisms, such as *Staphylococcus aureus* and *Streptococcus pneumoniae,* altered penicillin-binding proteins contribute to resistance as well and will be considered in the chapters devoted to those organisms. β-Lactamases existed long before antibiotics were used clinically. β-Lactamase-producing *Bacillus licheniformis* have been revived from soil attached to plants collected in the 17th century (119), while an antiquity of two billion years has been estimated based on the phylogeny and current diversity of the serine β-lactamases (49). Their ubiquity in diverse settings has led to much speculation concerning the relationship and transfer of these resistance determinants between the environment and humans. The ecology of such antibiotic resistance is a topic that has been a dominant theme throughout Stuart Levy's career (73).

Hundreds of β-lactamases have been characterized. They can be broadly divided into enzymes with a serine residue at the active site, similar to bacterial penicillin-binding proteins, from which they likely evolved (64), and metallo-enzymes with Zn^{2+} as cofactor and a separate heritage (42). Each group has been further subdivided by structure or function as described below. Based on substrate spectrum and reaction with inhibitors, the entire set can be divided into 11 or more functional groups (16) or into four molecular classes based on structure (1, 62, 102). β-Lactamase nomenclature is confusing and idiosyncratic at best. The enzymes have been named according to the substrates attacked, some biochemical property, the site of their discovery, the name of the patient providing the first sample, the bacterial host, the names of investigators, etc. Stuart Levy would surely find the naming of tetracycline resistance genes far more logical.

CLASSIFICATION SCHEMES

The β-lactamase community currently recognizes two major classification schemes (Table 1) based either on function or structure (16). Historically, these enzymes were separated according to biochemical activities, with classification based on hydrolysis of specific substrates and the sensitivity to selected inhibitors (130). However, when nucleotide sequencing became routinely available and inexpensive, investigators began to use molecular classification most frequently. This was probably because it was almost effortless to obtain a nucleotide sequence compared to the time and resources required to isolate and purify an enzyme for biochemical characterization. Unfortunately, molecular classification does not currently provide quantitative information about the function of the enzyme, thereby limiting its utility in understanding the role of the β-lactamase in resistance.

The functional classification of Bush, Jacoby, and Medeiros (16) attempted to link function with molecular structure as seen in Table 1. Twelve functional groups have currently been described, based primarily on substrate hydrolysis profiles, and are shown with their correspondence to the four molecular classes. Although inhibitor profiles are also listed, greater heterogeneity is seen among the various groups regarding inhibitory activities with clavulanic acid and tazobactam. Thus, inhibitor profiles are generally less predictable than are substrate profiles among each of the functional groups.

Reassuringly, relationships between functional groups and molecular classes are still reasonably well aligned, in spite of the doubling in the number of β-lactamases identified in the 10 years since the scheme was introduced (K. Bush, *104th Gen. Meet. Am. Soc. Microbiol.,* New Orleans, La.). Much of the increase

George Jacoby • Lahey Clinic, Burlington, MA 01805. **Karen Bush** • Johnson & Johnson Pharmaceutical Research & Development, L.L.C., Raritan, NJ 08869.

Table 1. Functional groups of β-lactamases based on Bush, Jacoby, and Medeiros scheme (16)

Functional group	Representative enzymes	Major substrates[a]	Inhibitor profile[b]	Molecular class
1	AmpC	Narrow-spectrum cephalosporins (cefazolin, cephalothin, cefamandole, cefuroxime, others), cephamycins (cefoxitin, cefotetan), oxyiminocephalosporins (cefotaxime, ceftazidime, ceftriaxone, others)	Clav (−), aztreonam (+)	C
2a	*S. aureus* PC1	Benzylpenicillin (penicillin G), aminopenicillins (amoxicillin and ampicillin), carboxypenicillins (carbenicillin and ticarcillin), ureidopenicillin (piperacillin)	Clav (+)	A
2b	TEM-1, SHV-1	Benzylpenicillin, aminopenicillins, carboxypenicillins, ureidopenicillin, narrow-spectrum cephalosporins, cefoperazone	Clav (+)	A
2be	TEM-ESBLs,[c] SHV-ESBLs	Penicillins, all cephalosporins except cephamycins, monobactam (aztreonam)	Clav (+)	A
2br	TEM-IRTs[d] SHV-10	Penicillin inhibitor combinations (ampicillin-sulbactam, piperacillin-tazobactam, ticarcillin-clavulanic acid)	Clav (−), TZB (+)	A
2c	PSE-1	Benzylpenicillin, aminopenicillins, carboxypencillins, ureidopenicillin	Clav (+)	A
2d	OXA-1, OXA-10	Penicillins, (cl)oxacillin, cefoperazone	Clav (+/−)	D
2de[e]	OXA-ESBLs	Penicillins, oxacillin, cephalosporins except cephamycins	Clav (−)	D
2df[e]	OXA-24, OXA-40	Penicillins, oxacillin, carbapenems (ertapenem, imipenem, meropenem)	Clav (+)	D
2f	SME-1, IMI-1, NMC-A, KPC-1	Penicillins, cephalosporins, carbapenems	Clav (+)	A
3	L1, CcrA, VIM-1, IMP-1	Penicillins, cephalosporins, carbapenems	Clav (−), EDTA(+)	B

[a] With exceptions for particular enzymes.
[b] IC_{50} of clavulanic acid that were >10 μM were considered to be Clav(−). IC_{50} that were <1 μM were considered to be Clav(+). IC_{50} that were <1 μM for tazobactam are TZB (+).
[c] ESBL, extended-spectrum β-lactamase.
[d] IRTs, inhibitor-resistant TEM β-lactamases.
[e] New group not included in reference 16. Note that functional group 4 enzymes are not included, as this group has not been fully characterized.

in numbers of enzymes is attributed to the identification of extended-spectrum β-lactamases (ESBLs). Thus, a new subgroup 2de (Table 1) was added to the 1995 scheme, based on the acquisition of cephalosporin-hydrolyzing activities by previously identified oxacillinases (50). In addition, a group 2df was added to acknowledge the recently described carbapenem-hydrolyzing activity of certain class D enzymes (12).

β-LACTAMASE ORIGINS

β-Lactamase genes seem originally to have been located on the bacterial chromosome, where genes for such enzymes as Bla1 and Bla2 β-lactamases of *Bacillus anthracis* (80), or their respective penicillinase and carbapenemase homologs in *Bacillus cereus* (148), are found. The CcrA metallo-enzyme of *Bacteroides fragilis* (126) is also chromosomally determined. Many gram-negative bacilli have a chromosomal *ampC* gene including *Acinetobacter calcoaceticus, Citrobacter freundii, Enterobacter cloacae, Morganella morganii, Proteus rettgeri, Pseudomonas aeruginosa,* and *Serratia marcescens. Escherichia coli* has an *ampC* gene as well, but it is normally expressed at such a low level that it contributes little to resistance (100). *Klebsiella* spp. lack an *ampC* gene, but *Klebsiella pneumoniae* usually has a chromosomal gene for a class A enzyme in the SHV, LEN, or related families (48), while *Klebsiella oxytoca* has chromosomal genes for OXY-1, OXY-2, OXY-3, or OXY-4 β-lactamase (47). Even *Mycobacterium tuberculosis* has several chromosomal-encoded β-lactamases (145).

Plasmids from gram-negative bacteria collected in the preantibiotic era lacked antibiotic resistance genes (25), but the penicillinases that appeared in *S. aureus* after the introduction of penicillin (e.g., PC1) or in *E. coli* following the clinical use of ampicillin (e.g., TEM-1 and TEM-2) were plasmid encoded. In both cases transposons were responsible: Tn*552* and related transposons carrying genes for a staphylococcal β-lactamase and its regulatory system and members of the Tn*3* family carrying a gene for TEM β-lactamase. Subsequently transposons were implicated in the spread of staphylococcal β-lactamase to *Enterococcus faecalis* (90, 128) and of TEM β-lactamase to plasmids in *Haemophilus influenzae* and *Neisseria gonorrhoeae* (32, 33).

Table 2. Known cassette-associated β-lactamase genes

β-Lactamase(s)	Reference(s)
CARB-3, CARB-4	51
CEF-1	36
GES-1, GES-2, GES-3	36, 146
IBC-1, IBC-2	36
IMP-1, IMP-2, IMP-3, IMP-4, IMP-5, IMP-6, IMP-7, IMP-8, IMP-10, IMP-12, IMP-13	24, 28, 36, 51, 58, 140, 151
LCR-1	21
OXA-1, OXA-2, OXA-3, OXA-4, OXA-5, OXA-7, OXA-8, OXA-9, OXA-10, OXA-11, OXA-13, OXA-14, OXA-15, OXA-16, OXA-17, OXA-18, OXA-19, OXA-20, OXA-21, OXA-28, OXA-31, OXA-32	35, 36, 51
PSE-1, PSE-4	51
VEB-1	36
VIM-1, VIM-2, VIM-3, VIM-4, VIM-7	36, 111, 141

Other β-lactamase genes are carried by multiresistance transposons (71), and many of the newly described β-lactamase genes are incorporated into plasmids as gene cassettes within integrons or integron-like structures. Table 2 lists *bla* genes identified in typical gene cassettes usually as components of class 1 integrons determining resistance to sulfonamide, often streptomycin or another aminoglycoside, or trimethoprim. Other *bla* genes, for example CTX-M-2 (3), CTX-M-9 (132), DHA-1 (144), and SPM-1 (117), lack the 59-base element of typical gene cassettes but have been found in class 1 integrons together with Orf513 or a related gene coding for a putative recombinase. In many cases the likely chromosomal origin of the β-lactamase gene is known (Table 3).

For other *bla* genes, insertion sequences IS*Ecp1* (116) or IS*26* (93, 121) have been involved in mobilization to a plasmid location. Sometimes the process captures adjacent chromosomal genes as well. For example, the SHV-5 gene in plasmid pACM1 is located on the remnant of a compound transposon surrounded by defective copies of IS*26* and genes unrelated to resistance that map next to the SHV-1 gene on the *K. pneumoniae* chromosome (120, 121). Sequence comparison of many SHV alleles suggests that such IS*26*-mediated plasmid mobilization has occurred at least twice (37).

Table 3. Likely origin of some plasmid-mediated *bla* genes

β-Lactamase	Origin	Reference(s)
ACC	*Hafnia alvei*	45
ACT	*Enterobacter asburiae*	131
CFE	*Citrobacter freundii*	93
CMY	*C. freundii* (some), ? *Aeromonas* spp.	8, 29
CTX-M	*Kluyvera* spp.	11
DHA	*Morganella morganii*	7
FOX	*Aeromonas caviae*	39
LAT	*C. freundii*	142
MIR	*Enterobacter cloacae*	104
MOX	? *Aeromonas* spp.	124
SHV	*Klebsiella pneumoniae*	48
TEM	Unknown	

INTRODUCTION OF NEW β-LACTAMS

New β-lactamases, or shifts in the prevalence of previously identified enzymes, have arisen over the past 50 years with the introduction of new classes of β-lactam-containing drugs. Figure 1 demonstrates the propagation of important β-lactamase populations that may be associated with the clinical use of novel β-lactam-containing antibiotics. The sources of these populations include previously unimportant enzymes such as the chromosomal SHV-1 β-lactamase in many strains of *K. pneumoniae,* or previously unrecognized mutant β-lactamases such as the ESBLs that were selected by the oxyimino-cephalosporins. The first documented change in background β-lactamase production was reported in the 1940s following the introduction of penicillin into clinical practice. After penicillin was used more frequently in post-World War II medicine, the incidence of penicillinase-producing *S. aureus* increased in one British hospital from ≤8% in 1945 to almost 60% of clinical isolates in less than five years (6, 83).

Other examples of apparent cause and effect relationships include the discovery of cephalosporin C and subsequent introduction of cephalosporin analogs, together with the introduction of broad-spectrum penicillins, such as ampicillin, leading to the identification of plasmid-encoded broad-spectrum penicillinases such as TEM-1. This ubiquitous β-lactamase among the *Enterobacteriaceae,* with the ability to hydrolyze not only many penicillins but also the early cephalosporins such as cephalothin (16, 26), became the most prevalent plasmid-encoded enzyme in epidemiological surveys of the late 1970s and early 1980s (81, 84). A similar plasmid-encoded enzyme, SHV-1, was also later recognized as a constitutive β-lactamase in *K. pneumoniae* (4, 82).

Attempts to identify new agents able to withstand hydrolysis by TEM-1 and SHV-1 led to a proliferation of innovative, resistance-selecting, β-lactam-containing agents beginning in 1978 with cefoxitin followed by the approval of cefotaxime in 1981 (84). Among these agents were the cephamycins (cefoxitin, cefotetan, cefmetazole), the oxyimino-cephalosporins

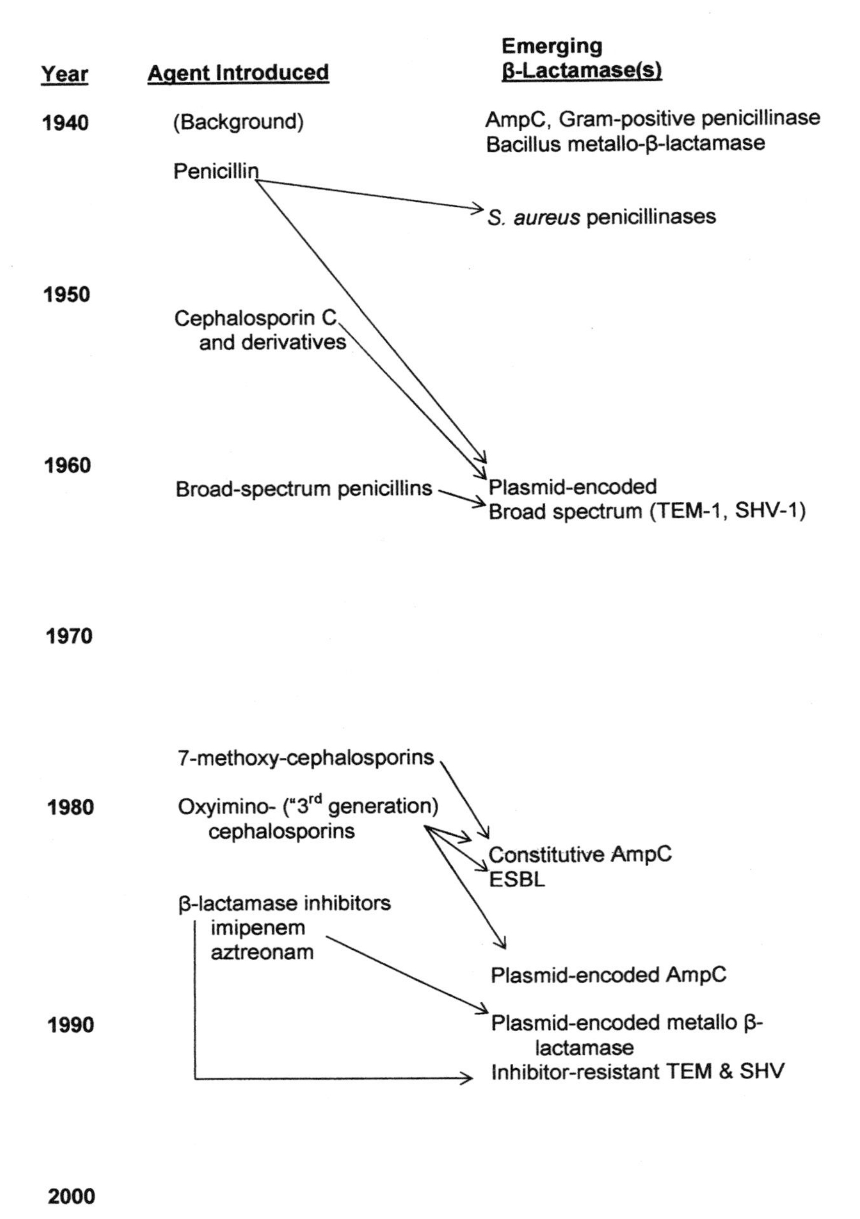

Figure 1. Effects of newly introduced β-lactam-containing agents on β-lactamase populations.

(cefotaxime, ceftazidime, ceftriaxone, cefepime), the β-lactamase inhibitor combinations (clavulanic acid-amoxicillin, clavulanic acid-ticarcillin, sulbactam-ampicillin, and tazobactam-piperacillin), the monobactam aztreonam, and the carbapenems imipenem and meropenem (15). As a result of this influx of novel semisynthetic chemical structures, an explosion of new β-lactamases began during the 1980s. Because new structures such as imipenem, aztreonam, and clavulanic acid, as well as the cephalosporins, were all natural products or close derivatives of natural products, it is tempting to speculate that these newly identified enzymes were lurking in nature but had not been recognized because of a lack of selective pressure.

Within two to three years following the introduction of the new cephalosporins and aztreonam, gram-negative bacilli with ESBLs and with derepressed AmpC β-lactamases were reported. Although the first high producing AmpC on a plasmid was recognized in 1987 (104), similar plasmid-encoded AmpC enzymes became more frequently identified by the mid-1990s (114). Imipenem use was met with increased numbers of carbapenem-hydrolyzing β-lactamases, with the

Zn^{2+}-containing enzymes more frequently identified than the serine enzymes (10). Perhaps the most disconcerting aspect of the carbapenem resistance profile was the identification and subsequent expansion of the plasmid-encoded enzymes in the IMP and VIM families (57, 69).

The slowest specific compound-related resistance to emerge was resistance to the β-lactamase inhibitor combinations, perhaps because resistance was required to two different agents, i.e., the inhibitor and its accompanying penicillin. The first resistance mechanism to the inhibitor combinations was the selection of gram-negative bacteria with overproduction of inhibitor-susceptible β-lactamases, such as the TEM-1 or TEM-2 enzymes. Enzyme overproduction remains the predominant resistance mechanism among the *Enterobacteriaceae* (112). In addition, multiple "inhibitor-resistant" TEM and SHV enzymes have now been identified (18), with one to three mutations conferring the resistance phenotype compared to the parent TEM-1 or SHV-1 enzyme (59). These inhibitor-resistant enzymes generally have slower hydrolysis rates for penicillins and cephalosporins, are more susceptible to inhibition by tazobactam than clavulanate and sulbactam, and do not confer resistance to cephalosporins (18).

MULTIPLICITY OF β-LACTAMASES

When β-lactamases were first described, it was thought that these enzymes would be species specific, with one major enzyme per organism. However, that view was challenged by the mid-1950s when *B. cereus* was shown to produce two β-lactamases with different substrate specificities, one a penicillinase and the other a Zn^{2+}-requiring cephalosporinase (a metallo-β-lactamase) (98). It is now known that almost every metallo-β-lactamase-producing organism also constitutively produces another penicillinase or cephalosporinase (125). This may be because metallo-β-lactamases often have lower catalytic efficiencies (k_{cat}/K_m) for most penicillins and cephalosporins than do traditional penicillinases or AmpC cephalosporinases. Thus, the production of a "pan-hydrolyzing" metallo-enzyme is probably sufficient to confer resistance to carbapenems, but the producing organism needs additional protection from other β-lactams.

In studies examining the occurrence of plasmid-encoded ESBLs, the number of enzymes per organism has been reported to be as high as five in *K. pneumoniae* (34, 52), with three or four enzymes commonly reported from other studies (143, 152). This β-lactamase multiplicity in klebsiellae is facilitated by the presence of the bla_{OXY} chromosomal enzymes in *K. oxytoca* (40) and SHV-1 and related enzymes in *K. pneumoniae* (4). However, some of these resistance determinants are not permanently maintained, as plasmids can be lost when selection pressure is relieved (143).

PREVALENCE OF ENZYMES

Bacterial Species

β-Lactamases have been identified in virtually every bacterial species with some key exceptions. Chromosomally encoded β-lactamases occur in most species, with a few exceptions noted below. Plasmid-mediated β-lactam resistance is frequently reported in *Enterobacteriaceae* and *P. aeruginosa*.

The enterococci or salmonellae do not produce a distinctive chromosomal β-lactamase. Although a small number of enterococcal strains have acquired a plasmid-encoded staphylococcal β-lactamase to serve this function (91, 157), few isolates are currently identified as β-lactamase producers (9, 56). Some salmonellae have become highly β-lactam resistant due to the acquisition of a variety of plasmid-encoded β-lactamases, including ESBLs and AmpC enzymes (105, 106). No traditional β-lactamase has ever been identified in *S. pneumoniae* or *Helicobacter pylori*. A cysteine-rich protein A in some strains of *H. pylori* has been described to hydrolyze penicillins and cephalosporins (87), but with rates four to five orders of magnitude slower than rates recorded for these substrates with common serine β-lactamases such as AmpC or TEM-1. Zoocin A, a streptococcolytic enzyme produced by *Streptococcus equi* subsp. Zooepidemicus, is another nontraditional penicillin-binding protein (PBP) with weak β-lactamase activity (53). Perhaps selected PBPs in β-lactam-resistant *H. pylori* or streptococci have a sufficiently rapid turnover of bound β-lactam so that an additional β-lactamase activity is unnecessary (27).

Geographical Distribution

Although β-lactamases enjoy global dissemination, certain enzymes, or enzyme families, have established themselves in preferred geographical areas. As indicated above, genes encoding AmpC cephalosporinases are present in virtually all *Enterobacteriaceae* and pseudomonads. However, a different scenario is found for the plasmid-encoded β-lactamases where localized differences prevail. In a survey of hospitals from seven countries, SHV-type ESBLs were more common than TEM-type ESBLs, even though the progenitor TEM-1 β-lactamase has been identified more frequently than SHV-1 in historical surveillance studies through the early 1990s (75, 84). In the compilation shown in Table 4, geographical areas reasonably close to each other sometimes exhibit different regional enzyme patterns. Notable outbreaks of strains producing TEM-24 in France (31, 44) have not been repeated in Italy, where TEM-24 was a minor ESBL in a 1999 Italian survey (113). Reports of TEM-derived ESBLs are common in the United States but are less frequently identified in Canada (89) and are exceedingly rare in South America (107, 123). The increasingly high global incidence of CTX-M enzymes in areas such as Canada, South America, and China has not yet reached the United States, although there have been isolated reports of their existence in these locations (88). Plasmid-encoded metallo-β-lactamases, originally identified in Japan, appear to have been localized there for almost a decade before they were identified in Europe. Interestingly, the IMP-family of enzymes has been centered in Japan (154), with a non-Japanese IMP enzyme, IMP-12, only recently identified in Italy (28). Instead, the family of VIM carbapenemases has predominated in Europe, with only a single VIM enzyme reported from Japan (153). These geographical hot-spots are probably due to the ease with which plasmids can be transferred within the local community and within a single hospital and the prevalence of the parent enzyme before selective antibiotic pressure was applied.

In many cases, localized antibiotic use has obviously contributed to the selection of specific sets of enzymes (110). For example, the emergence of ceftazidime-hydrolyzing ESBLs has been associated with increased use of ceftazidime in Massachusetts (129), New York (85), and California (94). The prevalence of plasmid-encoded metallo-β-lactamases in Japan must be driven by local carbapenem prescribing patterns (5). Additionally, the substitution of cefotaxime or ceftriaxone for ceftazidime is most likely associated with the selection of the CTX-M β-lactamases in diverse areas such as China, the United Kingdom, and South America (92, 110).

FACTORS MODIFYING EXPRESSION

The level of β-lactam resistance produced by an enzyme is determined by a number of factors besides its intrinsic hydrolytic activity. Resistance is generally proportional to the level of expression and hence to gene dosage and promoter efficiency. Despite the potential advantage in terms of gene number of carriage on small multicopy plasmids, most plasmid-acquired *bla* genes are found on large plasmids with low copy number in clinical isolates, the other genes on such plasmids presumably providing additional compensatory benefits. TEM- and SHV-type ESBLs are at particular disadvantage because the mutations that extend the spectrum of the enzymes also reduce the specific activity (122). Consequently, a successful ESBL will often have a more efficient promoter, which may explain why more ESBLs than would be expected are derived from TEM-2, which is less common but has a more efficient native promoter than TEM-1. Indeed, different promoters associated with TEM genes can be responsible for up to 10-fold variations in enzyme activity (68). Another TEM ESBL gene has acquired a 10-fold more efficient promoter by upstream insertion of an IS*1*-like element (46). Many chromosomal AmpC genes are inducible (97). The required regulatory elements sometimes accompany the structural genes when they are acquired by plasmids and sometimes not. For

Table 4. Reported regional occurrence of plasmid-encoded β-lactamases or β-lactamase families

Enzyme group	Geographical region	Predominant enzyme(s)	Reference(s)
2be (ESBL)	Canada	SHV-12, CTX-M family	89
	China	CTX-M family, SHV-12	19
	France	TEM-3, TEM-24, SHV-4, CTX-M family	31, 44, 156
	Italy	TEM-52, SHV-12	113
	Japan	Toho-1, SHV-5, SHV-12, TEM-26	55, 150
	Korea	SHV-2a, SHV-12, TEM-52, TEM-116	63, 66
	South Africa	TEM-26, SHV-2, SHV-5	115
	South America	CTX-M family PER-2	11, 123
	Turkey	OXA- family, PER-1	22, 23
	United Kingdom	SHV-5, some TEM enzymes, CTX-M family	14, 136
	United States	TEM-10, TEM-12, TEM-26, SHV-5, SHV-7, SHV-12	13, 60, 72, 143, 149
3 (metallo-β-lactamase)	Europe[a]	IMP, VIM	17, 28, 43, 69, 118, 134
	Japan	IMP family, VIM-2	137, 153, 154
	Singapore	IMP, VIM	67
	United States	VIM[b]	141

[a] Includes Croatia, France, Greece, Italy, and Portugal.
[b] Two different enzymes in geographically separated regions (141; Quinn and Queenan, personal communication).

example, plasmid-mediated ACT-1, CFE-1, CMY-13, DHA-1, and DHA-2 are inducible (38, 41, 86, 93, 127), but the majority of AmpC genes on plasmids are expressed constitutively.

Another important factor affecting resistance is the rate at which a β-lactam antibiotic traverses the outer cell membrane to enter the periplasmic space where β-lactamase is located in gram-negative bacteria. The rate of diffusion through the porin channels of the *E. coli* outer membrane is a function of the molecule's hydrophobicity, charge, and hindrance by bulky side chains (155). Expression of the porin channels is regulated by several systems and may be lost by mutation. For example, most clinical isolates of *K. pneumoniae* express two major porin proteins, OmpK35 and OmpK36 (54). Isolates expressing ESBLs, however, usually express only OmpK36 or lack both OmpK35 and OmpK36, because this loss facilitates resistance. In particular, loss of OmpK36 augments resistance to cefoxitin and to oxyimino- and zwitterionic cephalosporins in strains producing ESBLs and to carbapenems in stains producing plasmid-mediated AmpC-type enzymes (30, 61, 79).

Furthermore, porin channels are often components of multidrug efflux pumps that play an increasingly recognized role in antibiotic resistance (99) (chapter 19). The AcrAB-TolC system of *E. coli* and the MexAB-OprM system of *P. aeruginosa* are prime examples of active efflux systems that contribute to resistance to β-lactams as well as to other agents. For example, for *E. coli* lacking a plasmid-mediated β-lactamase the MIC of ampicillin falls from 2.5 to 0.6 μg/ml with deletion of AcrAB (101), and for *P. aeruginosa* the MIC of carbenicillin falls from 64 to 8 μg/ml in the absence of the MexAB-OprM system and rises to 512 μg/ml with its overexpression (74).

CLINICAL DETECTION OF β-LACTAMASES

Tests for the presence of β-lactamase are useful to guide therapy for such organisms as *H. influenzae, Neisseria* spp., and *Moraxella catarrhalis* where β-lactamase production does not occur in all clinical isolates. Enzyme activity can be detected with a chromogenic cephalosporin, such as nitrocefin, or by linkage of penicillin hydrolysis to a color change mediated by iodine or a pH indicator (76).

An AmpC inducible species can be recognized with a double disk test in which a cefoxitin disk is placed on a lawn of bacteria 25 mm apart from a disk containing the test cephalosporin (133). Induction by cefoxitin is indicated by blunting of the cephalosporin zone of inhibition (76). Such a test may be useful with *Enterobacter* spp., *C. freundii, Serratia* spp., *M. morganii*, or *Providencia* spp. for β-lactams that are good substrates for hydrolysis but poor inducers, such as many oxyimino-cephalosporins and piperacillin. The organisms may appear susceptible on routine testing yet give rise to mutants constitutively expressing AmpC enzymes at high levels with consequent resistance to most penicillins and cephalosporins, a phenomenon that has been observed in 19% of patients with apparently susceptible *Enterobacter* bacteremia treated with oxyimino-cephalosporins (20). Although the alternative of reporting all non-urinary tract isolates of *Enterobacter* spp, *C. freundii, Serratia* spp., *M. morganii*, or *Providencia* spp. as resistant to oxyimino-β-lactams has been recommended (77), this approach would severely limit the use of cephalosporins and other β-lactams for the treatment of infections caused by the *Enterobacteriaceae*.

Tests for ESBL detection in *E. coli, K. pneumoniae*, or *K. oxytoca* as defined by the Clinical and Laboratory Standards Institute (CLSI, formerly NCCLS) rely on an initial screening for reduced susceptibility followed by a confirmatory test of clavulanic acid inhibition of resistance (95, 96). The high MIC resistance breakpoints (and corresponding low resistance breakpoints for disk diffusion) that initially led many ESBL producing strains to be overlooked (60, 65) have been adjusted over the years to more realistic values that are still not uniform from country to country (Table 5). Table 5 also lists the current CLSI recommended screening criteria. Confirmation involves retesting with a range of ceftazidime and

Table 5. Susceptibility (S) and resistance (R) breakpoints for dilutional or disk susceptibility testing and recommended screening parameters for ESBL detection by the CLSI and British Society for Antimicrobial Chemotherapy (BSAC)

Antibiotic	CLSI MIC (S/R)	BSAC MIC (S/R)	CLSI zone diameter (mm) (R/S)	BSAC zone diameter (mm) (R/S)	CLSI screening MIC (μg/ml)	CLSI screening zone diameter (mm)
Aztreonam	≤8/≥32	≤1/≥2	≤15/≥22	≤23/≥24	≥2	≤27
Cefotaxime	≤8/≥64	≤1/≥2	≤14/≥23	≤29/≥30	≥2	≤27
Cefpodoxime	≤2/≥8	≤1/≥2	≤17/≥21	≤33/≥34	≥8	≤17
Ceftazidime	≤8/≥32	≤2/≥4	≤14/≥18	≤21/≥22[a]	≥2	≤22
Ceftriaxone	≤8/≥64	≤1/≥2	≤12/≥21	≤27/≥28	≥2	≤25

[a] For *E. coli* and *Klebsiella* spp. (2).

cefotaxime concentrations with and without 4 μg/ml clavulanic acid or by disks containing these antibiotics with and without 10 μg of clavulanic acid. A positive test is a ≥threefold decrease in MIC or ≥5 mm increase in zone of inhibition. Adaptations for Etest strips and various automated systems have been developed. Inclusion of both cefotaxime and ceftazidime is necessary to detect the CTX-M-type ESBLs conferring little or no ceftazidime resistance. Establishing exactly which ESBL is present is a job for a reference laboratory. Criteria for ESBL detection in other *Enterobacteriaceae* or in *P. aeruginosa* are lacking, but a recent survey indicated that few ESBL producers would be overlooked (135). Getting clinical laboratories to perform the necessary testing can be a problem (138, 139). An alternate view holds that if the screening breakpoint is set low enough, further testing is not necessary, i.e., that if an organism tests susceptible to a β-lactam, that antibiotic can be used for therapy even if an ESBL is present. However, clinical evidence to support this viewpoint is weak, based upon the use of United States breakpoints. When the outcomes of 32 patients with bacteremia caused by ESBL-producing *K. pneumoniae* who were treated with a cephalosporin to which the infecting organism tested nonresistant were reviewed, failure occurred in four of four with organisms testing intermediate and 18 of 33 patients with apparently susceptible isolates (108, 149). In contrast, only 1 of 27 patients with ESBL-producing *K. pneumoniae* bacteremia failed monotherapy with a carbapenem (mainly imipenem) (109).

The presence of an AmpC-type enzyme can be suspected by the appearance of cefoxitin resistance in an organism that normally tests susceptible. The distinction is important because organisms with a plasmid-encoded AmpC enzyme may test susceptible to some cephalosporins (114), yet, based on still limited results for patients with infections caused by plasmid-mediated AmpC producing *K. pneumoniae,* therapy with imipenem has produced better outcomes than cephalosporin treatment (103). In addition to acquisition of plasmid-encoded AmpC varieties, *E. coli* can become cefoxitin resistant by OmpF porin loss and up-mutations in the promoter for its otherwise poorly expressed chromosomal AmpC enzyme (78), and *K. pneumoniae* can acquire cefoxitin resistance by porin loss alone (79). Carbapenem resistance will usually raise concern about a carbapenemase in organisms other than *P. aeruginosa* or *Proteus mirabilis*. Tests combining EDTA and imipenem have been proposed for metallo-carbapenemase detection (70, 147), but false positives are common, especially in *S. marcescens* (K. Bush, unpublished data).

CONCLUSIONS

β-Lactamases are ancient enzymes, widespread in the microbial world, that in recent years have coevolved with β-lactam antibiotics as aminopenicillins, cephalosporins, cephamycins, oxyimino-cephalosporins, monobactams, and carbapenems have been developed to target new pathogens or to overcome existing resistance. β-Lactamase genes have been incorporated into plasmids, transposons, and gene cassettes that have facilitated spread to new hosts and now provide pathogens with hydrolytic activity toward a wide range of β-lactam substrates. Expanded activity has occurred as well by mutations near the active site of the enzyme that allow attack on what were designed to be nonhydrolyzable substrates. β-Lactamase inhibitor resistant enzymes have appeared more recently. Although β-lactamase nomenclature is chaotic, classification schemes allow the enzymes to be divided into 11 or more functional groups and 4 structural classes. Challenges remain for the clinical lab in how and how much to distinguish these enzymes and for the clinician and pharmaceutical world in how to deal with their ever-evolving variety.

REFERENCES

1. **Ambler, R. P.** 1980. The structure of β-lactamases. *Philos. Trans. R. Soc. Lond.* **B 289:**321–331.
2. **Andrews, J. M.** 2004. BSAC standardized disc susceptibility testing method (version 3). *J. Antimicrob. Chemother.* **53:**713–728.
3. **Arduino, S. M., P. H. Roy, G. A. Jacoby, B. E. Orman, S. A. Pineiro, and D. Centron.** 2002. $bla_{CTX\text{-}M\text{-}2}$ is located in an unusual class 1 integron (In35) which includes Orf513. *Antimicrob. Agents Chemother.* **46:**2303–2306.
4. **Babini, G. S., and D. M. Livermore.** 2000. Are SHV β-lactamases universal in *Klebsiella pneumoniae? Antimicrob. Agents Chemother.* **44:**2230.
5. **Bandoh, K., K. Ueno, K. Watanabe, and N. Kato.** 1993. Susceptibility patterns and resistance to imipenem in the *Bacteroides fragilis* group species in Japan: a 4-year study. *Clin. Infect. Dis.* **16 Suppl 4:**S382–S386.
6. **Barber, M., and J. E. M. Whitehead.** 1949. Bacteriophage types in penicillin-resistant staphylococcal infections. *Br. Med. J.* **2:**565–569.
7. **Barnaud, G., G. Arlet, C. Verdet, O. Gaillot, P. H. Lagrange, and A. Philippon.** 1998. *Salmonella enteritidis*: AmpC plasmid-mediated inducible β-lactamase (DHA-1) with an *ampR* gene from *Morganella morganii. Antimicrob. Agents Chemother.* **42:**2352–2358.
8. **Bauernfeind, A., I. Stemplinger, R. Jungwirth, and H. Giamarellou.** 1996. Characterization of the plasmidic β-lactamase CMY-2, which is responsible for cephamycin resistance. *Antimicrob. Agents Chemother.* **40:**221–224.
9. **Bertrand, X., M. Thouverez, P. Bailly, C. Cornette, and D. Talon.** 2000. Clinical and molecular epidemiology of hospital *Enterococcus faecium* isolates in eastern France. Members of Reseau Franc-Comtois de Lutte contre les Infections Nosocomiales. *J. Hosp. Infect.* **45:**125–134.

10. **Bonfiglio, G., G. Russo, and G. Nicoletti.** 2002. Recent developments in carbapenems. *Expert. Opin. Investig. Drugs* **11:**529–544.
11. **Bonnet, R.** 2004. Growing group of extended-spectrum β-lactamases: the CTX-M enzymes. *Antimicrob. Agents Chemother.* **48:**1–14.
12. **Bou, G., A. Oliver, and J. Martinez-Beltran.** 2000. OXA-24, a novel class D β-lactamase with carbapenemase activity in an *Acinetobacter baumannii* clinical strain. *Antimicrob. Agents Chemother.* **44:**1556–1561.
13. **Bradford, P. A., C. E. Cherubin, V. Idemyor, B. A. Rasmussen, and K. Bush.** 1994. Multiply resistant *Klebsiella pneumoniae* strains from two Chicago hospitals: identification of the extended-spectrum TEM-12 and TEM-10 ceftazidime-hydrolyzing β-lactamases in a single isolate. *Antimicrob. Agents Chemother.* **38:**761–766.
14. **Brenwald, N. P., G. Jevons, J. M. Andrews, J. H. Xiong, P. M. Hawkey, and R. Wise.** 2003. An outbreak of a CTX-M-type β-lactamase-producing *Klebsiella pneumoniae:* the importance of using cefpodoxime to detect extended-spectrum β-lactamases. *J. Antimicrob. Chemother.* **51:**195–196.
15. **Bush, K.** 2002. The impact of β-lactamases on the development of novel antimicrobial agents. *Curr. Opin. Investig. Drugs* **3:** 1284–1290.
16. **Bush, K., G. A. Jacoby, and A. A. Medeiros.** 1995. A functional classification scheme for β-lactamases and its correlation with molecular structure. *Antimicrob. Agents Chemother.* **39:**1211–1233.
17. **Cardoso, O., R. Leitao, A. Figueiredo, J. C. Sousa, A. Duarte, and L. V. Peixe.** 2002. Metallo-β-lactamase VIM-2 in clinical isolates of *Pseudomonas aeruginosa* from Portugal. *Microb. Drug Resist.* **8:**93–97.
18. **Chaibi, E. B., D. Sirot, G. Paul, and R. Labia.** 1999. Inhibitor-resistant TEM β-lactamases: phenotypic, genetic and biochemical characteristics. *J. Antimicrob. Chemother.* **43:**447–458.
19. **Chanawong, A., F. H. M'Zali, J. Heritage, J. H. Xiong, and P. M. Hawkey.** 2002. Three cefotaximases, CTX-M-9, CTX-M-13, and CTX-M-14, among *Enterobacteriaceae* in the People's Republic of China. *Antimicrob. Agents Chemother.* **46:**630–637.
20. **Chow, J. W., M. J. Fine, D. M. Shlaes, J. P. Quinn, D. C. Hooper, M. P. Johnson, R. Ramphal, M. M. Wagener, D. K. Miyashiro, and V. L. Yu.** 1991. Enterobacter bacteremia: clinical features and emergence of antibiotic resistance during therapy. *Ann. Intern. Med.* **115:**585–590.
21. **Couture, F., J. Lachapelle, and R. C. Levesque.** 1992. Phylogeny of LCR-1 and OXA-5 with class A and class D β-lactamases. *Mol. Microbiol.* **6:**1693–1705.
22. **Danel, F., L. M. Hall, D. Gur, and D. M. Livermore.** 1998. OXA-16, a further extended-spectrum variant of OXA-10 β-lactamase, from two *Pseudomonas aeruginosa* isolates. *Antimicrob. Agents Chemother.* **42:**3117–3122.
23. **Danel, F., L. M. C. Hall, D. Gur, and D. M. Livermore.** 1995. OXA-14, another extended-spectrum variant of OXA-10 (PSE-2) β-lactamase from *Pseudomonas aeruginosa. Antimicrob. Agents Chemother.* **39:**1881–1884.
24. **Da Silva, G. J., M. Correia, C. Vital, G. Ribeiro, J. C. Sousa, R. Leitao, L. Peixe, and A. Duarte.** 2002. Molecular characterization of *bla*$_{(IMP-5)}$, a new integron-borne metallo-β-lactamase gene from an *Acinetobacter baumannii* nosocomial isolate in Portugal. *FEMS Microbiol. Lett.* **215:**33–39.
25. **Datta, N., and V. M. Hughes.** 1983. Plasmids of the same Inc groups in Enterobacteria before and after the medical use of antibiotics. *Nature* (London) **306:**616–617.
26. **Datta, N., and M. H. Richmond.** 1966. The purification and properties of a penicillinase whose sythesis is mediated by an R-factor in *Escherichia coli. Biochem. J.* **98:**204–209.
27. **DeLoney, C. R., and N. L. Schiller.** 2000. Characterization of an in vitro-selected amoxicillin-resistant strain of *Helicobacter pylori. Antimicrob. Agents Chemother.* **44:**3368–3373.
28. **Docquier, J. D., M. L. Riccio, C. Mugnaioli, F. Luzzaro, A. Endimiani, A. Toniolo, G. Amicosante, and G. M. Rossolini.** 2003. IMP-12, a new plasmid-encoded metallo-β-lactamase from a *Pseudomonas putida* clinical isolate. *Antimicrob. Agents Chemother.* **47:**1522–1528.
29. **Doi, Y., N. Shibata, K. Shibayama, K. Kamachi, H. Kurokawa, K. Yokoyama, T. Yagi, and Y. Arakawa.** 2002. Characterization of a novel plasmid-mediated cephalosporinase (CMY-9) and its genetic environment in an *Escherichia coli* clinical isolate. *Antimicrob. Agents Chemother.* **46:**2427–2434.
30. **Doménech-Sánchez, A., L. Martínez-Martínez, S. Hernández-Allés, M. del Carmen Conejo, A. Pascual, J. M. Tomás, S. Albertí, and V. J. Benedí.** 2003. Role of *Klebsiella pneumoniae* OmpK35 porin in antimicrobial resistance. *Antimicrob. Agents Chemother.* **47:**3332–3335.
31. **Dumarche, P., C. De Champs, D. Sirot, C. Chanal, R. Bonnet, and J. Sirot.** 2002. TEM derivative-producing *Enterobacter aerogenes* strains: dissemination of a prevalent clone. *Antimicrob. Agents Chemother.* **46:**1128–1131.
32. **Elwell, L. P., J. De Graaff, D. Seibert, and S. Falkow.** 1975. Plasmid-linked ampicillin resistance in *Haemophilus influenzae* type b. *Infect. Immun.* **12:**404–410.
33. **Elwell, L. P., M. Roberts, L. W. Mayer, and S. Falkow.** 1977. Plasmid-mediated beta-lactamase production in *Neisseria gonorrhoeae. Antimicrob. Agents Chemother.* **11:**528–533.
34. **Essack, S. Y., L. M. Hall, D. G. Pillay, M. L. McFadyen, and D. M. Livermore.** 2001. Complexity and diversity of *Klebsiella pneumoniae* strains with extended-spectrum β-lactamases isolated in 1994 and 1996 at a teaching hospital in Durban, South Africa. *Antimicrob. Agents Chemother.* **45:**88–95.
35. **Fluit, A. C., and F. J. Schmitz.** 1999. Class 1 integrons, gene cassettes, mobility, and epidemiology. *Eur. J. Clin. Microbiol. Infect. Dis.* **18:**761–770.
36. **Fluit, A. C., and F. J. Schmitz.** 2004. Resistance integrons and super-integrons. *Clin. Microbiol. Infect.* **10:**272–288.
37. **Ford, P. J., and M. B. Avison.** 2004. Evolutionary mapping of the SHV β-lactamase and evidence for two separate IS26-dependent *bla*$_{SHV}$ mobilization events from the *Klebsiella pneumoniae* chromosome. *J. Antimicrob. Chemother.* **54:**69–75.
38. **Fortineau, N., L. Poirel, and P. Nordmann.** 2001. Plasmid-mediated and inducible cephalosporinase DHA-2 from *Klebsiella pneumoniae. J. Antimicrob. Chemother.* **47:**207–210.
39. **Fosse, T., C. Giraud-Morin, I. Madinier, and R. Labia.** 2003. Sequence analysis and biochemical characterisation of chromosomal CAV-1 (*Aeromonas caviae*), the parental cephalosporinase of plasmid-mediated AmpC 'FOX' cluster. *FEMS Microbiol. Lett.* **222:**93–98.
40. **Fournier, B., G. Arlet, P. H. Lagrange, and A. Philippon.** 1994. *Klebsiella oxytoca*: resistance to aztreonam by overproduction of the chromosomally encoded β-lactamase. *FEMS Microbiol. Lett.* **116:**31–36.
41. **Gaillot, O., C. Clement, M. Simonet, and A. Philippon.** 1997. Novel transferable β-lactam resistance with cephalosporinase characteristics in *Salmonella enteritidis. J. Antimicrob. Chemother.* **39:**85–87.
42. **Garau, G., I. García-Sáez, C. Bebrone, C. Anne, P. Mercuri, M. Galleni, J. M. Frère, and O. Dideberg.** 2004. Update of the standard numbering scheme for class B β-lactamases. *Antimicrob. Agents Chemother.* **48:**2347–2349.

43. **Giakkoupi, P., G. Petrikkos, L. S. Tzouvelekis, S. Tsonas, N. J. Legakis, and A. C. Vatopoulos.** 2003. Spread of integron-associated VIM-type metallo-β-lactamase genes among imipenem-nonsusceptible *Pseudomonas aeruginosa* strains in Greek hospitals. *J. Clin. Microbiol.* **41:**822–825.
44. **Giraud-Morin, C., and T. Fosse.** 2003. A seven-year survey of *Klebsiella pneumoniae* producing TEM-24 extended-spectrum β-lactamase in Nice University Hospital (1994–2000). *J. Hosp. Infect.* **54:**25–31.
45. **Girlich, D., T. Naas, S. Bellais, L. Poirel, A. Karim, and P. Nordmann.** 2000. Biochemical-genetic characterization and regulation of expression of an ACC-1-like chromosome-borne cephalosporinase from *Hafnia alvei. Antimicrob. Agents Chemother.* **44:**1470–1478.
46. **Goussard, S., W. Sougakoff, C. Mabilat, A. Bauernfeind, and P. Courvalin.** 1991. An IS1–like element is responsible for high-level synthesis of extended-spectrum β-lactamase TEM-6 in Enterobacteriaceae. *J. Gen. Microbiol.* **137:**2681–2687.
47. **Granier, S. A., V. Leflon-Guibout, F. W. Goldstein, and M. H. Nicolas-Chanoine.** 2003. New *Klebsiella oxytoca* β-lactamase genes bla_{OXY-3} and bla_{OXY-4} and a third genetic group of *K. oxytoca* based on bla_{OXY-3}. *Antimicrob. Agents Chemother.* **47:** 2922–2928.
48. **Hæggman, S., S. Löfdahl, A. Paauw, J. Verhoef, and S. Brisse.** 2004. Diversity and evolution of the class A chromosomal β-lactamase gene in *Klebsiella pneumoniae. Antimicrob. Agents Chemother.* **48:**2400–2408.
49. **Hall, B. G., and M. Barlow.** 2004. Evolution of the serine β-lactamases: past, present and future. *Drug Resist. Updates* **7:**111–123.
50. **Hall, L. M. C., D. M. Livermore, D. Gur, M. Akova, and H. E. Akalin.** 1993. OXA-11, an extended-spectrum variant of OXA-10(PSE-2) β-lactamase from *Pseudomonas aeruginosa. Antimicrob. Agents Chemother.* **37:**1637–1644.
51. **Hall, R. M., and C. M. Collis.** 1998. Antibiotic resistance in gram-negative bacteria: the role of gene cassettes and integrons. *Drug Resist. Updates* **1:**109–119.
52. **Hanson, N. D., K. S. Thomson, E. S. Moland, C. C. Sanders, G. Berthold, and R. G. Penn.** 1999. Molecular characterization of a multiply resistant *Klebsiella pneumoniae* encoding ESBLs and a plasmid-mediated AmpC. *J. Antimicrob. Chemother.* **44:**377–380.
53. **Heath, L. S., H. E. Heath, P. A. LeBlanc, S. R. Smithberg, M. Dufour, R. S. Simmonds, and G. L. Sloan.** 2004. The streptococcolytic enzyme zoocin A is a penicillin-binding protein. *FEMS Microbiol. Lett.* **236:**205–211.
54. **Hernández-Allés, S., S. Albertí, D. Álvarez, A. Doménech-Sánchez, L. Martínez-Martínez, J. Gil, J. M. Tomás, and V. J. Benedí.** 1999. Porin expression in clinical isolates of *Klebsiella pneumoniae. Microbiology* **145:**673–679.
55. **Hirakata, Y.** 2001. Extended-spectrum β-lactamases (ESBLs) producing bacteria. *Nippon Rinsho* **59:**694–700.
56. **Hsieh, S. R.** 2000. Antimicrobial susceptibility and species identification for clinical isolates of enterococci. *J. Microbiol. Immunol. Infect.* **33:**253–257.
57. **Ito, H., Y. Arakawa, S. Ohsuka, R. Wacharotayankun, N. Kato, and M. Ohta.** 1995. Plasmid-mediated dissemination of the metallo-β-lactamase gene bla_{IMP} among clinically isolated strains of *Serratia marcescens. Antimicrob. Agents Chemother.* **39:**824–829.
58. **Iyobe, S., H. Kusadokoro, A. Takahashi, S. Yomoda, T. Okubo, A. Nakamura, and K. O'Hara.** 2002. Detection of a variant metallo-β-lactamase, IMP-10, from two unrelated strains of *Pseudomonas aeruginosa* and an *Alcaligenes xylosoxidans* strain. *Antimicrob. Agents Chemother.* **46:**2014–2016.
59. **Jacoby, G., and K. Bush.** Amino acid sequences for TEM, SHV and OXA extended-spectrum and inhibitor resistant β-lactamases, http://www.lahey.org/studies/webt.htm.
60. **Jacoby, G. A., and P. Han.** 1996. Detection of extended-spectrum β-lactamases in clinical isolates of *Klebsiella pneumoniae* and *Escherichia coli. J. Clin. Microbiol.* **34:**908–911.
61. **Jacoby, G. A., D. M. Mills, and N. Chow.** 2004. Role of β-lactamases and porins in resistance to ertapenem and other β-lactams in *Klebsiella pneumoniae. Antimicrob. Agents Chemother.* **48:**3203–3206.
62. **Jaurin, B., and T. Grundström.** 1981. *ampC* cephalosporinase of *Escherichia coli* K-12 has a different evolutionary origin from that of β-lactamases of the penicillinase type. *Proc. Natl. Acad. Sci. USA* **78:**4897–4901.
63. **Jeong, S. H., I. K. Bae, J. H. Lee, S. G. Sohn, G. H. Kang, G. J. Jeon, Y. H. Kim, B. C. Jeong, and S. H. Lee.** 2004. Molecular characterization of extended-spectrum β-lactamases produced by clinical isolates of *Klebsiella pneumoniae* and *Escherichia coli* from a Korean nationwide survey. *J. Clin. Microbiol.* **42:**2902–2906.
64. **Joris, B., J. M. Ghuysen, G. Dive, A. Renard, O. Dideberg, P. Charlier, J. M. Frère, J. A. Kelly, J. C. Boyington, P. C. Moews, and J. R. Knox.** 1988. The active-site-serine penicillin-recognizing enzymes as members of the *Streptomyces* R61 DD-peptidase family. *Biochem. J.* **250:**313–324.
65. **Katsanis, G. P., J. Spargo, M. J. Ferraro, L. Sutton, and G. A. Jacoby.** 1994. Detection of *Klebsiella pneumoniae* and *Escherichia coli* strains producing extended-spectrum β-lactamases. *J. Clin. Microbiol.* **32:**691–696.
66. **Kim, Y. K., H. Pai, H. J. Lee, S. E. Park, E. H. Choi, J. Kim, J. H. Kim, and E. C. Kim.** 2002. Bloodstream infections by extended-spectrum β-lactamase-producing *Escherichia coli* and *Klebsiella pneumoniae* in children: epidemiology and clinical outcome. *Antimicrob. Agents Chemother.* **46:**1481–1491.
67. **Koh, T. H., G. S. Babini, N. Woodford, L. H. Sng, L. M. Hall, and D. M. Livermore.** 1999. Carbapenem-hydrolysing IMP-1 beta-lactamase in *Klebsiella pneumoniae* from Singapore. *Lancet* **353:**2162. (Letter.)
68. **Lartigue, M. F., V. Leflon-Guibout, L. Poirel, P. Nordmann, and M. H. Nicolas-Chanoine.** 2002. Promoters *P3, Pa/Pb, P4,* and *P5* upstream from bla_{TEM} genes and their relationship to β-lactam resistance. *Antimicrob. Agents Chemother.* **46:** 4035–4037.
69. **Lauretti, L., M. L. Riccio, A. Mazzariol, G. Cornaglia, G. Amicosante, R. Fontana, and G. M. Rossolini.** 1999. Cloning and characterization of bla_{VIM}, a new integron-borne metallo-β-lactamase gene from a *Pseudomonas aeruginosa* clinical isolate. *Antimicrob. Agents Chemother.* **43:**1584–1590.
70. **Lee, K., Y. S. Lim, D. Yong, J. H. Yum, and Y. Chong.** 2003. Evaluation of the Hodge test and the imipenem-EDTA double-disk synergy test for differentiating metallo-beta-lactamase-producing isolates of *Pseudomonas* spp. and *Acinetobacter* spp. *J. Clin. Microbiol.* **41:**4623–4629.
71. **Levesque, R. C., and G. A. Jacoby.** 1988. Molecular structure and interrelationships of multiresistance β-lactamase transposons. *Plasmid* **19:**21–29.
72. **Levison, M. E., Y. V. Mailapur, S. K. Pradham, G. A. Jacoby, P. D. Adams, C. L. Emery, P. L. May, and P. G. Pitsakis.** 2002. Regional occurrence of plasmid-mediated SHV-7, an extended-spectrum β-lactamase, in *Enterobacter cloacae* in Philadelphia teaching hospitals. *Clin. Infect. Dis.* **35:**1551–1554.
73. **Levy, S. B.** 2002. The 2000 Garrod lecture. Factors impacting on the problem of antibiotic resistance. *J. Antimicrob. Chemother.* **49:**25–30.
74. **Li, X. Z., L. Zhang, and K. Poole.** 2000. Interplay between the MexA-MexB-OprM multidrug efflux system and the outer

membrane barrier in the multiple antibiotic resistance of *Pseudomonas aeruginosa*. *J. Antimicrob. Chemother.* **45:**433–436.

75. **Liu, P. Y., D. Gur, L. M. Hall, and D. M. Livermore.** 1992. Survey of the prevalence of β-lactamases amongst 1000 gram-negative bacilli isolated consecutively at the Royal London Hospital. *J. Antimicrob. Chemother.* **30:**429–447.
76. **Livermore, D. M., and D. F. Brown.** 2001. Detection of β-lactamase-mediated resistance. *J. Antimicrob. Chemother.* **48 Suppl 1:**59–64.
77. **Livermore, D. M., T. G. Winstanley, and K. P. Shannon.** 2001. Interpretative reading: recognizing the unusual and inferring resistance mechanisms from resistance phenotypes. *J. Antimicrob. Chemother.* **48 Suppl 1:**87–102.
78. **Martínez-Martínez, L., M. C. Conejo, A. Pascual, S. Hernández-Allés, S. Ballesta, E. Ramírez De Arellano-Ramos, V. J. Benedí, and E. J. Perea.** 2000. Activities of imipenem and cephalosporins against clonally related strains of *Escherichia coli* hyperproducing chromosomal β-lactamase and showing altered porin profiles. *Antimicrob. Agents Chemother.* **44:**2534–2536.
79. **Martínez-Martínez, L., A. Pascual, S. Hernández-Allés, D. Alvarez-Díaz, A. I. Suárez, J. Tran, V. J. Benedí, and G. A. Jacoby.** 1999. Roles of β-lactamases and porins in activities of carbapenems and cephalosporins against *Klebsiella pneumoniae*. *Antimicrob. Agents Chemother.* **43:**1669–1673.
80. **Materon, I. C., A. M. Queenan, T. M. Koehler, K. Bush, and T. Palzkill.** 2003. Biochemical characterization of β-lactamases Bla1 and Bla2 from *Bacillus anthracis*. *Antimicrob. Agents Chemother.* **47:**2040–2042.
81. **Matthew, M.** 1979. Plasmid-mediated β-lactamases of gram-negative bacteria: properties and distribution. *J. Antimicrob. Chemother.* **5:**349–358.
82. **Matthew, M., R. W. Hedges, and J. T. Smith.** 1979. Types of β-lactamase determined by plasmids in gram-negative bacteria. *J. Bacteriol.* **138:**657–662.
83. **Medeiros, A.** 1997. Evolution and dissemination of β-lactamases accelerated by generations of β-lactam antibiotics. *Clin. Infect. Dis.* **24:**S19–S45.
84. **Medeiros, A. A.** 1984. β-lactamases. *Br. Med. Bull.* **40:**18–27.
85. **Meyer, K. S., C. Urban, J. A. Eagan, B. J. Berger, and J. J. Rahal.** 1993. Nosocomial outbreak of *Klebsiella* infection resistant to late-generation cephalosporins. *Ann. Intern. Med.* **119:**353–358.
86. **Miriagou, V., L. S. Tzouvelekis, L. Villa, E. Lebessi, A. C. Vatopoulos, A. Carattoli, and E. Tzelepi.** 2004. CMY-13, a novel inducible cephalosporinase encoded by an *Escherichia coli* plasmid. *Antimicrob. Agents Chemother.* **48:**3172–3174.
87. **Mittl, P. R., L. Luthy, P. Hunziker, and M. G. Grutter.** 2000. The cysteine-rich protein A from *Helicobacter pylori* is a β-lactamase. *J. Biol. Chem.* **275:**17693–17699.
88. **Moland, E. S., J. A. Black, A. Hossain, N. D. Hanson, K. S. Thomson, and S. Pottumarthy.** 2003. Discovery of CTX-M-like extended-spectrum β-lactamases in *Escherichia coli* isolates from five US States. *Antimicrob. Agents Chemother.* **47:**2382–2383.
89. **Mulvey, M. R., E. Bryce, D. Boyd, M. Ofner-Agostini, S. Christianson, A. E. Simor, and S. Paton.** 2004. Ambler class A extended-spectrum beta-lactamase-producing *Escherichia coli* and *Klebsiella* spp. in Canadian hospitals. *Antimicrob. Agents Chemother.* **48:**1204–1214.
90. **Murray, B. E.** 1992. β-lactamase-producing enterococci. *Antimicrob. Agents Chemother.* **36:**2355–2359.
91. **Murray, B. E., and B. D. Mederski-Samoraj.** 1983. Transferable β-lactamase: a new mechanism for in vitro penicillin resistance in *Streptococcus faecalis*. *J. Clin. Invest.* **72:**1168–1171.
92. **Mushtaq, S., N. Woodford, N. Potz, and D. M. Livermore.** 2003. Detection of CTX-M-15 extended-spectrum β-lactamase in the United Kingdom. *J. Antimicrob. Chemother.* **52:**528–529.
93. **Nakano, R., R. Okamoto, Y. Nakano, K. Kaneko, N. Okitsu, Y. Hosaka, and M. Inoue.** 2004. CFE-1, a novel plasmid-encoded AmpC β-lactamase with an *ampR* gene originating from *Citrobacter freundii*. *Antimicrob. Agents Chemother.* **48:**1151–1158.
94. **Naumovski, L., J. P. Quinn, D. Miyashiro, M. Patel, K. Bush, S. B. Singer, D. Graves, T. Palzkill, and A. M. Arvin.** 1992. Outbreak of ceftazidime resistance due to a novel extended-spectrum β-lactamase in isolates from cancer patients. *Antimicrob. Agents Chemother.* **36:**1991–1996.
95. **NCCLS.** 2003. *Methods for Dilution Antimicrobial Susceptibility Tests for Bacteria That Grow Aerobically*, 6th ed. *Approved standard M7-A6.* National Committee for Clinical Laboratory Standards, Wayne, Pa.
96. **NCCLS.** 2003. Performance standards for antimicrobial disk susceptibility tests—eighth edition. Approved standard. NCCLS document M2–A8. National Committee for Clinical Laboratory Standards, Wayne, Pa.
97. **Neu, H. C., and N. X. Chin.** 1985. A perspective on the present contribution of β-lactamases to bacterial resistance with particular reference to induction of β-lactamase and its clinical significance. *Chemioterapia* **4:**63–70.
98. **Newton, G. G. F., and E. P. Abraham.** 1956. Isolation of cephalosporin C, a penicillin-like antibiotic containing D-alpha aminoadipic acid. *Biochem. J.* **62:**651–658.
99. **Nikaido, H.** 1996. Multidrug efflux pumps of gram-negative bacteria. *J. Bacteriol.* **178:**5853–5859.
100. **Normark, S., S. Lindquist, and F. Lindberg.** 1986. Chromosomal β-lactam resistance in enterobacteria. *Scand. J. Infect. Dis. Suppl.* **49:**38–45.
101. **Okusu, H., D. Ma, and H. Nikaido.** 1996. AcrAB efflux pump plays a major role in the antibiotic resistance phenotype of *Escherichia coli* multiple-antibiotic-resistance (Mar) mutants. *J. Bacteriol.* **178:**306–308.
102. **Ouellette, M., L. Bissonnette, and P. H. Roy.** 1987. Precise insertion of antibiotic resistance determinants into Tn21-like transposons: nucleotide sequence of the OXA-1 β-lactamase gene. *Proc. Natl. Acad. Sci. USA* **84:**7378–7382.
103. **Pai, H., C. I. Kang, J. H. Byeon, K. D. Lee, W. B. Park, H. B. Kim, E. C. Kim, M. D. Oh, and K. W. Choe.** 2004. Epidemiology and clinical features of bloodstream infections caused by AmpC-type-β-lactamase-producing *Klebsiella pneumoniae*. *Antimicrob. Agents Chemother.* **48:**3720–3728.
104. **Papanicolaou, G. A., A. A. Medeiros, and G. A. Jacoby.** 1990. Novel plasmid-mediated β-lactamase (MIR-1) conferring resistance to oxyimino- and α-methoxy β-lactams in clinical isolates of *Klebsiella pneumoniae*. *Antimicrob. Agents Chemother.* **34:**2200–2209.
105. **Parry, C. M.** 2003. Antimicrobial drug resistance in *Salmonella enterica*. *Curr. Opin. Infect. Dis.* **16:**467–472.
106. **Paterson, D. L.** 2001. Extended-spectrum β-lactamases: the European experience. *Curr. Opin. Infect. Dis.* **14:**697–701.
107. **Paterson, D. L., K. M. Hujer, A. M. Hujer, B. Yeiser, M. D. Bonomo, L. B. Rice, and R. A. Bonomo.** 2003. Extended-spectrum β-lactamases in *Klebsiella pneumoniae* bloodstream isolates from seven countries: dominance and widespread prevalence of SHV- and CTX-M-type β-lactamases. *Antimicrob. Agents Chemother.* **47:**3554–3560.
108. **Paterson, D. L., W. C. Ko, A. Von Gottberg, J. M. Casellas, L. Mulazimoglu, K. P. Klugman, R. A. Bonomo, L. B. Rice, J. G. McCormack, and V. L. Yu.** 2001. Outcome of cephalosporin treatment for serious infections due to apparently susceptible organisms producing extended-spectrum

β-lactamases: implications for the clinical microbiology laboratory. *J. Clin. Microbiol.* **39:**2206–2212.

109. **Paterson, D. L., W. C. Ko, A. Von Gottberg, S. Mohapatra, J. M. Casellas, H. Goossens, L. Mulazimoglu, G. Trenholme, K. P. Klugman, R. A. Bonomo, L. B. Rice, M. M. Wagener, J. G. McCormack, and V. L. Yu.** 2004. Antibiotic therapy for *Klebsiella pneumoniae* bacteremia: implications of production of extended-spectrum β-lactamases. *Clin. Infect. Dis.* **39:** 31–37.
110. **Paterson, D. L., W. C. Ko, A. Von Gottberg, S. Mohapatra, J. M. Casellas, H. Goossens, L. Mulazimoglu, G. Trenholme, K. P. Klugman, R. A. Bonomo, L. B. Rice, M. M. Wagener, J. G. McCormack, and V. L. Yu.** 2004. International prospective study of *Klebsiella pneumoniae* bacteremia: implications of extended-spectrum β-lactamase production in nosocomial Infections. *Ann. Intern. Med.* **140:**26–32.
111. **Patzer, J., M. A. Toleman, L. M. Deshpande, W. Kaminska, D. Dzierzanowska, P. M. Bennett, R. N. Jones, and T. R. Walsh.** 2004. *Pseudomonas aeruginosa* strains harbouring an unusual bla_{VIM-4} gene cassette isolated from hospitalized children in Poland (1998–2001). *J. Antimicrob. Chemother.* **53:**451–456.
112. **Perez-Moreno, M. O., M. Perez-Moreno, M. Carulla, C. Rubio, A. M. Jardi, and J. Zaragoza.** 2004. Mechanisms of reduced susceptibility to amoxycillin-clavulanic acid in *Escherichia coli* strains from the health region of Tortosa (Catalonia, Spain). *Clin. Microbiol. Infect.* **10:**234–241.
113. **Perilli, M., E. Dell'Amico, B. Segatore, M. R. de Massis, C. Bianchi, F. Luzzaro, G. M. Rossolini, A. Toniolo, G. Nicoletti, and G. Amicosante.** 2002. Molecular characterization of extended-spectrum β-lactamases produced by nosocomial isolates of *Enterobacteriaceae* from an Italian nationwide survey. *J. Clin. Microbiol.* **40:**611–614.
114. **Philippon, A., G. Arlet, and G. A. Jacoby.** 2002. Plasmid-determined AmpC-type β-lactamases. *Antimicrob. Agents Chemother.* **46:**1–11.
115. **Pitout, J. D., K. S. Thomson, N. D. Hanson, A. F. Ehrhardt, E. S. Moland, and C. C. Sanders.** 1998. β-Lactamases responsible for resistance to expanded-spectrum cephalosporins in *Klebsiella pneumoniae, Escherichia coli,* and *Proteus mirabilis* isolates recovered in South Africa. *Antimicrob. Agents Chemother.* **42:**1350–1354.
116. **Poirel, L., J. W. Decousser, and P. Nordmann.** 2003. Insertion sequence IS*Ecp1B* is involved in expression and mobilization of a $bla_{(CTX-M)}$ β-lactamase gene. *Antimicrob. Agents Chemother.* **47:**2938–2945.
117. **Poirel, L., M. Magalhaes, M. Lopes, and P. Nordmann.** 2004. Molecular analysis of metallo-β-lactamase gene bla_{SPM-1}-surrounding sequences from disseminated *Pseudomonas aeruginosa* isolates in Recife, Brazil. *Antimicrob. Agents Chemother.* **48:**1406–1409.
118. **Poirel, L., T. Naas, D. Nicolas, L. Collet, S. Bellais, J. D. Cavallo, and P. Nordmann.** 2000. Characterization of VIM-2, a carbapenem-hydrolyzing metallo-β-lactamase and its plasmid- and integron-borne gene from a *Pseudomonas aeruginosa* clinical isolate in France. *Antimicrob. Agents Chemother.* **44:** 891–897.
119. **Pollock, M. R.** 1967. Origin and function of penicillinase: a problem in biochemical evolution. *Br. Med. J.* **4:**71–77.
120. **Preston, K. E., and R. A. Venezia.** 2002. Chromosomal sequences from *Klebsiella pneumoniae* flank the SHV-5 extended-spectrum β-lactamase gene in pACM1. *Plasmid* **48:** 73–76.
121. **Preston, K. E., R. A. Venezia, and K. A. Stellrecht.** 2004. The SHV-5 extended-spectrum β-lactamase gene of pACM1 is located on the remnant of a compound transposon. *Plasmid* **51:**48–53.
122. **Queenan, A. M., B. Foleno, C. Gownley, E. Wira, and K. Bush.** 2004. Effects of inoculum and β-lactamase activity in AmpC- and extended-spectrum β-lactamase (ESBL)-producing *Escherichia coli* and *Klebsiella pneumoniae* clinical isolates tested by using NCCLS ESBL methodology. *J. Clin. Microbiol.* **42:**269–275.
123. **Quinteros, M., M. Radice, N. Gardella, M. M. Rodriguez, N. Costa, D. Korbenfeld, E. Couto, and G. Gutkind.** 2003. Extended-spectrum β-lactamases in *Enterobacteriaceae* in Buenos Aires, Argentina, public hospitals. *Antimicrob. Agents Chemother.* **47:**2864–2867.
124. **Raskine, L., I. Borrel, G. Barnaud, S. Boyer, B. Hanau-Bercot, J. Gravisse, R. Labia, G. Arlet, and M. J. Sanson-Le-Pors.** 2002. Novel plasmid-encoded class C β-lactamase (MOX-2) in *Klebsiella pneumoniae* from Greece. *Antimicrob. Agents Chemother.* **46:**2262–2265.
125. **Rasmussen, B. A., and K. Bush.** 1997. Carbapenem-hydrolyzing β-lactamases. *Antimicrob. Agents Chemother.* **41:**223–232.
126. **Rasmussen, B. A., Y. Yang, N. Jacobus, and K. Bush.** 1994. Contribution of enzymatic properties, cell permeability, and enzyme expression to microbiological activities of beta-lactams in three *Bacteroides fragilis* isolates that harbor a metallo-beta-lactamase gene. *Antimicrob. Agents Chemother.* **38:**2116–2120.
127. **Reisbig, M. D., and N. D. Hanson.** 2002. The ACT-1 plasmid-encoded AmpC β-lactamase is inducible: detection in a complex β-lactamase background. *J. Antimicrob. Chemother.* **49:**557–560.
128. **Rice, L. B., and S. H. Marshall.** 1992. Evidence of incorporation of the chromosomal β-lactamase gene of *Enterococcus faecalis* CH19 into a transposon derived from staphylococci. *Antimicrob. Agents Chemother.* **36:**1843–1846.
129. **Rice, L. B., S. H. Willey, G. A. Papanicolaou, A. A. Medeiros, G. M. Eliopoulos, R. C. Moellering, Jr., and G. A. Jacoby.** 1990. Outbreak of ceftazidime resistance caused by extended-spectrum β-lactamases at a Massachusetts chronic-care facility. *Antimicrob. Agents Chemother.* **34:**2193–2199.
130. **Richmond, M. H., and R. B. Sykes.** 1973. The β-lactamases of gram-negative bacteria and their possible physiological roles. *Adv. Microb. Physiol.* **9:**31–88.
131. **Rottman, M., Y. Benzerara, B. Hanau-Bercot, C. Bizet, A. Philippon, and G. Arlet.** 2002. Chromosomal *ampC* genes in *Enterobacter* species other than *Enterobacter cloacae,* and ancestral association of the ACT-1 plasmid-encoded cephalosporinase to *Enterobacter asburiae. FEMS Microbiol. Lett.* **210:**87–92.
132. **Sabate, M., F. Navarro, E. Miro, S. Campoy, B. Mirelis, J. Barbe, and G. Prats.** 2002. Novel complex sul1-type integron in *Escherichia coli* carrying $bla_{CTX-M-9}$. *Antimicrob. Agents Chemother.* **46:**2656–2661.
133. **Sanders, C. C., and W. E. Sanders, Jr.** 1979. Emergence of resistance to cefamandole: possible role of cefoxitin-inducible beta-lactamases. *Antimicrob. Agents Chemother.* **15:**792–797.
134. **Sardelic, S., L. Pallecchi, V. Punda-Polic, and G. M. Rossolini.** 2003. Carbapenem-resistant *Pseudomonas aeruginosa*-carrying VIM-2 metallo-β-lactamase determinants, Croatia. *Emerg. Infect. Dis.* **9:**1022–1023.
135. **Schwaber, M. J., P. M. Raney, J. K. Rasheed, J. W. Biddle, P. Williams, J. E. McGowan, Jr., and F. C. Tenover.** 2004. Utility of NCCLS guidelines for identifying extended-spectrum β-lactamases in non-*Escherichia coli* and Non-*Klebsiella* spp. of *Enterobacteriaceae. J. Clin. Microbiol.* **42:**294–298.

136. **Shannon, K., P. Stapleton, X. Xiang, A. Johnson, H. Beattie, F. El Bakri, B. Cookson, and G. French.** 1998. Extended-spectrum β-lactamase-producing *Klebsiella pneumoniae* strains causing nosocomial outbreaks of infection in the United Kingdom. *J. Clin. Microbiol.* **36:**3105–3110.
137. **Shibata, N., Y. Doi, K. Yamane, T. Yagi, H. Kurokawa, K. Shibayama, H. Kato, K. Kai, and Y. Arakawa.** 2003. PCR typing of genetic determinants for metallo-β-lactamases and integrases carried by gram-negative bacteria isolated in Japan, with focus on the class 3 integron. *J. Clin. Microbiol.* **41:**5407–5413.
138. **Stevenson, K. B., M. Samore, J. Barbera, J. W. Moore, E. Hannah, P. Houck, F. C. Tenover, and J. L. Gerberding.** 2003. Detection of antimicrobial resistance by small rural hospital microbiology laboratories: comparison of survey responses with current NCCLS laboratory standards. *Diagn. Microbiol. Infect. Dis.* **47:**303–311.
139. **Tenover, F. C., M. J. Mohammed, T. S. Gorton, and Z. F. Dembek.** 1999. Detection and reporting of organisms producing extended-spectrum β-lactamases: survey of laboratories in Connecticut. *J. Clin. Microbiol.* **37:**4065–4070.
140. **Toleman, M. A., D. Biedenbach, D. Bennett, R. N. Jones, and T. R. Walsh.** 2003. Genetic characterization of a novel metallo-β-lactamase gene, bla_{IMP-13}, harboured by a novel Tn*5051*-type transposon disseminating carbapenemase genes in Europe: report from the SENTRY worldwide antimicrobial surveillance programme. *J. Antimicrob. Chemother.* **52:**583–590.
141. **Toleman, M. A., K. Rolston, R. N. Jones, and T. R. Walsh.** 2004. bla_{VIM-7}, an evolutionarily distinct metallo-β-lactamase gene in a *Pseudomonas aeruginosa* isolate from the United States. *Antimicrob. Agents Chemother.* **48:**329–332.
142. **Tzouvelekis, L. S., E. Tzelepi, and A. F. Mentis.** 1994. Nucleotide sequence of a plasmid-mediated cephalosporinase gene (bla_{LAT-1}) found in *Klebsiella pneumoniae. Antimicrob. Agents Chemother.* **38:**2207–2209.
143. **Urban, C., K. S. Meyer, N. Mariano, J. J. Rahal, R. Flamm, B. A. Rasmussen, and K. Bush.** 1994. Identification of TEM-26 β-lactamase responsible for a major outbreak of ceftazidime-resistant *Klebsiella pneumoniae. Antimicrob. Agents Chemother.* **38:**392–395.
144. **Verdet, C., G. Arlet, G. Barnaud, P. H. Lagrange, and A. Philippon.** 2000. A novel integron in *Salmonella enterica* serovar Enteritidis, carrying the bla_{DHA-1} gene and its regulator gene *ampR*, originated from *Morganella morganii. Antimicrob. Agents Chemother.* **44:**222–225.
145. **Voladri, R. K., D. L. Lakey, S. H. Hennigan, B. E. Menzies, K. M. Edwards, and D. S. Kernodle.** 1998. Recombinant expression and characterization of the major β-lactamase of *Mycobacterium tuberculosis. Antimicrob. Agents Chemother.* **42:**1375–1381.
146. **Wachino, J., Y. Doi, K. Yamane, N. Shibata, T. Yagi, T. Kubota, H. Ito, and Y. Arakawa.** 2004. Nosocomial spread of ceftazidime-resistant *Klebsiella pneumoniae* strains producing a novel class A β-lactamase, GES-3, in a neonatal intensive care unit in Japan. *Antimicrob. Agents Chemother.* **48:**1960–1967.
147. **Walsh, T. R., A. Bolmstrom, A. Qwarnstrom, and A. Gales.** 2002. Evaluation of a new Etest for detecting metallo-β-lactamases in routine clinical testing. *J. Clin. Microbiol.* **40:**2755–2759.
148. **Wang, W., P. S. Mezes, Y. Q. Yang, R. W. Blacher, and J. O. Lampen.** 1985. Cloning and sequencing of the β-lactamase I gene of *Bacillus cereus* 5/B and its expression in *Bacillus subtilis. J. Bacteriol.* **163:**487–492.
149. **Wong-Beringer, A., J. Hindler, M. Loeloff, A. M. Queenan, N. Lee, D. A. Pegues, J. P. Quinn, and K. Bush.** 2002. Molecular correlation for the treatment outcomes in bloodstream infections caused by *Escherichia coli* and *Klebsiella pneumoniae* with reduced susceptibility to ceftazidime. *Clin. Infect. Dis.* **34:**135–146.
150. **Yagi, T., H. Kurokawa, N. Shibata, K. Shibayama, and Y. Arakawa.** 2000. A preliminary survey of extended-spectrum β-lactamases (ESBLs) in clinical isolates of *Klebsiella pneumoniae* and *Escherichia coli* in Japan. *FEMS Microbiol. Lett.* **184:**53–56.
151. **Yan, J. J., W. C. Ko, and J. J. Wu.** 2001. Identification of a plasmid encoding SHV-12, TEM-1, and a variant of IMP-2 metallo-β-lactamase, IMP-8, from a clinical isolate of *Klebsiella pneumoniae. Antimicrob. Agents Chemother.* **45:**2368–2371.
152. **Yang, Y., N. Bhachech, P. A. Bradford, B. D. Jett, D. F. Sahm, and K. Bush.** 1998. Ceftazidime-resistant *Klebsiella pneumoniae* and *Escherichia coli* isolates producing TEM-10 and TEM-43 β-lactamases from St. Louis, Missouri. *Antimicrob. Agents Chemother.* **42:**1671–1676.
153. **Yatsuyanagi, J., S. Saito, S. Harata, N. Suzuki, Y. Ito, K. Amano, and K. Enomoto.** 2004. Class 1 integron containing metallo-β-lactamase gene bla_{VIM-2} in *Pseudomonas aeruginosa* clinical strains isolated in Japan. *Antimicrob. Agents Chemother.* **48:**626–628.
154. **Yomoda, S., T. Okubo, A. Takahashi, M. Murakami, and S. Iyobe.** 2003. Presence of *Pseudomonas putida* strains harboring plasmids bearing the metallo-β-lactamase gene bla_{IMP} in a hospital in Japan. *J. Clin. Microbiol.* **41:**4246–4251.
155. **Yoshimura, F., and H. Nikaido.** 1985. Diffusion of β-lactam antibiotics through the porin channels of *Escherichia coli* K-12. *Antimicrob. Agents Chemother.* **27:**84–92.
156. **Yuan, M., H. Aucken, L. M. C. Hall, T. L. Pitt, and D. M. Livermore.** 1998. Epidemiological typing of klebsiella with extended-spectrum β-lactamases from European intensive care units. *J. Antimicrob. Chemother.* **41:**527–539.
157. **Zscheck, K. K., and B. E. Murray.** 1991. Nucleotide sequence of the β-lactamase gene from *Enterococcus faecalis* HH22 and its similarity to staphylococcal β-lactamase genes. *Antimicrob. Agents Chemother.* **35:**1736–1740.

Frontiers in Antimicrobial Resistance: a Tribute to Stuart B. Levy
Edited by D. G. White, M. N. Alekshun, and P. F. McDermott

Chapter 6

Macrolide, Lincosamide, Streptogramin, Ketolide, and Oxazolidinone Resistance

MARILYN C. ROBERTS AND JOYCE SUTCLIFFE

Macrolides, lincosamides, streptogramins, ketolides, and oxazolidinones (MLSKO), though chemically distinct (Fig. 1), share overlapping binding sites on the 50S subunit of the ribosome (13, 37, 42, 44, 121, 122). Bacteria acquire resistance to one or more of these antibiotics by the acquisition of a gene(s) or with certain mutations (135, 150).

MACROLIDES

Erythromycin was discovered in 1952 and represented the first member of the macrolides. This class of drug was found to have good activity against gram-positive cocci, other gram-positive species (e.g., *Corynebacterium diphtheriae, Listeria monocytogenes*), and some gram-negative bacteria such as *Campylobacter, Bordetella pertusis* and *Neisseria*, as well as *Entamoeba histolytica* (49, 66, 67, 104, 105, 115). In contrast, *Enterobacteriaceae* and *Pseudomonas* sp. were thought to be innately nonsusceptible to erythromycin, though this assumption has changed recently (50, 118). By 1956, erythromycin resistance had been identified in French staphylococci and shortly thereafter was found in the United States (135). Resistance in these isolates was due to the presence of an acquired rRNA methylase (*erm*) gene. By the 1970s, these genes were found in a variety of gram-positive bacteria including *Streptococcus pneumoniae*. Today macrolide-resistant *S. pneumoniae* represent >80% of the isolates in Asia, 30 to 60% of the isolates in parts of Africa, and 30 to 40% of the isolates surveyed in parts of Europe and the United States and are usually multidrug resistant (17a).

In the 1980s, semisynthetic derivatives of erythromycin (e.g., azithromycin, clarithromycin, dirithromycin, and roxithromycin) were developed. These had improved intracellular and tissue penetration and improved pharmacodynamics and required fewer doses per day than erythromycin (6). These derivatives have a broader spectrum of activity and were therefore used for disease caused by *Haemophilus influenzae, Moraxella catarrhalis, Mycobacterium avium-intracellulare,* other rapid-growing mycobacteria, *Borrelia burgdorferi,* and the parasites *Plasmodium* sp., *Toxoplasa gondii,* and *Babesia microti* (6, 115). Clarithromycin and azithromycin have similar activity and became available just as the HIV era resulted in an increase in *M. avium-intracellulare* disease. More recently, long-term administration of azithromycin has been shown to be effective in patients with pulmonary *Pseudomonas aeruginosa* (50, 118). In the laboratory, azithromycin treatment can reduce adherence of *P. aeruginosa* to bronchial mucins and may also reduce expression of virulence factors in *P. aeruginosa* (22, 117).

KETOLIDES

The ketolides are macrolide derivatives that have recently become available for human therapy (19). These are semisynthetic derivatives of erythromycin A and are poor inducers of the rRNA methylase genes (*erm*), which are widespread in gram-positive and selected gram-negative bacteria (Tables 1 and 2) (53, 66, 67, 70, 75, 77, 82, 115, 135, 154). Telithromycin (Ketek) was the first ketolide for human therapy. This antibiotic is two to four times more active than clarithromycin against susceptible gram-positive cocci and nearly equivalent to azithromycin against *H. influenzae* in vitro. Telithromycin has notably improved activity against *S. pneumoniae* isolates carrying either *erm* or *mef*(A) genes (13, 19, 51, 121). The ketolides penetrate a variety of cells and achieve high intracellular concentrations, which may account for their good activity against intracellular pathogens (19, 89).

Marilyn C. Roberts • Department of Pathobiology, Box 357238, University of Washington, Seattle, WA 98195. **Joyce Sutcliffe** • Rib-X Pharmaceuticals, Inc., New Haven, CT 06511.

LINCOSAMIDES

The lincosamides (e.g., clindamycin and lincomycin) have activity against gram-positive cocci, a variety of gram-positive and gram-negative anaerobes, and *T. gondii* (20, 31, 33, 36, 59). The lincosamides are a distinct group of antibiotics that are structurally unrelated to macrolides. Lincomycin, a natural product, has been used for many years in the treatment of animal diseases. Clindamycin is a semisynthetic derivative of lincomycin and is used to treat a variety of human diseases, especially anaerobic infections (18a, 142). The sugar moiety of the lincosamide structure overlaps the amino sugar at the C5 position of the macrolides, which accounts for why some resistance mutations confer both macrolide and lincosamide resistance (122, 164).

STREPTOGRAMINS

The streptogramins used in the United States are composed of two chemically distinct components, streptogramin A and streptogramin B. The two components, which have very different chemical structures, when combined, prevent bacterial growth at a

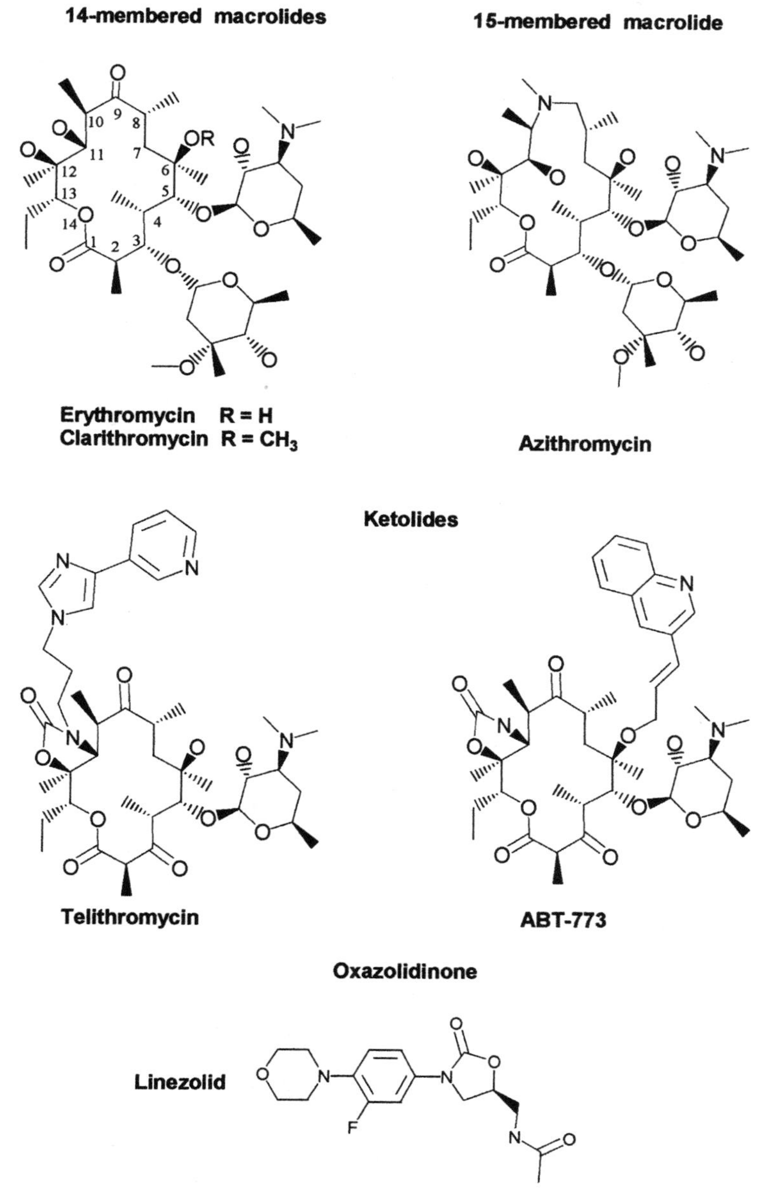

Figure 1. Structures of the various MLSKO antibiotics. (*Continued on following page.*)

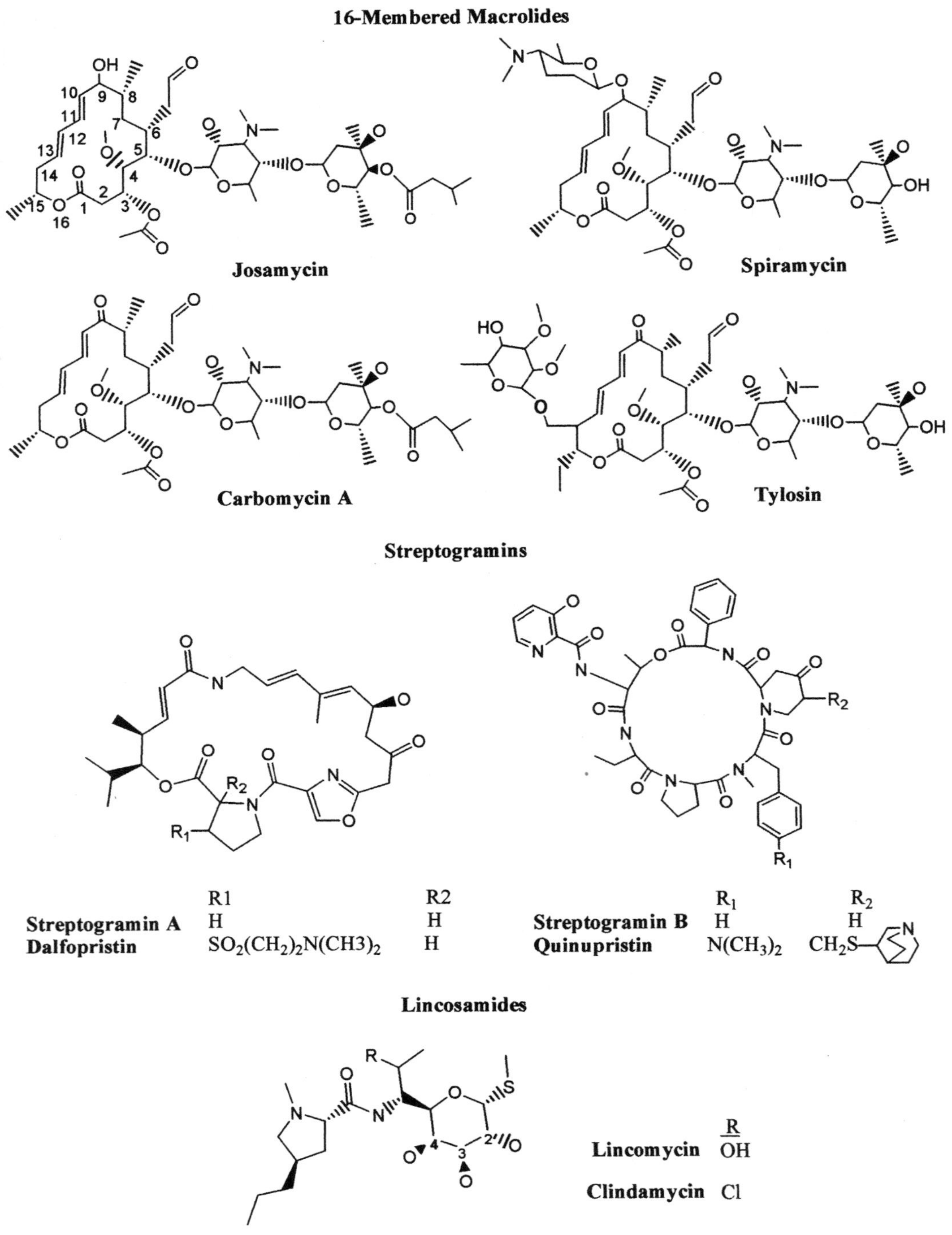

16-Membered Macrolides

Josamycin

Spiramycin

Carbomycin A

Tylosin

Streptogramins

	R1	R2
Streptogramin A	H	H
Dalfopristin	$SO_2(CH_2)_2N(CH3)_2$	H

	R_1	R_2
Streptogramin B	H	H
Quinupristin	$N(CH_3)_2$	CH_2S–

Lincosamides

	R
Lincomycin	OH
Clindamycin	Cl

Figure 1. *Continued*

lower concentration than does either component alone. The synergistic combination is used for therapy, and the most recent combination for human use is quinupristin/dalfopristin (Synercid) (9, 65, 74, 79–81, 132). Virginiamycin, one of the first streptogramins available, has been used as a feed additive for growth promotion and disease prevention in poultry, swine, and cattle for the last 30 years. The streptogramin B component shares an overlapping ribosomal binding site, on the 50S ribosomal subunit, with the macrolides and lincosamides (44, 144, 164) and is impacted by *erm* methylation at A2058 (*Escherichia coli* numbering). In contrast, the streptogramin A component binds nearer the peptidyl transferase center, blocking peptide bond

Table 1. Mechanism of resistance[a]

rRNA methylases (*erm* genes)	Efflux genes	Inactivating enzymes			
		Esterase	Lyase	Transferases	4 phosphorylases
(A), (B), (C), (D), (E), (F), (G), (H) (I), (N), (O), (Q), (R), (S), (T), (U) (V), (W), (X), (Y), (Z), (30), (31), (32), (33), (34), (35), (36), (37), (38), (39) (40)	*mef*(A), *msr*(A), *msr*(C), *msr*(D)[b] *car*(A), *lmr*(A), *ole*(B), *ole*(C) *srm*(B), *tlr*(C), *vga*(A), *vga*(B) *lsa*(A), *lsa*(B)	*ere*(A), *ere*(B)	*vgb*(A), *vgb*(B)	*lnu*(A), *lnu*(B), *lnu*(C)[c], *lnu*(F)[d] *vat*(A), *vat*(B), *vat*(C), *vat*(D), *vat*(E), *vat*(F)	*mph*(A), *mph*(B), *mph*(C), *mph*(D)

[a] http://faculty.washington.edu/marilynr.
[b] See reference 27.
[c] See reference 80a.
[d] See reference 46a.

Table 2. Bacteria carrying specific resistance genes[a]

Gene	Gram-positive genera[b]	Gram-negative genera[b]
rRNA methylases (MLS_B)[c]		
erm(A)	*Peptostreptococcus, Staphylococcus, Streptococcus*	*Actinobacillus, Bacteroides, Prevotella*
erm(B)	*Aerococcus, Arcanobacterium, Bacillus, Corynebacterium, Clostridium, Eubacterium, Enterococcus, Lactobacillus, Pediococcus, Peptostreptococcus, Rothia, Staphylococcus, Streptococcus*	*Acinetobacter, Actinobacillus, Bacteroides, Escherichia, Enterobacter, Fusobacterium, Haemophilus, Klebsiella, Neisseria, Pantoeae, Proteus, Pseudomonas, Porphyromonas, Serratia, Wolinella, Treponema, Citrobacter*
erm(C)	*Actinomyces, Bacillus, Corynebacterium, Enterococcus, Eubacterium, Lactobacillus, Micrococcus, Peptostreptococcus, Staphylococcus, Streptococcus*	*Actinobacillus, Bacteroides, Haemophilus, Neisseria, Prevotella, Wolinella*
erm(D), *erm*(34)	*Bacillus*	
erm(F)	*Actinomyces, Clostridium, Enterococcus, Gardnerella, Mobiluncus, Peptostreptococcus, Staphylococcus, Streptococcus*	*Actinobacillus, Bacteroides, Fusobacterium, Haemophilus, Neisseria, Porphyromonas, Prevotella, Selenomonas, Treponema, Veillonella, Wolinella*
erm(G)	*Bacillus, Lactobacillus,*	*Bacteroides, Prevotella, Porphyromonas*
erm(E), *erm*(H), *erm*(I), *erm*(N), *erm*(O), *erm*(S), *erm*(U), *erm*(V), *erm*(Z)	*Streptomyces*	
erm(30), *erm*(31), *erm*(32)	*Streptomyces*	
erm(Q)	*Clostridium, Staphylococcus, Streptococcus*	*Actinobacillus, Bacteroides, Wolinella*
erm(R)	*Arthrobacter*	
erm(T)	*Lactobacillus*	
erm(W)	*Micromonospora*	
erm(X)	*Arcanobacterium, Corynebacterium, Propionibacterium*	
erm(Y), *erm*(33)	*Staphylococcus*	
erm(35)		*Bacteroides*
erm(36)	*Micrococcus*	
erm(37), *erm*(38), *erm*(39)	*Mycobacterium*	
erm(40)	*Mycobacterium*	
ATP transporters[d]		
car(A), *ole*(B), *ole*(C), *srm*(B), *tlr*(C)	*Streptomyces*	

Continued on following page

Table 2. *Continued*

Gene	Gram-positive genera[b]	Gram-negative genera[b]
msr(A), *vga*(A), *vga*(B)	*Staphylococcus*	
msr(C)	*Enterococcus*	
msr(D)	*Corynebacterium, Enterococcus, Staphylococcus, Streptococcus*	*Acinetobacter, Citrobacter, Bacteroides*[e] *Enterobacter, Escherichia, Klebsiella, Morganella, Neisseria, Proteus, Providencia, Pseudomonas, Ralstonia, Serratia, Stenotrophomonas*
lsa(A)	*Enterococcus*	
lsa(B)	*Staphylococcus*	
Major facilitator transporters[f]		
lmr(A)	*Streptomyces*	
mef(A)	*Corynebacterium, Enterococcus, Micrococcus, Staphylococcus, Streptococcus*	*Acinetobacter, Citrobacter, Bacteroides,*[g] *Enterobacter, Escherichia, Klebsiella, Morganella, Neisseria, Pantoeae, Pseudomonas, Proteus, Providencia, Ralstonia, Serratia, Stenotrophomonas*
Esterases[h]		
ere(A)		*Citrobacter, Enterobacter, Escherichia, Klebsiella, Pantoeae, Providencia, Pseudomonas, Serratia, Stenotrophomonas, Vibrio, Staphylococcus*
ere(B)		*Acinetobacter, Citrobacter, Escherichia, Enterobacter, Klebsiella, Proteus, Pseudomonas, Staphylococcus*
Lyase[i]		
vgb(A)	*Enterococcus, Staphylococcus*	
vgb(B)	*Staphylococcus*	
Transferases[j]		
lnu(A)	*Staphylococcus*	
lnu(B)	*Enterococcus, Staphylococcus*	
lnu(C)	*Streptococcus*	
lnu(F)	*Escherichia*	
vat(A), *vat*(B), *vat*(C)	*Staphylococcus*	
vat(D)	*Enterococcus*	
vat(E)	*Enterococcus, Lactobacillus*	
vat(F)		*Yersinia*
Phosphorylases[k]		
mph(A)		*Citrobacter, Enterobacter, Escherichia, Klebsiella, Pantoeae, Pseudomonas, Proteus, Serratia, Stenotrophomonas*
mph(B)		*Escherichia, Enterobacter, Proteus, Pseudomonas*
mph(C)	*Staphylococcus*	*Stenotrophomonas*
mph(D)		*Escherichia, Klebsiella, Pantoeae, Pseudomonas, Proteus, Stenotrophomonas*

[a] The *erm*(B) gene is the only rRNA methylase screened for in enteric gram-negative bacteria. Thus the distribution of other *erm* genes in aerobic gram-negative genera is most likely underrepresented in the table.
[b] http://faculty.washington.edu/marilynr. This website is updated yearly.
[c] Macrolide-lincosamide-streptogramin B resistance.
[d] Resistance to either lincomycin, oleandomycin, spriamycin, tylosin, or streptogramin A or both erythromycin and streptogramin B.
[e] Reference 153, the *msr*(D) gene may not be functional.
[f] Resistance to either lincomycin or erythromycin.
[g] The *mef*(A) gene produces a truncated protein which is missing N-terminal amino acids and most likely is not functional.
[h] Resistance to erythromycin.
[i] Resistance to streptogramin B.
[j] Resistance to either lincomycin or streptogramin A.
[k] Resistance to macrolides.

formation (42, 43, 44, 105). There are a number of different acquired genes (*vat, vga*) that are able to modify the streptogramin A component and confer resistance (Tables 1 and 2). Most recently, the structures of the *vat*(D) gene with substrates streptogramin A and quinupristin-dalfopristin have been solved (62, 134).

OXAZOLIDINONES

Linezolid is the first oxazolidinone to be marketed for treatment of nosocomial gram-positive infections, including MRSA, penicillin-resistant pneumococci, and vancomycin-resistant enterococci (158). Linezolid binds to the 50S ribosomal subunit in a way that is not impacted by methylation at nucleotide A2058 within the 23S rRNA (*E. coli* numbering for all rRNA nucleotides). Its binding site may overlap with that of chloramphenicol, as some linezolid-resistant mutants are also chloramphenicol-resistant (32, 61). A plethora of mutations have been derived in the laboratory in different bacteria, such as *Halobacterium halobium, E. coli,* and *S. aureus* (59, 59a, 101), but clinical resistance in enterococci has been limited to the G2576U mutation. This mutation has also been described in clinical isolates of *S. aureus* (86, 147, 159), and more recently, a second mutation in 23S rRNA (A2500U) was described in staphylococcal isolates from a patient taking linezolid over a several year period (86). Linezolid is well absorbed into the CSF, bone, fat, and muscle and can be taken orally or intravenously. In addition, linezolid has good coverage against most gram-positive pathogens causing community-acquired and nosocomial pneumonia, bacteremia, skin and soft tissue infections, and *Nocardia brasiliensis* (72, 78, 149, 158) and has potential for treatment of a variety of anaerobic infections (162).

MECHANISMS OF ACTION

The Ribosome

The ribosome is a complex macromolecular machine made up of RNA and protein (87a, 159a). Because its function of converting genetic information encoded in mRNA into proteins is an essential step for all forms of life, it is a valuable target for antimicrobial drug development.

The ribosome has two subunits, a 30S subunit comprised of 16S rRNA and ~20 ribosomal proteins and a 50S subunit composed of two rRNA species, 5S and 23S, along with ~30 ribosomal proteins. tRNA molecules interact with mRNA and 30S at their 5′ end, carrying cognate amino acids attached to their 3′ end. The tRNA molecules sit between the two subunits in one of three locations, the A-site (which accommodates charged tRNAs), the P-site (where amino acids are added to the growing peptide), and the E-site (where uncharged tRNAs are positioned following peptide bond formation). Protein synthesis begins when an initiator tRNA (containing methionine) assembles in the P-site of the ribosome. A second amino acid is then added from another tRNA molecule that has now entered the A-site, and peptide bond formation is mediated solely by 23S rRNA of the 50S subunit. The protein grows within a tunnel of the 50S subunit that is large enough to accommodate ~35 amino acids. The protein is extruded on the backside of the 50S subunit (on the side opposite of the tRNA and 30S interactions), thereby allowing proper folding into a functional protein. The movement of the tRNA molecules between the two ribosomal subunits is called translocation. This cyclic process of amino acid addition and peptide bond formation continues until a stop codon permits the binding of release factors, resulting in the release of the full-length protein from the ribosome (87a, 159a).

The MLSKO antibiotics inhibit protein synthesis by binding to the 50S ribosomal subunit and blocking peptide bond formation and/or translation. The binding of 14-, 15-, and 16-membered macrolides, clindamycin, and two ketolides has been described at atomic detail in *Haloarcula marismortui* and *Deinococcus radiodurans* (43, 44, 121, 122, 147a). Binding of the 14- and 15-macrolides (erythromycin, azithromycin) enhances dissociation of peptidyl-tRNAs and result in an inhibition of protein synthesis (43, 87a, 122). Occlusion of the tunnel by binding of macrolides near the tunnel entrance, however, is not seen for all macrolides (147a). Thus, it is not completely clear how erythromycin blocks elongation of protein translation such that the predominant peptide dissociated from the P-site is seven amino acids in length (144, 147a). Recent data by Tenson et al. (144) suggest that ketolides allow the synthesis of peptides as long as 8 to 12 amino acids, whereas antibiotics closer to the peptidyl transfer (e.g., clindamycin and josamycin) block protein synthesis when peptidyl-tRNA only has a peptide of 2 to 3 amino acids. Ketolides have additional contacts along the tunnel wall, most notably at positions below the macrocyclic lactone. Although no direct contact is made between the side chain attached to the C11-C12 carbamate of Ketek or the C6 position of ABT-773, these drugs protect A752 in helix 35 from chemical modification (71, 147a). Clindamycin sits in a position that partially overlaps the amino sugar extending from the C5 position of the macrolides. The crystal structure of quinupristin, a streptogramin B analog, reveals that this compound binds at the entrance to the ribosomal tunnel, and these data are consistent with biochemical studies that place it in the same space as the macrolides.

The mechanism of action of oxazolidinones is not fully elucidated, but they have been described as inhibitors of tRNA binding. Linezolid is also known to disrupt elongation by blocking peptide bond formation. Linezolid was modeled recently into the space that chloramphenicol occupies in *D. radiodurans,* suggesting that linezolid is an A-site inhibitor. Mutations conferring chloramphenicol and linezolid resistance are consistent with the latter. Cross-linking data with an analog of linezolid, however, suggests that linezolid binds near ribosomal protein L27, a protein whose N terminus is near the P-site.

Enterobacteriaceae and *Pseudomonas* sp. have been considered to be nonsusceptible to erythromycin, due to innate multidrug-resistance transporters that have MLSKO compounds as substrates (104). In addition to these transporters, our recent work illustrated that 78% of randomly selected gram-negative bacteria actually carry known MLS genes (100). Thus, both gram-positive (1–4, 8, 10, 15, 16, 23, 24, 48, 52, 53, 57, 70, 75, 77, 82–85, 88, 111, 114, 120, 125, 130, 136, 146, 152, 156, 160) and gram-negative bacteria (5, 7, 14, 26, 40, 61, 63, 64, 76, 91, 97–100, 104, 126, 140, 145, 153, 154) become additionally resistant to MLSK antibiotics by acquisition of resistance genes (Table 2) (113, 115, 135, http://faculty.washington.edu/marilynr/). Acquired resistance mechanisms include specific efflux pumps for 14- and 15-membered macrolides, MS antibiotics, rRNA methylases that alter the 50S and reduce the binding of MLSK antibiotics, or by genes that inactivate MLS antibacterials (Table 1). Most of the genes appear to have a gram-positive origin based on G+C content and codon usage. The host range of the various genes varies widely among the genera of bacteria (Table 2).

ACQUIRED MLSKO RESISTANCE

rRNA Methylation

rRNA methylases, encoded by *erm* genes, were the first acquired genes to be identified that conferred resistance to macrolides, lincosamides, and streptogramin B (the MLS_B phenotype). These genes continue to play a dominant role in bacterial resistance to macrolides today (Table 2). These genes are found in aerobic and anaerobic gram-positive and gram-negative bacteria (Table 2). Currently, 32 different rRNA methylases have been characterized (Table 1). Each adds one or two methyl groups to a single adenine (A2058 in *E. coli*) in the 23S rRNA moiety (105). Many of these genes were identified in gram-positive species and found to occur naturally in *Streptomycetes,* where they perform a self-protective role for macrolide producers (Table 2). Recently, each macrolide-resistant species of *Mycobacterium* has also been found to carry an innate rRNA methylase (93, 94) (Table 2). The genes found in *Streptomycetes* and *Mycobacterium* have a high (>70%) G+C content, while those found in other bacteria have a low (≤40%) G+C content, suggesting an origin in a gram-positive ancestor.

Expression of the rRNA methylase genes can be induced or expressed constitutively. Induction generally occurs at a posttranscriptional level by translation attenuation and is the result of altered conformations of the *erm* mRNA containing one or more small peptides (referred to as leader peptides) upstream of the methylase gene. Under normal conditions, the conformation of the mRNA allows for translation of the leader peptide but prevents translation of the methylase coding sequence. In the presence of an inducing antibiotic, a conformational change in the mRNA occurs that allows the methylase gene to be translated and the host to expresses a resistant phenotype (125, 135). Initially, translation attenuation of *erm*(C) was described where 14- (e.g., clarithromycin and erythromycin) and 15-membered macrolides (e.g., azithromycin) were shown to act as inducers, and other macrolides (e.g., josamycin, spiramycin, and tylosin), lincosamides, and streptogramin B were not inducers. In the streptococci, however, the list of permissive inducers is longer (114, 135). The pattern of induction is not necessarily related to the class of *erm* gene but rather to the structure of the specific attenuator that controls *erm* expression or, for a specific attenuator, the structure of the MLS_BK antibiotic (60). Recently, Douthwaite's laboratory (32) found that *S. pneumoniae* strains carrying *erm*(B) are mostly monomethylated, whereas *erm* positive *Streptococcus pyogenes* and *S. aureus* strains are primarily dimethylated at A2058, which affects resistance to various macrolides. The *Mycobacterium erm* genes are regulated at the translation step. These genes confer resistance to macrolides and lincosamides but not to streptogramin B (94).

A recent article described the use of noninducer antibiotics to select for constitutive mutants of an *S. aureus* that was carrying an inducibly expressed *erm*(A) gene (123). All mutants were found to have structural alterations in the translational attenuator region. These included deletions of varying sizes (e.g., tandem duplications and point mutations) as well as an insertion of IS*256* in the upstream region. These alterations prevented the formation of the mRNA secondary structure in the *erm*(A) regulatory region and/or favored formation of secondary structures that allowed translation to be independent of the presence or absence of inducers. Thus, for those investigators who are interested in complete understanding of a resistance phenotype, mutation in the regulatory regions must also be considered.

The *erm*(33) gene is interesting because its sequence suggests that this gene was created by a recombination event between an *erm*(R) and an *erm*(A) gene (125). Because it has less than 80% amino acid identity with either the *erm*(C) or *erm*(A) genes, it was given its own designation. A similar recombination between two tetracycline resistance genes [*tet*(O) and *tet*(W)] has recently been identified, but in this case a new designation was not required (133).

Efflux Genes

Efflux of MLSKO antibiotics is accomplished by 12 members of the ATP binding cassette (ABC) family of proteins and two major facilitator superfamily (MFS) transporters. These genes confer a variety of resistance patterns, including resistance to lincomycin, olenadomycin, spiramycin, tylosin, streptogramin A, 14- and 15-membered macrolides, and streptogramin B (Table 2) (70, 83, 111, 135). All these proteins pump antibiotic(s) out of the cell or the cellular membrane. This action keeps the intracellular concentrations of the antibiotic low, thereby allowing the ribosomes to function. These genes have been found in *Enterococcus* spp. and *Staphylococcus* spp. where they reside on plasmids or in the chromosome. The *vga*(A) and *vga*(B) genes share 59% amino acid identity and have G+C contents of 29 to 36% and confer resistance to streptogramin A. Msr(A) confers inducible resistance to 14- and 15-membered macrolides and streptogramin B (the MS_B phenotype). The hydrophilic protein made from the *msr*(A) gene contains two ATP-binding motifs, which is a characteristic of the ABC transporters. A recent review details the mechanism of resistance and relationship to other ABC transporters including Msr(D) (111). Msr(A) tends to confer lower levels of erythromycin resistance than the rRNA methylases (111). Recent work suggests that both the *mef*(A) and the *msr*(D) genes are expressed as a single mRNA transcript in pneumococci and both contribute to resistance.

The two MFS-transporters are *lmr*(A), which codes for a lincomycin-specific efflux pump in *Streptomyces*, and the *mef*(A) gene, an efflux pump specific for 14- and 15-membered macrolides found in gram-positive and gram-negative bacteria (Table 1). The *mef*(A) is always found upstream of the *msr*(D) gene, and together these two genes have the widest host range of all the non-rRNA methylases (Table 2).

Inactivating Genes

There are currently 18 known inactivating enzymes involved in MLSKO resistance, including two esterases, two lyases, 10 transferases, and four phosphorylases (Table 1). There is little amino acid homology between the two erythromycin esterases produced by the *ere*(A) and *ere*(B) genes, though both hydrolyze the lactone ring, creating the same inactive products (135). These have been identified in both gram-negative and gram-positive bacteria. A truncated version with >30% amino acid homology to *ere*(A) has been found upstream of the erythromycin biosynthetic cluster in *Streptomyces erythraea* (135). The two lyases, *vgb*(A) and *vgb*(B), confer streptogramin B resistance by linearizing the cyclic peptide portion of the molecule. These two genes are found on plasmids in *Enterococcus* spp. and *Streptococcus* spp. (2, 88, 115, 135). *vgb* orthologs are also found in the chromosomes of other bacteria like *Bordetella pertussis* and *Streptomyces coelicolor* (88). The transferases confer resistance to either lincomycin or streptogramin A. The *lnu*(A) and *lnu*(B) genes code for lincosamide nucleotidyltransferases. These two enzymes adenylate the hydroxyl group in position 3 or 4 on the lincomycin and clindamycin, respectively, converting them to inactive drugs (16). Very recently a third gene, *lnu*(C), has been identified in *Streptococcus agalactiae,* which has characteristics similar to those of enzymes encoded by the *lnu*(A) and *lnu*(B) genes (168). A fourth gene, originally named *linF* but renamed *lnu*(F), has been reported from *E. coli* (46a). This represents the first lincosamide nucleotidyltransferase identified in gram-negative bacteria and suggests that lincosamide nucleotidyltransferases should be considered when examining lincosamide resistance in gram-negative bacteria (Table 2).

The *vat*(A), *vat*(B), *vat*(C), *vat*(D), and *vat*(E) genes encode streptogramin A acetyltransferases, which inactivate streptogramin A by acetylation. These enzymes, which share 50 to 60% amino acid identity, are members of the hexapeptide repeat family of acetyltransferases that include chloramphenicol acetyltransferase (45, 134, 157). The *vat*(A), *vat*(B), and *vat*(C) genes have only been identified in *Staphylococcus* spp., while *vat*(D) has been found in *Enterococcus* spp. and *vat*(E) in *Enterococcus* spp. and more recently in *Lactobacillus* spp. (Table 2). All four genes are usually associated with plasmids. A fifth streptogramin A acetyltransferase, encoded by *vat*(F), has been identified in the chromosome of the gram-negative *Yersinia enterocolitica*. This appears to be an innate gene (126), which does function when cloned into *E. coli*.

Four different macrolide phosphotransferases, specified by *mph*(A), *mph*(B), *mph*(C), and *mph*(D), have been identified. These enzymes add a phosphate group to the 2′-OH group of the desoamine or mycaminose sugar positioned at C5 on the macrocyclic lactone ring. These enzymes have limited homology with each other at either the DNA or amino acid sequence level and weak overall homology to other phosphotransferases (135). Much stronger homology is found at two conserved C-terminal regions (63, 135). The *mph*(A) gene is inducibly regulated and inactivates 14- and 15-membered macrolides. When *mph*(A) was cloned into an *E. coli* recipient, the MIC of erythromycin increased fourfold. When *mrx,* which is downstream of the *mph*(A) gene in the natural host, was also cloned into *E. coli,* higher levels of resistance occurred. The *mrx* gene product has homology with the MFS transporters, though there is no evidence to suggest that Mrx functions as an efflux pump (97).

The Mph(B) phosphotransferase inactivates 14-, 15-, and 16-membered macrolides in *E. coli* but has a more limited phenotype when cloned into *S. aureus* (confers resistance to 16-membered macrolides) (61, 99). Studies have identified four aspartic acids as essential for catalytic activity, and these are conserved among phosphotransferases (140). In *Stenotrophomonas,* the Mph(C) phosphotransferase has been found in a locus containing a cadmium efflux pump and its regulator (*cadC/cadA*), as well as a resolvase-invertase (*bin3*) (5). When the entire locus was cloned into an *E. coli* host bearing an inactive AcrAB multidrug resistance efflux protein, the MIC to erythromycin increased from 2 to 32 μg/ml (5). In contrast, when DNA carrying the *mph*(C) gene from *S. aureus* was cloned into a susceptible *S. aureus,* no increase in erythromycin MICs was found, nor was there evidence of erythromycin inactivation by a radiolabeled assay (23). This suggests that another gene is required along with the *mph*(C) for expression in *S. aureus.* Some *Nocardia* strains and *S. coelicolor* also phosphorylate macrolides (135). However, the corresponding gene(s) from these species have not been identified.

Mutational Resistance

In recent surveys of macrolide resistant gram-positive bacteria, between 1 and 4% of the isolates do not carry any of the known acquired macrolide resistance genes (137, 138). More recent work suggests that many of these isolates may have mutations in their rRNA genes and/or ribosomal protein genes which confer macrolide resistance (11, 12, 17, 20, 21, 25, 30, 34, 35, 47, 54–56, 137–139, 142, 163), instead of or in addition to MLS acquired genes (Table 3). Mutational changes usually lead to moderate changes in susceptibility. Mutations associated with resistance

Table 3. Species identified with 23S rRNA, L4, and/or L22 ribosomal protein mutations conferring resistance to MLSKO antibiotics[a]

Bacterium	23S rRNA[b]	L4	L22
Gram positive, cell wall-free			
Arcanobacterium pyogenes	Yes		
Enterococcus sp.	Yes		
Mycobacterium sp.	Yes		
Mycoplasma hominis	Yes		
Mycoplasma fermentans	Yes (innate)		
Mycoplasma pneumoniae	Yes	Yes	Yes
Propionibacterium sp.	Yes		
Staphylococcus aureus	Yes	Yes	Yes
Streptococcus pneumoniae	Yes	Yes	Yes
Streptococcus pyogenes	Yes	Yes	Yes
Gram negative			
Brachyspira hyodysenteriae	Yes		
Brachyspira pilosicoli	Yes		
Bordetella pertussis	Yes		
Campylobacter jejuni/coli	Yes		
Chlamydia trachomatis	Yes		
Escherichia coli	Yes	Yes	Yes
Haemophilus influenzae	Yes	Yes	Yes
Helicobacter pylori	Yes		
Neisseria gonorrhoeae	Yes		
Treponema sp.	Yes		

[a] Data from references 11, 18a, 30, 35, 54, 59a, 62, 79–80, 86, 95, 96, 102a, 103, 106, 107a, 135, 138, 139, 142, and 163.
[b] At either the A2058 or A2059 position with *E. coli* numbering.

to macrolides (59a, 80, 92, 102a, 107a, 110, 135, 137, 138, 150), ketolides (12, 21, 34a, 34b, 135), streptogramins (9, 21, 34a, 80), or oxazolidinones (17a, 34a, 39, 56, 81, 86, 101, 108, 135a, 147, 159, 161, 164) have been described.

Mutations in the 23S rRNA can confer on the host bacteria reduced susceptibility to macrolide, lincosamide and streptogramin B, macrolide and lincosamide, telithromycin and/or linezolid resistance. Various mutations have been identified in the 23S rRNA primarily at A2058 or A2059 position (*E. coli* numbering) (Table 3). Originally these mutations were found in pathogens that had one or two copies of the 23S rRNA like *Mycobacterium* or *Helicobacter* (92, 142). Clinical strains and laboratory mutants of *S. pneumoniae* that are heterozygous at these 23S rRNA loci, however, have been identified (30, 135, 139, 150). As the number of 23S rRNA loci acquire mutations, the resistance level in the host normally increases (81, 138). Mutations in 23S rRNA have been found in *Arcanobacterium pyogenes* (58), *H. influenzae* (103a), *Staphylococcus aureus* (17a, 79, 107, 107a, 135a), *S. pyogenes* (17, 80), *Propionibacterium* sp. (116), *Mycoplasma* sp. (35, 73), *Bordetella* (11), *Brachyspira* sp. (59, 59a), *Campylobacter* (54, 96), *Chlamydia trachomatis* (87), *E. coli* (135), *Neisseria gonorrhoeae*

(55, 95), and *Treponema* (68) (Table 3). 23S rRNA mutations may lead to clinically relevant changes in susceptibility as seen in both clinical and spontaneous mutants of *N. gonorrhoeae,* giving MICs for erythromycin of 32 to 64 μg/ml (95). Experiments in the laboratory have often been successful in predicting mutations seen in the clinic.

Several reviews on macrolide resistance, including resistance due to nucleotide(s) base substitutions in the 23S rRNA have been recently published (34b, 135) and are recommended for an in-depth coverage of the topic. The authors list the various mutations currently known and provide phenotypes and references. The level of resistance varies depending on the bacterium and the mutation(s). An interesting corollary to the work above is the recent sequencing of the 23S rRNA from intrinsically macrolide-resistant mycoplasma. These species have base pair changes in their 23S rRNA at positions that are mutated in other resistant isolates and are hypothesized to be the reason for their intrinsic resistance (102).

Mutations in ribosomal proteins L4 and/or L22 have been identified that confer reduced susceptibility to the newer agent telithromycin or linezolid and/or to the other members of the MLS group. Laboratory created mutants and recent clinical gram-positive isolates have been found with the same mutations (79, 102a, 107a, 137, 138) (Table 3). Mutant selection in vitro can be done by growing bacteria on increasing drug concentrations or exposing isolates to a plate containing a gradient of antibiotic. Many of the clinical isolates with mutations in either the 23S rRNA and/or ribosomal proteins were identified because they did not appear to carry one of the acquired genes known to mediate resistance. Thus, the prevalence of these mutations in conjunction with acquired macrolide-resistant mutations may be underestimated.

Intrinsic Efflux Systems

In the last 10 years the role of intrinsic efflux pumps in mediating the antibiotic resistance of gram-negative and gram-positive bacteria has been elucidated (69, 135, 155). These systems use the proton motive force or ATP as energy sources for transport. Overexpression of the efflux pumps may result in increased resistance to antibiotics, as well as some dyes, detergents, and disinfectants (25, 69, 163). A few of these efflux pumps have the MLS group antibiotics as substrates. In *N. gonorrhoeae* the MTR efflux pump is regulated by a transcriptional regulator from the *mtrR* gene. Missense, deletion, or insertions within the *mtrR* coding region result in increased resistance (25, 128, 135, 163). Upstream of *mtrR* a one base pair deletion of an adenine results in the loss of expression of *mtrR* and leads to a fourfold increase in resistance to erythromycin, penicillin, and tetracycline (25). A homolog of the *N. gonorrhoeae* MtrCD(E) system has been identified in *H. influenzae* (HI0894-HI0895). Expression of this gene produced a small decrease in susceptibility to erythromycin. When either of the genes was disrupted, the susceptibility to erythromycin increased (119). Numerous other innate efflux systems have been identified and are discussed in more detail in chapters 19 and 20.

ROLE OF MOBILE GENETIC ELEMENTS IN ACQUIRED RESISTANCE

A few bacteria, such as *S. pneumoniae,* have a combination of acquired resistance genes, mutations, and mosaic genes, the latter of which carry parts of genes found in related species (25, 26, 48, 65, 130, 137–139, 159). Yet in the majority of bacteria, mobile elements are associated with most of the clinically relevant resistance to this group of antibiotics. These elements are thought to play a major role in the movement of antibiotic resistance genes between strains, species, and ecosystems (113, 129, 153). The distribution of the various MLS resistance genes varies by genera and species and may be due to basic bacterial physiology and/or related to the type of mobile element with which particular genes are associated. A number of MLS genes with low G+C content, such as *erm*(B) and *mef*(A), are functional in both gram-positive and gram-negative species (7, 76, 100). Association with mobile elements allows the MLS resistance genes to be linked to a variety of other antibiotic resistance genes as illustrated in Table 4. This may improve the genes ability to survive in the bacterial population even when MLS antibiotics are not in the bacterial host environment (113).

rRNA Methylase Genes

Of the 32 different rRNA methylase genes, 12 genes are exclusively found in *Streptomyces* sp., another 13 genes are in single genera, and the remaining seven genes have been identified in multiple genera (Table 2). The distribution of the various *erm* genes in gram-positive cocci (streptococci, staphylococci, enterococci) varies by genus and by site of isolation (Table 1) (77). Seven *erm* genes [*erm*(A), *erm*(B), *erm*(C), *erm*(F), *erm*(G), *erm*(Q), and *erm*(X)] are found in multiple genera (Table 2). All but the *erm*(X) gene have been found in both gram-positive and gram-negative species. Many of these genes are associated with conjugative transposons and/or plasmids (Table 4). These genes have been identified in

Table 4. MLSKO antibiotic resistance genes linked to other antibiotic resistance genes or elements[a]

Gene	Linkage	Phenotype
rRNA methylases		
erm(A)	*lnu*(A)	Lincomycin and erythromycin
	msr(A)	Erythromycin and streptogramin B
	vga(A)	Streptogramin A
	vgb(A)	Streptogramin B
	Tn*554*	Transposon
erm(B)	*vat*(E)	Streptogramin A
	tet(M)	Tetracycline (Tn*916*-Tn*1545*)
	tet(Q)	Tetracycline
	erm(G)	MLS_B
	cat(pC221)	Chloramphenicol (Tn*916*-Tn*1545*)
	aphA-3	Kanamycin
	vat(B)	Streptogramin A
	lnu(A)	Lincomycin
	msr(A)	Erythromycin and streptogramin B
	vga(A)	Streptogramin A
	vgb(A)	Streptogramin B
	mef(A)	Tn*2010*, 14- and 15-membered macrolides
erm(C)	*lnu*(A)	Lincomycin
	msr(A)	Erythromycin and streptogramin B
	vga(A)	Streptogramin A
	vgb(A)	Streptogramin B
erm(F)	*tet*(X)	Tetracycline
	tet(Q)	Tetracycline
Efflux		
mef(A)	*msr*(D)[b]	Erythromycin, 14- and 15-membered macrolides
	tet(O)	Tetracycline *tet*(O)-*mef* element
	tet(M)	Tetracycline (Tn*2009*)
	tet(W)	Tetracycline
msr(A)	*mph*(C)	Macrolides
vga(A)	Tn*5406*	Transposon
Esterases		
ere(A)	Class 1 integron[c]	
	Class 2 integron	
	dfrA5	Trimethoprim
Lyases		
vgb(A)	*vga*(A)	Streptogramin A
	vat(A)	Streptogramin A
	erm(A)	MLS_B
	erm(B)	MLS_B
	erm(C)	MLS_B
vgb(B)	*vat*(C)	Streptogramin A
Transferases		
vat(A)	*vgb*(A)	Streptogramin B
vat(B)	*erm*(B)	MLS_B
vat(C)	*vgb*(B)	Streptogramin B
vat(D)	*erm*(B)	MLS_B
vat(E)	*tcrB*	Copper resistance
Phosphotransferases		
mph(A)	*mrx*[d]	Related to MSF transporters
mph(C)	*cadC*/*cadA*	Cadmium efflux

[a] This table is meant to illustrate the diversity of the genes and elements that the MLS genes have been linked to (2–5, 16, 27, 38, 40, 43, 45, 52, 53, 83, 85, 104, 112, 120, 145, 153).
[b] The *mef*(A) and the *msr(*D) genes appear to be together in all clinical isolates examined from any genera; however, when cloned separately into naïve strains of *S. pneumoniae,* both conferred macrolide resistance alone (27).
[c] Integron codes for an integrase (IntI) that mediates recombination between a recombination site (*attI*) and a target recombination sequence (*attC;* 59-base element.).
[d] Required for high level erythromycin resistance in *E. coli.*

various gram-positive and gram-negative populations isolated in the 1950s and 1960s (8, 24, 26). The *erm*(F) gene is often associated with conjugative transposons and has been linked to the *tet*(Q) gene, conferring tetracycline resistance by encoding for a ribosomal protection protein, or the *tet*(X) gene, coding for the first enzyme to use tetracycline as a substrate in the presence of oxygen (Table 3). Both linkages were first described in *Bacteroides* sp. (24, 129, 148). Conjugative transposons carrying the *erm*(F) gene have also been shown to mobilize coresident plasmids in *Bacteroides* spp. (46). Recently, the *erm*(B) gene has been associated with a conjugative transposon in *Bacteroides* spp. (40). The *erm*(B) gene is often linked with the *tet*(M) gene on Tn*916*-Tn*1545*-like conjugative transposons (85, 112).

Intrinsic macrolide resistance in *Mycobacterium tuberculosis* [*erm*(37)], *M. smegmatis* [*erm*(38)] and other macrolide-resistant *Mycobacterium* species [*erm*(39), *erm*(40)] results from innate rRNA methylases, which have the same mode of action as acquired *erm* genes and those found in the producers (Tables 1 and 2) (93, 94). These genes are found in the chromosome and appear to be unique to each different *Mycobacterium* species. The *M. smegmatis erm*(38) gene is on the chromosome and is not associated with known or putative mobile elements. Unlike other *erm* genes the *Mycobacterium* genes seem to confer resistance to macrolides and lincosamides but not streptogramin B (94). They also are regulated at the level of transcription rather than translation.

Efflux Genes (ATP-Transporters and MSF-Transporters)

Five of the ATP-transporters, *car*(A), *lmr*(A), *ole*(B), *ole*(C), and *srm*(B), are found only in *Streptomyces* spp.; three of the ATP-transporters, *msr*(A), *vga*(A), and *vga*(B), are found in *Staphylococcus* spp.; and one ATP-transporter, *msr*(C), is found in *Enterococcus* spp. (Table 2) (70, 83, 111, 115, 135). Among the MSF-transporters, the *lmr*(A) gene is found in *Streptomyces* spp. In contrast, the *mef*(A) gene has a broader host range and is found in six gram-positive and 14 gram-negative genera (25, 75–77, 100, 115) (Table 2). Recently, the *mef*(A) gene was found to be the most common MLS gene in randomly selected commensal gram-negative oral and urine bacteria from 12 genera (100).

The *mef*(A) gene is always linked to a gene downstream called *orf5*, which has homology with the *msr*(A) gene and has been named *msr*(D) (10, 28, 38, 120) (Table 1). This gene has been shown to function independently in *S. pneumoniae* to confer erythromycin resistance (27). The *mef*(A) gene is often associated with conjugative transposons but has been found in nonconjugative transposons such as Tn*1207.1* (120). Both conjugative and nonconjugative elements have been shown to be inserted into the chromosomal *comEC* gene in *S. pneumoniae* isolates, so that part of the gene is flanking each side of the element. These elements are also linked to a prophage in the chromosome (10). Recently, we have shown that the *tet*(O) gene, which codes for a ribosomal protection protein, is found upstream of *mef*(A) in Italian *S. pyogenes* isolates where it could be moved from one chromosome to another by conjugation (38). This is the first time that the *mef*(A) gene has been linked to the tetracycline resistant *tet*(O) gene. This gene was 99 to 100% identical to previously characterized *tet*(O) genes and is upstream of the *mef*(A) gene and part of a 60 kb element. We have found that the new *tet*(O)-*mef*(A) element was also integrated into a prophage in *S. pyogenes* but not flanked by the *comEC* gene (18).

The *mef*(A) gene associated with the mega element was inserted within the Tn*916* conjugative element upstream of the *tet*(M) gene in *S. pneumoniae*. Two different isolates had the *mef*(A) gene along with *msr*(D) plus five additional *orf*s inserted into the *orf6* Tn*916* gene to create Tn*2009*, a new composite element (28). This is the first description of the *mef*(A) gene linked to *tet*(M), a ribosomal protection protein. Recently, we have found that Tn*2009*-like elements are found in other genera (unpublished observation). At the recent ASM 2004 meeting, a report was presented which identified an oral *tet*(W) gene inserted into a *mef*(A) gene, creating a novel transposon element which linked the *mef*(A) gene to yet another ribosome protection protein (151).

The *msr*(A) gene is found in *Staphylococcus* spp. and is often associated with plasmids in *Staphylococcus* spp. (83). The *msr*(C) gene appears to be exclusively found in *Enterococcus* spp., while the *msr*(D) gene has a host range identical to the *mef*(A) gene, including both gram-positive and gram-negative genera (Table 2). The *vga*(A) and *vga*(B) genes are often associated with plasmids in *Staphylococcus* spp. (2, 3) (Table 4).

Inactivating Enzymes

Esterases, *ere*(A) and *ere*(B), are enzymes that hydrolyze the macrocyclic lactone of 14- and 15-membered macrolides. The *ere*(A) gene is unusual, because it has been associated with both class 1 and class 2 integrons in *Vibrio* sp. and *E. coli* (Table 3) (14, 104, 126). Although rare, either gene can be found in *S. aureus* (123a, 123b) Our recent work suggests that these two genes are more widely spread among gram-negative genera than the current literature indicates

(100). The two lyases *vgb*(A) and *vgb*(B) inactivate streptogramin B antibiotics, are found on plasmids, and are often linked to other antibiotic resistance genes in *Enterococcus* spp. and *Staphylococcus* sp. or *Staphylococcus* sp., respectively (2, 3, 23, 88) (Table 4).

Five of the *vat* genes found in *Enterococcus* spp. and/or *Staphylococcus* spp. are often associated with plasmids (1–4, 45, 53), while the recently described *vat*(F) gene was identified in *Y. enterocolitica* (126). These genes are acetyltransferases that inactivate streptogramin A by introducing an acetyl group at position C14 (4, 29), preventing a key hydrogen bond between C14-hydroxyl and G2505 in the 23S rRNA (44). Often other genes that confer resistance to streptogramin A or B antibiotics are linked to the *vat* genes (Table 4). Recently, the *vat*(D) and *erm*(B) genes were found as part of a mobile element *Enterococcus faecium* (53).

The *lnu*(A) and *lnu*(B) genes encode for 3-lincomyin, 4-clindamycin O-nucleotidyltransferases that use a nucleoside 5′-triphosphate as a nucleotidyl donor and magnesium as cofactor. These genes are often plasmid-borne in *Staphylococcus* spp. or *Enterococcus* spp. and *Staphylococcus* spp., respectively (15, 115), and have been linked to various MLS genes (Table 4). The *lnu*(F) gene, identified in an *E. coli* isolate, was part of a class I integron which also carried six different trimethoprim resistance genes and three aminoglycoside resistance genes (Table 4) (46a). Unfortunately, no other work has been done with this gene.

Among the four phosphorylases, the *mph*(A), *mph*(B), and *mph*(D) genes have been found exclusively in gram-negative species (Table 2) (61, 63, 97–100, 115). These are often associated with plasmids and can be linked to other genes (Table 4). When the *mph*(B) gene, from *E. coli*, was introduced into a macrolide susceptible *S. aureus*, the strain was able to inactivate a 16-membered macrolide (70). One of the *mph*(B) genes that resided on a plasmid in *E. coli* BM2506 was found on a 39-kb transposon similar to Tn*2610* (99). The *mph*(C) gene was originally found in *Staphylococcus* spp., but recently it has been identified in the gram-negative *Stenotrophomonas* spp. (5) where it was part of a transposon, suggesting that this gene may have the potential for a wider host distribution (Table 2).

THE FUTURE

Erythromycin was discovered over 50 years ago, but its use diminished over time with increased resistance development and the introduction of β-lactam antibiotics that could effectively treat β-lactamase-positive *S. aureus* infections. However, with the addition of the newer macrolides, streptogramins, ketolides, and oxazolidinones, use of this group of antibiotics has increased, and today they are important in the treatment of a variety of infectious diseases in both community and hospital environments. Increased frequency of antibiotic resistance often appears to be associated with increased consumption of macrolide antibiotics (127), especially if a robust clone becomes endemic.

Over the last few years there has continued to be an increasing level of MLSKO resistance. New acquired mechanisms of resistance have been identified (15, 16, 27) as well as ribosome-based mutations (Table 3), and identification of previously characterized acquired MLS genes in new genera have been made (5, 40, 64, 76, 100). Ketolide resistance in streptococci and oxazolidinone resistance in gram-positive enterococci or staphylococci are surfacing with increased therapeutic use. Ketolide resistance occurs in clinical pneumococcal isolates that carry certain truncated *erm*(B) genes with or without a mutated ribosomal protein L4 and in group A streptococci with certain *erm*(B) genes (109, 139). There is always the concern that additional acquired MLS genes will evolve to provide enhanced resistance to either ketolides. MLSK resistance can be selected in the laboratory from isolates carrying the inducible *erm*(A) genes (123). The spread of these isolates could significantly influence the utility of these newer antibiotics in therapy. Thus far, target-based mutations in certain 23S rRNA residues (G2576U, U2500A, G2505A, G2057A, and A2059G) or a 6-base pair deletion found in the gene which specifies ribosome protein L4 are the only known mechanisms of resistance that develop with linezolid selection (39, 56, 81, 86, 108, 147, 159). The only certainty is that this field is not stagnant, and with the number of genera resistant to MLSKO antibiotics increasing, resistance is likely to spread faster now than in the last 25 years.

REFERENCES

1. **Allignet, J., and N. El Solh.** 1995. Diversity among the gram-positive acetyltransferases inactivating streptogramin A and structurally related compounds and characterization of a new staphylococcal determinant, *vatB. Antimicrob. Agents Chemother.* **39:**2027–2036.
2. **Allignet, J., and N. El Solh.** 1999. Comparative analysis of staphylococcal plasmids carrying three streptogramin-resistance genes: *vat-vgb-vga. Plasmid* **42:**134–138.
3. **Allignet, J., N. Liassine, and N. El Solh.** 1998. Characterization of a staphylococcal plasmid related to pUB110 and carrying two novel genes, *vatC* and *vgbB,* encoding resistance to streptogramins A and B and similar antibiotics. *Antimicrob. Agents Chemother.* **42:**1794–1798.
4. **Allignet, J., V. Loncle, C. Simenel, M. Delepierre, and N. El Solh.** 1993. Sequence of a staphylococcal gene, *vat,* encoding

an acetyltransferase inactivating the A-type compounds of virginiamycin-like antibiotics. *Gene* **130:**91–98.

5. **Alonso, A., P. Sanchezm, and J. L. Marinez.** 2000. *Stenotrophomonas maltophilia* D457R contains a cluster of genes from gram-positive bacteria involved in antibiotic and heavy metal resistance. *Antimicrob. Agents Chemother.* **44:**1778–1782.
6. **Amsden, G. W.** 2001. Advanced-generation macrolides: tissue-directed antibiotics. *Inter. J. Antimicrob. Agents* **18:**S11–S15.
7. **Arthur, M., A. Andremont, and P. Courvalin.** 1987. Distribution of erythromycin esterase and rRNA methylase genes in members of the family *Enterobacteriaceae* highly resistant to erythromycin. *Antimicrob. Agents Chemother.* **31:**404–409.
8. **Atkinson, B. A., A. Abu-Al-Jaibat, and D. J. LeBlanc.** 1997. Antibiotic resistance among enterococci isolated from clinical specimens between 1953 and 1954. *Antimicrob. Agents Chemother.* **41:**1598–1600.
9. **Aumercier, M., S. Bouchallab, M. L. Capmau, F. Le Goffic.** 1992. RP59500: a proposed mechanism for its bacterial activity. *J. Antimicrob. Chemother.* **30:**9–14.
10. **Banks, D. J., S. F. Porcella, K. D. Barbian, J. M. Martin, and J. M. Musser.** 2003. Structure and distribution of an unusual chimeric genetic element encoding macrolide resistance in phylogenetically diverse clones of group A *Streptococcus. J. Infect. Dis.* **188:**1898–1908.
11. **Bartkus, J. M., A. Juni, K. Ehresmann, C. A. Miller, G. N. Sanden, P. K. Cassiday, M. Saubolle, B. Lee, J. Long, A. R. Harrison, Jr., and J. M. Besser.** 2003. Identification of a mutation associated with erythromycin resistance in *Bordetella pertussis*: implications for surveillance of antimicrobial resistance. *J. Clin. Microbiol.* **41:**1167–1172.
12. **Berisio, R., J. Harms, F. Schluenzen, R. Zarivach, H. A. Hansen, P. Fucini, and A. Yonath.** 2003. Structural insight into the antibiotic action of telithromycin against resistant mutants. *J. Bacteriol.* **185:**4276–4279. Erratum *In J. Bacteriol.* **185:**5027.
13. **Besier, S., K.-P. Hunfeld, I. Giesser, V. Schafer, V. Brade, and T. A. Wichelhaus.** 2003. Selection of ketolide resistance in *Staphylococcus aureus. Int. J. Antimicrob. Agents* **22:**87–88.
14. **Biskri, L., and D. Mazel.** 2003. Erythromycin esterase gene *ere*(A) is located in a functional gene cassette in an unusual class 2 integron. *Antimicrob. Agents Chemother.* **47:**3326–3331.
15. **Bozdogan, B., L. Berrezouga, M.-S. Kuo, D. A. Yurek, K. A. Farley, B. J. Stockman, and R. Leclercq.** 1999. A new resistance gene, *linB*, conferring resistance to lincosamides by nucleotidylation in *Enterococcus faecium* HM1025. *Antimicrob. Agents Chemother.* **43:**925–929.
16. **Bozdogan, B., S. Galopin, and R. Leclercq.** 2004. Characterization of a new *erm*-related macrolide resistance gene present in probiotic strains of *Bacillus clausii. Appl. Environ. Microbiol.* **70:**280–284.
17. **Bozdogan, B., P. C. Appelbaum, L. Ednie, I. N. Grivea, and G. A. Syrogiannopoulos.** 2003. Development of macrolide resistance by ribosomal protein L4 mutation in *Streptococcus pyogenes* during miocamycin treatment of an eight-year-old Greek child with tonsillopharyngitis. *Clin. Microbiol. Infect.* **9:**966–969.

17a. **Bozdogan, B., P. C. Appelbaum, L. M. Kelly, D. B. Hoellman, A. Tambic-Andrasevic, L. Drukalska, W. Hryniewicz, H. Hupkova, M. R. Jacobs, J. Kolman, M. Konkoly-Thege, J. Miciuleviciene, M. Pana, L. Setchanova, J. Trupl, and P. Urbaskova.** 2004. Activity of telithromycin compared with seven other agents against 1039 *Streptococcus pyogenes* pediatric isolates from ten centers in Central and Eastern Europe. *Clin. Microbiol. Infect.* **9:**741–745.

17b. **Bozdogan, B., M. V. Patel, S. V. Gupte, N. de Souza, H. Khorakiwala, K. Sreenivas.** 2002. Characterization of linezoid resistant mutants of *Staphylococcus aureus* and Streptococcus pneumoniae isolated from mouse in terms of 23S rRNA mutations, abstr. C1, p. 1609. *In Proceedings of the 42nd Interscience Conference on Antimicrobial Agents and Chemotherapy.* American Society for Microbiology, Washington, D.C.

18. **Brenciani, A., K. K. Ojo, A. Monachetti, S. Menzo, M. C. Roberts, P. E. Varaldo, and E. Giovanetti.** 2004. Distribution and molecular analysis of *mef*(A)-containing elements in tetracycline-susceptible and -resistant *Streptococcus pyogenes* clinical isolates with efflux-mediated erythromycin resistance. *J. Antimicrob. Chemother.* **54:**991–998.

18a. **Brook, I. and K. Shah.** 2003. Effect of amoxicillin or clindamycin on the adenoids bacterial flora. *Otolary. Head Neck Surg.* **129:**5–10.

19. **Bryskier, A., and A. Denis.** 2001. Ketolides: novel antibacterial agents designed to overcome resistance to erythromycin A within gram-positive cocci, p 97–140. *In* W. Schonfeld and H. A. Kirst (ed.), *Macrolide Antibiotics.* Birkhauser Verlag, Basel, Switzerland.
20. **Camps, M., G. Arrizabalaga, and J. Boothroyd.** 2002. An rRNA mutation identifies the apicoplast as the target for clindamycin in *Toxoplasma gondii. Mol. Microbiol.* **43:**1309–1328.
21. **Canu, A., B. Malbruny, M. Coquemont, T. A. Davies, P. C. Appelbaum, and R. Leclercq.** 2002. Diversity of ribosomal mutations conferring resistance to macrolides, clindamycin, streptogramin, and telithromycin in *Streptococcus pneumoniae. Antimicrob. Agents Chemother.* **46:**125–131.
22. **Carfartan, G., P. Gerardin, D. Turck, and M.-O. Husson.** 2004. Effect of subinhibitory concentrations of azithromycin on adherence of *Pseudomonas aeruginosa* to bronchial mucins collected from cystic fibrosis patients. *J. Antimicrob. Chemother.* **53:**686–688.
23. **Cheng, J., T. Grege, L. Wondrack, P. Courvalin, and J. Sutcliffe.** 1999. Characterization of genes involved in erythromycin resistance in a clinical strain of *Staphylococcus aureus. In* Abstract of the 39th Interscience Conference on Antimicrobial Agents and Chemotherapy, #83, p. 114, San Francisco, Calif.
24. **Chung, W. O., C. Werckenthin, S. Schwarz, and M. C. Roberts.** 1999. Host range of the *ermF* rRNA methylase gene in human and animal bacteria. *J. Antimicrob. Chemother.* **43:** 5–14.
25. **Cousin, S. L., Jr., W. L. Whittington, and M. C. Roberts.** 2003. Acquired macrolide resistance genes and the 1 bp deletion in the *mtrR* promoter in *Neisseria gonorrhoeae. J. Antimicrob. Chemother.* **51:**131–133.
26. **Cousin, S. L., Jr., W. L. Whittington, and M. C. Roberts.** 2003. Acquired macrolide resistance genes in pathogenic *Neisseria* spp. Isolated between 1940 and 1987. *Antimicrob. Agents Chemother.* **47:**3877–3880.
27. **Daly, M., R. Flamm, and V. Shortridge.** 2003. The prevalence of *mef*(A) vs *mef*(E) in *S. pneumoniae* and the characterization of associated *msr*(A) homolog element. *In* Abstract of the 43rd Interscience Conference on Antimicrobial Agents and Chemotherapy, C2-71, p. 112, Chicago Ill.
28. **Del Grosso, M., A. C. d'Abusco, F. Iannelli, G. Pozzi, and A. Panatosti.** 2004. Tn*2009*, a Tn*916*-like element containing *mef*(E) in *Streptococcus pneumoniae. Antimicrob. Agents Chemother.* **48:**2037–2042.
29. **De Meester, C., and J. Rondelet.** 1976. Microbial acetylation of M factor of virginiamycin. *J. Antibio.* **29:**1297–1305.
30. **Depardieu, F., and P. Courvalin.** 2001. Mutation in 23S rRNA responsible for resistance to 16-membered macrolides and

streptogramins in *Streptococcus pneumnoniae. Antimicrob. Agents Chemother.* **45:**319–323.

31. **Dixon, J.** 1967. Pneumococcus resistant to erythromycin and lincomycin. *Lancet* **1:**573.

32. **Douthwaite, S.** 2003. Differentiating macrolide and ketolide interactions: implications for resistance, abst. 1333, p. 524. *In Program and Abstracts of the 43rd Interscience Conference on Antimicrobial Agents and Chemotherapy,* American Society for Microbiology, Washington, D.C.

33. **Edlund, C., E. Sillerstrom, E. Wahlund, and C. E. Nord.** 1998. *In vitro* activity of HMR 3647 against anaerobic bacteria. *J. Chemoth.* **10:**280–284.

34. **Farrell, D. J., S. Douthwaite, I. Morrissey, S. Bakker, J. Poehlsgaard, L. Jakobsen, and D. Felmingham.** 2003. Macrolide resistance by ribosomal mutation in clinical isolates of *Streptococcus pneumoniae* from the PROTEKT 1999-2000 study. *Antimicrob. Agents Chemother.* **47:**1777–1783.

34a. **Farrell, D. J., I. Morrissey, S. Bakker, S. Buckridge, and D. Felmingham.** 2004. In vitro activities of telithromycin, linezolid, and quinupristin-dalfopristin against *Streptococcus pneumoniae* with macrolide resistance due to ribosomal mutations. *Antimicrob. Agents Chemother.* **48:**3169–3171.

34b. **Franceschi, F., Z. Kanyo, E. C. Sherer, and J. Sutcliffe.** Macrolide resistance from the ribosome perspective. *Curr. Drug Targets Infect. Disord.* **4:**177–191.

35. **Furneri, P. M., G. Rappazzo, M. P. Musumarra, P. D. Pietro, L. S. Catania, and L. S. Roccasalva.** 2001. Two new point mutations at A2062 associated with resistance to 16-membered macrolide antibiotics in mutant strains of *Mycoplasma hominis. Antimicrob. Agents Chemother.* **45:**2958–2960.

36. **Garcia-Rodriquez, J. A., J. E. Garcia-Sanchez, and J. L. Munoz-Bellido.** 1995. Antimicrobial resistance in anaerobic bacteria: current situation. *Anaerobe* **1:**69–80.

37. **Garza-Ramos, G., L. Xiong, P. Zhong, and A. Mankin.** 2001. Binding site of macrolide antibiotics on the ribosome: new resistance mutation identifies a specific interaction of ketolides with rRNA. *J. Bacteriol.* **184:**6898–6907.

38. **Giovanetti, E., A. Brenciani, R. Lupidi, M. C. Roberts, and P. E. Varaldo.** 2003. The presence of the *tet*(O) gene in erythromycin and tetracycline-resistant strains of *Streptococcus pyogenes. Antimicrob. Agents Chemother.* **47:**2844–2849.

39. **Gonzales, R. D., P. C. Schreckenberger, M. B. Graham, S. Kelkar, K. DenBesten, and J. P. Quinn.** 2001. Infections due to vancomycin-resistant *Enterococcus faecium* resistant to linezolid. *Lancet* **357:**1179.

40. **Gupta, A., H. Vlamakis, N. Shoemaker, and A. A. Salyers.** 2003. A new *Bacteroides* conjugative transposon that carries an *ermB* gene. *Appl. Environ. Microbiol.* **69:**6455–663.

41. **Hammerum, A. M., S. E. Flannagan, D. B. Clewell, and L. B. Jensen.** 2001. Indication of transposition of a mobile DNA element containing the *vat*(D) and *erm*(B) genes in *Enterococcus faecium. Antimicrob. Agents Chemother.* **45:**3223–3225.

42. **Hansen, J. L., P. B. Moore, and T. A. Steitz.** 2003. Structures of five antibiotics bound at the peptidyl transferase center of the large ribosomal subunit. *J. Mol. Biol.* **330:**1061–1075.

43. **Hansen, J. L., J. A. Ippolito, N. Ban, P. Nissen, P. B. Moore, and T. A. Steitz.** 2002. The structures of four macrolide antibiotics bound to the large ribosomal subunit. *Mol. Cell* **10:**117–128.

44. **Harms, J. M., F. Schluenzen, P. Fucini, H. Bartels, and A. E. Yonath.** 2004. Alterations at the peptidyl transferase centre of the ribosome induced by the synergistic action of the streptogramins dalfopristin and quinupristin. *BMC Biology* **2:**4.

45. **Haroche, J., J. Alligent, and N. El Solh.** 2002. Tn*5406*, a new staphylococcal transposon conferring resistance to streptogramin A and related compounds including dalfopristin. *Antimicrob. Agents Chemother.* **46:**2337–2343.

46. **Hecht, D. W., J. S. Thompson, and M. H. Malamy.** 1989. Characterization of the termini and transposition products of Tn*4399*, a conjugal mobilizing transposon of *Bacteroides fragilis. Proc. Natl. Acad. Sci.* **86:**5340–5344.

46a. **Heir, E., B.-A. Lindstedt, T. M. Leegaard, E. Gjernes, and G. Kapperud.** 2004. Prevalence and characterization of integrons in blood culture *Enterobacteriaceae* and gastrointestinal *Escherichia coli* in Norway and reporting of a novel class I integron-located lincosamide resistance gene. *Ann. Clin. Microbiol. Antimicrob.* **3:**12–20.

47. **Hsueh, P.-R., L.-J. Teng, C.-M. Lee, W.-K. Huang, T.-L. Wu, J.-H. Wan, D. Yang, J.-M. Shyr, Y.-C. Chuang, J.-J. Yan, J.-J. Lu, J.-J. Wu, W.-C. Ko, F.-Y. Chang, Y.-C. Yang, Y.-J. Lau, Y.-C. Liu, H.-S. Leu, C.-Y. Liu, and K.-T. Luh.** 2003. Telithromycin and quinupristin-dalfopristin resistance in clinical isolates of *Streptococcus pyogenes*: SMART Program 2001 Data. *Antimicrob. Agents Chemother.* **47:**2152–2157.

48. **Hyde, T. B., K. Gay, D. S. Stephens, D. J. Vugia, M. Pass, S. Johnson, N. L. Barrett, W. Schaffner, P. R. Cieslak, P. S. Maupin, E. R. Zell, J. H. Jorgensen, R. R. Facklam, and C. G. Whitney.** 2001. Macrolide resistance among invasive *Streptococcus pneumoniae* isolates. *JAMA* **286:**1857–1862.

49. **Iacoviello, V. R., and S. H. Zinner.** 2001. Macrolides: a clinical overview, 15–24. *In* W. Schonfeld and H. A. Kirst (ed.), *Macrolide Antibiotics.* Birkhauser Verlag, Basel, Switzerland.

50. **Jaffe, A., F. Francis, M. Rosenthal, M. Bush.** 1998. Long-term azithromycin may improve lung function in children with cystic fibrosis. *Lancet* **351:**420.

51. **Jalava, J., J. Kataja, H. Seppala, and P. Huovinen.** 2001. In vitro activities of the novel ketolide telithromycin (HMR 3647) against erythromycin-resistant *Streptococcus* species. *Antimicrob. Agents Chemother.* **45:**789–793.

52. **Jensen, L. B., A. M. Hammerum, R. M, and F. M. Aarestrup, A. E. van den Bogaard, and E. E. Stobberingh.** 1998. Occurrence of *satA* and *vgb* genes in streptogramin-resistant *Enterococcus faecium* isolates of animal and human origins in The Netherlands. *Antimicrob. Agents Chemother.* **42:**3330–3331.

53. **Jensen, L. B., A. M. Hammerum, R. M, and F. M. Aarestrup.** 2000. Linkage of *vat*(E) and *erm*(B) in streptogramin-resistant *Enterococcus faecium* isolates from Europe. *Antimicrob. Agents Chemother.* **44:**2231–2232.

54. **Jensen, L. B., and F. M. Aarestrup.** 2001. Macrolide resistance in *Campylobacter coli* of animal origin in Denmark. *Antimicrob. Agents Chemother.* **45:**371–372.

55. **Johnson, S. R., A. L. Sandul, M. Parekh, S. A. Wang, J. S. Knapp, and D. L. Trees.** 2003. Mutations causing in vitro resistance to azithromycin in *Neisseria gonorrhoeae. Int. J. Antimicrob. Agents* **21:**414–419.

56. **Jones, R. N., P. Della-Latta, L. V. Lee, and D. J. Biedenbach.** 2002. Linezolid-resistant *Enterococcus faecium* isolated from a patient without prior exposure to an oxazolidinone: report from the SENTRY antimicrobial surveillance program. *Diagn. Microbiol. Infect. Dis.* **42:**137–139.

57. **Jost, G. H., A. C. Field, H. T. Trinh, J. G. Songer, and S. J. Billington.** 2003. Tylosin resistance in *Arcanobacterium pyogenes* is encoded by an Erm X determinant. *Antimicrob. Agents Chemother.* **47:**3519–3524.

58. **Jost, G. H., H. T. Trinh, J. G. Songer, and S. J. Billington.** 2004. Ribosomal mutations in *Arcanobacterium pyogenes* confer a unique spectrum of macrolide resistance. *Antimicrob. Agents Chemother.* **48:**1021–1023.

59. **Karlsson, M., C. Fellstrom, M. U. Heldtander, K. E. Johansson, and A. Franklin.** 1999. Genetic basis of macrolide and lin-

cosamide resistance in *Brachyspira (Serpulina) hyodysenteriae*. *FEMS Microbiol. Lett.* **172**:255–260.

59a. **Karlsson, M., C. Fellstrom, K. E. Johansson, and A. Franklin.** 2004. Antimicrobial resistance in *Brachyspira pilosicoli* with special reference to point mutations in the 23S rRNA gene associated with macrolide and lincosamide resistance. *Microb. Drug Resist.* **10**:204–208.

60. **Kamimiya, S., and B. Weisblum.** 1997. Induction of *ermSV* by 16-membered-ring macrolide antibiotics. *Antimicrob. Agents Chemother.* **41**:530–534.

61. **Katayama, J., H. Okada, K. O'Hara, and N. Noguchi.** 1998. Isolation and characterization of two plasmids that mediate macrolide resistance in *Escherichia coli:* transferability and molecular properties. *Biol. Pharm. Bull.* **21**:326–329.

62. **Kehoe, L. E., J. Snidwongse, P. Courvalin, J. B. Rafferty, and I. A. Murray.** 2003. Structural basis of Synercid (quinupristin-dalfopristin) resistance in gram-positive bacterial pathogens. *J. Biol. Chem.* **278**:29963–29970.

63. **Kim, Y.-H., M.-C. Baek, S.-S. Choi, B.-K. Kim, and E.-C. Choi.** 1996. Nucleotide sequence, expression and transcriptional analysis of the *Escherichia coli mphK* gene encoding macrolide-phosphotransferase K. *Mol. Cell* **6**:153–160.

64. **Kim, Y.-H., C.-J. Cha, C.-E. Cerniglia.** 2002. Purification and characterization of an erythromycin esterase from an erythromycin-resistant *Pseudomonas* sp. (*ereA*). *FEMS Microbiol. Lett.* **210**:239–244.

65. **Kugler, K. C., G. A. Denys, M. L. Wilson, and R. N. Jones.** 2000. Serious streptococcal infections produced by isolates resistant to streptogramins (quinupristin-dalfopristin): Case reports from the SENTRY antimicrobial surveillance program. *Diagn. Microbiol. Infect. Dis.* **36**:269–272.

66. **Leclercq, R.** 2002. Mechanisms of resistance to macrolides and lincosamides: Nature of the resistance elements and their clinical implications. *Clin. Infect. Dis.* **34**:482–492.

67. **Leclercq, R., and P. Courvalin.** 2002. Resistance to macrolides and related antibiotics in *Streptococcus pneumoniae*. *Antimicrob. Agents Chemother.* **46**:2727–2734.

68. **Lee, S. Y., Y. Ning, and J. C. Fenno.** 2002. 23S rRNA point mutation associated with erythromycin resistance in *Treponema denticola*. *FEMS Microbiol. Lett.* **207**:39–42.

69. **Li, X. Z., and H. Nikaido.** 2004. Efflux-mediated drug resistance in bacteria. *Drugs* **64**:159–204.

70. **Lina, G., A. Quaglia, M.-E. Reverdy, R. Leclercq, R. Vandenesch, and J. Etienne.** 1999. Distribution of genes encoding resistance to macrolides, lincosamides, and streptogramins among staphylococci. *Antimicrob. Agents Chemother.* **43**:1062–1066.

71. **Liu, M., and D. S. Douthwaite.** 2002. Activity of the ketolide telithromycin is refractory to erm monomethylation of bacterial rRNA. *Antimicrob. Agents Chemother.* **46**:1629–1633.

72. **Livermore, D. M.** 2003. Linezolid *in vitro*: mechanism and antibacterial spectrum. *J. Antimicrob. Chemother.* **51**:ii9–ii16.

73. **Lucier, T. S., K. Heitzman, K.-K. Liu, and, P.-C. Hu.** 1995. Transition mutations in the 23S rRNA of erythromycin-resistant isolates of *Mycoplasma pneumoniae*. *Antimicrob. Agents Chemother.* **39**:2770–2773.

74. **Luh, K.-T., P.-R. Hsueh, L.-J. Teng, H.-J. Pan, Y.-C. Chen, J.-J. Lu, J.-J. Wu, and S.-W. Ho.** 2000. Quinupristin-dalfopristin resistance among gram-positive bacteria in Taiwan. *Antimicrob. Agents Chemother.* **44**:3374–3380.

75. **Luna, V. A., P. Coates, E. A. Eady, J. Cove, T. T. H. Nguyen, and M. C. Roberts.** 1999. A variety of gram-positive bacteria carry mobile *mef* genes. *J. Antimicrob. Chemother.* **44**:19–25.

76. **Luna, V. A., S. Cousin, Jr., W. L. H. Whittington, and M. C. Roberts.** 2000. Identification of the conjugative *mef* gene in clinical *Acinetobacter junii* and *Neisseria gonorrhoeae* isolates. *Antimicrob. Agents Chemother.* **44**:2503–2506.

77. **Luna, V. A., M. Heiken, K. Judge, C. Uelp, N. Van Kirk, H. Luis, M. Bernardo, J. Leitao, and M. C. Roberts.** 2002. Distribution of the *mef*(A) gene in gram-positive bacteria from healthy Portuguese children. *Antimicrob. Agents Chemother.* **46**:2513–2517.

78. **MacGowan, A. P.** 2003. Pharmacokinetic and pharmacodynamic profile of linezolid in healthy volunteers and patients with gram-positive infections. *J. Antimicrob. Chemother.* **51**:ii17–ii25.

79. **Malbruny, B., A. Canu, B. Bozdogan, B. Fantin, V. Zarrouk, S. Dutka-Malen, C. Feger, and R. Leclercq.** 2002. Resistance to quinupristin-dalfopristin due to mutations of L22 ribosomal protein in *Staphylococcus aureus*. *Antimicrob. Agents Chemother.* **46**:2200–2207.

80. **Malbruny, B., K. Nagai, M. Coquemont, B. Bozdogan, A. T. Andrasevic, H. Hupkova, R. Leclercq, and P. C. Appelbaum.** 2002. Resistance to macrolides in clinical isolates of *Streptococcus pyogenes* due to ribosomal mutations. *J. Antimicrob. Chemother.* **49**:935–939.

80a. **Malbruny B., A. M. Werno, T. P. Anderson, D. R. Murdoch, and R. Leclercq.** 2004. A new phenotype of resistance to lincosamide and streptogramin A-type antibiotics in *Streptococcus agalactiae* in New Zealand. *J. Antimicrob. Chemother.* **54**:1040–1044.

81. **Marshall, S. H., C. J. Donskey, R. Hutton-Thomas, R. A. Salata, and L. B. Rice.** 2002. Gene dosage and linezolid resistance in *Enterococcus faecium* and *Enterococcus faecalis*. *Antimicrob. Agents Chemother.* **46**:3334–3336.

82. **Martel, A., V. Meulenaere, L. A. Devriese, A. Decostere, and F. Haesebrouck.** 2003. Macrolide- and lincosamide-resistance in the gram-positive nasal and tonsillar flora of pigs. *Microbial Drug Res.* **9**:293–297.

83. **Matsuoka, M., K. Endou, H. Kobayashi, M. Inoue, and Y. Nakajima.** 1997. A dyadic plasmid that shows MLS and PMS resistance in *Staphylococcus aureus*. *FEMS Microbiol. Lett.* **148**:91–96.

84. **Matsuoka, M., M. Inoue, Y. Nakajima, and K. Endou.** 2002. New *erm* gene in *Staphylococcus aureus* clinical isolates. *Antimicrob. Agents Chemother.* **46**:211–215.

85. **McDougal, L. K., F. C. Tenover, L. N. Lee, J. K. Rasheed, J. E. Patterson, J. H. Jorgensen, and D. J. LeBlanc.** 1998. Detection of Tn*917*-like sequences within a Tn*916*-like conjugative transposon (Tn*3872*) in erythromycin-resistant isolates of *Streptococcus pneumoniae*. *Antimicrob. Agents Chemother.* **42**:2312–2318.

86. **Meka, V. G., S. K. Pillai, G. Sakoulas, C. Wennersten, L. Venkataraman, P. C. Degirolami, G. M. Eliopoulos, R. C. Moellering, Jr., and H. S. Gold.** 2004. Linezolid resistance in sequential *Staphylococcus aureus* isolates associated with a T2500A mutation in the 23S rRNA gene and loss of a single copy of rRNA. *J. Infect. Dis.* **190**:311–317.

87. **Misyurina, O. Y., E. V. Chipitsyna, Y. P. Finashutina, V. N. Lazarev, T. A. Akopian, A. M. Savicheva, and V. M. Govorun.** 2004. Mutations in a 23S rRNA gene of *Chlamydia trachomatis* associated with resistance to macrolides. *Antimicrob. Agents Chemother.* **48**:1347–1349.

87a. **Moore, P. B., and T. A Steitz.** 2003. The structural basis of large ribosomal subunit function. *Annu. Rev. Biochem.* **72**:813–850.

88. **Mukhtar, T. A., K. P. Koteva, D. W. Hughes, and G. D. Wright.** 2001. *Vgb* from *Staphylococcus aureus* inactivates streptogramin B antibiotics by an elimination mechanism not hydrolysis. *Biochem.* **40**:8877–8886.

89. **Muller-Serieys, C., J. Andrews, F. Vacheron, and C. Cantalloube.** 2004. Tissue kinetics of telithromycin, the first ketolide antibacterial. *J. Antimicrob. Chemother.* **53:**149–157.
90. **Mutnick, A. H., V. Enne, and R. N. Jones.** 2003. Linezolid resistance since 2001: SENTRY antimicrobial surveillance program. *Ann. Pharmacother.* **37:**769–774.
91. **Nakamura, A., I. Miyakozawa, K. Nakazawa, K. O'Hara, and T. Sawai.** 2000. Detection and characterization of a macrolide 2′-phosphotransferase from a *Pseudomonas aeruginosa* clinical isolated. *Antimicrob. Agents Chemother.* **44:** 3241–3242.
92. **Nash, K. A., and C. B. Inderlied.** 1996. Rapid detection of mutations associated with macrolide resistance in *Mycobacterium avium* complex. *Antimicrob. Agents Chemother.* **40:**1748–1750.
93. **Nash, K. A.** 2003. Intrinsic macrolide resistance in *Mycobacterium smegmatis* is conferred by a novel *erm* gene, *erm*(38). *Antimicrob. Agents Chemother.* **47:**3053–3060.
94. **Nash, K. A., Y. Zhang, B. A. Brown-Elliott, R. J. Wallace, Jr.** 2004. Identification of an rRNA methylase gene, *erm*(39), which confers inducible macrolide resistance to *M. fortuitum*, abstr. U-017. *In Program and Abstracts of the 104th General Meeting of the American Society for Microbiology.* American Society for Microbiology, Washington, D.C.
95. **Ng, L. K., I. Martin, G. Liu, and L. Bryden.** 2002. Mutations in 23S rRNA associated with macrolide resistance in *Neisseria gonorrhoeae. Antimicrob. Agents Chemother.* **46:**3020–5025.
96. **Niwa, H., T. Chuma, K. Okamoto, and K. Itoh.** 2001. Rapid detection of mutations associated with resistance to erythromycin *Campylobacter jejuni/coli* by PCR and line probe assay. *Int. J. Antimicrob. Agents* **18:**359–364.
97. **Noguchi, N., A. Emura, H. Matsuyama, K. O'Hara, M. Sasatsum, and M. Kono.** 1995. Nucleotide sequence and characterization of erythromycin resistance determinant that encodes macrolide 2′-Phosphotransferase I in *Escherichia coli. Antimicrob. Agents Chemother.* **39:**2359–2363.
98. **Noguchi, N., J. Katayama, and M. Sasatsu.** 2000. A transposon carrying the gene *mph*B for macrolide 2′-phosphotransferase II. *FEMS Microbiol. Lett.* **192:**175–178.
99. **Noguchi, N., Y. Tamura, J. Katayama, and K. Narui.** 1998. Expression of the *mphB* gene for macrolide 2′-phosphotransferase II from *Escherichia coli* in *Staphylococcus aureus. FEMS Microbiol. Lett.* **159:**337–342.
100. **Ojo, K. K., C. Ulep, N. Van Kirk, H. Luis, M. Bernardo, J. Leitao, and M. C. Roberts.** 2004. The *mef*(A) gene predominates among seven macrolide resistant genes identified in 13 gram-negative genera from healthy Portuguese children. *Antimicrob. Agents Chemother.* **48:**3451–3456.
101. **Paterson, D. L., A. W. Pasculle, and K. McCurry.** 2003. Linezolid: the first oxazolidinone antimicrobial. *Ann. Intern. Med.* **139:**863–864.
102. **Pereyre, S., P. Gonzalez, B. de Barbeyrac, A. Darnige, H. Renaudin, A. Charron, S. Raherison, C. Bebear, and C. M. Bebear.** 2002. Mutations in 23S rRNA account for intrinsic resistance to macrolides in *Mycoplasma hominis* and *Mycoplasma fermentans* and for acquire resistance to macrolides in *M. hominis. Antimicrob. Agents Chemother.* **46:**3142–3150.
103. **Pereyre, S., C. Guyot, H. Renaudin, A. Charron, C. Bebear, and C. M. Bebear.** 2004. In vitro selection and characterization of resistance to macrolides and related antibiotics in *Mycoplasma pneumoniae. Antimicrob. Agents Chemother.* **48:** 460–465.
103a. **Peric, M., B. Bozdogan, M. R. Jacobs, and P. C. Appelbaum.** 2003. Effects of an efflux mechanism and ribosomal mutations on macrolide susceptibility of *Haemophilus influenzae* clinical isolates. *Antimicrob. Agents Chemother.* **47:**1017–1022.
104. **Plante, I., D. Centron, and P. H. Roy.** 2003. An susceptible cassette encoding erythromycin esterase, *ere*(A), from *Providencia stuartii. Antimicrob. Agents Chemother.* **51:**787–790.
105. **Poehlsgaard, J., and S. Douthwaite.** 2003. Macrolide antibiotic interaction and resistance on the bacterial ribosome. *Cur. Opin. Invest. Drugs.* **4:**140–148.
106. **Porse, B. T., and R. A. Garrett.** 1999. Sites of interaction of streptogramin A and B antibiotics in the peptidyl transferase loop of 23S rRNA and the synergism of their inhibitory mechanisms. *J. Mol. Biol.* **286:**275–387.
107. **Prunier, A. L., B. Malbruny, D. Tande, B. Picard, and R. Leclercq.** 2002. Clinical isolates of *Staphylococcus aureus* with ribosomal mutations conferring resistance to macrolides. *Antimicrob. Agents Chemother.* **46:**3054–3056.
107a. **Prunier, A. L., H. N. Trong, D. Tande, C. Segond, and R. Leclercq.** 2005. Mutation of L4 ribosomal protein conferring unusual macrolide resistance in two independent clinical isolates of *Staphylococcus aureus. Microb. Drug Resist.* **11:**18–20.
107b. **Prystowsky, J., F. Siddiqui, J. Chosay, D. L. Shinabarger, J. Millichap, L. R. Peterson, and G. A. Noskin.** 2001. Resistance to linezolid: characterization of mutations in rRNA and comparison of their occurrences in vancomycin-resistant enterococci. *Antimicrob. Agents Chemother.* **45:**2154–2156.
108. **Rahim, S., S. K. Pillai, H. S. Gold, L. Venkataraman, K. Inglima, and R. A. Press.** 2003. Linezolid-resistant, vancomycin-resistant *Enterococcus faecium* infection in patients without prior exposure to linezolid. *Clin. Infect. Dis.* **36:**e146–e148.
109. **Reinert, R. R., R. Lutticken, J. A. Sutcliffe, A. Tait-Kamradt, M. Y. Cil, H. M. Schorn, A. Bryskier, and A. Al-Lahham.** 2004. Clonal relatedness of erythromycin-resistant *Streptococcus pyogenes* isolates in Germany. *Antimicrob. Agents Chemother.* **48:**1369–1373.
110. **Reinert, R. R., A. Wild, P. Appelbaum, R. Lutticken, M. Y. Cil, and A. Al-Lahham.** 2003. Ribosomal mutations conferring resistance to macrolides in *Streptococcus pneumoniae* clinical strains isolated in Germany. *Antimicrob. Agents Chemother.* **47:**2319–2322.
111. **Reynolds, E., J. I. Ross, and J. H. Cove.** 2003. *msr*(A) and related macrolide/streptogramin resistance determinants: incomplete transporters? *Int. J. Antimicrob. Agents.* **22:**228–236.
112. **Rice, L. B.** 1998. Tn*916* family conjugative transposons and dissemination of antimicrobial resistance determinants. *Antimicrob. Agents Chemother.* **42:**1871–1877.
113. **Roberts, M. C.** 1989. Gene transfer in the urogenital and respiratory tract, p. 347–375. *In* S. B. Levy and R. V. Miller (ed.), *Gene Transfer in the Environment.* McGraw-Hill Publishing Co., New York, N.Y.
114. **Roberts, M. C., and M. B. Brown.** 1994. Macrolide-lincosamide resistance determinants in streptococcal species isolated from the bovine mammary gland. *Vet. Microbiol.* **40:**253–261.
115. **Roberts, M. C., J. Sutcliffe, P. Courvalin, L. B. Jensen, J. Rood, and H. Seppala.** 1999. Nomenclature for macrolide and macrolide-lincosamide streptogramin B antibiotic resistance determinants. *Antimicrob. Agents Chemother.* **43:**2823–2830.
116. **Ross, J. I., A. M. Snelling, E. A. Eady, J. H. Cove, W. J. Cunliffe, J. J. Leyden, P. Collignon, B. Dreno, A. Reynaud, J. Fluhr, and S. Oshima.** 2001. Phenotypic and genotypic characterization of antibiotic-resistant *Propionibacterium acnes* isolated from acne patients attending dermatology clinics in Eu-

rope, the U.S.A., Japan and Australia. *Br. J. Dermatol.* **144:** 339–346.

117. **Saiman, L., Y. Chen, P. San Gabriel, C. Knirsch.** 2002. Synergistic activities of macrolide antibiotics against *Pseudomonas aeruginosa, Burkholderia cepacia, Stenotrophomonas maltophilia,* and *Alcaligenes xylosoxidans* isolated from patients with cystic fibrosis. *Antimicrob. Agents Chemother.* **46:**1105–1107.
118. **Saiman, L., B. C. Marshall, N. Mayer-Hamblett, J. L. Burns, A. L. Quittner, D. A. Cibene, S. Coquillette, A. Y. Fieberg, F. J. Accurso, P. W. Campbell, III.** 2003. Azithromycin in patients with cystic fibrosis chronically infected with *Pseudomonas aeruginosa*: a randomized controlled trial. *JAMA* **290:**1749–1756.
119. **Sanchez, L., W. Pan, M. Vinas, and H. Nicaido.** 1997. The *acrAB* homology of *Haemophilus influenzae* codes of a functional multidrug efflux pump. *J. Bacteriol.* **179:**6855–6857.
120. **Santagati, M., F. Iannelli, C. Cascone, F. Campanile, M. R. Oggioni, S. Sefani, and F. Pozzi.** 2003. The novel conjugative transposon Tn*1207.3* carries the macrolide efflux gene *mef*(A) in *Streptococcus pyogenes. Microb. Drug Res.* **9:**243–247.
121. **Schlunzen, F., J. M. Harms, F. Franceschi, H. A. Hansen, H. Bartels, E. Zarivach, and A. Yonath.** 2003. Structural basis for the antibiotic activity of ketolides and azalides. *Structure* **11:**329–338.
122. **Schlunzen, F., R. Zarivach, J. Harms, A. Bashan, A. Tocilj, R. Albrecht, A. Yonath, and F. Franceschi.** 2001. Structural basis for the interaction of antibiotics with the peptidyl transferase centre in eubacteria. *Nature* **413:**814–821.
123. **Schmitz, F.-J., J. Petridou, H. Jagusch, N. Astfalk, S. Scheuring, and S. Schwarz.** 2002. Molecular characterization of ketolide-resistance *erm*(A)-carrying *Staphylococcus aureus* isolates selected in vitro by telithromycin, ABT-773, quinupristin and clindamycin. *J. Antimicrob. Chemother.* **49:**611–617.

123a. **Schmitz, F. J., J. Petridou, A. C. Fluit, U. Hadding, G. Peters, and C. von Eiff.** 2000. Distribution of macrolide-resistance genes in *Staphylococcus aureus* blood-culture isolates from fifteen German university hospitals. M.A.R.S. Study Group. Multicentre Study on Antibiotic Resistance in Staphylococci. *Eur. J. Clin. Microbiol. Infect. Dis.* **19:**385–387.

123b. **Schmitz, F. J., R. Sadurski, A. Kray, M. Boos, R. Geisel, K. Kohrer, J. Verhoef, and A. C. Fluit.** 2000. Prevalence of macrolide-resistance genes in *Staphylococcus aureus* and *Enterococcus faecium* isolates from 24 European university hospitals. *J. Antimicrob. Chemother.* **45:**891–894.

124. **Schmitz, F.-J., W. Witte, G. Werner, J. Petridou, A. C. Fluit, and S. Schwarz.** 2001. Characterization of the translational attenuator of 20 methicillin-resistant, quinupristin/dalfopristin-resistant *Staphylococcus aureus* isolates with reduced susceptibility to glycopeptides. *J. Antimicrob. Chemother.* **48:**939–941.
125. **Schwarz, S., C. Kehrenberg, and K. K. Ojo.** 2002. *Staphylococcus sciuri* gene *erm*(33), encoding inducible resistance to macrolides, lincosamides, and streptogramin B antibiotics, is a product of recombination between *erm*(R) and *erm*(A). *Antimicrob. Agents Chemother.* **46:**3621–3623.
126. **Seoane, A., and J. M. G. Lobo.** 2000. Identification of a streptogramin A acetyltransferase gene in the chromosome of *Yersinia enterocolitica. Antimicrob. Agents Chemother.* **44:**905–909.
127. **Seppala, H., T. Klaukka, J. Vuopio-Varkila, A. Muotiala, H. Helenius, K. Lager, P. Huovinen.** 1997. The effect of changes in the consumption of macrolide antibiotics on erythromycin resistance in Group A streptococci in Finland. *N. Engl. J. Med.* **337:**441–446.
128. **Shafer, W. M., J. T. Balthazar, K. E. Hagman, and S. A. Morse.** 1999. Missense mutations that alter the DNA-binding domain of the MtrR protein occur frequently in rectal isolates of *Neisseria gonorrhoeae* that are resistant to faecal lipids. *Microbiology* **141:**907–911.
129. **Shoemaker, N. B., H. Vlamikis, K. Hayes, and A. A. Salyers.** 2001. Evidence for extensive resistance gene transfer among *Bacteroides* spp. and between *Bacteroides* and other genera in the human colon. *Appl. Environ. Microbiol.* **67:**561–568.
130. **Shortridge, V. D., G. V. Doern, A. B. Brueggmann, J. M. Beyer, and R. K. Flamm.** 1999. Prevalence of macrolide resistance mechanism in *Streptococcus pneumoniae* isolates from a multi-center susceptible resistance surveillance study conducted in the United States in 1994-1995. *Clin. Infect. Dis.* **29:**1186–1188.
131. **Sinclair, A., C. Arnold, and N. Woodford.** 2003. Rapid detection and estimation by pyrosequencing of 23S rRNA genes with a single nucleotide polymorphism conferring linezolid resistance in enterococci. *Antimicrob. Agents Chemother.* **47:** 3620–3622.
132. **Speciale, A., K. La Ferla, F. Caccamo, and G. Nicoletti.** 1999. Antimicrobial activity of quinupristin/dalfopristin, a new injectable streptogramin with wide gram-positive spectrum. *Int. J. Antimicrob. Agents* **13:**21–28.
133. **Stanton, T. B., and S. B. Humphrey.** 2003. Isolation of tetracycline-resistant *Megasphaera elsdenii* strains with novel mosaic gene combinations of *tet*(O) and *tet*(W) from swine. *Appl. Environ. Microbiol.* **69:**3874–3882.
134. **Sugantino, M., and S. L. Roderick.** 2002. Crystal structure of Vat(D): an acetyltransferase that inactivates streptogramin group A antibiotics. *Biochemistry* **41:**2209–2216.
135. **Sutcliffe, J. A., and R. Leclercq.** 2003. Mechanisms of resistance to macrolides, lincosamides and ketolides, p. 281–317. *In* W. Schonfeld and H. A. Kirst (ed.), *Macrolide Antibiotics.* Birkhauser Verlag, Basel, Switzerland.

135a. **Swaney, S. M., D. L. Shinabarger, R. D. Schaadt, J. H. Bock, J. L. Slightom, and G. E. Zurenko.** 1998. Oxazolidinone resistance is associated with a mutation in the peptidyl transferase region of 23S rRNA, abstr. C-104. *In Proceedings of the 38th Interscience Conference on Antimicrobial Agents and Chemotherapy* American Society for Microbiology, Washington, D.C.

136. **Tait-Kamradt, A., J. Clancy, M. Cronan, F. Dib-Hajj, L. Wondrack, W. Yan, and J. Sutcliffe.** 1997. *mefE* is necessary for the erythromycin-resistant M phenotype in *Streptococcus pneumoniae. Antimicrob. Agents Chemother.* **41:**2251–2255.
137. **Tait-Kamradt, A., T. Davies, P. C. Appelbaum, F. Depardieu, P. Courvalin, J. Pettipas, L. Wondrack, A. Walker, M. R. Jacobs, and J. Sutcliffe.** 2000. Two new mechanisms of macrolide resistance in clinical strains of *Streptococcus pneumoniae* from Eastern Europe and North America. *Antimicrob. Agents Chemother.* **44:**3395–3401.
138. **Tait-Kamradt, A., T. Davies, M. Cronan, M. R. Jacobs, P. C. Appelbaum, and J. Sutcliffe.** 2000. Mutations in 23S rRNA and L4 ribosomal protein account for resistance in pneumococcal strains selected in vitro by macrolide passage. *Antimicrob. Agents Chemother.* **44:**2118–2125.
139. **Tait-Kamradt, A., R. R. Reinert, A. Al-Lahham, D. E. Low, and J. A. Sutcliffe.** 2001. High-level ketolide-resistant streptococci, p. 101. *In Program and Abstracts of the 41st Interscience Conference on Antimicrobial Agents and Chemotherapy.* American Society for Microbiology, Washington, D.C.
140. **Taniguchi, L., A. Nakamura, K. Tsurubuchi, A. Ishii, K. O'Hara, and T. Sawai.** 1999. Identification of functional

amino acids in the macrolide 2″ phosphtransferase II. *Antimicrob. Agents Chemother.* **43**:2063–2065.

141. **Tateda, K., Y. Ishii, T. Matsumoto, N. Furuya, M. Nagashima, T. Matsunaga, A. Onho, S. Miyazaki, and K. Yamaguchi.** 1996. Direct evidence for antipseudomonal activity of macrolides: exposure-dependent bactericidal activity and inhibition of protein synthesis by erythromycin, clarithromycin, and azithromycin. *Antimicrob. Agents Chemother.* **40**:2271–2275.
142. **Taylor, D. E., Z. Ge, D. Purych, T. Lo, and K. Hiratsuka.** 1997. Cloning and sequence analysis of two copies of a 23S rRNA gene from *Helicobacter pylori* and association of clarithromycin resistance with 23S rRNA mutations. *Antimicrob. Agents Chemother.* **43**:2621–2628.
143. **Teng, L.-J., P.-R. Hsueh, J.-C. Tsai, S.-J. Liaw, S.-W. Ho, and K.-T. Luh.** 2002. High incidence of cefoxitin and clindamycin resistance among anaerobes in Taiwan. *Antimicrob. Agents Chemother.* **46**:2908–2913.
144. **Tenson, T., M. Lovmar, and M. Ehrenbert.** 2003. The mechanism of action of macrolides, lincosamides and streptogramin B reveals the nascent peptide exit path in the ribosome. *J. Mol. Biol.* **330**:1005–1014.
145. **Thungapathra, M., K. K. Amita, K. K. Sinha, S. R. Chaudhuri, P. Garg, T. Ramamurthy, G. B. Nair, and A. Ghosh.** 2002. Occurrence of antibiotic resistance gene cassettes *aac(6′)-Ib, dfrA5, dfrA12,* and *ereA2* in Class I integrons in Non-O1, Non-O139 *Vibrio cholerae* strains in India. *Antimicrob. Agents Chemother.* **46**:2948–2955.
146. **Tomasz, A.** 1999. New faces of an old pathogen: emergence and spread of a multidrug-resistant *Streptococcus pneumoniae. Am. J. Med.* **107**:55S–66S.
147. **Tsiodras, S., H. S. Gold, G. Sakoulas, G. M. Eliopoulos, C. Wennersten, L. Vendataraman, R. C. Moellering, and M. J. Ferraro.** 2001. Linezolid resistance in a clinical isolate of *Staphylococcus aureus. Lancet* **358**:207–208.

147a. **Tu, D., G. Blaha, P. B. Moore, and T. A. Steitz.** 2005. Structures of MLSBK antibiotics bound to mutated large ribosomal subunits provide a structural explanation for resistance. *Cell* **121**:257–270.

148. **Valentine, P. J., N. B. Shoemaker, and A. A. Salyers.** 1988. Mobilization of *Bacteroides* plasmids by *Bacteroides* conjugal elements. *J. Bacteriol.* **170**:1319–1324.
149. **Vera-Cabrera, L., A. Gomez-Flores, W. G. Escalante-Fuentes, and O. Welsh.** 2001. In vitro activity of PNU-100766 (linezolid), a new oxazolidinone antimicrobial, against *Norcardia brasiliensis. Antimicrob. Agents Chemother.* **45**:3629–2620.
150. **Vester, B., and S. Douthwaite.** 2001. Macrolide resistance conferred by base substitutions in 23S rRNA. *Antimicrob. Agents Chemother.* **45**:1–12.
151. **Villedieu, A., H. A. Hussain, M. Wilson, and P. Mullany.** 2004. Identification of an oral *tet*(W) gene contained within a novel transposon-like structure, abstr. H-114. *In Program and Abstracts of the 104th General Meeting of the American Society for Microbiology.* American Society for Microbiology, Washington, D.C.
152. **Waites, K., C. Johnson, B. Gray, K. Edwards, M. Crain, and W. Benjamin, Jr.** 2000. Use of clindamycin disks to detect macrolide resistance mediated by *ermB* and *mefE* in *Streptococcus pneumoniae* isolates from adults and children. *J. Clin. Microbiol.* **38**:1731–1734.
153. **Wang, Y., G. R. Wang, A. Shelby, N. B. Shoemaker, and A. A. Salyers.** 2003. A newly discovered *Bacteroides* conjugative transposon, CTnGERM1, contains genes also found in gram-positive bacteria. *Appl. Environ. Microbiol.* **69**:4594–4603.

153a. **Wareham, D. W., M. Wilks, D. Ahmed, J. S. Brazier, and M. Miller.** 2005. Anaerobic sepsis due to multidrug-resistant *Bacteroides fragilis*: microbiological cure and clinical response with linezolid therapy. *Clin. Infect. Dis.* **40**:e67–e68.

154. **Wasteson, Y., D. E. Roe, K. Falk, and M. C. Roberts.** 1996. Characterization of antibiotic resistance in *Actinobacillus pleuropneumoniae. Vet. Microbiol.* **48**:41–50.
155. **Webber, M. A., and L. J. V. Piddock.** 2003. The importance of efflux pumps in bacterial antibiotic resistance. *J. Antimicrob. Chemother.* **51**:9–11
156. **Werner, G., C. Cuny, F.-J. Schmitz, and W. Witte.** 2001. Methicillin-resistant, quinupristin-dalfopristin-resistant *Staphylococcus aureus* with reduced sensitivity to glycopeptides. *J. Clin. Microb.* **39**:3586–3590.
157. **Werner, G., I. Klare, and W. Witte.** 2002. Molecular analysis of streptogramin resistance in enterococci. *Int. J. Med. Microbiol.* **292**:81–94.
158. **Wilcox, M. H.** 2003. Efficacy of linezolid versus comparator therapies in gram-positive infections. *J. Antimicrob. Chemother.* **51**(Suppl. S2):ii27–ii35.
159. **Wilson, P., J. A. Andrews, R. Charlesworth, R. Walesby, M. Singer, D. J. Farrell, and M. Robbins.** 2003. Linezoid resistance in clinical isolates of *Staphylococcus aureus. J. Antimicrob. Chemother.* **51**:186–188.

159a. **Wimberly, B. D.** 2003. Crystal structure of the ribosome and ribosomal subunits, p. 237–246. *In* L. Brakier-Gingras and J. Lapointe (ed.), *Translation Mechanisms.* Landes Bioscience/Eurekah, Georgetown, Tex., and Kluwer Academic/Plenum Publishers.

160. **Wondrack, L., M. Massa, B. V. Yang, and J. Sutcliffe.** 1996. Clinical strain of *Staphylococcus aureus* inactivates and causes efflux of macrolides. *Antimicrob. Agents Chemother.* **40**:992–998.
161. **Woodford, N., L. Tysall, C. Auckland, M. W. Stockdale, A. J. Lawson, R. A. Walkerm, and D. M. Livermore.** 2002. Detection of oxazolidinone-resistant *Enterococcus faecalis* and *Enterococcus faecium* strains by real-time PCR and PCR-restriction fragment length polymorphism analysis. *J. Clin. Microbiol.* **40**:298–4300.
162. **Yagi, B. H., and G. E. Zurenko.** 2003. An in vitro time-kill assessment of linezolid and anaerobic bacteria. *Anaerobe* **9**:1–3.
163. **Zarantonelli, L., G. Borthagaray, E. H. Lee, W. Veal, and W. M. Shafer.** 2001. Decrease susceptibility to azithromycin and erythromycin mediated by a novel *mtr*(R) promoter mutation in *Neisseria gonorrhoeae. J. Antimicrob. Chemother.* **47**:651–654.
164. **Zurenko, G. E., W. M. Todd, B. Hafkin, B. Meyers, C. Kauffman, and J. Bock, et al.** 1999. Development of linezolid-resistant *Enterococcus faecium* in two compassionate use program patients treated with linezolid, abstr. C-848. *In Proceedings of the 39th Interscience Conference on Antimicrobial Agents and Chemotherapy.* American Society for Microbiology, Washington, D.C.

Frontiers in Antimicrobial Resistance: a Tribute to Stuart B. Levy
Edited by D. G. White, M. N. Alekshun, and P. F. McDermott

Chapter 7

Aminoglycoside Resistance Mechanisms

David D. Boehr, Ian F. Moore, and Gerard D. Wright

Inspired by the success of Rene Dubos in identifying antimicrobial substances produced by soil bacteria, Selman Waksman's group at Rutgers embarked on a search for antibiotics that resulted in the identification of the first antibiotic from an actinomycete, actinomycin, in 1940 (119). This success was to be the harbinger of what was to become the "Golden Age" of antibiotic discovery over the next two decades where the actinomycetes would dominate. Waksman then discovered streptomycin (Fig. 1), the first aminoglycoside-amincyclitol antibiotic, from a culture of *Streptomyces griseus* (89). Streptomycin was the first antibiotic to prove effective in the treatment of tuberculosis, and its introduction to clinical practice had a monumental effect on disease treatment. A reflection of the impact of this discovery was that Waksman was awarded the Nobel Prize in Medicine in 1952.

Five years after the discovery of streptomycin, Waksman reported the second aminoglycoside antibiotic, neomycin (Fig. 1) in 1949 (120). Toxicity issues associated with neomycin and emerging streptomycin resistance spurred additional aminoglycoside discovery in the 1950s and early 1960s with the group of Umezawa, in particular, in Japan reporting kanamycin (114) and gentamicin by Weinstein at Schering (123) (Fig. 1), both of which found clinical use. The neomycin and kanamycin/gentamicin classes of aminoglycoside represent the two most common structural classes of aminoglycoside antibiotic. They both share a central 2-deoxyaminocyclitol ring structure that is substituted with sugars at either position 4,5 (neomycin) or 4,6 (kanamycin/gentamicin).

Resistance to kanamycin was reported in the late 1960s and found to be primarily associated with enzymatic modification of the drugs (112, 113). The emergence of resistance essentially ended the clinical utility of kanamycin and launched a two decade search for new agents, natural and semisynthetic, that were not susceptible to resistance mechanisms, e.g., tobramycin, amikacin, and netilmicin (Fig. 1). However, the breadth of resistance has proven difficult to overcome with new agents, and the availability of newer antibacterial agents with a similar microbial spectrum, such as the fluoroquinolones in the 1970s, has served to marginalize the use of aminoglycosides in the clinic.

Despite the excellent spectrum of antibacterial activity and the fact that most aminoglycosides are bactericidal, there has not been a new aminoglycoside brought to market in North America for over 20 years. The history of the aminoglycosides therefore represents the dramatic impact that resistance can have on limiting therapeutic options and the power of natural selection in the selection for resistance that has been so well articulated by Stuart Levy (57). That said, a thorough understanding of the mechanism of action of antibiotics and their resistance mechanisms can be leveraged in the design of new agents and mechanisms to overcome resistance. This review focuses on these aspects of aminoglycoside biochemistry.

MODE OF ACTION AND ENTRY INTO CELLS

The major target of aminoglycosides is the bacterial ribosome, as first suggested by in vivo experiments demonstrating a marked decrease in protein synthesis following treatment of cells with aminoglycosides and in vitro experiments on bacterial extracts showing that aminoglycoside treatment resulted in repression of both initiation and elongation in protein synthesis (66, 99, 100, 121). Chemical footprinting studies and careful correlation analysis of ribosomal mutation with aminoglycoside resistance implicated specific ribosomal proteins and the tRNA binding site (A site) of the 16S rRNA as the most important determinants of aminoglycoside binding and action

David D. Boehr, Ian F. Moore, and Gerard D. Wright • Antimicrobial Research Centre, Department of Biochemistry, McMaster University, 1200 Main Street West, Hamilton, Ontario, Canada, L8N 3Z5.

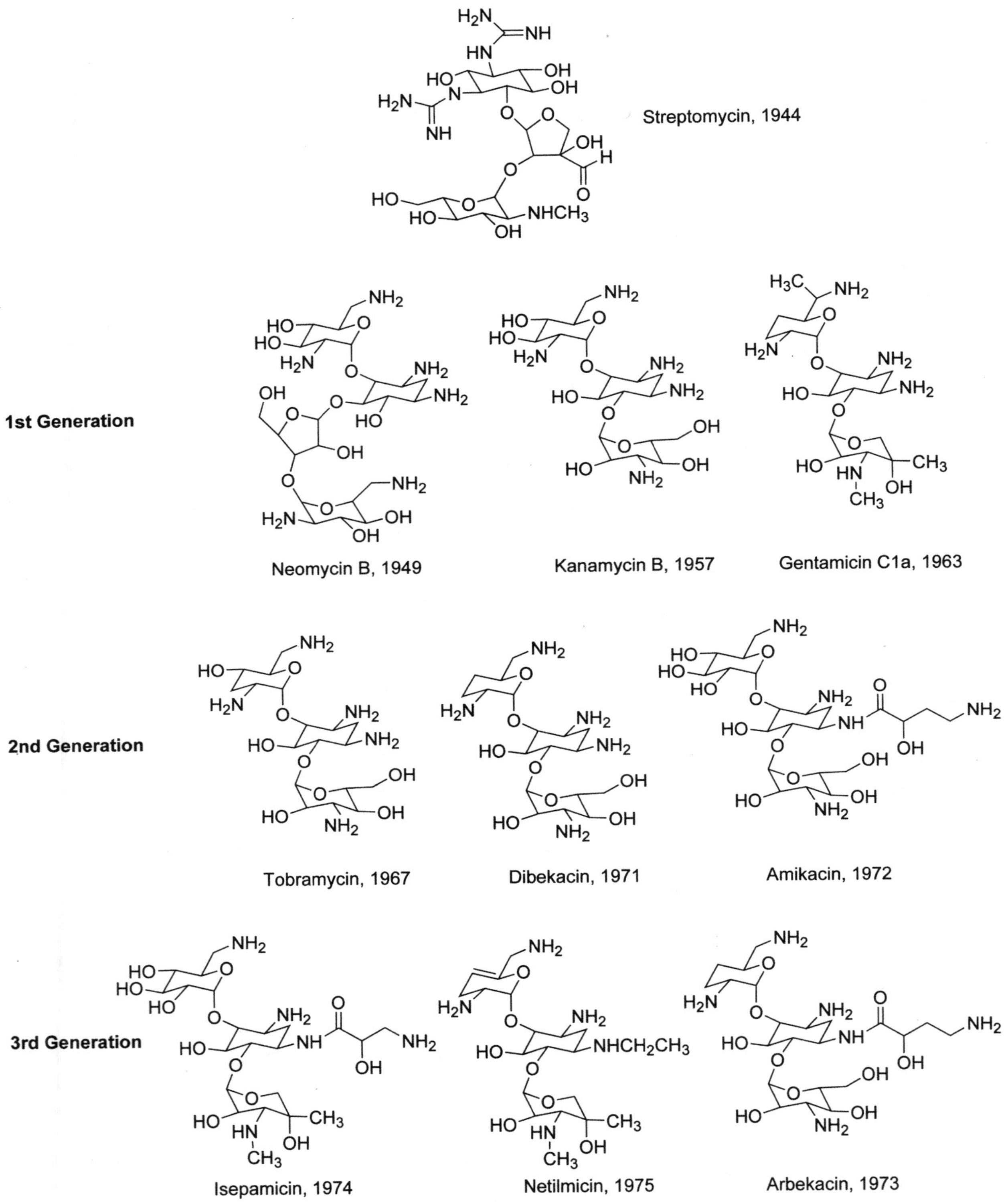

Streptomycin, 1944

1st Generation

Neomycin B, 1949

Kanamycin B, 1957

Gentamicin C1a, 1963

2nd Generation

Tobramycin, 1967

Dibekacin, 1971

Amikacin, 1972

3rd Generation

Isepamicin, 1974

Netilmicin, 1975

Arbekacin, 1973

Figure 1. Sixty years of aminoglycoside discovery.

(6, 7, 56, 70, 71, 75, 90, 92, 127). More recently, nuclear magnetic resonance (NMR) solution structures of aminoglycosides bound to a 27-nucleotide portion of the 16S rRNA (34–36, 131), and X-ray crystal structures of aminoglycosides bound to the 30S ribosomal subunit (15) have further pinpointed their site of action.

Aminoglycosides, belonging to the neomycin and kanamycin classes, bind to the major groove of the 16S rRNA where they make numerous contacts with the RNA, either directly or indirectly through intermediary water molecules. Binding of these antibiotics displaces the "proofreading" bases A1492 and A1493, signaling to the protein translation machinery that it has found a correct codon-anticodon match, regardless of the proper base pairing. This interaction leads to a decrease in protein translation fidelity. It has been suggested that mistranslated and misfolded membrane-directed proteins can compromise cellular integrity, eventually leading to cell lysis (26). This would help to explain why most aminoglycosides are bactericidal and not bacteriostatic as are other antibiotics that target the prokaryotic ribosome and impact translation but act with dissimilar molecular mechanism.

Thus, the major target of aminoglycoside action resides within the cell. However, considering that aminoglycosides are highly positively charged compounds, there are questions regarding how such a chemical species could cross the hydrophobic component of the cell membrane. There is some evidence that, at least in *Escherichia coli,* the oligopeptide binding protein, the periplasmic component of the major oligopeptide transport system, may play an important role in aminoglycoside uptake as mutants with reduced oligopeptide binding protein expression are resistant to aminoglycosides (1, 50, 51). However, although oligopeptide binding protein may be involved in some capacity, it is not known whether it has a direct or indirect effect on entering the chain elongation phase (99, 100, 121).

What is known about aminoglycoside entry is that it is a multiphasic event, where the highly positively charged antibiotic first crosses the outer membrane and periplasmic space in gram-negative organisms or the cell wall assembly of gram-positive organisms before it binds to the exterior of the cell membrane through electrostatic forces (26, 45). After the very rapid initial binding, there appear to be at least two additional energy-dependent phases (45). The first phase (EDPI) is a much slower phase where a small amount of aminoglycoside gains entry to the cells. This phase appears to be highly dependent on the bacterial electron transport system (12, 13, 45, 68). The proton motive force (PMF) generated by the electron transport system also appears to be critical, considering that uncouplers such as 2,4-dinitrophenol and carbonyl cyanide *m*-chlorophenol hydrazone that dissipate the PMF also inhibit aminoglycoside uptake (12, 13, 45, 68). The second phase, termed energy dependent phase II (EDPII), is unique to aminoglycoside sensitive bacteria. Resistant bacteria continue to show a low level of aminoglycoside uptake according to EDPI but do not show the much increased uptake characteristic of EDPII (45). It has been suggested that initial aminoglycoside entry leads to mistranslated membrane-directed proteins that form "transient pores" through which additional aminoglycoside can enter the cell. Eventually the protein translation machinery is overwhelmed and the membrane integrity is compromised sufficiently to cause cell death.

MECHANISMS OF RESISTANCE

Aminoglycoside Phosphotransferases

There are seven classes of aminoglycoside phosphotransferases, as dictated by differing regiospecificity of phosphate transfer to the aminoglycoside structure, including APH(3′), APH(2″), APH(3″), APH(6′), APH(9), APH(4) and APH(7″) (Fig. 2A). The most common aminoglycoside kinases are APH(3′)-IIIa and APH(2″)-Ia [C-terminal domain of the bifunctional aminoglycoside phosphotransferase-acetyltransferase AAC(6′)-APH(2″)] in gram-positive organisms, and APH(3′)-Ia and APH(3′)-IIa in gram-negative organisms.

APH(3′)-IIIa

The most thoroughly studied aminoglycoside kinase is APH(3′)-IIIa. It is especially important in gram-positive pathogens such as *Enterococcus, Staphylococcus,* and *Streptococcus,* but it has also been detected in the gram-negative organism *Campylobacter coli* (76, 101). The gene has been cloned from *Streptococcus faecalis* (108) and *Staphylococcus aureus* (43) and overexpressed in *E. coli* (63). The protein purifies as a mixture of dimer and monomer, although the two forms of the enzyme are indistinguishable with respect to in vitro enzyme kinetic activity (63).

The enzyme has a broad substrate profile, being able to efficiently modify both 4,5- and 4,6-disubstituted classes of aminoglycosides (63). NMR has definitively assigned the site of phosphoryl modification for the kanamycin class as the 3′ hydroxyl (63) and has confirmed the ability of APH(3′)-IIIa to dually modify many 4,5-disubstituted aminoglycosides, as initially suggested by the biphasic progress curves seen for members of the neomycin class (106).

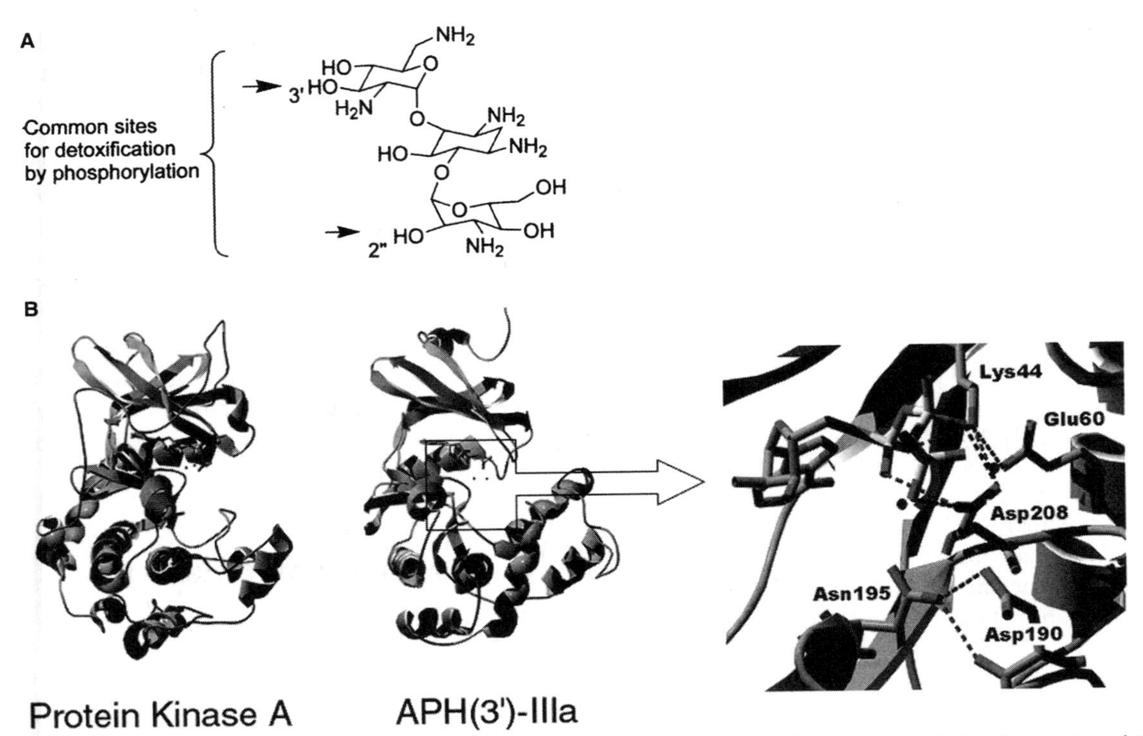

Figure 2. Structure and function of aminoglycoside phosphotransferases. (A) Sites of APH-catalyzed phosphorylation of 2-deoxystreptoamine aminoglycosides. (B) Structure of APH(3′)-IIIa, demonstrating the structural similarity eukaryotic Ser/Thr protein kinase A. Inset, blow up of active site region showing five conserved amino acids found in both APHs and protein kinases.

NMR analysis of the ternary complex of APH(3′)-IIIa bound with butirosin and Cr.ATP.APH(3′)-IIIa has demonstrated that both the 3′- and 5″-hydroxyls are in close proximity to the γ-phosphoryl group of ATP (21). Double modification occurs on the 3′ and 5″ hydroxyls, where initial phosphotransfer is dependent on the individual aminoglycoside (106).

A series of experiments has demonstrated that APH(3′)-IIIa follows a Theorell-Chance kinetic mechanism (64, 65). Theorell-Chance is a special case of an ordered BiBi kinetic mechanism where the reaction occurring in the ternary complex (i.e., chemistry) does not contribute significantly to the overall rate of catalysis. For APH(3′)-IIIa, ATP binds first, followed by the binding of aminoglycoside, phosphoryl transfer, and immediate release of phosphorylated antibiotic. The release of ADP is the primary rate-determining step in catalysis (64, 65).

The chemical mechanism of APH(3′)-IIIa has also been studied. The highly conserved residue His188 was initially thought to serve as a temporary docking site for a phosphate in a double displacement mechanism, but site-directed mutagenesis of this residue to Ala did not greatly affect the enzyme's activity, and further work using positional isotope labeled ATP established that there is a direct attack of the aminoglycoside hydroxyl on to the γ-phosphate of ATP (105).

Additional mutagenesis work has been aided by the determination of the X-ray crystal structures of APH(3′)-IIIa (14, 48). The structure shows remarkable similarity to the eukaryotic protein kinase (ePK) family, with a smaller N-terminal lobe composed mostly of β-sheets and a larger C-terminal lobe composed mostly of α-helices (48) (Fig. 2). The ATP-binding pocket is formed by a deep cleft between the two lobes, and a putative aminoglycoside binding site was also identified near the modeled terminal phosphate of ATP (48). The importance of this anionic depression in the binding of positively charged antibiotic has been confirmed with substrate modeling studies and directed mutagenesis of conserved residues in the region (107). These studies have also illuminated the capacity of the enzyme to accept a wide range of antibiotic conformations, suggested by its substrate profile and NMR studies modeling the bound conformations of the aminoglycoside butirosin A (21). More recently, structures of APH(3′)-IIIa bound with kanamycin A and neomycin have been solved, further delineating the antibiotic binding pocket (33). The flexible aminoglycoside binding region is divided into three subsites capable of accommodating structurally different components of the aminoglycoside substrates (33). The broad regiospecificity is also enhanced by the high number of potential electrostatic interactions in the region between Glu/Asp residues of the protein and the positively charged amino groups of the aminoglycoside. These studies have also illuminated the similarity of aminoglycoside binding to rRNA and resistance enzyme (20). Although the binding scaffold obviously differs, the conformations adopted by the aminoglycosides and the H-bonding scheme are similar, suggesting that APH(3′)-IIIa has successfully mimicked the aminoglycoside binding region of the 16S rRNA so as to compete efficiently for the antibiotic (33).

APH(3′)-IIIa is not only structurally similar to ePKs, but it is also functionally similar. Certain ePK inhibitors can act on APH(3′)-IIIa (25), and the enzyme demonstrates serine protein kinase activity (24). The molecular mechanisms of enzyme-catalyzed aminoglycoside and protein/peptide phosphorylation appear to be similar (11, 103). Although APH(3′)-IIIa only shares five absolutely conserved residues with ePKs [Lys44, Glu60, Asp190, Asn195, and Asp208 using APH(3′)-IIIa numbering, see Fig. 2B], kinetic analyses of a number of active-site directed mutants have helped to ascribe similar roles to these residues in both protein classes (11). In particular, it has been suggested that Asp190 in APH(3′)-IIIa acts as an active site base to help deprotonate the incoming aminoglycoside hydroxyl and facilitate phosphoryl transfer in an associative-type reaction. However, site directed mutagenesis of this residue was inconsistent with such a role, and deprotonation likely occurs simultaneously or after the phosphoryl transfer event in a dissociative-type reaction, analogous to what has been observed in the protein kinase field (11). The most important residue is Asp208, considering that mutation to Glu, Asn, or Ala completely abolishes kinase activity (11). This residue chelates the catalytic Mg^{2+} and is suggested to act as a sort of fulcrum over which the γ-P-O bond is elongated (103).

Other aminoglycoside kinases

APH(3′)-I is the most common class of aminoglycoside kinase in gram-negative bacteria. APH(3′)-Ia (94) and APH(3′)-IIa (93) have been cloned, overexpressed in *E. coli,* purified to homogeneity, and characterized. Similar to APH(3′)-IIIa, APH(3′)-Ia can also dimerize (94). Moreover, kinetic analysis has demonstrated that catalysis occurs through a ternary complex via a rapid equilibrium random mechanism, suggesting that phosphoryl transfer is direct (94). Bromoacetylated analogs of neamine were used to map the active site of APH(3′)-IIa (83, 129), and more recently, the 3D structure of APH(3′)-IIa in complex with kanamycin has been solved (74). In contrast to APH(3′)-IIIa, pre-steady-state analysis of APH(3′)-Ia and APH(3′)-IIa with 4′-deoxy-4′,4′-difluoroaminoglycosides has suggested that there is significant nucleophilic participation in the generation of the transition state (53). However, it should be stated that "associative" and "dissociative" represent extremes along a continuum of mechanistic possibilities, and the details may differ between individual kinases.

In contrast to the APH(3′) class of enzymes, APH(9) from *Legionella pneumophila* has very strict substrate specificity, being able to bind and modify only spectinomycin (104). The gene has been cloned, and the protein has been overexpressed, purified, and characterized (104). Site-directed mutagenesis has confirmed the conserved nature of important active site residues [Lys44 and Asp190 using APH(3′)-IIIa numbering], and the molecular mechanism of phosphoryl transfer appears to be similar to APH(3′)-IIIa (104).

Aminoglycoside Nucleotidyltransferases

There are four classes of aminoglycoside nucleotidyltransferases, including ANT(6), ANT(4′), ANT(3″), and ANT(2″). The best studied are ANT(4′) and ANT(2″), prominent in the aminoglycoside resistance profiles of gram-negative organisms (Fig. 3).

ANT(4′)

The ANT(4′) enzyme has been purified from *Staphylococcus epidermidis* (88), *S. aureus* (110), and overexpressing constructs of *E. coli* (86). The enzyme has a wide aminoglycoside substrate profile, being able to detoxify members of both the 4,5- and 4,6-disubstituted classes of aminoglycosides (16). Moreover, it is capable of using a number of nucleotide triphosphates as adenyl donors, including ATP, GTP, CTP, and UTP (16). These results suggest a measure of plasticity in both substrate and cofactor binding sites that has been revealed by crystal structures of ANT(4′) (Fig. 3) (77, 86). The regiospecificity of nucleotidyl transfer to the antibiotic has been confirmed by NMR approaches (41).

Kinetic studies have demonstrated that ANT(4′) operates through an ordered BiBi kinetic mechanism with aminoglycoside binding first, followed by nucleotide binding and adenyl transfer (16). Pyrophosphate leaves the enzyme first, before the nucleotidylated aminoglycoside exits the active site pocket (16). These results suggest that the reaction is catalyzed by a direct displacement without the involvement of a covalent enzyme intermediate. From these results, together with the crystal structure and pH dependence of the kinetic parameters, it has been suggested that Glu145 acts as a general base for the activation of the 4′-hydroxyl of the aminoglycoside to assist in an

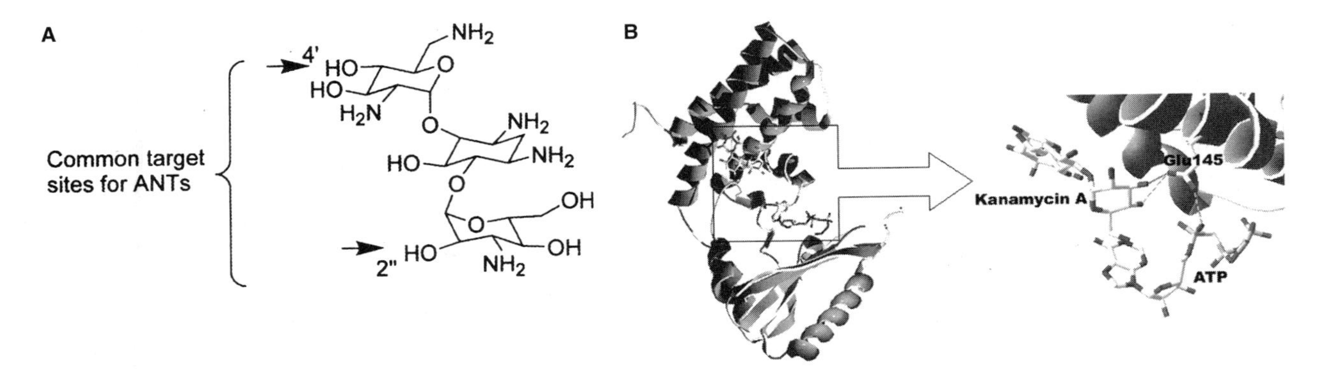

Figure 3. Structure and function of aminoglycoside nucleotidyltransferases. (A) Sites of ANT-catalyzed modification of 2-deoxystreptoamine aminoglycosides. (B) Structure of one monomer of ANT(4′) with inset showing active site residues and orientation of kanamycin and ATP, which lie in the active site formed at the dimer interface.

associative-type mechanism (16). Transition-state structure analysis based on heavy-atom isotope kinetic isotope effects has confirmed that the reaction is slightly associative, consistent with other solution and enzymatic phosphodiester reactions (42).

ANT(2″)

ANT(2″)-Ia is one of the most important determinants of aminoglycoside resistance in gram-negative organisms. It can modify a wide range of 4,6-disubstituted aminoglycosides (40). Steady-state kinetic analysis has indicated that the enzyme follows a Theorell-Chance mechanism, similar to APH(3′)-IIIa, with nucleotide binding first, followed by aminoglycoside binding, group transfer, release of pyrophosphate, and lastly, the release of nucleotidylated antibiotic (39, 40). The final step, release of modified aminoglycoside, is the rate-determining step in catalysis (39, 40).

Substrate binding to ANT(2″)-Ia has been studied by NMR methods (31). Similar to what has been found with APH(3′)-IIIa, the aminoglycosides can adopt multiple conformations depending on the antibiotic, and these conformations are similar when the aminoglycosides are bound to resistance protein and to rRNA (31).

Aminoglycoside Acetyltransferases

Aminoglycoside acetyltransferases (AACs) are the largest group of aminoglycoside inactivating enzymes with close to 50 unique enzymes identified from both gram-positive and gram-negative bacteria. AACs work with acetyl coenzyme A as a cosubstrate, transferring the acetyl group to amino groups of the aminoglycoside (*N*-acetyltransferases). *O*-Acetyltransferase activity has been shown in the AAC portion of the bifunctional enzyme AAC(6′)-APH(2″) (23) and the mycobacterial AAC(2′)-Ic (46). AACs are subdivided into four classes based on the regiospecificity of acetyl transfer. These include AAC(1), AAC(3), AAC(2′), and AAC(6′) (Fig. 4).

AAC(1) and AAC(3)

AAC(1) and AAC(3) acetylate either of the two amino groups found on the central deoxystreptamine ring of aminoglycosides (Fig. 4). There are only two members of the AAC(1) subclass, one from *E. coli* that can acetylate apramycin and 4,5-disubstituted aminoglycosides and the second from an *Actinomycetes* strain whose resistance profile is unknown (58, 97). Neither of the two *aac(1)* genes provides clinically relevant resistance. On the other hand, members of the AAC(3) subclass do provide clinically relevant resistance. There are 16 members of this subclass; most have been isolated from clinically resistant strains of gram-negative bacteria such as *Pseudomonas aeruginosa* (78, 80, 91). These genes are typically found on mobile genetic elements like integrons, plasmids, and transposons.

AAC(2′) and AAC(6′)

Members of the AAC(2′) and AAC(6′) subclass modify amino groups on the prime ring of AGs (Fig. 4). AAC(6′) is the largest subclass of AACs; there are 27 unique members of this class identified from a wide variety of bacteria including clinically resistant strains (19, 59). The majority of *aac(6′)* genes are found on transferable plasmids and transposons; there

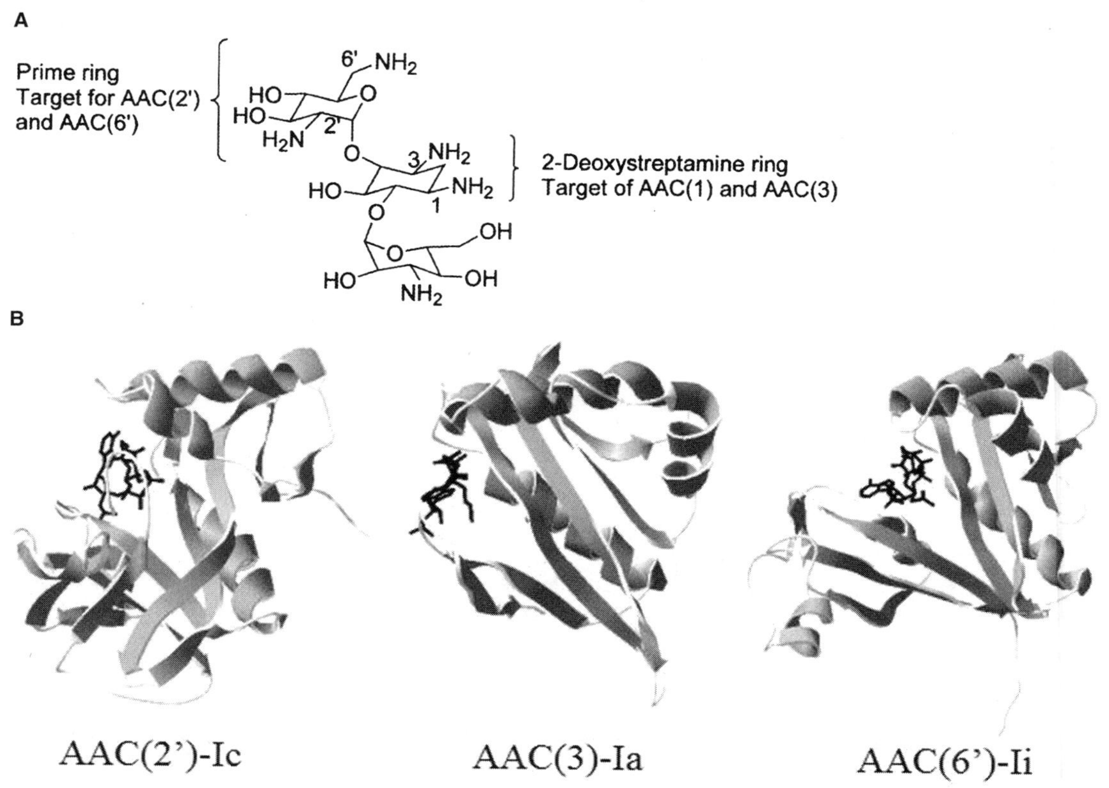

Figure 4. Structure and function of aminoglycoside acetyltransferases. (A) Sites of AAC-catalyzed acetylation of 2-deoxystreptoamine aminoglycosides. (B) Comparison of the 3D-structures of the AAC subclasses.

are, however, four members that are encoded by chromosomal genes. All AAC(6′) members have a broad specificity for aminoglycosides including the clinically important aminoglycosides gentamicin, tobramycin, and kanamycin. Included in this class is the bifunctional enzyme AAC(6′)-APH(2″), which can both acetylate and phosphorylate the same aminoglycoside (see below).

The five members of the AAC(2′) subclass are chromosomally encoded genes found in four species of *Mycobacteria* and one species of *Providencia*. These enzymes do not provide any clinically important resistance. AAC(2′)-Ic from *Mycobacterium tuberculosis* shows aminoglycoside acetylation activity in vitro (46); however, *M. tuberculosis* resistance to aminoglycosides is provided by point mutations in either the 16S rRNA gene or the S12 ribosomal protein gene (5). Blanchard and coworkers have asked what the function of this gene is if not to provide aminoglycoside resistance (116). Based on their crystal structure of AAC(2′)-Ic complexed with an aminoglycoside substrate (117), they postulate that the function of the gene in vivo is to acetylate one of the pseudodisaccharide precursors to mycothiol, an abundant thiol found in mycobacterial species.

Structures of AACs

There have been three crystal structures of AACs reported to date, one each from the AAC(3), AAC(2′), and AAC(6′) subclasses (Fig. 4) (117, 126, 128). All AAC enzymes share a common fold that classifies them as members of the GCN5-related N-acyltransferase (GNAT) superfamily. Other members of this superfamily include enzymes with such diverse substrates and acyl donors as histone acetyltransferases, N-myristoyltransferase, and serotonin N-acetyltransferase (30). There are no absolutely conserved residues or mechanisms of acetyl transfer among GNAT family members, only a common fold for binding coenzyme A (CoA). The structural similarity of AACs to enzymes with such a wide range of substrates has suggested that some AACs might have alternate physiological activity. As mentioned above, AAC(2′)-Ic has been postulated to have carbohydrate acetylating activity. The chromosomally encoded AAC(6′)-Ii from a clinically isolated strain of *Enterococcus faecium* has been shown to acetylate small basic proteins, including calf histones H3 and H4 (128). However, the physiologically relevant substrate from *E. faecium* has not been identified.

AAC(6′)-APH(2″)

AAC(6′)-APH(2″) was initially discovered in gram-positive cocci such as *Enterococcus* and *Staphylococcus* (47, 52, 102) but has also been detected in gram-negative bacteria (67). More recently, it has been discovered as part of the resistance cassette conferring vancomycin resistance to *S. aureus* (122). The protein has both ATP-dependent phosphorylation and acetyl-CoA-dependent acetylation activities and has the capacity to inactivate nearly all aminoglycoside antibiotics belonging to the 4,5- and 4,6-disubstituted classes (22).

The gene has been cloned from *S. aureus* (85) and *E. faecalis* (32), and the protein has been purified from *S. aureus* (85), *S. epidermidis* (110), and overexpressing constructs in *E. coli* (4) and *Bacillus subtilis* (23). Sequence homology studies and the construction of truncated proteins have demonstrated that the acetyltransferase [AAC(6′)-Ie] and phosphotransferase [APH(2″)-Ia] activities are associated with the N-terminal and C-terminal domains, respectively (32). These findings suggest that the resistance gene may have arisen from a gene fusion event between an AAC and an APH (32). In support of this hypothesis, homologues of both domains have been isolated, including APH(2″)-Ic from *Enterococcus gallinarum* (18) and APH(2″)-Id from *Enterococcus cassiflavus* (109) that show similarity to the APH domain, and AAC(6′)-Im from *E. coli* and *E. faecalis* that shows similarity to the AAC domain (17). Moreover, AAC(6′)-Im can be detected together with APH(2″)-Ib, where the genes are only 40 nucleotides apart, and consequently, the genes can be transferred simultaneously (17).

The two domains appear to be functionally distinct from one another, with each activity following a random rapid equilibrium steady-state kinetic mechanism independent from the other activity's required cofactor (60). The predicted regiospecificity of modification for both phosphoryl (2″-OH) and acetyl (6-NH_2) modification of kanamycin A (4) and arbekacin (55) has been confirmed, and furthermore, aminoglycoside bifunctionalization (O-phospho and N-acetyl derivatization of the same molecule) has been detected (4). The enzyme also has an even more startling array of activities, including alternate sites of phosphorylation for neomycin B (3′ and 3″) and lividomycin A (5′), as well as the ability to catalyze O-acetyl transfer (23). The unique ability of the acetyltransferase domain has been attributed to the ability of Asp99 to act as active site base (9) that may be absent in other aminoglycoside acetyltransferases (29). Thus, the APH(2″)-Ia activity, especially, appears to be impervious to the structure of the aminoglycoside, and together with the unique AAC(6′) domain, this enzyme may be the most threatening aminoglycoside resistance enzyme known to date.

Ribosomal Methylation

Ribosome methylation, a form of resistance previously known to occur only in gram-positive aminoglycoside-producing bacteria, has recently been found in clinically isolated strains of *P. aeruginosa* (130), *Klebsiella pneumoniae* (38), and *Serratia marcescens* (28). The methylation of specific bases in the ribosome by 16S rRNA methyltransferases blocks critical hydrogen bond interactions that provide affinity between the aminoglycoside and rRNA. The first gene reported, *rmtA*, from *P. aeruginosa* conferred high-level resistance to 4,6-disubstituted aminoglycosides in the host strain, a *P. aeruginosa* transconjugant and *E. coli* transformed with a plasmid harbouring *rmtA*. *rmtA* encodes a putative 251 amino acid protein found to be similar (30 to 35%) to 16S rRNA methylases from aminoglycoside-producing actinomycetes (130). Homologies include GrmB and Sgm from two sisomicin producers *Micromonospora rosea* and *M. zionensis* and Kmr from the kanamycin producing organism *Streptomyces kanamyceticus*. Indeed, a cell-free cytosolic preparation of *P. aeruginosa* containing RmtA was shown to accelerate incorporation of ^{3}H-methyl groups into *P. aeruginosa* 30S ribsomes.

Plasmid encoded *armA* from *K. pneumoniae* also confers high-level resistance to 4,6-disubstituted aminoglycosides (38). The deduced protein ArmA is 257 amino acids and also has close similarity to 16S rRNA methyltransferases from aminoglycoside producing species, including the enzymes mentioned above. 16S rRNA methyltransferases from aminoglycoside producing organisms are grouped into two classes, those that provide protection to 4,6-disubsituted aminoglycosides and those that provide protection to 4,5-disubstituted aminoglycosides. Based on the resistance profile of *armA*, it belongs to the 4,6-disubstituted class of rRNA methyltransferases that are known to methylate N-7 of guanine 1405 within the 16S rRNA.

rmtB identified in *S. marcescens* provides high-level resistance to the 4,6-disubstituted aminoglycosides and *E. coli* that are transformed with a plasmid bearing *rmtB* (28). RmtB is 251 amino acids and has 82% identity with RmtA and 29% identity with ArmA. It shares the same level of identity with GrmB, Sgm, and Kmr. The authors made a His-tagged version of RmtB and showed that the purified enzyme incorporated ^{3}H-methyl groups into 16S rRNA from *E.*

coli. Based on the resistance profile of RmtB, the authors predict that it belongs to the family of aminoglycoside resistance methyltransferases that methylate G1405 within the 16S rRNA.

Bacteria accepting genes from hereditarily distant organisms are not new, but this mechanism of resistance to aminoglycosides in human pathogens is very efficient and is of significant concern.

Aminoglycoside Efflux

Antibiotic efflux contributes significantly to multidrug resistance in a number of gram-negative bacteria (79). The efflux system relevant to antibiotic resistant bacteria is the resistance-nodulation-division (RND). In gram-negative bacteria, the inner membrane RND component works in conjunction with a periplasmic membrane fusion protein (MFP) and a channel forming outer membrane factor (OMF) to pump antibiotics across both membranes and out of the cell. The RND family uses PMF for export, and the range and number of molecules that are substrates for RND/MFP/OMF efflux pumps vary from organism to organism. The bulk of substrates are lipophilic or amphiphilic. There are, however, three examples of efflux genes that can export the hydrophilic cationic aminoglycosides. They are found in *Burkholderia pseudomallei* (72), *E. coli* (84), and *P. aeruginosa* (2, 69, 124). The *B. pseudomallei* and *P. aeruginosa* efflux pumps are tripartite with RND, MFP, and OMF components encoded by the *amrB*, *amrA*, and *oprA* genes, respectively. The *P. aeruginosa* efflux pump is encoded by the *mexY*, *mexX*, and *oprM* genes. MexY shares considerable similarity with AcrB, the RND component of the *E. coli* multidrug efflux transporter AcrAB-TolC, whose crystal structure has been solved (73). The *E. coli* efflux pump consists of only the RND component encoded by the *acrD* gene. Lipophilic molecules require export across the outer membrane since reentry across the cytoplasmic membrane is rapid. Hydrophilic molecules have a slow reentry rate across the cytoplasmic membrane and do not necessarily need to be exported across the outer membrane. Correspondingly, the *E. coli* pump, AcrD, which consists of only the RND component, was found to accommodate hydrophilic molecules not lipophilic molecules. The MexXY-OprM and AcrAB-OprA pumps accommodate lipophilic antibiotics in addition to aminoglycosides, so consequently, it is not known whether all three components are needed for export of aminoglycosides in these two systems. It is also not known whether the aminoglycosides have the same binding site as lipophilic antibiotics on MexXY-OprM and AcrAB-OprA.

A recent study looked at the contribution of the MexXY transporter to aminoglycoside resistance in clinical strains of *P. aeruginosa* (95). Expression of *mexXY* in 14 strains was determined by reverse transcription-PCR. It was found that most strains had elevated levels of *mexXY* expression; however, the levels of expression did not correlate with MICs for the resistant strains, suggesting that *mexXY* was not the sole factor contributing to resistance in these strains. Similarly, deletion of *mexXY* from the strains also had a varied effect on susceptibility. Although most efflux pumps expel antimicrobials, they also pump a range of other molecules (bile salts, fatty acids) found in the organism's environment. The *mexXY* system is a rare example where expression of the efflux genes has been found to be induced by antibiotics that are substrates for the pump; however, the exact gene that controls regulation is not known (61).

INHIBITION OF AMINOGLYCOSIDE RESISTANCE

Based on the abundance of mechanistic and structural information for aminoglycoside modifying enzymes, considerable work has been devoted to development of inhibitors of this class of enzymes. These inhibitors have the potential to reverse resistance to aminoglycoside antibiotics. Such a strategy has been employed in the clinically useful combinations of β-lactam–β-lactamase inhibitors (98). Work in the area of aminoglycoside resistance enzyme inhibitors can be divided into three areas: aminoglycoside-based inhibitors, mechanism-based inhibitors, and nonaminoglycoside-based inhibitors.

Use of aminoglycosides, which themselves are antibiotics but are not substrates of a particular resistance enzyme, has been successful. This strategy involves using aminoglycosides where the functional group that is modified by a particular enzyme is missing. Two examples are tobramycin and dibekacin, which lack a 3′ hydroxyl group, the site of phosphorylation of kanamycin by APH(3′) (Fig. 1). Tobramycin and dibekacin are competitive inhibitors of APH(3′) and are potent antibiotics (115). Unfortunately, tobramycin and dibekacin are substrates for other resistance enzymes, including APH(2″) and AAC(6′). A second strategy involves modifying aminoglycosides by removing interactions that are critical for affinity between the aminoglycoside and a particular enzyme. The enzyme ANT(2″) requires an equatorial hydroxyl group at position 5 of the 2-deoxystreptamine ring, and the 5-hydroxy group is not critical for antibiotic activity. Thus the semisynthetic molecules 5-*epi*-sisomicin and

5-*epi*-gentamicin, which have the 5 hydroxyl group in the axial position, are effective against a broad spectrum of bacteria that express ANT(2″), AAC(3), and AAC(2′) (Fig. 5) (37, 49, 118). The semisynthetics were substrates for the enzymes but were much poorer, with K_m values that were much higher than the parent compounds. Similarly, Roestamadji et al. synthesized a series of deaminated analogues of neamine and kanamycin A for use against APH(3′)-Ia and APH(3′)-IIa (81). The deaminated compounds were predicted and proved to be poorer substrates for the enzymes since the amino groups provide important binding energy to the enzyme active site, and these compounds showed activity against strains carrying APH(3′)-Ia and APH(3′)-IIa. These compounds were, however, phosphorylated by APH(3′)-IIIa (62). Sucheck et al. also synthesized a series of neamine derivatives modified at the 5 position, including neamine dimers linked through the 5 position. These dimers were found to be competitive inhibitors of the APH(2″) activity of the AAC(6′)-APH(2″) bifunctional enzyme as well as poor substrates for AAC(6′)-Ii and APH(3′)-IIIa (96).

Blocking the site of inactivation can also produce an effective inhibitor. Such is the example of the semisynthetic kanamycin derivatives 6′-N-methylkanamycin A and 3′,4′-dideoxy-6′-N-methylkanamycin B (111). In these two molecules, the site of modification is chemically blocked with a methyl group, thus preventing modification by 6′-N-acetylating enzymes. These two compounds do possess antibacterial activity against strains possessing 6′-N-acetylating enzymes. This type of approach is not broadly applicable, however, since bacteria often carry multiple modifying enzymes and many of the structural requirements of aminoglycosides for antibiotic activity are targeted by the resistance enzymes. Recently Disney et al. have reported the use of guanidinoglycosides as inhibitors of AAC(6′) activity (27). 1,3,3′-Triguanidinoglycosides of kanamycin A and B, neomycin, ribostamycin, paromomycin, and lividomycin were synthesized and linked to a solid support via the 6′-amino group. Detection of binding was accomplished with fluorescently labeled AAC(2′) and AAC(6′). 6′-β-ala-guanidinoribostamycin was the best inhibitor of both AAC(2′) and AAC(6′), with K_i values in the 10 to 200 μM range. This is the first example where microarraying has been used to detect inhibitors of aminoglycoside resistance enzymes.

Mechanism-based inhibitors of aminoglycoside resistance enzymes are an important class of inhibitors although their ability to reverse resistance has not been shown. A bisubstrate analogue inhibitor of AAC(3)-I was prepared enzymatically by reacting chloroacetyl coenzyme A with gentamicin (125). During the reaction, the chloroacetyl group was transferred to the 3 position of gentamicin. Subsequently, the sulfur atom of the CoA product displaced the chlorine atom on the transferred acyl group to generate the linked molecule. The bisubstrate inhibitor had a K_i value of 0.5 to 20 nM against AAC(3)-I. However, it did not show any effect on the antibiotic activity of gentamicin against an *E. coli* strain harbouring AAC(3)-I, presumably due to its inability to cross the cell membrane. Another example of mechanism-based inactivators are the 2′-deamino-2′-nitro derivatives of neamine and kanamycin B, which were synthesized to inactivate APH(3′)s (82). When the derivative is phosphorylated, the phosphoryl group spontaneously eliminates to produce a reactive inactivating species that traps an active site nucleophile. A covalent bond is formed between the enzyme and the derivative, and the enzyme is irreversibly inactivated. Although these compounds inactivated the enzyme, they did not reverse resistance in bacteria. Other reactive compounds shown to inactivate APH(3′) were 1,3,2′,6′-bromoacetylated analogues of neamine (83, 129).

A clever example of an aminoglycoside able to combat resistant bacteria by being able to regenerate itself is the 3′-oxo analogue of kanamycin A (Fig. 6) (44). Here the compound does not inhibit a resistance enzyme. The compound exists in equilibrium with its hydrated derivative and, when phosphorylated, generates an unstable intermediate that nonenzymatically

sisomicin 5-*epi*-sisomicin gentamicin B 5-*epi*-gentamicin B

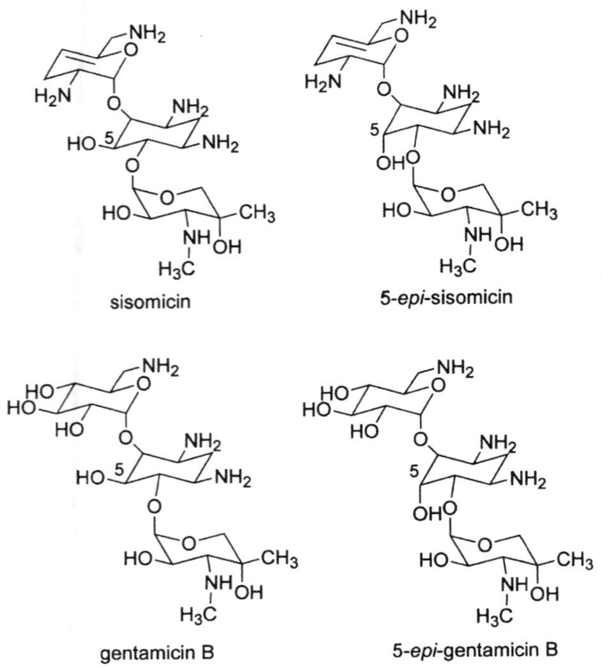

Figure 5. Semisynthetic aminoglycosides 5-*epi*-sisomicin and 5-*epi*-gentamicin retain antibiotic activity even against bacterial strains harboring ANT(2″), AAC(3), and AAC(2′).

eliminates phosphate to regenerate the ketone. Thus the molecule is able to maintain an active form of itself inside the cell. The compound is not as effective an antibiotic as kanamycin A, with a 30-fold higher MIC. It did have a four- to eightfold lower MIC than kanamycin A when tested against a resistant strain of *E. coli* bearing APH(3′)-Ia.

Identification of nonaminoglycoside-based inhibitors of aminoglycoside modifying enzymes has been less prevalent. The natural product 7-hydroxytropolone was identified and found to be a potent inhibitor of ANT(2″) and ANT(4′) (3, 54, 87). The molecule is competitive with ATP binding ($K_i = 10\ \mu M$) and shows potentiation of tobramycin activity against resistant *E. coli* strains that contain ANT(2″). The similarity in chemical mechanism and 3D structure between APHs and protein kinases lead to the testing of known inhibitors of protein kinases against APHs (25). Isoquinolinesulfonamides were found to be good inhibitors of APH(3′)-IIIa and AAC(6′)-APH(2″). These compounds, however, did not reverse resistance in strains of *Enterococcus* bearing these enzymes. Also, the enzyme AAC(6′)-APH(2″) is inactivated by the compound wortmannin, a known phosphatidyllinositol 3-kinase inhibitor (10), and 1-bromomethylphenanthrene and similar compounds (9).

The structures of AACs and APHs have shown that these enzymes have a negatively charged binding pocket to accommodate the positively charged aminoglycoside, and they have been shown to be active on peptide substrates. Therefore, Boehr et al. screened a series of small cationic peptides as inhibitors of APH(3′)-IIIa, AAC(6′)-Ii, and AAC(6′)-APH(2″) (8). They found two peptides that were able to inhibit both the APH and AAC. The peptide indolicidin was a competitive inhibitor of AAC with respect to aminoglycoside and a noncompetitive inhibitor of APH. The peptides screened were known antimicrobial agents that act on the outside of the cell. Consequently, none of the peptides showed synergistic action with aminoglycosides in bacteria containing the resistant genes.

CONCLUSIONS

Aminoglycosides have found over 60 years of continuous clinical use and been instrumental in the management of infectious disease during this time. This success, however, has resulted in full realization of the "antibiotic paradox" articulated by Stuart Levy where antibiotic use inevitably selects for the emergence of resistance mechanisms (57). In the case of aminoglycosides, this resistance has diminished the clinical utility of these compounds. However, molecular research over the last decade has resulted in an excellent understanding of the mode of action, interaction with target, and various resistance mechanisms. These efforts now have the potential to be applied in rational design and synthesis of new aminoglycosides or inhibitors of resistance enzymes that could result in a rejuvenation of the aminoglycoside field. The next decade therefore promises to be one of discovery and innovation in the aminoglycoside field.

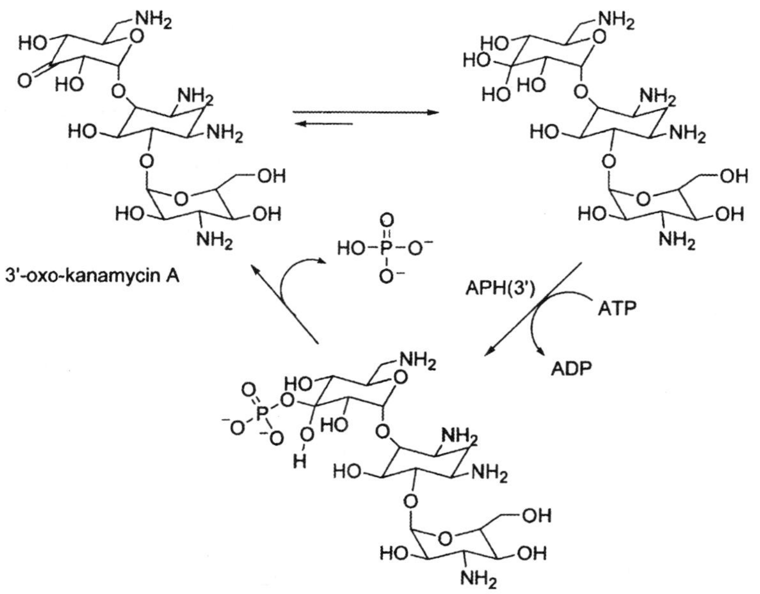

Figure 6. Evasion of APH-mediated resistance by 3′-oxo analogue of kanamycin A.

REFERENCES

1. **Acosta, M. B., R. C. Ferreira, G. Padilla, L. C. Ferreira, and S. O. Costa.** 2000. Altered expression of oligopeptide-binding protein (OppA) and aminoglycoside resistance in laboratory and clinical *Escherichia coli* strains. *J. Med. Microbiol.* **49:**409–413.
2. **Aires, J. R., T. Kohler, H. Nikaido, and P. Plesiat.** 1999. Involvement of an active efflux system in the natural resistance of *Pseudomonas aeruginosa* to aminoglycosides. *Antimicrob. Agents Chemother.* **43:**2624–2628.
3. **Allen, N. E., W. E. Alborn, Jr., J. N. Hobbs, Jr., and H. A. Kirst.** 1982. 7–Hydroxytropolone: an inhibitor of aminoglycoside-2″-O-adenylyltransferase. *Antimicrob. Agents Chemother.* **22:**824–831.
4. **Azucena, E., I. Grapsas, and S. Mobashery.** 1997. Properties of a bifunctional bacterial antibiotic resistance enzyme that catalyzes ATP-dependent 2″-phosphorylation and acetyl-CoA dependent 6′-acetylation of aminoglycosides. *J. Am. Chem. Soc.* **119:**2317–2318.
5. **Basso, L. A., and J. S. Blanchard.** 1998. Resistance to antitubercular drugs. *Adv. Exp. Med. Biol.* **456:**115–144.
6. **Birge, E. A., and C. G. Kurland.** 1969. Altered ribosomal protein in streptomycin-dependent *Escherichia coli. Science* **166:**1282–1284.
7. **Blanchard, S. C., D. Fourmy, R. G. Eason, and J. D. Puglisi.** 1998. rRNA chemical groups required for aminoglycoside binding. *Biochemistry* **37:**7716–7724.
8. **Boehr, D. D., K. A. Draker, K. Koteva, M. Bains, R. E. Hancock, and G. D. Wright.** 2003. Broad-spectrum peptide inhibitors of aminoglycoside antibiotic resistance enzymes. *Chem. Biol.* **10:**189–196.
9. **Boehr, D. D., S. I. Jenkins, and G. D. Wright.** 2003. The molecular basis of the expansive substrate specificity of the antibiotic resistance enzyme aminoglycoside acetyltransferase-6′-aminoglycoside phosphotransferase-2″. The role of ASP-99 as an active site base important for acetyl transfer. *J. Biol. Chem.* **278:**12873–12880.
10. **Boehr, D. D., W. S. Lane, and G. D. Wright.** 2001. Active site labeling of the gentamicin resistance enzyme AAC(6′)-APH(2″) by the lipid kinase inhibitor wortmannin. *Chem. Biol.* **8:**791–800.
11. **Boehr, D. D., P. R. Thompson, and G. D. Wright.** 2001. Molecular mechanism of aminoglycoside antibiotic kinase APH(3′)-IIIa: roles of conserved active site residues. *J. Biol. Chem.* **276:**23929–23936.
12. **Bryan, L. E., and H. M. Van den Elzen.** 1976. Streptomycin accumulation in susceptible and resistant strains of *Escherichia coli* and *Pseudomonas aeruginosa*. *Antimicrob. Agents Chemother.* **9:**928–938.
13. **Bryan, L. E., and H. M. Van Den Elzen.** 1977. Effects of membrane-energy mutations and cations on streptomycin and gentamicin accumulation by bacteria: a model for entry of streptomycin and gentamicin in susceptible and resistant bacteria. *Antimicrob. Agents Chemother.* **12:**163–177.
14. **Burk, D. L., W. C. Hon, A. K. Leung, and A. M. Berghuis.** 2001. Structural analyses of nucleotide binding to an aminoglycoside phosphotransferase. *Biochemistry* **40:**8756–8764.
15. **Carter, A. P., W. M. Clemons, D. E. Brodersen, R. J. Morgan-Warren, B. T. Wimberly, and V. Ramakrishnan.** 2000. Functional insights from the structure of the 30S ribosomal subunit and its interactions with antibiotics. *Nature* **407:**340–348.
16. **Chen-Goodspeed, M., J. L. Vanhooke, H. M. Holden, and F. M. Raushel.** 1999. Kinetic mechanism of kanamycin nucleotidyltransferase from *Staphylococcus aureus. Bioorg. Chem.* **27:**395–408.
17. **Chow, J. W., V. Kak, I. You, S. J. Kao, J. Petrin, D. B. Clewell, S. A. Lerner, G. H. Miller, and K. J. Shaw.** 2001. Aminoglycoside resistance genes aph(2″)-Ib and aac(6′)-Im detected together in strains of both *Escherichia coli* and *Enterococcus faecium*. *Antimicrob. Agents Chemother.* **45:**2691–2694.
18. **Chow, J. W., M. J. Zervos, S. A. Lerner, L. A. Thal, S. M. Donabedian, D. D. Jaworski, S. Tsai, K. J. Shaw, and D. B. Clewell.** 1997. A novel gentamicin resistance gene in *Enterococcus*. *Antimicrob. Agents Chemother.* **41:**511–514.
19. **Costa, Y., M. Galimand, R. Leclercq, J. Duval, and P. Courvalin.** 1993. Characterization of the chromosomal *aac(6′)-Ii* gene specific for *Enterococcus faecium*. *Antimicrob. Agents Chemother.* **37:**1896–1903.
20. **Cox, J. R., D. R. Ekman, E. L. DiGiammarino, A. Akal-Strader, and E. H. Serpersu.** 2000. Aminoglycoside antibiotics bound to aminoglycoside-detoxifying enzymes and RNA adopt similar conformations. *Cell Biochem. Biophys.* **33:**297–308.
21. **Cox, J. R., and E. H. Serpersu.** 1997. Biologically important conformations of aminoglycoside antibiotics bound to an aminoglycoside 3′-phosphotransferase as determined by transferred nuclear Overhauser effect spectroscopy. *Biochemistry* **36:**2353–2359.
22. **Culebras, E., and J. L. Martinez.** 1999. Aminoglycoside resistance mediated by the bifunctional enzyme 6′-N-aminoglycoside acetyltransferase-2″-O-aminoglycoside phosphotransferase. *Front. Biosci.* **4:**D1–D8.
23. **Daigle, D. M., D. W. Hughes, and G. D. Wright.** 1999. Prodigious substrate specificity of AAC(6′)-APH(2″), an aminoglycoside antibiotic resistance determinant in enterococci and staphylococci. *Chem. Biol.* **6:**99–110.
24. **Daigle, D. M., G. A. McKay, P. R. Thompson, and G. D. Wright.** 1999. Aminoglycoside antibiotic phosphotransferases are also serine protein kinases. *Chem. Biol.* **6:**11–18.
25. **Daigle, D. M., G. A. McKay, and G. D. Wright.** 1997. Inhibition of aminoglycoside antibiotic resistance enzymes by protein kinase inhibitors. *J. Biol. Chem.* **272:**24755–24758.
26. **Davis, B. D.** 1987. Mechanism of bactericidal action of aminoglycosides. *Microbiol. Rev.* **51:**341–350.
27. **Disney, M. D., S. Magnet, J. S. Blanchard, and P. H. Seeberger.** 2004. Aminoglycoside microarrays to study antibiotic resistance. *Angew Chem. Int. Ed. Engl.* **43:**1591–1594.
28. **Doi, Y., K. Yokoyama, K. Yamane, J. Wachino, N. Shibata, T. Yagi, K. Shibayama, H. Kato, and Y. Arakawa.** 2004. Plasmid-mediated 16S rRNA methylase in *Serratia marcescens* conferring high-level resistance to aminoglycosides. *Antimicrob. Agents Chemother.* **48:**491–496.
29. **Draker, K. A., and G. D. Wright.** 2004. Molecular mechanism of the enterococcal aminoglycoside 6′-N-acetyltransferase: role of GNAT-conserved residues in the chemistry of antibiotic inactivation. *Biochemistry* **43:**446–454.
30. **Dyda, F., D. C. Klein, and A. B. Hickman.** 2000. GCN5–related N-acetyltransferases: a structural overview. *Annu. Rev. Biophys. Biomol. Struct.* **29:**81–103.
31. **Ekman, D. R., E. L. DiGiammarino, E. Wright, E. D. Witter, and E. H. Serpersu.** 2001. Cloning, overexpression, and purification of aminoglycoside antibiotic nucleotidyltransferase (2″)-Ia: conformational studies with bound substrates. *Biochemistry* **40:**7017–7024.
32. **Ferretti, J. J., K. S. Gilmore, and P. Courvalin.** 1986. Nucleotide sequence analysis of the gene specifying the bifunctional 6′-aminoglycoside acetyltransferase 2″-aminoglycoside

phosphotransferase enzyme in *Streptococcus faecalis* and identification and cloning of gene regions specifying the two activities. *J. Bacteriol.* **167**:631–638.

33. **Fong, D. H., and A. M. Berghuis.** 2002. Substrate promiscuity of an aminoglycoside antibiotic resistance enzyme via target mimicry. *EMBO J.* **21**:2323–2331.
34. **Fourmy, D., M. I. Recht, S. C. Blanchard, and J. D. Puglisi.** 1996. Structure of the A site of *Escherichia coli* 16S ribosomal RNA complexed with an aminoglycoside antibiotic. *Science* **274**:1367–1371.
35. **Fourmy, D., M. I. Recht, and J. D. Puglisi.** 1998. Binding of neomycin-class aminoglycoside antibiotics to the A-site of 16 S rRNA. *J. Mol. Biol.* **277**:347–362.
36. **Fourmy, D., S. Yoshizawa, and J. D. Puglisi.** 1998. Paromomycin binding induces a local conformational change in the A-site of 16 S rRNA. *J. Mol. Biol.* **277**:333–345.
37. **Fu, K. P., and H. C. Neu.** 1978. Activity of 5–episisomicin compared with that of other aminoglycosides. *Antimicrob. Agents Chemother.* **14**:194–200.
38. **Galimand, M., P. Courvalin, and T. Lambert.** 2003. Plasmid-mediated high-level resistance to aminoglycosides in *Enterobacteriaceae* due to 16S rRNA methylation. *Antimicrob. Agents Chemother.* **47**:2565–2571.
39. **Gates, C. A., and D. B. Northrop.** 1988. Determination of the rate-limiting segment of aminoglycoside nucleotidyltransferase 2″-I by pH and viscosity-dependent kinetics. *Biochemistry* **27**: 3834–3842.
40. **Gates, C. A., and D. B. Northrop.** 1988. Substrate specificities and structure-activity relationships for the nucleotidylation of antibiotics catalyzed by aminoglycoside nucleotidyltransferase 2″-I. *Biochemistry* **27**:3820–3825.
41. **Gerratana, B., W. W. Cleland, and L. A. Reinhardt.** 2001. Regiospecificity assignment for the reaction of kanamycin nucleotidyltransferase from *Staphylococcus aureus*. *Biochemistry* **40**:2964–2971.
42. **Gerratana, B., P. A. Frey, and W. W. Cleland.** 2001. Characterization of the transition-state structure of the reaction of kanamycin nucleotidyltransferase by heavy-atom kinetic isotope effects. *Biochemistry* **40**:2972–2977.
43. **Gray, G. S., and W. M. Fitch.** 1983. Evolution of antibiotic resistance genes: the DNA sequence of a kanamycin resistance gene from *Staphylococcus aureus*. *Mol. Biol. Evol.* **1**: 57–66.
44. **Haddad, J., S. Vakulenko, and S. Mobashery.** 1999. An antibiotic cloaked by its own resistance enzyme. *J. Am. Chem. Soc.* **121**:11922–11923.
45. **Hancock, R. E.** 1981. Aminoglycoside uptake and mode of action—with special reference to streptomycin and gentamicin. I. Antagonists and mutants. *J. Antimicrob. Chemother.* **8**:249–276.
46. **Hegde, S. S., F. Javid-Majd, and J. S. Blanchard.** 2001. Overexpression and mechanistic analysis of chromosomally encoded aminoglycoside 2′-N-acetyltransferase (AAC(2′)-Ic) from *Mycobacterium tuberculosis*. *J. Biol. Chem.* **276**:45876–45881.
47. **Hodel-Christian, S. L., and B. E. Murray.** 1991. Characterization of the gentamicin resistance transposon Tn5281 from *Enterococcus faecalis* and comparison to staphylococcal transposons Tn4001 and Tn4031. *Antimicrob. Agents Chemother.* **35**:1147–1152.
48. **Hon, W. C., G. A. McKay, P. R. Thompson, R. M. Sweet, D. S. Yang, G. D. Wright, and A. M. Berghuis.** 1997. Structure of an enzyme required for aminoglycoside antibiotic resistance reveals homology to eukaryotic protein kinases. *Cell* **89**:887–895.
49. **Kabins, S. A., and C. Nathan.** 1978. In vitro activity of 5-episisomicin in bacteria resistant to other aminoglycoside antibiotics. *Antimicrob. Agents Chemother.* **14**:391–397.
50. **Kashiwagi, K., A. Miyaji, S. Ikeda, T. Tobe, C. Sasakawa, and K. Igarashi.** 1992. Increase of sensitivity to aminoglycoside antibiotics by polyamine-induced protein (oligopeptide-binding protein) in *Escherichia coli*. *J. Bacteriol.* **174**:4331–4333.
51. **Kashiwagi, K., M. H. Tsuhako, K. Sakata, T. Saisho, A. Igarashi, S. O. da Costa, and K. Igarashi.** 1998. Relationship between spontaneous aminoglycoside resistance in *Escherichia coli* and a decrease in oligopeptide binding protein. *J. Bacteriol.* **180**:5484–5488.
52. **Kaufhold, A., A. Podbielski, T. Horaud, and P. Ferrieri.** 1992. Identical genes confer high-level resistance to gentamicin upon *Enterococcus faecalis, Enterococcus faecium,* and *Streptococcus agalactiae*. *Antimicrob. Agents Chemother.* **36**:1215–1218.
53. **Kim, C., J. Haddad, S. B. Vakulenko, S. O. Meroueh, Y. Wu, H. Yan, and S. Mobashery.** 2004. Fluorinated aminoglycosides and their mechanistic implication for aminoglycoside 3′-phosphotransferases from gram-negative bacteria. *Biochemistry* **43**:2373–2383.
54. **Kirst, H. A., G. G. Marconi, F. T. Counter, P. W. Ensminger, N. D. Jones, M. O. Chaney, J. E. Toth, and N. E. Allen.** 1982. Synthesis and characterization of a novel inhibitor of an aminoglycoside-inactivating enzyme. *J. Antibiot.* (Tokyo) **35**:1651–1657.
55. **Kondo, S., A. Tamura, S. Gomi, Y. Ikeda, T. Takeuchi, and S. Mitsuhashi.** 1993. Structures of enzymatically modified products of arbekacin by methicillin resistant Staphylococcus aureus. *J. Antibiot.* **46**:310–315.
56. **Leclerc, D., P. Melancon, and L. Brakier-Gingras.** 1991. Mutations in the 915 region of *Escherichia coli* 16S ribosomal RNA reduce the binding of streptomycin to the ribosome. *Nucleic. Acids Res.* **19**:3973–3977.
57. **Levy, S. B.** 1992. *The Antibiotic Paradox*. Plenum Press, New York, N.Y.
58. **Lovering, A. M., L. O. White, and D. S. Reeves.** 1987. AAC(1): a new aminoglycoside-acetylating enzyme modifying the Cl aminogroup of apramycin. *J. Antimicrob. Chemother.* **20**:803–813.
59. **Magnet, S., P. Courvalin, and T. Lambert.** 1999. Activation of the cryptic *aac(6′)-Iy* aminoglycoside resistance gene of *Salmonella* by a chromosomal deletion generating a transcriptional fusion. *J. Bacteriol.* **181**:6650–6655.
60. **Martel, A., M. Masson, N. Moreau, and F. Le Goffic.** 1983. Kinetic studies of aminoglycoside acetyltransferase and phosphotransferase from *Staphylococcus aureus* RPAL. Relationship between the two activities. *Eur. J. Biochem.* **133**:515–521.
61. **Masuda, N., E. Sakagawa, S. Ohya, N. Gotoh, H. Tsujimoto, and T. Nishino.** 2000. Contribution of the MexX-MexY-oprM efflux system to intrinsic resistance in *Pseudomonas aeruginosa*. *Antimicrob. Agents Chemother.* **44**:2242–2246.
62. **McKay, G. A., J. Roestamadji, S. Mobashery, and G. D. Wright.** 1996. Recognition of aminoglycoside antibiotics by enterococcal-staphylococcal aminoglycoside 3′-phosphotransferase type IIIa: role of substrate amino groups. *Antimicrob. Agents Chemother.* **40**:2648–2650.
63. **McKay, G. A., P. R. Thompson, and G. D. Wright.** 1994. Broad spectrum aminoglycoside phosphotransferase type III from Enterococcus: overexpression, purification, and substrate specificity. *Biochemistry* **33**:6936–6944.
64. **McKay, G. A., and G. D. Wright.** 1995. Kinetic mechanism of aminoglycoside phosphotransferase type IIIa. Evidence for a

Theorell-Chance mechanism. *J. Biol. Chem.* **270:**24686–24692.

65. **McKay, G. A., and G. D. Wright.** 1996. Catalytic mechanism of enterococcal kanamycin kinase (APH(3′)-IIIa): viscosity, thio, and solvent isotope effects support a Theorell-Chance mechanism. *Biochemistry* **35:**8680–8685.
66. **Mehta, R., and W. S. Champney.** 2003. Neomycin and paromomycin inhibit 30S ribosomal subunit assembly in *Staphylococcus aureus*. *Curr. Microbiol.* **47:**237–243.
67. **Miller, G. H., F. J. Sabatelli, R. S. Hare, Y. Glupczynski, P. Mackey, D. Shlaes, K. Shimizu, and K. J. Shaw.** 1997. The most frequent aminoglycoside resistance mechanisms—changes with time and geographic area: a reflection of aminoglycoside usage patterns? Aminoglycoside Resistance Study Groups. *Clin. Infect. Dis.* **24**(Suppl. 1):S46–S62.
68. **Miller, M. H., S. C. Edberg, L. J. Mandel, C. F. Behar, and N. H. Steigbigel.** 1980. Gentamicin uptake in wild-type and aminoglycoside-resistant small-colony mutants of *Staphylococcus aureus*. *Antimicrob. Agents Chemother.* **18:**722–729.
69. **Mine, T., Y. Morita, A. Kataoka, T. Mizushima, and T. Tsuchiya.** 1999. Expression in *Escherichia coli* of a new multidrug efflux pump, MexXY, from *Pseudomonas aeruginosa*. *Antimicrob. Agents Chemother.* **43:**415–417.
70. **Moazed, D., and H. F. Noller.** 1987. Interaction of antibiotics with functional sites in 16S ribosomal RNA. *Nature* **327:**389–394.
71. **Montandon, P. E., R. Wagner, and E. Stutz.** 1986. *E. coli* ribosomes with a C912 to U base change in the 16S rRNA are streptomycin resistant. *EMBO J.* **5:**3705–3708.
72. **Moore, R. A., D. DeShazer, S. Reckseidler, A. Weissman, and D. E. Woods.** 1999. Efflux-mediated aminoglycoside and macrolide resistance in *Burkholderia pseudomallei*. *Antimicrob. Agents Chemother.* **43:**465–470.
73. **Murakami, S., R. Nakashima, E. Yamashita, and A. Yamaguchi.** 2002. Crystal structure of bacterial multidrug efflux transporter AcrB. *Nature* **419:**587–593.
74. **Nurizzo, D., S. C. Shewry, M. H. Perlin, S. A. Brown, J. N. Dholakia, R. L. Fuchs, T. Deva, E. N. Baker, and C. A. Smith.** 2003. The crystal structure of aminoglycoside-3′-phosphotransferase-IIa, an enzyme responsible for antibiotic resistance. *J. Mol. Biol.* **327:**491–506.
75. **Ozaki, M., S. Mizushima, and M. Nomura.** 1969. Identification and functional characterization of the protein controlled by the streptomycin-resistant locus in *E. coli*. *Nature* **222:**333–339.
76. **Papadopoulou, B., and P. Courvalin.** 1988. Dispersal in *Campylobacter* spp. of aphA-3, a kanamycin resistance determinant from gram-positive cocci. *Antimicrob. Agents Chemother.* **32:**945–948.
77. **Pedersen, L. C., M. M. Benning, and H. M. Holden.** 1995. Structural investigation of the antibiotic and ATP-binding sites in kanamycin nucleotidyltransferase. *Biochemistry* **34:**13305–13311.
78. **Poirel, L., T. Lambert, S. Turkoglu, E. Ronco, J. Gaillard, and P. Nordmann.** 2001. Characterization of Class 1 integrons from *Pseudomonas aeruginosa* that contain the bla(VIM-2) carbapenem-hydrolyzing beta-lactamase gene and of two novel aminoglycoside resistance gene cassettes. *Antimicrob. Agents Chemother.* **45:**546–552.
79. **Poole, K.** 2001. Multidrug resistance in gram-negative bacteria. *Curr. Opin. Microbiol.* **4:**500–508.
80. **Riccio, M. L., J. D. Docquier, E. Dell'Amico, F. Luzzaro, G. Amicosante, and G. M. Rossolini.** 2003. Novel 3-N-aminoglycoside acetyltransferase gene, *aac(3)-Ic,* from a *Pseudomonas aeruginosa* integron. *Antimicrob. Agents Chemother.* **47:**1746–1748.
81. **Roestamadji, J., I. Grapsas, and S. Mobashery.** 1995. Loss of individual electrostatic interactions between aminoglycoside antibiotics and resistance enzymes as an effective means to overcoming bacterial drug resistance. *J. Am. Chem. Soc.* **117:**11060–11069.
82. **Roestamadji, J., I. Grapsas, and S. Mobashery.** 1995. Mechanism-based inactivation of bacterial aminoglycoside 3′-phosphotransferases. *J. Am. Chem. Soc.* **117:**80–84.
83. **Roestamadji, J., and S. Mobashery.** 1998. The use of neamine as a molecular template: inactivation of bacterial antibiotic resistance enzyme aminoglycoside 3′-phosphotransferase type IIa. *Bioorg. Med. Chem. Lett.* **8:**3483–3488.
84. **Rosenberg, E. Y., D. Ma, and H. Nikaido.** 2000. AcrD of *Escherichia coli* is an aminoglycoside efflux pump. *J. Bacteriol.* **182:**1754–1756.
85. **Rouch, D. A., M. E. Byrne, Y. C. Kong, and R. A. Skurray.** 1987. The aacA-aphD gentamicin and kanamycin resistance determinant of Tn4001 from *Staphylococcus aureus*: expression and nucleotide sequence analysis. *J. Gen. Microbiol.* **133:**3039–3052.
86. **Sakon, J., H. H. Liao, A. M. Kanikula, M. M. Benning, I. Rayment, and H. M. Holden.** 1993. Molecular structure of kanamycin nucleotidyltransferase determined to 3.0Å-resolution. *Biochemistry* **32:**11977–11984.
87. **Saleh, N. A., A. Zwiefak, W. Peczynska-Czoch, M. Mordarski, and G. Pulverer.** 1988. New inhibitors for aminoglycoside-adenylyltransferase. *Zentralbl. Bakteriol. Mikrobiol. Hyg.* [*A*]. **270:**66–75.
88. **Santanam, P., and F. H. Kayser.** 1978. Purification and characterization of an aminoglycoside inactivating enzyme from *Staphylococcus epidermidis* FK109 that nucleotidylates the 4′- and 4″-hydroxyl groups of the aminoglycoside antibiotics. *J. Antibiot.* (Tokyo) **31:**343–351.
89. **Schatz, A., E. Bugie, and S. A. Waksman.** 1944. Streptomycin, a substance exibiting antibiotic activity against Gram-positive and Gram-negative bacteria. *Proc. Soc. Exp. Biol. Med.* **55:**66–69.
90. **Schatz, A., E. Bugie, and S. A. Waksman.** 1973. Selman Abraham Waksman, Ph.D. 22 July 1888—16 August 1973. Streptomycin reported. *Ann. Intern. Med.* **79:**678.
91. **Schwocho, L. R., C. P. Schaffner, G. H. Miller, R. S. Hare, and K. J. Shaw.** 1995. Cloning and characterization of a 3-*N*-aminoglycoside acetyltransferase gene, *aac(3)-Ib,* from *Pseudomonas aeruginosa*. *Antimicrob. Agents Chemother.* **39:**1790–1796.
92. **Sigmund, C. D., M. Ettayebi, and E. A. Morgan.** 1984. Antibiotic resistance mutations in 16S and 23S ribosomal RNA genes of *Escherichia coli*. *Nucleic Acids Res.* **12:**4653–63.
93. **Siregar, J. J., S. A. Lerner, and S. Mobashery.** 1994. Purification and characterization of aminoglycoside 3′-phosphotransferase type IIa and kinetic comparison with a new mutant enzyme. *Antimicrob. Agents Chemother.* **38:**641–647.
94. **Siregar, J. J., K. Miroshnikov, and S. Mobashery.** 1995. Purification, characterization, and investigation of the mechanism of aminoglycoside 3′-phosphotransferase type Ia. *Biochemistry* **34:**12681–12688.
95. **Sobel, M. L., G. A. McKay, and K. Poole.** 2003. Contribution of the MexXY multidrug transporter to aminoglycoside resistance in *Pseudomonas aeruginosa* clinical isolates. *Antimicrob. Agents Chemother.* **47:**3202–3207.
96. **Sucheck, S. J., A. L. Wong, K. M. Koeller, D. D. Boehr, K.-A. Draker, P. Sears, G. D. Wright, and C.-H. Wong.** 2000. Design of bifunctional antibiotics that target bacterial rRNA and in-

hibit resistance-causing enzymes. *J. Am. Chem. Soc.* **122**:5230–5231.

97. **Sunada, A., M. Nakajima, Y. Ikeda, S. Kondo, and K. Hotta.** 1999. Enzymatic 1-N-acetylation of paromomycin by an actinomycete strain #8 with multiple aminoglycoside resistance and paromomycin sensitivity. *J. Antibiot.* (Tokyo) **52**:809–814.
98. **Sutherland, R.** 1995. Beta-Lactam/beta-lactamase inhibitor combinations: development, antibacterial activity and clinical applications. *Infection* **23**:191–200.
99. **Tai, P. C., B. J. Wallace, and B. D. Davis.** 1978. Streptomycin causes misreading of natural messenger by interacting with ribosomes after initiation. *Proc. Natl. Acad. Sci. USA* **75**:275–279.
100. **Tai, P. C., B. J. Wallace, E. L. Herzog, and B. D. Davis.** 1973. Properties of initiation-free polysomes of *Escherichia coli*. *Biochemistry* **12**:609–615.
101. **Taylor, D. E., W. Yan, L. K. Ng, E. K. Manavathu, and P. Courvalin.** 1988. Genetic characterization of kanamycin resistance in *Campylobacter coli*. *Ann. Inst. Pasteur. Microbiol.* **139**:665–676.
102. **Thal, L. A., J. W. Chow, J. E. Patterson, M. B. Perri, S. Donabedian, D. B. Clewell, and M. J. Zervos.** 1993. Molecular characterization of highly gentamicin-resistant *Enterococcus faecalis* isolates lacking high-level streptomycin resistance. *Antimicrob. Agents Chemother.* **37**:134–137.
103. **Thompson, P. R., D. D. Boehr, A. M. Berghuis, and G. D. Wright.** 2002. Mechanism of aminoglycoside antibiotic kinase APH(3′)-IIIa: role of the nucleotide positioning loop. *Biochemistry* **41**:7001–7007.
104. **Thompson, P. R., D. W. Hughes, N. P. Cianciotto, and G. D. Wright.** 1998. Spectinomycin kinase from *Legionella pneumophila*. Characterization of substrate specificity and identification of catalytically important residues. *J. Biol. Chem.* **273**:14788–14795.
105. **Thompson, P. R., D. W. Hughes, and G. D. Wright.** 1996. Mechanism of aminoglycoside 3′-phosphotransferase type IIIa: His188 is not a phosphate-accepting residue. *Chem. Biol.* **3**:747–755.
106. **Thompson, P. R., D. W. Hughes, and G. D. Wright.** 1996. Regiospecificity of aminoglycoside phosphotransferase from Enterococci and Staphylococci (APH(3′)-IIIa). *Biochemistry* **35**:8686–8695.
107. **Thompson, P. R., J. Schwartzenhauer, D. W. Hughes, A. M. Berghuis, and G. D. Wright.** 1999. The COOH terminus of aminoglycoside phosphotransferase (3′)-IIIa is critical for antibiotic recognition and resistance. *J. Biol. Chem.* **274**:30697–30706.
108. **Trieu-Cuot, P., and P. Courvalin.** 1983. Nucleotide sequence of the *Streptococcus faecalis* plasmid gene encoding the 3′5″-aminoglycoside phosphotransferase type III. *Gene* **23**:331–341.
109. **Tsai, S. F., M. J. Zervos, D. B. Clewell, S. M. Donabedian, D. F. Sahm, and J. W. Chow.** 1998. A new high-level gentamicin resistance gene, *aph(2″)-Id*, in *Enterococcus spp*. *Antimicrob. Agents Chemother.* **42**:1229–1232.
110. **Ubukata, K., N. Yamashita, A. Gotoh, and M. Konno.** 1984. Purification and characterization of aminoglycoside-modifying enzymes from *Staphylococcus aureus* and *Staphylococcus epidermidis*. *Antimicrob. Agents Chemother.* **25**:754–759.
111. **Umezawa, H., Y. Nishimura, T. Tsuchiya, and S. Umezawa.** 1972. Syntheses of 6′-N-methyl-kanamycin and 3′,4′-dideoxy-6′-N-methylkanamycin B active against resistant strains having 6′-N-acetylating enzymes. *J. Antibiot.* (Tokyo) **25**:743–745.
112. **Umezawa, H., M. Okanishi, S. Kondo, K. Hamana, R. Utahara, K. Maeda, and S. Mitsuhashi.** 1967. Phosphorylative inactivation of aminoglycopside antibioics by *Escherichia coli* carrying R factor. *Science* **157**:1559–1561.
113. **Umezawa, H., M. Okanishi, R. Utahara, K. Maeda, and S. Kondo.** 1967. Isolation and structure of kanamycin inactivated by a cell free system of kanamycin-resistant *E. coli*. *J. Antibiot.* (Tokyo) **20**:136–141.
114. **Umezawa, H., M. Ueda, K. Maeda, K. Yagishita, S. Kando, Y. Okami, R. Utahara, Y. Osato, K. Nitta, and T. Kakeuchi.** 1957. Production and isolation of a new antibiotic kanamycin. *J. Antibiot.* **10**:181–189.
115. **Umezawa, H., S. Umezawa, T. Tsuchiya, Y. Okazaki, R. Muto, and Y. Nishimura.** 1971. 3′,4′-dideoxy-kanamycin B active against kanamycin-resistant *Escherichia coli* and *Pseudomonas aeruginosa*. *J. Antibiot.* (Tokyo) **24**:485–487.
116. **Vetting, M., S. L. Roderick, S. Hegde, S. Magnet, and J. S. Blanchard.** 2003. What can structure tell us about in vivo function? The case of aminoglycoside-resistance genes. *Biochem. Soc. Trans.* **31**:520–522.
117. **Vetting, M. W., S. S. Hegde, F. Javid-Majd, J. S. Blanchard, and S. L. Roderick.** 2002. Aminoglycoside 2′-N-acetyltransferase from *Mycobacterium tuberculosis* in complex with coenzyme A and aminoglycoside substrates. *Nat. Struct. Biol.* **9**:653–658.
118. **Waitz, J. A., G. H. Miller, E. Moss, Jr., and P. J. Chiu.** 1978. Chemotherapeutic evaluation of 5-episisomicin (Sch 22591), a new semisynthetic aminoglycoside. *Antimicrob. Agents Chemother.* **13**:41–48.
119. **Waksman, S., and H. B. Woodruff.** 1940. Bacteriostatic and bactericidal substances produced by soil Actinomyces. *Proc. Soc. Exp. Biol. Med.* **45**:609–614.
120. **Waksman, S. A., and H. A. Lechevalier.** 1949. Neomycin, a new antibiotic active against streptomycin-resistant bacteria, including tuberculosis organisms. *Science* **109**:305–307.
121. **Wallace, B. J., P. C. Tai, E. L. Herzog, and B. D. Davis.** 1973. Partial inhibition of polysomal ribosomes of *Escherichia coli* by streptomycin. *Proc. Natl. Acad. Sci. USA* **70**:1234–1237.
122. **Weigel, L. M., D. B. Clewell, S. R. Gill, N. C. Clark, L. K. McDougal, S. E. Flannagan, J. F. Kolonay, J. Shetty, G. E. Killgore, and F. C. Tenover.** 2003. Genetic analysis of a high-level vancomycin-resistant isolate of *Staphylococcus aureus*. *Science* **302**:1569–1571.
123. **Weinstein, M. J., G. M. Luedemann, E. M. Oden, G. H. Wagman, J. P. Rosselet, J. A. Marquez, C. T. Coniglio, W. Charney, H. L. Herzog, and J. Black.** 1963. Gentamicin, a new antibiotic complex from *Micromonospora*. *J. Med. Chem.* **6**:463–464.
124. **Westbrock-Wadman, S., D. R. Sherman, M. J. Hickey, S. N. Coulter, Y. Q. Zhu, P. Warrener, L. Y. Nguyen, R. M. Shawar, K. R. Folger, and C. K. Stover.** 1999. Characterization of a *Pseudomonas aeruginosa* efflux pump contributing to aminoglycoside impermeability. *Antimicrob. Agents Chemother.* **43**:2975–2983.
125. **Williams, J. W., and D. B. Northrop.** 1979. Synthesis of a tight-binding, multisubstrate analog inhibitor of gentamicin acetyltransferase I. *J. Antibiot.* (Tokyo) **32**:1147–1154.
126. **Wolf, E., A. Vassilev, Y. Makino, A. Sali, Y. Nakatani, and S. K. Burley.** 1998. Crystal structure of a GCN5-related N-acetyltransferase: *Serratia marcescens* aminoglycoside 3-N-acetyltransferase. *Cell* **94**:439–449.
127. **Woodcock, J., D. Moazed, M. Cannon, J. Davies, and H. F. Noller.** 1991. Interaction of antibiotics with A- and P-site-specific bases in 16S ribosomal RNA. *EMBO J.* **10**:3099–3103.
128. **Wybenga-Groot, L. E., K. Draker, G. D. Wright, and A. M. Berghuis.** 1999. Crystal structure of an aminoglycoside 6′-

N-acetyltransferase: defining the GCN5-related N-acetyltransferase superfamily fold. *Structure Fold. Des.* **7**:497–507.

129. **Yang, Y., J. Roestamadji, S. Mobashery, and R. Orlando.** 1998. The use of neamine as a molecular template: identification of active site residues in the bacterial antibiotic resistance enzyme aminoglycoside 3′-phosphotransferase type IIa by mass spectroscopy. *Bioorg. Med. Chem. Lett.* **8**:3489–3494.

130. **Yokoyama, K., Y. Doi, K. Yamane, H. Kurokawa, N. Shibata, K. Shibayama, T. Yagi, H. Kato, and Y. Arakawa.** 2003. Acquisition of 16S rRNA methylase gene in *Pseudomonas aeruginosa*. *Lancet* **362**:1888–1893.

131. **Yoshizawa, S., D. Fourmy, and J. D. Puglisi.** 1998. Structural origins of gentamicin antibiotic action. *EMBO J.* **17**:6437–6448.

Frontiers in Antimicrobial Resistance: a Tribute to Stuart B. Levy
Edited by D. G. White, M. N. Alekshun, and P. F. McDermott

Chapter 8

Glycopeptide Resistance in Enterococci

FLORENCE DEPARDIEU AND PATRICE COURVALIN

Although more than 20 *Enterococcus* species have been identified, only two are responsible for the majority of human infections. Until recently, *E. faecalis* had been the predominant enterococcal species, accounting for 80 to 90% of all clinical isolates while *E. faecium* accounted for 5 to 15% (66, 81, 94, 112, 128). Other *Enterococcus* species (*E. gallinarum, E. casseliflavus, E. flavescens, E. durans, E. avium,* and *E. raffinosus*) are isolated much less frequently and account for less than 5% of clinical isolates (66, 128). Enterococci have been recognized as an important cause of endocarditis and emerged as a common cause of hospital-acquired infections in the mid to late 1970s (91, 131). This was probably related, at least in part, to the increasing use of third generation cephalosporins, to which enterococci are naturally resistant (105). Enterococci are currently ascendant nosocomial pathogens, having become the second most common microorganisms recovered from nosocomial urinary tract and wound infections and the third most common cause of nosocomial bacteremia in the United States (36, 52, 81, 94, 131, 145).

Because most enterococci are tolerant to the bactericidal activity of β-lactams and glycopeptides, bactericidal synergy between one of these classes of antibiotics and an aminoglycoside is needed to treat most serious enterococcal infections such as endocarditis and meningitis (18, 84). Until the late 1980s, vancomycin was the only drug that could be consistently relied on for the treatment of infections caused by multidrug-resistant enterococci (63, 105). Enterococci are intrinsically resistant to low concentrations of lincosamides, aminoglycosides, and trimethoprim-sulfamethoxazole (63, 105) and of β-lactams due to the low affinity of several penicillin-binding proteins (PBP) for penicillins (67, 103). They have also acquired resistance to high levels of the penicillins by overproduction or alteration of PBP5 (58, 93, 129) or rarely by synthesis of a β-lactamase (104) and to aminoglycosides, chloramphenicol, macrolides-lincosamides-streptogramins, fluoroquinolones, tetracycline, and rifampin and more recently to the glycopeptides (63, 105).

Vancomycin, a glycopeptide antibiotic, was in clinical use for more than 30 years without the emergence of resistance. Teicoplanin is another glycopeptide not available in the United States but used in Europe. Because of their activity against methicillin-resistant staphylococci and other gram-positive bacteria, these drugs have been widely used for therapy against infections due to these organisms. In 1988, the first isolates of vancomycin-resistant *E. faecalis* and *E. faecium* were reported by investigators in France (86) and the United Kingdom (139); such strains were detected later in hospitals from the east coast of the United States (60). Vancomycin-resistant enterococci have spread with unanticipated rapidity and are now encountered in hospitals in most countries (23, 34, 44, 57, 81, 138, 140).

We shall first review the mode of action and the mechanism of resistance to glycopeptides, as exemplified by the VanA-type mediated by transposon Tn*1546*, which is widely spread in enterococci. The diversity, regulation, evolution, and origin of glycopeptide resistance will then be discussed.

THE TARGET OF GLYCOPEPTIDES

In susceptible enterococci, as in other genera, peptidoglycan synthesis requires various steps (Fig. 1). In the cytoplasm, a racemase converts L-alanine (L-Ala) to D-alanine (D-Ala), and then two molecules of D-Ala are joined by a ligase leading to the dipeptide D-Ala-D-Ala, which is then added to UDP-*N*-acetylmuramyl-tripeptide to form the UDP-*N*-acetylmuramyl-pentapeptide (Fig. 1). The latter is bound to the

Florence Depardieu and Patrice Courvalin • Unite des Agents Antibacteriens, Institut Pasteur, 25, rue du Docteur Roux, 75724 Paris Cedex 15, France.

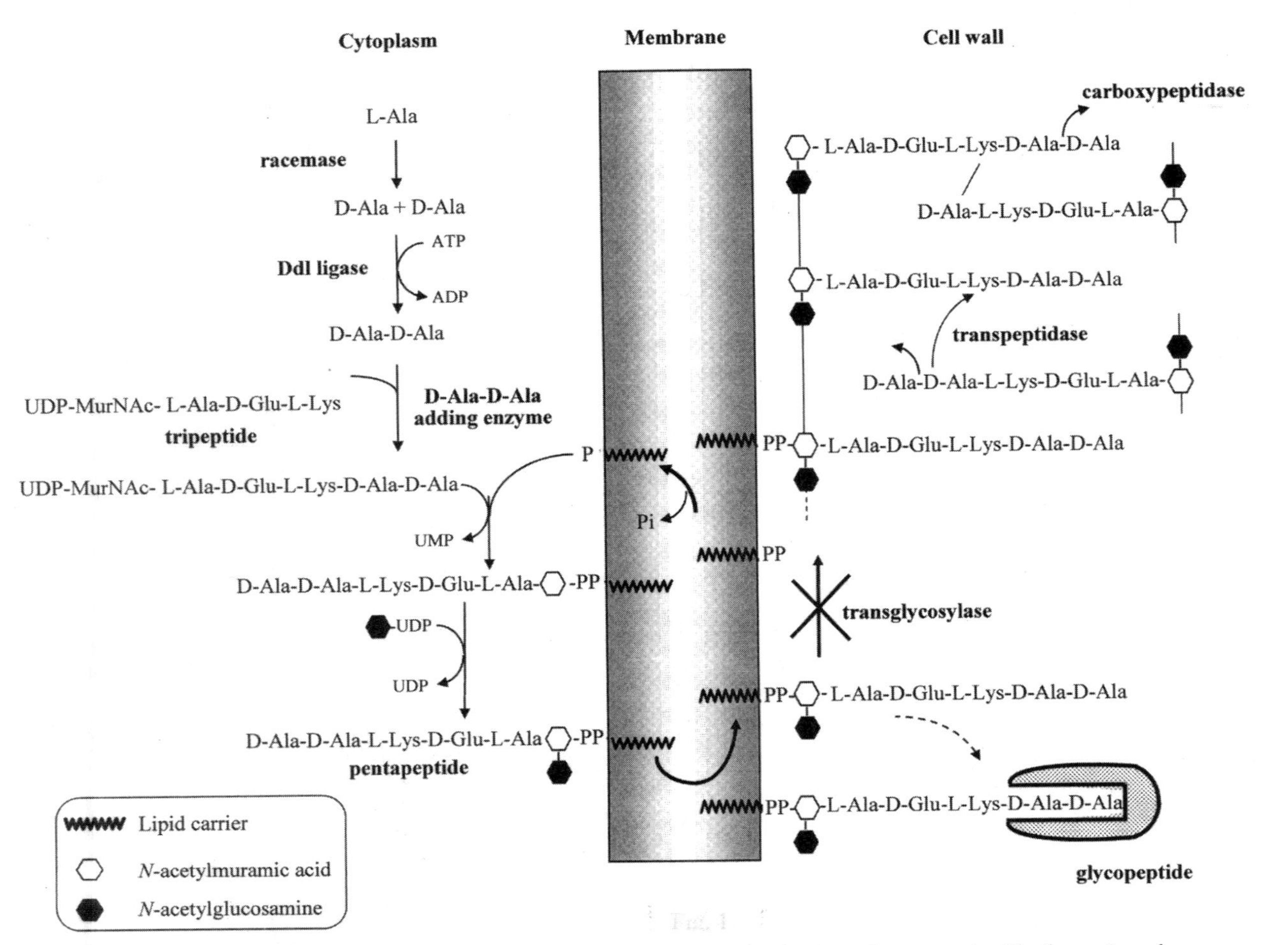

Figure 1. Schematic representation of peptidoglycan biosynthesis and mode of action of vancomycin. The formation of complexes between the antibiotic and the C-terminal D-alanyl-D-alanine moiety of peptidoglycan precursors prevents transfer of precursors from the lipid carrier to the peptidoglycan by transglycosylation. The reactions catalyzed by transpeptidases and D,D-carboxypeptidases are also inhibited.

undecaprenol lipid carrier that, following the addition of GlcNAc from UDP-GlcNAc, allows translocation of the precursors to the outer surface of the cytoplasmic membrane. *N*-acetylmuramyl-pentapeptide, when incorporated into nascent peptidoglycan by transglycosylation, allows the formation of cross bridges by penicillin-susceptible transpeptidation which contribute to the strength of the peptidoglycan layer (17).

The structure of glycopeptides, such as vancomycin and teicoplanin, is based on a heptapeptide domain in which five amino acid residues are common to all glycopeptides (Fig. 2) (22, 121). The drugs do not penetrate into the cytoplasm, and interaction with the target can only take place after translocation of the precursors to the outer surface of the membrane (Fig. 1) (22, 107). Gram-negative bacteria are insensitive to glycopeptides, since they possess an impermeable outer membrane. Glycopeptides bind by five hydrogen bonds to the D-Ala-D-Ala C terminus of precursors containing the pentapeptide moiety with high affinity (Fig. 2), blocking the addition of pentapeptide precursors by transglycosylation to the nascent peptidoglycan chain and preventing subsequent cross-linking catalyzed by D,D-transpeptidases (Fig. 1) (22, 121). Consequently, there is an accumulation of cytoplasmic precursors.

MECHANISMS OF RESISTANCE

Glycopeptides do not interact with cell wall biosynthetic enzymes but form complexes with peptidoglycan precursors and prevent their incorporation into the wall (Figs. 1 and 2) (22, 121). Consequently, the activity of glycopeptides is not determined by the affinity of target enzymes for the molecules but by the substrate specificity of the enzymes that determine the structure of peptidoglycan precursors. Glycopeptide resistance is due to the presence of gene clusters that

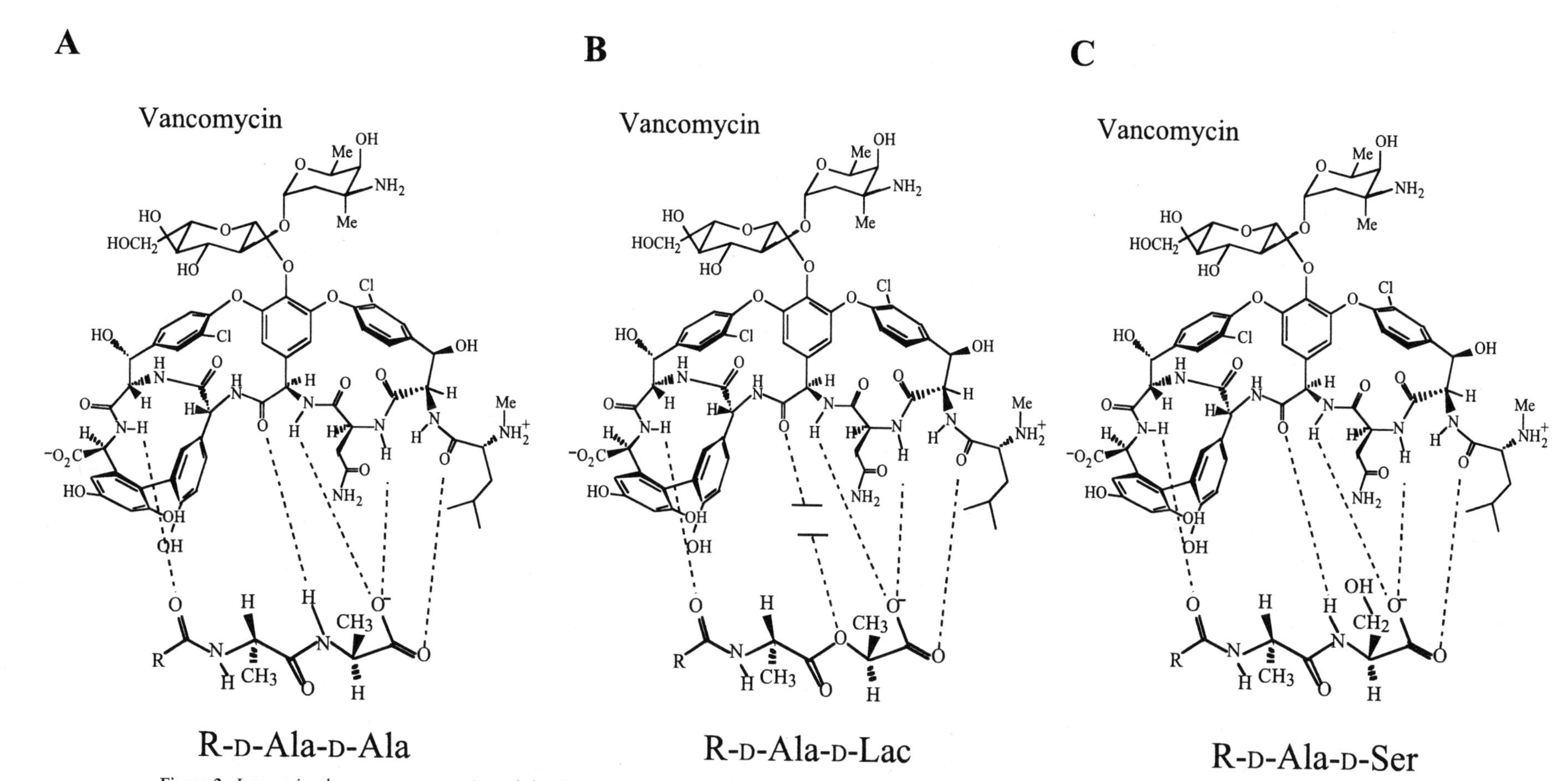

Figure 2. Interaction between vancomycin and the C-terminal end of peptidoglycan precursors. Binding of vancomycin to the (A) D-Ala-D-Ala extremity; (B) D-Ala-D-Lac extremity (substitution of the NH group by an oxygen prevents the formation of the central hydrogen bond); (C) D-Ala-D-Ser extremity (substitution of a CH_3 group by a CH_2OH group induces steric hindrance to binding).

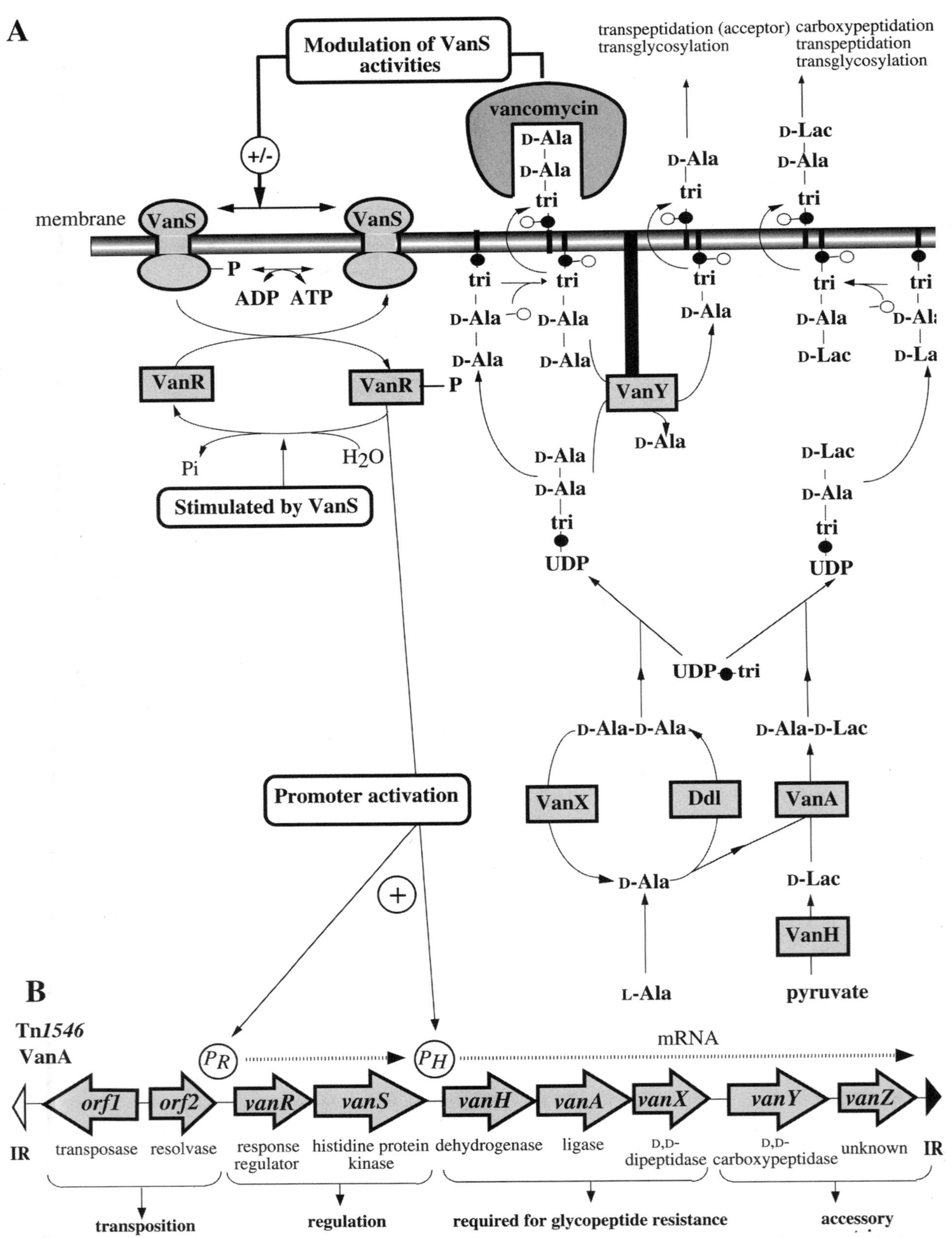

Figure 3. VanA-type glycopeptide resistance and map of Tn*1546*. (A) Schematic representation of the synthesis of peptidoglycan precursors in the VanA-type BM4147 resistant strain. (B) Organization of the *vanA* operon. Open arrows represent coding sequences and indicate the direction of transcription. Open and closed arrowheads labeled IR_L and IR_R indicate the left and right inverted repeats of the transposon. The regulator *vanR* and *vanS* and resistance genes *vanH, vanA, vanX, vanY,* and *vanZ* are cotranscribed from promoters P_R and P_H, respectively.

encode enzymes (i) for synthesis of low-affinity precursors in which the C-terminal D-Ala residues are replaced by a D-lactate (D-Lac) (16, 30) or D-serine (D-Ser) (124) and (ii) for elimination of the high-affinity precursors normally produced by the host (Fig. 3) (12).

Target Modification

VanA-type resistance, characterized by inducible resistance to high levels of vancomycin and teicoplanin, was the first to be described and is mediated by transposon Tn*1546* (Fig. 3B) (15) or closely related elements. The transposon encodes a dehydrogenase (VanH) that reduces pyruvate to D-Lac and a ligase of broad substrate specificity (VanA) that catalyzes the formation of an ester bond between D-Ala and D-Lac (Fig. 2B and Fig. 3A) (17). The resulting D-Ala-D-Lac depsipeptide replaces the dipeptide D-Ala-D-Ala in the pathway of peptidoglycan synthesis (Fig. 3A). The substitution eliminates a hydrogen bond critical for antibiotic binding (Fig. 2B) and considerably decreases the affinity for glycopeptides (30). The normal complement of cell wall biosynthetic enzymes encoded by the host chromosome tolerates the substitution of D-Lac for D-Ala (123). However, PBPs that catalyze the final transpeptidation reactions in the assembly pathway may display different activity for D-Ala and D-Lac-terminating precursors which could account for modification of the extent of cross-linking (27, 41). Induction of glycopeptide resistance is associated with an increase in susceptibility to β-lactams in certain strains of enterococci (2, 20).

The VanC resistance phenotype was described first in *E. gallinarum* (49, 88) and then in *E. casseliflavus* (106), which possess intrinsic low-level resistance to vancomycin and are susceptible to teicoplanin (Table 1). Three genes, *vanT, vanC,* and *vanXY*$_C$, are required for VanC-type resistance (Fig. 4) (4, 122). *vanT* encodes a membrane-bound serine racemase, VanT, which produces D-Ser (5, 6). The *vanC* gene product synthesizes D-Ala-D-Ser, which replaces D-Ala-D-Ala in late peptidoglycan precursors (124). Substitution of the ultimate D-Ala by a D-Ser does not alter hydrogen bonding of vancomycin to the target, but the replacement of the methyl side chain by hydroxylmethyl results in steric hindrance, which reduces slightly the affinity for vancomycin (Fig. 2C) (26).

Removal of the Susceptible Target

Coproduction of precursors ending in D-Ala or D-Lac does not result in resistance (12). Under these conditions, binding of glycopeptides to D-Ala-D-Ala-containing precursors bound to the lipid carrier at the external surface of the membrane would sequester the lipid carrier and thereby prevent translocation of additional precursors, including those terminated by D-Lac (Fig. 3). The interaction of a glycopeptide with its normal target is prevented by the removal of precursors terminating in D-Ala (120). Two enzymes are involved in this process (Fig. 3): a cytoplasmic D,D-dipeptidase (VanX) that hydrolyzes the dipeptide D-Ala-D-Ala synthesized by the host Ddl ligase (123)

Table 1. Glycopeptide resistance in enterococci

Resistance	Acquired					Intrinsic	
	High level	Variable level	Moderate level	Low level		Low level	High level
Type	VanA	VanB	VanD	VanG	VanE	VanC1/C2/C3	
MIC (mg/liter)							
Vancomycin	64–1,000	4–1,000	64–128	16	8–32	2–32	≥1,000
Teicoplanin	16–512	0.5–1	4–64	0.5	0.5	0.5–1	≥256
Conjugation	+	+	–	–	+	–	
Mobile element	Tn*1546*	Tn*1547* Tn*1549*					
Species	*E. faecium* *E. faecalis* *E. gallinarum* *E. casseliflavus* *E. avium* *E. durans* *E. mundtii* *E. raffinosus*	*E. faecium* *E. faecalis* *S. bovis*	*E. faecium* *E. faecalis*	*E. faecalis*	*E. faecalis*	*E. gallinarum* *E. casseliflavus* *E. flavescens*	*Leuconostoc* *Lactococcus* *Pediococcus*
Expression	Inducible	Inducible	Constitutive	Inducible	Inducible	Constitutive, inducible	Constitutive
Location	Plasmid, chromosome	Plasmid, chromosome	Chromosome	Chromosome	Chromosome	Chromosome	
Modified target	D-Ala-D-Lac	D-Ala-D-Lac	D-Ala-D-Lac	D-Ala-D-Ser	D-Ala-D-Ser	D-Ala-D-Ser	D-Ala-D-Lac

A

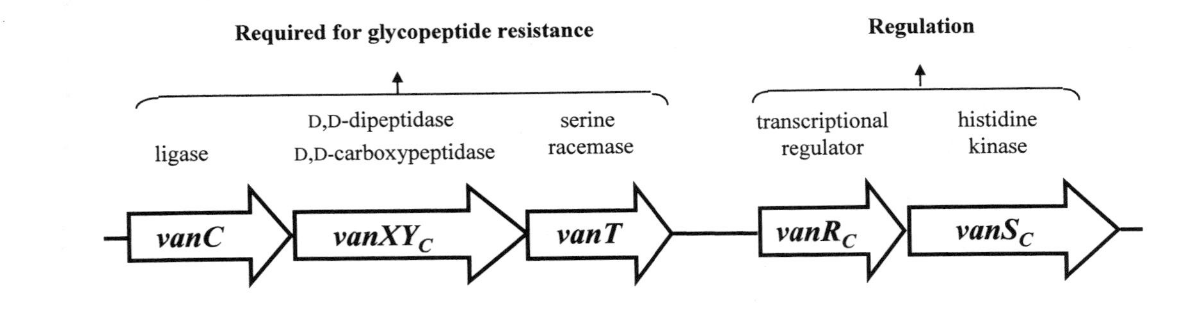

B

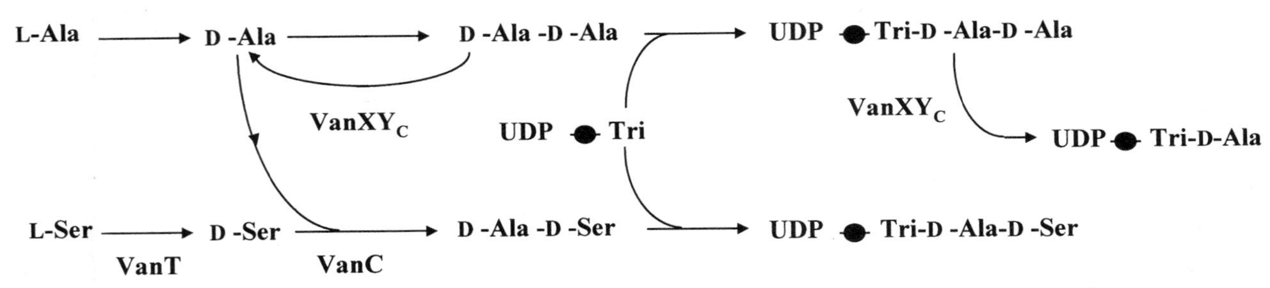

Figure 4. VanC-type glycopeptide resistance. (A) Organization of the *vanC* operon. Open arrows represent coding sequences and indicate the direction of transcription. (B) Schematic representation of the synthesis of peptidoglycan precursors in the VanC-type BM4174 strain. Tri, L-Ala-γ-D-Glu-L-Lys.

and a membrane-bound D,D-carboxypeptidase (VanY) that removes the C-terminal D-Ala residue of late peptidoglycan precursors when elimination of D-Ala-D-Ala by VanX is incomplete (7, 13).

In contrast to VanA-type resistance, in which the VanX and VanY activities are encoded by two genes (7), $VanXY_C$ has both D,D-dipeptidase and D,D-carboxypeptidase activity and thus catalyzes hydrolysis of the dipeptide D-Ala-D-Ala and removal of the terminal D-Ala from pentapeptide [D-Ala] (Fig. 4B) (122).

TYPES OF RESISTANCE

Six types of vancomycin-resistant enterococci have been characterized on phenotypic and genotypic bases as summarized in Table 1: five types possess acquired resistance (VanA, B, D, E, and G) and one type, VanC, is an intrinsic property of *E. gallinarum, E. casseliflavus,* and *E. flavescens.* Classification of glycopeptide resistance is now based on the primary sequence of the structural genes for the resistance ligases rather than on the levels of resistance to glycopeptides, since the MIC ranges of vancomycin and teicoplanin against the various types overlap (Table 1). VanA-type strains display high-level inducible resistance to both vancomycin and teicoplanin, whereas VanB-type strains have variable levels of inducible resistance to vancomycin only, since teicoplanin is not an inducer (Table 1) (12, 119). VanD-type strains are characterized by constitutive resistance to moderate levels of the two glycopeptides (Table 1) (45). VanC, VanE, and VanG are resistant to low levels of vancomycin but remain susceptible to teicoplanin (Table 1) (55, 88, 101).

Although all six types of resistance involve genes encoding related enzymatic functions, they can be distinguished by the location of the genes and by the various modes of regulation of gene expression (Fig. 5). The *vanA* and *vanB* operons are located on plasmids or in the chromosome (17), whereas the *vanD* (32, 45), *vanG* (42), *vanE* (1), and *vanC* (4) operons have so far been found exclusively in the chromosome.

VanA

This is the most frequently encountered type of glycopeptide resistance in enterococci (Table 1). The prototype VanA resistance element, Tn*1546*, was originally detected on plasmid pIP816 from *E. faecium* BM4147, and is a 10,851-bp Tn*3*-related transposon (Fig. 3B) (15). Tn*1546* encodes nine polypeptides

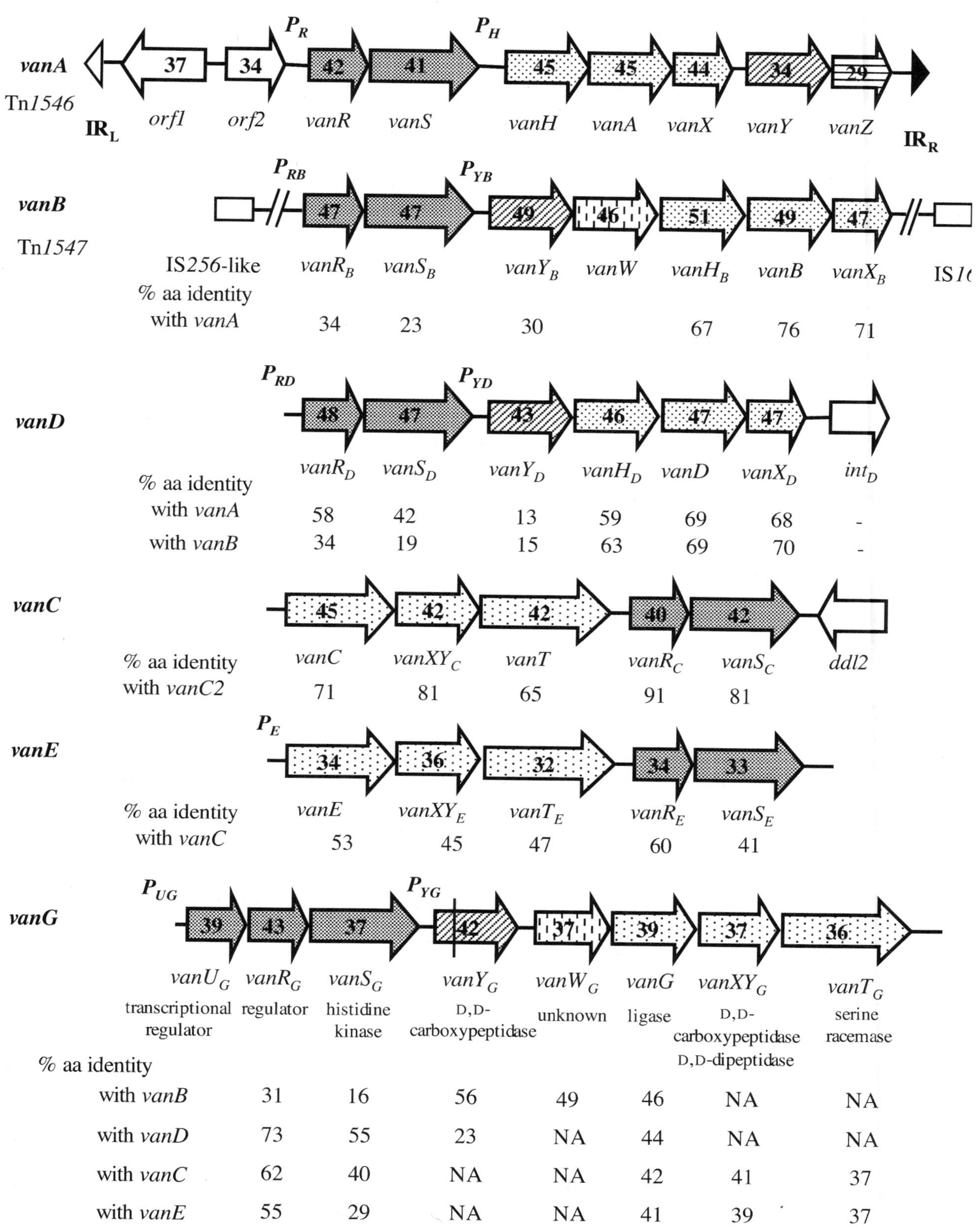

Figure 5. Comparison of the *van* gene clusters. Open arrows represent coding sequences and indicate the direction of transcription. The guanosine plus cytosine content (% GC) is indicated in the arrows. The percentage of amino acid (aa) identity between the deduced proteins of reference strains BM4147 (VanA) (15), V583 (VanB) (53), BM4339 (VanD) (32), BM4174 (VanC) (4), BM4405 (VanE) (1), and BM4518 (VanG) (42) is indicated under the arrows. Transposon Tn*1546* (10851 bp) carries the *vanA* operon and is delineated by 38-bp imperfect inverted repeats represented by arrowheads (IR_L and IR_R). Transposon Tn*1547* (65 kb) carries the *vanB* gene cluster and is delineated by insertion sequences IS*256*-like and IS*16* in direct orientation (represented by boxes). The vertical bar in $vanY_G$ indicates the frameshift mutation leading to a predicted truncated protein.

(Fig. 3B) that can be assigned to different functional groups: transposition functions (ORF1 and ORF2), regulation of vancomycin resistance genes (VanR and VanS), synthesis of depsipeptide D-Ala-D-Lac (VanH and VanA), and hydrolysis of precursors of peptidoglycan (VanX and VanY); the function of VanZ remains unknown. The *vanR, vanS, vanH, vanA,* and *vanX* genes are necessary and sufficient for inducible resistance (7, 9). Proteins VanY and VanZ are not essential, but their production increases the level of resistance to vancomycin and teicoplanin, respectively (10, 13). VanY is a penicillin-insensitive Zn^{2+}-dependent D,D-carboxypeptidase exhibiting a higher catalytic efficiency for hydrolysis of substrates ending in D-Ala-D-Ala than for D-Ala-D-Lac (7). VanZ alone confers teicoplanin resistance by an unknown mechanism that does not involve incorporation of a substituent of D-Ala-D-Ala into peptidoglycan precursors (10).

The VanR and VanS proteins constitute a two-component regulatory system that regulates the transcription of the *vanHAXYZ* gene cluster (Figs. 3 and 5) (14). This system is composed of a cytoplasmic VanR response regulator that acts as a transcriptional activator and a VanS histidine kinase which is associated with the membrane (Fig. 3) (148).

The *vanA* gene cluster has been reported mainly in *E. faecium* and *E. faecalis* but also in *E. avium* (127), *E. durans* (35, 70, 138), and *E. raffinosus* (36). Atypical isolates of *E. gallinarum* and *E. casseliflavus* that are highly resistant to both vancomycin and teicoplanin contained the *vanA* gene cluster in addition to the *vanC-1* or *vanC-2* operons characteristic of these species (25, 48).

VanB

As in VanA-type strains, acquired VanB-type resistance is due to the synthesis of peptidoglycan precursors ending in the depsipeptide D-Ala-D-Lac instead of the dipeptide D-Ala-D-Ala (Fig. 3A) (12, 102). The organization and functionality of the *vanB* cluster is similar to that of *vanA* (Fig. 5) but differs in its regulation, since vancomycin but not teicoplanin is an inducer (Table 1) (12, 119). The *vanB* operon (Fig. 5) contains the $vanH_BBX_B$ resistance genes encoding, respectively, a dehydrogenase, a ligase, and a dipeptidase that have a high level of amino acid sequence identity (67 to 76% identity) with the corresponding deduced proteins of the *vanA* operon and the $vanR_BS_B$ regulatory genes encoding a two-component regulatory system only distantly related to VanRS (34 and 24% identity) (53). The $VanY_B$ D,D-carboxypeptidase with low homology to VanY (30% identity) is present, but the position of its structural gene between the regulatory genes and $vanH_BBX_B$ differs from that for *vanY* in Tn*1546*. The function of the additional VanW protein found only in the *vanB* cluster between $VanY_B$ and $VanH_B$ is unknown (Fig. 5) (53). There is no gene related to *vanZ* (Fig. 5).

On the basis of sequence differences in the $vanS_B$-$vanY_B$ intergenic region and adjacent coding sequences, the *vanB* gene cluster can be divided into three subtypes: *vanB1* (54), *vanB2* (38, 64, 111), and *vanB3* (38, 111). There is no correlation between *vanB* subtype and the level of vancomycin resistance.

As in VanA-type strains, increased transcription of the $vanY_BWH_BBX_B$ operon is associated with increased incorporation of D-Ala-D-Lac into peptidoglycan precursors, to the detriment of D-Ala-D-Ala, and with a gradual increase in vancomycin resistance levels (12). Levels of D,D-dipeptidase activity ($VanX_B$) correlate with levels of vancomycin resistance (12). More complete elimination of D-Ala-D-Ala-containing precursors is required for teicoplanin resistance (12).

VanD

Acquired VanD-type resistance is due to constitutive production of peptidoglycan precursors with the majority ending in D-Ala-D-Lac (Table 1 and Fig. 6) (45, 113, 114). The organization of the *vanD* operon (Fig. 5), located exclusively in the chromosome in strains that have been studied, is similar to those of *vanA* and *vanB* (44, 45). However, no genes homologous to *vanZ* or *vanW* from the *vanA* and *vanB* operons, respectively, are present (Fig. 5). VanD-type strains share other characteristics that distinguish them from VanA- and VanB-type enterococci; in particular, resistance is constitutively expressed and is not transferable by conjugation to other enterococci (45, 113, 114).

VanD-type strains have negligible D,D-dipeptidase activity, encoded by $vanX_D$ alleles, despite the presence in $VanX_D$ of the critical residues implicated in the binding of Zn^{2+} and in catalysis (33, 45). Lack of such an activity should result in a glycopeptide-susceptible phenotype, since these bacteria are unable to eliminate peptidoglycan precursors ending in D-Ala-D-Ala, the target for glycopeptides. However, in VanD-type strains, the susceptible pathway does not function due to an inactive D-Ala:D-Ala ligase as the result of various mutations in the chromosomal *ddl* gene (Fig. 6) (45). The *ddl* gene is disrupted by a 5-bp insertion in BM4339 (32), by an insertion of the IS*19* (also called IS*Efm1*) element in BM4416 (29, 113), and by a single mutation $G_{184}S$ next to the serine involved in the binding of D-Ala1 in strain 10/96A (45). Consequently the strains should only grow in the presence of vancomycin since they rely on the inducible resistance

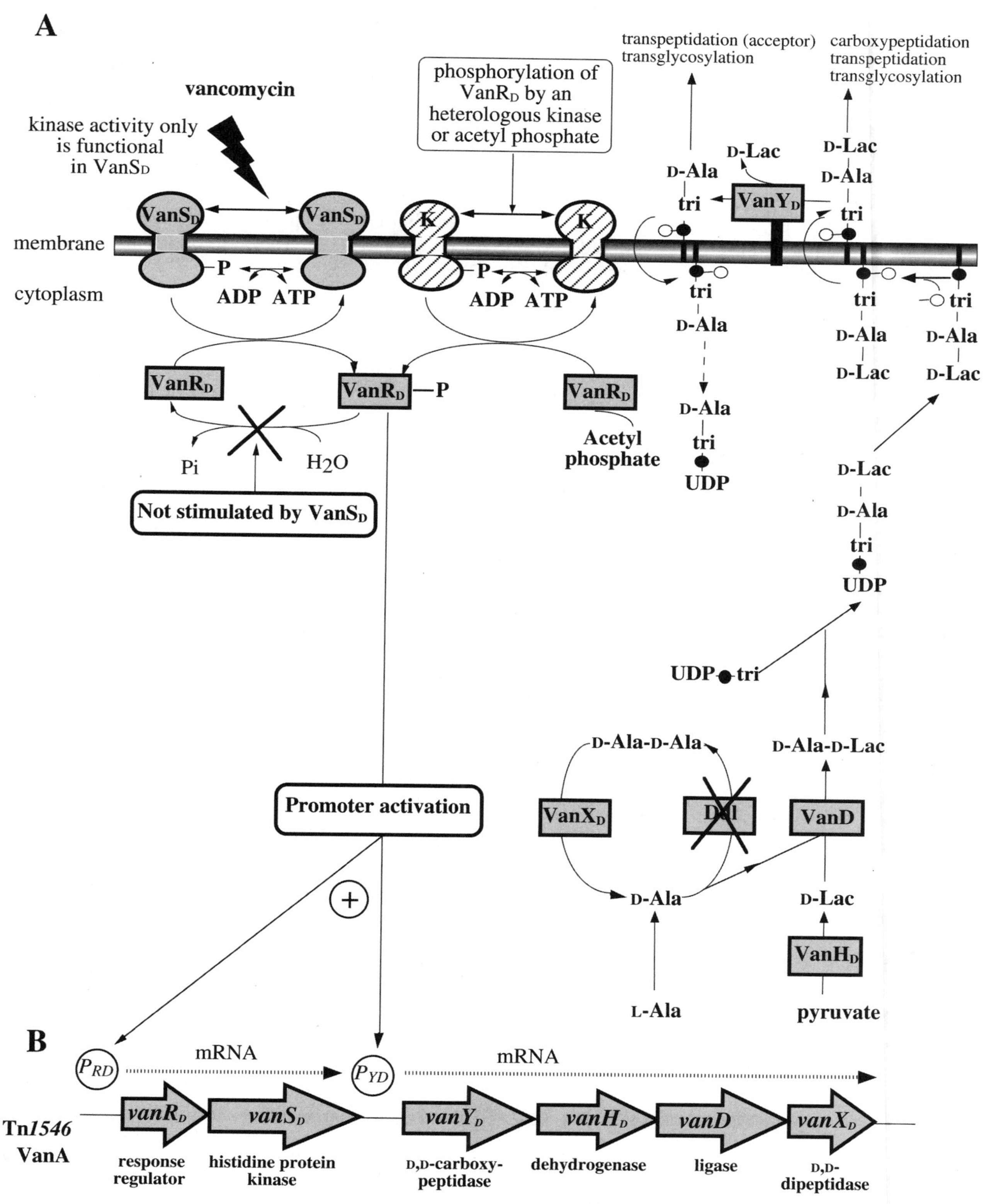

Figure 6. VanD-type glycopeptide resistance. (A) Schematic representation of the synthesis of peptidoglycan precursors in a VanD-type resistant strain. (B) Organization of the *vanD* operon. Open arrows represent coding sequences and indicate the direction of transcription. Regulatory $vanR_D$ and $vanS_D$ and resistance genes $vanY_D$, $vanH_D$, *vanD*, and $vanX_D$ are cotranscribed from promoters P_{RD} and P_{YD}, respectively. ●, N-acetylmuramic acid; ○, N-acetylglucosamine; tri, L-Ala-g-D-Glu-L-Lys; |, undecaprenylpyrophosphate (lipid carrier).

pathway for peptidoglycan synthesis. However, this is not the case since the *vanD* clusters are expressed constitutively due to mutations either in the $VanS_D$ sensor (Fig. 6) or, as determined recently, in the $VanR_D$ regulator (44). Strain BM4339 has a mutation at position 173 in the $VanS_D$ sensor leading to a Pro to Ser substitution in a critical region, since it alters a residue close to histidine 166, which corresponds to the putative autophosphorylation site of $VanS_D$ (45). N-97-330 (also designated BM4416) has suffered a 1-bp deletion at position 670 (BM4339 numbering), which results in a frameshift mutation presumably leading to the synthesis of a 233-amino acid truncated and nonfunctional sensor instead of a protein containing 381 amino acids as in BM4339 (29). Strain 10/96A contains a different type of mutational event since insertion sequence IS*Efa4* was found 45-bp downstream from the start site of the $vanS_D$ gene (45).

The $VanY_D$ D,D-carboxypeptidases belong to the PBP family of catalytic serine enzymes that are susceptible to benzylpenicillin (32, 45, 113) and are distinct from VanY and $VanY_B$, which are penicillin-insensitive Zn^{2+}-dependent D,D-carboxypeptidases. Another unusual feature of VanD-type strains is their only slightly diminished susceptibility to teicoplanin (MIC, 4 μg/ml) (Table 1) despite constitutive production of peptidoglycan precursors that terminate mainly in D-Ala-D-Lac. Recently, three new VanD-type strains from Australia have been characterized: one *E. faecium* and for the first time two *E. faecalis* (44). The three strains were distinguished by their varied assortment of mutations, confirming that all VanD-type strains isolated so far have suffered mutations in the *ddl* housekeeping and in the acquired $vanS_D$ or $vanR_D$ acquired genes that lead to constitutive resistance to vancomycin (44). Bacteria that constantly activate the *vanD* operon, by mutation in the two-component regulatory system, and that have eliminated the susceptible pathway, by inactivation of the Ddl ligase, do not require $VanX_D$ activity (Fig. 6). Whatever the actual sequence of mutational events may be, VanD-type strains provide a remarkable example of tinkering of both intrinsic and acquired genes to achieve higher levels of antibiotic resistance.

VanC

Enterococci belonging to the *E. gallinarum, E. casseliflavus,* and *E. flavescens* species are intrinsically resistant to low levels of vancomycin but remain susceptible to teicoplanin (Table 1). The VanC phenotype is expressed constitutively or inducibly (130), at least in some strains of *E. gallinarum,* due to the production of peptidoglycan precursors ending in D-Ser (26, 69, 124). Three *vanC* genes encoding D-Ala:D-Ser ligases have been described: *vanC-1* in *E. gallinarum, vanC-2* in *E. casseliflavus,* and *vanC-3* in *E. flavescens* (88, 106). The organization of the *vanC* operon (Fig. 5), which is chromosomally located and not transferable, is distinct from those of *vanA, vanB,* and *vanD.* Three proteins are required for VanC-type resistance: VanT, a membrane-bound serine racemase which produces D-Ser; VanC, a ligase which catalyzes the synthesis of D-Ala-D-Ser, and $VanXY_C$, which possesses both D,D-dipeptidase and D,D-carboxypeptidase activities and allows hydrolysis of precursors ending in D-Ala (Fig. 4) (4, 122).

$VanXY_C$ has a cytoplasmic location and contains consensus sequences for binding zinc, stabilizing the binding of the substrate and catalyzing hydrolysis that are present in both VanX- and VanY-type enzymes of VanA- and VanB-type strains (122). The protein has very low dipeptidase activity against D-Ala-D-Ser, unlike VanX, and no activity against UDP-MurNAc-pentapeptide [D-Ser] (122). Sequence comparison indicates that $VanXY_C$ is more closely related to VanY- than to VanX-type proteins but lacks the membrane spanning segment present in both VanY and $VanY_B$. VanT possesses serine and alanine racemase activities, and the transmembrane domain of this protein is probably also involved in the uptake of L-Ser from the external medium (6). In the *vanA, vanB,* and *vanD* clusters the genes encoding the two-component regulatory systems (VanRS, $VanR_BS_B$, or $VanR_DS_D$) are located upstream from the resistance genes, whereas in the *vanC* cluster they are downstream from *vanT* (Fig. 5) (4). An additional gene, *ddl2,* located downstream from the regulatory genes $vanR_CS_C$ and encoding a protein that has structural similarity with D-Ala:D-Ala ligases, was found in the VanC prototype strain BM4174 (Fig. 5) (3). Thus *E. gallinarum* produces at least three ligases: two for synthesis of D-Ala-D-Ala and one for synthesis of D-Ala-D-Ser. The deduced proteins of the *vanC-2* operon from *E. casseliflavus* display high degrees of identity (from 71 to 91%) with those encoded by the *vanC* operon (50), and those of the *vanC-3* gene cluster from *E. flavescens* display extensive identity with *vanC-2,* from 97 to 100%, including the intergenic regions (51). It is therefore difficult to distinguish *E. casseliflavus* and *E. flavescens* as two different species (51).

VanE

The VanE phenotype reported in two isolates of *E. faecalis* (28, 55) has low-level resistance to vancomycin and susceptibility to teicoplanin. This is due to synthesis of peptidoglycan precursors terminating in D-Ala-D-Ser as in intrinsically resistant *Enterococcus* spp. (Table 1). The *vanE* cluster has identical organization to that of the *vanC* operon (Fig. 5) (1, 28). The $vanR_ES_E$ regulatory genes are located downstream from the resistance genes which, in addition to the *vanE* ligase gene, include $vanXY_E$ and $vanT_E$, which encode a bifunctional D,D-dipeptidase-D,D-carboxypeptidase and a serine racemase, respectively (Fig. 5). Expression of vancomycin resistance is inducible in the VanE prototype strain BM4405, although the VanS sensor is likely to be inactive due to the presence of a stop codon in the 5′ portion of the gene (1). Cross talk with another kinase of the host could explain the resistance inducibility. Recently, four new VanE-type *E. faecalis* have been isolated in Australia (85).

VanG

Acquired VanG-type resistance was detected in four isolates of *E. faecalis* from Brisbane in Australia and is characterized by resistance to low levels of vancomycin (MIC, 16 μg/ml) but susceptibility to teicoplanin (MIC, 0.5 μg/ml) (101) and by the inducible synthesis of peptidoglycan precursors ending in D-Ala-D-Ser (42). The chromosomal *vanG* cluster is composed of seven genes recruited from various *van* operons (Fig. 5) (42, 101). The 3′ end encodes VanG, a D-Ala:D-Ser ligase; $VanXY_G$, a putative bifunctional D,D-peptidase; and $VanT_G$, a serine racemase. In deduced VanG, the EKY motif specific for D-Ala:D-Ser ligases was present (42). Upstream from the structural genes for these proteins are $vanW_G$, of unknown function, having 49% identity with *vanW* from the *vanB* operon (Fig. 5), and $vanY_G$, containing a frameshift mutation which results in premature termination and accounts for the lack of UDP-MurNAc-tetrapeptide in the cytoplasm (42). In the absence of the frameshift mutation, $VanY_G$ has homology with Zn^{2+} dependent D,D-carboxypeptidases, the highest identity (56%) being with $VanY_B$ D,D-carboxypeptidase (Fig. 5). In contrast to the other *van* operons, the 5′ end of the gene cluster contains three genes, $vanU_G$, $vanR_G$, and $vanS_G$, encoding a putative regulatory system (Fig. 5) (42). $vanR_G$ and $vanS_G$ have the highest homology with $vanR_D$ and $vanS_D$ (Fig. 5). Interestingly, the additional $vanU_G$ gene encodes a predicted transcriptional activator (42). A protein of this type has not previously been associated with glycopeptide resistance.

Glycopeptide-Dependent Strains

An interesting phenomenon that has developed in some VanA- and VanB-type enterococci is vancomycin dependence (Fig. 7). These strains are not only resistant to vancomycin, or to both vancomycin and teicoplanin, but also require their presence for growth. Variants of glycopeptide-resistant *E. faecalis* and *E. faecium* that grow only in the presence of glycopeptides have been isolated in vitro (20), in animal models (18), and from patients treated for long periods of time with vancomycin (46, 59, 68, 141, 145). A glycopeptide-dependent mutant of a strain of *E. avium* has also been reported (127, 134). These are all derivatives from enterococci of the VanA- (127, 134) or VanB-type (46, 59, 68, 141, 145). These glycopeptide-dependent strains are also able to grow in the absence of glycopeptides if supplied with the dipeptide D-Ala-D-Ala, suggesting that they are unable to produce the ligase encoded by the chromosomal *ddl* gene (Fig. 7). In the presence of vancomycin, the *vanA*- or *vanB*-encoded D-Ala:D-Lac ligase is induced, which overcomes the defect of synthesis of peptidoglycan precursors ending in D-Ala-D-Ala due to lack of a functional Ddl following various mutations and thus permits growth of the bacteria (20, 141). Since these strains require particular growth conditions, the prevalence of vancomycin-dependent enterococci is probably underestimated in routine laboratories. Reversion to vancomycin independence has been observed and occurs either by a mutation that leads to constitutive production of D-Ala-D-Lac and is consequently resistant to teicoplanin or to a mutation that restores the synthesis of D-Ala-D-Ala leading to VanB phenotype inducible by vancomycin (20, 46, 141).

GLYCOPEPTIDE RESISTANCE IN OTHER GRAM-POSITIVE BACTERIA

Lactic acid bacteria, including certain species belonging to the genera *Lactobacillus, Lactococcus, Leuconostoc,* and *Pediococcus* are intrinsically resistant to

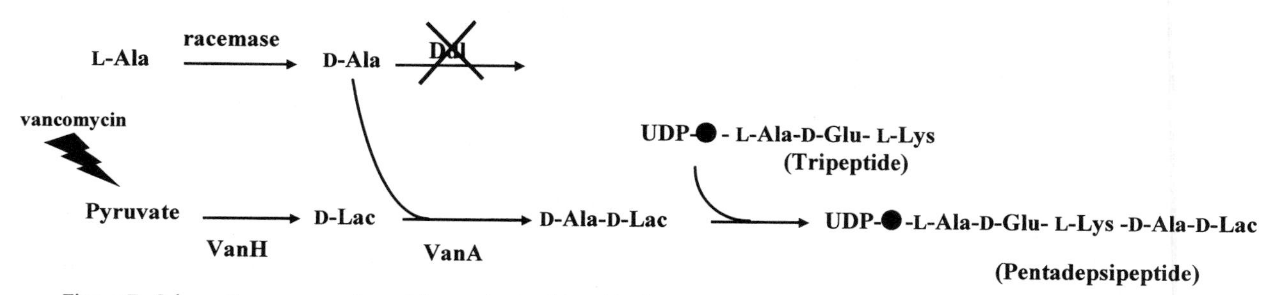

Figure 7. Schematic representation of the synthesis of peptidoglycan precursors in a vancomycin-dependent strain. Due to inactivation of the indigenous D-Ala:D-Ala ligase (Ddl), vancomycin is required in the culture medium to induce expression of the resistance pathway, thus allowing bacterial growth.

high levels of vancomycin and teicoplanin by synthesis of peptidoglycan precursors ending exclusively in D-Lac, as demonstrated in *Lactobacillus casei, Leuconostoc mesenteroides,* and *Pediococcus pentosaceus* (Table 1) (26, 73). These are opportunistic pathogens rarely responsible for infections. They contain a ligase that catalyzes the formation of D-Ala-D-Lac but do not produce D-Ala-D-Ala (62). Resistance to glycopeptides has been reported in clinical isolates of *Erysipelothrix rhusiopathiae,* an organism that may infect humans who have been exposed to animals or contaminated animal products (65). The *vanA* gene cluster was also found in a clinical isolate of *Bacillus circulans* (92).

In the laboratory, conjugal transfer of VanA-type vancomycin resistance genes from enterococci to other gram-positive cocci has been obtained. Recipient organisms in successful in vitro transfer have included group A and viridans group streptococci, and *Listeria monocytogenes* (24, 87). Transfer of resistance genes to *Staphylococcus aureus,* resulting in high levels of resistance to vancomycin, was obtained in vitro (108) and in an animal model. Most importantly, this transfer also occurs in vivo. Recently, three methicillin-resistant *S. aureus* strains with high or moderate levels of resistance to vancomycin and teicoplanin were isolated in Michigan, Pennsylvania, and New York after acquisition of the *vanA* gene cluster (82, 133, 137, 142). In each strain, the VanA-carrying genetic element Tn*1546* was found to be part of a plasmid (56, 137, 142). For the *S. aureus* with high resistance to vancomycin, it has been shown that vancomycin resistance was due to acquisition of a plasmid from an *E. faecalis* strain isolated from the same patient (56, 142).

A *van* cluster, designated *vanF* and homologous with the *vanA* cluster, was found in the vancomycin-resistant biopesticide *Paenibacillus popilliae* (110). This operon is composed of five genes encoding homologues of VanY, VanZ, VanH, VanA, and VanX (110). Orientation and alignment of the genes essential for resistance (*vanH/vanH$_F$, vanA/vanF,* and *vanX/vanX$_F$*) are identical in VanF and VanA. A *vanB*-related gene sequence has also been found in *Streptococcus bovis* (116). More recently, several anaerobic bacteria of the genus *Clostridium* carrying the *vanB* gene have been isolated from human stools (136).

REGULATION OF GLYCOPEPTIDE RESISTANCE

Expression of VanA-, VanB-, VanD-, VanE-, VanC-, and VanG-type resistance is regulated by a VanS/VanR-type two-component signal transduction system, composed of a membrane-bound histidine kinase (VanS, VanS$_B$, VanS$_D$, VanS$_E$, VanS$_C$, or VanS$_G$) and a cytoplasmic response regulator (VanR, VanR$_B$, VanR$_D$, VanR$_E$, VanR$_C$, or VanR$_G$) that acts as a transcriptional activator (Figs. 5 and 8) (1, 4, 14, 32, 42, 45, 53). In the *vanA, vanB, vanD,* and *vanG* operons, the genes encoding the two-component regulatory system (*vanRS, vanR$_B$S$_B$, vanR$_D$S$_D$, vanR$_G$S$_G$*) are present upstream from the structural genes for the resistance proteins (Fig. 5) (17, 32, 45), whereas in the *vanC* and *vanE* clusters, *vanR$_C$S$_C$* and *vanR$_E$S$_E$* are located downstream (Fig. 5) (1, 4).

VanS-type sensors are thought to comprise an N-terminal glycopeptide sensor domain, with two membrane-spanning segments and a C-terminal cytoplasmic kinase domain (Fig. 8A) (14, 148). Following a signal related to the presence of a glycopeptide in the culture medium, the cytoplasmic domain of VanS catalyzes ATP-dependent autophosphorylation on a specific histidine residue and transfers the phosphate group to an aspartate residue of VanR present in the effector domain (Fig. 8) (14, 148). VanS also stimulates dephosphorylation of VanR (8, 9, 148).

Purified VanS$_B$ autophorylates itself in the presence of ATP and acts as both a histidine protein kinase and a VanR$_B$ phosphoprotein phosphatase in vitro (43). The VanS sensor therefore modulates the level of phosphorylation of the VanR regulator (8, 9): it acts primarily as a phosphatase under noninducing conditions and as a kinase in the presence of glycopeptides, leading to phosphorylation of the response regulator and activation of the resistance genes (Fig. 8B) (8, 9, 43, 79). The phosphatase activity of VanS is required for negative regulation of resistance genes in the absence of glycopeptides preventing accumulation of phospho-VanR phosphorylated by acetyl phosphate or by kinases encoded by the host chromosome (Figs. 3 and 8B) (8).

The regulatory and resistance genes are transcribed from distinct promoters (Fig. 5) P_R, P_{RB}, P_{RD}, and P_H, P_{YB}, P_{YD}, respectively, that are coordinately regulated (Fig. 5) (8, 9, 33, 43, 53). Phosphorylation of VanR-type regulators enhances the affinity of the regulatory regions for the promoters and stimulates transcription of the regulatory and resistance genes of the *van* cluster (Fig. 3) (79). As an example, in VanA-type strains the promoters are not activated in the absence of VanR and VanS, are induced by glycopeptides when VanR and VanS are present, and are constitutively activated by VanR in the absence of VanS, due presumably to phosphorylation of VanR by host kinases (8, 9). The *vanC* cluster of *E. gallinarum* BM4174 is expressed constitutively, and two regions upstream from *vanC* and *vanR$_C$* were postulated as po-

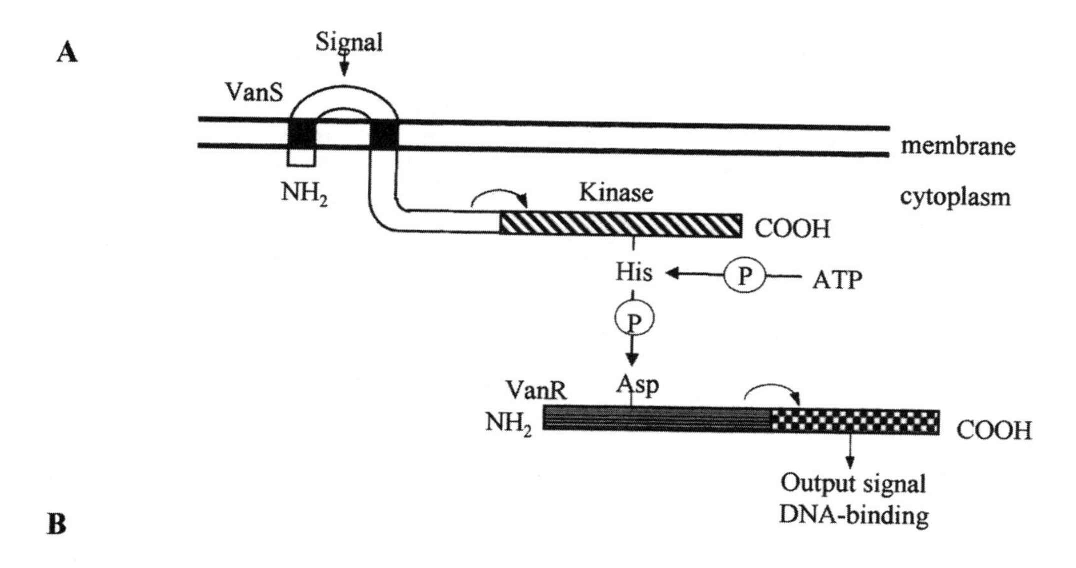

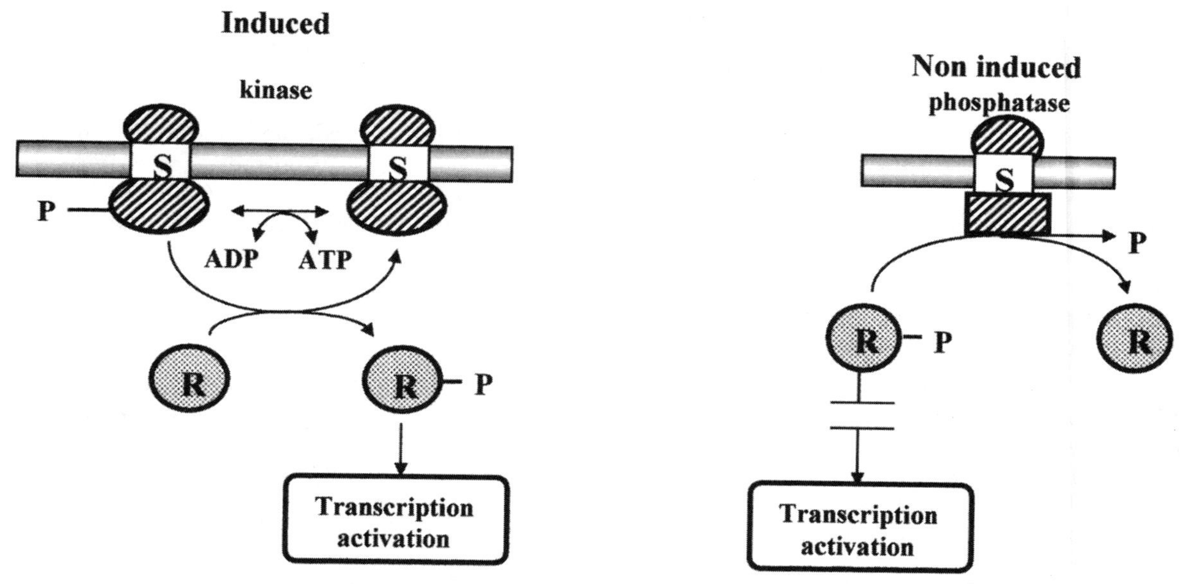

Figure 8. Schematic representation of, and regulation by, the VanRS two-component regulatory system. (A) Structure of VanR and VanS proteins. Asp, aspartate; His, histidine; P, phosphate. (B) Model for positive (phosphorylation) and negative (dephosphorylation) control of VanR by VanS.

tential promoters (Fig. 5) (4). However, the five genes of the *vanE* operon in *E. faecalis* BM4405 are cotranscribed from a single promoter upstream from *vanE* (Fig. 5) (1).

In contrast to the other *van* gene clusters, the *vanG* operon contains three genes, $vanU_G$, $vanR_G$, and $vanS_G$, encoding a putative regulatory system which are cotranscribed constitutively from the P_{UG} promoter, whereas transcription of $vanY_G$, W_G, G, XY_G, T_G is inducible and initiated from the P_{YG} promoter (Fig. 5) (42). Among Van-type strains, this is the first operon to be regulated in this manner.

ACQUISITION OF TEICOPLANIN RESISTANCE BY VanB-TYPE ENTEROCOCCI

Enterococci harboring clusters of the *vanB* class remain susceptible to teicoplanin, since this antibiotic is not an inducer. However, mutations in the $vanS_B$ sensor gene have been obtained in vitro, following selection on teicoplanin, that have led to three phenotypic classes (constitutive or teicoplanin-inducible expression of the resistance genes, or heterogeneous expression) due to three distinct alterations of $VanS_B$ function (20) and in vivo in animal models (18). Mutations leading to teicoplanin resistance also confer low-level resistance to the glycopeptide oritavancin (formerly LY333328) (11). Derivatives of VanB-type strains resistant to teicoplanin have been isolated from two patients following treatment with vancomycin (78) or teicoplanin (83), but the isolates were not studied further.

Inducible Phenotype

Mutations leading to inducible expression of resistance by vancomycin and by teicoplanin result from amino acid substitutions in the sensor domain of $VanS_B$ (Fig. 9) (20). A minority of the substitutions are located between the two putative transmembrane segments of $VanS_B$ (Fig. 9). This portion of the sensor is located at the outer surface of the membrane and may therefore interact directly with ligands, such as glycopeptides, which do not penetrate into the cytoplasm. The majority of the substitutions are located in the linker that connects the membrane-associated domain to the cytoplasmic catalytic domain (Fig. 9). The N-terminal domain of $VanS_B$ is involved in signal recognition and is involved in alteration of specificity that allows induction by teicoplanin but not by moenomycin (8, 19).

VanS and $VanS_B$ may sense the presence of glycopeptides by different mechanisms. VanA-type resistance is inducible by glycopeptides and moenomycin but not by drugs that inhibit the reactions preceding (such as ramoplanin) or following (such as bacitracin and penicillin G) transglycosylation (19, 71). This narrow specificity suggests that accumulation of lipid intermediate II, resulting from inhibition of transglycosylation, may be the signal recognized by the VanS sensor. This would account for induction by antibiotics that inhibit the same step of peptidoglycan synthesis but have different structures and modes of action (19, 71).

Constitutive Phenotype

In the wild-type VanS-type sensor, five blocks (H, N, G1, F, and G2) of the kinase domain are highly conserved (Fig. 9). The H block is responsible for both autophosphorylation and kinase/phosphatase activities, and G1 and G2 correspond to ATP binding blocks. Mutations responsible for constitutive expression of the *vanB* cluster lead to amino substitutions at two specific positions located on either side of the histidine at position 233, which corresponds to the putative autophosphorylation site of $VanS_B$ (Fig. 9) (20). Constitutive expression of glycopeptide resistance is most probably due to impaired dephosphorylation of $VanR_B$ by $VanS_B$, as similar substitutions affecting homologous residues of related sensor kinases impair the phosphatase but not the kinase activity of the proteins (20). These results confirm that dephosphorylation of $VanR_B$ is required to prevent transcription of the resistance genes and indicate that the phosphatase activity is negatively regulated by vancomycin (8).

VanB-type *E. faecium* BM4524 is resistant to vancomycin but susceptible to teicoplanin. The *E. faecium* derivative BM4525, isolated from the same patient two weeks later, was constitutively resistant to high levels of both vancomycin and teicoplanin (43). Increased glycopeptide resistance in BM4525 was shown to be due to the combination of a frameshift mutation leading to the loss of Ddl ligase activity and to constitutive synthesis of pentadepsipeptide precursors secondary to loss of $VanS_B$ phosphatase activity following deletion of a six amino acid moiety which partially overlaps the conserved G2 ATP-binding domain (Fig. 10) (43).

Heterogeneous Phenotype

The heterogeneous mutants most probably harbor null alleles of $vanS_B$ since the mutations introduced translation termination codons at various positions of the gene (Fig. 9) (21). The antibiotic disk diffusion assay revealed the presence of inhibition zones containing scattered colonies of resistant bacteria that grew predominantly in 48 h (18, 21).

GENETIC BACKGROUND OF THE *van* OPERONS

vanA Gene Cluster

The *vanA* gene cluster was detected originally on the nonconjugative transposon Tn*1546,* which belongs to the Tn3 family (Fig. 3B) (15). VanA-type resistance in clinical isolates of enterococci is mediated by genetic elements identical or closely related to Tn*1546* that are generally carried by self-transferable plasmids (36, 49, 72, 77, 86) and, occasionally, by the host chromosome as part of larger conjugative elements (74). Tn*1546*-like elements are highly conserved except for the presence of insertion sequences that have transposed into intergenic regions not essential for expression of glycopeptide resistance. The high degree of sequence conservation in the *vanRSHAX* cluster, although the isolates are geographically and epidemiologically unrelated, suggests that diversification of VanA elements occurred following the transfer of a progenitor Tn*1546* element to enterococci. Only a very few point mutations have been identified in this gene cluster (77, 143), with a single mutation reported in the actual *vanA* gene (77). More common is a G→T change in *vanX* (at position 8234 in Tn*1546*) (80).

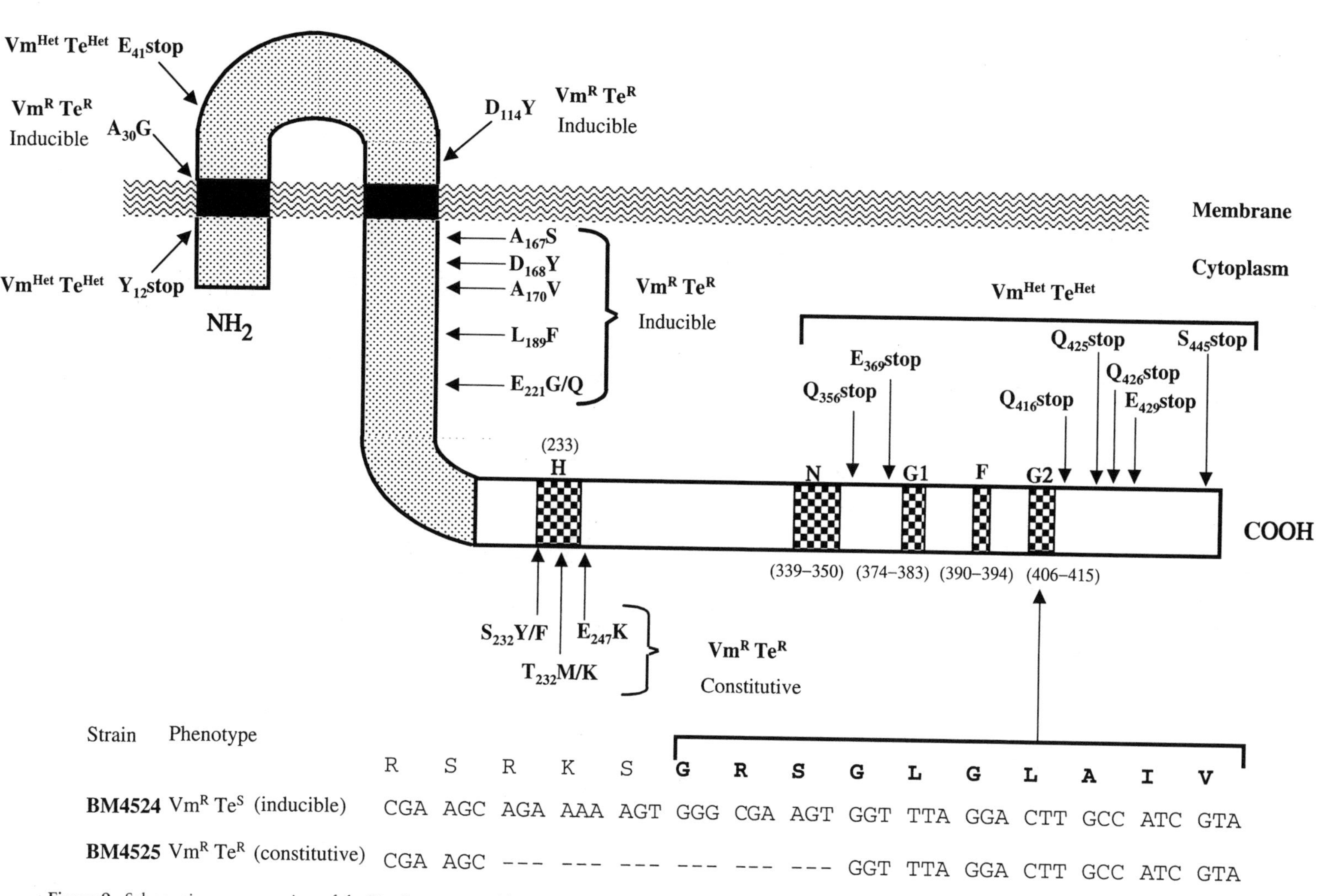

Figure 9. Schematic representation of the $VanS_B$ sensor and location of the amino acid substitutions in teicoplanin-resistant mutants. H, N, G1, F, and G2 refer to the motifs conserved in histidine protein kinases. The putative membrane-associated sensor domain (dotted black) containing transmembrane segments (black) and the putative cytoplasmic kinase domain (white) are indicated. Het, heterogeneously resistant; R, resistant; Vm, vancomycin; Te, teicoplanin.

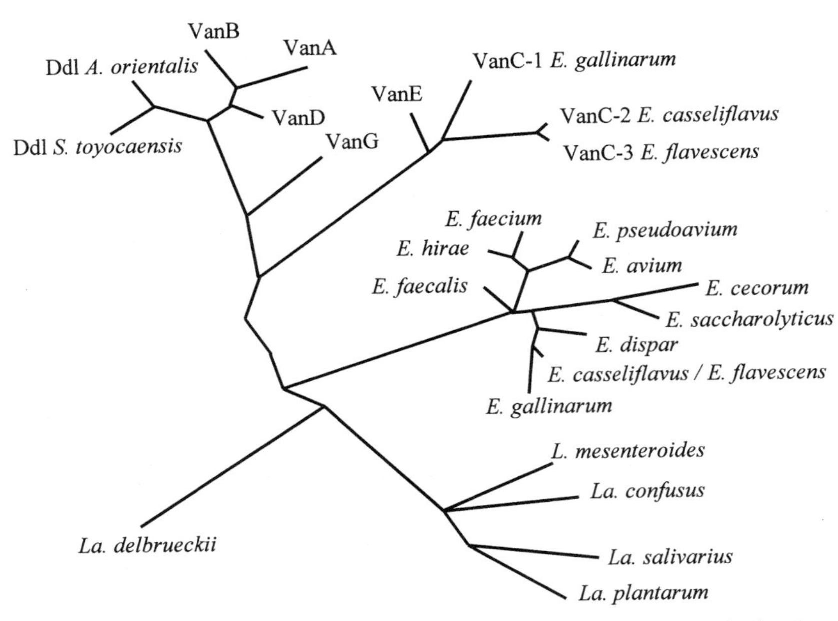

Figure 10. Phylogenetic tree derived from the alignment of D-Ala:D-Ala ligases and related enzymes.

Much greater diversity is found upstream from the *vanR* gene or downstream from *vanX* and results from the presence of deletions, rearrangements, and insertion sequences (IS) in genes not essential for glycopeptide resistance (*orf1, orf2, vanY,* and *vanZ*) and in the intergenic regions (109, 143).

IS*1251* has been found between *vanS* and *vanH,* especially in VanA elements from strains collected in the United States (75) but also in occasional isolates from Ireland and Norway (135). Less frequently, the element IS*1542* has been detected in strains isolated in the United Kingdom (40, 89, 144) and IS*1476* in the *vanY* gene from strains isolated in Canada (95). In contrast IS*1216V* appears to be ubiquitous, since insertions have been found in the *vanX-vanY* intergenic region, upstream from *vanR,* in *orf2* (109) and *vanS* (39), and complexed with an IS*3*-like element at the left terminus (74, 89, 109, 132, 143). The multiple insertion sites suggest that this element is actively mobile and indicate that the movement of IS elements is likely to be crucial in the evolution of VanA elements. Conjugal transfer of plasmids that have acquired Tn*1546*-like elements by transposition appears to be responsible for the spread of glycopeptide resistance in enterococci.

vanB Gene Cluster

Gene clusters related to *vanB* are generally carried by large (90 to 250 kb) elements that are transferable by conjugation from chromosome to chromosome (118). One of these elements (250 kb) was found to contain the composite transposon Tn*1547* (64 kb) delineated by insertion sequences belonging to the IS*256* and IS*16* families (Fig. 5) (117). Plasmid-borne *vanB*-related gene clusters have also been detected in clinical isolates of enterococci (126, 146). Much of the dissemination of VanB resistance appears to have resulted from the spread of *vanB2* clusters carried on Tn*916*-like conjugative transposons (37, 38, 100, 125). Two related elements (27- and 34-kb in size) have been characterized and designated Tn*5382* (31) and Tn*1549* (61). These elements have been identified widely in the United States and in Europe. In strains of *E. faecium* from the United States, Tn*5382* has been shown to insert immediately downstream from *pbp5,* which encodes the low affinity penicillin-binding protein 5 (76). In the United Kingdom and Ireland, the *vanB2* cluster was found in unrelated isolates of enterococci and was associated with Tn*5382*-like elements, although linkage to *pbp5* was not demonstrated (100). In two strains from the United Kingdom with plasmid-mediated *vanB2* resistance (147), the Tn*5382*-related transposon Tn*1549* was found inserted into the *traE1* and *uvrB* genes of pAD1-like pheromone-response plasmids (61). Recently, the organization of the chromosomal *vanB2* operon and flanking regions in an *E. faecium* clinical isolate was found to be similar to that in Tn*1549* (43). The *vanB1* cluster has been associated with the large transposon Tn*1547* but not with Tn*5382*-like elements (117).

Insertion sequences seem to be integrated in *vanB* clusters far less frequently than within VanA elements. An IS*Enfa200* was identified between $vanS_B$ and $vanY_B$

in certain isolates of *E. faecium* with *vanB2* clusters from the United States (37). An IS*150*-like element downstream from $vanX_B$ was described recently as being present in several *vanB2*-type *E. faecium* strains (43, 90).

vanG Gene Cluster

Transfer of VanG-type glycopeptide resistance to *E. faecalis* JH2-2 was associated with the movement, from chromosome to chromosome, of genetic elements of ca. 240 kb carrying also *ermB*-encoded erythromycin resistance (42). Sequence determination of the flanking regions of the *vanG* cluster in donor and transconjugants revealed the same 4-bp direct repeats and 22-bp imperfect inverted repeats that delineated the large element (42).

ORIGIN OF THE *van* RESISTANCE GENES

There is no homology between the *vanA* or *vanB* genes and DNA from glycopeptide-susceptible enterococci (49, 119). This suggests that resistance may have originated in another bacterial genus. The base composition (%GC) of the genes forming the *vanA* and *vanB* clusters (43 and 48%, respectively) is higher than that of chromosomal DNA from *E. faecium* (39%) or *E. faecalis* (38%) (15, 53). Furthermore the base composition differs greatly within a cluster (from 29 to 45% for the *vanA* operon and 46 to 51% for the *vanB* operon). These data provide strong evidence (i) that the glycopeptide resistance genes have an extrageneric origin and (ii) that the clusters may be composed of genes originating from various sources.

As already mentioned, several intrinsically vancomycin-resistant species, such as *Pediococcus* sp., *Leuconostoc* sp., and some lactobacilli, also produce peptidoglycan precursors that terminate in D-Lac (26, 73) and could potentially be the source of resistance ligases. Analysis of the genes encoding D-Ala:D-Ala ligase-related enzymes in strains of *L. mesenteroides, Lactobacillus plantarum, Lactobacillus salivarius,* and *Lactobacillus confusus* indicate that the enzymes of these glycopeptide-resistant species are more closely related to each other (47 to 63% amino acid homology) than to the enzyme of the glycopeptide-susceptible species *Lactobacillus leichmanii* (Fig. 10) (62). However, the D-Ala:D-Lac ligases found in these organisms are only distantly related to the VanA, VanB, and VanD ligases as demonstrated by the phylogenetic tree based on the alignment of the deduced sequences of D-Ala:D-Ala ligases and related enzymes (Fig. 10). Moreover, probes for *vanA* and *vanB* did not hybridize with DNA prepared from intrinsically glycopeptide-resistant lactic acid bacteria (49, 119).

Another more closely related group of *vanHAX* homologues, on the basis of deduced amino-acid sequence and G+C content, is found in the vancomycin-resistant biopesticide *P. popilliae* (110). The *vanF* operon is composed of five genes ($vanY_F$, $vanZ_F$, $vanH_F$, *vanF,* and $vanX_F$) encoding homologues of VanY, VanZ, VanH, VanA, and VanX, and the genes essential for resistance ($vanH_F$, *vanF,* and $vanX_F$) are organized and oriented as in VanA-type strains. The evolutionary lineage of these groups of homologous genes is not clear, but they may have a common ancestor or *vanF* could be an ancestor of the resistance operons acquired by enterococci.

The glycopeptide-producing organisms also represent a potential source for the resistance genes. Three genes encoding homologues of VanH, VanA, and VanX have been identified in organisms which produce glycopeptide antibiotics: the A47934 producer *Streptomyces toyocaensis* NRRL15009, the vancomycin producer *Amycolatopsis orientalis* C329.2, the chloro-eremomycin producer *A. orientalis* 18098, the ristocetin producer *A. orientalis* subsp. Lurida, and the teicoplanin-avoparcin producer *Amycolatopsis coloradensis* subsp. Labeda (96, 97). The A47934 biosynthetic gene cluster from *S. toyocaensis* NRRL15009 possesses the *vanRst* and *vanSst* genes that are homologues of *vanR* and *vanS,* which encode a two-component regulatory system (115). In vancomycin-resistant enterococci, these genes are in close proximity to the *vanHAX* operon, whereas in *S. toyocaensis,* they are separated by approximately 20 kb. However, the %GC of the corresponding genes (*vanRst, vanSst, vanH, ddlM, ddlN,* and *vanX*) in the organisms *S. toyocaensis* and *A. orientalis* (97–99, 115), is higher (60%) than those of *vanA, vanB,* and *vanD,* suggesting that acquisition of the resistance genes by *Enterococcus* is probably not a recent event.

Identification of a *vanB* gene cluster in *Clostridium* (136) or a *vanD* gene cluster in anaerobic bacteria of the human bowel flora (47) suggests that they may serve as a reservoir of vancomycin resistance *vanB* or *vanD* genes. The deduced amino acid sequences of $VanR_D$, $VanS_D$, $VanY_D$, $VanH_D$, VanD, and $VanX_D$ from an anaerobic bacterial strain are 97 to 100% identical with those of VanD-type *E. faecium* BM4339 (Fig. 5) (32). An *intD* gene with 99% identity to that of the *E. faecium* strain was found to be associated with the *vanD* gene cluster (32, 47).

No glycopeptide producers were found to synthesize peptidoglycan precursors ending in D-Ala-D-Ser, suggesting that the origin of the VanC, E, and G-type of resistance is different from that of VanA, B, and D (Fig. 10). The *vanC* and *vanE* operons have a high degree of identity (41 to 60%); acquired

VanE-type resistance in *E. faecalis* could be due to acquisition of a chromosomal operon from another species of *Enterococcus* such as *E. gallinarum* or *E. casseliflavus/flavescens* (Fig. 10) (1). The *vanG* cluster is composed of genes recruited from various *van* operons (42). $VanR_G$ and $VanS_G$ have the highest homology with the two-component regulatory system ($VanR_DS_D$) in VanD-type strains. $VanY_G$ has the highest identity with $VanY_B$ D,D-carboxypeptidase and $VanW_G$ with VanW, which is present only in the *vanB* operon. The 3′ end of the *vanG* cluster encodes VanG, $VanXY_G$, and $VanT_G$, as in VanC- and VanE-type strains. In spite of the presence in deduced VanG of the EKY motif specific for D-Ala:D-Ser ligases (42), the *vanG* gene is phylogenetically closer to the structural genes for the D-Ala:D-Lac ligases than to those for D-Ala:D-Ser ligases (Fig. 10).

CONCLUSIONS

In enterococci, major advances in the understanding of the mechanism of glycopeptide resistance have been achieved. However, the origin of both the resistance genes and the resistant bacteria themselves remains unclear. Glycopeptides, alone or in combination with aminoglycosides, often constitute the only therapy for multiresistant strains of staphylococci, streptococci, and enterococci. The emergence and dissemination of high-level resistance to glycopeptides in enterococci in the past two decades has resulted in clinical isolates resistant to all antibiotics of proven efficacy. Although enterococci are not highly pathogenic, the incidence of glycopeptide resistance among clinical isolates is increasing and the enterococci have become important as nosocomial pathogens and as a reservoir of resistance genes. Dissemination of glycopeptide resistance to more pathogenic bacteria such as staphylococci and streptococci has already occurred since there is no barrier to heterospecific expression or transfer of glycopeptide resistance genes among these bacteria. The mobility of the *vanA* and *vanB* gene clusters by conjugation and transposition is expected to facilitate such transfer in nature. Due to the expected dissemination of these phenotypes among gram-positive bacteria, there is a continued need to expand our antimicrobial armamentarium. The recent approvals of the oxazolidinone antimicrobial linezolid and the lipopeptide antibiotic daptomycin by the U.S. FDA has increased therapeutic options for treatment of glycopeptide resistant enterococcal infections. Also several second-generation glycopeptide antimicrobials (e.g., dalbavancin and oritavancin) as well as the minocycline derivative tigecycline are currently in clinical trials in adults. However, the appropriate indications and cost effectiveness of each of these new antimicrobials in our therapeutic strategies against gram-positive pathogens remains to be determined.

REFERENCES

1. **Abadía, L., P. Courvalin, and B. Périchon.** 2002. *vanE* gene cluster of vancomycin-resistant *Enterococcus faecalis* BM4405. *J. Bacteriol.* **184:**6457–6464.
2. **Al-Obeid, S., D. Billot-Klein, J. Van Heijenoort, E. Collatz, and L. Gutmann.** 1992. Replacement of the essential penicillin-binding protein 5 by high-molecular mass PBPs may explain vancomycin-b-lactam synergy in low-level vancomycin-resistant *Enterococcus faecium* D366. *FEMS Microbiol. Lett.* **91:**79–84.
3. **Ambur, O. H., P. E. Reynolds, and C. A. Arias.** 2002. D-Ala:D-Ala ligase gene flanking the *vanC* cluster: evidence for presence of three ligase genes in vancomycin-resistant *Enterococcus gallinarum* BM4174. *Antimicrob. Agents Chemother.* **46:**95–100.
4. **Arias, C. A., P. Courvalin, and P. E. Reynolds.** 2000. *vanC* cluster of vancomycin-resistant *Enterococcus gallinarum* BM4174. *Antimicrob. Agents Chemother.* **44:**1660–1666.
5. **Arias, C. A., M. Martin-Martinez, T. L. Blundell, M. Arthur, P. Courvalin, and P. E. Reynolds.** 1999. Characterization and modelling of VanT: a novel, membrane-bound, serine racemase from vancomycin-resistant *Enterococcus gallinarum* BM4174. *Mol. Microbiol.* **31:**1653–1664.
6. **Arias, C. A., J. Weisner, J. M. Blackburn, and P. E. Reynolds.** 2000. Serine and alanine racemase activities of VanT: a protein necessary for vancomycin resistance in *Enterococcus gallinarum* BM4174. *Microbiology* **146:**1727–1734.
7. **Arthur, M., F. Depardieu, L. Cabanié, P. Reynolds, and P. Courvalin.** 1998. Requirement of the VanY and VanX D,D-peptidases for glycopeptide resistance in enterococci. *Mol. Microbiol.* **30:**819–830.
8. **Arthur, M., F. Depardieu, and P. Courvalin.** 1999. Regulated interactions between partner and non partner sensors and responses regulators that control glycopeptide resistance gene expression in Enterococci. *Microbiology* **145:**1849–1858.
9. **Arthur, M., F. Depardieu, G. Gerbaud, M. Galimand, R. Leclercq, and P. Courvalin.** 1997. The VanS sensor negatively controls VanR-mediated transcriptional activation of glycopeptide resistance genes of Tn*1546* and related elements in the absence of induction. *J. Bacteriol.* **179:**97–106.
10. **Arthur, M., F. Depardieu, C. Molinas, P. Reynolds, and P. Courvalin.** 1995. The *vanZ* gene of Tn*1546* from *Enterococcus faecium* BM4147 confers resistance to teicoplanin. *Gene* **154:**87–92.
11. **Arthur, M., F. Depardieu, P. Reynolds, and P. Courvalin.** 1999. Moderate-level resistance to glycopeptide LY333328 mediated by genes of the *vanA* and *vanB* clusters in Enterococci. *Antimicrob. Agents Chemother.* **43:**1875–1880.
12. **Arthur, M., F. Depardieu, P. Reynolds, and P. Courvalin.** 1996. Quantitative analysis of the metabolism of soluble cytoplasmic peptidoglycan precursors of glycopeptide-resistant enterococci. *Mol. Microbiol.* **21:**33–44.
13. **Arthur, M., F. Depardieu, H. A. Snaith, P. E. Reynolds, and P. Courvalin.** 1994. Contribution of VanY D,D-carboxypeptidase to glycopeptide resistance in *Enterococcus faecalis* by hydrolysis of peptidoglycan precursors. *Antimicrob. Agents Chemother.* **38:**1899–1903.
14. **Arthur, M., C. Molinas, and P. Courvalin.** 1992. The VanS-VanR two-component regulatory system controls synthesis of

depsipeptide peptidoglycan precursors in *Enterococcus faecium* BM4147. *J. Bacteriol.* **174:**2582–2591.

15. **Arthur, M., C. Molinas, F. Depardieu, and P. Courvalin.** 1993. Characterization of Tn*1546*, a Tn3-related transposon conferring glycopeptide resistance by synthesis of depsipeptide peptidoglycan precursors in *Enterococcus faecium* BM4147. *J. Bacteriol.* **175:**117–127.
16. **Arthur, M., C. Molinas, S. Dutka-Malen, and P. Courvalin.** 1991. Structural relationship between the vancomycin resistance protein VanH and 2-hydroxycarboxylic acid dehydrogenases. *Gene* **103:**133–134.
17. **Arthur, M., P. Reynolds, and P. Courvalin.** 1996. Glycopeptide resistance in enterococci. *Trends Microbiol.* **4:**401–407.
18. **Aslangul, E., M. Baptista, B. Fantin, F. Depardieu, M. Arthur, P. Courvalin, and C. Carbon.** 1997. Selection of glycopeptide-resistant mutants of VanB-type *Enterococcus faecalis* BM4281 in vitro and in experimental endocarditis. *J. Infect. Dis.* **175:**598–605.
19. **Baptista, M., F. Depardieu, P. Courvalin, and M. Arthur.** 1996. Specificity of induction of glycopeptide resistance genes in *Enterococcus faecalis. Antimicrob. Agents Chemother.* **40:**2291–2295.
20. **Baptista, M., F. Depardieu, P. Reynolds, P. Courvalin, and M. Arthur.** 1997. Mutations leading to increased levels of resistance to glycopeptide antibiotics in VanB-type enterococci. *Mol. Microbiol.* **25:**93–105.
21. **Baptista, M., P. Rodrigues, F. Depardieu, P. Courvalin, and M. Arthur.** 1999. Single-cell analysis of glycopeptide resistance gene expression in teicoplanin-resistant mutants of a VanB-type *Enterococcus faecalis. Mol. Microbiol.* **32:**17–28.
22. **Barna, J. C. J., and D. H. Williams.** 1984. The structure and mode of action of glycopeptide antibiotics of the vancomycin group. *Ann. Rev. Microbiol.* **38:**339–357.
23. **Bell, J. M., J. C. Paton, and J. Turnidge.** 1998. Emergence of vancomycin-resistant enterococci in Australia: phenotypic and genotypic characteristics of isolates. *J. Clin. Microbiol.* **36:**2187–2190.
24. **Biavasco, E., E. Giovanetti, A. Miele, C. Vignaroli, B. Facinelli, and P. E. Varaldo.** 1996. In vitro conjugative transfer of VanA vancomycin resistance between *Enterococci* and *Listeriae* of different species. *Eur. J. Clin. Microbiol. Infect. Dis.* **15:**50–59.
25. **Biavasco, F., C. Paladini, C. Vignaroli, G. Foglia, E. Manso, and P. E. Varaldo.** 2001. Recovery from a single blood culture of two *Enterococcus gallinarum* isolates carrying both *vanC-1* and *vanA* cluster genes and differing in glycopeptide susceptibility. *Eur. J. Clin. Microbiol. Infect. Dis.* **20:**309–314.
26. **Billot-Klein, D., L. Gutmann, S. Sablé, E. Guittet, and J. van Heijenoort.** 1994. Modification of peptidoglycan precursors is a common feature of the low-level vancomycin-resistant VANB-type *Enterococcus* D366 and of the naturally glycopeptide-resistant species *Lactobacillus casei, Pediococcus pentosaceus, Leuconostoc mesenteroides,* and *Enterococcus gallinarum. J. Bacteriol.* **176:**2398–2405.
27. **Billot-Klein, D., D. Shlaes, D. Bryant, D. Bell, J. van Heijenoort, and L. Gutmann.** 1996. Peptidoglycan structure of *Enterococcus faecium* expressing vancomycin resistance of the VanB type. *Biochem. J.* **313:**711–715.
28. **Boyd, D. A., T. Cabral, P. Van Caeseele, J. Wylie, and M. R. Mulvey.** 2002. Molecular characterisation of the *vanE* gene cluster in vancomycin-resistant *Enterococcus faecalis* N00-410 isolated in Canada. *Antimicrob. Agents Chemother.* **46:**1977–1979.
29. **Boyd, D. A., J. Conly, H. Dedier, G. Peters, L. Robertson, E. Slater, and M. R. Mulvey.** 2000. Molecular characterization of the *vanD* gene cluster and a novel insertion element in a vancomycin-resistant enterococcus isolated in Canada. *J. Clin. Microbiol.* **38:**2392–2394.
30. **Bugg, T. D. H., G. D. Wright, S. Dutka-Malen, M. Arthur, P. Courvalin, and C. T. Walsh.** 1991. Molecular basis for vancomycin resistance in *Enterococcus faecium* BM4147: biosynthesis of a depsipeptide peptidoglycan precursor by vancomycin resistance proteins VanH and VanA. *Biochemistry* **30:**10408–10415.
31. **Carias, L. L., S. D. Rudin, C. J. Donskey, and L. B. Rice.** 1998. Genetic linkage and cotransfer of a novel, *vanB*-containing transposon (Tn5382) and a low-affinity penicillin-binding protein 5 gene in a clinical vancomycin-resistant *Enterococcus faecium* isolate. *J. Bacteriol.* **180:**4426–4434.
32. **Casadewall, B., and P. Courvalin.** 1999. Characterization of the *vanD* glycopeptide resistance gene cluster from *Enterococcus faecium* BM4339. *J. Bacteriol.* **181:**3644–3648.
33. **Casadewall, B., P. E. Reynolds, and P. Courvalin.** 2001. Regulation of expression of the *vanD* glycopeptide resistance gene cluster from *Enterococcus faecium* BM4339. *J. Bacteriol.* **183:**3436–3446.
34. **Centers for Disease Control and Prevention.** 1993. Nosocomial enterococci resistant to vancomycin. United States. 1989-1993. *Morb. Mortal. Wkly. Rep.* **42:**597–599.
35. **Cercenado, E., S. Unal, C. T. Eliopoulos, L. G. Rubin, H. D. Isenberg, R. C. Moellering, Jr., and G. M. Eliopoulos.** 1995. Characterization of vancomycin resistance in *Enterococcus durans. J. Antimicrob. Chemother.* **36:**821–825.
36. **Clark, N. C., R. C. Cooksey, B. C. Hill, J. M. Swenson, and F. C. Tenover.** 1993. Characterization of glycopeptide-resistant enterococci from U.S. hospitals. *Antimicrob. Agents Chemother.* **37:**2311–2317.
37. **Dahl, K. H., E. W. Lundblad, T. P. Rokenes, O. Olsvik, and A. Sundsfjord.** 2000. Genetic linkage of the *vanB2* gene cluster to Tn5382 in vancomycin-resistant enterococci and characterization of two novel insertion sequences. *Microbiology* **146:**1469–1479.
38. **Dahl, K. H., G. S. Simonsen, O. Olsvik, and A. Sundsfjord.** 1999. Heterogeneity in the *vanB* gene cluster of genomically diverse clinical strains of vancomycin-resistant enterococci. *Antimicrob. Agents Chemother.* **43:**1105–1110.
39. **Darini, A. L., M. F. Palepou, D. James, and N. Woodford.** 1999. Disruption of *vanS* by IS*1216V* in a clinical isolate of *Enterococcus faecium* with VanA glycopeptide resistance. *Antimicrob. Agents Chemother.* **43:**995–996.
40. **Darini, A. L., M. F. Palepou, and N. Woodford.** 1999. Nucleotide sequence of IS*1542*, an insertion sequence identified within VanA glycopeptide resistance elements of enterococci. *FEMS Microbiol. Lett.* **173:**341–346.
41. **De Jonge, B. L. M., S. Handwerger, and D. Gage.** 1996. Altered peptidoglycan composition in vancomycin-resistant *Enterococcus faecalis. Antimicrob. Agents Chemother.* **40:**863–869.
42. **Depardieu, F., M. G. Bonora, P. E. Reynolds, and P. Courvalin.** 2003. The *vanG* glycopeptide resistance operon from *Enterococcus faecalis* revisited. *Mol. Microbiol.* **50:**931–948.
43. **Depardieu, F., P. Courvalin, and T. Msadek.** 2003. A six amino acid deletion, partially overlapping the $VanS_B$ G2 ATP-binding motif, leads to constitutive glycopeptide resistance in VanB-type *Enterococcus faecium. Mol. Microbiol.* **50:**1069–1083.
44. **Depardieu, F., M. Kolbert, H. Pruul, J. Bell, and P. Courvalin.** 2004. Characterisation of new VanD-type Vancomycin Resistant *Enterococcus faecium* and *Enterococcus faecalis. Antimicrob. Agents Chemother.* **48:**3892–3904.
45. **Depardieu, F., P. E. Reynolds, and P. Courvalin.** 2003. VanD-type vancomycin-resistant *Enterococcus faecium* 10/96A. *Antimicrob. Agents Chemother.* **47:**7–18.

46. **Dever, L. L., S. M. Smith, S. Handwerger, and R. H. Eng.** 1995. Vancomycin-dependent *Enterococcus faecium* isolated from stool following oral vancomycin therapy. *J. Clin. Microbiol.* **33:**2770–2773.
47. **Domingo, M. C., A. Huletsky, R. Giroux, K. Boissinot, A. Bernal, F. J. Picard, and M. G. Bergeron.** 2003. Characterization of a *vanD* gene cluster from anaerobic bacterium of the normal flora of human bowel. 43rd Interscience Conference on Antimicrobial Agents and Chemotherapy (ICAAC, Chicago, Illinois), Abs. C2-2166.
48. **Dutka-Malen, S., B. Blaimont, G. Wauters, and P. Courvalin.** 1994. Emergence of high-level resistance to glycopeptides in *Enterococcus gallinarum* and *Enterococcus casseliflavus*. *Antimicrob. Agents Chemother.* **38:**1675–1677.
49. **Dutka-Malen, S., R. Leclercq, V. Coutant, J. Duval, and P. Courvalin.** 1990. Phenotypic and genotypic heterogeneity of glycopeptide resistance determinants in gram-positive bacteria. *Antimicrob. Agents Chemother.* **34:**1875–1879.
50. **Dutta, I., and P. E. Reynolds.** 2002. Biochemical and genetic characterization of the *van*C-2 vancomycin resistance gene cluster of *Enterococcus casseliflavus* ATCC 25788. *Antimicrob. Agents Chemother.* **46:**3125–3132.
51. **Dutta, I., and P. E. Reynolds.** 2003. The *vanC-3* vancomycin resistance gene cluster of *Enterococcus flavescens* CCM439. *J. Antimicrob. Chemother.* **51:**703–706.
52. **Eliopoulos, G. M.** 1997. Vancomycin-resistant enterococci. Mechanism and clinical relevance. *Infect. Dis. Clin. North Am.* **11:**851–865.
53. **Evers, S., and P. Courvalin.** 1996. Regulation of VanB-type vancomycin resistance gene expression by the $VanS_B$-$VanR_B$ two-component regulatory system in *Enterococcus faecalis* V583. *J. Bacteriol.* **178:**1302–1309.
54. **Evers, S., P. E. Reynolds, and P. Courvalin.** 1994. Sequence of the *vanB* and *ddl* genes encoding D-alanine:D-lactate and D-alanine:D-alanine ligases in vancomycin-resistant *Enterococcus faecalis* V583. *Gene* **140:**97–102.
55. **Fines, M., B. Perichon, P. Reynolds, D. F. Sahm, and P. Courvalin.** 1999. VanE, a new type of acquired glycopeptide resistance in *Enterococcus faecalis* BM4405. *Antimicrob. Agents Chemother.* **43:**2161–2164.
56. **Flannagan, S. E., J. W. Chow, S. M. Donabedian, W. J. Brown, M. B. Perri, M. J. Zervos, Y. Ozawa, and D. B. Clewell.** 2003. Plasmid content of a vancomycin-resistant *Enterococcus faecalis* isolate from a patient also colonized by *Staphylococcus aureus* with a VanA phenotype. *Antimicrob. Agents Chemother.* **47:**3954–3959.
57. **Fontana, R., M. Ligozzi, A. Mazzariol, G. Veneri, and G. Cornaglia.** 1998. Resistance of enterococci to ampicillin and glycopeptide antibiotics in Italy. The Italian Surveillance Group for Antimicrobial Resistance. *Clin. Infect. Dis.* **27** Suppl 1:S84–S86.
58. **Fontana, R., M. Ligozzi, F. Pittaluga, and G. Satta.** 1996. Intrinsic penicillin resistance in enterococci. *Microb. Drug Resist.* **2:**209–213.
59. **Fraimow, H. S., D. L. Jungkind, D. W. Lander, D. R. Delso, and J. L. Dean.** 1994. Urinary tract infection with an *Enterococcus faecalis* isolate that requires vancomycin for growth. *Ann. Intern. Med.* **121:**22–26.
60. **Frieden, T. R., S. S. Munsiff, D. E. Low, B. M. Willey, G. Williams, Y. Faur, W. Eisner, S. Warren, and B. Kreiswirth.** 1993. Emergence of vancomycin-resistant enterococci in New York City. *Lancet* **342:**76–79.
61. **Garnier, F., S. Taourit, P. Glaser, P. Courvalin, and M. Galimand.** 2000. Characterization of transposon Tn*1549*, conferring VanB-type resistance in *Enterococcus* spp. *Microbiology* **146:**1481–1489.
62. **Gay Elisha, B., and P. Courvalin.** 1995. Analysis of genes encoding D-alanine:D-alanine ligase-related enzymes in *Leuconostoc mesenteroides* and *Lactobacillus* spp. *Gene* **152:**79–83.
63. **Gold, H. S.** 2001. Vancomycin-resistant enterococci: mechanisms and clinical observations. *Clin. Infect. Dis.* **33:**210–219.
64. **Gold, H. S., S. Unal, E. Cercenado, C. Thauvin-Eliopoulos, G. M. Eliopoulos, C. B. Wennersten, and R. C. Moellering, Jr.** 1993. A gene conferring resistance to vancomycin but not teicoplanin in isolates of *Enterococcus faecalis* and *Enterococcus faecium* demonstrates homology with *vanB, vanA,* and *vanC* genes of enterococci. *Antimicrob. Agents Chemother.* **37:**1604–1609.
65. **Gorby, G. L., and J. E. Peacock, Jr.** 1988. *Erysipelothrix rhusiopathiae* endocarditis: microbiologic, epidemiologic, and clinical features of an occupational disease. *Rev. Infect. Dis.* **10:**317–325.
66. **Gordon, S., J. M. Swenson, B. C. Hill, N. E. Pigott, R. R. Facklam, R. C. Cooksey, C. Thornsberry, W. R. Jarvis, and F. C. Tenover.** 1992. Antimicrobial susceptibility patterns of common and unusual species of enterococci causing infections in the United States. *J. Clin. Microbiol.* **30:**2373–2378.
67. **Grayson, M. L., G. M. Eliopoulos, C. B. Wennersten, K. L. Ruoff, P. C. De Girolami, M. J. Ferraro, and R. C. Moellering, Jr.** 1991. Increasing resistance to b-lactam antibiotics among clinical isolates of *Enterococcus faecium:* a 22-year review at one institution. *Antimicrob. Agents Chemother.* **35:**2180–2184.
68. **Green, M., J. H. Shlaes, K. Barbadora, and D. M. Shlaes.** 1995. Bacteremia due to vancomycin-dependent *Enterococcus faecium*. *Clin. Infect. Dis.* **20:**712–714.
69. **Grohs, P., L. Gutmann, R. Legrand, B. Schoot, and J. L. Mainardi.** 2000. Vancomycin resistance is associated with serine-containing peptidoglycan in *Enterococcus gallinarum*. *J. Bacteriol.* **182:**6228–6232.
70. **Hall, L. M. C., H. Y. Chen, and R. J. Williams.** 1992. Vancomycin-resistant *Enterococcus durans*. *Lancet* **340:**1105.
71. **Handwerger, S., and A. Kolokathis.** 1990. Induction of vancomycin resistance in *Enterococcus faecium* by inhibition of transglycosylation. *FEMS Microbiol. Lett.* **70:**167–170.
72. **Handwerger, S., M. J. Pucci, and A. Kolokathis.** 1990. Vancomycin resistance is encoded on a pheromone response plasmid in *Enterococcus faecium* 228. *Antimicrob. Agents Chemother.* **34:**358–360.
73. **Handwerger, S., M. J. Pucci, K. J. Volk, J. Liu, and M. Lee.** 1994. Vancomycin-resistant *Leuconostoc mesenteroides* and *Lactobacillus casei* synthesize cytoplasmic peptidoglycan precursors that terminate in lactate. *J. Bacteriol.* **176:**260–264.
74. **Handwerger, S., and J. Skoble.** 1995. Identification of chromosomal mobile element conferring high-level vancomycin resistance in *Enterococcus faecium*. *Antimicrob. Agents Chemother.* **39:**2446–2453.
75. **Handwerger, S., J. Skoble, L. F. Discotto, and M. J. Pucci.** 1995. Heterogeneity of the *vanA* gene cluster in clinical isolates of enterococci from the northeastern United States. *Antimicrob. Agents Chemother.* **39:**362–368.
76. **Hanrahan, J., C. Hoyen, and L. B. Rice.** 2000. Geographic distribution of a large mobile element that transfers ampicillin and vancomycin resistance between *Enterococcus faecium* strains. *Antimicrob. Agents Chemother.* **44:**1349–1351.
77. **Hashimoto, Y., K. Tanimoto, Y. Ozawa, T. Murata, and Y. Ike.** 2000. Amino acid substitutions in the VanS sensor of the VanA-type vancomycin-resistant *Enterococcus* strains result in high-level vancomycin resistance and low-level teicoplanin resistance. *FEMS Microbiol. Lett.* **185:**247–254.

78. **Hayden, M. K., G. M. Trenholme, J. E. Schultz, and D. F. Sahm.** 1993. In vivo development of teicoplanin resistance in a VanB *Enterococcus faecium* isolate. *J. Infect. Dis.* **167:**1224–1227.
79. **Holman, T. R., Z. Wu, B. L. Wanner, and C. T. Walsh.** 1994. Identification of the DNA-binding site for the phosphorylated VanR protein required for vancomycin resistance in *Enterococcus faecium. Biochemistry* **33:**4625–4631.
80. **Jensen, L. B.** 1998. Internal size variations in Tn*1546*-like elements due to the presence of IS*1216V. FEMS Microbiol. Lett.* **169:**349–354.
81. **Jones, R. N., H. S. Sader, M. E. Erwin, and S. C. Anderson.** 1995. Emerging multiply resistant enterococci among clinical isolates. I. Prevalence data from 97 medical center surveillance study in the United States. Enterococcus Study Group. *Diagn. Microbiol. Infect. Dis.* **21:**85–93.
82. **Kacica, M., and L. C. McDonald.** 2004. Vancomycin-resistant *Staphylococcus aureus*-New York, 2004. *Morb. Mortal. Wkly. Rep.* **53:**322–323.
83. **Kawalec, M., M. Gniadkowski, J. Kedzierska, A. Skotnicki, J. Fiett, and W. Hryniewicz.** 2001. Selection of a teicoplanin-resistant *Enterococcus faecium* mutant during an outbreak caused by vancomycin-resistant enterococci with the VanB phenotype. *J. Clin. Microbiol.* **39:**4274–4282.
84. **Krogstad, D. J., and A. R. Pargwette.** 1980. Defective killing of enterococci: a common property of antimicrobial agents acting on the cell wall. *Antimicrob. Agents Chemother.* **17:**965–968.
85. **Lambert, E., C. McCullough, G. Coombs, J. Pearson, F. O'Brien, J. Bell, A. Berry, and K. Christiansen.** 2002. Multiple isolation of *Enterococcus faecalis* with VanE glycopeptide resistance in Australia. 42nd Interscience Conference on Antimicrobial Agents and Chemotherapy (ICAAC, San Diego, California), Abs. C2-1118.
86. **Leclercq, R., E. Derlot, J. Duval, and P. Courvalin.** 1988. Plasmid-mediated resistance to vancomycin and teicoplanin in *Enterococcus faecium. N. Engl. J. Med.* **319:**157–161.
87. **Leclercq, R., E. Derlot, M. Weber, J. Duval, and P. Courvalin.** 1989. Transferable vancomycin and teicoplanin resistance in *Enterococcus faecium. Antimicrob. Agents Chemother.* **33:**10–15.
88. **Leclercq, R., S. Dutka-Malen, J. Duval, and P. Courvalin.** 1992. Vancomycin resistance gene *vanC* is specific to *Enterococcus gallinarum. Antimicrob. Agents Chemother.* **36:**2005–2008.
89. **Lee, W. G., J. Y. Huh, S. R. Cho, and Y. A. Lim.** 2004. Reduction in glycopeptide resistance in vancomycin-resistant enterococci as a result of *vanA* cluster rearrangements. *Antimicrob. Agents Chemother.* **48:**1379–1381.
90. **Lee, W. G., and W. Kim.** 2003. Identification of a novel insertion sequence in *vanB2*-containing *Enterococcus faecium. Lett. Appl. Microbiol.* **36:**186–190.
91. **Lewis, C. M., and M. J. Zervos.** 1990. Clinical manifestations of enterococcal infection. *Eur. J. Clin. Microbiol. Infect. Dis.* **9:**111–117.
92. **Ligozzi, M., G. Lo Cascio, and R. Fontana.** 1998. *vanA* gene cluster in a vancomycin-resistant clinical isolate of *Bacillus circulans. Antimicrob. Agents Chemother.* **42:**2055–2059.
93. **Ligozzi, M., F. Pittaluga, and R. Fontana.** 1996. Modification of penicillin-binding protein 5 associated with high-level ampicillin resistance in *Enterococcus faecium. Antimicrob. Agents Chemother.* **40:**354–357.
94. **Low, D. E., N. Keller, A. Barth, and R. N. Jones.** 2001. Clinical prevalence, antimicrobial susceptibility, and geographic resistance patterns of enterococci: results from the SENTRY Antimicrobial Surveillance Program, 1997-1999. *Clin. Infect. Dis.* **32** Suppl 2:S133–S145.
95. **MacKinnon, M. G., M. A. Drebot, and G. J. Tyrrell.** 1997. Identification and characterization of IS*1476*, an insertion sequence-like element that disrupts VanY function in a vancomycin-resistant *Enterococcus faecium* strain. *Antimicrob. Agents Chemother.* **41:**1805–1807.
96. **Marshall, C. G., G. Broadhead, B. K. Leskiw, and G. D. Wright.** 1997. D-Ala-D-Ala ligases from glycopeptide antibiotic-producing organisms are highly homologous to the enterococcal vancomycin-resistance ligases VanA and VanB. *Proc. Natl. Acad. Sci. USA* **94:**6480–6483.
97. **Marshall, C. G., I. A. Lessard, I. Park, and G. D. Wright.** 1998. Glycopeptide antibiotic resistance genes in glycopeptide-producing organisms. *Antimicrob. Agents Chemother.* **42:**2215–2220.
98. **Marshall, C. G., and G. D. Wright.** 1998. DdlN from vancomycin-producing *Amycolatopsis orientalis* C329.2 is a VanA homologue with D-alanyl-D-lactate ligase activity. *J. Bacteriol.* **180:**5792–5795.
99. **Marshall, C. G., and G. D. Wright.** 1997. The glycopeptide antibiotic producer *Streptomyces toyocaensis* NRRL15009 has both D-alanyl-D-alanine and D-alanyl-D-lactate ligases. *FEMS Microbiol. Lett.* **157:**295–299.
100. **McGregor, K. F., C. Nolan, H. K. Young, M. F. Palepou, L. Tysall, and N. Woodford.** 2001. Prevalence of the *vanB2* gene cluster in VanB glycopeptide-resistant enterococci in the United Kingdom and the Republic of Ireland and its association with a Tn5382-like element. *Antimicrob. Agents Chemother.* **45:**367–368.
101. **McKessar, S. J., A. M. Berry, J. M. Bell, J. D. Turnidge, and J. C. Paton.** 2000. Genetic characterization of *vanG*, a novel vancomycin resistance locus of *Enterococcus faecalis. Antimicrob. Agents Chemother.* **44:**3224–3228.
102. **Meziane-Cherif, D., M.-A. Badet-Denisot, S. Evers, P. Courvalin, and B. Badet.** 1994. Purification and characterization of the VanB ligase associated with type B vancomycin resistance in *Enterococcus faecalis* V583. *FEBS Lett.* **354:**140–142.
103. **Moellering, R. C., Jr.** 1991. The *enterococcus:* a classic example of the impact of antimicrobial resistance on therapeutic options. *J. Antimicrob. Chemother.* **28:**1–12.
104. **Murray, B. E.** 1992. Beta-lactamase-producing enterococci. *Antimicrob. Agents Chemother.* **36:**2355–2359.
105. **Murray, B. E.** 1990. The life and times of the *Enterococcus. Clin. Microbiol. Rev.* **3:**46–65.
106. **Navarro, F., and P. Courvalin.** 1994. Analysis of genes encoding D-alanine:D-alanine ligase-related enzymes in *Enterococcus casseliflavus* and *Enterococcus flavescens. Antimicrob. Agents Chemother.* **38:**1788–1793.
107. **Nieto, M., and H. R. Perkins.** 1971. Modifications of the acyl-D-alanyl-D-alanine terminus affecting complex-formation with vancomycin. *Biochem. J.* **123:**789–803.
108. **Noble, W. C., Z. Virani, and R. G. A. Cree.** 1992. Co-transfer of vancomycin and other resistance genes from *Enterococcus faecalis* NCTC 12201 to *Staphylococcus aureus. FEMS Microbiol. Lett.* **93:**195–198.
109. **Palepou, M. F., A. M. Adebiyi, C. H. Tremlett, L. B. Jensen, and N. Woodford.** 1998. Molecular analysis of diverse elements mediating VanA glycopeptide resistance in enterococci. *J. Antimicrob. Chemother.* **42:**605–612.
110. **Patel, R., K. Piper, F. R. Cockerill, 3rd, J. M. Steckelberg, and A. A. Yousten.** 2000. The biopesticide *Paenibacillus popilliae* has a vancomycin resistance gene cluster homologous to the enterococcal VanA vancomycin resistance gene cluster. *Antimicrob. Agents Chemother.* **44:**705–709.
111. **Patel, R., J. R. Uhl, P. Kohner, M. K. Hopkins, J. M. Steckelberg, B. Kline, and F. R. Cockerill, 3rd.** 1998. DNA sequence variation within *vanA, vanB, vanC-1,* and *vanC-2/3* genes of

clinical *Enterococcus* isolates. *Antimicrob. Agents Chemother.* **42**:202–205.

112. **Patterson, J. E., A. H. Sweeney, M. Simm, N. Carley, R. Mangi, J. Sabetta, and R. W. Lyons.** 1995. Analysis of 110 serious enterococcal infections. Epidemiology, antibiotic susceptibility, and outcome. *Medicine* **74**:191–200.
113. **Périchon, B., B. Casadewall, P. Reynolds, and P. Courvalin.** 2000. Glycopeptide-resistant *Enterococcus faecium* BM4416 is a VanD-type strain with an impaired D-Alanine:D-Alanine ligase. *Antimicrob. Agents Chemother.* **44**:1346–1348.
114. **Périchon, B., P. E. Reynolds, and P. Courvalin.** 1997. VanD-type glycopeptide-resistant *Enterococcus faecium* BM4339. *Antimicrob. Agents Chemother.* **41**:2016–2018.
115. **Pootoolal, J., M. G. Thomas, C. G. Marshall, J. M. Neu, B. K. Hubbard, C. T. Walsh, and G. D. Wright.** 2002. Assembling the glycopeptide antibiotic scaffold: The biosynthesis of A47934 from *Streptomyces toyocaensis* NRRL15009. *Proc. Natl. Acad. Sci. USA* **99**:8962–8967.
116. **Poyart, C., C. Pierre, G. Quesne, B. Pron, P. Berche, and P. Trieu-Cuot.** 1997. Emergence of vancomycin resistance in the genus *Streptococcus:* characterization of a *vanB* transferable determinant in *Streptococcus bovis*. *Antimicrob. Agents Chemother.* **41**:24–29.
117. **Quintiliani, R., Jr. and P. Courvalin.** 1996. Characterization of Tn*1547*, a composite transposon flanked by the IS*16* and IS*256*-like elements, that confers vancomycin resistance in *Enterococcus faecalis* BM4281. *Gene* **172**:1–8.
118. **Quintiliani, R., Jr. and P. Courvalin.** 1994. Conjugal transfer of the vancomycin resistance determinant *vanB* between enterococci involves the movement of large genetic elements from chromosome to chromosome. *FEMS Microbiol. Lett.* **119**:359–364.
119. **Quintiliani, R., Jr., S. Evers, and P. Courvalin.** 1993. The *vanB* gene confers various levels of self-transferable resistance to vancomycin in enterococci. *J. Infect. Dis.* **167**:1220–1223.
120. **Reynolds, P. E.** 1998. Control of peptidoglycan synthesis in vancomycin-resistant enterococci: D,D-peptidases and D,D-carboxypeptidases. *Cell Mol. Life Sci.* **54**:325–331.
121. **Reynolds, P. E.** 1989. Structure, biochemistry and mechanism of action of glycopeptide antibiotics. *Eur. J. Clin. Microbiol. Infect. Dis.* **8**:943–950.
122. **Reynolds, P. E., C. A. Arias, and P. Courvalin.** 1999. Gene *vanXYC* encodes D,D-dipeptidase (VanX) and D,D-carboxypeptidase (VanY) activities in vancomycin-resistant *Enterococcus gallinarum* BM4174. *Mol. Microbiol.* **34**:341–349.
123. **Reynolds, P. E., F. Depardieu, S. Dutka-Malen, M. Arthur, and P. Courvalin.** 1994. Glycopeptide resistance mediated by enterococcal transposon Tn*1546* requires production of VanX for hydrolysis of D-alanyl-D-alanine. *Mol. Microbiol.* **13**:1065–1070.
124. **Reynolds, P. E., H. A. Snaith, A. J. Maguire, S. Dutka-Malen, and P. Courvalin.** 1994. Analysis of peptidoglycan precursors in vancomycin-resistant *Enterococcus gallinarum* BM4174. *Biochem. J.* **301**:5–8.
125. **Rice, L. B.** 2001. Emergence of vancomycin-resistant enterococci. *Emerg. Infect. Dis.* **7**:183–187.
126. **Rice, L. B., L. L. Carias, C. L. Donskey, and S. D. Rudin.** 1998. Transferable, plasmid-mediated *vanB*-type glycopeptide resistance in *Enterococcus faecium*. *Antimicrob. Agents Chemother.* **42**:963–964.
127. **Rosato, A., J. Pierre, D. Billot-Klein, A. Buu-Hoi, and L. Gutmann.** 1995. Inducible and constitutive expression of resistance to glycopeptides and vancomycin dependence in glycopeptide-resistant *Enterococcus avium*. *Antimicrob. Agents Chemother.* **39**:830–833.
128. **Ruoff, K. L., L. de la Maza, M. J. Murtagh, J. D. Spargo, and M. J. Ferraro.** 1990. Species identities of enterococci isolated from clinical specimens. *J. Clin. Microb.* **28**:435–437.
129. **Rybkine, T., J. L. Mainardi, W. Sougakoff, E. Collatz, and L. Gutmann.** 1998. Penicillin-binding protein 5 sequence alterations in clinical isolates of *Enterococcus faecium* with different levels of beta-lactam resistance. *J. Infect. Dis.* **178**:159–163.
130. **Sahm, D. F., L. Free, and S. Handwerger.** 1995. Inducible and constitutive expression of *vanC-1* encoded resistance to vancomycin in *Enterococcus gallinarum*. *Antimicrob. Agents Chemother.* **39**:1480–1484.
131. **Schaberg, D. R., D. H. Culver, and R. P. Gaynes.** 1991. Major trends in the microbial etiology of nosocomial infection. *Am. J. Med.* **91**:72–75.
132. **Schouten, M. A., R. J. Willems, W. A. Kraak, J. Top, J. A. Hoogkamp-Korstanje, and A. Voss.** 2001. Molecular analysis of Tn*1546*-like elements in vancomycin-resistant enterococci isolated from patients in Europe shows geographic transposon type clustering. *Antimicrob. Agents Chemother.* **45**:986–989.
133. **Sievert, D. M., M. L. Boulton, G. Stolman, D. Johnson, M. G. Stobierski, F. P. Downes, P. A. Somsel, J. T. Rudrik, W. J. Brown, W. Hafeez, T. Lundstrom, E. Flanagan, R. Johnson, J. Mitchell, and S. Chang.** 2002. *Staphylococcus aureus* resistant to vancomycin. *Morb. Mortal. Wkly. Rep.* **51**:565–567.
134. **Sifaoui, F., and L. Gutmann.** 1997. Vancomycin dependence in a *vanA*-producing *Enterococcus avium* strain with a nonsense mutation in the natural D-Ala:D-Ala ligase gene. *Antimicrob. Agents Chemother.* **41**:1409.
135. **Simonsen, G. S., M. R. Myhre, K. H. Dahl, O. Olsvik, and A. Sundsfjord.** 2000. Typeability of Tn*1546*-like elements in vancomycin-resistant enterococci using long-range PCRs and specific analysis of polymorphic regions. *Microb. Drug Resist.* **6**: 49–57.
136. **Stinear, T. P., D. C. Olden, P. D. Johnson, J. K. Davies, and M. L. Grayson.** 2001. Enterococcal *vanB* resistance locus in anaerobic bacteria in human faeces. *Lancet* **357**:855–856.
137. **Tenover, F. C., L. M. Weigel, P. C. Appelbaum, L. K. McDougal, J. Chaitram, S. McAllister, N. Clark, G. Killgore, C. M. O'Hara, L. Jevitt, J. B. Patel, and B. Bozdogan.** 2004. Vancomycin-resistant *Staphylococcus aureus* isolate from a patient in Pennsylvania. *Antimicrob. Agents Chemother.* **48**:275–280.
138. **Torres, C., J. A. Reguera, M. J. Sanmartin, J. C. Perez-Diaz, and F. Baquero.** 1994. *vanA*-mediated vancomycin-resistant *Enterococcus* spp. in sewage. *J. Antimicrob. Chemother.* **33**:553–561.
139. **Uttley, A. H., C. H. Collins, J. Naidoo, and R. C. Georges.** 1988. Vancomycin resistant enterococci. *Lancet* **i**:57–58.
140. **Uttley, A. H. C., R. C. George, J. Naidoo, N. Woodford, A. P. Johnson, C. H. Collins, D. Morrison, A. J. Gilfillan, L. E. Fitch, and J. Heptonstall.** 1989. High-level vancomycin-resistant enterococci causing hospital infections. *Epidem. Infect.* **103**:173–181.
141. **Van Bambeke, F., M. Chauvel, P. E. Reynolds, H. S. Fraimow, and P. Courvalin.** 1999. Vancomycin-dependent *Enterococcus faecalis* clinical isolates and revertant mutants. *Antimicrob. Agents Chemother.* **43**:41–47.
142. **Weigel, L. M., D. B. Clewell, S. R. Gill, N. C. Clark, L. K. McDougal, S. E. Flannagan, J. F. Kolonay, J. Shetty, G. E. Killgore, and F. C. Tenover.** 2003. Genetic analysis of a high-level vancomycin-resistant isolate of *Staphylococcus aureus*. *Science* **302**:1569–1571.
143. **Willems, R. J., J. Top, N. van den Braak, A. van Belkum, D. J. Mevius, G. Hendriks, M. van Santen-Verheuvel, and J. D. van**

Embden. 1999. Molecular diversity and evolutionary relationships of Tn*1546*-like elements in enterococci from humans and animals. *Antimicrob. Agents Chemother.* **43:**483–491.

144. **Woodford, N., A. M. Adebiyi, M. F. Palepou, and B. D. Cookson.** 1998. Diversity of VanA glycopeptide resistance elements in enterococci from humans and nonhuman sources. *Antimicrob. Agents Chemother.* **42:**502–508.
145. **Woodford, N., A. P. Johnson, D. Morrison, and D. C. E. Speller.** 1995. Current perspectives on glycopeptide resistance. *Clin. Microbiol. Rev.* **8:**585–615.
146. **Woodford, N., B. L. Jones, Z. Baccus, H. A. Ludlam, and D. F. J. Brown.** 1995. Linkage of vancomycin and high-level gentamicin resistance genes on the same plasmid in a clinical isolate of *Enterococcus faecalis. J. Antimicrob. Chemother.* **35:**179–184.
147. **Woodford, N., D. Morrison, A. P. Johnson, A. C. Bateman, J. G. Hastings, T. S. Elliott, and B. Cookson.** 1995. Plasmid-mediated *vanB* glycopeptide resistance in enterococci. *Microb. Drug. Resist.* **1:**235–240.
148. **Wright, G. D., T. R. Holman, and C. T. Walsh.** 1993. Purification and characterization of VanR and the cytosolic domain of VanS: a two-component regulatory system required for vancomycin resistance in *Enterococcus faecium* BM4147. *Biochemistry* **32:**5057–5063.

Frontiers in Antimicrobial Resistance: a Tribute to Stuart B. Levy
Edited by D. G. White, M. N. Alekshun, and P. F. McDermott

Chapter 9

Phenicol Resistance

STEFAN SCHWARZ AND DAVID G. WHITE

Chloramphenicol is produced synthetically but was originally isolated from a *Streptomyces* sp. isolated from a soil sample collected in a mulched field near Caracas, Venezuela (hence the species identification, *Streptomyces venezuelae*) (14, 69). The chemical structure of chloramphenicol and the methods used for synthesis were elucidated initially by Parke Davis Laboratories, who distributed the drug as chloromycetin (119). It has the chemical formula of D(-)-threo-p-nitrophenyl-2-dichloracetamido-1,3-propanediol and has a benzene ring in its structure (Fig. 1). Production of the oral preparation, chloromycetin palmitate, has been discontinued in the United States but is still available in other countries (14). Chloramphenicol is a highly stable antimicrobial which can pass through biological membranes to reach intracellular bacterial infections and can readily traverse the blood-brain barrier (14, 205, 211).

Chloramphenicol was once rightly considered as one of the most potent and useful antibiotics in human and veterinary medicine. If not for serious potent toxicities, it would most likely continue to be an important weapon in our current antimicrobial armamentarium. Adverse reactions include optic neuritis, gray baby syndrome, and gray toddler syndrome, which are associated with either prolonged use or increased serum concentrations of chloramphenicol (14, 119). The most devastating complication of chloramphenicol administration is bone marrow suppression and may occur by two very distinct mechanisms (14). The first and most common form is a dose-related suppression associated with inhibition of mitochondrial protein synthesis and is reversible when chloramphenicol use is discontinued (14). The second and often fatal form of suppression has been termed irreversible idiosyncratic aplastic anemia (14, 259). Interestingly, development of the disease does not appear to be dependent on dose or duration of exposure to the antimicrobial. A large body of indirect evidence favors a complex mechanism involving metabolic transformation of the *p*-NO_2 group of chloramphenicol by the predisposed subject, leading to the production of a toxic intermediate causing stem cell damage (259). However, the pathogenesis of aplastic anemia from chloramphenicol treatment remains unsure.

Recognition of the potential role of the *p*-NO_2 group in chloramphenicol toxicity prompted the design of analogues with various substitutes at this position. The substitution of a sulfomethyl group (-SO_2CH_3) for the *p*-NO_2 group at the 1-phenyl moiety led to the development of thiamphenicol and florfenicol (Fig. 1) (205). Thiamphenicol is an analogue of chloramphenicol in which the *p*-NO_2 group on the benzene ring is replaced by a sulfomethyl group (119). Thiamphenicol has been shown to possess good in vitro activity against difficult-to-treat multidrug resistant bacterial pathogens including *Streptococcus pneumoniae, Haemophilus influenzae, Staphylococcus aureus,* and *Escherichia coli* (137). However, it is also prone to ready inactivation by bacterial chloramphenicol acetyltransferase (55). The fluorinated phenicol derivative, florfenicol, is a synthetic drug which substitutes a fluorine atom for the hydroxyl group on the 3′ carbon of thiamphenicol and is considerably less affected by this enzymatic modification (157, 224). Florfenicol retains the broad spectrum and strong antibacterial activity of chloramphenicol and possesses the more favorable toxicity profile of thiamphenicol because it also lacks the aromatic nitro group (55). Besides the fluoro substitution at C-3 (in florfenicol), very few other group substitutions are tolerated without detrimental effects on antimicrobial activity (205).

Stefan Schwarz • Institut für Tierzucht, Bundesforschungsanstalt für Landwirtschaft (FAL), 31535 Neustadt-Mariensee, Germany.
David G. White • Division of Animal and Food Microbiology, Office of Research, Center for Veterinary Medicine, U. S. Food & Drug Administration, Laurel, MD 20708.

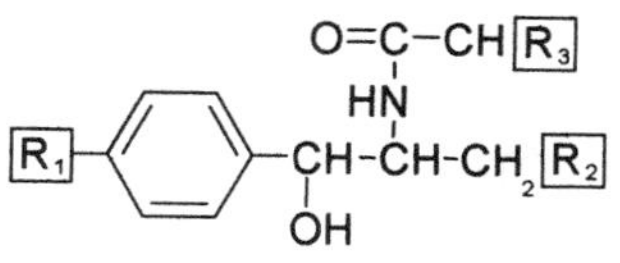

	R1	R2	R3
Chloramphenicol	$-NO_2$	$-OH$	$=Cl_2$
Thiamphenicol	$-SO_2CH_3$	$-OH$	$=Cl_2$
Florfenicol	$-SO_2CH_3$	$-F$	$=Cl_2$

Figure 1. Phenicol chemical structures.

SPECTRUM OF ACTIVITY AND MODE OF ACTION

The prokaryotic ribosome is sufficiently different from that of eukaryotic cells that a large number of antibacterial agents, including phenicols, function by interfering with various stages of bacterial protein synthesis (Fig. 2) (138). Chloramphenicol is a highly specific and potent inhibitor of protein synthesis and binds to the 70S ribosome and inhibits the peptidyl transferase reaction, which forms the peptide bond between amino acids (138, 198). Although the precise mode of action remains to be established, it is thought that phenicols occupy the position of the amino acid attached to the A-site tRNA, apparently decreasing the catalytic rate constant of peptidyl transferase, preventing peptide bond formation (43, 198).

Chloramphenicol is rarely used today in the United States due to the above mentioned adverse effects and the availability of many alternative antimicrobial agents that possess better safety profiles. It is highly effective against a variety of gram-positive and gram-negative bacteria, spirochetes, chlamydiae, mycoplasmas, and rickettsiae (14, 205). Chloramphenicol

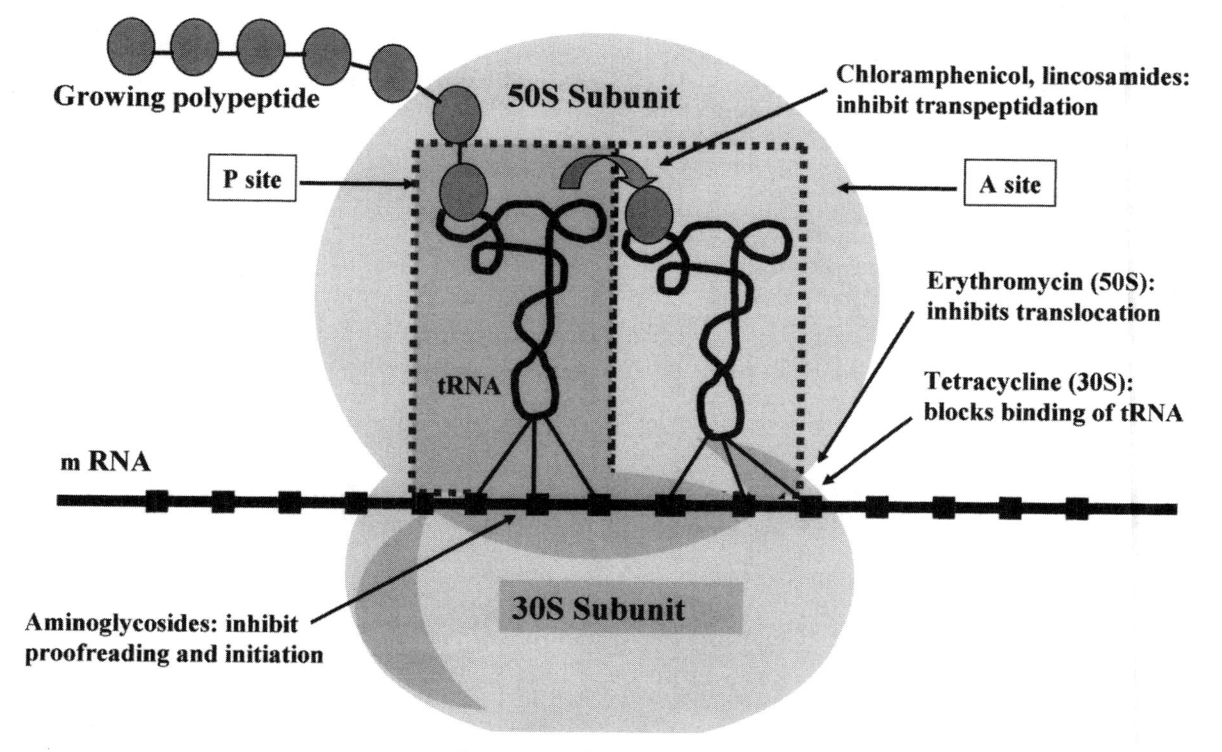

Figure 2. 70S bacterial ribosome

still remains as an alternative antimicrobial agent for treatment of bacterial meningitis caused by *Neisseria meningitidis, H. influenzae,* and *S. pneumoniae* due to its bactericidal nature against these organisms as well as its ability to achieve elevated concentrations in the cerebrospinal fluid (14, 205). It is estimated that bacterial meningitis causes 125, 000 deaths each year in infants and young children, and 96% of these occur in less developed countries where up to 50% of children with this disease die and 25 to 50% of survivors have neurological sequelae (76). Although third-generation cephalosporins are optimal empirical therapy for bacterial meningitis, they are neither obtainable nor affordable in many developing countries; therefore the majority of children worldwide are currently treated with cheaper alternatives, such as chloramphenicol (76).

Chloramphenicol is also being used in patients who have acquired immunodeficiency syndrome and VRE or methicillin-resistant *S. aureus* MRSA infections that do not respond to streptogramins or linezolid (14). It was also considered the universal treatment for typhoid fever in developing countries, until the emergence of chloramphenicol resistance worldwide limited its use (54, 166). However, increasing reports of *Salmonella enterica* serovar Typhi resistant to fluoroquinolones and other first line therapeutic agents, coupled with decreases in observed chloramphenicol resistance, may suggest its reevaluation as a treatment for typhoid fever (135, 167, 244).

Chloramphenicol was used extensively in veterinary medicine until concerns over its toxicity and association with aplastic anemia in humans emerged (82, 209). The possibility that trace residues of chloramphenicol in animal food products may induce potential adverse effects led the U.S. Food and Drug Administration to prohibit its use in food animals in the 1980s (82). The use of chloramphenicol in food-producing animals in the European Union (EU) was also banned in 1994 due to similar concerns (205). However, it is still used to some degree among pets and non-food-producing animals in both the United States (U.S.) and EU.

Thiamphenicol shares chloramphenicol's broad spectrum activity; however, it has not been associated with serious haematological side effects (e.g., fatal aplastic anemia) since the aromatic nitro group responsible is absent (21, 66, 259). Thiamphenicol is usually administered as the glycinate derivative due to its limited solubility in water and is rapidly absorbed by both the oral and the parenteral route (66). It is not inactivated by metabolic processes and diffuses freely into tissues as it is not bound to plasma proteins (66). Thiamphenicol glycinate acetylcysteinate (TGA) is a water soluble ester of thiamphenicol and has been employed for more than 30 years in the treatment of respiratory tract infections (66, 129). The combination of broad spectrum antimicrobial properties of thiamphenicol and mucolytic activity of *N*-acetylcysteine makes bronchial secretions more fluid and facilitates antibiotic absorption without altering its activity (3). TGA possesses effective pharmacokinetic characteristics, good tolerability, and a broad spectrum of effectiveness against several gram-positive and gram-negative organisms, especially those involved in upper and lower respiratory tract infections (including *S. pneumoniae, Streptococcus pyogenes, H. influenzae,* and *Moraxella catarrhalis*) (3, 137, 236). Thiamphenicol has also been shown to be a low cost, effective therapy for both uncomplicated gonococcal infection in males and *Haemophilus ducreyi* associated chancroid infections (22, 23, 252).

Florfenicol is a relatively new drug in the veterinary antimicrobial arsenal and is of great value against infectious diseases that formerly responded favorably to chloramphenicol. Florfenicol [*d*-threo-3-fluoro-2-dichloroacetamido-1-(4-methylsulfonylphenyl)-1-propanol] is a synthetic, broad-spectrum fluorinated analogue of thiamphenicol and chloramphenicol but with none of the chronic toxicity and associated health risks, such as depressed bone marrow function, leading to disorders such as aplastic anemia (205). Florfenicol is primarily bacteriostatic, and the mechanism of action is the same as that of chloramphenicol and thiamphenicol, that is, binding to the 50S ribosomal subunit and inhibiting protein synthesis (43, 224). No fluorinated chloramphenicol analogues are currently used in human medicine.

Due to its activity against many of the primary bacterial pathogens associated with bovine respiratory disease (BRD), e.g., *Mannheimia hemolytica, Pasteurella multocida,* and *Histophilus somni* (formerly *Haemophilus somnus*), florfenicol was approved in the U.S. in 1996, primarily to treat and control BRD in feedlot cattle (248). It was also approved for use in cattle and swine in the EU in 1995 and 2000, respectively (205). Florfenicol is approved for treatment of bovine interdigital phlegmon (foot rot, acute interdigital necrobacillosis, infectious pododermatitis) associated with *Fusobacterium necrophorum* and *Bacteroides melaninogenicus,* as well as swine respiratory disease (SRD) associated with *Actinobacillus pleuropneumoniae, P. multocida, Salmonella* serovar Choleraesuis and *Streptococcus suis* type 2 in the U.S. (46). Florfenicol has also been approved in numerous countries for the treatment and control of diseases in farmed fish and other aquatic species since the late 1980s (75, 93, 116).

IN VITRO SUSCEPTIBILITY TESTING PARAMETERS

There are a number of standards and guidelines currently available for antimicrobial susceptibility testing and interpretive criteria of certain phenicols (e.g., chloramphenicol and florfenicol). These include standards and guidelines published by the Clinical and Laboratory Standards Institute (CLSI; formerly NCCLS), the British Society for Antimicrobial Chemotherapy (BSAC), the Japanese Society for Chemotherapy (JSC), the Swedish Reference Group for Antibiotics (SRGA), the Comité de L'Antibiogramme de la Société Française de Microbiologie (CASFM), the Werkgroep richtlijnen gevoeligheidsbepalingen (WRG system, NL), and the Deutsches Institut für Normung (DIN), among others (247).

CLSI quality control zone diameters and MIC ranges exist for chloramphenicol to several bacteria including *E. coli* ATCC 25922, *S. aureus* ATCC 25923, *E. faecalis* ATCC 29212, *H. influenzae* ATCC 49247, and *S. pneumoniae* ATCC 49619 (156). Current CLSI susceptible and resistant MIC breakpoints for chloramphenicol against organisms other than streptococci are ≤ 8 μg/ml and ≥ 32 μg/ml, respectively (156). The one exception is *Haemophilus* spp. where susceptible and resistant breakpoints are ≤ 2 μg/ml and ≥ 8 μg/ml, respectively. Susceptible breakpoints for streptococci, including *S. pneumoniae* are ≤ 4 μg/ml; however, the resistant breakpoints differ by one doubling dilution (≥ 16 μg/ml versus ≥ 8 μg/ml, respectively). Chloramphenicol is not considered in the CLSI recommended primary test and report category A for either nonfastidious or fastidious organisms, with the exception of *Yersinia pestis* (156). It is, however, considered a group B agent which warrants primary testing but selective reporting for *Haemophilus* spp. and *Streptococcus* spp. other than *S. pneumoniae*. Chloramphenicol is most often included in the group C category, which represents susceptibility testing of alternative antimicrobial agents against certain bacteria, including *Staphylococcus* spp., *Enterococcus* spp., *S. pneumoniae*, *Enterobacteriaceae*, *Pseudomonas aeruginosa*, and *Acinetobacter* spp. (156).

A greater number of CLSI quality control strains and ranges exist for broth microdilution and disk diffusion testing of florfenicol. These include *S. aureus* (ATCC 25923), *E. faecalis* ATCC 29212, *E. coli* ATCC 25922, *S. pneumoniae* ATCC 49619, *H. somni* ATCC700025, *A. pleuropneumoniae* ATCC 27090, and more recently the aquatic pathogen *Aeromonas salmonicida* (153, 154, 156). Interpretive criteria for florfenicol exist for *P. multocida*, *M. haemolytica*, and *H. somni* from bovine respiratory diseases as well as *A. pleuropneumoniae*, *Bordetella bronchiseptica*, and *S. suis* from SRDs in pigs and *Salmonella* serovar Choleraesuis infections in pigs (153). With the exception of *Salmonella* serovar Choleraesuis, CLSI susceptible and resistant MIC breakpoints for florfenicol are ≤ 2 μg/ml and ≥ 8 μg/ml, respectively (*Salmonella* serovar Choleraesuis breakpoints are one doubling dilution greater for both) (155). Since florfenicol has veterinary specific interpretive criteria, it is included in the CLSI recommended primary test and report category A for both swine and cattle (153).

As thiamphenicol is not used in the U.S., there is no CLSI quality control or interpretive criteria available.

MECHANISMS OF RESISTANCE

A review of the literature shows that numerous mechanisms of chloramphenicol resistance exist among both gram-negative and gram-positive bacteria. Chloramphenicol resistance among bacteria is frequently due to the presence of the antibiotic inactivating enzyme chloramphenicol acetyltransferase (CAT), which catalyzes the acetyl-S-CoA-dependent acetylation of chloramphenicol at the 3-hydroxyl group (211). The product 3-acetoxy chloramphenicol does not bind to bacterial ribosomes and is not an inhibitor of peptidyltransferase. The synthesis of CAT is constitutive in *E. coli* and other gram-negative bacteria which harbor plasmids bearing the structural gene for the enzyme, whereas gram-positive bacteria such as staphylococci and streptococci synthesize CAT only in the presence of chloramphenicol and related compounds (211). Additionally, other chloramphenicol resistance mechanisms have been described, including target site mutations, permeability barriers, phosphotransferase inactivation, and a growing number of reports detailing efflux systems (26, 27, 50, 99, 149, 169, 205, 211, 248).

One unfortunate development related to the discovery and characterization of multiple chloramphenicol resistance genes is the lack of a uniform nomenclature system. This has led to some confusion and inconsistencies in assigning gene descriptions as in some cases identical designations have been published for different phenicol resistance genes, whereas other times different designations have been published for virtually the same resistance gene (205). Schwarz et al. have recently put forth a nomenclature grouping system of phenicol resistance genes based on percentage identity at the DNA and amino acid levels, which we will follow as well (205).

Enzymatic Inactivation

Bacterial resistance to chloramphenicol and thiamphenicol is most commonly mediated by mono- and diacetylation by CAT enzymes, preventing the subsequent binding of chloramphenicol to the 50S ribosomal subunit (149, 205, 211). Due to the replacement of the hydroxyl group at position C-3 with a fluorine residue, the acceptor site for acetyl groups was structurally altered in the veterinary phenicol, florfenicol. This modification rendered florfenicol resistant to inactivation by CAT enzymes, and consequently, chloramphenicol-resistant strains in which resistance is solely attributed to CAT activity are susceptible to florfenicol (43).

Chloramphenicol acetyltransferase (*cat*) genes are widespread among most genera of gram-positive and gram-negative bacteria and represent the best-understood mechanism of microbial resistance to chloramphenicol (205). All naturally occurring CAT variants reported to date are in the size range of 23 to 26 kDa, and the active forms in solution are homotrimers (149, 203, 205, 211). There are two defined types of CATs which differ distinctly in their structure: the classical CATs, referred to in this chapter as type A CATs and the novel CATs, also known as xenobiotic CATs (149, 205), but referred to in this chapter as type B CATs. Schwarz et al. also identified CAT variants within the whole genome sequences of *Bacillus cereus, Brucella melitensis, Mesorhizobium loti, Rhodobacter capsulatus,* and *Streptococcus agalactiae* strain 2306 (56, 98, 102, 229, 240). However, the potential CAT variants encoded by these five presumed *cat* genes do not exhibit structural features that allow their assignment to either type A or type B at this time.

Almost all of the *cat* genes identified in gram-positive bacteria are inducibly expressed, and chloramphenicol is the inducer (211). Extensive studies of two of these genes, one from *Bacillus pumilus* (89) and one from *S. aureus* (35), have demonstrated that regulation occurs via translational attenuation with chloramphenicol acting as an inducer (130). In brief, *cat* genes are continuously transcribed, but the transcripts are not translated because the ribosome binding site (RBS) for the *cat* coding sequence is sequestered in RNA secondary structure (130, 192). Immediately upstream from, and overlapping with, the region of secondary structure is a short open reading frame (ORF) known as the leader ORF. Ribosome stalling in the leader ORF alters the RNA secondary structure, exposing the RBS for the *cat* coding sequence, resulting in its translation. It appears that the sequence of the leader peptide and the antibiotic inducer chloramphenicol dictate the site of ribosome stalling in the leader (192). Inducible *catA* genes usually mediate high-level resistance to chloramphenicol displaying MICs of ≥ 128 μg/ml (205).

Translational attenuation has also been proposed as the regulatory mechanism for the Cm-inducible *catB1* gene from *A. tumefaciens* (192). Additionally, several *catB* genes exist on gene cassettes and thus are transcribed from a common promoter located in the integron. It has been shown that cassette-borne genes located closest to the promoter are more highly expressed than distal cassettes (88, 168, 195). One relevant example to this chapter was made by Rowe-Magnus et al., who investigated the level of Cm resistance mediated by the cassette-borne gene *catB9* in relation to its position within a multidrug resistance integron from *Vibrio cholerae* (195). When *catB9* was placed in the most distal seventh position of the integron, chloramphenicol MICs were < 1 μg/ml. However, when the *catB9* cassette was placed in one of the first four positions, chloramphenicol MICs were ≥ 25 μg/ml (195).

Type A chloramphenicol acetyltransferases

Type A CATs have been detected among a large number of gram-positve and gram-negative bacteria (149, 205, 212, 214). Despite amino acid differences, type A CATs do share some common properties. For example, all currently described type A CATs share conserved amino acids which are involved in either substrate binding, catalytic activities, folding of the monomers, or assembly of the monomers to a trimer (205). Some type A CATs possess specific properties, such as an enhanced susceptibility to inactivation by thiol-specific modifying reagents, or the ability to confer coresistance to fusidic acid (148, 241).

There are at least 16 distinct groups of *catA* genes, A-1 to A-16 (Table 1). The analogous type A CAT proteins assigned to each group exhibit amino acid sequence identities of more than 80%. The archetype *cat* gene of group A-1 (*catI*) was originally characterized as part of transposon Tn*9* in *E. coli* (7) and has been identified on multiple resistance plasmids from different gram-negative bacteria, including *Acinetobacter* spp., and *Photobacterium damselae* subsp. *piscicida,* formerly known as *Pasteurella piscicida* (71, 116). Group A-1 *cat* genes have more recently been identified in *S. enterica* serovar Typhi, *Serratia marcescens,* and *Shigella flexneri* (83, 131, 164). Additionally, enzymes resembling that encoded by *catI* have also been detected in the anaerobe *Bacteroides ochraceus* (211). The group A-2 *cat* genes (*catII*) have primarily been found on plasmids of *H. influenzae, Haemophilus parainfluenzae,* and *H. ducreyi,* which usually possess additional resistance genes (e.g., tetracycline) (148, 189, 191). Genes related to *catII* have also been detected in *E. coli* and *P. damselae* subsp. *piscicida* (144,

Table 1. Type A chloramphenicol acetyltransferases

Group	Gene designation(s)	Bacterium	Gene location[a]	Database accession no(s).	Reference(s)[b]
A-1	*cat*I	*Escherichia coli*	Tn*9*, R429	V00622	7
	*cat*I	*Acinetobacter baumannii*	Chrom. (Tn*2670*)	M62822	71
	cat	*Acinetobacter calcoaceticus*	Tn*2670*-like	M37690	71
	pp-cat	*Photobacterium damselae*	pSP9351	D16171	115
	cat	*Salmonella* serovar Typhi	pHCM1	AL513383, NC_003384	164
	cat	*Serratia marcescens*	R478	NC_005211, BX664015	83
	cat	*Shigella flexneri*	Chrom.	AF326777	131
A-2	*cat*II	*Haemophilus influenzae*	pRI234, pMR375	X53797	148
	*cat*II	*Escherichia coli*	pSa	X53796	148
	*cat*II	*Aeromonas salmonicida*	pAR-32	AJ517791	219
	cat	*Photobacterium damselae*	Plasmid	AB082569	144
A-3	*cat*III	*Shigella flexneri*	R387	X07848	147
	catA3	*Mannheimia* taxon 10	pMHSCS1	AJ249249	108
	catA3	*Mannheimia varigena*	pMVSCS1	AJ319822	109
	cat	Uncultured eubacterium	pIE1130	NC_004973, AJ271879	230
A-4	*cat*	*Proteus mirabilis*	Chrom.	M11587	47
A-5	*cat*	*Streptomyces acrimycini*	Chrom.	P20074	146
A-6	*cat86*	*Bacillus pumilus*	Chrom.	K00544	90
A-7	*cat* (pC221)	*Staphylococcus aureus*	pC221	X02529	33, 181
	*cat*C, *cat*	*Staphylococcus aureus*	pKH7	U38429	33, 181
	cat	*Staphylococcus aureus*	pUB112	X02872	143
	cat	*Staphylococcus intermedius*	pSCS1	M64281	35
	cat	*Staphylococcus aureus*	pSCS6	X60827	207
	cat	*Bacillus subtilis*	pTZ12	M16192	45
	cat	*Streptococcus agalactiae*	pGB354	U83488	9
	cat	*Streptococcus agalactiae*	pIP501	X65462	125
	cat	*Enterococcus faecalis*	pRE25	X92945	235 200
A-8	*cat* (pC223)	*Staphylococcus aureus*	pC223	NC_005243, AY355285	218
	cat	*Staphylococcus aureus*	pSCS7	M58516	202
	cat	*Listeria monocytogenes*	pWDB100	X68412	87
	cat	*Staphylococcus aureus*	pSBK203	M90091	41
	cat	*Lactococcus lactis*	pK214	X92946	173
	cat	*Staphylococcus haemolyticus*	pSCS5	M58515	201
	cat	*Enterococcus faecium*	pRUM	NC_005000, AF507977	184
A-9	*cat* (pC194)	*Staphylococcus aureus*	pC194	V01277	95
	cat-TC	*Lactobacillus reuteri*	pTC82	U75299	128
	cat	*Streptococcus suis*	TN*Ss1*	AB080798	225
	cat	*Staphylococcus aureus*	pMC524-MBM	AJ312056	24
A-10	*cat*	*Bacillus clausii*	Chrom.	AY238971	78
A-11	*catP*	*Clostridium perfringens*	pIP401:Tn*4451*	U15027	15
	catP	*Neisseria meningitidis*	Chrom.	AF031037	77
	*cat*D	*Clostridium difficile*	Chrom. (Tn*4453*)	X15100, AF226276	132
A-12	*catS*	*Streptococcus pyogenes*	Chrom.	X74948	234
A-13	*cat*	*Campylobacter coli*	pNR9589	M35190	243
A-14	*cat*	*Listonia anguillarum*	pJA7324	S48276	260
A-15	*catB*	*Clostridum butyricum*	Chrom.	M93113	97
A-16	*catQ*	*Clostridium perfringens*	Chrom.	M55620	16

[a] Plasmid and transposon designations if available; chrom., chromosome.
[b] Modified from reference 205.

148). A CAT enzyme similar to that encoded by *catII* was also reportedly identified in *Bacteroides fragilis* (211). The members of group A-3 are commonly found on plasmids in *Enterobacteriaceae* (147) and *Pasteurellaceae* (108, 109, 237), but a *cat* gene from this group has also been detected on a plasmid from an uncultured eubacterium plasmid (230). The CAT groups A-4 to A-6 are represented by unique *cat* genes, all of which are chromosomally located in either *Proteus mirabilis, Streptomyces acrimycini,* or *B. pumilus* (47, 90, 146).

Group A-7 to A-9 *cat* genes have a broad host range and have been found in a variety of gram-positive bacteria including *Staphylococcus* spp. (24, 33, 35, 41, 45, 95, 181, 201, 202, 207, 218), *Streptococcus* spp. (125, 170, 171, 225, 234, 235, 251), *Enterococcus* spp. (85, 171, 200, 234), *Bacillus subtilis* (9), *Listeria monocytogenes* (87), *Lactobacillus reuteri* (128), and *Lactococcus lactis* (172, 173). The prototype plasmids from which the respective *cat* genes sequences were deposited in the databases are pC221 (33, 181), pSCS7 (202), and pC194 (95). These three *cat* gene groups are commonly located on small multicopy plasmids which carry either the *cat* gene alone or in combination with a streptomycin or macrolide resistance gene (204, 206, 235). In some instances, such genes were associated with multidrug resistance plasmids or conjugative transposons (13, 173, 245).

The only *cat* representative of group A-10 was found in the chromosome of the probiotic bacterium *Bacillus clausii* (78). Group A-11 comprises two *cat* genes to date, *catP* and *catD,* identified in either *Clostridium* spp. or *N. meningitidis* isolates (15, 77, 97, 133, 216, 220, 253). The *catP* gene identified in *Clostridium perfringens* was shown to be a part of the chloramphenicol resistance transposon Tn*4451,* which is able to integrate into both plasmids and the chromosome (15). The spread of Tn*4451*- or Tn*4453*-associated *catP* genes from *Clostridium* spp. to *N. meningitidis* represents a serious health threat since chloramphenicol is still the standard therapy for meningococcal meningitis in developing countries (77).

Groups A-12 to A-16 only contain one gene each to date. The *catS* gene from *S. pyogenes* represents group A-12 and has been detected in streptococcal isolates of serogroups A, B, and G (234). Even though only a portion of this gene sequence is deposited in the NCBI database, analysis of the amino acid sequence deduced from this internal segment shows approximately 77% identity to the *catP* and *catD* gene products. The *cat* gene of group A-13 is located on plasmid pNR9589 of *Campylobacter coli,* which also harbors an *aphA*-3 gene coding for kanamycin resistance (243), whereas the group A-14 *cat* gene has been shown to be part of the multidrug resistance plasmid from *Vibrio anguillarum* (reclassified as *Listonia anguillarum*), exhibiting resistance to tetracycline, sulfonamides, and streptomycin (260). The representatives of the remaining *cat* gene groups of A-15 and A-16 include the *catB* gene from *Clostridium butyricum* and *catQ* gene from *C. perfringens,* which are both chromosomally located (16, 97).

NCBI GenBank database searches also revealed the presence of putative *catA* genes in several other bacteria, including *Bacillus anthracis* Ames (185), *Deinococcus radiodurans* R1 (249), *Clostridium acetobutylicum* ATCC 824 (161), *Clostridium tetani* E88 (36), *Zymomonas mobilis* (123), and *Bacteroides thetaiotaomicron* VPI-5482 (255). These putative *catA* genes were identified through whole genome sequencing and may be considered as separate new groups if CAT activity is confirmed in the future.

Type B chloramphenicol acetyltransferases

Type B CATs, also referred to as xenobiotic acetyltransferases, also inactivate chloramphenicol by acetylation. Type B CATs share some common properties with the type A CATs: native type B CATs are also homotrimers composed of monomers which are in the range of 209 to 212 amino acids (149). However, on the basis of their amino acid sequences, type B CATs differ distinctly from type A CATs in their structure and seem to be related to other acetylating enzymes identified in *Staphylococcus* and *Enterococcus* involved in resistance to streptogramin A compounds, such as Vat(D) (formerly known as SatA) (186), Vat(E) (formerly known as SatG) (246), Vat(A) (formerly known as Vat) (6), or Vat(B) (5).

There are at least five different groups of type B *cat* genes: B-1 to B-5 (Table 2). The first type *catB* gene, *catB1,* was detected in the chromosome of *Agrobacterium tumefaciens* (84, 227). The *catB2* gene was initially characterized on the *E. coli* multidrug resistance transposon Tn*2424* (163). Closely related group B-2 *catB2* genes have also been identified on plasmids from *Salmonella* serovar Enteritidis, *P. multocida, B. bronchiseptica,* or from an uncultured eubacterium (101, 205, 239, 254). The group B-3 comprises a number of genes so far designated as *catB3* through *catB6,* and *catB8* (Table 2). These genes have often been associated with either multidrug resistance transposons such as Tn*840* from *Morganella morganii* (120) or plasmid-borne multidrug resistance integrons from a number of bacterial species comprising the *Enterobacteriaceae, B. bronchiseptica* as well as *P. aeruginosa* (37, 96, 101, 117, 121, 124, 162, 233, 242, 257). Group B-4 only contains the *catB7* gene, which has only been found in the chromosomes of both *P. aeruginosa* PAO222 and PAO1 (222, 250), whereas

Table 2. Type B chloramphenicol acetyltransferases

Group	Gene designation(s)	Bacterium	Gene location[a]	Database accession no(s).	Reference(s)[b]
B-1	*cat, catB*	*Agrobacterium tumefaciens*	Chrom.	M58472	227
	cat	*Agrobacterium tumefaciens*	Chrom.	NC_003063	84
B-2	*catB2*	*Escherichia coli*	pNR79:Tn*2424*	AF047479	163
	catB2	*Salmonella* serovar Enteritidis	Plasmid	AJ487034	239
	catB2	Uncultured eubacterium	pSp39	AY139601	228
	catB2	*Pasteurella multocida*	pJR1	NC_004771, AY232670	254
	catB2	*Bordetella bronchiseptica*	pKBB958	AJ879564	101
B-3	*catB3*	*Salmonella* serovar Typhimurium	pWBH301	AJ009818	37
	catB3	*Salmonella* serovar Typhimurium	IncF1 plasmid	AJ310778	233
	catB3	*Acinetobacter baumannii*	Chrom.	AF445082	96
	catB3	*Escherichia coli*	pHSH2	AY259086	242
	catB3	*Bordetella bronchiseptica*	pkBB668	AJ844287	101
	catB4	*Enterobacter aerogenes*	pWBH301	U13880	37
	catB4	*Klebsiella pneumoniae*	pEKP0787-1	AF322577	257
	cat, catB5	*Morganella morganii*	Tn*840*	X82455	120
	catB6	*Pseudomonas aeruginosa*	pPAM-101	AJ223604	121
	catB8	*Klebsiella pneumoniae*	pKB42	AF227506	117
	catB8	*Salmonella* serovar Typhi	pST2301	AY123251	162
	catB8	*Pseudomonas aeruginosa*	Unknown	AF322577	124
B-4	*catB7*	*Pseudomonas aeruginosa*	Chrom.	AF036933	250
	catB7	*Pseudomonas aeruginosa*	Chrom.	AE004506	222
B-5	*catB9*	*Vibrio cholerae*	Chrom.	AF462019	195
	catB9	*Vibrio cholerae*	Chrom.	NC_002506	91

[a] Plasmid and transposon designations if available; chrom., chromosome.
[b] Modified from reference 205.

group B-5 solely consists of the *catB9* genes located as part of a super-integron in the chromosome of *V. cholerae* (91, 194).

During whole genome sequencing, a gene related to group B-2 *cat* genes was found in the chromosome of *Shewanella oneidensis* strain MR-1, and a putative *catB* gene was further detected in the chromosome of *Vibrio parahaemolyticus* (92, 134). Much like the story with the putative *catA* genes described above, CAT activity has yet to be determined. Incomplete amino acid sequences of additional CAT B proteins from *Serratia marcescens, P. aeruginosa, Bacillus sphaericus,* and *S. aureus* have also been reported (149) and suggest a wider distribution of *catB* genes among gram-positive and gram-negative bacteria than previously thought.

Chloramphenicol Drug Exporters

Energy-driven drug efflux systems are increasingly recognized as mechanisms of antibiotic resistance in a wide range of gram-positive and gram-negative bacteria. The first evidence that resistance to antibiotics was caused by an active efflux was put forward in 1980 by Levy and colleagues (139). Chromosomally located or acquired by bacteria, they can either be activated by environmental signals or by regulatory gene mutations. These pumps are classified in two groups according to their energy source, the secondary transporters using the proton motive force (PMF) and the ATP-binding cassette (ABC) transporters using ATP hydrolysis (182). These pumps extrude toxic compounds, limiting their accumulation in cells, and may have limited or broad substrates. Most of the bacterial MDR pumps use the PMF, whereas only two examples of ABC-type multidrug transporter have been well characterized in bacteria, LmrA in *Lactococcus lactis* and MsrA in *S. aureus* (68, 187).

Specific exporters and multidrug transporters have both been associated with chloramphenicol/florfenicol resistance (40, 179, 180, 182). Specific transporters have a substrate spectrum which is commonly limited to a small number of structurally closely related compounds and are often associated with mobile genetic elements which can easily be interchanged between bacterial species. Specific transporters also commonly mediate distinctly higher levels of resistance as compared to those of multidrug transporters (205). Specific efflux systems have been mainly identified with resistances to macrolides, lincosamides, and/or streptogramins, tetracyclines, as well as chloramphenicol/florfenicol in both gram-positive and gram-negative bacteria (40, 179, 180, 182, 205). The multidrug transporters, in contrast, are able to remove antimicrobials

Table 3. Drug exporters mediating resistance to chloramphenicol or chloramphenicol-florfenicol

Group	Gene designation(s)	Bacterium	Gene location[a]	Database accession no(s).	Reference(s)[b]
E-1	*cmlB*, *cmlA2*	*Enterobacter aerogenes*	pIP833	AF034958	174
	cmlA	*Salmonella* serovar Typhimurium	Plasmid	AJ487033	44
	cmlA5	*Escherichia coli*	R751 (Tn*2000*)	AF205943	177
	cmlA1	*Klebsiella pneumoniae*	pILT-3	AF458080	175
	cmlA1	*Pseudomonas aeruginosa*	RPL11 (Tn*1403*)	AF313472	168
	cmlA4	*Klebsiella pneumoniae*	pTK1	AF156486	176
	cmlA4	*Salmonella* serovar Agona	Plasmid	AJ867237	141
	cmlA5	Uncultured bacterium	pSp1	AY115475	228
	cmlA6	*Pseudomonas aeruginosa*	Plasmid	AF294653	12
	cmlA7	*Pseudomonas aeruginosa*	Chrom.	AJ511268	188
	cmlA, *cmlA1*	*Pseudomonas aeruginosa*	pR1033:Tn*1696*	U12338, M64556, AF078527	28, 221
E-2	*cml*	*Escherichia coli*	R26	M22614	60
E-3	*cmlA*-like	*Salmonella* serovar Typhimurium DT104	Chrom.	AF071555	34
	floR	*Salmonella* serovar Typhimurium DT104	Chrom.	AF118107	10
	flo	*Salmonella* serovar Typhimurium DT104	Chrom.	AJ251806	74
	floR	*Salmonella* serovar Typhimurium DT104	Chrom.	AF261825	32
	floR	*Salmonella* serovar Typhimurium DT104	Chrom.	AY339985	118
	flo	*Escherichia coli*	Plasmid	AF252855	248
	floR	*Escherichia coli*	Plasmid	AF231986	50
	floR	*Escherichia coli*	pMBSF1	AJ518835	29
	flo	*Escherichia coli*	pEC05	AY775258	215
	floR	*Escherichia coli*	Plasmids	AY499129, AY499130, AY517519	67
	floR	*Klebsiella pneumoniae*	R55	AF332662	49
	floR	*Vibrio cholerae*	Chrom. (SXT element)	AY034138	94
	floR	*Vibrio cholerae*	Chrom. (SXT element)	AY055428	20
	floR	*Pasteurella multocida*	pCCK381	AJ871969	112
	pp-flo	*Photobacterium damsela*	pSP92088	D37826	116
E-4	*fexA*	*Staphylococcus lentus*	pSCFS2	AJ549214	110
E-5	*cml*	*Streptomyces lividans*	Chrom.	X59968	58
E-6	*cmlv*	*Streptomyces venezuelae*	Chrom.	U09991	145
E-7	*cmrA*	*Rhodococcus erythropolis*	Tn*5561*	AF015087	150
	cmr	*Rhodococcus fascians*	pRF2	Z12001	57
E-8	*cmr*	*Corynebacterium glutamicum*	pXZ10145	U85507	262
	cmx	*Corynebacterium striatum*	pTP10:Tn*5564*	AF024666	226

[a] Plasmid and transposon designations if available; chrom., chromosome.
[b] Modified from reference 205.

of different classes and other toxic compounds from the bacterial cell and occasionally play a role in the intrinsic resistance of some bacteria to certain antimicrobials. These genes are usually located on the bacterial chromosome; however, a multidrug resistance efflux pump was recently identified on a conjugative plasmid in *E. coli* (89).

Specific exporters

Genes associated with either the export of chloramphenicol or florfenicol have been identified in a plethora of gram-negative and gram-positive bacteria. Short descriptions of mobile genes coding for specific efflux proteins, including those mediating the export of chloramphenicol and/or florfenicol, have recently been published by Butaye et al. and Schwarz et al. (40, 205). At least eight different groups of specific exporters, E-1 to E-8, are currently known (Table 3). Only the exporters assigned to groups E-3 and E-4 have been reported to mediate resistance to both chloramphenicol and florfenicol (205).

The first report of chloramphenicol resistance not attributed to enzyme inactivation was published in 1979 in *P. aeruginosa* and was associated with the Tn*1696* transposon residing on a P-1-type plasmid R1033 (196). Later studies demonstrated that this resistance phenotype was attributed to the presence of a gene conferring nonenzymatic resistance to chloramphenicol, termed *cmlA*, on the In4 integron of Tn*1696* (28). Sequence analysis of the *cmlA* gene showed a major hydrophobic polypeptide of 419 amino acids

(44, 228) which consisted of 12 predicted transmembrane domains, suggesting similarity to other transmembrane transport proteins of the major facilitator superfamily (28, 80, 221). It was further shown to be the first gene cassette to contain its own promoter, and regulation of *cmlA* expression was inducibly regulated via translational attenuation (221). The presence of the cloned *cmlA* gene in *E. coli* leads to, in addition to active pumping of the drug from the cell, a reduction in outer membrane permeability to chloramphenicol by repressing the synthesis of a major porin protein (28).

A number of closely related or identical *cmlA* genes have been identified from several gram-negative bacteria including *E. coli* (26, 248), *Salmonella* (42, 44, 79, 141), *Klebsiella pneumoniae* (175, 176), and *P. aeruginosa* (12, 168, 188) and have been assigned to group E-1 (Table 3). The least similar member of group E-1 is the *cmlA2* gene (e.g., *cmlB*) identified on the plasmid located In*40* integron of *Enterobacter aerogenes* BM2688 (174). The *cmlA2* gene had 83.7% nucleotide sequence identity with the previously described nonenzymatic chloramphenicol resistance *cmlA1* gene. The deduced CmlA2 protein of 409 amino acids, shared only 85.5% identity with CmlA1 and 50.3% identity with the polypeptide predicted from the *P. damselae* subsp. *piscicida pp-flo* gene, which confers resistance to florfenicol (174). However, much like *cmlA*, the *cmlA2* gene is also expressed from its own promoter.

Another nonenzymzatic chloramphenicol resistance gene, termed *cml*, was shown to be located on plasmid R26 and was inducible in the presence of chloramphenicol (59, 60). The CML protein does not share significant homology to other characterized chloramphenicol resistance proteins and only consists of 302 amino acids and five transmembrane segments. It is, however, in part similar to the distinctly larger CmlA protein of Tn*1696* and represents group E-2. Much like *cmlA* homologues, it does not mediate resistance to fluorinated chloramphenicol analogs, e.g., florfenicol (59).

A number of genes have recently been identified that confer resistance to chloramphenicol and florfenicol via efflux and have been combined in group E-3 (Table 3). These genes have been given several designations ranging from *pp-flo*, *cmlA*-like, to *flo* or *floR;* however, they are all closely related and show 91 to 100% identity in their nucleotide sequences and 89 to 100% identity in the amino acid sequences of their products (50, 62, 113, 116, 140, 205, 248). The first report detailing a member of this group was published in 1996 by researchers in Japan who identified a novel plasmid-encoded gene (*pp-flo*) from *P. damselae* subsp. *piscicida* that encoded resistance to both chloramphenicol and florfenicol (116). More recently, Bolton et al. described a gene with 97% homology to the *pp-flo* gene among *Salmonella enterica* serovar Typhimurium DT104 isolates, which they termed flo_{St} (30). Many isolates of serovar Typhimurium DT104 conferring the classic penta-resistance phenotype of ampicillin, chloramphenicol (florfenicol), streptomycin, sulfonamides, and tetracycline (ACSSuT) possess a similar genetic makeup comprising the *flo* and *tet*(G) genes bracketed by two class 1 integrons carrying the *pse-1* and *aadA2* cassettes clustered on a 14-kb region of the serovar Typhimurium DT104 genome (11, 30, 32, 34, 51, 74). This antibiotic resistance gene cluster is included in a chromosomal genomic island called *Salmonella* genomic island 1 (SGI1) (32). SGI1 is approximately 43 kb in size and is located between the *thdF* gene and a novel retron sequence in the chromosome of serovar Typhimurium DT104 (32). SGI1 variants have also been identified in serovar Typhimurium DT120 and other *Salmonella enterica* serovars, including Agona, Paratyphi B, Albany, and Newport, suggesting horizontal gene transfer of this region (31, 32, 52, 61, 63, 140, 205).

The *flo* gene has also been identified on plasmids and in the chromosome of chloramphenicol-resistant *E. coli* recovered from cattle, poultry, and pigs in the U.S., France, Germany, and more recently China (27, 29, 50, 65, 67, 113, 248). It has also been characterized on the IncC plasmid R55 from *K. pneumoniae* and a closely related plasmid from *Salmonella* serovar Newport (49, 140), and in the large conjugative elements (SXT constins) of *V. cholerae* E1 Tor O1 and O139 isolates (20, 94). Analysis of plasmid DNA sequences upstream and downstream of the *flo* gene and in the SXT elements showed regions with large homologies. Additionally, DNA sequence analysis revealed the presence of ORFs (*orfA*, *orfA5′Δ3′*, *orfA3′Δ5′*) whose gene products showed considerable homology to transposase proteins, suggesting that the *flo* gene may be located within a mobile genetic element (205). Recently the *flo* gene was found to be part of the novel 4284-bp transposon Tn*floR* from *E. coli.* Tn*floR* consists of the gene *floR*, a putative regulatory gene and the transposase gene *tnpA* (formerly known as *orfA*). A circular form of *TnfloR* was detected and suggested the mobility of this transposon (64).

Recently the first staphylococcal florfenicol-chloramphenicol efflux gene, *fexA,* was detected on the pSCFS2 plasmid in *Staphylococcus lentus* and represents group E-4 (110). The FexA protein consists of 475 amino acids and exhibits 14 transmembrane domains. Induction of *fexA* expression by chloramphenicol and florfenicol occurs via translational attenuation. However, the FexA protein differs from all previously known proteins involved in chloramphenicol and florfenicol efflux (110). The gene *fexA* was recently shown to be part of the novel transposon

Tn*558* from *S. lentus* (111). Similarities between Tn*558* and Tn*554* from *Staphylococcus aureus* included the arrangement of the transposase genes *tnpA* to *C* and an *att554*-like target sequence. Circular forms of Tn*558* were also detected and suggested that this transposon was functionally active. Although the members of this transposon family share the same overall structure and mode of transposition, they differed distinctly in their resistance gene regions. The macrolide-lincosamide-streptogramin B resistance gene *erm*(A) and the spectinomycin resistance gene *spc* in Tn*554* were replaced by a variant of the streptogramin A resistance gene *vga*(A) in Tn*5406* and by the florfenicol-chloramphenicol exporter gene *fexA* and a putative oxidoreductase gene in Tn*558* (111). The identification of *fexA* as part of a functionally active transposon is an important observation with regard to the mobility of *fexA* and the spread of combined resistance to florfenicol and chloramphenicol.

Additional chloramphenicol exporters have been identified among bacteria commonly associated with soil and the environment. Chloramphenicol efflux genes have been identified among *Streptomyces* (groups E-5 and E-6), *Rhodococcus* (group E-7), or *Corynebacterium* species (group E-8) (57, 58, 145, 150, 226, 262). The *cml* gene identified in the chromosome of *Streptomyces lividans* 1326 encodes a protein with 12 transmembrane segments (58), whereas the chromosomal *cmlv* gene of the chloramphenicol producer *S. venezuelae* ISP5230, encodes a protein of 10 transmembrane segments assumed to be involved in chloramphenicol export (145). Other chloramphenicol exporters consisting of 12 transmembrane domains include those encoded by the *cmr* and *cmrA* genes from *Rhodococcus fascians* and *Rhodococcus erythropolis* (57, 150). The *cmr* gene was found on the conjugative plasmid pRF2 from *Rhodococcus fascians* (163), while the related *cmrA* gene appeared to be associated with transposon Tn*5561* in *R. erythropolis* (164). A chloramphenicol efflux protein, termed *cmx*, has also been identified in a clinical multidrug resistant *Corynebacterium striatum* isolate and was associated with the Tn*5564* transposable element (226). A closely related chloramphenicol resistance gene, *cmr*, was found on a transposon-like element closely related to Tn*5564* on plasmid pXZ10145 in *Corynebacterium glutamicum* (262). Unlike the *cmlA* and *flo* homologues, the Cmx and Cmr proteins from *Corynebacterium* spp. appear to have only 10 transmembrane segments (205).

As previously seen with the *catA* and *catB* genes, annotations of florfenicol resistance genes have also been detected in whole genome sequences of *C. tetani* strain E88 (36), *B. melitensis* (56), *Yersinia pestis* (165), *Fusobacterium nucleatum*, and *F. nucleatum* subsp. *vincentii* (103, 104), *Legionella pneumophila* (183), *Heliobacillus mobilis* (184), and the photosynthetic green-sulfur bacterium *Chlorobium tepidum* TLS (70). However, the mechanisms of these corresponding proteins in florfenicol resistance have yet to be determined.

Multidrug transporters

In addition to specific exporters, a number of multidrug transporter systems have been identified whose substrate spectrum includes chloramphenicol and/or florfenicol. (Table 4) Generally speaking, chlor-

Table 4. Multidrug transporters associated with chloramphenicol efflux in gram-negative bacteria

Bacterium	Efflux pump	Regulator gene(s)	Gene location[a]	Database accession no(s).	Reference(s)
Escherichia coli	AcrAB-TolC, AcrEF-TolC	*acrR*, *marA*, *soxS*, *rob*, *sdiA*	Chrom.	M94248, U00734, AE000152	4, 126, 231
Salmonella serovar Typhimurium	AcrAB	?	Chrom.	U78314, AF209870	17, 18
Escherichia coli	OqxAB	?	pOLA52	AY241669	89
Pseudomonas aeruginosa	MexAB-OprM, MexCD-OprJ	*mexR* (AB); *nfxB* (CD)	Chrom.	L11616, U23763, AB011381	142, 180
Pseudomonas aeruginosa	MexEF-OprN	*mexT*	Chrom.	X99514, AJ007825	142, 180
Escherichia coli	MdfA (CMR)	?	Chrom.	Y08743, U44900	25, 159
Vibrio cholerae	VceABC	?	Chrom.	AF012101	53, 73
Burkholderia cepacia	CeoAB	*ceoR*	Chrom.	U97042, U38944	151
Pseudomonas putida	ArpABC	*arpR*	Chrom.	AF183959	114
Pseudomonas putida	TtgABC, TtgDEF, TtgGHI	*ttgR* (ABC)	Chrom.	AF031417, Y19106, AF299253	178, 193
Pseudomonas stutzeri	TbtABM	?	Chrom.	AF041464	100

[a] Plasmid and transposon designations if available; chrom, chromosome.

amphenicol and florfenicol MICs are usually lower in bacteria expressing multidrug transporter systems as compared to those specified by specific exporters (e.g., *acrAB* versus *flo*).

The multiple antibiotic resistance (*mar*) locus and *mar* regulon in *E. coli* and other members of the *Enterobacteriaceae* is a paradigm for a generalized response locus leading to increased expression of efflux pumps. One such efflux system, the AcrAB pump, extrudes biocides such as triclosan, chlorhexidine, and quaternary ammonium compounds as well as multiple antibiotics, including phenicols (126). Located in the inner membrane, the drug:proton antiporter AcrB captures its substrates within the phospholipid bilayer and transports them into the external medium via the outer membrane channel TolC (231). The cooperation between AcrB and TolC is presumably mediated by the periplasmic protein AcrA. All three components are required for efficient transport, because disruption of any of the three genes results in hypersusceptibility of *E. coli* to various substrates (231).

In *E. coli*, overexpression of the transcriptional activator *marA* results in the decreased expression of the OmpF porin protein in addition to increased expression of the AcrAB multidrug efflux pump (4). This mechanism confers decreased susceptibility to a large number of structurally unrelated antimicrobial agents, including chloramphenicol and florfenicol. The importance of this efflux system was reiterated recently by Baucheron et al., who showed that despite the presence of the *flo* efflux pump in *Salmonella* serovar Typhimurium DT104, chloramphenicol-florfenicol resistance appeared to be highly dependent on the presence of AcrAB-TolC, since the introduction of mutations in the respective *acrB* and *tolC* genes resulted in susceptible or intermediate resistance phenotypes (18). Another example of a chloramphenicol multidrug transporter in *E. coli* is MdfA (also termed Cmr), a 410 amino acid long membrane protein of the MFS family of secondary transporters (25, 159). Close homologues of MdfA have recently been identified in several pathogenic bacteria including *S. flexneri* (99% homology), *Salmonella* serovar Typhi (90% homology), and *Y. pestis* (73% homology) (127). Transport experiments have shown that MdfA is a multidrug resistance transporter that is driven by the proton electrochemical gradient and functions as a drug/proton antiporter (1).

Multidrug transporters whose substrate spectrum includes chloramphenicol have also been described in *P. aeruginosa* (178, 180). Similar to the AcrAB-TolC system in *E. coli*, these multidrug transporters are also composed of three components. To date, seven resistance-nodulation-division (RND) multidrug efflux systems have been described in *P. aeruginosa*, including MexAB-OprM, MexCD-OprJ, MexEF-OprN, MexXY-OprM, MexJK-OprM, and, most recently, MexHI-OpmD and MexVW-OprM (142). The genes encoding these systems are arranged as operons, with the first gene encoding a linker protein that is associated with the cytoplasmic membrane (MexA, MexC, MexE, MexX, MexJ, MexH, and MexV). The second gene encodes an efflux pump that exports substrates across the inner membrane (MexB, MexD, MexF, MexY, MexK, MexI, and MexW), and in several instances, a third gene encodes an outer membrane protein, which facilitates passage of the substrate across the outer membrane (OprM, OprJ, and OprN). The three components form a channel that traverses the inner membrane, periplasm, and outer membrane and allows the substrate to be effluxed directly from the cytoplasm or cytoplasmic membrane to the extracellular environment (178, 180). The MexAB-OprM was the first described MDR efflux system in *P. aeruginosa* and is constitutively expressed in wild-type strains. Expression contributes to intrinsic and acquired resistance to a number of antimicrobials including chloramphenicol (180).

Similar multidrug transporters which can also export chloramphenicol have been identified in the cystic fibrosis pathogen *Burkholderia cepacia* (CeoAB-OpcM) and *Pseudomonas putida* (ArpAB-ArpC; TtgAB-TtgC; TtgGH-TtgI) (151, 178, 193). Five ORFs have been identified in the *B. cepacia* antibiotic efflux gene cluster. There is apparently a single transcriptional unit, with *llpE* encoding a lipase-like protein, *ceoA* encoding a putative periplasmic linker protein, *ceoB* encoding a putative cytoplasmic membrane protein, and *opcM* encoding a previously described outer membrane protein (151). A putative LysR-type transcriptional regulatory gene, *ceoR*, is divergently transcribed upstream of the structural gene cluster. Nair et al. showed that when the three putative pump structural genes (*ceoAB-opcM*) were cloned into the plasmid construct pAM1, a 10-fold increase in chloramphenicol resistance was observed (151).

The *arpABC* operon of *P. putida* is involved in the active efflux of multiple antibiotics, such as tetracycline, carbenicillin, streptomycin, erythromycin, novobiocin, and chloramphenicol (114). The TtgGHI and TtgABC efflux pumps of *P. putida* play a key role in the innate and induced tolerance to organic solvents and antibiotics such as ampicillin and chloramphenicol (193). Tributyltin (TBT) is a toxic agent used in marine antifouling paints. TBT-resistant strains of *Pseudomonas stutzeri* have been isolated among the bacterial flora of a polluted harbor and associated with the presence of the *tbtABM* operon (100). TbtABM exhibited the greatest homology (60.9 to 84.9%) with the TtgDEF and SrpABC systems and conferred multidrug resistance to *n*-hexane, nalidixic acid, sulfamethoxazole, and chloramphenicol (100). The deduced amino acid sequences encoded by these efflux systems show a striking resemblance to proteins of the RND division family.

Nishino and Yamaguchi cloned all 37 putative drug transporter genes in the *E. coli* chromosome and evaluated their drug resistance phenotypes (160). Only three plasmid constructs conferred increased resistance to chloramphenicol compared to the control strain: pUCmdfA (16-fold), pUCydhE (twofold), and pU-CacrAB (eightfold). Additionally, the *yidY* ORF was cloned into the pTrc6His expression vector and increased chloramphenicol resistance twofold, suggesting potential involvement (160). None of the other cloned drug extrusion translocases contributed to elevated chloramphenicol MICs.

The situation is not so concrete with regard to the contribution of multidrug transporters from gram-positive bacteria, such as NorA from *S. aureus* or Bmr and Blt from *B. subtilis* with chloramphenicol resistance (2, 158, 205, 258). Though data show some association, it is difficult at this time to determine if these mechanisms have a significant impact on chloramphenicol resistance among these organisms.

The recent isolation of a conjugative plasmid from *E. coli* containing a multidrug efflux pump may signify the early stages of mobilization of these transport systems to DNA mobile elements (89). Hansen et al. showed that this new multidrug efflux system, termed *oqxAB,* conferred resistance to the swine growth enhancer, olaquindox, as well as to chloramphenicol (MICs > 64 μg/ml). This efflux system is homologous to several previously described resistance-nodulation-cell-division family efflux systems from several different bacterial species, and efflux was shown to be dependent on the host TolC outer membrane protein (89).

Through much of the published literature, it is unknown if florfenicol is a substrate of most of these multidrug transporter systems, as chloramphenicol has been the phenicol antimicrobial used most often in studies. However, unpublished data in our laboratory indicates that at least the AcrAB-TolC system in *E. coli* provides resistance to florfenicol (D. White, personnel communication). Utilizing strains from Dr. Stuart B. Levy and Dr. Laura McMurry (Tufts University School of Medicine, Boston, Mass.), it was determined that two high level Mar mutants, LM312.1 and LM313.1, demonstrated high level resistance to florfenicol (>256 μg/ml) in the absence of either the *flo* or *cmlA* genes, indicating that florfenicol maybe a substrate for the AcrAB efflux system in *E. coli* (Table 5). Subsequent research appears to support this hypothesis as increased chloramphenicol and florfenicol MICs have been associated with overexpression of the AcrAB-TolC efflux system in *E. aerogenes* (81).

Table 5. Effect of Mar mutations on chloramphenicol and florfenicol antimicrobial susceptibilities

Strain	Genotype	MIC (μg/ml)[a]		PCR result[b]	
		CML	FFC	*flo*	*cmlA*
AG100	Wild type	4.0	8.0	−	−
AG102	Mar mutant	32	8.0	−	−
LM312.1	Mar mutant	128	>256	−	−
LM313.1	Mar mutant	128	>256	−	−
CVM1804	Clinical strain	32	16	−	+
CVM1805	Clinical strain	>256	64	−	+
CVM1808	Clinical strain	>256	>256	+	−
CVM1831	Clinical strain	>256	>256	+	−

[a] Antimicrobial susceptibility was determined following NCCLS/CLSI broth microdilution methods (156). CML, chloramphenicol; FFC, florfenicol.
[b] PCR using *flo* and *cmlA* primers (27).

Lastly, it is important to remember that several types of multidrug transporters may coexist together in the same bacterium along with specific transporters on mobile genetic elements. For example, Lee et al. investigated the effects of simultaneous expression of several efflux pumps, including specific exporters such as CmlA and several multidrug transporters such as MdfA, AcrAB-TolC, or MexAB-OprM on antimicrobial resistance in *E. coli* and *P. aeruginosa* (122). They further demonstrated that when efflux pumps of different structural types were combined in the same cell, the observed antibiotic resistance was much higher than that conferred by each of the pumps expressed singly. However, simultaneous expression of pairs of single-component or multicomponent efflux pumps did not produce strong increases in antibiotic resistance (122).

Other Resistance Mechanisms

Although the CAT mechanism for chloramphenicol resistance is widespread through the bacterial kingdom, it is not used by the chloramphenicol-producing *Streptomyces* strains to protect itself from the toxic activity of this metabolite. Data show that the chloramphenicol-producing *S. venezuelae* modifies the primary (C-3) hydroxyl of the antibiotic by a novel inactivating enzyme, chloramphenicol 3-O-phosphotransferase (CPT) (99). Recently, the 3-O-phosphotransferase was crystallized and its X-ray structure was determined (99).

Decreased intracellular chloramphenicol accumulation and subsequent resistance development may result from mutations that cause reduced expression of major outer membrane proteins in *H. influenzae* and *B. cepacia* in the absence of chloramphenicol acetyltransferase activity (38, 39). The expression of the *cmlA* gene inTn*1696*-carrying strains also induces dramatic changes in the inner and outer membrane protein content, in particular diminishing the production of both OmpA and OmpC major porins (28). Increased chloramphenicol resistance was also associated with the lack of the OmpF protein in *Salmonella* serovar Typhi as well as in MAR mutants, since the MarA transcriptional activator is able to activate the gene *micF* which produces an antisense RNA that effectively inhibits *ompF* translation (4, 232).

Mutations in the major ribosomal protein gene cluster of *E. coli* and *B. subtilis* as well as the 23S rRNA gene of *E. coli* and the archaeon *Halobacterium halobium* have also been associated with chloramphenicol resistance (8, 19, 72, 136). However, unlike other ribosomal mutations that confer resistance to antimicrobial protein biosynthesis inhibitors, e.g., macrolide-lincosamide-streptogramin antibiotics (190, 238), chloramphenicol resistance as a consequence of target site mutation/modification has rarely been documented. This may be due to the incompatibility of structural changes at the peptidyl transferase center with satisfactory ribosome function (205).

Lastly, a novel gene, termed *cfr,* which mediates resistance to chloramphenicol and florfenicol has recently been identified on plasmid pSCFS1 from *Staphylococcus sciuri* (106, 208). The corresponding Cfr protein shows no homology to any of the currently described chloramphenicol resistance proteins and does not inactivate either chloramphenicol or florfenicol. Structural comparisons revealed a certain degree of similarity with members of the radical SAM protein family. A potential regulatory region which resembled a translational attenuator has been detected immediately upstream of the *cfr* reading frame; however, the phenicol resistance mechanism has yet to be ascertained (208).

PREVALENCE OF CHLORAMPHENICOL RESISTANCE AND ASSOCIATED RESISTANCE DETERMINANTS

A large number of type A and B *cat* genes, as well as numerous specific chloramphenicol exporters (e.g., *flo, cmlA*), exist on mobile DNA elements, such as plasmids, transposons, and integrons (Tables 1 through 3). Integrons are particularly unique systems for capturing and disseminating antibiotic resistance genes among gram-negative bacteria and perhaps gram-positive bacteria (152). Integrons act as receptors of antibiotic resistance cassettes, and these genes can be inserted into both chromosomes and plasmids via site-specific recombination. There have been more than 50 cassettes identified to date conferring resistance to beta-lactams, aminoglycosides, trimethoprim, chloramphenicol, streptomycin, and other antimicrobials. The location of particular chloramphenicol resistance genes on DNA mobile elements has led to a rapid dissemination among numerous gram-positive and gram-negative bacteria.

Since *catB* and *cmlA* genes are often associated with integrons containing other resistance genes, coselection pressures certainly must exist in the absence of chloramphenicol usage. The observation that virtually the same *cat* gene (e.g., those of groups B-2 or B-3) or exporter gene (e.g., those of group E-1) has been identified in numerous gram-negative bacteria such as *Salmonella, Escherichia, Klebsiella,* and *Pseudomonas* underlines the efficient dissemination of these cassette-borne genes.

In Japan, sulfonamide-, chloramphenicol-, and streptomycin-resistant isolates of *A. salmonicida* subsp. *salmonicida* were detected in 1970. In one of these strains, a 29-MDa plasmid, pAr-32, was found to be responsible for resistance to all three drugs and was later shown to contain a class 1 integron adjacent to the chloramphenicol resistance gene *catA2* (219). The pAr-32 integron is very similar to the In6 integron of pSa, except for the additional *aacA4* cassette in In6. Another study identified the *catB2* gene in 31 (10%) of 313 motile aeromonads from Danish rainbow trout farms. In all cases the *catB2* cassette was located in a class 1 integron that also contained gene cassettes conferring resistance to streptomycin and trimethoprim (199).

Plasmids mediating chloramphenicol resistance have been identified in both porcine and bovine *P. multocida* isolates, as well as bovine *"P. haemolytica"* isolates (107, 237, 256). Early work by Vassort-Bruneau et al. showed the presence of *catI* as well as *catIII* genes among bovine *P. multocida* and *"P. haemolytica"* isolates (237). Similar or identical *catI* genes have also been detected on plasmids or in the chromosomal DNA of *Acinetobacter baumannii* (M62822), *Acinetobacter calcoaceticus* (M37690), and *P. damselae* subsp. *piscicida* (D16171) (107). The *catIII* gene proved to be located on small plasmids of 5.1 kb, while the *catI* gene was located on plasmids of either 17.1 or 5.5 kb (237). Plasmid-borne *catIII* genes have also been detected in porcine *P. aerogenes* and bovine *Mannheimia* isolates by PCR (107). The *catIII* gene was detected on a 5.6 kb plasmid (pMVSCS1) in *M. varigena* which also conferred coresistance to sulfonamides and streptomycin (109). Several *catIII* genes virtually identical to that from *Mannheimia* have been found on plasmid R387 in *Shigella flexneri* and *E. coli* (X07848; P00484), and on plasmids obtained from uncultured eubacteria (AJ271879; AJ293027) (107). However, none of these chloramphenicol resistance plasmids from *Pasteurella* and *Mannheimia* isolates mediated resistance to the fluorinated chloramphenicol derivative, florfenicol. Although the *flo* gene was recently detected on a 10.8-kb plasmid from bovine *P. multocida* (112), florfenicol resistance has rarely been detected among bovine and porcine respiratory pathogens (105).

White et al. characterized 48 antimicrobial resistant strains of *E. coli* isolated from diarrheic calves for chloramphenicol resistance phenotypes and genotypes (248). Forty-two of 44 isolates that displayed decreased susceptibilities to florfenicol (MIC >16 μg/ml) were positive via PCR for the *flo* gene. All florfenicol susceptible (MIC <8 μg/ml) *E. coli* isolates were negative for the *flo* gene ($n = 4$). Twelve *E. coli* isolates were positive via PCR for the *cmlA* gene and all displayed resistance to chloramphenicol (MIC >32 μg/ml). Additionally, eight isolates were positive for both *flo* and *cmlA* genes and displayed resistance to both florfenicol and chloramphenicol (MICs >64 μg/ml). The *E. coli flo* gene was located on several large molecular weight plasmids approximately 225 kb in size (248). Keyes et al. previously identified the *flo* gene on high-molecular weight plasmids of 186 and 204 kb among clinical avian *E. coli* isolates (113). The presence of such large MW plasmids is not unprecedented in *E. coli*. Additionally, the original report of florfenicol resistance among *P. piscicida* was attributed to a transferable R plasmid (116). However, not all of the *flo* positive bovine *E. coli* isolates contained large *flo* carrying plasmids, indicating that some of the genes may also be chromosomally borne. Recent work by Singer et al. also showed that the *flo* gene was present in 8% ($n = 164/1{,}987$) of *E. coli* isolates recovered from dairy cattle in the U.S. (217). Fifteen of these *E. coli* isolates also possessed the *cmlA* efflux gene.

A similar study showed greater than 50% of beta-hemolytic clinical *E. coli* isolates from diseased swine in the U.S. were chloramphenicol resistant (27). The *cmlA* gene was detected by PCR in 47 of the 48 chloramphenicol-resistant isolates, and four of these also possessed the *catII* gene. The one chloramphenicol-resistant isolate that did not contain either *cmlA* or *catII* possessed the *flo* gene, and also demonstrated resistance to florfenicol at a MIC of 256 μg/ml (27). Interestingly, chloramphenicol MICs were not higher for the isolates with a *cmlA*$^+$ *catII*$^+$ genotype than for those with the *cmlA* gene alone, suggesting that there is no additive effect from the two resistance mechanisms. Subsequent work showed that the *cmlA* gene was present on large plasmids greater than 100 kbp (26). Fifty-two percent of the isolates were able to transfer chloramphenicol resistance to an *E. coli* recipient at conjugation frequencies ranging from 10^{-3} to 10^{-8} per recipient. Antimicrobial susceptibility tests on transconjugant strains demonstrated that resistance to sulfamethoxazole, tetracycline, and kanamycin frequently transferred along with chloramphenicol resistance. The transconjugant strains possessed at least two distinct class 1 integrons that linked *cmlA* to both aminoglycoside resistance genes *aadA1* and *aadA2* and either to *sul1* or to *sul3* sulphonamide resistance genes (26). These results suggest that in the absence of specific chloramphenicol selection pressure, the *cmlA* gene is maintained by virtue of gene linkage to genes encoding resistance to antimicrobials that are currently approved for use in food animals.

The situation appears similar in Europe and China, where investigators have reported chloramphenicol and florfenicol resistance among *E. coli* isolates recovered from cattle, swine, amd poultry (29, 65, 67, 86). Guerra et al. determined that approximately 9% ($n = 28/317$) of animal *E. coli* isolates from Germany demonstrated resistance to chloramphenicol (86). The most commonly identified chloramphenicol resistance genes were *catA* (68%) and *cmlA1*-like (36%); however, no *flo* genes were detected. The *flo* gene has also been detected among clinical bovine *E. coli* isolates from France and Germany and clinical swine *E. coli* isolates from Germany (29, 65). The *flo* gene has also been recently identified in *E. coli* isolates associated with calf diarrhea from China during 1982 through 1988, representing one of the earliest reports of this efflux mechanism in *E. coli* (67).

Conjugative plasmids containing the *flo* gene and the extended spectrum cephalosporin resistance gene *bla*$_{CMY-2}$ have also been detected in *E. coli* strains associated with nosocomial infections in dogs (197), as well as *Salmonella* serovar Typhimurium and Newport strains from animals, retail foods, and humans (48, 62, 261). The presence of more than one antimicrobial resistance gene on a plasmid allows for co-selection of multidrug resistance phenotypes. Circulation of linked resistance genes such as *bla*$_{CMY-2}$ and *floR* among bacteria in food animal production environments is significant in light of the possibility of contamination of food products with these organisms. These findings further highlight the potential ramifications of antimicrobial selective pressures in food animal production settings and the important interface between veterinary and human medicine.

SUMMARY AND CONCLUSIONS

Chloramphenicol was the first truly broad spectrum antimicrobial available for clinical use that demonstrated excellent tissue penetration. It was the drug of choice for numerous serious infections in both animals and humans, until better and safer alternatives became available (14). Even though the use of chloramphenicol has been reduced widely in human medicine in industrial countries, it is still extensively used worldwide in developing countries for the treatment of severe diarrhea, pneumonia, meningitis, and other infectious diseases, or in persons who are allergic to penicillin (14). There continues to be some interest in the phenicol family of antimicrobials due both to their broad spectrum of activity and the potential to modify this class to eliminate associated aplastic anemia.

Since the first reports of enzymatic acetylation in the early 1960s, bacteria have developed a number of mechanisms which enable them to circumvent the inhibitory effects of chloramphenicol (210, 213, 223). The first and still most frequently encountered mechanism of bacterial resistance is enzymatic inactivation by acetylation of the drug via different types of CATs, many of which are located on plasmids or transposons (149, 205). However, there are also reports on other mechanisms of chloramphenicol resistance, such as efflux systems, inactivation by phosphotransferases, mutations of the target site and permeability barriers (26, 27, 50, 99, 149, 169, 205, 211, 248). Since many of the phenicol resistance determinants reside on mobile genetic elements (Tables 1 through 3) that most often are associated with other resistance genes, coselection and persistence of phenicol resistance genes may occur even if there is no direct selective pressure imposed by the use of these antimicrobial agents.

REFERENCES

1. **Adler, J., O. Lewinson, and E. Bibi.** 2004. Role of a conserved membrane-embedded acidic residue in the multidrug transporter MdfA. *Biochemistry* **43:**518–525.
2. **Ahmed, M., L. Lyass, P. N. Markham, S. S. Taylor, N. Vazquez-Laslop, and A. A. Neyfakh.** 1995. Two highly similar multidrug transporters of *Bacillus subtilis* whose expression is differentially regulated. *J. Bacteriol.* **177:**3904–3910.
3. **Albini, E., G. Belluco, M. Berton, G. Schioppacassi, and D. Ungheri.** 1999. In vitro antibacterial activity of thiamphenicol glycinate acetylcysteinate against respiratory pathogens. *Arzneimittelforschung* **49:**533–537.
4. **Alekshun, M. N., and S. B. Levy.** 1999. The *mar* regulon: multiple resistance to antibiotics and other toxic chemicals. *Trends Microbiol.* **7:**410–413.
5. **Allignet, J., and N. El Solh.** 1995. Diversity among the gram-positive acetyltransferases inactivating streptogramin A and structurally related compounds and characterization of a new staphylococcal determinant, *vatB*. *Antimicrob. Agents Chemother.* **39:**2027–2036.
6. **Allignet, J., V. Loncle, C. Simenel, M. Delepierre, and N. El Solh.** 1993. Sequence of a staphylococcal gene, *vat*, encoding an acetyltransferase inactivating the A-type compounds of virginiamycin-like antibiotics. *Gene* **130:**91–98.
7. **Alton, N. K., and D. Vapnek.** 1979. Nucleotide sequence analysis of the chloramphenicol resistance transposon Tn9. *Nature* **282:**864–869.
8. **Anderson, L. M., T. M. Henkin, G. H. Chambliss, and K. F. Bott.** 1984. New chloramphenicol resistance locus in *Bacillus subtilis*. *J. Bacteriol.* **158:**386–388.
9. **Aoki, T., N. Noguchi, M. Sasatsu, and M. Kono.** 1987. Complete nucleotide sequence of pTZ12, a chloramphenicol-resistance plasmid of *Bacillus subtilis*. *Gene* **51:**107–111.
10. **Arcangioli, M. A., S. Leroy-Setrin, J. L. Martel, and E. Chaslus-Dancla.** 1999. A new chloramphenicol and florfenicol resistance gene flanked by two integron structures in *Salmonella typhimurium* DT104. *FEMS Microbiol. Lett.* **174:**327–332.
11. **Arcangioli, M. A., S. Leroy-Setrin, J. L. Martel, and E. Chaslus-Dancla.** 2000. Evolution of chloramphenicol resistance, with emergence of cross-resistance to florfenicol, in bovine *Salmonella typhimurium* strains implicates definitive phage type (DT) 104. *J. Med. Microbiol.* **49:**103–110.
12. **Aubert, D., L. Poirel, J. Chevalier, S. Leotard, J. M. Pages, and P. Nordmann.** 2001. Oxacillinase-mediated resistance to cefepime and susceptibility to ceftazidime in *Pseudomonas aeruginosa*. *Antimicrob. Agents Chemother.* **45:**1615–1620.
13. **Ayoubi, P., A. O. Kilic, and M. N. Vijayakumar.** 1991. Tn5253, the pneumococcal omega (cat tet) BM6001 element, is a composite structure of two conjugative transposons, Tn5251 and Tn5252. *J. Bacteriol.* **173:**1617–1622.
14. **Balbi, H. J.** 2004. Chloramphenicol: a review. *Pediatr. Rev.* **25:**284–288.
15. **Bannam, T. L., P. K. Crellin, and J. I. Rood.** 1995. Molecular genetics of the chloramphenicol-resistance transposon Tn4451 from *Clostridium perfringens:* the TnpX site-specific recombinase excises a circular transposon molecule. *Mol. Microbiol.* **16:**535–551.
16. **Bannam, T. L., and J. I. Rood.** 1991. Relationship between the *Clostridium perfringens catQ* gene product and chloramphenicol acetyltransferases from other bacteria. *Antimicrob. Agents Chemother.* **35:**471–476.
17. **Baucheron, S., H. Imberechts, E. Chaslus-Dancla, and A. Cloeckaert.** 2002. The AcrB multidrug transporter plays a major role in high-level fluoroquinolone resistance in *Salmonella enterica* serovar typhimurium phage type DT204. *Microb. Drug Resist.* **8:**281–289.
18. **Baucheron, S., S. Tyler, D. Boyd, M. R. Mulvey, E. Chaslus-Dancla, and A. Cloeckaert.** 2004. AcrAB-TolC directs efflux-mediated multidrug resistance in *Salmonella enterica* serovar typhimurium DT104. *Antimicrob. Agents Chemother.* **48:** 3729–3735.
19. **Baughman, G. A., and S. R. Fahnestock.** 1979. Chloramphenicol resistance mutation in *Escherichia coli* which maps in the major ribosomal protein gene cluster. *J. Bacteriol.* **137:** 1315–1323.
20. **Beaber, J. W., B. Hochhut, and M. K. Waldor.** 2002. Genomic and functional analyses of SXT, an integrating antibiotic resistance gene transfer element derived from *Vibrio cholerae*. *J. Bacteriol.* **184:**4259–4269.
21. **Beers, D. V., E. Schoutens, M. P. Vanderlinden, and E. Yourassowsky.** 1975. Comparative in vitro acitvity of chloramphenicol and thiamphenicol on common aerobic and anaerobic gram-negative bacilli (*Salmonella* and *Shigella* excluded). *Chemotherapy* **21:**73–81.
22. **Belda, J. W., L. F. Siqueira, and L. J. Fagundes.** 2000. Thiamphenicol in the treatment of chancroid. A study of 1, 128 cases. *Rev. Inst. Med. Trop. Sao Paulo* **42:**133–135.
23. **Belda, W., M. F. dos Santos Junior, L. J. Fagundes, L. F. Siqueira, C. Lombardi, and W. Francisco.** 1984. Minute treatment with thiamphenicol in water for acute gonococcal urethritis in male patients. *Sex Transm. Dis.* **11:**420–422.
24. **Bhakta, M., and M. Bal.** 2003. Identification and characterization of a shuttle plasmid with antibiotic resistance gene from *Staphylococcus aureus*. *Curr. Microbiol.* **46:**413–417.
25. **Bibi, E., J. Adler, O. Lewinson, and R. Edgar.** 2001. MdfA, an interesting model protein for studying multidrug transport. *J. Mol. Microbiol. Biotechnol.* **3:**171–177.
26. **Bischoff, K. M., D. G. White, M. E. Hume, T. L. Poole, and D. J. Nisbet.** 2005. The chloramphenicol resistance gene *cmlA*

is disseminated on transferable plasmids that confer multiple-drug resistance in swine *Escherichia coli. FEMS Microbiol. Lett.* **243**:285–291.

27. **Bischoff, K. M., D. G. White, P. F. McDermott, S. Zhao, S. Gaines, J. J. Maurer, and D. J. Nisbet.** 2002. Characterization of chloramphenicol resistance in beta-hemolytic *Escherichia coli* associated with diarrhea in neonatal swine. *J. Clin. Microbiol.* **40**:389–394.
28. **Bissonnette, L., S. Champetier, J. P. Buisson, and P. H. Roy.** 1991. Characterization of the nonenzymatic chloramphenicol resistance (*cmlA*) gene of the In4 integron of Tn1696: similarity of the product to transmembrane transport proteins. *J. Bacteriol.* **173**:4493–4502.
29. **Blickwede, M., and S. Schwarz.** 2004. Molecular analysis of florfenicol-resistant *Escherichia coli* isolates from pigs. *J. Antimicrob. Chemother.* **53**:58–64.
30. **Bolton, L. F., L. C. Kelley, M. D. Lee, P. J. Fedorka-Cray, and J. J. Maurer.** 1999. Detection of multidrug-resistant *Salmonella enterica* serotype typhimurium DT104 based on a gene which confers cross-resistance to florfenicol and chloramphenicol. *J. Clin. Microbiol.* **37**:1348–1351.
31. **Boyd, D., A. Cloeckaert, E. Chaslus-Dancla, and M. R. Mulvey.** 2002. Characterization of variant *Salmonella* genomic island 1 multidrug resistance regions from serovars Typhimurium DT104 and Agona. *Antimicrob. Agents Chemother.* **46**:1714–1722.
32. **Boyd, D., G. A. Peters, A. Cloeckaert, K. S. Boumedine, E. Chaslus-Dancla, H. Imberechts, and M. R. Mulvey.** 2001. Complete nucleotide sequence of a 43-kilobase genomic island associated with the multidrug resistance region of *Salmonella enterica* serovar Typhimurium DT104 and its identification in phage type DT120 and serovar Agona. *J. Bacteriol.* **183**:5725–5732.
33. **Brenner, D. G., and W. V. Shaw.** 1985. The use of synthetic oligonucleotides with universal templates for rapid DNA sequencing: results with staphylococcal replicon pC221. *EMBO J.* **4**:561–568.
34. **Briggs, C. E., and P. M. Fratamico.** 1999. Molecular characterization of an antibiotic resistance gene cluster of *Salmonella typhimurium* DT104. *Antimicrob. Agents Chemother.* **43**:846–849.
35. **Bruckner, R., and H. Matzura.** 1985. Regulation of the inducible chloramphenicol acetyltransferase gene of the *Staphylococcus aureus* plasmid pUB112. *EMBO J.* **4**:2295–2300.
36. **Bruggemann, H., S. Baumer, W. F. Fricke, A. Wiezer, H. Liesegang, I. Decker, C. Herzberg, R. Martinez-Arias, R. Merkl, A. Henne, and G. Gottschalk.** 2003. The genome sequence of *Clostridium tetani,* the causative agent of tetanus disease. *Proc. Natl. Acad. Sci. USA* **100**:1316–1321.
37. **Bunny, K. L., R. M. Hall, and H. W. Stokes.** 1995. New mobile gene cassettes containing an aminoglycoside resistance gene, *aacA7,* and a chloramphenicol resistance gene, *catB3,* in an integron in pBWH301. *Antimicrob. Agents Chemother.* **39**:686–693.
38. **Burns, J. L., L. A. Hedin, and D. M. Lien.** 1989. Chloramphenicol resistance in *Pseudomonas cepacia* because of decreased permeability. *Antimicrob. Agents Chemother.* **33**:136–141.
39. **Burns, J. L., P. M. Mendelman, J. Levy, T. L. Stull, and A. L. Smith.** 1985. A permeability barrier as a mechanism of chloramphenicol resistance in *Haemophilus influenzae. Antimicrob. Agents Chemother.* **27**:46–54.
40. **Butaye, P., A. Cloeckaert, and S. Schwarz.** 2003. Mobile genes coding for efflux-mediated antimicrobial resistance in gram-positive and gram-negative bacteria. *Int. J. Antimicrob. Agents* **22**:205–210.
41. **Byeon, W. H.** 1992. Genbank accession no. M90091. National Center for Biotechnology Information.
42. **Cabrera, R., J. Ruiz, F. Marco, I. Oliveira, M. Arroyo, A. Aladuena, M. A. Usera, M. T. Jimenez De Anta, J. Gascon, and J. Vila.** 2004. Mechanism of resistance to several antimicrobial agents in *Salmonella* clinical isolates causing traveler's diarrhea. *Antimicrob. Agents Chemother.* **48**:3934–3939.
43. **Cannon, M., S. Harford, and J. Davies.** 1990. A comparative study on the inhibitory actions of chloramphenicol, thiamphenicol and some fluorinated derivatives. *J. Antimicrob. Chemother.* **26**:307–317.
44. **Carattoli, A., F. Tosini, W. P. Giles, M. E. Rupp, S. H. Hinrichs, F. J. Angulo, T. J. Barrett, and P. D. Fey.** 2002. Characterization of plasmids carrying CMY-2 from expanded-spectrum cephalosporin-resistant *Salmonella* strains isolated in the United States between 1996 and 1998. *Antimicrob. Agents Chemother.* **46**:1269–1272.
45. **Cardoso, M., and S. Schwarz.** 1992. Nucleotide sequence and structural relationships of a chloramphenicol acetyltransferase encoded by the plasmid pSCS6 from *Staphylococcus aureus. J. Appl. Bacteriol.* **72**:289–293.
46. **Center for Veterinary Medicine.** FDA approved animal drug products online database system. U.S. Food and Drug Administration. 2005.
47. **Charles, I. G., J. W. Keyte, and W. V. Shaw.** 1985. Nucleotide sequence analysis of the *cat* gene of *Proteus mirabilis:* comparison with the type I (Tn9) *cat* gene. *J. Bacteriol.* **164**:123–129.
48. **Chen, S., S. Zhao, D. G. White, C. M. Schroeder, R. Lu, H. Yang, P. F. McDermott, S. Ayers, and J. Meng.** 2004. Characterization of multiple-antimicrobial-resistant salmonella serovars isolated from retail meats. *Appl. Environ. Microbiol.* **70**:1–7.
49. **Cloeckaert, A., S. Baucheron, and E. Chaslus-Dancla.** 2001. Nonenzymatic chloramphenicol resistance mediated by IncC plasmid R55 is encoded by a *floR* gene variant. *Antimicrob. Agents Chemother.* **45**:2381–2382.
50. **Cloeckaert, A., S. Baucheron, G. Flaujac, S. Schwarz, C. Kehrenberg, J. L. Martel, and E. Chaslus-Dancla.** 2000. Plasmid-mediated florfenicol resistance encoded by the *floR* gene in *Escherichia coli* isolated from cattle. *Antimicrob. Agents Chemother.* **44**:2858–2860.
51. **Cloeckaert, A., and S. Schwarz.** 2001. Molecular characterization, spread and evolution of multidrug resistance in *Salmonella enterica* typhimurium DT104. *Vet. Res.* **32**:301–310.
52. **Cloeckaert, A., B. K. Sidi, G. Flaujac, H. Imberechts, I. D'Hooghe, and E. Chaslus-Dancla.** 2000. Occurrence of a *Salmonella enterica* serovar typhimurium DT104-like antibiotic resistance gene cluster including the *floR* gene in *S. enterica* serovar agona. *Antimicrob. Agents Chemother.* **44**:1359–1361.
53. **Colmer, J. A., J. A. Fralick, and A. N. Hamood.** 1998. Isolation and characterization of a putative multidrug resistance pump from *Vibrio cholerae. Mol. Microbiol.* **27**:63–72.
54. **Cook, A. T., and D. E. Marmion.** 1949. Chloromycetin in the treatment of typhoid fever; 14 cases treated in the Middle East. *Lancet* **2**:975–979.
55. **De Craene, B. A., P. Deprez, E. D'Haese, H. J. Nelis, B. W. Van den, and P. De Leenheer.** 1997. Pharmacokinetics of florfenicol in cerebrospinal fluid and plasma of calves. *Antimicrob. Agents Chemother.* **41**:1991–1995.
56. **DelVecchio, V. G., V. Kapatral, R. J. Redkar, G. Patra, C. Mujer, T. Los, N. Ivanova, I. Anderson, A. Bhattacharyya, A. Lykidis, G. Reznik, L. Jablonski, N. Larsen, M. D'Souza, A. Bernal, M. Mazur, E. Goltsman, E. Selkov, P. H. Elzer, S. Hagius, D. O'Callaghan, J. J. Letesson, R. Haselkorn, N. Kyrpi-

des, and R. Overbeek. 2002. The genome sequence of the facultative intracellular pathogen *Brucella melitensis*. *Proc. Natl. Acad. Sci. USA* **99:**443–448.

57. **Desomer, J., D. Vereecke, M. Crespi, and M. Van Montagu.** 1992. The plasmid-encoded chloramphenicol-resistance protein of *Rhodococcus fascians* is homologous to the transmembrane tetracycline efflux proteins. *Mol. Microbiol.* **6:**2377–2385.
58. **Dittrich, W., M. Betzler, and H. Schrempf.** 1991. An amplifiable and deletable chloramphenicol-resistance determinant of *Streptomyces lividans* 1326 encodes a putative transmembrane protein. *Mol. Microbiol.* **5:**2789–2797.
59. **Dorman, C. J., and T. J. Foster.** 1982. Nonenzymatic chloramphenicol resistance determinants specified by plasmids R26 and R55-1 in *Escherichia coli* K-12 do not confer high-level resistance to fluorinated analogs. *Antimicrob. Agents Chemother.* **22:**912–914.
60. **Dorman, C. J., T. J. Foster, and W. V. Shaw.** 1986. Nucleotide sequence of the R26 chloramphenicol resistance determinant and identification of its gene product. *Gene* **41:**349–353.
61. **Doublet, B., P. Butaye, H. Imberechts, D. Boyd, M. R. Mulvey, E. Chaslus-Dancla, and A. Cloeckaert.** 2004. Salmonella genomic island 1 multidrug resistance gene clusters in *Salmonella enterica* serovar Agona isolated in Belgium in 1992 to 2002. *Antimicrob. Agents Chemother.* **48:**2510–2517.
62. **Doublet, B., A. Carattoli, J. M. Whichard, D. G. White, S. Baucheron, E. Chaslus-Dancla, and A. Cloeckaert.** 2004. Plasmid-mediated florfenicol and ceftriaxone resistance encoded by the *floR* and $bla_{(CMY-2)}$ genes in *Salmonella enterica* serovars Typhimurium and Newport isolated in the United States. *FEMS Microbiol. Lett.* **233:**301–305.
63. **Doublet, B., R. Lailler, D. Meunier, A. Brisabois, D. Boyd, M. R. Mulvey, E. Chaslus-Dancla, and A. Cloeckaert.** 2003. Variant *Salmonella* genomic island 1 antibiotic resistance gene cluster in *Salmonella enterica* serovar Albany. *Emerg. Infect. Dis.* **9:**585–591.
64. **Doublet, B., S. Schwarz, C. Kehrenberg, and A. Cloeckaert.** 2005. The florfenicol resistance gene *floR* is part of a novel transposon. *Antimicrob. Agents Chemother.* **49:**2106–2108.
65. **Doublet, B., S. Schwarz, E. Nussbeck, S. Baucheron, J. L. Martel, E. Chaslus-Dancla, and A. Cloeckaert.** 2002. Molecular analysis of chromosomally florfenicol-resistant *Escherichia coli* isolates from France and Germany. *J. Antimicrob. Chemother.* **49:**49–54.
66. **Drago, L., E. De Vecchi, M. C. Fassina, B. Mombelli, and M. R. Gismondo.** 2000. Serum and lung levels of thiamphenicol after administration of its glycinate N-acetylcysteinate ester in experimentally infected guinea pigs. *Int. J. Antimicrob. Agents* **13:**301–303.
67. **Du, X., C. Xia, J. Shen, B. Wu, and Z. Shen.** 2004. Characterization of florfenicol resistance among calf pathogenic *Escherichia coli. FEMS Microbiol. Lett.* **236:**183–189.
68. **Ecker, G. F., K. Pleban, S. Kopp, E. Csaszar, G. J. Poelarends, M. Putman, D. Kaiser, W. N. Konings, and P. Chiba.** 2004. A three-dimensional model for the substrate binding domain of the multidrug ATP binding cassette transporter LmrA. *Mol. Pharmacol.* **66:**1169–1179.
69. **Ehrlich, J. Q., Q. R. Bartz, R. M. Smith, D. A. Joslyn, and P. R. Burkholder.** 1947. Chlormycetin, a new antibiotic from a soil actinomycete. *Science* **106:**417.
70. **Eisen, J. A., K. E. Nelson, I. T. Paulsen, J. F. Heidelberg, M. Wu, R. J. Dodson, R. Deboy, M. L. Gwinn, W. C. Nelson, D. H. Haft, E. K. Hickey, J. D. Peterson, A. S. Durkin, J. L. Kolonay, F. Yang, I. Holt, L. A. Umayam, T. Mason, M. Brenner, T. P. Shea, D. Parksey, W. C. Nierman, T. V. Feldblyum, C. L. Hansen, M. B. Craven, D. Radune, J. Vamathevan, H. Khouri, O. White, T. M. Gruber, K. A. Ketchum, J. C. Venter, H. Tettelin, D. A. Bryant, and C. M. Fraser.** 2002. The complete genome sequence of *Chlorobium tepidum* TLS, a photosynthetic, anaerobic, green-sulfur bacterium. *Proc. Natl. Acad. Sci. USA* **99:**9509–9514.
71. **Elisha, B. G., and L. M. Steyn.** 1991. Identification of an *Acinetobacter baumannii* gene region with sequence and organizational similarity to Tn2670. *Plasmid* **25:**96–104.
72. **Ettayebi, M., S. M. Prasad, and E. A. Morgan.** 1985. Chloramphenicol-erythromycin resistance mutations in a 23S rRNA gene of *Escherichia coli. J. Bacteriol.* **162:**551–557.
73. **Federici, L., D. Du, F. Walas, H. Matsumura, J. Fernandez-Recio, K. S. McKeegan, M. I. Borges-Walmsley, B. F. Luisi, and A. R. Walmsley.** 2005. The crystal structure of the outer membrane protein VCEC from the bacterial pathogen vibrio cholerae at 1.8 A resolution. *J. Biol. Chem.* **280:**15307–15314.
74. **Frech, G., C. Kehrenberg, and S. Schwarz.** 2003. Resistance phenotypes and genotypes of multiresistant *Salmonella enterica* subsp. enterica serovar Typhimurium var. Copenhagen isolates from animal sources. *J. Antimicrob. Chemother.* **51:** 180–182.
75. **Fukui, H., Y. Fujihara, and T. Kano.** 1987. In vitro and in vivo antibacterial activities of florfenicol, a new fluorinated analog of thiamphenicol, against fish pathogens. *Fish. Pathol.* **22:**201–207.
76. **Fuller, D. G., T. Duke, F. Shann, and N. Curtis.** 2003. Antibiotic treatment for bacterial meningitis in children in developing countries. *Ann. Trop. Paediatr.* **23:**233–253.
77. **Galimand, M., G. Gerbaud, M. Guibourdenche, J. Y. Riou, and P. Courvalin.** 1998. High-level chloramphenicol resistance in *Neisseria meningitidis. N. Engl. J. Med.* **339:**868–874.
78. **Galopin, S., and Leclercq R.** GenBank accession no. AY238971. National Center for Biotechnology Information. 2003.
79. **Gebreyes, W. A., and S. Thakur.** 2005. Multidrug-resistant *Salmonella enterica* serovar Muenchen from pigs and humans and potential interserovar transfer of antimicrobial resistance. *Antimicrob. Agents Chemother.* **49:**503–511.
80. **George, A. M., and R. M. Hall.** 2002. Efflux of chloramphenicol by the CmlA1 protein. *FEMS Microbiol. Lett.* **209:** 209–213.
81. **Ghisalberti, D., M. Masi, J. M. Pages, and J. Chevalier.** 2005. Chloramphenicol and expression of multidrug efflux pump in *Enterobacter aerogenes. Biochem. Biophys. Res. Commun.* **328:**1113–1118.
82. **Gilmore, A.** 1986. Chloramphenicol and the politics of health. *CMAJ.* **134:**423, 426–5.
83. **Gilmour, M. W., N. R. Thomson, M. Saunders, J. Parkhill, and D. E. Taylor.** 2003. Genbank accession no. NC_005211. National Center for Biotechnology Information.
84. **Goodner, B., G. Hinkle, S. Gattung, N. Miller, M. Blanchard, B. Qurollo, B. S. Goldman, Y. Cao, M. Askenazi, C. Halling, L. Mullin, K. Houmiel, J. Gordon, M. Vaudin, O. Iartchouk, A. Epp, F. Liu, C. Wollam, M. Allinger, D. Doughty, C. Scott, C. Lappas, B. Markelz, C. Flanagan, C. Crowell, J. Gurson, C. Lomo, C. Sear, G. Strub, C. Cielo, and S. Slater.** 2001. Genome sequence of the plant pathogen and biotechnology agent *Agrobacterium tumefaciens* C58. *Science* **294:**2323–2328.
85. **Grady, R., and F. Hayes.** 2003. Axe-Txe, a broad-spectrum proteic toxin-antitoxin system specified by a multidrug-resistant, clinical isolate of *Enterococcus faecium. Mol. Microbiol.* **47:**1419–1432.
86. **Guerra, B., E. Junker, A. Schroeter, B. Malorny, S. Lehmann, and R. Helmuth.** 2003. Phenotypic and genotypic characterization of antimicrobial resistance in German *Escherichia coli*

isolates from cattle, swine and poultry. *J. Antimicrob. Chemother.* **52**:489–492.

87. **Hadorn, K., H. Hachler, A. Schaffner, and F. H. Kayser.** 1993. Genetic characterization of plasmid-encoded multiple antibiotic resistance in a strain of *Listeria monocytogenes* causing endocarditis. *Eur. J. Clin. Microbiol. Infect. Dis.* **12**:928–937.
88. **Hall, R. M., C. M. Collis, M. J. Kim, S. R. Partridge, G. D. Recchia, and H. W. Stokes.** 1999. Mobile gene cassettes and integrons in evolution. *Ann. N.Y. Acad. Sci.* **870**:68–80.
89. **Hansen, L. H., E. Johannesen, M. Burmolle, A. H. Sorensen, and S. J. Sorensen.** 2004. Plasmid-encoded multidrug efflux pump conferring resistance to olaquindox in *Escherichia coli. Antimicrob. Agents Chemother.* **48**:3332–3337.
90. **Harwood, C. R., D. M. Williams, and P. S. Lovett.** 1983. Nucleotide sequence of a *Bacillus pumilus* gene specifying chloramphenicol acetyltransferase. *Gene* **24**:163–169.
91. **Heidelberg, J. F., J. A. Eisen, W. C. Nelson, R. A. Clayton, M. L. Gwinn, R. J. Dodson, D. H. Haft, E. K. Hickey, J. D. Peterson, L. Umayam, S. R. Gill, K. E. Nelson, T. D. Read, H. Tettelin, D. Richardson, M. D. Ermolaeva, J. Vamathevan, S. Bass, H. Qin, I. Dragoi, P. Sellers, L. McDonald, T. Utterback, R. D. Fleishmann, W. C. Nierman, O. White, S. L. Salzberg, H. O. Smith, R. R. Colwell, J. J. Mekalanos, J. C. Venter, and C. M. Fraser.** 2000. DNA sequence of both chromosomes of the cholera pathogen *Vibrio cholerae. Nature* **406**:477–483.
92. **Heidelberg, J. F., I. T. Paulsen, K. E. Nelson, E. J. Gaidos, W. C. Nelson, T. D. Read, J. A. Eisen, R. Seshadri, N. Ward, B. Methe, R. A. Clayton, T. Meyer, A. Tsapin, J. Scott, M. Beanan, L. Brinkac, S. Daugherty, R. T. DeBoy, R. J. Dodson, A. S. Durkin, D. H. Haft, J. F. Kolonay, R. Madupu, J. D. Peterson, L. A. Umayam, O. White, A. M. Wolf, J. Vamathevan, J. Weidman, M. Impraim, K. Lee, K. Berry, C. Lee, J. Mueller, H. Khouri, J. Gill, T. R. Utterback, L. A. McDonald, T. V. Feldblyum, H. O. Smith, J. C. Venter, K. H. Nealson, and C. M. Fraser.** 2002. Genome sequence of the dissimilatory metal ion-reducing bacterium *Shewanella oneidensis. Nat. Biotechnol.* **20**:1118–1123.
93. **Ho, S. P., T. Y. Hsu, M. H. Che, and W. S. Wang.** 2000. Antibacterial effect of chloramphenicol, thiamphenicol and florfenicol against aquatic animal bacteria. *J. Vet. Med. Sci.* **62**:479–485.
94. **Hochhut, B., Y. Lotfi, D. Mazel, S. M. Faruque, R. Woodgate, and M. K. Waldor.** 2001. Molecular analysis of antibiotic resistance gene clusters in vibrio cholerae O139 and O1 SXT constins. *Antimicrob. Agents Chemother.* **45**:2991–3000.
95. **Horinouchi, S., and B. Weisblum.** 1982. Nucleotide sequence and functional map of pC194, a plasmid that specifies inducible chloramphenicol resistance. *J. Bacteriol.* **150**:815–825.
96. **Houang, E. T., Y. W. Chu, W. S. Lo, K. Y. Chu, and A. F. Cheng.** 2003. Epidemiology of rifampin ADP-ribosyltransferase (*arr-2*) and metallo-beta-lactamase (*blaIMP-4*) gene cassettes in class 1 integrons in *Acinetobacter* strains isolated from blood cultures in 1997 to 2000. *Antimicrob. Agents Chemother.* **47**:1382–1390.
97. **Huggins, A. S., T. L. Bannam, and J. I. Rood.** 1992. Comparative sequence analysis of the *catB* gene from *Clostridium butyricum. Antimicrob. Agents Chemother.* **36**:2548–2551.
98. **Ivanova, N., A. Sorokin, I. Anderson, N. Galleron, B. Candelon, V. Kapatral, A. Bhattacharyya, G. Reznik, N. Mikhailova, A. Lapidus, L. Chu, M. Mazur, E. Goltsman, N. Larsen, M. D'Souza, T. Walunas, Y. Grechkin, G. Pusch, R. Haselkorn, M. Fonstein, S. D. Ehrlich, R. Overbeek, and N. Kyrpides.** 2003. Genome sequence of *Bacillus cereus* and comparative analysis with *Bacillus anthracis. Nature* **423**:87–91.
99. **Izard, T.** 2001. Structural basis for chloramphenicol tolerance in *Streptomyces venezuelae* by chloramphenicol phosphotransferase activity. *Protein Sci.* **10**:1508–1513.
100. **Jude, F., C. Arpin, C. Brachet-Castang, M. Capdepuy, P. Caumette, and C. Quentin.** 2004. TbtABM, a multidrug efflux pump associated with tributyltin resistance in *Pseudomonas stutzeri. FEMS Microbiol. Lett.* **232**:7–14.
101. **Kadlec, K., C. Kehrenberg, and S. Schwarz.** 2005. GenBank accession nos. AJ844287 and AJ879564. National Center for Biotechnology Information.
102. **Kaneko, T., Y. Nakamura, S. Sato, E. Asamizu, T. Kato, S. Sasamoto, A. Watanabe, K. Idesawa, A. Ishikawa, K. Kawashima, T. Kimura, Y. Kishida, C. Kiyokawa, M. Kohara, M. Matsumoto, A. Matsuno, Y. Mochizuki, S. Nakayama, N. Nakazaki, S. Shimpo, M. Sugimoto, C. Takeuchi, M. Yamada, and S. Tabata.** 2000. Complete genome structure of the nitrogen-fixing symbiotic bacterium *Mesorhizobium loti. DNA Res.* **7**:331–338.
103. **Kapatral, V., I. Anderson, N. Ivanova, G. Reznik, T. Los, A. Lykidis, A. Bhattacharyya, A. Bartman, W. Gardner, G. Grechkin, L. Zhu, O. Vasieva, L. Chu, Y. Kogan, O. Chaga, E. Goltsman, A. Bernal, N. Larsen, M. D'Souza, T. Walunas, G. Pusch, R. Haselkorn, M. Fonstein, N. Kyrpides, and R. Overbeek.** 2002. Genome sequence and analysis of the oral bacterium *Fusobacterium nucleatum* strain ATCC 25586. *J. Bacteriol.* **184**:2005–2018.
104. **Kapatral, V., N. Ivanova, I. Anderson, G. Reznik, A. Bhattacharyya, W. L. Gardner, N. Mikhailova, A. Lapidus, N. Larsen, M. D'Souza, T. Walunas, R. Haselkorn, R. Overbeek, and N. Kyrpides.** 2003. Genome analysis of *F. nucleatum* sub spp *vincentii* and its comparison with the genome of *F. nucleatum* ATCC 25586. *Genome Res.* **13**:1180–1189.
105. **Kehrenberg, C., J. Mumme, J. Wallmann, J. Verspohl, R. Tegeler, T. Kuhn, and S. Schwarz.** 2004. Monitoring of florfenicol susceptibility among bovine and porcine respiratory tract pathogens collected in Germany during the years 2002 and 2003. *J. Antimicrob. Chemother.* **54**:572–574.
106. **Kehrenberg, C., K. K. Ojo, and S. Schwarz.** 2004. Nucleotide sequence and organization of the multiresistance plasmid pSCFS1 from *Staphylococcus sciuri. J. Antimicrob. Chemother.* **54**:936–939.
107. **Kehrenberg, C., G. Schulze-Tanzil, J. L. Martel, E. Chaslus-Dancla, and S. Schwarz.** 2001. Antimicrobial resistance in *Pasteurella* and *Mannheimia:* epidemiology and genetic basis. *Vet. Res.* **32**:323–339.
108. **Kehrenberg, C., and S. Schwarz.** 2001. Occurrence and linkage of genes coding for resistance to sulfonamides, streptomycin and chloramphenicol in bacteria of the genera *Pasteurella* and *Mannheimia. FEMS Microbiol. Lett.* **205**:283–290.
109. **Kehrenberg, C., and S. Schwarz.** 2002. Nucleotide sequence and organization of plasmid pMVSCS1 from *Mannheimia varigena:* identification of a multiresistance gene cluster. *J. Antimicrob. Chemother.* **49**:383–386.
110. **Kehrenberg, C., and S. Schwarz.** 2004. fexA, a novel *Staphylococcus lentus* gene encoding resistance to florfenicol and chloramphenicol. *Antimicrob. Agents Chemother.* **48**:615–618.
111. **Kehrenberg, C., and S. Schwarz.** 2005. Florfenicol-chloramphenicol exporter gene *fexA* is part of the novel transposon Tn558. *Antimicrob. Agents Chemother.* **49**:813–815.
112. **Kehrenberg, C., and S. Schwarz.** 2005. Plasmid-borne florfenicol resistance in *Pasteurella multocida. J. Antimicrob. Chemother.* **55**:773–775.
113. **Keyes, K., C. Hudson, J. J. Maurer, S. Thayer, D. G. White, and M. D. Lee.** 2000. Detection of florfenicol resistance genes

in *Escherichia coli* isolated from sick chickens. *Antimicrob. Agents Chemother.* **44:**421–424.

114. **Kieboom, J., and J. de Bont.** 2001. Identification and molecular characterization of an efflux system involved in *Pseudomonas putida* S12 multidrug resistance. *Microbiology* **147:**43–51.
115. **Kim, E., and T. Aoki.** 1993. The structure of the chloramphenicol resistance gene on a transferable R plasmid from the fish pathogen, *Pasteurella piscicida. Microbiol. Immunol.* **37:** 705–712.
116. **Kim, E., and T. Aoki.** 1996. Sequence analysis of the florfenicol resistance gene encoded in the transferable R-plasmid of a fish pathogen, *Pasteurella piscicida. Microbiol. Immunol.* **40:**665–669.
117. **Kim, J.** 2003. Genbank accession no. AF227506. National Center for Biotechnology Information.
118. **Kim, S.** 2003. GenBank accession no. AY339985. National Center for Biotechnology Information.
119. **Kucers, A., and N. Bennett.** 1987. Chloramphenicol and Thiamphenicol, p. 757–807. *In* A. Kucers and N. Bennett (ed.), The use of antibiotics. J. B. Lippincott Co., Philadelphia, PA.
120. **Kupzig, S., and P. M. Bennett.** 1994. Genbank accession no. X82455. National Center for Biotechnology Information.
121. **Laraki, N., M. Galleni, I. Thamm, M. L. Riccio, G. Amicosante, J. M. Frere, and G. M. Rossolini.** 1999. Structure of In31, a bla_{IMP}-containing *Pseudomonas aeruginosa* integron phyletically related to In5, which carries an unusual array of gene cassettes. *Antimicrob. Agents Chemother.* **43:**890–901.
122. **Lee, A., W. Mao, M. S. Warren, A. Mistry, K. Hoshino, R. Okumura, H. Ishida, and O. Lomovskaya.** 2000. Interplay between efflux pumps may provide either additive or multiplicative effects on drug resistance. *J. Bacteriol.* **182:**3142–3150.
123. **Lee, H. J., and H. S. Kang.** 1999. GenBank accession no. AF124757. National Center for Biotechnology Information.
124. **Lee, K., J. H. Yum, D. Yong, K. H. Roh, Y. Chong, and G. M. Rossolini.** 2001. GenBank accession no. AF418284. National Center for Biotechnology Information.
125. **Lee, M. H., A. Nittayajarn, and C. E. Rubens.** 1997. GenBank accession no. U83488. National Center for Biotechnology Information.
126. **Levy, S. B.** 2002. Active efflux, a common mechanism for biocide and antibiotic resistance. *Symp. Ser. Soc. Appl. Microbiol.* 65S–71S.
127. **Lewinson, O., J. Adler, G. J. Poelarends, P. Mazurkiewicz, A. J. Driessen, and E. Bibi.** 2003. The *Escherichia coli* multidrug transporter MdfA catalyzes both electrogenic and electroneutral transport reactions. *Proc. Natl. Acad. Sci. USA* **100:**1667–1672.
128. **Lin, C. F., Z. F. Fung, C. L. Wu, and T. C. Chung.** 1996. Molecular characterization of a plasmid-borne (pTC82) chloramphenicol resistance determinant (cat-TC) from *Lactobacillus reuteri* G4. *Plasmid* **36:**116–124.
129. **Lombardi, A., L. Drago, E. De Vecchi, B. Mombelli, and M. R. Gismondo.** 2001. Antimicrobial activity of thiamphenicol-glycinate-acetylcysteinate and other drugs against *Chlamydia pneumoniae. Arzneimittelforschung* **51:**264–267.
130. **Lovett, P. S.** 1990. Translational attenuation as the regulator of inducible cat genes. *J. Bacteriol.* **172:**1–6.
131. **Luck, S. N., S. A. Turner, K. Rajakumar, H. Sakellaris, and B. Adler.** 2001. Ferric dicitrate transport system (Fec) of *Shigella flexneri* 2a YSH6000 is encoded on a novel pathogenicity island carrying multiple antibiotic resistance genes. *Infect. Immun.* **69:**6012–6021.
132. **Lyras, D., and J. I. Rood.** 2000. Transposition of Tn4451 and Tn4453 involves a circular intermediate that forms a promoter for the large resolvase, TnpX. *Mol. Microbiol.* **38:**588–601.
133. **Lyras, D., C. Storie, A. S. Huggins, P. K. Crellin, T. L. Bannam, and J. I. Rood.** 1998. Chloramphenicol resistance in *Clostridium difficile* is encoded on Tn4453 transposons that are closely related to Tn4451 from *Clostridium perfringens. Antimicrob. Agents Chemother.* **42:**1563–1567.
134. **Makino, K., K. Oshima, K. Kurokawa, K. Yokoyama, T. Uda, K. Tagomori, Y. Iijima, M. Najima, M. Nakano, A. Yamashita, Y. Kubota, S. Kimura, T. Yasunaga, T. Honda, H. Shinagawa, M. Hattori, and T. Iida.** 2003. Genome sequence of *Vibrio parahaemolyticus:* a pathogenic mechanism distinct from that of V cholerae. *Lancet* **361:**743–749.
135. **Mandal, S., M. D. Mandal, and N. K. Pal.** 2004. Reduced minimum inhibitory concentration of chloramphenicol for *Salmonella enterica* serovar Typhi. *Indian J. Med. Sci.* **58:**16–23.
136. **Mankin, A. S., and R. A. Garrett.** 1991. Chloramphenicol resistance mutations in the single 23S rRNA gene of the archaeon *Halobacterium halobium. J. Bacteriol.* **173:**3559–3563.
137. **Marchese, A., E. A. Debbia, E. Tonoli, L. Gualco, and A. M. Schito.** 2002. In vitro activity of thiamphenicol against multiresistant *Streptococcus pneumoniae, Haemophilus influenzae* and *Staphylococcus aureus* in Italy. *J. Chemother.* **14:**554–561.
138. **McDermott, P. F., R. D. Walker, and D. G. White.** 2003. Antimicrobials: modes of action and mechanisms of resistance. *Int. J. Toxicol.* **22:**135–143.
139. **McMurry, L., R. E. Petrucci, Jr., and S. B. Levy.** 1980. Active efflux of tetracycline encoded by four genetically different tetracycline resistance determinants in *Escherichia coli. Proc. Natl. Acad. Sci. USA* **77:**3974–3977.
140. **Meunier, D., S. Baucheron, E. Chaslus-Dancla, J. L. Martel, and A. Cloeckaert.** 2003. Florfenicol resistance in *Salmonella enterica* serovar Newport mediated by a plasmid related to R55 from *Klebsiella pneumoniae. J. Antimicrob. Chemother.* **51:**1007–1009.
141. **Michael, G. B., M. Cardoso, and S. Schwarz.** Class 1 integron-associated gene cassettes in *Salmonella enterica* subsp. *enterica* serovar Agona isolated from pig carcasses in Brazil. *J. Antimicrob. Chemother.,* in press.
142. **Middlemiss, J. K., and K. Poole.** 2004. Differential impact of MexB mutations on substrate selectivity of the MexAB-OprM multidrug efflux pump of *Pseudomonas aeruginosa. J. Bacteriol.* **186:**1258–1269.
143. **Moon, K. H., B. R. Lee, S. J. Yoon, C. H. Koh, H. D. Jeong, and D. S. Lee.** 1995. Nucleotide sequences of the REP and CAT proteins encoded by the chloramphenicol resistance plasmid pKH7. *Yakhak Hoeji* **39:**676–680.
144. **Morii, H., N. Hayashi, and K. Uramoto.** 2003. Cloning and nucleotide sequence analysis of the chloramphenicol resistance gene on conjugative R plasmids from the fish pathogen *Photobacterium damselae* subsp. *piscicida. Dis. Aquat. Organ* **53:**107–113.
145. **Mosher, R. H., D. J. Camp, K. Yang, M. P. Brown, W. V. Shaw, and L. C. Vining.** 1995. Inactivation of chloramphenicol by O-phosphorylation. A novel resistance mechanism in *Streptomyces venezuelae* ISP5230, a chloramphenicol producer. *J. Biol. Chem.* **270:**27000–27006.
146. **Murray, I. A., J. A. Gil, D. A. Hopwood, and W. V. Shaw.** 1989. Nucleotide sequence of the chloramphenicol acetyltransferase gene of *Streptomyces acrimycini. Gene* **85:**283–291.
147. **Murray, I. A., A. R. Hawkins, J. W. Keyte, and W. V. Shaw.** 1988. Nucleotide sequence analysis and overexpression of the gene encoding a type III chloramphenicol acetyltransferase. *Biochem. J.* **252:**173–179.

148. **Murray, I. A., J. V. Martinez-Suarez, T. J. Close, and W. V. Shaw.** 1990. Nucleotide sequences of genes encoding the type II chloramphenicol acetyltransferases of *Escherichia coli* and *Haemophilus influenzae*, which are sensitive to inhibition by thiol-reactive reagents. *Biochem. J.* **272:**505–510.

149. **Murray, I. A., and W. V. Shaw.** 1997. O-Acetyltransferases for chloramphenicol and other natural products. *Antimicrob. Agents Chemother.* **41:**1–6.

150. **Nagy, I., G. Schoofs, J. Vanderleyden, and R. De Mot.** 1997. Transposition of the IS21-related element IS1415 in *Rhodococcus erythropolis*. *J. Bacteriol.* **179:**4635–4638.

151. **Nair, B. M., K. J. Cheung, Jr., A. Griffith, and J. L. Burns.** 2004. Salicylate induces an antibiotic efflux pump in *Burkholderia cepacia* complex genomovar III (*B. cenocepacia*). *J. Clin. Invest.* **113:**464–473.

152. **Nandi, S., J. J. Maurer, C. Hofacre, and A. O. Summers.** 2004. Gram-positive bacteria are a major reservoir of Class 1 antibiotic resistance integrons in poultry litter. *Proc. Natl. Acad. Sci. USA* **101:**7118–7122.

153. **NCCLS.** 2002. Performance standards for antimicrobial disk and dilution susceptibility tests for bacteria isolated from animals; Approved Standard. NCCLS document M31-A2. National Committee for Clinical Laboratory Standards, Wayne, Pa.

154. **NCCLS.** 2003. Methods for antimicrobial disk susceptibility testing of bacteria isolated from aquatic animals. M42-R. NCCLS, Wayne, Pa.

155. **NCCLS.** 2004. Performance standards for antimicrobial disk and dilution susceptibility tests for bacteria isolated from animals; Informational supplement (May 2004). NCCLS document M31-S1. NCCLS, Wayne, Pa.

156. **NCCLS.** 2004. Performance standards for antimicrobial susceptibility testing. NCCLS document M100-S14. National Committee for Clinical Laboratory Standards, Wayne, Pa.

157. **Neu, H. C., and K. P. Fu.** 1980. In vitro activity of chloramphenicol and thiamphenicol analogs. *Antimicrob. Agents Chemother.* **18:**311–316.

158. **Neyfakh, A. A., V. E. Bidnenko, and L. B. Chen.** 1991. Efflux-mediated multidrug resistance in *Bacillus subtilis:* similarities and dissimilarities with the mammalian system. *Proc. Natl. Acad. Sci. USA* **88:**4781–4785.

159. **Nilsen, I. W., I. Bakke, A. Vader, O. Olsvik, and M. R. El Gewely.** 1996. Isolation of *cmr,* a novel *Escherichia coli* chloramphenicol resistance gene encoding a putative efflux pump. *J. Bacteriol.* **178:**3188–3193.

160. **Nishino, K., and A. Yamaguchi.** 2001. Analysis of a complete library of putative drug transporter genes in *Escherichia coli*. *J. Bacteriol.* **183:**5803–5812.

161. **Nolling, J., G. Breton, M. V. Omelchenko, K. S. Makarova, Q. Zeng, R. Gibson, H. M. Lee, J. Dubois, D. Qiu, J. Hitti, Y. I. Wolf, R. L. Tatusov, F. Sabathe, L. Doucette-Stamm, P. Soucaille, M. J. Daly, G. N. Bennett, E. V. Koonin, and D. R. Smith.** 2001. Genome sequence and comparative analysis of the solvent-producing bacterium *Clostridium acetobutylicum*. *J. Bacteriol.* **183:**4823–4838.

162. **Pai, H., J. H. Byeon, S. Yu, B. K. Lee, and S. Kim.** 2003. *Salmonella enterica* serovar typhi strains isolated in Korea containing a multidrug resistance class 1 integron. *Antimicrob. Agents Chemother.* **47:**2006–2008.

163. **Parent, R., and P. H. Roy.** 1992. The chloramphenicol acetyltransferase gene of Tn2424: a new breed of cat. *J. Bacteriol.* **174:**2891–2897.

164. **Parkhill, J., G. Dougan, K. D. James, N. R. Thomson, D. Pickard, J. Wain, C. Churcher, K. L. Mungall, S. D. Bentley, M. T. Holden, M. Sebaihia, S. Baker, D. Basham, K. Brooks, T. Chillingworth, P. Connerton, A. Cronin, P. Davis, R. M. Davies, L. Dowd, N. White, J. Farrar, T. Feltwell, N. Hamlin, A. Haque, T. T. Hien, S. Holroyd, K. Jagels, A. Krogh, T. S. Larsen, S. Leather, S. Moule, P. O'Gaora, C. Parry, M. Quail, K. Rutherford, M. Simmonds, J. Skelton, K. Stevens, S. Whitehead, and B. G. Barrell.** 2001. Complete genome sequence of a multiple drug resistant *Salmonella enterica* serovar Typhi CT18. *Nature* **413:**848–852.

165. **Parkhill, J., B. W. Wren, N. R. Thomson, R. W. Titball, M. T. Holden, M. B. Prentice, M. Sebaihia, K. D. James, C. Churcher, K. L. Mungall, S. Baker, D. Basham, S. D. Bentley, K. Brooks, A. M. Cerdeno-Tarraga, T. Chillingworth, A. Cronin, R. M. Davies, P. Davis, G. Dougan, T. Feltwell, N. Hamlin, S. Holroyd, K. Jagels, A. V. Karlyshev, S. Leather, S. Moule, P. C. Oyston, M. Quail, K. Rutherford, M. Simmonds, J. Skelton, K. Stevens, S. Whitehead, and B. G. Barrell.** 2001. Genome sequence of *Yersinia pestis,* the causative agent of plague. *Nature* **413:**523–527.

166. **Parry, C. M.** 2004. The treatment of multidrug-resistant and nalidixic acid-resistant typhoid fever in Viet Nam. *Trans. R. Soc. Trop. Med. Hyg.* **98:**413–422.

167. **Parry, C. M.** 2004. Typhoid Fever. *Curr. Infect. Dis. Rep.* **6:**27–33.

168. **Partridge, S. R., G. D. Recchia, H. W. Stokes, and R. M. Hall.** 2001. Family of class 1 integrons related to In4 from Tn1696. *Antimicrob. Agents Chemother.* **45:**3014–3020.

169. **Paulsen, I. T.** 2003. Multidrug efflux pumps and resistance: regulation and evolution. *Curr. Opin. Microbiol.* **6:**446–451.

170. **Pepper, K., G. de Cespedes, and T. Horaud.** 1988. Heterogeneity of chromosomal genes encoding chloramphenicol resistance in streptococci. *Plasmid* **19:**71–74.

171. **Pepper, K., C. Le Bouguenec, G. de Cespedes, and T. Horaud.** 1986. Dispersal of a plasmid-borne chloramphenicol resistance gene in streptococcal and enterococcal plasmids. *Plasmid* **16:**195–203.

172. **Perreten, V., G. Moschetti, and M. Teuber.** 1995. GenBank accession no. X92945. National Center for Biotechnology Information.

173. **Perreten, V., F. Schwarz, L. Cresta, M. Boeglin, G. Dasen, and M. Teuber.** 1997. Antibiotic resistance spread in food. *Nature* **389:**801–802.

174. **Ploy, M. C., P. Courvalin, and T. Lambert.** 1998. Characterization of In40 of *Enterobacter aerogenes* BM2688, a class 1 integron with two new gene cassettes, *cmlA2* and *qacF. Antimicrob. Agents Chemother.* **42:**2557–2563.

175. **Poirel, L., J. W. Decousser, and P. Nordmann.** 2003. Insertion sequence ISEcp1B is involved in expression and mobilization of a $bla_{(CTX-M)}$ beta-lactamase gene. *Antimicrob. Agents Chemother.* **47:**2938–2945.

176. **Poirel, L., T. Le, I. T. Naas, A. Karim, and P. Nordmann.** 2000. Biochemical sequence analyses of GES-1, a novel class A extended-spectrum beta-lactamase, and the class 1 integron In52 from *Klebsiella pneumoniae*. *Antimicrob. Agents Chemother.* **44:**622–632.

177. **Poirel, L., T. Naas, M. Guibert, E. B. Chaibi, R. Labia, and P. Nordmann.** 1999. Molecular and biochemical characterization of VEB-1, a novel class A extended-spectrum beta-lactamase encoded by an *Escherichia coli* integron gene. *Antimicrob. Agents Chemother.* **43:**573–581.

178. **Poole, K.** 2001. Multidrug efflux pumps and antimicrobial resistance in *Pseudomonas aeruginosa* and related organisms. *J. Mol. Microbiol. Biotechnol.* **3:**255–264.

179. **Poole, K.** 2001. Multidrug resistance in gram-negative bacteria. *Curr. Opin. Microbiol.* **4:**500–508.

180. **Poole, K.** 2004. Efflux-mediated multiresistance in gram-negative bacteria. *Clin. Microbiol. Infect.* **10:**12–26.

181. **Projan, S. J., J. Kornblum, S. L. Moghazeh, I. Edelman, M. L. Gennaro, and R. P. Novick.** 1985. Comparative sequence and functional analysis of pT181 and pC221, cognate plasmid replicons from *Staphylococcus aureus*. *Mol. Gen. Genet.* **199:**452–464.
182. **Putman, M., H. W. van Veen, and W. N. Konings.** 2000. Molecular properties of bacterial multidrug transporters. *Microbiol. Mol. Biol. Rev.* **64:**672–693.
183. **Rankin, S., Z. Li, and R. R. Isberg.** 2002. Macrophage-induced genes of *Legionella pneumophila:* protection from reactive intermediates and solute imbalance during intracellular growth. *Infect. Immun.* **70:**3637–3648.
184. **Raymond, J., O. Zhaxybayeva, J. P. Gogarten, S. Y. Gerdes, and R. E. Blankenship.** 2002. Whole-genome analysis of photosynthetic prokaryotes. *Science* **298:**1616–1620.
185. **Read, T. D., S. N. Peterson, N. Tourasse, L. W. Baillie, I. T. Paulsen, K. E. Nelson, H. Tettelin, D. E. Fouts, J. A. Eisen, S. R. Gill, E. K. Holtzapple, O. A. Okstad, E. Helgason, J. Rilstone, M. Wu, J. F. Kolonay, M. J. Beanan, R. J. Dodson, L. M. Brinkac, M. Gwinn, R. T. DeBoy, R. Madpu, S. C. Daugherty, A. S. Durkin, D. H. Haft, W. C. Nelson, J. D. Peterson, M. Pop, H. M. Khouri, D. Radune, J. L. Benton, Y. Mahamoud, L. Jiang, I. R. Hance, J. F. Weidman, K. J. Berry, R. D. Plaut, A. M. Wolf, K. L. Watkins, W. C. Nierman, A. Hazen, R. Cline, C. Redmond, J. E. Thwaite, O. White, S. L. Salzberg, B. Thomason, A. M. Friedlander, T. M. Koehler, P. C. Hanna, A. B. Kolsto, and C. M. Fraser.** 2003. The genome sequence of *Bacillus anthracis* Ames and comparison to closely related bacteria. *Nature* **423:**81–86.
186. **Rende-Fournier, R., R. Leclercq, M. Galimand, J. Duval, and P. Courvalin.** 1993. Identification of the *satA* gene encoding a streptogramin A acetyltransferase in *Enterococcus faecium* BM4145. *Antimicrob. Agents Chemother.* **37:**2119–2125.
187. **Reynolds, E., J. I. Ross, and J. H. Cove.** 2003. Msr(A) and related macrolide/streptogramin resistance determinants: incomplete transporters? *Int. J. Antimicrob. Agents* **22:**228–236.
188. **Riccio, M. L., J. D. Docquier, E. Dell'Amico, F. Luzzaro, G. Amicosante, and G. M. Rossolini.** 2003. Novel 3-N-aminoglycoside acetyltransferase gene, *aac(3)-Ic*, from a *Pseudomonas aeruginosa* integron. *Antimicrob. Agents Chemother.* **47:**1746–1748.
189. **Roberts, M., A. Corney, and W. V. Shaw.** 1982. Molecular characterization of three chloramphenicol acetyltransferases isolated from *Haemophilus influenzae*. *J. Bacteriol.* **151:**737–741.
190. **Roberts, M. C.** 2004. Resistance to macrolide, lincosamide, streptogramin, ketolide, and oxazolidinone antibiotics. *Mol. Biotechnol.* **28:**47–62.
191. **Roberts, M. C., L. A. Actis, and J. H. Crosa.** 1985. Molecular characterization of chloramphenicol-resistant *Haemophilus parainfluenzae* and *Haemophilus ducreyi*. *Antimicrob. Agents Chemother.* **28:**176–180.
192. **Rogers, E. J., M. S. Rahman, R. T. Hill, and P. S. Lovett.** 2002. The chloramphenicol-inducible *catB* gene in *Agrobacterium tumefaciens* is regulated by translation attenuation. *J. Bacteriol.* **184:**4296–4300.
193. **Rojas, A., A. Segura, M. E. Guazzaroni, W. Teran, A. Hurtado, M. T. Gallegos, and J. L. Ramos.** 2003. In vivo and in vitro evidence that TtgV is the specific regulator of the TtgGHI multidrug and solvent efflux pump of *Pseudomonas putida*. *J. Bacteriol.* **185:**4755–4763.
194. **Rowe-Magnus, D. A., A. M. Guerout, L. Biskri, P. Bouige, and D. Mazel.** 2003. Comparative analysis of superintegrons: engineering extensive genetic diversity in the Vibrionaceae. *Genome Res.* **13:**428–442.
195. **Rowe-Magnus, D. A., A. M. Guerout, and D. Mazel.** 2002. Bacterial resistance evolution by recruitment of super-integron gene cassettes. *Mol. Microbiol.* **43:**1657–1669.
196. **Rubens, C. E., W. F. McNeill, and W. E. Farrar, Jr.** 1979. Transposable plasmid deoxyribonucleic acid sequence in *Pseudomonas aeruginosa* which mediates resistance to gentamicin and four other antimicrobial agents. *J. Bacteriol.* **139:**877–882.
197. **Sanchez, S., M. A. McCrackin Stevenson, C. R. Hudson, M. Maier, T. Buffington, Q. Dam, and J. J. Maurer.** 2002. Characterization of multidrug-resistant *Escherichia coli* isolates associated with nosocomial infections in dogs. *J. Clin. Microbiol.* **40:**3586–3595.
198. **Schlunzen, F., R. Zarivach, J. Harms, A. Bashan, A. Tocilj, R. Albrecht, A. Yonath, and F. Franceschi.** 2001. Structural basis for the interaction of antibiotics with the peptidyl transferase centre in eubacteria. *Nature* **413:**814–821.
199. **Schmidt, A. S., M. S. Bruun, I. Dalsgaard, and J. L. Larsen.** 2001. Incidence, distribution, and spread of tetracycline resistance determinants and integron-associated antibiotic resistance genes among motile aeromonads from a fish farming environment. *Appl. Environ. Microbiol.* **67:**5675–5682.
200. **Schwarz, F. V., V. Perreten, and M. Teuber.** 2001. Sequence of the 50-kb conjugative multiresistance plasmid pRE25 from *Enterococcus faecalis* RE25. *Plasmid* **46:**170–187.
201. **Schwarz, S., and M. Cardoso.** 1991. Molecular cloning, purification, and properties of a plasmid-encoded chloramphenicol acetyltransferase from *Staphylococcus haemolyticus*. *Antimicrob. Agents Chemother.* **35:**1277–1283.
202. **Schwarz, S., and M. Cardoso.** 1991. Nucleotide sequence and phylogeny of a chloramphenicol acetyltransferase encoded by the plasmid pSCS7 from *Staphylococcus aureus*. *Antimicrob. Agents Chemother.* **35:**1551–1556.
203. **Schwarz, S., and E. Chaslus-Dancla.** 2001. Use of antimicrobials in veterinary medicine and mechanisms of resistance. *Vet. Res.* **32:**201–225.
204. **Schwarz, S., and S. Grolz-Krug.** 1991. A chloramphenicol-streptomycin-resistance plasmid from a clinical strain of *Staphylococcus sciuri* and its structural relationships to other staphylococcal resistance plasmids. *FEMS Microbiol. Lett.* **66:**319–322.
205. **Schwarz, S., C. Kehrenberg, B. Doublet, and A. Cloeckaert.** 2004. Molecular basis of bacterial resistance to chloramphenicol and florfenicol. *FEMS Microbiol. Rev.* **28:**519–542.
206. **Schwarz, S., and W. C. Noble.** 1994. Structure and putative origin of a plasmid from *Staphylococcus hyicus* that mediates chloramphenicol and streptomycin resistance. *Lett. Appl. Microbiol.* **18:**281–284.
207. **Schwarz, S., U. Spies, and M. Cardoso.** 1991. Cloning and sequence analysis of a plasmid-encoded chloramphenicol acetyltransferase gene from *Staphylococcus intermedius*. *J. Gen. Microbiol.* **137:**977–981.
208. **Schwarz, S., C. Werckenthin, and C. Kehrenberg.** 2000. Identification of a plasmid-borne chloramphenicol-florfenicol resistance gene in *Staphylococcus sciuri*. *Antimicrob. Agents Chemother.* **44:**2530–2533.
209. **Settepani, J. A.** 1984. The hazard of using chloramphenicol in food animals. *J. Am. Vet. Med. Assoc.* **184:**930–931.
210. **Shaw, W. V.** 1967. The enzymatic acetylation of chloramphenicol by extracts of R factor-resistant *Escherichia coli*. *J. Biol. Chem.* **242:**687–693.
211. **Shaw, W. V.** 1983. Chloramphenicol acetyltransferase: enzymology and molecular biology. *CRC Crit Rev. Biochem.* **14:**1–46.
212. **Shaw, W. V.** 1992. Chemical anatomy of antibiotic resistance: chloramphenicol acetyltransferase. *Sci. Prog.* **76:**565–580.

213. **Shaw, W. V., and R. F. Brodsky.** 1967. Chloramphenicol resistance by enzymatic acetylation: comparative aspects. *Antimicrobial. Agents Chemother.* **7:**257–263.
214. **Shaw, W. V., and A. G. Leslie.** 1991. Chloramphenicol acetyltransferase. *Annu. Rev. Biophys. Biophys. Chem.* **20:**363–386.
215. **Sherwood, K. J., and B. Wiedemann.** 2004. GenBank accession no. AY775258. National Center for Biotechnology Information.
216. **Shultz, T. R., J. W. Tapsall, P. A. White, C. S. Ryan, D. Lyras, J. I. Rood, E. Binotto, and C. J. Richardson.** 2003. Chloramphenicol-resistant *Neisseria meningitidis* containing *catP* isolated in Australia. *J. Antimicrob. Chemother.* **52:**856–859.
217. **Singer, R. S., S. K. Patterson, A. E. Meier, J. K. Gibson, H. L. Lee, and C. W. Maddox.** 2004. Relationship between phenotypic and genotypic florfenicol resistance in *Escherichia coli. Antimicrob. Agents Chemother.* **48:**4047–4049.
218. **Smith, M. C., and C. D. Thomas.** 2003. GenBank accession no. NC_005243. National Center for Biotechnology Information.
219. **Sorum, H., T. M. L'Abee-Lund, A. Solberg, and A. Wold.** 2003. Integron-containing IncU R plasmids pRAS1 and pAr-32 from the fish pathogen *Aeromonas salmonicida. Antimicrob. Agents Chemother.* **47:**1285–1290.
220. **Steffen, C., and H. Matzura.** 1989. Nucleotide sequence analysis and expression studies of a chloramphenicol-acetyltransferase-coding gene from *Clostridium perfringens. Gene* **75:**349–354.
221. **Stokes, H. W., and R. M. Hall.** 1991. Sequence analysis of the inducible chloramphenicol resistance determinant in the Tn1696 integron suggests regulation by translational attenuation. *Plasmid* **26:**10–19.
222. **Stover, C. K., X. Q. Pham, A. L. Erwin, S. D. Mizoguchi, P. Warrener, M. J. Hickey, F. S. Brinkman, W. O. Hufnagle, D. J. Kowalik, M. Lagrou, R. L. Garber, L. Goltry, E. Tolentino, S. Westbrock-Wadman, Y. Yuan, L. L. Brody, S. N. Coulter, K. R. Folger, A. Kas, K. Larbig, R. Lim, K. Smith, D. Spencer, G. K. Wong, Z. Wu, I. T. Paulsen, J. Reizer, M. H. Saier, R. E. Hancock, S. Lory, and M. V. Olson.** 2000. Complete genome sequence of *Pseudomonas aeruginosa* PA01, an opportunistic pathogen. *Nature* **406:**959–964.
223. **Suzuki, Y., and S. Okamoto.** 1967. The enzymatic acetylation of chloramphenicol by the multiple drug-resistant *Escherichia coli* carrying R factor. *J. Biol. Chem.* **242:**4722–4730.
224. **Syriopoulou, V. P., A. L. Harding, D. A. Goldmann, and A. L. Smith.** 1981. In vitro antibacterial activity of fluorinated analogs of chloramphenicol and thiamphenicol. *Antimicrob. Agents Chemother.* **19:**294–297.
225. **Takamatsu, D., M. Osaki, and T. Sekizaki.** 2003. Chloramphenicol resistance transposable element TnSs1 of *Streptococcus suis*, a transposon flanked by IS6-family elements. *Plasmid* **49:**143–151.
226. **Tauch, A., Z. Zheng, A. Puhler, and J. Kalinowski.** 1998. Corynebacterium striatum chloramphenicol resistance transposon Tn5564: genetic organization and transposition in *Corynebacterium glutamicum. Plasmid* **40:**126–139.
227. **Tennigkeit, J., and H. Matzura.** 1991. Nucleotide sequence analysis of a chloramphenicol-resistance determinant from *Agrobacterium tumefaciens* and identification of its gene product. *Gene* **98:**113–116.
228. **Tennstedt, T., R. Szczepanowski, S. Braun, A. Puehler, and A. Schlueter.** 2003. Occurrence of integron-associated resistance gene cassettes located on antibiotic resistance plasmids isolated from a wastewater treatment plant. *FEMS Microbiol. Ecol.* **45:**239–252.
229. **Tettelin, H., V. Masignani, M. J. Cieslewicz, J. A. Eisen, S. Peterson, M. R. Wessels, I. T. Paulsen, K. E. Nelson, I. Margarit, T. D. Read, L. C. Madoff, A. M. Wolf, M. J. Beanan, L. M. Brinkac, S. C. Daugherty, R. T. DeBoy, A. S. Durkin, J. F. Kolonay, R. Madupu, M. R. Lewis, D. Radune, N. B. Fedorova, D. Scanlan, H. Khouri, S. Mulligan, H. A. Carty, R. T. Cline, S. E. Van Aken, J. Gill, M. Scarselli, M. Mora, E. T. Iacobini, C. Brettoni, G. Galli, M. Mariani, F. Vegni, D. Maione, D. Rinaudo, R. Rappuoli, J. L. Telford, D. L. Kasper, G. Grandi, and C. M. Fraser.** 2002. Complete genome sequence and comparative genomic analysis of an emerging human pathogen, serotype V *Streptococcus agalactiae. Proc. Natl. Acad. Sci. USA* **99:**12391–12396.
230. **Tietze, E., and K. Smalla.** 2000. Genbank accession no. NC_004973. National Center for Biotechnology Information.
231. **Tikhonova, E. B., and H. I. Zgurskaya.** 2004. AcrA, AcrB, and TolC of *Escherichia coli* form a stable intermembrane multidrug efflux complex. *J. Biol. Chem.* **279:**32116–32124.
232. **Toro, C. S., S. R. Lobos, I. Calderon, M. Rodriguez, and G. C. Mora.** 1990. Clinical isolate of a porinless *Salmonella typhi* resistant to high levels of chloramphenicol. *Antimicrob. Agents Chemother.* **34:**1715–1719.
233. **Tosini, F., P. Visca, I. Luzzi, A. M. Dionisi, C. Pezzella, A. Petrucca, and A. Carattoli.** 1998. Class 1 integron-borne multiple-antibiotic resistance carried by IncFI and IncL/M plasmids in *Salmonella enterica* serotype Typhimurium. *Antimicrob. Agents Chemother.* **42:**3053–3058.
234. **Trieu-Cuot, P., G. de Cespedes, F. Bentorcha, F. Delbos, E. Gaspar, and T. Horaud.** 1993. Study of heterogeneity of chloramphenicol acetyltransferase (CAT) genes in streptococci and enterococci by polymerase chain reaction: characterization of a new CAT determinant. *Antimicrob. Agents Chemother.* **37:**2593–2598.
235. **Trieu-Cuot, P., G. de Cespedes, and T. Horaud.** 1992. Nucleotide sequence of the chloramphenicol resistance determinant of the streptococcal plasmid pIP501. *Plasmid* **28:**272–276.
236. **Tullio, V., A. Cuffini, N. Mandras, J. Roana, G. Banche, D. Ungheri, and N. Carlone.** 2004. Influence of thiamphenicol on the primary functions of human polymorphonuclear leucocytes against *Streptococcus pyogenes. Int. J. Antimicrob. Agents* **24:**381–385.
237. **Vassort-Bruneau, C., M. C. Lesage-Descauses, J. L. Martel, J. P. Lafont, and E. Chaslus-Dancla.** 1996. CAT III chloramphenicol resistance in *Pasteurella haemolytica* and *Pasteurella multocida* isolated from calves. *J. Antimicrob. Chemother.* **38:**205–213.
238. **Vester, B., and S. Douthwaite.** 2001. Macrolide resistance conferred by base substitutions in 23S rRNA. *Antimicrob. Agents Chemother.* **45:**1–12.
239. **Villa, L., C. Mammina, V. Miriagou, L. S. Tzouvelekis, P. T. Tassios, A. Nastasi, and A. Carattoli.** 2002. Multidrug and broad-spectrum cephalosporin resistance among *Salmonella enterica* serotype Enteritidis clinical isolates in southern Italy. *J. Clin. Microbiol.* **40:**2662–2665.
240. **Vlcek, C., V. Paces, N. Maltsev, J. Paces, R. Haselkorn, and M. Fonstein.** 1997. Sequence of a 189-kb segment of the chromosome of *Rhodobacter capsulatus* SB1003. *Proc. Natl. Acad. Sci. USA* **94:**9384–9388.
241. **Volker, T. A., S. Iida, and T. A. Bickle.** 1982. A single gene coding for resistance to both fusidic acid and chloramphenicol. *J. Mol. Biol.* **154:**417–425.
242. **Wang, M., J. H. Tran, G. A. Jacoby, Y. Zhang, F. Wang, and D. C. Hooper.** 2003. Plasmid-mediated quinolone resistance in clinical isolates of *Escherichia coli* from Shanghai, China. *Antimicrob. Agents Chemother.* **47:**2242–2248.
243. **Wang, Y., and D. E. Taylor.** 1990. Chloramphenicol resistance in *Campylobacter coli:* nucleotide sequence, expression, and cloning vector construction. *Gene* **94:**23–28.

244. **Wasfy, M. O., R. Frenck, T. F. Ismail, H. Mansour, J. L. Malone, and F. J. Mahoney.** 2002. Trends of multiple-drug resistance among *Salmonella* serotype Typhi isolates during a 14-year period in Egypt. *Clin. Infect. Dis.* **35:**1265–1268.
245. **Werckenthin, C., M. Cardoso, J. L. Martel, and S. Schwarz.** 2001. Antimicrobial resistance in staphylococci from animals with particular reference to bovine *Staphylococcus aureus,* porcine *Staphylococcus hyicus,* and canine *Staphylococcus intermedius. Vet. Res.* **32:**341–362.
246. **Werner, G., and W. Witte.** 1999. Characterization of a new enterococcal gene, *satG,* encoding a putative acetyltransferase conferring resistance to streptogramin A compounds. *Antimicrob. Agents Chemother.* **43:**1813–1814.
247. **White, D. G., J. Acar, F. Anthony, A. Franklin, R. Gupta, T. Nicholls, Y. Tamura, S. Thompson, E. J. Threlfall, D. Vose, M. van Vuuren, H. C. Wegener, and M. L. Costarrica.** 2001. Antimicrobial resistance: standardisation and harmonisation of laboratory methodologies for the detection and quantification of antimicrobial resistance. *Rev. Sci. Tech.* **20:**849–858.
248. **White, D. G., C. Hudson, J. J. Maurer, S. Ayers, S. Zhao, M. D. Lee, L. Bolton, T. Foley, and J. Sherwood.** 2000. Characterization of chloramphenicol and florfenicol resistance in *Escherichia coli* associated with bovine diarrhea. *J. Clin. Microbiol.* **38:**4593–4598.
249. **White, O., J. A. Eisen, J. F. Heidelberg, E. K. Hickey, J. D. Peterson, R. J. Dodson, D. H. Haft, M. L. Gwinn, W. C. Nelson, D. L. Richardson, K. S. Moffat, H. Qin, L. Jiang, W. Pamphile, M. Crosby, M. Shen, J. J. Vamathevan, P. Lam, L. McDonald, T. Utterback, C. Zalewski, K. S. Makarova, L. Aravind, M. J. Daly, and C. M. Fraser.** 1999. Genome sequence of the radioresistant bacterium *Deinococcus radiodurans* R1. *Science* **286:**1571–1577.
250. **White, P. A., H. W. Stokes, K. L. Bunny, and R. M. Hall.** 1999. Characterisation of a chloramphenicol acetyltransferase determinant found in the chromosome of *Pseudomonas aeruginosa. FEMS Microbiol. Lett.* **175:**27–35.
251. **Widdowson, C. A., P. V. Adrian, and K. P. Klugman.** 2000. Acquisition of chloramphenicol resistance by the linearization and integration of the entire staphylococcal plasmid pC194 into the chromosome of *Streptococcus pneumoniae. Antimicrob. Agents Chemother.* **44:**393–395.
252. **Widy-wirski, A. Meheus, B. Lala, P. Piot, and J. D'costa.** 1984. Thiamphenicol in treatment of genital gonorrhoeae in Central African Republic. *Afr. J. Sex Transmi. Dis.* **1:**11–13.
253. **Wren, B. W., P. Mullany, C. Clayton, and S. Tabaqchali.** 1989. Nucleotide sequence of a chloramphenicol acetyl transferase gene from *Clostridium difficile. Nucleic Acids Res.* **17:**4877.
254. **Wu, J., H. K. Shieh, J. Shien, S. Gong, and P. Chang.** 2003. GenBank accession no. NC_004771. National Center for Biotechnology Information.
255. **Xu, J., M. K. Bjursell, J. Himrod, S. Deng, L. K. Carmichael, H. C. Chiang, L. V. Hooper, and J. I. Gordon.** 2003. A genomic view of the human-*Bacteroides thetaiotaomicron* symbiosis. *Science* **299:**2074–2076.
256. **Yamamoto, J., T. Sakano, and M. Shimizu.** 1990. Drug resistance and R plasmids in *Pasteurella multocida* isolates from swine. *Microbiol. Immunol.* **34:**715–721.
257. **Yan, J. J., W. C. Ko, and J. J. Wu.** 2001. Identification of a plasmid encoding SHV-12, TEM-1, and a variant of IMP-2 metallo-beta-lactamase, IMP-8, from a clinical isolate of *Klebsiella pneumoniae. Antimicrob. Agents Chemother.* **45:**2368–2371.
258. **Yoshida, H., M. Bogaki, S. Nakamura, K. Ubukata, and M. Konno.** 1990. Nucleotide sequence and characterization of the *Staphylococcus aureus norA* gene, which confers resistance to quinolones. *J. Bacteriol.* **172:**6942–6949.
259. **Yunis, A. A.** 1988. Chloramphenicol: relation of structure to activity and toxicity. *Annu. Rev. Pharmacol. Toxicol.* **28:**83–100.
260. **Zhao, J., and T. Aoki.** 1992. Cloning and nucleotide sequence analysis of a chloramphenicol acetyltransferase gene from *Vibrio anguillarum. Microbiol. Immunol.* **36:**695–705.
261. **Zhao, S., S. Qaiyumi, S. Friedman, R. Singh, S. L. Foley, D. G. White, P. F. McDermott, T. Donkar, C. Bolin, S. Munro, E. J. Baron, and R. D. Walker.** 2003. Characterization of *Salmonella enterica* serotype Newport isolated from humans and food animals. *J. Clin. Microbiol.* **41:**5366–5371.
262. **Zheng, Z., S. Shi, X. Jiang, Z. Wang, and L. Caro.** 1997. GenBank accession no. U85507. National Center for Biotechnology Information.

PART III. MULTIPLE ANTIMICROBIAL RESISTANCE MECHANISMS

SECTION EDITOR: Patrick F. McDermott

THE ABILITY OF BACTERIA TO DEVELOP ANTIMICROBIAL RESISTANCE was recognized by Fleming with the discovery of penicillin in 1928. For the next 25 years, as other classes of antimicrobials were discovered, the dynamics of resistance development were generally understood as incremental changes from low levels to high and from single resistances to multiple resistances, although the latter was relatively infrequent. The situation today is much different. After 60 years of widespread antimicrobial use in clinical and veterinary medicine and in agriculture, multidrug resistance has become common. This development can be explained by the accumulation of linked genes on large transmissible plasmids, expression of multidrug efflux pumps, and the temporal accumulation of mutations selected by extensive antimicrobial exposure. After much research emphasis on plasmid-mediated multidrug resistance, seminal advances made by Dr. Levy's laboratory showed that such multiple antimicrobial resistance (Mar) mechanisms also reside on the bacterial chromosome and constitute part of the native cellular defenses to a variety of environmental insults. This section provides a discussion on bacterial multiresistance to environmental and chemotherapeutic compounds, with a special emphasis on the role of the Mar regulon and analogous systems.

Frontiers in Antimicrobial Resistance: a Tribute to Stuart B. Levy
Edited by D. G. White, M. N. Alekshun, and P. F. McDermott

Chapter 10

Multiple Antimicrobial Resistance

ANTHONY M. GEORGE

This chapter is an overview of multiple antimicrobial resistance mechanisms in bacteria. It begins, however, with a brief history of my beginnings in this field, under the tutelage of Stuart B. Levy at Tufts University School of Medicine in Boston, Mass. Multidrug resistance was a relatively new phenomenon in 1981, with only preliminary observations in bacteria and the identification of P-glycoprotein in human cancer. I went to Boston in 1981 as a postdoctoral fellow with an elementary background in biochemistry and microbiology and came away in 1983 as a reasonably accomplished scientist. In those days, investigative molecular genetics was more classical than modern in that DNA cloning and sequencing was very new, laborious, and technically difficult to perform (ready-use kits had not come onto the market yet). Nevertheless, by using bacterial genetic methods, an open mind, and the searching, eager, and driving force of Stuart Levy's scientific appetite for discovery, we managed to identify the *mar* locus in *Escherichia coli* (25, 26). The intervening two decades have now seen the reporting of many other multidrug resistance mechanisms in all three kingdoms.

The Mar system had its beginnings in Stuart Levy's serendipitous observation that cells of an *E. coli* K-12 laboratory strain could be selected, grown, and maintained on a very low level of tetracycline (~1 μ/ml) on Luria-Bertani (LB) agar plates. We went on to observe that cells from the original population could be "trained or adapted"—at low frequency ($\sim 10^{-8}$)—to grow on LB agar containing successively higher concentrations of tetracycline (up to 200 μg/ml) and that this process could be mimicked or paralleled by using chloramphenicol (from 5 μg to >400 μg/ml). This observation could also be accomplished by subculturing low-level mutants with tetracycline or chloramphenicol at only 5 μ/ml, whereupon cells from these cultures could be plated on LB agar containing very high concentrations of either tetracycline or chloramphenicol. When the antibiotics were removed, the high-level resistances reverted to basal levels in just several days. The tandem tetracycline-chloramphenicol resistance phenotypes led us to test for—and find—cross-resistance between these and other antibiotics such as nalidixic acid, rifampin, puromycin, and β-lactams (25). We went on to demonstrate that this complex phenotype was due—at least in part—to an efflux mechanism for tetracycline.

The genetic site of this multiple antibiotic resistance phenotype was mapped (by a single Tn*5* insertion) to a locus within the large cotransduction gap of the *E. coli* chromosome (26). Interestingly, the Tn*5* mutant was even more susceptible to all antibiotics than the original wild-type strain, suggesting that a basal level of resistance was present even in wild-type cells. This new locus was designated *mar* for multiple antibiotic resistance, and it was at this point that my fellowship concluded. The next generation of researchers in Stuart's laboratory took this study to the molecular and mechanistic level by cloning and sequencing the *mar* locus; identifying the *marR* and *marA* genes as transcriptional repressor and activator, respectively, of the *marRAB operon;* and identifying the mar regulon as a global effector in the *E. coli* chromosome. The *mar* system is discussed at length by many of Stuart's "extended family" of colleagues and peers in other chapters of this book and in the references cited therein.

From these beginnings comes the central theme of this chapter, multiple antimicrobial resistance mechanisms. By definition, this is intended to mean multidrug resistance (MDR) that is encoded by single chromosomal or plasmid genes or operons. The chief mechanism by which the phenotype is displayed is active efflux (extrusion) of structurally diverse compounds. The energy for the translocation process is

Anthony M. George • Department of Cell and Molecular Biology, University of Technology Sydney, P.O. Box 123, Broadway, N.S.W., Australia 2007.

derived from the proton motive force or ATP hydrolysis. The context of this definition of MDR is not intended to include R-plasmid resistance genes, transposons, or integron cassettes, which collectively imbue the host bacterium with a multiple resistance phenotype—a collage of genes of which each confers resistance to a single antibiotic or single class of antibiotics. This overview will provide a brief, descriptive compendium of genes and systems and then discuss two of the chief conundrums of MDR: what does a multidrug site look like, and how do single proteins flaunt the "one-substrate, one-enzyme" rule?

MULTIDRUG RESISTANCE EFFLUX SYSTEMS

The number of MDR systems now known in bacteria is extensive, having been amassed from genetic, microbiological, biochemical, and phylogenetic investigations and, more recently, from bioinformatic sequence analysis of complete bacterial chromosomes. Multidrug efflux systems have been reviewed more extensively and knowledgeably than is possible in this overview, and the reader is directed to a number of excellent reviews (28, 30, 53, 54, 57, 64, 66, 67, 70, 71, 74, 77, 89, 103). The chief intention of this review is to provide a brief update and outline of multidrug systems in bacteria and then discuss in more detail current knowledge of the basis of the MDR phenomenon.

Multidrug transporters may be classified into two divisions: proton motive force-dependent secondary transporters (67) and ABC transporters (24, 100). The first category is divided into four subclasses—the major facilitator (MF), resistance-nodulation-cell division (RND), small multidrug resistance (SMR), and multidrug and toxic compound extrusion (MATE) families (Fig. 1). MF, RND, and SMR transporters are H^+/drug antiporters, MATE pumps are Na^+/drug antiporters, and ABC transporters are energized by ATP hydrolysis (Table 1). Listings for all five families are given in Tables 1–5. Putative and extant multidrug transporters of the five families occur in the *E. coli* genome (68). MF transporters are responsible for most of the MDR in gram-positive bacteria, though SMR transporters are also found in some species. In gram-negative bacteria, most MF and SMR pumps are not expressed under normal conditions (84) and RND systems are the main contributors. RND and MATE MDR pumps (Table 2) have only been found in gram-negative organisms (71). The MF and ABC superfamilies are large and ancient and are found in all three kingdoms (74). Although RND MDR pumps have been located only in gram-negative bacteria, an extensive sequence and phylogenetic study revealed that RND permeases are ubiquitous among the archeae, eubacteria, and eukaryia (89). Though these RNDs are involved mostly in physiological functions, the authors predicted that multidrug efflux will be identified as an ancillary function.

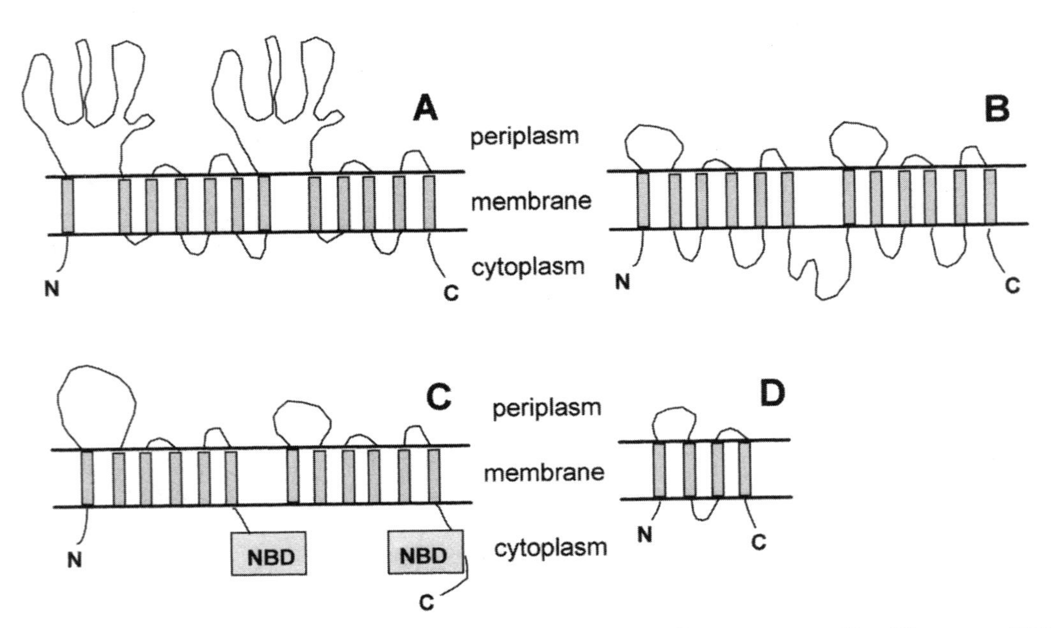

Figure 1. Transmembrane (TM) topologies of the major classes of multidrug efflux transporters. (A to D) represent RND, MF, ABC, and SMR families, respectively. The MATE family is not depicted, as its topology resembles the MF superfamily in having 12 TM segments. Some MF pumps have an extra 2 TM segments, that is, 14 in all. Vertical rectangles delineate TM spans of about 18 residues each. Intra- and extracytoplasmic connecting loops represent approximate scale lengths of residues, except for ABC transporters (C), whose large ATP-binding cassette domains (NBDs) are drawn as boxes. Note also that ABC transporters in bacteria are not usually contiguous polypeptides but are composites of four separate subunits (two TMDs and two NBDs) or dimers of two half transporters, as depicted in C.

Table 1. RND multidrug transporters

Efflux pump	Bacterial host	Substrates[a]	Regulator gene(s)
AcrAB-TolC, AcrEF-TolC	*Escherichia coli*	CML, TET, ERY, NOV, TMP, SDS, pine oil, CV, FUS, ACR, EB, ACF, FAs, BSs, BLs, OSs	*acrR, marA, soxS, rob, sdiA*
SmeABC, SmeCDE	*Stenotrophomonas maltophilia*	AGs, BLs FQs (ABC); ERY, TET, EB (CDE)	*smrR* (ABC)
TtgABC, TtgDEF, TtgGHI	*Pseudomonas putida*	CML, AMP, TET, NAL, TOL (ABC); AHs DEF); AMP, CAB, AHs (GHI)	*ttgR* (ABC)
MexAB-OprM, MexCD-OprJ	*Pseudomonas aeruginosa*	CML, TET, NOV, TMP, MLs, EB, ACR, SDS, BLs, AHs, FQs, TSN, CV, (SULs, TOL; AB only)	*mexR* (AB); *nfxB* (CD)
MexEF-OprN	*Pseudomonas aeruginosa*	CML, TMP, TCN, AHs, FQs	*mexT*
MexXY-OprM	*Pseudomonas aeruginosa*	TET, ERY, FQs, AGs	*mexZ*
TbtABM	*Pseudomonas stutzeri*	Tributyltin, *n*-hexane, NAL, CML, SMX	
MdtABC	*Escherichia coli*	NOV, BSs	*baeR*
YhiUV	*Escherichia coli*	DOX, ERY, R6G, BZK, SDS, DEO, CV	*evgA*
CmeABC	*Campylobacter jejuni*	TET, RIF, BLs, BSs	
CzcCBA	*Pseudomonas aeruginosa*	IPM, heavy metals	*czcRS*
MtrCDE	*Neisseria gonorrhoeae*	FAs, BSs, peptides	*mtrA*
CeoAB	*Burkholderia cepacia*	CML, CIP, TMP	
BpeAB-OprB	*Burkholderia pseudomallei*	AGs, BLs, MLs, PMs	*bpeR*
ArpABC	*Pseudomonas putida*	CML, TET, NOV, ERY, CAB, STR	
MepABC	*Pseudomonas putida*	TET, NOV, ERY, AHs	

[a] ACF, acriflavine; ACR, acridine; AGs, aminoglycosides; AHs, aromatic hydrocarbons; AMP, ampicillin; BLs, β-lactams; BSs, bile salts; BZK, benzylkonium; CAB, carbenicillin; CCCP, carbonyl cyanide *m*-chlorophenylhydrazone; CIP, ciprofloxacin; CML, chloramphenicol; CV, crystal violet; DAPI, 4′,6-diamidino-2-phenylindole; DAU, daunomycin; DEO, deoxycholate; DOX, doxorubicin; EB, ethidium bromide; ERY, erythromycin; FAs, fatty acids; FQs, fluoroquinolones; FUS, fusidic acid; IPM, imipenem; KAN, kanamicin; LCs, lincosamides; MLs, macrolides; MV, methyl viologen; NAL, nalidixic acid; NOR, norfloxacin; NOV, novobiocin; OSs, organic solvents; OTC, oxytetracycline; OXO, oxacillin; PFN, proflavin; PMs, polymyxins; PUR, puromycin; QACs, quaternary ammonium compounds; QLs, quinolones; RIF, rifampin; R6G, rhodamine 6G; SDS, sodium dodecyl sulfate; SGs, streptogramins; SMX, sulfamethoxalone; STR, streptomycin; SUL, sulfonamide; TCN, triclosan; TET, tetracycline; TMP, trimethoprim; TOL, toluene; TPP$^+$, tetraphenylphosphonium; VGN, virginiamycin.

There are at least five, six, and seven putative SMR proteins in *E. coli, P. aeruginosa,* and *B. subtilis,* respectively, and one of the *P. aeruginosa* proteins shows close identity to EmrE from *E. coli* (47). This transporter, named $EmrE_{Pae}$, is now known to make a contribution to the intrinsic resistance of *P. aeruginosa* to aminoglycosides and cationic dyes. This SMR diversity is reflected within the other families and their bacterial hosts. The substrate spectra of many of these transporters overlap, which poses the intriguing question of why a bacterium with an economically organized genome harbors such a large set of multidrug efflux pumps (5). Moreover, why are some systems subjected to strict regulatory activation and repression, while others such as the SMRs are expressed constitutively? Most systems are subject to some form of transcriptional and/or translational control, perhaps against the deleterious effects of overexpression (30).

Table 2. MF multidrug transporters

Efflux pump	Bacterial host	Substrates[a]	Regulator gene(s)
MdeA	*Staphylococcus aureus*	QACs, EB, NOV, FUS, VGN	
MdfA	*Escherichia coli*	EB, TPP$^+$, DAU, R6G, RIF, TET PUR, CML, ERY, AGs, FQs, BZK	
EmeA	*Enterococcus faecalis*	EB, CLN, NOR, TPP+, DAPI, Hoechst 33342, BZK	
LmrP	*Lactococcus lactis*	MLs, LCs, SGs, TETs	
VceAB	*Vibrio cholerae*	DEO, NAL, CML, CCCP	
QacA (pl)[b]	*Staphylococcus aureus*	Antiseptics, disinfectants, intercalating dyes	*qacR*
NorA	*Staphylococcus aureus*	QACs, QLs, acridine dyes, Hoechst 3334	*norR, agr, sarA*
Bmr	*Bacillus subtilis*	EB, ACR, CML, PUR, TPP+, FQs, QACs, R6G	
PmrA	*Streptococcus pneumoniae*	EB, ACF, FQs	
Mdt(A)	*Lactococcus lactis*	MLs, LCs, tetracyclines	
FarAB	*Neisseria gonorrhoeae*	Long-chain fatty acids	*farR, mtrR*
LfrA	*Mycobacterium smegmatis*	EB, ACR, FQs(?)	

[a] See Table 1, footnote *a*, for abbreviations for substrates.
[b] pl, plasmid encoded.

Table 3. SMR multidrug transporters

Efflux pump	Bacterial host	Substrates[a]
EmrE	*Escherichia coli*	EB, TET, TPP$^+$, QACs, MV, CV
Smr, QacG, QacH, QacJ (pl)[b]	*Staphylococcus aureus*	QACs, EB, cationic biocides
QacE (pl)	*Staphylococcus aureus*	EB, QACs, CV, PFN
TehA	*Escherichia coli*	TPP$^+$, EB, PFN, CV

[a] See Table 1, footnote *a*, for abbreviations for substrates.
[b] pl, plasmid encoded.

GLOBAL REGULATION SYSTEMS

Multidrug transporters are often expressed under very precise and elaborate transcriptional control, in response to environmental signals, substrates, or xenobiotics. For example, in *E. coli,* EmrR derepresses transcription of the MDR efflux genes *emrAB* in response to a variety of chemicals or substrates (48). The AcrAB pump is regulated on several levels, the first of which is by the contiguous AcrR repressor (50). The *acrRAB* operon is also regulated globally by the transcriptional activator MarA (3, 6), in response to anionic lipophilic compounds such as salicylate that bind to a MarR dimer, causing its release from the *marO* site (27). SoxS is another global regulator that activates *acrAB* and a group of other genes in response to oxidative stress (28). RobA is a third global activator of *acrAB* (85, 96). MarA, SoxS, and RobA are small proteins belonging to the XylA/AraC family of transcriptional regulators but are different in that they do not bind ligands and bind as monomers to asymmetric operators, rather than as dimers to palindromic DNA. To this group may be added PqrA from *P. vulgaris* (35), whose amino acid sequence has a high percentage of similarity with SoxS and MarA and was able to elicit a multidrug phenotype when cloned into *E. coli.* SdiA (31) is a global activator of *acrAB* and three other *E. coli* MDR loci, namely *tolC, acrEF,* and *acrD.* SdiA, a homologue of the LuxR quorum-sensing regulator, also acts as a positive regulator of genes involved in septation, motility, and chemotaxis (94, 95).

Why global regulation in bacteria? Clearly, this phenomenon is based on the unicellular nature of bacteria and their environmental exposure to a raft of agents, toxins, drugs, organic solvents, heavy metals, oxidants, secondary metabolites, and other chemicals (30, 63). It has the advantage of putting the host on a "war footing"—the constitutive expression of MarA in *E. coli,* for instance, causes the differential expression, by activation or repression, of over 60 chromosomal genes (6, 75). The *E. coli mar* locus is also found among other enteric bacteria (12). Interestingly, the complex choreography of the global response to relatively simple inducer signals requires that the entire process be effectively shut down when the inducer is removed. The first stage in this process involves the repression of further transcription of *soxS* and *marA* genes by SoxR (15) and MarR (3), respectively. Next, extant copies of SoxS and MarA—that are intrinsically unstable—are removed by rapid protease degradation in the absence of inducer ligands (29). It is possible that SoxS and MarA—and other regulators—exist in two conformations that correspond to the DNA bound and unbound states and that it is the latter free conformation that is particularly susceptible to degradation (13). Regulation at the protein stability level for transcriptional regulators is relatively uncommon in *E. coli.*

Homologues of MDR genes or operons are widespread among bacteria (12, 74). Phylogenetic analysis had shown that there are similar numbers of multidrug efflux systems in pathogenic and nonpathogenic bacteria, suggesting that these transporters have not arisen recently in response to antimicrobial therapy (74). Somewhat oddly, drug efflux systems appear to have evolved infrequently in evolutionary time yet have been maintained as genetic traits of the hosts. Here, the difficulty of "signposting" the habitats of bacterial species from primordial times to the present could be offset by presuming that constant exposure to metabolic toxins and xenobiotics provided sufficient selective pressure. Or, do we need to consider

Table 4. MATE multidrug transporters

Efflux pump	Bacterial host	Substrates[a]
VcrM	*Vibrio cholerae*	ACR, DAPI, TPP+, Hoechst 33342, R6G, EB
NorM	*Vibrio parahaemolyticus*	FQs, AGs, dyes
MorE / YdhE	*Escherichia coli*	FQs, AGs, ACR, TPP$^+$
VcmA	*Vibrio parahaemolyticus*	DAU, DOX, STR, KAN, ACR, DAPI, FQs, Hoechst 33342, EB
VmrA	*Vibrio cholerae*	DAPI, EB, ACR, TPP$^+$

[a] See Table 1, footnote *a*, for abbreviations for substrates.

Table 5. ABC multidrug transporters

Efflux pump	Bacterial host	Substrates[a]
MsrA (pl)[b]	*Staphylococcus epidermidis*	MLs, SGs
MsrB (pl)	*Staphylococcus xylosus*	MLs, SGs
MsrC	*Enterococcus faecium*	MLs, SGs
EfrAB	*Enterococcus faecalis*	DAU, DOX, ACR, DAPI, TPP$^+$
YvcC	*Bacillus subtilis*	Hoechst 33342, others
LmrA	*Lactococcus lactis*	R6G, EB, DAU, TPP$^+$
TetAB (pl)	*Corynebacterium striatum*	TET, OTC, OXA
VcaM	*Vibrio cholerae*	TET, DOX, DAU, FQs

[a] See Table 1, footnote *a*, for abbreviations for substrates.
[b] pl, plasmid encoded.

and weigh the relative contributions of clinical, environmental, and physiological signals?

Though MDR global systems either impinge upon or elaborate multidrug transporters, it is probable that many of the transported drugs or compounds are opportunistic passengers, and the physiological substrate(s) are often unknown. This could help to explain the occurrence of several MDR systems within a single bacterial genome. The search for natural substrates of multidrug transporters is an odyssey that is seldom rewarded, perhaps because these exporter ferrymen have such a diversity of passengers, but some examples have been identified. The remarkable substrate spectrum of the promiscuous human P-glycoprotein (MDR1) is perhaps the most noteworthy example since it has many dozens of substrates, modulators, and chemosensitizers (80), yet its natural substrate range is believed to be limited to short-chain cellular lipids. The *E. coli* AcrAB system is inducible by toxic fatty acids and exports bile salts (86). The natural substrate of the AcrEF transporter might be the tryptophan metabolite indole, which is known to impair bacterial growth and hence must be exported (40). These examples suggest that the translocation of multiple drugs is a fortuitous consequence of a natural efflux function or that bacteria use multidrug transporters for functions related to both normal physiology and toxin extrusion (53). The failure to identify any significant correlation between clinical drug resistance and number and distribution of multidrug transporters among pathogenic bacteria lends argument to notions of the coupling of MDR between physiological functions and environmental signals and stresses (53).

SOME WELL-KNOWN SINGLE AND TRIPARTITE PUMPS

In *P. aeruginosa,* a combination of low outer membrane permeability and the presence of multiple efflux pumps renders this organism intrinsically resistant to a wide range of compounds (12), including aminoglycosides, fluoroquinolones, β-lactams, aromatic hydrocarbons, and cationic dyes (Table 1). In wild-type cells, these resistances are due mainly to the MexAB-OprM pump and to a lesser extent to the MexXY-OprM RND and EmrE$_{Pae}$ SMR pumps (47). The MexCD-OprJ and MexEF-OprN pumps only appear to be expressed in *nfxB* and *nfxC* MDR mutants, respectively (41). The main pump, MexAB-OprM, is regulated by an upstream repressor gene, *mexR,* that belongs to the MarR family. A MexR dimer binds to two sites near overlapping promoters for *mexR* and *mexAB-OprM.* Expression of MexAB-OprM in wild-type cells may be constitutive from a proximal promoter, while hyperexpression in MDR mutants is due to additional expression from a MexR-regulated distal promoter (23).

P. aeruginosa thrives in environments polluted with organic matter, and it is also an opportunistic pathogen of particular importance in immunocompromised and cystic fibrosis patients. This bacterium is intrinsically resistant to antibiotics, solvents, and heavy metals, which is accounted for by the combination of low outer membrane permeability and multiple efflux pumps (63). Cross-resistance between carbapenem and heavy metals in the *P. aeruginosa* reference strain PT5 was demonstrated recently (69). Heavy metal resistance was correlated with increased expression of the metal efflux pump CzcCBA and its cognate two-component regulator genes *czcRS* and with reduced expression of the OprD porin (Table 1). The *czcRS* region is located upstream and divergent to the *czcCBA* operon, whose transcription is activated by the CzcR regulator. Sequencing of the *czcRS* region in eight independent zinc- and imipenem-resistant mutants revealed the same V194L mutation in the CzcS sensor protein. CzcS, like EnvZ in *E. coli,* is a class I histidine kinase that contains separate catalytic and ATP-binding domains, with the former involved in dimerization, phosphotransfer, and phosphatase activities. The sensor protein transfers a phosphoryl group to an aspartate residue on the regulator protein. The activated regulator then triggers expression of target genes. Overexpression of an efflux pump and down-regulation of a porin is an efficient means of preventing the influx of xenobiotics, with this type of coregulation being reported for *nfxC* mutants of *P. aeruginosa* in which the MexEF-OprN pump is up-regulated and OprD expression decreased (42). It has also been noted in *E. coli* in which the micF regulatory RNA—in response to internal and external stress signals—simultaneously increases expression of the AcrAB pump and decreases expression of the OmpF porin (14, 85).

A very interesting and potentially important mechanism of resistance has been reported in biofilm microbial communities, within which antibiotic resistance can be up to 1,000-fold greater than in planktonic cells (34). A genetic determinant of high-level biofilm-specific resistance to several classes of antibiotics was identified recently in *P. aeruginosa* (57). The *ndvB* locus is required for the synthesis of periplasmic cyclic glucans that are abundant in *P. aeruginosa* biofilm communities. These glucans are able to bind and sequester antibiotics in the periplasm, leading to high-level resistance. An *ndvB* mutant, which still grew within a biofilm but did not produce glucans, was susceptible to a range of antibiotics. The high-level resistance is therefore proposed to result solely from the glucan-specific sequestration of antibiotics in the periplasm, slowing the passage of the antibiotics to their sites of action within the cell (51).

MdfA is a multidrug transporter in *E. coli* and belongs to the MFS family (17, 65). Homologues have now been found in *Shigella, Salmonella,* and *Yersinia.* MdfA is a drug/proton antiporter that recognizes a diverse range of charged and uncharged lipophilic drugs (Table 2). Competition assays between TPP^+ and chloramphenicol suggested that both drugs bind to distinct but interacting sites. The transport of neutral substrates is electrogenic—that is, it is driven by both components ($\Delta\psi$ and ΔpH) of the proton electrochemical gradient, whereas the extrusion of cationic substrates is electroneutral and driven by the ΔpH only (46). Thus, discrete transport reactions are governed by different components of the *pmf.*

Within the SMR family, there are two subclasses. The first contains multidrug transporters representing about 75% of the total and including such members as EmrE from *E. coli,* Smr (QacC) from *E. coli,* and QacE from *S. aureus* (Table 3). There are at least 60 homologues of EmrE among gram-positive and gram-negative bacteria (76). The second group, representing 25% of the total, contains members that have been termed SUGs (82). This second group has poorly defined activities, even though they are predicted to deploy the same topology as the SMRs and contain the signature SMR sequence. EmrE—the most studied of the SMRs—typically catalyzes the electrogenic efflux of cationic aromatic hydrocarbons of varying size, structure, and charge in exchange for two or more protons (56). FTIR and NMR studies support the prediction of a four-helix bundle within the membrane (53). Early studies that attempted to determine the topology and oligomeric state of EmrE were inconclusive, but these have been bolstered recently by a high-resolution crystal structure (49) and an in vitro synthesis of a fully functional EmrE (19). The authors of the crystal structure study concluded that EmrE is a tetramer or dimer of dimers. This structure does not explain all of the previous biochemical evidence and requires significant conformational change to undergo the transport cycle. The in vitro synthesised EmrE protein was functional in proteoliposomes and formed a dimer that was very difficult to dissociate. However, like the crystal structure, a working model of dimer or tetramer could not be easily decided. Most groups conclude that EmrE must assemble into some oligomeric complex in order to function, since it is otherwise difficult to explain how a compact monomeric four-helix bundle would be capable of extruding large hydrophobic drugs. Another recent study takes advantage of the singular observation that detergent-solubilized EmrE binds substrates with high affinity in a pH-dependent manner, thereby enabling the mechanism of proton exchange to be examined (83). It is argued that substrates and protons compete for the same binding site and that this site encompasses the highly conserved Glu-14 residue in the center of the first transmembrane segment. Against this is the observation that at least one substrate—methyl viologen—is not recognized by detergent-solubilized EmrE and that Lys-22, within the first external loop, is as conserved as Glu-14 and might also be involved in some way in the translocation mechanism.

The *norA* gene of *S. aureus* was first isolated on the basis of its resistance to quinolones (99) and was shown subsequently to elicit resistance to a broad range of lipophilic and monocationic compounds, including some quinolones, acridine dyes, and quaternary ammonium compounds (62). NorA belongs to the MF secondary transporter superfamily and is a close homologue of Bmr from *B. subtilis* (61) that confers a similar broad resistance phenotype (Table 2). Another homologue, named PmrA, has been identified in *S. pneumoniae* (27). NorA in *E. coli* everted vesicles or reconstituted in proteoliposomes (102) was able to transport the hydrophobic Hoechst 33342, which was inhibited competitively by hydrophilic fluoroquinolones but noncompetitively by the hydrophobic fluoroquinolone sparfloxacin. These data suggested that NorA is a self-sufficient transporter with two functionally distinct drug-binding sites. Virulence factors in *S. aureus* are controlled by two global regulatory loci, *agr* and *sarA*. The *norR* gene encodes an activator of the *norA* gene promoter, but only in the presence of the *agr* and *sarA* global transcriptional effectors (88). NorR was shown additionally to be a multifunctional regulator that affected cell surface properties as well as NorA expression and likely other multidrug efflux pumps in *S. aureus*. NorR is unlikely to be autoregulatory in the same way as MarR is for the Mar regulon.

The *qacA-qacR* locus is prevalent on MDR plasmids in many clinical *S. aureus* strains (68, 73) and confers resistance to more than 30 distinct compounds including mono- and bivalent cationic lipophilic antimicrobials, antiseptics, disinfectants, and intercalating dyes (Table 2). It is one of a very few known extrachromosomal MDRs specified by a single gene. QacA is an MF transporter that is negatively regulated by the divergently encoded QacR, a member of the TetR family. QacR binds as a pair of dimers to the *qacA* operator sequence, a large inverted repeat immediately upstream from the promoter (30). QacR also represses QacB, a close homologue of QacA that differs in only seven nucleotides and contains two fewer charged residues, resulting in little or no resistance to divalent cationic ligands (67). QacR was shown to bind in vivo to the natural plant alkaloids berberine, palmitine, and nitidine and in vitro to chelerythrine and avicin (31). Though the ability of QacA to efflux these is still to be tested, berberine, at least, is a substrate for the QacA and NorA *S. aureus* pumps and is therefore likely to be a natural MDR substrate (31).

Neisseria gonorrhoeae is a strictly human pathogen that contains at least two MDR systems, the MtrCDE (33) and FarAB (45) efflux pumps that belong to the RND and MFS families, respectively (Tables 1 and 2). MtrCDE exports host-derived antimicrobial hydrophobic agents such as bile salts, peptides, and free fatty acids. The *mtrCDE* operon is controlled by the upstream *mtrR* gene that encodes the MtrR transcriptional repressor belonging to the TetR family and by a positive regulator MtrA, belonging to the AraC family. FarAB is a second pump in *N. gonorrhoeae* that elaborates resistance to host-derived long-chain fatty acids. It requires the MtrE protein as the outer membrane porin and was thought previously to be regulated directly by binding of MtrR to the *farAB* promoter. However, the recent discovery of the *farR* gene has clarified the role of MtrR in both the *mtr* and *far* systems (44). FarR is the negative regulatory binding protein of the *farAB* promoter, but unusually in *Neisseria,* MtrR binds to the *farR* promoter and thereby indirectly up-regulates the *farAB* operon. This makes the MtrR modulation of both *mtr* and *far* global systems somewhat akin to the *emrAB-emrR* and *marRAB* in *E. coli* and the *mexAB-OprM-mexR* system in *P. aeruginosa.* Indeed, FarR, EmrR, MarR, and MexR all belong to the MarR family of transcriptional regulators.

WHAT DOES A MULTIDRUG SITE LOOK LIKE? REGULATOR VERSUS TRANSPORTER

The DNA-binding repressor TetR employs a traditional drug-binding mechanism in which the tetracycline-$Mg2^{+}$ complex is bound by a precise and dense network of hydrogen bonds and van der Waals interactions between the ligand, the TetR binding pocket, and the surrounding water molecules (41). The tetracycline efflux transporter presumably forms a similar closed network of specific interactions with substrate. This ligand-transporter interaction obeys the so-called one enzyme, one-substrate rule, a tenet of protein biochemistry. But can it be broken? One major difference between specific enzyme-substrate binding and multidrug binding is that the former involves many hydrogen bonds and the latter mainly hydrophobic interactions. Multidrug transporters and regulators certainly flout the one-enzyme, one-substrate rule in having many substrates that are largely nonpolar and that bind to hydrophobic residues with the protein binding pocket. Similar rule-breaking has been observed in some cytosolic enzymes. For example, the porcine-odorant-binding protein (93) binds the same aromatic ligand in more than one orientation, demonstrating the looseness and flexibility of the site. Another is naphthalene 1,2-dioxygenase (8) whose large binding cavity accommodates hundreds of different substrates, including those with planar and nonplanar phenyl rings—biphenyl, for instance, has its phenyl rings in perpendicular planes. A related enzyme, toluene 2-monooxygenase (59), recognizes both aromatic and small, chlorinated aliphatic compounds, which bind within a presumably large, loose pocket.

Many transcriptional factors—activators and repressors—are able to bind the same array of drugs that are effluxed by the transporters they regulate. The recently solved resolution structures of four transcriptional regulators, MarR (4), BmrR (104), TipA (38), and QacR (78), furnishes us a first look at a multidrug-binding site. Can such sites be mimicked in membrane transporters, with sites having the plasticity to open, close, and bind in the same way as regulator sites, with multiple hydrophobic stacking interactions from aromatic residues, electrostatic neutralizing interactions, and conformational movements or even interconversions of loops, helices, or strands? Recent evidence seems to be moving in that direction, but there is still a strong school of thought and body of evidence that puts multidrug-binding sites within the transporter membrane channels, yet within which the required flexibility of the binding sites might not be possible. This has been argued most resolutely for perhaps the greatest multidrug transporter of them all, P-glycoprotein, an ABC ATP-binding cassette protein. Here, the abiding notion is that the membrane-embedded helices of the transporter are able to tilt or rotate to adopt the flexibility required to bind multiple ligands; not only is this still to be proven, but it should pertain also to other MDR transporters for which similar ideas have yet to be proffered. The recent crystal structure of the ABC transporters MsbA and BtuCD provide only equivocal support to the channel membrane site model, chiefly because neither protein was crystallized with bound substrate, nor was activity demonstrated for either transporter. The juxtaposition of multidrug regulator and transporter binding sites is covered in recent reviews and will be elaborated in more detail below (30, 37, 55, 77).

The solution structures of transcriptional regulators support a "hydrophobic basket" model of multidrug binding (60) that might have relevance to transporters as the multidrug portfolios are similar. Transcriptional regulators BmrR, MarR, TipA, and QacR all bind drugs in hydrophobic pockets. For BmrR and QacR, modeling and structural studies have shown that ligands penetrate a deep hydrophobic pocket. QacR has been crystallised with six different drugs (ethidium, rhodamine 6G, dequalinium, crystal violet, malachite green, and berberine) and is presently the benchmark for multidrug binding models (78, 79). In the QacR-rhodamine 6G and QacR-ethidium bromide complexes, the drug molecules are much smaller than the binding pocket, whose unoccupied space is filled with water molecules. Other larger substrates are capable of straddling the binding footprints of smaller drugs, with spaces filled in by water molecules. It is tempting to postulate that transporter multidrug-binding sites might have similar architectures to QacR and BmrR and that these sites may be sculpted from the loops at either end of the TMDs. Many transporters are crowned with large intra- and/or extracytosolic loop domains, and it is difficult to reconcile their complete omission from substrate site construction or filtering—as recognised in membrane proteins with gated channels, in porins, and in RND transporters.

The *B. subtilis* BmrR regulator binds TPP^+ in a pocket lined by hydrophobic residues and one glutamate (104). Ligand binding could only occur following a helix-coil transition that widened the pocket. In the *S. aureus* regulator, QacR, the large binding pocket was expanded by the expulsion of tyrosine side chains, had at least two differential binding sites, and involved many hydrophobic and three electrostatic (glutamate) interactions. A direct comparison of the binding cavities of QacR and AcrB (see below) showed that there are more residue side chains closer to the bound ligand in QacR than in AcrB. In the latter, there are only three residues near bound ethidium, and the other side of the ethidium molecule floats freely in a large cavity. A change in only one ligand-ethidium interaction (Phe to Ala) abolishes resistance to ethidium and other cationic dyes. In QacR, a bound ethidium molecule is near three glutamates that are involved in ligand binding, in addition to other hydrophobic interactions. This difference of looser binding in AcrB and tighter binding in QacR is explained as being due possibly to mobile anionic phospholipid heads that move into the AcrB site and contact the cationic ligand (101).

Though both QacR and AcrB require large, flexible hydrophobic binding pockets, a more precise binding network of interactions might be required in regulators in order to trigger the conformational change needed for a regulator to bind to its cognate DNA site. On the other hand, transporter composite binding sites need to be more pliant, as binding is only a transient step in the translocation process that requires perhaps a series of rapid, reversible interactions between the cavity and ligand as the latter is ferried into and out of the binding pocket. In support of this notion is the observation of differential selectivity of ligand types by mostly electrostatic interactions as they enter the vestibules of transporters (101). A recent study investigated the binding of QacR to pentamidine and hexamidine that differ by only one methylene carbon (58). These two substrates had dissimilar binding modes, whereas the positively charged benzamidine groups of hexamidine were neutralized by two glutamate residues in the QacR binding pocket, only one of pentamidine's charges was neutralized by a glutamate that, surprisingly, was a different one than the two that interacted with hexamidine. The second pentamidine charge was neutralized by carbonyl and side chain oxygen atoms, demonstrating that a formal negative charge is not a prerequisite for binding positively charged drugs, underscoring the versatility/flexibility of binding pockets (58).

The clearest indication of an MDR transporter efflux mechanism has been afforded by the AcrAB-TolC tripartite system (57, 103), consisting of the RND AcrB transporter, AcrA membrane fusion protein, and TolC outer membrane porin. The assembled complex is proposed to consist of an elongated TolC homotrimer β-barrel conduit that protrudes from the outer membrane into the periplasmic space where it docks with the oppositely protruding AcrB headpiece, held in place by an AcrA collar. The docked funnel-headpiece structure formed from TolC and AcrB trimers contains a vestibule or cavity that is proposed to open and close like an iris and forms the substrate-binding site (21, 101). Drugs are captured from the periplasmic side of the inner membrane and translocated via the central AcrB-TolC vestibule to the outside of the cell. A diverse range of ligand structures may be accommodated within the central vestibule in much the same way as has been described for the binding pockets of the transcriptional regulators QacR (78) and BmrR (104). However, there is a problem in transposing this notion to transporters that do not have large periplasmic or extracellular domains. Most transporters have enough residues in flexible loops on either side of the membrane that could combine with largely hydrophobic collars at the membrane portals to form substrate pockets or baskets. Even the small SMRs are predicted to associate into larger functional multimers. How important are the transmembrane channels versus the collars and loops on either side? Why would there be parallel and duplicate evolution of two different kinds of multidrug binding sites in efflux pumps? Indeed, why do some transporters bind only single substrates while others are multidrug transporters? These questions are the subject of the next section.

WHERE ARE THE MULTIDRUG BINDING SITES IN EFFLUX PUMPS?

The SUG protein SugE from *E. coli* was back-mutated to an SMR by substituting four conserved residues (82). Surprisingly, the SugE quadruple mutant had the opposite effect of making the host even more susceptible to quaternary ammonium compounds and ethidium, including the observation of an increase in ethidium uptake, leading the authors to conclude that SugE may be an importer rather than exporter. Three of the mutated residues are in the middle of TM2 and the fourth is on the first external loop; importantly, the conserved glutamate in TM1—believed to be critical in the exporter function of SMRs—was present in the SugE mutant. Of course, it could be that this study simply omitted to make the substitutions (whichever they might be) that would have effected the actual interconversion of SugE to an SMR. Nevertheless, the authors were puzzled by the observation that two proteins with identical topologies, namely EmrE and mutated SugE, could transport similar substrates in opposite directions. This confusing result also suggests that the critical glutamate residue in TM1 is insufficient of itself to effect efflux (see below). It is noteworthy that non-MDR SUGs also contain the SMR signature sequence and its role in MDR efflux must therefore be benign.

There are several examples of efflux transporters that rely on membrane-embedded acidic residues for recognition of cationic drugs (2). The list includes QacA in *S. aureus,* LmrP in *L. lactis,* and EmrE and MdfA in *E. coli.* These last two have been studied closely and have yielded conflicting data on the critical importance of a buried glutamate residue. A highly conserved and membrane-embedded glutamate residue (Glu-14) in the SMR transporter EmrE has been implicated in proton coupling and substrate-binding (32, 98). This glutamate residue is conserved in more than 50 SMRs. In two substitution mutants, G14C and G14D, resistance was abolished, but G14D was able to transport drug at 30 to 50% of the wild type (84). Cysteine substitutions of all other charged residues had no effect on resistance, and it was concluded that Glu-14—the only membrane-embedded charged residue in EmrE—was critical for resistance and transport. Another highly conserved residue (Lys-22) lies just outside TM1 in the periplasmic space, and though not important alone, it is possible that this lysine might form a salt bridge with the oppositely charged glutamate, which might explain why transport is still observed in the G14D substitution mutant but not for G14C.

MdfA contains a single glutamate residue in the middle of TM1 (of 12 TMs) and changes to this residue affect substrate recognition (18). A recent study concluded that the glutamate residue does not play an essential role in the multidrug transport mechanism of MdfA. Neutral substitutions of the glutamate residue partially preserved transport of cationic substrates and did not affect greatly the transport of neutral drugs (1). The conclusion is that—like increasing evidence elsewhere—there is structural and chemical promiscuity in multidrug recognition pockets. The importance of a glutamate residue near the selectivity filter of the KcsA K^+ selectivity filter has recently been questioned (9). This glutamate forms a carboxyl-carboxylate salt bridge with a nearby aspartate residue. When the glutamate was substituted with a valine, the channel still displayed conductance and K^+/Na^+ selectivity, suggesting that the glutamate was not a primary determinant of ion selectivity or conductance either directly or through the salt bridge. It is now clear that not only do many efflux transporters lack membrane-embedded acidic residues but some pumps suspected of using these residues may not require them absolutely, as the examples discussed above seem to imply. This revives the question of studies that demonstrated a crucial role for the binding of cationic substrates to acidic membrane-embedded residues. Indeed, in the case of AcrB (101), it was proposed that substrates interacted directly with the acidic headgroups of lipids in the absence of a transporter acidic residue.

We have just seen that charged residues within the membrane channels of efflux pumps might not be as critical as originally thought, yet the evidence either way is still not strong enough to exclude them from all aspects of proton and/or drug flux. Indeed, regulator binding pockets often include charged residues, but they are not as important as the cumulative binding capacity of hydrophobic interactions between substrates and binding sites. So, where are the multidrug recognition pockets in efflux pumps? The drug-binding pocket is believed to encompass a large volume (78, 101) and has the flexibility to allow binding of dissimilar substrates by both different and shared binding determinants. As mentioned in the previous section and in examples discussed below, evidence is mounting that drug-binding sites in multidrug efflux pumps are fashioned from the loops on one or both sides of the membrane and may include the vestibule collars of the membrane channels. This notion fulfils the requirements for multidrug binding: flexibility and charged residues in the loops, and hydrophobic stacking forces within and between the folded loops and membrane vestibules.

A recent study identified three binding site regions in the human ABC Multidrug Resistance Protein (MRP1) transporter (39). The three regions were the large intracytoplasmic loop domain separating TMD1 and TMD2, the cytosolic linker domain separating TMD2 and TMD3, and the last two TM segments. Three conserved tryptophan, one phenylalanine, and one nonconserved Tyr residue appear to form an aromatic basket or "net" proximal to the cytosolic membrane bilayer interface in MRP1 (7). Although this site is on the inner side of the membrane, it is consistent with other reports that predict low-affinity substrate-binding sites in this position in human P-glycoprotein (36, 91) and in the *E. coli* MF pump MdfA (1).

Human P-glycoprotein (MDR1) has many dozens of substrates, modulators, and chemosensitizers (80), whereas the other human P-glycoprotein, MDR3, translocates phosphatidylcholine only (90). It comes as something of a surprise, therefore, to find that MDR3 might be capable of effluxing a suite of MDR1 drug substrates (81) and that the chief difference between MDR1 and MDR3 is not the configuration of substrate-binding sites but the ubiquitous (MDR1) or limited (MDR3) location of each in humans. Another recent example is afforded by the LmrA and MsbA ABC pumps. LmrA is an *L. lactis* multidrug transporter (92) and MsbA is an *E. coli* Lipid A transporter (105). It has been reported recently that LmrA is capable of transporting the MsbA substrate, Lipid A, and that MsbA is capable of transporting several LmrA drug substrates (72). What do these examples tell us of the nature of specific versus multidrug substrate recognition and transport? How many other examples are there that flout the one-enzyme, one-substrate rule? How can transporters—with more or less identical transmembrane channel topologies—filter, bind, and translocate multiple substrates that often differ in polarity, size, and shape?

Secondary transporters vary in size from the small SMRs (~100 residues) to medium MF and MATE proteins (~420 residues) to large RNDs (~1,000 residues), yet most are predicted to deploy a 12-helix membrane channel. The shortfall in SMRs is met by putative multimers of 4-helix bundles, and the surfeit of residues in RNDs is taken up in two large periplasmic loop domains. The one certainty is that these channels provide the proton translocation pathway for symporters and antiporters alike. But do the channels contain the substrate-binding sites? Some examples already discussed above go against this notion, particularly the RND pumps. A phylogenic comparison of 93 integral membrane proteins with multiple TMDs (87) and a screening project of the exons and flanking intronic regions of 24 human membrane transporter genes (43) found that amino acid diversity in TMDs was significantly lower than in the loop domains, suggesting that TMDs have special functional constraints that cannot account for the functional diversity of these proteins. That we might look, therefore, to the loop domains is reinforced by the next and final example.

The RND-type MexB and MexY *pmf* transporters from *P. aeruginosa* export β-lactams and aminoglycosides, respectively (16). MexB and MexY each have two large periplasmic loops that extend out of the TMDs. When the MexY loops were replaced by the MexB loops, the hybrid now exhibited MexB-like β-lactam selectivity and failed to recognize aminoglycosides. When the TM segments of MexB were replaced one-by-one with TM segments from MexY for all 12 segments, all of these hybrids still showed aminoglycoside selectivity. These data suggested that RND transporters select and efflux drugs within the large periplasmic domains and that the TMDs are unlikely to participate in substrate selectivity (16). An earlier study of the RND-type MexD transporter used mutagenesis and variation in substrate recognition to show that the large external periplasmic loop domains contain multiple sites of interaction for structurally diverse compounds (52). The same conclusion was reached for the AcrB and AcrD transporter periplasmic loops (20). More research is required to determine if this example also applies to the other transporter families.

QUO VADIS?

In the purely clinical sense, antibiotic resistance is a major problem for several reasons. First, the development of new antibiotics is much slower than the emergence of resistance to these antibiotics. For example, linezolid, an oxazolidinone, reached the market in 2000 and resistant bacteria were identified within two years of this release; this augurs badly for the development of new antimicrobial drugs (10). During antibiotic treatments, bacteria may avoid cell death in the usual way of being resistant by one or more mechanisms, but additionally, there is the problem of nonmultiplying bacteria that can assume roles as persistors and/or mutators that survive antibiotic treatments and may even proliferate later with newly acquired low- or high-level resistance phenotypes (10, 11) gained from horizontal gene transfer or from activation of intrinsic multidrug systems, which of themselves are an innate advantage for many pathogens. Two new approaches to tackling the drug-resistance problem are receiving increasing attention. The first involves targeting nonmultiplying latent bacteria, which could reduce the duration of chemotherapy and the rate of development of resistance (10). The second method, which shows even more promise, involves activating chromosomal suicide modules that trigger programmed cell death of bacteria. Such modules have been found recently in *E. coli* and some pathogens (22). Some common antibiotics that inhibit transcription/translation or folic acid metabolism activate these suicide modules, and perhaps future research could be directed at identifying compounds that exclusively activate programmed cell death modules.

REFERENCES

1. **Adler, J., and E. Bibi.** 2004. Determinants of substrate recognition by the *Escherichia coli* multidrug transporter MdfA identified on both sides of the membrane. *J. Biol. Chem.* **279:** 8957–8965.
2. **Adler, J., O. Lewinson, and E. Bibi.** 2004. Role of a conserved membrane-embedded acidic residue in the multidrug transporter MdfA. *Biochemistry* **43:**518–525.
3. **Alekshun, M. N., and S. B. Levy.** 1999. The mar regulon: multiple resistance to antibiotics and other toxic chemicals. *Trends Microbiol.* **7:**410–413.
4. **Alekshun, M. N., S. B. Levy, T. R. Mealy, B. A. Seaton, and J. F. Head.** 2001. The crystal structure of MarR, a regulator of multiple antibiotic resistance, at 2.3 Å resolution. *Nat. Struct. Biol.* **8:**710–714.
5. **Baranova, N., and H. Nikaido.** 2002. The BaeSR two-component regulatory system activates transcription of the *yegMNOB* (*mdtABCD*) transporter gene cluster in *Escherichia coli* and increases its resistance to novobiocin and deoxycholate. *J. Bacteriol.* **184:**4168–4176.
6. **Barbosa, T. M., and S. B. Levy.** 2000. Differential expression of over 60 chromosomal genes in *Escherichia coli* by constitutive expression of MarA. *J. Bacteriol.* **182:**3467–3474.
7. **Campbell, J. D., K. Koike, C. Moreau, M. S. P. Sansom, R. G. Deeley, and S. P. C. Cole.** 2004. Molecular modeling correctly predicts the functional importance of Phe^{594} in transmembrane helix 11 of the multidrug resistance protein, MRP1 (ABCC1). *J. Biol. Chem.* **279:**463–468.
8. **Carredano, E., A. Karlsson, B. Kauppi, D. Choudhury, R. E. Parales, J. V. Parales, K. Lee, D. T. Gibson, H. Eklund, and S. Ramaswamy.** 2000. Substrate binding site of naphthalene 1, 2-dioxygenase: functional implications of indole binding. *J. Mol. Biol.* **321:**621–632.
9. **Choi, H., and L. Heginbotham.** 2004. Functional influence of the pore helix glutamate in the KcsA K^+ channel. *Biophys. J.* **86:**2137–2144.
10. **Chopra, I., A. J. O'Neill, and K. Miller.** 2003. The role of mutators in the mergence of antibiotic-resistant bacteria. *Drug Res. Updates* **6:**137–145.
11. **Coates, A., Y. Hu, R. Bax, and C. Page.** 2002. The future challenges facing the development of new antimicrobial drugs. *Nature Rev.* **1:**895–910.
12. **Cohen, S. P., W. Yan, and S. B. Levy.** 1993. A multidrug resistance regulatory chromosomal locus is widespread among enteric bacteria. *J. Infect. Dis.* **168:**484–488.
13. **Danghi, B., P. Pelupessey, R. G. Martin, J. L. Rosner, J. M. Louis, and A. M. Gronenborn.** 2001. Structure and dynamics of MarA-DNA complexes: an NMR investigation. *J. Mol. Biol.* **313:**1067–1081.
14. **Delihas, N., and S. Forst.** 2001. *MicF:* an antisense RNA gene involved in response of *Escherichia coli* to global stress factors. *J. Mol. Biol.* **313:**1–12.
15. **Demple, B.** 1996. Redox signaling and gene control in the *Escherichia coli soxRS* oxidative stress regulon—a review. *Gene* **179:**53–57.
16. **Eda, S., H. Maseda, and T. Nakae.** 2003. An elegant means of self-protection in gram-negative bacteria by recognizing and extruding xenobiotics from the periplasmic space. *J. Biol. Chem.* **278:**2085–2088.
17. **Edgar, R., and E. Bibi.** 1997. MdfA, an *Escherichia coli* multidrug resistance protein with an extraordinarily broad spectrum of drug recognition. *J. Bacteriol.* **179:**2274–2280.
18. **Edgar, R., and E. Bibi.** 1999. A single membrane-embedded negative charge is critical for recognizing positively charged drugs by the *Escherichia coli* multidrug resistance protein MdfA. *EMBO J.* **18:**822–832.
19. **Elbaz, Y., S. Steiner-Mordoch, T. Danieli, and S. Schuldiner.** 2004. *In vitro* synthesis of fully functional EmrE, a multidrug transporter, and study of its oligomeric state. *Proc. Natl. Acad. Sci. USA* **101:**1519–1524.
20. **Elkins, C. A., and H. Nikaido.** 2002. Substrate specificity of the RND-type multidrug efflux pumps AcrB and AcrD of *Escherichia coli* is determined predominantly by two large periplasmic loops. *J. Bacteriol.* **184:**6490–6498.
21. **Elkins, C. A., and H. Nikaido.** 2003. Chimeric analysis of AcrA function reveals the importance of its C-terminal domain in its interaction with the AcrB multidrug efflux pump. *J. Bacteriol.* **185:**5349–5356.
22. **Engelberg-Kulka, H., B. Sat, M. Reches, S. Amitai, and R. Hazan.** 2004. Bacterial programmed cell death systems as targets for antibiotics. *Trends Microbiol.* **12:**66–71.
23. **Evans, K., L. Adewoye, and K. Poole.** 2001. MexR repressor of the *mexAB-oprM* multidrug efflux operon of *Pseudomonas aeruginosa:* identification of MexR binding sites in the *mexA-mexR* intergenic region. *J. Bacteriol.* **183:**807–812.
24. **Fath, M. J., and R. Kolter.** 1993. ABC transporters: bacterial exporters. *Microbiol. Rev.* **57:**995–1017.
25. **George, A. M., and S. B. Levy.** 1983a. Amplifiable resistance to tetracycline, chloramphenicol, and other antibiotics in *Escherichia coli:* involvement of a non-plasmid-determined efflux of tetracycline. *J. Bacteriol.* **155:**531–540.
26. **George, A. M., and S. B. Levy.** 1983b. Gene in the major cotransduction gap of the *Escherichia coli* K-12 linkage map

required for the expression of chromosomal resistance to tetracycline and other antibiotics. *J. Bacteriol.* **155:**541–548.

27. **Gill, M. J., N. P. Brenwald, and R. Wise.** 1999. Identification of an efflux pump gene, *pmrA*, associated with fluoroquinolone resistance in *Streptococcus pneumoniae*. *Antimicrob. Agents Chemother.* **43:**187–189.
28. **Godsey, M. H., E. E. Zheleznova, and R. G. Brennan.** 2002. Structural biology of bacterial multidrug resistance gene regulators. *J. Biol. Chem.* **277:**40169–40172.
29. **Griffith, K. L., I. M. Shah, and R. E. Wolf, Jr.** 2004. Proteolytic degradation of *Escherichia coli* transcription activators SoxS and MarA as the mechanism for reversing the induction of the superoxide (SoxRS) and multiple antibiotic resistance (Mar) regulons. *Mol. Microbiol.* **51:**1801–1816.
30. **Grkovic, S., M. H. Brown, and R. A. Skurray.** 2002. Regulation of bacterial drug export systems. *Microbiol. Mol. Biol. Rev.* **66:**671–701.
31. **Grkovic, S., K. M. Hardie, M. H. Brown, and R. A. Skurray.** 2003. Interactions of the QacR multidrug-binding protein with structurally diverse ligands: implications for the evolution of the binding pocket. *Biochemistry* **42:**15226–15236.
32. **Gutman, N., S. Steiner-Mordoch, and S. Schuldiner.** 2003. An amino acid cluster around the essential Glu-14 is part of the substrate- and proton-binding domain of EmrE, a multidrug transporter from *Escherichia coli*. *J. Biol. Chem.* **278:**16082–16087.
33. **Hagman, K. E., W. Pan, B. G. Spratt, J. T. Balthazar, R. C. Judd, and W. M. Shafer.** 1995. Resistance of *Neisseria gonorrhoeae* to antimicrobial hydrophobic agents is modulated by the *mtrCDE* efflux system. *Microbiology* **141:**611–622.
34. **Hoyle, B. D., and W. J. Costerton.** 1991. Bacterial resistance to antibiotics: the role of biofilms. *Prog. Drug Res.* **37:**91–105.
35. **Ishida, H., H. Fuziwara, Y. Kaibori, T. Horiuchi, K. Sato, and Y. Osada.** 1995. Cloning of multidrug resistance gene *pqrA* from *Proteus vulgaris*. *Antimicrob. Agents Chemother.* **39:**453–457.
36. **Jones, P. M., and A. M. George.** 2000. Symmetry and structure in P-glycoprotein and ABC transporters. *Eur. J. Biochem.* **267:**5298–5305.
37. **Jones, P. M., and A. M. George.** 2004. The ABC transporter structure and mechanism: perspectives on recent research. *Cell. Mol. Life Sci.* **61:**682–699.
38. **Kahmann, J. D., H.-J. Sass, M. G. Allan, H. Seto, C. J. Thompson, and S. Grzesiek.** 2003. Structural basis for antibiotic recognition by the TipA class of multidrug-resistance transcriptional regulators. *EMBO J.* **22:**1824–1834.
39. **Karwatsky, J., R. Daoud, J. Cai, P. Gros, and E. Georges.** 2003. Binding of a photoaffintiy analogue of glutathione to MRP1 (ABCC1) within two cytoplasmic regions (L0 and L1) as well as transmembrane domains 10-11 and 16-17. *Biochemistry* **42:**3286–3294.
40. **Kawamura-Sato, K., K. Shibayama, T. Horii, Y. Iimuma, Y. Arakawa, and M. Ohta.** 1999. Role of multiple efflux pumps in *Escherichia coli* in indole expulsion. *FEMS Microbiol. Lett.* **179:**345–352.
41. **Kisker, C., W. Hinrichs, K. Tovar, W. Hillen, and W. Saenger.** 1995. The complex formed between Tet repressor and tetracycline-Mg^{2+} reveals mechanism of antibiotic resistance. *J. Mol. Biol.* **247:**260–280.
42. **Köhler, T., M. Michea-Hamzehpour, U. Henze, N. Gotoh, L. Kocjancic-Curty, and J. C. Pechère.** 1997. Characterization of MexE-MexF-OprN, a positively regulated multidrug efflux system of *Pseudomonas aeruginosa*. *Mol. Microbiol.* **23:**345–254.
43. **Leabman, M. K., C. C. Huang, J. DeYoung, E. J. Carlson, T. R. Taylor, M. de la Cruz, S. J. Johns, D. Stryke, M. Kawamoto, T. J. Urban, D. L. Kroetz, T. E. Ferrin, A. G. Clark, N. Risch, I. Herskowitz, and K. M. Giacomini.** 2003. Natural variation in human membrane transporter genes reveals evolutionary and functional constraints. *Proc. Natl. Acad. Sci. USA* **100:**5896–5901.
44. **Lee, E.-H., C. Rouquette-Loughlin, J. P. Folster, and W. M. Shafer.** 2003. FarR regulates the *farAB*-encoded efflux pump of *Neisseria gonorrhoeae* via an MtrR regulatory mechanism. *J. Bacteriol.* **185:**7145–7152.
45. **Lee, E.-H., and W. M. Shafer.** 1999. The *farAB* efflux pump mediates resistance of gonococci to long-chained antibacterial fatty acids. *Mol. Microbiol.* **33:**839–845.
46. **Lewinson, O., J. Adler, G. J. Poelarends, P. Mazurkiewicz, A. J. M. Driessen, and E. Bibi.** 2003. The *Escherichia coli* multidrug transporter MdfA catalyzes both electrogenic and electroneutral transport reactions. *Proc. Natl. Acad. Sci. USA* **100:**1667–1672.
47. **Li, X.-Z., K. Poole, and H. Nikaido.** 2003. Contributions of MexAB-OprM and an EmrE homolog to intrinsic resistance of *Pseudomonas aeruginosa* to aminoglycosides and dyes. *Antimicrob. Agents Chemother.* **47:**27–33.
48. **Lomovskaya, O., K. Lewis, and A. Martin.** 1995. EmrR is a negative regulator of the *Escherichia coli* multidrug resistance pump EmrAB. *J. Bacteriol.* **177:**2328–2334.
49. **Ma, C., and G. Chang.** 2004. Structure of the multidrug resistance efflux transporter EmrE from *Escherichia coli*. *Proc. Natl. Acad. Sci. USA* **101:**2852–2857.
50. **Ma, D., M. Alberti, C. Lynch, H. Nikaido, and J. E. Hearst.** 1996. The local repressor AcrR plays a modulating role in the regulation of *acrAB* genes of *Escherichia coli* by global stress signals. *Mol. Microbiol.* **19:**101–112.
51. **Mah, T.-F., B. Pitts, B. Pellock, G. C. Walker, P. S. Stewart, and G. A. O'Toole.** 2003. A genetic basis for *Pseudomonas aeruginosa* biofilm antibiotic resistance. *Nature* **426:**306–310.
52. **Mao, W., M. S. Warren, D. S. Black, T. Satou, T. Murata, T. Nishino, N. Gotoh, and O. Lomovskaya.** 2002. On the mechanism of substrate specificity by resistance nodulation division (RND)-type multidrug resistance pumps: the large periplasmic pools of MexD from *Pseudomonas aeruginosa* are involved in substrate recognition. *Mol. Microbiol.* **46:**889–901.
53. **Markham, P. N., and A. A. Neyfakh.** 2001. Efflux-mediated resistance in gram-positive bacteria. *Curr. Opin. Microbiol.* **4:**509–514.
54. **Martin, R. G., and J. L. Rosner.** 1995. Binding of lurified multiple antibiotic-resistance repressor protein (MarR) to *mar* operator sequences. *Proc. Natl. Acad. Sci. USA* **92:**5456–5460.
55. **McKeegan, K. S., M. I. Borges-Walmsley, and A. R. Walmsley.** 2003. The structure and function of drug pumps: an update. *Trends Microbiol.* **11:**21–29.
56. **Mordoch, S. S., D. Granot, M. Lebendiker, and S. Schuldiner.** 1999. Scanning cysteine accessibility of EmrE, an H^+-coupled multidrug transporter from *Escherichia coli,* reveals a hydrophobic pathway for solutes. *J. Biol. Chem.* **274:**19480–19486.
57. **Murakami, S., and A. Yamaguchi.** 2003. Multidrug-exporting secondary transporters. *Curr. Opin. Struct. Biol.* **13:**443–452.
58. **Murray, D. S., M. A. Schumacher, and R. G. Brennan.** 2004. Crystal structures of QacR-diamidine complexes reveal additional multidrug-binding modes and a novel mechanism of drug charge neutralization. *J. Biol. Chem.* **279:**14365–14371.
59. **Newman, L. M., and L. P. Wackett.** 1997. Trichloromethylene oxidation by purified toluene 2-monooxygenase: products, kinetics, and turnover-dependent inactivation. *J. Bacteriol.* **179:**90–96.
60. **Neyfakh, A. A.** 2002. Mystery of multidrug transporters: the answer can be simple. *Mol. Microbiol.* **44:**1123–1130.
61. **Neyfakh, A. A., V. E. Bidnenko, and L. B. Chen.** 1991. Efflux-mediated multidrug resistance in *Bacillus subtilis:* similarities

and dissimilarities with the mammalian system. *Proc. Natl. Acad. Sci. USA* **88:**478–8785.
62. **Neyfakh, A. A., C. M. Borsch, and G. W. Kaatz.** 1993. Fluoroquinolone resistance protein NorA of *Staphylococcus aureus* is a multidrug efflux transporter. *Antimicrob. Agents Chemother.* **37:**128–129.
63. **Nikaido, H.** 1994. Prevention of drug access to bacterial targets: permeability barriers and active efflux. *Science* **264:**382–388.
64. **Nikaido, H.** 1998. Multiple antibiotic resistance and efflux. *Curr. Opin. Microbiol.* **1:**516–523.
65. **Nilsen, I. W., I. Baake, A. Vader, Ø. Olsvik, and M. R. El-Gewely.** 1996. Isolation of *cmr,* a novel *Escherichia coli* chloramphenicol resistance gene encoding a putative efflux pump. *J. Bacteriol.* **178:**3188–3193.
66. **Paulsen, I. T., M. H. Brown, and R. A. Skurray.** 1996. Proton-dependent multidrug efflux systems. *Microbiol. Rev.* **60:**575–608.
67. **Paulsen, I. T., J. Chen, K. E. Nelson, and M. H. Saier, Jr.** 2001. Comparative genomics of microbial drug efflux systems. *J. Mol. Microbiol. Biotechnol.* **3:**145–150.
68. **Paulsen, I. T., M. K. Silwinski, and M. H. Saier, Jr.** 1998. Microbial genomic analysis: global comparisons of transport capabilities based on phylogenies, bioenergetics and substrate specificities. *J. Mol. Biol.* **277:**573–592.
69. **Perron, K., O. Caille, C. Rossier, C. van Delden, J.-L. Dumas, and T. Köhler.** 2004. CzcR-CzcS, a two-component system involved in heavy metal and carbenpenem resistance in *Pseudomonas aeruginosa. J. Biol. Chem.* **279:**8761–8768.
70. **Poole, K.** 2002. Outer membranes and efflux: the path to multidrug resistance in gram-negative bacteria. *Curr. Pharm. Biotech.* **3:**77–98.
71. **Putman, M., H. W. van Veen, J. E. Degener, and W. N. Konings.** 2000. Antibiotic resistance: era of the multidrug pump. *Mol. Microbiol.* **36:**772–773.
72. **Reuter, G., T. Janvilisri, H. Venter, S. Shahi, L. Balakrishnan, and H. W. van Veen.** 2003. The ATP binding cassette multidrug transporter LmrA and lipid transporter MsbA have overlapping substrate specificities. *J. Biol. Chem.* **278:**35193–35198.
73. **Rouch, D. A., D. S. Cram, D. DiBerardino, T. G. Littlejohn, and R. A. Skurray.** 1990. Efflux-mediated antiseptic resistance gene *qacA* from *Staphylococcus aureus:* common ancestry with tetracycline- and sugar-transport proteins. *Mol. Microbiol.* **4:**2051–2062.
74. **Saier, M. H. Jr., I. T. Paulsen, M. K. Sliwinski, S. S. Pao, R. A. Skurray, and H. Nikaido.** 1998. Evolutionary origins of multidrug and drug-specific efflux pumps in bacteria. *FASEB J.* **12:**265–274.
75. **Schneiders, T., T. M. Barbosa, L. M. McMurry, and S. B. Levy.** 2004. The *Escherichia coli* transcriptional regulator MarA directly represses transcription of *purA* and *hdeA. J. Biol. Chem.* **279:**9037–9042.
76. **Schulinder, S., D. Granot, S. S. Mordoch, S. Nino, D. Rotem, M. Soskin, C. G. Tate, and H. Yerushalmi.** 2001. Small is mighty: EmrE, a multidrug transporter as an experimental paradigm. *News Physiol. Sci.* **16:**130–134.
77. **Schumacher, M. A., and R. G. Brennan.** 2002. Structural mechanisms of multidrug recognition and regulation by bacterial multidrug transcription factors. *Mol. Microbiol.* **45:**885–893.
78. **Schumacher, M. A., M. C. Miller, S. Grkovic, M. H. Brown, R. A. Skurray, and R. G. Brennan.** 2001. Structural mechanisms of QacR induction and multidrug recognition. *Science* **294:**2158–2163.
79. **Schumacher, M. A., M. C. Miller, S. Grkovic, M. H. Brown, R. A. Skurray, and R. G. Brennan.** 2002. Structural basis for cooperative DNA binding by two dimers of the multidrug-binding protein QacR. *EMBO J.* **21:**1210–1218.
80. **Sharom, F. J.** 1997. The P-glycoprotein efflux pump: how does it transport drugs? *J. Membr. Biol.* **160:**161–175.
81. **Smith, A. J., A. van Helvoort, G. van Meer, K. Szabó, E. Welker, G. Szakács, A. Váradi, B. Sarkadi, and P. Borst.** 2000. MDR3 P-glycoprotein, a phosphatidylcholine translocase, transports several cytotoxic drugs and directly interacts with drugs as judged by interference with nucleotide trapping. *J. Biol. Chem.* **275:**23530–23539.
82. **Son, M. S., C. Del Castilho, K. A. Duncalf, D. Carney, J. H. Weiner, and R. J. Turner.** 2003. Mutagenesis of SugE, a small multidrug resistance protein. *Biochem. Biophys. Res. Commun.* **312:**914–921.
83. **Soskine, M., Y. Adam, and S. Schuldiner.** 2004. Direct evidence for substrate-induced proton release in detergent-solubilized EmrE, a multidrug transporter. *J. Biol. Chem.* **279:**9951–9955.
84. **Sulavik, M. C., C. Houseweart, C. Cramer, N. Jiwani, N. Murgolo, J. Greene, B. DiDomenico, K. J. Shaw, G. H. Miller, R. Hare, and G. Shimer.** 2001. Antibiotic susceptibility profiles of *Escherichia coli* strains lacking multidrug efflux pump genes. *Antimicrob. Agents Chemother.* **45:**1126–1136.
85. **Tanaka, T., T. Horli, K. Shibayama, K. Sato, S. Ohsuka, Y. Arawaka, K.-I. Yamaka, K. Takagi, and M. Ohta.** 1997. RobA-induced multiple antibiotic resistance largely depends on the activation of the AcrAB efflux. *Microbiol. Immunol.* **41:**697–702.
86. **Thanassi, D. G., L. W. Cheng, and H. Nikaido.** 1997. Active efflux of bile salts by *Escherichia coli. J. Bacteriol.* **179:**2515–2518.
87. **Tourasse, N. J., and W. H. Li.** 2000. Selective constraints, amino acid composition, and the rate of protein evolution. *Mol. Biol. Evol.* **17:**656–664.
88. **Truong-Bolduc, Q. C., X. Zhang, and D. C. Hooper.** 2003. Characterization of NorR protein, a multifunctional regulator of *norA* expression in *Staphylococcus aureus. J. Bacteriol.* **185:**3127–2138.
89. **Tseng, T.-T., K. S. Gratwick, J. Kollman, D. Park, D. H. Nies, A. Goffeau, and M. H. Saier, Jr.** 1999. The RND permease superfamily: an ancient, ubiquitous and diverse family that includes human disease and development proteins. *J. Mol. Microbiol. Biotechnol.* **1:**107–125.
90. **van Helvoort, A., A. J. Smith, H. Sprong, I. Fritzsche, A. H. Schinkel, P. Borst, and G. van Meer.** 1996. MDR1 P-Glycoprotein is a lipid translocase of broad specificity, while MDR3 P-glycoprotein specifically translocates phosphatidylcholine. *Cell* **87:**507–517.
91. **van Veen, H. W., A. Margolles, M. Muller, C. F. Higgins, and W. N. Konings.** 2000. The homodimeric ATP-binding cassette transporter LmrA mediates multidrug transport by an alternating two-site (two-cylinder engine) mechanism. *EMBO J.* **19:**2503–2514.
92. **van Veen, H. W., K. Vanema, H. Bolhuis, I. Oussenko, J. Kok, B. Poolman, A. J. M. Driessen, and W. N. Konings.** 1996. Multidrug resistance mediated by a bacterial homolog of the human multidrug transporter MDR1. *Proc. Natl. Acad. Sci. USA* **93:**10668–10672.
93. **Vincent, F., S. Spinelli, R. Ramoni, S. Grolli, P. Pelosi, C. Cambillau, and M. Tegoni.** 2000. Complexes of porcine odorant binding protein with odorant molecules belonging to different chemical classes. *J. Mol. Biol.* **305:**459–469.
94. **Wang, X. D., P. A. de Boer, and L. I. Rothfield.** 1991. A factor that positively regulates cell division by activating transcription of the major cluster of essential cell division genes of *Escherichia coli. EMBO J.* **10:**3363–3372.
95. **Wei, Y., J.-M. Lee, D. R. Smulski, and R. A. LaRossa.** 2001. Global impact of sdiA amplification revealed by comprehen-

sive gene expression profiling of *Escherichia coli*. *J. Bacteriol.* **183:**2265–2272.

96. **White, D. G., J. D. Goldman, B. Demple, and S. B. Levy.** 1997. Role of the *acrAB* locus in organic solvent tolerance mediated by expression of *marA*, *soxS*, or *robA* in *Escherichia coli*. *J. Bacteriol.* **179:**6122–6126.
97. **Yerushalmi, H., M. Lebendiker, and S. Schuldiner.** 1996. Negative dominance studies demonstrate the oligomeric structure of Emr, a multidrug antiporter from *Escherichia coli*. *J. Biol. Chem.* **271:**31044–31048.
98. **Yerushalmi, H., and S. Schuldiner.** 2000. An essential glutamyl residue in EmrE, a multidrug transporter from *Escherichia coli*. *J. Biol. Chem.* **275:**5264–5269.
99. **Yoshida, H., M. Bogaki, S. Nakamura, K. Ubukata, and M. Konno.** 1990. Nucleotide sequence and characterization of the *Staphylococcus aureus norA* gene, which confers resistance to quinolones. *J. Bacteriol.* **172:**6942–6949.
100. **Young, J., and I. B. Holland.** 1999. ABC transporters: bacterial exporters—revisited five years on. *Biochim. Biophys. Acta* **1461:**177–200.
101. **Yu, E. W., J. R. Aires, and H. Nikaido.** 2003. AcrB multidrug efflux pump of *Escherichia coli:* composite substrate-binding cavity of exceptional flexibility generates its extremely wide substrate specificity. *J. Bacteriol.* **185:**5657–5664.
102. **Yu, J.-L., L. Grinius, and D. C. Hooper.** 2002. NorA functions as a multidrug efflux protein in both cytoplasmic membrane vesicles and reconstituted proteoliposomes. *J. Bacteriol.* **184:**1370–1377.
103. **Zgurskaya, H. I., and H. Nikaido.** 2000. Multidrug resistance mechanisms: drug efflux across two membranes. *Mol. Microbiol.* **37:**219–225.
104. **Zheleznova, E. E., P. M. Markham, A. A. Neyfakh, and R. G. Brennan.** 1999. Structural basis of multidrug recognition by BmrR, a transcription activator of a multidrug transporter. *Cell* **96:**353–362.
105. **Zhou, Z., K. A. White, A. Polissi, C. Georgopoulos, and C. R. H. Raetz.** 1998. Function of *Escherichia coli* MsbA, an essential ABC transporter, in lipid A and phospholipid biosynthesis. *J. Biol. Chem.* **273:**12466–12475.

Frontiers in Antimicrobial Resistance: a Tribute to Stuart B. Levy
Edited by D. G. White, M. N. Alekshun, and P. F. McDermott

Chapter 11

Metal Resistance Loci of Bacterial Plasmids

Anne O. Summers

Like genes for antibiotic resistance, xenobiotic degradation, and virulence factors among other traits carried by plamids, the metal resistance genes are not intrinsic to the replication, maintenance, or transfer of plasmids. They are part of the genetic baggage that plasmids ferry around the ocean of prokaryotic genomes. Plasmids carry a huge variety of genes, and humans have selected out of this variety the above noted traits to study because of their immediate importance to us for control of diseases or pollution. This chapter provides an introduction to metallobiology, brief descriptions of the well-studied plasmid-coded metal resistances, and a short survey of the genetic and biochemical connections between metals and antibiotic resistances.

METALS IN BIOLOGY

It is now widely recognized that inorganic elements, especially those called metals, were essential for the origin of life and are presently a major aspect of all living systems (37). This fact also means living systems are vulnerable to the toxic effects of chemically related, but nonidentical metallic elements. Note that the term "metal" is generally applied to elements characterized by ductility and luster in their bulk form, a certain electronic plasticity (i.e., several stable valence states) at standard temperatures, and, in some, the ability to form covalent bonds.

The subset of metallic elements essential for life (Table 1) is relatively abundant in the Earth's surface. These elements occupy a distinct region near the center of the Periodic Table, referred to as the "transition," occurring mostly in the fourth period (row). There is also one essential element, selenium, among the so-called metalloid elements of the main group on the right-hand side of the table. Adjacent to the beneficial metal/metalloid elements in the Periodic Table are metal and metalloid elements sufficiently similar to the essential ones so as to interact readily with the organic molecules of life, but sufficiently different in their properties that they are dangerous to life rather than beneficial to it. In the course of evolving to occupy myriad niches on this planet, single cell organisms have been exposed very directly to the dangers of these toxic elements and have evolved the appropriate survival mechanisms (Table 2).

All metals, beneficial or toxic, exhibit some preferences for their ligands (17). The so-called soft or thiophilic metals (Ag, Cd, Hg) prefer sulfur; the hard (oxophilic) transition metals (Mn, Mo) and the metalloids (As, Sb, Bi) prefer oxygen ligands; in its lower 3+ oxidation state, arsenic can also take sulfur ligands. Lead, iron, cobalt, nickel, copper, and zinc can take oxygen or sulfur ligands, and the latter four often take nitrogen ligands. The metals also vary in their preferred number and arrangment of ligands (referred to as their "coordination geometry"), and these preferences are important in how proteins distinguish one metal from another and in how they can tune a metal's oxidation potential (17). As to covalent bonds, Hg and Pb can form alkylated compounds (although only the former are stable in water) as can the metalloid elements As, Se, and Te.

Their electronic plasticity makes metals valuable for doing chemistry in living systems, especially redox chemistry; of all the essential metals (and the single essential metalloid), only zinc is not redox active. But this very redox flexibility also means that even the essential metals can generate dangerous reactive oxygen species (ROS) unless the intracellular concentrations of their "free" hydrated forms are controlled. So, apart from specific resistance systems for the toxic elements, all organisms have evolved systems for balancing the intracellular availability of the redox active essential metals (36). These "metal ion homeostasis" systems are chromosomally encoded and typically consist of two

Anne O. Summers • Department of Microbiology, 527 Biological Sciences Building, University of Georgia, Athens, GA 30602-2605.

Table 1. Biologically useful heavy metal and metalloid elements[a]

Class		Metal (common valences)	Examples of cellular roles
Transition metal	First period	Manganese, Mn(II to VII)	Peroxidases, dismutases, photosynthesis
		Iron, Fe(II/III)	Respiratory enzymes
		Cobalt, Co(I/II)	Carbon-1 metabolism
		Nickel, Ni(I/II)	Hydrogenases
		Copper, Cu(I/II)	Electron transport, dioxygen transport and activation
		Zinc, Zn(II)	Enzyme cofactor, regulatory protein structural element
Transition metal	Second period	Molybdenum, Mo(II/IV)	Nitrogen fixation[b]
Metalloid	First period	Selenium, Se(III/V)	Selenocysteine (21st amino acid)

[a] Data from references 17 and 37.
[b] Vanadium and tungsten are also essential for nitrogen fixation in some bacteria.

subsystems: one to scavenge the essential metal when environmental concentrations are low and another to pump out the excess metal when such occurs. Many genes in bacterial plasmid-borne metal resistance systems are homologous to those of the chromosomal homeostasis systems (see below).

THE METAL RESISTANCE SYSTEMS

Mercury Resistance

Mercury is ubiquitous in our external and internal environments because of its occurrence in fossil fuels, its widespread use in industrial processes and products, and its use in approximately 80% of dental restorations (Table 2) (3). Bacteria can exhibit resistance just to inorganic Hg(II) or to organomercurials, RHg(I), as well (3). The former, called narrow spectrum resistance, is provided by a transport system consisting of a periplasmic protein (MerP) and an inner membrane protein (MerT) that brings ionic Hg(II) into the cytosol where it is reduced by MerA, a homolog of glutathione reductase with some additional domains. Specifically, MerA has an aminoterminal domain (a homolog of MerP) that captures incoming Hg(II) and conveys it to the active site with the assistance of a cysteine pair in MerA's signature C-terminal domain. The active site, located in the MerA "core" region just beyond the amino-terminal domain, is characteristic of the homodimeric flavin disulfide oxidoreductase family and includes two cysteines which receive the incoming Hg(II) from the two cysteines of the C-terminus of the other monomer. An FAD mounted near the active site cysteines delivers electons from NADPH to the Hg(II) bound to that cysteine pair. The resulting monoatomic, lipophilic Hg(0) simply volatilizes out of the cell. Although chemically simple, this process is energetically costly; each NADPH consumed might otherwise be used in anabolism or, after transhydrogenation to NADH, produce 3 ATPs via the aerobic respiratory chain.

The other enzyme associated with Hg resistance, MerB is found only in broad spectrum *mer* loci. MerB is a unique 22.4-kDa monomeric enzyme that protonolyzes the covalent carbon-mercury bond in compounds such as the disinfectant, merthiolate, or the biocide, phenylmercuric acetate. The regulator of *mer* transcriptional expression, the repressor-activator protein, MerR, was the first example of an entirely novel type of procaryotic regulator (3, 7); other members of the MerR-family were subsequently found to be regulators for chromosomal genes encoding cobalt, (CoaR), copper (CueR), and zinc (ZntR) efflux systems and for part of the oxidative shock response (SoxR) (7). MerD is the MerR antagonist and is a unique protein whose structure and mechanism are not yet understood. The periplasmic protein MerP (72 resides in secreted form) is the archetype of a small heavy metal binding domain with the central motif, GMTCxxC, that has been observed in free-standing proteins or as a subdomain (often in repeats) of larger metal trafficking proteins (8, 16). The inner membrane protein MerT (116 residues, three transmembrane domains, and two cysteine pairs) has no obvious relatives. It is presumed that the MerP cysteines hand off Hg(II) to MerT's cysteines and the latter move the metal into the cytosol, perhaps directly to MerA, but the details of Hg(II) uptake are presently unknown.

Naturally occurring *mer* operons are mosaics based on the above core genes. A typical arrangement in gram-negative bacteria is a two operon locus consisting of divergently transcribed *merR* and *merTP(C, F)A(B)DEG*. Genes in parentheses are those that provide accessory or unknown functions in suppport of the core functions and are not present in all examples of the *mer* operon. These include, in addition to MerB, the inner membrane Hg(II) transporter, MerC; the membrane-bound MerF, thought to be involved in Hg(II) uptake; the periplasmic MerG, apparently involved in organomercurial resistance; and MerE, a putative membrane protein which has never been

Table 2. Plasmid-determined heavy metal and metalloid resistance systems

Class	Metal (valence)	Cellular targets	Resistance mechanism (Model organisms)	Representative functional proteins	Regulation (family)	Reference(s)
Cation	Ag(I)	Thiols, imines	Efflux (*Salmonella enterica*)	SilE, periplasmic Ag(I) binding protein; SilP, P-type ATPase pump; SilCBA, chemiosmotic RND cation/anion antiporter	SilRS (2-component, sensor-responder)	49
	Cd(II), Co(II), Cr(VI), Pb(II), Ni(II), Zn(II)	Thiols, carboxyls	Efflux (*Staphylococcus aureus*)	CadA, Cpx-type ATPase pump, expels Cd(II), Pb(II), Zn(II)	CadC, repressor (SmtB/ArsR family)	9, 39, 55
			Efflux (*Ralsonia metallidurens*)	ChrA, chromate efflux; ChrC, superoxide dismutase; CnrCBA, RND superfamily, expels Co(II); and Ni(II); CzcCBA, RND superfamily, expels Co(II), Zn(II), and Cd(II); CzcD, CDF-mediated transporter, expels Co(II), Zn(II) and Cd(II); CadA, Cpx-type ATPase pump, expels Cd(II), Pb(II), Zn(II); PbrDCBAT, P-type ATPase superfamily	ChrIBF (unknown), CnrYXH (novel family operating via σ^{ECF}), CzcYXH (novel family operating via σ^{ECF}), CadC (SmtB/ArsR family), PbrR (MerR-family)	19, 23 6, 30, 36
	Hg(II)/RHg(I)	Thiols	Reduction/protonolysis (homologs occur in many eubacteria)	MerP, periplasmic protein; MerT (or MerC), inner membrane transport protein(s); MerA, Hg(II) reductase; MerB, organomercurial lyase; MerD, antagonist of MerR; MerE, MerF putative membrane proteins of unknown function; MerG, periplasmic protein providing organomercury resistance	MerR	3
	Cu(I/II)	Thiols, imines	Efflux (porcine *E. coli*, plant pathogenic *Pseudomonas* and *Xanthomonas*)	*pcoABCDE*	*pcoRS* on plasmid *cusRS* on chromosome (2 component, sensor-responder) chromosome	7
Oxoanion	As(III/V), Sb(III/V)	Thiols, phosphate	Efflux/reduction (homologs occur in many eubacteria)	ArsC, arsenate reductase; ArsA, ATP-dependent efflux pump; ArsB, membrane translocase	ArsR, repressor (SmtB/ArsR family); ArsD, a repressor with a glutaredoxin-like subdomain	9, 34, 58
	Te(III)	PMF (thiols?)	Efflux/reduction?	Plasmid-encoded TerZABCDEF, functions unknown. Four other unique loci in gram-negative bacterial plasmids or chromosomes. Some TeR may be epiphenomena of sulfur metabolism.	Unknown	52

isolated (3). Perhaps surprisingly, all *mer* proteins studied have highly conserved cysteine residues through which they interact with Hg(II). However, unlike the cysteines of other proteins that are rendered inactive by binding to Hg(II), some presently undefined aspect(s) of the *mer* proteins allow(s) their cysteines to function in signal transduction, transport, and enzyme catalysis with this most toxic of common metals.

Silver Resistance

Silver has also long been used as an antiseptic preparation and is still used in photography (with consequent waste processing issues) despite the advent of digital imaging (49). It is also the second most abundant metal in amalgam dental resotrations (after mercury). The silver resistance (*sil*) locus (49) is controlled by a two-component sensor-responder pair, SilRS, whose genes reside in an operon adjacent to, but transcribed divergently from, the structural gene operon, *silCBAP*. This latter operon encodes two biochemically distinct efflux pumps. SilP is a P-type ATPase, and SilCBA is a member of the resistance-nodulation-cell division-(RND) family of chemiosmotically driven efflux pumps. A periplasmic binding protein, SilE, is encoded by a transcript just downstream of *silRS*, and two short open reading frames (ORFs) of unknown function are interpolated into the *sil* operon for a total of nine proteins in three operons to constitute this resistance. Homologs of these genes occur on the *E. coli* chromosome and may be involved in copper homeostasis. This is not surprising since Cu(I), the form of copper that occurs in the reduced cytosolic environment, and Ag(I) are chemically similar and can take either sulfur or nitrogen ligands. Indeed, the periplasmic binding protein, SilE, which operates in an aerobic environment, has 10 histidines but no cysteines and can bind 5 Ag(I), Cu(II), or Cd(II) ions.

The first described silver resistance locus lies on the 180 kb, Inc H1 conjugative plasmid pMG101 and was originally found in a *Salmonella* strain from a hospital burn unit where silver sulfadiazine had been in use (49). This conjugative plasmid also carries genes encoding resistance to mercury, to tellurite, and to several antibiotics. Silver resistance is less widely found than mercury resistance, but it can generally be found where Ag(I) is in use. Note that although silver resistant cells growing on silver-containing medium turn black as a result of silver reduction, reduction is not part of the silver resistance mechanism and apparently occurs adventiously as cells enter stationary phase.

Copper Resistance

Copper is an essential element with redox-related metabolic roles; for this reason, in excess it can be dangerous to the cell in fostering the generation of ROS. Its microbicidal properties at high concentrations are exploited in its common uses in food animal production (15, 57) and horticulture (5, 18). In bacteria, relatively high tolerance for excess copper is afforded by chromosomal homeostasis loci in both gram-negative (11, 35, 42) and gram-positive (2, 50) model systems that have been examined in detail. Although genes for copper uptake in gram-negative bacteria are not yet defined, the efflux half of chromosomal copper homeostasis systems consists of various copper-sensors controlling several types of copper efflux pumps (44). The CPx-type ATPase, CopA, pumps Cu(I) from the cytoplasm, and the 3-component RND-type *cusABC* system pumps copper from the periplasm as well; indeed, that may be its more important role. The Cop system appears to function under aerobic conditions and the Cus system under anaerobic conditions (41). In addition, the chromosomally encoded multicopper oxidase *cueO*, a periplasmic enzyme, protects periplasmic proteins from Cu(II) damage by an as yet undefined process. The regulators and pumps of these chromosomally encoded Cu(I) efflux systems will also sense and pump Ag(I), respectively.

In bacteria isolated from metal contaminated environments, these chromosomal copper efflux systems are often assisted by a distinct but homologous plasmid-encoded, two-operon system, called *pcoABCDRS* and *pcoE* in the porcine *E. coli* plasmid pRJ1004 (aka *copABCDRS* and *copE* in the plant pathogenic *Pseudomonas syringae*) (Table 2) (44). The details of the additional protection provided by the *pco/cop* systems are just beginning to be established, and the current model suggests that the 605 residue PcoA, a periplasmic multicopper oxidase homologous to chromosomally encoded CueO, may catalyze oxidation of catechols such as enterobactin concommitant with their sequestration of Cu(II). The resulting formation of a brown precipitate in the periplasm would prevent Cu(II) from being taken up by the cell. PcoB is a 296 residue outer membrane protein, which apparently interacts with PcoA to provide nearly maximal copper resistance even without PcoC and PcoD. These latter two proteins are a 126 residue periplasmic copper binding protein and a 309 residue inner membrane protein, respectively. They are thought to take Cu(I) into the cell for incorporation into the PcoA multicopper oxidase. The *pcoABCD* operon is under control of the plasmid-borne two component sensor-response regulator, *pcoRS*, and also of the chromosomally encoded CusRS. The second operon *pcoE* is under control of the chromosomal response regulators, *cusRS*. PcoE appears to be a periplasmic copper sequestration protein, homologous to SilE, which performs the same role in silver resistance.

Cadmium, Cobalt, Nickel, Zinc, and Lead Resistances

Cobalt, nickel, and zinc are all essential metals, and all but zinc are redox active. Lead and cadmium have no beneficial biological functions and are both quite toxic. Both of these latter elements have long been used in paints, and lead was also used through much of the last century as a gasoline additive, leading to widespread environmental dispersal. Operons conferring single and multiple resistances to various subsets of these four transition metals and the main group heavy metal, lead, are based on several distinct types of efflux pumps. Two nonconjugative megaplasmids of *Ralstonia metallidurens* (30, 36) and a large metal- and antibiotic-resistant plasmid of *S. aureus* have been rich resources for study of these efflux mediated resistances (Table 2).

In *R. metallidurens* there are three metal resistance loci on pMOL28 (180 kb): a transposon conferring resistance to mercury, *merTPADE,* functioning as described above and two nontransposable loci conferring resistance to chromate (*chrI, chrBACEF,* a novel chromate/sulfate antiporter) and to cobalt and nickel (*cnrYXHCBA,* a proton-motive-force-driven RND-type metal-proton antiporter that can efflux substrates from the cytoplasm of the periplasm).

The larger *R. metallidurens* plasmid pMOL30 (250 kb) carries, in addition to a distinct *mer* transposon, three other nontransposable metal resistance loci conferring resistance to cobalt, zinc, and cadmium (*czcNICBADE,* an RND-type antiporter and related metallochaperones), to lead (*pbrTRABCD,* a CPx-type ATPase and related lead metallochaperones), and to copper (*copTKABCDIGFLH,* a CPx-type ATPase and several metallochaperones). Plasmid pMOL30 and the chromosome of the *Ralstonia* strain also carry CadA-like CPx-type ATPases, related to the archetypal Cd(II)/Pb(II)/Zn(II)-efflux pump carried by the plasmid pI258 of *Staphylococcus aureus* (39). Both of these plasmids and the *R. metallidurens* chromsome are littered with homologs of *czc, cnr,* and *cus* loci and operonic fragments of other resistance loci including silver and arsenic resistances (30, 36).

The regulatory elements associated with the various resistance loci in *R. metalllidurens* fall into the same three groups as those seen with metal resistance loci in other bacteria: the MerR-like activator-repressors (e.g., PbrR), the two-component histidine kinase sensor-respondors (e.g., CopRS), and the SmtB/ArsR family (e.g., CadC). The extensive redundancy of these resistance loci and their regulators in *R. metallidurens* makes it quite a challenge to sort out their respective contributions to resistance. However, it also suggests that there are robust mechanisms in cells under heavy metal stress to freely multiply and disperse these survival systems throughout the cell's replicon "real estate."

Arsenic and Antimony Resistance

Arsenic is a fairly abundant toxic element that is leached from natural minerals and accumulated by some plants and animals as organoarsenicals. It is also a major component of several pesticides used in agriculture (34). The metalloid elements arsenic and antimony are soluble oxoanions in their 3+ and 5+ formal oxidation states. Their uptake in the 3+ state is not well defined, but arsenic in the 5+ state (arsenate) enters the cell on phosphate transporters and, acting as an analog of phosphate, it thwarts the formation of high energy phosphoryl bonds. However, when the bacterial plasmid encoded *ars* operon is present, the intracellular 5+ metalloids are reduced to the 3+ state by ArsC, the arsenate reductase, one of three main components of the *ars* operon. In the 3+ state these metalloid elements have a high affinity for the sulfhydryl groups of the ATP-dependent ArsAB efflux pump which pumps them out of the cell (58). The synthesis of the *ars* structural genes is inducible by arsenate, arsenite, or antimonite, which provoke an allosteric change in the transcriptional repressors ArsR and/or ArsD, resulting in their release from the *ars* operator (26). At very low concentrations of these metalloids, ArsR leaves the operator allowing transcription of the efflux pump subunits and ArsD. The concentration of ArsD gradually rises and represses the operon again but only when intracellular concentrations of the metalloids begin to fall. This tandem regulation is thought to be necessary to avoid overexpression of ArsAB, which can be toxic to the cells.

In addition to the ArsC reductase associated with the *ars* operon of *E. coli,* there are at least two other phylogenetically distinct families of arsenate reductases found in prokaryotes and eukaryotes (33), all of which have similar analogous arrangements of the key active site sulfur ligands for the 5+ metalloid substrate but otherwise have little similarity. All also use glutaredoxin and glutathione as electron donors for metalloid reduction, and it is generally agreed that these enzymes arose by convergent evolution.

Anaerobic bacteria from several genera found in arsenic contaminated environments are capable of reducing arsenate as a source of metabolic energy through a periplasmic enzyme called ArrAB (34, 47). This process is not a resistance mechanism per se, although the bacteria that use it have relatively high tolerance for arsenite and arsenate (46). Whether the relevant genes are chromosomal or plasmid borne has not been established, although the phylogenetic distribution of this property suggests that it might be horizontally transferred (47).

Tellurite Resistance

Tellurium is not an abundant element, and its limited industrial use in catalysts and electronics has not resulted in extensive environmental dissemination. As the oxoanion tellurite, Te(III) is used in several diagnostic media since this readily reduced form of the metalloid makes black deposits of elemental Te(0) inside bacteria allowing ready visualization of those bacteria able to grow in its presence (e.g., enterococci). Gram-negative bacteria display varying degrees of tolerance to this metalloid. Five operons, two of them on plasmids and the rest on chromosomes of three unrelated gram-negative bacteria, have been shown to confer resistance to tellurite (52). The greatest level of resistance (MIC of 1,024 μg/ml) is conferred by the *terZABCDEFI* locus carried by the large, narrow host range conjugative IncH12 plasmid R478 and its homologues. Unfortunately, it has been difficult to study this locus; its coding regions have few obvious relatives, and cloning the genes for expression studies has been thwarted by their toxicity when separated. Knockout mutations suggest that this locus may also be involved in bacteriophage restriction. The other plasmid locus *klaABC* (aka *kilAtelBC*) carried by plasmid RP4 actually confers a postsegregational killing mechanism and only confers tellurite resistance when there is a specific nonsense mutation in the *telB* gene. Interestingly, the chromsomal tellurite resistance loci are unrelated to either plasmid locus and to each other, suggesting that the resistance the latter provides is an epiphenomenon of some undefined cellular function. Recent findings on interactions of the chromsomally encoded *teh* proteins with glutathione and S-adenosylmethionein suggest a role for Teh proteins in defense against redox damage to key cellular metabolites (14, 28). The ubiquity of these chromosomal tellurite resistance loci as well as the occurrence of the IncH12 plasmids in gram-negative pathogens suggests that what is observed as tellurite resistance arises from one or more undefined cellular processes that provide the host bacteria with some other advantage, perhaps in colonizing their animal hosts (53). This idea is also consistent with the very limited natural abundance of tellurite and the fact that it has never been used as a biocide as compared to As, Hg, and Cu.

ANTIBIOTICS AND METALS: GENETIC AND BIOCHEMICAL CONNECTIONS

Although the ubiquity of bacterial plasmids first became apparent in the context of antibiotic resistance (31, 32, 38, 45), it soon became obvious that resistance to widely used mercurial antiseptics and disinfectants was also a plasmid-encoded phenotype, often genetically linked to antibiotic resistances. The use of mercury as a disinfectant is quite ancient (10), and its use in dental "silver" fillings began in the early 19th century, so it is likely that plasmid-encoded genes for resistance to this toxic metal antedated the widespread use of antibiotics begun in the mid-1940s. Hospital and community use of mercurial disinfectants continued for at least four decades after the introduction of antibiotics, and large amounts of mercury are still used in dentistry, ensuring opportunity for simultaneous selection of Hg and antibiotic resistances in the commensal microbiota of humans.

Arsenic compounds also have a long history as anti-infectives, notably the 19th century introduction of arsenic (Salvarsan) for the treatment of syphilis (54). Although no longer used in human medicine, arsenic compounds are still widely used to control coccidia in animal husbandry and nematodes in horticulture (34). In both settings, particularly the former, antibiotics are also employed, again affording simultaneous coselection for both kinds of resistance (4, 20).

The essential metal copper is added as a growth supplement for swine production (15, 57) and is used as a pesticide in horticulture (5, 18). As antibiotics are also used in both settings (29, 56), it is not surprising to isolate bacteria exhibiting plasmid-mediated coresistance to copper and certain antimicrobials (1, 4, 13).

The above three examples concern selective regimens that simultaneously enrich for bacterial strains with genes for both antibiotic and specific metal resistances. Frequently these traits are physically linked by carriage on the same plasmids (48) or transposable elements (27). But are there also direct interactions between metal ions and antibiotics in biological systems? Several recent studies demonstrate that antibiotics with oxygen or nitrogen atoms available as multidentate ligands (chelates) for metal cations may give rise to very reactive chemical complexes. Copper derivatives of aminoglycoside antibiotics are very effective as in vitro chemical nucleases (12, 22, 25, 51).

Copper and zinc adducts of cephalexin significantly increased its in vivo activity (21). Although there are not many studies bearing on the in vivo biological effects of metal-antibiotic complexes, at present it seems that in vitro results are more dramatic than the in vivo effects, possibly because of the low availability of Cu(II) and Zn(II) in the cytosol (40). Lastly, since there is regulatory overlap between stress responses, exposure to metals that induce ROS formation may trigger broad responses (43) that diminish general susceptibility to antibiotics.

GENOMIC CONSIDERATIONS

Although heavy metal resistances in bacteria first appeared to be largely plasmid encoded, genome sequencing has revealed that homologs of transition metal resistance genes and their regulators abound in all prokaryotic chromosomes, most likely for managing the cell's use of their beneficial and essential metal relatives. These chromosomally encoded homeostasis systems for essential transition metals appear to be ancestral to the plasmid-encoded genes which subsequently evolved to handle the toxic transition metals. The origins of the metalloid resistances, whether they appear on plasmids or the chromosome, are less obvious. Selenium, the only essential metalloid element, is incorporated into proteins via covalent bonds, so the concept of homeostasis does not apply in the same way as it does for the transition metals. The proteins involved in resistance to arsenic or tellurium compounds are not phylogenetically related to the chromosomal genes involved in uptake or metabolism of selenium but rather to various chromsomal xenobiotic efflux pumps and/or redox defense proteins. An origin in both transition metal homeostasis and redox defense systems seems to apply to mercury resistance; the regulator MerR is a homolog of copper, cobalt, and zinc regulators, but the MerA gene is a homolog of glutathione reductase. The common theme, like that of resistance to antibiotics, is that metal/metalloid resistance genes can be recruited from among metabolic genes responsible for managing the toxicity of a normal metabolite or from the genes targeted by the toxic agent.

Is there anything special about having metal/metalloid resistance genes on a plasmid rather than on a chromosome? Of course, the same question applies to antibiotic resistance genes, since several of them (e.g., beta-lactamases) are chromosomally located in many taxa and many others are transposon-borne and can reside on any replicon. We do know that bacteria isolated from metal-impacted environments typically carry several metal resistances on large conjugative plasmids, suggesting that this genetic arrangement has selective advantage beyond that afforded by expression of related chromosomal loci. In general, plasmids allow duplication of the metal homeostasis genes and operons and consequent diversification without interrupting essential spatial relationships on the chromosome. Regulation of these plasmid-borne systems is often shared between chromsomal and plasmid regulatory elements, as noted above. Moreover, since plasmids are adventitious genetic elements, they can be abandoned when they are no longer necessary (although the existence of postsegregational killing mechanisms constrains this flexibility [24]). However, we are still a long way from understanding how any plasmid-borne function interdigitates with the physiology of even a single host, much less how these loci are able to fit into the complex biology and regulation of the many hosts they may occupy once they become plasmid borne.

As for the genomics of the plasmids themselves, presently we lack sufficient information on the phylogeny and population biology of these large plasmids to discern the basis for their host preferences and/or environmental distribution; only 12% of the ca. 700 recorded plasmid genome sequences are over 90 kb. Moreover, very recent comparisons of several genomes of the promiscuous IncP-1β group (48) led the authors to speculate that conjugation and recombination are sufficiently rife in this group that the genetic load (resistance or xenobiotic metabolism genes) carried by any given plasmid will reflect the selective pressure most recently experienced by its host. In other words, the evolutionary clock of the "floating genome" (plasmids, phages, and transposons [27]) is marked not in millenia or even centuries but perhaps in weeks and days. Increasing efforts at sequencing the agents of horizontal gene exchange, including plasmids, will soon fill in the picture of where, how, and how often plasmid-borne metal resistances arise and become linked with antibiotic resistance and/or virulence genes and what differences that makes ecologically and epidemiologically to the bacteria that harbor them and the corresponding human, plant, and animal hosts of these bacteria. Such information will doubtless provide insights useful for intervening in the spread of antibiotic resistance genes.

IN APPRECIATION

In the early 1980s, Stuart afforded my group the opportunity to "run the Periodic Table" on freshly collected human and animal fecal bacterial isolates. He thereby set the stage for our subsequent discovery that Hg exposure from the installation of dental "silver" fillings was as effective as a course of antibiotics in provoking proliferation of antibiotic multiresistant bacteria in the commensal microbiota. Over the years Stuart's pioneering perspective has moved us all toward a global model of the problem of antibiotic resistance. Since his early days as the only physician chicken-farmer in Boston, his deeply ecological vision has continuously informed and inspired my science and that of many others. And beyond his scientific legacy, a testimonial to his insight, energy, and eloquence is the accelerating movement toward change in public policy in this increasingly critical area.

Acknowledgments. Thanks to several colleagues who provided background and updates on their current work. Space limitations and the nonspecialist audience for this brief review constrained the information and citations to more general publications. Work in my own lab is supported by the U.S. Department of Energy program in Natural and Accelerated Bioremediation (NABIR) and the U.S. Department of Agriculture National Research Initiative.

REFERENCES

1. **Aarestrup, F. M., H. Hasman, L. B. Jensen, M. Moreno, I. A. Herrero, L. Dominguez, M. Finn, and A. Franklin.** 2002. Antimicrobial resistance among enterococci from pigs in three European countries. *Appl. Environ. Microbiol.* **68:**4127–4129.
2. **Banci, L., I. Bertini, R. Del Conte, J. Markey, and F. J. Ruiz-Duenas.** 2001. Copper trafficking: the solution structure of *Bacillus subtilis* CopZ. *Biochemistry* **40:**15660–15668.
3. **Barkay, T., S. M. Miller, and A. O. Summers.** 2003. Bacterial mercury resistance from atoms to ecosystems. *FEMS Microbiol. Rev.* **27:**355–384.
4. **Bender, C. L., and D. A. Cooksey.** 1986. Indigenous plasmids in *Pseudomonas syringae* pv. *tomato:* conjugative transfer and role in copper resistance. *J. Bacteriol.* **165:**534–541.
5. **Besnard, E., C. Chenu, and M. Robert.** 2001. Influence of organic amendments on copper distribution among particle-size and density fractions in Champagne vineyard soils. *Environ. Pollut.* **112:**329–337.
6. **Borremans, B., J. L. Hobman, A. Provoost, N. L. Brown, and D. van de Lelie.** 2001. Cloning and functional analysis of the *pbr* lead resistance determinant of *Ralstonia metallidurans* CH34. *J. Bacteriol.* **183:**5651–5658.
7. **Brown, N. L., J. V. Stoyanov, S. P. Kidd, and J. L. Hobman.** 2003. The MerR family of transcriptional regulators. *FEMS Microbiol. Rev.* **27:**145–163.
8. **Bull, P. C., and D. W. Cox.** 1994. Wilson disease and Menkes disease: new handles on heavy-metal transport. *Trends Genet.* **10:**246–252.
9. **Busenlehner, L. S., M. A. Pennella, and D. P. Giedroc.** 2003. The SmtB/ArsR family of metalloregulatory transcriptional repressors: structural insights into prokaryotic metal resistance. *FEMS Microbiol. Rev.* **27:**131–143.
10. **Canady, R. A., C. S. Rabe, and K. Gan.** 1996. Toxicological Profile for Mercury (update). U.S. Dept. of Health and Human Services, Public Health Service Agency for Toxic Substances and Disease Registry, Atlanta, Ga.
11. **Canovas, D., I. Cases, and V. de Lorenzo.** 2003. Heavy metal tolerance and metal homeostasis in *Pseudomonas putida* as revealed by complete genome analysis. *Environ. Microbiol.* **5:**1242–1256.
12. **Chen, C. A., and J. A. Cowan.** 2002. In vivo cleavage of a target RNA by copper kanamycin A. Direct observation by a fluorescence assay. *Chem. Commun.* (Camb)**:**196–197.
13. **Cooksey, D. A., and H. R. Azad.** 1992. Accumulation of copper and other metals by copper-resistant plant pathogenic and saprophytic pseudomonads. *Appl. Environ. Microbiol.* **58:**274–278.
14. **Dyllick-Brenzinger, M., M. Liu, T. L. Winstone, D. E. Taylor, and R. J. Turner.** 2000. The role of cysteine residues in tellurite resistance mediated by the TehAB determinant. *Biochem. Biophys. Res. Commun.* **277:**394–400.
15. **Edmonds, M. S., O. A. Izquierdo, and D. H. Baker.** 1985. Feed additive studies with newly weaned pigs: efficacy of supplemental copper, antibiotics and organic acids. *J. Anim. Sci.* **60:**462–469.
16. **Finney, L. A., and T. V. O'Halloran.** 2003. Transition metal speciation in the cell: insights from the chemistry of metal ion receptors. *Science* **300:**931–936.
17. **Frausto de Silva, J. J. R., and R. J. P. Williams.** 1991. *The Biological Chemistry of the Elements: the Inorganic Chemistry of Life.* Clarendon Press, Oxford, United Kingdom.
18. **Gallagher, D. L., K. M. Johnston, and A. M. Dietrich.** 2001. Fate and transport of copper-based crop protectants in plasticulture runoff and the impact of sedimentation as a best management practice. *Water Res.* **35:**2984–2994.
19. **Goldberg, M., T. Pribyl, S. Juhnke, and D. H. Nies.** 1999. Energetics and topology of CzcA, a cation/proton antiporter of the resistance-nodulation-cell division protein family. *J. Biol. Chem.* **274:**26065–26070.
20. **Hasman, H., and F. M. Aarestrup.** 2002. *tcrB*, a gene conferring transferable copper resistance in Enterococcus faecium: occurrence, transferability, and linkage to macrolide and glycopeptide resistance. *Antimicrob. Agents Chemother.* **46:**1410–1416.
21. **Iqbal, M. S., A. R. Ahmad, M. Sabir, and S. M. Asad.** 1999. Preparation, characterization and biological evaluation of copper(II) and zinc(II) complexes with cephalexin. *J. Pharm. Pharmacol.* **51:**371–375.
22. **Jezowska-Bojczuk, M., W. Szczepanik, W. Lesniak, J. Ciesiolka, J. Wrzesinski, and W. Bal.** 2002. DNA and RNA damage by Cu(II)-amikacin complex. *Eur. J. Biochem.* **269:**5547–5556.
23. **Juhnke, S., N. Peitzsch, N. Hubener, C. Grosse, and D. H. Nies.** 2002. New genes involved in chromate resistance in *Ralstonia metallidurans* strain CH34. *Arch. Microbiol.* **179:**15–25.
24. **Lehnherr, H., and M. Yarmolinsky.** 1995. Addiction protein Phd of plasmid prophage P1 is a substrate of the ClpXp serine-protease of *Escherichia coli. Proc. Natl. Acad. Sci. USA* **92:**3274–3277.
25. **Lesniak, W., W. R. Harris, J. Y. Kravitz, J. Schacht, and V. L. Pecoraro.** 2003. Solution chemistry of copper(II)-gentamicin complexes: relevance to metal-related aminoglycoside toxicity. *Inorg. Chem.* **42:**1420–1429.
26. **Li, S., B. P. Rosen, M. I. Borges-Walmsley, and A. R. Walmsley.** 2002. Evidence for cooperativity between the four binding sites of the dimeric ArsD, and As(III)-responsive transcriptional regulator. *J. Biol. Chem.* **277:**25992–256002.
27. **Liebert, C. A., R. M. Hall, and A. O. Summers.** 1999. Transposon Tn*21*, flagship of the floating genome. *Microbiol. Mol. Biol. Rev.* **63:**507–522.
28. **Liu, M., R. J. Turner, T. L. Winstone, A. Saetre, M. Dyllick-Brenzinger, G. Jickling, L. W. Tari, J. H. Weiner, and D. E. Taylor.** 2000. *Escherichia coli* TehB requires S-adenosylmethionine as a cofactor to mediate tellurite resistance. *J. Bacteriol.* **182:**6509–6513.
29. **McEwen, S. A., and P. J. Fedorka-Cray.** 2002. Antimicrobial use and resistance in animals. *Clin. Infect. Dis.* **34:**S93–S106.

30. **Mergeay, M., S. Monchy, T. Vallaeys, V. Auquier, A. Benotmane, P. Bertin, S. Taghavi, J. Dunn, D. van der Lelie, and R. Wattiez.** 2003. *Ralstonia metallidurans,* a bacterium specifically adapted to toxic metals: towards a catalogue of metal-responsive genes. *FEMS Microbiol. Rev.* **27:**385–410.
31. **Miller, M. A., and S. A. Harmon.** 1967. Genetic association of determinants controlling resistance to mercuric chloride, production of penicillinase and synthesis of methionine in *Staphylococcus aureus. Nature* **215:**531–532.
32. **Moore, B.** 1960. A new screen test and selective medium for the rapid detection of epidemic strains of *Staphylococcus aureus. Lancet* **ii:**453–458.
33. **Mukhopadhyay, R., and B. P. Rosen.** 2002. Arsenate reductases in prokaryotes and eukaryotes. *Environ. Health Perspect.* **110:** 745–748.
34. **Mukhopadhyay, R., B. P. Rosen, L. T. Phung, and S. Silver.** 2002. Microbial arsenic: from geocycles to genes and enzymes. *FEMS Microbiol. Rev.* **26:**311–325.
35. **Munson, G. P., D. L. Lam, F. W. Outten, and T. V. O'Halloran.** 2000. Identification of a copper-responsive two-component system on the chromosome of *Escherichia coli* K-12. *J. Bacteriol.* **182:**5864–5871.
36. **Nies, D. H.** 2003. Efflux-mediated heavy metal resistance in prokaryotes. *FEMS Microbiol. Rev.* **27:**313–339.
37. **Nies, D. H.** 2004. Essential and toxic effects of elements on microorganisms, p. 257–276. *In* K. Anke, M. Ihnat, and M. Stoeppler (ed.), *Metals and Their Compounds in the Environment.* Wiley-VCH, Weinheim, Germany.
38. **Novick, R. P., and C. Roth.** 1968. Plasmid resistance to inorganic salts in *Staphylococcus aureus. J. Bacteriol.* **95:**1335–1342.
39. **Nucifora, G., L. Chu, T. K. Misra, and S. Silver.** 1989. Cadmium resistance from *Staphylococcus aureus* plasmid pI258 *cadA* gene results from a cadmium-efflux ATPase. *Proc. Natl. Acad. Sci. USA* **86:**3544–3548.
40. **Outten, C. E., and T. V. O'Halloran.** 2001. Femtomolar sensitivity of metalloregulatory proteins controlling zinc homeostasis. *Science* **292:**2488–2492.
41. **Outten, F. W., D. L. Huffman, J. A. Hale, and T. V. O'Halloran.** 2001. The independent *cue* and *cus* systems confer copper tolerance during aerobic and anaerobic growth in *Escherichia coli. J. Biol. Chem.* **276:**30670–30677.
42. **Outten, F. W., C. E. Outten, J. Hale, and T. V. O'Halloran.** 2000. Transcriptional activation of an *Escherichia coli* copper efflux regulon by the chromosomal MerR homolog, CueR. *J. Biol. Chem.* **275:**31024–31029.
43. **Randall, L. P., and M. J. Woodward.** 2002. The multiple antibiotic resistance (mar) locus and its significance. *Res. Vet. Sci.* **72:**87–93.
44. **Rensing, C., and G. Grass.** 2003. *Escherichia coli* mechanisms of copper homeostasis in a changing environment. *FEMS Microbiol. Rev.* **27:**197–213.
45. **Richmond, M. H., and M. John.** 1964. Co-transduction by a staphylococcal phage of the genes responsible for penicillinase synthesis and resistance to mercury salts. *Nature* **202:**1360–1361.
46. **Saltikov, C. W., A. Cifuentes, K. Venkateswaran, and D. K. Newman.** 2003. The *ars* detoxification system is advantageous but not required for As(V) respiration by the genetically tractable *Shewanella* species strain ANA-3. *Appl. Environ. Microbiol.* **69:**2800–2809.
47. **Saltikov, C. W., and D. K. Newman.** 2003. Genetic identification of a respiratory arsenate reductase. *Proc. Natl. Acad. Sci. USA* **100:**10983–10988.
48. **Schluter, A., H. Heuer, R. Szczepanowski, L. J. Forney, C. M. Thomas, A. Puhler, and E. M. Top.** 2003. The 64,508 bp IncP-1β antibiotic multiresistance plasmid pB10 isolated from a waste-water treatment plant provides evidence for recombination between members of different branches of the IncP-1β group. *Microbiology* **149:**3139–3153.
49. **Silver, S.** 2003. Bacterial silver resistance: molecular biology and uses and misuses of silver compounds. *FEMS Microbiol. Rev.* 27:341-353.
50. **Solioz, M., and J. V. Stoyanov.** 2003. Copper homeostasis in *Enterococcus hirae. FEMS Microbiol. Rev.* **27:**183–195.
51. **Szczepanik, W., E. Dworniczek, J. Ciesiolka, J. Wrzesinski, J. Skala, and M. Jezowska-Bojczuk.** 2003. In vitro oxidative activity of cupric complexes of kanamycin A in comparison to in vivo bactericidal efficacy. *J. Inorg. Biochem.* **94:**355–364.
52. **Taylor, D. E.** 1999. Bacterial tellurite resistance. *Trends Microbiol.* **7:**111–115.
53. **Taylor, D. E., M. Rooker, M. Keelan, L. K. Ng, I. Martin, N. T. Perna, N. T. Burland, and F. R. Blattner.** 2002. Genomic variability of O islands encoding tellurite resistance in enterohemorrhagic *Escherichia coli* O157:H7 isolates. *J. Bacteriol.* **184:**4690–4698.
54. **Thorburn, A. L.** 1983. Paul Ehrlich: pioneer of chemotherapy and cure by arsenic (1854-1915). *Br. J. Vener. Dis.* **59:**404–405.
55. **Tsai, K. J., Y. F. Lin, M. D. Wong, H. H. C. Yang, H. L. Fu, and B. P. Rosen.** 2002. Membrane topolocy of the pI258 CadA Cd(II)/Pb(II)/Zn(II)-translocating P-type ATPase. *J. Bioenerg. Biomembr.* **34:**147–156.
56. **Vidaver, A. K.** 2002. Uses of antimicrobials in plant agriculture. *Clin. Infect. Dis.* **34:**S107–S110.
57. **Williams, J. R., A. G. Morgan, D. A. Rouch, N. L. Brown, and B. T. Lee.** 1993. Copper-resistant enteric bacteria from United Kingdom and Australian piggeries. *Appl. Environ. Microbiol.* **59:**2531–2537.
58. **Wong, M. D., B. Fan, and B. P. Rosen.** 2003. Bacterial transport ATPases for monovalent, divalent and trivalent soft metal ions, p. 159–178. *In* J. Kaplan, Y. Wada, and M. Futai (ed.), *Ion-Pumping ATPases: Biochemisry, Cell Biology and Pathophysiology.* Wiley-VCH, Weinhein, Germany.

Frontiers in Antimicrobial Resistance: a Tribute to Stuart B. Levy
Edited by D. G. White, M. N. Alekshun, and P. F. McDermott

Chapter 12

Biocides and Resistance

BONNIE M. MARSHALL AND LAURA M. MCMURRY

In 1945, not long after his serendipitous and historic discovery of penicillin, Alexander Fleming uttered the first words of caution concerning penicillin abuse. This followed on the heels of discovery in his laboratory that penicillin-resistant mutants could grow in the presence of increasingly higher amounts of the drug. Nonetheless, his words did nothing to stem the tide of frenzied use of penicillin and other "miracle drugs" for all manner of illnesses and conditions, irrespective of its appropriateness. In some ways, the present "antibacterial craze" ominously mirrors the early days of antibiotic use and ironically stems in part from a need to overcome the consequences of antibiotic abuse.

Now that antibiotic resistance is escalating to alarming proportions and the discovery of new antibiotics is waning, a greater reliance on infection prevention has evolved—from fear as well as necessity. In the 1990s, the Centers for Disease Control and Prevention (CDC) launched a much-heralded educational campaign that made remarkable strides in the arena of community hygiene (30). This increased awareness, however, coupled with the reports of menacing, antibiotic-resistant "superbugs," has fueled a certain germ phobia that, taken to the extreme, resulted in a deluge of consumer healthcare and cleaning products touting "antibacterial" uses. The environmental impact of using products containing these antibacterial agents has not been studied in great detail. Stuart Levy was one of the earliest investigators to question the wisdom of this uncontrolled, burgeoning use of such chemicals outside the clinical setting (66, 67).

Those who argue that triclosan and other biocides have been used for multiple decades without difficulty often fail to note that their usage has changed from specific, controlled application in hospital and other health care settings to nonprescribed, indiscriminate application in households and the community. It is only recently that these agents have been incorporated into a vast array of consumer and health care products. The claims stated or implied by such products tend to quell the public's insecurities and may actually encourage a sense of complacency toward prudent hygiene. In recent years, much research has highlighted the shortcomings of some of these biocides, noting in particular that their efficacy and benefits are more elusive under home use conditions than in the hospital where their use has a proven track record of efficacy.

BACKGROUND

Disinfecting agents used in the hospital, household, and industry environments are sometimes variably referred to as antibacterials, antiseptics, disinfectants, or, in broader terms, germicides or biocides. Here, the more inclusive term "biocides" will be used for the agents that include the antiseptics (agents applied to skin surfaces) as well as the disinfectants (compounds applied to inanimate surfaces for cleaning and microbial reduction). Biocides can be divided into two major categories (Table 1). The nonresidue-producers, represented primarily by the chlorines, peroxides, and alcohols, are either unstable or volatile and do not persist in an active form in the environment. In contrast, the residue-producing biocides possess remanent qualities, which presumably make them long acting. The latter are the focus of this discussion, since their purported ability to act like antibiotics, including a potential for selecting resistant mutants, may compromise the efficacy of either themselves or antibiotics.

TRICLOSAN AND TRICLOCARBAN

The number of "antibacterial" products on the market has grown from about two dozen in the early 1990s to 1,000 or more. Among personal hygiene prod-

Bonnie M. Marshall and Laura M. McMurry • Department of Molecular Biology and Microbiology, Tufts University School of Medicine, Boston, MA 02111.

Table 1. Biocides and their mechanisms of action[a]

Biocide group	Biocide	Mode/target of action
Nonresidue Producers		
Alcohols	Ethanol, isopropanol	Membrane permeability
Aldehydes	Gluteraldhyde, formaldehyde	Cytoplasmic amino group
Oxidizing agents	Chlorine compounds, iodine compounds (betadine, povidone iodine), hydrogen peroxide, peracetic acid, ozone, ethylene oxide	Oxidation of cell components with leakage of cellular constituents
Residue Producers		
Anilides	Triclocarban	Destruction of cytoplasmic membrane
Biguanides	Chlorhexidine, alexidine, polymeric biguanides	Low concentrations: affect membrane integrity; high concentrations: congealing of cytoplasm
Diamidines	Propamidine isethionate	Inhibition of oxygen uptake; amino acid leakage
Phenols and cresols		
bis-phenols	Triclosan, hexachlorophene, fentichlor	High concentrations: disrupt cell wall; precipitate cell proteins. Low concentrations: inactivate enzyme systems; leakage of metabolites from cell wall
Halophenols	Chloroxylenol (PCMX), 2-phenylphenol	
Other	Pine oil disinfectants	
Heavy metals	Silver compounds, mercury compounds	Electron transport chain thiol groups
Quaternary ammonium compounds (cationic detergents)	Cetrimide, benzalkonium chloride, cetylpyridinium chloride	Inactivation of energy-producing enzymes; denaturation of cell proteins

[a] Compiled from references 7, 14, and 79.

ucts, the agents triclocarban and triclosan are the most common biocide components (http://householdproducts.nlm.nih.gov/index.htm). The former, an anilide, is active primarily against gram-positive bacteria and is found in bar soaps and deodorants. This compound has significantly less persistence (substantivity) on the skin and is less well studied than the more prevalent agent, triclosan, which has a long history as a fabric preservative (15). Since its early days as a remarkable life-extender for U.S. Army boots, which deteriorated rapidly in the high heat and humidity of the Korean war, its use has expanded markedly as a component of many textile-based products in the home environment, including socks, sheets, pillowcases, mattresses, sporting gear, etc. (30).

Compared to older agents (e.g., alcohol, chlorhexidine, chloroxylenol, and povidone iodine), triclosan reduces microbial counts less rapidly but possesses a more sustained activity because it binds to the stratum corneum of the skin and therefore acts to reduce both the resident as well as the transient microflora (15). Assessed to be beneficial in high-risk clinical environments, triclosan is effectively employed in surgical scrubs and hand and body washes, largely for the control of pathogenic staphylococci. Because of its mildness, substantivity, and sustained microbial action, triclosan became a popular component in liquid hand soaps, deodorants, and oral care products, all of which are marketed for public consumption.

More recently, triclosan has been melt-mixed into a myriad of plastic products ranging from kitchen cutting boards and utensils to children's toys, spas, and outdoor furniture. Published experiments suggest that leaching does not occur from the embedded portion of the agent (56, 57). Thus, only surface-level quantities are actually exposed, and biological activity occurs only in the presence of moisture. Many of these products came to bear "antibacterial" labels, implying health benefits as a result of their use. Concluding that their efficacy in "disease prevention" was unfounded, the Environmental Protection Agency (EPA) took action and banned this labeling practice in 1997 (30). Nonetheless, the incorporation of biocides still continues for its original "preservative" reasons—intended for control of odor, mold and mildew stains, and the prevention of material degradation of products as diverse as carpeting, adhesives, paints, plumbing fixtures, and laminate flooring (http://www.microban.com).

QUATERNARY AMMONIUM COMPOUNDS (QACs)

Another major group of commonly used residue-producing biocides consists of the quaternary ammonium compounds (QACs) that are also known as cationic detergents (Table 1). Of this rather large group, benzalkonium chloride and cetrimide are the more common and recognizable. These products have been used for many years, chiefly as preservatives for cosmetics, and in hospitals and other clinical settings as skin and surface disinfectants. QACs are also widely used in the food industry and in veterinary applications, particularly to control staphylococci. First introduced as biocides in the 1930s, QAC use has expanded significantly over the last two decades, and these compounds have become the primary antimicrobial agent in a large number of household cleaning compounds (http://householdproducts.nlm.nih.gov/index.htm).

BIOCIDE ACTIVITY

Unlike antibiotics, biocide behavior varies with its concentration. At low concentrations, single cell targets may be affected (as with antibiotics), and point mutations are possible in these target sites, which can result in demonstrable tolerance to the biocide (see below). At higher concentrations (i.e., those applied in practical applications, which commonly range from 600 to 20,000 μg/ml) (53), however, multiple targets are affected, leading to a series of cumulative cell disruptions that culminate in rapid cell death (79). These differences in activity complicate an understanding of resistance to biocides because the concentration which inhibits cell growth (i.e., the MIC) is lower than that which rapidly kills them over a short contact time (i.e., minimum bactericidal concentration [MBC]). Whereas antibiotics function effectively at their MIC by suppressing bacterial growth and allowing host immune systems to eliminate the infection, biocides lack this advantage and must be used at much higher concentrations. Typically, it is the MBC, or "killing dose," which is used in evaluating biocide efficacy. While some investigators define "resistance" as lack of effectiveness in vivo, others have demonstrated and reported effects at the level of the MIC, which is 100 to 10,000 times lower than concentrations used in practice (52). Low-level resistance, as measured by MIC testing, has become of greater interest in recent years, but data are limited and the significance is yet unclear. Because the concept of resistance is more nebulous for biocides than for antibiotics and a clear-cut definition is lacking for most species, the terms "tolerance" or "reduced susceptibility" are deemed more appropriate for describing the low levels of resistance that do not manifest as biocide failure in practice (14, 36).

MECHANISMS OF RESISTANCE

Features of the outer cell wall and membrane (particularly among gram-negative bacteria), the spore coats of certain spore-forming species, and the lipid-rich waxy cell coat of mycobacteria provide especially high natural (intrinsic) biocide resistance, which renders the cell impermeable to uptake. The remarkable biocide resistance demonstrated by biofilms is linked to the unique organizational structure of a specialized consortium of organisms in which certain members undergo nutrient deprivation and physiological change, rendering them more tolerant. In addition, as a result of decreasing diffusion, the bottom layers become inaccessible to biocides. Although not truly "resistant," such cells are "protected" and can sometimes persist in the presence of high concentrations of biocide formulations (79). These mechanisms differ from the powerful (intrinsic) efflux pumps that also confer high-level resistance to toxic compounds (22) (see below). Some organisms (e.g., *Bacillus subtilis*) are inherently resistant to triclosan via a resistant enoyl reductase (FabL) (42); others can resist biocides by inactivation of the compounds (40, 83, 86, 87, 91).

In contrast, acquired resistance is achieved through genetic changes, typically a mutation in a biocide's target site (see below). Mutational resistance may be acquired through exposure to low-level (subbactericidal) biocide concentrations and can sometimes be increased by stepwise exposure to higher concentrations (Table 2). Alternatively, resistance may be acquired by uptake of plasmids or transposons, which encode transporters or other biocide resistance determinants. The acquisition of resistance through these mechanisms varies by species and may or may not be maintained stably. The finding of biocide resistance genes on plasmids is particularly important, since plasmid transfer by conjugation (bacterial mating) is a promiscuous form of genetic exchange (93).

A rather large number of plasmid-borne genes have been identified that encode resistance to the QACs. These *qac* genes (variously designated and reported as *qacA-E, G, H,* and *J*) (16, 45, 47, 70, 79, 92) all reside in gram-positive bacteria (with the exception of QacE, which is found in gram-negative organisms) and code for an energy-dependent efflux mechanism (see below). The resistance determinants are differentiated by their origins of derivation (e.g., from hospital, food, or veterinary sources) but demonstrate marked degrees of homology. Thus, they have been structurally grouped into two families, the QacA/B and Smr (small multidrug resistance) families (63). Of note, the *qac* genes have been found on plasmids, including large conjugative multiresistance plasmids (primarily pSK1 and its derivatives), those specifying β-lactamases, and on elements conferring heavy metal resistance (79). The most notable of these is the QacA/B system, which expresses a high level of resistance not only to QACs but also to biocide dyes (e.g., acridine and ethidium bromide), diamidines, and chlorhexidine. QacA plasmids may also carry transposons that code for gentamicin, trimethoprim, and kanamycin resistance (120). Some plasmids that bear *qac* genes in commensal and pathogenic staphylococci are similar, suggesting there is a common pool of shared resistance determinants (63). Moreover, trans-

Table 2. In vitro biocide adaptation/mutation and cross-resistance to unrelated biocides and antibiotics

Species	Adapted to	Cross-resistance[a]	Reference
Gram-positive species			
Staphylococcus aureus	Pine oil cleaner	Oxacillin	94
S. aureus	Biocide dyes (acriflavin, acrinol), QACs[b] (benzethonium and benzalkonium chloride), chlorhexidine	Biocide dyes, QACs, fluoroquinolones (norfloxacin, ofloxacin)	89
MRSA[c]	QAC (benzalkonium chloride)	Ofloxacin, oxacillin, cloxacillin, moxolactam, flomoxef, cefmetazole	4
Leuconostoc	QAC (benzalkonium chloride)	Gentamicin	107
Gram-negative species			
Pseudomonas aeruginosa	QAC (N-dodecylpyridinium chloride)	Benzalkonium chloride	118
P. aeruginosa	QAC (benzalkonium chloride)	Chloramphenicol, polymixin B, tobramycin	71
P. stutzeri	Chlorhexidine	Triclosan, cetylpyridinium chloride, erythromycin ampicillin, nalidixic acid, polymixin B	100
Stenotrophomonas maltophilia	Triclosan	Tetracycline, chloramphenicol, ciprofloxacin	101
E. coli	Chloroxylenol, chlorhexidine + QAC (cetrimide), glutaraldehyde, iodine	Tetracycline, ampicillin, streptomycin, chloramphenicol	111
E. coli O157	Triclosan, QAC (benzalkonium chloride)	Chlorhexidine, benzalkonium chloride, amoxicillin, amoxicillin-clavulanate, chloramphenicol, erythromycin, imipenem, tetracycline, trimethoprim, triclosan	19, 20
E. coli O157	Clorhexidine	Triclosan	
E. coli O55	Triclosan	Trimethoprim	
E. coli K-12	Triclosan	Chloramphenicol	
E. coli	Pine oil	Tetracycline, chloramphenicol, nalidixic acid	85
E. coli	QAC (benzalkonium chloride)	Ampicillin, penicillin, norfloxacin, nalidixic acid, kanamycin, gentamicin, chloramphenicol, tetracycline, erythromycin, ethidium bromide, proflavine	60
E. coli	QAC (cetyltrimethylammonium bromide	Ampicillin, amoxicillin, sulbenicillin, chloramphenicol, erythromycin, kanamycin, tetracycline, nalidixic acid, trimethoprim, other QACs, heavy metals, oxidants and organic compounds	51
E. coli	QAC (didecyldimethylammonium chloride	Chloramphenicol	123
Serratia marcescens	Chlorhexidine	QAC	34
Salmonella serovar Typhimurium	QAC (benzalkonium chloride	Chlorhexidine	19
Salmonella serovar Virchow	Chlorhexidine, benzalkonium chloride	Triclosan, amikacin, amoxicillin, chlorhexidine, chloramphenicol, imipenem, triclosan, trimethoprim	19
Salmonella enterica wild-type/ MAR mutants	Phenolic farm disinfectant	Tetracycline, ampicillin, ciprofloxacin, cyclohexane	95
	Triclosan	Ampicillin, cyclohexane	
Other			
Mycobacterium smegmatis	Triclosan	Diazaborine, isoniazid	80

[a] Partially or fully resistant to one or more of the stated agents.
[b] QAC, quaternary ammonium compounds.
[c] MRSA, methicillin-resistant *S. aureus*.

fer of biocide genes, both by conjugation and by transduction (i.e., via bacterial viruses), has been observed in the presence of low-level biocide concentrations (6, 93).

Both intrinsic and acquired resistance mechanisms act by reducing the cellular uptake of biocides. While impermeability features (described above) act as external barriers, many bacterial species exhibit tolerance via efflux pumps, which are membrane-based structures that actively efflux toxic substances by pumping them out across the cell membranes (124). These are generally plasmid encoded in gram-positive bacteria and chromosomally encoded among gram-negative organisms. Four efflux pump systems operating in gram-negative bacteria (MexAB, MexCD, and MexEF in *Pseudomonas aeruginosa* and the AcrAB system in *E. coli*) are members of the resistance-nodulation-cell division (RND) family and have been shown to facilitate transport of a large number of biocides (including triclosan and QACs), antibiotics, and toxic dyes (124). The MexAB efflux system is responsible for the very high intrinsic resistance to triclosan (1000 μg/ml) observed in *P. aeruginosa* (25, 103). Other efflux pumps are silent but may be expressed under selective pressure when exposed to a biocide (25). Many clinical strains express multiple RND efflux pumps. Since efflux is a resistance mechanism shared by both antibiotic and biocide-resistant strains, mutations or gene acquisitions that lead to this activity could result in the expression of resistance to both antibiotics and biocides—the two antimicrobial groups employed to control infection (65).

LABORATORY STUDIES OF BIOCIDE RESISTANCE: ADAPTATION TO BIOCIDE TOLERANCE

Just as bacteria can develop resistance to antibiotics, they can mutate to tolerate biocides by exposing them to incrementally higher concentrations of these chemicals. Various attempts at laboratory-induced tolerance have resulted in mixed findings. Some report that low-level adaptations are possible but unstable (113), while others have shown that biocide-tolerant mutants of both clinical and nonclinical isolates are easy to obtain in vitro and are quite stable (19, 32, 44, 118) but exhibit a marked difference in the speed with which they can build tolerance (19). In addition, a growing number of species have demonstrated cross-resistance, not only to other biocides but to antibiotics and other toxic compounds as well (Table 2). For example, Moken et al. reported that *E. coli* mutants that were resistant to pine oil were cross-resistant to tetracycline, chloramphenicol, and nalidixic acid (see appendix). This was followed by other laboratory examinations of triclosan-selected mutations in *E. coli* that turned out not to express cross-resistance but instead led to the finding of a single cell target for triclosan activity (see below). The pathogens *Salmonella enterica* and *E. coli* O157 could readily adapt to stable biocide tolerance (30 or more days), and very high triclosan tolerance could be attained in *E. coli* (1024 μg/ml, which is about half the concentration found in commercial handsoaps). Whereas *E. coli* O157 showed cross-resistance to multiple antibiotics and biocides, cross resistance in *Salmonella* was more limited and varied with the serotype (20) (Table 2).

Recently, additional evidence for the development of biocide and antibiotic cross-resistance in vitro was shown in three serovars of *S. enterica*. Isolates which were selected using a sub-bactericidal concentration of a phenolic farm disinfectant demonstrated higher (10 to 100-fold) frequencies of mutation to tetracycline, ampicillin, and ciprofloxacin resistance. Those exposed to triclosan exhibited higher mutation frequencies to ampicillin resistance only (95). However, both exposures resulted in increased cyclohexane tolerance, which is indicative of AcrAB efflux pump overexpression (125). Importantly, resistance to ciprofloxacin was expressed at clinically significant levels (95). The reverse phenomenon has also been reported (i.e., strains adapted to antibiotic resistance can demonstrate tolerance to biocides) (55, 81, 88, 94).

In contrast to the above findings, no change was reported in cell biotype or antibiotic resistance profiles of *S. aureus* or *S. epidermidis* following long-term (105 days) growth in triclosan (24). Studies of triclosan exposure among anaerobic bacteria from dental plaque likewise demonstrated that the phenomenon of adaptability is not universal. These and other studies suggested that additional genetic targets besides FabI (see below) may be involved in triclosan resistance (21, 78).

MECHANISM OF ACTION OF TRICLOSAN

Note: This section gives major emphasis to studies on triclosan done by the Levy group. Subsequent work on this topic from other laboratories is also described briefly.

Initial experiments by Merri Moken, a high school student, showed that mutants selected for resistance to the household disinfectant pine oil were also multiply antibiotic resistant (Mar) (85) (see appendix). These findings inspired the Levy group to examine another household antimicrobial, triclosan. That work (82) led accidentally to the previously unknown cellular target of triclosan, which was found to be an essential enzyme involved in lipid biosynthesis. This enzyme has homologs in many other microorganisms, including those causing tuberculosis and malaria, and triclosan has now been shown to have activity against these homologs (see below). More recently, triclosan has been shown to prevent infections in at least two mouse models (104, 116).

Triclosan was first described as a degerming agent for soaps in 1968 to control the growth of both gram-negative and gram-positive organisms (33). The late 1990s saw a burgeoning use of triclosan in soaps, toys, toothpaste, cutting boards, etc. to kill household germs. Although use of antibiotics has historically led to self-resistance and sometimes to cross-resistance (68), it was thought that such resistance for triclosan would be unlikely, since triclosan appeared not to have a single target but rather to physically lyse the membranes of microorganisms (97). The motive of the Levy group was to test the presumption about resistance.

Studies in *Escherichia coli*

Mar mutants (overexpressing the *E. coli mar* locus with attendant overexpression of the activator protein MarA) were found to be resistant to triclosan by a factor of two or so. This was presumably because MarA induced the gene for the AcrAB multidrug efflux pump, of which triclosan (and pine oil) were substrates (81).

When trying to select Mar mutants on plates containing triclosan (0.2 μg/ml) in a manner analogous to that using tetracycline (35) and pine oil (85), the Levy group found surviving colonies on triclosan-containing media at a frequency of about 1 in 10^8. However, to our surprise, these mutants were not resistant to any antibiotics, and deletion of the *mar* locus did not lower the level of triclosan resistance. Therefore the triclosan-resistant colonies were not *acrRAB* or Mar mutants.

Suspecting that we had selected a mutation in a gene for an unknown triclosan/drug efflux pump, we cloned a *Sau*IIIA genomic digest of the mutant with the highest resistance. Triclosan-resistant clones were found at the frequency expected for a single locus, and all promoted the same level of triclosan resistance. The clone contained only two genes, *ycjD* and *envM*. The first, completely unknown, definitely did not specify an efflux pump. The second, *envM*, was also called *fabI* and encoded enoyl reductase involved in fatty acid biosynthesis (hence the *fab* designation).

Deletion of each gene separately clearly demonstrated that triclosan resistance was associated only with *fabI*. When the mutant *fabI* was replaced with the wild-type gene, resistance was much lower (although, likely due to a plasmid multicopy effect, resistance was still higher than that for wild-type *E. coli*). Therefore, a mutated *fabI* gene was causing the triclosan resistance of the highest level mutant. The sequence of this entire mutant *fabI* gene differed by only one base pair from the wild type sequence, giving rise to a FabI protein with a Gly93Val mutation (82). With considerable satisfaction, we noted that a Gly93Ser mutation, at the identical residue, had been reported to cause resistance to diazaborine, a known inhibitor of FabI protein (12). Four of the five triclosan-resistant mutants tested were also found to be resistant to diazaborine.

Four other independently isolated mutants with different levels of triclosan were studied. P1 transductions showed that resistance for all four mutants mapped to the *fabI* region of the chromosome. The sequence of the *fabI* gene of two of these mutants revealed FabI to be mutated to give Met159Thr or Phe203Leu. Since the crystal structure and cofactor NADH binding site of the FabI protein had been determined (9), we were able to wander through the virtual protein using a graphics program and see that all of our mutations were near the NADH binding site. Presumably all of the mutations prevented binding/action of triclosan at this site (see below).

It looked therefore as if triclosan did have a single target after all. That no Mar mutants had been selected on triclosan, even though they have a low-level triclosan resistance, may have been because the selective levels were too high or the frequency of *fabI* mutations was greater than that of *mar* mutations.

Since FabI is an essential enzyme catalyzing one of the steps in lipid synthesis, triclosan would be expected to inhibit lipid synthesis—and it did so (the control was diazaborine, the previously described FabI inhibitor) (82). Triclosan did not inhibit either protein or DNA synthesis (82).

Studies in Mycobacteria

Since the FabI equivalent in *Mycobacterium tuberculosis* was InhA, similar genetic methods were used to see if this kind of enzyme could also be targeted by triclosan, using the related InhA in the nonpathogenic species *Mycobacterium smegmatis*. Just as with *E. coli*, resistant colonies appeared when cultures of *M. smegmatis* were spread on plates containing triclosan. Some of these colonies had cross-resistance to isoniazid (an InhA inhibitor), and an isoniazid-resistant mutant with a known mutation in *inhA* turned out to be triclosan resistant. The triclosan-selected mutants proved to have point mutations in the active site of InhA (80). Since *M. smegmatis* and *M. tuberculosis* InhA were conserved at the residues where our mutations had been found, it was puzzling to learn that the latter organism had been tallied as triclosan resistant (122). More recently, *M. tuberculosis*, as well as its purified InhA, has been reported to be sensitive to triclosan (90). Triclosan was bound in the active site as shown in the crystal structure of InhA in *M. tuberculosis* (58).

Killing by Triclosan

An early study had shown that a release of UV light-absorbing material from *E. coli* cells and a loss of cell absorbance and viability resulted from treatment with triclosan at levels above those which halted growth

(97); the conclusion was that the triclosan caused lesions in the cell membrane which led to death by leakage. Although triclosan caused no cell leakage for *Porphyromonas gingivalis* (121), it did release potassium from *S. aureus,* which correlated with killing (114).

MICs determined on agar plates cannot distinguish between mere growth inhibition and loss of viability. Using liquid cultures, we found that the growth rate of wild-type cells was inhibited 50% at 0.15 μg/ml triclosan, while loss of optical density and viability required 8 μg/ml (81, 82). These values suggested that plate MICs of 0.3 to 0.8 μg/ml had reflected inhibition of growth (81, 82). Growth in liquid culture for the Gly93Val mutant was inhibited 50% at 13 μg/ml, while viability was retained at quite high levels (>256 μg/ml), as though the Gly93Val mutation decreased both the inhibition of growth and the killing of cells by triclosan (81, 82). Protection of cells from killing by the Gly93Val mutation was dependent on the presence of the AcrAB efflux pump (81). These results suggest the possibility that triclosan has two modes of action, both involving inhibition of FabI and one involving direct action on the membrane. At relatively low concentrations of triclosan, growth would be inhibited because not enough lipids were made. At higher concentrations of triclosan, the low-lipid membrane resulting from FabI inhibition might be more easily permeabilized by triclosan acting physically on the membrane itself, more acutely if the AcrAB pump were not removing the drug from the membrane. More work will be needed to resolve this issue.

Studies in Various Organisms on Purified FabI and Use of Triclosan as a Curative Drug

Soon after our 1998 publication (81), another group proved that triclosan inhibited purified FabI protein (43). That triclosan did in fact bind near the NADH binding site in *E. coli* FabI and that the residues affected in our mutants were in contact with the drug was shown by the cocrystal structure (64). FabI of other organisms was found to be targeted by triclosan, including that of *Staphylococcus aureus* (41), *Bacillus subtilis* (42), and *Pseudomonas aeruginosa* (48). Strikingly, the *Plasmodium falciparum* (malaria) FabI was also inhibited by triclosan (116). Moreover, triclosan was found to be effective in mouse models against *P. berghei* (116) and *E. coli* O55:B5 (104).

BIOCIDE RESISTANCE IN THE ENVIRONMENT

Following the investigations of pine oil and triclosan resistance (see above), Stuart Levy began openly expressing concerns over the rising number of biocide-containing products and the indiscriminate ways in which they were being applied outside of the controlled, health care environment. These concerns noted that community applications are more casual and for shorter exposures (usually just seconds not minutes) than those practiced in clinical environments. The impacts of these products on the environmental flora, in particular the potential to select for resistant mutants in the community environment, to promote cross-resistance to antibiotics, and to permit proliferation of intrinsically resistant strains, were also questioned (66, 69). Following is an overview of biocide tolerance as it presents in health care, the food industry, and agricultural environments.

Retrospective Assays

Biocide resistance in the clinical arena

A serious compromise to health care would be the appearance of pathogens which are not only resistant to antibiotics but also fail eradication with antiseptics and/or disinfectants. Presently, the most highly biocide-resistant organisms are those expressing high-level, intrinsic resistance. Bacterial genera such as *Burkholderia, Serratia, Alcaligenes,* and *Pseudomonas,* which can thrive within concentrated biocide solutions (possibly as biofilms), have proved to be the culprits in a number of troublesome hospital outbreaks of antibiotic resistant infection (99). Clinically significant biocide resistance in *Staphylococcus* has been reported when therapy for methicillin-resistant *Staphylococcus aureus* (MRSA) eradication (through application of mupirocin ointment and triclosan baths) failed to eliminate the organism from colonized patients (10, 28). Such strains were resistant to both the antibiotic mupirocin and to triclosan (MIC = 2 to 4 μg/ml), and these traits were cotransferable (28). Nonetheless, in a recent screen of nosocomial pathogens, all disinfectants were still found to be potent at in-use concentrations, with the exception of chlorhexidine, against which many strains of *Serratia marcescens, P. aeruginosa,* and *Klebsiella pneumoniae* demonstrated resistance. It was postulated that the loss of efficacy might be due to an efflux system that is controlled by a QAC gene (105).

Since the finding in 1978 of two *P. aeruginosa* plasmids expressing increased tolerance to hexachlorophene (as well as transferable coresistance to sulfadiazine and gentamicin) (117), reports have cited low-level tolerance to a variety of biocides among diverse species of clinical origin (52, 99). MRSA strains are among the most scrutinized, primarily because their plasmids are so promiscuous and they commonly carry a gentamicin resistance plasmid, which also confers a somewhat increased tolerance for biocides such as chlorhexidine and QACs (79). The *qacA* gene was prevalent in French isolates of *S. aureus* (11), and wide dissemination of four or more biocide resistance genes has been reported among Japanese MRSA (88).

Biocide testing of bacteria at MIC levels, which measure subtle rises in tolerance, has historically been regarded as irrelevant, since commercial products contain concentrations that are 100 to 10,000-fold higher than those used in MIC assays and both sensitive and "tolerant" strains will succumb to these high levels (52). However, since the finding of a single target site (FabI) for triclosan (82), MIC susceptibility testing has received more attention. In the late 1990s, surveys reported that 2.2 to 7.5% of MRSA had elevated MICs of triclosan (1 to 8 μg/ml). These studies demonstrated that MRSA were not universally susceptible to the biocide and suggested that tolerance could be increasing and should be tracked (5, 10, 113).

Other investigators have made focused comparisons of antibiotic-susceptible strains and their resistant counterparts in order to specifically determine whether antibiotic resistance might be linked with greater biocide tolerance. Some MRSA have shown more tolerance (1.5- to 4-fold) to chlorhexidine, propamidine, and QACs than methicillin-susceptible *S. aureus* (MSSA), even though their bactericidal efficacy in time kill assays remained relatively unaffected (113). Recently a majority of GISA (glycopeptide-resistant *S. aureus*) isolates from a French hospital showed ~100-fold higher MICs to triclosan than either MRSA or MSSA; both GISA and MRSA exhibited 2 to 64-fold more tolerance to other biocides (benzalkonium chloride and chlorhexidine) (38). In contrast, others have failed to detect consistent differences between any comparisons of susceptible and resistant isolates (52). At present, vancomycin-resistant enterococci (VRE) are no more resistant to biocides than their susceptible counterparts (VSE) in either MIC or time-kill assays (39, 113).

Biocide tolerance to QACs has been also been reported in the coagulase-negative staphylococci (CNS). While CNS were once considered almost exclusively as "commensals," they are now emerging as important opportunistic pathogens. QAC genes were found in more than one-third of Australian clinical CNS in the 1980s (40% of these bore two different *qac* genes residing on two separate plasmids) (63) and over 90% of Norwegian clinical CNS in the 1990s (106). The latter also provided the first evidence for actual linkage of antibiotic and biocide resistance genes on the same multiresistance plasmid in clinical isolates. A positive correlation was found between frequencies of tolerance to the QAC benzalkonium chloride (3 to 8 μg/ml) and resistance to antibiotics, including penicillin, oxacillin, cephalothin, cefuroxime, imipenem, trimethoprim, gentamicin, doxycycline, and fusidic acid. Nearly 25% of the isolates were also tolerant to chlorhexidine and other unrelated biocides. The QAC genes were found to be closely linked physically with beta-lactamase genes (coding for ampicillin resistance) and with a tetracycline resistance gene [*tet*(K)] but were not linked to the other antibiotic resistances. These studies revealed a number of genetic rearrangements that probably resulted from recombination and cointegration activities—all of which were maintained on stable plasmids. (106). The findings provide further credence to the hypothesis that selective pressure from either antibiotic or biocide agents could lead to coselection of both resistance types.

Biocide resistance in the food and agricultural industries

QACs are widely used as disinfectants in the food industry, and there is concern over the increasing incidence of biocide resistance in food-associated bacteria. Biocide rotation of chemically unrelated compounds has been recommended to control the problem (59), but cross-resistance to unrelated compounds could undermine this strategy. Here, as in clinical venues, the findings differ depending on the concentration used (MIC versus MBC) and the species tested. An examination of seven species of CNS, which were derived from poultry plants, red meat plants, baguettes, and chicken carcasses, showed a potential for widespread distribution of QAC resistance determinants in the food industry, and some were also cross-resistant to ampicillin, penicillin, tetracycline, erythromycin, and trimethoprim (44, 46). Like hospital *S. aureus* isolates, food-associated CNS isolates bearing the *qacA/B* determinants also expressed beta-lactam resistance, and half bore the beta-lactamase (*blaZ*) gene on the same plasmid (109). Among *Lactobacillus* species, which can also manifest as opportunistic pathogens, nearly 18% of isolates recovered from food or treated processing surfaces were found to be either tolerant (10 to 29 ug/ml) or resistant (>30 ug/ml) to the QAC disinfectant benzalkonium chloride (BC). BC-resistant strains systematically exhibited cross-resistance (2- to 3-fold higher MIC) to gentamicin and chlorhexidine but not to other antibiotics or biocides (108). In contrast, BC tolerance among gram-negatives and enterococci from food remains low (<1%), and no systematic cross-resistance to other antimicrobials is evident (109).

In agriculture, biocides are used extensively in animal husbandry. A novel plasmid-borne resistance determinant (*qacJ*) has recently been recovered from staphylococci in equines treated with a QAC for chronic skin infections. It expressed a higher level of tolerance to BC (6 μg/ml) but more variable stability than did previous staphylococcal *qac* genes of the Smr family and was found among multiple *Staphylococcus* species, suggesting extensive transfer (16). Isolates of *S. aureus,* derived from cows receiving daily QAC (cetrimide) treatment for 10 years for mastitis control, bore coresistance to cetrimide, tetracycline, and penicillin, but at low frequencies (<5%). Significantly, however, these carried a QAC-bearing plasmid similar to that recovered from human and food isolates (17).

Prospective Assays

The biocide controversy has sparked an increased interest in the undertaking of prospective in situ or microcosm-based studies that would help shed greater insight into the impacts of biocide application as practiced in the community environment. Of particular focus have been household environments in which biocides have been extensively applied, i.e., kitchens and bathrooms, and the oral cavity—a site of long-term intensive biocide use for the control of dental diseases such as caries and periodontitis.

Household studies

Prompted by concerns over the laboratory findings of resistant mutants to pine oil and triclosan, coupled with a dearth of information on antibacterial impacts on antibiotic susceptibility in the home environment, Stuart Levy's laboratory undertook the first evaluation of households that compared the antibiotic susceptibility of bacterial flora from those using and not using biocide-containing hygiene and cleaning products. A snapshot survey was conducted of 38 "normal" households in the Boston and Cincinnati areas to evaluate whether household use of residue-producing biocides affected the frequency and levels of antibiotic resistance in these homes. In this examination of nearly 500 kitchen and bathroom samples, the gram-negative, gram-positive, and total bacterial flora (including potential pathogens) were screened for antibiotic resistance phenotypes. High numbers of bacteria were recovered from the surfaces of both user and nonuser groups, especially from sponges and dish cloths, and from kitchen and bathroom sink drains (73), corroborating the findings of previous investigators (27, 54). In this one-point-in-time survey, no significant differences were noted in the overall recoveries of the different bacterial types or pathogens, despite reported regular biocide use. While single and multiple antibiotic resistances were common in both kitchen and bathroom environments, no consistent trends could be identified that signaled significant differences between the two test groups (B.M. Marshall et al., ms in preparation).

A subsequent collaboration with Elaine Larson's group at Columbia University in New York examined the biocide tolerance of speciated, bacterial isolates cultured from the hands of New York City homemakers participating in a randomized trial of biocide-containing and control products for a one-year period. This provided an ideal opportunity to expand the limited available baseline data on tolerance to biocides in the household environment and to determine any relationships with antibiotic susceptibility. Most of the species, which included *K. pneumoniae, Pseudomonas fluorescens, P. putida, Acinetobacter* spp., *Enterobacter* spp., *S. aureus,* and other commensal *Staphylococcus* spp., exhibited a bimodal distribution of MICs, demonstrating a heterogeneity of tolerance to triclosan. The susceptibilities of gram-positive bacteria reflected a range of MICs (0.03 to 2.0 μg/ml) that corroborated those of earlier reports (5, 10, 28, 113), but there was no correlation between triclosan tolerance and user status, nor was any relationship found with the expression of antibiotic resistance (3). In addition, biocide-tolerant strains could be found both before and after biocide exposure, suggesting that selection may have occurred from biocide usage prior to the start of the study.

These studies have been followed by others of household surfaces and skin flora in which no notable correlations were found between antibiotic susceptibility and tolerance to multiple biocide types (27, 76, 77). Kitchen sink drains constitute a major site of biofilm formation and also receive a high degree of biocide exposure. Working with 17 characterized isolates from drain biofilms, McBain et al. found species-dependent alterations involving both increases and decreases in QAC susceptibility when the isolates were exposed to QACs by multiple passage in vitro. However, these changes were not necessarily reflected in the complex microbial communities of sink-drain biofilms in situ. (77). Biofilm microcosms exposed to triclosan-containing detergent for a total of 6 months exhibited a marked stability of microcosm flora, and there were no significant alterations in microbial susceptibilities to any of several biocides or to six antibiotics tested (76).

Studies of the oral cavity

The environment of the mouth has been the site of a number of studies which have examined the impacts of biocide-containing (primarily triclosan and chlorhexidine) hygiene formulations on the oral supra- and subgingival flora. These encompass both prospective, in situ studies and microcosms mimicking dental caries or periodontal disease states. In general, these studies (which spanned as much as 5 years) demonstrated either a marked suppression, or considerable fluctuation, of putative dental pathogens but no observed increase in tolerance to either the biocides themselves or to antibiotics (29, 112, 115). In a 2003 survey of facultative supragingival microflora from 10 subjects, only 0.0014% demonstrated elevated triclosan MICs (>25 μg/ml), and these data were consistent with those found in a 1991 survey of 144 subjects (98). Interestingly, while the more recent studies confirm the lack of deleterious effects on mouth flora, they do report population shifts that stem from suppression of sensitive populations (including the putative dental pathogens) and clonal expansion of more (intrinsically) tolerant species (74, 75). Saunders observed significant ecological shifts following just 6 hours of exposure to triclosan in which the more sensitive gram-negative anaerobes were markedly suppressed and the more tolerant anaerobic streptococci became dominant (102).

Failure to demonstrate resistance

A number of difficulties inherent in studies performed in situ could account for the failure to find any significant accumulation of resistance to either biocides or antibiotics in the prospective environmental studies. One problem is that there is no clear indication of the length of biocide exposure time required before concluding a positive or negative impact. The New York study, with one year of regular biocide application, is the longest but did not detect a significant increase in risk of biocide tolerance or cross-resistance to antibiotics (3). The selection of test isolates is often subjective, and surveys are frequently limited by the number of isolates that can be examined. Phenotypically identical populations may be genotypically different. Thus the selection of single isolates based on plate phenotypes may miss small subpopulations of resistant strains. Consequently, resistance could be lurking below the levels of detection in these studies. Moreover, detecting significant differences in antibiotic resistance is often difficult because the baseline frequencies of such resistance in the environment are already so high. Findings may also be clouded by the complexities of bacterial dynamics operating in situ but not found in vitro. Theoretically, the use of residue-producing biocides might be expected to result in a prolonged reduction of bacterial counts, particularly if residues accumulated over time. However, the evidence does not support this. Both commensals and opportunistic pathogens readily repopulate household surfaces that are regularly treated with biocides, presumably due to recontamination by heavily contaminated sponges and cloths (27, 54). One possible explanation may reside in the ability of certain bacteria and fungi to degrade or inactivate triclosan and QACs (50, 76, 83, 86). Zones of triclosan clearing have been reported around highly resistant species growing on plates with supersaturated levels of triclosan (76, 83). Triclosan-degrading species have been isolated from sludge and soil (40, 83) and QAC-degraders are abundant in sink drain biofilms (76). The implication is that, under the right conditions, certain organisms may be actively degrading triclosan in situ. The removal of biocide (or reduction to subinhibitory levels) can allow recolonization of the niche by alternate species. For example, MRSA could colonize triclosan-laden plastic that had been precolonized with a triclosan-degrading *Alcaligenes* species (83). It is conceivable that the complexities of biocide flux in the environment, i.e., the build up and subsequent degradation of residues, are in part responsible for the failure thus far to detect statistically significant biocide resistance or antibiotic cross-resistance in in situ evaluations.

CONTINUING CONCERNS SURROUNDING BIOCIDE USE

Environmental Residues

Despite the evidence for biocide degradation by certain species, this process is slow and incomplete, even under optimal conditions (76). Profiles of vertical cores extracted from lake sediment reflect the increases in triclosan popularity and build up between the early 1960s and the present, and point to a marked stability in sediment (110). Due to the great numbers of products that continue to be marketed, used, and ultimately discarded as waste, a concern remains over the generation of biocide gradients in the environment (15, 66). Trace quantities of triclosan have been detected in rivers, streams, wastewater, seawater (110), fish bile, aquatic biota (84), breast milk (1), and blood plasma (49). In theory, these subbactericidal levels have the potential to generate mutations in exposed bacteria, much as those demonstrated in the laboratory. With continued selective pressure exerted by low levels of relatively ubiquitous biocide and antibiotic residues, resistance could theoretically continue to escalate. Moreover, it has been suggested that the selective pressures of both these agents together produce the greatest likelihood of biocide and antibiotic resistance genes appearing jointly on the same plasmid (107). In addition, it has been noted that because of the shared mechanism of efflux pumps which can be triggered by both antibiotics and biocides, removing the selective pressure of either chemical from the environment would be insufficient to decrease resistance, since either one can select for resistance to the other (69).

Like antibiotics, biocides can depress susceptible populations and allow the subsequent expansion of tolerant ones (75). As most healthy environments are characterized by a heterogeneity of species, a disruption and shifting of niche populations results in the reduction of population diversification (74, 76) and is cause for concern, particularly in the proliferation of troublesome pathogens such as *Pseudomonas* and *Stenotrophomonas,* which are difficult to eradicate (69). Selection of biocide-tolerant strains that happen to contain antibiotic resistance genes further compounds the concern. It has been suggested that the overzealous application of biocides in the home environment may be an underlying cause for the emergence of MRSA in the community (67). It was hypothesized that changes occurring as a result of biocide pressures on environmental bacteria permit MRSA to become more dominant in the flora and that recent laboratory findings support this concept (4). When MRSA were subjected to low-level QAC selection, strains emerged with cross-resistance to a variety of beta-lactam antibiotics but not to other antibiotics. These are remarkably similar to the MRSA phenotype which has appeared in the community (8).

Elevated Biocide Tolerance

Recent surveys of large numbers of species isolated from a variety of environments have revealed a range in biocide susceptibility, with small frequencies of outliers that exhibit significantly elevated tolerance (3, 5). Some argue that these increased MICs are of no concern because these strains still succumb to in-use biocide concentrations and therefore present no problems clinically (30). However, among critics of widespread biocide use, elevated MICs are recognized as a "first step" in the progression toward a genuine resistance problem (31, 68, 96). From a retrospective look at the antibiotic era, we now know that rising MIC levels are a signal for the potential development of even greater resistance, which can ultimately manifest as treatment failure. While the available temporal surveys of biocide susceptibilities do not yet point to clear increases in MICs, the question still remains as to whether rising biocide MICs have the potential to attain levels high enough to thwart biocide efficacy in practice.

Biocide Use in the Home

The conflicting evidence emanating from biocide studies leaves unresolved questions concerning the casual use of biocides outside of health care environments. A clear causal link exists between hand hygiene and the transmission of infection (2, 23). While there is ample evidence that the addition of biocide agents to hand soaps is helpful in controlling nosocomial infections in the clinical environment (53), the relationships between hygiene and infectious disease in the home are still debated (37). In purchasing biocide-containing products, the public anticipates an added health benefit. However, at present, the evidence suggests that this casual use does not result in any significant reduction in symptoms of infectious illness (62) and consequently lends support to the argument for reducing home use of residue-producing biocides (119). For general hand hygiene, the available evidence suggests that frequent and thorough washing with regular (i.e., nonbiocidal) soap is sufficient (61, 72). However, for home cleaning, the conclusions are less clear. Hot water and detergent alone have not demonstrated significant microbial reduction in this environment (18, 26, 54). Although use of a "targeted" approach to household cleaning has not yet been clearly defined for the public, the International Scientific Forum on Home Hygiene (IFH) suggests that such an approach would reserve biocide use for preventing bacterial transfer in high risk activities involving probable cross-contamination and infection, for example, between food items, from hands to food, and vice versa, with particular attention to the decontamination of sponges and dishcloths, which are known to harbor and spread large numbers of bacteria (14). Levy has suggested that biocides should preferably be reserved for the care of at-home patients in extended recovery situations, particularly those who have overly taxed or depleted immune systems and thus are more vulnerable to infection (66). Guidelines for biocide use in the home are currently under development (http://ifh-homehygiene.org/2003/index.html).

Recent findings have called into question the efficacy of biocide that is incorporated into plastics. A study of control and triclosan-incorporated plastic showed no significant differences in the biofilm populations which formed after 3 weeks exposure to drinking water. While the actual triclosan concentration at the biofilm level could not be measured, it was determined that trace quantities (0.012 μg/ml) were desorbed into the immediate surrounding medium (56). This concentration would be ineffective against all but the most sensitive bacteria, and therefore could potentially act, not only to select and promote the growth of mutants, but also to select for proliferation of the more tolerant species present in the environment.

CONCLUSIONS

Presently, the most significant biocide resistance is that expressed by intrinsic protective mechanisms afforded by biofilms and strains expressing high-level efflux pumps (25, 79, 103). The accumulated laboratory evidence shows a range of capacities for the development of low-level biocide tolerance, which in some species bears cross-resistance to unrelated biocide chemicals and to clinically important antibiotics (Table 2). Given these indicators and the wide dissemination of biocide genes on transferable plasmids in diverse environments, it is somewhat surprising that biocide resistance is not more evident in the natural environment, particularly with the finding of ubiquitous residue levels of biocide contamination.

Of the surveys performed in situ which have examined a relationship between biocide and antibiotic resistance, no clear pattern of linkage has yet emerged in either the clinical or the household settings. For the moment, this is reassuring and points to antibiotic use and misuse as the major contributors to the present antibiotic resistance crisis.

As a result of the inconsistencies in the total accumulated evidence on biocide resistance, experts have suggested that clinical isolates should be regularly evaluated for rising tolerance (31), but it would seem equally prudent to watch these levels among household and other community environments where sublethal concentrations are likely and where conditions may be optimal for resistance selection. Early detection may thwart a more serious problem later. To this end, the FDA concluded that, while there is no evidence that biocides have contributed significantly to the problem of antibiotic resistance, a surveillance program is deemed advisable (52).

To avoid compromising the efficacy of antimicrobials, experts have concluded that biocides should not be used indiscriminately or frivolously and suggest application under conditions which promote rapid and effective kill and in a manner that avoids the build-up of residues that could potentially select for resistant strains (14). It is not clear, however, how this build-up is to be avoided. To circumvent the problems posed by these agents, the use of nonresidue biocides (chlorine, alcohols, peroxides) is preferable (13). The American Medical Association has urged further evaluation of consumer biocides and regulation of those for which acquired resistance mechanisms have been demonstrated (119).

While most antibiotic resistance developed rather quickly after introduction of each new drug, nearly 40 years passed before vancomycin resistance surfaced as a clinical problem. It has been proposed that, given sufficient exposure to biocides over time, biocide resistance of the acquired type may eventually emerge (65). Complex factors operating in the environmental arena may be complicating the development of biocide resistance and obscuring or delaying its detection. If time is a critical element in this process, then perhaps "only time will tell." In the meantime, it would seem that surveillance and prudent use are worthwhile endeavors.

IN APPRECIATION

When I arrived as a newly graduated Medical Technologist in 1976, little did I know the depth and range of challenges I would encounter in the coming 29 years of working in the laboratory of Stuart Levy. From scatological assays of farm animals and of humans in clinical trials to molecular tracking of antibiotic resistance genes, surveying food and household bacteria, writing, editing and creating intricate graphics, the stimulating opportunities afforded from working alongside this brilliant and visionary mind have kept me commuting 155 miles each day to find what exciting experimental venture awaits next on the horizon. Stuart, thank you for providing me a rewarding and fascinating journey of inquiry and learning! (B. M. M.)

Acknowledgements. This chapter was written by B.M.M., with the exception of the section on Mechanisms of Triclosan Action (L.M.M.) and the appendix (M.C.M.)

APPENDIX

High School Student Discovered Mutants Resistant to Pine Oil, Which Were Also Resistant to Multiple Antibiotics

Merri C. Moken

As a freshman in high school in 1992, I was searching for an innovative and exciting scientific investigation to perform for my high school science fair. Sparked by a Dow Scrubbing Bubbles commercial, it began relatively simply. Seeing all of those little cartoon brushes running around on the television, cleaning and sanitizing everything and magically leaving sparkling paths behind them, prompted me to question how we know that disinfectants really work. Little did I realize at the time that this research would balloon into a more than 4-year study and would ultimately be the first to prove that a disinfectant can select for bacterial mutants that are resistant to antibiotics because of the increased expression of *marA*.

In the process of evaluating 10 different disinfectants against both gram-positive and gram-negative bacteria via the agar diffusion method, I found that bacterial colonies grew in the zones of inhibition caused by some of the disinfectants. I thought that they might be mutants that had increased resistance to disinfectants. This proved to be true; the astonishing fact was that people all over the world were using products that could select for resistant bacteria.

I wondered whether these disinfectant-"induced" bacterial mutants could also be resistant to other chemical agents, in particular antibiotics, and whether disinfectants might be actually exacerbating the antibiotic resistance problem which had come to the forefront in the news. During the second year of my research, I found that mutants selected on each of several disinfectants exhibited a marked increase in antibiotic resistance when compared with the wild type of the same bacteria. Gram-negative bacterial mutants were on average 50% more resistant to antibiotics, and gram-positive bacterial mutants were on average 28% more resistant to antibiotics than the wild-type bacteria. This finding was exhilarating!

In the third year of my research, I set about finding how the disinfectants caused this antibiotic resistance. I concentrated on *E. coli* bacteria and pine oil. At that time there was an increased awareness in the press about *E. coli* bacteria and the sicknesses attributable thereto. Many of my antibiotic-resistant mutants resulted from selection on disinfectants containing pine oil as an active ingredient. Such pine oil-based disinfectants produced bacterial mutants that were resistant to multiple antibiotics. At this point, I had read publications on outer membrane proteins and *mar* by Drs. Stuart Levy and Laura McMurry and decided to contact them. I called Laura, explaining my research and hypotheses and asked her to send me several strains of bacteria expressing different outer membrane proteins, which I could use as standards of comparison in my tests and on electrophoresis gels. I was thrilled to get Laura's kind response and the bacterial strains. The mutants were analyzed using polyacrylamide electrophoresis, and this analysis showed that they seemed to lack *ompF* and had additional proteins (overexpressed) that corresponded to those present in some of the *mar* strains. These findings and the antibiotic resistances themselves led me to think that the mutants selected by the disinfectants might also overexpress *marA*.

To see whether the *mar* gene was the cause of this antibiotic resistance, I used P1 bacteriophage, sent to me by Drs. Levy and McMurry, to delete the *mar* locus by transduction of a kanamycin cassette that replaced the locus. The P1 transductants no longer were resistant to either pine oil or antibiotics, confirming that *mar* was responsible for the resistances. Proof of this hypothesis was thrilling, especially since this had never been previously discovered.

During this last year of my research, I visited with Drs. Levy and McMurry. It was exciting to finally meet and talk to the kind people with whom I had been corresponding. Dr. Levy, Dr. McMurry, and I met for hours, discussing my research and findings. Dr. Levy was gracious, enthusiastic, warm, and welcoming. For a high school student to receive this type of a welcome from someone of his stature was overwhelming and very fulfilling. Dr. Levy kindly offered to have his group try to replicate my research, following my protocols and procedures, and if successful, coauthor a paper with me. I eagerly agreed. To have such a prestigious team of scientists provide their time, consultation, and guidance was an honor, and their validation enabled my research to be taken seriously in the scientific community. Until such a time, this work by a high school student was viewed with skepticism at best and, in general, with disbelief. My research was ultimately recognized and rewarded in both the Intel International Science and Engineering Fair (Grand Winner—Microbiology) and the Westinghouse Science Talent Search Competition (Finalist). Dr. McMurry did reproduce my results with pine oil, and after several additional tests, our findings were published a year later in *Antimicrobial Agents and Chemotherapy* (85).

Note: Work on the disinfectant pine oil was begun by high school student Merri Moken in Morristown, N.J. The role of the Levy laboratory was to confirm and extend her findings. We were indeed able to select pine oil-resistant mutants of a different strain of *E. coli*, and as she predicted, they were also resistant to low levels of multiple antibiotics. Known Mar mutants (selected on tetracycline) from the Levy lab in turn were resistant to pine oil, and when the *mar* locus was inactivated in the new pine oil mutants, the resistances were all lost (85). All but one of the pine oil mutants overexpressed the *marRAB* transcript, as expected for a Mar mutant. As was the case for the antibiotic resistances, the resistance to pine oil depended on the multidrug efflux pump AcrAB (up-regulated by MarA).

REFERENCES

1. **Adolfsson-Erici, M., M. Pettersson, J. Parkkonen, and J. Sturve.** 2002. Triclosan, a commonly used bactericide found in human milk and in the aquatic environment in Sweden. *Chemosphere* **46:**1485–1489.
2. **Aiello, A. E., and E. L. Larson.** 2002. What is the evidence for a causal link between hygiene and infections? *Lancet Infect. Dis.* **2:**103–110.
3. **Aiello, A. E., B. Marshall, S. B. Levy, P. Della-Latta, and E. Larson.** Submitted for publication.
4. **Akimitsu, N., H. Hamamoto, R. Inoue, M. Shoji, A. Akamine, K. Takemori, N. Hamasaki, and K. Sekimizu.** 1999. Increase in resistance of methicillin-resistant *Staphylococcus aureus* to beta-lactams caused by mutations conferring resistance to benzalkonium chloride, a disinfectant widely used in hospitals. *Antimicrob. Agents Chemother.* **43:**3042–3043.
5. **Al-Doori, Z., D. Morrison, G. Edwards, and C. Gemmell.** 2003. Susceptibility of MRSA to triclosan. *J. Antimicrob. Chemother.* **51:**185–186.
6. **al-Masaudi, S. B., M. J. Day, and A. D. Russell.** 1991. Effect of some antibiotics and biocides on plasmid transfer in *Staphylococcus aureus*. *J. Appl. Bacteriol.* **71:**239–243.
7. **Anon.** 2001. Disinfectants in Consumer Products Publication #2001/05E. Health Council of the Netherlands.
8. **Anon.** 1999. Four pediatric deaths from community-acquired methicillin-resistant *Staphylococcus aureus:* Minnesota and North Dakota 1997-1999. *Morb. Mortal. Wkly. Rep.* **48:**707–710.
9. **Baldock, C., J. B. Rafferty, S. E. Sedelnikova, P. J. Baker, A. R. Stuitje, A. R. Slabas, T. R. Hawkes, and D. W. Rice.** 1996. A mechanism of drug action revealed by structural studies of enoyl reductase. *Science* **274:**2107–2110.
10. **Bamber, A. I., and T. J. Neal.** 1999. An assessment of triclosan susceptibility in methicillin-resistant and methicillin-sensitive *Staphylococcus aureus*. *J. Hosp. Infect.* **41:**107–109.
11. **Behr, H., M. E. Reverdy, C. Mabilat, J. Freney, and J. Fleurette.** 1994. [Relationship between the level of minimal inhibitory concentrations of five antiseptics and the presence of qacA gene in *Staphylococcus aureus*]. *Pathol. Biol.* (Paris) **42:**438–444.
12. **Bergler, H., P. Wallner, A. Ebeling, B. Leitinger, S. Fuchsbichler, H. Aschauer, G. Kollenz, G. Hogenauer, and F. Turnowsky.** 1994. Protein EnvM is the NADH-dependent enoyl-ACP reductase (FabI) of *Escherichia coli*. *J. Biol. Chem.* **269:**5493–5496.
13. **Beumer, R., S. Bloomfield, M. Exner, G. Fara, K. Nath, and E. Scott** 2003, posting date. Biocide usage and antimicrobial resistance in home settings: an update. International Scientific Forum on Home Hygiene (IFH). [Online.]
14. **Beumer, R., S. Bloomfield, M. Exner, G. Fara, K. Nath, and E. Scott** 2000, posting date. Microbial Resistance and Biocides. International Scientific Forum on Home Hygiene (IFH). [Online.]
15. **Bhargava, H. N., and P. A. Leonard.** 1996. Triclosan: applications and safety. *Am. J. Infect. Control* **24:**209–218.
16. **Bjorland, J., T. Steinum, M. Sunde, S. Waage, and E. Heir.** 2003. Novel plasmid-borne gene qacJ mediates resistance to quaternary ammonium compounds in equine *Staphylococcus aureus, Staphylococcus simulans,* and *Staphylococcus intermedius*. *Antimicrob. Agents Chemother.* **47:**3046–3052.

17. **Bjorland, J., M. Sunde, and S. Waage.** 2001. Plasmid-borne smr gene causes resistance to quaternary ammonium compounds in bovine *Staphylococcus aureus. J. Clin. Microbiol.* **39:**3999–4004.
18. **Bloomfield, S. F.** 2002. Significance of biocide usage and antimicrobial resistance in domiciliary environments. *J. Appl. Microbiol.* **92:**144S–157S.
19. **Braoudaki, M., and A. C. Hilton.** 2004. Adaptive resistance to biocides in *Salmonella enterica* and *Escherichia coli* O157 and cross-resistance to antimicrobial agents. *J. Clin. Microbiol.* **42:**73–78.
20. **Braoudaki, M., and A. C. Hilton.** 2004. Low level of cross-resistance between triclosan and antibiotics in *Escherichia coli* K-12 and *E. coli* O55 compared to *E. coli* O157. *FEMS Microbiol. Lett.* **235:**305–309.
21. **Brenwald, N. P., and A. P. Fraise.** 2003. Triclosan resistance in methicillin-resistant *Staphylococcus aureus* (MRSA). *J. Hosp. Infect.* **55:**141–144.
22. **Brooun, A., S. Liu, and K. Lewis.** 2000. A dose-response study of antibiotic resistance in *Pseudomonas aeruginosa* biofilms. *J. Antimicrob. Chemother.* **44:**640–646.
23. **Bryan, J. L., J. Cohran, and E. L. Larson.** 1994. Handwashing: a ritual revisited., p. 163–178. *In* W. A. Rutala (ed.), *Chemical Germicides in Healthcare.* Association for Professionals in Infection Control and Epidemiology, Inc., Washington, D.C.
24. **Cantrell, F.** 2001. The effect of prolonged exposure to triclosan on staphylococci. Presented at the 101st Annual Meeting of the American Society for Microbiology.
25. **Chuanchuen, R., R. R. Karkhoff-Schweizer, and H. P. Schweizer.** 2003. High-level triclosan resistance in *Pseudomonas aeruginosa* is solely a result of efflux. *Am. J. Infect. Control* **31:**124–127.
26. **Cogan, T. A., J. Slader, S. F. Bloomfield, and T. J. Humphrey.** 2002. Achieving hygiene in the domestic kitchen: the effectiveness of commonly used cleaning procedures. *J. Appl. Microbiol.* **92:**885–892.
27. **Cole, E. C., R. M. Addison, J. R. Rubino, K. E. Leese, P. D. Dulaney, M. S. Newell, J. Wilkins, D. J. Gaber, T. Wineinger, and D. A. Criger.** 2003. Investigation of antibiotic and antibacterial agent cross-resistance in target bacteria from homes of antibacterial product users and nonusers. *J. Appl. Microbiol.* **95:**664–676.
28. **Cookson, B. D., H. Farrelly, P. Stapleton, R. P. Garvey, and M. R. Price.** 1991. Transferable resistance to triclosan in MRSA. *Lancet* **337:**1548–1549.
29. **Cullinan, M. P., S. M. Hamlet, B. Westerman, J. E. Palmer, M. J. Faddy, and J. G. Seymour.** 2003. Acquisition and loss of *Porphyromonas gingivalis, Actinobacillus actinomycetemcomitans* and *Prevotella intermedia* over a 5-year period: effect of a triclosan/copolymer dentifrice. *J. Clin. Periodontol.* **30:**532–541.
30. **Favero, M. S.** 2002. Products containing biocides: perceptions and realities. *J. Appl. Microbiol.* **92** (Suppl.):72S–77S.
31. **Fraise, A. P.** 2002. Biocide abuse and antimicrobial resistance—a cause for concern? *J. Antimicrob. Chemother.* **49:**11–12.
32. **Fraise, A. P.** 2002. Susceptibility of antibiotic-resistant cocci to biocides. *J. Appl. Microbiol.* **31:**158S–162S.
33. **Furia, T. E., and A. G. Schenkel.** 1968. New Broad Spectrum Bacteriostat. *Soap and Chemical Specialties* **44:**47–50,116–122.
34. **Gandhi, I., A. D. Sawant, L. A. Wilson, and D. G. Ahearn.** 1993. Adaptation and growth of *Serratia marcescens* in contact lens disinfectant solutions containing chlorhexidine gluconate. *Appl. Environ. Microbiol.* **59:**183–188.
35. **George, A. M., and S. B. Levy.** 1983. Amplifiable resistance to tetracycline, chloramphenicol, and other antibiotics in *Escherichia coli:* involvement of a non-plasmid-determined efflux of tetracycline. *J. Bacteriol.* **155:**531–540.
36. **Gerba, C. P., and P. Rusin.** 2001. Relationship between the use of antiseptics/disinfectants and the development of antimicrobial resistance, p. 187–194. *In* W. A. Rutala (ed.), *Disinfection, Sterilization and Antisepsis: Principles and Practices in Healthcare Facilities.* Association for Professionals in Infection Control and Epidemiology, Washington, D.C.
37. **Gilbert, P., A. J. McBain, and S. F. Bloomfield.** 2002. Biocide abuse and antimicrobial resistance: being clear about the issues. *J. Antimicrob. Chemother.* **50:**137–139; author reply 139–40.
38. **Goldstein, F. W., C. Soria, A. Ly, and M. D. Kitzis.** 2004. Very high prevalence of resistance to tricolsan and other antiseptics among French glycopeptide-intermediate *S. aureus* (GISA). Presented at the 44th Interscience Conference on Antimicrobial Agents and Chemotherapy, Washington, D.C.
39. **Haines, K. A., D. A. Klein, G. McDonnell, and D. Pretzer.** 1997. Could antibiotic-resistant pathogens be cross-resistant to hard-surface disinfectants? *Am. J. Infect. Control* **25:**439–441.
40. **Hay, A. G., P. M. Dees, and G. S. Sayler.** 2001. Growth of a bacterial consortium on triclosan. *FEMS Microbiol. Ecol.* **36:**105–112.
41. **Heath, R. J., J. Li, G. E. Roland, and C. O. Rock.** 2000. Inhibition of the Staphylococcus aureus NADPH-dependent enoyl-acyl carrier protein reductase by triclosan and hexachlorophene. *J. Biol. Chem.* **275:**4654–4659.
42. **Heath, R. J., N. Su, C. K. Murphy, and C. O. Rock.** 2000. The enoyl-[acyl-carrier-protein] reductases FabI and FabL from *Bacillus subtilis. J. Biol. Chem.* **275:**40128–40133.
43. **Heath, R. J., Y. T. Yu, M. A. Shapiro, E. Olson, and C. O. Rock.** 1998. Broad spectrum antimicrobial biocides target the FabI component of fatty acid synthesis. *J. Biol. Chem.* **273:** 30316–30320.
44. **Heir, E., G. Sundheim, and A. L. Holck.** 1999. Identification and characterization of quaternary ammonium compound resistant staphylococci from the food industry. *Int J. Food Microbiol.* **48:**211–219.
45. **Heir, E., G. Sundheim, and A. L. Holck.** 1999. The *qacG* gene on plasmid pST94 confers resistance to quaternary ammonium compounds in staphylococci isolated from the food industry. *J. Appl. Microbiol.* **86:**378–388.
46. **Heir, E., G. Sundheim, and A. L. Holck.** 1995. Resistance to quaternary ammonium compounds in *Staphylococcus* spp. isolated from the food industry and nucleotide sequence of the resistance plasmid pST827. *J. Appl. Bacteriol.* **79:**149–156.
47. **Heir, E., G. Sundheim, and A. L. Holck.** 1998. The *Staphylococcus qacH* gene product: a new member of the SMR family encoding multidrug resistance. *FEMS Microbiol. Lett.* **163:**49–56.
48. **Hoang, T. T., and H. P. Schweizer.** 1999. Characterization of *Pseudomonas aeruginosa* enoyl-acyl carrier protein reductase (FabI): a target for the antimicrobial triclosan and its role in acylated homoserine lactone synthesis. *J. Bacteriol.* **181:**5489–5497.
49. **Hovander, L., T. Malmberg, M. Athanasiadou, I. Athanassiadis, S. Rahm, A. Bergman, and E. K. Wehler.** 2002. Identification of hydroxylated PCB metabolites and other phenolic halogenated pollutants in human blood plasma. *Arch. Environ. Contam. Toxicol.* **42:**105–117.
50. **Hundt, K., D. Martin, E. Hammer, U. Jonas, M. K. Kindermann, and F. Schauer.** 2000. Transformation of triclosan by *Trametes versicolor* and *Pycnoporus cinnabarinus. Appl. Environ. Microbiol.* **66:**4157–4160.

51. **Ishikawa, S., F. Matsumura, F. Yoshizako, and T. Tsuchido.** 2002. Characterization of a cationic surfactant-resistant mutant isolated spontaneously from *Escherichia coli*. *J. Appl. Microbiol.* **92:**261–268.
52. **Jones, R. D.** 1999. Bacterial resistance and topical antimicrobial wash products. *Am. J. Infect. Control* **27:**351–363.
53. **Jones, R. D., H. B. Jampani, J. L. Newman, and A. S. Lee.** 2000. Triclosan: a review of effectiveness and safety in health care settings. *Am. J. Infect. Control* **28:**184–196.
54. **Josephson, K. L., J. R. Rubino, and I. L. Pepper.** 1997. Characterization and quantification of bacterial pathogens and indicator organisms in household kitchens with and without the use of a disinfectant cleaner. *J. Appl. Microbiol.* **83:**737–750.
55. **Joynson, J. A., B. Forbes, and R. J. Lambert.** 2002. Adaptive resistance to benzalkonium chloride, amikacin and tobramycin: the effect on susceptibility to other antimicrobials. *J. Appl. Microbiol.* **93:**96–107.
56. **Junker, L. M., and A. G. Hay.** 2004. Effects of triclosan incorporation into ABS plastic on biofilm communities. *J. Antimicrob. Chemother.* **53:**989–996.
57. **Kalyon, B. D., and U. Olgun.** 2001. Antibacterial efficacy of polymers containing triclosan and other antimicrobial additives: REPLY. *Amer. J. Infect. Cont.* **29:**429–430.
58. **Kuo, M. R., H. R. Morbidoni, D. Alland, S. F. Sneddon, B. B. Gourlie, M. M. Staveski, M. Leonard, J. S. Gregory, A. D. Janjigian, C. Yee, J. M. Musser, B. Kreiswirth, H. Iwamoto, R. Perozzo, W. R. Jacobs, Jr., J. C. Sacchettini, and D. A. Fidock.** 2003. Targeting tuberculosis and malaria through inhibition of enoyl reductase: compound activity and structural data. *J. Biol. Chem.* **278:**20851–20859.
59. **Langsrud, S., G. Sundheim, and R. Borgmann-Strahsen.** 2003. Intrinsic and acquired resistance to quaternary ammonium compounds in food-related *Pseudomonas* spp. *J. Appl. Microbiol.* **95:**874–882.
60. **Langsrud, S., G. Sundheim, and A. L. Holck.** 2004. Cross-resistance to antibiotics of *Escherichia coli* adapted to benzalkonium chloride or exposed to stress-inducers. *J. Appl. Microbiol.* **96:**201–208.
61. **Larson, E., A. Aiello, L. V. Lee, P. Della-Latta, C. Gomez-Duarte, and S. Lin.** 2003. Short- and long-term effects of handwashing with antimicrobial or plain soap in the community. *J. Commun. Hlth.* **28:**139–150.
62. **Larson, E. L., S. X. Lin, C. Gomez-Pichardo, and P. Della-Latta.** 2004. Effect of antibacterial home cleaning and handwashing products on infectious disease symptoms: a randomized, double-blind trial. *Ann. Intern. Med.* **140:**321–329.
63. **Leelaporn, A., I. T. Paulsen, J. M. Tennent, T. G. Littlejohn, and R. A. Skurray.** 1994. Multidrug resistance to antiseptics and disinfectants in coagulase-negative staphylococci. *J. Med. Microbiol.* **40:**214–220.
64. **Levy, C. W., A. Roujeinikova, S. Sedelnikova, P. J. Baker, A. R. Stuitje, A. R. Slabas, D. W. Rice, and J. B. Rafferty.** 1999. Molecular basis of triclosan activity. *Nature* **398:**383–384.
65. **Levy, S. B.** 2002. Active efflux, a common mechanism for biocide and antibiotic resistance. *J. Appl. Microbiol.* **92:**65S–71S.
66. **Levy, S. B.** 2001. Antibacterial household products: cause for concern. *Emerg. Infect. Dis.* **7:**512–515.
67. **Levy, S. B.** 2000. Antibiotic and antiseptic resistance: impact on public health. *Pediatr. Infect. Dis. J.* **19:**S120–S122.
68. **Levy, S. B.** 2002. The Antibiotic Paradox: How the Misuse of Antibiotics Destroys Their Curative Powers. Perseus Publishing, Cambridge, Mass.
69. **Levy, S. B.** 2002. Antimicrobial consumer products: where's the benefit? What's the risk? *Arch. Dermatol.* **138:**1087–1088.
70. **Littlejohn, T. G., I. T. Paulsen, M. T. Gillespie, J. M. Tennent, M. Midgley, I. G. Jones, A. S. Purewal, and R. A. Skurray.** 1992. Substrate specificity and energetics of antiseptic and disinfectant resistance in *Staphylococcus aureus*. *FEMS Microbiol. Lett.* **74:**259–265.
71. **Loughlin, M. F., M. V. Jones, and P. A. Lambert.** 2002. *Pseudomonas aeruginosa* cells adapted to benzalkonium chloride show resistance to other membrane-active agents but not to clinically relevant antibiotics. *J. Antimicrob. Chemother.* **49:**631–639.
72. **Luby, S. P., M. Agboatwalla, J. Painter, A. Altaf, W. L. Billhimer, and R. M. Hoekstra.** 2004. Effect of intensive handwashing promotion on childhood diarrhea in high-risk communities in Pakistan: a randomized controlled trial. *J. Am. Med. Assoc.* **291:**2547–2554.
73. **Marshall, B. M., E. Robleto, T. Dumont, W. Billhimer, K. Wiandt, B. Keswick, and S. B. Levy.** 2003. The frequency of bacteria and antibiotic resistance in homes that use and do not use surface antibacterial agents. Presented at the Annual 103rd Meeting of the American Society for Microbiology, Washington, D.C.
74. **McBain, A. J., R. G. Bartolo, C. E. Catrenich, D. Charbonneau, R. G. Ledder, and P. Gilbert.** 2003. Effects of a chlorhexidine gluconate-containing mouthwash on the vitality and antimicrobial susceptibility of in vitro oral bacterial ecosystems. *Appl. Environ. Microbiol.* **69:**4770–4776.
75. **McBain, A. J., R. G. Bartolo, C. E. Catrenich, D. Charbonneau, R. G. Ledder, and P. Gilbert.** 2003. Effects of triclosan-containing rinse on the dynamics and antimicrobial susceptibility of in vitro plaque ecosystems. *Antimicrob. Agents Chemother.* **47:**3531–3538.
76. **McBain, A. J., R. G. Bartolo, C. E. Catrenich, D. Charbonneau, R. G. Ledder, B. B. Price, and P. Gilbert.** 2003. Exposure of sink drain microcosms to triclosan: population dynamics and antimicrobial susceptibility. *Appl. Environ. Microbiol.* **69:**5433–5422.
77. **McBain, A. J., R. G. Ledder, L. E. Moore, C. E. Catrenich, and P. Gilbert.** 2004. Effects of quaternary-ammonium-based formulations on bacterial community dynamics and antimicrobial suseptibility. *Appl. Environ. Microbiol.* **70:**3449–3456.
78. **McBain, A. J., R. G. Ledder, P. Sreenivasan, and P. Gilbert.** 2004. Selection for high-level resistance by chronic triclosan exposure is not universal. *J. Antimicrob. Chemother.* **53:**772–777.
79. **McDonnell, G., and A. D. Russell.** 1999. Antiseptics and disinfectants: activity, action, and resistance. *Clin. Microbiol. Rev.* **12:**147–179.
80. **McMurry, L. M., P. F. McDermott, and S. B. Levy.** 1999. Genetic evidence that InhA of Mycobacterium smegmatis is a target for triclosan. *Antimicrob. Agents. Chemother.* **43:**711–713.
81. **McMurry, L. M., M. Oethinger, and S. B. Levy.** 1998. Overexpression of marA, soxS, or acrAB produces resistance to triclosan in laboratory and clinical strains of *Escherichia coli*. *FEMS Microbiol. Lett.* **166:**305–309.
82. **McMurry, L. M., M. Oethinger, and S. B. Levy.** 1998. Triclosan targets lipid synthesis. *Nature* **394:**531–532.
83. **Meade, M. J., R. L. Waddell, and T. M. Callahan.** 2001. Soil bacteria *Pseudomonas putida* and *Alcaligenes xylosoxidans* subsp. denitrificans inactivate triclosan in liquid and solid substrates. *FEMS Microbiol. Lett.* **204:**45–48.
84. **Miyazaki, T., T. Yamagishi, and M. Matsumoto.** 1984. Residues of 4-chloro-1-(2,4-dichlorophenoxy)-2 methoxy-

benzene (triclosan methyl) in aquatic biota. *Contam. Toxicol.* **32:**227–232.

85. **Moken, M. C., L. M. McMurry, and S. B. Levy.** 1997. Selection of multiple-antibiotic-resistant (mar) mutants of *Escherichia coli* by using the disinfectant pine oil: roles of the mar and acrAB loci. *Antimicrob. Agents Chemother.* **41:**2770–2772.
86. **Nishihara, T., T. Okamoto, and N. Nishiyama.** 2000. Biodegradation of didecyldimethylammonium chloride by *Pseudomonas fluorescens* TN4 isolated from activated sludge. *J. Appl. Microbiol.* **88:**641–647.
87. **Nishiyama, N., Y. Toshima, and Y. Ikeda.** 1995. Biodegradation of alkyltrimethylammonium salts in activated sludge. *Chemosphere* **30:**593–603.
88. **Noguchi, N., M. Hase, M. Kitta, M. Sasatsu, K. Deguchi, and M. Kono.** 1999. Antiseptic susceptibility and distribution of antiseptic-resistance genes in methicillin-resistant *Staphylococcus aureus. FEMS Microbiol. Lett.* **172:**247–253.
89. **Noguchi, N., M. Tamura, K. Narui, K. Wakasugi, and M. Sasatsu.** 2002. Frequency and genetic characterization of multidrug-resistant mutants of *Staphylococcus aureus* after selection with individual antiseptics and fluoroquinolones. *Biolog. Pharmaceut. Bull.* **25:**1129–1132.
90. **Parikh, S. L., G. Xiao, and P. J. Tonge.** 2000. Inhibition of InhA, the enoyl reductase from *Mycobacterium tuberculosis,* by triclosan and isoniazid. *Biochemistry* **39:**7645–7650.
91. **Patrauchan, M. A., and P. J. Oriel.** 2003. Degradation of benzyldimethylalkylammonium chloride by *Aeromonas hydrophila* sp. *J. Appl. Microbiol.* **94:**266–272.
92. **Paulsen, I. T., T. G. Littlejohn, P. Radstrom, L. Sundstrom, O. Skold, G. Swedberg, and R. A. Skurray.** 1993. The 3′ conserved segment of integrons contains a gene associated with multidrug resistance to antiseptics and disinfectants. *Antimicrob. Agents Chemother.* **37:**761–768.
93. **Pearce, H., S. Messager, and J. Y. Maillard.** 1999. Effect of biocides commonly used in the hospital environment on the transfer of antibiotic-resistance genes in *Staphylococcus aureus. J. Hosp. Infect.* **43:**101–107.
94. **Price, C. T., V. K. Singh, R. K. Jayaswal, B. J. Wilkinson, and J. E. Gustafson.** 2002. Pine oil cleaner-resistant *Staphylococcus aureus:* reduced susceptibility to vancomycin and oxacillin and involvement of SigB. *Appl. Environ. Microbiol.* **68:**5417–5421.
95. **Randall, L. P., S. W. Cooles, L. J. Piddock, and M. J. Woodward.** 2004. Effect of triclosan or a phenolic farm disinfectant on the selection of antibiotic-resistant *Salmonella enterica. J. Antimicrob. Chemother.* **54:**621–627.
96. **Randall, L. P., and M. J. Woodward.** 2002. The multiple antibiotic resistance (mar) locus and its significance. *Res. Vet. Sci.* **72:**87–93.
97. **Regos, J., and H. R. Hitz.** 1974. Investigations on the mode of action of Triclosan, a broad spectrum antimicrobial agent. *Zbl. Bakt. Hyg.* **226:**390–401.
98. **Reynolds, H. S., P. K. Sreenivasan, R. Subramanyam, V. I. Haraszthy, D. Cummins, and J. J. Zambon.** 2003. Media and method-dependent variations in minimal inhibitory concentrations of triclosan and 2-t-butyl-5-(4-t-butylphenyl)-phenol. Presented at the Annual Meeting of the American Society for Microbiology, Washington D.C.
99. **Russell, A. D.** 2002. Introduction of biocides into clinical practice and the impact on antibiotic-resistant bacteria. *Symp. Ser. Soc. Appl. Microbiol.* **31:**121S–135S.
100. **Russell, A. D., U. Tattawasart, J. Y. Maillard, and J. R. Furr.** 1998. Possible link between bacterial resistance and use of antibiotics and biocides. *Antimicrob. Agents Chemother.* **42:**2151.
101. **Sanchez, P., E. Moreno, and J. L. Martinez.** 2005. The biocide triclosan selects *Stenotrophomonas maltophilia* mutants that overproduce the SmeDEF multidrug efflux pump. *Antimicrob. Agents Chemother.* **49:**781–782.
102. **Saunders, K. A., J. Greenman, and C. McKenzie.** 2000. Ecological effects of triclosan and triclosan monophosphate on defined mixed cultures of oral species grown in continuous culture. *J. Antimicrob. Chemother.* **45:**447–452.
103. **Schweizer, H. P.** 1998. Intrinsic resistance to inhibitors of fatty acid biosynthesis in *Pseudomonas aeruginosa* is due to efflux: application of a novel technique for generation of unmarked chromosomal mutations for the study of efflux systems. *Antimicrob. Agents Chemother.* **42:**394–398.
104. **Sharma, S., T. N. Ramya, A. Surolia, and N. Surolia.** 2003. Triclosan as a systemic antibacterial agent in a mouse model of acute bacterial challenge. *Antimicrob. Agents Chemother.* **47:**3859–3866.
105. **Shimizu, M., K. Okuzumi, A. Yoneyama, T. Kunisada, M. Araake, H. Ogawa, and S. Kimura.** 2002. In vitro antiseptic susceptibility of clinical isolates from nosocomial infections. *Dermatol.* **204** (Suppl. 1):21–27.
106. **Sidhu, M. S., E. Heir, T. Leegaard, K. Wiger, and A. Holck.** 2002. Frequency of disinfectant resistance genes and genetic linkage with beta-lactamase transposon Tn552 among clinical staphylococci. *Antimicrob. Agents Chemother.* **46:**2797–2803.
107. **Sidhu, M. S., E. Heir, H. Sorum, and A. Holck.** 2001. Genetic linkage between resistance to quaternary ammonium compounds and beta-lactam antibiotics in food-related *Staphylococcus* spp. *Microb. Drug Resist.* **7:**363–71.
108. **Sidhu, M. S., S. Langsrud, and A. Holck.** 2001. Disinfectant and antibiotic resistance of lactic acid bacteria isolated from the food industry. *Microb. Drug. Resist.* **7:**73–83.
109. **Sidhu, M. S., H. Sorum, and A. Holck.** 2002. Resistance to quaternary ammonium compounds in food-related bacteria. *Microb. Drug Resist.* **8:**393–399.
110. **Singer, H., S. Muller, C. Tixier, and L. Pillonel.** 2002. Triclosan: occurrence and fate of a widely used biocide in the aquatic environment: field measurements in wastewater treatment plants, surface waters, and lake sediments. *Environ. Sci. Technol.* **36:**4998–5004.
111. **Sivaji, Y., and A. Mandal.** 1987. Antibiotic resistance & serotypic variations in *Escherichia coli* induced by disinfectants. *Indian J. Med. Res.* **86:**165–172.
112. **Sreenivasan, P., and A. Gaffar.** 2002. Antiplaque biocides and bacterial resistance: a review. *J. Clin. Periodon.* **29:**965–974.
113. **Suller, M. T., and A. D. Russell.** 1999. Antibiotic and biocide resistance in methicillin-resistant *Staphylococcus aureus* and vancomycin-resistant enterococcus. *J. Hosp. Infect.* **43:**281–291.
114. **Suller, M. T., and A. D. Russell.** 2000. Triclosan and antibiotic resistance in *Staphylococcus aureus. J. Antimicrob. Chemother.* **46:**11–8.
115. **Sullivan, A., B. Wretlind, and C. E. Nord.** 2003. Will triclosan in toothpaste select for resistant oral streptococci? *Clin. Microbiol. Infect.* **9:**306–309.
116. **Surolia, N., and A. Surolia.** 2001. Triclosan offers protection against blood stages of malaria by inhibiting enoyl-ACP reductase of *Plasmodium falciparum. Nat. Med.* **7:**167–173.
117. **Sutton, L., and G. A. Jacoby.** 1978. Plasmid-determined resistance to hexachlorophene in Pseudomonas aeruginosa. *Antimicrob. Agents Chemother.* **13:**634–636.
118. **Tabata, A., H. Nagamune, T. Maeda, K. Murakami, Y. Miyake, and H. Kourai.** 2003. Correlation between resistance of *Pseudomonas aeruginosa* to quaternary ammonium

compounds and expression of outer membrane protein OprR. *Antimicrob. Agents Chemother.* **47:**2093–2099.

119. **Tan, L., N. H. Nielsen, D. C. Young, Z. Trizna, and A. M. A. Council on Scientific Affairs.** 2002. Use of antimicrobial agents in consumer products. *Arch. Dermatol.* **138:**1082–1086.
120. **Tennent, J. M., B. R. Lyon, M. T. Gillespie, J. W. May, and R. A. Skurray.** 1985. Cloning and expression of *Staphylococcus aureus* plasmid-mediated quaternary ammonium resistance in *Escherichia coli. Antimicrob. Agents Chemother.* **27:**79–83.
121. **Villalain, J., C. R. Mateo, F. J. Aranda, S. Shapiro, and V. Micol.** 2001. Membranotropic effects of the antibacterial agent Triclosan. *Arch. Biochem. Biophys.* **390:**128–136.
122. **Vischer, W. A., and J. Regos.** 1974. Antimicrobial spectrum of Triclosan, a broad-spectrum antimicrobial agent for topical application. *Zentralbl. Bakteriol.* **226:**376–389.
123. **Walsh, S. E., J. Y. Maillard, A. D. Russell, C. E. Catrenich, D. L. Charbonneau, and R. G. Bartolo.** 2003. Development of bacterial resistance to several biocides and effects on antibiotic susceptibility. *J. Hosp. Infect.* **55:**98–107.
124. **Webber, M. A., and L. J. Piddock.** 2003. The importance of efflux pumps in bacterial antibiotic resistance. *J. Antimicrob. Chemother.* **51:**9–11.
125. **White, D. G., J. D. Goldman, B. Demple, and S. B. Levy.** 1997. Role of the acrAB locus in organic solvent tolerance mediated by expression of marA, soxS, or robA in *Escherichia coli. J. Bacteriol.* **179:**6122–6126.

Frontiers in Antimicrobial Resistance: a Tribute to Stuart B. Levy
Edited by D. G. White, M. N. Alekshun, and P. F. McDermott

Chapter 13

The Nexus of Oxidative Stress Responses and Antibiotic Resistance Mechanisms in *Escherichia coli* and *Salmonella*

BRUCE DEMPLE

This chapter discusses the mechanistic overlap between seemingly separate regulatory systems, one responding to oxidative stress, the other governing antibiotic resistance mechanisms encoded in the bacterial chromosome. The realization of this overlap came as I developed a lasting relationship with Stuart Levy as a collaborator, colleague, and friend. Thus, the scientific story parallels the personal one, and I will interweave the two narratives here.

FROM ANNOYANCE TO INSIGHT

When I first established my own laboratory at Harvard in 1984, one of our major interests was to understand the signaling and defense mechanisms employed by *Escherichia coli* to respond to oxidative stress. For our purposes, we had defined oxidative stress as a condition in which the levels of reactive species such as superoxide radical and hydrogen peroxide temporarily overwhelm cellular defenses (41). These defenses would include front-line activities such as catalase and superoxide dismutase, as well as corrective systems to rebalance metabolism and repair damage to key macromolecules such as DNA. My own prior work had demonstrated the existence of a distinct response in *E. coli* to defend against H_2O_2 (14), and a later study identified the *oxyR* gene as the main regulator responsible for that response (9). Our work had shown the existence of other responses to superoxide-generating agents such as menadione (MD) (19), and one of my first graduate students, Jean T. Greenberg, set about looking for the presumed novel regulatory systems.

Jean's approach was to isolate mutants with increased resistance to redox-related agents other than H_2O_2. She addressed this goal in a typically thorough way by carrying out a large-scale experiment in which she selected *E. coli* strains able to survive on lethal levels of compounds such as MD, paraquat, various metals, etc. Jean eventually identified the *soxR* gene through studies of MD-resistant (MD^R) strains (20), but her initial effort had generated a very large number of other variants. The *soxR* work, which eventually connected to antibiotic resistance, took all our attention, so we never did revisit the Petri plates piled high in our freezer room. I always wondered what other interesting regulatory systems might have been there awaiting discovery.

The *soxR* mutants Jean isolated fit nicely with our ideas of how such strains should behave. They showed increased activities of many antioxidant defense and repair functions and exhibited increased resistance to agents like MD and paraquat. These compounds cause greatly increased fluxes of superoxide in the cell, and in *soxR* mutants the potential toxicity is offset by increased activity of Mn-containing superoxide dismutase and other activities. The MD^R strains did not have elevated resistance to H_2O_2 or to nonoxidative damaging agents such as UV light or alkylating compounds, which indicated the specificity of the regulatory system. The *soxR* locus was discovered independently in the laboratory of Bernard Weiss (43).

We needed to characterize the MD^R strains genetically and map the *soxR* locus to facilitate its cloning (the completed *E. coli* genome was still a few years off [5]). A common approach to genetic analysis involves using bacterial recombination to put something easy to select, typically an antibiotic resistance marker, near the new mutation of interest so that the two alleles segregate together. Such approaches were greatly facilitated through the use of transposons, especially engineered versions (44), which we employed in our studies. During the genetic analysis, Jean noticed that the *soxR* strains produced a high background growth in antibiotic-containing media, even without the introduction of resistance markers. In short order

Bruce Demple • Department of Genetics and Complex Diseases, Harvard School of Public Health, Boston, MA 02115.

she quantified this resistance using the classic minimum inhibitory concentration assay. We were quite impressed the morning after her first such test when there was no question of the result: the resistance of the MD^R (*soxR*) strains was several times higher than that of the controls, and this was true for several antibiotics.

Although the MD^R strains had clearly elevated resistance to chloramphenicol, ampicillin, and tetracycline, some of the most commonly used antibiotics in *E. coli* bacterial genetics, the resistance was modest and could be overcome by employing somewhat higher antibiotic concentrations for our selections. In this way, we were able to link a minitransposon to the locus of MD^R and map this on the *E. coli* chromosome near 92 min. The ensuing genetic analysis demonstrated that both the MD^R and the antibiotic resistance phenotypes were due to mutations mainly at the same genetic locus and that these were accompanied by elevated expression of a set of proteins inducible by MD and paraquat in wild-type bacteria. We named this locus *soxR*, for *s*uper*ox*ide *r*esponse, and showed that it acts in an overall positive manner: *soxR*-deletion mutants lost inducibility of protective activities and all the corresponding resistance phenotypes.

A REGULATORY CROSSROADS: A FAMILY OF CONTROL PROTEINS FOR OXIDATIVE STRESS AND ANTIBIOTIC RESISTANCE GENES

The genetic analysis of the MD^R strains revealed one with a mutation located far from *soxR*, at a locus we initially named *soxQ*. The *soxQ1* mutation was mapped near 34 minutes on the *E. coli* chromosome, near two other mutations known to affect multiple antibiotic resistance, *marA1* (17) and *cfxB* (23). Not only did the *soxQ1* strain exhibit multiple antibiotic resistance, but the *cfxB* and *marA1* mutations conferred oxidant-resistance phenotypes (e.g., MD^R) similar to those associated with *soxQ1*. These phenotypes were reflected in elevated levels of several oxidative stress enzymes (e.g., Mn-containing superoxide dismutase and glucose-6-phosphate dehydrogenase) associated with the *soxQ1*, *marA1*, and *cfxB* mutations. In addition, the overall protein expression patterns, determined by two-dimensional gel electrophoresis, were very similar between the *soxQ1* and *cfxB* strains, with increased expression of six proteins in common with *soxR*^{C} strains, as well as seven others. It appeared that *soxQ1*, *cfxB*, and *marA1* affected the same regulatory locus and were possibly alleles of the same gene (18).

It soon became apparent that the observed regulatory overlap was embodied in the similar proteins encoded at the *soxR* and *soxQ/mar* loci. Cloning and DNA sequencing showed that the *soxR* locus included a second gene, *soxS*, which acted as the direct positive regulator of target genes (2, 47). The function of *soxR* was activation of the *soxS* gene in a redox-dependent fashion, to give an overall two-stage system of positive regulation (35, 48). Figure 1 summarizes the overall regulatory arrangement of the *soxRS* system. The *mar* locus also turned out to contain several genes, including *marA*, encoding a close homolog of *soxS*, and *marR*, encoding a negative regulator of *marA*. Thus, the coregulation of genes such as *sodA*, *zwf*, etc. was apparently due to the structural similarity of the predicted SoxS and MarA (SoxQ) proteins.

We carried out a more detailed analysis of the *soxQ1* and *cfxB* mutations and the *mar* locus in collaboration with the Levy laboratory. That work (3) showed that both mutations affected the same gene in the *mar* locus, designated *marR*, in which several previously isolated mar mutations were also located (10). The *cfxB* allele was a large deletion in *marR* and evidently a null mutation (3), which provided important evidence consistent with the suggestion that *marR* encodes a repressor (10).

The emerging models for regulation in the *soxRS* and *marRAB* systems pulled together diverse observations for a unified view of the inducible mechanism of antibiotic resistance in *E. coli*. Oxidative stress signals activate SoxR, leading to transcriptional induction of the *soxS* gene (13). The expressed SoxS protein binds the promoters of the regulon genes, which we demonstrated by showing that the purified protein recruits RNA polymerase to the target promoters in vitro (29). For the *marRAB* system, environmental signals would inactivate the MarR repressor, boosting the transcription of the *marRAB* operon and increasing MarA synthesis. In each case, antibiotic resistance depended on both the repression of the OmpF outer membrane protein by the *micF*-encoded antisense RNA (8, 12) and the activation of the *acrAB*-encoded efflux pump (46). The combination of limited uptake and increased efflux is a particularly potent mechanism for generalized antibiotic resistance (33).

The overlap in specificity between the two systems (activation of both oxidative stress and antibiotic resistance genes) is embodied in the second-stage regulators, SoxS and MarA. These small proteins are homologous to the C-terminal domain of the AraC/XylS family of transcription activators (2, 10, 47). It is the C-terminal domain of these larger proteins that contains the specific DNA binding and transcription-activating functions (16). As our analysis was underway, a new *soxS/marA*-related sequence named *rob* appeared in the public databases (42). Unlike AraC and its relatives, the predicted Rob protein had its SoxS/MarA-like domain located at the N terminus, with a C-terminal region not clearly related to known

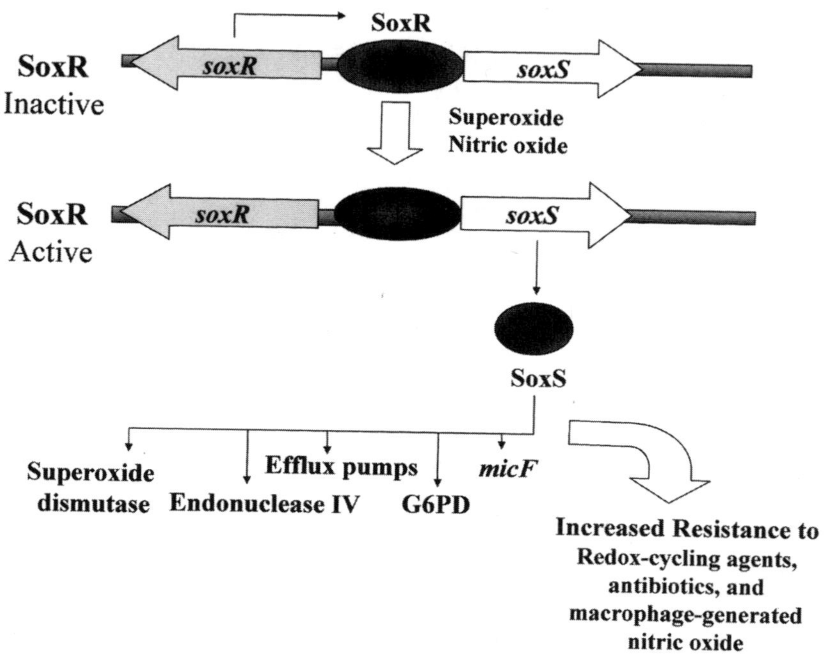

Figure 1. Control and phenotypic output in the *soxRS* system. Both latent and activated forms of SoxR protein bind tightly to a site between the −10 and −35 motifs of the *soxS* promoter. Through its 2Fe-2S iron-sulfur centers, the transcription activating function of SoxR is triggered either by oxidative stress signals generated by redox-cycling agents such as MD or paraquat (resulting in oxidation) or by exposure to nitric oxide (resulting in nitrosylation of the centers). The increased expression of SoxS protein increases its occupancy of binding sites in the target promoters of the regulon, where it recruits σ70 RNA polymerase to activate transcription. The activated genes increased resistance to oxidative stress (e.g., superoxide dismutase) and antibiotics (e.g., micF RNA). Adapted from reference 37.

proteins (Fig. 2). Although the Rob N terminus shared greater identity with SoxS and MarA than did other known AraC/XylS family members protein, Rob had been isolated as a binding activity for the rightward replication origin of the *E. coli* chromosome (42). Nevertheless, we showed that overexpression of Rob, or expression of only the N-terminal domain, activated some oxidative stress and antibiotic resistance genes (4). This activation resulted from specific Rob binding of the target promoters at similar DNA sites to those recognized by SoxS and MarA (28).

A seemingly disparate observation emerged from studies from a laboratory in Japan: both multicopy *rob* (31) and a *soxR*-constitutive mutation (32) conferred resistance to organic solvents in *E. coli.* A collaboration with the Levy laboratory (46) confirmed this new phenotype and extended it to MarA expression and showed that the resistance depends on the activity of the AcrAB efflux pump. Thus, antibiotic resistance and solvent tolerance in *E. coli* intersect through a set of genes regulated in common by MarA, SoxS, and Rob proteins (see also chapters 15 and 17).

The signaling processes that activate the various systems described here have been only partly revealed. For the *soxRS* regulon (Fig. 1), oxidative stress imposed by superoxide-generating agents (35) or macrophage-generated nitric oxide (15) are the predominant known activation signals (22). Both in vivo and in vitro studies indicate that *marRAB* derepression can occur through binding of salicylate and other compounds to the MarR protein (1, 11, 30). However, the structural basis for selective drug binding and altered interaction with the *mar* operator remains to be established.

A structure for Rob protein complexed with the *micF* promoter (27) indicated a rather different DNA binding mode than found for a MarA-DNA complex (38). In addition, for Rob protein, the additional domain unrelated to AraC proteins faced away from the DNA. A comparison of this domain structure revealed similarity to the drug-binding domain of BmrR, in which it also faces away from the DNA in the cocomplex (21), which suggests that these structures form a family of effector domains (Alexey G. Murzin, personal communication; see http://scop.mrc-lmb.cam.ac.uk/scop/data/scop.b.e.ig.b.html). It seems likely that small molecule binding to this domain activates Rob for DNA binding or transcription activation, which would be required to govern the activity of this abundant (42) protein. In this regard, Rob-mediated transcription in vivo is activated by synthetic dipyridyl compounds, a response that is dependent on the

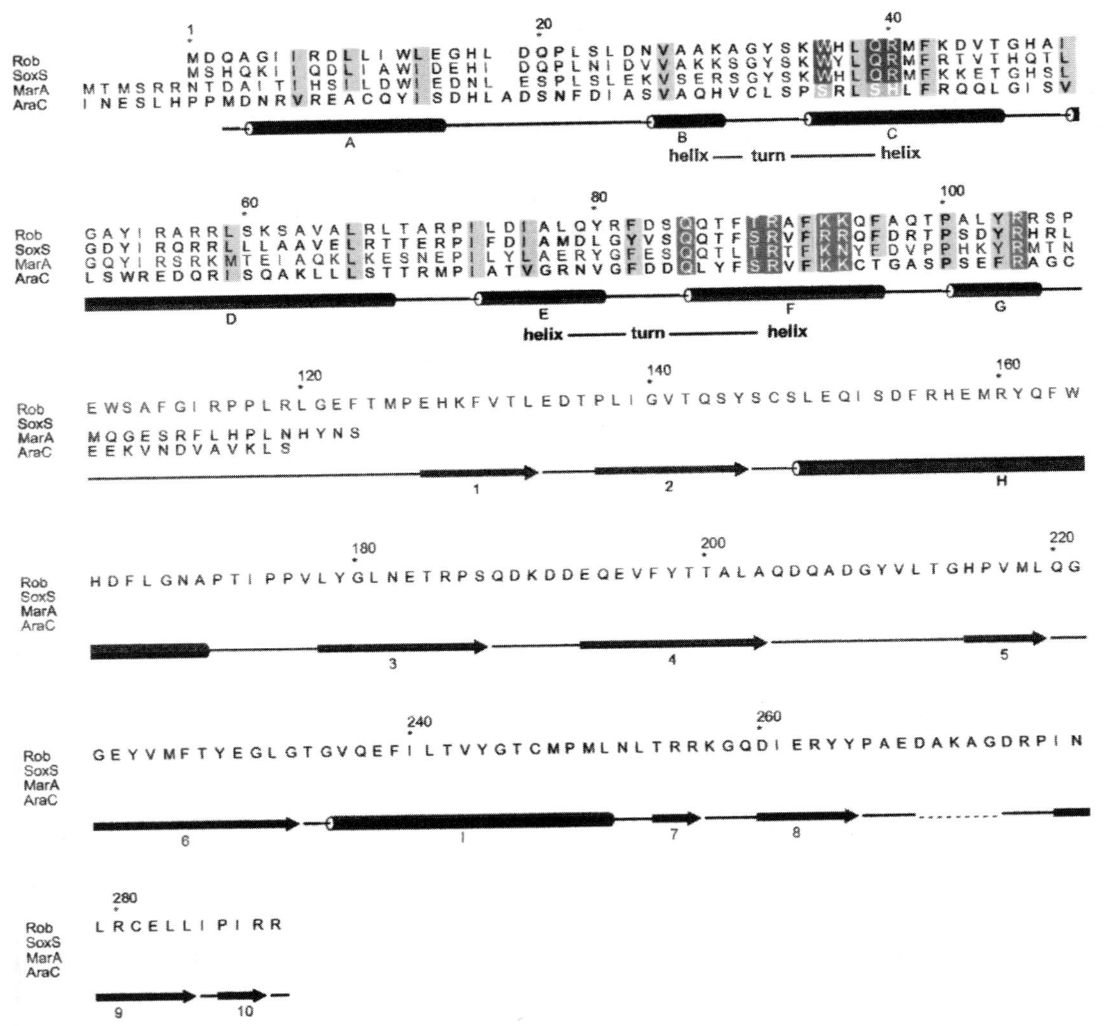

Figure 2. A conserved DNA binding domain in the Rob, SoxS, MarA, and AraC proteins. Rob protein secondary structure (27) is shown below an alignment of the *E. coli* Rob, SoxS, MarA, and AraC proteins. A common domain of ~100 residues is shared in all the proteins, while Rob has an additional 170 C-terminal residues, probably involved in modulating its transcription-activating function (see text). The first helix-turn-helix motif directly contacts specific DNA sequences; the second shows different secondary contacts in the Rob-DNA structure (27) compared to a MarA-DNA complex (38). The shaded residues correspond to conserved buried hydrophobic residues or contact residues in the DNA binding motifs (27). Adapted from reference 27.

C-terminal domain (40). The in vitro DNA binding activity was not affected (40), consistent with the idea of the allosteric activation of Rob by ligand (effector) binding. A more recent study identified likely physiological ligands for Rob protein: bile salts and decanoate, a fatty acid (39). Such regulation indicates how the overall antibiotic resistance of bacteria such as *E. coli* is in tune with multiple environmental cues.

CLINICAL ANTIBIOTIC RESISTANCE AND THE ROLE OF *soxRS*

The isolation of antibiotic-resistant regulatory mutations such as *mar* or *soxR* in the laboratory suggested that they might also occur in a clinical setting. To address this question, Levy and colleagues focused on *E. coli* strains resistant to fluoroquinolones, thus avoiding the frequent plasmid- and transposon-based mechanisms that provide powerful resistance to other classes of antibiotics. They cleverly added the additional screen of organic solvent tolerance (46), which was not expected to be affected by quinolone-resistance mutations in topoisomerase genes. That work (36) was the first to reveal the elevated expression of MarA or SoxS protein in a subset of resistant strains isolated from patients. Moreover, the strains overexpressing MarA all had point mutations in the *marR* gene, consistent with inactivation of the *mar* operon repressor (36). Thus, mutations in a regulatory locus appeared to be involved in the development of at least this type of clinical antibiotic resistance.

In another collaboration with Stuart Levy, we examined the properties of a *Salmonella enterica* (serovar Typhimurium) strain (St46) that developed ciprofloxacin resistance during antibiotic therapy (24). In this case, a quinolone-sensitive isolate (St45) was available from the same patient before the antibiotic treatment (24). Compared to several other clinical strains or St45, the quinolone-resistant St46 harbored elevated levels of Mn-containing superoxide dismutase, indicative of *soxRS* activation. Indeed, St46 had strongly elevated expression of both *soxS* mRNA and SoxS protein (26). The cloned *soxRS* genes from St46 conferred resistance to multiple antibiotics in a laboratory strain of *S. enterica,* while the St45 genes did not, and comparison of the DNA sequences revealed just one difference, a point mutation in the *soxR* gene of St46 (26). The most compelling experiment involved the introduction of a plasmid with the wild-type *S. enterica soxRS* genes, which suppressed the antibiotic resistance of St46 but not of St45 (26). In the absence of oxidative stress, competition of the mutant SoxR in St46 by the nonactivated wild-type protein effectively shut down *soxS* expression and thus antibiotic resistance. This result was direct evidence for the contribution of *soxRS* to the resistance of the clinical strain.

Additional work (26a) now extends these observations to *E. coli* strains exhibiting quinolone-resistance (36, 45). The new studies also hint at additional mutations outside *soxRS* that influence its activity, perhaps affecting the system(s) that maintain SoxR in the reduced and inactive state (25). The *soxR* mutations identified in clinical strains, together with genetic analyses conducted in laboratory strains (6, 7), have begun to help reveal the structural basis for redox signal transduction in this unusual transcription regulator.

RATIONALIZING THE REGULATORY OVERLAP

The convergence of different regulatory systems on both oxidative stress and antibiotic resistance genes was unexpected. Although each regulon independently controls additional genes, the overlap seems to have been preserved in both *E. coli* and *S. enterica,* in both laboratory and clinical strains. Evidently there is an evolutionary advantage to this arrangement, perhaps by allowing the integration of signals from various types of environmental threats to elicit expression of a core set of defense genes. In the case of *rob,* perhaps the environmental selection is for response to bile salts and fatty acids, which the bacteria would certainly encounter. For *soxRS,* there are numerous environmental compounds that can generate superoxide through redox cycling (41) which, together with nitric oxide (15, 34) generated by inflammatory processes, may constitute naturally occurring activators. In a similar vein, binding of environmental toxins or other compounds to MarR would explain the activity of *marRAB.*

POSTSCRIPT

Analyzing the regulatory mechanisms of *soxRS* and *marRAB* and their roles in resistance to multiple antibiotics proved to be an exciting adventure. The clear relevance of this work to human health was a bonus, as was the chance to learn about whole new areas of research. I have frequently cited the story for my students to illustrate how scientific research can lead in unexpected directions and how one needs to take the initiative to approach other researchers. My reward in that respect was to meet Stuart Levy and his research group, to enjoy the productive and stimulating collaborative efforts, and to develop a friendship built on this experience. I am grateful for this chance to contribute to the effort to honor my friend.

REFERENCES

1. **Alekshun, M. N., and S. B. Levy.** 1999. Alteration of the repressor activity of MarR, the negative regulator of the *Escherichia coli* marRAB locus, by multiple chemicals in vitro. *J. Bacteriol.* **181:**4669–4672.
2. **Amábile-Cuevas, C. F., and B. Demple.** 1991. Molecular characterization of the soxRS genes of *Escherichia coli:* two genes control a superoxide stress regulon. *Nucleic Acids Res.* **19:**4479–4484.
3. **Ariza, R. R., S. P. Cohen, N. Bachhawat, S. B. Levy, and B. Demple.** 1994. Repressor mutations in the marRAB operon that activate oxidative stress genes and multiple antibiotic resistance in *Escherichia coli. J. Bacteriol.* **176:**143–148.
4. **Ariza, R. R., Z. Li, N. Ringstad, and B. Demple.** 1995. Activation of multiple antibiotic resistance and binding of stress-inducible promoters by *Escherichia coli* Rob protein. *J. Bacteriol.* **177:**1655–1661.
5. **Blattner, F. R., G. Plunkett, 3rd, C. A. Bloch, N. T. Perna, V. Burland, M. Riley, J. Collado-Vides, J. D. Glasner, C. K. Rode, G. F. Mayhew, J. Gregor, N. W. Davis, H. A. Kirkpatrick, M. A. Goeden, D. J. Rose, B. Mau, and Y. Shao.** 1997. The complete genome sequence of *Escherichia coli* K-12. *Science* **277:**1453–1474.
6. **Chander, M., and B. Demple.** 2004. Functional analysis of SoxR residues affecting transduction of oxidative stress signals into gene expression. *J. Biol. Chem.* **279:**41603–41610.
7. **Chander, M., L. Raducha-Grace, and B. Demple.** 2003. Transcription-defective soxR mutants of *Escherichia coli:* isolation and in vivo characterization. *J. Bacteriol.* **185:**2441–2450.
8. **Chou, J. H., J. T. Greenberg, and B. Demple.** 1993. Posttranscriptional repression of *Escherichia coli* OmpF protein in response to redox stress: positive control of the micF antisense RNA by the soxRS locus. *J. Bacteriol.* **175:**1026–1031.
9. **Christman, M. F., R. W. Morgan, F. S. Jacobson, and B. N. Ames.** 1985. Positive control of a regulon for defenses against oxidative stress and some heat-shock proteins in *Salmonella typhimurium. Cell* **41:**753–762.

10. **Cohen, S. P., H. Hächler, and S. B. Levy.** 1993. Genetic and functional analysis of the multiple antibiotic resistance (mar) locus in *Escherichia coli. J. Bacteriol.* **175:**1484–1492.
11. **Cohen, S. P., S. B. Levy, J. Foulds, and J. L. Rosner.** 1993. Salicylate induction of antibiotic resistance in *Escherichia coli:* activation of the mar operon and a mar-independent pathway. *J. Bacteriol.* **175:**7856–7862.
12. **Cohen, S. P., L. M. McMurry, and S. B. Levy.** 1988. marA locus causes decreased expression of OmpF porin in multiple-antibiotic-resistant (Mar) mutants of *Escherichia coli. J. Bacteriol.* **170:**5416–5422.
13. **Demple, B.** 1996. Redox signaling and gene control in the *Escherichia coli* soxRS oxidative stress regulon—a review. *Gene* **179:**53–57.
14. **Demple, B., and J. Halbrook.** 1983. Inducible repair of oxidative DNA damage in *Escherichia coli. Nature* **304:**466–468.
15. **Ding, H., and B. Demple.** 2000. Direct nitric oxide signal transduction via nitrosylation of iron-sulfur centers in the SoxR transcription activator. *Proc. Natl. Acad. Sci. USA* **97:**5146–5150.
16. **Gallegos, M. T., R. Schleif, A. Bairoch, K. Hofmann, and J. L. Ramos.** 1997. Arac/XylS family of transcriptional regulators. *Microbiol. Mol. Biol. Rev.* **61:**393–410.
17. **George, A. M., and S. B. Levy.** 1983. Gene in the major cotransduction gap of the *Escherichia coli* K-12 linkage map required for the expression of chromosomal resistance to tetracycline and other antibiotics. *J. Bacteriol.* **155:**541–548.
18. **Greenberg, J. T., J. H. Chou, P. A. Monach, and B. Demple.** 1991. Activation of oxidative stress genes by mutations at the soxQ/cfxB/marA locus of *Escherichia coli. J. Bacteriol.* **173:**4433–4439.
19. **Greenberg, J. T., and B. Demple.** 1989. A global response induced in Escherichia coli by redox-cycling agents overlaps with that induced by peroxide stress. *J. Bacteriol.* **171:**3933–3939.
20. **Greenberg, J. T., P. Monach, J. H. Chou, P. D. Josephy, and B. Demple.** 1990. Positive control of a global antioxidant defense regulon activated by superoxide-generating agents in *Escherichia coli. Proc. Natl. Acad. Sci. USA* **87:**6181–6185.
21. **Heldwein, E. E., and R. G. Brennan.** 2001. Crystal structure of the transcription activator BmrR bound to DNA and a drug. *Nature* **409:**378–382.
22. **Hidalgo, E., H. Ding, and B. Demple.** 1997. Redox signal transduction via iron-sulfur clusters in the SoxR transcription activator. *Trends Biochem. Sci.* **22:**207–210.
23. **Hooper, D. C., J. S. Wolfson, K. S. Souza, E. Y. Ng, G. L. McHugh, and M. N. Swartz.** 1989. Mechanisms of quinolone resistance in *Escherichia coli:* characterization of nfxB and cfxB, two mutant resistance loci decreasing norfloxacin accumulation. *Antimicrob. Agents Chemother.* **33:**283–290.
24. **Howard, A. J., T. D. Joseph, L. L. Bloodworth, J. A. Frost, H. Chart, and B. Rowe.** 1990. The emergence of ciprofloxacin resistance in *Salmonella typhimurium. J. Antimicrob. Chemother.* **26:**296–298.
25. **Koo, M. S., J. H. Lee, S. Y. Rah, W. S. Yeo, J. W. Lee, K. L. Lee, Y. S. Koh, S. O. Kang, and J. H. Roe.** 2003. A reducing system of the superoxide sensor SoxR in *Escherichia coli. EMBO J.* **22:**2614–2622.
26. **Koutsolioutsou, A., E. A. Martins, D. G. White, S. B. Levy, and B. Demple.** 2001. A soxRS-constitutive mutation contributing to antibiotic resistance in a clinical isolate of *Salmonella enterica* (Serovar *typhimurium*). *Antimicrob. Agents Chemother.* **45:**38–43.

26a. **Koutsolioutsou, A., S. Peña-Llopis, and B. Demple.** 2005. Constitutive *soxR* mutations contribute to multiple antibiotic resistance in clinical *Escherichia coli* isolates. *Antimicrob. Agents Chemother.*, in press.

27. **Kwon, H. J., M. H. Bennik, B. Demple, and T. Ellenberger.** 2000. Crystal structure of the *Escherichia coli* Rob transcription factor in complex with DNA. *Nat. Struct. Biol.* **7:**424–430.
28. **Li, Z., and B. Demple.** 1996. Sequence specificity for DNA binding by *Escherichia coli* SoxS and Rob proteins. *Mol. Microbiol.* **20:**937–945.
29. **Li, Z., and B. Demple.** 1994. SoxS, an activator of superoxide stress genes in *Escherichia coli.* Purification and interaction with DNA. *J. Biol. Chem.* **269:**18371–18377.
30. **Martin, R. G., and J. L. Rosner.** 1995. Binding of purified multiple antibiotic-resistance repressor protein (MarR) to mar operator sequences. *Proc. Natl. Acad. Sci. USA* **92:**5456–5460.
31. **Nakajima, H., K. Kobayashi, M. Kobayashi, H. Asako, and R. Aono.** 1995. Overexpression of the robA gene increases organic solvent tolerance and multiple antibiotic and heavy metal ion resistance in *Escherichia coli. Appl. Environ. Microbiol.* **61:**2302–2307.
32. **Nakajima, H., M. Kobayashi, T. Negishi, and R. Aono.** 1995. soxRS gene increased the level of organic solvent tolerance in *Escherichia coli. Biosci. Biotechnol. Biochem.* **59:**1323–1325.
33. **Nikaido, H.** 1994. Prevention of drug access to bacterial targets: permeability barriers and active efflux. *Science* **264:**382–388.
34. **Nunoshiba, T., T. deRojas-Walker, J. S. Wishnok, S. R. Tannenbaum, and B. Demple.** 1993. Activation by nitric oxide of an oxidative-stress response that defends *Escherichia coli* against activated macrophages. *Proc. Natl. Acad. Sci. USA* **90:**9993–9997.
35. **Nunoshiba, T., E. Hidalgo, C. F. Amabile Cuevas, and B. Demple.** 1992. Two-stage control of an oxidative stress regulon: the *Escherichia coli* SoxR protein triggers redox-inducible expression of the soxS regulatory gene. *J. Bacteriol.* **174:**6054–6060.
36. **Oethinger, M., I. Podglajen, W. V. Kern, and S. B. Levy.** 1998. Overexpression of the marA or soxS regulatory gene in clinical topoisomerase mutants of *Escherichia coli. Antimicrob. Agents Chemother.* **42:**2089–2094.
37. **Pomposiello, P. J., and B. Demple.** 2001. Redox-operated genetic switches: the SoxR and OxyR transcription factors. *Trends Biotechnol.* **19:**109–114.
38. **Rhee, S., R. G. Martin, J. L. Rosner, and D. R. Davies.** 1998. A novel DNA-binding motif in MarA: the first structure for an AraC family transcriptional activator. *Proc. Natl. Acad. Sci. USA* **95:**10413–10418.
39. **Rosenberg, E. Y., D. Bertenthal, M. L. Nilles, K. P. Bertrand, and H. Nikaido.** 2003. Bile salts and fatty acids induce the expression of *Escherichia coli* AcrAB multidrug efflux pump through their interaction with Rob regulatory protein. *Mol. Microbiol.* **48:**1609–1619.
40. **Rosner, J. L., B. Dangi, A. M. Gronenborn, and R. G. Martin.** 2002. Posttranscriptional activation of the transcriptional activator Rob by dipyridyl in *Escherichia coli. J. Bacteriol.* **184:**1407–1416.
41. **Sies, H.** 1991. Oxidative stress: introduction, p. xv-xxii. *In* H. Sies (ed.), *Oxidative Stress: Oxidants and Antioxidants.* Academic Press, London, United Kingdom.
42. **Skarstad, K., B. Thony, D. S. Hwang, and A. Kornberg.** 1993. A novel binding protein of the origin of the *Escherichia coli* chromosome. *J. Biol. Chem.* **268:**5365–5370.
43. **Tsaneva, I. R., and B. Weiss.** 1990. soxR, a locus governing a superoxide response regulon in *Escherichia coli* K-12. *J. Bacteriol.* **172:**4197–4205.
44. **Way, J. C., M. A. Davis, D. Morisato, D. E. Roberts, and N. Kleckner.** 1984. New Tn10 derivatives for transposon muta-

genesis and for construction of lacZ operon fusions by transposition. *Gene* **32:**369–379.

45. **Webber, M. A., and L. J. Piddock.** 2001. Absence of mutations in marRAB or soxRS in acrB-overexpressing fluoroquinolone-resistant clinical and veterinary isolates of *Escherichia coli. Antimicrob. Agents Chemother.* **45:**1550–1552.
46. **White, D. G., J. D. Goldman, B. Demple, and S. B. Levy.** 1997. Role of the acrAB locus in organic solvent tolerance mediated by expression of marA, soxS, or robA in *Escherichia coli. J. Bacteriol.* **179:**6122–6126.
47. **Wu, J., and B. Weiss.** 1991. Two divergently transcribed genes, soxR and soxS, control a superoxide response regulon of *Escherichia coli. J. Bacteriol.* **173:**2864–2871.
48. **Wu, J., and B. Weiss.** 1992. Two-stage induction of the soxRS (superoxide response) regulon of *Escherichia coli. J. Bacteriol.* **174:**3915–3920.

Frontiers in Antimicrobial Resistance: a Tribute to Stuart B. Levy
Edited by D. G. White, M. N. Alekshun, and P. F. McDermott

Chapter 14

The *mar* Locus

THAMARAI SCHNEIDERS, HERBERT HAECHLER, AND WILLIAM YAN

Antibiotic resistance in bacteria can arise due to the acquisition of extrachromosomal genetic elements, and/or loss or reduction in normal gene function through mutations and enzymatic modifications (9). The drug resistance phenotypes conferred by these genetic determinants are usually drug and mechanism specific. Hence, it is common that the acquisition of these elements result in high-level antibiotic resistance and, in some cases, therapeutic failure (70, 80, 81). In contrast, low-level antibiotic resistance that occurs after challenge with subinhibitory drug concentrations can result from a broader variety of biochemical and adaptive mechanisms, i.e., stress response, and emerge and evolve under nonantibiotic pressure (9, 61). Thus genetic adaptations that occur due to low-level resistance mechanisms have important implications and relevance. The first such adaptive mechanism discovered was mapped to the multiple antibiotic resistance *mar* locus at 34.05 min of the *Escherichia coli* linkage map (16, 28, 29, 37). As such, the discovery of the *mar* locus is significant for two reasons: first, it confers a multidrug resistance phenotype (18, 28, 29, 37) and second, recent findings have demonstrated that this locus controls genes in cellular metabolism, physiology, and virulence (10, 71).

E. coli selected on subinhibitory concentrations of chloramphenicol and tetracycline were isolated at frequencies of 10^{-7} and 10^{-8} (28) and produced *mar* mutants that are cross-resistant to a wide range of antibiotics which include the penicillins, cephalosporins, nalidixic acid, and rifampin (28, 29, 37). This resistance phenotype has expanded to include the fluoroquinolones, agents imposing oxidative stress, organic solvents, and disinfectants such as pine oil and triclosan (65, 89). The level of resistance achieved by *mar* mutants can be enhanced by repeated subculture on higher antibiotic concentrations (28). However, high-levels of resistance can only be partially attributed to the *mar* locus as transduction from high- or low-level mutants accounts for low-level resistance only (4, 5). The *mar* resistance phenotype is inducible by salicylate, acetaminophen, sodium benzoate, 2-4 dinitrophenol, cinnamate as well as the redox cycling compounds menadione, plumbagin, and aryloxoalcanoic acids (5, 8, 84). In addition, the *mar* phenotype is able to confer protection against rapid cell killing by fluoroquinolones as inactivation of the locus in Mar mutants results in wild-type susceptibility to these drugs (32). The *mar* phenotype is completely reversed by the insertion of the Tn*5* transposon into the locus or via the deletion of the *mar* locus, which leads to hypersensitivity to various antibiotics (28, 29)

Although mar-mediated resistance is not fully understood, the mechanisms attributed to this phenotype involve the ATP driven drug specific efflux system AcrAB (69, 89) and decreased influx by the reduction of the outer membrane protein F (OmpF) (19).

CHARACTERIZATION OF *mar* MUTANTS

The *mar* locus was defined by an inactivating Tn*5* insertion, *marA*::Tn*5*, into *marA* at 34.05 min of the *E. coli* K-12 chromosome (28, 29). Initial studies cloned the junction fragment containing DNA from both *marA* and Tn*5*, which was then used as a probe for screening a plasmid library of wild-type *E. coli* K-12 for unmutated *mar* sequences (16, 37). Smaller fragments were subsequently subcloned on plasmid vectors and checked for their ability to complement *mar* in trans in a strain harboring a 39 kb deletion which includes the *mar* locus and surrounding re-

Thamarai Schneiders • Department of Molecular Microbiology and Microbiology, Tufts University, School of Medicine, Boston, MA 02111. **Herbert Haechler** • Institute of Veterinary Bacteriology, University of Berne, Laenggassstrasse 122, 3012 Berne, Switzerland. **William Yan** • Food Directorate, Health Canada, Sir Frederick Banting Research Centre, Address Locator 2204A1, Tunney's Pasture, Ottawa, Ontario K1A 0L2, Canada.

gions (37). In this way, a 7.8 kb *HpaI/PstI* fragment was found to complement the deleted phenotype and also found to be sufficient for multiple antibiotic resistance to develop (37). Further studies were able to reduce the size of the functional fragment to less than 2 kb (54, 86).

Extrachromosomal Mar resistance was found to be equally stable, and the plasmid-borne Mar phenotype reached higher levels of resistance than that provided by the chromosome, due to the higher gene dose (16, 37). The results were reproducible and demonstrated that the phenotype could be restored in trans (26, 37).

The sequence of the entire functional 7.8 kb *HpaI-PstI* fragment was determined twice, from both the wild type and a *mar* mutant where 28 open reading frames (ORFs) with a potential coding capacity of greater than 60 amino acids were detected. The wild type and *mar* mutant sequences were found to differ by a single point mutation which led to an amino acid substitution within an ORF consisting of 144 amino acids, later designated MarR (16, 37). At that time, sequence analysis of MarR did not show any significant identity with any proteins in the database. Further analysis of six other *mar* mutants revealed either a point mutation, a short tandem duplication, or an insertion of an IS-element in either *marO* or *marR* (Fig. 1) (16). Since multiple copies of *marO* were able to titrate the repressory effect of MarR and confer the antibiotic resistance phenotype, it was postulated that MarR is the repressor of the *marRAB* locus and does so by acting on *marO*. These observations—coupled with the fact that an increase in *mar*-specific transcripts was demonstrated only in *mar* mutants or after relevant antibiotic induction, but were normally down regulated in the wild type—confirmed that MarR was indeed the negative regulator of the *marRAB* operon (16).

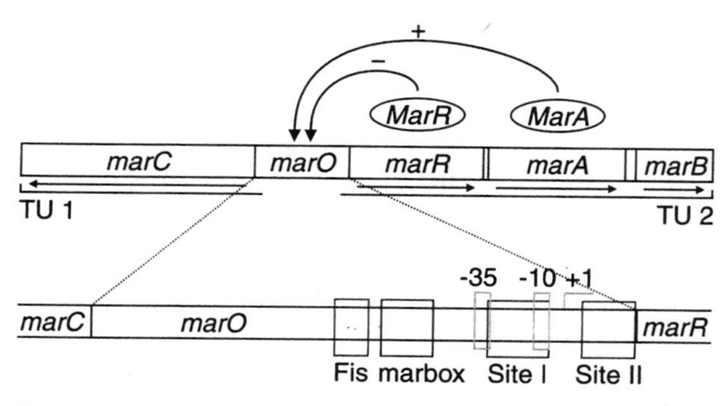

Figure 1. Genetic organization of the *mar* locus. *marC* and *marRAB* are divergently transcribed from the central operator region *marO* (16, 19, 29). MarR is the repressor of the *marRAB* operon and does so by binding to sites within *marO*, depicted in the figure as Site I and Site II (16, 19). MarA (127AA) activates the transcription of the *marRAB* locus by binding to the marbox (53). This activation is enhanced by the binding of the DNA bending protein Fis at the binding site also known as the "accessory" marbox (58). *marB* encodes for a protein (72AA) that has no known function (16). *marC* encodes for a putative transmembrane protein sized at 221AA with no known function (62).

The gene immediately downstream of MarR, identified as the original insertion site for the Tn*5* transposon was called *marA*. MarA encodes for 127 amino acids and shares homology with the AraC/XylS family of transcriptional regulators, which are involved in a wide range of cellular activities (23, 25, 56). Based on the size of the transcripts expressed from the *mar* locus, *marR, marA,* and a third ORF (MarB) encoding a 72 amino acid protein were considered to be part of the operon (Fig. 1). Like MarR, MarB shared no significant homology with other proteins in the database when compared initially; however, recent analysis has shown that homologous proteins exist in *Shigella flexneri* (11), *Enterobacter aerogenes,* and *Klebsiella pneumoniae.* The divergently transcribed ORF upstream of the *marRAB* operon, later called MarC, was found to be induced in the presence of tetracycline and chloramphenicol (16), but its role in mediating a MAR (multiple antibiotic resistance) phenotype is still unclear (62).

GENE ARRANGEMENT AND REGULATION OF THE *marRAB* LOCUS

The *E. coli mar* locus comprises two transcriptional units that are divergently transcribed from a common operator region, *marO* (Fig. 1). Transcriptional unit 1 (TU 1) encodes MarC, which has a hydropathy profile suggestive of a putative integral transmembrane protein with six transmembrane helices (62). The other transcriptional unit, unit 2 (TU 2), encodes three ORFs, which translate to MarR (the repressor), MarA (the activator), and MarB (a protein with no known function) (Fig. 1). Initial attempts at mapping the transcriptional start site of the *marRAB* locus were unsuccessful. Primer extension analysis identified the transcriptional start point as nucleotide 1418 (according to the Cohen sequence numbering [16]) within the central operator region *marO* (5). Northern blot analysis sized *mar*-specific transcripts at ~1.1kb and ~0.9kb, where the bigger transcript correlates well with the position of the transcriptional start site and the size of the *mar* operon (5). The reason for the smaller ~0.9kb transcript, however, is still unclear. It is possible that an internal promoter located within *marR* could produce a smaller transcript. Such a promoter has already been described in the *emr* locus (4, 48), which encodes a multidrug efflux pump (47). This promoter is growth phase regulated and found not to be under the control of the operon's repressor (4, 48).

Increased or constitutive transcription of *marA* can be seen in both nonmutated strains induced with antibiotics (like tetracycline and chloramphenicol) or nonantibiotics (like salicylate) or strains harboring mutations within *marO* and *marR*, where the RNA levels can increase up to 16-fold (16, 17). Analysis of the *marO* region revealed the presence of two pentameric subelements (inverted repeats) named site I and site II, which were found to be essential for MarR binding and repression (57). Other transcriptional factors like MarA, SoxS, and Rob are able to enhance transcription via *cis*-acting element(s) within *marO* identified as putative MarA binding sites (4). These binding sites have a degenerate consensus sequence (AYnGCACnnWnnRYYAAAYn [51]), and based on this sequence, several putative marboxes were found within *marO* and *marR* (5). However, footprinting experiments have clearly demonstrated that the actual MarA binding site is situated 59 to 64 nucleotides upstream of the −35 hexamer (58). The positions of the binding sites of MarA and MarR imply a role for autoregulation of the *marRAB* locus by both proteins (53). Despite MarR mediated repression, low-levels of *marA* are still detectable in unmutated strains, where this basal constitutive expression of *marA* may be needed for some expression of the *marRAB* locus and other members of the regulon (53, 64). Phenotypic evidence supports this, as deletion of the wild-type *mar* locus results in increased sensitivity to antibiotics and oxidative stress agents (16, 33, 37). Another binding site, originally called the "accessory" marbox, was actually found to bind Fis (Fig. 1), a small regulatory and DNA-bending protein. The simultaneous presence of Fis and MarA at their respective binding sites has been shown to increase transcription from the *marRAB* promoter in contrast to when only MarA is bound to the marbox (58). To date there is no experimental data demonstrating the mechanism of transcription termination at the *marRAB* operon; it possibly involves a putative RNA stem-loop at nucleotides 2542 to 2559 (16) located immediately downstream of *marB* (4), where these sequences may resemble a rho-independent structure and cause transcriptional pausing.

The gene arrangement of the *marRAB* locus appears to be conserved in 14 out of 53 species of *Enterobacteriaceae*, as was shown by Southern hybridization (20). The 14 species belong to the genera *Citrobacter, Enterobacter, Escherichia, Hafnia, Klebsiella, Salmonella,* and *Shigella,* and Mar mutants could be derived from *Salmonella* sp. and *Enterobacter agglomerans* (20). In addition, the *mar* operon in *Salmonella enterica* serotype Typhimurium was later shown to be highly similar, both structurally and functionally, to the one in *E. coli,* except for *marB* (85).

GENETIC ELEMENTS OF THE *mar* LOCUS

MarR

MarR is the prototype of a newly described family of proteins in *E. coli.* Since the discovery and characterization of MarR coupled with the surge of bacterial genome sequencing, an ever increasing number of proteins with the family signature [AspXArgX5(Leu/Ile)ThrX2Gly (4)] are being reported (47, 72, 79). The MarR family of proteins are characterized by the ability to sense phenolic compounds (87). These proteins also bind palindromic sequences as dimers and generally function as repressors (4, 36, 83, 87). Homologs of MarR include proteins involved in a myriad of cellular functions, such as virulence regulators like RovA in *Yersinia enterocolitica,* which is involved in the regulation of invasin in mammalian cells in response to the environment (76); ScoC required for peptide transport as well as sporulation initiation in *Bacillus subtilis* (44); PecS in *Erwinia chrysanthemi,* which is involved in the synthesis of virulence factors involved in plant infection (79); and lastly, two equally well studied systems of clinical significance, EmrR, which regulates the efflux pump *emrAB* (47) in *E. coli* and MexR, which controls the *mexAB*-OprM efflux pump in *Pseudomonas aeruginosa* (72, 90). Studies involving EmrR have shown that it shares significant amino acid and biochemical similarity with MarR (87). Unlike MexR, both EmrR and MarR are inducible by salicylate and EmrR is also able to repress the *marRAB* operon (4, 5, 87).

Microarray studies analyzing gene expression in Dam (DNA adenine methylase) deficient cells observed a small but reproducible increase in *marR* expression. Although the exact significance of this increase is not clear (46), this finding suggests that *marR* itself is subject to regulation and can also control the expression of other proteins. There are examples of MarR family members involved in global gene regulation, e.g., PecS, which has been shown to regulate at least 11 virulence genes in *E. chrysanthemi* (79). Therefore it is likely that MarR mediated regulation is not limited to the *marRAB* locus, and the above-mentioned reasons suggest a larger regulatory role for this protein.

That MarR is the repressor of the *mar* operon was demonstrated early on in investigations of the locus (16, 84). In the absence of inducers or mutations within either *marO* or *marR,* a functional MarR is able to repress effectively by binding to two pentameric sites within *marO* (site I and site II) (Fig. 1), where the *marR-marO* interaction is highly specific and has an apparent K_d of 10^{-9}M (84). Site I overlaps both the −35 and −10 hexamers and lies between these sequences, and site II lies downstream of site I between the transcriptional start and the translational start of the protein (Fig. 1) (16, 57, 58). Footprinting studies with DNAase I have shown that MarR binds *marO* on both strands of DNA (53). In support of this observation, the crystal structure of MarR does reveal a single winged helix DNA-binding motif which would result in a MarR monomer bound to each of the DNA binding sites on opposite strands of the DNA helix (6). Despite initial observations that site I alone may suffice for the repression of *marRAB* (57), recent data show that the deletions of either site I or site II alone are not sufficient to completely abolish repression (60). Thus complete repression appears to require the presence of both sites (60). Therefore the arrangement of both site I and site II allows MarR to function as an efficient repressor of the *marRAB* operon and of itself. The translational efficiency of MarR is appreciably lower than the rate of *marR* transcription, and the initial hypothesis suggested that the alternative start codon (GTG) may be responsible; however, recent data have now demonstrated that the low translational efficiency of MarR is due to a poor Shine-Dalgarno sequence (60). Inefficient MarR translation suggests leaky regulation of the *marRAB* operon and may result in enhanced cell sensitivity to environmental signals.

The most efficient antagonist of the MarR-DNA interaction is salicylate (2, 17, 84). In addition, MarR has also been found to interact efficiently with other aromatic compounds such as plumbagin, menadione, 2,4-dinitro-phenol, and aryloxoalcanoic acids (2, 8, 84). Derepression of the *marRAB* locus has been demonstrated with both tetracycline and chloramphenicol, but as no interaction with these antibiotics and the MarR-DNA complex can be demonstrated by binding studies, this effect appears to be mediated through an indirect mechanism (4).

Consistent with its role as the repressor of the *marRAB* locus, mutations which result in an elevation of *marA* transcription have been mapped to the MarR protein (Fig. 2) (1, 43, 49, 67, 68). Mutational and crystallographic analyses demonstrate that MarR has distinctive domains divided into a single oligomerization domain situated at the amino terminus and two helix-turn-helix domain(s) (HTH) at the carboxy terminus (Fig. 2) (1, 6). The oligomerization domain is involved in the dimerization of the protein, which is presumed to be a precursor to DNA binding and repressor function; the HTH domain is known to be involved in DNA binding (1, 3, 6).

Mutations within the oligomerization domain (e.g., Gln42Amber) were not able to negatively complement in the *marR* expressing host strain and exhibited no repressor activity in the Δ*marRAB* background, supporting the role of the oligomerization domain in making essential protein-protein contacts

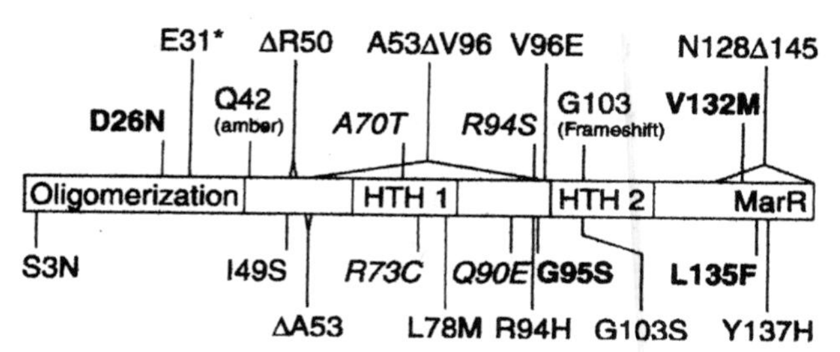

Figure 2. Mutations within *marR*, the repressor of the *marRAB* locus. Mutations described in MarR in both laboratory and clinical mutants which result in a multidrug resistance phenotype (43, 49, 67, 68) are shown. All the mutations have been shown to reduce or abolish the repressory effect by MarR and are localized all over the protein. The mutations shown in italics (*A70T, R73C, Q90E,* and *R94S*) are clustered in the HTH domains and result in defective proteins with little or no DNA binding activity (1). The superrepressor mutations shown in bold (**D26N, G95S, V132M,** and **L135F**) all exhibit increased DNA binding with a variable response to salicylate (3). The mutant protein Q42Amber demonstrates no repressor activity (1). The asterisk indicates a stop codon at amino acid E31 (68).

(1). In addition, the deletion of the first 19 amino acids of MarR within this domain completely eliminates repressor activity (84), confirming the role of this region in MarR dimerization and repression.

Studies of negatively complementing mutations clustered within the HTH domains at positions Arg73Cys, Ala70Thr, and Gln90Glu demonstrated that these mutations resulted in decreased or no DNA binding (1). Other mutants that encode for a truncated protein also result in a negatively complementing phenotype which substantiates the role of the HTH in DNA binding (1). In fact, the crystal structure of MarR in complex with salicylate supports the genetic and biochemical analysis outlined here; for example, the mutation Arg73, which abolishes DNA binding, maps to the α-helix 4 within the HTH domains, which has been predicted to be involved in DNA-binding (1, 6). Superrepressor mutations at positions Asp26Asn, Gly95Ser, Val132Met, and Leu135Phe result in proteins that display appreciably greater DNA-binding activity (via increased nonspecific interactions with MarR and DNA [2]) but with a variable response to the inducer salicylate (3). The mutational analysis of the MarR protein with both the negatively complementing phenotypes and superrepressor mutants shows that these lesions are spread throughout the protein and are not localized to a particular site (Fig. 2).

Studies with laboratory mutants and clinically resistant *E. coli* isolates have also revealed that mutations occur all over the protein and not only in mutational hotspots (1, 3, 49, 66–68) (Fig 2). Perhaps the more instructive example of MarR function comes from the study of a fluoroquinolone resistant *E. coli* isolate that harbors a mutation in MarR resulting in an 18 amino acid deletion in the carboxy terminal domain (CTD), which results in a MarR protein that exhibits greatly reduced DNA binding and partially affected dimerization (66). As HTH motifs in the DNA binding domain remained unaffected by the mutation, it is possible that this CTD deletion affects protein binding and dimerization functions. This suggests that dimerization is a prerequisite for DNA binding (66). However, in order to understand the MarR-DNA interactions completely, we must await the structural determination of this complex.

MarA

MarA belongs to a group of transcriptional regulators typified by the proteins AraC from *E. coli* and XylS from *Pseudomonas putida* collectively known as the AraC/XylS family (23, 25, 56). The prototypic family signature is a 100 amino acid region of sequence similarity that forms an independently folding domain containing two HTH DNA binding motifs (23, 25, 56). A recent database search has identified 830 family members to date (23), although the functional activities for all these proteins have not been deduced. This group of regulators have been shown to be involved in an array of cell associated functions, which include virulence regulation in pathogens (e.g., ToxT in *Vibrio cholerae* [40] and ExsA in *P. aeruginosa* [39]), metabolic functions, cell physiology (e.g., Gad X, [88]) and stress response (e.g., MarA, SoxS, Rob [4, 7, 10, 25, 56, 71]). Structural analysis of MarA has shown that it also shares similarity with the integrase family of site-specific recombinases (31).

MarA is the second gene in the *marRAB* locus, and its expression is negatively regulated by MarR. Despite this control, basal levels of *marA* transcription are demonstrable through autoactivation of its own expression via a marbox (asymmetric binding site) that lies upstream of the MarR sites of repression (Fig. 1) (53, 58). This site is accessible to the other members of the MarA subfamily, and it has been shown that both SoxS and Rob can activate *marRAB* expression. The deletion of this marbox abrogates the positive regulation by the three regulators MarA, SoxS, and Rob (53) . The binding of Fis, a DNA bending protein, to an accessory marbox augments MarA mediated activation, and the DNA binding of either protein to its cognate binding site occurs independently of either protein (58). The number of MarA molecules present in the cell in the repressed and derepressed state differ by about 10-fold (50). In fact the level of MarA protein is controlled by an active process and has been recently shown to be subject to proteolytic control by the Lon protease (34). Thus, MarA is subject to regulation at both the transcriptional and posttranslational level.

The antibiotic resistant phenotype observed in *mar* mutants arises primarily through the induction of the multidrug efflux pump, AcrAB, in *E. coli* (69). This positive regulation is mediated via the binding of MarA to the marbox sequence in the *acrAB* promoter and is one of the first examples of genes that are regulated by MarA.

Microarray studies have shown that MarA controls approximately 60 genes in the *E. coli* genome when expression profiles of a *marA* deleted strain are compared with that of a *marA* constitutive expresser (10), and in addition 153 genes were found to be affected when the expression profiles of salicylate and paraquat-treated samples were compared (71). Further analyses by standard molecular biology techniques and bioinformatics report that the true regulon size of directly activated genes by MarA may be around 55 (55, 59). The MarA regulon shares considerable overlap with genes regulated by SoxS, although the levels of activation differ (10, 52, 55, 71). This promoter discrimination by SoxS and MarA has been attributed to the marbox sequence itself, regardless of the different promoter arrangements, although the exact involvement of the different DNA bases is unclear (52). Perhaps an analysis of electrochemical interactions between activator proteins such as MarA/SoxS/Rob and the different DNA bases within the marbox may provide a clue.

Like most of its family members, MarA was thought to be an exclusive transcriptional activator, but microarray data have shown that a subset of genes are negatively regulated (10, 71). Extensive genetic studies have shown that positively regulated genes contain MarA binding sites in two promoter arrangements known as the class I and class II promoters. Class I promoters lie upstream of the −35 hexamer and in the "forward" orientation, such that the carboxyl terminus of MarA lies proximal to the RNA polymerase (RNAP) binding site. In the class II promoters, the marbox overlaps the −35 hexamer and lies in the "backward" orientation, such that the amino terminus of MarA lies proximal to RNAP. At both class I and class II promoters, contacts between RNAP and MarA result in transcriptional activation (41, 51). It has been demonstrated in vitro that the alpha-CTD of RNAP interactions with MarA is crucial for transcriptional activation of class I promoters, e.g., *zwf* (42). In vivo analysis to determine the role of alpha-NTD of RNAP in transcriptional activation by MarA at both class I and II promoters found that mutant alpha-NTD derivatives conferred no significant defects in transcriptional activation (24). Thus alpha-NTD is not a target for MarA mediated activation.

Recent analysis of MarA and RNAP interactions using NMR-based chemical shift mapping demonstrates that the alpha-CTD RNAP uses a "265-like determinant" to contact MarA at a site distant from DNA, which is in contrast to the interactions of RNAP with activators like Fis and CRP (21). In these cases, the "265-like determinant" interacts directly with DNA. Mutations within MarA at positions His14, Ser15, Asp18, Trp19, Asp22, Arg36, and Arg37 that abolish transcriptional activation are those contacts made with the 265-like determinant of RNAP (Fig. 3) (21). This result indicates that the RNAP alpha-CTD docking site on MarA facilitates the interaction of both proteins at the two different promoter configurations. (This is reviewed in more detail by Martin and Rosner in chapter 17). An examination of two negatively regulated genes has shown that the MarA binding site has a similar DNA sequence to the positively regulated genes, and spatially it overlaps the −35 hexamer and lies with the amino terminus of MarA proximal to RNAP (82). It remains to be seen how the MarA-RNAP interactions play out at the repressed promoters, if at all. Current bioinformatics approaches estimate that at least 10,000 binding sites are present in an exponentially growing *E. coli* cell

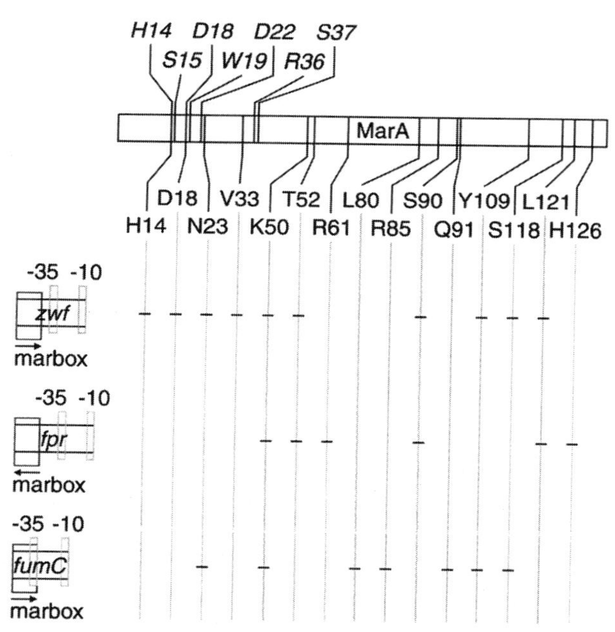

Figure 3. MarA residues important for transcriptional activation and DNA binding. Alanine mutations within the MarA protein, which result in promoter specific defects in vivo, are shown. Chemical shift mapping of the MarA-DNA complexes and β-galactosidase assays measuring the activation of the different promoter classes regulated by MarA demonstrate the effects on transcriptional activation (30). The − indicates the negative effect on transcriptional activation by MarA at the different promoters. The MarA residues shown in italics have been demonstrated to be critical for interactions with RNA polymerase at all three promoter configurations (21). Both *zwf* and *fpr* are representative of class I promoters and *fumC* is a class II promoter (51).

(50); hence, one of the fundamental questions surrounding MarA mediated regulation is "How does MarA locate its cognate binding site given the degeneracy of these marbox sequences?" Studies have shown that MarA is able to interact with both the RNAP core enzyme and holoenzyme in the absence of DNA (50). Therefore, MarA and RNAP can form a binary complex in solution and subsequently scan the chromosome for binding sites, which would enhance the probability of finding a "true" MarA and RNAP binding site.

The crystal structure of MarA has demonstrated that it is the first prokaryotic transcriptional activator with a bipartite HTH motif consisting of seven α-helices which fold to form two HTH subdomains connected by a long linker α-helix (77). This linker helix interacts with the DNA sugar phosphate groups and controls the orientation and the tandem binding of the HTH motifs (22, 77). Generally both HTH domains contribute to binding by inserting into the major groove of the DNA (i.e., marbox) and bending the DNA by ≈35° (77). MarA interacts with the *mar* marbox by inserting into the recognition helices (helix 3 and 6), which protrude from the same side of the protein into the binding site. Therefore, MarA binds as a monomer to one face of the helix at the *mar* marbox (77).

X-ray crystallographic analysis of MarA predicted both base specific and nonspecific contacts with the *mar* marbox (77). Alanine mutagenesis of MarA residues positioned at Arg46 (helix-3) and Arg96 (helix-6) severely disrupt activation at both class I and II promoters (30) despite the fact that the DNA bases contacting these amino acids are not conserved between the different promoter classes (class I and II) tested.

Several other contacts (i.e., Thr93, Glu31, and Thr97), predicted to be important for MarA-marbox interaction based on the crystallographic structure, were found not to hinder MarA activity in vivo (30, 77). The cocrystal structure predicted that the protruding side chains of helix 3 and helix 6 are docked into the major groove of the *mar* marbox where five amino acid residues in the two helices make van der Waals contacts with the DNA bases: Trp42 and Gln45 in helix-3 and Gln91, Gln92, and Thr95 in helix-6 (77). Mutagenesis of the helix-3 residues in particular disrupted MarA activity, whereas substitutions within helix-6 at Gln91 and Thr95 resulted in variable activity at different promoter classes (Fig. 3). The substitution at Gln91 appeared to have a promoter-specific defect primarily at the class II promoters, *fumC* and *micF*.

Alanine substitutions outside the MarA DNA binding regions, especially in the hydrophobic core within the HTH regions and solvent exposed amino acid positions (Trp19, Glu21, Leu28, Pro78, and Arg110), severely disrupt MarA activity (30). Generally, amino acid substitutions located throughout the protein surface are able to affect MarA activity (Fig. 3) at the different promoters to variable levels. However, alanine substitutions on MarA at positions Glu77, Leu80, and Gln91 result in class II promoter-specific defects (30). Like the CAP protein (14), MarA appears to have a class II promoter-specific region. In contrast, no single class I specific region on MarA could be deduced from the current data, as the substitutions affecting this class of promoters were found all over the protein surface (Fig. 3) (22, 30).

The activating potential and binding affinities of the MarA alanine mutants are severely affected when substitutions within the amino-terminal region, helix-3, and the hydrophobic core are present in contrast to alterations within the carboxy-terminal region, helix-6. This effect is attributed to the greater sequence conservation of the marbox sequence to which helix-3 binds and strongly suggests that the critical point of contact is between helix-3 of MarA and the first 10 nucleotides of the marbox sequence (30, 35).

Studies investigating protein-protein interactions and regulation by other transcription factors will be necessary to further the understanding of MarA in the cell. Evidence for such cellular activities is emerging; for example, MarA-mediated repression levels of *hdeA* in vivo are appreciably greater in comparison to in vitro levels, suggesting a coordinate regulation with another protein in vivo (82). Secondly, the negative regulation of MarA by the periplasmic protein MppA and another mutated locus (12, 45) provides the first example of non-MarR mediated repression.

MarB AND MarC

MarB

MarB is transcribed as part of the *marRAB* locus and possesses its own ribosome binding site (16). In defining the minimal regulatory sequences required to induce, repress, and raise constitutive *mar* mutants, it was found in all cases that *marB* was not required (54). The overexpression of *marB* alone does not have any effect on drug resistance (5), although a 10-fold increase has been reported in the protein production of MalE-MarA fusion when extracted from a *malE-marAB* construct as compared to one without *marB* (5). Therefore it is possible that *marB* may have a role in stabilizing protein production of MarA in *E. coli.* When the rates of cell killing by the fluoroquinolones was investigated in a 39kb deletion mutant (Δ *marRAB* and flanking regions), both transcriptional units (TU1 and TU2 [Fig. 1]) of the *mar* locus (which included *marB*) were required to confer a protective effect (32). This requirement mirrors another finding in which complementation by both transcriptional units was required to restore the organic solvent tolerance phenotype in the same 39-kb deletion mutant (Δ *marRAB*) (89). Thus, the requirements of *marB* to confer the *mar* phenotype appear to be strain specific. Despite its apparent lack of function, *marB* is conserved in other members of the family *Enterobacteriaceae* where a recent database search demonstrated homologous proteins of at least 50% identity in *Shigella flexneri, S. enterica* serotypes Typhimurium and Enteritidis, and *Klebsiella pneumoniae.*

MarC

MarC is positioned 211 nucleotides upstream of *marR* and encodes a putative integral membrane protein of 221 amino acids with six predicted transmembrane domains (Fig. 1) (62). Two paralogs of MarC, YchE (65%), and YhgN (63%), which possess similar hydrophobicity profiles and size are present in *E. coli* (4, 62), although both of these proteins have no assigned metabolic function in the cell. Stringent Southern hybridization analysis with an *E. coli marC* probe found *marC*-like sequences in several members of the *Enterobacteriaceae* family like *Salmonella* spp., *S. dysenteriae, Klebsiella* spp., *Enterobacter cloacae,* and *Yersinia enterocolitica* (62). Blast searches of the NCBI database for *marC* homologs found a large number located in other species, which include *Bacillus subtilis* (65% similarity) and *Aquifex aeolicus* (67%) (62). However, the search for contiguous *marA*-like sequences did not reveal any in most species, suggesting that *marC* may be regulated independently of *marO* (62).

The transcriptional start site of *marC* has been mapped to nucleotide 1266 (62) based on the sequence numbering by Cohen et al. (16). The expression of *marC* is relatively low in the absence of inducing agents like chloramphenicol and tetracycline (16, 62), and this is attributed to suboptimal spacing between the −10 and −35 hexamers in the RNA polymerase binding sites. Chloramphenicol and tetracycline, but not salicylate, can induce *marC*-specific transcription although the exact mechanism is not apparent. This finding is in contrast to that demonstrated in *Salmonella* serotype Typhimurium, where *marC* is inducible by salicylate (85). Like MarB, MarC does not appear to have a role in the *mar* phenotype, and accordingly, no contribution to drug resistance, organic solvent tolerance, or bile resistance could be linked to the overexpression nor the deletion of *marC* or its paralogs *ychE* and *yhgN* (62).

MOLECULAR EVOLUTION OF MarA

Within the AraC/XylS family of proteins, MarA belongs to a subgroup comprised of TetD, SoxS, PqrA, RamA, AarP, and YgcK (25, 38), all of which are shorter than 166 amino acids. This means that MarA, which is 127 amino acids (4, 5, 26) in length, contains barely more than the 100 amino acid consensus making up the conserved DNA-binding domain of the AraC/XylS family. This is unique, primarily because most XylS/AraC family members contain 250 to 350 some even up to 500 amino acids within the DNA binding domains, where these additional sequences usually form an effector-binding and/or dimerizing domain fused to the DNA-binding domains by a short flexible linker (25, 56). This domain is thought to sense specific environmental signals and to trigger the regulator accordingly. By analogy, protein families involved in secretion (27) and sensing specificity in two-domain XylS/AraC regulators might evolve through the modular exchange of sensor domains. This poses the questions of whether MarA has a sensor domain and whether sequence translations up- and downstream of MarA are related to known sensor domains. Alignments with the AraC/XylS regulators of protein translations in all three forward reading frames of sequences downstream of MarA showed no homology. However, 20.5 to 26.9% amino acid identity with the XylS sensor domain was detected within a stretch of 140 amino acids immediately upstream of MarA and in the same reading frame. This finding indicated that the sequence homology between MarA and XylS theoretically spanned 270 rather than just 129 of the XylS protein's 321 amino acids; however, the upstream region was untranslatable due to several stop codons.

Therefore, it is possible that regulator genes might have been translocated to distant operons during evolution and that sensor domains might have been replaced or deleted or evolved to new function(s). In the case of MarA one might speculate that (i) the sensor domain lost association with the DNA-binding domain, (ii) the sensor-coding DNA eroded through several amber/ochre/opal mutations, and (iii) this DNA finally evolved into coding for a protein with a new function, i.e., MarR, in another reading frame. Consequently, the lack of a sensor domain for environmental signals has resulted in an arrangement where *marA* expression is tightly down-regulated. This repressory effect is only removed if MarR is antagonized by an inducer or mutated. A striking coincidence is that both MarR and the XylS sensor domain both respond to small aromatic compounds (87).

These observations hint at an evolutionary link of MarR and A to a distant group of two component signal-transduction systems which include EnvZ/OmpR (78). This supports the view that evolution of all prokaryotic regulators, repressors, and activators follow basic principles (73).

CONSERVATION OF THE *mar* LOCUS IN OTHER BACTERIA

Southern hybridization analysis with an *E. coli marRAB* probe has demonstrated that the *marRAB* locus is widespread and highly conserved among members of *Enterobacteriaceae* such as *S. flexneri, Citrobacter freundii, K. pneumoniae, Klebsiella oxytoca,* and *Enterobacter cloacae* (20). Apart from illustrating the ubiquitousness of the *mar* locus, this finding underscores the possibility that the functional and regulatory aspects may also be conserved.

The overexpression of *E. coli marA* in *Enterobacter aerogenes* (15) and *Mycobacterium smegmatis* (63) demonstrates that the gene is functional in both species, implying that a *mar* operon or a *mar*-like regulatory system exists in these organisms. Studies describing the effect of salicylate induction in bacteria demonstrate that there is an increase in antibiotic resistance to a wide spectrum of agents (13, 74). This shows that many *mar*-like systems may be present in bacteria across all genera. Furthermore, salicylate has been shown to increase the frequency at which fusidic acid resistant *Staphylococcus aureus* (74) can be obtained, suggesting that a *mar*-like gene or locus provides protection from antibiotic challenge, an effect previously observed in *E. coli* (32). When grown in the presence of sodium salicylate, *Pseudomonas cepacia* exhibited a multiple antibiotic resistance phenotype attributed to the loss of an outer membrane protein (13), a phenomenon also observed in *E. coli* overexpressing *marA* (19). Database searches inform us that MarA homologs are present in nearly all bacterial species; however, the functional roles of these proteins in conferring "multiple antibiotic resistance" or "multiple adaptive response" are unknown. It is important to note that in most species the genetic arrangement of the *mar* operon is not conserved and that individual genes that represent the members of the MarR and MarA families are in fact distant from one another in the chromosome. It remains to be shown whether these genes are chromosomal elements that specify intrinsic antibiotic resistance or regulatory components in conferring the relevant phenotype.

CLINICAL IMPLICATIONS AND CONCLUSIONS

Reports of multidrug resistance in clinical and environmental isolates which overexpress *marA* or *soxS* are increasing (43, 49, 67, 68, 75). Although the levels of resistance reached either by induction or by relevant regulatory gene mutations causing overexpression of either *marA* or *soxS* are low to moderate, it staves off cell death until other mechanisms develop. This property has been clearly demonstrated in previous experiments (32). Thus MarA and other such regulatory proteins could provide a stepping stone to higher levels of resistance. It is accepted that the overexpression of MarA upregulates the expression of the drug efflux pump AcrAB and causes a decrease in outer membrane protein F production (18, 69). When considering the current data on the Mar regulon (10, 71), it is possible that other *marA*-regulated genes directly or indirectly affecting antibiotic resistance may be discovered.

IN APPRECIATION

It has been an honor to contribute to this book celebrating Stuart Levy's work. The opportunity of working with him has taught us so much, and the research has been nothing short of exciting. He has been a constant source of encouragement and support throughout, and we thank him for everything.

Acknowledgments. We thank Johannes Schneiders for drawing the figures. We thank Dr. S. B. Levy for providing results prior to publication and Bonnie Marshall for helpful comments on the manuscript.

REFERENCES

1. **Alekshun, M. N., Y. S. Kim, and S. B. Levy.** 2000. Mutational analysis of MarR, the negative regulator of *marRAB* expression in *Escherichia coli,* suggests the presence of two regions required for DNA binding. *Mol. Microbiol.* **35:**1394–1404.
2. **Alekshun, M. N., and S. B. Levy.** 1999. Alteration of the repressor activity of MarR, the negative regulator of the *Escherichia coli marRAB* locus, by multiple chemicals in vitro. *J. Bacteriol.* **181:**4669–4672.
3. **Alekshun, M. N., and S. B. Levy.** 1999. Characterization of MarR superrepressor mutants. *J. Bacteriol.* **181:**3303–3306.
4. **Alekshun, M. N., and S. B. Levy.** 1999. The *mar* regulon: multiple resistance to antibiotics and other toxic chemicals. *Trends Microbiol.* **7:**410–413.
5. **Alekshun, M. N., and S. B. Levy.** 1997. Regulation of chromosomally mediated multiple antibiotic resistance: the *mar* regulon. *Antimicrob. Agents Chemother.* **41:**2067–2075.
6. **Alekshun, M. N., S. B. Levy, T. R. Mealy, B. A. Seaton, and J. F. Head.** 2001. The crystal structure of MarR, a regulator of multiple antibiotic resistance, at 2.3 Å resolution. *Nat. Struct. Biol.* **8:**710–714.
7. **Ariza, R. R., Z. Li, N. Ringstad, and B. Demple.** 1995. Activation of multiple antibiotic resistance and binding of stress-inducible promoters by *Escherichia coli* Rob protein. *J. Bacteriol.* **177:**1655–1661.
8. **Balague, C., and E. G. Vescovi.** 2001. Activation of multiple antibiotic resistance in uropathogenic *Escherichia coli* strains by aryloxoalcanoic acid compounds. *Antimicrob. Agents Chemother.* **45:**1815–1822.
9. **Baquero, F.** 2001. Low-level antibacterial resistance: a gateway to clinical resistance. *Drug Resist. Update* **4:**93–105.
10. **Barbosa, T. M., and S. B. Levy.** 2000. Differential expression of over 60 chromosomal genes in *Escherichia coli* by constitutive expression of MarA. *J. Bacteriol.* **182:**3467–3474.
11. **Barbosa, T. M., and S. B. Levy.** 1999. Genetic analysis of the *Shigella flexneri mar* locus, abstr. A42:9. *In 99th General Meeting of the American Society for Microbiology.* American Society for Microbiology, Washington, D.C.
12. **Bina, X., V. Perreten, and S. B. Levy.** 2003. The periplasmic protein MppA requires an additional mutated locus to repress *marA* expression in *Escherichia coli. J. Bacteriol.* **185:**1465–1469.
13. **Burns, J. L., and D. K. Clark.** 1992. Salicylate-Inducible antibiotic resistance in *Pseudomonas cepacia* associated with absence of a pore-forming outer membrane protein. *Antimicrob. Agents Chemother.* **36:**2280–2285.
14. **Busby, S., and R. H. Ebright.** 1999. Transcription activation by catabolite activator protein (CAP). *J. Mol. Biol.* **293:**199–213.
15. **Chollet, R., C. Bollet, J. Chevalier, M. Mallea, J. M. Pages, and A. Davin-Regli.** 2002. *mar* operon involved in multidrug resistance of *Enterobacter aerogenes. Antimicrob. Agents Chemother.* **46:**1093–1097.
16. **Cohen, S. P., H. Hachler, and S. B. Levy.** 1993. Genetic and functional analysis of the multiple antibiotic resistance (*mar*) locus in *Escherichia coli. J. Bacteriol.* **175:**1484–1492.
17. **Cohen, S. P., S. B. Levy, J. Foulds, and J. L. Rosner.** 1993. Salicylate induction of antibiotic resistance in *Escherichia coli:* activation of the *mar* operon and a *mar*-independent pathway. *J. Bacteriol.* **175:**7856–7862.
18. **Cohen, S. P., L. M. McMurry, D. C. Hooper, J. S. Wolfson, and S. B. Levy.** 1989. Cross-resistance to fluoroquinolones in multiple-antibiotic-resistant (Mar) *Escherichia coli* selected by tetracycline or chloramphenicol: decreased drug accumulation associated with membrane changes in addition to OmpF reduction. *Antimicrob. Agents Chemother.* **33:**1318–1325.
19. **Cohen, S. P., L. M. McMurry, and S. B. Levy.** 1988. *marA* locus causes decreased expression of OmpF porin in multiple-antibiotic-resistant (Mar) mutants of *Escherichia coli. J. Bacteriol.* **170:**5416–5422.
20. **Cohen, S. P., W. Yan, and S. B. Levy.** 1993. A multidrug resistance regulatory chromosomal locus is widespread among enteric bacteria. *J. Infect. Dis.* **168:**484–488.
21. **Dangi, B., A. M. Gronenborn, J. L. Rosner, and R. G. Martin.** 2004. Versatility of the carboxy-terminal domain of the α subunit of RNA polymerase in transcriptional activation: use of the DNA contact site as a protein contact site for MarA. *Mol. Microbiol.* **54:**45–59.
22. **Dangi, B., P. Pelupessey, R. G. Martin, J. L. Rosner, J. M. Louis, and A. M. Gronenborn.** 2001. Structure and dynamics of MarA-DNA complexes: an NMR investigation. *J. Mol. Biol.* **314:**113–127.
23. **Egan, S. M.** 2002. Growing repertoire of AraC/XylS activators. *J. Bacteriol.* **184:**5529–5532.
24. **Egan, S. M., A. J. Pease, J. Lang, X. Li, V. Rao, W. K. Gillette, R. Ruiz, J. L. Ramos, and R. E. Wolf, Jr.** 2000. Transcription activation by a variety of AraC/XylS family activators does not depend on the class II-specific activation determinant in the N-terminal domain of the RNA polymerase alpha subunit. *J. Bacteriol.* **182:**7075–7077.

25. **Gallegos, M. T., R. Schleif, A. Bairoch, K. Hofmann, and J. L. Ramos.** 1997. Arac/XylS family of transcriptional regulators. *Microbiol. Mol. Biol. Rev.* **61:**393–410.
26. **Gambino, L., S. J. Gracheck, and P. F. Miller.** 1993. Overexpression of the MarA positive regulator is sufficient to confer multiple antibiotic resistance in *Escherichia coli. J. Bacteriol.* **175:**2888–2894.
27. **Genin, S., and C. A. Boucher.** 1994. A superfamily of proteins involved in different secretion pathways in gram-negative bacteria: modular structure and specificity of the N-terminal domain. *Mol. Gen. Genet.* **243:**112–118.
28. **George, A. M., and S. B. Levy.** 1983. Amplifiable resistance to tetracycline, chloramphenicol, and other antibiotics in *Escherichia coli:* involvement of a non-plasmid-determined efflux of tetracycline. *J. Bacteriol.* **155:**531–540.
29. **George, A. M., and S. B. Levy.** 1983. Gene in the major cotransduction gap of the *Escherichia coli* K-12 linkage map required for the expression of chromosomal resistance to tetracycline and other antibiotics. *J. Bacteriol.* **155:**541–548.
30. **Gillette, W. K., R. G. Martin, and J. L. Rosner.** 2000. Probing the *Escherichia coli* transcriptional activator MarA using alanine-scanning mutagenesis: residues important for DNA binding and activation. *J. Mol. Biol.* **299:**1245–1255.
31. **Gillette, W. K., S. Rhee, J. L. Rosner, and R. G. Martin.** 2000. Structural homology between MarA of the AraC family of transcriptional activators and the integrase family of site-specific recombinases. *Mol. Microbiol.* **35:**1582–1583.
32. **Goldman, J. D., D. G. White, and S. B. Levy.** 1996. Multiple antibiotic resistance (*mar*) locus protects *Escherichia coli* from rapid cell killing by fluoroquinolones. *Antimicrob. Agents Chemother.* **40:**1266–1269.
33. **Greenberg, J. T., J. H. Chou, P. A. Monach, and B. Demple.** 1991. Activation of oxidative stress genes by mutations at the *soxQ/cfxB/marA* locus of *Escherichia coli. J. Bacteriol.* **173:** 4433–4439.
34. **Griffith, K. L., I. M. Shah, and R. E. Wolf, Jr.** 2004. Proteolytic degradation of *Escherichia coli* transcription activators SoxS and MarA as the mechanism for reversing the induction of the superoxide (SoxRS) and multiple antibiotic resistance (Mar) regulons. *Mol. Microbiol.* **51:**1801–1816.
35. **Griffith, K. L., and R. E. Wolf, Jr.** 2001. Systematic mutagenesis of the DNA binding sites for SoxS in the *Escherichia coli zwf* and *fpr* promoters: identifying nucleotides required for DNA binding and transcription activation. *Mol. Microbiol.* **40:**1141–1154.
36. **Grkovic, S., M. H. Brown, and R. A. Skurray.** 2002. Regulation of bacterial drug export systems. *Microbiol. Mol. Biol. Rev.* **66:**671–701, table of contents.
37. **Hächler, H., S. P. Cohen, and S. B. Levy.** 1991. *marA*, a regulated locus which controls expression of chromosomal multiple antibiotic resistance in *Escherichia coli. J. Bacteriol.* **173:**5532–5538.
38. **Hächler, H., S. P. Cohen, and S. B. Levy.** 1996. Untranslated sequence upstream of MarA in the multiple antibiotic resistance locus of *Escherichia coli* is related to the effector-binding domain of the XylS transcriptional activator. *J. Mol. Evol.* **42:**409–413.
39. **Hovey, A. K., and D. W. Frank.** 1995. Analyses of the DNA-binding and transcriptional activation properties of ExsA, the transcriptional activator of the *Pseudomonas aeruginosa* exoenzyme S regulon. *J. Bacteriol.* **177:**4427–4436.
40. **Hulbert, R. R., and R. K. Taylor.** 2002. Mechanism of ToxT-dependent transcriptional activation at the *Vibrio cholerae tcpA* promoter. *J. Bacteriol.* **184:**5533–5544.
41. **Jair, K. W., W. P. Fawcett, N. Fujita, A. Ishihama, and R. E. Wolf, Jr.** 1996. Ambidextrous transcriptional activation by SoxS: requirement for the C-terminal domain of the RNA polymerase alpha subunit in a subset of *Escherichia coli* superoxide-inducible genes. *Mol. Microbiol.* **19:**307–317.
42. **Jair, K. W., R. G. Martin, J. L. Rosner, N. Fujita, A. Ishihama, and R. E. Wolf, Jr.** 1995. Purification and regulatory properties of MarA protein, a transcriptional activator of *Escherichia coli* multiple antibiotic and superoxide resistance promoters. *J. Bacteriol.* **177:**7100–7104.
43. **Kern, W. V., M. Oethinger, A. S. Jellen-Ritter, and S. B. Levy.** 2000. Non-Target gene mutations in the development of fluoroquinolone resistance in *Escherichia coli. Antimicrob. Agents Chemother.* **44:**814–820.
44. **Koide, A., M. Perego, and J. A. Hoch.** 1999. ScoC regulates peptide transport and sporulation initiation in *Bacillus subtilis. J. Bacteriol.* **181:**4114–4117.
45. **Li, H., and J. T. Park.** 1999. The periplasmic murein peptide-binding protein MppA is a negative regulator of multiple antibiotic resistance in *Escherichia coli. J. Bacteriol.* **181:**4842–4847.
46. **Lobner-Olesen, A., M. G. Marinus, and F. G. Hansen.** 2003. Role of SeqA and Dam in *Escherichia coli* gene expression: a global/microarray analysis. *Proc. Natl. Acad. Sci. USA* **100:** 4672–4677.
47. **Lomovskaya, O., and K. Lewis.** 1992. Emr, an *Escherichia coli* locus for multidrug resistance. *Proc. Natl. Acad. Sci. USA* **89:** 8938–8942.
48. **Lomovskaya, O., K. Lewis, and A. Matin.** 1995. EmrR is a negative regulator of the *Escherichia coli* multidrug resistance pump EmrAB. *J. Bacteriol.* **177:**2328–2334.
49. **Maneewannakul, K., and S. B. Levy.** 1996. Identification of *mar* mutants among quinolone-resistant clinical isolates of *Escherichia coli. Antimicrob. Agents Chemother.* **40:**1695–1698.
50. **Martin, R. G., W. K. Gillette, N. I. Martin, and J. L. Rosner.** 2002. Complex formation between activator and RNA polymerase as the basis for transcriptional activation by MarA and SoxS in *Escherichia coli. Mol. Microbiol.* **43:**355–370.
51. **Martin, R. G., W. K. Gillette, S. Rhee, and J. L. Rosner.** 1999. Structural requirements for marbox function in transcriptional activation of *mar/sox/rob* regulon promoters in *Escherichia coli:* sequence, orientation and spatial relationship to the core promoter. *Mol. Microbiol.* **34:**431–441.
52. **Martin, R. G., W. K. Gillette, and J. L. Rosner.** 2000. Promoter discrimination by the related transcriptional activators MarA and SoxS: differential regulation by differential binding. *Mol. Microbiol.* **35:**623–634.
53. **Martin, R. G., K. W. Jair, R. E. Wolf, Jr., and J. L. Rosner.** 1996. Autoactivation of the *marRAB* multiple antibiotic resistance operon by the MarA transcriptional activator in *Escherichia coli. J. Bacteriol.* **178:**2216–2223.
54. **Martin, R. G., P. S. Nyantakyi, and J. L. Rosner.** 1995. Regulation of the multiple antibiotic resistance (*mar*) regulon by *mar*ORA sequences in *Escherichia coli. J. Bacteriol.* **177:**4176–4178.
55. **Martin, R. G., and J. L. Rosner.** 2003. Analysis of microarray data for the *marA, soxS,* and *rob* regulons of *Escherichia coli. Methods Enzymol.* **370:**278–280.
56. **Martin, R. G., and J. L. Rosner.** 2001. The AraC transcriptional activators. *Curr. Opin. Microbiol.* **4:**132–137.
57. **Martin, R. G., and J. L. Rosner.** 1995. Binding of purified multiple antibiotic-resistance repressor protein (MarR) to *mar* operator sequences. *Proc. Natl. Acad. Sci. USA* **92:**5456–5460.
58. **Martin, R. G., and J. L. Rosner.** 1997. Fis, an accessorial factor for transcriptional activation of the *mar* (multiple antibiotic resistance) promoter of *Escherichia coli* in the presence of the activator MarA, SoxS, or Rob. *J. Bacteriol.* **179:**7410–7419.

59. **Martin, R. G., and J. L. Rosner.** 2002. Genomics of the *marA/soxS/rob* regulon of *Escherichia coli:* identification of directly activated promoters by application of molecular genetics and informatics to microarray data. *Mol. Microbiol.* **44:**1611–1624.
60. **Martin, R. G., and J. L. Rosner.** 2004. Transcriptional and translational regulation of the *marRAB* multiple antibiotic resistance operon in *Escherichia coli. Mol. Microbiol.* **53:**183–191.
61. **Martinez, J. L., and F. Baquero.** 2000. Mutation frequencies and antibiotic resistance. *Antimicrob. Agents Chemother.* **44:**1771–1777.
62. **McDermott, P. F., I. Podglajen, M. P. Draper, J. L. Dzink-Fox, and S. B. Levy.** Unpublished data.
63. **McDermott, P. F., D. G. White, I. Podglajen, M. N. Alekshun, and S. B. Levy.** 1998. Multidrug resistance following expression of the *Escherichia coli marA* gene in *Mycobacterium smegmatis. J. Bacteriol.* **180:**2995–2998.
64. **Miller, P. F., L. F. Gambino, M. C. Sulavik, and S. J. Gracheck.** 1994. Genetic relationship between *soxRS* and *mar* loci in promoting multiple antibiotic resistance in *Escherichia coli. Antimicrob. Agents Chemother.* **38:**1773–1779.
65. **Moken, M. C., L. M. McMurry, and S. B. Levy.** 1997. Selection of multiple-antibiotic-resistant (*mar*) mutants of *Escherichia coli* by using the disinfectant pine oil: roles of the *mar* and *acrAB* loci. *Antimicrob. Agents Chemother.* **41:**2770–2772.
66. **Notka, F., H. J. Linde, A. Dankesreiter, H. H. Niller, and N. Lehn.** 2002. A C-terminal 18 amino acid deletion in MarR in a clinical isolate of *Escherichia coli* reduces MarR binding properties and increases the MIC of ciprofloxacin. *J. Antimicrob. Chemother.* **49:**41–47.
67. **Oethinger, M., W. V. Kern, A. S. Jellen-Ritter, L. M. McMurry, and S. B. Levy.** 2000. Ineffectiveness of topoisomerase mutations in mediating clinically significant fluoroquinolone resistance in *Escherichia coli* in the absence of the AcrAB efflux pump. *Antimicrob. Agents Chemother.* **44:**10–13.
68. **Oethinger, M., I. Podglajen, W. V. Kern, and S. B. Levy.** 1998. Overexpression of the *marA* or *soxS* regulatory gene in clinical topoisomerase mutants of *Escherichia coli. Antimicrob. Agents Chemother.* **42:**2089–2094.
69. **Okusu, H., D. Ma, and H. Nikaido.** 1996. AcrAB efflux pump plays a major role in the antibiotic resistance phenotype of *Escherichia coli* multiple-antibiotic-resistance (Mar) mutants. *J. Bacteriol.* **178:**306–308.
70. **Perez-Trallero, E., J. M. Marimon, L. Iglesias, and J. Larruskain.** 2003. Fluoroquinolone and macrolide treatment failure in pneumococcal pneumonia and selection of multidrug-resistant isolates. *Emerg. Infect. Dis.* **9:**1159–1162.
71. **Pomposiello, P. J., M. H. Bennik, and B. Demple.** 2001. Genome-wide transcriptional profiling of the *Escherichia coli* responses to superoxide stress and sodium salicylate. *J. Bacteriol.* **183:**3890–3902.
72. **Poole, K., K. Tetro, Q. Zhao, S. Neshat, D. E. Heinrichs, and N. Bianco.** 1996. Expression of the multidrug resistance operon *mexA-mexB-oprM* in *Pseudomonas aeruginosa: mexR* encodes a regulator of operon expression. *Antimicrob. Agents Chemother.* **40:**2021–2028.
73. **Prag, G., S. Greenberg, and A. B. Oppenheim.** 1997. Structural principles of prokaryotic gene regulatory proteins and the evolution of repressors and gene activators. *Mol. Microbiol.* **26:**619–620.
74. **Price, C. T., and J. E. Gustafson.** 2001. Increases in the mutation frequency at which fusidic acid-resistant *Staphylococcus aureus* arise with salicylate. *J. Med. Microbiol.* **50:**104–106.
75. **Randall, L. P., A. M. Ridley, S. W. Cooles, M. Sharma, A. R. Sayers, L. Pumbwe, D. G. Newell, L. J. Piddock, and M. J. Woodward.** 2003. Prevalence of multiple antibiotic resistance in 443 *Campylobacter* spp. isolated from humans and animals. *J. Antimicrob. Chemother.* **52:**507–510.
76. **Revell, P. A., and V. L. Miller.** 2000. A chromosomally encoded regulator is required for expression of the *Yersinia enterocolitica inv* gene and for virulence. *Mol. Microbiol.* **35:**677–685.
77. **Rhee, S., R. G. Martin, J. L. Rosner, and D. R. Davies.** 1998. A novel DNA-binding motif in MarA: the first structure for an AraC family transcriptional activator. *Proc. Natl. Acad. Sci. USA* **95:**10413–10418.
78. **Ronson, C. W., B. T. Nixon, and F. M. Ausubel.** 1987. Conserved domains in bacterial regulatory proteins that respond to environmental stimuli. *Cell* **49:**579–581.
79. **Rouanet, C., S. Reverchon, D. A. Rodionov, and W. Nasser.** 2004. Definition of a consensus DNA-binding site for PecS, a global regulator of virulence gene expression in *Erwinia chrysanthemi* and identification of new members of the PecS regulon. *J. Biol. Chem.* **279:**30158–30167.
80. **Rupali, P., O. C. Abraham, M. V. Jesudason, T. J. John, A. Zachariah, S. Sivaram, and D. Mathai.** 2004. Treatment failure in typhoid fever with ciprofloxacin susceptible *Salmonella enterica* serotype Typhi. *Diagn. Microbiol. Infect. Dis.* **49:**1–3.
81. **Rzeszutek, M., A. Wierzbowski, D. J. Hoban, J. Conly, W. Bishai, and G. G. Zhanel.** 2004. A review of clinical failures associated with macrolide-resistant *Streptococcus pneumoniae. Int. J. Antimicrob. Agents* **24:**95–104.
82. **Schneiders, T., T. M. Barbosa, L. M. McMurry, and S. B. Levy.** 2004. The *Escherichia coli* transcriptional regulator MarA directly represses transcription of *purA* and *hdeA. J. Biol. Chem.* **279:**9037–9042.
83. **Schumacher, M. A., and R. G. Brennan.** 2002. Structural mechanisms of multidrug recognition and regulation by bacterial multidrug transcription factors. *Mol. Microbiol.* **45:**885–893.
84. **Seoane, A. S., and S. B. Levy.** 1995. Characterization of MarR, the repressor of the multiple antibiotic resistance (*mar*) operon in *Escherichia coli. J. Bacteriol.* **177:**3414–3419.
85. **Sulavik, M. C., M. Dazer, and P. F. Miller.** 1997. The *Salmonella typhimurium mar* locus: molecular and genetic analyses and assessment of its role in virulence. *J. Bacteriol.* **179:**1857–1866.
86. **Sulavik, M. C., L. F. Gambino, and P. F. Miller.** 1994. Analysis of the genetic requirements for inducible multiple-antibiotic resistance associated with the *mar* locus in *Escherichia coli. J. Bacteriol.* **176:**7754–7756.
87. **Sulavik, M. C., L. F. Gambino, and P. F. Miller.** 1995. The MarR repressor of the multiple antibiotic resistance (*mar*) operon in *Escherichia coli:* prototypic member of a family of bacterial regulatory proteins involved in sensing phenolic compounds. *Mol. Med.* **1:**436–446.
88. **Tramonti, A., P. Visca, M. De Canio, M. Falconi, and D. De Biase.** 2002. Functional characterization and regulation of *gadX*, a gene encoding an AraC/XylS-like transcriptional activator of the *Escherichia coli* glutamic acid decarboxylase system. *J. Bacteriol.* **184:**2603–2613.
89. **White, D. G., J. D. Goldman, B. Demple, and S. B. Levy.** 1997. Role of the *acrAB* locus in organic solvent tolerance mediated by expression of *marA, soxS,* or *robA* in *Escherichia coli. J. Bacteriol.* **179:**6122–6126.
90. **Xiong, A., A. Gottman, C. Park, M. Baetens, S. Pandza, and A. Matin.** 2000. The EmrR protein represses the *Escherichia coli emrRAB* multidrug resistance operon by directly binding to its promoter region. *Antimicrob. Agents Chemother.* **44:**2905–2907.

Frontiers in Antimicrobial Resistance: a Tribute to Stuart B. Levy
Edited by D. G. White, M. N. Alekshun, and P. F. McDermott

Chapter 15

The *mar* Regulon

TERESA M. BARBOSA AND PABLO J. POMPOSIELLO

Escherichia coli is a natural inhabitant of diverse environmental niches, surviving well in spite of the multitude of external cellular stresses to which it is exposed. Its success is largely due to the presence of regulatory loci and control switches that quickly sense the environment and alter the expression of a network of genes on the bacterial chromosome. These responses, which act to reestablish cellular homeostasis, allow the bacteria to deal with the external hazard, be it superoxide generating agents, antibiotics, nutrient or oxygen limitation, organic solvents, bile salts, or a variety of other insults. The *E. coli mar* regulon, which is under the control of the global transcriptional regulator MarA, mediates one such adaptational response.

The chromosomal multiple antibiotic resistance *marRAB* operon was first characterized genetically in *E. coli* (14) but has more recently been characterized in *Salmonella enterica* serovar Typhimurium (87) and *Shigella flexneri* (10). In all three genera, the *marRAB* operon (described in detail elsewhere in this volume) specifies two regulatory proteins and a third protein, MarB, whose function remains unknown. The first regulatory protein, MarR, is an autorepressor of the operon. It acts by binding to sequences in the operator *marO* and negatively regulates expression of *marRAB* (2, 85). The second protein is a transcriptional factor (MarA) that induces its own transcription in the absence of MarR repression (14, 42, 57).

MarA, in addition to inducing its own transcription, governs the differential expression of the *mar* regulon members. The resultant Mar phenotype includes resistance to structurally unrelated antibiotics, organic solvents, oxidative stress agents, and household disinfectants (4) (see chapter 14).

In this chapter we will review our current understanding of the *mar* regulon, focusing on some of the more interesting features of regulation and highlighting the considerable homology and cross-regulation that exists between MarA and the related transcriptional regulators SoxS and Rob. The extensive overlap observed in the regulons and phenotypes associated with these proteins will be discussed.

MarA, SoxS, AND Rob REGULATORY CIRCUITS

The *mar* regulon consists of a large group of chromosomal genes directly or indirectly regulated by MarA. This regulon is also frequently referred to as the *mar/sox/rob* regulon because the two *E. coli* MarA homologues, SoxS and Rob, recognize the same regulatory DNA element in the promoter of regulated genes. As a consequence they all appear to control an overlapping set of genes.

All three proteins are members of the AraC/XylS family of bacterial transcriptional regulators (26, 59), a general feature of which includes the presence of a conserved DNA binding domain with two HTH motifs. It is also predicted that a carboxyl- or amino-terminal effector binding domain will be part of the structure of many of the family members (26).

MarA and SoxS, which are small proteins of 127 (54) and 107 amino acid residues, respectively, are limited to a DNA binding domain with two HTH binding motifs (see chapter 17). These bind and bend the DNA by 35° in order to establish contacts at two adjacent major grooves (78). Rob, on the other hand, is a larger protein of 289 amino acid residues and, in addition to its homologous HTH binding motifs, contains a carboxyl-terminal effector-binding domain (43, 82). A more detailed structural and functional analysis of the MarA protein subfamily is given elsewhere in this volume (see chapter 17).

Teresa M. Barbosa • Instituto de Tecnologia Química e Biológica, Universidade Nova de Lisboa, Avenida da República, Apartado 127, 2781-901 Oeiras Codex, Portugal. **Pablo J. Pomposiello** • Department of Microbiology, University of Massachusetts, Amherst, MA 01003.

The exact molecular mechanisms responsible for overlapping regulation by MarA, SoxS, and Rob are still poorly understood. For example, SoxS mediated a multiple-resistance phenotype that is essentially indistinguishable from the MarA-mediated response (13, 34, 64) and appeared partly dependent on an intact *mar* locus (64), while MarA appears to operate independently of SoxS (9, 73).

In spite of the sequence homology shared between these three proteins, the ways in which they are regulated differ significantly. Nevertheless, the cross-regulation that evidently occurs between them could ultimately play a role in optimizing the cell's response to different stresses through the differential regulation of specific promoters (Fig. 1).

Regulation of MarA

Since MarA does not possess an effector-binding domain, the *marRAB* system responds to inducers by altering the production of MarA (Fig. 1). In general, activation of *marRAB* transcription occurs when the repressing function of MarR is disrupted, either as a result of mutations in *marO* or *marR* or by direct interaction between the MarR repressor and a chemical inducer, such as phenolic compounds or oxidative

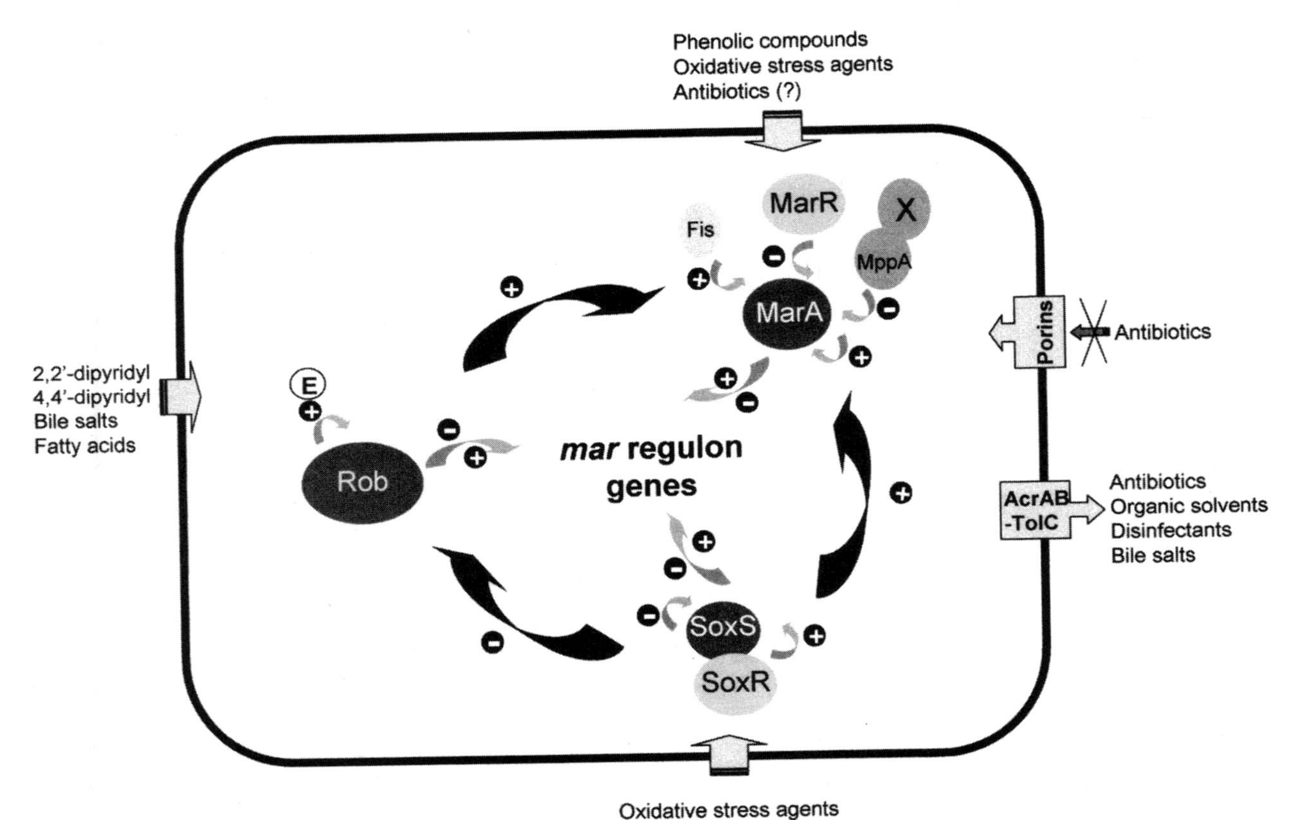

Figure 1. MarA, SoxS, and Rob regulatory circuits. MarA, SoxS, and Rob can mediate a global cellular stress response to different toxic compounds by governing the expression of a common network of chromosomal genes, the *mar* regulon, which is involved in a variety of different cell functions. This includes active efflux of toxic compounds (e.g., through the activation of the AcrAB-TolC complex), reduction of cell permeability (e.g., through the decreased expression of porins, such as OmpF, and modulation of expression of other membrane proteins), detoxification (e.g., through the increased expression of cytoprotective and repair enzymes), and others. A multitude of external stresses are sensed by the MarRAB, SoxRS, and Rob systems, which ultimately result in the modulation through different pathways of the expression of the three transcriptional factors. Some of these stimuli result in the activation of more than one sensory system, while other stimuli display restricted activation of only one system. For example, MarA is produced when MarR is inactivated either by mutations or by interaction with inducing agents, such as phenolic compounds and certain oxidative stress agents. *marA* expression can also be induced by SoxS and Rob and enhanced by Fis. Additionally, MppA, in combination with mutations at an unknown locus in *E. coli*, is capable of influencing *marA* expression through what appears to be a MarR-independent pathway. In contrast, oxidative stress agents oxidize SoxR, which in turn activates expression of SoxS. No other signals are known to result in increased levels of SoxS. Rob is produced constitutively, but recently it has been shown that its expression can be repressed in a SoxS-dependent manner. Although Rob is known to accumulate to high concentrations in the cell, its activation in vivo is thought to be mediated by inducing agents that bind to the carboxyl-terminus effector-binding domain of this protein, such as 2,2′- and 4,4′- dipyridyl, bile salts, and fatty acids (represented as E in the figure).

stress agents, e.g., menadione, plumbagin, and sodium salicylate (1, 4, 15). Most of these compounds abrogate MarR repression by disrupting its ability to bind DNA, allowing for derepression of the *marRAB* operon to occur and the de novo synthesis of MarA to take place (1) (see chapter 18).

Expression of *marA* is also under positive regulation by SoxS and Rob (3, 11) and is enhanced by Fis—an accessory factor that binds to *marO* sequences located upstream from the MarA, SoxS, and Rob binding site (60). In contrast, an effect of MarA on *soxS* or *rob* expression has not been detected (Fig. 1).

Several interesting twists to the standard mechanism of regulation of *marRAB* expression have been described. For example, we have observed induction of the operon by sodium salicylate in a MarR-deficient strain that should in theory be fully derepressed (15, 53, 57, 60) (Barbosa and Levy, unpublished). Induction was also achieved by concentrations of 2,2′-dipyridyl that are below the 5 mM threshold reported to be necessary for the activation of Rob and, therefore, for Rob dependent activation of regulon members (82) (see below). Together, these results suggest that induction of *marRAB* transcription can bypass MarR, but additional studies are required to better understand the molecular basis for this regulation.

Another unusual feature of *marA* expression has been described in a recent report by Bina et al. where the periplasmic murein tripeptide binding protein MppA appeared to be capable of repressing *marA* transcription in a MarR-deficient strain (12) (Fig. 1). This repression was not solely dependent on MppA and required additional, as yet unspecified, mutation(s). Again, further studies will be required to gain a deeper understanding of this pattern of regulation.

While these studies indicate that transcriptional regulation represents the principal control pathway for MarA expression, recent studies also suggest that activity can be regulated posttranslationally by protease regulation of MarA stability (36) (see chapter 17). Future studies will need to address the contribution of this regulation to the overall development and persistence of the Mar phenotype.

Regulation of SoxS

Like MarA, SoxS mediates cellular adaptation to stress via the coordinated transcriptional modulation of gene expression. In particular, SoxS mediates a genetic response to increased intracellular levels of superoxide (21, 28) and nitric oxide (20, 68). The activation of SoxS expression renders the cell resistant to superoxide-generating agents, like the redox-cycling compounds paraquat and menadione, nitric oxide but also to heavy metals, organic solvents, chlorine, and multiple antibiotics.

The activity of SoxS is regulated exclusively by its intracellular concentration, and both the transcription rate of the *soxS* gene and the stability of the SoxS protein are under active regulation (5, 36, 95). The transcriptional regulation of SoxS by SoxR constitutes a paradigm for redox-sensitive genetic switches (76, 97) and is revisited in this volume (chapter 13). A *soxRS* system has also been characterized in *S. enterica* (75).

The SoxR protein is a transcriptional activator of the *soxS* gene that belongs to the MerR family of DNA-binding proteins (Fig. 1). It is a redox-sensing homodimer consisting of 17-kD subunits, each one of which contains a redox-active [2Fe-2S] center. The DNA binding activity of SoxR does not depend on the presence (39) or oxidation state (28) of the iron-sulfur clusters, but only SoxR with oxidized [2Fe-2S] centers activates transcription (21, 28). Activation of SoxR through the one-electron oxidation of its [2Fe-2S] centers therefore corresponds to an allosteric transition in the promoter DNA-SoxR complex. This transition does not increase binding by RNA polymerase, but rather stimulates the formation of the “open” complex essential for initiating transcription (39). The reduction of the [2Fe-2S] clusters occurs rapidly after cessation of the oxidative signal, and the *soxS* transcriptional rate decreases to basal levels (22). This rapid deactivation of SoxR indicated that the rereduction of oxidized SoxR must be an active process. Indeed, a recent study linked the products of the *rsx* and *rseC* genes to a reducing system for SoxR (45). Interestingly, SoxR can be activated by exposure of cells to nitric oxide (NO) in vitro, or by coculture with activated murine macrophages (68). The activation of SoxR by NO depends on the reversible nitrosylation of the [2Fe-2S] clusters and has been shown both in vivo and in vitro (20). Nitrosylated SoxR is stable in vitro but in vivo is rapidly destabilized, suggesting an active mechanism for reconstitution of iron-sulfur clusters (20). The nature of this denitrosylating mechanism remains unknown.

The ability of reactive oxygen species (ROS) to damage virtually all biological macromolecules is reflected in the diversity of genes activated by redox-sensitive regulatory systems. SoxS-activated mechanisms seem to be geared not only toward prevention and repair of oxidative damage, but also for protection against xenobiotics that damage iron-sulfur clusters.

Curiously, although SoxS up-regulates *marA* transcription (57), it appears to repress *rob* expression (63), but neither of the two affects the levels of *soxS* (Fig. 1). Instead, and like many other regulatory proteins, both SoxS and SoxR limit their own expression by binding the promoters of their own structural genes (40, 69).

Regulation of Rob

Rob, in contrast to MarA and SoxS, is produced constitutively and accumulates to about 5,000 to 10,000 copies per cell (43, and references therein). Yet despite this apparent excess, its function remains somewhat cryptic. While it has been reported to induce the expression of *mar* regulon promoters in vivo and to be associated with increased resistance to antibiotics and organics solvents (7, 52, 66), it can only do so when overexpressed from a strong promoter in a high-copy-number plasmid. In contrast, normal levels of Rob in wild-type cells appear incapable of influencing the expression of *mar* regulon genes to any great extent, and the effects of its deletion appear to be limited to small decreases in organic solvent tolerance and in the expression of a relatively small number of genes (11, 92,). Nevertheless, this apparent lack of intrinsic activity in vivo contrasts with the high-binding affinity shown for the promoter sequences of many regulon members in vitro. In many cases this can be even higher than that of MarA and SoxS (46, 54) and has been reported to result in transcriptional activation of a number of *mar* regulon promoters in vitro (43).

While contradictions between in vivo and in vitro regulation have proven confusing, recent studies on the previously uncharacterized carboxyl-terminus of Rob are providing new insights into its regulation and function and are helping to shed some light on these apparent inconsistencies. Two independent studies have now proposed that this carboxyl-terminus region encodes an effector binding domain that plays a role in the posttranscriptional activation of the protein and consequently in the expression of *mar* regulon genes (81, 82) (Fig. 1). For example, Rob activation of several regulon gene promoters could be significantly enhanced in the presence of the effector 2,2′-dipyridyl (82). Insights into the molecular basis for this regulation come from NMR studies, which suggest that direct interactions between the effector molecules and the carboxyl-terminus domain of Rob would produce conformational changes that could generate an active protein in vivo. Besides 2,2′-dipyridyl, 4,4′-dipyridyl, bile salts and fatty acids also appear to induce the *mar* regulon via direct interaction with Rob (81, 82) (Fig. 1). How Rob is active in vitro or when overexpressed in the absence of an effector molecule in vivo remains unclear, particularly when endogenous Rob is inactive in the absence of such an effector molecule.

DEFINING THE *mar* REGULON

Since MarA was first reported as a global regulator, there has been debate concerning the size of the *mar* regulon, and successive studies have aimed to identify its constituent members. Basic genetic and proteomic approaches have been used to help identify a number of bona fide *mar* regulon members, including *micF, ompF, inaA, fumC, sodA, zwf, acrAB,* and *marRAB* (4) (Table 1). Other elements such as the *mlr* genes and *slp* have also been identified but have not been so exhaustively studied (86).

More recently, with the advent of genome-wide transcriptome analysis, two independent macroarray studies have provided new insights into the multitude of genes that constitute the transcriptional network of the regulon. Barbosa and Levy reported the differential expression of 62 genes (47 up-regulated and 15 down-regulated) by comparing the expressing profiles of *mar*-deleted cells and cells constitutively expressing MarA (9). In a similar, yet experimentally different study, Pomposiello and colleagues examined the transcriptome of wild-type cells exposed to the MarA-inducing agent sodium salicylate (73). They reported the altered expression of 144 genes (84 up-regulated and 60 down-regulated), of which 19 had been identified in the Barbosa and Levy study. In a separate set of experiments, Pomposiello et al. also looked at gene expression in a MarA deficient strain that had been transformed to express MarA from an inducible promoter on a multicopy plasmid. A total of 88 genes were differentially expressed following induction (67 up-regulated and 21 down-regulated), of which 20 (15 up- and 5 down-regulated) were in common with the sodium salicylate exposure study, and 21 were in common with the Barbosa and Levy study. In all cases the regulated genes were dispersed throughout the chromosome, either individually or located as clusters of polycistronic units. The reader is referred to both references 9 and 73 for a complete list of the differentially regulated genes under the described experimental parameters.

Although the overlap between the regulons described in these studies was not as large as might have been anticipated, it would be naïve to assume that the same repertoire of responsive genes would be identified given that distinct experimental conditions were used. If nothing else, the results from these studies attest to the sensitivity and specificity of the cellular response mechanisms that are involved in the development of the Mar phenotype, and it is increasingly evident that no single model will simultaneously provide the full dimension of the regulon. For example, while many of the bona fide *mar* regulon members were identified by these studies, others such as *slp* and *poxB* that are expressed during the stationary phase (86, 89) were not detected in any of the arrays studies, where cells were harvested during the exponential growth phase.

Table 1. Bona fide *mar/sox* regulon genes[a]

Unique identifier	Gene name	Function	Promoter class[b]	Regulation[c]		References[d]
				SoxS	MarA	
b0096	*lpxC*	Lipid A biosynthesis	ND	+	+	73, Kenyon and Pomposiello (unpublished)
b0168	*map*	Methionine aminopeptidase	I*	+	+	9, 61, 73
b0463-0462	*acrAB*	Drug efflux pump	I	+	+	9, 56, 61, 73, 92
b0578	*nfnB*	Oxygen-insensitive NADPH nitroreductase	II	+	++	8, 9, 73
b0684-0683	*fldA-fur*	Flavodoxin A-transcriptional repressor of iron uptake genes	I	+	No effect	9, 27, 73, 96, Barbosa and Levy (unpublished)
b0850-0851-0852	*ybjC-nfsA-rimK*	Unknown-nitrodeductase A-modifier of ribosomal protein S6	II	+	+	9, 51, 61, 72, 73
b0871	*poxB*	Pyruvate dehydrogenase/oxidase	I	++	+	56, 61
b0929	*ompF*	Outer membrane protein 1a	NA	−	−	9, 16, 17, 73
b0950	*pqiA*	Paraquat-inducible protein A	II	++	+	56, 61
b1101	*ptsG*	Glucose transporter, PTS system	ND	+	+	73, Fig. 2 (this chapter), Liu and Pomposiello (unpublished)
b1276	*acnA*	Aconitase A	I	+	+	9, 38, 61
b1277	*ribA*	GTP-cyclase	I	+	+	9, 44, 61
b1415	*aldA*	Aldehyde dehydrogenase A	II	+	+	9, 61
b1452			II	+	+	61, 73
b1530-1531-1532	*marRAB*	Multiple drug resistance operon	I	+	++	15, 56, 57, 60, 61, 73
1611	*fumC*	Fumarase C	II	++	+	9, 49, 56, 61
b1852	*zwf*	Glucose-6-P dehydrogenase	I*	+	+	9, 34, 56, 61, 73
b2159	*nfo*	Endonuclease IV	II	++	+	34, 56, 61, 88
b2237	*inaA*	Unknown, induced by acid stress	II	+	++	9, 56, 61, 83
b2414	*cysK*	Cysteine synthase A	ND	+	No effect	9, 73, Pomposiello (unpublished)
b2895	*fldB*	Flavodoxin B	I	+	+?	27, 61
b2947	*gshB*	Glutathione synthase B	ND	No effect	+	9, 73, Fig. 2 (this chapter)
b2962-2963	*yggX-mltC*	Iron trafficking?/peptidoglycan hydrolase	ND	+	No effect	73, 77, Fig. 2 (this chapter)
b3028	*mdaB*	Flavoprotein modulator of drug activity B	II	+	++	9, 61
b3035	*tolC*	Outer membrane protein	II	+	+	6, 9, 61
b3160	*yhbW*	Putative enzyme	II	+	++	9, 61
b3510	*hdeA*	Periplasmic protein involved in acid resistance	− − −	−?	−	73, 84
b3908	*sodA*	Superoxide dismutase (Mn)	II	++	+	9, 34, 56, 61
b3924	*fpr*	NADP-dependent ferredoxin reductase	I	++	+	50, 56, 61
b4025	*pgi*	Glucosephosphate isomerase	II	++	+	9, 61, 73
b4062	*soxS*	Transcription factor, modulates response to O_2^- and NO	ND	−	No effect	9, 69
b4177	*purA*	Adenylosuccinate synthase	− − −	−?	−	9, 84
b4396	*rob*	Transcription factor	ND	−	No effect	9, 63
EG30063	*micF*	Antisense regulator of porin OmpF	II	+	++	13, 56, 61

[a] Regulon members whose differential expression by MarA and/or SoxS has been confirmed by more than one experimental approach.
[b] For specific marbox sequences and respective configurations, see references 8, 55, and 61 and references therein. ND, not determined; NA, not applied, indirect regulation.
[c] +, activation; ++, relatively larger degree of activation; −, repression; ?, no available comparative data.
[d] Due to space restriction, we are only able to refer to selected publications. The reader is strongly advised to consult original research studies referenced therein.

In addition to the technical limitations, there are also interpretive subjectivities such as the establishment of threshold values, which have to be taken into account specifically when macroarray studies are used (9, 73, 79, 80). For example, of the 144 and 88 genes identified as differentially regulated in the sodium salicylate and MarA induction models, 22 and 19, respectively, showed similar patterns of expression in the Barbosa and Levy study. However, they were deemed to be below the chosen experimental threshold and were thus omitted from the final calculations. Clearly, these limitations restrict the comprehensive interpretation of the macroarray data and may be responsible for the misidentification or exclusion of some *mar* regulon members. Nevertheless, even if controversy still persists as to the real size of this system, there can be no doubt as to the usefulness of this approach for the identification of novel regulon members. Classical analysis has by now confirmed that several of the macroarray identified genes are indeed legitimate members of the regulon (8, 61, 72, 77) (Table 1).

While the macroarray data cannot be viewed as either definitive or all-answering in its own right, one of its major strengths lies in its ability to perceive trends and identify families of genes that are regulated. One very insightful application of the data has been described by Martin et al. who have employed bioinformatics approaches to analyze the promoter sequences of macroarray identified genes in order to try to discriminate between genes that are directly controlled by the transcriptional regulators and those that are indirect targets of cascade responses (58, 61). While the use of computer approaches to search for degenerate binding sites such as the MarA recognition sequence is not without problems and may fail to identify sequences that differ considerably from the consensus but are biologically important (see below), the approach has proven very informative (61) (Table 1).

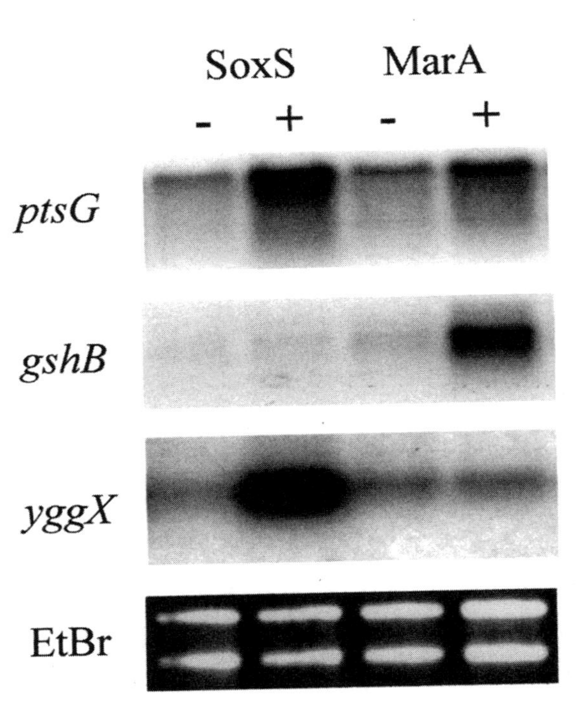

Figure 2. Northern blotting reveals differential regulation of gene transcription by SoxS and MarA. The SoxS and MarA proteins were expressed in *E. coli* in the absence of stress from IPTG-inducible constructs. Total RNA was extracted, purified, separated by electrophoresis in an agarose gel, transferred to a Nytran membrane, and hybridized sequentially with gene-specific probes. The bottom panel shows the EtBr stain of a gel run in parallel with the same amounts of total RNA.

MECHANISMS FOR TRANSCRIPTIONAL ACTIVATION BY MarA

Promoter *cis*-Acting Region for MarA Transcriptional Regulation: the marbox

Direct activation of *mar* regulon genes occurs through binding of MarA as a monomer to an asymmetric, degenerate, 20 bp recognition sequence known as the marbox, which is located in the promoter of controlled genes. The marbox sequence is also recognized by the two MarA homologues, SoxS and Rob, although the relative levels of activation appear to vary greatly between the regulatory proteins (Fig. 2) (see below).

Promoters of activated genes have been divided into two classes, class I and class II, depending on the configuration (i.e., location and orientation) of the marbox in relation to the RNA polymerase (RNAP) signatures. Class I promoters have marboxes with a backward orientation and are located upstream from the −10 hexamer (38–40 or 50 bp), e.g., *mar, acrAB* (55, 93) (Fig. 3, Table 1). Exceptions to this orientation are the Class I* promoters of *zwf* and *map,* which have a forward orientation but are only 30 bp upstream from the −10 (Fig. 3, Table 1) (55, 61, 93). The promoter of *acnA* also constitutes a deviation from this rule since its proposed marbox is only 29 nt from the −10 but it retains the backward orientation (61) (Table 1). On the other hand, in all class II promoters, the marbox overlaps the −35 hexamer (17–20 bp from the −10) in a forward orientation (55, 61) (Fig. 3). Examples of this kind of marbox are found in promoters of *fumC, inaA,* and *sodA* (Table 1). It is of interest to note that the functional activity of both classes of promoters depends on both the location and orientation of the respective marbox, and if either parameter is changed the activation can be lost (55, 93).

As an increasing number of MarA-binding promoter elements are characterized, the marbox consen-

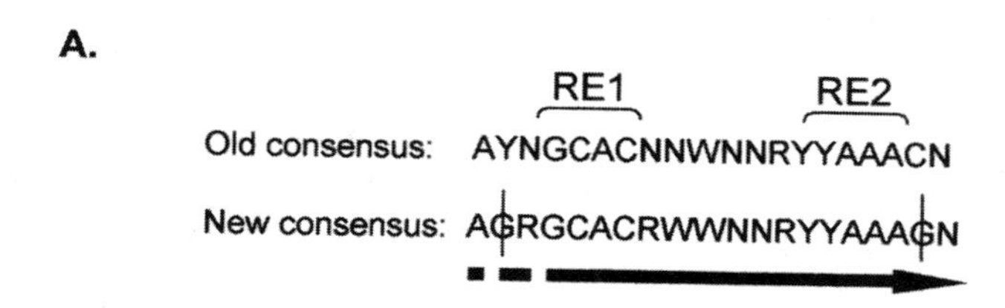

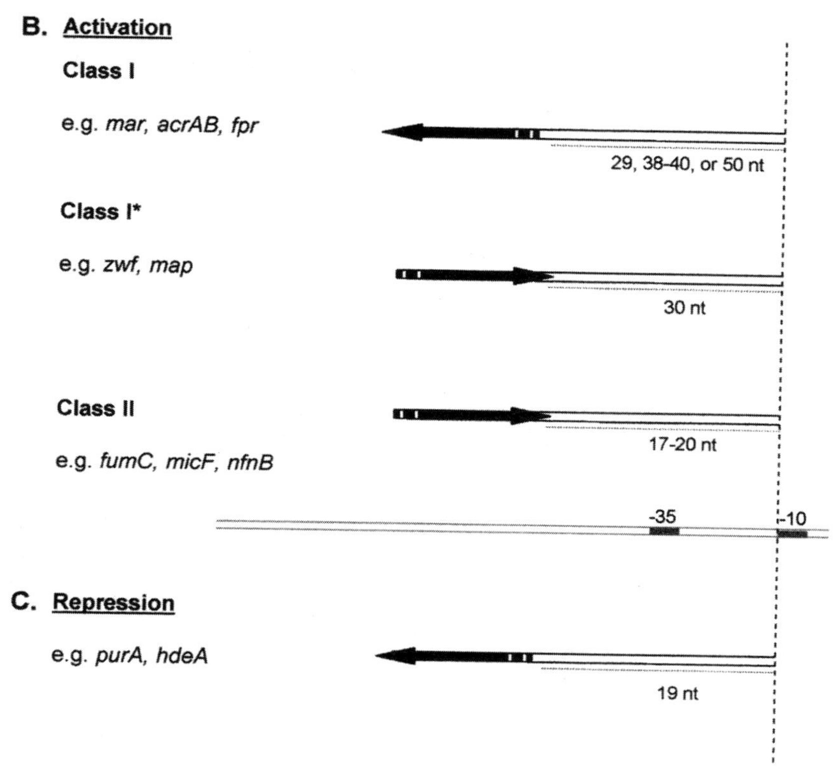

Figure 3. MarA, SoxS, and Rob recognition sequence within the promoter of controlled genes. (A) Comparison of the most recently defined consensus for the 20 bp degenerate marbox sequences in the forward orientation: "old consensus" (55) and "new consensus" (61); N, any base; R = A/G; W = A/T; Y = C/T; G̸ = any base but G. The location of the most conserved recognition elements within the marbox sequence, RE1 and RE2, is indicated. (B) Location and orientation of the marbox in class I (backward), class I* (forward), and class II *mar* regulon-activated promoters (55, 61). (C) Location and orientation of the marbox in the promoters of down-regulated genes (84). Arrows depict the marbox, while the direction of the arrow-head represents the functional orientation of the marbox relative to the −10 and −35 RNAP recognition sequences (gray rectangles). Distances between the marbox and the −10 hexamer are indicated.

sus sequence has progressively been refined (Fig. 3) (55, 61). Nevertheless, the sequence remains degenerate (17 defined nucleotides out of 20) to an extent whereby a computer search of the *E. coli* genome for sequences that match the marbox in 13 nucleotides out of the 17 identifies ~5,000 positions, while a match of 12 occurs ~13,000 times (61). Considering that a marbox with as many as nine mismatches from the consensus is still functional (8), these numbers could even be an underestimation of the real possibilities. The importance of these observations comes from the realization that other very divergent marboxes may be important in the cell's response to environmental hazards. On the other hand, these figures include a large number of false positives, such as marboxes with a nonfunctional configuration that are of little importance in a physiological context.

Importantly, although multiple marboxes can be found in the promoter regions of *mar* regulon genes, in vivo studies have demonstrated that it is the one closest to the promoter signatures that plays the major role in transcriptional control (55, 61). In general, MarA-mediated transcription does not require the activity of additional transcription factors or auxiliary enhancer proteins. One reported exception to this is FIS, which enhances transcription of the *marRAB* operon (60) (see above).

It has been proposed that the unique nucleotide sequence of each marbox may be the basis for the differential binding properties and/or levels of control that the three transcriptional regulators exhibit at specific promoter sequences (8, 18, 37, 47, 56). These unique properties should then allow the cell to fine tune its response in a manner that would most appropriately reflect the type of stress to which it was being exposed.

Despite the general degeneracy observed in marbox sequences, two areas of high conservation can be found. Following the designation of Griffith and Wolf (37), these elements are referred to as recognition element 1, RE1 (nt 4–7, consensus GCAC), and RE2 (nt 15–18, consensus YAAA) (37, 55, 61) (Fig. 3). These two sites, which are located on the same side of the DNA helix, have been shown to be particularly important for the establishment of essential contacts with the two HTH binding motifs of MarA, SoxS, or Rob (18, 37, 46, 47, 78). Therefore, while some early studies indicated that the first part of the marbox containing RE1 was the most important for activation (46–48, 59), it is now recognized that both elements are essential and perhaps of equal importance for MarA-mediated transcription activation to occur (18, 29, 37, 61, 78). Importantly, even for the *nfnB* marbox, which shares limited similarity to the consensus marbox sequence (9 nt mismatches out of 17 defined nucleotides) and is the only one known to contain a RE2 element that is completely different from the consensus, both elements were shown to be important for MarA activation in vivo (8). The structure of MarA and the specific protein-DNA interactions that are proposed to occur during activation are discussed in more detail later in this volume (chapter 17).

TRANSCRIPTIONAL REPRESSION BY MarA

MarA has long been described exclusively as an activator. Only expression of the outer membrane porin OmpF was known to be repressed in a MarA-dependent fashion, and this down-regulation was shown to result from MarA induction of the small antisense RNA *micF,* which then negatively regulates the translation of *ompF* (17, 19). Nevertheless, both the macroarray studies described above reported that a considerable subgroup of genes are actually repressed by MarA (9, 73), although it was not possible to ascertain the exact nature of this repression.

Two such MarA repressed genes, *purA* (which codes for adenylosuccinate synthase) and *hdeA* (which encodes a periplasmic protein involved in acid resistance) (25, 84, 91), have recently been analyzed in order to specifically address the question of whether repression is direct or indirect (84). These studies demonstrated that repression, like activation, can be mediated by the binding of MarA to a marbox, or a marbox homologous sequence in the promoter region of repressed genes. In these cases, however, the marbox overlapped the −35 promoter motif in the backward orientation (Fig. 3, Table 1). Compared with class II activated promoters, therefore, it appears that repression is associated with the orientation of the marbox rather than necessarily with its sequence or location. Presumably, the two different orientations relative to the −35 hexamer result in distinct interactions between MarA, the RNAP, and the DNA, one leading to activation and the other to repression. Binding sites that overlap or lie between the hexamers of RNAP have been described for several repressors (31).

Curiously, the repression of both *purA* and *hdeA* genes appeared to be independent of the presence of a complete marbox. In the case of *purA,* for example, the RE1 region but not the RE2 was important for repression, while in the case of *hdeA,* RE2 was the element required for repression. The physiological significance of *purA* and *hdeA* repression by MarA is unknown, and additional studies will be required to fully understand the roles that decreased expression might play in development of the Mar phenotype.

While it is evident that MarA can function as a dual regulator and that both activation and repression can be mediated by binding directly to a related *cis*-regulatory sequence with a specific configuration at the promoter of controlled genes, it is important to note that repression of gene expression is not unique to MarA. SoxS has been shown to repress its own transcription as well as that of *rob* in vivo (63, 69). However, while in vitro binding of purified SoxS to promoters sequences of *soxS* and *rob* has been demonstrated, the specific regulatory sequences that mediate this interaction have not been identified. Proof for direct interaction and for the functional and physiological relevance of this binding in vivo have yet to be presented. Although additional studies are required to understand the molecular basis for MarA repressive function, it remains possible that features of these mechanisms might be relevant and applicable to negative regulation by SoxS and Rob proteins.

REGULATION OF REGULON MEMBERS: DNA-MarA-RNAP INTERACTIONS

The different configurations of marboxes imply that the regulators must be able to interact with the RNAP in different orientations and suggest that different DNA-regulator-RNAP interactions can be established depending on the promoter. However, despite the obvious truth of this statement, the specific interactions that influence regulator-RNAP complex assembly at the different promoters only now start to be unravelled (see chapter 17). Earlier work suggested that activation of class I promoters depended on interactions between the regulators and the carboxyl-terminus domain of the α-subunit of RNAP (41–43, 55), while activation of class II promoters seemed independent of both carboxyl-and amino-terminal interactions (23, 41). However, more recent evidence indicates that interactions between the regulatory protein and the α-CTD are required for the activation of all different type regulon promoters (29, 54) (chapter 17).

One of the biggest challenges is to develop a model for transcription regulation that explains how MarA is able to recognize true *mar* regulon marboxes from the large number of putative MarA-binding sites present in the *E. coli* genome. To help explain this phenomenon, two studies have independently developed and tested what has been described as the scanning or prerecruitment theory (35, 54). This theory hypothesizes that the regulator (MarA or SoxS) and RNAP form a binary complex, and it is only after formation of this complex has occurred that the chromosome will be scanned for functional promoters where the marbox and the RNAP hexamers are appropriately spaced and orientated (35, 54). This theory would help explain why transcriptional activation does not necessarily occur at sites that are predicted by computer analysis to represent marboxes and why biological activity in vivo does not always correlate with in vitro binding (61). How such a binary protein complex is able to assemble and function at the different types of *mar* regulon promoters remains to be fully characterized.

CONTRIBUTION OF INDIVIDUAL GENES FOR THE DEVELOPMENT OF THE Mar PHENOTYPE

Given the broad Mar phenotype, it is of little surprise to find that MarA-controlled genes are involved in a multitude of cellular processes. This includes carbon metabolism and catabolism, biosynthesis of cofactors, amino acid metabolism and biosynthesis, fatty acid and nucleotide metabolism, cell wall biosynthesis, transport and binding, acid adaptation, protection responses such as antibiotic resistance, cell envelope, macromolecule synthesis, sugar transport, glycolysis, and others (9, 73, 74). While the physiological role of several *mar* regulon encoded proteins has been experimentally validated, for others it is predicted on the basis of their functional annotation and homologies with known proteins. Examples of these include the b0447 gene that encodes a putative LRP-like transcriptional regulator and *yadG,* which encodes a putative ATP-binding component of a transport system. Genes b1448 and *yggJ* on the other hand are examples of regulated genes that have no known homologues, and their functions are unknown.

While macroarray analysis has provided information regarding the identity of regulon members, the challenge facing us in this postgenomic era is to understand how the entirety of regulon encoded proteins can interact and work together to build the Mar phenotype—to better understand the proteomics of the stress response. In the next section we will attempt to bring together some of the knowledge regarding regulon members, be they individual proteins or families of related proteins, and try to build a more comprehensive picture of how the phenotype develops. Given the very significant overlap between the *mar* and *sox* regulons and their associated phenotypes, we will reference the information that is available regarding both MarA- and/or SoxS-mediated antibiotic resistance and oxidative stress responses when possible.

Reduction of Cell Permeability

MarA and SoxS activate transcription of the small, antisense RNA *micF,* which down-regulates translation of the outer membrane porin OmpF (13, 19). This porin is a major component of the outer envelope and regulates the intake of small, hydrophilic molecules. The *micF*-mediated down-regulation of OmpF results in a decrease in cellular permeability with a resulting increase in resistance to toxic compounds, including several antibiotics (16, 17) (Fig. 1).

Interestingly, the antibiotic resistance mediated by a *soxR*-constitutive mutation was only partially reversed upon deletion of *micF.* Accordingly it has been predicted that the *sox* regulon includes other components that contribute to general antibiotic resistance (13). Recently, the *lpxC* gene, coding for the rate-limiting step in lipid A biosynthesis, was shown to be under MarA and SoxS control (Kenyon and Pomposiello, unpublished) (73). Since *lpxC* mutations result in increased cellular permeability and susceptibility to antibiotics, it was suggested that the MarA/SoxS-mediated activation of *lpxC* might result in a decrease in cellular permeability. Indeed, a partially defective *lpxC* mutant showed reduced ability to decrease permeability in response to oxidative stress (Kenyon and Pomposiello, unpublished). These observations suggest that the *sox* regulon, and likely the *mar* regulon, modulates cellular permeability in response to stress through the composition of the outer envelope, and that at least two pathways contribute to this defense mechanism.

Besides OmpF, changes in the expression of other outer membrane proteins, e.g., increased OmpX, have also been reported in *E. coli marR* mutants and strains overexpressing MarA (6, 9).

Efflux of Toxic Compounds

The membrane-spanning AcrAB-TolC complex is capable of the active extrusion of toxic xenobiotics, and increased expression of the genes that encode this efflux system represents a key step in the development of the Mar and Sox phenotypes (see chapter 19). The complex itself is a broad specificity pump, recognizing and extruding lipophilic and amphiphilic compounds with a range of different chemical structures. The only requirement for a compound to be a substrate of the AcrAB-TolC pump is a hydrophobic moiety that allows partition into the membrane (67). Thus, SoxS and MarA-mediated activation of the *acrAB-tolC* genes triggers a general mechanism of resistance to xenobiotics. The essential role of the AcrAB proteins in MarA and SoxS-mediated resistance to multiple antibiotics is supported by the observation that *acrAB* mutant strains display increased sensitivity to antibiotics, even in the presence of constitutive MarA or SoxS expression (70). The pump is also known to play a role in the *E. coli* tolerance to organic solvents such as cyclohexane (92). TolC appears to be equally important since inactivation of *tolC* reverses the phenotype associated with AcrAB (24).

Increased activation of *acrAB* expression in a MarA-dependent fashion is observed also when agents such as pine oil (65) and household disinfectants such as the widely used biocide triclosan are used to select *mar* mutants (62). Induction of AcrAB also occurs in response to bile salts and fatty acids, but in this case it seems to occur via Rob and independently of both MarA and SoxS (81). Here, as described previously, activation of *acrAB* transcription occurs as a result of binding of these compounds to the effector binding domain of Rob.

Besides the predominant role played by the AcrAB-TolC system, there is evidence that other as yet unidentified drug-specific efflux pumps (e.g., to chloramphenicol, fluoroquinolones, tetracyclines) appear to be concomitantly involved in the development of the Mar phenotype (4, 70).

Detoxification

Many genes of the *mar/sox* regulon are thought to be involved in dealing with the toxic effects resulting from exposure to toxic compounds such as oxidative stress agents. This protection can be mediated through a variety of different measures, which may include scavenging of reactive species (*sodA*, Mn-containing superoxide dismutase), synthesis of reducing species (*acnA*, aconitase, and *zwf*, glucose-6-phosphate dehydrogenase), and repair of oxidative damage (*nfo*, endonuclease IV and fpr, NADPH-ferredoxin oxidoreductase). Some of these are discussed below in more detail.

Isoenzyme switch

Cells have evolved strategies for meeting metabolic requirements under stress. The adaptation to changing environmental conditions is frequently achieved by differential regulation of isoenzymes. The transcription of genes coding for two groups of biosynthetic isoenzymes is differentially regulated by SoxS. *E. coli* possesses two aconitases, A and B, encoded by the *acnA* and *acnB* genes (38). Both enzymes catalyze the reversible formation of isocitrate from citrate, with *cis*-aconitate as an enzyme-bound intermediary. The activity of both aconitases depends on a 4Fe-4S cluster at the active site. Early work showed that during aerobic growth, aconitase B provides most of the activity but decreases under oxidative stress (38). In contrast, aconitase A activity is low under normal aerobic growth but is induced under oxidative stress in a SoxRS-dependent manner. As result of this differential activity, the total aconitase activity shows only a slight increase under oxidative stress (38). This observation was interpreted as the replacement of an oxido-sensitive enzyme by an oxido-resistant one. This suggestion was confirmed by recent studies that measured the stability of aconitase activity in cell extracts prepared from *acnA*- and *acnB*-deficient strains. In cell extracts, aconitase A was more resistant to atmospheric oxygen, superoxide, and hydrogen peroxide than aconitase B (90). Interestingly, the activities of the purified aconitase A and aconitase B were equally labile in vitro, showing that the differential resistance to oxidants is not intrinsic to the protein structure. Moreover, addition of cell extract increased the stability of aconitase A but not that of aconitase B. The factor responsible for the stabilization of aconitase A was DNAse and RNAse resistant but did not resist boiling or dialysis (90). These results suggested that a protein is responsible for the protection of aconitase A from oxidative damage, and it has been proposed that the YggX is the specific factor involved (77, 90). Interestingly, while *acnA* is up-regulated by both SoxS and MarA (9, 73) (Table 1), the *yggX* gene is regulated by SoxS (77) but not by MarA and has been proposed to encode an intracellular iron-trafficking protein (33) (Fig. 2) (see below).

Fumarase, another TCA cycle activity, can be carried by three isoenzymes, fumarase A, B, and C. While fumarase A and B are heat-labile enzymes that require iron for activity, fumarase C is resistant to heat and does not require iron for activity (94). Fumarase A and B seem to be the aerobic fumarases, since both enzymes are down-regulated during anaerobic growth (71). Transcription of *fumC* is activated by SoxS and MarA (49, 71).

Sugar transport and catabolism

Redox-cycling agents produce superoxide at the expense of NADPH, causing a simultaneous loss of reducing power and increased level of oxidative damage. The pentose-phosphate pathway replenishes NADPH, and flow of carbon through this pathway is supposed to be enhanced by the SoxS/MarA-mediated activation of *zwf,* which codes for glucose-6-P-phosphate dehydrogenase (34). Additionally, MarA and SoxS activate transcription of the *ptsG* gene, which codes for the glucose-specific, transmembrane PTS transporter EIIBC (73) (Liu and Pomposiello, unpublished) (Fig. 2). The increase in sugar transport and catabolism might be a conserved antioxidant response, since proteomic studies revealed that the yeast *S. cerevisiae* also redirects carbohydrate fluxes towards regeneration of NADPH at the expense of glycolysis (30).

Reduction of redox-active prosthetic groups

Two genes coding for flavodoxins and their NADPH-dependent reductase are activated by SoxS (27, 50). Flavodoxin A and B, encoded by *fldA* (an essential gene) and *fldB,* respectively, are small flavin mononuclotide-containing reductases that participate in biosynthetic processes, such as synthesis of cobalamin and biotin. The NADPH-dependent flavodoxin (ferredoxin) oxidoreductase is encoded by *fpr.* Activation of the *fpr-fldAB* pathway should increase the shuttling of electrons from NADPH into biosynthetic processes and thus assist in maintaining the redox homeostasis of the cell under stress.

Iron-Associated Genes

Iron is an essential element for the bacterial cell and plays an important role in bacterial pathogenesis. On the other hand, it also catalyzes the production of hydroxyl ions via the Fenton reaction, which can cause damage to cellular components and lead to cell death.

SoxS activates expression of the *fur* gene, which encodes an iron-binding repressor of iron uptake genes that represses its own transcription (73, 96). The metabolic relevance of this activation is uncertain. It has been shown that oxidative stress causes oxidation of iron in metalloproteins, leading to disassembly of iron-sulfur clusters and an increase in free intracellular iron. Since an increase in free iron can be deleterious to the cell, an increased level of Fur repressor could result in sequestration of the free iron and increased repression of iron uptake.

Alternatively, it has been suggested that oxidative stress might damage Fur-Fe complexes through Fenton chemistry and that maintenance of repression of iron uptake might require increased synthesis of Fur. This suggested loss of Fur-mediated repression is supported by the observation that certain Fur-repressed genes are up-regulated under superoxide stress. In contrast to the situation with SoxS, there is no evidence for an increase in the expression levels of the *fur* or *fldA-fur* transcripts by MarA, either by macroarray (9) or by Northern analysis (Barbosa and Levy, unpublished).

In addition to Fur, SoxS activates the expression of YggX, another protein that has been linked to iron metabolism. In *Salmonella enterica,* overexpression of YggX suppresses the hypersensitive phenotype of a glutathione-deficient (*gshA*) strain, which grows poorly in aerobic minimal medium. Moreover, overproduction of YggX in *gshA* strains suppresses the decreased activity of aconitase, rescues the elevated level of spontaneous mutagenesis, and protects against exogenous redox-cycling compounds (32). Purified YggX binds iron with low affinity and protects DNA from oxidative damage from Fenton reactions (33). As mentioned previously, in *E. coli, yggX* is under transcriptional control by SoxS (77). An *E. coli yggX* deletion mutant grows aerobically at the same rate as the isogenic wild-type strain but is hypersensitive to superoxide-producing agents. Moreover, the expression of YggX seems to be a limiting factor for aconitase activity under superoxide stress (77). These observations have led to a model for YggX function, in which YggX mediates intracellular trafficking of iron between uptake systems and biosynthetic reactions, such as the formation of iron-sulfur clusters.

CONCLUDING REMARKS

The *marRAB* and *soxRS* systems translate specific environmental stresses into a coordinated cellular response by converting a damaging signal into the activation of a transcriptional factor, MarA or SoxS. The subsequent regulation of the *mar* or *sox* regulon members results in a global modification that allows the cell to overcome the insults being experienced. The key strength of these generalized responses lies in the fact that different environmental challenges can govern the expression of a common repertoire of adaptive genes.

The direct activation and repression of genes by MarA can occur through binding of the protein to a

20 bp degenerate marbox sequence located in the promoter of controlled genes. This sequence can exist in diverse configurations and with different levels of degeneracy, suggesting that unique interactions can be formed with transcriptional complexes that contain the regulatory proteins and the RNAP. Although SoxS and Rob recognize the same sequence, the degree of transcriptional regulation of any given promoter by the three proteins can vary substantially. This is thought to arise from sequence variation in the 20 bp cognate marbox site. A comprehensive understanding of the molecular switches that control these interactions remains a key objective of future studies, as it underlies both the specificity and flexibility of the adaptive response.

Although gene macroarrays have identified a plethora of genes that are regulated by MarA and SoxS, one striking feature of the accumulated data is that MarA and SoxS can activate genes that are not part of operons. However, many of these genes code for keystone metabolic functions that limit the flux of intermediaries through whole pathways. Examples of this selective activation include *zwf* of the pentose-phosphate pathway, *lpxC* which is involved in LPS biosynthesis, and *ptsG* involved in glucose transport.

While understanding the molecular cross-talk that underlies regulon expression and how this is elicited by natural stresses remains an important goal for the future, the prevailing long-term objective is to understand the physiological relevance of these changes within the cell and to more accurately map the development of the phenotype. This will offer new possibilities for identifying targets for novel antimicrobial therapies to better deal with the growing problem of antimicrobial resistance. With such an extended regulon, the most obvious target might be the regulator itself, be it MarA or its homologues. However more selective targeting in the future could provide the clinician with a range of stronger and more specific tools for therapeutic treatments.

IN APPRECIATION

The personal contribution of Stuart Levy to this work cannot be overstated. And yet, his contribution has extended far beyond the design and interpretation of important experiments. He continues to be the great communicator and champion for this cause, taking the significant progress being made in the lab and challenging society's perspective on the global threat represented by antibiotic resistance. As a mentor, he has impassioned a whole new generation of researchers from across the world to take up this challenge, encouraging them to push the limits of knowledge and understanding and providing them with the intellectual freedom and vibrant innovative environment in which to develop. Stuart, thank you.

Acknowledgments. We thank L. M. McMurry and most particularly J. L. Rosner for helpful comments and suggestions on the manuscript.

REFERENCES

1. **Alekshun, M. N., and S. B. Levy.** 1999. Alteration of the repressor activity of MarR, the negative regulator of the *Escherichia coli marRAB* locus, by multiple chemicals in vitro. *J. Bacteriol.* **181:**4669–4672.
2. **Alekshun, M. N., and S. B. Levy.** 1999. Characterization of MarR superrepressor mutants. *J. Bacteriol.* **181:**3303–3306.
3. **Alekshun, M. N., and S. B. Levy.** 1999. The *mar* regulon: multiple resistance to antibiotics and other toxic chemicals. *Trends Microbiol.* 7:410–413.
4. **Alekshun, M. N., and S. B. Levy.** 1997. Regulation of chromosomally mediated multiple antibiotic resistance: the *mar* regulon. *Antimicrob. Agents Chemother.* **41:**2067–2075.
5. **Amabile-Cuevas, C. F., and B. Demple.** 1991. Molecular characterization of the *soxRS* genes of *Escherichia coli:* two genes control a superoxide stress regulon. *Nucleic Acids Res.* **19:**4479–4484.
6. **Aono, R., N. Tsukagoshi, and M. Yamamoto.** 1998. Involvement of outer membrane protein TolC, a possible member of the *mar-sox* regulon, in maintenance and improvement of organic solvent tolerance of *Escherichia coli* K-12. *J. Bacteriol.* **180:**938–944.
7. **Ariza, R. R., Z. Li, N. Ringstad, and B. Demple.** 1995. Activation of multiple antibiotic resistance and binding of stress-inducible promoters by *Escherichia coli* Rob protein. *J. Bacteriol.* **177:**1655–1661.
8. **Barbosa, T. M., and S. B. Levy.** 2002. Activation of the *Escherichia coli nfnB* gene by MarA through a highly divergent marbox in a class II promoter. *Mol. Microbiol.* **45:**191–202.
9. **Barbosa, T. M., and S. B. Levy.** 2000. Differential expression of over 60 chromosomal genes in *Escherichia coli* by constitutive expression of MarA. *J. Bacteriol.* **182:**3467–3474.
10. **Barbosa, T. M., and S. B. Levy.** 1999. Presented at the 99th General Meeting of the American Society for Microbiology, Chicago, USA. Abstract A42, p 9.
11. **Bennik, M. H., P. J. Pomposiello, D. F. Thorne, and B. Demple.** 2000. Defining a *rob* regulon in *Escherichia coli* by using transposon mutagenesis. *J. Bacteriol.* **182:**3794–3801.
12. **Bina, X., V. Perreten, and S. B. Levy.** 2003. The periplasmic protein MppA requires an additional mutated locus to repress *marA* expression in *Escherichia coli. J. Bacteriol.* **185:**1465–1469.
13. **Chou, J. H., J. T. Greenberg, and B. Demple.** 1993. Posttranscriptional repression of *Escherichia coli* OmpF protein in response to redox stress: positive control of the *micF* antisense RNA by the *soxRS* locus. *J. Bacteriol.* **175:**1026–1031.
14. **Cohen, S. P., H. Hachler, and S. B. Levy.** 1993. Genetic and functional analysis of the multiple antibiotic resistance (*mar*) locus in *Escherichia coli. J. Bacteriol.* **175:**1484–1492.
15. **Cohen, S. P., S. B. Levy, J. Foulds, and J. L. Rosner.** 1993. Salicylate induction of antibiotic resistance in *Escherichia coli:* activation of the *mar* operon and a mar-independent pathway. *J. Bacteriol.* **175:**7856–7862.
16. **Cohen, S. P., L. M. McMurry, D. C. Hooper, J. S. Wolfson, and S. B. Levy.** 1989. Cross-resistance to fluoroquinolones in mul-

tiple-antibiotic-resistant (Mar) *Escherichia coli* selected by tetracycline or chloramphenicol: decreased drug accumulation associated with membrane changes in addition to OmpF reduction. *Antimicrob. Agents Chemother.* **33:**1318–1325.

17. **Cohen, S. P., L. M. McMurry, and S. B. Levy.** 1988. *marA* locus causes decreased expression of OmpF porin in multiple-antibiotic-resistant (Mar) mutants of *Escherichia coli. J. Bacteriol.* **170:**5416–5422.
18. **Dangi, B., P. Pelupessey, R. G. Martin, J. L. Rosner, J. M. Louis, and A. M. Gronenborn.** 2001. Structure and dynamics of MarA-DNA complexes: an NMR investigation. *J. Mol. Biol.* **314:**113–127.
19. **Delihas, N., and S. Forst.** 2001. MicF: an antisense RNA gene involved in response of *Escherichia coli* to global stress factors. *J. Mol. Biol.* **313:**1–12.
20. **Ding, H., and B. Demple.** 2000. Direct nitric oxide signal transduction via nitrosylation of iron-sulfur centers in the SoxR transcription activator. *Proc. Natl. Acad. Sci. USA* **97:**5146–5150.
21. **Ding, H., E. Hidalgo, and B. Demple.** 1996. The redox state of the [2Fe-2S] clusters in SoxR protein regulates its activity as a transcription factor. *J. Biol. Chem.* **271:**33173–33175.
22. **Ding, H. G., and B. Demple.** 1997. In vivo kinetics of a redox-regulated transcriptional switch. *Proc. Natl. Acad. Sci. USA* **94:**8445–8449.
23. **Egan, S. M., A. J. Pease, J. Lang, X. Li, V. Rao, W. K. Gillette, R. Ruiz, J. L. Ramos, and R. E. Wolf, Jr.** 2000. Transcription activation by a variety of AraC/XylS family activators does not depend on the class II-specific activation determinant in the N-terminal domain of the RNA polymerase alpha subunit. *J. Bacteriol.* **182:**7075–7077.
24. **Fralick, J. A.** 1996. Evidence that TolC is required for functioning of the Mar/AcrAB efflux pump of *Escherichia coli. J. Bacteriol.* **178:**5803–5805.
25. **Gajiwala, K. S., and S. K. Burley.** 2000. HDEA, a periplasmic protein that supports acid resistance in pathogenic enteric bacteria. *J. Mol. Biol.* **295:**605–612.
26. **Gallegos, M. T., R. Schleif, A. Bairoch, K. Hofmann, and J. L. Ramos.** 1997. Arac/XylS family of transcriptional regulators. *Microbiol. Mol. Biol. Rev.* **61:**393–410.
27. **Gaudu, P., and B. Weiss.** 2000. Flavodoxin mutants of *Escherichia coli* K-12. *J. Bacteriol.* **182:**1788–1793.
28. **Gaudu, P., and B. Weiss.** 1996. SoxR, a [2Fe-2S] transcription factor, is active only in its oxidized form. *Proc. Natl. Acad. Sci. USA* **93:**10094–10098.
29. **Gillette, W. K., R. G. Martin, and J. L. Rosner.** 2000. Probing the *Escherichia coli* transcriptional activator MarA using alanine-scanning mutagenesis: residues important for DNA binding and activation. *J. Mol. Biol.* **299:**1245–1255.
30. **Godon, C., G. Lagniel, J. Lee, J. M. Buhler, S. Kieffer, M. Perrot, H. Boucherie, M. B. Toledano, and J. Labarre.** 1998. The H_2O_2 stimulon in *Saccharomyces cerevisiae. J. Biol. Chem.* **273:**22480–22489.
31. **Gralla, J. D., and J. Collado-Vides.** 1996. Organization and function of transcriptional regulatory elements, p. 1232–1245. *In* F. C. Neidhardt (ed.), Escherichia coli *and* Salmonella: *Cellular and Molecular Biology.* ASM Press, Washington, D.C.
32. **Gralnick, J., and D. Downs.** 2001. Protection from superoxide damage associated with an increased level of the YggX protein in *Salmonella enterica. Proc. Natl. Acad. Sci. USA* **98:**8030–8035.
33. **Gralnick, J. A., and D. M. Downs.** 2003. The YggX protein of *Salmonella enterica* is involved in Fe(II) trafficking and minimizes the DNA damage caused by hydroxyl radicals: residue CYS-7 is essential for YggX function. *J. Biol. Chem.* **278:**20708–20715.
34. **Greenberg, J. T., P. Monach, J. H. Chou, P. D. Josephy, and B. Demple.** 1990. Positive control of a global antioxidant defense regulon activated by superoxide-generating agents in *Escherichia coli. Proc. Natl. Acad. Sci. USA* **87:**6181–6185.
35. **Griffith, K. L., I. M. Shah, T. E. Myers, M. C. O'Neill, and R. E. Wolf, Jr.** 2002. Evidence for "pre-recruitment" as a new mechanism of transcription activation in *Escherichia coli:* the large excess of SoxS binding sites per cell relative to the number of SoxS molecules per cell. *Biochem. Biophys. Res. Commun.* **291:**979–986.
36. **Griffith, K. L., I. M. Shah, and R. E. Wolf, Jr.** 2004. Proteolytic degradation of *Escherichia coli* transcription activators SoxS and MarA as the mechanism for reversing the induction of the superoxide (SoxRS) and multiple antibiotic resistance (Mar) regulons. *Mol. Microbiol.* **51:**1801–1816.
37. **Griffith, K. L., and R. E. Wolf, Jr.** 2001. Systematic mutagenesis of the DNA binding sites for SoxS in the Escherichia coli *zwf* and *fpr* promoters: identifying nucleotides required for DNA binding and transcription activation. *Mol. Microbiol.* **40:**1141–1154.
38. **Gruer, M. J., and J. R. Guest.** 1994. Two genetically-distinct and differentially-regulated aconitases (AcnA and AcnB) in *Escherichia coli. Microbiology* **140:**2531–2541.
39. **Hidalgo, E., and B. Demple.** 1994. An iron-sulfur center essential for transcriptional activation by the redox-sensing SoxR protein. *EMBO J.* **13:**138–146.
40. **Hidalgo, E., V. Leautaud, and B. Demple.** 1998. The redox-regulated SoxR protein acts from a single DNA site as a repressor and an allosteric activator. *EMBO J.* **17:**2629–2636.
41. **Jair, K. W., W. P. Fawcett, N. Fujita, A. Ishihama, and R. E. Wolf, Jr.** 1996. Ambidextrous transcriptional activation by SoxS: requirement for the C-terminal domain of the RNA polymerase alpha subunit in a subset of *Escherichia coli* superoxide-inducible genes. *Mol. Microbiol.* **19:**307–317.
42. **Jair, K. W., R. G. Martin, J. L. Rosner, N. Fujita, A. Ishihama, and R. E. Wolf, Jr.** 1995. Purification and regulatory properties of MarA protein, a transcriptional activator of *Escherichia coli* multiple antibiotic and superoxide resistance promoters. *J. Bacteriol.* **177:**7100–7104.
43. **Jair, K. W., X. Yu, K. Skarstad, B. Thony, N. Fujita, A. Ishihama, and R. E. Wolf, Jr.** 1996. Transcriptional activation of promoters of the superoxide and multiple antibiotic resistance regulons by Rob, a binding protein of the *Escherichia coli* origin of chromosomal replication. *J. Bacteriol.* **178:**2507–2513.
44. **Koh, Y. S., J. Choih, J. H. Lee, and J. H. Roe.** 1996. Regulation of the *ribA* gene encoding GTP cyclohydrolase II by the *soxRS* locus in *Escherichia coli. Mol. Gen. Genet.* **251:**591–598.
45. **Koo, M. S., J. H. Lee, S. Y. Rah, W. S. Yeo, J. W. Lee, K. L. Lee, Y. S. Koh, S. O. Kang, and J. H. Roe.** 2003. A reducing system of the superoxide sensor SoxR in *Escherichia coli. EMBO J.* **22:**2614–2622.
46. **Kwon, H. J., M. H. Bennik, B. Demple, and T. Ellenberger.** 2000. Crystal structure of the *Escherichia coli* Rob transcription factor in complex with DNA. *Nat. Struct. Biol.* **7:**424–430.
47. **Li, Z., and B. Demple.** 1996. Sequence specificity for DNA binding by Escherichia coli SoxS and Rob proteins. *Mol. Microbiol.* **20:**937–945.
48. **Li, Z., and B. Demple.** 1994. SoxS, an activator of superoxide stress genes in *Escherichia coli.* Purification and interaction with DNA. *J. Biol. Chem.* **269:**18371–18377.
49. **Liochev, S. I., and I. Fridovich.** 1992. Fumarase C, the stable fumarase of *Escherichia coli,* is controlled by the *soxRS* regulon. *Proc. Natl. Acad. Sci. USA* **89:**5892–5896.

50. **Liochev, S. I., A. Hausladen, W. F. Beyer, Jr., and I. Fridovich.** 1994. NADPH: ferredoxin oxidoreductase acts as a paraquat diaphorase and is a member of the *soxRS* regulon. *Proc. Natl. Acad. Sci. USA* **91:**1328–1331.
51. **Liochev, S. I., A. Hausladen, and I. Fridovich.** 1999. Nitroreductase A is regulated as a member of the soxRS regulon of *Escherichia coli. Proc. Natl. Acad. Sci. USA* **96:**3537–3579.
52. **Ma, D., M. Alberti, C. Lynch, H. Nikaido, and J. E. Hearst.** 1996. The local repressor AcrR plays a modulating role in the regulation of *acrAB* genes of *Escherichia coli* by global stress signals. *Mol. Microbiol.* **19:**101–112.
53. **Maneewannakul, K., and S. B. Levy.** 1996. Identification for *mar* mutants among quinolone-resistant clinical isolates of *Escherichia coli. Antimicrob. Agents Chemother.* **40:**1695–1698.
54. **Martin, R. G., W. K. Gillette, N. I. Martin, and J. L. Rosner.** 2002. Complex formation between activator and RNA polymerase as the basis for transcriptional activation by MarA and SoxS in *Escherichia coli. Mol. Microbiol.* **43:**355–370.
55. **Martin, R. G., W. K. Gillette, S. Rhee, and J. L. Rosner.** 1999. Structural requirements for marbox function in transcriptional activation of *mar/sox/rob* regulon promoters in *Escherichia coli:* sequence, orientation and spatial relationship to the core promoter. *Mol. Microbiol.* **34:**431–441.
56. **Martin, R. G., W. K. Gillette, and J. L. Rosner.** 2000. Promoter discrimination by the related transcriptional activators MarA and SoxS: differential regulation by differential binding. *Mol. Microbiol.* **35:**623–634.
57. **Martin, R. G., K. W. Jair, R. E. Wolf, Jr., and J. L. Rosner.** 1996. Autoactivation of the *marRAB* multiple antibiotic resistance operon by the MarA transcriptional activator in *Escherichia coli. J. Bacteriol.* **178:**2216–2223.
58. **Martin, R. G., and J. L. Rosner.** 2003. Analysis of microarray data for the marA, soxS, and *rob* regulons of *Escherichia coli. Methods Enzymol.* **370:**278–280.
59. **Martin, R. G., and J. L. Rosner.** 2001. The AraC transcriptional activators. *Curr. Opin. Microbiol.* **4:**132–137.
60. **Martin, R. G., and J. L. Rosner.** 1997. Fis, an accessorial factor for transcriptional activation of the *mar* (multiple antibiotic resistance) promoter of *Escherichia coli* in the presence of the activator MarA, SoxS, or Rob. *J. Bacteriol.* **179:**7410–7419.
61. **Martin, R. G., and J. L. Rosner.** 2002. Genomics of the *marA/soxS/rob* regulon of *Escherichia coli:* identification of directly activated promoters by application of molecular genetics and informatics to microarray data. *Mol. Microbiol.* **44:**1611–1624.
62. **McMurry, L. M., M. Oethinger, and S. B. Levy.** 1998. Overexpression of *marA, soxS* or *acrAB* produces resistance to triclosan in *Escherichia coli. FEMS Microbiol. Lett.* **166:**305–309.
63. **Michan, C., M. Manchado, and C. Pueyo.** 2002. SoxRS down-regulation of *rob* transcription. *J. Bacteriol.* **184:**4733–4738.
64. **Miller, P. F., L. F. Gambino, M. C. Sulavik, and S. J. Gracheck.** 1994. Genetic relationship between *soxRS* and *mar* loci in promoting multiple antibiotic resistance in *Escherichia coli. Antimicrob. Agents Chemother.* **38:**1773–1779.
65. **Moken, M. C., L. M. McMurry, and S. B. Levy.** 1997. Selection of multiple-antibiotic-resistant (*mar*) mutants of *Escherichia coli* by using the disinfectant pine oil: roles of the *mar* and *acrAB* loci. *Antimicrob. Agents Chemother.* **41:**2770–2772.
66. **Nakajima, H., K. Kobayashi, M. Kobayashi, H. Asako, and R. Aono.** 1995. Overexpression of the *robA* gene increases organic solvent tolerance and multiple antibiotic and heavy metal ion resistance in *Escherichia coli. Appl. Environ. Microbiol.* **61:**2302–2307.
67. **Nikaido, H., M. Basina, V. Nguyen, and E. Y. Rosenberg.** 1998. Multidrug efflux pump AcrAB of *Salmonella typhimurium* excretes only those beta-lactam antibiotics containing lipophilic side chains. *J. Bacteriol.* **180:**4686–4692.
68. **Nunoshiba, T., T. de Rojas-Walker, J. S. Wishnok, S. R. Tannenbaum, and B. Demple.** 1993. Activation by nitric oxide of an oxidative-stress response that defends *Escherichia coli* against activated macrophages. *Proc. Natl. Acad. Sci. USA* **90:**9993–9997.
69. **Nunoshiba, T., E. Hidalgo, Z. Li, and B. Demple.** 1993. Negative autoregulation by the *Escherichia coli* SoxS protein: a dampening mechanism for the *soxRS* redox stress response. *J. Bacteriol.* **175:**7492–7494.
70. **Okusu, H., D. Ma, and H. Nikaido.** 1996. AcrAB efflux pump plays a major role in the antibiotic resistance phenotype of *Escherichia coli* multiple-antibiotic-resistance (Mar) mutants. *J. Bacteriol.* **178:**306–308.
71. **Park, S. J., and R. P. Gunsalus.** 1995. Oxygen, iron, carbon, and superoxide control of the fumarase *fumA* and *fumC* genes of *Escherichia coli:* role of the *arcA*, *fnr*, and *soxR* gene products. *J. Bacteriol.* **177:**6255–6262.
72. **Paterson, E. S., S. E. Boucher, and I. B. Lambert.** 2002. Regulation of the *nfsA* gene in *Escherichia coli* by SoxS. *J. Bacteriol.* **184:**51–58.
73. **Pomposiello, P. J., M. H. Bennik, and B. Demple.** 2001. Genome-wide transcriptional profiling of the *Escherichia coli* responses to superoxide stress and sodium salicylate. *J. Bacteriol.* **183:**3890–3902.
74. **Pomposiello, P. J., and B. Demple.** 2002. Global adjustment of microbial physiology during free radical stress. *Adv. Microb. Physiol.* **46:**319–341.
75. **Pomposiello, P. J., and B. Demple.** 2000. Identification of SoxS-regulated genes in *Salmonella enterica* serovar *typhimurium. J. Bacteriol.* **182:**23–29.
76. **Pomposiello, P. J., and B. Demple.** 2001. Redox-operated genetic switches: the SoxR and OxyR transcription factors. *Trends Biotechnol.* **19:**109–114.
77. **Pomposiello, P. J., A. Koutsolioutsou, D. Carrasco, and B. Demple.** 2003. SoxRS-regulated expression and genetic analysis of the *yggX* gene of *Escherichia coli. J. Bacteriol.* **185:**6624–6632.
78. **Rhee, S., R. G. Martin, J. L. Rosner, and D. R. Davies.** 1998. A novel DNA-binding motif in MarA: the first structure for an AraC family transcriptional activator. *Proc. Natl. Acad. Sci. USA* **95:**10413–10418.
79. **Rhodius, V. A., and R. A. LaRossa.** 2003. Uses and pitfalls of microarrays for studying transcriptional regulation. *Curr. Opin. Microbiol.* **6:**114–119.
80. **Richmond, C. S., J. D. Glasner, R. Mau, H. Jin, and F. R. Blattner.** 1999. Genome-wide expression profiling in *Escherichia coli* K-12. *Nucleic Acids Res.* **27:**3821–3835.
81. **Rosenberg, E. Y., D. Bertenthal, M. L. Nilles, K. P. Bertrand, and H. Nikaido.** 2003. Bile salts and fatty acids induce the expression of *Escherichia coli* AcrAB multidrug efflux pump through their interaction with Rob regulatory protein. *Mol. Microbiol.* **48:**1609–1619.
82. **Rosner, J. L., B. Dangi, A. M. Gronenborn, and R. G. Martin.** 2002. Posttranscriptional activation of the transcriptional activator Rob by dipyridyl in *Escherichia coli. J. Bacteriol.* **184:**1407–1416.
83. **Rosner, J. L., and J. L. Slonczewski.** 1994. Dual regulation of *inaA* by the multiple antibiotic resistance (*mar*) and superoxide (*soxRS*) stress response systems of *Escherichia coli. J. Bacteriol.* **176:**6262–6269.

84. **Schneiders, T., T. M. Barbosa, L. M. McMurry, and S. B. Levy.** 2004. The *Escherichia coli* transcriptional regulator MarA directly represses transcription of *purA* and *hdeA*. *J. Biol. Chem.* **279:**9037–9042.
85. **Seoane, A. S., and S. B. Levy.** 1995. Characterization of MarR, the repressor of the multiple antibiotic resistance (*mar*) operon in *Escherichia coli*. *J. Bacteriol.* **177:**3414–3419.
86. **Seoane, A. S., and S. B. Levy.** 1995. Identification of new genes regulated by the *marRAB* operon in *Escherichia coli*. *J. Bacteriol.* **177:**530–535.
87. **Sulavik, M. C., M. Dazer, and P. F. Miller.** 1997. The *Salmonella typhimurium mar* locus: molecular and genetic analyses and assessment of its role in virulence. *J. Bacteriol.* **179:**1857–1866.
88. **Tsaneva, I. R., and B. Weiss.** 1990. *soxR*, a locus governing a superoxide response regulon in *Escherichia coli* K-12. *J. Bacteriol.* **172:**4197–4205.
89. **Van Dyk, T. K., B. L. Ayers, R. W. Morgan, and R. A. Larossa.** 1998. Constricted flux through the branched-chain amino acid biosynthetic enzymes acetolactate synthase triggers elevated expression of genes reglated by *rpoS* and internal acidification. *J. Bacteriol.* **180:**785–792.
90. **Varghese, S., Y. Tang, and J. A. Imlay.** 2003. Contrasting sensitivities of *Escherichia coli* Aconitases A and B to oxidation and iron depletion. *J. Bacteriol.* **185:**221–230.
91. **Waterman, S. R., and P. L. Small.** 1996. Identification of sigma S-dependent genes associated with the stationary-phase acid-resistance phenotype of *Shigella flexneri*. *Mol. Microbiol.* **21:** 925–940.
92. **White, D. G., J. D. Goldman, B. Demple, and S. B. Levy.** 1997. Role of the *acrAB* locus in organic solvent tolerance mediated by expression of *marA*, *soxS*, or *robA* in *Escherichia coli*. *J. Bacteriol.* **179:**6122–6126.
93. **Wood, T. I., K. L. Griffith, W. P. Fawcett, K. W. Jair, T. D. Schneider, and R. E. Wolf, Jr.** 1999. Interdependence of the position and orientation of SoxS binding sites in the tran scriptional activation of the class I subset of *Escherichia coli* superoxide-inducible promoters. *Mol. Microbiol.* **34:**414–430.
94. **Woods, S. A., S. D. Schwartzbach, and J. R. Guest.** 1988. Two biochemically distinct classes of fumarase in *Escherichia coli*. *Biochem. Biophys. Acta* **954:**14–26.
95. **Wu, J., and B. Weiss.** 1991. Two divergently transcribed genes, soxR and soxS, control a superoxide response regulon of *Escherichia coli*. *J. Bacteriol.* **173:**2864–2871.
96. **Zheng, M., B. Doan, T. D. Schneider, and G. Storz.** 1999. OxyR and SoxRS regulation of *fur*. *J. Bacteriol.* **181:**4639–4643.
97. **Zheng, M., and G. Storz.** 2000. Redox sensing by prokaryotic transcription factors. *Biochem. Pharmacol.* **59:**1–6.

Frontiers in Antimicrobial Resistance: a Tribute to Stuart B. Levy
Edited by D. G. White, M. N. Alekshun, and P. F. McDermott

Chapter 16

Identification of Mar Mutants among Clinical Bacterial Isolates

JOANN DZINK-FOX AND MARGRET OETHINGER

Since the discovery of antibacterial products in the late 1800s, the search for an agent(s) to treat infectious diseases continues into the present decade. Although the search for antibiotics did not intensify until the 1940s, when penicillin became available to the public to treat infections, the rapid development and production of not only natural but new synthetic antimicrobials continues. As a result of widespread usage and sometimes misusage of antimicrobials, the emergence of resistant bacteria has occurred. Bacteria that had once been susceptible to most antibiotics are now resistant to these and other structurally unrelated antibiotics. Studies have shown that the newly acquired resistances were attributed to the expression of intrinsic resistance plasmids and/or their transfer to other bacterial species. While these resistance genes have been present in bacteria in the environment, the widespread usage of antibiotics therapeutically, agriculturally, and/or in animal husbandry has created a selective pressure that has provided more resistant strains within bacterial populations. The result is an evolutionary advantage with widespread emergence of resistant isolates within the gastrointestinal tract of humans.

Aside from resistance found on plasmids, integrons, and/or transposons, high-level antibiotic resistance has been linked to mutations in chromosomal genes or to expression of genes on the chromosome that are not usually expressed but may be switched on under selective pressure of antibiotics.

In recent years, a great deal of interest has focused on chromosomal multidrug resistance such as the multiple antibiotic resistance (*mar*) locus in *Escherichia coli* and *Salmonella* spp. Studies in Stuart Levy's laboratory (Tufts University School of Medicine) discovered that plasmid-free *E. coli,* when grown on low levels of tetracycline, not only demonstrated resistance to tetracycline, but also showed pleiotropic resistance to several other unrelated antibiotics (chloramphenicol, rifampin, nalidixic acid, puromycin, ampicillin, cephalosporins, penicillin G, and minocycline) (19, 20). In addition, these mutants showed increasingly higher levels of resistance with continued exposure to the antibiotics. The resistance locus mapped to 34 min on the *E. coli* K-12 linkage map and the implicated gene was designated Multiple Antibiotic Resistance (*mar*) locus (20). The *mar* operon was characterized, and the increased antibiotic resistance of the original Mar mutant strain (AG102) was shown to be due to a point mutation in the *marR* gene, resulting in overexpression of *marA*. This overexpression resulted in the differential expression of other ancillary genes, such as *soxS* and *micF* among others (11) and in decreased susceptibilities to structurally unrelated antibiotics including the fluoroquinolones. Overexpression of *marA* also decreases the susceptibilities to organic solvents, oxidative stress agents, and disinfectants such as pine oil and triclosan (1, 39). Although much of the initial work was done with *E. coli,* evidence of the *mar* operon and Mar mutants has been found on the chromosomes of other bacterial species within the *Enterobacteriaceae,* as well as *Neisseria* spp., *Listeria* spp., *Yersinia* spp., *Mycobacterium tuberculosis,* and *Staphylococcus aureus*. This chapter will address the involvement of the *mar* (or *mar*–like) operon in various bacterial species that may cause infections in humans and the potential impact of this resistance locus on modern therapeutics.

ESCHERICHIA COLI AND *SALMONELLA*

Although considered a nonpathogenic commensal present in feces, in developing countries pathogenic *E. coli* is a major cause of intestinal infections as-

JoAnn Dzink-Fox • Novartis Institute for BioMedical Research, Inc., 100 Technology Square, Room 4204, Cambridge, MA 02139.
Margret Oethinger • Yale University School of Medicine, Yale-New Haven Hospital, Department of Laboratory Medicine, 333 Cedar Street, New Haven, CT 06504.

sociated with high childhood mortality and morbidity. In contrast, *E. coli* in developed countries is often associated with community-acquired and nosocomial extraintestinal infections. Treatment usually involves β-lactam antibiotics, fluoroquinolones, and, if necessary, an aminoglycoside. *Salmonella* spp., on the other hand, are not normally found in the human bowel and when introduced by ingesting contaminated food or water cause gastroenteritis or dysentery, except for *Salmonella enterica* serovar Typhi, which causes a generalized infection. Antibiotic treatment is usually withheld except in severe cases, because shedding of enteric salmonella can be extended.

Escherichia coli

The *mar* locus was discovered when the genetic basis of tetracycline resistance in *E. coli* was investigated (19, 20). Briefly, the *marRAB* locus consists of an operator (*marO*) from which two divergent transcripts are expressed (*marC* and *marRAB*). Proteins encoded by the operon are affected by the negative regulator, MarR (1, 9). MarA, a positive transcriptional activator, when overexpressed (due to either mutations in *marO* or *marR*) results in multidrug resistance by controlling the expression of at least 60 genes in *E. coli* (4). MarA up-regulates its own production and a total of 47 genes including AcrA and TolC, both components of the AcrAB efflux pump, which pumps out structurally unrelated antibiotics, dyes, and disinfectants (4). MarA also down-regulates 15 genes including OmpF, an outer membrane protein through which hydrophilic substances enter the cell (4, 10, 11). The level of resistance that is typically seen upon induction of the *mar* locus is low (two- to fourfold) and does not reach the MIC breakpoint for clinical resistance (19). For instance, the MIC of ofloxacin for the wild-type *E. coli* K-12 strain AG100 is 0.03 μg/ml, that of *marA* overexpressing AG112 (derived from AG100 by selection on tetracycline) is 0.125 μg/ml. The breakpoint for clinical intermediate resistance is 1 μg/ml. The function of *marC* or *marB* is not known at this time. It was the extensive work done with *E. coli* and the increasing incidence of antibiotic resistances in other species of bacteria that led to the search for the *mar* operon in other bacterial species.

Initially, the role of *mar* in clinically significant, high-level resistance was unclear. However, the first indication that the *mar* locus was involved in the emergence of clinical resistance to fluoroquinolones came from the observation that clinical fluoroquinolone resistance in *E. coli* was almost always associated with pleiotropic resistance, i.e., with resistance to structurally unrelated antibiotics, termed the "Mar phenotype" (52). The spectrum of resistance was reminiscent of that conferred by overexpression of the *mar* locus (10). Possible mechanisms, as described above, were decreased expression of the OmpF porin (11) and overexpression of the AcrAB efflux pump (47), both resulting in a decreased intracellular drug concentration. Second, the most important mechanism of fluoroquinolone resistance is mutations in the structural genes coding for topoisomerases II (*gyrA*) and topoisomerase IV (*parC*) (13, 26, 50, 58, 59). Point mutations in the so-called quinolone-resistance determining regions (QRDR) affect the affinity of the fluoroquinolone to the gyrase or topoisomerase IV, respectively. Prior studies showed that overall the level of fluoroquinolone resistance correlated with the number of mutations that had accumulated in the *E. coli* genome. Yet, in a substantial number of strains, identical mutations did not lead to the same level of resistance. In other words, mutations in the QRDR regions did not account for the entire level of resistance in some strains. Additional mutations to one or more of the regulatory genes in *marRAB* (9), *soxRS* (2, 63), or *robA* (3) were likely to contribute to the resistance phenotype. Third, the relatively rapid increase of fluoroquinolone resistance rates among clinical *E. coli* isolates has been surprising since wild-type *E. coli* is intrinsically highly susceptible to second generation fluoroquinolones such as ciprofloxacin or ofloxacin (25). In the laboratory, fluoroquinolone resistance was difficult to attain by selection in the presence of fluoroquinolones, with mutation rates between $1{:}10^{-10}$ and $1{:}10^{-11}$ (45). Early experiments in Stuart Levy's laboratory (Tufts University School of Medicine) showed that high-level fluoroquinolone resistant mutants could be readily selected from Mar mutants but not from wild-type *E. coli* isolates (10). The most likely explanation for this observation was the lower drug concentrations noted in Mar mutants as compared to wild-type strains. This could provide Mar mutants with an advantage for survival.

Hence the group of fluoroquinolone-resistant *E. coli* with pleiotropic resistance to structurally unrelated antimicrobial agents was the group of *E. coli* most likely to harbor mutations affecting the *mar* operon or *mar*-like systems. The first study exploring the role of the *mar* operon in clinical antibiotic resistance in *E. coli* was reported by Maneewannakul and Levy (36). Of 23 fluoroquinolone-resistant clinical *E. coli* isolates from three geographically different laboratories, three constitutively expressed *marA* and were found to have mutations in the *marR* gene (Table 1). Inactivation of the *mar* locus by insertion of a kanamycin resistance cassette significantly decreased levels of many antibiotics (Table 2). Five additional strains showed variable responses to induction by salicylate or tetracyclines but had wild-type *marR* (36).

Table 1. Gene sequences of mutant *marR* in fluoroquinolone-resistant *Escherichia coli* strains from humans and animals reported in the literature to date

Study	Strain	Fluoroquinolone MIC (μg/ml)	Cyclohexane tolerance[e]	Mutation in *marR* (nucleotide change) at amino acid:						
				Ser 3	26-31	Glu 31	Leu 36	Cys 47	Ile 49	Arg 50
Maneewannakul (36)	KM-D	1.5[a]	–	–	–	–	–	–	–	–
	KM-F	>256[a]	–	–	–	–	–	–	–	ΔCCG
	J28	4[a]	–	Asn (1452G →A)	–	–	–	–	–	–
Oethinger (46)	S20	0.25[b]	+	–	–	–	–	–	–	–
	M19	2[b]	+	–	–	Stop codon inserted (1535G →T)	–	–	–	–
	NH52	8[b]	–	–	–	–	–	–	–	–
	HO17	32[b]	+	–	–	–	–	–	–	–
	HO99	32[b]	+	–	–	–	–	–	–	–
	E22	64[b]	+	–	–	–	–	–	Ser (1590T →G)	–
Linde (33)	EP1	16[c]	–	–	–	–	–	–	–	–
	EP2	256[c]	+	–	–	–	–	–	–	–
Webber (61)	I276	32[c]	ND	–	–	–	–	–	–	–
	I278	8[c]	ND	–	–	–	–	–	–	–
	I279	8[c]	ND	–	–	–	–	–	–	–
	I280	32[c]	ND	–	–	–	–	–	–	–
	I283	32[c]	ND	–	–	–	–	–	–	–
Komp (31)	C1171	>32[c]	+	NA[g]	–	–	–	–	–	–
	C1181	>32[c]	–	NA[g]	–	–	–	–	–	–
	C1203	>32[c]	+	NA[g]	–	–	–	–	–	–
	C1168	>32[c]	+	NA[g]	–	–	–	–	–	–
	C1162	32[c]	+	NA[g]	–	–	–	–	–	–
	C1172	24[c]	+	NA[g]	–	–	–	–	–	–
	C1186	32[c]	+	NA[g]	–	–	–	Gly	–	–
	C1200	16[c]	+	NA[g]	–	–	–	–	–	–
	C1180	3[c]	–	NA[g]	–	–	–	–	–	–
	C47	0.05[c]	+	NA[g]	+17 nt (duplication)	–	–	–	–	–
	C25	0.01[c]	–	NA[g]	–	–	–	–	–	–
Sáenz et al. (55)	Co1		ND	–	–	–	–	–	–	–
	Co19	N[d]	ND	–	–	–	–	–	–	–
	Co45	N, C[d]	ND	–	–	–	Gln	–	–	–
	Co71	N, C[d]	ND	–	–	–	–	–	–	–
	Co80	N[d]	ND	–	–	–	–	–	–	–
	Co82	N, C[d]	ND	–	–	–	–	–	–	–
	Co110	N[d]	ND	–	–	–	–	–	–	–
	Co125	N, C[d]	ND	–	–	–	–	–	–	–
	Co177	N[d]	ND	–	–	–	–	–	–	–
	Co201	N[d]	ND	–	–	–	–	–	–	–
	Co227	N[d]	ND	–	–	–	–	–	–	–
	Co228	N[d]	ND	–	–	–	–	–	–	–
	Co232	N, C[d]	ND	–	–	–	–	–	–	–
	Co279	N[d]	ND	–	–	–	–	–	–	–
	Co354	N, C[d]	ND	–	–	–	–	–	–	–

[a] Fluoroquinolone tested was norfloxacin.
[b] Fluoroquinolone tested was ofloxacin.
[c] Fluoroquinolone tested was ciprofloxacin.
[d] Resistance phenotype; indicates resistance to nalidixic acid (N) or ciprofloxacin (C) (55).
[e] Cyclohexane tolerance was tested on LB agar overlaid by the organic solvent and grown for 24 h at 30°C (44).
[f] Δ, deletion.
[g] Several strains carried reportedly a Ser 3 Asn mutation, but strains are not identified in reference 31.

Mutation in *marR* (nucleotide change) at amino acid:											
Ala 52	Ala 53	Lys 62	Asp 76	Leu 78	Val 79	Arg 94	Val 96	Gly 103		Glu 110	Tyr 137
–	Δ[f]1602-1731	–	–	–	–	–	–	Ser (1751G →A)	–	–	–
–	–	–	–	–	–	–	Glu (1731T →A)	–	–	–	–
–	Δ1601-1603	–	–	–	–	–	–	–	–	–	–
–	–	–	–	–	–	Ser (1724C →A)	–	Ser (1751G →A)	–	–	His (1853T →C)
–	–	–	–	–	–	–	–	Ser (1751G →A)	–	–	His (1853T →C)
–	–	–	–	–	–	–	–	frame-shift Δ1751	–	–	–
–	–	–	–	–	–	His (1725G →A	–	Ser (1751G →A)	–	–	His (1853T →C)
–	–	–	–	Met (1676C →A)	–	–	–	–	–	–	–
–	–	–	–	–	–	–	–	–	–	–	–
–	–	–	–	–	–	–	–	–	–	–	–
–	–	–	–	–	–	–	–	–	frameshift Δ1821	–	–
–	–	–	–	–	–	–	–	–	–	–	His (1853T →C)
–	–	–	–	–	–	–	–	–	–	–	His (1853T →C)
–	–	–	–	–	–	–	–	–	–	–	His (1853T →C)
–	–	–	–	–	–	–	–	–	–	–	His (1853T →C)
–	–	–	–	–	–	–	–	–	–	–	His (1853T →C)
–	–	–	–	–	–	–	–	–	–	Stop codon inserted	–
–	–	Arg	–	–	–	–	–	–	–	–	–
–	–	Arg	–	–	–	–	–	–	–	–	–
Arg (ΔG)	–	–	–	–	–	–	–	Asn	–	–	–
–	–	–	–	Met	–	–	–	–	–	–	–
–	–	–	–	Met	–	–	–	–	–	–	–
–	–	–	–	–	–	–	–	–	–	–	–
–	–	–	Gly	–	–	–	–	–	–	–	–
–	–	–	–	–	Ile	–	–	–	–	–	–
–	–	–	–	–	–	–	–	–	–	–	–
–	–	Arg	–	–	–	–	–	–	–	–	–
–	–	Ser	–	–	–	–	–	–	–	–	His
–	–	–	–	–	–	–	–	Ser	–	–	His
–	–	–	–	–	–	–	–	Ser	–	–	His
–	–	–	–	–	–	–	–	Ser	–	–	His
–	–	–	–	–	–	–	–	Ser	–	–	His
–	–	–	–	–	–	–	–	Ser	–	–	His
–	–	–	–	–	–	–	–	Ser	–	–	His
–	–	–	–	–	–	–	–	Ser	–	–	His
–	–	–	–	–	–	–	–	Ser	–	–	His
–	–	–	–	–	–	–	–	Ser	–	–	His
–	–	–	–	–	–	–	–	Ser	–	–	His
–	–	–	–	–	–	–	–	Ser	–	–	His
–	–	–	–	–	–	–	–	Ser	–	–	His
–	–	–	–	–	–	–	–	Ser	–	–	His
–	–	–	–	–	–	–	–	Ser	–	–	His

Table 2. Effect of inactivation of the *mar* locus (*Bsp*HI fragment) by a kanamycin resistance gene replacement on antibiotic susceptibility[a]

Strain	MIC (μg/ml)						
	Gradient plate method			E-test			
	Tetracycline	Chloramphenicol	Ampicillin	Rifampin	Norfloxacin	Cephalothin	Fleroxacin
AG100	8.3	6.6	2.3	12.0	0.023	2.0	0.064
AG100/Kan	7.0	2.2	1.7	12.0	0.023	2.0	0.064
AG102	13.3	37.7	20.0	24.0	1.25	16.0	1.5
AG102/Kan	7.5	2.2	12.1	12.0	0.028	2.0	0.064
KM-D	7.8	17.8	12.2	24.0	1.5	12.0	8.0
KM-D/Kan	7.8	11.1	11.1	12.0	1.5	8.0	8.0
KM-F	>100	44.4	>100	8.0	>256	>256	>256
KM-F/Kan	>100	8.3	>100	4.0	>256	>256	≤256
J28	9.4	47.2	21.7	12.0	4.0	41.5	16.0
J28/Kan	5.6	33.3	6.0	6.0	1.0	8.0	4.0

[a] Results are representative of duplicate experiments. Reprinted with kind permission from reference 36.

Among European cancer patients who were treated prophylactically with ofloxacin during profound neutropenia, the rate of fluoroquinolone-resistant *E. coli* isolated from blood cultures had increased by more than fourfold within a 5 year period (12). Fluoroquinolone-resistant strains were analyzed from 10 cancer centers for epidemiological relatedness and for mechanisms of resistance and compared with fluoroquinolone-resistant *E. coli* isolated from hospitalized patients without cancer, as well as with fluoroquinolone-susceptible isolates from cancer patients at a single center (13, 28, 42). Screening a total of 138 independently obtained *E. coli* isolates (comprising the above described fluoroquinolone-resistant *E. coli* strains from European cancer centers) with a recently described organic solvent tolerance test (62) yielded 25 strains that grew under 100% cyclohexane or 75% cyclohexane (cyclohexane:hexane 3:1) (44). Fluoroquinolone-resistant *E. coli* exhibited organic solvent (cyclohexane) tolerance at a significantly higher frequency (30%) than fluoroquinolone-susceptible strains (2%; $P < 0.001$) (Fig. 1). We also observed a statistically significant trend of higher cyclohexane tolerance with increasing ofloxacin MICs (Fig. 1). Northern blot analysis and subsequent functional reporter gene assay showed that 6 of 25 cyclohexane-tolerant strains were Mar mutants (Fig. 2) (46). Analysis of the *soxRS* operon that was also shown to be associated with increased antibiotic resistance and increased organic solvent tolerance (62) identified four Sox mutants (46). Among *E. coli* strains from cancer patients, three Mar (Table 1) and three Sox mutants were identified for which the fluoroquinolone MICs were higher than those for *E. coli* strains with identical structural gene mutations, i.e., in *gyrA* and *parC* genes (Table 3). These Mar and Sox mutants were also resistant to triclosan (40).

It appears that the resistance acquired in Mar mutants and "Mar-like" mutants—such as Sox mutants—acts as an important "stepping stone" to higher drug resistance (1). In an attempt to dissect the time course of the events from wild-type to high-level fluoroquinolone resistance in vivo, i.e., within a given patient, a prospective study on colonization with fluoroquinolone-resistant *E. coli* among cancer patients was performed in adult patients with neutropenia (<1,000 neutrophils/μl) secondary to cancer or cytotoxic therapy who received oral ofloxacin as an an-

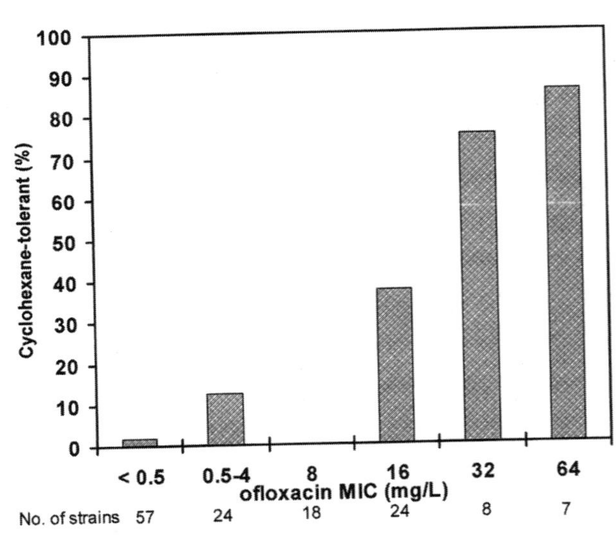

Figure 1. Percentage of cyclohexane-tolerant strains among 138 clinical *E. coli* isolates, grouped according to their ofloxacin MIC. The number of strains in each group is given at the bottom of the figure. Strains with an ofloxacin MIC of < 0.5 mg/liter are a control group of 57 fluoroquinolone-susceptible strains. Of the 24 fluoroquinolone-intermediate-resistant strains, three were cyclohexane tolerant. They had ofloxacin MICs of 1, 2, and 4 mg/liter, respectively.

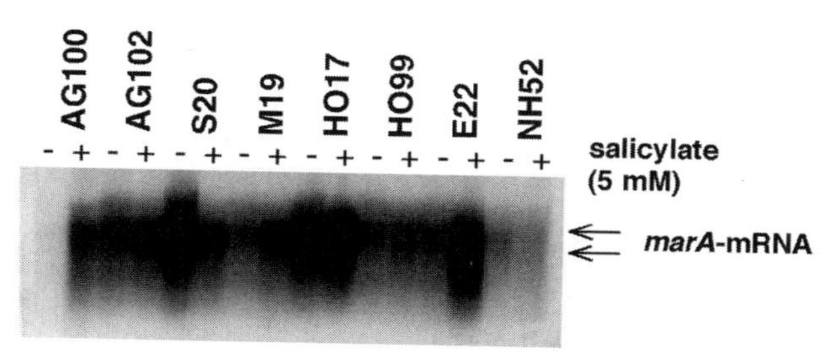

Figure 2. Northern blot analysis of *marRAB* mRNAs prepared from clinical *E. coli* strains incubated without (−) and with (+) 5 mM sodium salicylate for 45 min. RNA samples were transferred to Hybond-N$^+$ membranes and probed with radioactively labeled *marA*. Arrows point to prominent transcripts of ~1.1 and ~0.9 kbp. [Reprinted with kind permission from reference 46].

tibacterial prophylaxis (43). Surveillance cultures from oropharyngeal washings and feces were obtained once or twice weekly from the patients over a period of 18 months. Cultures from blood or other sites were obtained as clinically indicated (43). Fluoroquinolone-resistant *E. coli* were cultured from at least one body site in 16 cancer patients. Population analysis in 11 patients indicated that a homogenous population of high-level fluoroquinolone-resistant *E. coli* emerged after a median of 34 days (range: 12-102 days). In the cancer patients, fluoroquinolone-resistant *E. coli* was the only species of aerobic gram-negative bacilli present. Of note, we did not observe *E. coli* which exhibited low or intermediate levels of fluoroquinolone resistance during drug administration, but we found an apparently homogenous population with roughly identical, high ofloxacin MICs. We concluded that low-level resistant mutants, thus, may be transient and rapidly mutate to high-level resistance (43). Alternatively, the *mar* operon or Mar-like systems may be up-regulated only transiently (43).

In contrast, in vitro passage with ofloxacin of the clinical fluoroquinolone-susceptible isogenic *E. coli* strain of two selected isolates yielded a typical pattern of step-wise increasing fluoroquinolone resistance (29). Laboratory *E. coli* K-12 strains AG100 (wild-type *mar*) and AG112 (Mar mutant) were used as controls. Interestingly, passage of both laboratory and clinical *E. coli* strains with fluoroquinolones first led to selection of *gyrA* mutants and subsequently to Mar phenotype mutants which generally overexpressed MarA or SoxS (29). Ciprofloxacin MICs for first step mutants associated with *gyrA* mutations were in the range of 0.06 to 0.5 μg/ml. Second and third step mutants had a Mar phenotype including cyclohexane resistance and exhibited ciprofloxacin MICs of 0.5 to 32 μg/ml. They also showed decreased accumulation of ciprofloxacin compared to wild-type or first step mutants (Fig. 3). The addition of carbonyl cyanide *m*-chlorophenylhydrazone (CCCP) abolished this effect, suggesting that an efflux pump was important in the phenotype of these second and higher step mutants (29).

Linde and Notka (33, 41) reported on two sequential, isogenic *E. coli* isolates from one patient who had different MICs of ciproflocaxin (isolate EP1 = 16 μg/ml versus isolate EP2 = 256 μg/ml) but identical mutations in structural genes in the quinolone resistance determining regions. Isolate EP2 was also more resistant to chloramphenicol, tetracyclines, cefuroxime, and organic solvents. A deletion of adenine (ΔA1821) was found in *marR* of isolate EP2, which resulted in an 18-amino-acid C-terminal deletion in the MarR protein (Table 1). The causative relationship

Table 3. Overexpression of *marA* and *soxS* in clinical fluoroquinolone-resistant *E. coli* strains with identical mutations in the regions determining quinolone resistance in *gyrA* and *parC*[a]

Strain[b]	Ofloxacin MIC (μg/ml)	Cyclohexane tolerance[c]	Mutation in[d]: *gyrA* at amino acid: Ser 83	Mutation in[d]: *gyrA* at amino acid: Asp 87	Mutation in[d]: *parC* at amino acid: Ser 80	Mutation in[d]: *parC* at amino acid: Glu 84	Overexpression of *marA*	Overexpression of *soxS*
E5	8	−	Leu	Asn		Lys	−	−
E10[e]	64	−	Leu	Asn		Lys	−	−
NH1	8	−	Leu	Gly	Ile		−	−
HO17	32	+	Leu	Gly	Ile		+	−
HO12	8	−	Leu	Asn	Ile		−	−
HO13[e]	32	−	Leu	Asn	Ile		−	−
E7	32	+	Leu	Asn	Ile		−	−
HO99	32	+	Leu	Asn	Ile		+	−
E3	64	+	Leu	Asn	Ile		−	+
E19	64	+	Leu	Asn	Ile		−	+
E22	64	+	Leu	Asn	Ile		+	−

[a] Reprinted from reference 46 with permission.
[b] The strains, including HO strains, are genotypically unrelated (42), and they are all bloodstream isolates except for NH1, which is a urinary tract isolate.
[c] Cyclohexane tolerance was tested on LB agar overlaid by the organic solvent and grown for 24 h at 30°C (44).
[d] See reference 13.
[e] Increased fluoroquinolone resistance not associated with cyclohexane tolerance.

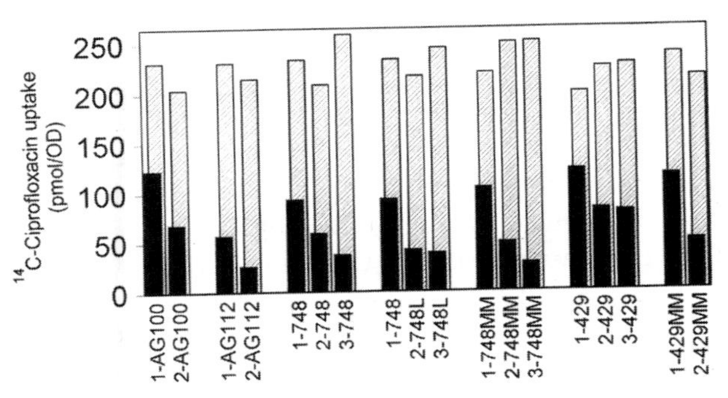

Figure 3. Radiolabeled ciprofloxacin uptake in energized cells with (hatched bars) and without (solid bars) CCCP (200 μM). First-, second-, and third-step fluoroquinolone-resistant mutants, isolated in vitro, are designated by the prefixes 1-, 2-, and 3-, respectively. MM means selection of in vitro mutants on minimal medium. Results are expressed as picomoles of ciprofloxacin per A_{600} unit (OD, optical density) and represent the mean of duplicate steady-state accumulation measurements. Experiments were repeated three times, with values for the accumulation relative to that of the parental strains differing by <10%. The results shown here are the results of one representative experiment. (Reprinted from reference 29 with kind permission.)

between ΔA1821 and the Mar phenotype was demonstrated both by the replacement of the wild-type *marR* by *marR* ΔA1821 in isolate EP1 and by complementation with wild-type *marR* in trans in isolate EP2. In isolate EP2 complemented with wild-type *marR*, susceptibility to chloramphenicol was restored completely, whereas susceptibility to ciprofloxacin was restored only partially. Northern blotting demonstrated increased expression of *marA* and *acrAB* but not of *soxS* in isolate EP2 compared to EP1. Thus, the deletion of A1821 in *marR* in the clinical isolate EP2 appeared to be an important contributory cause of the increase in the MIC of ciprofloxacin by four dilutions and of decreased susceptibility to unrelated antibiotics (33).

The genetic basis for fluoroquinolone resistance was examined in 30 high-level fluoroquinolone-resistant *E. coli* clinical isolates from Beijing, China (60). These isolates showed multiple chromosomal mutations, including mutations in the topoisomerase genes (*gyrA* and *parC*). By sequence analysis of the QRDR, all strains showed double point mutations in *gyrA* and *parC*. Although those were the most frequent cause of fluoroquinolone resistance, higher levels of resistance in *E. coli* resulted from double mutations in *gyrA* with a single *parC* mutation; the highest level of resistance resulted from an additional mutation in *parC*. Sixty-three percent of the strains (19 of 30) were resistant to organic solvents and overexpressed AcrA, an obligatory component of the AcrAB multidrug efflux system. Surprisingly, none of the strains overexpressed either the MarA protein or *soxS* RNA. As a result, we then went on to look for possible mutations in *acrR*, the regulator of the *acrAB*, in a subset of strains. We found that six of eight strains overexpressing AcrA had amino acid substitutions, deletions, or duplications in the AcrR repressor. However, two strains lacked mutations in *acrR* or the *acrAB* promoter/operator region, yet they overexpressed AcrA, were cyclohexane tolerant, and were highly resistant to multiple antimicrobials (60). The role of the AcrAB efflux pump was investigated in 15 clinical *E. coli* strains that had ciprofloxacin MICs ≤ 1 μg/ml, and in 10 high-level fluoroquinolone-resistant *E. coli* strains (ciprofloxacin MICs ≥ 32 μg/ml) (37). AcrA was overexpressed by ≥ 170% in 9 of the 10 high-level resistant strains but not in any of the 15 susceptible or low-level fluoroquinolone-resistant isolates (37). The *mar* operon, however, was not analyzed in this study.

Webber et al. studied 36 fluoroquinolone-resistant *E. coli* from humans and animals, 27 of which displayed a Mar phenotype (61). They determined the amount of *acrB*, *marA*, and *soxS* mRNA in these strains and showed that 11 of the 36 strains had elevated *acrB* mRNA. In 6 of these 11 strains, overexpression of *acrB* was due to an Arg 45 Cys point mutation in the repressor protein AcrR. Ten of the 36 isolates appeared to overexpress *soxS*, and five appeared to overexpress *marA*. A number of mutations were found in the *marR* (Table 1) and *soxR* repressor genes, correlating with greater amounts of *marA* and *soxS* mRNA, respectively (61). A similar study on 54 clinical *E. coli* strains from patients with uncomplicated urinary tract infections analyzed the DNA sequences of *gyrA*, *gyrB*, *parC*, *parE*, *marOR*, and *acrR* (31). Mutations in the *marOR* sequence were identified in 11 strains (Table 1). Three strains had mutations that would cause premature termination of *marR* translation. In one case this was caused directly by a nonsense codon, while the other two carried frame shift mutations. Eight strains had single-amino-acid alterations in the *marR*-coding sequence (Table 1). Two of the three strains with the Lys 62 Arg mutation were organic solvent susceptible, thus confirming our previous observation that this mutation does not contribute to organic solvent tolerance. A newly described mutation, Val 79 Ile, was also not associated with an organic solvent tolerance phenotype (Table 1). Strain C1186 carried a Cys 47 Gly mutation and, in addition, a 5-nt deletion in the *marO* region. The deleted sequence (nt -72 to -68 upstream of the *marR* GTG start codon) is one-half of a tandem repeat sequence located between the -35 region of the *marR* promoter and the Mar box sequence to which the MarA activator binds (1). C1186 also carried an IS*1* insertion in *acrR*. In addition to the alterations described above, many strains had either Ser or Asn at codon 3. Both amino acids were seen in strains that were susceptible to norfloxacin and ciprofloxacin. Hence, these genotypes appear to be neutral with respect to fluoroquinolone resistance (31), as previously described (46). The authors observed a correlation between the fluoroquinolone MIC and the number of genes that were affected by mutations, with resistant strains having mutations in two to five of the genes examined. They concluded that the evolution of fluoroquinolone resistance involves the accumulation of multiple mutations in several genes.

A recent study analyzed 17 nonpathogenic *E. coli* strains of human, animal, and food origins that showed a pleiotropic resistance phenotype (55). The strains showed a wide variety of antibiotic resistance genes, many of them carried by class 1 and class 2 integrons. Amino acid changes in *marR* and mutations in *marO* were identified for 15 and 14 *E. coli* strains, respectively (55). All 15 strains with *marR* mutations, however, showed the Gly 103 Ser and Tyr 137 His mutations (Table 1) that have previously been shown to represent genotypic variations without loss of repressor activity (46). One additional amino acid change in *marR*, Leu 36 Gln, was found only in strain Co45 (Table 1).

Salmonella

Several reports described the isolation of *Salmonella* spp. with a Mar phenotype from human clinical infections during antibiotic therapy (21, 27, 48, 51, 53). Recent evidence showed that active efflux via the AcrAB-TolC multidrug efflux system is the primary mechanism of resistance to ciprofloxacin in *Salmonella enterica* serovar Typhimurium (5, 22). Eaves et al. (16) studied three post-therapy serovar Typhimurium strains along with the pretherapy parental strain previously described (51, 53). Compared with the pretherapy isolate (isolate L3), the levels of expression of one or more of the *acrB*, *acrF*, and *acrD* genes were increased in all three posttherapy clinical isolates. These isolates also accumulated less ciprofloxacin; the level of accumulation could be increased by the addition of CCCP. No change in the level of *soxS* expression was detected, but the level of *marA* expression was increased in the posttherapy clinical isolates (16).

Randall et al. (54) investigated the relationship between organic solvent tolerance and increased resistance to multiple antibiotics, particularly to ciprofloxacin, for *Salmonella* isolates. Ciprofloxacin resistance was greatest when *gyrA* mutations were associated with cyclohexane tolerance (32). Cyclohexane-tolerant *Salmonella* isolates in this study were not tested for up-regulation of *acrAB*, *marRAB*, *robA*, or *soxRS* genes; thus, the Mar phenotype could have been due to up-regulation of these or as yet undefined genes.

KLEBSIELLA AND *ENTEROBACTER*

Klebsiella spp. and *Enterobacter* spp. are opportunistic pathogens commonly isolated from patients with urinary tract, respiratory, and/or wound infections. *Enterobacter aerogenes* is emerging as the third leading cause of nosocomial respiratory tract infections, exhibiting increased resistance to many broad-spectrum antibiotics. Antimicrobial treatment varies from β-lactam antibiotics to aminoglycosides, quinolones, trimethoprim, and chloramphenicol. Resistance to these antimicrobials is usually plasmid mediated, frequently resulting in multiresistance. Recently, there has been an increase of multiply resistant strains due to chromosomal gene mutations in the DNA gyrase and topoisomerase IV genes, or mutations resulting in decreased outer membrane (OmpF) permeability, and/or drug accumulation (AcrAB). The latter two have been reported to be due to the overexpression of chromosomal regulatory loci of *marRAB* and *soxRS*.

Klebsiella

A recent study (56) demonstrated that fluoroquinolone-resistant *Klebsiella pneumoniae* strains had mutations in structural genes *gyrA* and *parC* in agreement with similar studies in *E. coli* (13, 24, 29, 58, 59, 60, 64, 65). In contrast, however, no overexpression of *marA* or *soxS* was detected in these strains of *Klebsiella*. Interestingly, this study also found increased expression of the *Klebsiella* transcriptional regulator *ramA* (40% sequence homology to *marA*) in a few strains (56). The RamA protein is related to the *E. coli* MarA and SoxS proteins in that they all belong to the AraC family of transcriptional activators, are similar in length, and have a helix-turn-helix DNA-binding domain near the N terminus (1, 2, 9, 17, 63). When *ramA* was cloned and overexpressed in *E. coli*, a multidrug resistance phenotype was observed. Although not completely understood, RamA, a transcriptional activator, is thought to be part of an operon that comprises a gene specifying a putative channel-forming outer membrane protein, RomA (18, 30). In addition, increased expression of *ramA* is directly associated with the multidrug efflux pump, AcrAB.

Even though MarA and RamA are not interchangeable, these transcriptional activators may be able to recognize similar regulatory signals, which in turn activate the expression of other diverse genes.

Enterobacter

Antibiotic resistance in *Enterobacter* species is largely attributed to changes in or deletions of certain outer membrane proteins (15, 23) and/or decreased accumulation across the inner membrane (6, 14, 35), thus conferring a multiple antibiotic resistance phenotype. The *romA* gene responsible for this multiresistant phenotype in *Enterobacter cloacae* was previously identified by Komatsu (30); however, these changes were later shown to be the result of expression of a *ramA*-like gene. Clinical resistance in these strains could be seen with β-lactams, chloramphenicol, tetracyclines, quinolones, and the carbapenems. The gene encoding RamA was recently identified in a clinical multiresistant *E. aerogenes* strain (8). Overexpression of RamA induced a Mar phenotype, a decrease in porin production, and increased production of AcrA, a component of the AcrAB-TolC drug efflux pump. RamA enhanced transcription of the *marRAB* operon but was also able to induce the Mar phenotype in a *mar*-deleted strain (8).

The mechanism of fluoroquinolone resistance in two isogenic isolates of *E. cloacae* from the same patient with different ciprofloxacin MICs (Ecl#1 = 0.25 μg/ml and Ecl#2 = 1 μg/ml) was analyzed (34). Ecl#2 accumulated significantly less norfloxain and displayed higher levels of expression of *marR* and *acrB* than Ecl#1. Hence increased fluoroquinolone efflux contributed to the increased resistance, underscoring the sequential development of fluoroquinolone resistance (34).

Chollet et al. identified and sequenced a homologue of the *mar* operon in *E. aerogenes* (7). When compared to known homologues in *E. coli, Salmonella* serovar Typhimurium, *E. cloacae,* and *K. pneumoniae,* the predicted amino acid alignments of MarR, MarA, and MarB showed >80% similarity. MarA-*Ea* showed greater amino acid similarity to MarA-*Kp* (96.1%) than MarA-*Eco* (88.4%). Similarity between MarR-*Ea* and MarR-*Kp* was 91%, and between MarR-*Ea* and MarR-*Eco* it was 81.2%. This greater similarity between Mar-*Ea* and Mar-*Kp* is obviously due to their phylogenetic proximity. By characterizing the involvement of the *marA* gene with increased outer membrane impermeability and drug efflux demonstrating a multidrug resistance phenotype, they also determined that the *E. aerogenes marRAB* operon was not only structurally but also functionally analogous to the *E. coli* and *Salmonella* operons (7, 57).

Mar IN OTHER SPECIES

Mycobacterium

Although the majority of multidrug resistance studies have been focused primarily on the *Enterobacteriaceae,* multidrug resistance in *Mycobacterium* spp. is increasing at an alarming rate. On a worldwide basis, spread of *M. tuberculosis* among various human populations is a major public health concern, contributing to 8 to 10 million new cases of tuberculosis annually, with 2 to 3 million deaths resulting from the disease (data according to the World Health Organization). Although the greatest morbidity and mortality occur in developing countries, tuberculosis is also found at a lower but significant rate in developed countries due to reactivation of old infections or recent infections in immunocompromised individuals. Recommended treatment of tuberculosis requires a multiple drug regimen (isoniazid, rifampin, pyrazinamide, and ethambutol) for approximately 6 months to prevent the emergence of resistant bacilli.

Resistance in *Mycobacterium* spp. was previously thought to be due to the accumulation of independent chromosomal mutations; however, studies using *M. smegmatis* (38) suggested that a *mar*-like regulatory system is also present in *M. tuberculosis.* When *E. coli marA* was cloned into *M. smegmatis,* increased resistance to multiple antimicrobials, including rifampin, isoniazid, ethambutol, chloramphenicol, and tetracycline, was observed at 37°C. Furthermore, this resistance was directly related to MarA, as evidenced by insertional inactivation of the MarA helices (38). Although the mechanism of MarA-mediated resistance is currently unknown, the observed phenotype may be the result of an indirect interaction of MarA with endogenous mycobacterial (efflux) proteins. In support of this notion, two proteins have been identified in *M. tuberculosis* that are similar to MarA (49).

CONCLUSIONS

A substantial body of evidence shows that a Mar phenotype of resistance can develop in clinical isolates of *E. coli, Salmonella* spp., and other *Enterobacteriaceae.* The presence of cyclohexane-tolerant veterinary isolates of *Salmonella* spp. also suggests that similar multiple resistance can develop in animals. Studies indicate that mutations giving rise to increased expression of the transcriptional activators *marA* and *soxS* or others affect the expression of a variety of different genes, including *ompF* and *acrAB.* The net result is that expression of OmpF is reduced and less drug is able to enter the cell. Importantly, expression of *acrAB* is increased, enhancing efflux from the cell. Research in the last decade has tremendously increased our understanding of these systems of adaptive response.

REFERENCES

1. **Alekshun, M. N., and S. B. Levy.** 1997. Regulation of chromosomally mediated multiple antibiotic resistance: the *mar* regulon. *Antimicrob. Agents Chemother.* **41:**2067–2075.

2. **Amábile-Cuevas, C. F., and B. Demple.** 1991. Molecular characterization of the *soxRS* genes of *Escherichia coli:* two genes control a superoxide stress regulon. *Nucleic Acids Res.* **19:**4479–4484.
3. **Ariza, R. R., Z. Li, N. Ringstad, and B. Demple.** 1995. Activation of multiple antibiotic resistance and binding of stress-inducible promoters by *Escherichia coli* Rob protein. *J. Bacteriol.* **177:**1655–1661.
4. **Barbosa, T. M., and S. B. Levy.** 2000. Differential expression of over 60 chromosomal genes in *Escherichia coli* by constitutive expression of MarA. *J. Bacteriol.* **182:**3467–3474.
5. **Baucheron, S., H. Imberechts, E. Chaslus-Dancla, and A. Cloeckaert.** 2002. The AcrB multidrug transporter plays a major role in high-level fluoroquinolone resistance in *Salmonella enterica* serovar Typhimurium phage type DT204. *Microb. Drug Resist.* **8:**281–289.
6. **Charrel, R. N., J. M. Pages, P. De Micco, and M. Malléa.** 1996. Prevalence of outer membrane porin alteration in beta-lactam-antibiotic-resistant *Enterobacter aerogenes. Antimicrob. Agents Chemother.* **40:**2854–2858.
7. **Chollet, R., C. Bollet, J. Chevalier, M. Malléa, J. M. Pages, and A. Davin-Regli.** 2002. Mar Operon involved in multidrug resistance of *Enterobacter aerogenes. Antimicrob. Agents Chemother.* **46:**1093–1097.
8. **Chollet, R., J. Chevalier, C. Bollet, J. M. Pages, and A. Davin-Regli.** 2004. RamA is an alternate activator of the multidrug resistance cascade in *Enterobacter aerogenes. Antimicrob. Agents Chemother.* **48:**2518–2523.
9. **Cohen, S. P., H. Hächler, and S. B. Levy.** 1993. Genetic and functional analysis of the multiple antibiotic resistance (*mar*) locus in *Escherichia coli. J. Bacteriol.* **175:**1484–1492.
10. **Cohen, S. P., L. M. McMurry, D. C. Hooper, J. S. Wolfson, and S. B. Levy.** 1989. Cross-resistance to fluoroquinolone in multiple-antibiotic-resistant (Mar) *Escherichia coli* selected by tetracycline or chloramphenicol: decreased drug accumulation associated with membrane changes in addition to OmpF reduction. *Antimicrob. Agents Chemother.* **33:**1318–1325.
11. **Cohen, S. P., L. M. McMurry, and S. B. Levy.** 1988. *marA* locus causes decreased expression of OmpF porin in multiple-antibiotic-resistant (Mar) mutants of *Escherichia coli. J. Bacteriol.* **170:**5416–5422.
12. **Cometta, A., T. Calandra, J. Bille, and M. P. Glauser.** 1994. *Escherichia coli* resistant to fluoroquinolones in patients with cancer and neutropenia. *N. Engl. J. Med.* **330:**1240–1241.
13. **Conrad, S., M. Oethinger, K. Kaifel, G. Klotz, R. Marre, and W. V. Kern.** 1996. *gyrA* mutations in high-level fluoroquinolone-resistant clinical isolates of *Escherichia coli. J. Antimicrob. Chemother.* **38:**443–455.
14. **Dang, P., L. Gutmann, C. Quentin, R. Williamson, and E. Collatz.** 1998. Some properties of *Serratia marcescens, Salmonella paratyphi A,* and *Enterobacter cloacae* with non-enzyme-dependent multiple resistance to beta-lactam antibiotics, aminoglycosides, and quinolones. *Rev. Infect. Dis.* **10:**899–904.
15. **de Champs, C., C. Henquell, D. Guelon, D. Sirot, N. Gazuy, and J. Sirot.** 1993. Clinical and bacteriological study of nosocomial infections due to *Enterobacter aerogenes* resistant to imipenem. *J. Clin. Microbiol.* **31:**123–127.
16. **Eaves, D. J., V. Ricci, and L. J. V. Piddock.** 2004. Expression of *acrB, acrF, acrD, marA,* and *soxS* in *Salmonella enterica* serovar typhimurium: role in multiple antibiotic resistance. *Antimicrob. Agents Chemother.* **48:**1145–1150.
17. **Gambino, L., S. J. Gracheck, and P. F. Miller.** 1993. Overexpression of the MarA positive regulator is sufficient to confer multiple antibiotic resistance in *Escherichia coli. J. Bacteriol.* **175:**2888–2894.
18. **George, A. M., R. M. Hall, and H. W. Stokes.** 1995. Multidrug resistance in *Klebsiella pneumoniae:* a novel gene, *ramA,* confers a multidrug resistance phenotype in *Escherichia coli. Microbiology* **141:**1909–1920.
19. **George, A. M., and S. B. Levy.** 1983a. Amplifiable resistance to tetracycline, chloramphenicol, and other antibiotics in *Escherichia coli:* involvement of a non-plasmid-determined efflux of tetracycline. *J. Bacteriol.* **155:**531–540.
20. **George, A. M., and S. B. Levy.** 1983b. Gene in the major cotransduction gap of the *Escherichia coli* K-12 linkage map required for the expression of chromosomal resistance to tetracycline and other antibiotics. *J. Bacteriol.* **155:**541–548.
21. **Gibb, A. P., C. S. Lewin, and O. J. Garden.** 1991. Development of quinolone resistance and multiple antibiotic resistance in *Salmonella bovismorbificans* in a pancreatic abscess. *J. Antimicrob. Chemother.* **28:**318–321.
22. **Giraud, E., A. Cloeckaert, D. Kerboeuf, and E. Chaslus-Dancla.** 2000. Evidence for active efflux as the primary mechanism of resistance to ciprofloxacin in *Salmonella enterica* serovar Typhimurium. *Antimicrob. Agents Chemother.* **44:**1223–1228.
23. **Gutmann, L., R. Williamson, N. Moreau, M. D. Kitzis, E. Collatz, J. F. Acar, and F. W. Goldstein.** 1985. Cross-resistance to nalidixic acid, trimethoprim, and chloramphenicol associated with alterations in outer membrane proteins of *Klebsiella, Enterobacter,* and *Serratia. J. Infect. Dis.* **151:**501–507.
24. **Heisig, P.** 1996. Genetic evidence for a role of *parC* mutation in the development of high-level fluoroquinolone resistance in *Escherichia coli. Antimicrob. Agents Chemother.* **40:**879–885.
25. **Hooper, D. C.** 1995. Quinolone mode of action. *Drugs* **49** (Suppl. 2):10–15.
26. **Hooper, D. C.** 1999. Mechanisms of fluoroquinolone resistance. *Drug Resist. Updates* **2:**38–55.
27. **Howard, A. J., T. D. Joseph, L. L. Bloodworth, J. A. Frost, H. Chart, and B. Rowe.** 1990. The emergence of ciprofloxacin resistance in *Salmonella* Typhimurium. *J. Antimicrob. Chemother.* **26:**296–298.
28. **Kern, W. V., E. Andriof, M. Oethinger, P. Kern, J. Hacker, and R. Marre.** 1994. Emergence of fluoroquinolone-resistant *Escherichia coli* at a cancer center. *Antimicrob. Agents Chemother.* **38:**681–687.
29. **Kern, W. V., M. Oethinger, A. S. Jellen-Ritter, and S. B. Levy.** 2000. Non-target gene mutations in the development of fluoroquinolone resistance in *Escherichia coli. Antimicrob. Agents Chemother.* **44:**814–820.
30. **Komatsu, T., M. Ohta, N. Kido, Y. Arakawa, H. Ito, and N. Kato.** 1991. Increased resistance to multiple drugs by introduction of the *Enterobacter cloacae romA* gene into OmpF porin-deficient mutants of *Escherichia coli* K-12. *Antimicrob. Agents Chemother.* **35:**2155–2158.
31. **Komp Lindgren, P., A. Karlsson, and D. Hughes.** 2003. Mutation rate and evolution of fluoroquinolone resistance in *Escherichia coli* isolates from patients with urinary tract infections. *Antimicrob. Agents Chemother.* **47:**3222–3232.
32. **Liebana, E., C. Clouting, C. A. Cassar, L. P. Randall, R. A. Walker, E. J. Threlfall, F. A. Clifton-Hadley, A. M. Ridley, and R. H. Davies.** 2002. Comparison of *gyrA* mutations, cyclohexane resistance, and the presence of class I integrons in *Salmonella enterica* from farm animals in England and Wales. *J. Clin. Microbiol.* **40:**1481–1486.
33. **Linde, H. J., F. Notka, M. Metz, B. Kochanowski, P. Heisig, and N. Lehn.** 2000. In vivo increase in resistance to ciprofloxacin in *Escherichia coli* associated with deletion of the C-terminal part of MarR. *Antimicrob. Agents Chemother.* **44:**1865–1868.
34. **Linde, H. J., F. Notka, C. Irtenkauf, J. Decker, J. Wild, H. H. Niller, P. Heisig, and N. Lehn.** 2002. Increase in MICs of

ciprofloxacin in vivo in two closely related clinical isolates of *Enterobacter cloacae. J. Antimicrob. Chemother.* **49:**625–630.
35. **Malléa, M., J. Chevalier, C. Bornet, A. Eyraud, A. Davin-Regli, C. Bollet, and J. M. Pages.** 1998. Porin alteration and active efflux: two in vivo drug resistance strategies used by *Enterobacter aerogenes. Microbiology* **144:**3003–3009.
36. **Maneewannakul, K., and S. B. Levy.** 1996. Identification for mar mutants among quinolone-resistant clinical isolates of *Escherichia coli. Antimicrob. Agents Chemother.* **40:**1695–1698.
37. **Mazzariol, A., Y. Tokue, T. M. Kanegawa, G. Cornaglia, and H. Nakaido.** 2000. High-level fluoroquinolone-resistant clinical isolates of *Escherichia coli* overproduce multidrug efflux protein AcrA. *Antimicrob. Agents Chemother.* **44:**3441–3443.
38. **McDermott, P. F., D. G. White, I. Podglajen, M. N. Alekshun, and S. B. Levy.** 1998. Multidrug resistance following expression of the *Escherichia coli marA* gene in *Mycobacterium smegmatis. J. Bacteriol.* **180:**2995–2998.
39. **McMurry, L. M., M. Oethinger, and S. B. Levy.** 1998a. Triclosan targets lipid synthesis. *Nature* **394:**531–532.
40. **McMurry, L. M., M. Oethinger, and S. B. Levy.** 1998b. Overexpression of *marA*, *soxS*, or *acrAB* produces resistance to triclosan in laboratory and clinical strains of *Escherichia coli. FEMS Microbiol. Lett.* **166:**305–309.
41. **Notka, F., H. J. Linde, A. Dankesreiter, H. H. Niller, and N. Lehn.** 2002. A C-terminal 18 amino acid deletion in MarR in a clinical isolate of *Escherichia coli* reduces MarR binding properties and increases the MIC of ciprofloxacin. *J. Antimicrob. Chemother.* **49:**41–47.
42. **Oethinger, M., S. Conrad, K. Kaifel, A. Cometta, J. Belle, G. Klotz, M. P. Glauser, R. Marre, The International Antimicrobial Therapy Cooperative Group of the European Organization for Research and Treatment of Cancer, and W. V. Kern.** 1996. Molecular epidemiology of fluoroquinolone-resistant *Escherichia coli* bloodstream isolates from patients admitted to European cancer centers. *Antimicrob. Agents Chemother.* **40:**387–392.
43. **Oethinger, M., A. S. Jellen-Ritter, S. Conrad, R. Marre, and W. V. Kern.** 1998a. Colonization and infection with fluoroquinolone-resistant *Escherichia coli* among cancer patients: clonal analysis. *Infection* **26:**379–384.
44. **Oethinger, M., W. V. Kern, J. D. Goldman, and S. B. Levy.** 1998b. Association of organic solvent tolerance and fluoroquinolone resistance in clinical isolates of *Escherichia coli. J. Antimicrob. Chemother.* **41:**111–114.
45. **Oethinger, M., W. V. Kern, A. S. Jellen-Ritter, L. M. McMurry, and S. B. Levy.** 2000. Ineffectiveness of topoisomerase mutations in mediating clinically significant fluoroquinolone resistance in *Escherichia coli* in the absence of the AcrAB efflux pump. *Antimicrob. Agents Chemother.* **44:**10–13.
46. **Oethinger, M., I. Podglajen, W. V. Kern, and S. B. Levy.** 1998c. Overexpression of the *marA* or *soxS* regulatory gene in clinical topoisomerase mutants of *Escherichia coli. Antimicrob. Agents Chemother.* **42:**2089–2094.
47. **Okusu, H., D. Ma, and H. Nikaido.** 1996. AcrAB efflux pump plays a major role in the antibiotic resistance phenotype of *Escherichia coli* multiple-antibiotic-resistance (Mar) mutants. *J. Bacteriol.* **178:**306–308.
48. **Pers, C., P. Sogaard, and L. Pallesen.** 1996. Selection of multiple resistance in *Salmonella enteritidis* during treatment with ciprofloxacin. *Scand. J. Infect. Dis.* **28:**529–531.
49. **Philipp, W. J., S. Poulet, K. Eiglmeier, L. Pascopella, V. Balasubramanian, B. Heym, S. Bergh, B. B. R. Bloom, W. R. J. Jacobs, and S. T. Cole.** 1996. An integrated map of the genome of the tubercle bacillus, *Mycobacterium tuberculosis* H37Rv, and comparison with *Mycobacterium leprae. Proc. Natl. Acad. Sci. USA* **93:**3132–3137.
50. **Piddock, L. J.** 1999. Mechanisms of fluoroquinolone resistance: an update 1994–1998. *Drugs* **58** (Suppl. 2):11–18.
51. **Piddock, L. J., D. J. Griggs, M. C. Hall, and Y. F. Jin.** 1993. Ciprofloxacin resistance in clinical isolates of *Salmonella* Typhimurium obtained from two patients. *Antimicrob. Agents Chemother.* **37:**662–666.
52. **Piddock, L. J., M. C. Hall, and R. N. Walters.** 1991. Phenotypic characterization of quinolone-resistant mutants of Enterobacteriaceae selected from wild type, *gyrA* type and multiply-resistant (*marA*) type strains. *J. Antimicrob. Chemother.* **28:**185–198.
53. **Piddock, L. J., D. G. White, K. Gensberg, L. Pumbwe, and D. J. Griggs.** 2000. Evidence for an efflux pump mediating multiple antibiotic resistance in *Salmonella enterica* serovar Typhimurium. *Antimicrob. Agents Chemother.* **44:**3118–3121.
54. **Randall, L. P., and M. J. Woodward.** 2001. Multiple antibiotic resistance (mar) locus in *Salmonella enterica* serovar Typhimurium DT104. *Appl. Environ. Microbiol.* **67:**1190–1197.
55. **Sáenz, Y., L. Brinas, E. Dominguez, J. Ruiz, M. Zarazaga, J. Vila, and C. Torres.** 2004. Mechanisms of resistance in multiple-antibiotic-resistant *Escherichia coli* strains of human, animal, and food origins. *Antimicrob. Agents Chemother.* **48:**3996–4001.
56. **Schneiders, T., S. G. B. Amyes, and S. B. Levy.** 2003. Role of *acrR* and *ramA* in fluoroquinolone resistance in clinical *Klebsiella pneumoniae* isolates from Singapore. *Antimicrob. Agents Chemother.* **47:**2831–2837.
57. **Sulavik, M. C., M. Dazer, and P. F. Miller.** 1997. The *Salmonella typhimurium mar* locus: molecular and genetic analyses and assessment of its role in virulence. *J. Bacteriol.* **179:**1857–1866.
58. **Vila, J., J. Ruiz, P. Goñi, and M. T. J. de Anta.** 1996. Detection of mutations in *parC* in quinolone-resistant clinical isolates of *Escherichia coli. Antimicrob. Agents Chemother.* **40:**491–493.
59. **Vila, J., J. Ruiz, F. Marco, A. Barcelo, P. Goín, E. Giralt, and T. J. de Anta.** 1994. Association between double mutations in *gyrA* gene of ciprofloxacin-resistant clinical isolates of *Escherichia coli* and MICs. *Antimicrob. Agents Chemother.* **38:**2477–2479.
60. **Wang, H., J. L. Dzink-Fox, M. Chen, and S. B. Levy.** 2001. Genetic characterization of highly fluoroquinolone-resistant clinical *Escherichia coli* strains from China: role of acrR mutations. *Antimicrob. Agents Chemother.* **45:**1515–1521.
61. **Webber, M. A., and L. J. V. Piddock.** 2001. Absence of mutations in *marRAB* or *soxRS* in *acrB*-overexpressing fluoroquinolone-resistant clinical and veterinary isolates of *Escherichia coli. Antimicrob. Agents Chemother.* **45:**1550–1552.
62. **White, D. G., J. D. Goldman, B. Demple, and S. B. Levy.** 1997. Role of the *acrAB* locus in organic solvent tolerance mediated by expression of *marA*, *soxS*, or *robA* in *Escherichia coli. J. Bacteriol.* **179:**6122–6126.
63. **Wu, J., and B. Weiss.** 1991. Two divergently transcribed genes, *soxR* and *soxS*, control a superoxide response regulon of *Escherichia coli. J. Bacteriol.* **173:**2864–2871.
64. **Yoshida, H., T. Kojima, J. Yamagishi, and S. Nakamura.** 1988. Quinolone-resistant mutations of the *gyrA* gene of *Escherichia coli. Mol. Gen. Genet.* **211:**1–7.
65. **Yoshida, H., M. Bogaki, M. Nakamura, L. M. Yamanaka, and S. Nakamura.** 1991. Quinolone resistance-determining region in the DNA gyrase *gyrB* gene of *Escherichia coli. Antimicrob. Agents Chemother.* **35:**1647–1650.

Frontiers in Antimicrobial Resistance: a Tribute to Stuart B. Levy
Edited by D. G. White, M. N. Alekshun, and P. F. McDermott

Chapter 17

Structure and Function of MarA and Its Homologs

ROBERT G. MARTIN AND JUDAH L. ROSNER

In the early 1960s, multiply antibiotic resistant bacteria were being discovered in increasing numbers. The resistances were often identified as being transferable by mating and, subsequently, found to be borne on a variety of plasmids and transposons. These resistances were usually due to genes that encoded activities specific to a single class of antibiotics. Thus, several genes were responsible for the "multiple" resistances.

In 1983, George and Levy (21, 22) found that single chromosomal mutations in *Escherichia coli* could engender low-level resistance to a wide range of antibiotics. What is important about this discovery is that it showed that normal bacteria possess the ability to become multiply antibiotic resistant without acquiring foreign DNA. To map the resistance, they isolated mutants which lost the multiple antibiotic resistance due to the integration of a transposon (Tn*5*). The transposon was then mapped to a gene which they called *marA* (multiple antibiotic resistance) and which they localized to 34 minutes on the *E. coli* chromosome. They soon found that MarA increased multiple antibiotic resistance by two mechanisms: (i) the enhancement of drug efflux, later shown to be due to AcrAB-mediated efflux (40); and (ii) reduction of antibiotic influx due to increased transcription of *micF*, which in turn reduced translation of an outer membrane porin encoded by *ompF* (13).

Our laboratory became involved in this work since we had observed that salicylate increased multiple antibiotic resistance phenotypically by increasing the expression of *micF* and reducing *ompF* translation (20, 55, 57, 61). Together with the Levy laboratory, we showed that salicylate and related compounds activate expression of the *mar* operon (14). As explored in chapter 18 of this volume, salicylate was later shown to interact with MarR, thereby derepressing expression of the *mar* operon with attendant synthesis of MarA. Subsequently, the entire *marRAB* region was cloned (31) and sequenced (12) by the Levy group, leading to the conclusion that MarA was a transcriptional activator in the AraC family (53) and very closely related to SoxS (1, 12, 70). We have spent the last decade studying the molecular basis of transcriptional activation by MarA, the activator first identified by Stuart Levy and coworkers.

STRUCTURE

MarA was originally annotated as being a 129 amino acid protein since that is the longest open reading frame in the sequence which begins Met1-Thr2-Met3. . . (12). However, from the disparity between the size of the in vivo MarA and the cloned 129 amino acid MarA, we suggested that MarA probably starts at Met3 in vivo and is 127 amino acids long (46). Proper spacing between the ribosome binding site and the fMet also suggests starting at Met3. However, at this time we choose not to alter the numbering system for MarA amino acids since this would produce much confusion.

A ribbon diagram of MarA showing 116 of the 127 amino acids (the 6 amino-terminal and 5 carboxy-terminal residues are disordered in the crystal structure) is illustrated in Fig. 1 (53). Although AraC was the first transcriptional activator ever identified (17), no crystal structure for any member of this large family had been obtained in nearly 40 years. Subsequently, the structure was solved (37) for a second member of the MarA subfamily, Rob (which has an additional ~120 amino acid carboxy-terminal domain), confirming the structural similarity of these molecules. There are two main features of the structures: the proteins bind DNA as monomers and there are two helix-turn-helix (HTH) motifs connected by a 27Å-long, rigid helix. In the case of MarA, helices 3 and 6 are inserted into adjacent portions of the major groove of the

Robert G. Martin and Judah L. Rosner • Laboratory of Molecular Biology, NIDDK, National Institutes of Health, Bethesda, MD 20892.

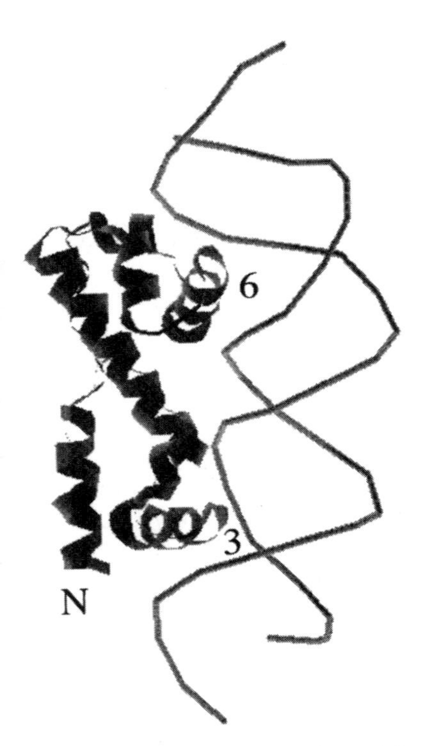

Figure 1. Ribbon diagram of MarA interacting with its DNA binding site (53). The DNA shows an overall bend of ~35°.

1 2 3 4 5 6 7 8 9 10 11 12 13 14 15 16 17 18 19 20
AHRGCACRWWNNRYYAAAHN

Figure 2. Consensus sequence for MarA/SoxS/Rob DNA binding site. R = A or G, W = A or T, Y = T or C, H = a, T, or C, and N = any nucleotide. Note that the binding site is usually indicated as being 20 bp long since any additional bp at either end strongly enhances binding (46).

DNA, inducing a ~35° bend in the DNA. The surface of the molecule facing DNA is highly positively charged, whereas the surface facing away from the DNA is highly negatively charged with the exception of one small positively charged region and a narrow strip that is neutral.

SPECIFICITY OF BINDING

MarA, SoxS, and Rob are now known to transcriptionally activate overlapping families of promoters referred to as the *marA, soxS,* and *rob* regulons or as the *marA/soxS/rob* regulon since the profile of promoters is so similar (discussed in more detail below). A 20 bp long MarA or SoxS binding site has been identified for nearly all of the 25 known regulon promoters, and detailed consensus sequences have been identified for MarA (46) and SoxS (28). In addition, a generalized consensus for all three proteins which differs in only minor ways from the individual consensuses has been proposed (42) (Fig. 2).

Two properties of this sequence are immediately apparent: it is asymmetric (the sequence as illustrated has been arbitrarily defined as in the "forward" orientation) and so highly degenerate that it should be found with a frequency of ~10,000 copies per *E. coli* chromosome (45). We have experimentally verified by gel-shift analysis that these "random sites" found within known genes of *E. coli* do bind MarA, SoxS, and Rob (46).

As might be expected from the high degree of degeneracy, there is a wide range of binding affinities exhibited among the binding sites for the different promoters. In general, binding occurs with moderate affinity, with dissociation constants ranging from ~10 nM^{-1} to >1 μM^{-1} (23, 28, 29, 45, 46). Although there is considerable variation, in general Rob binds more tightly than SoxS, which in turn binds more tightly than MarA. However, binding affinity is a poor predictor of activation in vivo. There is no apparent correlation for a given activator between the binding affinity and the extent of activation for the different promoters of its regulon in vivo, e.g., the binding of MarA to *inaA* is hard to detect, yet MarA activates *inaA* to a greater extent than other promoters that it binds tightly. Nonetheless, we have noted that certain promoters discriminate between the different activators. For example, MarA activates *fpr* and *fumC* to a much lesser extent than does SoxS in vivo, and this correlates, albeit weakly, with the relative binding affinities of the activators for the promoters (44). This latter correlation does not extend to Rob. It has also been shown that many of the "random sites" bind all three activators as tightly or more tightly than the bona fide promoters, ~10 nM^{-1} (45). Taken together, these observations suggest that binding is not the only factor involved in activation.

POSITION AND ORIENTATION OF BINDING SITES

Because the MarA/SoxS/Rob binding site is asymmetric, its orientation relative to the RNA polymerase (RNAP) binding sites can differ at different promoters. Transcriptionally activated promoters are generally categorized as class I or II depending upon whether the activator binding site lies upstream of or overlaps, respectively, the −35 binding signal for RNAP. The functional binding sites for MarA, SoxS, and Rob activated promoters have been found only at specific distances and orientations from the RNAP −10 signal, and experimental alteration of either the distance or the orientation leads to loss of activation (43, 46, 69).

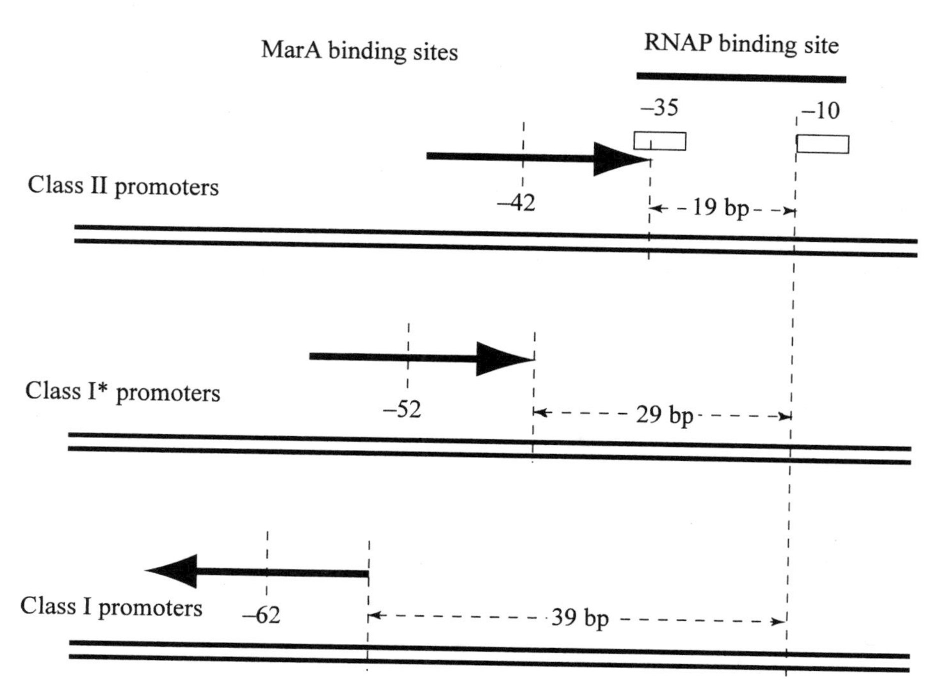

Figure 3. Diagram showing functional configurations of the MarA binding sites relative to the RNAP binding site (43, 46). The MarA binding sites centered at −42 and −52 nt (relative to the transcription start site) are in the forward orientation while those centered at −62 and −72 (not shown) are in the backward orientation. The distances of the MarA binding sites from the −10 site indicate that MarA is essentially on the same surface of the DNA relative to RNAP in each configuration.

Thus, particular configurations are required so that activator and RNAP can interact while bound to their respective sites on the promoter.

Depending on the distance from the RNAP binding site, the MarA binding sites can be functional in two orientations (Fig. 3). In class II promoters, the binding sites are centered at nt −42 (relative to the start site of transcription) and are oriented in the forward direction as indicated above. A potential interaction with the σ−subunit of RNAP is possible at this position but has not been studied. Two orientations are functional for class I promoters. A minority of such promoter binding sites, designated class I*, are centered at nt −52 and oriented in the forward direction like class II binding sites. As a consequence, MarA is bound on the same surface of the DNA with its carboxy-terminus closest to RNAP in both classes II and I*. By contrast, the binding sites of class I promoters centered at nt −62 or −72 are in the backward orientation so that the amino terminus of MarA is closest to RNAP, although it is still bound to the same DNA surface. Because of this geometry, there is a position on the surface of MarA that has precisely the same spatial relationship to RNAP bound at both class I* and class I promoters (Fig. 4). That position is between amino acids D18 and D22. The relative positions of two other important amino acids (discussed below), W19 and R36, are also indicated.

INTERACTION BETWEEN MarA AND THE CARBOXY-TERMINAL DOMAIN OF THE α-SUBUNIT OF RNAP

Although early work in vitro suggested that the carboxy-terminal domain (CTD) of the α-subunit of RNAP was required only for class I promoters (33–36), it is now clear that interaction of MarA with the α-CTD is required for the activation of all *marA* and *soxS* regulon promoters. For example, some mutants of *marA* (23) and *soxS* (29) obtained by exhaustive alanine scanning mutagenesis are defective for both class I and class II promoter activation. MarA mutants defective only for class I promoters have not been found, suggesting that contact of MarA with α-CTD is required for both class I and class II promoters. NMR spectroscopy of a MarA-DNA-α-CTD complex has now provided the basis for a molecular model of the interaction between MarA and α-CTD (15, 16) and has been confirmed by mutational analysis and chromatography (Fig. 5).

The principal interactions predicted by the model between MarA and α-CTD amino acids, respectively, are W19:**N294**; W19:**R265**; R36:**N294**; R36:L295; S37:P293; S37:**N294**; D18:**R265**; D18:**N268**; D22:**V264**; D22:**R265**; D22:**K298**; H14:**N268**; S15:**N294**. These interactions were unexpected since the α-CTD amino acids (in bold) involved in binding to MarA are part of the "265-determinant." At other promoters,

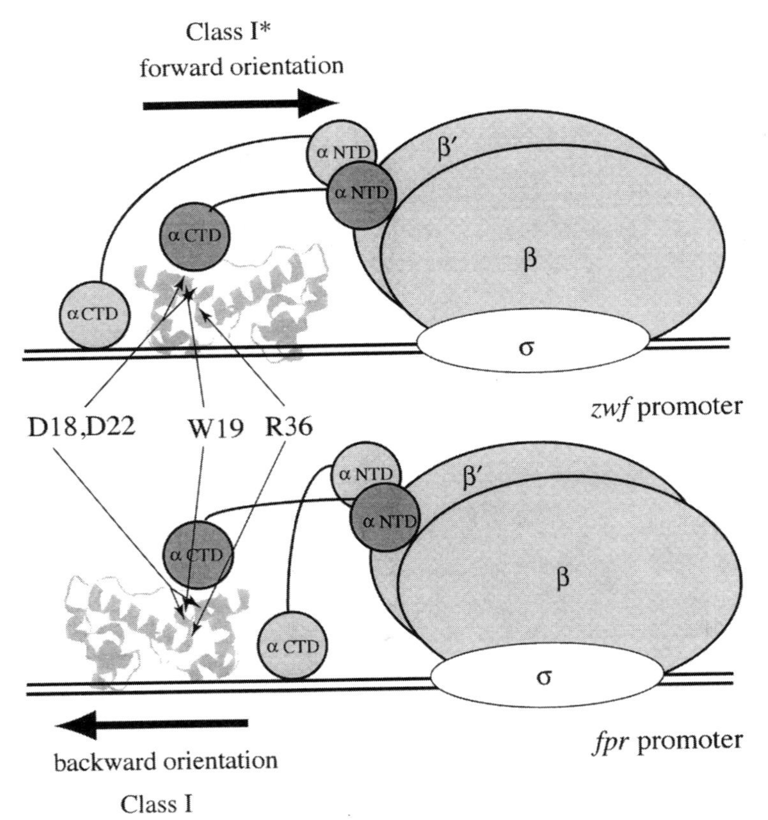

Figure 4. Representations of RNAP interacting via α-CTD with MarA at two promoter configurations (not to scale). The α, β, β′, and σ subunits of RNAP are shown. Each α-CTD is connected to the α-NTD by a flexible tether (curved lines). Significant MarA amino acids involved at the interface with α-CTD are indicated. The α-CTD domain that does not contact MarA is shown in contact with the DNA, but this is speculative. The model predicts that the spatial relationship between α-CTD and MarA will be similar at both class I* (*zwf*) and class I (*fpr*) promoters.

the 265-determinant of α-CTD is used exclusively to contact DNA (UP elements or DNA sequences adjacent to the transcriptional activators, CRP, and Fis) and does not contact the activator (4, 8, 9, 18, 19, 25, 48, 58–60, 72). Instead, other surfaces of α-CTD contact the activator at the CRP and Fis activated promoters (Fig. 5).

A second feature of the model that may be significant is that conformational changes at the MarA surface resulting from interaction with α-CTD appear to be propagated internally through the molecule to amino acids of helix 3, causing small conformational changes at Q45 and W42. These amino acids are critical for DNA binding and recognition of the consensus sequence. Thus, the binding of MarA by α-CTD may enhance MarA's ability to recognize its DNA binding site. Finally, it is worth noting that the principal contacts between MarA and α-CTD are hydrophobic, and most involve backbone atoms of the peptide chains rather than the amino acid side groups. Thus, amino acid substitutions at these sites frequently do not alter transcriptional activation.

MarA-RNAP SCANNING COMPLEXES

Transcriptional activators are commonly believed to function by binding to DNA in the promoter region and then recruiting RNAP thereby increasing the occupancy of RNAP at the promoter (51). Several lines of reasoning suggest that MarA may not function via this recruitment model (45).

(i) The general problem is that there are too many potential MarA/SoxS/Rob binding sites on the chromosome and too few molecules of MarA to saturate them. As pointed out above, each chromosome contains ~10,000 potential binding sites for MarA. Since a rapidly growing cell has >3 chromosomes per cell (7), saturation would require that the cell contain >30,000 copies of MarA. However, active induction occurs when the cell has fewer than 2,300 molecules per cell; even when MarA is produced from a high copy number plasmid with strong transcription and translation signals, the cell contains only ~15,000 molecules (45) because of the high turnover rate of MarA (31).

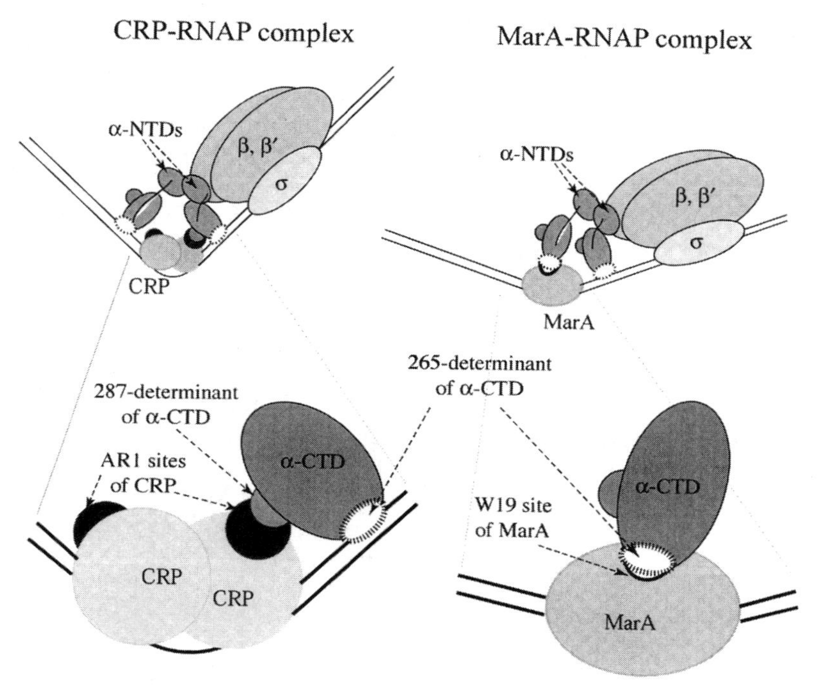

Figure 5. Two different types of interaction of α-CTD with activators at class I promoters. RNAP is indicated as in Fig. 4. In the CRP (or FIS) complex, one α-CTD contacts the AR1 site of CRP (black balls) via the 287-determinant (gray protrusion) and DNA via the 265-determinant (white ovals with dashed borders), whereas the other α-CTD makes contact only with DNA. In the MarA complex, the model envisions the 265-determinant of one α-CTD as making contact with the W19 site of MarA only.

(ii) Increasing the number of MarA binding sites, which might be expected to bind the limited number of MarA molecules, does not decrease activation. Multiple high affinity binding sites for MarA lacking RNAP binding sites were introduced into cells on a high copy number plasmid. This did not reduce activation of a *marA* regulon promoter even though it provided many "random sites" (45). As a positive control, the addition of high affinity binding sites for MarR were shown to derepress the regulon, indicating that they bound sufficient MarR to lower its effective concentration in the cell.

(iii) Rob, present in the cell at 5,000 to 10,000 molecules per cell (65, 67), binds the promoters of most members of the *marA* regulon more tightly than MarA, but Rob functions poorly as a transcriptional activator unless posttranscriptionally activated (54, 56). Moreover, the binding by Rob does not interfere with the activation of the regulon when MarA is derepressed, e.g., by treatment with salicylate (14, 45). Indeed, the kinetics and profiles of activation are similar for all of the promoters of the regulon that have been tested although the MarA binding constants vary widely from ~20 nM^{-1} to >1 μM^{-1}. Thus, activation by MarA is not blocked by the stronger binding of many molecules of Rob.

To explain these results, we suggested that MarA does not activate transcription by recruitment but by a "scanning mechanism" (45). This hypothesis proposes that MarA first interacts with RNAP to form a complex which then scans the chromosome for the appropriate combination of signals: the −10 and −35 RNAP signals and an appropriately configured MarA binding site.

In support of this model, we have shown that MarA and SoxS form stable complexes with RNAP (both holoenzyme and apoenzyme lacking the α-subunit) in the absence of DNA, and that Rob–RNAP complexes are less stable and therefore less likely to interfere with the MarA- (or SoxS-)RNAP complexes (45). Furthermore, overexpression of mutant MarA molecules that can bind DNA but not RNAP (due to an alanine substitution at W19) (16), do not interfere with induction of regulon promoters (unpublished results). This contrasts with CRP- or Fis-regulated promoters where the overexpression of "positive control" mutants which cannot bind RNAP inhibits induction by wild-type, presumably by occupying the activator binding site and blocking access of the wild-type activator (3, 24). Finally, the NMR observations that interaction with RNAP engenders a conformational shift in MarA at those amino acids involved in

DNA recognition (see previous section) may account for the enhanced binding of RNAP observed on titration with MarA or SoxS (45). A similar model, termed "prerecruitment" rather than "scanning," was subsequently proposed for SoxS (27).

A NOVEL COMPLEX FORMED BETWEEN THE MarA ACTIVATOR AND RNAP

We have proposed that the α-CTD of RNAP interacts with MarA in a manner that differs from how it interacts with CRP or Fis (16), so that the transcriptional complexes are structurally and functionally novel. Activation of class I promoters by CRP and MarA is illustrated in Fig. 5. In the CRP (or FIS) complex, a single α-CTD subunit simultaneously interacts with both activator and DNA. In the MarA complex, that is not possible since the α-CTD contacts the MarA at a site distant from the DNA. Thus, if α contacts the DNA, it would have to be via a second α-CTD, as indicated (Fig. 5). This has implications for how DNA UP elements might affect activation.

UP elements, a stretch of 5 to 6 adenines located ~5 nt beyond the 3′ end of the activator binding site (8, 18, 19, 25, 47, 60), are known to stimulate transcription by enhancing the interaction of the 265-determinant of α-CTD with DNA. We reasoned that an UP element located at the site of interaction between activator and α-CTD (see Fig. 5) might have different effects for CRP and MarA stimulated promoters. In the case of MarA, an UP element so located would be expected to behave additively with regard to activation, since an individual α-CTD either makes contact with MarA or with DNA but not both at the same time. Thus, an UP element in the absence of MarA would activate transcription by the binding of one subunit of α-CTD. However, this α-CTD subunit cannot bind to MarA since the surface of α-CTD used to bind DNA is now unavailable for binding MarA. For MarA to be effective on this promoter, a second subunit of α-CTD would have to be bound so that the rate of activation upon the addition of MarA should be no greater than the rate of activation by MarA in the absence of the UP element, i.e., there can be no cooperative interaction between an α-CTD subunit and both the UP element and MarA. On the other hand, the presence of an UP element appropriately located in a CRP-stimulated promoter should result in the cooperative binding of α-CTD to the AR1 site of CRP via the 287-determinant and to the UP element via the 265-determinant (8).

Such behavior has been observed in an initial study (Fig. 6). Addition of an UP element raised the basal activity of both a *lacZ* (CRP-stimulated) and an *fpr* (MarA-stimulated) promoter. When CRP expression was induced by growth in fructose, the rate of activation of the *lacZ* promoter was 1.8-fold greater when the promoter contained an UP element (Fig. 6). But when the *fpr* promoter was further stimulated by MarA, the rate of activation was not substantially increased (1.2-fold; Fig. 6). These results are consistent with different modes of transcriptional activation by MarA and CRP and may explain why no UP elements are found at any of the *marA* regulon promoters.

It has been demonstrated that two members of the AraC family, RhaS and MelR, interact with domain 4 of the σ subunit of RNAP at class II promoters (5, 26). We previously observed that σ is not needed to bind RNAP to MarA in the absence of DNA in vitro (45). Thus, if MarA also interacts with σ at class II promoters, perhaps it does so only after the scanning complex has bound to the DNA.

ROLE OF DNA BENDING IN THE FORMATION OF ACTIVATED COMPLEXES

Lawson et al. (38) have recently proposed a structural model for the activated complex formed by CRP with RNAP and DNA and have stressed the importance of DNA bending in the formation of this complex. Although we have no additional structural data on the MarA complexes, we have noted (unpublished data) that a DNase I hypersensitive site 5 nt downstream of the MarA binding region appears in footprints when the activated complex is formed. This suggests that the DNA bending induced by MarA (Fig. 1 and 5) may be increased in the complex formed with RNAP.

PROMOTERS ACTIVATED BY MarA (SoxS AND Rob)

In an attempt to define all of the promoters activated by MarA and SoxS, microarray analyses have been carried out under a variety of conditions including constitutive expression of *marA* on a plasmid, derepression of *marA* by salicylate, and induction of *soxS* with paraquat (2, 50). Of course, it is difficult to distinguish primary effects from secondary effects in such experiments. In addition, salicylate or paraquat treatments would be expected to result in the induction of salicylate- and paraquat-specific promoters in addition to *marA/soxS/rob*-specific promoters. It is therefore not surprising that of the 150 genes found to be elevated in these studies, only 25 promoters have thus far been definitively documented as activated by MarA, SoxS, or Rob (42). These are *acnA, acrAB, fldA, fldB, fpr, mar, poxB, ribA, map, zwf, aldA, b1452, fumC,*

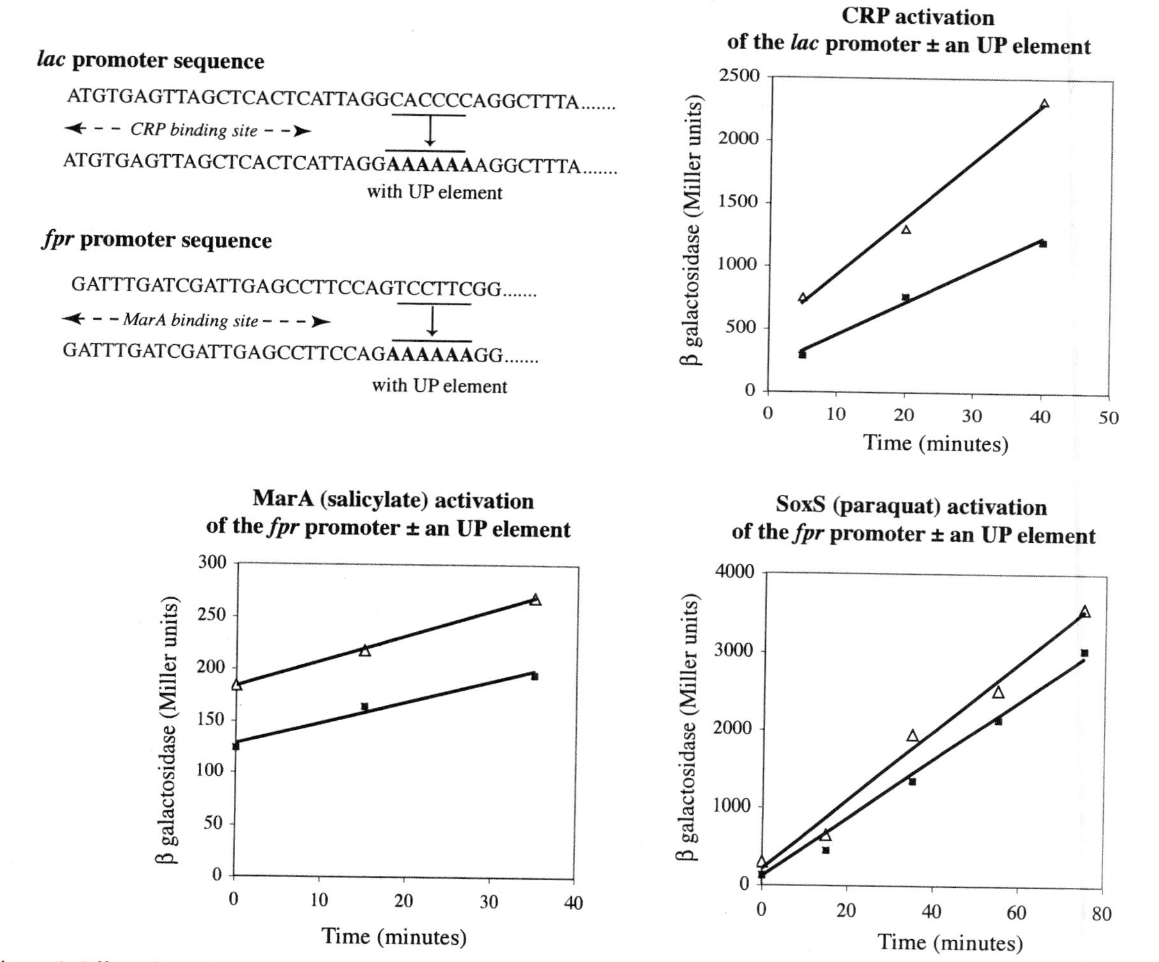

Figure 6. Effect of UP elements on the induction of the *lac* and *fpr* promoters. Variants of the *lac* and *fpr* promoters were constructed with or without the indicated UP element. The cells were treated with fructose, salicylate or paraquat to express CRP, MarA or SoxS, respectively. The effects on promoter activity were then assayed using a *lacZ* reporter gene fused to each construct (β-galactosidase specific activity expressed in Miller Units). Additive effects of the UP element were seen with MarA and SoxS, but a cooperative effect was seen with CRP.

inaA, mdaB, micF, nfnB, nfo, pgi, pqiA, sodA, tolC, ybjC, yggX (49), and *ybbW*. Statistical analysis suggests that about 30 more regulon promoters may exist.

MarA AS A REPRESSOR

Since DNA binding proteins can be both activators and repressors, it was reasonable to wonder whether MarA could also be a repressor. The microarray studies of Barbosa and Levy (2) and Pomposiello et al. (50) suggested that many promoters might be repressed by MarA and SoxS, respectively, but it was not clear whether this was a direct or indirect effect. Recently, work from the Levy laboratory has shown that two promoters (*purA* and *hdeA*) are repressed by MarA in vivo and in vitro (62). The in vitro assays, containing only promoter DNA, RNAP, and MarA, demonstrated a direct role for MarA in the repression. Apparently, the MarA binds to a marbox that overlaps the −35 signal for RNAP as in transcriptional activation, but the marbox is in the opposite ("backward") orientation than found for activation of class II promoters. It will be of interest to determine whether MarA binding alone interferes with the transcription or whether MarA must also interact with RNAP at these sites to affect the repression.

MarA PARALOGS

The sequence of the *E. coli* chromosome (6) indicates 23 paralogs of AraC may be encoded. We have noted that there is a high degree of sequence homology at 14 positions within the first 50 amino acids of MarA (16). Of these "strong consensus" positions, only three are surface exposed and are not involved in DNA binding. The consensus amino acids at these three positions are aromatic at position 19 and aliphatic at positions 33 and 37. Positions 19 and 37 were identified as crucial in the MarA-α-CTD interface (16), and the location of V33 suggests an important role for positioning of the aromatic ring of W19.

Unlike the "strong consensus" amino acids which are mostly uncharged, 8 of the 12 less well conserved "weak consensus" amino acids that are surface exposed are polar (positions 14, 15, 21–23, 25, 36, and 42) and seven of these are charged. These weak consensus amino acids are not randomly distributed and may define subclasses within the AraC family. For example, although only nine of the paralogs have a basic amino acid at position 36 (of importance in MarA binding to α-CTD), these include all four members of the MarA/SoxS/Rob/TetD subfamily.

MarA ORTHOLOGS

Evidence that MarA-like systems exist in other enterobacteria has been adduced by Northern blotting (11). Eight MarA orthologs have been identified in several bacterial species where they seem to play a similar role in activating multiple antibiotic resistance (10, 23, 32, 39, 41, 63, 66, 68, 71) as indicated by phenotypes such as induction of resistance by salicylate, reduced synthesis of outer membrane porins, increased efflux of xenobiotics, or increased expression of efflux pumps and increased organic solvent tolerance. We have sequence data for seven of the eight orthologs, which show varying similarity to MarA from ~30% identity (~50% homology) to ~75% identity. Like MarA, all contain a basic amino acid at position 36 and a tryptophan at position 19.

PHYSIOLOGY OF THE *marA* REGULON

We now have a wealth of mechanistic information on the workings of the *marRAB* operon, MarR repression and derepression (chapter 18), the regulon promoters, and transcriptional activation and repression. The challenge ahead will be to integrate this information in terms of how the system is used to sense and adapt to the environment. What natural signals are involved in sensing and what are the roles of the various genes that are regulated? For example, the *mar* operon is unusual in that its transcription is temperature-sensitive (64). Is this a clue that the *mar* system originated as an adaptation to survival outside of a warm-blooded host, perhaps when in contact with plant materials likely to contain salicylates? How does the particular profile of gene activation and repression contribute to resistance to drugs and oxidative stress? In addition to the control of influx and efflux of xenobiotics, and in addition to antioxidative stress functions, what roles do the remaining genes play in the response? Finally, what is the global role of *mar* in the genesis of clinical multiple antibiotic resistance?

ADDENDUM

Do Antibiotics Induce *marRAB* Transcription in *Escherichia coli?*

Given that derepression of the *marRAB* operon results in resistance to antibiotics, it is reasonable to ask whether antibiotics induce transcription of the operon since that would be highly adaptive. Indeed, the amount of *marRAB* mRNA that was detected by Northern blots was elevated when the cells were treated with sublethal doses of either chloramphenicol or tetracycline but not when untreated or treated with sublethal doses of ampicillin, nalidixic acid, or norfloxacin (1a). This result has been repeatedly observed with a variety of laboratory and clinical strains (1a, 14a, 41a). However, the increased amounts of transcripts do not appear to be due to antibiotic-induced derepression of *marR* since the antibiotic-specific increase was seen (and is even more pronounced) in many *marR* mutants. Furthermore, wild-type *marA* is not necessary for this effect either (1a).

Attempts to demonstrate antibiotic-induced expression of the operon by in vivo measurements have not been successful. Tetracycline, over a wide range of concentrations, did not induce the expression of the MarA transcriptionally activated genes *acrRAB::lacZ* or *inaA::lacZ* (40a; Rosner and Martin, unpublished results). Therefore, we have undertaken double disc diffusion experiments (47a) to test whether treatment of cells with various concentrations of ampicillin, chloramphenicol, or tetracycline renders them more resistant to these and other antibiotics. A paper disc with the presumptive inducer was placed at the center of a Luria agar plate that had been overlayed with 10^8 cells. At varying distances from this disc, other discs with antibiotics (ampicillin, chloramphenicol, cephalexin, nalidixic acid, or tetracycline) were placed. Diffusion from the discs exposes the cells to a wide range of transient inducer and antibiotic concentrations which intersect at points determined by the placement of the discs (Fig. 7). In addition to chloramphenicol and tetracycline, ampicillin was tested as an inducer because microarray experiments indicated that *marA* and *soxS* levels were increased following treatment with ampicillin even though the expression of other regulon members was not significantly affected (36a). Salicylate, which derepresses the *marRAB* operon, was included as a control inducer (14b).

Resistance to cephalexin was promoted by salicylate, as seen by the reduced zone of inhibition surrounding the cephalexin disc closest to the salicylate disc (Fig. 7, bottom right). However, ampicillin, chloramphenicol, and tetracycline did not increase resistance to cephalexin; at some concentrations, enhanced sensitivity was seen (Fig 7, top left, top right, bottom left). Only the well-known, *mar*-independent effect of protein synthesis inhibition on quinolone resistance was seen (data not shown) (31a). Thus, unlike salicylate, which increases MarA expression, the antibiotics did not induce resistance to diverse antibiotics.

One caution in interpreting these experiments is that both reporter gene expression and antibiotic resistance require protein synthesis for their manifestation. However, if induction of *marRAB*

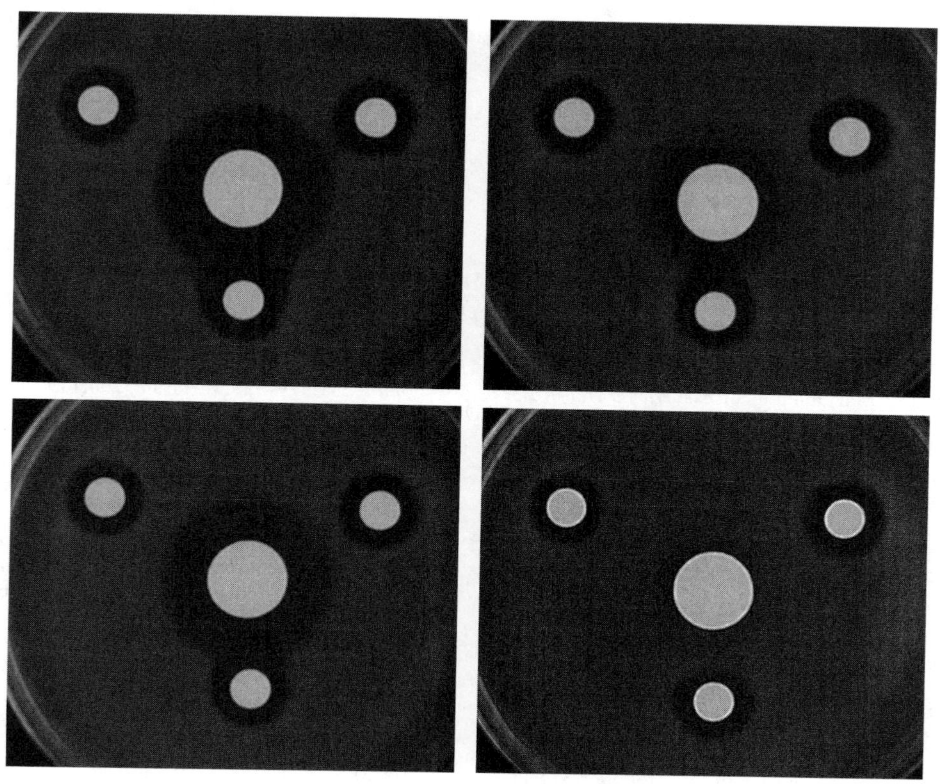

Figure 7. Double disk diffusion assays (47a) showing effects of several antibiotics on resistance to cephalexin. About 2×10^8 strain AG100 cells (14b) were plated on Luria agar, and a 12.8 mm paper disk containing either (A) 87.5 μg of ampicillin, (B) 25 μg of chloramphenicol, (C) 87.5 μg of tetracycline, or (D) 13.2 mg of sodium salicylate was placed on the center of the plate. Three 6.4 mm paper disks containing 50 μg of cephalexin were added to the periphery of the plates at different distances from the central disk, and the plates were incubated at 30°C overnight. A reduced zone of inhibition around the cephalexin disk facing the central disk indicates induction of resistance, as seen at the bottom of (D).

transcription by sublethal levels of protein synthesis inhibitors like chloramphenicol or tetracycline is not followed by protein synthesis, it is hard to understand how the transcription would be adaptive.

These observations raise the possibility that the increased *marRAB* mRNA found after treatment with chloramphenicol or tetracycline is not due to greater expression of *marRAB* but to selective recovery of the mRNA from cells in which protein-synthesis has been inhibited. Since effects of protein-synthesis inhibitors on mRNA have been reported (39a), this alternative possibility must be seriously considered.

REFERENCES

1. **Amábile-Cuevas, C. F., and B. Demple.** 1991. Molecular characterization of the *soxRS* genes of *Escherichia coli:* two genes control a superoxide stress regulon. *Nucleic Acids Res.* **19:**4479–4484.

1a. **Ariza, R. R., S. P. Cohen, N. Bachhawat, S. B. Levy, and B. Demple.** 1994. Repressor mutations in the *marRAB* operon that activate oxidative stress genes and multiple antibiotic resistance in *Escherichia coli. J. Bacteriol.* **176:**143–148.

2. **Barbosa, T. M., and S. B. Levy.** 2000. Differential expression of over 60 chromosomal genes in *Escherichia coli* by constitutive expression of MarA. *J. Bacteriol.* **182:**3467–3474.

3. **Bell, A. I.** 1990. Regulation of transcription by FNR and CRP: two homologous transcription activators of *Escherichia coli.* Ph. D. Thesis, Faculty of Science of the University of Birmingham.

4. **Benoff, B., H. Yang, C. L. Lawson, G. Parkinson, J. Liu, E. Blatter, Y. W. Ebright, H. M. Berman, and R. H. Ebright.** 2002. Structural basis of transcription activation: the CAP-CTD-DNA complex. *Science* **297:**1562–1566.

5. **Bhende, P. M., and S. M. Egan.** 2000. Genetic evidence that transcription activation by RhaS involves specific amino acid contacts with sigma 70. *J. Bacteriol.* **182:**4959–4969.

6. **Blattner, F. R., G. Plunkett, III, C. A. Bloch, N. T. Perna, V. Burland, M. Riley, J. Collado-Vides, J. D. Glasner, C. K. Rode, G. F. Mayhew, J. Gregor, N. W. Davis, H. A. Kirkpatrick, M. A. Goeden, D. J. Rose, B. Mau, and Y. Shao.** 1997. The complete genome sequence of *Escherichia coli* K-12. *Science* **277:**1453–1474.

7. **Bremer, H., and P. P. Dennis.** 1987. Modulation of chemical composition and other parameters of the cell by growth rate, p. 1553–1569. *In* F. C. Neidhardt (ed.), Escherichia coli *and* Salmonella: *Cellular and Molecular Biology,* vol. 2 American Society for Microbiology Press, Washington, D.C.

8. **Busby, S., and R. Ebright.** 1999. Transcription activation by catabolite activator protein (CAP). *J. Mol. Biol.* **293:**199–213.

9. **Cheng, Y. S., W. Z. Yang, R. C. Johnson, and H. S. Yuan.** 2002. Structural analysis of the transcriptional activation on Fis: crys-

tal structures of six Fis mutants with different activation properties. *J. Mol. Biol.* **302:**1139–1151.
10. **Chollet, R., C. Bollet, J. Chevalier, M. Mallea, J. M. Pages, and A. Davin-Regli.** 2002. *mar* Operon involved in multidrug resistance of *Enterobacter aerogenes*. *Antimicrob. Agents Chemother.* **46:**1093–1097.
11. **Cohen, S. P., W. Yan, and S. B. Levy.** 1993. A multidrug resistance regulatory chromosomal locus is widespread among enteric bacteria. *J. Infect. Dis.* **168:**484–488.
12. **Cohen, S. P., H. Hächler, and S. B. Levy.** 1993. Genetic and functional analysis of the multiple antibiotic resistance (*mar*) locus in *Escherichia coli*. *J. Bacteriol.* **175:**1484–1492.
13. **Cohen, S. P., L. M. McMurry, and S. B. Levy.** 1988. *marA* locus causes decreased expression of OmpF porin in multiple-antibiotic-resistant (Mar) mutants of *Escherichia coli*. *J. Bacteriol.* **170:**5416–5422.
14. **Cohen, S. P., S. B. Levy, J. Foulds, and J. L. Rosner.** 1993. Salicylate induction of antibiotic resistance in Escherichia coli: activation of the *mar* operon and a *mar*-independent pathway. *J. Bacteriol.* **175:**7856–7862.
14a. **Cohen, S. P., H. Hächler, and S. B. Levy.** 1993. Genetic and functional analysis of the multiple antibiotic resistance (*mar*) locus in *Escherichia coli*. *J. Bacteriol.* **175:**1484–1492.
14b. **Cohen, S. P., S. B. Levy, J. Foulds, and J. L. Rosner.** 1993. Salicylate induction of antibiotic resistance in *Escherichia coli:* activation of the *mar* operon and a *mar*-independent pathway. *J. Bacteriol.* **175:**7856–7862.
15. **Dangi, B., P. Pelupessey, R. G. Martin, J. L. Rosner, J. M. Louis, and A. M. Gronenborn.** 2001. Structure and dynamics of MarA-DNA complexes: an NMR investigation. *J. Mol. Biol.* **314:**113–127.
16. **Dangi, B., A. M. Gronenborn, J. L. Rosner, and R. G. Martin.** Versatility of the carboxy-terminal domain of the α subunit of RNA polymerase in transcriptional activation: use of the DNA contact site as a protein contact site for MarA. *Mol. Microbiol.*, in press.
17. **Englesberg, E.** 1961. Enzymatic characterization of 17 L-arabinose negative mutants of *E. coli*. *J. Bacteriol.* **81:**996–1006.
18. **Estrem, S. T., T. Gaal, W. Ross, and R. L. Gourse.** 1998. Identification of an UP element consensus sequence for bacterial promoters. *Proc. Natl. Acad. Sci. USA* **95:**9761–9766.
19. **Estrem, S. T., W. Ross, T. Gaal, Z. W. Chen, W. Niu, R. H. Ebright, and R. L. Gourse.** 1999. Bacterial promoter architecture: subsite structure of UP elements and interactions with the carboxy-terminal domain of the RNA polymerase alpha subunit. *Genes Dev.* **13:**2134–2147.
20. **Foulds, J., D. M. Murray, T. Chai, and J. L. Rosner.** 1989. Decreased permeation of cephalosporins through the outer membrane of *Escherichia coli* grown in salicylates. *Antimicrob. Agents Chemother.* **33:**412–417.
21. **George, A. M., and S. B. Levy.** 1983. Amplifiable resistance to tetracycline, chloramphenicol, and other antibiotics in *Escherichia coli:* involvement of a non-plasmid-determined efflux of tetracycline. *J. Bacteriol.* **155:**531–540.
22. **George, A. M., and S. B. Levy.** 1983. Gene in the major cotransduction gap of the *Escherichia coli* K-12 linkage map required for the expression of chromosomal resistance to tetracycline and other antibiotics. *J. Bacteriol.* **155:**541–548.
23. **Gillette, W. K., R. G. Martin, and J. L. Rosner.** 2000. Probing the *Escherichia coli* transcriptional activator MarA using alanine-scanning mutagenesis: residues important for DNA binding and activation. *J. Mol. Biol.* **299:**1245–1255.
24. **Gosink, K. K., T. Gaal, A. J. Bokal IV, and R. L. Gourse.** 1996. A positive control mutant of the transcription activator protein FIS. *J. Bacteriol.* **178:**5182–5187.
25. **Gourse, R. L, W. Ross, and T. Gaal.** 2000. UPs and downs in bacterial transcription initiation: the role of the alpha subunit of RNA polymerase in promoter recognition. *Mol. Microbiol.* **37:**687–695.
26. **Grainger, D. C., C. L. Webster, T. A. Belyaeva, E. I. Hyde, and S. J. Busby.** 2004. Transcription activation at the *Escherichia coli melAB* promoter: interactions of MelR with its DNA target site and with domain 4 of the RNA polymerase sigma subunit. *Mol. Microbiol.* **51:**1297–1309.
27. **Griffith, K. L, I. M. Shah, T. E. Myers, M. C. O'Neill, and R. E. Wolf, Jr.** 2002. Evidence for "pre-recruitment" as a new mechanism of transcription activation in *Escherichia coli:* the large excess of SoxS binding sites per cell relative to the number of SoxS molecules per cell. *Biochem. Biophys. Res. Commun.* **291:**979–986. (Erratum, **294:**1191.)
28. **Griffith, K. L., and R. E. Wolf, Jr.** 2001. Systematic mutagenesis of the DNA binding sites for SoxS in the *zwf* and *fpr* promoters of *Escherichia coli:* identifying nucleotides required for DNA binding and transcription activation. *Mol. Microbiol.* **40:**1141–1154.
29. **Griffith, K. L., and R. E. Wolf, Jr.** 2002. A comprehensive alanine scanning mutagenesis of the *Escherichia coli* transcriptional activator SoxS: identifying amino acids important for DNA binding and transcription activation. *J. Mol. Biol.* **322:**237–257.
30. **Griffith, K. L., I. M. Shah, and R. E. Wolf, Jr.** 2004. Proteolytic degradation of *Escherichia coli* transcription activators SoxS and MarA as the mechanism for reversing the induction of the superoxide (SoxRS) and multiple antibiotic resistance (Mar) regulons. *Mol. Microbiol.* **51:**1801–1816.
31. **Hächler, H., S. P. Cohen and S. B. Levy.** 1991. *marA*, a regulated locus which controls expression of chromosomal multiple antibiotic resistance in *Escherichia coli*. *J. Bacteriol.* **173:**5532–5538.
31a. **Hooper, D. C., and Wolfson J. S.** 1993. Mechanism of quinolone action and bacterial killing, p. 53–75. *In* D. C. Hooper and J. S. Wolfson (ed.), *Quinolone Antimicrobial Agents*. American Society for Microbiology, Washington D.C.
32. **Ishida, H., H. Fuziwara, Y. Kaibori, T. Horiuchi, K. Sato, and Y. Osada.** 1995. Cloning of multidrug resistance gene *pqrA* from *Proteus vulgaris*. *Antimicrobiol. Agents Chemother.* **39:**453–457.
33. **Ishihama, A.** 1992. Role of the RNA polymerase alpha subunit in transcription activation. *Mol. Microbiol.* **6:**3283–3288.
34. **Jair, K.-W., R. G. Martin, J. L. Rosner, N. Fujita, A. Ishihama, and R. E. Wolf, Jr.** 1995. Purification and regulatory properties of MarA protein, a transcriptional activator of *Escherichia coli* multiple antibiotic and superoxide resistance promoters. *J. Bacteriol.* **17:**7100–7104.
35. **Jair, K.-W., W. P. Fawcett, N. Fujita, A. Ishihama, and R. E. Wolf, Jr.** 1996. Ambidextrous transcriptional activation by SoxS: requirement for the C-terminal domain of the RNA polymerase alpha subunit in a subset of *Escherichia coli* superoxide-inducible genes. *Mol. Microbiol.* **19:**307–317.
36. **Jair, K.-W., X. Yu, K. Skarstad, B. Thöny, N. Fujita, A. Ishihama, and R. E. Wolf, Jr.** 1996. Transcriptional activation of promoters of the superoxide and multiple antibiotic resistance regulons by Rob, a binding protein of the *Escherichia coli* origin of chromosomal replication. *J. Bacteriol.* **178:**2507–2513.
36a. **Kaldalu, N., R. Mei, and K. Lewis.** 2004. Killing by ampicillin an ofloxacin induces overlapping changes in *Escherichia coli* transcription profile. *Antimicrob. Agents Chemother.* **48:**890–896.
37. **Kwon, H. J., M. H. J. Bennik, B. Demple, and T. Ellenberger.** 2000. Crystal structure of the *Escherichia coli* Rob transcription factor in complex with DNA. *Nature Struct. Biol.* **7:**424–430.

38. **Lawson, C. L, D. Swigon, K. S. Murakami, S. A. Darst, H. M. Berman, and R. H. Ebright.** 2004. Catabolite activator protein: DNA binding and transcription activation. *Curr. Opin. Struct. Biol.* **14:**10–20.
39. **Linde, H. J., F. Notka, C. Irtenkauf, J. Decker, J. Wild, H. H. Niller, P. Heisig, and N. Lehn.** 2002. Increase in MICs of ciprofloxacin in vivo in two closely related clinical isolates of *Enterobacter cloacae*. *J. Antimicrobiol. Chemother.* **49:**625–630.

39a. **Lopez, P. J., I. Marchand, O. Yarchuk, and M. Dreyfus.** 1998. Translation inhibitors stabilize *Escherichia coli* mRNAs independently of ribosome protection. *Proc. Natl. Acad. Sci. USA* **95:**6067–6072.

40. **Ma, D., D. N. Cook, M. Alberti, N. G. Pon, H. Nikaido, and J. E. Hearst.** 1995. Genes *acrA* and *acrB* encode a stress-induced efflux system of *Escherichia coli*. *Mol. Microbiol.* **16:**45–55.

40a. **Ma, D., D. N. Cook, M. Alberti, N. G. Pon, H. Nikaido, and J. E. Hearst.** 1995. Genes *acrA* and *acrB* encode a stress-induced efflux system of *Escherichia coli*. *Mol. Microbiol.* **16:**45–55.

41. **Macinga, D. R., M. M. Parojcic, and P. N. Rather.** 1995. Identification and analysis of *aarP*, a transcriptional activator of the 2′-N-acetyltransferase in *Providencia stuartii*. *J. Bacteriol.* **177:**3407–3413.

41a. **Maneewannakul, K., and S. B. Levy.** 1996. Identification for *m* mutants among quinolone-resistant clinical isolates of *Escherichia coli*. *Antimicrob. Agents Chemother.* **40:**1695–1698.

42. **Martin, R. G., and J. L. Rosner.** 2002. Genomics of the *marA/soxS/rob* regulon of *Escherichia coli:* identification of directly activated promoters by application of molecular genetics and informatics to microarray data. *Mol. Microbiol.* **44:**1611–1624.
43. **Martin, R. G., and J. L. Rosner.** 1997. Fis, an accessorial factor for transcriptional activation of the *mar* (multiple antibiotic resistance) promoter of *Escherichia coli* in the presence of the activator MarA, SoxS or Rob. *J. Bacteriol.* **179:**7410–7419.
44. **Martin, R. G., W. K. Gillette, and J. L. Rosner.** 2000. Promoter discrimination by the related transcriptional activators MarA and SoxS: differential regulation by differential binding. *Mol. Microbiol.* **35:**623–634.
45. **Martin, R. G., W. K. Gillette, N. I. Martin, and J. L. Rosner.** 2002. Complex formation between activator and RNA polymerase as the basis for transcriptional activation by MarA and SoxS in *Escherichia coli*. *Mol. Microbiol.* **43:**355–370.
46. **Martin, R. G., W. K. Gillette, S. Rhee, and J. L. Rosner.** 1999. Structural requirements for marbox function in transcriptional activation of *mar/sox/rob* regulon promoters in *Escherichia coli:* sequence, orientation and spatial relationship to the core promoter. *Mol. Microbiol.* **34:**431–441.
47. **McLeod, S. M., S. E. Aiyar, R. L. Gourse, and R. C. Johnson.** 2002. The C-terminal domains of the RNA polymerase subunits: contact site with Fis and localization during co-activation with CRP at the *Escherichia coli proP* P2 promoter. *J. Mol. Biol.* **316:**517–519.

47a. **Möller, O., and J. Holmgren.** 1969. A paper disc technique for studying antibacterial synergism. *Acta Pathol. Microbiol. Scand.* **76:**141–145.

48. **Pan, C. Q., S. E. Finkel, S. E. Cramton, J.-A. Feng, D. S. Sigman, and R. C. Johnson.** 1996. Variable structures of Fis-DNA complexes determined by flanking DNA—protein contacts. *J. Mol. Biol.* **264:**675–695.
49. **Pomposiello, P. J., A. Koutsolioutsou, D. Carrasco, and B. Demple.** 2003. SoxRS-regulated expression and genetic analysis of the *yggX* gene of *Escherichia coli*. *J. Bacteriol.* **185:**6624–6632.
50. **Pomposiello, P. J., M. H. J. Bennik and B. Demple.** 2001. Genome-wide transcriptional profiling of the *Escherichia coli* responses to superoxide stress and sodium salicylate. *J. Bacteriol.* **183:**3890–3902.
51. **Ptashne, M., and A. Gann.** 1997. Transcriptional activation by recruitment. *Nature* **386:**569–577.
52. **Ramos, J. L., F. Rojo, L. Zhou, and K. N. Timmis.** 1990. A family of positive regulators related to the *Pseudomonas putida* TOL plasmid XylS and the *Escherichia coli* AraC activators. *Nucleic Acids Res.* **18:**2149–2152.
53. **Rhee, S., R. G. Martin, J. L. Rosner, and D. R. Davies.** 1998. A novel DNA-binding motif in MarA: the first structure for an AraC family transcriptional activator. *Proc. Natl. Acad. Sci. USA* **95:**10413–10418.
54. **Rosenberg, E. Y., D. Bertenthal, M. L. Nilles, K. P. Bertrand, and H. Nikaido.** 2003. Bile salts and fatty acids induce the expression of *Escherichia coli* AcrAB multidrug efflux pump through their interaction with Rob regulatory protein. *Mol. Microbiol.* **48:**1609–1619.
55. **Rosner, J. L.** 1985. Nonheritable resistance to chloramphenicol and other antibiotics induced by salicylates and other chemotactic repellents in *Escherichia coli* K-12. *Proc. Natl. Acad. Sci. USA* **82:**8771–8774.
56. **Rosner, J. L., B. Dangi, A. M. Gronenborn, and R. G. Martin.** 2002. Posttranscriptional activation of the transcriptional activator Rob by dipyridyl in *Escherichia coli*. *J. Bacteriol.* **184:**1407–1416.
57. **Rosner, J. L., T. J. Chai, and J. Foulds.** 1991. Regulation of *ompF* porin expression by salicylate in *Escherichia coli*. *J. Bacteriol.* **173:**5631–5638.
58. **Ross, W., K. K. Gosink, J. Salomon, K. Igarashi, C. Zou, A. Ishihama, K. Severinov, and R. L. Gourse.** 1993. A third recognition element in bacterial promoters: DNA binding by the alpha subunit of RNA polymerase. *Science* **262:**1407–1413.
59. **Savery, N. J., G. S. Lloyd, M. Kainz, T. Gaal , W. Ross, R. H. Ebright, R. L. Gourse, and S. J. Busby.** 1998. Transcription activation at Class II CRP-dependent promoters: identification of determinants in the C-terminal domain of the RNA polymerase alpha subunit. *EMBO J.* **17:**3439–3447.
60. **Savery, N. J., G. S. Lloyd, S. J. Busby, M. S. Thomas, R. H. Ebright, and R. L. Gourse.** 2002. Determinants of the C-terminal domain of the *Escherichia coli* RNA polymerase alpha subunit important for transcription at class I cyclic AMP receptor protein-dependent promoters. *J. Bacteriol.* **184:**2273–2280.
61. **Sawai, T., S. Hirano, and A. Yamaguchi.** 1989. Repression of porin synthesis by salicylate in *Escherichia coli, Klebsiella pneumoniae* and *Serratia marcescens*. *FEMS Lett.* **40:**233–237.
62. **Schneiders, T., T. M. Barbosa, L. M. McMurry, and S. B. Levy.** 2004. The *Escherichia coli* transcriptional regulator MarA directly represses transcription of *purA* and *hdeA*. *J. Biol. Chem.* **279:**9037–9042.
63. **Schneiders, T., S. G. Amyes, and S. B. Levy.** 2003. Role of AcrR and *ramA* in fluoroquinolone resistance in clinical *Klebsiella pneumoniae* isolates from Singapore. *Antimicrobiol. Agents Chemother.* **47:**2831–2837.
64. **Seoane, A. S., and S. B. Levy.** 1995. Characterization of MarR, the repressor of the multiple antibiotic resistance (*mar*) operon in *Escherichia coli*. *J. Bacteriol.* **177:**3414–3419.
65. **Skarstad, K., B. Thöny, D. S. Hwang, and A. Kornberg.** 1993. A novel binding protein of the origin of the *Escherichia coli* chromosome. *J. Biol. Chem.* **268:**5365–5370.
66. **Sulavik, M. C., M. Dazer, and P. F. Miller.** 1997. The *Salmonella typhimurium mar* locus: molecular and genetic analyses and assessment of its role in virulence. *J. Bacteriol.* **179:**1857–1866.
67. **Talukder, A. A., A. Iwata, A. Nishimura, S. Ueda, and A. Ishihama.** 1999. Growth-phase-dependent variation in protein composition of the *Escherichia coli* nucleoid. *J. Bacteriol.* **181:**6361–6370.

68. **Tibbetts, R. J., T. L. Lin, and C. C. Wu.** 2003. Phenotypic evidence for inducible multiple antimicrobial resistance in *Salmonella choleraesuis. FEMS Microbiol. Lett.* **218:**333–338.
69. **Wood, T. I., K. L. Griffith, W. P. Fawcett, K.-W., Jair, T. D. Schneider, and R. E. Wolf, Jr.** 1999. Interdependence of the position and orientation of SoxS binding sites in the transcription activation of the class I subset of *Escherichia coli* superoxide-inducible promoters. *Mol. Microbiol.* **34:**414–430.
70. **Wu, J., and B. Weiss.** 1991. Two divergently transcribed genes, *soxR* and *soxS*, control a superoxide response regulon of *Escherichia coli. J. Bacteriol.* **173:**2864–28671.
71. **Yassien, M. A., H. E. Ewis, C. D. Lu, and A. T. Abdelal.** 2002. Molecular cloning and characterization of the *Salmonella enterica* serovar Paratyphi B *rma* gene, which confers multiple drug resistance in *Escherichia coli. Antimicrobiol. Agents Chemother.* **46:**360–366.
72. **Yasuno, K., T. Yamazaki, Y. Tanaka, T. S. Kodama, A. Matsugami, M. Katahira, A. Ishihama, and K. Y. Kyogoku.** 2001. Interaction of the C-terminal domain of the *E. coli* RNA polymerase subunit with the UP element: recognizing the backbone structure in the minor groove surface. *J. Mol. Biol.* **306:**213–225.

Frontiers in Antimicrobial Resistance: a Tribute to Stuart B. Levy
Edited by D. G. White, M. N. Alekshun, and P. F. McDermott

Chapter 18

Function and Structure of MarR Family Members

MICHAEL N. ALEKSHUN AND JAMES F. HEAD

Members of the MarR protein family are found in a number of bacteria important in both the medical and environmental settings and are also distributed throughout the Archaea, Eukaryotes, and bacteriophage (6, 93). Here they function generally as both transcription repressors and activators (as monomeric, homodimeric, or heterodimeric complexes) to regulate a host of bacterial phenotypes (Table 1).

Many MarR family members were identified originally based on their ability to regulate multiple antibiotic resistance (Mar) (34, 35, 85). More recently, a number of MarR proteins have been shown to be important to the survival of a number of bacterial pathogens during infection. Included are proteins from *Yersinia enterocolitica* (79), *Salmonella enterica* serovars Choleraesuis and Typhimurium (42, 49), and *Staphylococcus aureus* (14).

The activities of a number of MarR family members are responsive to small organic molecules. *Escherichia coli* MarR (4, 60, 93) and EmrR (71) are modulated by compounds containing phenolic rings including salicylate and its derivatives. The DNA binding activity of *Butyrivibrio fibrisolvens* CinR is modulated by cinnamic acid sugar esters (salicylate precursors) (18). OhrR from *Xanthomonas campestris* pv. *phaseoli* is regulated by organic peroxides and hydrogen peroxide in vitro (71).

From a structural standpoint, MarR proteins are members of the helix-turn-helix (HTH) superfamily (69) and are further classified as members of the winged-helix subfamily (28) of HTH proteins. There has been a surge of 3-dimensional structures reported recently for a number of MarR family members. Crystallographic or NMR structures are now available for *E. coli* MarR (7), MexR from *Pseudomonas aeruginosa* (50), an *Enterococcus faecalis* SlyA-like protein (97), YusO from *Bacillus subtilis,* and BlaI, MecI, SarA, SarR, and SarS from *S. aureus* (32, 33, 48, 51, 64, 89).

Given the widespread conservation of MarR family members in diverse bacterial genera, and the roles that they play in processes fundamental to the survival of the microbe, these proteins can be useful tools for understanding and exploring bacterial physiology. That the activities of many MarR family members are modulated by small organic molecules makes them attractive targets for new anti-infection therapeutics.

FUNCTIONS OF MarR FAMILY MEMBERS

Regulation of Multiple Antibiotic Resistance

Gram-negative bacteria

Enterobacteriaceae. Regulation of the *E. coli* multiple antibiotic resistance (Mar) phenotype (34, 35) is dependent on two transcription factors. MarR (multiple antibiotic resistance repressor) negatively regulates and MarA (multiple antibiotic resistance activator) positively controls Mar (17, 92).

Under wild-type or noninducing conditions, the activity of MarR prevents expression of the *marRAB* cistron in *E. coli* and other *Enterobacteriaceae* by binding to a site upstream of *marR* (the *mar* operator [*marO*]) (17). *marO* consists of two MarR binding sites (Fig. 1); site I is located between the −35 and −10 RNAP promoter sequences, and site II is positioned immediately distal to the MarR start codon (60). Each site is composed of two palindromic half-sites that are separated by a 2-base pair (bp) spacer (Fig. 1), and thus the half-sites would not lie precisely on the same face of the DNA duplex (7). Martin and Rosner have shown recently that both sites I and II are required for full repression by MarR and each presumably functions independently (61). These studies have also

Michael N. Alekshun • Paratek Pharmaceuticals, Inc., 75 Kneeland Street, Boston, MA 02111. **James F. Head** • Boston University School of Medicine, Department of Physiology and Biophysics, 715 Albany Street, Boston, MA 02118.

Table 1. A representative collection of MarR family members and their functions

Organism	Protein	Function[a]
Gram-negative bacteria		
E. coli	MarR	Mar
Y. enterocolitica	RovA	Virulence
S. enterica serovar Typhimurium	SlyA	Virulence, OXS
P. aeruginosa	MexR	Mar
N. gonorrhoeae	FarR	Mar
Gram-positive bacteria		
S. aureus	SarA, SarR, SarT, SarH1 (SarS), SarU, and SarV	Virulence
	Rot	Virulence
	NorR (MgrA/Rat)	Mar and virulence
	BlaI and MecI	β-Lactam resistance
	TcaR	Teicoplanin resistance
Other bacteria		
M. thermoautotrophicum	MTH313	Unknown
Eukaryotes	HNF-3	Cell differentiation
Bacteriophage		
Mu	MuR	transposition

[a] Abbreviations: Mar, multiple antibiotic resistance; OST, organic solvent tolerance; OXS, resistance to oxidative stress.

shown that MarR translation is inefficient, and this property is attributed to a nonconsensus Shine-Dalgarno sequence (61).

MarR can be rendered inactive either through genetic mutation (43, 56) or via its direct interaction with a small molecule inducer (4, 60). Once MarR is inactive, the expression of MarA increases, and MarA then functions to activate the Mar phenotype. This is achieved through the up-regulation of the genes encoding the AcrAB multidrug efflux system.

A mutational approach was used to study regions of MarR important for both its interaction with salicylate and DNA (3, 5). In the first study, MarR superrepressors, mutant proteins selected for a lack of response to salicylate in a whole screen using a lethal phenotype, were characterized (5). Further analysis of these mutants demonstrated that all exhibited an increased affinity for *marO* (5). In the second set of experiments, a more extensive mutagenic approach was performed to identify MarR mutants that functioned in a negative complementing manner in a whole cell screen using a β-galactosidase reporter fusion (3). Twenty-two individual mutations were identified that abolished or reduced the DNA

```
MarR        Site I     5′ TTATACTTGCCTGGGCAATATTAT 3′
            Site II    5′ AATTACTTGCCAGGGCAACTAAT 3′
MexR        Site I     5′ AAATGTGGTTGATCCAGTCAACTATTTTG 3′
            Site II    5′ ATTTTAGTTGACCTTATCAACCTTGTTT 3′
SlyA                   5′ TTAGCAAGCTAA 3′
PecS                   5′ CGANWTCGTATATTACGANNNCG 3′
MecI/BlaI              5′ RNATTACAYNTGTARZNT 3′

hRFX1       Co X-tal   5′ CGTTACCATGGTAACG 3′
            physiol    5′ GTTGCCCGGCAAC 3′
HSF                    5′ GGTTCTAGAACC 3′
```

Figure 1. DNA binding sites of MarR family members. Each protein binding site is symmetric and is composed of two palindromic half sites (underlined sequences). The MarR (60) and MexR (24) binding sites were determined using footprinting experiments. The SlyA (91), PecS (82), and MecI/BlaI (32) binding sites were determined using both molecular and biochemical approaches. The physiological (physiol) and DNA-protein cocrystal (Co X-tal) hRFX1 (a eukaryotic winged-helix protein) binding sites are shown for comparison (29). Abbreviations: N, any nucleotide; W, A, or T; Z, G, or T; R, purine; Y, pyrimidine.

binding activity of MarR or reduced levels of protein expression (3).

E. coli contains an additional MarR paralog, MprA (also called EmrR) (53), that regulates multiple drug resistance (MDR) in this host. Unlike MarR, MprA (EmrR) directly regulates the expression of a multidrug efflux system called EmrAB (53). EmrAB confers resistance to carbonyl cyanide *m*-chlorophenylhydrazone (CCCP), nalidixic acid, and a number of other toxic agents (52). MprA (EmrR) also repressed expression of a *marR-lacZ* fusion and this repressor activity was affected negatively by salicylate (93).

***Pseudomonas aeruginosa*.** Although the *P. aeruginosa* genome specifies 13 MarR orthologs, only one (MexR) has been characterized extensively. MexR is encoded divergently and positioned directly proximal to the genes specifying the MexAB-OprM MDR efflux system (73), an efflux pump with an extraordinarily broad substrate profile. MexR negatively regulates its own expression and that of *mexAB-oprM* (24, 73, 90).

The MexR binding site has been characterized (24), and its overall organization is similar to that of the MarR binding sites (Fig. 1). There is an important difference, however, between the MexR binding site and that of other MarR family members (Fig. 1). The palindromic MexR half-sites are separated by a distance of 5 bp (24), and this would place each on opposite faces of the DNA duplex (Fig. 1). Thus, DNA recognition by MexR may be different structurally from other MarR proteins (50).

Mutations in *mexR,* which alter the repressor activity of the protein, occur in laboratory-derived strains (1, 84) and in clinical isolates (10, 70, 100). The latter presumably occur during both single and dual antibiotic treatment regimens (100).

***Neisseria gonorrhoeae*.** In a genome-wide search for potential transcription regulators of the FarAB MDR efflux pump, which confers resistance to toxic long chain fatty acids (47), the genes specifying two putative MarR-like proteins were identified (46). One (FarR) encoded a protein of ~21 kDa that bound in a sequence specific manner to regions upstream of both *farAB* and *farR* in electrophoretic mobility shift assays (EMSA) (46). Additional experiments with gene fusions demonstrated that FarR autogenously regulated its own expression and functioned as a repressor of FarAB synthesis (46). *N. gonorrhoeae* possesses another MDR efflux system, MtrCDE, which confers resistance to antibiotics and noxious fatty acids (37) and is required for infection of the murine genital tract (41). While expression of MtrDCE is under the control of MtrR, a TetR family member (38), MtrR also binds specifically to a sequence upstream of *farR* and acts as a repressor of FarR expression (46). In this manner, MtrR increases expression of FarAB through its inhibitory effects on FarR synthesis (46).

Gram-positive bacteria

While many reports regarding the activity of MarR-family members regulating MDR have been described in gram-negative bacteria, there are examples of this function in gram-positive organisms. In *S. aureus,* overexpression of the multidrug efflux pump NorA engenders resistance to particular quinolones, disinfectants, and dyes (96, 99). A biochemical approach was used to identify potential regulators of *norA* expression, and these experiments resulted in the discovery of NorR (also termed Rat (39) and MgrA (54) [see below]) (95). *norR* specifies a protein of ~18 kDa that in EMSA binds an operator upstream of *norA* but not to sequences proximal to *norR* (95). Unlike the other well-characterized gram-negative MarR-family members that negatively regulate MDR efflux pump expression, NorR appears to have a positive effect (acts as an activator) on NorA expression (95). Recent studies with *S. aureus* SarA demonstrated that inactivation of the gene specifying this protein affected susceptibility to antibiotics, ethidium, and household cleaners (67).

Regulation of Single Antibiotic Resistances

β-Lactam resistance in *Staphylococcus aureus* and *Bacillus licheniformis*

Susceptibility to the β-lactams in *S. aureus* and *Bacillus licheniformis* are in part controlled by the activity of MarR-like (winged-helix) proteins. BlaI and MecI regulate the expression of the BlaZ β-lactamase and staphylococcal cassette chromosome *mec* (SC-C*mec*), also termed *mecA* and which specifies penicillin binding protein PBP2′ or PBP2a and affords protection against all β-lactams in *S. aureus* (9, 64). Synthesis of the BlaP β-lactamase in *B. licheniformis* is also under the control of BlaI (25). BlaI and MecI repress expression of *blaZ* and *mecA* in the absence of a β-lactam, but when the cell is exposed to an antibiotic within this class a series of cytoplasmic events occur. A sensory membrane protein, either BlaR1 or MecR1, which contains both extracellular sensory and intracellular metalloprotease domains, recognizes the presence of a β-lactam, and this results in the proteolytic cleavage of BlaI or MecI, thereby relieving repression (reviewed in reference 9).

The MecI and BlaI binding sites are very similar to MarR operator (32). Garcia-Castellanos et al. have produced a consensus binding site for these proteins (Fig. 1).

Teicoplanin resistance in *Staphylococcus aureus*

In the laboratory, resistance to teicoplanin (a glycopeptide) is easier to achieve than is resistance to vancomycin (another glycopeptide antibiotic), and while studying the stepwise increase in teicoplanin resistance in laboratory derived mutants, a novel putative sensory mechanism was identified (11). Brandenberger et al., found a chromosomal locus specifying three proteins: TcaR (MarR family member), TcaA (transmembrane sensory protein), and TcaB (hypothetical membrane protein). Inactivation of the *tcaRAB* operon in the *S. aureus* COL (MRSA) background led to a fourfold decrease in susceptibility to teicoplanin and a twofold increase in methicillin susceptibility by an unknown mechanism(s) (11). When individual components of the *tcaRAB* locus were supplied in trans to the deletion mutant, *tcaA* by itself was sufficient in restoring wild-type antibiotic susceptibility (55). It is tempting to suggest that the TcaRAB system either senses the presence of teicoplanin or a teicoplanin-lipid II complex or that it somehow modifies the structure of the cell wall to confer resistance (11). Whether or not TcaR participates directly in sensing glycopeptide stress or plays a more fundamental role in TcaRAB expression is currently unknown.

Roles in Bacterial Pathogenesis and Global Gene Regulation

Gram-negative bacteria

Yersinia enterocolitica. The *Yersinia enterocolitica* RovA is involved in regulating the expression of *inv,* which specifies an invasion required for the early stages of infection of this host and a number of other virulence genes (23). Deletion of the gene specifying RovA in *Y. enterocolitica* resulted in a 70- to 500-fold increase in the oral 50% lethal dose (LD_{50}) in murine models of infection (21, 79). This virulence defect is in part dependent on the route of infection since the *rovA* mutant exhibits only an 11-fold increase in the LD_{50} when mice are infected intraperitoneally (21). The virulence defect of the *rovA* strain appears to be related to the inability of the mutant to produce inflammation, since the mutant does not elicit IL-1α production in the Peyer's patches (areas of lymphoid tissue on the intestinal mucosa) of infected mice (21).

Salmonella. SlyA was originally identified as a regulator of a cryptic hemolysin in *E. coli* K-12 (49, 68). Deletion of the gene specifying SlyA in either *S. enterica* serovars Typhimurium or Choleraesuis severely attenuates the virulence of these organisms and results in an 1,000- to 100,000-fold increase in the LD_{50} following oral, intraperitoneal, or intravenous routes of infection (19, 42, 49). SlyA mediates susceptibility to oxidative stress but is not required for a response to metabolites of nitric oxide synthase (13). A *slyA::lacZ* fusion is induced in serovar Typhimurium grown in macrophages (13), and SlyA is required for the destruction of Peyer's patch membranous epithelial (M) cells and the survival of the bacterium within the Peyer's patches of infected mice (19).

A cursory investigation of the SlyA regulon (a collection of genes regulated by the transcription factor) has been attempted (91). These results demonstrated that SlyA either directly or indirectly participates in the regulation of proteins that play roles in motility (FliC), iron acquisition (IroN), membrane permeability (OmpA, OmpC, and OmpF), and other virulence associated functions (PagC) (91). These experiments also resulted in the identification of a consensus SlyA binding sequence (Fig. 1) (91).

Erwinia chrysanthemi. *E. chrysanthemi* is a pathogen that infects plants, and the virulence of this organism is in part mediated by PecS (80). This protein regulates the expression of pectate lyase, cellulase, polygalacturonase enzymes, and indigoidine (a blue pigment that affords protection against oxidative stress) (80). Deletion of *pecS* in *E. chrysanthemi* conferred resistance to hydrogen peroxide (H_2O_2), due to increased synthesis of indigoidine, and relative to a wild-type strain the mutant is more virulent (81). These data confirm that PecS contributes to virulence in both a positive and negative manner (81).

A molecular biochemical approach was used to characterize a consensus PecS binding sequence (Fig. 1) (82). These studies also identified genes involved in flagellar biosynthesis as new members of the PecS regulon (82).

Gram-positive bacteria

Staphylococcus aureus. *S. aureus* regulates the expression of numerous virulence factors in a complex and highly coordinated manner, and two of the most extensively characterized regulatory loci are *agr* and *sar* (14, 16, 66). More recently, myriad *S. aureus* winged-helix proteins that regulate virulence in this pathogen have also been described, including members of the SarA family, Rot, and NorR (MgrA/Rat) (14, 16, 66) (Table 2).

SarA and its family members. SarA is an ~15 kDa protein that positively and negatively regulates a number of virulence associated genes (66). The multitude of functions regulated directly and indirectly by

Table 2. MarR orthologs from *Staphylococcus aureus* that regulate virulence

Protein	Function
SarA	Pleiotropic regulator of gene expression
SarR	Regulator of SarA expression
SarS (SarH1)	Controls cell wall and extracellular protein expression
SarT	Repressor of α-hemolysin
SarU	Regulates RNAIII and other virulence gene expression
SarV	Controls protease and autolysin expression
Rot	Repressor of toxins
TcaR	Regulates biofilm formation

SarA and the complex regulatory nature of the *sarA* locus have been described in full detail elsewhere (16, 66).

A custom Affymetrix GeneChip, encompassing 86% of the *S. aureus* COL genome, was used recently to characterize the SarA regulon (22). SarA had a negative effect on the expression of 44 genes, including those encoding lipase (*lip*), nuclease (*nuc*), protein A (*spa*), and *sarT* and *sarH1* (*sarS*) (see below) and a positive effect on the expression of 76 genes such as the *agr* locus, fibronectin binding protein A and B (*fnbA* and *fnbB*), and α-(*hla*) and δ-toxins (*hld*) (22).

Since mutations in *sarA*, and *agr*, incompletely attenuate virulence in animal models of infection (8), it was surmised that other regulatory factors contribute to *S. aureus* virulence. This hypothesis led to the identification of other SarA family members, including SarR, SarH1 (SarS), SarT, SarU, SarV, SarX, SarY, and SarZ (14). SarR (13.6 kDa) binds *sar* promoter fragments in EMSA and down-regulates *sarA* expression, which in turn affects the expression of many SarA regulated genes (57). SarH1 (also called SarS) is larger (29.9 kDa) and functions as an activator of *spa* and a repressor of *hla*, and its own expression is negatively regulated by *agr* and SarA (15, 94). The expression of *sarT*, specifying a 16.1 kDa protein, is negatively regulated by SarA (87). While SarT acts as a repressor of *agr*, *hla*, and *sarU* expression and an activator of SarS synthesis, its overall effects in the cell may lie downstream of those mediated by *agr* and SarA (86, 87). *sarU* is located contiguous to but is transcribed divergently from *sarT*, and its gene product (29 kDa) has a positive effect on *agr* expression (58). SarV (12.8 kDa) affects *agr* RNA expression as well as genes specifying proteases and autolysins (59). Its own expression is regulated by SarA ad MgrA (59). The functions of SarX, SarY, and SarZ are not yet known (14).

ROT. While deciphering the regulatory mechanisms involved in the production of α-toxin in an *agr* mutant, transposon mutagenesis identified a locus that partially restored expression of this virulence factor (63). This locus was subsequently termed *rot* (repressor of toxins) since it was found to regulate both *hla* (α-toxin) and protease expression in a growth dependent manner (63).

Using an Affymetrix GeneChip (see above), Rot was found to function in a global manner as both a positive and negative regulator of transcription (83). These data confirm the earlier findings with *hla* and add *hlb* (β-toxin), the *spl* and *ssp* operons (specifying staphylococcal protease activity), *geh* (lipase), and a number of other virulence factors to the list of negatively regulated genes (83). Rot also positively modulated the expression of *spa* (protein A), other cell surface proteins, and SarS (83).

NORR (MGRA/RAT). *norR* (*mgrA/rat*) null mutants were initially found to exhibit a cell-clumping phenotype in liquid media (95). Subsequent experimental data suggested that *norR* (*mgrA/rat*) null mutants were also defective in regulating autolysis (39) and that *norR* (*mgrA/rat*) overexpressing strains synthesized aberrant levels of a number of virulence factors (54). NorR (MgrA/Rat) was eventually found to regulate genes involved in autolysis (*lytRS*, *arlRS*, *lytM*, and *lytN*), cell wall metabolism (*scdA* and *sspA*), and pathogenesis (*cap8*, *spa*, *hla*, and *nuc*) (39).

TCAR. In addition to its role in regulating antibiotic susceptibility, TcaR has been shown recently to participate in the complex *S. aureus* virulence cascade. Jefferson et al. used a biochemical approach to identify regulatory factors of the *ica* (intracellular adhesion) locus, which is involved in biofilm formation (40). These studies demonstrated that while TcaR binds specifically to *ica* and negatively regulates its expression in vitro, the overall effect on biofilm formation in a strain deleted for *tcaR* is modest presumably because of other regulatory factors in the cell (40).

McCallum et al. used microarrays to characterize the transcriptome of wild type *S. aureus* COL and a strain bearing a >6-kb chromosomal deletion, which encompassed the *tcaRAB* operon, and found altered expression of only three transcripts (*sarS*, *spa*, and *sasF* [specifying a cell-wall associated protein]) (62). Subsequent experiments directly demonstrated that TcaR was responsible for altered expression of *sarS* and *sasF* in the mutant (62).

It is of interest to note that *S. aureus* NCTC8325, the parent strain of a number of laboratory mutants that have been used to decipher virulence regulatory networks in this host, bears a naturally mutant *tcaR* locus (62). Thus, the information derived on the regulatory properties of TcaR and SarS (see above) in NCTC8325 genetic backgrounds should be reevaluated with this new finding in mind (62).

Listeria monocytogenes. Approximately 7% of the genomes of *L. monocytogenes* (a human pathogen) and *Listeria innocua* (a nonpathogenic bacterium) are devoted to specifying a variety of transcription factors (2, 36). A proteomic approach was used to study the response of *L. monocytogenes* to both lethal (pH 3.5) and nonlethal (pH 5.5) acidic growth conditions (72). In these experiments, two MarR-family members of ~15 kDa were identified; one (YfiV) was induced at pH 5.5, while the other (YxaD) was induced at both pH 3.5 and 5.5 (72). Deletion of the gene specifying YxaD in *L. monocytogenes* affected the growth of this organism in media with an alkaline pH or media containing ethanol, but a virulence defect in mice was not noted (78).

Other bacteria

Mycobacterium marinum has been used as a surrogate for *Mycobacterium tuberculosis* infection because its course of infection and the resulting pathology in its natural host resembles that of *M. tuberculosis* infection in humans (31). Transposon mutagenesis was used to identify CrtR, a MarR family member in *M. marinum* that when mutated affected carotenoid biosynthesis in this host (31). Carotenoids are thought to protect *M. marinum* from the damaging effects of singlet oxygen and other toxic radicals (76). Further experiments suggested that CrtR acts as a repressor of *crt*, the carotenoid biosynthetic locus (31). Further experiments will be needed to confirm an in vivo role in infection for this protein.

Interaction of MarR Family Members with Small Organic Molecules

Escherichia coli

Until recently (4, 7, 60), it was unknown whether the inducibility of the *E. coli* Mar phenotype was attributed to a direct MarR-small molecule interaction or through other means (6). While data derived from experiments in vitro initially suggested a direct interaction (4, 60), the crystal structure of MarR in the presence of salicylate has been determined recently (7) (see below). MarR is also responsive to 2,4-dinitrophenol (DNP), menadione, and plumbagin (4). EmrR binds CCCP, carbonyl cyanide *p*-(trifluoromethoxy)-phenylhydrazone (FCCP), and DNP (12) and is responsive to CCCP, DNP, tetrachlorosalicylanilide, and nalidixic acid in vitro (98). More recently, the activity in vitro of MarR from *S. enterica* serovar Typhimurium has been shown to be affected negatively by deoxycholate, a bile salt (74). These data suggest that MarR and EmrR interact with a number of small molecules, which may have growth-inhibitory effects on the cell, and that this interaction results in the activation of microbial defense mechanisms, e.g., MDR efflux pumps, etc., allowing the cell to deal with the stress.

HpaR, a MarR protein from *E. coli* W, regulates the expression of the *hpa* operon, which specifies proteins involved in the catabolism of 4-hydroxyphenylacetic (4-HPA) acid (30). Purified HpaR was eluted from a phenyl Sepharose column with a number of 4-HPA analogs, including 3-HPA and 3,4-HPA but not other structurally similar aromatic compounds (30). 4-HPA, 3-HPA, and 3,4-HPA all induce expression of a HpaR responsive reporter in whole cells (30).

Comamonas testosteroni

The *C. testosteroni* CbaR is a transcription factor that regulates the catabolism of chlorinated benzoic acid (75). The DNA binding activity of CbaR in vitro was antagonized by 3-chlorobenzoate, protocatechuate, and benzoate but enhanced by 3-hydroxyl- and 4-carboxylbenzoate (75).

Butyrivibrio fibrisolvens

Expression of *cinB* (specifying a cinnamoyl ester hydrolase) in *B. fibrisolvens* is regulated by CinR, which binds in a sequence specific manner to the *cinRB* intergenic region (18). The DNA binding activity of CinR was negatively affected by cinnamic acid sugar esters but not by other metabolites of arabinoxylan catabolism (18).

Bacillus subtilis and *Xanthomonas campestris* pv. *phaseoli*

The bacterial organic hydroperoxide resistance (Ohr) proteins protect the host from organic peroxide stress (65). Expression of Ohr in a number of species is regulated by OhrR, and paralogs and orthologs of this protein are found in a variety of gram-negative and gram-positive bacteria (26). Experiments demonstrated that the DNA binding activity of the *B. subtilis* and *X. campestris* pv. *phaseoli* OhrR proteins in vitro is negatively affected by cumene hydroperoxide, *tert*-butyl hydroperoxide, and H_2O_2 (27, 71). That this negative affect was also observed with a thiol oxidizing agent and reversed in the presence of a thiol reducing compound (27) suggested that OhrR's responsiveness to peroxides was mediated by a cysteine residue (27, 71). Subsequent experiments using site directed mutagenesis of a cysteine residue conserved among OhrR proteins confirmed this hypothesis (27, 71). A mechanism whereby the OhrR cysteine residue is reversibly converted to a sulfenic acid was then proposed (27).

Use of small molecule MarR family members to inhibit virulence

Recent experiments have demonstrated that treatment with acetylsalicylate is protective in a rabbit model of *S. aureus* endocarditis (45). It has also been shown that salicylate attenuated virulence in both laboratory and clinically derived isolates of *S. aureus* (44). The salicylate mediated attenuation of virulence was dependent on SarA, and these data tie together a virulence modulating effect of salicylate in vitro with a protective effect in the infection model (44).

STRUCTURES OF MarR FAMILY MEMBERS

Winged-helix proteins share a common topology (28), and although the overall topology differs from protein to protein, e.g., MarR lacks a second wing structure, the HTH motifs in all cases is flanked by a wing structure(s). The HTH motif of winged-helix proteins also exhibits slight structural variations from other similar HTH elements (28). Winged-helix transcription factors use the wing (29), the HTH motif, or both elements together (32) to interact with DNA. The salient structural features of a number of MarR orthologs are described below.

Gram-Negative Bacteria

E. coli and *P. aeruginosa*

The structures of two gram-negative MarR proteins have been described: *E. coli* MarR (7) and *P. aeruginosa* MexR (50). Crystals of *E. coli* MarR were produced in the presence of salicylate. The structure shows MarR as a homodimer formed by a perfect twofold rotation of the subunit (Fig. 2). The dimer is held together by interactions between the overlapping N and C termini of each subunit, which form a closely intertwined domain. The globular DNA-binding domain of each subunit is connected to this region by a pair of long helices. The two DNA-binding domains themselves interact with each other only through a pair of salt bridges and thus appear likely to be able to reorient to accommodate binding to DNA. Mobility in one of the helices connecting the DNA-binding domain to the dimerization domain gives further indication that the molecule can "flex" to permit reorientation of the DNA-binding regions. In the salicylate complex, two

A

B

Figure 2. Ribbon representation of MarR with bound salicylate (PDB ID 1JGS). (A) The MarR dimer with one chain in black and the other in white. Bound salicylates are shown as space-filled models. (B) View of the MarR dimer from below (relative to top figure) that shows a 25Å slab of the structure centered on the winged-helix motifs and with the bound salicylates shown as space-filled models.

molecules of salicylate appear to be bound to each DNA-binding domain, one on either side of the H3–DNA-binding helix. This location would be consistent with a potential steric blockade or interference with the interaction of the helix with DNA, in keeping with the inhibitory effects of salicylate.

The crystal structure of MexR gives additional insight into the range of possible conformational flexibility in these dimeric structures. The overall structure of MexR resembles that of MarR, but the MexR crystal structure shows the dimer in four different conformational states (Fig. 3). It seems likely that these states may represent different physiologically relevant structures in which the DNA-binding domains are held in different orientations. Changes in the orientation of the helices within the dimerization domain appear to contribute to these reorientations, as well as additional interactions of part of one MexR C-terminal tail between the DNA-binding domains in another dimer. Differences in these interactions and possible flexibilities of MarR and MexR may indicate distinct modes of regulation as well as differences in the organization of the DNA sites to which the proteins must bind.

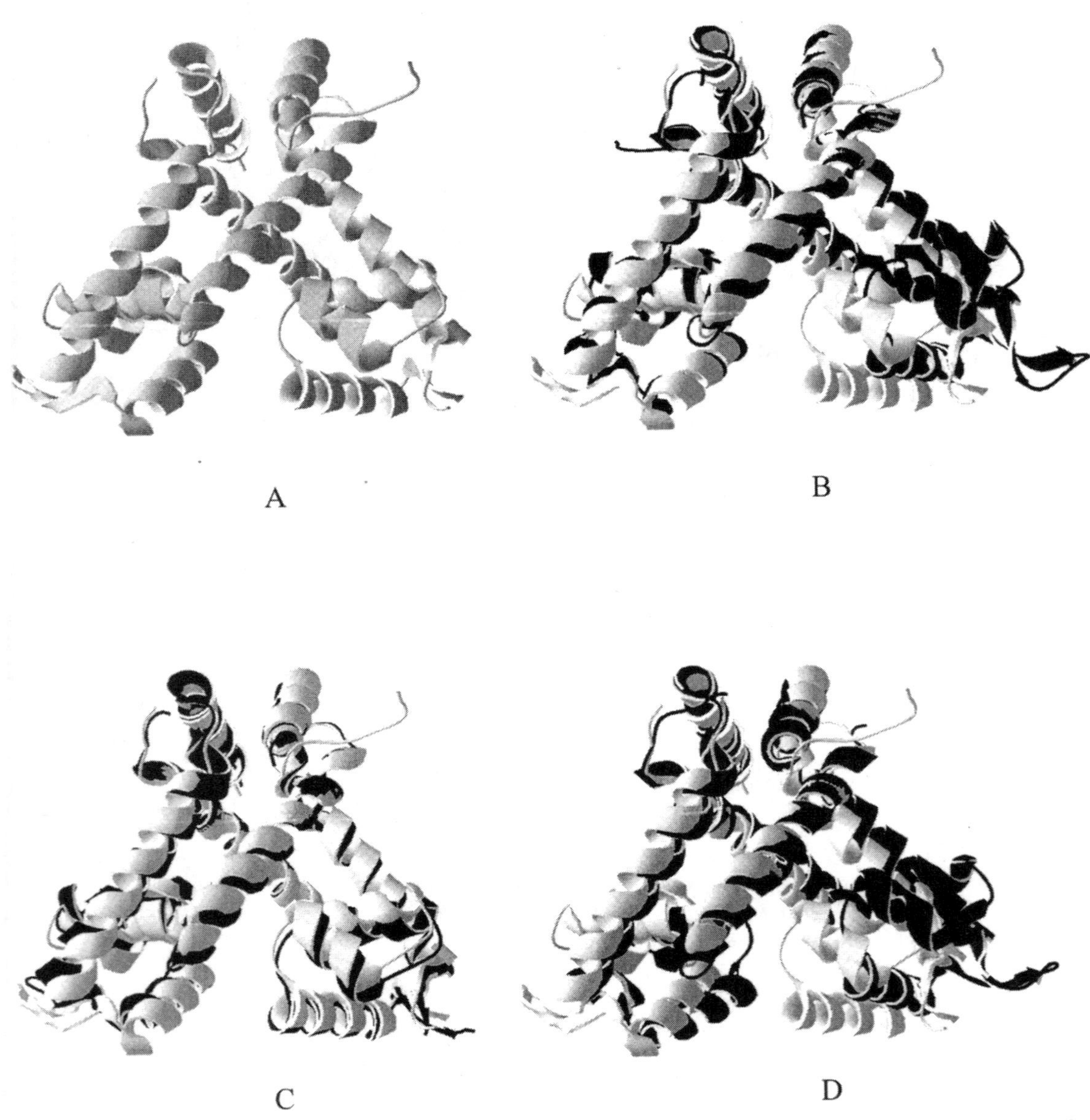

Figure 3. Ribbon representations of MexR (PDB ID 1LNW). Four structures from the same crystal, aligned with a best fit on residues 4-16 and 116-139, showing the relative differences in the positions of the DNA binding domains. (A) A/B chains shown in white. (B-D) the C/D, E/F, and G/H chains (black), respectively, are superimposed on the A/B chains (white).

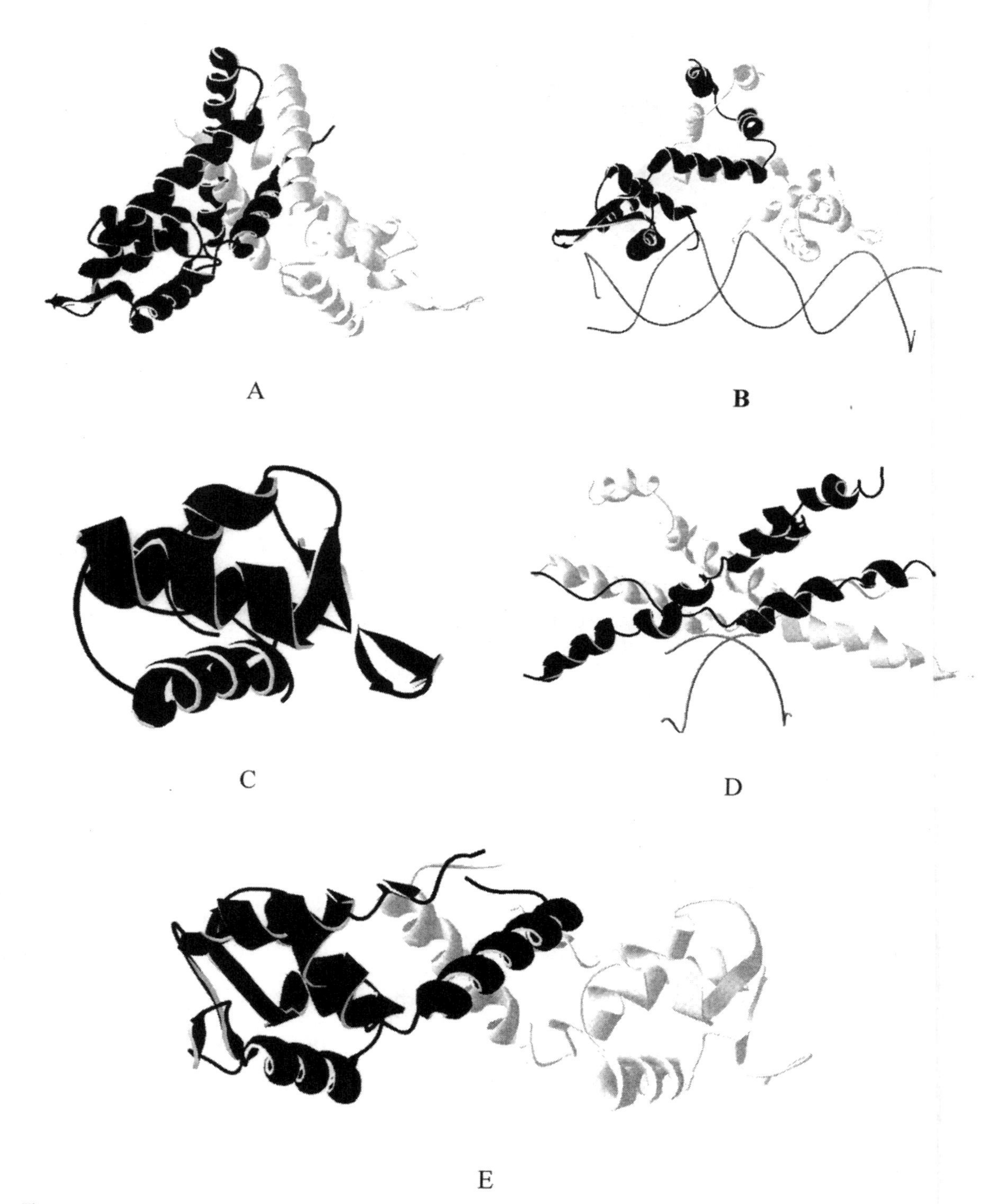

Figure 4. Structures of gram-positive MarR orthologs (PDB ID). The proteins are oriented so that the (putative) DNA binding domains are facing down. (A) *E. faecalis* SlyA-like protein (1LJ9). (B) *S. aureus* MecI in the presence of DNA (represented as a wireframe model) (1SAX). (C) *S. aureus* BlaI (1P6R). (D) *S. aureus* SarA in the presence of DNA (represented as a wireframe model) (1FZP). (E) *S. aureus* SarR (1HSJ) (SarR was crystallized as a fusion with maltose binding protein [MBP] [51], but the MBP structure was omitted from this figure). (F) *S. aureus* SarS (1P4X). (G) *B. subtilis* YusO (1S3J). (*Continued on following page.*)

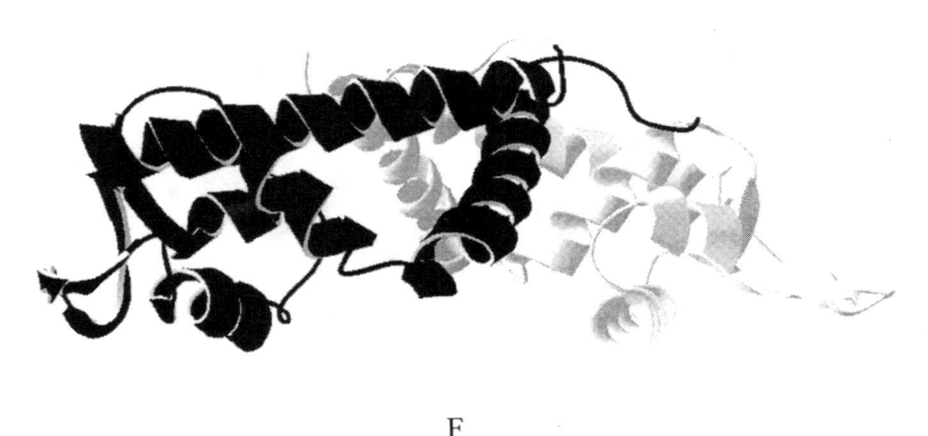

F

G

Figure 4. *Continued*

Gram-Positive Bacteria

S. aureus

There are five structures available for *S. aureus* MarR orthologs (Fig. 4 B–F), and Manna et al. have proposed a tentative classification for these proteins (59). Accordingly, one family would encompass single domain proteins (e.g., SarR and SarV), another would consist of two-domain proteins (e.g., SarS and SarU), and a third would be comprised of single-domain proteins with high homology to MarR (e.g., MgrA and others) (59) (Table 3). As noted previously (88), the structure of SarA (Fig. 4D), although 27% identical to SarR, appears to vary the most from the other MarR orthologs. The structure of MecI has been solved in the presence (32) and absence (33) of its operator. Although MecI and its operator closely resemble MarR and *marO,* respectively, the orientation of the DNA-binding domains is distinct between unbound MecI and MarR (as seen in complex with salicylate). The MecI dimer undergoes little conformational change in forming the DNA complex. The structure of MecI with DNA shows the repressor binding the palindromic operator with half-sites on the same face of the DNA duplex and inducing a convex bend in the DNA (Fig. 4B).

Enterococcus faecalis and *Bacillus subtilis*

During efforts to characterize structurally (http://www.mcsg.anl.gov) the *E. faecalis* genome, a SlyA-like protein was identified (Fig. 4A). Analysis of the protein's structure demonstrates a twofold symmetry of the putative DNA-binding domain, and this is con-

Table 3. Classification of SarA family members in *Staphylococcus aureus* (14)

Class 1	Class 2	Class 3
SarA	SarS	SarZ
SarR	SarU	NorR (MgrA/Rat)
SarT	SarY	SA2256[a]
SarV		SA2531[a]
SarX		

[a] Based in the *Staphylococcus aureus* COL genome (http://www.tigr.org).

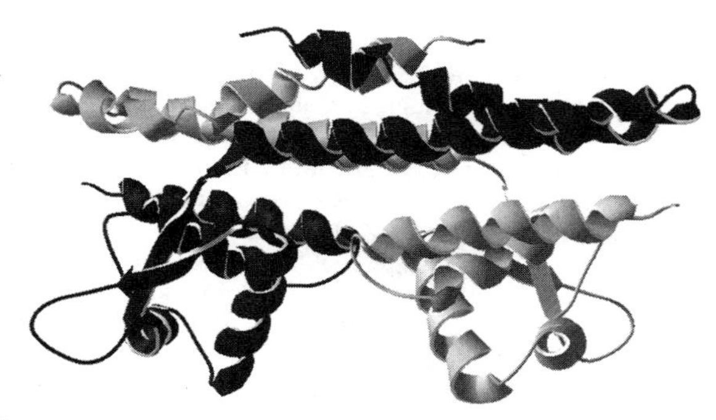

Figure 5. Ribbon representation of *M. jannaschii* Mj223 (PDB ID 1KU9).

sistent with the palindromic nature of the DNA targets to which SlyA and related proteins bind (97). The structure of the *B. subtilis* YusO has been deposited recently (Fig. 4G).

Archaea

Methanococcus jannaschii

There are a number of genes in *S. aureus* that contribute to methicillin resistance in this organism (20). *M. jannaschii* Mj223 was characterized originally by its similarity to a protein from *S. aureus* RUSA223, which plays a role in methicillin resistance (20). Mj223 is a dimeric structure that contains a basic and hydrophobic putative DNA binding domain, and these data suggest that it and RUSA223 function as transcription factors (77) (Fig. 5). Whether Mj223 and RUSA223 function as activators or repressors is currently unknown.

CONCLUDING REMARKS

The plethora of MarR paralogs and orthologs makes this group one of the most widely distributed protein families. Although once known only to play a role in regulating multiple antibiotic resistance, their contributions toward microbial virulence seem to be equally important and thus far underappreciated. The availability of crystal structures of the proteins themselves as well as small molecule- or DNA-co-crystal complexes may prove rewarding in subsequent structure-based drug design efforts.

REFERENCES

1. **Adewoye, L., A. Sutherland, R. Srikumar, and K. Poole.** 2002. The *mexR* repressor of the *mexAB-oprM* multidrug efflux operon in *Pseudomonas aeruginosa:* characterization of mutations compromising activity. *J. Bacteriol.* **184:**4308–4312.
2. **Alekshun, M. N.** 2001. Beyond comparison-antibiotics from genome data? *Nat. Biotechnol.* **19:**1124–1125.
3. **Alekshun, M. N., Y. S. Kim, and S. B. Levy.** 2000. Mutational analysis of MarR, the negative regulator of *marRAB* expression in *Escherichia coli,* suggests the presence of two regions required for DNA binding. *Mol. Microbiol.* **35:**1394–1404.
4. **Alekshun, M. N., and S. B. Levy.** 1999. Alteration of the repressor activity of MarR, the negative regulator of the *Escherichia coli marRAB* locus, by multiple chemicals in vitro. *J. Bacteriol.* **181:**4669–4672.
5. **Alekshun, M. N., and S. B. Levy.** 1999. Characterization of MarR superrepressor mutants. *J. Bacteriol.* **181:**3303–3306.
6. **Alekshun, M. N., and S. B. Levy.** 1997. Regulation of chromosomally mediated multiple antibiotic resistance: the *mar* regulon. *Antimicrob. Agents Chemother.* **41:**2067–2075.
7. **Alekshun, M. N., S. B. Levy, T. R. Mealy, B. A. Seaton, and J. F. Head.** 2001. The crystal structure of MarR a regulator of multiple antibiotic resistance at 2.3 Å resolution. *Nat. Struct. Biol.* **8:**710–714.
8. **Alksne, L. E., and S. J. Projan.** 2000. Bacterial virulence as a target for antimicrobial chemotherapy. *Curr. Opin. Biotechnol.* **11:**625–636.
9. **Berger-Bachi, B., and S. Rohrer.** 2002. Factors influencing methicillin resistance in staphylococci. *Arch. Microbiol.* **178:** 165–171.
10. **Boutoille, D., S. Corvec, N. Caroff, C. Giraudeau, E. Espaze, J. Caillon, P. Plesiat, and A. Reynaud.** 2004. Detection of an IS21 insertion sequence in the *mexR* gene of *Pseudomonas aeruginosa* increasing beta-lactam resistance. *FEMS Microbiol. Lett.* **230:**143–146.
11. **Brandenberger, M., M. Tschierske, P. Giachino, A. Wada, and B. Berger-Bachi.** 2000. Inactivation of a novel three-cistronic operon *tcaR-tcaA-tcaB* increases teicoplanin resistance in *Staphylococcus aureus. Biochim. Biophys. Acta* **1523:**135–139.
12. **Brooun, A., J. J. Tomashek, and K. Lewis.** 1999. Purification and ligand binding of EmrR, a regulator of a multidrug transporter. *J. Bacteriol.* **181:**5131–5133.
13. **Buchmeier, N., S. Bossie, C. Y. Chen, F. C. Fang, D. G. Guiney, and S. J. Libby.** 1997. SlyA, a transcriptional regulator of *Salmonella typhimurium,* is required for resistance to oxidative stress and is expressed in the intracellular environment of macrophages. *Infect. Immun.* **65:**3725–3730.
14. **Cheung, A. L., A. S. Bayer, G. Zhang, H. Gresham, and Y. Q. Xiong.** 2004. Regulation of virulence determinants in vitro and in vivo in *Staphylococcus aureus. FEMS Immunol. Med. Microbiol.* **40:**1–9.
15. **Cheung, A. L., K. Schmidt, B. Bateman, and A. C. Manna.** 2001. SarS, a SarA homolog repressible by *agr,* is an activator of protein A synthesis in *Staphylococcus aureus. Infect. Immun.* **69:**2448–2455.
16. **Cheung, A. L., and G. Zhang.** 2002. Global regulation of virulence determinants in *Staphylococcus aureus* by the SarA protein family. *Front. Biosci.* **7:**d1825–d1842.
17. **Cohen, S. P., H. Hächler, and S. B. Levy.** 1993. Genetic and functional analysis of the multiple antibiotic resistance (*mar*) locus in *Escherichia coli. J. Bacteriol.* **175:**1484–1492.
18. **Dalrymple, B. P., and Y. Swadling.** 1997. Expression of a *Butyrivibrio fibrisolvens* E14 gene (*cinB*) encoding an enzyme with cinnamoyl ester hydrolase activity is negatively regulated by the product of an adjacent gene (*cinR*). *Microbiology* **143:** 1203–1210.
19. **Daniels, J. J., I. B. Autenrieth, A. Ludwig, and W. Goebel.** 1996. The gene *slyA* of *Salmonella typhimurium* is required for destruction of M cells and intracellular survival but not for invasion or colonization of the murine small intestine. *Infect. Immun.* **64:**5075–5084.

20. **De Lencastre, H., S. W. Wu, M. G. Pinho, A. M. Ludovice, S. Filipe, S. Gardete, R. Sobral, S. Gill, M. Chung, and A. Tomasz.** 1999. Antibiotic resistance as a stress response: complete sequencing of a large number of chromosomal loci in *Staphylococcus aureus* strain COL that impact on the expression of resistance to methicillin. *Microb. Drug Resist.* 5:163–175.
21. **Dube, P. H., S. A. Handley, P. A. Revell, and V. L. Miller.** 2003. The rovA mutant of Yersinia enterocolitica displays differential degrees of virulence depending on the route of infection. *Infect. Immun.* **71:**3512–3520.
22. **Dunman, P. M., E. Murphy, S. Haney, D. Palacios, G. Tucker-Kellogg, S. Wu, E. L. Brown, R. J. Zagursky, D. Shlaes, and S. J. Projan.** 2001. Transcription profiling-based identification of *Staphylococcus aureus* genes regulated by the *agr* and/or *sarA* loci. *J. Bacteriol.* **183:**7341–7353.
23. **Ellison, D. W., M. B. Lawrenz, and V. L. Miller.** 2004. Invasin and beyond: regulation of *Yersinia* virulence by RovA. *Trends Microbiol.* **12:**296–300.
24. **Evans, K., L. Adewoye, and K. Poole.** 2001. MexR repressor of the *mexAB-oprM* multidrug efflux operon of *Pseudomonas aeruginosa:* identification of MexR binding sites in the *mexA-mexR* intergenic region. *J. Bacteriol.* **183:**807–812.
25. **Filee, P., K. Benlafya, M. Delmarcelle, G. Moutzourelis, J. M. Frere, A. Brans, and B. Joris.** 2002. The fate of the BlaI repressor during the induction of the *Bacillus licheniformis* BlaP beta-lactamase. *Mol. Microbiol.* **44:**685–694.
26. **Fuangthong, M., S. Atichartpongkul, S. Mongkolsuk, and J. D. Helmann.** 2001. OhrR is a repressor of *ohrA*, a key organic hydroperoxide resistance determinant in *Bacillus subtilis. J. Bacteriol.* **183:**4134–4141.
27. **Fuangthong, M., and J. D. Helmann.** 2002. The OhrR repressor senses organic hydroperoxides by reversible formation of a cysteine-sulfenic acid derivative. *Proc. Natl. Acad. Sci. USA* **99:**6690–6695.
28. **Gajiwala, K. S., and S. K. Burley.** 2000. Winged helix proteins. *Curr. Opin. Str. Biol.* **10:**110–116.
29. **Gajiwala, K. S., H. Chen, F. Cornille, B. P. Roques, W. Reith, B. Mach, and S. K. Burley.** 2000. Structure of the winged-helix protein hRFX1 reveals a new mode of DNA binding. *Nature* **403:**916–921.
30. **Galan, B., A. Kolb, J. M. Sanz, J. L. Garcia, and M. A. Prieto.** 2003. Molecular determinants of the *hpa* regulatory system of *Escherichia coli:* the HpaR repressor. *Nucleic Acids Res.* **31:**6598–6609.
31. **Gao, L. Y., R. Groger, J. S. Cox, S. M. Beverley, E. H. Lawson, and E. J. Brown.** 2003. Transposon mutagenesis of *Mycobacterium marinum* identifies a locus linking pigmentation and intracellular survival. *Infect. Immun.* **71:**922–929.
32. **Garcia-Castellanos, R., G. Mallorqui-Fernandez, A. Marrero, J. Potempa, M. Coll, and F. X. Gomis-Ruth.** 2004. On the transcriptional regulation of methicillin resistance: MecI repressor in complex with its operator. *J. Biol. Chem.* **11:**11.
33. **Garcia-Castellanos, R., A. Marrero, G. Mallorqui-Fernandez, J. Potempa, M. Coll, and F. X. Gomis-Ruth.** 2003. Three-dimensional structure of MecI. Molecular basis for transcriptional regulation of staphylococcal methicillin resistance. *J. Biol. Chem.* **278:**39897–39905.
34. **George, A. M., and S. B. Levy.** 1983. Amplifiable resistance to tetracycline, chloramphenicol, and other antibiotics in *Escherichia coli:* Involvement of a non-plasmid-determined efflux of tetracycline. *J. Bacteriol.* **155:**531–540.
35. **George, A. M., and S. B. Levy.** 1983. Gene in the major cotransduction gap of the *Escherichia coli* K-12 linkage map required for the expression of chromosomal resistance to tetracycline and other antibiotics. *J. Bacteriol.* **155:**541–548.
36. **Glaser, P., L. Frangeul, C. Buchrieser, C. Rusniok, A. Amend, F. Baquero, P. Berche, H. Bloecker, P. Brandt, T. Chakraborty, A. Charbit, F. Chetouani, E. Couve, A. de Daruvar, P. Dehoux, E. Domann, G. Dominguez-Bernal, E. Duchaud, L. Durant, O. Dussurget, K.-D. Entian, H. Fsihi, F. G.-D. Portillo, P. Garrido, L. Gautier, W. Goebel, N. Gomez-Lopez, T. Hain, J. Hauf, D. Jackson, L.-M. Jones, U. Kaerst, J. Kreft, M. Kuhn, F. Kunst, G. Kurapkat, E. Madueno, A. Maitournam, J. M. Vicente, E. Ng, H. Nedjari, G. Nordsiek, S. Novella, B. de Pablos, J.-C. Perez-Diaz, R. Purcell, B. Remmel, M. Rose, T. Schlueter, N. Simoes, A. Tierrez, J.-A. Vazquez-Boland, H. Voss, J. Wehland, and P. Cossart.** 2001. Comparative genomics of Listeria species. *Science* **294:**849–852.
37. **Hagman, K. E., W. Pan, B. G. Spratt, J. T. Balthazar, R. C. Judd, and W. M. Shafer.** 1995. Resistance of *Neisseria gonorrhoeae* to antimicrobial hydrophobic agents is modulated by the *mtrRCDE* efflux system. *Microbiology* **141:**611–622.
38. **Hagman, K. E., and W. M. Shafer.** 1995. Transcriptional control of the *mtr* efflux system of *Neisseria gonorrhoeae. J. Bacteriol.* **177:**4162–4165.
39. **Ingavale, S. S., W. Van Wamel, and A. L. Cheung.** 2003. Characterization of RAT, an autolysis regulator in *Staphylococcus aureus. Mol. Microbiol.* **48:**1451–1466.
40. **Jefferson, K. K., D. B. Pier, D. A. Goldmann, and G. B. Pier.** 2004. The teicoplanin-associated locus regulator (TcaR) and the intercellular adhesin locus regulator (IcaR) are transcriptional inhibitors of the *ica* locus in *Staphylococcus aureus. J. Bacteriol.* **186:**2449–2456.
41. **Jerse, A. E., N. D. Sharma, A. N. Simms, E. T. Crow, L. A. Snyder, and W. M. Shafer.** 2003. A gonococcal efflux pump system enhances bacterial survival in a female mouse model of genital tract infection. *Infect. Immun.* **71:**5576–5582.
42. **Kaneko, A., M. Mita, K. Sekiya, H. Matsui, K. Kawahara, and H. Danbara.** 2002. Association of a regulatory gene, *slyA* with a mouse virulence of *Salmonella enterica* serovar Choleraesuis. *Microbiol. Immunol.* **46:**109–113.
43. **Kern, W. V., M. Oethinger, A. S. Jellen-Ritter, and S. B. Levy.** 2000. Non-target gene mutations in the development of fluoroquinolone resistance in *Escherichia coli. Antimicrob. Agents Chemother.* **44:**814–820.
44. **Kupferwasser, L. I., M. R. Yeaman, C. C. Nast, D. Kupferwasser, Y. Q. Xiong, M. Palma, A. L. Cheung, and A. S. Bayer.** 2003. Salicylic acid attenuates virulence in endovascular infections by targeting global regulatory pathways in Staphylococcus aureus. *J. Clin. Invest.* **112:**222–233.
45. **Kupferwasser, L. I., M. R. Yeaman, S. M. Shapiro, C. C. Nast, P. M. Sullam, S. G. Filler, and A. S. Bayer.** 1999. Acetylsalicylic acid reduces vegetation bacterial density, hematogenous bacterial dissemination, and frequency of embolic events in experimental *Staphylococcus aureus* endocarditis through antiplatelet and antibacterial effects. *Circulation* **99:**2791–2797.
46. **Lee, E. H., C. Rouquette-Loughlin, J. P. Folster, and W. M. Shafer.** 2003. FarR regulates the farAB-encoded efflux pump of Neisseria gonorrhoeae via an MtrR regulatory mechanism. *J. Bacteriol.* **185:**7145–7152.
47. **Lee, E. H., and W. M. Shafer.** 1999. The *farAB*-encoded efflux pump mediates resistance of gonococci to long-chained antibacterial fatty acids. *Mol. Microbiol.* **33:**839–845.
48. **Li, R., A. C. Manna, S. Dai, A. L. Cheung, and G. Zhang.** 2003. Crystal structure of the SarS protein from *Staphylococcus aureus sarU*, a *sarA* homolog, is repressed by SarT and regulates virulence genes in *Staphylococcus aureus. J. Bacteriol.* **185:**4219–4225.
49. **Libby, S. J., W. Goebel, A. Ludwig, N. Buchmeier, F. Bowe, F. C. Fang, D. G. Guiney, J. G. Songer, and F. Heffron.** 1994.

A cytolysin encoded by Salmonella is required for survival within macrophages. *Proc. Natl. Acad. Sci. USA* **91:**489–493.

50. **Lim, D., K. Poole, and N. C. Strynadka.** 2002. Crystal structure of the MexR repressor of the *mexRAB-oprM* multidrug efflux operon of *Pseudomonas aeruginosa. J. Biol. Chem.* 277:29253–29259.
51. **Liu, Y., A. Manna, R. Li, W. E. Martin, R. C. Murphy, A. L. Cheung, and G. Zhang.** 2001. Crystal structure of the SarR protein from *Staphylococcus aureus. Proc. Natl. Acad. Sci. USA* **98:**6877–6882.
52. **Lomovskaya, O., and K. Lewis.** 1992. Emr, an *Escherichia coli* locus for multidrug resistance. *Proc. Natl. Acad. Sci. USA* **89:**8938–8942.
53. **Lomovskaya, O., K. Lewis, and A. Matin.** 1995. EmrR is a negative regulator of the *Escherichia coli* multidrug resistance pump EmrAB. *J. Bacteriol.* **177:**2328–2334.
54. **Luong, T. T., S. W. Newell, and C. Y. Lee.** 2003. Mgr, a novel global regulator in *Staphylococcus aureus. J. Bacteriol.* **185:**3703–3710.
55. **Maki, H., N. McCallum, M. Bischoff, A. Wada, and B. Berger-Bachi.** 2004. *tcaA* inactivation increases glycopeptide resistance in *Staphylococcus aureus. Antimicrob. Agents Chemother.* **48:**1953–1959.
56. **Maneewannakul, K., and S. B. Levy.** 1996. Identification of *mar* mutants among quinolone-resistant clinical isolates of *Escherichia coli. Antimicrob. Agents Chemother.* **40:**1695–1698.
57. **Manna, A., and A. L. Cheung.** 2001. Characterization of *sarR,* a modulator of *sar* expression in *Staphylococcus aureus. Infect. Immun.* **69:**885–896.
58. **Manna, A. C., and A. L. Cheung. 2003.** *sarU,* a *sarA* homolog, is repressed by SarT and regulates virulence genes in *Staphylococcus aureus. Infect. Immun.* **71:**343–353.
59. **Manna, A. C., S. S. Ingavale, M. Maloney, W. van Wamel, and A. L. Cheung.** 2004. Identification of *sarV* (SA2062), a new transcriptional regulator, is repressed by SarA and MgrA (SA0641) and involved in the regulation of autolysis in *Staphylococcus aureus. J. Bacteriol.* **186:**5267–5280.
60. **Martin, R. G., and J. L. Rosner.** 1995. Binding of purified multiple antibiotic-resistance repressor protein (MarR) to *mar* operator sequences. *Proc. Natl. Acad. Sci. USA* **92:**5456–5460.
61. **Martin, R. G., and J. L. Rosner.** 2004. Transcriptional and translational regulation of the marRAB multiple antibiotic resistance operon in *Escherichia coli. Mol. Microbiol.* **53:**183–191.
62. **McCallum, N., M. Bischoff, H. Maki, A. Wada, and B. Berger-Bachi.** 2004. TcaR, a putative MarR-like regulator of *sarS* expression. *J. Bacteriol.* **186:**2966–2972.
63. **McNamara, P. J., K. C. Milligan-Monroe, S. Khalili, and R. A. Proctor.** 2000. Identification, cloning, and initial characterization of *rot,* a locus encoding a regulator of virulence factor expression in *Staphylococcus aureus. J. Bacteriol.* **182:**3197–3203.
64. **Melckebeke, H. V., C. Vreuls, P. Gans, P. Filee, G. Llabres, B. Joris, and J. P. Simorre.** 2003. Solution structural study of BlaI: implications for the repression of genes involved in beta-lactam antibiotic resistance. *J. Mol. Biol.* **333:**711–720.
65. **Mongkolsuk, S., W. Praituan, S. Loprasert, M. Fuangthong, and S. Chamnongpol.** 1998. Identification and characterization of a new organic hydroperoxide resistance (*ohr*) gene with a novel pattern of oxidative stress regulation from *Xanthomonas campestris pv. phaseoli. J. Bacteriol.* **180:**2636–2643.
66. **Novick, R. P.** 2000. Pathogenicity factors and their regulation, p. 392–407. *In* V. A. Fischetti, R. P. Novick, J. J. Ferretti, D. A. Portnoy, and J. I. Rood (ed.), *Gram-Positive Pathogens.* ASM Press, Washington, D.C.
67. **O'Leary, J. O., M. J. Langevin, C. T. Price, J. S. Blevins, M. S. Smeltzer, and J. E. Gustafson.** 2004. Effects of *sarA* inactivation on the intrinsic multidrug resistance mechanism of *Staphylococcus aureus. FEMS Microbiol. Lett.* **237:**297–302.
68. **Oscarsson, J., Y. Mizunoe, B. E. Uhlin, and D. J. Haydon.** 1996. Induction of haemolytic activity in *Escherichia coli* by the *slyA* gene product. *Mol. Microbiol.* **20:**191–199.
69. **Pabo, C. O., and R. T. Sauer.** 1984. Protein-DNA recognition. *Annu. Rev. Biochem.* 53:293–321.
70. **Pai, H., J. Kim, J. H. Lee, K. W. Choe, and N. Gotoh.** 2001. Carbapenem resistance mechanisms in *Pseudomonas aeruginosa* clinical isolates. *Antimicrob. Agents Chemother.* **45:**480–484.
71. **Panmanee, W., P. Vattanaviboon, W. Eiamphungporn, W. Whangsuk, R. Sallabhan, and S. Mongkolsuk.** 2002. OhrR, a transcription repressor that senses and responds to changes in organic peroxide levels in *Xanthomonas campestris* pv. *phaseoli. Mol. Microbiol.* **45:**1647–1654.
72. **Phan-Thanh, L., and F. Mahouin.** 1999. A proteomic approach to study the acid response in *Listeria monocytogenes. Electrophoresis* **20:**2214–2224.
73. **Poole, K., K. Tetro, Q. Zhao, S. Neshat, D. E. Heinrichs, and N. Bianco.** 1996. Expression of the multidrug resistance operon *mexA-mexB-oprM* in *Pseudomonas aeruginosa: mexR* encodes a regulator of operon expression. *Antimicrob. Agents Chemother.* **40:**2021–2028.
74. **Prouty, A. M., I. E. Brodsky, S. Falkow, and J. S. Gunn.** 2004. Bile-salt-mediated induction of antimicrobial and bile resistance in *Salmonella typhimurium. Microbiology* **150:**775–783.
75. **Providenti, M. A., and R. C. Wyndham.** 2001. Identification and functional characterization of CbaR, a MarR-like modulator of the *cbaABC*-encoded chlorobenzoate catabolism pathway. *Appl. Environ. Microbiol.* **67:**3530–3541.
76. **Ramakrishnan, L., H. T. Tran, N. A. Federspiel, and S. Falkow.** 1997. A *crtB* homolog essential for photochromogenicity in *Mycobacterium marinum:* isolation, characterization, and gene disruption via homologous recombination. *J. Bacteriol.* **179:**5862–5868.
77. **Ray, S. S., J. B. Bonanno, H. Chen, H. de Lencastre, S. Wu, A. Tomasz, and S. K. Burley.** 2003. X-ray structure of an *M. jannaschii* DNA-binding protein: implications for antibiotic resistance in *S. aureus. Proteins* **50:**170–173.
78. **Rea, R. B., C. G. Gahan, and C. Hill.** 2004. Disruption of putative regulatory loci in *Listeria monocytogenes* demonstrates a significant role for Fur and PerR in virulence. *Infect. Immun.* **72:**717–727.
79. **Revell, P. A., and V. L. Miller.** 2000. A chromosomally encoded regulator is required for expression of the *Yersinia enterocolitica inv* gene and for virulence. *Mol. Microbiol.* **35:**677–685.
80. **Reverchon, S., W. Nasser, and J. Robert-Baudouy.** 1994. *pecS:* a locus controlling pectinase, cellulase and blue pigment production in *Erwinia chrysanthemi. Mol. Microbiol.* **11:**1127–1139.
81. **Reverchon, S., C. Rouanet, D. Expert, and W. Nasser.** 2002. Characterization of indigoidine biosynthetic genes in *Erwinia chrysanthemi* and role of this blue pigment in pathogenicity. *J. Bacteriol.* **184:**654–665.
82. **Rouanet, C., S. Reverchon, D. A. Rodionov, and W. Nasser.** 2004. Definition of a consensus DNA-binding site for PecS, a global regulator of virulence gene expression in *Erwinia chrysanthemi* and identification of new members of the PecS regulon. *J. Biol. Chem.* **279:**30158–30167.

83. **Said-Salim, B., P. M. Dunman, F. M. McAleese, D. Macapagal, E. Murphy, P. J. McNamara, S. Arvidson, T. J. Foster, S. J. Projan, and B. N. Kreiswirth.** 2003. Global regulation of *Staphylococcus aureus* genes by Rot. *J. Bacteriol.* **185:**610–619.
84. **Saito, K., H. Akama, E. Yoshihara, and T. Nakae.** 2003. Mutations affecting DNA-binding activity of the MexR repressor of *mexR-mexA-mexB-oprM* operon expression. *J. Bacteriol.* **185:**6195–6198.
85. **Saito, K., H. Yoneyama, and T. Nakae.** 1999. *nalB*-type mutations causing the overexpression of the MexAB-OprM efflux pump are located in the *mexR* gene of the *Pseudomonas aeruginosa* chromosome. *FEMS Microbiol. Lett.* **179:**67–72.
86. **Schmidt, K. A., A. C. Manna, and A. L. Cheung.** 2003. SarT influences *sarS* expression in *Staphylococcus aureus. Infect. Immun.* **71:**5139–5148.
87. **Schmidt, K. A., A. C. Manna, S. Gill, and A. L. Cheung.** 2001. SarT, a repressor of alpha-hemolysin in *Staphylococcus aureus. Infect. Immun.* **69:**4749–4758.
88. **Schumacher, M. A., B. K. Hurlburt, and R. G. Brennan.** 2001. Crystal structures of SarA, a pleiotropic regulator of virulence genes in *Staphylococcus aureus. Nature* **414:**85.
89. **Schumacher, M. A., B. K. Hurlburt, R. G. Brennan, T. M. Rechtin, A. F. Gillaspy, and M. S. Smeltzer.** 2001. Crystal structures of SarA, a pleiotropic regulator of virulence genes in *S. aureus:* Characterization of the SarA virulence gene regulator of *Staphylococcus aureus. Nature* **409:**215–219.
90. **Srikumar, R., C. J. Paul, and K. Poole.** 2000. Influence of mutations in the *mexR* repressor gene on expression of the MexA-MExB-OprM multidrug efflux system of *Pseudomonas aeruginosa. J. Bacteriol.* **182:**1410–1414.
91. **Stapleton, M. R., V. A. Norte, R. C. Read, and J. Green.** 2002. Interaction of the *Salmonella typhimurium* transcription and virulence factor SlyA with target DNA and identification of members of the SlyA regulon. *J. Biol. Chem.* **277:**17630–17637.
92. **Sulavik, M. C., L. F. Gambino, and P. F. Miller.** 1994. Analysis of the genetic requirements for inducible multiple-antibiotic resistance associated with the *mar* locus in *Escherichia coli. J. Bacteriol.* **176:**7754–7756.
93. **Sulavik, M. C., L. F. Gambino, and P. F. Miller.** 1995. The MarR repressor of the multiple antibiotic resistance (*mar*) operon in *Escherichia coli:* prototypic member of a family of bacterial regulatory proteins involved in sensing phenolic compounds. *Mol. Med.* **1:**436–446.
94. **Tegmark, K., A. Karlsson, and S. Arvidson.** 2000. Identification and characterization of SarH1, a new global regulator of virulence gene expression in *Staphylococcus aureus. Mol. Microbiol.* **37:**398–409.
95. **Truong-Bolduc, Q. C., X. Zhang, and D. C. Hooper.** 2003. Characterization of NorR protein, a multifunctional regulator of *norA* expression in *Staphylococcus aureus. J. Bacteriol.* **185:**3127–3138.
96. **Ubukata, K., N. Itoh-Yamashita, and M. Konno.** 1989. Cloning and expression of the *norA* gene for fluoroquinolone resistance in *Staphylococcus aureus. Antimicrob. Agents Chemother.* **33:**1535–1539.
97. **Wu, R. Y., R. G. Zhang, O. Zagnitko, I. Dementieva, N. Maltzev, J. D. Watson, R. Laskowski, P. Gornicki, and A. Joachimiak.** 2003. Crystal structure of *Enterococcus faecalis* SlyA-like transcriptional factor. *J. Biol. Chem.* **278:**20240–20244.
98. **Xiong, A., A. Gottman, C. Park, M. Baetens, S. Pandza, and A. Matin.** 2000. The EmrR protein represses the *Escherichia coli emrRAB* multidrug resistance operon by directly binding to its promoter region. *Antimicrob. Agents Chemother.* **44:**2905–2907.
99. **Yoshida, H., H. Bogaki, S. Nakamura, K. Ubukata, and H. Konno.** 1990. Nucleotide sequence and characterization of the *Staphylococcus aureus norA* gene, which confers resistance to quinolones. *J. Bacteriol.* **172:**6942–6949.
100. **Ziha-Zarifi, I., C. Llanes, T. Köhler, J.-C. Pechère, and P. Plésiat.** 1999. In vivo emergence of multidrug-resistant mutants of *Pseudomonas aeruginosa* overexpressing the active efflux system MexA-MexB-OprM. *Antimicrob. Agents Chemother.* **43:**287–291.

Frontiers in Antimicrobial Resistance: a Tribute to Stuart B. Levy
Edited by D. G. White, M. N. Alekshun, and P. F. McDermott

Chapter 19

Role, Structure, and Function of Multidrug Efflux Pumps in Gram-Negative Bacteria

Hiroshi Nikaido

My association with Stuart Levy began in the mid-1960s. It was such a long time ago, and the details of how we met each other are not clear in my mind, especially because I was not working on antibiotic resistance at that time. Possibly our meeting resulted from the fact that we both worked in the laboratory of the late Tsutomu Watanabe, at Keio University, Tokyo, who made a decisive contribution in recognizing that the transferable multidrug resistance in enterics was due to the self-transferring R plasmids. Stuart often claims that in those days I did not believe in the active efflux of tetracyclines. Of course this is not true. What I understood as a biochemist, and what perhaps could have presented some resistance to Stuart, was the notion that the intracellular accumulation of tetracycline at high levels did not necessarily imply the presence of a specific tetracycline influx pump in spite of the claims of some researchers; I remember spending some time in Stuart's laboratory discussing the way the pH difference between the cytoplasm and the medium affected the passive distribution of lipophilic, charged drugs like tetracycline. I had no idea that I would be drawn into the world of antibiotics many years later.

OVERVIEW OF MULTIDRUG EFFLUX PUMPS IN GRAM-NEGATIVE BACTERIA

Genomes of gram-negative bacteria usually contain multiple copies of genes belonging to each of the families of multidrug transporters, such as SMR (small multidrug resistance), MFS (major facilitator superfamily), MATE (multidrug and toxic compound extrusion), and RND (resistance nodulation division); (for an update on the transport gene classification, Milton Saier's transport website [http://www.biology.ucsd.edu/~msaier/transport/] is most useful; for the genome-wide analysis, Ian Paulsen's website [http://www.membranetransport.org] should be consulted). Some of the non-RND transporters pump out dyes, disinfectants, and sometimes antibiotics, but the range of compounds excreted tends to be limited outside the RND family. In contrast, RND pumps often transport an extremely wide range of compounds. Importantly, the substrates of RND transporters often include β-lactams and aminoglycosides, which are rarely subjected to efflux by other classes of transporters. The reported substrates of various RND transporters from gram-negative bacteria are listed in a recent review (31). In addition to the wide substrate range, RND transporters form a multiprotein complex together with the outer membrane (OM) channel of the outer membrane factor (OMF) family (such as TolC in *E. coli*) and the periplasmic linker protein of the membrane fusion protein (MFP) family (such as AcrA in *E. coli*). This construction allows the pump complex to excrete drugs directly into the medium (Fig. 1). Since the drugs must reenter the cell through the OM, an effective permeability barrier, the pump acts synergistically with the OM barrier and this synergy makes RND pumps exceedingly effective in creating drug resistance.

In 1993–1994, we came to the discovery of the *E. coli* AcrAB pump through collaboration with Dzwokai Ma in the laboratory of John Hearst in the Chemistry department of our campus (37) and also learned (32) that the multidrug efflux Xian-Zhi Li found as a phenomenon in *Pseudomonas aeruginosa* (29, 30) was mediated by the MexAB-OprM complex discovered earlier by Keith Poole's group (75). However, we realized only later how such transport complexes work through the extension of the pioneering work from S. Levy's laboratory. McMurry, Levy, and others not only

Hiroshi Nikaido • Department of Molecular and Cell Biology, Room 426 Barker Hall, University of California, Berkeley, CA 94720-3202.

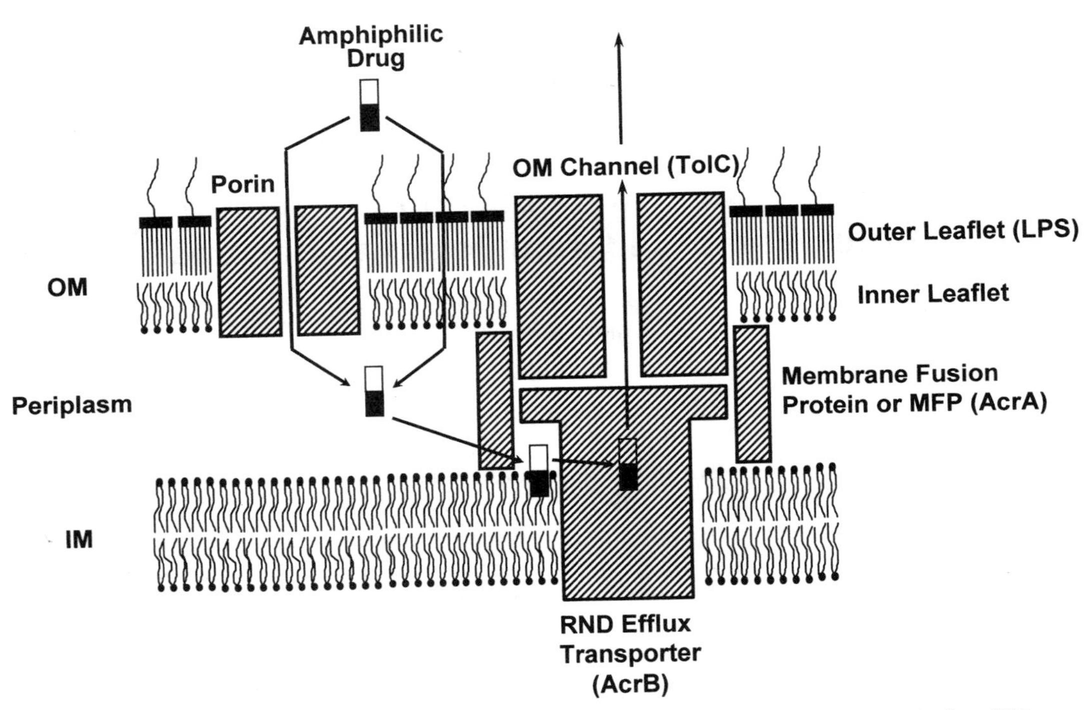

Figure 1. RND pump complex is likely to capture its amphiphilic substrates from the outer leaflet of IM.

discovered that wild-type *E. coli* cells carried out tetracycline efflux (48) but also found that the loss of a large-channel porin OmpF increased the tetracycline resistance (49). Our efforts to understand this porin effect by accumulation experiments and by computer simulation led to the conclusion that simple efflux pumps, such as Tet, excrete the drugs into the periplasmic space, whereas the endogenous pump operating in the susceptible *E. coli* extrude drugs directly into the media (95). This endogenous pump, which turned out to be AcrB (72), thus works synergistically with the OM permeability barrier (60, 62), a characteristic that is shared by other RND pumps of gram-negative bacteria. (Thus permeabilization of the OM exerts a sensitization effect, very similar to that caused by the inactivation of the AcrB pump, on the intact cell of gram-negative bacteria [64]). This explains why the multidrug-efflux-based resistance can become a major mechanism of resistance in *P. aeruginosa* with its OM of exceptionally low permeability. When carbenicillin began to be used for the treatment of *P. aeruginosa* infection, almost 80% of the carbenicillin-resistant clinical isolates from the British Isles did not produce carbenicillin-hydrolyzing β-lactamase and presumably owed their resistance to increased efflux (105). In a more recent example, a survey of levofloxacin-resistant isolates of *P. aeruginosa* from Japan revealed that the increased activity of the MexAB-OprM system plays a significant role in resistance in 96% of the strains (D. Cho, J. Blais, K. Tangen et al., *Progr. Abstr. 39th Intersci. Conf. Antimicrob. Agents Chemother. 1999,* p. 327).

Because of the predominant role RND pumps play in the resistance (both intrinsic and acquired) to commonly used antibiotics, this chapter will devote more space for the pumps of this type. Also because of the restriction in the length of this chapter, the discussion will be limited to *E. coli* and *P. aeruginosa,* as representative organisms. For a more extensive treatment, a recent review (31) should be consulted. A brief description of non-RND efflux pumps in these species will be given below.

A number of non-RND multidrug pumps are present in *E. coli.* EmrAB is an MFS efflux system that operates with the TolC channel and contributes to the intrinsic resistance of *E. coli* to nalidixic acid and carbonyl cyanide *m*-chlorophenylhydrazone (CCCP), a proton conductor (17, 34). Another chromosomally encoded MFS transporter, MdfA (also known as CmlA, Cmr), is a multidrug pump, but its substrate range is limited. It provides resistance to chloramphenicol (66), a compound that is often inactivated enzymatically by a specific chloramphenicol acetlytransferase (CAT), coded by a plasmid gene. MdfA-mediated chloramphenicol/H^+ antiport has been experimentally demonstrated (51). An SMR transporter, EmrE of *E. coli* is a well-studied efflux pump that will be discussed in detail below. Besides EmrE, *E. coli* contains at least four other SMR transporters. Of them, SugE needs to be overexpressed to reveal its export function (11).

Non-RND efflux pumps in *P. aeruginosa* have received less attention, presumably because of the very dominant roles RND pumps play in this organism. However, an Smr transporter has been shown to catalyze the efflux of not only dyes but also, rather surprisingly, aminoglycosides (33).

ROLE OF RND PUMPS IN ANTIBIOTIC RESISTANCE

Intrinsic Resistance

It is well-known that most gram-negative bacteria are inherently more resistant to antimicrobials than gram-positive bacteria. When Vaara (99) examined the natural antibiotics discovered in recent years, he found that about 90% of those had activity only against gram-positive bacteria, showing no or little activity against *E. coli*. The situation is not much different among chemically synthesized agents. The most recent of these compounds, oxazolidinones, show a strong activity against gram-positive bacteria but not against gram-negatives (85).

Most of this characteristic intrinsic resistance of gram-negative bacteria is caused by multidrug efflux through RND pumps, which are widely distributed among these bacteria and function in synergy with the OM barrier, as mentioned already. When the *acrAB* genes, coding for the constitutively expressed RND pump that contributes most of the efflux activity to important antibiotics in *E. coli* (72), are inactivated, compounds like cloxacillin, whose efficacy used to be thought as limited to gram-positives, suddenly become quite effective against *E. coli* and *Salmonella enterica* serovar Typhimurium as shown by the precipitous decrease of their MIC from 256 to 512 μg/ml to 2 μg/ml (47, 65). (Intrinsic resistance to oxazolidinones, mentioned above, is also due largely to the efflux by AcrB and its homologs [J. M. Buysse et al., *Progr. Abstr. 36th Intersci. Conf. Antimicrob. Agents Chemother. 1996*, abstract C42.]). Similar, large decreases in the MIC of novobiocin, erythromycin, fusidic acid, SDS, and cationic dyes also occur (36, 37), showing that efflux through AcrB is almost entirely responsible for the intrinsic resistance of enterics against these compounds. Similarly, in *P. aeruginosa*, inactivation of the main, constitutively expressed RND pump complex MexAB-OprM results in a strong decrease in MIC to such compounds as novobiocin, norfloxacin, tetracycline, carbenicillin, and azlocillin (30).

One often-asked question is the physiological function of RND multidrug efflux pumps, especially those that are expressed constitutively. For *E. coli*, in vitro reconstitution of AcrB (112) into proteoliposomes allowed us to make an estimate of affinity of AcrB to various substrates. This study showed that conjugated bile acids, such as taurocholic acid and glycocholic acids, have an apparent affinity at least one order of magnitude higher than antibiotics such as erythromycin and cloxacillin, suggesting that AcrB was optimized, during evolution, to pump out mainly bile acids, the major toxic chemical in its natural habitat. For soil and water dwellers, it is difficult to pinpoint the "natural" substrate for their main efflux pumps. However, gram-negative bacteria have always faced toxic chemicals in the environment, and OM cannot completely shut out lipophilic compounds, which can diffuse, albeit slowly (73), across the asymmetric bilayer of OM. Furthermore, in any environment there is probably more than one kind of toxic chemical. Thus it is reasonable to assume that unicellular organisms had to develop wide-specificity (multidrug) exporters in the very early stage of evolutionary history. The presence of RND transporters in all major kingdoms of life (97) strongly supports this hypothesis.

The OM of *P. aeruginosa* (and probably the OM of other fluorescent pseudomonads) is a very effective permeability barrier for hydrophilic compounds because of the low permeability of their porin channel (61). However, even in this case most drug molecules can reach half-equilibration across OM in a few minutes or less (61). Thus the presence of multidrug pumps is needed for the high intrinsic level of resistance in this organism, and the RND pumps, which work synergistically with the OM barrier, are especially effective in the organisms of this group (62, 63).

Examples of RND Pumps

E. coli and its relatives

Living in a natural habitat surrounded by high concentrations of bile salts and other antimicrobial inhibitors such as fatty acids, *E. coli* cells are armed with the OM as well as a wide range of efflux pumps. A survey of the *E. coli* genome revealed the presence of at least 37 either single-drug or multidrug, putative or proven, efflux transporters, which include 7ABC, 19 MFS, 1 MATE, 5 SMR, and 7 RND transporters (67). Nevertheless, the tripartite RND-type AcrAB-TolC system is the predominant pump in terms of efflux of commonly used antimicrobial agents (60).

In addition to the AcrAB pump, which is constitutively expressed, *E. coli* also possesses other RND transporters such as AcrEF, AcrD, YhiUV, and MdtABC that were demonstrated to extrude antibiotics. All these systems appear to require TolC as the OM component (67). (Although we reported that AcrD does not require TolC [81], a later study showed that TolC was needed for drug extrusion by AcrD [15]). Inactivation of the *acrEF, yhiUV,* and *mdtABCD* (*yegMNOB*) genes does not change drug susceptibility of wild-type *E. coli* under the standard laboratory growth conditions, indicating that these pumps are not expressed to a significant extent in wild-type cells (8, 38, 67, 91). Deletion of *acrD* renders mutants hypersusceptible to aminoglycosides (81). YhiUV-TolC overexpression is associated with resistance to doxorubicin, erythromycin, deoxycholate, and crystal violet, and MdtABC mediates resistance to bile salts and novobiocin (8, 56). Intriguingly, the MdtABC system contains two RND transporters, MdtB and MdtC, which are both required for efflux.

P. aeruginosa

Six RND-type multidrug efflux systems have been identified in *P. aeruginosa:* MexAB-OprM (75), MexCD-OprJ (74), MexEF-OprN (24), MexXY (also referred as MexGH- or AmrAB)-OprM (3, 52, 104), MexJK-OprM (10), and MexGHI-OpmD (2). Efflux operon for each of these encodes for an RND transporter (MexB, MexD, MexF, MexX, MexK, or MexI) in the inner (cytoplasmic) membrane (IM), a periplasmic MFP (MexA, MexC, MexE, MexY, MexJ, or MexH), and in many cases an OM channel protein (OprM, OprJ, OprN, or OpmD). In addition, the *P. aeruginosa* genome sequence reveals the presence of several additional RND-type systems (90). Of these, MexAB-OprM is expressed constitutively and has a major role in the intrinsic resistance of *P. aeruginosa* to commonly used antimicrobial agents. Since these pumps will be described in detail in chapter 26, only a brief outline will be given below.

MexAB-OprM. The overexpression of MexAB-OprM, causing acquired, elevated levels of multidrug resistance, is seen in at least two types of mutants, *nalB* and *nalC,* selected either in vitro or in vivo (42, 43, 89, 114). The overproduction or deletion of the MexAB system, however, does not strongly affect the MIC of imipenem, a carbapenem (46). Imipenem is likely a poor substrate of the pump. Possibly the presence of the OprD channel in the OM, which permits rapid penetration of imipenem (96), further weakens the effect of efflux. The antipseudomonal activity of other carbapenems such as panipenem and meropenem can, however, be strongly influenced by the MexAB-OprM pump (42, 70, 71).

MexCD-OprJ. MexCD-OprJ is apparently not expressed in wild-type *P. aeruginosa* under laboratory conditions (74, 88). Overexpression of this operon in *nfxB*-type mutants significantly increases resistance to quinolones, tetracycline, chloramphenicol, and fourth-generation cephems such as cefepime and cefpirome (46, 74). The *nfxB*-type mutants show variability in antibiotic resistance patterns, which can be classified into type A and type B. Type A mutants are resistant to ofloxacin, erythromycin, and new zwitterionic cephems (i.e., cefpirome, cefclidin, cefozopran, and cefoselis), and type B mutants are resistant to these agents as well as to tetracycline and chloramphenicol. Type B mutants, however, are four to eight times more susceptible to many conventional penicillins (e.g., carbenicillin), atypical β-lactams (e.g., moxalactam and aztreonam), carbapenems (e.g., imipenem and biapenem), and aminoglycosides (e.g., gentamicin and kanamycin) than the wild-type PAO1 (41). This hypersusceptibility is probably due to the down-regulation of the MexAB-OprM system (18, 28) and the AmpC β-lactamase (44) in the MexCD-OprJ-overproducing mutants. MexCD-OprJ expression can also be induced by some nonantibiotic agents, including ethidium bromide, acriflavin, tetraphenylphophonium and rhodamine 6G, all substrates of MexCD-OprJ (53).

MexEF-OprN. The *mexEF-oprN* system also is not expressed in wild-type strains of *P. aeruginosa* (24). MexEF-OprN is highly expressed in *nfxC* mutants, which show increased resistance to chloramphenicol, quinolones, and trimethoprim (24). Imipenem resistance is also seen in *nfxC* mutants, though this is apparently due to the down-regulation of OprD expression (69).

MexXY-OprM. Without a gene for an OM protein linked to the *mexXY* operon, the MexXY system utilizes OprM of the MexAB-OprM system (3, 52). Initial expression study in *E. coli* showed resistance to fluoroquinolones and macrolides (52); then MexXY, together with OprM, was shown to provide natural resistance to aminoglycoside antibiotics in *P. aeruginosa* (3). Overexpression of the MexXY-OprM pump is likely responsible for aminoglycoside resistance in "impermeability type" clinical isolates (104).

MexJK-OprM and MexGHI-OpmD. MexJK pump is not expressed in wild-type cells (10). While MexJK requires OprM for extruding ciprofloxacin, erythromycin, and tetracycline, it functions apparently independently of OprM for triclosan resistance. MexGHI-OpmD contains an MFP, MexH (mistakenly called an RND transporter); an RND transporter, MexI (mistakenly called an MFP member); an OM channel, OpmD; and, in addition, a small integral membrane protein of unknown function, MexG (2). This system is operative in wild-type cells and mediates resistance to vanadium (2) (vanadyl cation [$VO2^{2+}$] was added to the medium). Interestingly, *mexGHI-opmD* null mutants show increased resistance to tetracycline, netilmicin, and ticarcillin plus clavulanic acid, and this may be due to the compensating overexpression of other MDR pumps (2, 28). Possibly the same explanation applies to the lowered production in the growing culture of *N-acyl* homoserine lactones (2), and the data do not support the authors' conclusion that this system is involved in the transport of these signaling compounds.

Enhanced Resistance Due to Overexpression of RND Pumps

When the expression level of RND pumps becomes increased, this creates increased resistance over the level mediated by their normal level of expression. Such increases are frequently due to regulatory mutations. Spontaneous antibiotic-resistant mutants can often be selected in vitro on plates containing either a single antibiotic or multiple antibiotics. At least four types of multidrug resistant mutants of *P. aeruginosa* overexpressing efflux systems were selected in vitro, including the *nalB* or *nalC* type mutants overproducing MexAB-OprM, the *nfxB* type overproducing MexCD-OprJ, and the *nfxC* type overproducing MexEF-OprN. As is discussed below, similar mutations are seen among resistant isolates from clinical sources. Importantly, MDR mutants can also be selected by antiseptics or other toxic chemicals. For example, efflux-based MDR mutants of *E. coli* and *P. aeruginosa* were selected by triclosan, an antiseptic used in soaps, toothpastes, plastics, and even included in a commercially available selective growth medium for *P. aeruginosa* (9, 50). The relationship between biocide usage and antibiotic resistance is discussed in chapter 12.

A very brief outline of regulatory mechanisms for the pump expression is given here. A more detailed description can be found in a recent review (19) as well as in chapters 13 to 17. One consideration is that multidrug pumps of very broad specificity may even pump out normal metabolites, a possibility recently borne out with AcrAB (20). The overexpression of AcrAB was indeed found to be toxic for *E. coli* already in the first cloning attempt of the corresponding genes (37). Perhaps this toxicity explains why the expression of many pumps is controlled by an elaborate mechanism.

The expression of many multidrug pumps is controlled by local regulators, mostly repressors. These regulators play a predominant role in the expression of the RND multidrug efflux system of *P. aeruginosa.* Thus the gene *mexR,* located upstream of *mexAB-oprM,* encodes a repressor of the MarR family. "*nalB*" strains overproducing MexAB-OprM often carry mutations in *mexR* (1, 89). MexCD-OprJ expression is regulated by the NfxB repressor, encoded by a gene located upstream of the *mexCD-oprJ* operon (74). Expression of the *mexXY* operon is controlled by the TetR-like repressor MexZ, which is encoded by *mexZ,* which lies upstream of the *mexXY* genes (3). (MexXY is a rare drug efflux system that is known to be induced by the presence of its substrates in the medium [45]; the mechanism is not clear at present). The *mexEF-oprN* operon is positively regulated by a transcriptional activator, MexT (23). Intriguingly, the *mexT* gene is not altered in the *nfxC* mutants that overproduce MexEF-OprN; thus it is likely that the *nfxC* gene is located far away from the *mexT/mexEF-oprN* complex (23).

Regulatory mutations leading to the overexpression of RND pumps often play a decisive role in the development of resistant *P. aeruginosa* strains in a clinical setting. In an elegant study, 21 pre- and post-therapy pairs of *P. aeruginosa* isolates were analyzed (114). Ten posttherapy isolates with resistance to only β-lactams overexpressed AmpC β-lactamase, and the other 11 isolates, which had increased resistance to both β-lactams and non-β-lactams, overexpressed the MexAB-OprM system. Of the latter class, 10 out of 11 had mutations in the *mexR* gene. In a similar study of *P. aeruginosa* clinical isolates showing increased levels of resistance to aminoglycosides, all strains were found to overproduce MexXY (100), and many of them were found to contain mutations in the repressor gene *mexZ.* A few, however, did not have mutations in *mexZ* or its promoter, suggesting that the MexXY expression is also regulated by a currently unidentified gene elsewhere.

In *E. coli, acrR,* which is divergently transcribed from the *acrAB* genes, encodes a repressor (35). Perhaps because the inactivation of AcrR results in a too high level of production of AcrAB, it seems to serve only a subsidiary role in the regulation of AcrAB expression. However, analyses of highly fluoroquinolone-resistant clinical isolates of *E. coli* (101, 103) led to the discovery of *acrR* mutations causing AcrAB overexpression in these strains.

The AcrAB pump expression is predominately regulated by global regulators, especially by the XylS/AraC-type activators, MarA, SoxS, and Rob. These systems, whose study was pioneered by the Levy laboratory (12) and which play a large role in the overexpression of AcrAB pumps in resistant strains of clinical origin, will be discussed in detail in chapters 13 to 17 of this book. I note only that MarA and SoxR are unusually small-sized members of the XylS/AraC family that contain only the DNA-binding domains. Thus the regulation through these two proteins must occur at the level of their expression. The MarR repressor controls the expression of MarA, and the binding of MarR to DNA is inhibited by some inducers such as salicylate (5, 40). Although MarA regulates many genes, its involvement in antibiotic resistance, often referred as Mar phenotype, is mostly explained by the increased expression of AcrAB (72). Interestingly, *E. coli* strains carrying a null mutation in *mppA*, a gene encoding a periplasmic murein peptide-binding protein, overproduced MarA, and thus displayed a Mar phenotype (27).

Another small MarA homologue, SoxS, also increases the transcription of *acrAB* and thus mediates elevated multidrug resistance (36). Again, because SoxS contains only the DNA-binding domain, regulation via the *sox* system takes place only by the alteration of the steady-state levels of SoxS in the cell, and increases in SoxS occur by oxidation of the SoxR repressor (containing FeS centers) caused by the presence of superoxide in the environment (21). This mechanism allows *E. coli* to increase its antibiotic resistance in the presence of superoxide radicals, for example in tissues with the migration of polymorphonuclear leukocytes.

In contrast to MarA and SoxS, Rob is twice as large and contains a domain in addition to the DNA-binding domain. Rob was found to be involved in the regulation of the *acrAB* operon, because overproduction of plasmid-coded Rob resulted in resistance to organic solvents (6, 57). However, Rob appears to be synthesized constitutively, unlike MarA and SoxS. Recently, the transcriptional activation of *mar-sox* regulon genes (including *acrAB*) by Rob was shown to occur through the binding of inducers such as dipyridyl (82) and medium-chain fatty acids and some bile acids (80). In the latter case, the lipophilic inhibitors that are present in the normal habitat of *E. coli* were indeed shown to increase the MIC of amphiphilic antibiotics only in those cells producing the intact Rob protein.

SdiA, an *E. coli* protein that is homologous to the receptor of acyl homoserine lactone quorum-sensing signal, was found to regulate positively the AcrAB expression (76). However, *E. coli* K-12 lacks the genes for the production of acyl homoserine lactones, and it is not clear what signal SdiA is responding to in pure cultures of K-12.

Some multidrug pumps in *E. coli* are regulated by two-component systems. Thus BaeSR regulates the expression of RND transporter MdtABC (8, 56), and EvgAS regulates RND transporter YhiUV (68). In both cases we do not know the nature of the stimuli recognized by the sensor protein of the two-component systems.

Inhibitors of RND Pumps

Potent inhibitors of RND efflux pumps have been developed by the collaborative program between Microcide Pharmaceuticals and Daiichi Pharmaceuticals. Some of the compounds developed by using the initial lead compound Phe-Arg-β-naphthylamide (MC 207,110) could decrease the levofloxacin MICs for RND-pump-overproducing *P. aeruginosa* up to 64-fold and were nearly equally effective in inhibiting the MexABOprM, MexCDOprJ, and MexEFOprN systems (78) and RND pumps from some other gram-negative bacteria. Optimization produced stable compounds with decreased acute toxicity (77, 79), but the diamine structure, required for activity, resulted in tissue accumulation (102) of this class of broad-spectrum inhibitors. Inhibitors specific for MexAB-OprM system have also been discovered (58).

STRUCTURAL BASIS OF EFFLUX FUNCTION

SMR Transporters

SMR transporters typically extrude lipophilic cations, such as methylviologen, ethidium, acriflavin, benzalkonium, and tetraphenylphosphonium (TPP^+). Because they are small proteins (12 kD in EmrE of *E. coli*) each containing only four transmembrane (TM) helices, they have been studied intensively in terms of mechanism and structure especially by the laboratory of Schuldiner (for a review, see reference 84). There is a highly conserved glutamate residue (Glu 14 in EmrE) in the middle of the TM helix 1, and this residue was shown to be the only acidic residue needed for the function (107). This residue has an unusually high pK_a of about 7.5 (55) and is involved in the binding of the cationic substrates when deprotonated (108). Recently, the binding of substrates to this binding site was shown to result in the release of a proton in a detergent-solubilized EmrE preparation (86). Thus the action mechanism of this transporter can be described in a simplified (or oversimplified) hypothesis as follows. At the cytoplasmic side, the high cytoplasmic pH deprotonates Glu 14, and the cationic substrate is bound to this site. This binding produces a conformational change in EmrE (as shown experimentally in two-dimensional crystals [93]), and the Glu 14 (with the bound ligand) now becomes exposed to the periplasm. The acidic pH in the periplasm results in the protonation of Glu 14, simultaneously releasing the ligand.

Obviously the specific binding of ligands, conformational alteration of the protein, and the release of ligands would not be easily achieved with a protein with only four TM helices. Thus EmrE functions as an oligomer, and the necessity for the oligomeric structure was shown, for example, by the construction of heterooligomers in which the nonfunctional mutant protomer inhibits the transport by the functional, wild-type protein (83, 106). The projection (92) and the three-dimensional (98) structures of two-dimensional crystals of EmrE indeed show that the basic unit is an asymmetric dimer. Very interestingly, six of the eight TM helices are tilted in relation to the membrane normal, and they appear to form a funnel-like structure, with TPP^{+} bound in the center. Although the resolution of the model does not permit the identification of individual helices, TM1 must be close to the TM1 of the other protomer, as determined by the cross-linking (87) and site-directed spin labeling (26) studies.

RND Transporters

RND transporters are remarkable in several respects: (i) They appear to form a multiprotein complex containing a periplasmic protein, an MFP, and an OM channel, belonging to the OMF family. Thus the *E. coli* AcrB transporter occurs with an MFP, AcrA (to which it was chemically cross-linked [113]) and with an OMF, TolC (16) (Fig. 1). The *P. aeruginosa* MexB transporter occurs in an operon coding in addition to an MFP, MexA, and an OMF, OprM (75). (ii) Furthermore, synergy with the OM permeability barrier as well as the absence of accumulation of drugs in the periplasm (95) led to the conclusion that the pump complex was designed to extrude drugs directly into the external medium, bypassing the periplasm. These data suggested the construction of the efflux complex depicted in Fig. 1. (iii) These transporters pump out compounds of diverse chemical structures, and some of them display a truly remarkable versatility. For example, the substrates of the AcrB pump include most of the currently used antibiotics, antiseptics, disinfectants, detergents, dyes, and even solvents, which include uncharged as well as anionic, cationic, and zwitterionic compounds (60). MexB transports only a slightly more limited range of compounds (29, 30, 32, 60). However, most of the known substrates of these pumps contain large lipophilic domains and thus are at least amphiphilic (60). (iv) The range of substrates pumped out by RND transporters include β-lactams, which inhibit targets located in the periplasm and some of which cannot cross IM, as well as aminoglycosides, which are very hydrophilic and not expected to cross IM easily. Non-RND pumps rarely catalyze the efflux of these compounds.

The AcrB transporter was purified by using the histidine-rich sequence at its C terminus as a natural His-tag and was reconstituted into proteoliposomes (112). This was not an easy task, as AcrB protein was unstable in octyl-β-D-glucoside, which was needed for reconstitution. Furthermore, conventional drug accumulation assays did not produce positive results, presumably because most of the known drug substrates of AcrB spontaneously diffused across the lipid bilayer membrane thanks to their lipophilicity. A novel assay, in which the translocation of fluorescence-labeled phospholipids into "acceptor liposomes" was followed, was devised and showed that AcrB was indeed a proton/drug antiporter capable of moving even phospholipids (112). Finally, the phospholipid extrusion reaction was inhibited, presumably competitively, by various known substrates of AcrB pump, for example conjugated bile acids such as taurocholate and glycocholate and antibiotics such as cloxacillin and erythromycin. Half-maximal inhibition required only 10 to 20 μM concentrations of the conjugated bile salts, although several hundred micromolar concentrations of the antibiotics were needed for a similar inhibition (112). These results indicated that the AcrB pump was optimized during evolution to extrude bile salts, the major toxic compound *E. coli* faces in its natural habitat, the intestinal tract of higher animals.

Zgurskaya and Nikaido (112) also showed that in the AcrB-containing proteoliposomes, the flux of proton from the more acidic interior to the external medium was accelerated in the presence of known substrates of the transporter, presumably as a consequence of its antiporter mechanism. More recently, Aires and Nikaido (3a) used the proteoliposome reconstitution of the *E. coli* AcrD pump to examine the transport of aminoglycosides, the substrates of this pump (81). Because aminoglycosides are polycationic and hydrophilic, and therefore do not diffuse spontaneously across the lipid bilayer, we could show that the efflux of proton from the vesicles was accompanied by the influx and intravesicular accumulation of aminoglycosides. Although precise determination of molar ratios was not possible, it appeared that the antiport reaction involved a proton to substrate ratio that is not very far from 1.

One surprising feature of RND pumps was the active efflux of β-lactam compounds detected both for AcrB and MexB (30, 72). This is because the targets of action of β-lactams, PBPs, reside in the periplasm and at least the dianionic β-lactams were shown not to cross IM, thus remaining in the periplasm (30, 65). This finding indicated that some compounds are captured by the RND transporters from the periplasm, rather than the intuitively appealing location, the cytoplasm. Even these compounds, however, have to contain lipophilic side-chains for an efficient efflux (65). These results, together with the knowledge that most other substrates of RND pumps were lipophilic or amphiphilic, described above, led to the hypothesis that the substrates at least partially partitioned into the outer leaflet of IM and that the substrate capture occurred preferentially from this location (30, 60) (Fig. 1).

This hypothesis received strong support from the X-ray crystallographic structure of AcrB. Murakami et al. (54) made a major advance in the field by crystallizing His-tagged AcrB as homotrimers and then solving the structure of this large complex, despite limited resolution (3.5 Å) inherent in the crystal structures of these secondary transporters (the structure of the protein is similar to that of the liganded protein, shown in Fig. 2). Each of the protomers consists of a transmembrane domain with 12 membrane-spanning helices and a large periplasmic domain that protrudes by about 70 Å into the periplasm. The latter is composed of the two large periplasmic loops, each containing more than 300 amino acid residues, between transmembrane helices 1 and 2, as well as 7 and 8. The external end of the periplasmic domains contained a funnel-like cavity, and moreover had the dimension to fit the internal end of the TolC channel, which was earlier shown to consist of a β-barrel, which traverses the OM and a 100-Å-long, periplasmic, α-tunnel (25). The authors (54) pointed out that the large central cavity in the transmembrane domain of AcrB is connected to the exterior through vestibules that extend outward at the level of the external leaflet of the bilayer (Fig. 2) and suggested that the drugs enter from the periplasm into the central cavity through these vestibules while still associated with the external leaflet of the membrane bilayer, in accordance with our earlier hypothesis (30, 60). E. W. Yu and others (110) confirmed this idea further by showing that in the cocrystals of AcrB protein with four different ligands, ciprofloxacin, rhodamine 6G, ethidium, and dequalinium, the ligands were bound to the walls of the central cavity, as predicted (see "ciprofloxacin in central cavity" in Fig. 2). Furthermore, the extremely loose association of the ligands with the cavity wall and the frequent absence of acidic amino acid residues, usually expected to neutralize the cationic groups on the lig-

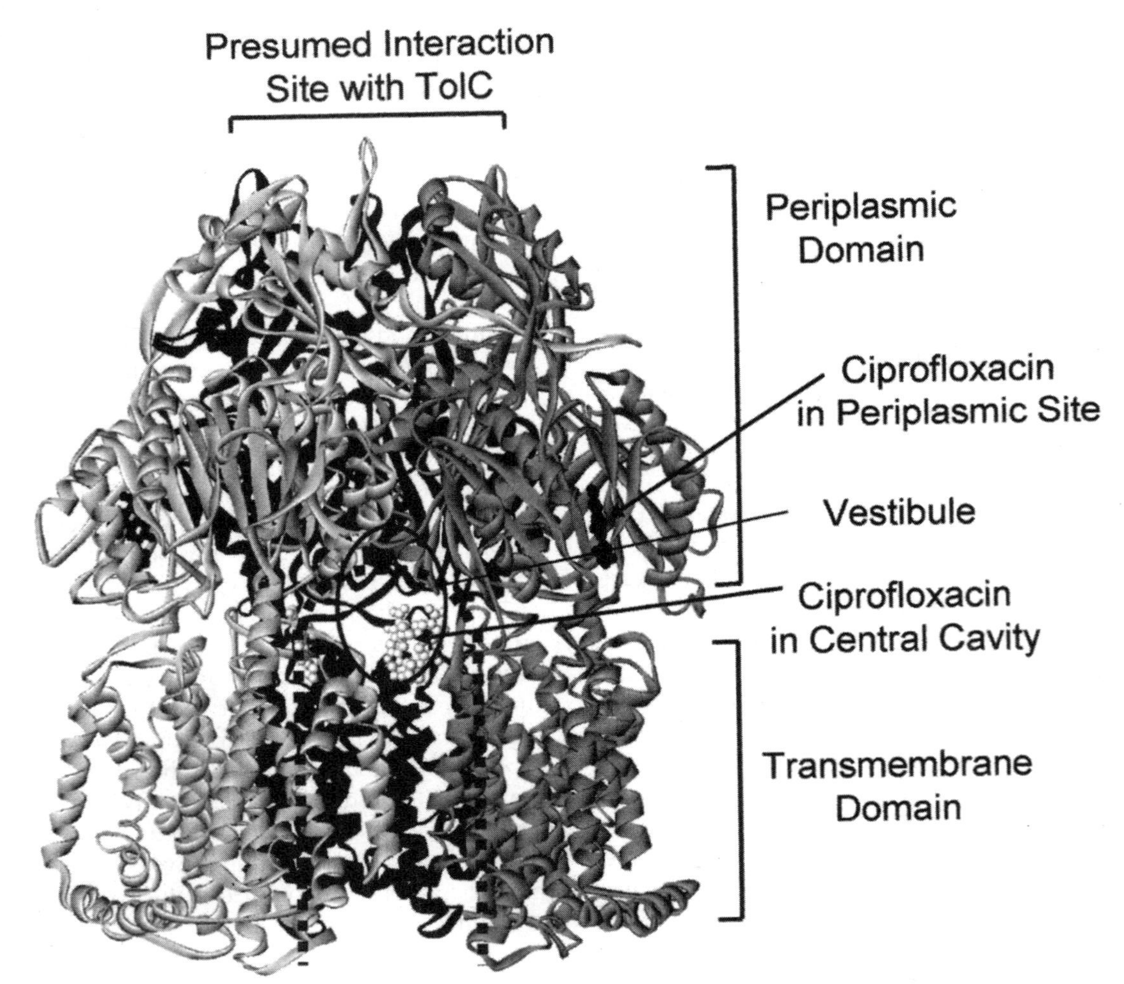

Figure 2. Three-dimensional structure of the AcrB trimer with liganded ciprofloxacin molecules. This structure shows the N109A mutant AcrB trimer, with six molecules of bound ciprofloxacin (Yu, McDermott, and Nikaido, submitted for publication). The three protomers of AcrB are shown as ribbon diagrams in different shades of gray. The ciprofloxacin molecules are shown in ball-and-stick models, those in the central cavity in light gray, and those in the periplasmic site in black. The drawing was made with DS Viewer Pro (Accelrys, San Diego, Calif.).

ands, led to the idea that the drugs are still partially partitioned into the outer leaflet of the lipid bilayer within the cavity and that the charges are neutralized by the acidic head groups of phospholipids (109). The binding site discovered can thus be assumed to be a composite site, involving not only the residues of AcrB but also phospholipids in the cavity. It thus seems most likely that most lipophilic drugs are indeed captured laterally via diffusion through the vestibule (Fig. 3). Aminoglycosides, although hydrophilic, are expected to become adsorbed to the head groups of acidic phospholipids and to diffuse into the central cavity again through the vestibule (Fig. 3).

The demonstration of the ligand binding to the central cavity (110), however, left open the question of how the ligands are transported to the funnel at the outer end of the periplasmic domain in preparation for their extrusion through the TolC channel tunnel. Furthermore, the binding to the central cavity, which is surrounded by transmembrane helices, would suggest that the transmembrane domain would determine the substrate specificity of these pumps. However, earlier genetic experiments (15) showed that the specificity of AcrB and AcrD was nearly entirely determined by their periplasmic loop domain. Moreover, when Mao et al. (39) mapped spontaneous mutations that altered the substrate specificity of MexD and allowed it now to pump out dianionic β-lactams such as carbenicillin, all of the mutations modified residues in the periplasmic domain and none affected residues in the transmembrane domain. These results suggest that the binding to the wall of the central cavity represents only the first step of the drug export process, and drugs then bind to second (and possibly third) binding sites within the periplasmic domain in their passage through the large AcrB protein to reach the opening of the TolC tunnel. Such subsequent binding events would have a large influence on the substrate specificity of the transporter.

Our recent results (E. W. Yu, G. McDermott, and H. Nikaido, submitted for publication) appear to be relevant to the issue just discussed. We purified and crystallized a single amino-acid-substitution mutant of AcrB (N109A), which is largely functional. The X-ray crystallographic structure of the mutant protein is similar to that of the wild-type protein, although there are subtle differences. Most interestingly, when the mutant protein was cocrystallized with five different ligands (ciprofloxacin, rhodamine 6G, and ethidium, which were also used with the wild-type AcrB, together with nafcillin and the pump inhibitor MC 207,110 [Phe-Arg-β-naphthylamide] [78]), we found that the drugs bound not only to the wall of the central cavity, but also to the large depression in the periplasmic domain of the mutant AcrB (see "ciprofloxacin in periplasmic site" in Fig. 2). This depression is the area that was suggested by Murakami et al. (54) as the site of binding of the periplasmic helper protein AcrA; indeed the X-ray crystallography data on our AcrA-AcrB cocrystal suggest that AcrA acts as a

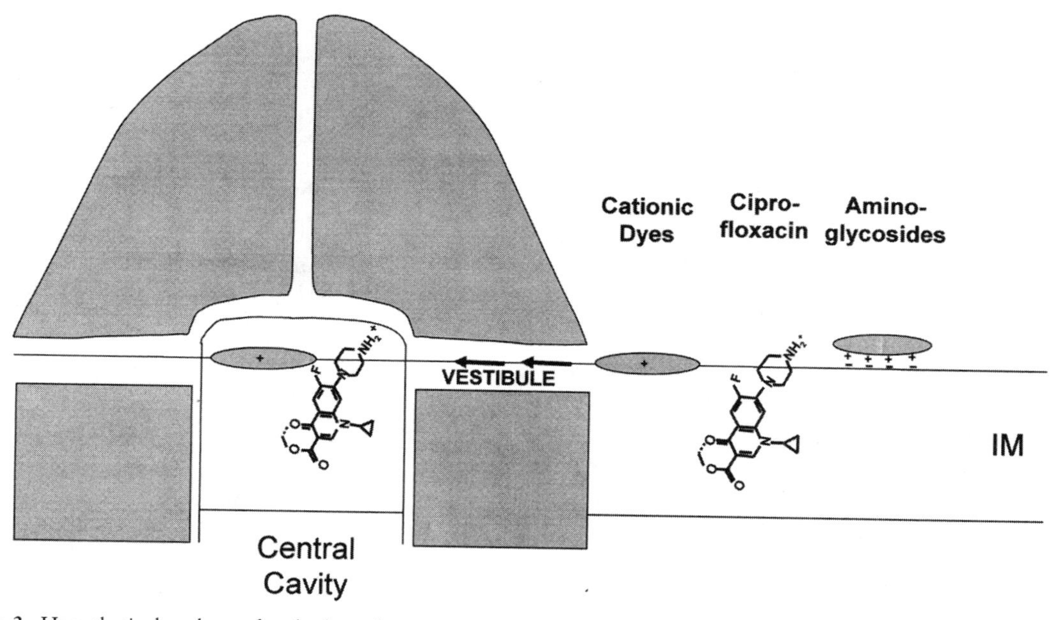

Figure 3. Hypothetical pathway for the lateral capture of substrates by RND group transporters. Cationic dyes and aminoglycosides become adsorbed at the interface region and on the surface of the outer leaflet of IM. Ciprofloxacin is partitioned into IM, with the protonated piperazine amino group near the interface. These substrates diffuse laterally through vestibules into the central cavity, where they become bound to its wall.

"cover" for bound ligands, closing the cavity (E. W. Yu, G. McDermott, H. I. Zgurskaya, and H. Nikaido, unpublished results). Furthermore, some of the spontaneous mutations that altered the substrate specificity of MexD (39) alter amino acid residues right in this area. It thus seems likely that the mutant AcrB protein mimics the transient conformation of the wild-type protein during the movement of substrates, which presumably bind to this new site before their export into the funnel and eventually into the TolC tunnel.

Another interesting feature of the N109A structure is that periplasmic binding of only the pump inhibitor MC 207,110 induces special conformational alterations in the periplasmic domain. We may thus understand eventually how this inhibitor works, and this knowledge may lead us to the design of more effective inhibitors in the future.

Studies on the other components of the RND transporter complex will be mentioned briefly. The X-ray crystallographic structure of TolC was elucidated by Koronakis et al. in 2000 (25). TolC exists as a trimer. It traverses the OM as a 12-stranded β-barrel, each protomer contributing four strands. As mentioned already, the channel in the β-barrel is continued as the central channel in a 100-Å-long α-helical tunnel extending into the periplasm. It has been proposed that the end of the tunnel, which is closed in the crystal structure, could become open through its interaction with the outer end of AcrB. Hydrodynamic (111) and electron diffraction (7) studies showed that AcrA, an MFP, is an elongated protein. Recently, an X-ray crystallographic structure of MexA, an AcrA homolog, has appeared (4). However, the C-terminal domain containing about 100 residues, which is thought to interact with the RND transporter (14, 59), is unfortunately disordered. Site-directed spin labeling study of AcrA (22) not only confirmed the oligomeric nature of this protein but also revealed a flexible conformation affected by an acidic pH.

CONCLUSIONS AND PERSPECTIVES

Among the different classes of multidrug efflux pumps found in gram-negative bacteria, the RND pumps appear to represent the ultimate level of sophistication in their ability to act synergistically with the OM barrier, to extrude the drugs directly into the medium, and to capture the drugs from the periplasm (or the outer leaflet of IM) before their entry into the cytoplasm. The initial binding of the drugs in the vast central cavity of the transporter, using a composite binding site including phospholipids, is probably essential in the unprecedented wide range of substrates that are pumped out by these transporters. Furthermore, the need for any unicellular organism to survive in the presence of many lipophilic toxic compounds means that all species of gram-negative bacteria are equipped with the pumps of this type, allowing them to develop increased resistance to most antimicrobial agents by simply overproducing these pumps.

Another alarming possibility in the future is the dissemination of plasmids coding for the pumps of this type. Already such plasmids have been reported (13, 94). Furthermore, although the substrate specificity of the existing RND pumps may not be optimized for the extrusion of antibiotics (112), selection in the laboratory has shown that single amino acid substitutions can easily lead to the broadening of the range of substrates (39). These findings suggest that a rapid dissemination of improved RND transporters could occur in the future, aided by selection with better, more efficient antimicrobial agents or even with simple disinfectants.

Acknowledgments. I thank the past and present members of the laboratory for their inspired contributions. Thanks are especially due to Xian-Zhi Li, who not only discovered the importance of the multidrug efflux process in antibiotic resistance of *P. aeruginosa*, but authored a major part of an extensive review (31), which was of great help in the writing of this chapter. Research in our laboratory has been supported by a grant from the U.S. Public Health Service (AI-09644 from the National Institute for Allergy and Infectious Diseases).

REFERENCES

1. **Adewoye, L., A. Sutherland, R. Srikumar, and K. Poole.** 2002. The *mexR* repressor of the *mexAB-oprM* multidrug efflux operon in *Pseudomonas aeruginosa:* characterization of mutations compromising activity. *J. Bacteriol.* **184:**4308–4312.
2. **Aendekerk, S., B. Ghysels, P. Cornelis, and C. Baysse.** 2002. Characterization of a new efflux pump, MexGHI-OpmD, from *Pseudomonas aeruginosa* that confers resistance to vanadium. *Microbiology* **148:**2371–2381.
3. **Aires, J. R., T. Kohler, H. Nikaido, and P. Plesiat.** 1999. Involvement of an active efflux system in the natural resistance of *Pseudomonas aeruginosa* to aminoglycosides. *Antimicrob. Agents Chemother.* **43:**2624–2628.

3a. **Aires, J. R., and H. Nikaido.** 2005. Aminoglycosides are captured from both periplasm and cytoplasm by the AcrD multidrug efflux transporter of *Escherichia coli. J. Bacteriol.* **187:** 1923–1929.

4. **Akama, H., T. Matsuura, S. Kashiwagi, H. Yoneyama, S. Narita, T. Tsukihara, A. Nakagawa, and T. Nakae.** 2004. Crystal structure of the membrane fusion protein, MexA, of the multidrug transporter in *Pseudomonas aeruginosa. J. Biol. Chem.* **279:**25939–25942.
5. **Alekshun, M. N., S. B. Levy, T. R. Mealy, B. A. Seaton, and J. F. Head.** 2001. The crystal structure of MarR, a regulator of multiple antibiotic resistance, at 2.3 Å resolution. *Nat. Struct. Biol.* **8:**710–714.
6. **Ariza, R. R., Z. Li, N. Ringstad, and B. Demple.** 1995. Activation of multiple antibiotic resistance and binding of stress-inducible promoters by *Escherichia coli* Rob protein. *J. Bacteriol.* **177:**1655–1661.

7. **Avila-Sakar, A. J., S. Misaghi, E. M. Wilson-Kubalek, K. H. Downing, H. Zgurskaya, H. Nikaido, and E. Nogales.** 2001. Lipid-layer crystallization and preliminary three-dimensional structural analysis of AcrA, the periplasmic component of a bacterial multidrug efflux pump. *J. Struct. Biol.* **136:**81–88.
8. **Baranova, N., and H. Nikaido.** 2002. The *baeSR* two-component regulatory system activates transcription of the *yegMNOB* (*mdtABCD*) transporter gene cluster in *Escherichia coli* and increases its resistance to novobiocin and deoxycholate. *J. Bacteriol.* **184:**4168–4176.
9. **Chuanchuen, R., K. Beinlich, T. T. Hoang, A. Becher, R. R. Karkhoff-Schweizer, and H. P. Schweizer.** 2001. Cross-resistance between triclosan and antibiotics in *Pseudomonas aeruginosa* is mediated by multidrug efflux pumps: exposure of a susceptible mutant strain to triclosan selects nfxB mutants overexpressing MexCD-OprJ. *Antimicrob. Agents Chemother.* **45:**428–432.
10. **Chuanchuen, R., C. T. Narasaki, and H. P. Schweizer.** 2002. The MexJK efflux pump of *Pseudomonas aeruginosa* requires OprM for antibiotic efflux but not for efflux of triclosan. *J. Bacteriol.* **184:**5036–5044.
11. **Chung, Y. J., and M. H. Saier, Jr.** 2002. Overexpression of the *Escherichia coli sugE* gene confers resistance to a narrow range of quaternary ammonium compounds. *J. Bacteriol.* **184:**2543–2545.
12. **Cohen, S. P., H. Hachler, and S. B. Levy.** 1993. Genetic and functional analysis of the multiple antibiotic resistance (*mar*) locus in *Escherichia coli. J. Bacteriol.* **175:**1484–1492.
13. **Dröge, M., A. Pühler, and W. Selbitschka.** 2000. Phenotypic and molecular characterization of conjugative antibiotic resistance plasmids isolated from bacterial communities of activated sludge. *Mol. Gen. Genet.* **263:**471–482.
14. **Elkins, C. A., and H. Nikaido.** 2003. Chimeric analysis of AcrA function reveals the importance of its C-terminal domain in its interaction with the AcrB multidrug efflux pump. *J. Bacteriol.* **185:**5349–5356.
15. **Elkins, C. A., and H. Nikaido.** 2002. Substrate specificity of the RND-type multidrug efflux pumps AcrB and AcrD of *Escherichia coli* is determined predominantly by two large periplasmic loops. *J. Bacteriol.* **184:**6490–6498.
16. **Fralick, J. A.** 1996. Evidence that TolC is required for functioning of the Mar/AcrAB efflux pump of *Escherichia coli. J. Bacteriol.* **178:**5803–5805.
17. **Furukawa, H., J. T. Tsay, S. Jackowski, Y. Takamura, and C. O. Rock.** 1993. Thiolactomycin resistance in *Escherichia coli* is associated with the multidrug resistance efflux pump encoded by emrAB. *J. Bacteriol.* **175:**3723–3729.
18. **Gotoh, N., H. Tsujimoto, M. Tsuda, K. Okamoto, A. Nomura, T. Wada, M. Nakahashi, and T. Nishino.** 1998. Characterization of the MexC-MexD-OprJ multidrug efflux system in Δ*mexA-mexB-oprM* mutants of *Pseudomonas aeruginosa. Antimicrob. Agents Chemother.* **42:**1938–1943.
19. **Grkovic, S., M. H. Brown, and R. A. Skurray.** 2002. Regulation of bacterial drug export systems. *Microbiol. Mol. Biol. Rev.* **66:**671–701.
20. **Helling, R. B., B. K. Janes, H. Kimball, T. Tran, M. Bundesmann, P. Check, D. Phelan, and C. Miller.** 2002. Toxic waste disposal in *Escherichia coli. J. Bacteriol.* **184:**3699–3703.
21. **Hidalgo, E., H. Ding, and B. Demple.** 1997. Redox signal transduction via iron-sulfur clusters in the SoxR transcription activator. *Trends Biochem. Sci.* **22:**207–210.
22. **Ip, H., K. Stratton, H. Zgurskaya, and J. Liu.** 2003. pH-induced conformational changes of AcrA, the membrane fusion protein of *Escherichia coli* multidrug efflux system. *J. Biol. Chem.* **278:**50474–50482.
23. **Köhler, T., S. F. Epp, L. K. Curty, and J. C. Pechère.** 1999. Characterization of MexT, the regulator of the MexE-MexF-OprN multidrug efflux system of *Pseudomonas aeruginosa. J. Bacteriol.* **181:**6300–6305.
24. **Köhler, T., M. Michea-Hamzehpour, U. Henze, N. Gotoh, L. K. Curty, and J. C. Pechère.** 1997. Characterization of MexE-MexF-OprN, a positively regulated multidrug efflux system of *Pseudomonas aeruginosa. Mol. Microbiol.* **23:**345–354.
25. **Koronakis, V., A. Sharff, E. Koronakis, B. Luisi, and C. Hughes.** 2000. Crystal structure of the bacterial membrane protein TolC central to multidrug efflux and protein export. *Nature* **405:**914–919.
26. **Koteiche, H. A., M. D. Reeves, and H. S. McHaourab.** 2003. Structure of the substrate binding pocket of the multidrug transporter EmrE: site-directed spin labeling of transmembrane segment 1. *Biochemistry* **42:**6099–6105.
27. **Li, H., and J. T. Park.** 1999. The periplasmic murein peptide-binding protein MppA is a negative regulator of multiple antibiotic resistance in *Escherichia coli. J. Bacteriol.* **181:**4842–4847.
28. **Li, X. Z., N. Barre, and K. Poole.** 2000. Influence of the MexA-MexB-OprM multidrug efflux system on expression of the MexC-MexD-OprJ and MexE-MexF-OprN multidrug efflux systems in *Pseudomonas aeruginosa. J. Antimicrob. Chemother* **46:**885–893.
29. **Li, X. Z., D. M. Livermore, and H. Nikaido.** 1994. Role of efflux pump(s) in intrinsic resistance of *Pseudomonas aeruginosa:* resistance to tetracycline, chloramphenicol, and norfloxacin. *Antimicrob. Agents Chemother.* **38:**1732–1741.
30. **Li, X. Z., D. Ma, D. M. Livermore, and H. Nikaido.** 1994. Role of efflux pump(s) in intrinsic resistance of *Pseudomonas aeruginosa:* active efflux as a contributing factor to beta-lactam resistance. *Antimicrob. Agents Chemother.* **38:**1742–1752.
31. **Li, X. Z., and H. Nikaido.** 2004. Efflux-mediated drug resistance in bacteria. *Drugs* **64:**159–204.
32. **Li, X. Z., H. Nikaido, and K. Poole.** 1995. Role of *mexA-mexB-oprM* in antibiotic efflux in *Pseudomonas aeruginosa. Antimicrob. Agents Chemother.* **39:**1948–1953.
33. **Li, X.-Z., K. Poole, and H. Nikaido.** 2003. Contributions of MexAB-OprM and an EmrE homolog to intrinsic resistance of *Pseudomonas aeruginosa* to aminoglycosides and dyes. *Antimicrob. Agents Chemother.* **47:**27–33.
34. **Lomovskaya, O., and K. Lewis.** 1992. Emr, an *Escherichia coli* locus for multidrug resistance. *Proc. Natl. Acad. Sci. USA* **89:**8938–8942.
35. **Ma, D., M. Alberti, C. Lynch, H. Nikaido, and J. E. Hearst.** 1996. The local repressor AcrR plays a modulating role in the regulation of *acrAB* genes of *Escherichia coli* by global stress signals. *Mol. Microbiol.* **19:**101–112.
36. **Ma, D., D. N. Cook, M. Alberti, N. G. Pon, H. Nikaido, and J. E. Hearst.** 1995. Genes *acrA* and *acrB* encode a stress-induced efflux system of *Escherichia coli. Mol. Microbiol.* **16:**45–55.
37. **Ma, D., D. N. Cook, M. Alberti, N. G. Pon, H. Nikaido, and J. E. Hearst.** 1993. Molecular cloning and characterization of *acrA* and *acrE* genes of *Escherichia coli. J. Bacteriol.* **175:** 6299–6313.
38. **Ma, D., D. N. Cook, J. E. Hearst, and H. Nikaido.** 1994. Efflux pumps and drug resistance in gram-negative bacteria. *Trends Microbiol.* **2:**489–493.
39. **Mao, W., M. S. Warren, D. S. Black, T. Satou, T. Murata, T. Nishino, N. Gotoh, and O. Lomovskaya.** 2002. On the mechanism of substrate specificity by resistance nodulation division (RND)-type multidrug resistance pumps: the large periplasmic

loops of MexD from *Pseudomonas aeruginosa* are involved in substrate recognition. *Mol. Microbiol.* **46:**889–901.

40. **Martin, R. G., and J. L. Rosner.** 1995. Binding of purified multiple antibiotic-resistance repressor protein (MarR) to mar operator sequences. *Proc. Natl. Acad. Sci. USA* **92:**5456–5460.
41. **Masuda, N., N. Gotoh, S. Ohya, and T. Nishino.** 1996. Quantitative correlation between susceptibility and OprJ production in NfxB mutants of *Pseudomonas aeruginosa. Antimicrob. Agents Chemother.* **40:**909–913.
42. **Masuda, N., and S. Ohya.** 1992. Cross-resistance to meropenem, cephems, and quinolones in *Pseudomonas aeruginosa. Antimicrob. Agents Chemother.* **36:**1847–1851.
43. **Masuda, N., E. Sakagawa, and S. Ohya.** 1995. Outer membrane proteins responsible for multiple drug resistance in *Pseudomonas aeruginosa. Antimicrob. Agents Chemother.* **39:**645–649.
44. **Masuda, N., E. Sakagawa, S. Ohya, N. Gotoh, and T. Nishino.** 2001. Hypersusceptibility of the *Pseudomonas aeruginosa nfxB* mutant to β-lactams due to reduced expression of the *ampC* β-lactamase. *Antimicrob. Agents Chemother.* **45:**1284–1286.
45. **Masuda, N., E. Sakagawa, S. Ohya, N. Gotoh, H. Tsujimoto, and T. Nishino.** 2000. Contribution of the MexX-MexY-oprM efflux system to intrinsic resistance in *Pseudomonas aeruginosa. Antimicrob. Agents Chemother.* **44:**2242–2246.
46. **Masuda, N., E. Sakagawa, S. Ohya, N. Gotoh, H. Tsujimoto, and T. Nishino.** 2000. Substrate specificities of MexAB-OprM, MexCD-OprJ, and MexXY-OprM efflux pumps in *Pseudomonas aeruginosa. Antimicrob. Agents Chemother.* **44:**3322–3327.
47. **Mazzariol, A., G. Cornaglia, and H. Nikaido.** 2000. Contributions of the AmpC beta-lactamase and the AcrAB multidrug efflux system in intrinsic resistance of *Escherichia coli* K-12 to beta-lactams. *Antimicrob. Agents Chemother.* **44:**1387–1390.
48. **McMurry, L. M., D. A. Aronson, and S. B. Levy.** 1983. Susceptible *Escherichia coli* cells can actively excrete tetracyclines. *Antimicrob. Agents Chemother.* **24:**544–551.
49. **McMurry, L. M., J. C. Cullinane, and S. B. Levy.** 1982. Transport of the lipophilic analog minocycline differs from that of tetracycline in susceptible and resistant *Escherichia coli* strains. *Antimicrob. Agents Chemother.* **22:**791–799.
50. **McMurry, L. M., M. Oethinger, and S. B. Levy.** 1998. Overexpression of *marA, soxS,* or *acrAB* produces resistance to triclosan in laboratory and clinical strains of *Escherichia coli. FEMS Microbiol. Lett.* **166:**305–309.
51. **Mine, T., Y. Morita, A. Kataoka, T. Mizushima, and T. Tsuchiya.** 1998. Evidence for chloramphenicol/H^+ antiport in Cmr (MdfA) system of *Escherichia coli* and properties of the antiporter. *J. Biochem.* (Tokyo) **124:**187–193.
52. **Mine, T., Y. Morita, A. Kataoka, T. Mizushima, and T. Tsuchiya.** 1999. Expression in *Escherichia coli* of a new multidrug efflux pump, MexXY, from *Pseudomonas aeruginosa. Antimicrob. Agents Chemother.* **43:**415–417.
53. **Morita, Y., Y. Komori, T. Mima, T. Kuroda, T. Mizushima, and T. Tsuchiya.** 2001. Construction of a series of mutants lacking all of the four major mex operons for multidrug efflux pumps or possessing each one of the operons from *Pseudomonas aeruginosa* PAO1: MexCD-OprJ is an inducible pump. *FEMS Microbiol. Lett.* **202:**139–143.
54. **Murakami, S., R. Nakashima, E. Yamashita, and A. Yamaguchi.** 2002. Crystal structure of bacterial multidrug efflux transporter AcrB. *Nature* **419:**587–593.
55. **Muth, T. R., and S. Schuldiner.** 2000. A membrane-embedded glutamate is required for ligand binding to the multidrug transporter EmrE. *EMBO J.* **19:**234–240.
56. **Nagakubo, S., K. Nishino, T. Hirata, and A. Yamaguchi.** 2002. The putative response regulator BaeR stimulates multidrug resistance of *Escherichia coli* via a novel multidrug exporter system, MdtABC. *J. Bacteriol.* **184:**4161–4167.
57. **Nakajima, H., K. Kobayashi, M. Kobayashi, H. Asako, and R. Aono.** 1995. Overexpression of the *robA* gene increases organic solvent tolerance and multiple antibiotic and heavy metal ion resistance in *Escherichia coli. Appl. Environ. Microbiol.* **61:**2302–2307.
58. **Nakayama, K., Y. Ishida, M. Ohtsuka, H. Kawato, K. Yoshida, Y. Yokomizo, S. Hosono, T. Ohta, K. Hoshino, H. Ishida, T. E. Renau, R. Leger, J. Z. Zhang, V. J. Lee, and W. J. Watkins.** 2003. MexAB-OprM-specific efflux pump inhibitors in *Pseudomonas aeruginosa.* Part 1: Discovery and early strategies for lead optimization. *Bioorg. Med. Chem. Lett.* **13:**4201–4204.
59. **Nehme, D., X. Z. Li, R. Elliot, and K. Poole.** 2004. Assembly of the MexAB-OprM multidrug efflux system of *Pseudomonas aeruginosa:* identification and characterization of mutations in *mexA* compromising MexA multimerization and interaction with MexB. *J. Bacteriol.* **186:**2973–2983.
60. **Nikaido, H.** 1996. Multidrug efflux pumps of gram-negative bacteria. *J. Bacteriol.* **178:**5853–5859.
61. **Nikaido, H.** 1989. Outer membrane barrier as a mechanism of antimicrobial resistance. *Antimicrob. Agents Chemother.* **33:**1831–1836.
62. **Nikaido, H.** 2001. Preventing drug access to targets: cell surface permeability barriers and active efflux in bacteria. *Semin. Cell Dev. Biol.* **12:**215–223.
63. **Nikaido, H.** 1994. Prevention of drug access to bacterial targets: permeability barriers and active efflux. *Science* **264:**382–388.
64. **Nikaido, H.** 1998. The role of outer membrane and efflux pumps in the resistance of gram-negative bacteria. Can we improve drug access? *Drug Res. Updates* **1:**93–98.
65. **Nikaido, H., M. Basina, V. Nguyen, and E. Y. Rosenberg.** 1998. Multidrug efflux pump AcrAB of *Salmonella typhimurium* excretes only those β-lactam antibiotics containing lipophilic side chains. *J. Bacteriol.* **180:**4686–4692.
66. **Nilsen, I. W., I. Bakke, A. Vader, O. Olsvik, and M. R. El-Gewely.** 1996. Isolation of cmr, a novel *Escherichia coli* chloramphenicol resistance gene encoding a putative efflux pump. *J. Bacteriol.* **178:**3188–3193.
67. **Nishino, K., and A. Yamaguchi.** 2001. Analysis of a complete library of putative drug transporter genes in *Escherichia coli. J. Bacteriol.* **183:**5803–5812.
68. **Nishino, K., and A. Yamaguchi.** 2002. EvgA of the two-component signal transduction system modulates production of the *yhiUV* multidrug transporter in *Escherichia coli. J. Bacteriol.* **184:**2319–2323.
69. **Ochs, M. M., M. P. McCusker, M. Bains, and R. E. Hancock.** 1999. Negative regulation of the *Pseudomonas aeruginosa* outer membrane porin OprD selective for imipenem and basic amino acids. *Antimicrob. Agents Chemother.* **43:**1085–1090.
70. **Okamoto, K., N. Gotoh, and T. Nishino.** 2002. Extrusion of penem antibiotics by multicomponent efflux systems MexAB-OprM, MexCD-OprJ, and MexXY-OprM of *Pseudomonas aeruginosa. Antimicrob. Agents Chemother.* **46:**2696–2699.
71. **Okamoto, K., N. Gotoh, and T. Nishino.** 2001. *Pseudomonas aeruginosa* reveals high intrinsic resistance to penem antibiotics: penem resistance mechanisms and their interplay. *Antimicrob. Agents Chemother.* **45:**1964–1971.
72. **Okusu, H., D. Ma, and H. Nikaido.** 1996. AcrAB efflux pump plays a major role in the antibiotic resistance phenotype of

Escherichia coli multiple-antibiotic-resistance (Mar) mutants. *J. Bacteriol.* **178:**306–308.

73. **Plésiat, P., and H. Nikaido.** 1992. Outer membranes of gram-negative bacteria are permeable to steroid probes. *Mol. Microbiol.* **6:**1323–1333.
74. **Poole, K., N. Gotoh, H. Tsujimoto, Q. Zhao, A. Wada, T. Yamasaki, S. Neshat, J. Yamagishi, X. Z. Li, and T. Nishino.** 1996. Overexpression of the *mexC-mexD-oprJ* efflux operon in *nfxB*-type multidrug-resistant strains of *Pseudomonas aeruginosa. Mol. Microbiol.* **21:**713–724.
75. **Poole, K., K. Krebes, C. McNally, and S. Neshat.** 1993. Multiple antibiotic resistance in *Pseudomonas aeruginosa:* evidence for involvement of an efflux operon. *J. Bacteriol.* **175:**7363–7372.
76. **Rahmati, S., S. Yang, A. L. Davidson, and E. L. Zechiedrich.** 2002. Control of the AcrAB multidrug efflux pump by quorum-sensing regulator SdiA. *Mol. Microbiol.* **43:**677–685.
77. **Renau, T. E., R. Leger, L. Filonova, E. M. Flamme, M. Wang, R. Yen, D. Madsen, D. Griffith, S. Chamberland, M. N. Dudley, V. J. Lee, O. Lomovskaya, W. J. Watkins, T. Ohta, K. Nakayama, and Y. Ishida.** 2003. Conformationally-restricted analogues of efflux pump inhibitors that potentiate the activity of levofloxacin in *Pseudomonas aeruginosa. Bioorg. Med. Chem. Lett.* **13:**2755–2758.
78. **Renau, T. E., R. Leger, E. M. Flamme, J. Sangalang, M. W. She, R. Yen, C. L. Gannon, D. Griffith, S. Chamberland, O. Lomovskaya, S. J. Hecker, V. J. Lee, T. Ohta, and K. Nakayama.** 1999. Inhibitors of efflux pumps in *Pseudomonas aeruginosa* potentiate the activity of the fluoroquinolone antibacterial levofloxacin. *J. Med. Chem.* **42:**4928–4931.
79. **Renau, T. E., R. Leger, R. Yen, M. W. She, E. M. Flamme, J. Sangalang, C. L. Gannon, S. Chamberland, O. Lomovskaya, and V. J. Lee.** 2002. Peptidomimetics of efflux pump inhibitors potentiate the activity of levofloxacin in *Pseudomonas aeruginosa. Bioorg. Med. Chem. Lett.* **12:**763–766.
80. **Rosenberg, E. Y., D. Bertenthal, M. L. Nilles, K. P. Bertrand, and H. Nikaido.** 2003. Bile salts and fatty acids induce the expression of *Escherichia coli* AcrAB multidrug efflux pump through their interaction with Rob regulatory protein. *Mol. Microbiol.* **48:**1609–1619.
81. **Rosenberg, E. Y., D. Ma, and H. Nikaido.** 2000. AcrD of *Escherichia coli* is an aminoglycoside efflux pump. *J. Bacteriol.* **182:**1754–1756.
82. **Rosner, J. L., B. Dangi, A. M. Gronenborn, and R. G. Martin.** 2002. Posttranscriptional activation of the transcriptional activator Rob by dipyridyl in *Escherichia coli. J. Bacteriol.* **184:**1407–1416.
83. **Rotem, D., N. Sal-man, and S. Schuldiner.** 2001. In vitro monomer swapping in EmrE, a multidrug transporter from *Escherichia coli,* reveals that the oligomer is the functional unit. *J. Biol. Chem.* **276:**48243–48249.
84. **Schuldiner, S., D. Granot, S. S. Mordoch, S. Ninio, D. Rotem, M. Soskin, C. G. Tate, and H. Yerushalmi.** 2001. Small is mighty: EmrE, a multidrug transporter as an experimental paradigm. *News Physiol. Sci.* **16:**130–134.
85. **Shinabarger, D. L., K. R. Marotti, R. W. Murray, A. H. Lin, E. P. Melchior, S. M. Swaney, D. S. Dunyak, W. F. Demyan, and J. M. Buysse.** 1997. Mechanism of action of oxazolidinones: effects of linezolid and eperezolid on translation reactions. *Antimicrob. Agents Chemother.* **41:**2132–2136.
86. **Soskine, M., Y. Adam, and S. Schuldiner.** 2004. Direct evidence for substrate-induced proton release in detergent-solubilized EmrE, a multidrug transporter. *J. Biol. Chem.* **279:**9951–9955.
87. **Soskine, M., S. Steiner-Mordoch, and S. Schuldiner.** 2002. Crosslinking of membrane-embedded cysteines reveals contact points in the EmrE oligomer. *Proc. Natl. Acad. Sci. USA* **99:**12043–12048.
88. **Srikumar, R., X. Z. Li, and K. Poole.** 1997. Inner membrane efflux components are responsible for β-lactam specificity of multidrug efflux pumps in *Pseudomonas aeruginosa. J. Bacteriol.* **179:**7875–7881.
89. **Srikumar, R., C. J. Paul, and K. Poole.** 2000. Influence of mutations in the *mexR* repressor gene on expression of the MexA-MexB-oprM multidrug efflux system of *Pseudomonas aeruginosa. J. Bacteriol.* **182:**1410–1414.
90. **Stover, C. K., X. Q. Pham, A. L. Erwin, S. D. Mizoguchi, P. Warrener, M. J. Hickey, F. S. Brinkman, W. O. Hufnagle, D. J. Kowalik, M. Lagrou, R. L. Garber, L. Goltry, E. Tolentino, S. Westbrock-Wadman, Y. Yuan, L. L. Brody, S. N. Coulter, K. R. Folger, A. Kas, K. Larbig, R. Lim, K. Smith, D. Spencer, G. K. Wong, Z. Wu, I. T. Paulsen, J. Reizer, M. H. Saier, R. E. Hancock, S. Lory, and M. V. Olson.** 2000. Complete genome sequence of *Pseudomonas aeruginosa* PA01, an opportunistic pathogen. *Nature* **406:**959–964.
91. **Sulavik, M. C., C. Houseweart, C. Cramer, N. Jiwani, N. Murgolo, J. Greene, B. DiDomenico, K. J. Shaw, G. H. Miller, R. Hare, and G. Shimer.** 2001. Antibiotic susceptibility profiles of *Escherichia coli* strains lacking multidrug efflux pump genes. *Antimicrob. Agents Chemother.* **45:**1126–1136.
92. **Tate, C. G., E. R. Kunji, M. Lebendiker, and S. Schuldiner.** 2001. The projection structure of EmrE, a proton-linked multidrug transporter from *Escherichia coli,* at 7 Å resolution. *EMBO J.* **20:**77–81.
93. **Tate, C. G., I. Ubarretxena-Belandia, and J. M. Baldwin.** 2003. Conformational changes in the multidrug transporter EmrE associated with substrate binding. *J. Mol. Biol.* **332:**229–242.
94. **Tauch, A., A. Schluter, N. Bischoff, A. Goesmann, F. Meyer, and A. Puhler.** 2003. The 79,370-bp conjugative plasmid pB4 consists of an IncP-1beta backbone loaded with a chromate resistance transposon, the *strA-strB* streptomycin resistance gene pair, the oxacillinase gene *bla* (NPS-1), and a tripartite antibiotic efflux system of the resistance-nodulation-division family. *Mol. Gen. Genom.* **268:**570–84.
95. **Thanassi, D. G., G. S. Suh, and H. Nikaido.** 1995. Role of outer membrane barrier in efflux-mediated tetracycline resistance of *Escherichia coli. J. Bacteriol.* **177:**998–1007.
96. **Trias, J., and H. Nikaido.** 1990. Outer membrane protein D2 catalyzes facilitated diffusion of carbapenems and penems through the outer membrane of *Pseudomonas aeruginosa. Antimicrob. Agents Chemother.* **34:**52–57.
97. **Tseng, T. T., K. S. Gratwick, J. Kollman, D. Park, D. H. Nies, A. Goffeau, and M. H. Saier, Jr.** 1999. The RND permease superfamily: an ancient, ubiquitous and diverse family that includes human disease and development proteins. *J. Mol. Microbiol. Biotechnol.* **1:**107–125.
98. **Ubarretxena-Belandia, I., J. M. Baldwin, S. Schuldiner, and C. G. Tate.** 2003. Three-dimensional structure of the bacterial multidrug transporter EmrE shows it is an asymmetric homodimer. *EMBO J.* **22:**6175–6181.
99. **Vaara, M.** 1993. Antibiotic-supersusceptible mutants of *Escherichia coli* and *Salmonella typhimurium. Antimicrob. Agents Chemother.* **37:**2255–2260.
100. **Vogne, C., J. R. Aires, C. Bailly, D. Hocquet, and P. Plesiat.** 2004. Role of the multidrug efflux system MexXY in the emergence of moderate resistance to aminoglycosides among *Pseudomonas aeruginosa* isolates from patients with cystic fibrosis. *Antimicrob. Agents Chemother.* **48:**1676–1680.
101. **Wang, H., J. L. Dzink-Fox, M. Chen, and S. B. Levy.** 2001. Genetic characterization of highly fluoroquinolone-resistant

clinical *Escherichia coli* strains from China: role of *acrR* mutations. *Antimicrob. Agents Chemother.* **45:**1515–1521.

102. **Watkins, W. J., Y. Landaverry, R. Leger, R. Litman, T. E. Renau, N. Williams, R. Yen, J. Z. Zhang, S. Chamberland, D. Madsen, D. Griffith, V. Tembe, K. Huie, and M. N. Dudley.** 2003. The relationship between physicochemical properties, in vitro activity and pharmacokinetic profiles of analogues of diamine-containing efflux pump inhibitors. *Bioorg. Med. Chem. Lett.* **13:**4241–4244.
103. **Webber, M., and L. J. Piddock.** 2001. Quinolone resistance in *Escherichia coli. Vet. Res.* **32:**275–284.
104. **Westbrock-Wadman, S., D. R. Sherman, M. J. Hickey, S. N. Coulter, Y. Q. Zhu, P. Warrener, L. Y. Nguyen, R. M. Shawar, K. R. Folger, and C. K. Stover.** 1999. Characterization of a *Pseudomonas aeruginosa* efflux pump contributing to aminoglycoside impermeability. *Antimicrob. Agents Chemother.* **43:**2975–2983.
105. **Williams, R. J., D. M. Livermore, M. A. Lindridge, A. A. Said, and J. D. Williams.** 1984. Mechanisms of beta-lactam resistance in British isolates of *Pseudomonas aeruginosa. J. Med. Microbiol.* **17:**283–293.
106. **Yerushalmi, H., M. Lebendiker, and S. Schuldiner.** 1996. Negative dominance studies demonstrate the oligomeric structure of EmrE, a multidrug antiporter from *Escherichia coli. J. Biol. Chem.* **271:**31044–31048.
107. **Yerushalmi, H., S. S. Mordoch, and S. Schuldiner.** 2001. A single carboxyl mutant of the multidrug transporter EmrE is fully functional. *J. Biol. Chem.* **276:**12744–12748.
108. **Yerushalmi, H., and S. Schuldiner.** 2000. A model for coupling of H(+) and substrate fluxes based on "time-sharing" of a common binding site. *Biochemistry* **39:**14711–14719.
109. **Yu, E. W., J. R. Aires, and H. Nikaido.** 2003. AcrB multidrug efflux pump of *Escherichia coli:* composite substrate-binding cavity of exceptional flexibility generates its extremely wide substrate specificity. *J. Bacteriol.* **185:**5657–5664.
110. **Yu, E. W., G. McDermott, H. I. Zgurskaya, H. Nikaido, and D. E. Koshland, Jr.** 2003. Structural basis of multiple drug-binding capacity of the AcrB multidrug efflux pump. *Science* **300:**976–980.
111. **Zgurskaya, H. I., and H. Nikaido.** 1999. AcrA is a highly asymmetric protein capable of spanning the periplasm. *J. Mol. Biol.* **285:**409–420.
112. **Zgurskaya, H. I., and H. Nikaido.** 1999. Bypassing the periplasm: reconstitution of the AcrAB multidrug efflux pump of *Escherichia coli. Proc. Natl. Acad. Sci. USA* **96:**7190–7195.
113. **Zgurskaya, H. I., and H. Nikaido.** 2000. Cross-linked complex between oligomeric periplasmic lipoprotein AcrA and the inner-membrane-associated multidrug efflux pump AcrB from *Escherichia coli. J. Bacteriol.* **182:**4264–4267.
114. **Ziha-Zarifi, I., C. Llanes, T. Kohler, J. C. Pechere, and P. Plesiat.** 1999. In vivo emergence of multidrug-resistant mutants of *Pseudomonas aeruginosa* overexpressing the active efflux system MexA-MexB-OprM. *Antimicrob. Agents Chemother.* **43:**287–291.

Frontiers in Antimicrobial Resistance: a Tribute to Stuart B. Levy
Edited by D. G. White, M. N. Alekshun, and P. F. McDermott

Chapter 20

Role of Multidrug Efflux Pumps in Gram-Positive Bacteria

GLENN W. KAATZ

Gram-positive organisms are significant human pathogens. Organisms of particular importance include *Staphylococcus aureus, Staphylococcus epidermidis,* and other coagulase-negative staphylococci; *Enterococcus faecalis* and *Enterococcus faecium; Streptococcus pneumoniae;* and, at this time of heightened concern regarding bioterrorism, *Bacillus anthracis.* Drug resistance among these pathogens complicates therapy of infections caused by them, and a considerable component of that resistance can be ascribed to the activity of membrane-based efflux proteins, hereafter referred to as "pumps." Of course, cellular transport processes unrelated to drug resistance, including efflux, are utilized by all bacteria to acquire nutrients, to establish the proper charge and pH gradient across the cytoplasmic membrane, and to extrude metabolic by-products (13, 40). These natural processes are integral to survival of the organism. It is possible that proteins mediating such processes may contribute to drug resistance if an antibacterial agent resembles its natural substrate(s) or substrate selectivity of the pump protein is low.

All transport proteins belong to one of five families on the basis of structural characteristics, mechanism(s) of action, and energy source used for substrate translocation. Primary active transporters utilize ATP energy to expel substrates, and secondary active transporters couple substrate efflux with the transport of an ion (most commonly H^+) down an electrochemical concentration gradient (13, 54, 60). Secondary active transporters are thus dependent on the proton motive force (pmf), which consists of the transmembrane proton gradient and the electrical potential across the membrane, for their function. Pump protein families consist of the ATP binding cassette (ABC), the major facilitator superfamily (MFS), the multidrug and toxin extrusion (MATE), the resistance-nodulation-division (RND), and the small multidrug resistance (SMR) families. The general structural characteristics of each of these pump families, which are similar between gram-positive and gram-negative organisms, is depicted in schematic fashion in Fig. 1. To date, only ABC, MFS, and SMR family MDR pump proteins have been described in gram-positive organisms.

Drug efflux pumps possessed by gram-positive organisms are encoded either by plasmid- or chromosomally based genes (54, 60). The expression of plasmid-based genes often is sufficient for relevant resistance to be observed without the need for additional mutations owing to the multicopy nature of these genetic elements. However, drug resistance due to chromosomally encoded MDR pump genes most often occurs because of increased gene expression, which can take place as a consequence of substrate-induced transcriptional activation, gene amplification, or the occurrence of regulatory mutations (25). Generally, drug-specific efflux pumps tend to be located on plasmids and thus are readily transmissible, whereas multidrug (MDR) efflux pumps are usually encoded on the chromosome and are not easily donated to another organism. For the purposes of this chapter, our discussion will focus entirely on MDR-type pumps.

The available genome sequence data for gram-positive organisms suggests the existence of numerous potential drug pumps in all of them. Using the comprehensive transporter database available at http://www.membranetransport.org and the search terms "efflux," "drug efflux," or "multidrug efflux," the list in Table 1 was compiled. The reason(s) for a single organism to possess such a large number of these proteins is not known. MDR pumps often have a rather extensive overlap in their substrate profiles, a characteristic that on the surface appears redundant and perhaps even wasteful with respect to resource utilization. There is no experimental data that explains, or even suggests, why this redundancy exists. It is possible that MDR pumps evolved multiple times from a common primordial ancestor, maintaining the

Glenn W. Kaatz • John D. Dingell VA Medical Center and Wayne State University School of Medicine, Detroit, MI 48201.

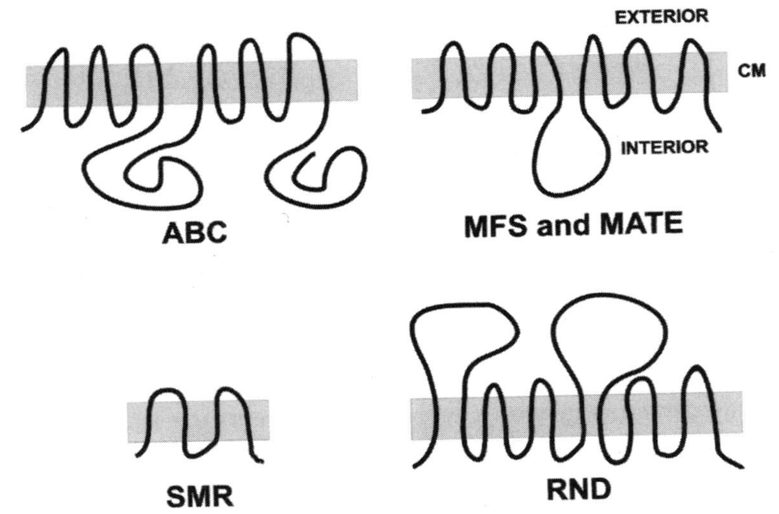

Figure 1. Schematic representation of the general structural characteristics of each family of MDR efflux proteins. ABC, ATP-binding cassette; MFS, major facilitator superfamily; MATE, multidrug and toxin extrusion; SMR, small multidrug resistance; RND, resistance-nodulation-division. The large central loop that is characteristic of MFS proteins is not necessarily as large as that depicted for MATE proteins. The cytoplasmic membrane (CM) is depicted in gray, and the interior and exterior of the cell are as indicated.

recognition of a specific set of substrates but acquiring the ability to transport a new toxin(s). It also is possible that they have specific physiologic functions and transport drugs only fortuitously. Continued research on MDR pumps will help us to understand these issues more completely.

The clinical relevance of MDR pump-related drug resistance in gram-positive bacteria has been questioned. With respect to *S. aureus,* an association between the NorA MDR pump and antiseptic resistance was shown to exist for approximately 20% of a series of methicillin- and antiseptic-resistant strains isolated in Japan in 1992 (52). The QacA and QacB MDR pumps of *S. aureus,* which encode proteins that are nearly identical and transport many different biocides and antiseptic agents, appear to be more widely distributed owing to their genes being carried on plasmids. A recent Japanese study demonstrated that *qacA/B* was present in 33 and 8% of methicillin-resistant and -susceptible strains, respectively, establishing the relevance of this resistance mechanism (5).

The greatest amount of information regarding the mechanism(s) of pump function and multidrug recognition as well as the regulation of pump gene expression in gram-positive organisms is available for MFS MDR efflux pumps. It is for this reason that the majority of our discussion will be devoted to pumps of this type. MDR pumps belonging to other families will be discussed briefly. A brief review of the status of the search for inhibitors of gram-positive drug pumps as a means to recover clinically relevant activity of substrates will conclude this chapter.

Table 1. Drug transport proteins predicted from analysis of genome data

Organism	No. of predicted pump proteins				
	ABC	MFS	SMR	MATE	RND
B. subtilis 168	3	33	3	4	1
B. anthracis Ames	21	54	19	4	4
E. faecalis V583	10	12	3	4	1
L. lactis IL1403	10	19	2	2	0
S. aureus COL	4	18	0	1	2
S. aureus N315	4	22	1	1	2
S. epidermidis 12228	8	18	2	1	1
S. pneumoniae TIGR4	7	2	0	1	0

MAJOR FACILITATOR SUPERFAMILY MDR EFFLUX PUMPS

The largest amount of research on MFS MDR efflux pumps has been conducted on QacA/B and NorA of *S. aureus,* Bmr and Blt of *B. subtilis,* and LmrP of *Lactococcus lactis.* Substrates for all MDR pumps, including those within the MFS, are typically amphipathic cations and can include biocides and antiseptics such as benzalkonium chloride and chlorhexidine, dyes such as ethidium bromide and acriflavine, and antimicrobial agents such as tetracycline and fluoroquinolones (13). With respect to NorA and Bmr, MDR occurs only when the chromosomally located genes for these proteins are either amplified (Bmr) or overexpressed as a result of regulatory mutations (NorA) (38, 49).

NorA

NorA was the first chromosomally encoded *S. aureus* MDR pump to be identified. The *norA* gene was cloned from the chromosome of a fluoroquinolone-resistant clinical strain in 1989 and based on its nucleotide sequence was predicted to encode a typical MFS-type protein with 12 membrane-spanning alpha helices (transmembrane segments, or TMSs) (67, 72). The expression of *norA* from a plasmid in either *Escherichia coli* or *S. aureus* resulted in fluoroquinolone resistance, which was noted to be more pronounced for fluoroquinolones that had hydrophilic characteristics. The uptake of enoxacin, which is hydrophilic, was reduced significantly in a *S. aureus* strain expressing *norA,* whereas the uptake of sparfloxacin, a more hydrophobic fluoroquinolone, was not affected. In the presence of carbonyl cyanide *m*-chlorophenyl hydrazone (CCCP), a protonophore that disrupts the pmf, the NorA-mediated enoxacin uptake defect was corrected and suggested that the pmf was important for NorA function. It was later shown that nigericin, which collapses the transmembrane proton gradient but leaves the electrical potential component of the pmf intact, abolished NorA-mediated efflux in an *E. coli* everted vesicle system (51). This established the pmf as the necessary energy source for NorA function. Further work using purified NorA reconstituted into proteoliposomes established that the protein is a drug::proton antiporter (73).

NorA has broad substrate specificity, including biocides, dyes, and hydrophilic FQs (50). In general NorA substrates are typical of those for MDR pumps, namely, amphipathic cations. NorA has modest homology with several tetracycline-specific efflux proteins of gram-negative bacteria (20 to 25%), but it has the highest degree of homology with Bmr, the product of the *Bacillus subtilis bmr* gene (44%) (see below). NorA activity is inhibited by reserpine, a compound known to interfere with the function of many MDR efflux proteins including mammalian p-glycoprotein. Other compounds are also capable of blocking NorA function, and these will be discussed subsequently.

An active area of research is the study of the regulation of *norA* expression. Recent work has identified an 18 kDa protein that binds upstream of the −35 motif of the *norA* promoter. The identity of this protein, initially named NorR, recently was established (66). Analysis of *S. aureus* genome data in the public domain reveals that *norR* is a conserved gene and lies 7.6 kb away from and on the opposite strand to *norA*. NorR is a protein of 147 residues that has modest homology with other regulatory proteins such as MarR and SarR. Experimental data suggest that repeats consisting of the consensus sequence TTAATT may be involved in the binding of NorR. Four such hexamers are located upstream of the −35 motif of the *norA* promoter (17). Subsequent work has revealed that NorR is not a specific regulator of *norA* expression but rather is a global regulator that, in addition to altering *norA* transcription when overexpressed, affects the transcription of several known autolytic regulators (30, 42). This protein, which independently has been named Rat and MgrA, is transcribed optimally from two promoters, positively regulates its own expression, and acts at the transcriptional level to enhance the expression of numerous genes whose products negatively impact the expression of murein hydrolases.

NorR/Rat/MgrA (hereafter referred to as MgrA, the name agreed upon by the codiscoverers of the protein) appears to augment *norA* expression in the presence of a transposon-mediated disruption of the *arlS* chromosomal locus (17). The *arlR-arlS* locus encodes a two-component regulatory system involved in adhesion, autolysis, and extracellular proteolytic activity of *S. aureus* (18). The exact mechanism by which an *arlS* disruption and MgrA interact to alter *norA* expression has not been established. One group has shown that the binding of MgrA to the *norA* promoter is modified in an *arlS*− strain such that increased *norA* expression is observed. The same group also has shown that overexpression of *mgrA* from an uncharacterized plasmid-based promoter in an *arlS*+ background also appears to result in an increase in the level of *norA* mRNA and reduced susceptibility to various NorA substrates (17, 66). In fact, these investigators have found that only when *mgrA* is overexpressed is an effect on *norA* transcript level observed in a wild-type background. These data suggest that a mutation in *arlS* augments, but is not a requirement for, an MgrA effect on the *norA* promoter and that wild-type levels of MgrA have little effect on *norA* expression. Highly fluoroquinolone-resistant strains of *S. aureus* have been described in which *norA* expression is enhanced in the absence of any modification of the *arlR-arlS* locus or change in *mgrA* expression, indicating that loci other than *arlR-arlS* and *mgrA* must be involved in the regulation of *norA* expression (18, 51, 65, 66). Work performed in our laboratory also has demonstrated that substrate exposure can augment *norA* expression, but the mediator(s) of this effect are unknown (35, 36).

Several years ago we identified a strain of *S. aureus* in which *norA* expression was enhanced in the presence of a T → A point mutation 11 bp downstream of the −10 promoter motif (*flqB* mutation) (38). The association of this mutation with augmented *norA* transcription was supported by its introduction into another *S. aureus* strain using plasmid integration (37). Other investigators showed that T → G or T → C *flqB* mutations also resulted in an apparent increase in *norA* expression (51, 53). The *flqB* mutation lies within the 5′ untranslated region of *norA* mRNA and could affect secondary structure and possibly mRNA half-life. Employing a Northern analysis approach, we previously found that in the presence of the T → A *flqB* mutation the half-life of the *norA* message was not increased, but using an RT-PCR approach others found a fivefold prolongation of half-life in a strain having a T → G *flqB* mutation (19 and unpublished data). The reason(s) for the differences in these data remains unknown.

Bmr and Blt

The *bmr* gene was cloned from a mutant of *Bacillus subtilis* selected in the laboratory for resistance to rhodamine 6G (49). In this mutant the expression of *bmr* was enhanced as a result of gene amplification. The *bmr* gene encodes a typical 12 TMS MFS-family protein that has a substrate profile very similar to that of NorA (48-50). As noted previously, NorA and Bmr share 44% sequence identity and thus are closely related in evolutionary terms. Bmr utilizes the pmf as its energy source, and its activity is inhibited by compounds that also inhibit the function of p-glycoprotein, including reserpine and verapamil.

The expression of *bmr* is affected by BmrR, a MerR-type activator protein encoded by a gene immediately downstream from *bmr* (3). The presence of Bmr substrates facilitates the binding to the *bmr* promoter by BmrR and subsequent activation of gene expression; thus, the expression of *bmr* can be considered inducible by its substrates (Fig. 2). The crystal structure BmrR at 2.8 Å resolution in the presence of the Bmr substrate tetraphenylphosphonium bromide (TPP) has been solved, and this work has revealed much about the binding site of a multidrug recognizing protein. The presence of TPP causes a conformational change in BmrR such that a large hydrophobic binding pocket is exposed (75). This pocket is lined with hydrophobic residues that can interact with hydrophobic portions of drug substrates via stacking and van der Waals contacts. There also is a glutamate residue within the binding pocket that forms an electrostatic interaction with the positively charged drug, which is an essential component of the cation selectivity of BmrR. The binding pocket is capable of accommodating drugs of various sizes and shapes, with the only requirements seeming to be an amphipathic nature and a positive charge. It is thought that Bmr, and perhaps most MDR transporters, may recognize substrates in a similar fashion. Confirmation of this hypothesis awaits the production of high-quality crystals of an MFS MDR transporter in the presence of substrates.

Blt is another *B. subtilis* MFS MDR transporter whose gene was isolated from a library of *B. subtilis* genomic fragments cloned into an expression vector (4). Unlike *bmr, blt* is not normally transcribed in wild-type strains. Blt is predicted to consist of 12 TMSs, is pmf-dependent, is highly homologous with Bmr (52%) and NorA (39%), and has a similar substrate profile to these pumps. The expression of *blt* is enhanced by the binding of BltR (encoded by *bltR,* found immediately upstream of *blt*) to the *blt* promoter region. This binding is thought to be affected by the interaction of substrates with BltR, although specific inducers have yet to be identified and no studies have been done with respect to determining the crystal structure of BltR. BmrR and BltR have homologous N-terminal DNA binding domains but differ significantly in their C-terminal inducer binding domains. These data suggest that Bmr and Blt probably have different physiologic functions. In fact, this position is supported by the fact that Blt has been shown to transport the natural polyamine spermidine, whereas Bmr does not have this capability (71).

In addition to the specific regulators of *bmr* and *blt* transcription just described, the expression of these genes is affected by MtaN, a global transcriptional regulator that interacts with their promoters, inducing transcription (7). MtaN consists of the N-terminal 109 residues of a larger protein, Mta (257 residues); the intact parent protein does not activate *bmr* or *blt* transcription. It is hypothesized that upon interacting with an inducer (as yet unidentified), the N- and C-terminal domains of Mta are functionally separated, allowing it to function as a transcriptional activator (Fig. 2). The crystal structure of MtaN at 2.75 Å res-

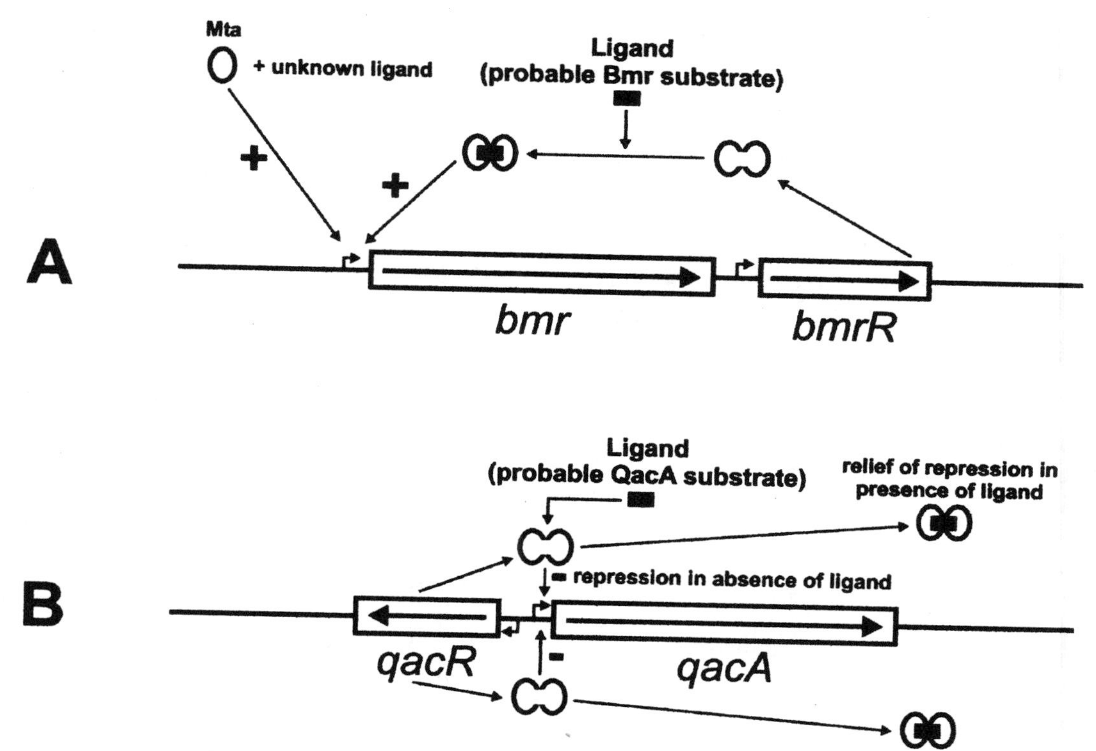

Figure 2. Regulation of *bmr* and *qacA* expression. (A) *bmr*. A BmrR dimer binds to a ligand, most likely a Bmr substrate, and in turn binds to the *bmr* promoter activating transcription. The ligand(s) for Mta are currently unknown. (B) *qacA*. In the absence of a ligand, most likely a QacA substrate, QacR binds as a pair of dimers to the *qacA* promoter and blocks transcription. Binding of a ligand to each QacR dimer results in a release of its repression on *qacA*, allowing transcription to proceed.

olution reveals it to be a winged helix-turn-helix protein, but the mechanism by which it activates transcription remains unknown (21).

QacA

QacA and the nearly identical QacB are pmf-dependent *S. aureus* MDR pumps that differ from NorA, Bmr, and Blt in that they are plasmid-encoded and thus relatively easy to transfer between strains. These proteins possess 14 rather than 12 TMSs and confer resistance to clinically relevant biocides, some of which are also transported by NorA (46, 54, 55). QacA and QacB differ by seven amino acid residues, which results in a difference in substrate specificity. QacA can transport mono- and divalent cations, whereas QacB only transports monovalent substrates. A negatively charged residue in the 10th TMS, present in QacA but absent in QacB, is responsible for this difference in substrate specificity (54, 55).

All *qacA/B* determinants are regulated by the divergently transcribed QacR repressor protein, which is a member of the MarR family of transcriptional regulators (25). Similar to BmrR, QacR interacts with QacA/B substrates (23, 26, 61). However, QacR is a repressor that normally is bound to the *qacA/B* promoter as a pair of dimers, whereas BmrR is an activator that only binds to the *bmr* promoter as a single dimer in the presence of substrate (3, 24, 62). The interaction of QacR with substrates is followed by its dissociation from the promoter and subsequent relief of its repression of *qacA/B* (Fig. 2).

QacR has been crystallized in the presence of multiple substrates, and the data generated were quite similar to that found for BmrR. QacR possesses a large hydrophobic drug binding pocket (1,100 Å^3) that is lined with aromatic residues as well as several glutamate residues that interact with positively charged substrates (47, 61, 62). Hydrophobic interactions occur with hydrophobic portions of substrate molecules, and the positive charge(s) of substrates is stabilized by electrostatic interactions with the glutamates. All substrates bind within this pocket, but the actual binding sites within the rather expansive drug binding pocket differ and depend on the particular drug. Thus, the presence of multiple drug binding sites within the same pocket explains how QacR can accommodate structurally different mono- and divalent compounds.

LmrP

Lactococcus lactis is an organism utilized in food manufacturing (28, 29). Konings and colleagues selected *L. lactis* mutants resistant to high concentrations of ethidium bromide, daunomycin, and rhodamine 6G, respectively, and found that these mutants were resistant to a broad array of additional drugs and actively effluxed ethidium (9). In one of these mutants the resistance profile was dependent on the proton motive force. Using a chromosomal DNA library created from this strain and expressed in *E. coli,* a clone that conferred multidrug resistance was discovered (10). The determinant was named *lmrP,* and it encoded a typical 12 TMS MFS-type protein. LmrP has 19 and 21% identity with NorA and Bmr, respectively, which is typical in that MFS MDR transporters often have higher homology to drug-specific pumps within the MFS than they do to each other (54). Deletion of *lmrP* from the chromosome resulted in augmented ethidium accumulation compared to wild-type strains, which was further increased by exposure to the ATPase inhibitor ortho-vanadate. This observation suggested the existence of an ATP-dependent drug extrusion system, which was later identified as LmrA (see below).

LmrP has been studied extensively with respect to its energetics and mechanism of substrate recognition and transport in the absence of crystal structure data. LmrP transport is dependent on both components of the pmf and appears to occur by an electrogenic nH+::drug exchange, where $n \geq 2$ (11). Competition for transport between the LmrP substrate Hoechst 33342 and other pump substrates occurred by competitive, noncompetitive, and uncompetitive processes, indicating that the protein likely contains multiple drug interaction sites (59). Drug acquisition by the transporter occurs at the inner leaflet of the cytoplasmic membrane followed by expulsion into the external medium (11). Similar observations have been made for QacA and LmrA, suggesting that this may be a general mechanism employed by many MDR efflux proteins regardless of the energy coupling mechanism used (12, 46).

Membrane-embedded charged residues are energetically unfavorable, and their presence in this location suggests that they may be important for transporter function. There are three membrane-embedded acidic residues in LmrP. Site-directed mutagenesis was used to change these residues, and from these studies it was concluded that Asp142 and Glu327 were important for substrate recognition (45). A multiple sequence alignment between LmrP, Bmr, and NorA reveals that a negative charge at a position homologous to Glu327 is conserved and suggests that this residue may have a similar function in these other MFS MDR transporters.

Other Gram-Positive MFS MDR Pumps

Several studies have suggested the presence of an efflux-mediated mechanism for low-level fluoroquinolone resistance in *S. pneumoniae* (8, 15, 16, 74). The genetic basis for this resistance mechanism was discovered by searching the *S. pneumoniae* genome for *norA* homologues, which led to the identification of *pmrA* (20). The contribution of PmrA to efflux was confirmed by insertionally inactivating the *pmrA* gene on the chromosome of a strain having an efflux-related phenotype and observing the reversion of this phenotype to wild-type drug susceptibility. Similar to NorA and Bmr, the activity of PmrA, which is predicted to be a 12 TMS protein, is inhibited by reserpine. Increased expression of *pmrA* has been associated with MDR in some instances, but not all MDR strains are found to overexpress this gene. This observation indicates that *S. pneumoniae* must possess other drug pumps, which also is suggested by analysis of genome data (Table 1) (14, 57, 58).

A genomics approach was used to identify EmeA, a NorA homolog found in *Enterococcus faecalis* (32). Deletion of *emeA* from the chromosome of the commonly employed laboratory strain *E. faecalis* OG1RF resulted in twofold MIC reductions for ethidium bromide and norfloxacin, both of which are also NorA substrates. From these data it can be concluded that EmeA contributes to the poor antimicrobial activity of norfloxacin in this strain and that overexpression of *emeA* in a clinical isolate may confer broad MDR similar to that observed for overexpression of NorA in *S. aureus*. At the time of this writing, no clinical isolates of *E. faecalis* that overexpress *emeA* have been identified.

A novel MFS *S. aureus* MDR efflux pump, MdeA, was recently described (27). Sequence analysis predicts that MdeA has a 14 TMS secondary structure, and its function is susceptible to reserpine as well as CCCP, indicating a dependence on the pmf for function. When overexpressed in *S. aureus* MdeA confers resistance to a broad array of substrates, including some of those transported by NorA and QacA/B (ethidium bromide, benzalkonium chloride) but also drugs not recognized by other *S. aureus* MDR efflux pumps, including virginiamycin, novobiocin, and fusidic acid. In contrast to NorA, MdeA does not appear to transport fluoroquinolones. Expression of *mdeA* in wild-type strains is low, but spontaneous mutants having increased transcription are selectable in vitro. These mutants, which have raised minimum inhibitory concentrations to MdeA substrates, have mutations in the *mdeA* promoter. No further information regarding the regulation of *mdeA* expression is available.

SMALL MULTIDRUG RESISTANCE FAMILY MDR EFFLUX PUMPS

SMR pumps are unique in that they are the smallest known drug transporters, consisting of only 4 TMSs (Fig. 1) (22, 56). These proteins are dependent on the pmf for activity, and it has been proposed that they may function as oligomeric complexes. They are found in both gram-positive and -negative organisms, and typical substrates include biocides and disinfectants; no SMR pump has been described that is capable of transporting antimicrobial agents in clinical use. The Smr protein of *S. aureus,* which is plasmid-based and thus easily transmissible, has been proposed to function as a homotrimer as the association of three Smr monomers would result in the formation of a typical 12 TMS pump. This hypothesis has not been verified experimentally. However, other members of this family have been found that clearly function as heterodimers, with the EbrAB MDR pump of *B. subtilis* being one example (44). This pump is composed of two SMR-like subunits, neither of which is sufficient to confer MDR alone, but the expression of *ebrA* and *ebrB* together results in an MDR phenotype. The same phenomenon has been described for another pair of *B. subtilis* SMR-type proteins, YkkC and YkkD (31). It thus appears that a functional pmf-dependent MDR efflux pump can consist of as few as eight TMSs.

ABC FAMILY MDR EFFLUX PUMPS

Typical ABC pumps consist of four domains, two membrane-embedded hydrophobic regions that carry out substrate recognition and translocation and two hydrophilic nucleotide binding domains (NBDs) that are the sites at which energy for the translocation process is generated via ATP hydrolysis (Fig. 1) (6). The best-characterized gram-positive ABC-type MDR pump is the LmrA protein of *L. lactis* (70). The *lmrA* gene was recovered from a genomic DNA library constructed in *E. coli* similar to that described previously for *lmrP.* The gene was found to encode a 590 residue polypeptide which consisted of only one hydrophobic domain (6 TMSs) and one nucleotide binding domain. Human p-glycoprotein is a typical four-domain ABC pump having two homologous halves, each consisting of six TMS regions followed by large hydrophilic domains containing the NBDs. Thus, LmrA appeared to be a "half-transporter" based on this typical ABC pump structure. LmrA and each half of human p-glycoprotein share 34% identity, and they have a significant overlap in their substrate specificities (70). In fact, when LmrA is expressed in a eukaryotic cell line, the efflux of cancer chemotherapeutic agents is observed (68). A comparison between p-glycoprotein and LmrA drug binding and export in this system revealed that the two proteins performed in highly similar manners. Subsequently it was established that LmrA functions as a homodimer, which generates a protein having typical ABC pump structural features (69).

As for LmrP, LmrA gathers substrates from the inner leaflet of the cytoplasmic membrane (12). The protein has two drug binding sites and two ATP binding sites, one each for each monomer. It is proposed to mediate multidrug transport by an alternating two-site mechanism termed the "two-cylinder engine" model, in which drug and ATP alternatively bind to their respective binding sites on each monomer (69). ATP binding at one NBD results in the appearance of a high-affinity drug binding site at the inner membrane surface of the transporter, whereas the ADP-bound state is associated with a low-affinity substrate binding site on the exterior surface of the protein. Hydrolysis of ATP results in translocation of the inner-leaflet bound substrate to the exterior and its release. ATP can then bind to the alternate NBD, thus allowing the cycle to repeat. It is possible that this mechanism of substrate transport is applicable to most, if not all, ABC-type export proteins. High-resolution crystals of LmrA or other bacterial or eukaryotic ABC-type MDR pumps with and without bound substrate will be required to establish or refute this mechanism of transport.

Other gram-positive ABC-type MDR pumps have been described, but none has been characterized as completely as LmrA. The BmrA protein of *B. subtilis* is an LmrA and p-glycoprotein homologue that probably also functions as a homodimer (63). The *E. faecalis* EfrAB pump is composed of two highly similar proteins, each containing a 6 TMS hydrophobic domain and a hydrophilic NBD (39). Neither protein is capable of conferring MDR by itself, but expression of *efrA* and *efrB* together does. This is the first example of an ABC MDR pump in a gram-positive bacterium that functions as a heterodimer.

MATE AND RND MDR EFFLUX PUMPS

To date, no MDR pumps belonging to either the MATE or RND family have been discovered in gram-positive organisms, but analysis of genome data suggests that they do exist (Table 1). Candidate genes can be amplified using PCR and cloned into multicopy plasmids under the control of an inducible promoter. This approach is most likely to yield genes encoding proteins that confer an MDR phenotype, as uncontrolled overexpression of pump proteins can be toxic for a host cell. Such research is anxiously awaited.

Inhibitors of Gram-Positive MDR Efflux Pumps

A concerted effort has been made in recent years to identify compounds capable of inhibiting the NorA MDR pump. This has included a massive screen of over 9,000 inorganic compounds, the study of naturally occurring plant alkaloids, and the study of specific compounds already in clinical use for other indications (1, 2, 33, 34, 43, 64). Drugs belonging to this latter class that have been shown to be capable of NorA inhibition include reserpine, verapamil, omeprazole, prochlorperazine, and paroxetine. Unfortunately, the concentrations required for significant pump inhibition by any of these compounds is beyond that which is clinically achievable. Using these drugs as a starting point, it may be possible to design derivatives that are less toxic and more active. It would be most desirable to develop an efflux pump inhibitor with activity against both pmf- and ATP-dependent pumps, but this seems unlikely on the basis of the fundamental structural and functional differences between these two protein families.

CONCLUSIONS

Gram-positive organisms possess a myriad of MDR efflux pumps belonging to several transporter protein families. The general consensus is that these pumps function as "hydrophobic vacuum cleaners" that detoxify the bacterial membrane. Some probably evolved from drug-specific pumps, whereas others probably have always had the capability to recognize multiple drugs. In both cases the accumulation of mutations in response to environmental stresses likely has led to the appearance of multiple pumps with overlapping substrate profiles. Some of these MDR pumps may have natural, physiologic functions as exemplified by the transport of spermidine by Blt of *B. subtilis*.

The study of the mechanism of multidrug recognition and transport is hampered by our general inability to produce high-resolution crystals of MDR pump proteins. The recent successful production of such crystals of the AcrB RND pump at 3.8 Å is a breakthrough in this area and has been discussed in chapter 19. However, for the most part our understanding of how MDR pumps work has been achieved by extension of conclusions reached in the study of multidrug binding proteins that regulate the expression of MDR pumps. It is reasonable to predict that the mechanisms of drug recognition by these proteins will be similar, if not identical, to that employed by the pumps themselves.

Bacterial MDR pumps contribute to the growing problem of antimicrobial resistance in bacteria, and in fact the activity of such pumps forms the basis for the innate multidrug resistance observed in some genera (41). Further research on bacterial MDR efflux pumps is warranted in an effort to improve our understanding of how these systems work, because such an understanding may allow the development of means to overcome this resistance mechanism and to recover useful activity of their antibiotic substrates. This may occur by the development of drugs designed to avoid efflux or by the discovery of broad-spectrum efflux pump inhibitors.

IN APPRECIATION

I first met Stuart Levy in the latter part of the 1980s when he came to Detroit to deliver a lecture to the faculty and students of the Department of Biochemistry at Wayne State University. Steve Lerner, a colleague of mine and also a friend of Stuart's, informed me of his visit, and I attended his lecture. Afterwards I was introduced to Dr. Levy and gave him a brief presentation concerning my own research, and to this day I remain grateful to him for his genuine interest in what I was doing despite my very junior status and relative naiveté at the time within the field of bacterial efflux transporters. Over the ensuing years we have continued to see each other or communicate by telephone, letter, or electronic mail, sharing ideas and strains in an effort to advance each of our own projects. Stuart has always taken the time to help me with problems, review grant proposals, and offer comments, and he never fails to stop by to say hello and discuss data at presentations I make at meetings. He truly is a great mentor, teacher, and colleague who is well-deserving of acclaim for all he has accomplished.

REFERENCES

1. **Aeschlimann, J. R., L. D. Dresser, G. W. Kaatz, and M. J. Rybak.** 1999. Effects of NorA inhibitors on the in vitro antibacterial activities and postantibiotic effects of levofloxacin, ciprofloxacin, and norfloxacin in genetically related strains of *Staphylococcus aureus*. *Antimicrob. Agents Chemother.* **43:**335–340.
2. **Aeschlimann, J. R., G. W. Kaatz, and M. J. Rybak.** 1999. The effects of NorA inhibition on the activities of levofloxacin, ciprofloxacin, and norfloxacin against two genetically related strains of *Staphylococcus aureus* in an in-vitro infection model. *Antimicrob. Agents Chemother.* **44:**343–349.
3. **Ahmed, M., C. M. Borsch, S. S. Taylor, N. Vazquez-Laslop, and A. A. Neyfakh.** 1994. A protein that activates expression of a multidrug efflux transporter upon binding the transporter substrates. *J. Biol. Chem.* **269:**1–8.
4. **Ahmed, M., L. Lyass, P. N. Markham, S. S. Taylor, N. Vazquez-Laslop, and A. A. Neyfakh.** 1995. Two highly similar multidrug transporters of Bacillus subtilis whose expression is differentially regulated. *J. Bacteriol.* **177:**3904–3910.

5. **Alam, M. M., N. Kobayashi, N. Uechara, and N. Wantanabe.** 2003. Analysis on distribution and genomic diversity of high-level antiseptic resistance genes *qacA* and qacB in human clinical isolates of *Staphylococcus aureus*. *Microb. Drug Resist.* **9:**109–21.
6. **Ambudkar, S. V., I. H. Lelong, J. Zhang, C. O. Cardarelli, M. M. Gottesman, and I. Pastan.** 1992. Partial purification and reconstitution of the human multidrug-resistance pump: characterization of the drug-stimulatable ATP hydrolysis. *Proc. Natl. Acad. Sci. USA* **89:**8472–8476.
7. **Baranova, N. N., A. Danchin, and A. A. Neyfakh.** 1999. Mta, a global MerR-type regulator of the *Bacillus subtilis* multidrug-efflux transporters. *Mol. Microbiol.* **31:**1549–1559.
8. **Baranova, N. N., and A. A. Neyfakh.** 1997. Apparent involvement of a multidrug transporter in the fluoroquinolone resistance of *Streptococcus pneumoniae*. *Antimicrob. Agents Chemother.* **41:**1396–1398.
9. **Bolhuis, H., D. Molenaar, G. J. Poelarends, H. W. van Veen, B. Poolman, A. J. M. Driessen. and W. N. Konings.** 1994. Proton motive force-driven and ATP-dependent extrusion systems in multidrug-resistant *Lactococcus lactis*. *J. Bacteriol.* **176:**6957–6964.
10. **Bolhuis, H., G. J. Poelarends, H. W. van Veen, B. Poolman, A. J. M. Driessen, and W. N. Konings.** 1995. The lactococcal *lmrP* gene encodes a proton motive force-dependent drug transporter. *J. Biol. Chem.* **270:**26092–26098.
11. **Bolhuis, H., H. W. van Veen, J. R. Brands, M. Putman, B. Poolman, A. J. M. Driessen, and W. N. Konings.** 1996. Energetics and mechanism of drug transport mediated by the lactococcal multidrug transporter LmrP. *J. Biol. Chem.* **271:**24123–24128.
12. **Bolhuis, H. H. W. van Veen, D. Molenaar, B. Poolman, A. J. M. Driessen, and W. N. Konings.** 1996. Multidrug resistance in *Lactococcus lactis:* evidence for ATP-dependent drug extrusion from the inner leaflet of the cytoplasmic membrane. *EMBO J.* **15:**4239–4245.
13. **Borges-Walmsley, M. I., K. S. McKeegan, and A. R. Walmsley.** 2003. Structure and function of efflux pumps that confer resistance to drugs. *Biochem. J.* **376:**313–338.
14. **Brenwald, N. P., P. Appelbaum, T. Davies, and M. J. Gill.** 2003. Evidence for efflux pumps, other than PmrA, associated with fluoroquinolone resistance in *Streptococcus pneumoniae*. *Clin. Microbiol. Infect.* **9:**140–143.
15. **Brenwald, N. P., M. J. Gill, and R. Wise.** 1997. The effect of reserpine, an inhibitor of multidrug efflux pumps, on the in-vitro susceptibilities of fluoroquinolone-resistant strains of *Streptococcus pneumoniae* to norfloxacin. *J. Antimicrob. Chemother.* **40:**458–460.
16. **Brenwald, N. P., M. J. Gill, and R. Wise.** 1998. Prevalence of a putative efflux mechanism among fluoroquinolone-resistant clinical isolates of *Streptococcus pneumoniae*. *Antimicrob. Agents Chemother.* **42:**2032–2035.
17. **Fournier, B., R. Aras, and D. C. Hooper.** 2000. Expression of the multidrug resistance transporter NorA from *Staphylococcus aureus* is modified by a two-component regulatory system. *J. Bacteriol.* **182:**664–671.
18. **Fournier, B., and D. C. Hooper.** 2000. A new two-component regulatory system involved in adhesion, autolysis, and extracellular proteolytic activity of *Staphylococcus aureus*. *J. Bacteriol.* **182:**3955–3964.
19. **Fournier, B., Q.-C. Truong-Bolduc, X. Zhang, and D. C. Hooper.** 2001. A mutation in the 5′ untranslated region increases stability of *norA* mRNA, encoding a multidrug resistance transporter of *Staphylococcus aureus*. *J. Bacteriol.* **183:**2367–2371.
20. **Gill, M. J., N. P. Brenwald, and R. Wise.** 1999. Identification of an efflux pump gene, *pmrA*, associated with fluoroquinolone resistance in *Streptococcus pneumoniae*. *Antimicrob. Agents Chemother.* **43:**187–189.
21. **Godsey, M. H., N. N. Baranova, A. A. Neyfakh, and R. G. Brennan.** 2001. Crystal structure of MtaN, a global multidrug transporter gene activator. *J. Biol. Chem.* **276:**47178–47184.
22. **Grinius, L. L., G. Dreguniene, E. B. Goldberg, C. H. Liao, and S. J. Projan.** 1992. A staphylococcal multidrug resistance gene product is a member of a new protein family. *Plasmid* **27:**119–129.
23. **Grkovic, S., M. H. Brown, N. J. Roberts, I. T. Paulsen, and R. A. Skurray.** 1998. QacR is a repressor protein that regulates expression of the *Staphylococcus aureus* multidrug efflux pump QacA. *J. Biol. Chem.* **273:**18665–18673.
24. **Grkovic, S., M. H. Brown, M. A. Schumacher, R. G, Brennan, and R. A. Skurray.** 2001. The staphylococcal QacR multidrug regulator binds a correctly spaced operator as a pair of dimers. *J. Bacteriol.* **183:**7102–7109.
25. **Grkovic, S., M. H. Brown, and R. A. Skurray.** 2002. Regulation of bacterial drug export systems. *Microbiol. Mol. Biol. Rev.* **66:**671–701.
26. **Grkovic, S., K. M. Hardie, M. H. Brown, and R. A. Skurray.** 2003. Interactions of the QacR multidrug-binding protein with structurally diverse ligands: implications for the evolution of the binding pocket. *Biochemistry* **42:**15226–15236.
27. **Huang, J., P. W. O'Toole, W. Shen, H. Amrine-Madsen, X. Jiang, N. Lobo, L. M. Palmer, L. Voelker, F. Fan, M. N. Gwynn, and D. McDevitt.** 2004. Novel chromosomally-encoded multidrug efflux transporter MdeA in *Staphylococcus aureus*. *Antimicrob. Agents Chemother.* **48:**909–917.
28. **Hugenholtz, J., and M. Kleerebezum. 1999.** Metabolic engineering of lactic acid bacteria: overview of the approaches and results of pathway rerouting involved in food fermentations. *Curr. Opin. Biotechnol.* **10:**492–497.
29. **Hugenholtz, J., and E. J. Smid.** 2002. Nutraceutical production with food-grade microorganisms. *Curr. Opin. Biothecnol.* **13:** 497–507.
30. **Ingavale, S. S., W. van Wamel, and A. L. Cheung.** 2003. Characterization of Rat, an autolysis regulator in *Staphylococcus aureus*. *Mol. Microbiol.* **48:**1451–1466.
31. **Jack, D. L., M. L. Storms, J. H. Tchieu, I. T. Paulsen, and M. H. Saier.** 2000. A broad-specificity multidrug efflux pump requiring a pair of homologous SMR-type proteins. *J. Bacteriol.* **182:**2311–2313.
32. **Jonas, B. M., B. E. Murray, and G. M. Weinstock.** 2001. Characterization of *emeA*, a *norA* homologue and multidrug resistance efflux pump, in *Enterococcus faecalis*. *Antimicrob. Agents Chemother.* **45:**3574–3579.
33. **Kaatz, G. W., V. V. Moudgal, S. M. Seo, J. Bondo Hansen, and J. E. Kristiansen.** 2003. Phenylpiperidine selective serotonin reuptake inhibitors interfere with multidrug efflux pump activity in *Staphylococcus aureus*. *Int. J. Antimicrob. Agents* **22:**254–261.
34. **Kaatz, G. W., V. V. Moudgal, S. M. Seo, and J. E. Kristiansen.** 2003. Phenothiazines and thioxanthenes inhibit multidrug efflux pump activity in *Staphylococcus aureus*. *Antimicrob. Agents Chemother.* **47:**719–726.
35. **Kaatz, G. W., and S. M. Seo.** 1995. Inducible NorA-mediated multidrug resistance in *Staphylococcus aureus*. *Antimicrob. Agents Chemother.* **39:**2650–2655.
36. **Kaatz, G. W., and S. M. Seo.** 2004. Effect of substrate exposure and other growth condition manipulations on *norA* expression. *J. Antimicrob. Chemother.* **54:**364–369.
37. **Kaatz, G. W., S. M. Seo, and T. J. Foster.** 1999. Introduction of a *norA* promoter region mutation into the chromosome of fluoroquinolone-susceptible strain of *Staphylococcus aureus* using plasmid integration. *Antimicrob. Agents Chemother.* **43:** 2222–2224.

38. **Kaatz, G. W., S. M. Seo, and C. A. Ruble.** 1993. Efflux-mediated fluoroquinolone resistance in *Staphylococcus aureus. Antimicrob. Agents Chemother.* **37:**1086–1094.
39. **Lee, E.-W., M. Nazmul Huda, T. Kuroda, T. Mizushima, and T. Tsuchiya.** 2003. EfrAB, an ABC multidrug efflux pump in *Enterococcus faecalis. Antimicrob. Agents Chemother.* **47:**3733–3738.
40. **Lolkema, J. S., B. Poolman, and W. N. Konings.** 1998. Bacterial solute uptake and efflux systems. *Curr. Opin. Microbiol.* **1:**248–253.
41. **Lomovskaya, O., M. S. Warren, A. Lee, J. Galazzo, R. Fronko, M. Lee, J. Blais, D. Cho, S. Chamberland, T. Renau, R. Leger, S. Hecker, W. Watkins, K. Hishino, H. Ishida, and V. J. Lee.** 2001. Identification and characterization of inhibitors of multidrug resistance efflux pumps in *Pseudomonas aeruginosa:* novel agents for combination therapy. *Antimicrob. Agents Chemother.* **45:**105–116.
42. **Luong, T. T., S. W. Newell, and C. Y. Lee.** 2003. *mgrA,* a novel global regulator in *Staphylococcus aureus. J. Bacteriol.* **185:** 3703–3710.
43. **Markham, P. N., E. Westhaus, K. Klyatchko, M. E. Johnson, and A. A. Neyfakh.** 1999. Multiple novel inhibitors of the NorA multidrug transporter of *Staphylococcus aureus. Antimicrob. Agents Chemother.* **43:**2404–2408.
44. **Masaoka, Y., Y. Ueno, Y. Morita, T. Kuroda, T. Mizushima, and T. Tsuchiya.** 2000. A two-component multidrug efflux pump, EbrAB, in *Bacillus subtilis. J. Bacteriol.* **182:**2307–2310.
45. **Mazurkiewicz, P., W. N. Konings, and G. J. Poelarends.** 2002. Acidic residues in the lactococcal multidrug efflux pump LmrP play critical roles in transport of lipophilic cationic compounds. *J. Biol. Chem.* **277:**26081–26088.
46. **Mitchell, B. A., I. T. Paulsen, M. H. Brown, and R. A. Skurray.** 1999. Bioenergetics of the staphylococcal multidrug export protein QacA. *J. Biol. Chem.* **274:**3541–3548.
47. **Murray, D. S., M. A. Schumacher, and R. G. Brennan.** 2004. Crystal structures of QacR-diamidine complexes reveal additional multidrug-binding modes and a novel mechanism of drug charge neutralization. *J. Biol. Chem.* **279:**14365–14371.
48. **Neyfakh, A. A.** 1992. The multidrug efflux transporter of *Bacillus subtilis* is a structural and functional homologue of the *Staphylococcus aureus* NorA protein. *Antimicrob. Agents Chemother.* **36:**484–485.
49. **Neyfakh, A. A., V. E. Bidnenko, and L. B. Chen.** 1991. Efflux-mediated multidrug resistance in *Bacillus subtilis:* similarities and dissimilarities with the mammalian system. *Proc. Natl. Acad. Sci. USA* **88:**4781–4785.
50. **Neyfakh, A. A., C. M. Borsch, and G. W. Kaatz.** 1993. Fluoroquinolone resistance protein NorA of *Staphylococcus aureus* is a multidrug efflux transporter. *Antimicrob. Agents Chemother.* **37:**128–129.
51. **Ng, E. Y. W., M. Trucksis, and D. C. Hooper.** 1994. Quinolone resistance mediated by *norA:* physiologic characterization and relationship to *flqB,* a quinolone resistance locus on the *Staphylococcus aureus* chromosome. *Antimicrob. Agents Chemother.* **38:**1345–1355.
52. **Noguchi N., M. Hase, M. Kitta, M. Sasatsu, K. Deguchi, and M. Kono.** 1999. Antiseptic susceptibility and distribution of antiseptic-resistance genes in methicillin-resistant *Staphylococcus aureus. FEMS Microbiol. Lett.* **172:**247–253.
53. **Noguchi, N., M. Tamura, K. Narui, K. Wakasugi, and M. Sasatu.** 2002. Frequency and genetic characterization of multidrug-resistant mutants of *Staphylococcus aureus* after selection with individual antiseptics and fluoroquinolones. *Biol. Pharm. Bull.* **25:**1129–1132.
54. **Paulsen, I. T., M. H. Brown, and R. A. Skurray.** 1996. Proton-dependent multidrug efflux systems. *Microbiol. Rev.* **60:**575–608.
55. **Paulsen, I. T., M. H. Brown, T. G. Littlejohn, B. A. Mitchell, and R. A. Skurray.** 1996. Multidrug resistance proteins QacA and QacB from *Staphylococcus aureus:* membrane topology and identification of residues involved in substrate specificity. *Proc. Natl. Acad. Sci. USA* **93:**3630–3635.
56. **Paulsen, I. T., R. A. Skurray, R. Tam, M. H. saier, R. J. Turner, J. H. weiner, E. B. Goldberg, and L. L. Grinius.** 1996. The SMR family: a novel family of multidrug efflux proteins involved with the efflux of lipophilic drugs. *Mol. Microbiol.* **19:**1167–1175.
57. **Pestova, E., J. J. Millichap, F. Siddiqui, G. A. Noskin, and L. R. Peterson.** 2002. Non-PmrA-mediated multidrug resistance in *Streptococcus pneumoniae. J. Antimicrob. Chemother.* **49:**553–556.
58. **Piddock, L. J. V., M. M. Johnson, S. Simjee, and L. Pumbwe.** 2002. Expression of efflux pump gene *pmrA* in fluoroquinolone-resistant and -susceptible clinical isolates of *Streptococcus pneumoniae. Antimicrob. Agents Chemother.* **46:**808–812.
59. **Putman, M., L. A. Koole, H. W. van Veen, and W. N. Konings.** 1999. The secondary multidrug transporter LmrP contains multiple drug interaction sites. *Biochemistry* **38:**13900–13905.
60. **Putman, M., H. W. van Veen, and W. N. Konings.** 2000. Molecular properties of bacterial multidrug transporters. *Microbiol. Mol. Biol. Rev.* **64:**672–693.
61. **Schumacher, M. A., M. C. Miller, S. Grkovic, M. H. Brown, R. A. Skurray, and R. G. Brennan.** 2001. Structural mechanisms of QacR induction and multidrug recognition. *Science* **294:**2158–2163.
62. **Schumacher, M. A., M. C. Miller, S. Grkovic, M. H. Brown, R. A. Skurray, and R. G. Brennan.** 2002. Structural basis for cooperative DNA binding by two dimers of the multidrug-binding protein QacR. *EMBO J.* **21:**1210–1218.
63. **Steinfels, E., C. Orelle, J.-R. Fantino, O. Dalmas, J.-L. Rigaud, F. Denizot, A. Di Pietro, and J.-M. Jault.** 2004. Characterization of YvcC (BmrA), a multidrug ABC transporter constitutively expressed in *Bacillus subtilis. Biochemistry* **43:**7491–7502.
64. **Stermitz, F. R., P. Lorenz, J. N. Tawara, L. A. Zenewicz, and K. Lewis.** 2000. Synergy in a medicinal plant: antimicrobial action of berberine potentiated by 5′-methoxyhydnocarpin, a multidrug pump inhibitor. *Proc. Natl. Acad. Sci. USA* **97:**1433–1437.
65. **Trucksis, M., J. S. Wolfson, and D. C. Hooper.** 1991. A novel locus conferring fluoroquinolone resistance in *Staphylococcus aureus. J. Bacteriol.* **173:**5854–5860.
66. **Truong-Bolduc, Q.-C., X. Zhang, and D. C. Hooper.** 2003. Characterization of NorR protein, a multifunctional regulator of *norA* expression in *Staphylococcus aureus. J. Bacteriol.* **185:**3127–3138.
67. **Ubukata, K., N. Itoh-Yamashita, and M. Konno.** 1989. Cloning and expression of the *norA* gene for fluoroquinolone resistance in *Staphylococcus aureus. Antimicrob. Agents Chemother.* **33:**1535–1539.
68. **van Veen, H. W., R. Callaghan, L. Soceneantu, A. Sardini, W. N. Konings, and C. F. Higgins.** 1998. A bacterial antibiotic-resistance gene that complements the human multidrug-resistance P-glycoprotein gene. *Nature* **391:**291–295.
69. **van Veen, H. W., A. margolles, M. Muller, C. F. Higgins, and W. N. Konings.** 2000. The homodimeric ATP-binding cassette transporter LmrA mediates multidrug transport by an alternating two-site (two-cylinder engine) mechanism. *EMBO J.* **19:**2503–2514.
70. **van Veen, H. W., K. Venema, H. Bolhuis, I. Oussenko, J. Kok, B. Poolman, A. J. M. Driessen, and W. N. Konings.** 1996. Multidrug resistance mediated by a bacterial homolog of the human multidrug transporter MDR1. *Proc. Natl. Acad. Sci. USA* **93:**10668–10672.

71. **Woolridge, D. P., N. Vazquez-Laslop, P. N. Markham, M. S. Chevalier, E. W. Garner, and A. A. Neyfakh.** 1997. Efflux of the natural polyamine spermidine facilitated by the *Bacillus subtilis* multidrug transporter Blt. *J. Biol. Chem.* **272:**8864–8868.
72. **Yoshida, H., M. Bogaki, S. Nakamura, K. Ubukata, and M. Konno.** 1990. Nucleotide sequence and characterization of the *Staphylococcus aureus norA* gene, which confers resistance to quinolones. *J. Bacteriol.* **172:**6942–6949.
73. **Yu, J.-L., L. Grinius, and D. C. Hooper.** 2002. NorA functions as a multidrug efflux protein in both cytoplasmic membrane vesicles and reconstituted proteoliposomes. *J. Bacteriol.* **184:**1370–1377.
74. **Zeller, V., C. Janoir, M. D. Kitzis, L. Gutmann, and N. J. Moreau.** 1997. Active efflux as a mechanism of resistance to ciprofloxacin in *Streptococcus pneumoniae*. *Antimicrob. Agents Chemother.* **41:**1973–1978.
75. **Zheleznova, E. E., P. N. Markham, A. A. Neyfakh, and R. G. Brennan.** 1999. Structural basis of multidrug recognition by BmrR, a transcriptional activator of a multidrug transporter. *Cell* **96:**353–362.

PART IV. ANTIMICROBIAL RESISTANCE IN SELECT PATHOGENS

SECTION EDITOR: David G. White

BACTERIAL ANTIMICROBIAL RESISTANCE has become a global problem in both the medical and agricultural fields. Antibiotic-resistant strains of bacteria are an increasing threat to animal and human health, with resistance mechanisms having been identified and reported for all known antimicrobials currently available for clinical use in human and veterinary medicine. The predicament of antibiotic resistance is worsened by the facts that many of the more virulent bacterial strains have acquired resistance to multiple, structurally unrelated antimicrobials and that few new antimicrobials are likely to be available in the next few years.

The contributions within this section provide an overview of the resistance situation among select pathogens, including *Mycobacterium, Enterococcus, Staphylococcus, Pseudomonas,* and *Plasmodium,* as well as a description of the progress made to establish both standardized susceptibility testing methods and interpretive criteria (i.e., breakpoints) for potential bacterial agents of bioterrorism.

Frontiers in Antimicrobial Resistance: a Tribute to Stuart B. Levy
Edited by D. G. White, M. N. Alekshun, and P. F. McDermott

Chapter 21

Advances in Vancomycin Resistance: Research in *Staphylococcus aureus*

Keiichi Hiramatsu, Maria Kapi, Yutaka Tajima, Longzhu Cui, Suwanna Trakulsomboon, and Teruyo Ito

Staphylococcus aureus is a part of normal bacterial flora of healthy humans. Since the advent of antibiotic chemotherapy in the last century, bacteria inhabiting our body have gone through extensive assaults by various classes of antibiotics. However, *S. aureus*, and probably other members of our flora as well, did not stop inhabiting the human body. It persisted by successfully altering its genetic traits.

We have seen in the history of chemotherapy that *S. aureus* acquired penicillinase production soon after humans started using penicillin G in the 1940s, followed by the worldwide dissemination of multidrug-resistant *S. aureus* strains in the 1950s, which bore additional resistance to tetracycline, erythromycin, and streptomycin. The successful development of methicillin and cephem antibiotics in the 1960s was again quickly followed by the emergence and worldwide dissemination of methicillin- and cephem-resistant *S. aureus* (MRSA) in the 1970s. Now, MRSA is prevailing both inside and outside hospitals, gradually replacing methicillin-susceptible *S. aureus* (MSSA), which is part of the normal flora of healthy humans (48).

Until recently, vancomycin was the only antibiotic that escaped being registered in the list of antibiotics conquered by MRSA. Vancomycin is not a drug recently developed for MRSA but is an old drug discovered in 1956. Why did vancomycin become the drug of choice for MRSA infections? No past or current antibiotic has ever been developed with better anti-MRSA activity and fewer side effects. Therefore, the emergence of vancomycin resistance has posed a significant health-care problem in modern medicine, and it still does in countries where MRSA prevails. In this section of the book, we shall concentrate on the mechanisms and biological features of vancomycin resistance in MRSA. Readers interested in the emergence and evolution of MRSA and mechanism of methicillin resistance are referred to other review articles (30).

EMERGENCE OF VANCOMYCIN RESISTANCE IN MRSA

Two classes of vancomycin resistance have been identified in MRSA. One has low-level resistance (L-VRSA, low-level vancomycin-resistant *S. aureus*) and the other high level (H-VRSA, high-level vancomycin-resistant *S. aureus*). L-VRSA is well known as VISA (vancomycin-intermediate *S. aureus*) in the United States (U.S.) since its vancomycin MIC is either 8 or 16 mg/liter, which, according to the National Committee for Clinical Laboratory Standards (NCCLS) of the U.S., are designated "intermediate" (between "susceptible" and "resistant"). On the other hand, the MICs for H-VRSA (so far three strains have been independently identified in the U.S.) are reported to be 32, 64, and 1,024 mg/liter, and they are called "VRSA" (vancomycin-resistant *S. aureus*) (2, 4, 5).

However, NCCLS breakpoints for vancomycin has not correlated well with the clinical therapeutic efficacy of vancomycin. According to clinical experience, VISA infection is refractory to vancomycin therapy, and even MRSA strains with MIC of 4 mg/liter are known to cause infection (17).

The authors first isolated clinical VISA strain Mu50 in 1996 from a 4-month old infant patient who underwent heart surgery on pulmonary atresia (25, 32). The postoperative MRSA wound infection resisted 29 days of vancomycin therapy (45 mg/kg per

Keiichi Hiramatsu, Maria Kapi, Yutaka Tajima, Longzhu Cui, and Teruyo Ito • Department of Bacteriology, Juntendo University, 2-1-1 Bunkyo-Ku, Tokyo, Japan 113-8421. **Suwanna Trakulsomboon** • Department of Medicine, Faculty of Medicine Siriraj Hospital, Mahidol University, Bangkoknoi, Bangkok 10700, Thailand.

day). However, the wound healed after an additional 12 days of vancomycin therapy combined with arbekacin (30 mg/kg per day), an aminoglycoside with potent anti-MRSA activity. The wound infection relapsed after 12 days of cessation of the combination therapy, and complete resolution required 23 days' combination therapy of sulbactam/ampicillin (300 mg/kg per day) and arbekacin (30 mg/kg per day) (25). This clinical course of the infection showed that therapeutic failure of vancomycin was caused by the vancomycin resistance of strain Mu50 rather than due to other factors such as the immunocompromised status of the patient or insufficient dosing of vancomycin.

Isolation of similar strains from clinical settings was subsequently reported from 10 countries, including the U.S. (Table 1). On the other hand, VRSA strains were first isolated in 2002 from a diabetic patient in Michigan (2, 6), and so far all three strains in the world were isolated from the U.S. (Table 1).

THE MECHANISM OF VANCOMYCIN RESISTANCE

Horizontal Transfer and Mutation

VRSA strains isolated from three American patients contained plasmids carrying the *vanA* gene complex. The gene complex is carried on a transposon which seems to have been transferred by conjugation from vancomycin-resistant enterococcus (VRE) that coexisted in the patients' bodies (2, 6). This interspecies conjugation has been known to occur since 1992, when Noble demonstrated it by experimental transfer of vancomycin resistance from VRE to *S. aureus* in the laboratory (47).

The enzyme encoded by the *vanA* gene, together with those encoded by the adjacent genes *vanH*, *vanX*, and *vanY*, replaces D-alanyl-D-alanine residues of *S. aureus* peptidoglycan by D-alanyl-D-lactate (52). Since D-alanyl-D-alanine is the target of vancomycin, and given that it cannot bind to D-alanyl-D-lactate with a

Table 1. Isolation of hetero-VISA, VISA, and VRSA strains from various countries

Year	Category of resistance	Country (state)[a]	Reference
1997	VISA	Japan	32
1997	Hetero-VISA	Japan	29
1998	VISA	France	53
1998	Hetero-VISA	UK	34
1999	VISA	US (MI, NJ)	60
1999	VISA	US (NY)	54
1999	VISA	Germany	7
1999	Hetero-VISA	Germany	18
1999	Hetero-VISA	Japan	23
1999	Hetero-VISA	Hong Kong	68
2000	Hetero-VISA	Italy	44
2000	VISA	US (IL)	3
2000	VISA and hetero-VISA	France	19
2000	VISA	Korea	38
2000	VISA	UK	33
2000	VISA	South Africa	16
2001	VISA	Brazil	50
2001	Hetero-VISA	Australia	66
2001	VISA	US	20
2001	VISA	UK	51
2001	VISA	France	8
2001	Hetero-VISA	Thailand	64
2002	VISA and hetero-VISA	Belgium	12
2002	VISA	Greece	65
2002	Hetero-VISA	Korea	37
2002	VRSA	US (MI)	2, 6
2002	VRSA	US (PA)	5
2003	VISA	Brazil	1
2003	VISA	US (MN)	46
2003	VISA and SA-RVS[b]	US	17
2003	VISA and hetero-VISA	France	13
2004	VISA and hetero-VISA	France	9
2004	VRSA	US (NY)	4

[b] UK, United Kingdom; MI, Michigan; NJ, New Jersey; NY, New York; IL, Illinois; PA, Pennsylvania; and MN, Minnesota.

[b] *S. aureus* with reduced vancomycin susceptibility.

high affinity as it does to D-alanyl-D-alanine, the bacterial cells under *vanA* gene expression become highly resistant to vancomycin (52). The mechanism seems essentially the same as the one elaborated with VRE (56), so interested readers are recommended to read chapters 8 and 22.

S. aureus with low-level vancomycin resistance, VISA, and its precursor hetero-VISA (see below), on the other hand, emerge through accumulation of mutations in the genome of vancomycin susceptible *S. aureus*.

Historical Consideration on the Emergence of Vancomycin Resistance in *S. aureus*

So far only three VRSA strains have been identified in the world, and all are clinical isolates from the U.S. It is evident that high prevalence of VRE in U.S. hospitals paved the way for the emergence of *vanA* gene transfer to MRSA. Rapid and steady rise in the number of MRSA and VRE in U.S. hospitals in the past two decades is another contributing factor, increasing the chance of encounter of the two species in patients' bodies (15). In parallel with this, the increase of multiresistant cocci is due to a significant increase in vancomycin use during the last two decades, which evidently served as a selective pressure for the emergence of VISA and VRSA in U.S. hospitals (24, 39).

The authors suspect that the significant rise of MRSA and VRE in U.S. hospitals is correlated with the generous use of broad-spectrum cephalosporins such as ceftriaxone and ceftazidime based on historical experience in Japan. The nation-wide use of third-generation cephalosporins in Japan in the early 1980s coincided with the abrupt and dramatic rise in MRSA detection rates (62).

After the mid-1980s, Japanese physicians, alerted by the abrupt increase of MRSA infections, started treating MRSA with the more potent beta-lactam antibiotics. Imipenem and several other beta-lactam antibiotics were frequently used for this purpose. This practice was inevitable, because the injectable form of vancomycin was not available in Japan until 1991. Introduction of another anti-MRSA antibiotic, arbekacin, was not launched until 1990.

As a result of this unique nation-wide use of antibiotics, the most prevalent MRSA in the early 1980s, clonotype 2B-30, defined by carriage of type-2B staphylococcal cassette chromosome *mec* (SCC*mec;* 36) and chromosome multilocus sequence type (MLST) 30 (14), disappeared. It was replaced by MRSA clone 2A-5, previously designated clonotype II-A (26, 62).

MRSA strains of clonotype 2A-5 were widespread in the U.S. in the 1980s but not in Europe (Ito et al., unpublished). Seven of eight American VISA strains as well as Japanese and Korean VISA strains belong to this MRSA clonotype. Two recently isolated VRSA strains bear this clonotype as well (45). This indicates that this MRSA clonotype is particularly suited for the generation of vancomycin resistance (14, 29).

Although clonotype 2A-5 is apparently suited for yielding VISA strains, it would be pertinent to point out that the level of vancomycin resistance of these isolates may be acquired practically from any MRSA genotype. An even much greater level of resistance could be achieved by repeated in vitro selection with vancomycin on strain COL of clonotype 1B-8 (58). However, it is possible that the genetic alteration causing vancomycin resistance (whether it is a mutation or acquisition of an exogenous gene [see below]) may be extremely costly in certain genetic backgrounds. The case of COL is illustrative: although it generated a highly vancomycin-resistant mutant (MIC > 100 mg/liter) by repeated selection, the mutant's success was greatly flawed by its extremely slow growth rate and loss of methicillin resistance. Other than vancomycin, the most important feature for MRSA strains is their ability to survive in the hospital environment (59).

Mechanism of Resistance of VISA

In the following section, we will discuss the mechanisms of vancomycin resistance. To understand this, knowledge of peptidoglycan (murein) synthesis is essential. Since assigned space precludes the detailed explanation, readers who are not familiar with the subject are recommended to read a more detailed account elsewhere (28).

Unlike VRSA, the common feature of resistance in VISA is a thickened cell wall (10). It was demonstrated by measuring peptidoglycan thickness using transmission electron microscopy. Since vancomycin has to bind to the D-alanyl-D-alanine residues of murein monomers present on the cytoplasmic membrane, thickened peptidoglycan layers outside the cytoplasmic membrane serve as an obstacle for access of vancomycin to its targets (11). This is because *S. aureus* peptidoglycan layers (usually about 20 layers thick) are known to have as many as 6×10^6 D-alanyl-D-alanine residues that serve as "false binding targets" for vancomycin (27). Vancomycin is thus "affinity trapped" inside the peptidoglycan layers by the false targets that are present in greater amounts in the cell wall of VISA strains relative to vancomycin susceptible *S. aureus* (VSSA) (11).

In the final step of peptidoglycan synthesis, the D-alanyl-D-alanine residues of murein monomers that constitute nascent peptidoglycan chains are split by the transpeptidase function of penicillin-binding proteins (PBPs). PBPs then catalyze cross-linking of the penultimate D-alanine residue of the nascent peptidoglycan chain to the pentaglycine chain of the preexisting peptidoglycan layer (28). Thus, PBP action reduces the number of false targets, D-alanyl-D-alanine residues, from the peptidoglycan. However, PBPs tend to skip the reaction and do not process all the murein monomer units in the cell wall, thus leaving 20 to 30% of the D-alanyl-D-alanine residues uncleaved (61). As a result, a suitable number of false targets remain in the peptidoglycan layers.

In the cell wall of Mu50, peptidoglycan cross-linking is significantly decreased as compared to hetero-VISA strain Mu3 or VSSA strains (22). That is, there are more D-alanyl-D-alanine false targets in the cell-wall peptidoglycan layers of Mu50 than in those of control strains. In fact, purified peptidoglycan of the Mu50 cell absorbs 2.4 times more vancomycin molecules than that of vancomycin-susceptible strain FDA209P (21).

Not only does the thickened cell wall absorb vancomycin and reduce its effective concentration from the infected tissue, it also retards the timing and action of vancomycin on the cytoplasmic target (Fig. 1). A thickened cell wall compared to a thin cell wall delays full saturation of vancomycin targets on the cytoplasmic membrane by as long as 30 to 60 min (Fig. 1). We consider this delay a manifestation of the "clogging" of the outer layers of peptidoglycan mesh structure by bound vancomycin molecules (Cui, L. et al., in preparation). This effect of a thickened cell wall further consumes vancomycin since the cell can continue production of more cell-wall peptidoglycan from beneath the preexisting layers. Moreover, the organism might even complete a round of cell division, since the time required to stop cell wall synthesis is comparable to or even longer than the doubling time of *S. aureus* (usually 20 to 40 min).

Genetic Basis for Vancomycin Resistance in VISA

The vancomycin resistance of VISA is not due to the acquisition of a resistance gene from another bacterial species. It is generated spontaneously from VSSA strains in vitro, though the development of the VISA phenotype does not occur through a single-step selection process (27, 67). Starting with VSSA strains, 19 cycles of passage on increasing concentrations of vancomycin are required to obtain VISA mutants. This prompted Schaaff et al. to hypothesize a contribution of a mutator genotype in the emergence of VISA strains (55). However, there have been no apparent mutator phenotypes or *mutS* mutations demonstrated in clinical MRSA isolates or VISA strains (50). Therefore it is more probable that *S. aureus* has gone through re-

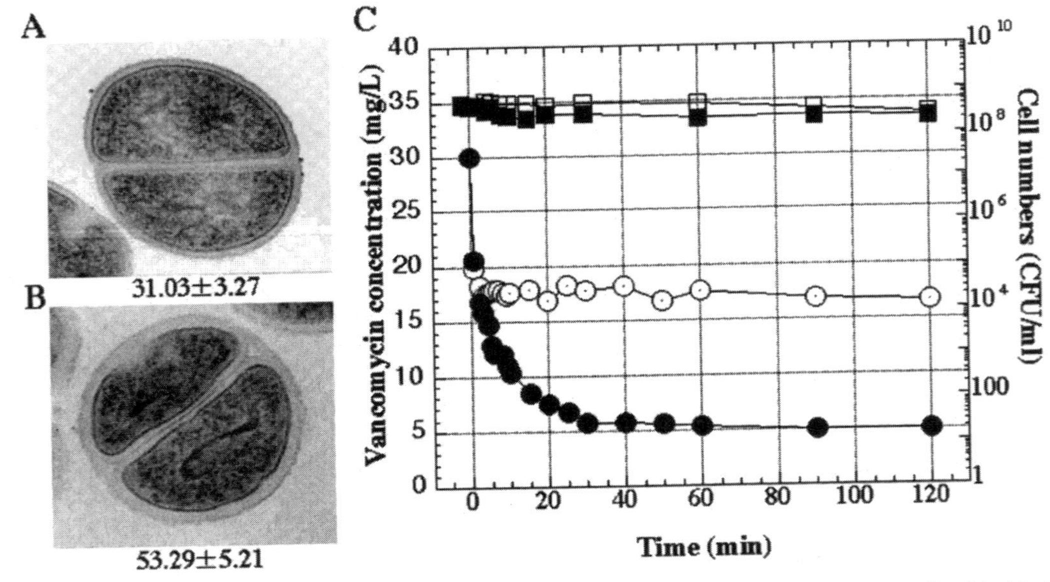

Figure 1. Thick cell wall prevents the saturation of peptidoglycan layers with vancomycin. The Mu50 cells with thin (A) and thick cell wall (B) were prepared by incubating the cells in resting media (RMs) with different nutrient compositions (11). RM does not support the cell growth. The numbers under the panels are mean and SD of cell-wall thickness in nm. (C) The cells were incubated in RM containing 30 mg/liter of vancomycin, and the time course of vancomycin consumption was monitored by HPLC. Symbols: squares, the cell densities of thin (open) and thick (closed) cell wall. Circles, vancomycin concentration in the culture supernatant of the cells with thin (open) and thick (closed) cell wall. Note that the cells with thick cell walls consume twice as much vancomycin as those with thin cell walls. Consumption of vancomycin by the cells with thick cell walls does not reach maximum before 60 min, whereas that by the latter cells reaches maximum before 5 min.

peated exposure to vancomycin and gradually accumulated many mutations to finally achieve VISA status. Hetero-VISA Mu3 shares remarkably many of the unusual phenotypes of Mu50 such as increased PBP2 expression, increased uptake of *N*-acetylglucosamine, and increased murein monomer pool sizes (21). This agrees well with this hypothesis that Mu3 has gone through many steps of genetic alterations but is still one or two more mutations away from those found in Mu50 (29). Hetero-VISA also expresses raised resistance (MIC, 8~32 mg/liter) to teicoplanin (another glycopeptide) (21). Thus, prevalence of hetero-VISA is in the "prelude" status for the emergence of VISA in hospitals (27).

Hetero-VISA can be selected in vitro from certain VSSA strains of clonotype 2A-5. Though modest in the rise of the level of resistance when measured by MIC assays, the mutants obtained are quite distinct from VSSA in that VISA strains frequently emerge (Fig. 2). Figure 2 shows the population analysis of *S. aureus* strains with various degrees of vancomycin susceptibilities. The growths of two VSSA strains, FDA209P (MSSA) and H1 (MRSA), are inhibited completely (inoculum sizes, 10^6~10^7 colony forming units [CFU]) by the presence of 4 mg/liter of vancomycin. On the other hand, almost all inoculated Mu50 cells grow in the presence of 4 mg/liter of vancomycin.

Mu3 gives rise to resistant subpopulations that are capable of growing in greater than 4 mg/liter of vancomycin. The subpopulations are mutants with increased vancomycin resistance, the level of which depends on the concentrations of vancomycin used for selection—the greater the concentration, the greater the resistance level of the selected mutants (29).

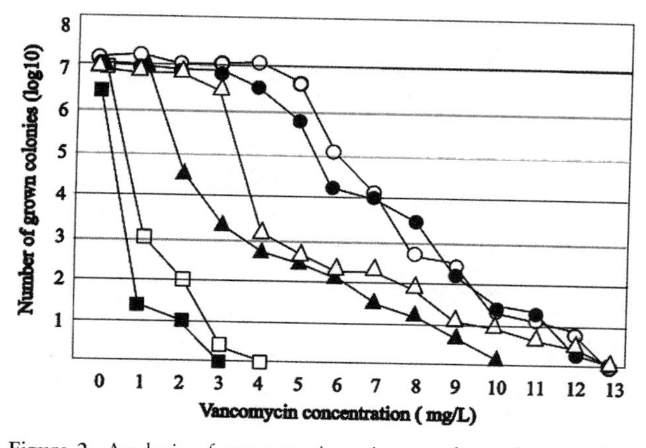

Figure 2. Analysis of vancomycin-resistant subpopulations of *S. aureus* strains. Plating efficiency test is performed with maximum inoculum size of about 10^7 CFU onto BHI agar plates of 8.5 cm in diameter containing various concentrations of vancomycin. Strains analyzed are Mu50 (open circle), PC27-14 (closed circle), PC27 (open triangle), Mu3 (closed triangle), H1 (open square), FDA209P (closed square). Note that PC27 and PC27-14 were obtained from Mu3 cell population based on their colony sizes alone without recourse to drug selection.

A remarkable feature of hetero-VISA is that it can be obtained from the VSSA strain by selection with beta-lactam antibiotics. Imipenem, for example, can be used to obtain hetero-VISA mutants. This is shown by using ΔIP, a heterogeneously methicillin-resistant *S. aureus* (hetero-MRSA) strain that still retains susceptibility to such potent beta-lactam antibiotics as imipenem (imipenem and oxacillin MICs, 8 and 32 mg/liter, respectively). By imipenem selection of ΔIP, homogeneously and highly methicillin-resistant *S. aureus* (homo-MRSA) mutants whose imipenem and oxacillin MICs are 128 mg/liter and 512 mg/liter, respectively, are obtained at (high) frequency of about 1×10^{-4}. A part of these mutant strains (5 to 10% in repeated experiments) are found to have raised resistance to glycopeptide antibiotics as well (28).

One of the imipenem-selected mutant strains, designated ΔIP-H14, has greatly increased transcription of *vraS* that encodes for the sensor kinase of a two-component system (*vraSR*) (40). We identified *vraSR* as the genes whose transcriptions were enhanced in Mu50 as compared to a vancomycin-susceptible strain Mu50ω (MIC, 1 mg/liter) (41). The latter strain was isolated from the patient from whom VISA strain Mu50 was isolated. They share identical pulsed-field gel electrophoresis (PFGE) patterns but exhibit dramatic differences in their susceptibilities to vancomycin. Therefore, *vraSR* is considered to be associated with the raised expression of the vancomycin resistance of Mu50.

By gene-specific deletion and complementation experiments, we found that *vraSR* is the two-component system that is required for beta-lactam- and vancomycin-induced transcription of *murZ, pbp2,* and *sgtB,* the genes encoding important enzymes of cell-wall peptidoglycan synthesis (40). MurZ is UDP-*N*-acetylglucosamine enolpyruvyl transferase, which is essential in murein monomer synthesis. PBP2 and SgtB (monofunctional glycosyltransferase) are involved in polymerization of murein monomers to complete cross-linked peptidoglycan layers. PBP2 is a peptidoglycan transpeptidase, and its artificial overexpression causes increased resistance to vancomycin and teicoplanin (21). Thus, *vraS*'s overexpression seems to have caused glycopeptide and imipenem resistance through the up-regulation of *S. aureus* cell-wall synthesis (40).

In other VISA strains, different genetic alterations seem to cause vancomycin resistance by producing a thickened cell wall. Recently Maki et al. showed via gene knockout experiments that inactivation of *tcaA* caused increased teicoplanin resistance, a concomitant marginal increase of vancomycin resistance, and a heteroresistant subpopulation profile (43). Truncation of *tcaA* has been demonstrated in vitro and in a clinical VISA strain from Michigan (43). The mechanism(s) of how the *tcaA* gene causes vancomycin hetero-resistance remains to be seen.

So far *vraSR* and *tcaA* have been the genes most clearly associated with the vancomycin hetero-resistance phenotype. Further studies are likely to uncover alternative genes that contribute to the production of a thickened cell wall. However, it is also clear that these gene alterations alone cannot achieve such a drastic change in the cell physiology without fatally affecting other physiological functions of the cell. Definitely, another mutation or combination of mutations is necessary to occur besides the alteration of *vraSR* or *tcaA* for VSSA cells to achieve VISA status. The second step mutation that converts hetero-VISA into VISA is still unknown. Alteration of sugar metabolism may be involved since Mu50 incorporates a high proportion of glucose taken from the medium into cell-wall peptidoglycan, whereas Mu3 does not. Whole genome sequencing of strain Mu3 is now underway. Comparison of the Mu3 genome with that of Mu50 will give us a clue regarding the cycle of genetic events in the acquisition of vancomycin resistance in the Mu3-Mu50 genetic backgrounds.

HETERO-VISA: A UNIQUE PARADIGM OF VANCOMYCIN RESISTANCE

On the High Incidence of Mutations in Hetero-VISA

Mu3 generates various mutants with varied degrees of reduced susceptibility to vancomycin. The appearance of resistant colonies on BHI agar with a vancomycin MIC of 4 mg/liter fluctuates, indicating that they are generated spontaneously in a subpopulation prior to the exposure to vancomycin. The mutation rate is on the order of 10^{-7}, which is higher than the mutation rate for spontaneous rifampin resistance in Mu3 ($\sim 1 \times 10^{-8}$). Therefore, we cannot assume a mutator phenotype to explain the high incidence of VISA mutants in Mu3 cell populations. Instead we hypothesized a "regulator mutation" to account for the high incidence of phenotypic conversions concentrated in the cell-wall-associated physiology (31).

An illustrative example for this concept is the mutational inactivation of *mecI,* whose gene product plays a role in regulating methicillin resistance (42). MecI is a repressor of *mecA*-gene transcription. Since *mecA* encodes PBP2′, which is responsible for methicillin resistance, the strain carrying the intact *mecI* gene does not express methicillin resistance. Such strains are represented by N315 and are thus called pre-MRSA (42). However, N315 spontaneously produces MRSA mutants at a high frequency (1×10^{-4}) by mutational inactivation of *mecI* (42).

A high-thoroughput sequencing strategy for regulator genes in combination with microarray transcription profiling in many isogenic VISA and hetero-VISA combinations might reveal several alternative series of regulator mutations.

Another Unique Feature of Hetero-VISA

Besides the heterogeneity in the degrees of resistance, hetero-VISA (and VISA to a lesser degree) exhibits extreme heterogeneity in the colony sizes when streaked onto drug-free agar plates (Fig. 3). This is a characteristic phenotype of hetero-VISA and is considered to reflect a resistance mechanism(s) underlying their vancomycin resistance phenotype. The VISA phenotype is closely associated with increased cell-wall thickness and decreased doubling time (i.e., growth rate) of the strains (10). Therefore, the observed heterogeneity in the colony sizes is a logical consequence of heterogeneity in vancomycin resistance.

Table 2 shows the list of subpopulations of Mu3 classified based on the colony size and timing of the colony formation on BHI agar plates. It is noticeable that Mu3 contains at least five distinct colony size classes (CSC): RC, SC, PC, C2, and C3. The latter three may be regarded as "small colony variants (SCV)," though they have no discreet difference from parent strain Mu3 in terms of auxotrophy known to characterize extant SCV strains observed in association with resistance acquisition against other antibiotics.

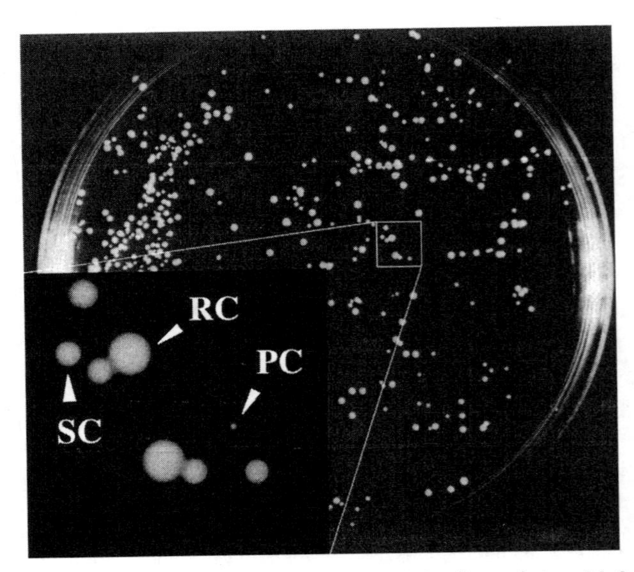

Figure 3. Heterogeneous colony sizes in Mu3 cell population. Mu3 cells were spread on brain heart infusion agar plate and incubated for 24 h at 37°C before the photo was taken. Inset: magnification of the area in the square. RC, regular sized colony; SC, small colony; PC, pin-point colony.

Table 2. Hetero-VISA strain Mu3 is composed of cells of various colony size classes

CSC on BHI agar at 37°C[a]	Frequency (%)	Mean doubling time ± SD (min)	No. of strains tested	No. of strains with vancomycin MIC (mg/liter) as measured at:											
				24 h			48 h			72 h					
				2	3	4	2	3	4	2	3	4	5	6	7
RC	29.3	28.84 ± 1.49	15	14	1	0	6	9	0	4	11	0	0	0	0
SC	69.5	33.34 ± 0.96	16	16	0	0	3	13	0	1	15	0	0	0	0
PC	0.96	36.57 ± 1.31	29	26	3	0	4	25	0	1	25	1	1	0	1
C2	0.097	50.6[b]	1	1	0	0	0	1	0	0	1	0	0	0	0
C3	0.143	69.65[b]	1	1	0	0	1	0	0	0	1	0	0	0	0

[a] CSC, colony size class: RC, regular colony formed after 24 h; SC, small colony after 24 h; PC, pinpoint colony formed after 24 h; C2, colony formed after 48 h; C3, colony formed after 72h.
[b] C2 and C3 are mostly unstable. One stable strain each was analyzed in this study.

As shown in Table 2, some of the strains belonging to PC have relatively high vancomycin resistance compared to those in other CSCs. PC27 shows a typical heterogeneous population profile (Fig. 2), and VISA mutants are selected at a high frequency (1 x 10^{-4}) in this genetic background.

PC constitutes less than 1% of the Mu3 cell population (Table 2), whereas RC and SC, having lower levels of resistance, dominate the Mu3 population. Mu3, since the initial colony isolation, seems to have accumulated significant number of "degraded" subpopulations with reduced vancomycin resistance in exchange for better growth capabilities (fitter in the drug-free condition). We also assume, however, that the Mu3 cell population still contains the "original Mu3" from which all the degraded subpopulations were derived. The original Mu3 cell subpopulation should satisfy the following requirements: (i) it has a typical heterogeneous type population curve; (ii) it can generate all kinds of "degraded" cells that are found in the Mu3 population, including those belonging to RC and SC; and (iii) it can generate VISA cells within its cell population at a high frequency (presumably at a greater frequency than the Mu3 used here). Strain PC27 satisfies all of these requirements.

As Figure 2 shows, PC27 has a typical heterogeneous population profile, and vancomycin exposure selects VISA mutants from PC27 at much higher frequency than from Mu3. In addition, one of the C2 colonies, PC27-14, found in the cell population of PC27 turned out to be very similar to Mu50 (Fig. 2). Thus, it is clear that PC27 spontaneously generates VISA within its cell population prior to exposure to vancomycin.

Clinical Significance of Hetero-VISA

The clinical significance of hetero-VISA is twofold. First, it generates VISA at high frequencies, thus hetero-VISA initiates refractory infection that eventually yields VISA after prolonged vancomycin therapy (29). Recently, this sequential event was proven to occur in vivo in an MRSA septicemia case (57). It is noteworthy that the patient was initially infected with hetero-VISA not with VSSA before the vancomycin treatment started. This again confirms our prediction that hetero-VISA is the essence of the problem, and an increase of hetero-VISA in a hospital is an ominous sign for emergence of VISA infections (27–29).

Another clinical significance of hetero-VISA is that it causes refractory infection in certain clinical circumstances. From our experience, the lower respiratory tract is one of the most vulnerable tissues for refractory infection caused by hetero-VISA. This probably is correlated with the limited concentration of vancomycin in tissue (2 to 3 mg/liter as measured in sputa), achieved with a regular drip-infusion regimen (29). In fact, a retrospective chart review of pneumonia cases in Juntendo Hospital suggested a correlation between the hetero-VISA phenotype of causative agents and vancomycin therapeutic failure (27). Our subsequent prospective study in collaboration with the infection control team at Mahidole University, Thailand, showed that the lower respiratory tract was the preferred site of hetero-VISA infection, and that hetero-VISA pneumonia cases resisted vancomycin therapy, resulting in greater mortality rate than with VSSA pneumonia cases (Trakulsomboon et al., in preparation). Therefore, if we perform more detailed analyses of the clinical efficacy of vancomycin therapy, the clinical impact of hetero-VISA would turn out to be more significant than we expected.

In this regard, it is noteworthy that an unique response to vancomycin therapy may occur with hetero-VISA infection since greater than 99.9% of its cell subpopulations are susceptible to 4 mg/liter of vancomycin (31). In fact, the patient infected with Mu3 was illustrative in this regard (25). Mu3 was isolated from the sputum of a 64-year-old male patient who developed MRSA pneumonia. The pneumonia responded favorably to vancomycin therapy (2 g/day) for 7 days, and the patient's condition improved gradually. His infection deteriorated during the next 4 days in spite of continued vancomycin therapy (25). This double-phased response of pneumonia was also observed with another Japanese patient infected with hetero-VISA strain (35).

DEVELOPING A NOVEL DETECTION METHOD OF VISA AND HETERO-VISA STRAINS

Antibiotic susceptibility tests performed in clinical microbiology laboratories are limited in efficient detection of VISA strains (63). Detection of hetero-VISA is utterly hopeless as long as we only stick to extant antibiotic susceptibility tests. Hetero-resistance presents a novel concept in antibiotic resistance; therefore, its detection requires a novel detection method. Population analysis is the gold standard for the detection of hetero-resistance (29, 69), but the method is not suited for testing many samples on routine basis. Therefore, quicker and simpler method is definitely needed for the detection of VISA and hetero-VISA clinical isolates.

NCCLS-based MIC determination is the "gold standard" for the identification of VISA (63). However, since the resistance of VISA is dependent on the production of a thick cell wall and MH recommended by NCCLS does not fully support the cell-wall thickening as BHI does, the method is not sensitive enough to detect VISA strains (11). This may explain why MRSA with vancomycin MIC 4 mg/liter measured in MH broth resisted vancomycin therapy like VISA does in actual infection (17). Therefore, MRSA strains with MIC 4 mg/liter in MH broth should be tested with a population analysis method.

In any case, some VISA strains exhibit borderline resistance to vancomycin with a MIC between 4 and 5 mg/liter (growth in 4 mg/liter but not in 5 mg/liter of the drug). Therefore, we need to develop more sensitive methods for the detection of VISA. Figure 4 illustrates the results of a novel method that detects metabolic activity of the viable cells in the presence of various concentrations of vancomycin. After 3 h incubation with vancomycin, menadione is added to reflect the reducing power of the viable cells, which is then detected by chemiluminescence based on alkaline-Lucigenin method after 1 h of incubation (70). Using this method, 12 VISA strains from seven countries could be clearly distinguished from 30 VSSA strains of various genetic backgrounds (Tajima et al., in preparation). As expected, hetero-VISA strains are not detectable using the same protocol. Adjustment of protocol, however, for hetero-VISA detection is underway.

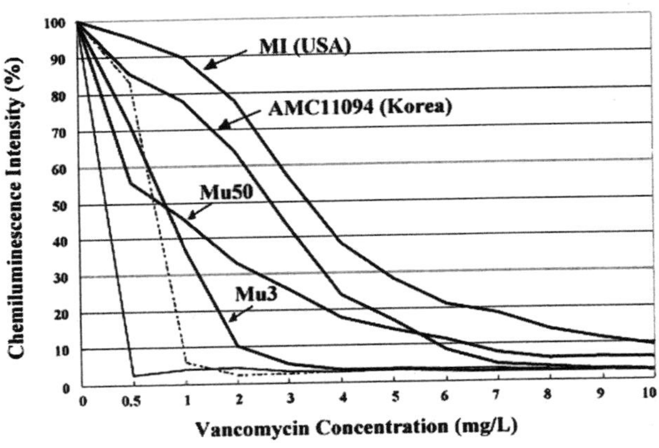

Figure 4. Novel detection method of VISA strains. The chemiluminescence method to detect metabolism of viable cells after 3 h exposure of the cells to vancomycin. Three representative VISA strains, Mu50, MI, and AMC11094, are clearly distinguished from VSSA strains FDA209P (thin line) and NCTC8325 (dotted line). Hetero-VISA strain Mu3 has a marginal pattern and is not well differentiated from VSSA with this protocol.

CONCLUSION

Considering that MRSA is still a ubiquitous problem in hospitals in every country, and vancomycin is continuously used as the first line antibiotic in most hospitals for gram-positive infections, hetero-VISA will stay as a significant threat to hospital infection control. Now, MRSA is a part of our natural flora and will continue to be so in the future. Therefore, elucidation of the secret of this organisms's genetic plasticity and its "dangerous neighbors" is of primary importance for us to protect ourselves by recognizing and outwitting its glaring strategies.

Acknowledgments. Molecular microbiological study on MRSA was supported by a Grant-in-Aid for the 21st Century Center of Excellence (COE) Research Program and for Scientific Research on Priority Areas (13226114) of Japanese Ministry of Education, Culture, Sports, Science and Technology. The prospective study of hetero-VISA was supported by Core University System Exchange Program under the Japan Society for the Promotion of Science (JSPS), coordinated by the University of Tokyo Graduate School of Medicine and Mahidol University.

REFERENCES

1. **Andrade-Baiocchi, S., M. C. Tognim, O. C. Baiocchi, and H. S. Sader.** 2003. Endocarditis due to glycopeptide-intermediate *Staphylococcus aureus:* case report and strain characterization. *Diagn. Microbiol. Infect. Dis.* **45:**149–152.
2. **Anonymous.** 2002. *Staphylococcus aureus* resistant to vancomycin—United States, 2002. *Morb. Mortal. Wkly. Rep.* **51:** 565–567.
3. **Anonymous.** 2000. *Staphylococcus aureus* with reduced susceptibility to vancomycin—Illinois, 1999. *Morb. Mortal. Wkly. Rep.* **48:**1165–1167.
4. **Anonymous.** 2004. Vancomycin-resistant *Staphylococcus aureus*—New York, 2004. *Morb. Mortal. Wkly. Rep.* **53:**322–323.

5. **Anonymous.** 2002. Vancomycin-resistant *Staphylococcus aureus*—Pennsylvania, 2002. *Morb. Mortal. Wkly. Rep.* **51:**902.
6. **Bartley, J.** 2002. First case of VRSA identified in Michigan. *Infect. Control Hosp. Epidemiol.* **23:**480.
7. **Bierbaum, G., K. Fuchs, W. Lenz, C. Szekat, and H. G. Sahl.** 1999. Presence of Staphylococcus aureus with reduced susceptibility to vancomycin in Germany. *Eur. J. Clin. Microbiol. Infect. Dis.* **18:**691–696.
8. **Bobin-Dubreux, S., M. E. Reverdy, C. Nervi, M. Rougier, A. Bolmstrom, F. Vandenesch, and J. Etienne.** 2001. Clinical isolate of vancomycin-heterointermediate *Staphylococcus aureus* susceptible to methicillin and in vitro selection of a vancomycin-resistant derivative. *Antimicrob. Agents Chemother.* **45:**349–352.
9. **Cartolano, G. L., M. Cheron, D. Benabid, M. Leneveu, and A. Boisivon.** 2004. Methicillin-resistant *Staphylococcus aureus* (MRSA) with reduced susceptibility to glycopeptides (GISA) in 63 French general hospitals. *Clin. Microbiol. Infect.* **10:**448–451.
10. **Cui, L., X. Ma, K. Sato, K. Okuma, F. C. Tenover, E. M. Mamizuka, C. G. Gemmell, M. N. Kim, M. C. Ploy, N. El-Solh, V. Ferraz, and K. Hiramatsu.** 2003. Cell wall thickening is a common feature of vancomycin resistance in *Staphylococcus aureus*. *J. Clin. Microbiol.* **41:**5–14.
11. **Cui, L., H. Murakami, K. Kuwahara-Arai, H. Hanaki, and K. Hiramatsu.** 2000. Contribution of a thickened cell wall and its glutamine nonamidated component to the vancomycin resistance expressed by *Staphylococcus aureus* Mu50. *Antimicrob. Agents Chemother.* **44:**2276–2285.
12. **Denis, O., C. Nonhoff, B. Byl, C. Knoop, S. Bobin-Dubreux, and M. J. Struelens.** 2002. Emergence of vancomycin-intermediate *Staphylococcus aureus* in a Belgian hospital: microbiological and clinical features. *J. Antimicrob. Chemother.* **50:**383–391.
13. **El Solh, N., M. Davi, A. Morvan, H. A. Damon, and N. Marty.** 2003. Characteristics of French methicillin-resistant *Staphylococcus aureus* isolates with decreased susceptibility or resistance to glycopeptides. *J. Antimicrob. Chemother.* **52:**691–694.
14. **Enright, M. C., D. A. Robinson, G. Randle, E. J. Feil, H. Grundmann, and B. G. Spratt.** 2002. The evolutionary history of methicillin-resistant *Staphylococcus aureus* (MRSA). *Proc. Natl. Acad. Sci. USA* **99:**7687–7692.
15. **Farr, B. M., C. D. Salgado, T. B. Karchmer, and R. J. Sherertz.** 2001. Can antibiotic-resistant nosocomial infections be controled? *Lancet Infect. Dis.* **1:**38–45.
16. **Ferraz, V., A. G. Duse, M. Kassel, A. D. Black, T. Ito, and K. Hiramatsu.** 2000. Vancomycin-resistant *Staphylococcus aureus* occurs in South Africa. *S. Afr. Med. J.* **90:**1113.
17. **Fridkin, S. K., J. Hageman, L. K. McDougal, J. Mohammed, W. R. Jarvis, T. M. Perl, and F. C. Tenover.** 2003. Epidemiological and microbiological characterization of infections caused by *Staphylococcus aureus* with reduced susceptibility to vancomycin, United States, 1997–2001. *Clin. Infect. Dis.* **36:**429–439.
18. **Geisel, R., F. J. Schmitz, L. Thomas, G. Berns, O. Zetsche, B. Ulrich, A. C. Fluit, H. Labischinsky, and W. Witte.** 1999. Emergence of heterogeneous intermediate vancomycin resistance in *Staphylococcus aureus* isolates in the Dusseldorf area. *J. Antimicrob. Chemother.* **43:**846–848.
19. **Guerin, F., A. Buu-Hoi, J. L. Mainardi, G. Kac, N. Colardelle, S. Vaupre, L. Gutmann, and I. Podglajen.** 2000. Outbreak of methicillin-resistant *Staphylococcus aureus* with reduced susceptibility to glycopeptides in a Parisian hospital. *J. Clin. Microbiol.* **38:**2985–2988.
20. **Hageman, J. C., D. A. Pegues, C. Jepson, R. L. Bell, M. Guinan, K. W. Ward, M. D. Cohen, J. A. Hindler, F. C. Tenover, S. K. McAllister, M. E. Kellum, and S. K. Fridkin.** 2001. Vancomycin-intermediate *Staphylococcus aureus* in a home health-care patient. *Emerg. Infect. Dis.* **7:**1023–1025.
21. **Hanaki, H., K. Kuwahara-Arai, S. Boyle-Vavra, R. S. Daum, H. Labischinski, and K. Hiramatsu.** 1998. Activated cell-wall synthesis is associated with vancomycin resistance in methicillin-resistant *Staphylococcus aureus* clinical strains Mu3 and Mu50. *J. Antimicrob. Chemother.* **42:**199–209.
22. **Hanaki, H., H. Labischinski, Y. Inaba, N. Kondo, H. Murakami, and K. Hiramatsu.** 1998. Increase in glutamine-non-amidated muropeptides in the peptidoglycan of vancomycin-resistant *Staphylococcus aureus* strain Mu50. *J. Antimicrob. Chemother.* **42:**315–320.
23. **Haraga, I., S. Nomura, and A. Nagayama.** 1999. The effects of vancomycin and beta-lactam antibiotics on vancomycin-resistant *Staphylococcus aureus*. *N. Engl. J. Med.* **341:**1624.
24. **Harbarth, S., W. Albrich, D. A. Goldmann, and J. Huebner.** 2001. Control of multiply resistant cocci: do international comparisons help? *Lancet Infect. Dis.* **1:**251–261.
25. **Hiramatsu, K.** 1998. The emergence of Staphylococcus aureus with reduced susceptibility to vancomycin in Japan. *Am. J. Med.* **104:**7S–10S.
26. **Hiramatsu, K.** 1995. Molecular evolution of MRSA. *Microbiol. Immunol.* **39:**531–543.
27. **Hiramatsu, K.** 1998. Vancomycin resistance in staphylococci. *Drug Resist. Updates* **1:**135–150.
28. **Hiramatsu, K.** 2001. Vancomycin-resistant *Staphylococcus aureus:* a new paradigm of antibiotic resistance. *Lancet Infect. Dis.* **1:**147–155.
29. **Hiramatsu, K., N. Aritaka, H. Hanaki, S. Kawasaki, Y. Hosoda, S. Hori, Y. Fukuchi, and I. Kobayashi.** 1997. Dissemination in Japanese hospitals of strains of Staphylococcus aureus heterogeneously resistant to vancomycin. *Lancet* **350:** 1670–1673.
30. **Hiramatsu, K., L. Cui, M. Kuroda, and T. Ito.** 2001. The emergence and evolution of methicillin-resistant *Staphylococcus aureus*. *Trends Microbiol.* **9:**486–493.
31. **Hiramatsu, K., and H. Hanaki.** 1998. Glycopeptide resistance in staphylococci. *Current Opin. Infect. Dis.* **11:**653–658.
32. **Hiramatsu, K., H. Hanaki, T. Ino, K. Yabuta, T. Oguri, and F. C. Tenover.** 1997. Methicillin-resistant *Staphylococcus aureus* clinical strain with reduced vancomycin susceptibility. *J. Antimicrob. Chemother.* **40:**135–136.
33. **Hood, J., G. F. S. Edwards, B. Cosgrove, E. Curran, D. Morrison, and C. G. Gemmell.** 2000. Vancomycin-intermediate *Staphylococcus aureus* at a Scottish hospital. *J. Infect.* **40:** A11–w01.
34. **Howe, R. A., K. E. Bowker, T. R. Walsh, T. G. Feest, and A. P. MacGowan.** 1998. Vancomycin-resistant Staphylococcus aureus. *Lancet* **351:**602.
35. **Kashii, Y., Y. Arakawa, M. Y. Momoi, and K. Hiramatsu.** 2002. Treatment of a pneumonia caused by heterogeneously vancomycin-resistant *Staphylococcus aureus* (hetero-VRSA) with a combination of arbekacin and ampicillin/sulbactam, abstr. 155-02. *In 10th International Symposium on Staphylococci and Staphylococcal Infections.*
36. **Katayama, Y., T. Ito, and K. Hiramatsu.** 2000. A new class of genetic element, Staphylococcus Cassette Chromosome *mec*, encodes methicillin resistance in *Staphylococcus aureus*. *Antimicrob. Agents Chemother.* **44:**1549–1555.
37. **Kim, M. N., S. H. Hwang, Y. J. Pyo, H. M. Mun, and C. H. Pai.** 2002. Clonal spread of Staphylococcus aureus heterogeneously resistant to vancomycin in a university hospital in Korea. *J. Clin. Microbiol.* **40:**1376–1380.
38. **Kim, M. N., C. H. Pai, J. H. Woo, J. S. Ryu, and K. Hiramatsu.** 2000. Vancomycin-intermediate Staphylococcus aureus in Korea. *J. Clin. Microbiol.* **38:**3879–3881.

39. **Kirst, H. A., D. G. Thompson, and T. I. Nicas.** 1998. Historical yearly usage of vancomycin. *Antimicrob. Agents Chemother.* **42:**1303–1304.
40. **Kuroda, M., H. Kuroda, T. Oshima, F. Takeuchi, H. Mori, and K. Hiramatsu.** 2003. Two-component system VraSR positively modulates the regulation of cell-wall biosynthesis pathway in Staphylococcus aureus. *Mol. Microbiol.* **49:**807–821.
41. **Kuroda, M., K. Kuwahara-Arai, and K. Hiramatsu.** 2000. Identification of the up- and down-regulated genes in vancomycin-resistant *Staphylococcus aureus* strains Mu3 and Mu50 by cDNA differential hybridization method. *Biochem. Biophys. Res. Commun.* **269:**485–490.
42. **Kuwahara-Arai, K., N. Kondo, S. Hori, E. Tateda-Suzuki, and K. Hiramatsu.** 1996. Suppression of methicillin resistance in a *mecA*-containing pre-methicillin-resistant *Staphylococcus aureus* strain is caused by the *mecI*-mediated repression of PBP2′ production. *Antimicrob. Agents Chemother.* **40:**2680–2685.
43. **Maki, H., N. McCallum, M. Bischoff, A. Wada, and B. Berger-Baechi.** 2004. *tcaA* inactivation increases glycopeptide resistance in *Staphylococcus aureus*. *Antimicrob. Agents Chemother.* **48:**1953–1959.
44. **Marchese, A., G. Balistreri, E. Tonoli, E. A. Debbia, and G. C. Schito.** 2000. Heterogeneous vancomycin resistance in methicillin-resistant Staphylococcus aureus strains isolated in a large Italian hospital. *J. Clin. Microbiol.* **38:**866–869.
45. **McDougal, L. K., C. D. Steward, G. E. Killgore, J. M. Chaitram, S. K. McAllister, and F. C. Tenover.** 2003. Pulsed-field gel electrophoresis typing of oxacillin-resistant *Staphylococcus aureus* isolates from the United States: establishing a national database. *J. Clin. Microbiol.* **41:**5113–5120.
46. **Naimi, T. S., D. Anderson, C. O'Boyle, D. J. Boxrud, S. K. Johnson, F. C. Tenover, and R. Lynfield.** 2003. Vancomycin-intermediate *Staphylococcus aureus* with phenotypic susceptibility to methicillin in a patient with recurrent bacteremia. *Clin. Infect. Dis.* **36:**1609–1612.
47. **Noble, W. C., Z. Virani, and R. G. Cree.** 1992. Co-transfer of vancomycin and other resistance genes from Enterococcus faecalis NCTC 12201 to Staphylococcus aureus. *FEMS Microbiol. Lett.* **72:**195–198.
48. **Okuma, K., K. Iwakawa, J. D. Turnidge, W. B. Grubb, J. M. Bell, F. G. O'Brien, G. W. Coombs, J. W. Pearman, F. C. Tenover, M. Kapi, C. Tiensasitorn, T. Ito, and K. Hiramatsu.** 2002. Dissemination of new methicillin-resistant Staphylococcus aureus clones in the community. *J. Clin. Microbiol.* **40:**4289–4294.
49. **Oliveira, G. A., A. M. Dell'Aquila, R. L. Masiero, C. E. Levy, M. S. Gomes, L. Cui, K. Hiramatsu, and E. M. Mamizuka.** 2001. Isolation in Brazil of nosocomial Staphylococcus aureus with reduced susceptibility to vancomycin. *Infect. Control Hosp. Epidemiol.* **22:**443–448.
50. **O'Neill, A. J., and I. Chopra.** 2003. Lack of evidence for involvement of hypermutability in emergence of vancomycin-intermediate *Staphylococcus aureus*. *Antimicrob. Agents Chemother.* **47:**1484–1485. (Letter.)
51. **Paton, R., T. Snell, F. X. Emmanuel, and R. S. Miles.** 2001. Glycopeptide resistance in an epidemic strain of methicillin-resistant *Staphylococcus aureus*. *J. Antimicrob. Chemother.* **48:**941–942.
52. **Perichon, B., and P. Courvalin.** 2000. Update on vancomycin resistance. *Int. J. Clin. Pract.* **54:**250–254.
53. **Ploy, M. C., C. Grelaud, C. Martin, L. de Lumley, and F. Denis.** 1998. First clinical isolate of vancomycin-intermediate *Staphylococcus aureus* in a French hospital. *Lancet* **351:**1212.
54. **Rotun, S. S., V. McMath, D. J. Schoonmaker, P. S. Maupin, F. C. Tenover, B. C. Hill, and D. M. Ackman.** 1999. *Staphylococcus aureus* with reduced susceptibility to vancomycin isolated from a patient with fatal bacteremia. *Emerg. Infect. Dis.* **5:**147–149.
55. **Schaaff, F., A. Reipert, and G. Bierbaum.** 2002. An elevated mutation frequency favors development of vancomycin resistance in *Staphylococcus aureus*. *Antimicrob. Agents Chemother.* **46:**3540–3548.
56. **Severin, A., K. Tabei, F. Tenover, M. Chung, N. Clarke, and A. Tomasz.** 2004. High level oxacillin and vancomycin resistance and altered cell wall composition in *Staphylococcus aureus* carrying the staphylococcal mecA and the enterococcal vanA gene complex. *J. Biol. Chem.* **279:**3398–3407.
57. **Sieradzki, K., T. Leski, J. Dick, L. Borio, and A. Tomasz.** 2003. Evolution of a vancomycin-intermediate *Staphylococcus aureus* strain in vivo: multiple changes in the antibiotic resistance phenotypes of a single lineage of methicillin-resistant *S. aureus* under the impact of antibiotics administered for chemotherapy. *J. Clin. Microbiol.* **41:**1687–1693.
58. **Sieradzki, K., and A. Tomasz.** 1996. A highly vancomycin-resistant laboratory mutant of *Staphylococcus aureus*. *FEMS Microbiol. Lett.* **142:**161–166.
59. **Sieradzki, K., S. W. Wu, and A. Tomasz.** 1999. Inactivation of the methicillin resistance gene mecA in vancomycin-resistant *Staphylococcus aureus*. *Microb. Drug Resist.* **5:**253–257.
60. **Smith, T. L., M. L. Pearson, K. R. Wilcox, C. Cruz, M. V. Lancaster, B. Robinson-Dunn, F. C. Tenover, M. J. Zervos, J. D. Band, E. White, and W. R. Jarvis.** 1999. Emergence of vancomycin resistance in *Staphylococcus aureus*. Glycopeptide-Intermediate *Staphylococcus aureus* Working Group. *N. Engl. J. Med.* **340:**493–501.
61. **Snowden, M. A., and H. R. Perkins.** 1990. Peptidoglycan cross-linking in *Staphylococcus aureus*. *Eur. J. Biochem.* **191:**373–377.
62. **Tanaka, T., K. Okuzumi, A. Iwamoto, and K. Hiramatsu.** 1995. A retrospective study on methicillin-resistant *Staphylococcus aureus* clinical strains in Tokyo University Hospital. *J. Infect. Chemother.* **1:**40–49.
63. **Tenover, F. C., M. V. Lancaster, B. C. Hill, C. D. Steward, S. A. Stocker, G. A. Hancock, C. M. O'Hara, N. C. Clark, and K. Hiramatsu.** 1998. Characterization of staphylococci with reduced susceptibility to vancomycin and other glycopeptides. *J. Clin. Microbiol.* **36:**1020–1027.
64. **Trakulsomboon, S., S. Danchaivijitr, Y. Rongrungruang, C. Dhiraputra, W. Susaemgrat, T. Ito, and K. Hiramatsu.** 2001. First report of methicillin-resistant *Staphylococcus aureus* with reduced susceptibility to vancomycin in Thailand. *J. Clin. Microbiol.* **39:**591–595.
65. **Tsakris, A., E. Papadimitriou, J. Douboyas, F. Stylianopoulou, and E. Manolis.** 2002. Emergence of vancomycin-intermediate *Staphylococcus aureus* and *S. sciuri*, Greece. *Emerg. Infect. Dis.* **8:**536–537.
66. **Ward, P. B., P. D. Johnson, E. A. Grabsch, B. C. Mayall, and M. L. Grayson.** 2001. Treatment failure due to methicillin-resistant *Staphylococcus aureus* (MRSA) with reduced susceptibility to vancomycin. *Med. J. Aust.* **175:**480–483.
67. **Watanakunakorn, C.** 1990. In-vitro selection of resistance of *Staphylococcus aureus* to teicoplanin and vancomycin. *J. Antimicrob. Chemother.* **25:**69–72.
68. **Wong, S. S., P. L. Ho, P. C. Woo, and K. Y. Yuen.** 1999. Bacteremia caused by staphylococci with inducible vancomycin heteroresistance. *Clin. Infect. Dis.* **29:**760–767.
69. **Wootton, M., R. A. Howe, R. Hillman, T. R. Walsh, P. M. Bennett, and A. P. MacGowan.** 2001. A modified population analysis profile (PAP) method to detect hetero-resistance to vancomycin in *Staphylococcus aureus* in a UK hospital. *J. Antimicrob. Chemother.* **47:**399–403.
70. **Yamashoji, S., I. Manome, and M. Ikedo.** 2001. Menadione-catalyzed O_2^- production by *Escherichia coli* cells: application of rapid chemiluminescent assay to antimicrobial susceptibility testing. *Microbiol. Immunol.* **45:**333–340.

Frontiers in Antimicrobial Resistance: a Tribute to Stuart B. Levy
Edited by D. G. White, M. N. Alekshun, and P. F. McDermott

Chapter 22

Enterococcus

ROLAND LECLERCQ AND PATRICE COURVALIN

Enterococci are normal hosts of the intestinal flora of humans and animals. *Enterococcus faecalis* is the species most frequently isolated in humans, accounting for 80 to 90% of enterococcal isolates, whereas *Enterococcus faecium* represents between 5 and 10% of clinical isolates. Nearly 15 other enterococcal species may be occasionally associated with infections, including *Enterococcus gallinarum, Enterococcus casseliflavus, Enterococcus avium, Enterococcus durans,* and *Enterococcus hirae.* Enterococci are facultative anaerobes which can survive in various extreme environmental conditions, being able to cope with oxidative stress, desiccation, high and low temperatures, and pH extremes. This explains why they are commonly isolated from food, soils, sewage, and water, frequently through fecal contamination.

Enterococci have a prolonged survival on the surfaces in hospital settings and have been isolated from the hands of health care workers. This ability to adapt to various and hostile environmental conditions is complementary to a large set of intrinsic and acquired resistances to antimicrobials characteristic of this species. Infections caused by enterococci include urinary tract infections, wound infections, intra-abdominal infections, endocarditis, and bacteremia. These infections may be acquired in the community or at the hospital. Enterococci are opportunistic pathogens now recognized as a major cause of nosocomial infections, being the second leading cause of nosocomial infection overall and the third leading cause of nosocomial bacteremia in National Nosocomial Infections Surveillance (NNIS) surveys in the United States (115). Often, the infecting strain originates from the patient's intestinal flora, and it seems that spread of resistant enterococci in hospitals results in colonization and replacement of the normal flora of the patients then followed by serious infections in those who are under severe underlying conditions and/or have been exposed to broad spectrum antibiotics (third generation cephalosporins and fluoroquinolones). Emergence of enterococci as nosocomial pathogens is mostly related to their multiple resistance to antimicrobials, which renders them well suited to survive in hospital settings where antibiotic selective pressure is heavy.

INTRINSIC RESISTANCE

The limited choice of antibiotics and the poor results of single-agent antimicrobial therapy in the treatment of enterococcal infections which require bactericidal activity of antimicrobials (e.g., endocarditis) are mostly related to intrinsic resistance in enterococci (Table 1). Enterococci are intrinsically resistant to numerous antimicrobial agents. In particular, they are intrinsically resistant to several β-lactam antibiotics, including cephalosporins and penicillinase-resistant penicillins and to clinically achievable levels of aminoglycosides. Additional resistances to clindamycin, amikacin, and vancomycin are characteristics of certain enterococcal species. The intrinsic resistances are related to low affinity antibiotic targets or to drug efflux. Multidrug efflux pumps have been characterized in *E. faecalis,* such as EfrAB, an ABC transporter which has a broad spectrum of substrates, including quinolones, tetracyclines, and antiseptics (77).

NON-SPECIES-RELATED INTRINSIC RESISTANCE

Beta-Lactams and Aminoglycosides

Compared with most streptococci, *E. faecalis* is 10 to 100 times less susceptible to penicillin G, with MICs generally ranging from 2 to 8 mg/liter. Similarly,

Roland Leclercq • Service de Microbiologie, CHU Côte de Nacre, Caen, France. **Patrice Courvalin** • Unité des Agents Antibacteriens, Institut Pasteur, Paris, France.

Table 1. Intrinsic resistance in enterococci

Antibiotic	Species	Mechanism of resistance
β-lactams	All enterococci	Low affinity PBPs (in particular PBP5 and homologues)
Aminoglycosides	All enterococci	Inefficient uptake
Trimethoprim-sulfamethoxazole	All enterococci	Inefficacy in vivo due to assimilation of exogenous folates
Lincosamides and streptogramins A-type	*E. faecalis, E. avium, E. gallinarum, E. casseliflavus*	Putative efflux
Aminoglycosides (except streptomycin and gentamicin)	*E. faecium*	Production of chromosomal AAC(6′)Ii enzyme
Vancomycin	*E. gallinarum, E. casseliflavus/flavescens*	Production of D-Ala-D-Ser ending peptidoglycan precursors

ampicillin (MIC, 1 to 4 mg/liter), amoxicillin, piperacillin, and imipenem are moderately active. The MICs of all available cephalosporins are high, and the drugs are ineffective in inhibiting enterococci in vivo. In *E. faecium* and also *E. raffinosus* and *E. hirae*, resistance is even more pronounced, with MICs of penicillin G generally ranging from 8 to 32 mg/liter with full resistance to piperacillin and imipenem. Although less is known for other enterococcal species, activity of penicillin G and ampicillin against *E. avium* and *E. gallinarum* seems to be similar to that against *E. faecalis*, whereas piperacillin is less active (MIC >16 mg/liter) (51).

The profile of low susceptibility to penicillins and resistance to cephalosporins has been associated with the presence of penicillin-binding proteins (PBPs) with low affinity for beta-lactams. Most species of enterococci have at least five PBPs (141). More specifically, intrinsic resistance of *E. faecium* and *E. hirae* to β-lactams is due to the presence of PBP5, which is highly expressed and has lower affinity for β-lactams than the other PBPs (47, 142). The strains grow normally in the presence of penicillin concentration which saturates all PBPs, except PBP5, which suggests that PBP5 can substitute for the functions of all the other PBPs of the cell. Interestingly, the amino acid sequence of PBP5 is 33% homologous to PBP2a, which confers methicillin resistance in staphylococci (45). Most of the work on PBP5 has been done on *E. hirae* ATCC 9790. PBPs related in their amino acid sequence to PBP5 have been identified in several enterococcal species. Structural similarities between those of *E. faecalis, E. faecium,* and *E. hirae* have been shown on the basis of cross reaction with antibodies directed against PBP5 of *E. hirae* (78).

In addition, even penicillins that have activity against enterococci are not bactericidal. This poor bactericidal activity is also observed with other cell-wall active agents such as the glycopeptides. The mechanism of tolerance of enterococci to these agents is not known. Killing enterococci can be achieved only when a penicillin or a glycopeptide is combined with an aminoglycoside, although the MICs of aminoglycosides generally range from 8 to 64 mg/liter. Uptake of aminoglycosides by the bacteria requires oxidatively generated energy, and the intrinsic resistance of enterococci to aminoglycosides is the consequence of inefficient active transport of the antibiotic due to poor membrane energization (20). Facilitation of aminoglycoside uptake that results from disruption of the cell-wall by the cell-wall active agents accounts for the synergism between cell-wall synthesis inhibitors and aminoglycosides (93).

TRIMETHOPRIM-SULFAMETHOXAZOLE (CO-TRIMOXAZOLE)

Enterococci appear susceptible in vitro to co-trimoxazole when tested in appropriate conditions (in media devoided of thymidine, which reverses the activity of co-trimoxazole). However, clinical failures during treatment of urinary tract infections due to enterococci have raised concern about the in vivo activity of the combination (50). In addition, co-trimoxazole is not efficacious in the treatment of experimental endocarditis due to *E. faecalis* in a rabbit model (52) and lacks in vitro bactericidal activity against *E. faecalis* (97). This might be due to the fact that enterococci can use exogenous folates in biological fluids containing these compounds, urine for example. Therefore, they can escape inhibition by antifolate agents, which could thus be ineffective in vivo. However, the reversal of the antifolate effect seems strikingly more important for *E. faecium* than for *E. faecalis* (53) and there is still controversy on the lack of efficacy of co-trimoxazole in the treatment of urinary tract infections or prostatitis due to *E. faecalis*.

SPECIES-RELATED INTRINSIC RESISTANCE

Lincosamides and Streptogramins A

Resistance to lincosamides (lincomycin and clindamycin) and streptogramins A (dalfopristin, pristinamycin II, and virginiamycin M), which defines the LS_A phenotype, is intrinsic in *E. faecalis*. Resistance to factor A of streptogramins is responsible for the loss of synergism between factors A and B composing the streptogramin mixture (quinupristin-dalfopristin, pristinamycin, and virginiamycin), which explains the intrinsic resistance of *E. faecalis* to this class of antibiotics. This resistance has been related, at least in part, to the expression of the chromosomal *lsa* gene in *E. faecalis* OG1RF, which appears to be species-specific (129). The *lsa*-like genes of two clinical isolates of *E. faecalis* susceptible to lincosamides contained mutations which produced premature termination codons, confirming the role of the gene in lincosamide resistance (41). In addition, differences in the region surrounding the *lsa* gene influence the expression of lincosamide resistance (128). The Lsa protein shows similarities to members of a superfamily of transport-related proteins known as ABC transporters. However, no transmembrane partner has been found associated with the Lsa protein, and the efflux mechanism has not been proven. The mechanism of intrinsic resistance to lincosamides and streptogramins A remains therefore incompletely elucidated. Other species of enterococci, including *E. avium, E. gallinarum,* and *E. casseliflavus,* also display the LS_A phenotype. The genetic basis for resistance in these species is unknown. By contrast, *E. faecium, E. durans,* and *E. hirae* do not exhibit this intrinsic resistance.

Aminoglycosides

All strains of *E. faecium* produce an aminoglycoside 6′-N-acetyltransferase encoded by the *aac(6′)-Ii* chromosomal gene (33). Production of the enzyme renders this organism resistant to moderate levels of kanamycin, tobramycin, amikacin, and netilmicin. The ubiquitous presence of this enzyme in *E. faecium* complicates the choice of antibiotics for treatment of infections caused by this organism since it abolishes the bactericidal activity of cell-wall inhibitors (penicillins and glycopeptides) with aminoglycosides except for streptomycin and gentamicin, which are not substrates for the enzyme (93). Surprisingly, synergism of amikacin combined with a penicillin seems to be retained, although this aminoglycoside is a substrate for the enzyme (33, 93). This feature could be related to the properties of the enzyme, in particular k_{cat}/K_m values, which are low compared to those of other aminoglycoside-modifying enzymes (143). This enzyme is thus not particularly well suited to aminoglycoside inactivation and may have an alternate physiological function (143). The X-ray structure of the enzyme has been recently determined (21).

Glycopeptides

Vancomycin has low activity (MIC = 2 to 32 mg/liter) against *E. gallinarum* and *E. casseliflavus/flavescens.* Although these species are rarely isolated in clinical settings, they may provoke bacteremia, endocarditis, and intra-abdominal infections (29). The lower MIC breakpoint of vancomycin (4 mg/liter) divides the bacterial population, with certain isolates being categorized as susceptible. The mechanisms of resistance to vancomycin of these species is detailed in chapter 8.

ACQUIRED RESISTANCE

Penicillins

Acquired resistance to penicillins can be due to production of a penicillinase or to modification of the PBP targets. Both mechanisms can be found in *E. faecium* and *E. faecalis,* but their respective frequency in these species differ. Penicillinase production was first reported in strains of *E. faecalis* from U.S. hospitals in 1981 and subsequently from Argentina, Canada, and Lebanon (96, 133). Only rare strains of *E. faecium* have been reported to synthesize the enzyme. Resistance is generally plasmid mediated and is often linked with high-level resistance to gentamicin. The penicillinases of *E. faecalis* are identical or highly homologous (10-amino acid difference) to that encoded by the *blaZ* gene of *Staphylococcus aureus* (133). In staphylococci, the production of β-lactamase is inducible and controlled by the regulatory genes *blaI* and *blaR1*. In enterococci, the regulatory genes may be present, or partially or totally absent, but all strains produce the penicillinase constitutively although at relatively low levels (133). For this reason, the penicillinase-producing enterococci may appear susceptible by the standard routine test with ampicillin (96). Use of a heavy inoculum is required to reveal resistance. Although a number of clones have been reported, this type of resistance remains for the most part confined to a few hospitals and does not account for the global increase in resistance of enterococci to penicillins (134).

This increase is mostly due to the spread of penicillin-resistant *E. faecium* that do not produce a β-lactamase. As previously mentioned, studies in

E. hirae, and subsequently in *E. faecium,* have shown that the presence of low affinity PBP5 in these species accounts for increased resistance to penicillins, as compared to *E. faecalis*. A number of strains belonging to these species may express higher levels of resistance. Penicillin-resistant mutants can also be readily selected in vitro among other enterococcal species, including *E. faecalis* (11). However, penicillin-resistant clinical isolates of *E. faecalis* are extremely rare, and only a limited number of species with resistance have been reported, including *E. faecium* and to a lesser extent *E. hirae* and *E. raffinosus*.

In *E. faecium* and *E. hirae,* two types of mechanisms account for increased resistance to the penicillins. The first mechanism is related to the observation that there is a direct correlation between the amount of PBP5 produced and the penicillin MIC (79). A laboratory mutant of *E. hirae* ATCC 9790 with an MIC of ampicillin equal to 80 mg/liter and overproducing its PBP5 had a deletion in the upstream *psr* regulator of the *pbp5* gene (79). Expression of PBP5 of *E. faecium* is also regulated, probably by the *psrfm* gene, which has 74% similarity with *psr* (147). Therefore, among strains of *E. faecium* with similar MICs of ampicillin, the mechanism of penicillin resistance is also attributed to overproduction of PBP5.

A second mechanism accounts for a high level of resistance to penicillins in *E. faecium* strains that do not overproduce PBP5. Substitutions in the primary sequence of PBP5 lead to an additional decrease in penicillin-binding capacity and appear responsible for higher levels of resistance (124). Mutations have been found in conserved motifs which are part of the active site of the enzyme (122, 147), and combinations of mutations yield even higher levels of resistance to β-lactams (114).

The mechanisms affecting chromosomal PBP5 are, by definition, not transferable. However, a *pbp5* gene colocated with a vancomycin resistance *vanB* gene cluster on a transposon has been described (24). Several strains bearing this genetic element have been reported in the United States (54). In addition, a *pbp3r* gene was found, together with genes for resistance to macrolides and aminoglycosides, on a plasmid in a strain of *E. hirae* isolated from a pig (111). Since the gene was identical to the *pbp3r* gene of *E. faecium,* the authors hypothesized transfer of the *pbp3r* gene from the chromosome of *E. faecium* to a plasmid. No other plasmid of this type has been found in clinical isolates.

Finally, a novel mechanism of resistance to β-lactams which does not involve PBPs has been reported in a laboratory mutant of *E. faecium* (86). A strain of *E. faecium* in which the *pbp5* gene had been deleted was used for selection of a mutant with a very high level of resistance to ampicillin (MIC > 2,000 mg/liter). It was shown that cross-linking during cell wall elongation occurred by a LD transpeptidation [L-Lys(3) - D-Asx-Lys(3)] which bypasses the usual β-lactam-susceptible DD-transpeptidation (86, 87). LD-transpeptidation was found to be completely insensitive to β-lactams. Although no clinical isolate was detected with this type of resistance, this mechanism might represent a future way for *E. faecium* to acquire an ultimate level of ampicillin resistance.

Aminoglycosides (High Level)

Enterococci can acquire resistance to aminoglycosides by three mechanisms: modification of the ribosomal target, alteration of antibiotic transport, and enzymic modification of the drugs. The first two mechanisms are due to chromosomal mutations, whereas the third one is generally mediated by plasmids or transposons and accounts for the vast majority of clinical isolates resistant to high levels of aminoglycosides (35, 36). However, in certain *E. faecalis* strains, high-level resistance to streptomycin can also be due to ribosomal mutations (44).

As in gram-negative organisms and staphylococci, inactivation of the aminoglycosides can occur through *N*-acetylation, *O*-phosphorylation, or *O*-nucleotidylation. The enzymes are named according to the reaction catalyzed, which is designated by a three-letter acronym (ANT, aminoglycoside nucleotidyltransferase; AAC, aminoglycoside acetyltransferase; APH, aminoglycoside phosphotransferase), and to the site modified, which is indicated by a number in parentheses. Table 2 provides a list of the enzymes reported in enterococci. The enzymes vary in their substrate ranges, which are often broad and overlap, and several of them are encoded by genes similar to those detected in staphylococci and borne by similar transposons (36).

In general, presence of an enzyme confers to the host high-level resistance toward the antibiotics that are modified in vitro (MICs > 1,000 mg/liter). As expected, the synergistic activity of the combination of a cell wall active agent, penicillin, or glycopeptide with an aminoglycoside is abolished in case of production of a modifying enzyme. It is important to note that this suppression may also occur when the aminoglycoside is a poor substrate for the enzyme and when, consequently, its bacteriostatic activity is not markedly affected. In particular, the rate of modification of amikacin by the enzymes which phosphorylate, adenylate, or acetylate kanamycin [in particular APH(3′)-III, ANT(4′), and APH(2″)-AAC(6′)] is generally not sufficient to confer high-level resistance to the host (MICs generally between 64 and 256 mg/liter) but can prevent synergism or even yield antagonism between these an-

Table 2. Aminoglycoside-modifying enzymes in enterococci

Enzyme	Organism	Location	Phenotype[a]	Reference(s)
Phosphotransferases				
APH(3′)-III	*E. faecalis, E. faecium, E. gallinarum, E. casseliflavus*	Plasmid, transposon	KAN, AMI, ISE, NEO, BUT, RIB LIV	23, 136
APH(2″)-Ia[b]	*E. faecalis, E. faecium, E. gallinarum, E. casseliflavus*	Plasmid, transposon	GENC1, NET, SIS, KAN, TOB, AMI, ISE, NEO, BUT, RIB, LIV	42, 46
APH(2″)-Ib	*E. faecium, E. durans*		GENC1, NET, KAN, TOB, DIB, AMI, ISE	71
APH(2″)-Ic	*E. gallinarum, E. faecalis, E. faecium*	Plasmid	GENC1, NET, KAN, TOB	30, 42
APH(2″)-Id	*E. casseliflavus, E. faecalis, E. faecium*	Chromosome	GENC1, NET, KAN, TOB	42, 138
Nucleotidyltransferases				
ANT(3′)-Ia	*E. faecalis*	Plasmid	STR	61
ANT(6)-Ia	*E. faecalis*	Plasmid	STR	31
ANT(9)-Ia	*E. faecalis, E. durans*	Plasmid	SPE	103
ANT(9)-Ib	*E. faecalis*	Plasmid	SPE	76
ANT(4′)-Ia	*E. gallinarum, E. faecalis*	—	KAN, TOB, AMI, NEO, DIB	25, 38
Acetyltransferases				
AAC(6′)-Ie[b]	*E. faecalis, E. faecium, E. casseliflavus, E. gallinarum*	Plasmid, transposon	KAN, TOB, AMI, NEO, FOR	42, 46
AAC(6′)-Ii[c]	*E. faecium*	Chromosome	KAN, TOB, AMI, NEO	33
AAC(6′)-Im				29

[a] AMI, amikacin; BUT, butirosin; FOR, fortimicin; GENC1, gentamicin C1; KAN, kanamycin; ISE, isepamycin; KM, kanamycin; LIV, lividomycin; NEO, neomycin; NET, netilmicin; RIB, ribostamycin; SIS, sisomicin; SPE, spectinomycin; STR, streptomycin; TOB, tobramycin.
[b] Part of the bifunctional enzyme AAC(6′)-APH(2″).
[c] Intrinsic resistance in *E. faecium.*

tibiotics and a penicillin or a glycopeptide in vitro (132) and in animal models (12). Similarly, high-level resistance to gentamicin due to production of the bifunctional enzyme APH(2″)-AAC(6′) prevents synergy between netilmicin and cell-wall inhibitors in the absence of high-level resistance to this aminoglycoside.

Aminoglycoside *O*-Phosphotransferases (APH)

Several members of the APH(2″) family have been characterized in enterococci. APH(2″)-Ia is remarkable since it is combined with the AAC(6′)-Ie enzyme in a single fusion protein, probably resulting from a fusion of two ancestral genes (46). This bifunctional enzyme possesses dual enzymic activity of acetylation and phosphorylation (46). The presence of this enzyme confers high-level resistance to gentamicin, kanamycin, and tobramycin and abolishes synergism between cell-wall inhibitors and all aminoglycosides except streptomycin. This enzyme is widespread in various species of enterococci both from human and animal origin and accounts for most isolates resistant to high levels of gentamicin (42, 103).

Other APH(2″) enzymes more recently characterized are less frequently encountered in enterococci from humans, whereas they can be found more commonly in enterococci from food animals. The APH(2″) family includes APH(2″)-Ib, APH(2″)-Ic, and APH(2″)-Id, which are related: APH(2″)-Ib has 32, 33, and 25% identity with APH(2″)-Id, APH(2″)-Ia, and APH(2″)-Ic, respectively (71).

The APH(2″)-Ib mediates high-level resistance to gentamicin, tobramycin kanamycin, netilmicin, and dibekacin and has activity against amikacin, isepamicin, and arbekacin (29, 81). This enzyme has been detected in vancomycin-resistant *E. faecium* isolates. It seems to be rare in other species of enterococci and has been detected so far in a single strain of *E. durans* isolated from milk (82) where it did not confer high-level resistance to gentamicin.

The APH(2″)-Ic enzyme, initially detected in a veterinary isolate of *E. gallinarum,* was borne by a plasmid (30). The *aph(2″)-Ic* gene has also been found in *E. faecium* and *E. faecalis* from humans and mostly from food animals and meat (42). The gene mediates a moderate level of resistance to gentamicin (MICs from 128 to 1,024 μg/ml), tobramycin, and kanamycin but not to amikacin or netilmicin (30). Despite intermediate resistance to gentamicin, the synergy between ampicillin and gentamicin is suppressed. This resistance may be missed by clinical laboratories, which routinely use gentamicin levels of 500 to 2,000 μg/ml to detect high-level resistance to gentamicin.

The APH(2″)-Id enzyme is responsible for high-level resistance to gentamicin, tobramycin, kanamycin, netilmicin, and dibekacin (138). The corresponding gene, *aph(2″)-Ic,* was first detected in an *E. casseliflavus* isolate from blood and has subsequently been found in *E. faecium* isolates from humans and food animals (42).

The gene encoding APH(3′)-III in *E. faecalis* was originally detected on plasmid pJH1 (136) and has also been found in staphylococci and *Streptococcus pneumoniae,* where it is borne by transposon Tn*1545* (23, 34). This enzyme has been purified and characterized extensively (88). The three-dimensional structure of the enzyme shows strong similarity with Ser/Thr and Tyr kinases, although the proteins have low amino acid sequence identity (<10%) (62). Both enzymes have an N-terminal domain consisting mostly of β-sheets and a C-terminal domain with α-helices. The active site of the enzymes lies between the two domains. This suggests that protein and aminoglycoside kinases may have evolved from a common ancestor.

Aminoglycoside *N*-Acetyltransferases (AAC)

In addition to the chromosomal aac(*6′*)-*Ii* gene, intrinsic to *E. faecium,* two other *aac*(*6′*) genes have been characterized. The *aac*(*6′*)-*Ie* has already been mentioned as part of the *aac*(*6′*)*Ie-aph*(*2″*)-*Ia* gene for the bifunctional enzyme. The *aac*(*6′*)-*Im* gene encodes an enzyme which was originally detected in *E. coli* 40 bp upstream from an *aph*(*2″*)-*Ia* gene. It was then shown that in the strain of *E. faecium* SF11770 where the *aph*(*2″*)-*Ia* gene had been initially characterized, the *aac*(*6′*)-*Im* gene was also present and that the two genes were linked in a similar fashion (29). The two genes appear to be independently translated and do not form a bifunctional enzyme. The AAC(6′)-Im enzyme inactivates kanamycin A, tobramycin, amikacin, dibekacin, and netimicin.

Aminoglycoside *O*-Nucleotidyltransferases (ANT)

The ANT(4′)-Ia enzyme is responsible for high-level resistance to kanamycin and tobramycin and inactivates amikacin and dibekacin because of a 4″ nucleotidyltransferase activity. It was found originally in a veterinary isolate of *E. gallinarum* initially identified as *E. faecium* (25). This enzyme is commonly found in staphylococci (104) but seems to be rare in enterococci.

Other ANT enzymes have a narrow spectrum of activity since they inactivate streptomycin [ANT(6)-Ia and ANT(3″)-Ia] or spectinomycin [(ANT(9)-Ib] only (31, 61, 103). The *ant*(*6′*)-*Ia* gene was found on a large 80.7-kb conjugative plasmid of *E. faecalis* and was associated with an *apha*(*3*)-*III* gene (103). This gene has also been detected in staphylococci (39). The ANT(3″)-Ia enzyme is more commonly found in *S. aureus* and gram-negative bacilli (31), whereas the ANT(9)-Ib is homologous to but distinct from the ANT(9)-Ia enzyme found in staphylococci (95).

Macrolides, Lincosamides, and Streptogramins

Table 3 provides a list of resistance phenotypes and of the corresponding genes of resistance to macrolides, lincosamides, and streptogramins. A common way for enterococci to resist macrolides is to produce a methylase, which modifies the ribosomal target site of these protein synthesis inhibitors. The precise site of methylation is an adenine at position 2058 (*E. coli* numbering) in the 23S RNA of the large 50S ribosomal subunit. The residue is located within a conserved region of domain V of 23S RNA and plays a key role in the binding of macrolide-lincosamide-streptogramin B antibiotics (MLS_B). As a consequence of 23S rRNA methylation, binding of erythromycin to

Table 3. Macrolide-lincosamide-streptogramin resistance genes in enterococci

Resistance phenotype[a]	Resistance gene	Resistance mechanism	Enterococcal species	Reference(s)
MLS_B	*erm*(B)	Ribosomal methylation	*E. faecalis, E. faecium, E. gallinarum, E. durans, E. hirae, E. casseliflavus, E. avium*	80, 83, 108
	erm(C)	Ribosomal methylation	*Enterococcus* sp.	83
	erm(F)	Ribosomal methylation	*Enterococcus* sp.	83
	erm(A)	Ribosomal methylation	*E. faecium, Enterococcus* sp.	83, 108
M	*mef*(A)/*msr*(D)	Efflux	*E. faecalis Enterococcus* sp.	37, 80, 83
MS_B	*msr*(C)	Putative efflux	*E. faecalis*	108, 127, 139
MS_B	*msr*(A)	Putative efflux	*E. faecium*	108
L	*lnu*(B)	Nucleotidylation	*E. faecium*	16
LS_A	Unknown	*E. faecium*		17
S_A	*vat*(D), vat(E)	Acetylation	*E. faecium, E faecalis*	112, 140
S_B	*vgb*(A)	Streptogramin B lyase	*E. faecium*	18, 67

[a] MLS_B, macrolides-lincosamides-streptogramins B; M, 14-, 15-membered-ring macrolides; MS_B, macrolides-streptogramins B; L, lincosamides; LS_A, lincosamides-streptogramins A; S_A, streptogramins A-type; S_B, streptogramins B type.

its target is impaired. There is cross-resistance between all macrolides, lincosamides, and streptogramins B (pristinamycin I, virginiamycin S, and quinupristin) because of overlapping binding sites of the drugs in 23S rRNA, which gave rise to the designation for the MLS_B phenotype.

Macrolide resistance methylases constitute a family of highly related proteins encoded by *erm* (Erythromycin Resistance Methylase) genes. In enterococci, resistance is mediated mostly by *erm*(B) genes [previously known as *ermB, erm*(AM), or *erm*(AMR)], and sparsely by *erm*(A), *erm*(C), or *erm*(F) genes (80, 84, 101, 108). These genes are shared with streptococci and, to a lesser extent, with staphylococci. Expression of resistance may be constitutive or inducible under the translational control of an attenuator located upstream from the *erm*(B) gene (64, 119). Inducible resistance is expressed as a cross-resistance between most members of the MLS_B group, including clindamycin and 16-membered macrolides, which are inducers at various degrees of ErmB methylase production. Whether inducible or constitutive, erythromycin resistance is usually expressed at a high level (MIC > 128 mg/liter).

Another, but less common, mechanism of resistance to erythromycin is acquisition of a *mef*(A) gene, which encodes a protein related to the Major Facilitator Superfamily of efflux proteins that pumps erythromycin out of the bacterial cell. This gene along with another gene, *msr*(D), codes for an ABC-transporter of a large conjugative transposon (37, 84). Similar elements are found in streptococci. The resistance is limited to the 14- and 15-membered ring macrolides (erythromycin, roxithromycin, clarithromycin, azithromycin) and is expressed at moderate levels (MIC of erythromycin = 2 to 16 mg/liter) (83).

The *msr*(A) gene from *Staphylococcus epidermidis,* encoding an ABC-transporter type protein and responsible for inducible resistance to 14- and 15-membered ring macrolides and streptogramins B, has rarely been found in enterococci (84, 120). More common in *E. faecium* is the *msr*(C) gene, which possesses 53 to 62% identity with *msr(A)*. Contribution of this gene to erythromycin resistance is probably modest in the absence of other macrolide resistance gene since its disruption in *E. faecium* lowers MIC of erythromycin by only one dilution. Although the gene was initially thought to be intrinsic in *E. faecium* (108, 127), it seems to be only highly prevalent (139). Isolated resistance to lincosamides has been reported in *E. faecium.* The *lnu*(B) gene (previously *linB*) encodes a lincosamide O-nucleotidyltransferase, which modifies an hydroxyl group of clindamycin and lincomycin at position 3 (16, 118). In the initial isolate, *lnu*(B) was located on a large conjugative plasmid.

In practice, macrolides and lincosamides are not used for the treatment of enterococcal infections because of their weak activity. As previously mentioned, the streptogramins have poor activity against *E. faecalis,* which is intrinsically resistant to the A-component of the streptogramin combination. However, the *E. faecium* species does not display this resistance, and the injectable streptogramin combination quinupristin-dalfopristin is used in the treatment of infections due to multidrug-resistant *E. faecium.* Synergistic binding of the two factors to the 50S ribosomal subunit causes conformational change in the peptidyltransferase center and inhibition of protein synthesis (55). The mechanisms of resistance to A and B types of streptogramin are different. In *E. faecium,* acetyltransferases encoded by the *vat*(D) (previously *satA*) or *vat*(E) (previously *satG*) genes inactivates streptogramins A (113, 140). Both genes are related to the acetyltransferase genes *vat*(A) (10), *vat*(B) (7), and *vatC* (8) reported in staphylococci. Resistance to streptogramins B is due either to hydrolysis of the antibiotic mediated by the *vgb* gene (18, 67) initially reported in *S. aureus* (9) or to modification of the ribosomal target by a 23S rRNA methylase encoded by the *erm*(B) gene as discussed above. The presence of a single *vat* or a *vgb* gene may only result in a moderate increase in MICs. Full resistance occurs when the genes are combined (17). Resistance is more common in isolates recovered from food animals than from humans (59). This has been related to the use of virginiamycin as a feed additive in animals. Finally, some isolates of *E. faecium* may be resistant to lincosamides and streptogramins A, displaying a LS_A phenotype (17). The mechanism of this resistance is unknown.

Fluoroquinolones

Mechanism of action of quinolones has been extensively studied in *E. coli, S. aureus,* and *S. pneumoniae* (63). In *E. coli,* quinolones inhibit bacteria by interacting with the DNA gyrase and topoisomerase IV complexed with bacterial DNA. The enzymes are responsible for DNA negative supercoiling and are essential for DNA replication. The DNA gyrase is a heterotetramer composed of two A (GyrA) and B (GyrB) subunits which are encoded by the *gyrA* and *gyrB* genes, respectively. Similarly, the topoisomerase IV is a heterotetramer composed of two C (ParC) and E (ParE) subunits which are encoded by the *parC* and *parE* genes, respectively. Resistance to quinolones may be due to mutational alteration of the target or efflux of the antimicrobials.

Resistance of enterococci to first generation quinolones is related to the low affinity of the drugs for their target (98). In addition, it has been shown that *E. faecalis*, and probably *E. faecium*, possess endogenous pumps which excrete fluoroquinolones (85). This might explain the weak activity of ciprofloxacin against enterococci. Newer fluoroquinolones, levofloxacin, gatifloxacin, and moxifloxacin, have an increased activity, but this activity remains moderate as compared to that against pneumococci. In addition, isolates resistant to ciprofloxacin are generally resistant to these fluoroquinolones.

Little is known on the mechanisms of action of quinolones and of resistance to these antimicrobials in enterococci. However, Nakanishi et al. showed that the DNA gyrase from *E. faecalis* ATCC 19433 has also two A and B subunits with properties similar to those of DNA gyrase from *Bacillus subtilis* and *Micrococcus luteus* (98). Inhibition of the supercoiling activity of the gyrase by the fluoroquinolones correlated with their antibacterial activities, whereas gyrases prepared from fluoroquinolone-resistant isolates were less sensitive to inhibition of supercoiling by the fluoroquinolones. In addition, when a GyrA subunit purified from a fluoroquinolone-resistant *E. faecalis* was combined with a wild-type GyrB subunit, the supercoiling reaction remained insensitive to the fluoroquinolones, showing the importance of the *gyrA* mutation. Similar findings were reported later for topoisomerase IV of *E. faecalis* (102). In addition, several studies have shown that resistance to fluoroquinolones was associated with mutations in the GyrA and ParC proteins of clinical and environmental isolates of *E. faecalis* (69, 75, 107, 131). *E. faecalis* resistant to low-levels of ciprofloxacin, norfloxacin, and ofloxacin have mutations of Serine at position 80 of ParC (*E. coli* coordinates) (69). Isolates with high-level resistance to ciprofloxacin combine mutations in GyrA (at Ser83 and Glu87) and ParC (at Ser80 and Glu84). These findings are similar to those for other gram-positive bacteria and *E. coli*. No clinical isolates with only *gyrA* mutation have been reported, and second step mutants have gyrA mutations (75). In the laboratory, a single *gyrA* mutant was obtained after a one-step mutation (102). Taken together, these data suggest that mutation in the *parC* gene is a prerequisite for mutations in *gyrA* to occur and that topoisomerase IV is the primary target of fluoroquinolones in *E. faecalis* as well as in other gram-positive cocci such as *S. aureus* and *S. pneumoniae*. However, this hypothesis remains to be demonstrated for enterococci. Also, this might not hold true for all fluoroquinolones, since it has been shown that in *S. pneumoniae* either gyrase or topoisomerase IV can be the primary target, depending on the structure of the fluoroquinolone (105). Resistance to fluoroquinolones is often combined with high-level resistance to gentamicin (113, 131). For fluoroquinolone-resistant *E. faecium*, findings were similar to those for *E. faecalis* with mutations in GyrA at positions 83 and 87 and in ParC with a mutation at position 80 (19, 43, 135). Finally, no *gyrB* or *parE* mutations or increased quinolone efflux have been reported in enterococci.

Tetracyclines

Tetracyclines inhibit protein synthesis by preventing binding of aminoacyl-tRNA molecules to the 30S ribosomal subunit (123). Despite the fact that enterococcal infections are not treated with tetracyclines, more than 60% of enterococci are resistant to tetracyclines (113). Tetracycline resistance is mediated by two major mechanisms, protection of the ribosomes by large cytoplasmic proteins similar to elongation factors and energy-dependent efflux of tetracycline (117). The ribosomal protection mechanisms identified so far fall into six classes, three of which have been reported in enterococci, TetM, TetO, and TetS, (2, 14, 26, 65, 146), after they were previously found in streptococci, *Listeria*, and *Campylobacter*, respectively (26, 91, 130). The *tet*(M) determinant is, by far, the most common resistance determinant in enterococci and is typically borne by conjugative transposons of the Tn*916* family located in the chromosome of the bacteria or on conjugative plasmids (14, 48, 106). In addition, a *tet*(U) determinant of unknown function but with similarity to *tet*(M) has been reported in *E. faecium* (116).

Whereas ribosomal protection confers cross resistance to tetracyclines and minocycline, isolates with an efflux mechanism display resistance to tetracyclines but remain susceptible to minocycline (106). The active efflux of tetracyclines is due to large proteins with 14 transmembrane domains encoded by the *tet*(L) and *tet*(K) genes. The *tet*(K) gene was first reported in *S. aureus* and is rare in enterococci. The *tet*(L) gene is more common and can be associated with *tet*(M) (65).

Trimethoprim

Trimethoprim inhibits the dihydrofolate reductase (DHFR) which catalyzes the reduction of dihydrofolate to tetrahydrofolate in prokaryotic and eukaryotic cells (60). The most frequent mechanism of resistance to trimethoprim is the production of an additional drug-resistant DHFR, often borne by mobile genetic elements (plasmids, transposons, and gene cassettes). In contrast to gram-negative bacteria, few DHFRs conferring high-level resistance to trimethoprim have been identified in gram-positive organisms. In a strain of *E. faecalis* highly resistant to trimethoprim (MIC $> 1,024$ μg/ml), a *dfrF* gene that also conferred high-level resistance to trimethoprim after cloning in *E. coli* was identified (32). The predicted amino acid sequence had 38 to 64% similarity with other dihydrofolate reductases from gram-positive and gram-negative organisms. The gene was apparently chromosomally located and could not be transferred by conjugation.

Oxazolidinones

The oxazolidinones bind to the ribosomal peptidyltransferase center, domain V of 23S rRNA, and prevent formation of the initiation complex formed by *N*-formyl-methionyl-tRNA, ribosomes, mRNA, and initiation factors IF2 and IF3, therefore blocking protein synthesis at an early stage (73, 143). Soon after the introduction in therapy of linezolid, the first commercially available oxazolidinone, resistance was reported in a few isolates of *E. faecalis* and *E. faecium*, mostly after prolonged courses of therapy (13, 49, 58, 68, 69). Rapidly, mutations in domain V of 23S rRNA were identified as responsible for oxazolidinone resistance.

In vitro mutants of *E. coli*, *Halobacterium halobium*, *E. faecium*, and *E. faecalis* resistant to linezolid were generated before resistant clinical isolates had been reported (81, 109, 144). Mutations selected in vitro have been predictive of those found in clinical isolates of gram-positive pathogens. All mutations in clinical isolates of linezolid-resistant *E. faecalis* and *E. faecium* were found at G2576 in the central loop of domain V (40, 68, 81, 110).

The *rrl* gene encoding 23S rRNA is usually present at several copies in bacteria. *E. faecium* has four copies of the gene, whereas *E. faecalis* contains six copies. Several studies have shown that there is a correlation between the number of mutated copies and the MICs of linezolid (81, 90, 121, 126). Resistant clinical isolates generally have MICs equal to 8 to 64 μg/ml (MICs $\leq$ 4 μg/ml for the susceptible strains) and have at least half of the gene copies mutated.

Chloramphenicol, Heavy Metals, Bacitracin, and Evernimicin

The renewed interest in the 1990s for chloramphenicol as an alternative therapeutic agent for infections due to enterococci came from the increasing prevalence of multidrug resistant *E. faecium*. However, in addition to poor efficacy of the drug against susceptible enterococci, possibly due to active efflux of the drug (85), 1 to 22 % of the strains have acquired resistance (145). In a large number of isolates, resistance is due to diacetylation of an hydroxyl group of the molecule by chloramphenicol-acetyltransferases (CAT) encoded by *cat* genes. Similar *cat* genes are found in enterococci, streptococci, and staphylococci, confirming active exchange of resistance determinants between these species. The *cat* genes from enterococcal plasmid pIP501and staphylococcal plasmids pSC57 and pC194 have been identified in both *E. faecalis* and *E. faecium* (106, 137)

Among gram-positive organisms, resistance to toxic heavy metals has been mostly studied in *S. aureus*. Much less is known in enterococci, although resistance to antiseptics, including heavy metals, could have a role in the remarkable persistence of enterococci in hospitals and in the environment.

The first resistance to a heavy metal was that to mercury, identified in 11 of 52 clinical isolates of *E. faecalis* studied by Zscheck and Murray (148). Resistance was due to acquisition of the *merA* and *merB* genes previously described associated to penicillin G resistance on the large penicillinase plasmids widespread in *S. aureus* (99). Resistance is due to mercury detoxification, as in *S. aureus*. Although mercury resistance was not initially associated with penicillinase plasmids in enterococci, it was detected later on transposon Tn*5385*, which is a composite of several smaller transposable elements originating in *Enterococcus* and *Staphylococcus*, which encode resistance to penicillin G by β-lactamase production, to erythromycin, gentamicin, mercuric chloride, streptomycin, and tetracycline (15).

Copper homeostasis is mediated in *E. hirae* by a *cop* operon (125). Acquired resistance to copper has been recently identified in a pig isolate of *E. faecium* (56). In this isolate, the *tcrB* gene located on a large conjugative plasmid (ca. 175 kb) confers a sevenfold increase in copper MIC. It encodes a putative heavy metal transporter protein belonging to the CPx-type ATPase family, which has similarity (46%) to the CopB protein from the *cop* operon of *E. hirae* (100). This gene was found in all copper-resistant *E. faecium* and *E. faecalis* pig isolates from three European countries and also from a few human isolates (4, 56). In the original *E. faecium*, the gene is colocated with the *vanA* gene cluster for glycopeptide resistance and the *erm*(B) macrolide resistance gene on the same conjugative plasmid, and these resistance markers are also frequently associated in the other copper-resistant isolates from pigs (58). The physical link between the resistance determinants has lead Hasman et al. to suggest that the use of copper as a food additive in piglets and also in slaughter pigs, broilers, and calves might be in part responsible for the persistence of vancomycin-resistant *E. faecium* in caeca of these food animals despite the ban of avoparcin (57).

The antiseptic resistance gene *qacE*Δ1, originally isolated from gram-negative bacteria, was found by Japanese authors in 9 of 48 strains of clinical isolates of *Enterococcus* (72). This gene was found identical to that of an integron of *Pseudomonas aeruginosa,* showing genetic exchange between these bacteria.

Zinc bacitracin is used as an antimicrobial in poultry, and a recent report from New Zealand showed that nearly all poultry enterococcal studied were highly resistant to the drug (89). Resistance to bacitracin is mediated in *E. faecalis* by the *bcrA* and *bcrB* genes located on a conjugative plasmid and organized in an operon with a third *bcrD* gene (89). The BcrA protein, which contains an ATP-binding domain, and the BrcB protein, which contains a membrane-spanning domain, may constitute a functional homodimeric ABC transporter that actively pumps bacitracin from the cell. Upstream from the genes was identified a *bcrR* gene encoding a transcriptional activator of the resistance genes.

Avilamycin is an oligosaccharidic antibiotic used as a food additive in animals (22). A related antibiotic, evernimicin, has been developed for therapy in humans but withdrawn for toxicity reasons. Both antibiotics display cross resistance (1). Resistance to avilamycin has been reported in enterococci from animals and rarely from human feces (1) and is due to alterations in the ribosomal protein L16 (6) or to acquisition of an *emtA* gene coding for a ribosomal methyltransferase (74, 88).

CONCLUSION

Evolution of enterococci toward resistance to multiple antimicrobials is a major cause of concern. Because of their presence in high quantities in the gut and the diversity of mobile genetic elements that they harbor, these microorganisms act as a hub for the dissemination of resistance genes. The fact that these organisms are less virulent than others is not reassuring, since acquisition of virulence genes by bacteria is common.

REFERENCES

1. **Aarestrup, F. M.** 1998. Association between decreased susceptibility to a new antibiotic for treatment of human diseases, everninomicin (SCH 27899), and resistance to an antibiotic used for growth promotion in animals, avilamycin. *Microb. Drug Resist.* **4:**137–141.
2. **Aarestrup, F. M., Y. Agerso, P. Gerner-Smidt, M. Madsen, and L. B. Jensen.** 2000. Comparison of antimicrobial resistance phenotypes and resistance genes in *Enterococcus faecalis* and *Enterococcus faecium* from humans in the community, broilers, and pigs in Denmark. *Diagn. Microbiol. Infect. Dis.* **37:**127–137.
3. **Aarestrup, F. M., and L. B. Jensen.** 2000. Presence of variations in ribosomal protein L16 corresponding to susceptibility of enterococci to oligosaccharides (avilamycin and evernimicin). *Antimicrob. Agents Chemother.* **44:**3425–3427.
4. **Aarestrup, F. M., H. Hasman, L. B. Jensen, M. Moreno, I. A. Herrero, L. Dominguez, M. Finn, and A. Franklin.** 2002. Antimicrobial resistance among enterococci from pigs in three European countries. *Appl. Environ. Microbiol.* **68:**4127–4129.
5. **Adrian, P. V., C. Mendrick, D. Loebenberg, P. McNicholas, K. J. Shaw, K. P. Klugman, R. S. Hare, and T. A. Black.** 2000. Evernimicin (SCH27899) inhibits a novel ribosome target site: analysis of 23S ribosomal DNA mutants. *Antimicrob. Agents Chemother.* **44:**3101–3106.
6. **Adrian, P. V., W. Zhao, T. A. Black, K. J. Shaw, R. S. Hare, and K. P. Klugman.** 2000. Mutations in ribosomal protein L16 conferring reduced susceptibility to evernimicin (SCH27899): implications for mechanism of action. *Antimicrob. Agents Chemother.* **44:**732–738.
7. **Allignet, J., and N. El Solh.** 1995. Diversity among the gram-positive acetyltransferases inactivating streptogramin A and structurally related compounds and characterization of a new staphylococcal determinant, *vatB*. *Antimicrob. Agents. Chemother.* **39:**2027–2036.
8. **Allignet, J., N. Liassine, and N. El Solh.** 1998. Characterization of a staphylococcal plasmid related to pUB110 and carrying two novel genes, *vatC* and *vgbB,* encoding resistance to streptogramins A and B and similar antibiotics. *Antimicrob. Agents. Chemother.* **42:**1794–1798.
9. **Allignet, J., V. Loncle, P. Mazodier, and N. El Solh.** 1988. Nucleotide sequence of a staphylococcal plasmid gene, *vgb,* encoding a hydrolase inactivating the B components of virginiamycin-like antibiotics. *Plasmid* **20:**271–275.
10. **Allignet, J., V. Loncle, C. Simenel, M. Delepierre, and N. El Solh.** 1993. Sequence of a staphylococcal gene, vat, encoding an acetyltransferase inactivating the A-type compounds of virginiamycin-like antibiotics. *Gene* **130:**91–98.
11. **Al-Obeid, S., L. Gutmann, and R. Williamson.** 1990. Modification of penicillin-binding proteins of penicillin-resistant mutants of different species of enterococci. *J. Antimicrob. Chemother.* **26:**613–618.
12. **Asseray, N., J. Caillon, N. Roux, C. Jacqueline, R. Bismuth, M. F. Kergueris, G. Potel., and D. Bugnon.** 2002. Different aminoglycoside-resistant phenotypes in a rabbit *Staphylococcus aureus* endocarditis infection model. *Antimicrob. Agents Chemother.* **46:**1591–1593.
13. **Auckland, C., L. Teare, F. Cooke, M. E. Kaufmann, M. Warner, G. Jones, K. Bamford, H. Ayles, and A. P. Johnson.** 2002. Linezolid-resistant enterococci: report of the first isolates in the United Kingdom. *J. Antimicrob Chemother.* **50:**743–746.
14. **Bentorcha, F., G. De Cespedes, and T. Horaud.** 1991. Tetracycline resistance heterogeneity in *Enterococcus faecium. Antimicrob. Agents Chemother.* **35:**808–812.
15. **Bonafede, M. E., L. L. Carias, and L. B. Rice.** 1997. Enterococcal transposon Tn5384: evolution of a composite transposon through cointegration of enterococcal and staphylococcal plasmids. *Antimicrob. Agents Chemother.* **41:**1854–1858.
16. **Bozdogan, B., L. Berrezouga, M. S. Kuo, D. A. Yurek, K. A. Farley, B. J. Stockman, and R. Leclercq.** 1999. A new resistance gene, *linB,* conferring resistance to lincosamides by nucleotidylation in *Enterococcus faecium* HM1025. *Antimicrob. Agents Chemother.* **43:**925–929.
17. **Bozdogan, B., and R. Leclercq.** 1999. Effects of genes encoding resistance to streptogramins A and B on the activity of quinupristin-dalfopristin against *Enterococcus faecium. Antimicrob. Agents Chemother.* **43:**2720–2725.
18. **Bozdogan, B., R. Leclercq, A. Lozniewski, and M. Weber.** 1999. Plasmid-mediated coresistance to streptogramins and vancomycin in *Enterococcus faecium* HM1032. *Antimicrob. Agents Chemother.* **43:**2097–2098.

19. **Brisse, S., A. C. Fluit, U. Wagner, P. Heisig, D. Milatovic, J. Verhoef, S. Scheuring, K. Kohrer, and F. J. Schmitz.** 1999. Association of alterations in ParC and GyrA proteins with resistance of clinical isolates of *Enterococcus faecium* to nine different fluoroquinolones. *Antimicrob. Agents Chemother.* **43:** 2513–2516.
20. **Bryan, L. E., and H. M. Van Den Elzen.** 1977. Effects of membrane-energy mutations and cations on streptomycin and gentamicin accumulation by bacteria: a model for entry of streptomycin and gentamicin in susceptible and resistant bacteria. *Antimicrob. Agents Chemother.* **12:**163–177.
21. **Burk, D. L., N. Ghuman, L. E. Wybenga-Groot, and A. M. Berghuis.** 2003. X-ray structure of the AAC(6′)-Ii antibiotic resistance enzyme at 1.8 A resolution; examination of oligomeric arrangements in GNAT superfamily members. *Protein Sci.* **12:**426–437.
22. **Butaye, P., L. A. Devriese, and F. Haesebrouck.** 2003. Antimicrobial growth promoters used in animal feed: effects of less well known antibiotics on gram-positive bacteria. *Clin. Microbiol. Rev.* **16:**175–188.
23. **Caillaud, F., P. Trieu-Cuot, C. Carlier, and P. Courvalin.** 1987. Nucleotide sequence of the kanamycin resistance determinant of the pneumococcal transposon Tn*1545:* evolutionary relationships and transcriptional analysis of *aphA-3* genes. *Mol. Gen. Genet.* **207:**509–513.
24. **Carias, L. L., S. D. Rudin, C. J. Donskey, and L. B. Rice.** 1998. Genetic linkage and cotransfer of a novel, *vanB*-containing transposon (Tn5382) and a low-affinity penicillin-binding protein 5 gene in a clinical vancomycin-resistant *Enterococcus faecium* isolate. *J. Bacteriol.* **180:**4426–4434.
25. **Carlier, C., and P. Courvalin.** 1990. Emergence of 4′,4″-aminoglycoside nucleotidyltransferase in enterococci. *Antimicrob. Agents Chemother.* **34:**1565–1569.
26. **Charpentier, E., G. Gerbaud, and P. Courvalin.** 1993. Characterization of a new class of tetracycline-resistance gene *tet*(S) in *Listeria monocytogenes* BM4210. *Gene* **131:**27–34.
27. **Charpentier, E., G. Gerbaud, and P. Courvalin.** 1994. Presence of the Listeria tetracycline resistance gene *tet*(S) in *Enterococcus faecalis. Antimicrob. Agents Chemother.* **38:**2330–2335.
28. **Choi, S. H., S. O Lee, T. H. Kim, J. W. Chung, E. J. Choo, Y. G. Kwak, M. N. Kim, Y. S. Kim, J. H. Woo, J. Ryu, and N. J. Kim.** 2004. Clinical features and outcomes of bacteremia caused by *Enterococcus casseliflavus* and *Enterococcus gallinarum:* analysis of 56 cases. *Clin. Infect. Dis.* **38:**53–61.
29. **Chow, J. W., V. Kak, I. You, S. J. Kao, J. Petrin, D. B. Clewell, S. A. Lerner, G. H. Miller, and K. J. Shaw.** 2001. Aminoglycoside resistance genes *aph*(2″)*-Ib* and *aac*(6′)-Im detected together in strains of both *Escherichia coli* and *Enterococcus faecium. Antimicrob. Agents Chemother.* **45:**2691–2694.
30. **Chow, J. W., M. J. Zervos, S. A. Lerner, L. A. Thal, S. M. Donabedian, D. D. Jaworski, S. Tsai, K. J Shaw, and D. B. Clewell.** 1997. A novel gentamicin resistance gene in *Enterococcus. Antimicrob. Agents Chemother.* **41:**511–514.
31. **Clark, N. C., O. Olsvik, J. M. Swenson, C. A. Spiegel, and F. C. Tenover.** 1999. Detection of a streptomycin/spectinomycin adenylyltransferase gene (aadA) in *Enterococcus faecalis. Antimicrob. Agents Chemother.* **43:**157–160.
32. **Coque, T. M., K. V. Singh, G. M. Weinstock, and B. E. Murray.** 1999. Characterization of dihydrofolate reductase genes from trimethoprim-susceptible and trimethoprim-resistant strains of *Enterococcus faecalis. Antimicrob. Agents Chemother.* **43:**141–147.
33. **Costa, Y., M. Galimand, R. Leclercq, J. Duval, and P. Courvalin.** 1993. Characterization of the chromosomal *aac(6′)-Ii* gene specific for *Enterococcus faecium. Antimicrob. Agents Chemother.* **37:**1896–1903.
34. **Courvalin, P., and C. Carlier.** 1987. Tn*1545:* a conjugative shuttle transposon. *Mol. Gen. Genet.* **206:**259–264.
35. **Courvalin, P., C. Carlier, and E. Collatz.** 1980. Plasmid-mediated resistance to aminocyclitol antibiotics in group D streptococci. *J. Bacteriol.* **143:**541–551.
36. **Courvalin, P., C. Carlier, and E. Collatz.** 1980. Structural and functional relationships between aminoglycoside-modifying enzymes from streptococci and staphylococci, p. 309–320. *In* S. Mitsuhashi, I. Rosival, and V. Kremery (ed.), *Medical and Biological Aspects of Resistant Strains.* Springer-Verlag, Berlin, Germany.
37. **Daly, M. M., S. Doktor, R. Flamm, and D. Shortridge.** 2004. Characterization and prevalence of MefA, MefE, and the associated *msr*(D) gene in *Streptococcus pneumoniae* clinical isolates. *J. Clin. Microbiol.* **42:**3570–3574.
38. **Del Campo, R., C. Tenorio, C. Rubio, J. Castillo, C. Torres, and R. Gomez-Lus.** 2000. Aminoglycoside-modifying enzymes in high-level streptomycin and gentamicin resistant *Enterococcus* spp. in Spain. *Int. J. Antimicrob. Agents.* **15:**221–226.
39. **Derbise, A., S. Aubert, and N. El Solh.** 1997. Mapping the regions carrying the three contiguous antibiotic resistance genes *aadE, sat4,* and *aphA-3* in the genomes of staphylococci. *Antimicrob. Agents Chemother.* **41:**1024–1032.
40. **Dibo, I., S. K. Pillai, H. S. Gold, M. R. Baer, M. Wetzler, J. L. Slack, P. A. Hazamy, D. Ball, C. B. Hsiao, P. L. McCarthy Jr., and B. H. Segal.** 2004. Linezolid-resistant *Enterococcus faecalis* isolated from a cord blood transplant recipient. *J. Clin. Microbiol.* **42:**1843–1845.
41. **Dina, J., B. Malbruny, and R. Leclercq.** 2003. Nonsense mutations in the lsa-like gene in *Enterococcus faecalis* isolates susceptible to lincosamides and Streptogramins A. *Antimicrob. Agents Chemother.* **47:**2307–2309.
42. **Donabedian, S. M., L. A. Thal, E. Hershberger, M. B. Perri, J. W. Chow, P. Bartlett, R. Jones, K. Joyce, S. Rossiter, K. Gay, J. Johnson, C. Mackinson, E. Debess, J. Madden, F. Angulo, and M. J. Zervos.** 2003. Molecular characterization of gentamicin-resistant enterococci in the United States: evidence of spread from animals to humans through food. *J. Clin. Microbiol.* **41:**1109–1113.
43. **El Amin, N. A., S. Jalal, and B. Wretlind.** 1999. Alterations in GyrA and ParC associated with fluoroquinolone resistance in *Enterococcus faecium. Antimicrob. Agents Chemother.* **43:**947–949.
44. **Eliopoulos, G. M., B. F. Farber, B. E. Murray, C. Wennersten, and R. C. Moellering, Jr.** 1984. Ribosomal resistance of clinical enterococcal to streptomycin isolates. *Antimicrob. Agents Chemother.* **25:**398–399.
45. **El Kharroubi, A., P. Jacques, G. Piras, J. Van Beeumen, J. Coyette, and J. M. Ghuysen.** 1991. The *Enterococcus hirae* R40 penicillin-binding protein 5 and the methicillin-resistant *Staphylococcus aureus* penicillin-binding protein 2′ are similar. *Biochem. J.* **280:**463–469.
46. **Ferretti, J. J., K. S. Gilmore, and P. Courvalin.** 1986. Nucleotide sequence analysis of the gene specifying the bifunctional 6′-aminoglycoside acetyltransferase 2″-aminoglycoside phosphotransferase enzyme in *Streptococcus faecalis* and identification and cloning of gene regions specifying the two activities. *J. Bacteriol.* **167:**631–638.
47. **Fontana, R., R. Cerini, P. Longoni, A. Grossato, and P. Canepari.** 1983. Identification of a streptococcal penicillin-binding protein that reacts very slowly with penicillin. *J. Bacteriol.* **155:**343–1350.
48. **Franke, A. E., and D. B. Clewell.** 1981. Evidence for a chromosome-borne resistance transposon (Tn*916*) in *Streptococcus faecalis* that is capable of "conjugal" transfer in the absence of a conjugative plasmid. *J. Bacteriol.* **145:**494–502.

49. **Gonzales, R. D., P. C. Schreckenberger, M. B. Graham, S. Kelkar, K. DenBesten, and J. P. Quinn.** 2001. Infections due to vancomycin-resistant *Enterococcus faecium* resistant to linezolid. *Lancet* **357:**1179.
50. **Goodhart, G. L.** 1984. In vivo versus in vitro susceptibility of enterococcus to trimethoprim-sulfamethoxazole. A pitfall. *JAMA* **252:**2748–2749.
51. **Gordon, S., J. M. Swenson, B. C. Hill, N. E. Pigott, R. R. Facklam, R. C. Cooksey, C. Thornsberry, W. R. Jarvis, and F. C. Tenover.** 1992. Antimicrobial susceptibility patterns of common and unusual species of enterococci causing infections in the United States. Enterococcal Study Group. *J. Clin. Microbiol.* **30:**2373–2378.
52. **Grayson, M. L., C. Thauvin-Eliopoulos, G. M. Eliopoulos, J. D. Yao, D. V. DeAngelis, L. Walton, J. L. Woolley, and R. C. Moellering, Jr.** 1990. Failure of trimethoprim-sulfamethoxazole therapy in experimental enterococcal endocarditis. *Antimicrob. Agents Chemother.* **34:**1792–1794.
53. **Hamilton-Miller, J. M.** 1988. Reversal of activity of trimethoprim against gram-positive cocci by thymidine, thymine and folates. *J. Antimicrob. Chemother.* **22:**35–39.
54. **Hanrahan, J., C. Hoyen, and L. B. Rice.** 2000. Geographic distribution of a large mobile element that transfers ampicillin and vancomycin resistance between *Enterococcus faecium* strains. *Antimicrob. Agents Chemother.* **44:**1349–1351.
55. **Harms, J. M., F. Schlunzen, P. Fucini, H. Bartels, and A. Yonath.** 2001. Alterations at the peptidyl transferase centre of the ribosome induced by the synergistic action of the streptogramins dalfopristin and quinupristin. *BMC Biol.* **2:**4.
56. **Hasman, H., and F. M. Aarestrup.** 2002. *tcrB*, a gene conferring transferable copper resistance in *Enterococcus faecium:* occurrence, transferability, and linkage to macrolide and glycopeptide resistance. *Antimicrob. Agents Chemother.* **46:**1410–1416.
57. **Hasman H, and F. M. Aarestrup.** 2005. Relationship between copper, glycopeptide, and macrolide resistance among *Enterococcus faecium* strains isolated from pigs in Denmark between 1997 and 2003. *Antimicrob. Agents Chemother.* **49:**454–456.
58. **Herrero, I. A., N. C. Issa, and R. Patel.** 2002. Nosocomial spread of linezolid-resistant, vancomycin-resistant *Enterococcus faecium. N. Engl. J. Med.* **346:**867–869.
59. **Hershberger, E., S. Donabedian, K. Konstantinou, and M. J. Zervos.** 2004. Quinupristin-dalfopristin resistance in gram-positive bacteria: mechanism of resistance and epidemiology. *Clin. Infect. Dis.* **38:**92–98.
60. **Hitchins, G. H.** 1973. Mechanism of action of trimethoprim-sulfamethoxazole. *J. Infect. Dis.* **128**(Suppl.):433–436.
61. **Hollingshead, S., and D. Vapnek.** 1985. Nucleotide sequence analysis of a gene encoding a streptomycin/spectinomycin adenylyltransferase. *Plasmid* **13:**17–30.
62. **Hon, W. C., G. A. McKay, P. R. Thompson, R. M. Sweet, D. S. Yang, G. D. Wright, and A. M. Berghuis.** 1997. Structure of an enzyme required for aminoglycoside antibiotic resistance reveals homology to eukaryotic protein kinases. *Cell* **89:**887–895.
63. **Hooper, D. C.** 2000. Mechanisms of fluoroquinolone resistance, p. 685–693. *In* V. A. Fischetti, R. P. Novick, J. J. Ferretti, D. A. Portnoy, and J. I. Rood (ed.), *Gram-Positive Pathogens.* ASM Press, Washington, D.C.
64. **Horinouchi, S., W. H. Byeon, and B. Weisblum.** 1983. A complex attenuator regulates inducible resistance to macrolides, lincosamides, and streptogramin type B antibiotics in *Streptococcus sanguis. J. Bacteriol.* **154:**1252–1262.
65. **Huys, G., K. D'Haene, J. M. Collard, and J. Swings.** 2004. Prevalence and molecular characterization of tetracycline resistance in *Enterococcus* isolates from food. *Appl. Environ. Microbiol.* **70:**1555–1562.
66. **Jackson, C. R., P. J. Fedorka-Cray, J. B. Barrett, and S. R. Ladely.** 2004. Genetic relatedness of high-level aminoglycoside-resistant enterococci isolated from poultry carcasses. *Avian Dis.* **48:**100–107.
67. **Jensen, L. B., A. M. Hammerum, F. M. Aerestrup, A. E. Van Den Bogaard, and E. E. Stobberingh.** 1998. Occurrence of *satA* and *vgb* genes in streptogramin-resistant *Enterococcus faecium* isolates of animal and human origins in The Netherlands. *Antimicrob. Agents Chemother.* **42:**3330–3331.
68. **Johnson, A. P., L. Tysall, M. V. Stockdale, N. Woodford, M. E. Kaufmann, M. Warner, D. M. Livermore, F. Asboth, and F. J. Allerberger.** 2002. Emerging linezolid-resistant *Enterococcus faecalis* and *Enterococcus faecium* isolated from two Austrian patients in the same intensive care unit. *Eur. J. Clin. Microbiol. Infect. Dis.* **21:**751–754.
69. **Jones, R. N., P. Della-Latta, L. V. Lee, and D. J. Biedenbach.** 2002. Linezolid-resistant *Enterococcus faecium* isolated from a patient without prior exposure to an oxazolidinone: report from the SENTRY Antimicrobial Surveillance Program. *Diagn. Microbiol. Infect Dis.* **42:**137–139.
70. **Kanematsu, E., T. Deguchi, M. Yasuda, T. Kawamura, Y. Nishino, and Y. Kawada.** 1998. Alterations in the GyrA subunit of DNA gyrase and the ParC subunit of DNA topoisomerase IV associated with quinolone resistance in *Enterococcus faecalis. Antimicrob. Agents Chemother.* **42:**433–435.
71. **Kao, S. J., I. You, D. B. Clewell, S. M. Donabedian, M. J. Zervos, J. Petrin, K. J. Shaw, and J. W. Chow.** 2000. Detection of the high-level aminoglycoside resistance gene *aph(2″)-Ib* in *Enterococcus faecium. Antimicrob. Agents Chemother.* **44:**2876–2879.
72. **Kazama, H., H. Hamashima, M. Sasatsu, and T. Arai.** 1998. Distribution of the antiseptic-resistance gene *qacE*Δ1 in gram-positive bacteria. *FEMS Microbiol. Lett.* **165:**295–299.
73. **Kloss, P., L. Xiong, D. L. Shinabarger, and A. S. Mankin.** 1999. Resistance mutations in 23 S rRNA identify the site of action of the protein synthesis inhibitor linezolid in the ribosomal peptidyl transferase center. *J. Mol. Biol.* **294:**93–101.
74. **Kofoed, C. B., and B. Vester.** 2002. Interaction of avilamycin with ribosomes and resistance caused by mutations in 23S rRNA. *Antimicrob. Agents Chemother.* **46:**3339–3342.
75. **Korten, V., W. M. Huang, and B. E. Murray.** 1994. Analysis by PCR and direct DNA sequencing of gyrA mutations associated with fluoroquinolone resistance in *Enterococcus faecalis. Antimicrob. Agents Chemother.* **38:**2091–2094.
76. **LeBlanc, D. J., L. N. Lee., and J. M. Inamine.** 1991. Cloning and nucleotide base sequence analysis of a spectinomycin adenyltransferase AAD(9) determinant from *Enterococcus faecalis. Antimicrob. Agents Chemother.* **35:**1804–1810.
77. **Lee, E. W., M. N. Huda, T. Kuroda, T. Mizushima, and T. Tsuchiya.** 2003. EfrAB, an ABC multidrug efflux pump in *Enterococcus faecalis. Antimicrob. Agents Chemother.* **47:**3733–3738.
78. **Ligozzi, M., M. Aldegheri, S. C. Predari, and R. Fontana.** 1991. Detection of penicillin-binding proteins immunologically related to penicillin-binding protein 5 of *Enterococcus hirae* ATCC 9790 in *Enterococcus faecium* and *Enterococcus faecalis. FEMS Microbiol. Lett.* **67:**335–339.
79. **Ligozzi, M., F. Pittaluga, and R. Fontana.** 1993. Identification of a genetic element (psr) which negatively controls expression of *Enterococcus hirae* penicillin-binding protein 5. *J. Bacteriol.* **175:**2046–2051.
80. **Lim, J. A., A. R. Kwon, S. K. Kim, Y. Chong, K. Lee, and B. C. Choi.** 2002. Prevalence of resistance to macrolide, lincosamide and streptogramin antibiotics in Gram-positive cocci

isolated in a Korean hospital. *J. Antimicrob. Chemother.* **49:** 489–495.

81. **Lobritz, M., R. Hutton-Thomas, S. Marshall, and L. B. Rice.** 2003. Recombination proficiency influences frequency and locus of mutational resistance to linezolid in *Enterococcus faecalis. Antimicrob. Agents Chemother.* **47:**3318–3320.
82. **Lopes, M. de F., T. Ribeiro, M. P. Martins, R. Tenreiro, and M. T. Crespo.** 2003. Gentamicin resistance in dairy and clinical enterococcal isolates and in reference strains. *J. Antimicrob. Chemother.* **52:**214–219.
83. **Luna, V. A., P. Coates, E. A. Eady, J. H. Cove, T. T. Nguyen, and M. C. Roberts.** 1999. A variety of gram-positive bacteria carry mobile *mef* genes. *J. Antimicrob. Chemother.* **44:**19–25.
84. **Luna, V. A., M. Heiken, K. Judge, C. Ulep, N. Van Kirk, H. Luis, M. Bernardo, J. Leitao, and M. C. Roberts.** 2002. Distribution of mef(A) in gram-positive bacteria from healthy Portuguese children. *Antimicrob. Agents Chemother.* **46:**2513–2517.
85. **Lynch, C., P. Courvalin, and H. Nikaido.** 1997. Active efflux of antimicrobial agents in wild-type strains of enterococci. *Antimicrob. Agents Chemother.* **41:**869–871.
86. **Mainardi, J. L, R. Legrand, M. Arthur, B. Schoot, J. van Heijenoort, and L. Gutmann.** 2000. Novel mechanism of β-lactam resistance due to by-pass of DD-transpeptidation in *Enterococcus faecium. J Biol. Chem.* **275:**16490–16496.
87. **Mainardi, J. L., V. More, M. Fourgeaud, J. Cremniter, D. Blanot, R. Legrand, C. Frehel, M. Arthur, J. Van Heijenoort, and L. Gutmann.** 2002. Balance between two transpeptidation mechanisms determines the expression of beta-lactam resistance in *Enterococcus faecium. J. Biol. Chem.* **277:**35801–35807.
88. **Mann, P. A., L. Xiong, A. S. Mankin, A. S. Chau, C. A. Mendrick, D. J. Najarian, C. A. Cramer, D. Loebenberg, E. Coates, N. J. Murgolo, F. M. Aarestrup, R. V. Goering, T. A. Black, R. S. Hare, and P. M. McNicholas.** 2001. *emtA*, a rRNA methyltransferase conferring high-level evernimicin resistance. *Mol. Microbiol.* **41:**1349–1356.
89. **Manson, J. M., S. Keis, J. M. Smith, and G. M. Cook.** 2004. Acquired bacitracin resistance in *Enterococcus faecalis* is mediated by an ABC transporter and a novel regulatory protein, BcrR. *Antimicrob. Agents Chemother.* **48:**3743–3748.
90. **Marshall, S. H., C. J. Donskey, R. Hutton-Thomas, R. A. Salata, and L. B. Rice.** 2002. Gene dosage and linezolid resistance in *Enterococcus faecium* and *Enterococcus faecalis. Antimicrob. Agents Chemother.* **46:**3334–3336.
91. **Martin, P., P. Trieu-Cuot, and P. Courvalin.** 1986. Nucleotide sequence of the *tetM* tetracycline resistance determinant of the streptococcal conjugative shuttle transposon Tn*1545*. *Nucleic Acids Res.* **14:**7047–7058.
92. **McKay, G. A., P. R. Thompson, and G. D. Wright.** 1994. Broad spectrum aminoglycoside phosphotransferase type III from *Enterococcus:* overexpression, purification, and substrate specificity. *Biochemistry* **33:**6936–6944.
93. **Moellering, R. C. Jr., O. M. Korzeniowski, M. A. Sande, and C. B. Wennersten.** 1979. Species-specific resistance to antimicrobial synergism in *Streptococcus faecium* and *Streptococcus faecalis. J. Infect. Dis.* **140:**203–208.
94. **Moellering, R. C. Jr., C. Wennersten, and A. N. Weinberg.** 1971. Studies on antibiotic synergism against enterococci. I. Bacteriologic studies. *J. Lab. Clin Med.* **77:**821–828.
95. **Murphy, E.** 1985. Nucleotide sequence of a spectinomycin adenyltransferase AAD(9) determinant from *Staphylococcus aureus* and its relationship to AAD(3″) (9). *Mol. Gen. Genet.* **200:**33–39.
96. **Murray, B. E.** 1992. Beta-lactamase-producing enterococci. *Antimicrob. Agents Chemother.* **36:**2355–2359.
97. **Najjar, A., and B. E. Murray.** 1987. Failure to demonstrate a consistent in vitro bactericidal effect of trimethoprim-sulfamethoxazole against enterococci. *Antimicrob. Agents Chemother.* **3:**808–810.
98. **Nakanishi, N., S. Yoshida, H. Wakebe, M. Inoue, and S. Mitsuhashi.** 1991. Mechanisms of clinical resistance to fluoroquinolones in *Enterococcus faecalis. Antimicrob. Agents Chemother.* **35:**1053–1059.
99. **Novick, R. P., and Roth C.** 1968 Plasmid-linked resistance to inorganic salts in *Staphylococcus aureus. J. Bacteriol.* **95:**1335–1342.
100. **Odermatt, A., H. Suter, R. Krapf, and M. Solioz.** 1993. Primary structure of two P-type ATPases involved in copper homeostasis in *Enterococcus hirae. J. Biol. Chem.* **268:**12775–12779.
101. **Oh, T. G., A. R. Kwon, and E. C. Choi.** 1998. Induction of *ermAMR* from a clinical strain of *Enterococcus faecalis* by 16-membered-ring macrolide antibiotics. *J. Bacteriol.* **180:**5788–5791.
102. **Onodera, Y., J. Okuda, M. Tanaka, and K. Sato.** 2002. Inhibitory activities of quinolones against DNA gyrase and topoisomerase IV of *Enterococcus faecalis. Antimicrob. Agents Chemother.* **46:**1800–1804.
103. **Ounissi, H., and P. Courvalin.** 1987. Nucleotide sequence of streptococcal genes, p. 275. *In* J. Feretti and R. Curtiss III (ed.), *Streptococcal Genetics.* American Society for Microbiology, Washington, D.C.
104. **Ounissi, H., E. Derlot, C. Carlier, and P. Courvalin.** 1990. Gene homogeneity for aminoglycoside-modifying enzymes in gram-positive cocci. *Antimicrob. Agents Chemother.* **34:**2164–2168.
105. **Pan, X.-S., and L. M. Fisher.** 1997. Targeting of DNA gyrase in *Streptococcus pneumoniae* by sparfloxacin: selective targeting of gyrase or topoisomerase IV by quinolones. *Antimicrob. Agents Chemother.* **41:**471–474.
106. **Pepper, K., T. Horaud, C. Le Bouguenec, and G. de Cespedes.** 1987. Location of antibiotic resistance markers in clinical isolates of *Enterococcus faecalis* with similar antibiotypes. *Antimicrob Agents Chemother.* **31:**1394–1402.
107. **Petersen, A., and L. B. Jensen.** 2004. Analysis of *gyrA* and *parC* mutations in enterococci from environmental samples with reduced susceptibility to ciprofloxacin. *FEMS Microbiol. Lett.* **231:**73–76.
108. **Portillo, A., F. Ruiz-Larrea, M. Zarazaga, A. Alonso, J. L. Martinez, and C. Torres.** 2000. Macrolide resistance genes in *Enterococcus* spp. *Antimicrob. Agents Chemother.* **44:**967–971.
109. **Prystowsky, J., F. Siddiqui, J. Chosay, D. L. Shinabarger, J. Millichap, L. R. Peterson, and G. A. Noskin.** 2001. Resistance to linezolid: characterization of mutations in rRNA and comparison of their occurrences in vancomycin-resistant enterococci. *Antimicrob. Agents Chemother.* **45:**2154–2156.
110. **Rahim, S., S. K. Pillai, H. S. Gold, L. Venkataraman, K. Inglima, and R. A. Press.** 2003. Linezolid-resistant, vancomycin-resistant *Enterococcus faecium* infection in patients without prior exposure to linezolid. *Clin. Infect. Dis.* **36:**E146–E148.
111. **Raze, D., O. Dardenne, S. Hallut, M. Martinez-Bueno, J. Coyette, and J. M. Ghuysen.** 1998. The gene encoding the low-affinity penicillin-binding protein 3r in *Enterococcus hirae* S185R is borne on a plasmid carrying other antibiotic resistance determinants. *Antimicrob. Agents Chemother.* **42:**534–539.
112. **Rende-Fournier, R., R. Leclercq, M. Galimand, J. Duval, and P. Courvalin.** 1993 Identification of the *satA* gene encoding a streptogramin A acetyltransferase in *Enterococcus faecium* BM4145. *Antimicrob. Agents Chemother.* **37:**2119–2125.

113. **Reynolds, R., N. Potz, M. Colman, A. Williams, D. Livermore, A. MacGowan A, and BSAC Extended Working Party on Bacteraemia Resistance Surveillance.** 2004. Antimicrobial susceptibility of the pathogens of bacteraemia in the UK and Ireland 2001-2002: the BSAC Bacteraemia Resistance Surveillance Programme. *J. Antimicrob. Chemother.* **53:**1018–1032.
114. **Rice, L. B., S. Bellais, L. L. Carias, R. Hutton-Thomas, R. A. Bonomo, P. Caspers, M. G. Page, and L. Gutmann.** 2004. Impact of specific *pbp5* mutations on expression of beta-lactam resistance in *Enterococcus faecium*. *Antimicrob. Agents Chemother.* **48:**3028–3032.
115. **Richards, M. J., J. R. Edwards, D. H. Culver, and R. P. Gaynes.** 2000. Nosocomial infections in combined medical-surgical intensive care units in the United States. *Infect. Control Hosp. Epidemiol.* **21:**510–515.
116. **Ridenhour, M. B., H. M. Fletcher, J. E. Mortensen, and L. Daneo-Moore.** 1996. A novel tetracycline-resistant determinant, *tet*(U), is encoded on the plasmid pKq10 in *Enterococcus faecium*. *Plasmid* **35:**71–80.
117. **Roberts, M. C.** 1996. Tetracycline resistance determinants: mechanisms of action, regulation of expression, genetic mobility, and distribution. *FEMS Microbiol. Rev.* **19:**1–24.
118. **Roberts, M. C., J. Sutcliffe, P. Courvalin, L. B. Jensen, J. Rood, and H. Seppala.** 1999. Nomenclature for macrolide and macrolide-lincosamide-streptogramin B resistance determinants. *Antimicrob. Agents Chemother.* **43:**2823–2830.
119. **Rosato, A., H. Vicarini, and R. Leclercq.** 1999. Inducible or constitutive expression of resistance in clinical isolates of streptococci and enterococci cross-resistant to erythromycin and lincomycin. *J. Antimicrob. Chemother.* **43:**559–562.
120. **Ross, J. I., E. A. Eady, J. H. Cove, W. J. Cunliffe, S. Baumberg, and J. C. Wootton.** 1990. Inducible erythromycin resistance in staphylococci is encoded by a member of the ATP-binding transport super-gene family. *Mol. Microbiol.* **4:**1207–1214.
121. **Ruggero, K. A., L. K. Schroeder, P. C. Schreckenberger, A. S. Mankin, and J. P. Quinn.** 2003. Nosocomial superinfections due to linezolid-resistant *Enterococcus faecalis:* evidence for a gene dosage effect on linezolid MICs. *Diagn. Microbiol. Infect. Dis.* **47:**511–513.
122. **Rybkine, T., J. L. Mainardi, W. Sougakoff, E. Collatz, and L. Gutmann.** 1998. Penicillin-binding protein 5 sequence alterations in clinical isolates of *Enterococcus faecium* with different levels of beta-lactam resistance. *J. Infect. Dis.* **178:**159–163.
123. **Schnappinger, D., and W. Hillen.** 1996. Tetracyclines: antibiotic action, uptake, and resistance mechanisms. *Arch. Microbiol.* **165:**359–369.
124. **Sifaoui, F., M. Arthur, L. Rice, and L. Gutmann.** 2001. Role of penicillin-binding protein 5 in expression of ampicillin resistance and peptidoglycan structure in *Enterococcus faecium*. *Antimicrob. Agents Chemother.* **45:**2594–2597.
125. **Silver, S., and L. T. Phung.** 1996. Bacterial heavy metal resistance: new surprises. *Annu. Rev. Microbiol.* **50:**753–789.
126. **Sinclair, A., C. Arnold, and N. Woodford.** 2003. Rapid detection and estimation by pyrosequencing of 23S rRNA genes with a single nucleotide polymorphism conferring linezolid resistance in enterococci. *Antimicrob. Agents Chemother.* **47:**3620–3622.
127. **Singh, K. V., K. Malathum, and B. E. Murray.** 2001. Disruption of an *Enterococcus faecium* species-specific gene, a homologue of acquired macrolide resistance genes of staphylococci, is associated with an increase in macrolide susceptibility. *Antimicrob. Agents Chemother.* **45:**263–266.
128. **Singh, K. V., and B. E. Murray.** 2005. Differences in the *Enterococcus faecalis lsa* locus that influence susceptibility to quinupristin-dalfopristin and clindamycin. *Antimicrob. Agents Chemother.* **49:**32–39.
129. **Singh, K. V., G. M. Weinstock, and B. E. Murray.** 2002. An *Enterococcus faecalis* ABC homologue (Lsa) is required for the resistance of this species to clindamycin and quinupristin-dalfopristin. *Antimicrob. Agents Chemother.* **46:**1845–1850.
130. **Sougakoff, W., B. Papadopoulou, P. Nordman, and P. Courvalin.** 1987. Nucleotide sequence and distribution of *tet*(O) gene encoding tetracycline resistance in *Campylobacter coli*. *FEMS Microbiol. Lett.* **44:**153–159.
131. **Tankovic, J., F. Mahjoubi, P. Courvalin, J. Duval, and R. Leclercq.** 1996. Development of fluoroquinolone resistance in *Enterococcus faecalis* and role of mutations in the DNA gyrase *gyrA* gene. *Antimicrob. Agents Chemother.* **40:**2558–2561.
132. **Thauvin, C., G. M. Eliopoulos, C. Wennersten, and R. C. Moellering, Jr.** 1985. Antagonistic effect of penicillin-amikacin combinations against enterococci. *Antimicrob. Agents Chemother.* **28:**78–83.
133. **Tomayko, J. F., K. K. Zscheck, K. V. Singh, and B. E. Murray.** 1996. Comparison of the beta-lactamase gene cluster in clonally distinct strains of *Enterococcus faecalis*. *Antimicrob. Agents Chemother.* **40:**1170–1174.
134. **Tomayko, J. F., and B. E. Murray.** 1995. Analysis of *Enterococcus faecalis* isolates from intercontinental sources by multilocus enzyme electrophoresis and pulsed-field gel electrophoresis. *J. Clin. Microbiol.* **33:**2903–2907.
135. **Torell, E., I. Kuhn, B. Olsson-Liljequist, S. Haeggman, B. M. Hoffman, C. Lindahl, and L. G. Burman.** 2003. Clonality among ampicillin-resistant *Enterococcus faecium* isolates in Sweden and relationship with ciprofloxacin resistance. *Clin. Microbiol. Infect.* **9:**1011–1019.
136. **Trieu-Cuot, P., and P. Courvalin.** 1983. Nucleotide sequence of the *Streptococcus faecalis* plasmid gene encoding the 3′5″-aminoglycoside phosphotransferase type III. *Gene* **23:**331–341.
137. **Trieu-Cuot, P., G. de Cespedes, F. Bentorcha, Delbos F, E. Gaspar, and T. Horaud.** 1993 Study of heterogeneity of chloramphenicol acetyltransferase (CAT) genes in streptococci and enterococci by polymerase chain reaction: characterization of a new CAT determinant. *Antimicrob. Agents Chemother.* **37:**2593–2598.
138. **Tsai, S. F., M. J. Zervos, D. B. Clewell, S. M. Donabedian, D. F. Sahm, and J. W. Chow.** 1998. A new high-level gentamicin resistance gene, *aph(2″)-Id*, in *Enterococcus* spp. *Antimicrob. Agents Chemother.* **42:**1229–1232.
139. **Werner, G., B. Hildebrandt, and W. Witte.** 2001. The newly described *msrC* gene is not equally distributed among all isolates of *Enterococcus faecium*. *Antimicrob. Agents Chemother.* **45:**3672–3673.
140. **Werner, G., and W. Witte.** 1999. Characterization of a new enterococcal gene, *satG*, encoding a putative acetyltransferase conferring resistance to streptogramin A compounds. *Antimicrob. Agents Chemother.* **43:**1813–1814.
141. **Williamson, R., L. Gutmann, T. Horaud, F. Delbos, and J. F. Acar.** 1986. Use of penicillin-binding proteins for the identification of enterococci. *J. Gen. Microbiol.* **132:**1929–1937.
142. **Williamson, R., C. le Bouguenec, L. Gutmann, and T. Horaud.** 1985. One or two low affinity penicillin-binding proteins may be responsible for the range of susceptibility of *Enterococcus faecium* to benzylpenicillin. *J. Gen. Microbiol.* **131:**1933–1940.
143. **Wright, G. D., and P. Ladak.** 1997. Overexpression and characterization of the chromosomal aminoglycoside 6′-N-acetyltransferase from *Enterococcus faecium*. *Antimicrob. Agents Chemother.* **41:**956–960.

144. **Xiong, L., P. Kloss, S. Douthwaite, N. M. Andersen, S. Swaney, D. L. Shinabarger, and A. S. Mankin.** 2000. Oxazolidinone resistance mutations in 23S rRNA of *Escherichia coli* reveal the central region of domain V as the primary site of drug action. *J. Bacteriol.* **182:**5325–5331.

145. **Zhanel, G. G., N. M. Laing, K. A. Nichol, L. P. Palatnick, A. Noreddin, T. Hisanaga, J.L. Johnson, D.J. Hoban, and NAVRESS Group.** 2003 Antibiotic activity against urinary tract infection (UTI) isolates of vancomycin-resistant enterococci (VRE): results from the 2002 North American Vancomycin Resistant Enterococci Susceptibility Study (NAVRESS). *J. Antimicrob. Chemother.* **52:**382–388.

146. **Zilhao, R., B. Papadopoulou, and P. Courvalin.** 1988. Occurrence of the *Campylobacter* resistance gene *tet*O in *Enterococcus* and *Streptococcus* spp. *Antimicrob. Agents Chemother.* **32:**1793–1796.

147. **Zorzi, W., X. Y. Zhou, O. Dardenne, J. Lamotte, D. Raze, J. Pierre, L. Gutmann, and J. Coyette.** 1996. Structure of the low-affinity penicillin-binding protein 5 PBP5fm in wild-type and highly penicillin-resistant strains of *Enterococcus faecium*. *J. Bacteriol.* **178:**4948–4957.

148. **Zscheck, K. K., and B. E. Murray.** 1990 Evidence for a staphylococcal-like mercury resistance gene in *Enterococcus faecalis*. *Antimicrob. Agents Chemother.* **34:**1287–1289.

Frontiers in Antimicrobial Resistance: a Tribute to Stuart B. Levy
Edited by D. G. White, M. N. Alekshun, and P. F. McDermott

Chapter 23

Streptococcus pneumoniae

JOYCE SUTCLIFFE AND MARILYN C. ROBERTS

Streptococcus pneumoniae is the bacterial pathogen most frequently implicated in community-acquired respiratory tract infections (RTI) (35, 107, 110) and also has the distinction of being the leading cause of invasive bacterial disease in the very young and the elderly (178). The introduction of a protein-polysaccharide conjugate vaccine (PncCV) targeting seven pneumococcal serotypes has led to a reduction in the rate of invasive disease in the United States, especially in children under the ages of 2 years (94, 177). It was recently shown that a 9-valent PncCV not only reduces invasive pneumococcal disease in vaccine recipients but also reduces pneumonia associated with seven different respiratory viral infections, presumably by preventing superimposed bacterial coinfection (94, 177). It has been well documented that respiratory viruses can contribute to bacterial infections through their destruction of respiratory epithelium, up-regulation of bacterial adhesion molecules, and by impacting bacterial adhesion through viral neuraminidase activity. The introduction of the vaccine, along with targeted patient and physician education of judicious use of antibiotics, has been shown to reduce antibiotic prescriptions in the community (64), but an overall reduction in antibiotic resistance has not been apparent (16). In fact the incidence of multidrug resistance in pneumococci unfortunately continues to rise (16, 77, 178).

The gram-positive pathogen *S. pneumoniae* shares the same ecological niche as the gram-negative bacteria *Haemophilus influenzae* and *Neisseria meningitidis*. All three are carried in the nasopharynx and can target the lung, ear, blood, and brain for infection. These three organisms also share some basic physiology in that all three are naturally transformable and have a tendency to lyse in stationary phase (52). Autolysis does appear important to the virulence phenotype as autolytic-deficient *S. pneumoniae* mutants have reduced virulence when compared to wild-type strains. As a result of autolysis, cell wall fragments are released and these serve as potent inducers of inflammation (52, 170). When such proinflammatory signals precede bacterial challenge, there is a significant reduction (10,000-fold) in the number of bacteria required for invasion (71). Additionally, autolytic-deficient strains may be less virulent due to the lack of release of pneumolysin, a cytoplasmically located exotoxin (52).

The medical importance of pneumococci is underscored by its ability to be both a community- and hospital-acquired pathogen. The high mortality rate and the serious permanent sequelae in survivors of invasive disease along with the concern of increasing antimicrobial resistance in isolates from both the community and the hospital make *S. pneumoniae* a key pathogen for targeted antimicrobial drug discovery (17, 73, 138).

ANTIMICROBIAL RESISTANCE

Prior to 1967, *S. pneumoniae* isolates were considered universally susceptible to antibiotics. This changed with the report of the first penicillin-resistant *S. pneumoniae* in 1967 (57). Seven years later penicillin-resistant *S. pneumoniae* was reported in the United States (111), and between 1974 and 1984, rates of penicillin-resistant *S. pneumoniae* of ≥10% were reported in a number of areas across the world (4). Today penicillin-resistant *S. pneumoniae* is found on every continent, with rates as high as 45 to 70% in some countries (46, 86, 97, 152). Parallel with the increasing spread of penicillin resistance has been the increasing resistance to other antibiotics used for treatment of pneumococcal infections. Penicillin-resistant *S. pneumoniae* are more likely to be resistant to clindamycin, chloramphenicol, macrolides, tetracyclines,

Joyce Sutcliffe • Rib-X Pharmaceuticals, Inc., New Haven, CT 06511. **Marilyn C. Roberts** • Department of Pathobiology, Box 357238, University of Washington, Seattle, WA 98195.

and trimethoprim/sulfamethoxazole (17, 36, 65, 74, 169, 178). Though over 90 serotypes of *S. pneumoniae* have been identified, only a few are found associated with multiple drug resistance (78).

Clinical strains of *S. pneumoniae* may become resistant to antibiotics by mutations of chromosomal genes, as is the case for resistance to fluoroquinolones, macrolides and ketolides, and trimethoprim and sulfamethoxazole; development of mosaic genes via natural transformation as seen in β-lactam and trimethoprim resistance; or by acquisition of new genes which may confer resistance to aminoglycosides, chloramphenicol, macrolides, lincosamides, streptogramins, and/or tetracyclines (78, 155). Mutations and mosaic genes are transferred to daughter cells during replication, while acquired genes may be transferred as a result of replication or by conjugal transfer to other species and genera in the bacterial population. A small number of multiply drug-resistant clones have successfully spread internationally due to clonal expansion (78, 101, 156). There are currently 26 international clones recognized; molecular typing and distribution of clones can be viewed at http://www.sph.emory.edu/PMEN/.

Unlike many other species, transmission of acquired antibiotic resistance genes by plasmids in pneumococci is uncommon. The association of a plasmid from the pDP1 family (related to pC194) with a small number of isolates has been seen, but there have been no specific functions ascribed to those carrying a member of this plasmid family (184). Most acquired resistance genes in pneumococci are found on transposons and are generally shuttled by either conjugation between live cells or transformation of DNA fragments from lysed cells. The two classes of conjugative transposons found in pneumococci are transferred as single-stranded DNA into recipient cells where the transposons become integrated into the chromosome. Tn*1545* is a 25.3-kb conjugative transposon and is related to the Tn*916* transposon, an 18-kb element. Together the Tn*916*–Tn*1545* family of elements represents one class found in pneumococci. These transposons normally carry a *tet*(M) gene which codes for a ribosomal protection protein and confers resistance to tetracycline, doxycycline, and minocycline (172). These transposons may also carry an *erm*(B) gene, which confers resistance to macrolide-lincosamide-streptogramin B antibiotics (MLS_B), and/or an *aphA-3* gene, which confers resistance to kanamycin (172). The second class of transposons found in *S. pneumoniae* is larger, and Tn*5253* exemplifies this class. Tn*5253* is really a composite transposon of Tn*5252* (47.5-kb) that encodes resistance to chloramphenicol and Tn*5251* (18-kb) that carries the *tet*(M) gene inserted into the central region of Tn5252 (5). The two components of the Tn*5253* can be conjugated independently or together; in the latter case, integration occurs within a single 72-bp target site. Pneumococci can serve as donors or recipients in either intraspecies or intergeneric matings of either transposon class. In addition, nonconjugative plasmids like pMV158 can be mobilized in streptococci, largely through MobM, the plasmid-encoded nicking-closing protein involved in the initiation and termination of the conjugative transfer, and the function(s) provided by auxiliary plasmids (31).

AMINOGLYCOSIDE RESISTANCE

Aminoglycosides bind to the 30S subunit of bacterial ribosomes via interactions with the universally conserved 16S rRNA residues A1492 and A1493 (all nucleotide residues throughout are numbered as described in *Escherichia coli* rRNA genes) (20). These latter residues interact intimately with the minor groove of the first two base pairs between the codon and anticodon, thus sensing Watson-Crick base-pairing geometry and discriminating against near-cognate tRNAs. Aminoglycoside antibiotics can facilitate binding of near-cognate tRNAs, leading to missense and nonsense mutations and reduced functionality of proteins (112).

Bacterial resistance to aminoglycoside antibiotics is almost exclusively accomplished through phosphorylation, adenylylation, or acetylation of the aminoglycoside substrate. Resistance is manifested in *S. pneumoniae* through an acquired phosphotransferase APH(3′)-IIIa that transfers a phosphate group from ATP to the 3′-hydroxyl of its substrates (kanamycin, ribostamycin, neomycin, amikacin, isepamicin, and butirosin) (171). Lividomycin lacks a 3′-hydroxyl group, and phosphorylation takes place at the 5″-hydroxyl group of the pentose. APH(3′)-IIIa is carried on Tn*1545*, often in combination with genes that confer resistance to tetracycline [*tet*(M)] and MLS_B antibiotics [*erm*(B)](153). Although there are seven phenotypic classes of APH(3′), APH(3′)-IIIa is the one most associated with gram-positive pathogens and the only one associated with *S. pneumoniae*.

The three-dimensional structure of APH(3′)-IIIa complexed to one of its products, ADP, has been determined to a resolution of 2.2 Å (66). Interestingly, the enzyme has overall three-dimensional similarity to eukaryotic kinases, and purified APH(3′)-IIIa has low-level, selective protein serine kinase activity (27). Further, known inhibitors of protein kinases (isoquinoline sulfonamides and flavanoids quercetin and genistein) inhibit APH(3′)-IIIa (28). Although the aminoglycoside detoxifying enzyme and protein kinases share less than 5% primary sequence conservation, they have structural homology in the nucleotide-positioning loop (NPL) that closes down on the enzyme active site upon binding of ATP. Further, a conserved serine [Ser27 in APH(3′)-IIIa] in the NPL directly interacts with the β-phosphate of ATP through its hydroxymethyl group and amide hydrogen, facilitating bond breakage with the γ-phosphate during formation of the metaphosphate-like transition. The amide hydrogen of Met26 helps to stabilize this interaction. Asp190 is located in the enzyme active site where it serves to deprotonate the reactive hydroxyl group of the aminoglycoside and is conserved. The common elements of structure and function are consistent with a common ancestry for protein and aminoglycoside kinases.

Aminoglycoside resistance in pneumococci is not usually reported in surveillance studies, largely because antibiotics from this class are rarely used to treat pneumococcal infections. Although APH(3′)-IIIa is carried on Tn*1545*, recent investigations of MLS_B-resistant pneumococci reveal that many resident Tn*1545* transposons no longer carry the *aphA-3* sequence as part of this resistance genetic element (100, 150).

MACROLIDE-LINCOSAMIDE-STREPTOGRAMIN B (MLS_B) RESISTANCE

Macrolides, lincosamides, and streptogramins, though chemically distinct, are usually considered together because they inhibit protein synthesis by binding to overlapping sites in the 50S bacterial ribosomal subunit (38, 45, 60, 61, 63, 128, 145, 146, 167). Macrolides bind about 15 Å from the peptidyl-transferase center, inhibiting peptide progression and resulting in the dissociation of peptidyl tRNA from the ribosomes. The hydrophobic face of the macrocyclic lactone of 14-, 15-, or 16-membered macrolides associates with the tunnel wall formed by positions 2057-2059. Amino sugars attached to the C5 position of the lactone ring point toward the peptidyl-transferase center. The presence of a monosaccharide versus a disaccharide at C5 permits longer peptides to be synthesized (61, 129, 167).

The crystal structures of telithromycin or ABT-773 complexed to *Deinococcus radiodurans* 50S have been solved, and these molecules bind at a roughly similar location to other macrolides (9, 145). Additional interactions seen for ABT-773 include a hydrogen bond between the nitrogen in the cyclic carbamate and U2609 and another hydrogen bond between the quinoline nitrogen and U790 (C803 in *D. radiodurans*) in domain II. The quinoline ring also interacts with U1782 (C1773 in *D. radiodurans*) from domain IV via hydrophobic contacts. Although telithromycin appeared to interact with U2609 in footprinting studies (47), there is no evidence in the crystal structure of a direct interaction (9). The macrocyclic lactone ring of telithromycin adopts a more extended conformation when bound to *D. radiodurans* 50S, thereby allowing a hydrogen bond between the 3-keto oxygen of telithromycin and the O2′ hydroxyl of C2610 of domain V (9, 145). There is no interaction between domain IV and telithromycin observed, but the imidazole moiety of the aryl-alkyl extension does stack with U790 from domain II.

Clindamycin complexed to *D. radiodurans* 50S forms hydrogen bonds with A2058, A2059, and G2505, explaining the partially overlapping binding sites of macrolides and clindamycin (146). This compound also interacts with C2452, a nucleotide shown to be involved in P-site tRNA binding. Thus, this compound has both an A- and P-site character to it and, not surprisingly, inhibits peptide bond formation. Quinupristin, the streptogramin B portion of Synercid, is bound at the entrance to the ribosomal tunnel in space that clearly overlaps where macrolides bind (63). There is an extensive network of hydrophobic interactions with portions of domains II, IV, and V and hydrogen bonds between A2062 and C2586. Thus, these three different antibiotic classes have partially overlapping binding sites.

Resistance to all three classes of antibiotics is normally due to the presence of an rRNA methylase gene, although mutations in 23S rRNA or ribosomal proteins can also confer a MLS_B-resistant phenotype (19, 45, 84, 159). There have been 32 different rRNA methylase genes described to date (http://faculty.washington.edu/marilynr/). Each methylase adds one or two methyl groups to a single adenine (A2058 in *E. coli*) in the 23S rRNA moiety of the ribosome (96). This modification prevents the binding of macrolide, lincosamide, and streptogramin B antibiotics to the 50S ribosomal subunit (45, 61, 128, 146). The *erm*(B) gene has been identified in *S. pneumoniae* since the 1970s and is often found on transposons of the Tn*916*-Tn*1545* family where it is upstream of the *tet*(M) gene (150, 172). In a recent study (161), MLS_B resistance could be induced by erythromycin, miocamycin and clindamycin in an *S. pneumoniae* strain carrying the *erm*(B) gene; the promiscuity of inducers in pneumococci has been previously described (84).

Recently, strains of *S. pneumoniae* carrying an *erm*(A) gene, coding for another rRNA methylase, were identified from 2,448 carriers of ≤2 years of age isolated between 1997 and 1999 in Greece (162). A number of different *erm* genes are found in other streptococci (134, 135), but this was the first description of a second *erm* gene in *S. pneumoniae*. Isolates with the *erm*(A) gene had uninduced MICs to erythromycin of 1 to 4 μg/ml, which were generally lower then *erm*(B)$^+$ *S. pneumoniae,* collected from the same population (161). Four of the *erm*(A)$^+$ *S. pneumoniae* isolates were shown to be induced by erythromycin but not to be induced by miocamycin or clindamycin (161). General information on the different MLS_B determinants can be found in chapter 6 of this book.

In the 1990s *S. pneumoniae* were isolated which were resistant to macrolides but not lincosamides or streptogramin B antibiotics (154, 160). These strains had lower levels of macrolide resistance than most isolates carrying the *erm*(B) gene (11, 134, 154, 160). These isolates with an M resistance phenotype carried a gene which produced a protein that pumps 14- and 15-membered macrolides outside the cell wall. The *mef*(A) gene [originally called *mef*(E) in pneumococci] was identified in all strains with an M phenotype (23, 163). This gene was shown to be adequate to confer macrolide resistance in *S. pneumoniae* (163). The *mef*(A) gene was later shown to be linked to an ABC transporter that has homology with *msr*(A) (48, 130, 141). Recently, this second gene has been shown to independently confer macrolide resistance in *S. pneumoniae* and has been named *msr*(D) (30, 70, 130). There is some discrepancy in the level of macrolide resistance observed in *mef*(A)$^+$ *msr*(D)$^-$ versus *mef*(A)$^-$ *msr*(D)$^+$ strains (30, 70). However, earlier work showed that both genes are expressed as a single mRNA transcript in pneumococci (48, 158). In the absence of induction, very little mRNA transcript is made, as shown by northern analysis. Erythromycin was a strong inducer of the *mef*(A) *msr*(D) transcript, while induction in the presence of ketolides telithromycin and ABT-773 was significantly less (158). When *msr*(D) was inactivated by insertion mutagenesis and confirmed by both DNA and RNA analysis, the *mef*(A)$^+$ *msr*(D)$^-$ strain could still efflux radiolabeled erythromycin (158) but not to the full extent seen in the isogenic *mef*(A)$^+$ *msr*(D)$^+$ strain. Thus, it appears that both genes contribute to resistance.

The *mef*(A) *msr*(D) genes have been associated with both conjugative and nonconjugative transposons. The mega genetic element is common among clinical strains of *S. pneumoniae* but has also been found in *Streptococcus salivarius* (26, 157), *Neisseria spp.* (92), and *Acinetobacter jejuni* (92). More recently it was the most commonly acquired MLS_B gene in a survey of 176 gram-negative commensal oral and urine isolates (113). In addition to the *mef*(A) *msr*(D) genes, mega has five additional open reading frames (ORFs), all highly homologous to Tn*1207.1*. Tn*1207.1* was described in pneumococci from Italy, and it has six ORFs in addition to *mef*(A) *msr*(D) genes (141). This element is integrated into the chromosomal *celB* gene, and this disruption impairs the ability of the host strain to be rendered competent for transformation. It is likely that Tn*1207.1* is a defective form of Tn*1207.3,* a conjugative transposon described in *Streptococcus pyogenes* (140), or as a chimeric element in which Tn*1207.3* has integrated into a prophage found in *S. pyogenes* 2812A (7). Tn*1207.3* has not yet been described in pneumococci.

Recently, we have shown that the *tet*(O) gene, which codes for a ribosomal protection protein, has been linked upstream of the *mef*(A) gene in Italian *S. pyogenes* isolates (50), though it is not clear if this element is also present in *S. pneumoniae.* Just recently, a new composite transposon composed of the mega element integrated into a Tn*916*-like element upstream of the structural *tet*(M) gene has been described (32). Integration of the new element, Tn*2009,* resulted in a 9.5-kb deletion in the pneumococcal chromosome such that the ends of the element were flanked by the chromosomal *spr1206* gene upstream and the *spr1199* gene downstream (32). The Tn*2009* element was not conjugative even though it was inserted within the very promiscuous Tn*916* conjugative element but could be transferred by transformation.

Mutations in either 23S rRNA alleles or ribosomal proteins L4 and L22 that confer resistance to one or more of the MLS_BK antibiotics have been found in clinical strains of *S. pneumoniae* (42, 45, 174). Often there is a measurable change in susceptibility as more of the 23S rRNA loci are modified (42, 164). Mutations have been identified frequently in one or more of the four 23S rRNA alleles at either the A2058 or A2059 position (42, 45, 174). A review on macrolide resistance including tables of nucleotide(s) base substitutions in 23S rRNA and ribosomal protein mutations in L4 and L22 was recently published (45) and is recommended for in-depth coverage of the topic.

Macrolide resistance varies geographically, but most countries are reporting double-digit frequencies of resistance in pneumococci (72, 138). Isolates from the Far East, in particular, are highly resistant to macrolides, with the majority of strains carrying the *erm*(B) methylase (48.8%) and a large percentage (13.2%) with both *erm*(B) and *mef*(A) (72). In the United States, 31% of the *S. pneumoniae* respiratory isolates from 2000 to 2001 were erythromycin-resistant, but 99.6% of the strains remained susceptible to telithromycin (138). In Europe, France had the highest percentage of macrolide resistance (>53%), with the majority of resistance occurring in children (144). In Europe, *erm*(B) is the predominant macrolide resistance determinant, while in the United States, resistance is mediated primarily via *mef*(A) (73, 144).

TETRACYCLINE RESISTANCE

Tetracyclines bind to the 30S subunit and primarily block the binding of the incoming aminoacyl-tRNA to the A site (14, 126). Consistent with this mechanism of action, tetracycline also inhibits the binding of both release factors RF-1 and RF-2 that occur during termination (15). More than one binding site has been seen within the 30S in both biochemical experiments and crystal structures (14, 40, 80, 126); however, the primary binding site appears to be located near the acceptor site of aminoacyl-tRNA (A site) between the head and body of the 30S. The binding pocket is ~20 Å long and 7 Å deep and is primarily formed by an irregular minor groove of helix H34 in combination with residues 964–967 from the H31 stem-loop (14). Tetracycline makes hydrogen bonds between oxygen atoms on its hydrophilic side with backbone phosphate oxygen atoms of H34, including important salt bridges mediated by magnesium. Another portion of the drug-binding pocket is formed by bases 1196 and 1054 that form a clamp that holds tetracycline by hydrophobic interactions.

Tetracycline resistance can be mediated by one of several different mechanisms. There have been 23 different tetracycline efflux genes, 13 ribosomal protection genes, 3 genes that inactivate tetracycline, and 1 unknown acquired tetracycline gene characterized to date (see http://faculty.washington.edu/marilynr/). However, until recently tetracycline-resistant *S. pneumoniae* have carried a single ribosomal protection gene, the *tet*(M) gene (22, 102). The *tet*(M) gene is usually associated with conjugative transposons and has recently been found in Tn*2009* where it is linked to *mef*(A)-*msr*(D) macrolide resistance genes. A few tetracycline-resistant *S. pneumoniae* have been identified which do not carry the *tet*(M) gene; isolates from South Africa and Washington State in the United States were found to harbor another ribosomal protection gene, *tet*(O) (93, 181). The *tet*(O)$^{+}$ *S. pneumoniae* were not able to conjugally transfer this gene, suggesting that the *tet*(O) is not associated with conjugative elements. However, conjugation may be seen in the future, especially if the *tet*(O)-*mef*(A) element in *S. pyogenes* (50) becomes part of the *S. pneumoniae* armamentarium. General information on tetracycline resistance due to ribosomal protection proteins can be found in chapter 2.

Recent clinical isolates of *S. pneumoniae* were collected from 33 medical centers in the United States during the winter of 1999-2000, and tetracycline resistance was 16.3% (37). Pneumococci from 26 countries isolated from adults with community-acquired respiratory tract infections were 71.3% susceptible to doxycycline (74). However, the Far East eclipses other parts of the world, recording 81.6% resistance to tetracycline in pneumococcal isolates (72).

β-LACTAM RESISTANCE

β-lactam antibiotics target one or more of the penicillin-binding proteins (PBPs) involved in a late step in peptidoglycan biosynthesis. The PBPs are responsible for transglycosylation of interconnecting glycan chains (polymers of *N*-acetylmuramic acid and *N*-acetylglucosamine) and cross-linking, or transpeptidation, of short peptides that are attached to the glycan chains. PBPs fall into two groups: the low-molecular-mass (LMM) PBPs are monofunctional enzymes acting mainly as D-alanyl D-alanine carboxypeptidases (DD-peptidase), while the high-molecular mass (HMM) PBPs are multimodular enzymes that have DD-peptidase activity and often transglycosylase activity. HMM-PBPs are subdivided into classes A and B based on whether one of their domains has transglycosylase activity (class A) or not (class B). *S. pneumoniae* has five HMM-PBPs (1a, 1b, 2a, 2b, 2x); mutations in PBP2x and PBP2b are primary resistance determinants, and their modification is required for the development of high level penicillin resistance (58, 132). The mutations that are responsible for the low-affinity PBPs are acquired by transformation of PBPs from viridans streptococci or penicillin-resistant pneumococci and homologous recombination resulting in mosaic PBP genes in pneumococci (58). Penicillin-resistant strains of pneumococci can harbor as many as 100 mutations spread throughout the PBP gene sequence in one or more of the HMM-PBPs.

Structural information on class B HMM-PBPs has been published for PBP2x from *S. pneumoniae* (34, 51, 106, 119), PBP5fm from *Enterococcus faecium* (142), and PBP2a from methicillin-resistant *Staphylococcus aureus* (89). PBP2x from a β-lactam-sensitive pneumococcal strain revealed a three-domain macromolecule composed of an elongated N-terminal unit, a central transpeptidase domain, and a C-terminal domain (51, 119). The transpeptidase domain is composed of an α/β domain formed by a central five-stranded anti-parallel β sheet supported by two helices (α1 and α11) on one side and a single helix (α8) on the other, with an α helical domain. The active site is located in a groove that parallels the five-stranded β sheet. There are three highly conserved sequence motifs for the transpeptidase domain (337Ser*XX*Lys, 395Ser*X*Asn, 547LysSerGly). The nucleophilic Ser337 is found as the N terminus of an α-helix (α2) in close proximity to Lys547 in both the sensitive and resistant forms of PBP2x (34, 51, 119). The conserved active site residues form a rigid network of hydrogen bonds; the residues that participate are Ser337, Lys340, Ser395, Asn397, and Lys547. In the apo form, a sulfate ion is bound above Ser337 and is coordinated by Ser395, Thr550, and a water molecule. Ser337 and Ser395 are critical for stabilization of the substrate or antibiotic in the active site cavity. In the sensitive PBP2x, Ser395 points into the active site, making close contact with Ser337, while Asn514 points away from the active site region and does not interfere with the position of Ser389. Mutation of Met339Phe results in a 4- to 10-fold reduction of the reaction rate with beta-lactams and a 3-fold faster release of the inactivated antibiotic (21).

Both structural evidence and site-directed mutagenesis have confirmed why certain mutations confer resistance (106, 119). β-lactam antibiotics interfere with transpeptidation because they are pseudosubstrates for the D-alanyl-D-alanine C terminus of the peptidoglycan stem peptides. As pseudosubstrates, they acylate the active site serine but are very slowly deacylated. A common mutation that is next to the active site serine in *S. pneumoniae* PBP2x, Thr338Ala, appears to weaken the local hydrogen-bonding network and destabilizes the buried water molecule crucial for deacylation (34). The Asn514His mutation generates a more open active site by causing Ser395 to point away from the active site such that this residue is no longer able to participate in the enzymatic reaction. The consequence of these two common mutations is a deficiency in ester bond formation and substrate acylation for both natural substrate and β-lactam antibiotics (34). The crystal structure of the double mutant T338A/M339F has been solved to a resolution of 2.4 Å and a distorted active site is seen, with a reorientation of the hydroxyl group of the active site Ser337. More recently a group of penicillin-resistant *S. pneumoniae* clinical isolates were found to have a Gln552Glu substitution; transformation studies were consistent with this single mutation being responsible for most of the resistance (122). The crystal structure of PBP2x containing Gln552Glu in the active site shows that this region is in a closed conformation (122).

Several other mechanisms of resistance to β-lactam antibiotics have been described in *S. pneumoniae.* In cefotaxime-resistant laboratory-derived mutants, mutations were found in *ciaH,* a histidine protein kinase involved in the early steps of competence regulation (56). The phenotype included deficiency in competence along with cefotaxime, but not piperacillin resistance. Piperacillin-resistant mutants were found associated with mutations in *cpoA,* a gene that is located downstream and in the same operon of the primary σ factor in pneumococci (54). In addition to piperacillin resistance, these mutants had a defect in competence development and reduced amounts of PBP1a. However, neither mechanism has been found in clinical strains. Since the major mechanism of resistance appears to be acquisition of mosaic PBPs from bacteria in their environment, resistance was manifested later in pneumococci than in other gram-positive species that acquired plasmids or transposons encoding β-lactamases (enzymes that hydrolyze the β-lactam bond) or novel PBPs (e.g., PBP2a in *S. aureus* that encodes methicillin resistance).

The frequency of resistance to penicillin is very high in many countries. Penicillin resistance of pneumococcal isolates from adults from 1998 to 2000 in the United States was 37.1%, with 25% being classified as high-level resistance (MIC of penicillin of ≥2 μg/ml) versus 12.1% having intermediate resistance (MIC between 0.12 and 1 μg/ml) (73). From the same study, isolates from Asia, Eastern Europe, and Western Europe had 54.7%, 14.6%, and 22.2% penicillin resistance, respectively. In comparing isolates from children, the incidence of penicillin resistance is higher than in adults and higher for strains isolated from respiratory tract infections versus invasive disease at any age. The vast majority of penicillin-resistant pneumococci are also coresistant to one or two other antibiotics, most notably trimethoprim-sulfamethoxazole and macrolides. The higher the level of penicillin resistance, the greater the proportion of isolates that are coresistant to either of these antibiotics. Thus, this increase in multidrug-resistant pneumococci (>20%) is now driving the charge for new antibiotics.

FLUOROQUINOLONE RESISTANCE

Fluoroquinolones interact with eubacterial type 2 topoisomerases, DNA gyrase, and topoisomerase IV (68). Both of these enzymes are essential for DNA replication, and both function as tetrameric heterodimers (A_2B_2). DNA gyrase and topoisomerase IV are encoded by sets of homologous genes, *gyrA* and *gyrB* and *parC* and *parE,* respectively. Both enzymes can relax positive superhelical coils, but only DNA gyrase can introduce negative superhelical twists. Topoisomerase IV requires the energy of ATP hydrolysis to relax superhelices or for decatenation, but DNA gyrase only uses ATP hydrolysis for its nonredundant function.

Resistance to quinolones is often due to mutational changes in the subunits of the two target enzymes. Mutations are clustered in a quinolone-resistance-determining region (QRDR) in the amino terminus of the A subunit of either gyrase or topoisomerase IV, usually occurring at positions 79, 81, or 83 (Ser81Phe and Tyr83Phe in *gyrA* or Ser79Phe in *parC*) (18, 68). Mutations in the B subunits are in the C-terminal half of the 70-kd protein, generally regarded as the portion that interacts with the A subunit. The mutations occur frequently in *gyrB* at positions 435 and 474 (Glu474Lys; Asp435Asn) and in *parE* at analogous positions (Glu422Asp or Asp435Asn) (18, 121, 176). In general, mutations in the B subunit must be combined with a mutation in one of the A subunits to give high-level fluoroquinolone resistance.

For some compounds there is a preference for mutational changes to occur first in either DNA gyrase or topoisomerase IV. For example, ciprofloxacin targets topoisomerase IV as its primary target in *S. pneumoniae* and other gram-positive bacteria (116). However, dimers of ciprofloxacin in which ciprofloxacin is tethered to itself or to pipemidic acid by linkage of C-7 piperazinyl rings appear to target DNA gyrase (53). When serially resistant mutants are evaluated, sparfloxacin and grepafloxacin initially select for mutations in DNA gyrase (105), while clinafloxacin appears to target both topoisomerases equally (117). However, when ciprofloxacin, sparfloxacin, and clinafloxacin were tested as inhibitors of the purified enzymes, all three were found to be more potent inhibitors of topoisomerase IV than DNA gyrase (118).

X-ray crystal structures and biochemistry studies have led to a two-gate model of type II topoisomerase function (76, 136). The A subunits of either type II topoisomerase are brought together when they bind a DNA segment (the gate or G segment). The N-terminal domains of GyrB or ParC form a protein clamp that dimerizes in the presence of ATP, trapping another DNA duplex segment (the transported or T segment). Binding of the second duplex fragment triggers DNA cleavage through nucleophilic attack on the DNA backbone and the formation of 5′ phosphotyrosyl enzyme-DNA linkages in the A subunit, forming a transient opening. The T segment is passed through the opening, the G segment is religated, and conformational changes occur such that the T segment exits through a gate formed by the opening of the GyrA domain. The enzyme is reset to its original conformation upon ATP hydrolysis. If the G and T segments are part of the same molecule, strand passage usually results in DNA relaxation or, sometimes in the case of gyrase, introduction of a negative supercoil. If the two segments are part of different DNA molecules, then catenation/decatenation occurs.

Crystal structures of fragments of GyrA and GyrB subunits from *E. coli* or other bacterial species are available (25, 29, 55, 82, 83, 104, 173, 182) as well as crystal structures of portions of ParC and ParE (8, 69). The GyrA59 N-terminal fragment of the A subunit is the minimal fragment that will function with the GyrB subunit to cleave DNA (131). When crystallized, the breakage-reunion fragment of GyrA forms a dimer composed of an N-proximal head and a C-proximal tail (104). The active-site of the breakage-reunion region resides in the "head" portion of the dimer and is a cluster of conserved residues formed by Tyr122 and Arg121 from one monomer and His80, Arg32, and Lys42 from the other monomer. The location of the quinolone-resistant mutations at the head-dimer interface is consistent with the finding that quinolone drugs bind strongly to the gyrase-DNA complex but weakly to either gyrase or DNA alone (183).

Recently the C-terminal 35-kd domain of GyrA (GyrA CTD) from *Borrelia burgdorferi* has been crystallized (25). This portion of the GyrA subunit is where the DNA duplex binds in a right-handed wrap, thereby introducing directionality and permitting the introduction of negative supercoils. The GyrA CTD adopts an unusual β-pinwheel fold with a large basic patch that is the likely site for DNA binding and bending. This fragment (as well as the C-terminal domain of ParC) is capable of bending DNA by ≥180° over a 40-bp region. The mechanistic model constructed by Corbett and colleagues along with the substrate preferences of topoisomerase IV account for its ability to act principally on positively supercoiled and catenated DNAs and explain the ability of DNA gyrase to introduce negative supercoils (25). A preliminary X-ray crystallographic analysis of the CTD of ParC from *Bacillus stearothermophilus* has also been published (69).

The crystal structures of the N-terminal 43-kDa fragments of GyrB and ParE are quite similar in function and structure (8, 182). The fragments bind ATP as a dimer, with the very N terminus of one monomer wrapping around the other subunit in the dimer. The N-terminal 24-kDa GyrB is mostly responsible for binding ATP (and coumarin antibiotics), while the C-terminal subdomain of the 43-kDa fragment forms the sides of a large hole in which the T-segment is captured. This cavity is lined with arginine residues that likely facilitate the transport of the T-segment DNA to the DNA gate that is located at the heterotetrameric interface. Both of these proteins belong to the GHKL phosphotransferase superfamily (39).

In addition to target mutations that confer quinolone resistance, there is an efflux pump in *S. pneumoniae* called PmrA (13, 49, 124). PmrA has 24% amino acid identity to NorA, an efflux pump from the major facilitator superfamily that contributes to antibacterial resistance in *S. aureus* (67, 68). Interestingly, it is the more hydrophilic fluoroquinolones that are substrates for PmrA, although hydrophobicity may not be the only factor in differentiating substrates as noted for NorA substrates in *S. aureus* (165). Inactivation of *pmrA* or inhibition by reserpine in clinical strains that are fluoroquinolone-resistant without any mutations in gyrase or topoisomerase IV resulted in four- to eightfold increases in susceptibility to norfloxacin (13). Increased expression of *pmrA* in pneumococcal strains resistant to ciprofloxacin did not always correlate with a phenotype consistent with an efflux mutant (125). Further, others have noted that other multidrug transporter proteins appear to be associated with resistance to fluoroquinolones, as can be seen in strains that have had their *pmrA* genes inactivated (12, 123).

Surveillance of pneumococcal isolates from the United States and Canada indicate that resistance to fluoroquinolones remains low. Levofloxacin-resistant respiratory tract isolates from the United States (2000-2001) and Canada (1999-2000) were found at frequencies of 0.8 and 1.4%, respectively (18, 138). In invasive isolates from the United States, resistance to levofloxacin increased from 0.1% in 1998 to 0.6% in 2001 ($P = 0.008$); however, a decrease to 0.4% was seen in 2002 (127). Resistance rose from 1998 to 2002, even in isolates covered by the conjugate vaccine. In this population, 16% of the strains had evidence of active efflux, while 94% had multiple mutations in the QRDRs of the *gyrA, gyrB, parC,* and *parE* genes. Levofloxacin-resistant invasive isolates were predominantly clonal; 58% of the isolates were related to five international clones, and 36 and 44% were coresistant to penicillin and macrolides, respectively. Further, 28% were multidrug-resistant to levofloxacin, macrolides, and penicillin.

The frequency of levofloxacin resistance increased from 1999–2000 to 2001–2002 from 0.3 to 0.7% in Europe and from 3.0 to 3.2% in Asia (75). Hong Kong at 14.3% observed incidence appeared to be the worldwide winner in frequency of fluoroquinolone-resistant isolates; furthermore, the majority of isolates were coresistant to penicillin, macrolides, and tetracycline (18, 72). Certain regions, like Brazil, South Africa, and Australia, had no levofloxacin-resistant mutants (18, 75).

TRIMETHOPRIM-SULFAMETHOXAZOLE (CO-TRIMOXAZOLE) RESISTANCE

The genes for the folate pathway in pneumococci are encoded by an operon containing four genes, *sulA, sulB, sulC, and sulD* (81). *sulB* encodes dihydrofolate reductase (DHFR), the last enzyme in the pathway that catalyzes the NADPH-dependent reduction of dihydrofolate to tetrahydrofolate. This protein is essential for the synthesis of thymidylate, purines, and several amino acids. Inhibition of the enzyme's activity leads to arrest of DNA synthesis and cell death. Trimethoprim (TMP) is a diaminopyrimidine that selectively inhibits bacterial dihydrofolate reductase (180). The enzyme dihydropteroate synthase (DHPS) is encoded by *sulA* and acts earlier in the same pathway, catalyzing the condensation of *para*-aminobenzoic acid (PABA) with 6-hydroxymethyl-7, 8-dihydropterin-pyrophosphate to form 7,8-dihydropteroate and pyrophosphate. DHPS is essential for the de novo synthesis of folate in prokaryotes, lower eukaryotes, and plants but is absent in mammals (6). Inhibition of this enzyme's activity by sulfonamide and sulfone drugs depletes the folate pool, resulting in growth inhibition and cell death. For example, sulfamethoxazole competes with the PABA binding site within DHPS (6, 180).

The crystal structures for dihydrofolate reductase from *E. coli, Lactobacillus casei,* and other bacterial species have been published (10, 44, 85, 88, 143, 148). All of these structurally determined, chromosomally encoded DHFRs are single domain proteins containing an eight-stranded mixed β-sheet, flanked by two α-helices. The DHFRs are organized into two subdomains: the adenosine-binding domain that binds the adenosine portion of NADPH, and the loop domain, which is dominated by three loops. The space between these subdomains forms the active site where dihydrofolate and inhibitors like trimethoprim bind.

In *S. pneumoniae,* the majority of trimethoprim resistance is mediated by a single amino acid change in the DHFR protein (*sulB*), Ile100Leu (2, 137, 147, 180). This mutation is analogous to the Ile94 position in *E. coli* DHFR, a residue that that forms a key hydrogen bond to the 4-amino group of trimethoprim. DHFRs in trimethoprim-resistant pneumococci have significantly more variability in their DNA sequences than trimethoprim-susceptible pneumococci (2, 137, 147, 180). Therefore, it appears that trimethoprim resistance in pneumococci could well be the result of heterologous recombination, resulting in mosaic *sulB* genes (2) akin to what is seen in lower affinity PBPs in pneumococci (58, 132, 155, 156). However, recently, several other mutations have been shown to modulate the resistance to trimethoprim (99). Spontaneous mutants with intermediate levels of resistance (64 μg/ml) containing a mutation resulting in Trp or Ser for Leu31 have been observed in the laboratory (99). A substantial increase in the trimethoprim IC_{50} was observed when Met53Ile was found in combination with Ile100Leu (99). Met53 and Leu31 are both in contact with trimethoprim in analogy to residues in the *E. coli* structure (10). Enzymatic characterization of the mutant DHFRs revealed not only an increase in the trimethoprim IC_{50}, but often considerably reduced K_m values for dihydrofolate (99). Thus, trimethoprim resistance can be costly, resulting in a reduction in the affinity for its natural substrates. Maskell et al. did note that some of the other mutations in DHFR appeared to be compensatory in terms of kinetic properties for natural substrates (99).

Crystal structures of bacterial DHPS have also been published (1, 6, 59). The DHPS monomer has a single domain that forms an eight-stranded α/β barrel characteristic of TIM-barrel proteins (59). The protein crystallizes as a noncrystallographic dimer; the dimer interface involves three α-helices. The active site is found at the C-terminal end of the central β-barrel fold. However, mutations that confer resistance to sulfonamides are spread over the surface of the enzyme, with some mutations occurring at the dimer interface. Despite the accessibility of the active sites of both monomers being similar, the substrate 6-hydroxymethyl-7, 8-dihydropterin-pyrophosphate appears to be bound to only one monomer. The binding model for PABA, sulfonamides, and sulfones can be inferred based on the *M. tuberculosis* DHPS structure, and it appears that these substrates/inhibitors bind in the active site near the tip of loop 2, which contains the Arg58 to Tyr63 region where the majority of amino acid insertions or duplications conferring sulfonamide resistance have been found (6, 57, 90, 115, 147, 180).

The first laboratory-derived sulfonamide-resistant *S. pneumoniae* isolates had a 6-bp repeat leading to a repetition of the amino acids Ile66 and Glu67 in SulA (originally termed the *sul*-d mutation) (90). More recently a number of sulfonamide-resistant clinical strains were analyzed, and a number of 3-bp and 6-bp insertions resulting in the repetition of one or two amino acids were observed (57, 98, 115, 147). These included repeats of Ile66-Glu67, Ser61-Ser62, Ser62-Tyr63, and Arg58-Pro59. In addition, single amino acid repeats (Ser61, Ser62, or Tyr63) were found (57, 98, 115, 147). There are also insertion mutations (Arg between Gly60 and Ser61, Glu-Ser between Ser61 and Ser62, Ser-Ser between Ser62 and Tyr63, and Asn or Ser between Ser62 and Tyr63, and SerTyr between 63Tyr-Val64) (115, 147). A number of the insertions or duplications were shown to be sufficient for sulfonamide resistance (57, 98, 115). When the enzymatic parameters of the DHPS enzymes containing the Ile66-Glu67, the Ser61, or the Ser61-Ser62 repeats were determined, resistance was explained by an increased K_i for sulfathiazole with very little effect on the K_m for either substrate (57). Thus, these mutations do not appear to affect the biosynthetic reaction rate significantly. Finally, sulfonamide resistance is likely more complicated as there are isolates that appear to have no mutations in DHPS (115).

There is a strong correlation between trimethoprim and sulfamethoxazole-trimethoprim (co-trimoxazole) resistance, and there appears to be little difference in the efficacy of trimethoprim as monotherapy versus its use in combination with sulfamethoxazole in treatment of pneumococcal diseases (3). The incidence of co-trimoxazole resistance in pneumococci among strains with decreased susceptibility to penicillin and other antibiotics is quite high (73, 147, 181). The incidence of co-trimoxazole resistance in *S. pneumoniae* isolates from the United States collected in 2001–2002 was 34.8% (16). In Europe the resistance rates varied from 0% in Austria to ~20% in Spain and France (147). In Asian adults, the incidence of co-trimoxazole resistance is very high, 74%, even higher than penicillin resistance (73).

CHLORAMPHENICOL RESISTANCE

Chloramphenicol is a small molecule that has been described as a structural analog of puromycin, an aminoacylated nucleoside. Although this antibiotic is considered a prototypic inhibitor of peptide bond synthesis in prokaryotes and early studies revealed one binding site (43), there have been other reports consistent with chloramphenicol having two binding sites, a high affinity site with a $K_d = 2$ μM and a low affinity site with a $K_d = 200$ μM (24, 87). Additionally, footprinting (protection of 23S rRNA residues A2058, A2059, A2062, A2451, and G2505) and mutational studies (G1057, G2061, A2062, G2447, A2451, C2452, A2503, and U2504 in 23S rRNA) are consistent with chloramphenicol binding to two sites (41, 95, 103, 175). These two binding sites are revealed by the two structures of 50S subunits complexed with chloramphenicol that have been published (62, 146). The A-site crevice formed by the bases of A2451 and C2452 is the high affinity site in which chloramphenicol binds in *Deinococcus radiodurans* 50S (146). This is likely the same site in which chloramphenicol binds independently of erythromycin (114). In *Haloarcula marismortui* 50S, chloramphenicol binds in the lower affinity site, within a hydrophobic crevice created by A2058 and A2059, at the entrance to the peptide exit channel (62). The latter binding site overlaps part of the macrolide binding site and helps explain why erythromycin can displace chloramphenicol (weak binding at site 2) but not the reverse (60, 61, 114, 145). Also consistent with different binding sites is one study that suggests that oligopeptides accumulate in the presence of chloramphenicol (133). Accumulation of short peptides also occurs with macrolide antibiotics (166, 167). Other studies have shown that chloramphenicol affects translational fidelity, causing frameshifting and nonsense suppression (168). Thus, the mechanism of action of chloramphenicol appears more complex than as a simple competitive inhibitor of A-site substrates.

Chloramphenicol resistance appears to be mediated by antibiotic inactivation in streptococci (5, 33, 108, 151, 179, 180). Resistance by the enzyme chloramphenicol acetyltransferase (*cat*) is generally inducible via a mechanism of translational attenuation (91). The enzyme uses acetyl coenzyme A as the acetyl donor, adding an acetyl group to the 3OH group of chloramphenicol. Diacetylation can also occur, following slow, nonenzymatic transfer of the acetyl group from the 3OH to the 1OH position, followed by enzyme-mediated acetylation at the 3OH (109). A single acetylation confers resistance, blocking key interactions of the 3OH group to G2505 and U2506, hydrogen-bonds that are mediated via a putative magnesium (146).

The *cat* gene in pneumococci appears to be 100% identical to the *cat* gene in plasmid pC194, originally described in *S. aureus* (151, 179). However, in pneumococci, the *cat* gene (and sometimes the entire pC194) has been found integrated into a conjugative element found on the chromosome (179). As mentioned earlier, this transposon is large (65.5 kb), and Tn*5253* is responsible for carrying both *cat* and *tet*(M) genes in pneumococci (5). The smaller Tn*5252* is homologous to the staphylococcal plasmid pC194 and can conjugate separately from the other component of Tn*5253*, the Tn*916*-like conjugal element (Tn*5251*) that encodes tetracycline resistance.

Chloramphenicol is still used for empiric therapy for respiratory tract infections in developing countries (33, 180). In addition, it is recommended in the *Sanford Guide to Antimicrobial Therapy* for bacterial meningitis caused by *S. pneumoniae*, meningococci, and *H. influenzae*, especially for patients displaying penicillin allergies (1). However, its use has decreased due to other antibiotics that are deemed safer. The level of resistance varies geographically, with pneumococcal isolates from 1999 having 10.6 and 12.7% resistance in the United States and Europe, respectively. A high rate of 38.5% resistance has been noted in Barcelona, Spain (33). In community-acquired pneumonia, penicillin-resistant pneumococcal isolates were 26.1% resistant to chloramphenicol in a recent study in Thailand (139). Earlier studies in Singapore (1995) showed that 20.1% of the strains were resistant to chloramphenicol (79).

SUMMARY

S. pneumoniae is an important community and hospital pathogen that is becoming increasingly multidrug resistant (17, 36, 65, 74, 169, 178). For community-acquired respiratory tract infections, therapy is generally empiric and drugs are needed to cover the common clones that are becoming more difficult to treat. In invasive situations and even in very highly penicillin-resistant pneumococci, the use of conjugate vaccines has significantly lessened incidence in children (120, 149, 177). However, understanding more about the pathogenesis and weak links in the pneumococcal armamentarium will provide new approaches for therapy against the leading cause of invasive and noninvasive disease in infants and young children.

REFERENCES

1. **Achari, A., D. O. Somers, J. N. Champness, P. K. Bryant, J. Rosemond, and D. K. Stammers.** 1997. Crystal structure of the anti-bacterial sulfonamide drug target dihydropteroate synthase. *Nat. Struct. Biol.* **4:**490–497.
2. **Adrian, P. V., and K. P. Klugman.** 1997. Mutations in the dihydrofolate reductase gene of trimethoprim-resistant isolates of *Streptococcus pneumoniae. Antimicrob. Agents Chemother.* **41:**2406–2413.
3. **Antimicrobial Therapy, Inc.** 2003. Clinical approach to initial choice of antimicrobial therapy, p. 5. *In* D. N. Gilbert, R. C. Moellering, Jr., and M. A. Sande (ed.), *The Sanford Guide to Antimicrobial Therapy,* 33rd ed. Antimicrobial Therapy, Inc., Hyde Park, Vt.
4. **Appelbaum, P. C.** 1987. World-wide development of antibiotic resistance in pneumococci. *Eur. J. Clin. Microbiol.* **6:**367–377.
5. **Ayoubi, P., A. O. Kilic, and M. N. Vijayakumar.** 1991. Tn*5253*, the pneumococcal omega (*cat tet*) BM6001 element, is a composite structure of two conjugative transposons, Tn*5251* and Tn*5252*. *J. Bacteriol.* **173:**1617–1622.
6. **Baca, A. M., R. Sirawaraporn, S. Turley, W. Sirawaraporn, and W. G. Hol.** 2000. Crystal structure of *Mycobacterium tuberculosis* 7,8-dihydropteroate synthase in complex with pterin monophosphate: new insight into the enzymatic mechanism and sulfa-drug action. *J. Mol. Biol.* **302:**1193–1212.
7. **Banks, D. J., S. F. Porcella, K. D. Barbian, J. M. Martin, and J. M. Musser.** 2003. Structure and distribution of an unusual chimeric genetic element encoding macrolide resistance in phylogenetically diverse clones of group A *Streptococcus. J. Infect. Dis.* **188:**1898–1908.
8. **Bellon, S., J. D. Parsons, Y. Wei, K. Hayakawa, L. L. Swenson, P. S. Charifson, J. A. Lippke, R. Aldape, and C. H. Gross.** 2004. Crystal structures of *Escherichia coli* topoisomerase IV ParE subunit (24 and 43 kilodaltons): a single residue dictates differences in novobiocin potency against topoisomerase IV and DNA gyrase. *Antimicrob. Agents Chemother.* **48:**1856–1864.
9. **Berisio, R., J. Harms, F. Schluenzen, R. Zarivach, H. A. Hansen, P. Fucini, and A. Yonath.** 2003. Structural insight into the antibiotic action of telithromycin against resistant mutants. *J. Bacteriol.* **185:**4276–4279.
10. **Bolin, J. T., D. J. Filman, D. A. Matthews, R. C. Hamlin, and J. Kraut.** 1982. Crystal structures of *Escherichia coli* and *Lactobacillus casei* dihydrofolate reductase refined at 1.7 Å resolution. I. General features and binding of methotrexate. *J. Biol. Chem.* **257:**13650–13662.
11. **Bozdogan, B., T. Bogdanovich, K. Kosowska, M. R. Jacobs, and P. C. Appelbaum.** 2004. Macrolide resistance in *Streptococcus pneumoniae:* clonality and mechanisms of resistance in 24 countries. *Curr. Drug Topics—Infect. Disorders* **4:**169–176.
12. **Brenwald, N. P., P. Appelbaum, T. Davies, and M. J. Gill.** 2003. Evidence for efflux pumps, other than PmrA, associated with fluoroquinolone resistance in *Streptococcus pneumoniae. Clin. Microbiol. Infect.* **9:**140–143.
13. **Brenwald, N. P., M. J. Gill, and R. Wise.** 1998. Prevalence of a putative efflux mechanism among fluoroquinolone-resistant

clinical isolates of *Streptococcus pneumoniae. Antimicrob. Agents Chemother.* **42:**2032–2035.

14. **Brodersen, D. E., W. M. Clemons, Jr., A. P. Carter, R. J. Morgan-Warren, B. T. Wimberly, and V. Ramakrishnan.** 2000. The structural basis for the action of the antibiotics tetracycline, pactamycin, and hygromycin B on the 30S ribosomal subunit. *Cell* **103:**1143–1154.
15. **Brown, C. M., K. K. McCaughan, and W. P. Tate.** 1993. Two regions of the *Escherichia coli* 16S ribosomal RNA are important for decoding stop signals in polypeptide chain termination. *Nucleic Acids Res.* **21:**2109–2115.
16. **Brown, S. D., and M. J. Rybak.** 2004. Antimicrobial susceptibility of *Streptococcus pneumoniae, Streptococcus pyogenes* and *Haemophilus influenzae* collected from patients across the USA, in 2001-2002, as part of the PROTEKT US study. *J. Antimicrob. Chemother.* **54** (Suppl 1.):I7–I15.
17. **Campbell, G. D., Jr., and R. Silberman.** 1998. Drug-resistant *Streptococcus pneumoniae. Clin. Infect. Dis.* **26:**1188–1195.
18. **Canton, R., M. Morosini, M. C. Enright, and I. Morrissey.** 2003. Worldwide incidence, molecular epidemiology and mutations implicated in fluoroquinolone-resistant *Streptococcus pneumoniae:* data from the global PROTEKT surveillance programme. *J. Antimicrob. Chemother.* **52:**944–952.
19. **Canu, A., B. Malbruny, M. Coquemont, T. A. Davies, P. C. Appelbaum, and R. Leclercq.** 2002. Diversity of ribosomal mutations conferring resistance to macrolides, clindamycin, streptogramin, and telithromycin in *Streptococcus pneumoniae. Antimicrob. Agents Chemother.* **46:**125–131.
20. **Carter, A. P., W. M. Clemons, D. E. Brodersen, R. J. Morgan-Warren, B. T. Wimberly, and V. Ramakrishnan.** 2000. Functional insights from the structure of the 30S ribosomal subunit and its interactions with antibiotics. *Nature* **407:**340–348.
21. **Chesnel, L., L. Pernot, D. Lemaire, D. Champelovier, J. Croize, O. Dideberg, T. Vernet, and A. Zapun.** 2003. The structural modifications induced by the M339F substitution in PBP2x from *Streptococcus pneumoniae* further decreases the susceptibility to beta-lactams of resistant strains. *J. Biol. Chem.* **278:**44448–44456.
22. **Chopra, I., and M. Roberts.** 2001. Tetracycline antibiotics: mode of action, applications, molecular biology, and epidemiology of bacterial resistance. *Microbiol. Mol. Biol. Rev.* **65:**232–260.
23. **Clancy, J., J. Petitpas, F. Dib-Hajj, W. Yuan, M. Cronan, A. V. Kamath, J. Bergeron, and J. A. Retsema.** 1996. Molecular cloning and functional analysis of a novel macrolide- resistance determinant, *mefA,* from *Streptococcus pyogenes. Mol. Microbiol.* **22:**867–879.
24. **Contreras, A., and D. Vazquez.** 1977. Cooperative and antagonistic interactions of peptidyl-tRNA and antibiotics with bacterial ribosomes. *Eur. J. Biochem.* **74:**539–547.
25. **Corbett, K. D., R. K. Shultzaberger, and J. M. Berger.** 2004. The C-terminal domain of DNA gyrase A adopts a DNA-bending beta-pinwheel fold. *Proc. Natl. Acad. Sci. USA* **101:**7293–7298.
26. **Cousin, S., Jr., W. L. Whittington, and M. C. Roberts.** 2003. Acquired macrolide resistance genes in pathogenic *Neisseria spp.* isolated between 1940 and 1987. *Antimicrob. Agents Chemother.* **47:**3877–3880.
27. **Daigle, D. M., G. A. McKay, P. R. Thompson, and G. D. Wright.** 1999. Aminoglycoside antibiotic phosphotransferases are also serine protein kinases. *Chem. Biol.* **6:**11–18.
28. **Daigle, D. M., G. A. McKay, and G. D. Wright.** 1997. Inhibition of aminoglycoside antibiotic resistance enzymes by protein kinase inhibitors. *J. Biol. Chem.* **272:**24755–24758.
29. **Dale, G. E., D. Kostrewa, B. Gsell, M. Stieger, and A. D'Arcy.** 1999. Crystal engineering: deletion mutagenesis of the 24 kDa fragment of the DNA gyrase B subunit from *Staphylococcus aureus. Acta Crystallogr. D. Biol. Crystallogr.* **55** (Pt 9):1626–1629.
30. **Daly, M. M., S. Doktor, R. Flamm, and D. Shortridge.** 2004. Characterization and prevalence of MefA, MefE, and the associated *msr*(D) gene in *Streptococcus pneumoniae* clinical isolates. *J. Clin. Microbiol.* **42:**3570–3574.
31. **de Antonio, C., M. E. Farias, M. G. de Lacoba, and M. Espinosa.** 2004. Features of the plasmid pMV158-encoded MobM, a protein involved in its mobilization. *J. Mol. Biol.* **335:**733–743.
32. **Del Grosso, M., A. Scotto d'Abusco, F. Iannelli, G. Pozzi, and A. Pantosti.** 2004. Tn*2009*, a Tn*916*-like element containing *mef*(E) in *Streptococcus pneumoniae. Antimicrob. Agents Chemother.* **48:**2037–2042.
33. **Deshpande, L. M., R. N. Jones, and M. A. Pfaller.** 2001. Accuracy of broth microdilution and E test methods for detecting chloramphenicol acetyl transferase mediated resistance in *Streptococcus pneumoniae:* geographic variations in the prevalence of resistance in The SENTRY Antimicrobial Surveillance Program (1999). *Diagn. Microbiol. Infect. Dis.* **39:**267–269.
34. **Dessen, A., N. Mouz, E. Gordon, J. Hopkins, and O. Dideberg.** 2001. Crystal structure of PBP2x from a highly penicillin-resistant *Streptococcus pneumoniae* clinical isolate: a mosaic framework containing 83 mutations. *J. Biol. Chem.* **276:**45106–45112.
35. **Dever, L. L., K. Shashikumar, and W. G. Johanson, Jr.** 2002. Antibiotics in the treatment of acute exacerbations of chronic bronchitis. *Expert Opin. Investig. Drugs* **11:**911–925.
36. **Doern, G. V., A. B. Brueggemann, H. Huynh, and E. Wingert.** 1999. Antimicrobial resistance with *Streptococcus pneumoniae* in the United States, 1997-98. *Emerg. Infect. Dis.* **5:**757–765.
37. **Doern, G. V., K. P. Heilmann, H. K. Huynh, P. R. Rhomberg, S. L. Coffman, and A. B. Brueggemann.** 2001. Antimicrobial resistance among clinical isolates of *Streptococcus pneumoniae* in the United States during 1999-2000, including a comparison of resistance rates since 1994-1995. *Antimicrob. Agents Chemother.* **45:**1721–1729.
38. **Douthwaite, S., L. H. Hansen, and P. Mauvais.** 2000. Macrolide-ketolide inhibition of MLS-resistant ribosomes is improved by alternative drug interaction with domain II of 23S rRNA. *Mol. Microbiol.* **36:**183–193.
39. **Dutta, R., and M. Inouye.** 2000. GHKL, an emergent ATPase/kinase superfamily. *Trends Biochem. Sci.* **25:**24–28.
40. **Epe, B., P. Woolley, and H. Hornig.** 1987. Competition between tetracycline and tRNA at both P and A sites of the ribosome of *Escherichia coli. FEBS Lett.* **213:**443–447.
41. **Ettayebi, M., S. M. Prasad, and E. A. Morgan.** 1985. Chloramphenicol-erythromycin resistance mutations in a 23S rRNA gene of *Escherichia coli. J. Bacteriol.* **162:**551–557.
42. **Farrell, D. J., I. Morrissey, S. Bakker, S. Buckridge, and D. Felmingham.** 2004. In vitro activity of telithromycin, linezolid and quinupristin-dalfopristin in *Streptococcus pneumoniae* with macrolide resistance due to ribosomal mutations. *Antimicrob. Agents Chemother.* **48:**3169–3171.
43. **Fernandez-Munoz, R., R. E. Monro, R. Torres-Pinedo, and D. Vazquez.** 1971. Substrate- and antibiotic-binding sites at the peptidyl-transferase centre of *Escherichia coli* ribosomes. Studies on the chloramphenicol, lincomycin and erythromycin sites. *Eur. J. Biochem.* **23:**185–193.
44. **Filman, D. J., J. T. Bolin, D. A. Matthews, and J. Kraut.** 1982. Crystal structures of *Escherichia coli* and *Lactobacillus casei* dihydrofolate reductase refined at 1.7 Å resolution. II. Environment of bound NADPH and implications for catalysis. *J. Biol. Chem.* **257:**13663–13672.

45. **Franceschi, F., Z. Kanyo, E. C. Sherer, and J. Sutcliffe.** 2004. Macrolide resistance from the ribosome perspective. *Curr. Drug Topics—Infect. Disorders* **4:**177–191.
46. **Friedland, I. R., and K. P. Klugman.** 1992. Antibiotic-resistant pneumococcal disease in South African children. *Am. J. Dis. Child* **146:**920–923.
47. **Garza-Ramos, G., L. Xiong, P. Zhong, and A. Mankin.** 2001. Binding site of macrolide antibiotics on the ribosome: new resistance mutation identifies a specific interaction of ketolides with rRNA. *J. Bacteriol.* **183:**6898–6907.
48. **Gay, K., and D. S. Stephens.** 2001. Structure and dissemination of a chromosomal insertion element encoding macrolide efflux in *Streptococcus pneumoniae. J. Infect. Dis.* **184:**56–65. Epub May 31, 2001.
49. **Gill, M. J., N. P. Brenwald, and R. Wise.** 1999. Identification of an efflux pump gene, *pmrA*, associated with fluoroquinolone resistance in *Streptococcus pneumoniae. Antimicrob. Agents Chemother.* **43:**187–189.
50. **Giovanetti, E., A. Brenciani, R. Lupidi, M. C. Roberts, and P. E. Varaldo.** 2003. Presence of the *tet*(O) gene in erythromycin- and tetracycline-resistant strains of *Streptococcus pyogenes* and linkage with either the *mef*(A) or the *erm*(A) gene. *Antimicrob. Agents Chemother.* **47:**2844–2849.
51. **Gordon, E., N. Mouz, E. Duee, and O. Dideberg.** 2000. The crystal structure of the penicillin-binding protein 2x from *Streptococcus pneumoniae* and its acyl-enzyme form: implication in drug resistance. *J. Mol. Biol.* **299:**477–485.
52. **Gosink, K., and E. Tuomanen.** 2000. *Streptococcus pneumoniae:* invasion and inflammation, p. 214–224. *In* V. Fischetti (ed.), *Gram-Positive Pathogens.* ASM Press, Washington, D.C.
53. **Gould, K. A., X. S. Pan, R. J. Kerns, and L. M. Fisher.** 2004. Ciprofloxacin dimers target gyrase in *Streptococcus pneumoniae. Antimicrob. Agents Chemother.* **48:**2108–2115.
54. **Grebe, T., J. Paik, and R. Hakenbeck.** 1997. A novel resistance mechanism against beta-lactams in *Streptococcus pneumoniae* involves CpoA, a putative glycosyltransferase. *J. Bacteriol.* **179:**3342–3349.
55. **Gross, C. H., J. D. Parsons, T. H. Grossman, P. S. Charifson, S. Bellon, J. Jernee, M. Dwyer, S. P. Chambers, W. Markland, M. Botfield, and S. A. Raybuck.** 2003. Active-site residues of *Escherichia coli* DNA gyrase required in coupling ATP hydrolysis to DNA supercoiling and amino acid substitutions leading to novobiocin resistance. *Antimicrob. Agents Chemother.* **47:**1037–1046.
56. **Guenzi, E., A. M. Gasc, M. A. Sicard, and R. Hakenbeck.** 1994. A two-component signal-transducing system is involved in competence and penicillin susceptibility in laboratory mutants of *Streptococcus pneumoniae. Mol. Microbiol.* **12:**505–515.
57. **Haasum, Y., K. Strom, R. Wehelie, V. Luna, M. C. Roberts, J. P. Maskell, L. M. Hall, and G. Swedberg.** 2001. Amino acid repetitions in the dihydropteroate synthase of *Streptococcus pneumoniae* lead to sulfonamide resistance with limited effects on substrate K(m). *Antimicrob. Agents Chemother.* **45:**805–809.
58. **Hakenbeck, R., A. Konig, I. Kern, M. van der Linden, W. Keck, D. Billot-Klein, R. Legrand, B. Schoot, and L. Gutmann.** 1998. Acquisition of five high-Mr penicillin-binding protein variants during transfer of high-level beta-lactam resistance from *Streptococcus mitis* to *Streptococcus pneumoniae. J. Bacteriol.* **180:**1831–1840.
59. **Hampele, I. C., A. D'Arcy, G. E. Dale, D. Kostrewa, J. Nielsen, C. Oefner, M. G. Page, H. J. Schonfeld, D. Stuber, and R. L. Then.** 1997. Structure and function of the dihydropteroate synthase from *Staphylococcus aureus. J. Mol. Biol.* **268:**21–30.
60. **Hansen, H. L., P. Mauvais, and S. Douthwaite.** 1999. The macrolide-ketolide antibiotic binding site is formed by structures in domain II and V of 23S ribosomal RNA. *Mol. Microbiol.* **31:**623–631.
61. **Hansen, J. L., J. A. Ippolito, N. Ban, P. Nissen, P. B. Moore, and T. A. Steitz.** 2002. The structures of four macrolide antibiotics bound to the large ribosomal subunit. *Mol. Cell.* **10:** 117–128.
62. **Hansen, J. L., P. B. Moore, and T. A. Steitz.** 2003. Structures of five antibiotics bound at the peptidyl transferase center of the large ribosomal subunit. *J. Mol. Biol.* **330:**1061–1075.
63. **Harms, J. M., F. Schlunzen, P. Fucini, H. Bartels, and A. Yonath.** 2004. Alterations at the peptidyl transferase centre of the ribosome induced by the synergistic action of the streptogramins dalfopristin and quinupristin. *BMC Biol.* **2:**4.
64. **Hennessy, T. W., K. M. Petersen, D. Bruden, A. J. Parkinson, D. Hurlburt, M. Getty, B. Schwartz, and J. C. Butler.** 2002. Changes in antibiotic-prescribing practices and carriage of penicillin-resistant *Streptococcus pneumoniae:* a controlled intervention trial in rural Alaska. *Clin. Infect. Dis.* **34:**1543–1550.
65. **Hoban, D. J., G. V. Doern, A. C. Fluit, M. Roussel-Delvallez, and R. N. Jones.** 2001. Worldwide prevalence of antimicrobial resistance in *Streptococcus pneumoniae, Haemophilus influenzae,* and *Moraxella catarrhalis* in the SENTRY Antimicrobial Surveillance Program, 1997-1999. *Clin. Infect. Dis.* **32** (Suppl. 2):S81–S93.
66. **Hon, W. C., G. A. McKay, P. R. Thompson, R. M. Sweet, D. S. Yang, G. D. Wright, and A. M. Berghuis.** 1997. Structure of an enzyme required for aminoglycoside antibiotic resistance reveals homology to eukaryotic protein kinases. *Cell* **89:**887–895.
67. **Hooper, D. C.** 2000. Mechanisms of action and resistance of older and newer fluoroquinolones. *Clin. Infect. Dis.* **31** (Suppl. 2):S24–S28.
68. **Hooper, D. C.** 2000. Mechanisms of fluoroquinolone resistance, p. 685–693. *In* V. Fischetti (ed.), *Gram-Positive Pathogens.* ASM Press, Washington, D.C.
69. **Hsieh, T. J., and N. L. Chan.** 2004. Crystallization and preliminary X-ray crystallographic analysis of the C-terminal domain of ParC protein from *Bacillus stearothermophilus. Acta Crystallogr. D. Biol. Crystallogr.* **60:**564–566.
70. **Iannelli, F., M. Santagati, J. D. Doquier, M. Cassone, M. R. Oggioni, G. Rossolini, S. Stefani, and G. Pozzi.** 2004. Type M Resistance to macrolides in streptococci is not due to the *mef*(A) gene, but to *mat*(A) encoding an ATP-dependent efflux pump, abstr. C1-1188. *Program and Abstracts of the 104th Interscience Conference on Antimicrobial Agents and Chemotherapy,* Washington, D.C.
71. **Idanpaan-Heikkila, I., P. M. Simon, D. Zopf, T. Vullo, P. Cahill, K. Sokol, and E. Tuomanen.** 1997. Oligosaccharides interfere with the establishment and progression of experimental pneumococcal pneumonia. *J. Infect. Dis.* **176:**704–712.
72. **Inoue, M., N. Y. Lee, S. W. Hong, K. Lee, and D. Felmingham.** 2004. PROTEKT 1999-2000: a multicentre study of the antibiotic susceptibility of respiratory tract pathogens in Hong Kong, Japan and South Korea. *Int. J. Antimicrob. Agents* **23:** 44–51.
73. **Jacobs, M. R.** 2004. *Streptococcus pneumoniae:* epidemiology and patterns of resistance. *Am. J. Med.* **117** (Suppl. 3A):3S–15S.
74. **Jacobs, M. R., D. Felmingham, P. C. Appelbaum, and R. N. Gruneberg.** 2003. The Alexander Project 1998-2000: susceptibility of pathogens isolated from community-acquired respiratory tract infection to commonly used antimicrobial agents. *J. Antimicrob. Chemother.* **52:**229–246.
75. **Jones, M. E., R. S. Blosser-Middleton, C. Thornsberry, J. A. Karlowsky, and D. F. Sahm.** 2003. The activity of levofloxacin

and other antimicrobials against clinical isolates of *Streptococcus pneumoniae* collected worldwide during 1999-2002. *Diagn. Microbiol. Infect. Dis.* **47**:579–586.

76. **Kampranis, S. C., A. D. Bates, and A. Maxwell.** 1999. A model for the mechanism of strand passage by DNA gyrase. *Proc. Natl. Acad. Sci. USA* **96**:8414–8419.
77. **Karlowsky, J. A., D. C. Draghi, C. Thornsberry, M. E. Jones, I. A. Critchley, and D. F. Sahm.** 2002. Antimicrobial susceptibilities of *Streptococcus pneumoniae, Haemophilus influenzae* and *Moraxella catarrhalis* isolated in two successive respiratory seasons in the US. *Int. J. Antimicrob. Agents* **20**:76–85.
78. **Klugman, K. P.** 2002. The successful clone: the vector of dissemination of resistance in *Streptococcus pneumoniae. J. Antimicrob. Chemother.* **50** (Suppl. S2):1–5.
79. **Koh, T. H., and R. V. Lin.** 1997. Increasing antimicrobial resistance in clinical isolates of *Streptococcus pneumoniae. Ann. Acad. Med. Singapore* **26**:604–608.
80. **Kolesnikov, I. V., N. Y. Protasova, and A. T. Gudkov.** 1996. Tetracyclines induce changes in accessibility of ribosomal proteins to proteases. *Biochimie* **78**:868–873.
81. **Lacks, S. A., B. Greenberg, and P. Lopez.** 1995. A cluster of four genes encoding enzymes for five steps in the folate biosynthetic pathway of *Streptococcus pneumoniae. J. Bacteriol.* **177**:66–74.
82. **Lafitte, D., V. Lamour, P. O. Tsvetkov, A. A. Makarov, M. Klich, P. Deprez, D. Moras, C. Briand, and R. Gilli.** 2002. DNA gyrase interaction with coumarin-based inhibitors: the role of the hydroxybenzoate isopentenyl moiety and the 5′-methyl group of the noviose. *Biochemistry* **41**:7217–7223.
83. **Lamour, V., L. Hoermann, J. M. Jeltsch, P. Oudet, and D. Moras.** 2002. Crystallization of the 43 kDa ATPase domain of *Thermus thermophilus* gyrase B in complex with novobiocin. *Acta Crystallogr. D. Biol. Crystallogr.* **58**:1376–1378. Epub, 20 July 2002.
84. **Leclercq, R., and P. Courvalin.** 2002. Resistance to macrolides and related antibiotics in *Streptococcus pneumoniae. Antimicrob. Agents Chemother.* **46**:2727–2734.
85. **Lee, H., V. M. Reyes, and J. Kraut.** 1996. Crystal structures of *Escherichia coli* dihydrofolate reductase complexed with 5-formyltetrahydrofolate (folinic acid) in two space groups: evidence for enolization of pteridine O4. *Biochemistry* **35**:7012–7020.
86. **Lee, H. J., J. Y. Park, S. H. Jang, J. H. Kim, E. C. Kim, and K. W. Choi.** 1995. High incidence of resistance to multiple antimicrobials in clinical isolates of *Streptococcus pneumoniae* from a university hospital in Korea. *Clin. Infect. Dis.* **20**:826–835.
87. **Lessard, J. L., and S. Pestka.** 1972. Studies on the formation of transfer ribonucleic acid-ribosome complexes. 23. Chloramphenicol, aminoacyl-oligonucleotides, and *Escherichia coli* ribosomes. *J. Biol. Chem.* **247**:6909–6912.
88. **Li, R., R. Sirawaraporn, P. Chitnumsub, W. Sirawaraporn, J. Wooden, F. Athappilly, S. Turley, and W. G. Hol.** 2000. Three-dimensional structure of *M. tuberculosis* dihydrofolate reductase reveals opportunities for the design of novel tuberculosis drugs. *J. Mol. Biol.* **295**:307–323.
89. **Lim, D., and N. C. Strynadka.** 2002. Structural basis for the beta lactam resistance of PBP2a from methicillin-resistant *Staphylococcus aureus. Nat. Struct. Biol.* **9**:870–876.
90. **Lopez, P., M. Espinosa, B. Greenberg, and S. A. Lacks.** 1987. Sulfonamide resistance in *Streptococcus pneumoniae:* DNA sequence of the gene encoding dihydropteroate synthase and characterization of the enzyme. *J. Bacteriol.* **169**:4320–4326.
91. **Lovett, P. S.** 1996. Translation attenuation regulation of chloramphenicol resistance in bacteria—a review. *Gene* **179**:157–162.
92. **Luna, V. A., S. Cousin, Jr., W. L. Whittington, and M. C. Roberts.** 2000. Identification of the conjugative *mef* gene in clinical *Acinetobacter junii* and *Neisseria gonorrhoeae* isolates. *Antimicrob. Agents Chemother.* **44**:2503–2506.
93. **Luna, V. A., and M. C. Roberts.** 1998. The presence of the *tetO* gene in a variety of tetracycline-resistant *Streptococcus pneumoniae* serotypes from Washington State. *J. Antimicrob. Chemother.* **42**:613–619.
94. **Madhi, S. A., K. P. Klugman, and The Vaccine Trialist Group.** 2004. A role for *Streptococcus pneumoniae* in virus-associated pneumonia. *Nat. Med.* **10**:811–813.
95. **Mankin, A. S., and R. A. Garrett.** 1991. Chloramphenicol resistance mutations in the single 23S rRNA gene of the archaeon *Halobacterium halobium. J. Bacteriol.* **173**:3559–3563.
96. **Maravic, G.** 2004. Macrolide resistance based on the Erm-mediated rRNA methylation. *Curr. Drug Topics—Infect. Disorders* **4**:193–202.
97. **Marton, A.** 1992. Pneumococcal antimicrobial resistance: the problem in Hungary. *Clin. Infect. Dis.* **15**:106–111.
98. **Maskell, J. P., A. M. Sefton, and L. M. Hall.** 1997. Mechanism of sulfonamide resistance in clinical isolates of *Streptococcus pneumoniae. Antimicrob. Agents Chemother.* **41**:2121–2126.
99. **Maskell, J. P., A. M. Sefton, and L. M. Hall.** 2001. Multiple mutations modulate the function of dihydrofolate reductase in trimethoprim-resistant *Streptococcus pneumoniae. Antimicrob. Agents Chemother.* **45**:1104–1108.
100. **McDougal, L. K., F. C. Tenover, L. N. Lee, J. K. Rasheed, J. E. Patterson, J. H. Jorgensen, and D. J. LeBlanc.** 1998. Detection of Tn*917*-like sequences within a Tn*916*-like conjugative transposon (Tn*3872*) in erythromycin-resistant isolates of *Streptococcus pneumoniae. Antimicrob. Agents Chemother.* **42**:2312–2318.
101. **McGee, L., L. McDougal, J. Zhou, B. G. Spratt, F. C. Tenover, R. George, R. Hakenbeck, W. Hryniewicz, J. C. Lefevre, A. Tomasz, and K. P. Klugman.** 2001. Nomenclature of major antimicrobial-resistant clones of *Streptococcus pneumoniae* defined by the pneumococcal molecular epidemiology network. *J. Clin. Microbiol.* **39**:2565–2571.
102. **McMurry, L. M., and S. B. Levy.** 2000. Tetracycline resistance in Gram-positive bacteria, p. 660–677. *In* V. Fischetti (ed.), *Gram-Positive Pathogens.* ASM Press, Washington, D.C.
103. **Moazed, D., and H. F. Noller.** 1987. Chloramphenicol, erythromycin, carbomycin and vernamycin B protect overlapping sites in the peptidyl transferase region of 23S ribosomal RNA. *Biochimie* **69**:879–884.
104. **Morais Cabral, J. H., A. P. Jackson, C. V. Smith, N. Shikotra, A. Maxwell, and R. C. Liddington.** 1997. Crystal structure of the breakage-reunion domain of DNA gyrase. *Nature* **388:** 903–906.
105. **Morris, J. E., X. S. Pan, and L. M. Fisher.** 2002. Grepafloxacin, a dimethyl derivative of ciprofloxacin, acts preferentially through gyrase in *Streptococcus pneumoniae:* role of the C-5 group in target specificity. *Antimicrob. Agents Chemother.* **46**:582–585.
106. **Mouz, N., E. Gordon, A. M. Di Guilmi, I. Petit, Y. Petillot, Y. Dupont, R. Hakenbeck, T. Vernet, and O. Dideberg.** 1998. Identification of a structural determinant for resistance to beta-lactam antibiotics in gram-positive bacteria. *Proc. Natl. Acad. Sci. USA* **95**:13403–13406.
107. **Murphy, T. F., and S. Sethi.** 2002. Chronic obstructive pulmonary disease: role of bacteria and guide to antibacterial selection in the older patient. *Drugs Aging* **19**:761–775.
108. **Murray, I. A.** 2000. Chloramphenicol resistance, p. 678–684. *In* V. Fischetti (ed.), *Gram-Positive Pathogens.* ASM Press, Washington, D.C.

109. **Murray, I. A., and W. V. Shaw.** 1997. O-Acetyltransferases for chloramphenicol and other natural products. *Antimicrob. Agents Chemother.* **41:**1–6.
110. **Musher, D. M.** 2000. *Streptococcus pneumoniae,* p. 2128–2147. *In* G. L. Mandell, J. E. Bennett, and R. Dolin (ed.), *Principles and Practices of Infectious Disease,* 5th ed. Churchill Livingstone, Edinburgh, United Kingdom.
111. **Naraqi, S., G. P. Kirkpatrick, and S. Kabins.** 1974. Relapsing pneumococcal meningitis: isolation of an organism with decreased susceptibility to penicillin G. *J. Pediatr.* **85:**671–673.
112. **Ogle, J. M., D. E. Brodersen, W. M. Clemons, Jr., M. J. Tarry, A. P. Carter, and V. Ramakrishnan.** 2001. Recognition of cognate transfer RNA by the 30S ribosomal subunit. *Science* **292:**897–902.
113. **Ojo, K. K., C. Ulep, N. Van Kirk, H. Luis, M. Bernardo, J. Leitao, and M. C. Roberts.** 2004. The *mef*(A) gene predominates among seven macrolide resistance genes identified in gram-negative strains representing 13 genera, isolated from healthy Portuguese children. *Antimicrob. Agents Chemother.* **48:**3451–3456.
114. **Oleinick, N. L., J. M. Wilhelm, and J. W. Corcoran.** 1968. Nonidentity of the site of action of erythromycin A and chloramphenicol on *Bacillus subtilis* ribosomes. *Biochim. Biophys. Acta* **155:**290–292.
115. **Padayachee, T., and K. P. Klugman.** 1999. Novel expansions of the gene encoding dihydropteroate synthase in trimethoprim-sulfamethoxazole-resistant *Streptococcus pneumoniae. Antimicrob. Agents Chemother.* **43:**2225–2230.
116. **Pan, X. S., and L. M. Fisher.** 1996. Cloning and characterization of the *parC* and *parE* genes of *Streptococcus pneumoniae* encoding DNA topoisomerase IV: role in fluoroquinolone resistance. *J. Bacteriol.* **178:**4060–4069.
117. **Pan, X. S., and L. M. Fisher.** 1998. DNA gyrase and topoisomerase IV are dual targets of clinafloxacin action in *Streptococcus pneumoniae. Antimicrob. Agents Chemother.* **42:**2810–2816.
118. **Pan, X. S., and L. M. Fisher.** 1999. *Streptococcus pneumoniae* DNA gyrase and topoisomerase IV: overexpression, purification, and differential inhibition by fluoroquinolones. *Antimicrob. Agents Chemother.* **43:**1129–1136.
119. **Pares, S., N. Mouz, Y. Petillot, R. Hakenbeck, and O. Dideberg.** 1996. X-ray structure of *Streptococcus pneumoniae* PBP2x, a primary penicillin target enzyme. *Nat. Struct. Biol.* **3:**284–289.
120. **Pelton, S. I., R. Dagan, B. M. Gaines, K. P. Klugman, D. Laufer, K. O'Brien, and H. J. Schmitt.** 2003. Pneumococcal conjugate vaccines: proceedings from an interactive symposium at the 41st Interscience Conference on Antimicrobial Agents and Chemotherapy. *Vaccine* **21:**1562–1571.
121. **Perichon, B., J. Tankovic, and P. Courvalin.** 1997. Characterization of a mutation in the *parE* gene that confers fluoroquinolone resistance in *Streptococcus pneumoniae. Antimicrob. Agents Chemother.* **41:**1166–1167.
122. **Pernot, L., L. Chesnel, A. Le Gouellec, J. Croize, T. Vernet, O. Dideberg, and A. Dessen.** 2004. A PBP2x from a clinical isolate of *Streptococcus pneumoniae* exhibits an alternative mechanism for reduction of susceptibility to beta-lactam antibiotics. *J. Biol. Chem.* **279:**16463–16470.
123. **Pestova, E., J. J. Millichap, F. Siddiqui, G. A. Noskin, and L. R. Peterson.** 2002. Non-PmrA-mediated multidrug resistance in *Streptococcus pneumoniae. J. Antimicrob. Chemother.* **49:**553–556.
124. **Piddock, L. J., M. Johnson, V. Ricci, and S. L. Hill.** 1998. Activities of new fluoroquinolones against fluoroquinolone-resistant pathogens of the lower respiratory tract. *Antimicrob. Agents Chemother.* **42:**2956–2960.
125. **Piddock, L. J., M. M. Johnson, S. Simjee, and L. Pumbwe.** 2002. Expression of efflux pump gene *pmrA* in fluoroquinolone-resistant and -susceptible clinical isolates of *Streptococcus pneumoniae. Antimicrob. Agents Chemother.* **46:**808–812.
126. **Pioletti, M., F. Schlunzen, J. Harms, R. Zarivach, M. Gluhmann, H. Avila, A. Bashan, H. Bartels, T. Auerbach, C. Jacobi, T. Hartsch, A. Yonath, and F. Franceschi.** 2001. Crystal structures of complexes of the small ribosomal subunit with tetracycline, edeine and IF3. *EMBO J.* **20:**1829–1839.
127. **Pletz, M. W., L. McGee, J. Jorgensen, B. Beall, R. R. Facklam, C. G. Whitney, and K. P. Klugman.** 2004. Levofloxacin-resistant invasive *Streptococcus pneumoniae* in the United States: evidence for clonal spread and the impact of conjugate pneumococcal vaccine. *Antimicrob. Agents Chemother.* **48:**3491–3497.
128. **Poehlsgaard, J., and S. Douthwaite.** 2003. Macrolide antibiotic interaction and resistance on the bacterial ribosome. *Curr. Opin. Investig. Drugs* **4:**140–148.
129. **Poulsen, S. M., C. Kofoed, and B. Vester.** 2000. Inhibition of the ribosomal peptidyl transferase reaction by the mycarose moiety of the antibiotics carbomycin, spiramycin and tylosin. *J. Mol. Biol.* **304:**471–481.
130. **Pozzi, G., F. Iannelli, M. R. Oggioni, M. Santagati, and S. Stefani.** 2004. Genetic elements carrying macrolide efflux genes in streptococci. *Curr. Drug Topics—Infect. Disorders* **4:**203–206.
131. **Reece, R. J., and A. Maxwell.** 1991. Probing the limits of the DNA breakage-reunion domain of the *Escherichia coli* DNA gyrase A protein. *J. Biol. Chem.* **266:**3540–3546.
132. **Reichmann, P., A. Konig, A. Marton, and R. Hakenbeck.** 1996. Penicillin-binding proteins as resistance determinants in clinical isolates of *Streptococcus pneumoniae. Microb. Drug Resist.* **2:**177–181.
133. **Rheinberger, H. J., and K. H. Nierhaus.** 1990. Partial release of AcPhe-Phe-tRNA from ribosomes during poly(U)-dependent poly(Phe) synthesis and the effects of chloramphenicol. *Eur. J. Biochem.* **193:**643–650.
134. **Roberts, M. C.** 2004. Distribution of macrolide, lincosamide, streptogramin, ketolide and oxazolidinone (MLSKO) resistance genes in gram-negative bacteria. *Curr. Drug Topics—Infect. Disorders* **4:**207–215.
135. **Roberts, M. C., J. Sutcliffe, P. Courvalin, L. B. Jensen, J. Rood, and H. Seppala.** 1999. Nomenclature for macrolide and macrolide-lincosamide-streptogramin B resistance determinants. *Antimicrob. Agents Chemother.* **43:**2823–2830.
136. **Roca, J., J. M. Berger, S. C. Harrison, and J. C. Wang. 1996.** DNA transport by a type II topoisomerase: direct evidence for a two-gate mechanism. *Proc. Natl. Acad. Sci. USA* **93:**4057–4062.
137. **Rudolph, K. M., A. J. Parkinson, and M. C. Roberts.** 2001. Mechanisms of erythromycin and trimethoprim resistance in the Alaskan *Streptococcus pneumoniae* serotype 6B clone. *J. Antimicrob. Chemother.* **48:**317–319.
138. **Rybak, M. J.** 2004. Increased bacterial resistance: PROTEKT US—an update. *Ann. Pharmacother.* **38:**S8–S13.
139. **Sangthawan, P., S. Chantaratchada, N. Chanthadisai, and A. Wattanathum.** 2003. Prevalence and clinical significance of community-acquired penicillin-resistant pneumococcal pneumonia in Thailand. *Respirology* **8:**208–212.
140. **Santagati, M., F. Iannelli, C. Cascone, F. Campanile, M. R. Oggioni, S. Stefani, and G. Pozzi.** 2003. The novel conjugative transposon Tn*1207.3* carries the macrolide efflux gene *mef*(A) in *Streptococcus pyogenes. Microb. Drug Resist.* **9:**243–247.
141. **Santagati, M., F. Iannelli, M. R. Oggioni, S. Stefani, and G. Pozzi.** 2000. Characterization of a genetic element carrying the

macrolide efflux gene *mef*(A) in *Streptococcus pneumoniae*. *Antimicrob. Agents Chemother.* **44**:2585–2587.

142. **Sauvage, E., F. Kerff, E. Fonze, R. Herman, B. Schoot, J. P. Marquette, Y. Taburet, D. Prevost, J. Dumas, G. Leonard, P. Stefanic, J. Coyette, and P. Charlier.** 2002. The 2.4-Å crystal structure of the penicillin-resistant penicillin-binding protein PBP5fm from *Enterococcus faecium* in complex with benzylpenicillin. *Cell. Mol. Life Sci.* **59**:1223–1232.
143. **Sawaya, M. R., and J. Kraut.** 1997. Loop and subdomain movements in the mechanism of *Escherichia coli* dihydrofolate reductase: crystallographic evidence. *Biochemistry* **36**:586–603.
144. **Schito, G. C., A. Marchese, D. Elkharrat, and D. J. Farrell.** 2004. Comparative activity of telithromycin against macrolide-resistant isolates of *Streptococcus pneumoniae:* results of two years of the PROTEKT surveillance study. *J. Chemother.* **16**:13–22.
145. **Schlunzen, F., J. M. Harms, F. Franceschi, H. A. Hansen, H. Bartels, R. Zarivach, and A. Yonath.** 2003. Structural basis for the antibiotic activity of ketolides and azalides. *Structure* (Camb.) **11**:329–338.
146. **Schlunzen, F., R. Zarivach, J. Harms, A. Bashan, A. Tocilj, R. Albrecht, A. Yonath, and F. Franceschi.** 2001. Structural basis for the interaction of antibiotics with the peptidyl transferase centre in eubacteria. *Nature* **413**:814–821.
147. **Schmitz, F. J., M. Perdikouli, A. Beeck, J. Verhoef, and A. C. Fluit.** 2001. Resistance to trimethoprim-sulfamethoxazole and modifications in genes coding for dihydrofolate reductase and dihydropteroate synthase in European *Streptococcus pneumoniae* isolates. *J. Antimicrob. Chemother.* **48**:935–936.
148. **Schneider, P., S. Hawser, and K. Islam.** 2003. Iclaprim, a novel diaminopyrimidine with potent activity on trimethoprim sensitive and resistant bacteria. *Bioorg. Med. Chem. Lett.* **13**:4217–4221.
149. **Schrag, S. J., L. McGee, C. G. Whitney, B. Beall, A. S. Craig, M. E. Choate, J. H. Jorgensen, R. R. Facklam, and K. P. Klugman.** 2004. Emergence of *Streptococcus pneumoniae* with very-high-level resistance to penicillin. *Antimicrob. Agents Chemother.* **48**:3016–3023.
150. **Seral, C., F. J. Castillo, C. Garcia, M. C. Rubio-Calvo, and R. Gomez-Lus.** 2000. Presence of conjugative transposon Tn*1545* in strains of *Streptococcus pneumoniae* with *mef*(A), *erm*(B), *tet*(M), *cat*pC194 and *aph3'*-III genes. *Enferm. Infecc. Microbiol. Clin.* **18**:506–511.
151. **Seral, C., F. J. Castillo, M. C. Rubio-Calvo, A. Fenoll, C. Garcia, and R. Gomez-Lus.** 2001. Distribution of resistance genes *tet*(M), *aph3'*-III, *cat*pC194 and the integrase gene of Tn*1545* in clinical *Streptococcus pneumoniae* harbouring *erm*(B) and *mef*(A) genes in Spain. *J. Antimicrob. Chemother.* **47**:863–866.
152. **Setchanova, L., and A. Tomasz.** 1999. Molecular characterization of penicillin-resistant *Streptococcus pneumoniae* isolates from Bulgaria. *J. Clin. Microbiol.* **37**:638–648.
153. **Shaw, K., and G. D. Wright.** 2000. Aminoglycoside resistance in gram-positive bacteria, p. 635–646. *In* V. Fischetti (ed.), *Gram-Positive Pathogens*. ASM Press, Washington, D.C.
154. **Shortridge, V. D., R. K. Flamm, N. Ramer, J. Beyer, and S. K. Tanaka.** 1996. Novel mechanism of macrolide resistance in *Streptococcus pneumoniae. Diagn. Microbiol. Infect. Dis.* **26**:73–78.
155. **Sibold, C., J. Henrichsen, A. Konig, C. Martin, L. Chalkley, and R. Hakenbeck.** 1994. Mosaic *pbpX* genes of major clones of penicillin-resistant *Streptococcus pneumoniae* have evolved from *pbpX* genes of a penicillin-sensitive *Streptococcus oralis. Mol. Microbiol.* **12**:1013–1023.
156. **Sibold, C., J. Wang, J. Henrichsen, and R. Hakenbeck.** 1992. Genetic relationships of penicillin-susceptible and -resistant *Streptococcus pneumoniae* strains isolated on different continents. *Infect. Immun.* **60**:4119–4126.
157. **Stadler, C., and M. Teuber.** 2002. The macrolide efflux genetic assembly of *Streptococcus pneumoniae* is present in erythromycin-resistant *Streptococcus salivarius. Antimicrob. Agents Chemother.* **46**:3690–3691.
158. **Sutcliffe, J.** 2001. MLS_BK Resistance Update, Presented at the 5th International Antibacterial Drug Discovery and Development Summit, Princeton, N.J.
159. **Sutcliffe, J., and R. Leclercq.** 2002. Mechanisms of resistance to macrolides, lincosamides, and ketolides, p. 281–317. *In* W. Schonfeld and H. A. Kirst (ed.), *Macrolide Antibiotics*. Birkhauser Verlag, Basel, Germany.
160. **Sutcliffe, J., A. Tait-Kamradt, and L. Wondrack.** 1996. *Streptococcus pneumoniae* and *Streptococcus pyogenes* resistant to macrolides but sensitive to clindamycin: a common resistance pattern mediated by an efflux system. *Antimicrob. Agents Chemother.* **40**:1817–1824.
161. **Syrogiannopoulos, G. A., I. N. Grivea, L. M. Ednie, B. Bozdogan, G. D. Katopodis, N. G. Beratis, T. A. Davies, and P. C. Appelbaum.** 2003. Antimicrobial susceptibility and macrolide resistance inducibility of *Streptococcus pneumoniae* carrying *erm*(A), *erm*(B), or *mef*(A). *Antimicrob. Agents Chemother.* **47**:2699–2702.
162. **Syrogiannopoulos, G. A., I. N. Grivea, A. Tait-Kamradt, G. D. Katopodis, N. G. Beratis, J. Sutcliffe, P. C. Appelbaum, and T. D. Davies.** 2000. Identification of *erm*(A) erythromycin resistance methylase gene in *Streptococcus pneumoniae* isolated in Greece. *Antimicrob. Agents Chemother.* **45**:342–344.
163. **Tait-Kamradt, A., J. Clancy, M. Cronan, F. Dib-Hajj, L. Wondrack, W. Yuan, and J. Sutcliffe.** 1997. *mefE* is necessary for the erythromycin-resistant M phenotype in *Streptococcus pneumoniae. Antimicrob. Agents Chemother.* **41**:2251–2255.
164. **Tait-Kamradt, A., T. Davies, P. C. Appelbaum, F. Depardieu, P. Courvalin, J. Petitpas, L. Wondrack, A. Walker, M. R. Jacobs, and J. Sutcliffe.** 2000. Two new mechanisms of macrolide resistance in clinical strains of *Streptococcus pneumoniae* from Eastern Europe and North America. *Antimicrob. Agents Chemother.* **44**:3395–3401.
165. **Takenouchi, T., F. Tabata, Y. Iwata, H. Hanzawa, M. Sugawara, and S. Ohya.** 1996. Hydrophilicity of quinolones is not an exclusive factor for decreased activity in efflux-mediated resistant mutants of *Staphylococcus aureus. Antimicrob. Agents Chemother.* **40**:1835–1842.
166. **Tenson, T., and M. Ehrenberg.** 2002. Regulatory nascent peptides in the ribosomal tunnel. *Cell* **108**:591–594.
167. **Tenson, T., M. Lovmar, and M. Ehrenberg.** 2003. The mechanism of action of macrolides, lincosamides and streptogramin B reveals the nascent peptide exit path in the ribosome. *J. Mol. Biol.* **330**:1005–1014.
168. **Thompson, J., M. O'Connor, J. A. Mills, and A. E. Dahlberg.** 2002. The protein synthesis inhibitors, oxazolidinones and chloramphenicol, cause extensive translational inaccuracy in vivo. *J. Mol. Biol.* **322**:273–279.
169. **Thornsberry, C., D. F. Sahm, L. J. Kelly, I. A. Critchley, M. E. Jones, A. T. Evangelista, and J. A. Karlowsky.** 2002. Regional trends in antimicrobial resistance among clinical isolates of *Streptococcus pneumoniae, Haemophilus influenzae,* and *Moraxella catarrhalis* in the United States: results from the TRUST Surveillance Program, 1999-2000. *Clin. Infect. Dis.* **34** (Suppl. 1):S4–S16.

170. **Tomasz, A., and W. Fischer.** 2000. The cell wall of *Streptococcus pneumoniae,* p. 191–200. *In* V. Fischetti (ed.), *Gram-Positive Pathogens.* ASM Press, Washington, D.C.
171. **Trieu-Cuot, P., and P. Courvalin.** 1983. Nucleotide sequence of the *Streptococcus faecalis* plasmid gene encoding the 3′5″-aminoglycoside phosphotransferase type III. *Gene* **23:**331–341.
172. **Trieu-Cuot, P., C. Poyart-Salmeron, C. Carlier, and P. Courvalin.** 1990. Nucleotide sequence of the erythromycin resistance gene of the conjugative transposon Tn*1545*. *Nucleic Acids Res.* **18:**3660.
173. **Tsai, F. T., O. M. Singh, T. Skarzynski, A. J. Wonacott, S. Weston, A. Tucker, R. A. Pauptit, A. L. Breeze, J. P. Poyser, R. O'Brien, J. E. Ladbury, and D. B. Wigley.** 1997. The high-resolution crystal structure of a 24-kDa gyrase B fragment from *E. coli* complexed with one of the most potent coumarin inhibitors, clorobiocin. *Proteins* **28:**41–52.
174. **Vester, B., and S. Douthwaite.** 2001. Macrolide resistance conferred by base substitutions in 23S rRNA. *Antimicrob. Agents Chemother.* **45:**1–12.
175. **Vester, B., and R. A. Garrett.** 1988. The importance of highly conserved nucleotides in the binding region of chloramphenicol at the peptidyl transfer centre of *Escherichia coli* 23S ribosomal RNA. *EMBO J.* **7:**3577–3587.
176. **Weigel, L. M., G. J. Anderson, R. R. Facklam, and F. C. Tenover.** 2001. Genetic analyses of mutations contributing to fluoroquinolone resistance in clinical isolates of *Streptococcus pneumoniae. Antimicrob. Agents Chemother.* **45:**3517–3523.
177. **Whitney, C. G., M. M. Farley, J. Hadler, L. H. Harrison, N. M. Bennett, R. Lynfield, A. Reingold, P. R. Cieslak, T. Pilishvili, D. Jackson, R. R. Facklam, J. H. Jorgensen, and A. Schuchat.** 2003. Decline in invasive pneumococcal disease after the introduction of protein-polysaccharide conjugate vaccine. *N. Engl. J. Med.* **348:**1737–1746.
178. **Whitney, C. G., M. M. Farley, J. Hadler, L. H. Harrison, C. Lexau, A. Reingold, L. Lefkowitz, P. R. Cieslak, M. Cetron, E. R. Zell, J. H. Jorgensen, and A. Schuchat.** 2000. Increasing prevalence of multidrug-resistant *Streptococcus pneumoniae* in the United States. *N. Engl. J. Med.* **343:**1917–1924.
179. **Widdowson, C. A., P. V. Adrian, and K. P. Klugman.** 2000. Acquisition of chloramphenicol resistance by the linearization and integration of the entire staphylococcal plasmid pC194 into the chromosome of *Streptococcus pneumoniae. Antimicrob. Agents Chemother.* **44:**393–395.
180. **Widdowson, C. A., and K. P. Klugman.** 1999. Molecular mechanisms of resistance to commonly used non-betalactam drugs in *Streptococcus pneumoniae. Semin. Respir. Infect.* **14:**255–268.
181. **Widdowson, C. A., K. P. Klugman, and D. Hanslo.** 1996. Identification of the tetracycline resistance gene, *tet*(O), in *Streptococcus pneumoniae. Antimicrob. Agents Chemother.* **40:**2891–2893.
182. **Wigley, D. B., G. J. Davies, E. J. Dodson, A. Maxwell, and G. Dodson.** 1991. Crystal structure of an N-terminal fragment of the DNA gyrase B protein. *Nature* **351:**624–629.
183. **Willmott, C. J., and A. Maxwell.** 1993. A single point mutation in the DNA gyrase A protein greatly reduces binding of fluoroquinolones to the gyrase-DNA complex. *Antimicrob. Agents Chemother.* **37:**126–127.
184. **Yother, J.** 2000. Genetics of *Streptococcus pneumoniae,* p. 232–243. *In* V. Fischetti (ed.), *Gram-Positive Pathogens.* ASM Press, Washington, D.C.

Frontiers in Antimicrobial Resistance: a Tribute to Stuart B. Levy
Edited by D. G. White, M. N. Alekshun, and P. F. McDermott

Chapter 24

Helicobacter and *Campylobacter*

PATRICK F. MCDERMOTT AND DIANE E. TAYLOR

Helicobacter and *Campylobacter* are related organisms that colonize and cause disease in the gastrointestinal tract. The genus *Campylobacter* was proposed in 1963 to differentiate *Vibrio fetus* from other less related *Vibrio* species. Originally named *Campylobacter pylori, Helicobacter* was distinguished as a separate genus in 1989 based on structural and biochemical features.

Helicobacter pylori is a major cause of gastroduodenal diseases, including ulcers and gastric cancer. Its discovery has led to use of eradication therapy to eliminate *H. pylori* from the stomach. Triple therapy, comprising two antimicrobials (clarithromycin and amoxicillin) and a proton pump inhibitor, is recommended as standard therapy (40). If this treatment fails, possibly because of antimicrobial resistance, other regimens include metronidazole, tetracycline, rifampin or, more rarely, fluoroquinolones. The resistance mechanisms described in *H. pylori* for these antimicrobials will be outlined in the following sections.

Campylobacter jejuni is a leading cause of bacterial gastroenteritis, a disease characterized by diarrhea, abdominal pain and fever. *Campylobacter* enteritis is usually self-limiting, with infrequent cases of severe or invasive illness for which antimicrobial therapy may be required. Most *Campylobacter* infections are sporadic single cases resulting from the consumption of contaminated food, milk, or water. Undercooked or mishandled poultry appear to be the most important sources of infection (22, 37). Large outbreaks are rare and have been linked to ingestion of contaminated milk and water (22). Even though the infective dose for humans has been estimated as low as 500 bacterial cells, foodborne outbreaks are not common. This is likely due to the fact that *Campylobacter* do not survive long when exposed to air and drying.

Treatment with erythromycin is effective early in the course of infection (97), and is currently considered the drug of choice for treating culture-confirmed cases (5). However, because *Campylobacter* gastroenteritis symptoms are indistinguishable from salmonellosis, a fluoroquinolone is often given pending (or in the absence of) culture results, and it is the treatment of choice for traveler's diarrhea (33). In some countries, tetracycline or doxycycline have been used as alternative therapeutics. Gentamicin, meropenem, clindamycin, telithromycin and azithromycin all show potent in vitro activity, and may prove valuable for controlling infections. The major resistance mechanisms are highlighted.

ANTIMICROBIAL SUSCEPTIBILITY TESTING

Campylobacter

While much has been reported on the resistance of *Campylobacter* to different antimicrobial agents, intralaboratory comparison of data has been hampered by the lack of a standardized and validated susceptibility testing method. Given the fastidious growth requirements of *Campylobacter,* previously established quality control (QC) organisms and testing methods are not suitable for testing bacteria in this genus (35). To address the need for a *Campylobacter* method, the Clinical and Laboratory Standards Institute (CLSI, formerly NCCLS) recently accepted an agar dilution method using *C. jejuni* ATCC 33560 as the QC organism, and QC ranges for five antimicrobials (59). Based on this reference method, a broth microdilution method was subsequently developed, with QC ranges for fourteen antimicrobials. The method involves testing on Mueller-Hinton broth supplemented with 2 to 5% lysed horse blood, at 36 + 1°C for

Patrick F. McDermott • Office of Research, Center for Veterinary Medicine, U.S. Food and Drug Administration, 8401 Muirkirk Road, Laurel, MD 20708. **Diane E. Taylor** • Department of Medical Microbiology & Immunology, 1-28 Medical Sciences Bldg., University of Alberta, Edmonton, T6G 2 H7, Alberta, Canada.

48 h and 42°C for 24 h in an humid atmosphere of 10% CO_2, 5% O_2, 85% N_2. It is essential that testing be done in a stable environment: incubation temperatures of 35 and 43°C both result in many instances of no growth. In addition, sealable plastic bags used in primary isolation are not adequate for maintaining the atmosphere. An incubator with a pressurized gas supply, or sealed plastic containers with Oxoid or Remel gas-generating sachets are preferable. This method shows very good day-to-day reproducibility within and between laboratories and should facilitate surveillance and laboratory testing of these organisms. The details of the broth method will be published in the CLSI document M100-S16.

Even with a robust testing method, there are currently no validated breakpoints for interpreting susceptibility testing results for *Campylobacter.* Establishing such interpretive criteria requires a large amount of data on the population distribution of MICs, information of pharmacokinetic/pharmacodynamic properties at the site of infection, and clinical outcome studies. Lacking clinical trial data, the CLSI is making an effort to set tentative breakpoints based mainly on properties of the bacterium. Similar approaches have been taken by other standards organizations. For example, the British Society for Antimicrobial Chemotherapy (BSAC) and the Comite de L'Antibiogramme de la Societe Francaise de Microbiologie (http://www.sfm.asso.fr/), among others, have proposed interpretative criteria for organisms belonging to this genus. These and other resistance breakpoints (15) are based largely on the population distribution of MICs. Many reports use the interpretative criteria generated for the *Enterobacteriaceae,* or in the case of erythromycin, those established for *Staphylococcus* spp. For ciprofloxacin, the use of 1 μg/ml as the resistant breakpoint is logical, given the pharmacokinetics of this class of agents, along with the broad bimodal nature of fluoroquinolone MICs (11, 60). For other antimicrobials, the appropriate breakpoints are less clear. Pending more complete information, current tentative interpretative criteria should be used with caution.

Helicobacter

The current method for *Helicobacter* susceptibility testing is based on agar dilution (13). The method involves testing on Mueller-Hinton agar containing 5% sheep blood and incubation for 72 h in a microaerobic atmosphere at 35 ± 2°C. *H. pylori* ATCC 43504 is the QC organism. Currently, there are QC ranges for amoxicillin, clarithromycin, metronidazole, telithromycin and tetracycline. Interpretive criteria are established only for clarithromycin (resistant MIC breakpoint, ≥1 μg/ml) (13).

ANTIMICROBIAL RESISTANCE MECHANISMS IN *HELICOBACTER PYLORI*

Clarithromycin Resistance

Erythromycin (Ery) and clarithromycin (Cla) are macrolide antimicrobials which bind to the 50S subunit of bacterial ribosomes and interfere with protein synthesis by inhibiting the elongation of peptide chains (96). Point mutations in the 23S rRNA gene were associated with clarithromycin resistance (Cla^r) in clinical isolates of *H. pylori* (90). Substitutions of adenine (A) to guanine (G) at either of two nucleotides (corresponding to positions 2058 and 2059 using *Escherichia coli* coordinates) within domain IV of the peptidyl transferase loop of 23S rRNA are responsible for Cla^r. The positions of the two nucleotides were demonstrated to be 2142 and 2143 in *H. pylori* coordinates, based on the determination of the transcription start site of *H. pylori* 23S rRNA (83).

Binding experiments using ^{14}C-labelled erythromycin demonstrated loss of drug binding to ribosomes of Cla^R *H. pylori* strains compared to susceptible parental strains (67). Point mutations in 23SrRNA likely alter the ribosome structure to inhibit binding of macrolide antimicrobials.

Usually mutations at position 2142 confer a higher level of resistance than those at 2143, with Cla MICs of 1 to 4 μg/ml for A2143G and 8 to 16 μg/ml for A2142G compared with 0.004 μg/ml for susceptible isolates (92). Moreover, mutations at 2142 were associated with high level cross resistance to macrolide, lincosamide and streptogramin B (MLS phenotype). In contrast, mutations at position 2143 conferred an intermediate level resistance to macrolides and lincomycin but lacked resistance to streptogramin B (92).

H. pylori possess two copies of 23SrRNA genes (83), which was confirmed by whole genome sequencing (3, 86). Clinical isolates, as well as in vitro created mutants, usually contain mutations in both gene copies (homozygous) (92), probably reflecting a high efficiency of DNA recombination in *H. pylori.* Natural transformation of Ery^R using the mutated 23S rDNA genes, followed by homologous recombination into the *H. pylori* genome, has been used extensively to study Ery^r (83, 92, 93).

While A-to-G transitions were observed in the majority of Cla^r clinical isolates, a few examples of A-to-C transversions at position A2142C have been described (80). No A-to-T mutations have ever been identified. In vitro site directed mutagenesis (16, 92) demonstrated that A-to-C mutations confer similar levels of resistance compared to A-to-G mutations, whereas A2142-to-T mutation also confers an intermediate level of resistance. Further examination of the in vitro site-directed mutants demonstrated that

A-to-G mutants grow significantly faster that A-to C or A-to T mutants, no matter whether the transition is at position 2142 or 2143. Moreover, multiplex sequence analysis demonstrated a quantitative competitive growth advantage of the A-to-G mutants over the A-to-C mutants (91). The competitive growth advantage was determined to be in the following order: A2142G>A2143G>>>A2142C>A2143C(A2142T), thus providing a rational observation for the mutation pattern observed in clinical isolates.

Amoxicillin Resistance

Amoxicillin (Amx) is a beta-lactam antimicrobial with MICs for *H. pylori* that range from <0.01 to 0.1 μg/ml (84). Amoxicillin inhibits cell wall synthesis, with resulting lysis of replicating bacterial cells. Amoxicillin resistance in *H. pylori* is rare, although it is very common in other bacterial species, frequently depending on the production of β-lactamases.

Studies on high level amoxicillin resistance (Amx^r) in a clinical isolate of *H. pylori* from Holland (MIC, 8 μg/ml) demonstrated that another resistance mechanism, modification of the bacterial cell wall by replacement of a penicillin binding protein (PBP), called PBP 1A (27), was responsible. In another study Amx^r strains of *H. pylori* were selected in vitro with MICs of 4 to 8 μg/ml. Comparative binding studies of PBP 1 for several beta-lactams demonstrated that PBP 1 was altered, and that the reduced uptake of bata-lactam antimicrobials could be selected for by prolonged exposure to Amx (17).

More recently, four clinical isolates from Italy with high level Amx^r (MIC, 64 μg/ml) were associated with alterations in PBP 1A, showing 36 nucleotides mutations resulting in 10 amino acid changes in the C-terminal portion of the protein (the putative-penicillin binding domain). These strains were associated with resistance to several other antimicrobials, some of which (tetracycline and chloramphenicol resistance) could also be acquired by natural transformation of Amx^r. This led the authors to propose that multidrug resistance depended on a combination of PBP 1A alteration and decreased membrane permeability (45).

Tetracycline Resistance

In an early report of tetracycline resistance (Tc^r) *H. pylori* from Australia (62) two isolates were cultured from a patient who failed therapy with colloidal bismuth subcitrate, metronidazole and tetracycline (Tc). These isolates were highly resistant to Tc with MICs of ≥258 μg/ml (antrum isolate − 108) and 32 μg/ml (corpus isolate − 109) compared with an MIC of 0.5 μg/ml for *H. pylori* 26695. Attempts to characterize these Tc^r *H. pylori* by shotgun cloning, DNA-DNA hybridization and PCR screening identified no evidence of the well-known Tc resistance determinates: efflux (see chapter 1) or ribosomal protection proteins (see chapter 2). Subsequently P. D. Midolo and J. Rood kindly sent DET strains 108 and 109 [see reference (88)], so that additional studies on Tc^r could be carried out.

Previously, a rare mechanism of Tc^r had identified a mutation in 16S rRNA in several clinical isolates of *Propionibacterium acnes* (a causative agent of acne vulgaris) in which was found a G-to-C mutation at position 1058 (*Escherichia coli* numbering) (77). This mutation facilitated a variable level of Tc^r with an MIC ranging from 2 to 64 μg/ml, compared with 0.12 to 1 μg/ml for susceptible *P. acnes*. The DNA sequence of the 16S rRNA genes from the two Tc^r isolates from Australia were compared with the sequences of 16S rRNA genes from susceptible isolates, including the two *H. pylori* genomes of 26695 and J99 which had already been sequenced (3, 86). Using sequence comparisons several mutations were identified in *H. pylori* strains 108 and 109, which both exhibited MICs of 64 μg/ml. Mutagenesis of specific 16S rRNA genes using recombinant PCR techniques, in combination with natural transformation of *H. pylori* DNA from Tc^r resistant strains, were used to identify the particular mutations involved in Tc^r for *H. pylori*. A triple mutation AGA965-967TTC, present in both 16S rRNA copies in strains 108 and 109, was solely responsible for Tc^r (88).

Tetracycline inhibits protein synthesis by binding to the 30S subunit of the ribosome and blocks the binding of aminoacyl-tRNA, thus stalling the synthesis of the growing peptide chains. Crystal structures of the 30S ribosome have identified one primary binding site and one to five weaker secondary binding sites (9, 71). The mutations AGA 965-967 UUC (h31 loop) are located in domain III of 16S rRNA and overlap both the primary tetracycline binding site and part of the P site. It has been suggested that the substitution of the middle G residue is primarily responsible for resistance, although DNA transformation experiments with various mutant DNA sequences from *H. pylori* did not confirm this hypothesis (26, 88). The most likely explanation for Tc^r is the loss or weakening of tetracycline binding due to alteration of the functional binding site. A decrease in the affinity of the ribosome for binding likely increases the ability of the rRNA to compete for binding.

The link between Tc^r and the mutation from the wild type AGA at positions 965-967 to TTC, in the two 16S rRNA genes of *H. pylori* was confirmed by Gerrits et al. (27) and by Dalidiene et al. (14). Although the later study used the 965-967 nomenclature, equivalent to that used in *E. coli,* the former study renumbered the mutations as AGA 926-928 TTC as the authors considered this conformed more closely to the alignment of bases in *H. pylori* 16S rRNA genes. We have chosen to continue to use the *E. coli* nomenclature here because, in subsequent work, we have introduced the *H. pylori* mutations into *E. coli* 16S rRNA genes to study tetracycline binding to *E. coli* ribosomes (see below).

To date, strains from Australia (88), The Netherlands (26), El Salvador, Lithuania (14) and Brazil (76), carrying high levels of Tc^r, contained the AGA956-TTC967 triplet mutation in the 16S rRNA genes. Nevertheless the tetracycline MICs for all these strains covers a wide range, which has not been clearly explained. The original Australian *H. pylori* strain 108 early after isolation, specified higher level resistance than that observed later (256 μg/ml compared with 64 μg/ml). Some transformants of strain 26695 prepared using 109 16S rDNA exhibited MICs of 4 to 16 μg/ml and this was attributed to a situation where only one copy of the 16S rRNA gene had been subjected to homologous recombination. The resulting mixture of susceptible wild-type and Tc^r mutant ribosomes could then account for intermediate levels of Tc^r (88). This explanation does not account for the lower level of resistance observed in other studies. Dailidiene et al. (14) have suggested that the range of Tc^r MICs in *H. pylori* may result from accumulations of changes that could "affect tetracycline-ribosome affinity or other functions (perhaps porins or efflux pumps)." However, other studies have shown no evidence for efflux systems (88) and excluded a variety of other genes that could be involved in Tc^r (26). In none of the studies of *H. pylori* was a mutation similar to that observed in *P. acnes* (G1058C) found, yet this mutation in *P. acnes* also produced a wide range of Tc^r levels from 2 to 64 μg/ml (77). It is possible that the interaction of tetracycline with the ribosome, which is modified by these 16S rRNA mutations, leads to unstable interactions, and these may influence the MIC values, perhaps depending on growth medium or other variables.

Site-directed mutagenesis in *H. pylori* using limited (seven) examples of substitutions within the triplet mutation suggested that single-and double-base-pair mutations mediate only low level Tc^r (MIC, 1 to 2 μg/ml) and also decreased growth rates in the presence of tetracycline (26). This study offers a possible explanation for the prevalence of the TTC mutation observed in clinical Tc^r isolates of *H. pylori.*

In a larger study, 32 different mutations at positions 965-967 were introduced into 16S rDNA in a strain of *E. coli* containing only a single copy of 16S rDNA on a multicopy plasmid (6). Working in *E. coli,* the majority of strains with equivalent to three changes at AGA 965-967 encoded a higher MIC of 8 μg/ml, compared with 0.25 μg/ml for AGA, although a few strains had MICs of 1 or 2 μg/ml. No obvious pattern emerged which could relate any particular nucleotide change to the increased level of Tc^r. It is of interest that *H. pylori* wild-type AGA specifies an MIC of 0.25 μg/ml in *E. coli,* whereas *E. coli* wild type is TGC, which specifies an MIC of 2 μg/ml (65). In contrast to Gerrits et al. (26) we found no difference in growth rate in the presence of tetracycline for *E. coli* strains with one or two mutations at 965-967. Perhaps *H. pylori* growth with these mutations responds differently from E. coli to tetracycline. Further studies are necessary to understand exactly how the mutations found in the Tc^r *H. pylori* are selected in vivo and to understand why such a wide variation in MIC is possible in *H. pylori* isolates which contain similar 16S rRNA mutations.

Metronidazole Resistance

Metronidazole (Mtz) was used in early combination therapy to eradicate *H. pylori,* although the MIC of susceptible isolates varies from 0.1 to 4 μg/ml. Mtz is a pro-drug and must be reduced in the stomach to a hydroxylamine derivative which damages DNA and appears to cause cell death by nicking DNA. Several mechanisms have been proposed to explain Mtz^r in *H. pylori.* Goodwin et al. (32) convincingly demonstrated that mutations in *rdxA,* which encodes an oxygen-insensitive NADPH nitroreductase, resulted in Mtz^r. Mutations in *rdxA* were confirmed by others to correlate with Mtz^r in *H. pylori* (42, 82). Mutations occurred spontaneously throughout the *rdxA* gene, resulting in MICs of 16 to 128 μg/ml (93). In *H. pylori* clinical isolates there is marked heterogeneity in the MIC of Mtz ranging from 8 to 256 μg/ml, implicating the involvement of additional genes.

Kwon et al. (46) and Jeong et al. (43) independently demonstrated that *frxA,* which codes for an NAD(P)H-flavin oxidoreductase, that is a paralog of *rdxA,* can also be involved in Mtz^r. They showed that inactivation of *rdxA* alone resulted in moderate Mtz^r (MIC, 16 to 32 μg/ml), whereas double mutants in *rdxA* and *frxA* conferred higher level resistance (MIC > 64 μg/ml). There is still controversy as to the roles of *rdxA* and *frxA* in Mtz^r (61). Moreover the *recA* gene (12) and the *fdxB* (encoding a ferredoxin-like protein) (46) have also been implicated in Mtz^r in *H. pylori.*

Rifampin Resistance

Rifampin (Rif^r) in *E. coli* and other bacteria, including *H. pylori,* results from the mutation of the *rpoB* gene, which encodes the β subunit of RNA polymerase to which rifampin binds (38, 39, 93). The MIC of spontaneous Rif^r mutations range from 32 to 256 μg/ml, whereas the MIC for the wild-type *H. pylori* strain was 0.1 μg/ml (93). Mutations were found at many positions in the *rpoB* gene. As yet Rif^r has been reported very infrequently in *H. pylori* (39), as it has been used in a limited number of patients.

Fluoroquinolone Resistance

Resistance to the fluoroquinolone ciprofloxacin (Cip) was acquired very rapidly by *H. pylori* isolates in a clinical trial in which ciprofloxacin was used to eradicate the microorganism (64). Since resistance to fluoroquinolones has also been acquired easily by *Campylobacter jejuni*, a close relative of *H. pylori* (19), this group of antimicrobials should be used with caution for *H. pylori* eradication therapy.

Cip resistance (Cip^r) is due to a mutation in the gene (*gyrA*) which encodes the A subunit of DNA gyrase (64, 93). This enzyme, which introduces negative superhelical turns into DNA, and is required both for DNA replication and RNA transcription, consists of two A and two B subunits. A quinolone resistance-determining region (QRDR) is located at the amino-terminus (amino acid residue 67-106) of the GyrA protein. In Cip^r *H. pylori* isolates, several types of base substitutes resulting in a single amino acid change in the QRDR of *gyrA*, were associated with an increase in MIC of Cip from <0.25 μg/ml 4 to 8 μg/ml, with cross-resistance to other fluoroquinolones (93).

ANTIMICROBIAL RESISTANCE IN *CAMPYLOBACTER*

Ciprofloxacin Resistance

While *Campylobacter* are generally considered susceptible to most antimicrobials of clinical importance, multidrug resistance is becoming more common. Of particular concern is the increased incidence of fluoroquinolone-resistant *C. jejuni* in recent years (34). This rise has been attributed in part to the use of fluoroquinolones in poultry medicine. Endtz et al. reported that the emergence of fluoroquinolone resistance in human *C. jejuni* infections in The Netherlands coincided with the approval of the veterinary fluoroquinolone, enrofloxacin, in poultry in 1987 (18). In Minnesota from 1992-1998, the number of quinolone-resistant infections increased from 1.3 to 10.2%. Two years after approval of the poultry fluoroquinolone sarafloxacin in 1995, Cip^r among *Campylobacter* in Minnesota doubled. Part of this increase was attributed to the acquisition of resistant strains from poultry meats (79). Among human *Campylobacter* isolates submitted to the CDC, ciprofloxacin resistance rose from 0% in 1989-1990 (34) to 21% in 2002 (11). The associations of Cip^r *Campylobacter* in humans to selection in the poultry production environment prompted the Food and Drug Administration to propose withdrawing approval of the new animal drug applications for use of the fluoroquinolones in poultry (21).

In food-producing animals such as cattle, poultry and swine, fecal *C. jejuni/C. coli* is regarded as a commensal organism. *C. jejuni* is routinely recovered from retail chicken products and poultry processing plants. Recent data examining retail chicken meats from around the U.S. indicate that ca. 57% of chicken breast samples were contaminated with *Campylobacter*, of which 13.8% were Cip^r (MIC, ≥4 μg/ml)(20), whereas *E. coli* and *Salmonella* from the same samples are largely susceptible. This can be explained in part by the fact that fluoroquinolone resistance in *Campylobacter* results from a single topoisomerase mutation, similar to that seen in *H. pylori*, and unlike *Salmonella* and *E. coli*, in which two mutations are considered necessary (see chapter 4).

As with other bacteria, fluoroquinolone resistance is due mainly to point mutations in DNA gyrase (*gyrA*). The most common mutation associated with high-level ciprofloxacin MICs (≥32 μg/ml) is a substitution of Ile for Thr at position 86 (homologous to Ser83 in *E. coli*) (4, 8, 36, 94). *gyrA* mutations conferring intermediate ciprofloxacin MICs (1 to 4 μg/ml) include Asp90 and Ala70 (55, 94). Additional *gyrA* mutations have been reported in resistant isolates (25), but their respective contributions to quinolone resistance have not been measured. No mutations in *gyrB* have been associated with FQ resistance, and *Campylobacter* lacks *parC* (68). Recently, Luo et al. (55) reported the interesting observation that fluoroquinolone-resistant *Campylobacter* with a Thr86Ile (C257T) mutation in *gyrA* exhibited a fitness advantage in the chicken intestine over isogenic susceptible strains. The requirement of only a single base change for high-level MICs, and the competitive advantage of these mutants in vivo, may help explain the rapid evolution of fluoroquinolone resistance in chickens exposed to sarafloxacin or enrofloxacin (56, 60, 89) and the widespread nature of Cip^r in retail raw meats (20) and human clinical isolates (11).

Multidrug efflux pumps are also operative in *Campylobacter* and contribute to fluoroquinolone susceptibility. Wild type susceptible isolates of *Campylobacter* display higher ciprofloxacin MICs (0.125 to 0.5 μg/ml) relative to other gram-negative enterics such as *E. coli* and *Salmonella* (MIC, 0.015 to 0.06 μg/ml). This intrinsic resistance appears to result from the constitutively-expressed efflux pump, *cmeB* (52, 73). When *cmeB* was inactivated by site-directed mutagenesis, ciprofloxacin MICs dropped to levels in the range for *E. coli* and *Salmonella* (52, 73). Expression of *cmeB,* perhaps *cmeF* as well as other loci (74), also contributes to acquired quinolone/multidrug resistance. Inactivation of the *cmeABC* operon in resistant clinical isolates decreased ciprofloxacin MICs near to that of wild-type isolates (56), showing that *cmeABC* is required to maintain acquired high-level fluoroquinolone MICs in *Campylobacter.* In addition to multiple antimicrobial resistance, *cmeABC* also confers bile resistance and is consequently required for intestinal colonization in chickens (53).

Macrolide Resistance

Macrolide resistance in *C. jejuni* is relatively rare in the U.S., with about 1% of human *C. jejuni* isolates showing erythromycin MICs $\geq$ 8μg/ml (11); however, increasing resistance has been reported in some regions (75). Resistance is generally higher in isolates from swine and poultry production environments (1), where macrolides are used routinely. It is also generally higher in *C. coli* than *C. jejuni* (2).

While macrolide resistance is incompletely understood in *Campylobacter,* most resistance appears to result from mutations in the ribosome at positions in the peptidyl transferase region of the *C. jejuni* 23S rRNA genes, corresponding in *E. coli* to nucleotide positions 2058 and 2059 (19), with a A2059G transition being most common. Resistance is generally higher in *C. coli,* where G2230A and T2268C base substitution have been reported (25). Ribosomal mutations in *Campylobacter* can confer cross-resistance to azithromycin and clarithromycin.

Although there are no known extra-chromosomal macrolide resistance determinants in *C. jejuni* (98), there is evidence of a role for efflux in erythromycin resistance (57, 69). The exposure of *Campylobacter* to an efflux pump inhibitor increased erythromycin susceptibility to wild type levels in intermediately-susceptible strains and to a lesser degree in resistant strains (57). In addition, inactivation of *cmeABC* yielded a 4- to 16-fold reduction in erythromycin MICs in wild-type susceptible strains (52, 74).

Tetracycline Resistance

Worldwide, Tc^r is frequent among human isolates of *C. jejuni.* In Canada, Tc^r has increased from approximately 7 to 9% in 1980-1981 (23) to 43 to 68% in 1998-2001 (24), with more recent resistant strains also showing higher MICs (31). In the U.S. from 1997 to 2002, tetracycline resistance has ranged from 38 to 47% (11), with comparable levels in Finnish isolates (36). In other regions, the proportion of resistant isolates is much higher (51, 78).

Tc^r is due to ribosomal protection mediated by the *tet*(O) gene (58), which confers resistance by blocking the primary tetracycline ribosomal binding site (87). *tet*(O) is widespread in *Campylobacter* worldwide, and no other *tet* alleles genes have been found to confer resistance. *tet*(O) is usually plasmid-borne (85) but may be located on the chromosome (72). *tet*(O) alleles in *C. jejuni* usually impart tetracycline resistance levels ranging from 32 μg/ml to 128 μg/ml; however, mutations in *tet*(O) can lead to MICs as high as 512 μg/ml (31).

Recently, two large self-transmissible Tc^r plasmids have been sequenced (7), one from *C. jejuni* and one from *C. coli,* isolated on separate continents about 20 years apart. Both plasmids had mosaic sequence structures, with gene signatures suggesting origins in various commensal and pathogenic bacteria, including *H. pylori.* Remarkably, the two plasmids were 94.3% identical at DNA sequence level and are widespread in plasmid-containing, Tc^r *Campylobacter* isolates. Other plasmid vehicles, ranging in size up to 100 kb, also carry Tc^r determinants (85).

Aminoglycoside Resistance

Aminoglycoside resistance is fairly well characterized at the gene level in various organisms. In *Campylobacter,* kanamycin resistance has been associated with the presence of the *aphA-3* gene (49), usually located on large plasmids (40 to >100 kb) that often carry *tet*(O) as well (30). Integrons also have been identified in *Campylobacter* (50, 54), which in one report were found to be common (16.4%) in isolates from different sources and to contain the aminoglycoside-modifying enzyme encoded by *aadA2* (66). In addition, spectinomycin/streptomycin resistance due to adenylyltranferases encoded by *aadA* and *aadE* has been identified with plasmids from human clinical isolates (70). Resistance to the aminoglycoside streptothricin has been linked to the *sat4* gene product in animal and clinical isolates from Europe (41). This agent has been used as an animal feed additive in Germany but is not used in the U.S.

Gentamicin has been recommended for the treatment of campylobacteriosis. Resistance to gentamicin is very rare in *Campylobacter* and has not been characterized extensively at the genetic level. In isolates recovered from a poultry production house, integrons carrying the *aacA-4* gene were detected in isolates resistant to tobramycin and gentamicin (50). It is not known if this mechanism is present in human clinical isolates.

Beta-Lactam Resistance

β-Lactamases are very common in *Campylobacter,* with over 80% of *C. jejuni* strains being producers (81); however, phenotypic resistance is variable. *Campylobacter* generally respond poorly to narrow-spectrum β-lactams, independent of β-lactamase production, presumably due to low-affinity PBPs. Early studies indicate that *C. jejuni* are inherently resistant to cefamandole, cefoxitin and cefoperazone. For this reason, cefoperazone is incorporated into some selective media used in primary isolation (44). Most isolates also are resistant to cephalothin and cefazolin, and resistance is variable for cefotaxime, moxalactam, piperacillin and ticarcillin (48). The most active beta-lactam agents include ampicillin, amoxicillin, cefpirome and imipenem (81). Meropenem also shows good activity against *Campylobacter* (47) and has been recommended as a treatment option (10, 63).

Other Resistances

Campylobacter are generally resistant to trimethoprim and sulfonamides, through mechanisms common to other bacteria. Trimethoprim resistance in *C. jejuni* is due to the chromosomal presence of acquired trimethoprim-resistant dihydrofolate reductase gene cassettes (*dfr1, dfr9*) (29). These elements were found associated with degenerate integron and transposon sequences, respectively. Sulfonamide resistance in *Campylobacter,* as in other organisms, results from mutations in dihydropteroate synthase (28). Chloramphenicol resistance is rare in *Campylobacter* and results from acetylase activity encoded by *cat* genes similar to those in the Clostridia (95). To our knowledge, no other genes for trimethoprim, sulfonamide or chloramphenicol resistance have been characterized to date.

SUMMARY

While *Campylobacter* and *Helicobacter* have been recognized as human pathogens for over 100 years, only recent microbiological advances have made detailed study of these fastidious pathogens possible. Much more work remains to be done to understand the evolution of resistance and the genetic bases underlying acquired and intrinsic resistance and to more fully investigate potential new therapies for clinical use.

REFERENCES

1. **Aarestrup, F. M., and J. Engberg.** 2001. Antimicrobial resistance of thermophilic *Campylobacter. Vet. Res.* **32:**311–321.
2. **Aarestrup, F. M., E. M. Nielsen, M. Madsen, and J. Engberg.** 1997. Antimicrobial susceptibility patterns of thermophilic *Campylobacter* spp. from humans, pigs, cattle, and broilers in Denmark. *Antimicrob. Agents Chemother.* **41:**2244–2250.
3. **Alm, R. A., L. S. Ling, D. T. Moir, B. L. King, E. D. Brown, P. C. Doig, D. R. Smith, B. Noonan, B. C. Guild, B. L. deJonge, G. Carmel, P. J. Tummino, A. Caruso, M. Uria-Nickelsen, D. M. Mills, C. Ives, R. Gibson, D. Merberg, S. D. Mills, Q. Jiang, D. E. Taylor, G. F. Vovis, and T. J. Trust.** 1999. Genomic-sequence comparison of two unrelated isolates of the human gastric pathogen *Helicobacter pylori. Nature* **397:**176–180.
4. **Alonso, R., E. Mateo, C. Girbau, E. Churruca, I. Martinez, and A. Fernandez-Astorga.** 2004. PCR-restriction fragment length polymorphism assay for detection of *gyrA* mutations associated with fluoroquinolone resistance in *Campylobacter coli. Antimicrob. Agents Chemother.* **48:**4886–4888.
5. **Anonymous.** 2003. The Sanford Guide to Antimicrobial Therapy. Antimicrobial Therapy, Inc., Hyde Park, Vt.
6. **Asai, T., C. Condon, J. Voulgaris, D. Zaporojets, B. Shen, M. Al Omar, C. Squires, and C. L. Squires.** 1999. Construction and initial characterization of *Escherichia coli* strains with few or no intact chromosomal rRNA operons. *J. Bacteriol.* **181:**3803–3809.
7. **Batchelor, R. A., B. M. Pearson, L. M. Friis, P. Guerry, and J. M. Wells.** 2004. Nucleotide sequences and comparison of two large conjugative plasmids from different *Campylobacter* species. *Microbiology* **150:**3507–3517.
8. **Beckmann, L., M. Muller, P. Luber, C. Schrader, E. Bartelt, and G. Klein.** 2004. Analysis of *gyrA* mutations in quinolone-resistant and -susceptible *Campylobacter jejuni* isolates from retail poultry and human clinical isolates by non-radioactive single-strand conformation polymorphism analysis and DNA sequencing. *J. Appl. Microbiol.* **96:**1040–1047.
9. **Brodersen, D. E., W. M. Clemons, Jr., A. P. Carter, R. J. Morgan-Warren, B. T. Wimberly, and V. Ramakrishnan.** 2000. The structural basis for the action of the antibiotics tetracycline, pactamycin, and hygromycin B on the 30S ribosomal subunit. *Cell* **103:**1143–1154.
10. **Burch, K. L., K. Saeed, A. D. Sails, and P. A. Wright.** 1999. Successful treatment by meropenem of *Campylobacter jejuni* meningitis in a chronic alcoholic following neurosurgery. *J. Infect.* **39:**241–243.
11. **CDC.** 2002 National Antimicrobial Resistance Monitoring System (NARMS) For Enteric Bacteria. Available at: http://www.cdc.gov/ncidod/dbmd/narms/. NARMS.2005.
12. **Chang, K. C., S. W. Ho, J. C. Yang, and J. T. Wang.** 1997. Isolation of a genetic locus associated with metronidazole resistance in *Helicobacter pylori. Biochem. Biophys. Res. Commun.* **236:**785–788.
13. **Clinical and Laboratory Standards Institute/NCCLS.** 2005. *Performance Standards for Antimicrobial Susceptibility Testing; Fifteenth Informational Supplement. CLSI/NCCLS document M100-S15.* Clinical and Laboratory Standards Institute, Wayne, Pa.
14. **Dailidiene, D., M. T. Bertoli, J. Miciuleviciene, A. K. Mukhopadhyay, G. Dailide, M. A. Pascasio, L. Kupcinskas, and D. E. Berg.** 2002. Emergence of tetracycline resistance in *Helicobacter pylori:* multiple mutational changes in 16S ribosomal DNA and other genetic loci. *Antimicrob. Agents Chemother.* **46:**3940–3946.
15. **DANMAP.** DANMAP 2001—Consumption of antimicrobial agents and occurrence of antimicrobial resistance in bacteria from food animals, foods, and humans in Denmark. Available at: http://www.dfvf.dk/.
16. **Debets-Ossenkopp, Y. J., A. B. Brinkman, E. J. Kuipers, C. M. Vandenbroucke-Grauls, and J. G. Kusters.** 1998. Explaining

the bias in the 23S rRNA gene mutations associated with clarithromycin resistance in clinical isolates of *Helicobacter pylori. Antimicrob. Agents Chemother.* **42:**2749–2751.

17. **DeLoney, C. R., and N. L. Schiller.** 2000. Characterization of an in vitro-selected amoxicillin-resistant strain of *Helicobacter pylori. Antimicrob. Agents Chemother.* **44:**3368–3373.
18. **Endtz, H. P., G. J. Ruijs, B. van Klingeren, W. H. Jansen, T. van der Reyden, and R. P. Mouton.** 1991. Quinolone resistance in *Campylobacter* isolated from man and poultry following the introduction of fluoroquinolones in veterinary medicine. *J. Antimicrob. Chemother.* **27:**199–208.
19. **Engberg, J., F. M. Aarestrup, D. E. Taylor, P. Gerner-Smidt, and I. Nachamkin.** 2001. Quinolone and macrolide resistance in *Campylobacter jejuni* and *C. coli:* resistance mechanisms and trends in human isolates. *Emerg. Infect. Dis.* **7:**24–34.
20. **FDA.** 2004. National Antimicrobial Resistance Monitoring System for Enteric Bacteria (NARMS): NARMS Retail Meat Annual Report, 2002. Rockville, Md.: U.S. Department of Health and Human Services, FDA.
21. **Federal Register.** 2000. 65 Fed. Reg. 64954 (Oct 31, 2000).
22. **Friedman, C. R., J. Neimann, H. C. Wegener, and R. V. Tauxe.** 2000. Epidemiology of *Campylobacter jejuni* infections in the United States and other industrialized nations, p. 130. *In* I. Nachamkin and M. J. Blaser (ed.), Campylobacter. American Society for Microbiology, Washington, D.C.
23. **Gaudreau, C., and H. Gilbert.** 1998. Antimicrobial resistance of clinical strains of *Campylobacter jejuni* subsp. *jejuni* isolated from 1985 to 1997 in Quebec, Canada. *Antimicrob. Agents Chemother.* **42:**2106–2108.
24. **Gaudreau, C., and H. Gilbert.** 2003. Antimicrobial resistance of *Campylobacter jejuni* subsp. *jejuni* strains isolated from humans in 1998 to 2001 in Montreal, Canada. *Antimicrob. Agents Chemother.* **47:**2027–2029.
25. **Ge, B., D. G. White, P. F. McDermott, W. Girard, S. Zhao, S. Hubert, and J. Meng.** 2003. Antimicrobial-resistant *Campylobacter* species from retail raw meats. *Appl. Environ. Microbiol.* **69:**3005–3007.
26. **Gerrits, M. M., M. Berning, A. H. van Vliet, E. J. Kuipers, and J. G. Kusters.** 2003. Effects of 16S rRNA gene mutations on tetracycline resistance in *Helicobacter pylori. Antimicrob. Agents Chemother.* **47:**2984–2986.
27. **Gerrits, M. M., D. Schuijffel, A. A. van Zwet, E. J. Kuipers, C. M. Vandenbroucke-Grauls, and J. G. Kusters.** 2002. Alterations in penicillin-binding protein 1A confer resistance to beta-lactam antibiotics in *Helicobacter pylori. Antimicrob. Agents Chemother.* **46:**2229–2233.
28. **Gibreel, A., and O. Skold.** 1999. Sulfonamide resistance in clinical isolates of *Campylobacter jejuni:* mutational changes in the chromosomal dihydropteroate synthase. *Antimicrob. Agents Chemother.* **43:**2156–2160.
29. **Gibreel, A., and O. Skold.** 2000. An integron cassette carrying *dfr1* with 90-bp repeat sequences located on the chromosome of trimethoprim-resistant isolates of *Campylobacter jejuni. Microb. Drug Resist.* **6:**91–98.
30. **Gibreel, A., O. Skold, and D. E. Taylor.** 2004. Characterization of plasmid-mediated *aphA-3* kanamycin resistance in *Campylobacter jejuni. Microb. Drug Resist.* **10:**98–105.
31. **Gibreel, A., D. M. Tracz, L. Nonaka, T. M. Ngo, S. R. Connell, and D. E. Taylor.** 2004. Incidence of antibiotic resistance in *Campylobacter jejuni* isolated in Alberta, Canada, from 1999 to 2002, with special reference to *tet*(O)-mediated tetracycline resistance. *Antimicrob. Agents Chemother.* **48:**3442–3450.
32. **Goodwin, A., D. Kersulyte, G. Sisson, S. J. Veldhuyzen van Zanten, D. E. Berg, and P. S. Hoffman.** 1998. Metronidazole resistance in *Helicobacter pylori* is due to null mutations in a gene (*rdxA*) that encodes an oxygen-insensitive NADPH nitroreductase. *Mol. Microbiol.* **28:**383–393.
33. **Guerrant, R. L., T. Van Gilder, T. S. Steiner, N. M. Thielman, L. Slutsker, R. V. Tauxe, T. Hennessy, P. M. Griffin, H. DuPont, R. B. Sack, P. Tarr, M. Neill, I. Nachamkin, L. B. Reller, M. T. Osterholm, M. L. Bennish, and L. K. Pickering.** 2001. Practice guidelines for the management of infectious diarrhea. *Clin. Infect. Dis.* **32:**331–351.
34. **Gupta, A., J. M. Nelson, T. J. Barrett, R. V. Tauxe, S. P. Rossiter, C. R. Friedman, K. W. Joyce, K. E. Smith, T. F. Jones, M. A. Hawkins, B. Shiferaw, J. L. Beebe, D. J. Vugia, T. Rabatsky-Ehr, J. A. Benson, T. P. Root, and F. J. Angulo.** 2004. Antimicrobial resistance among *Campylobacter* strains, United States, 1997-2001. *Emerg. Infect. Dis.* **10:**1102–1109.
35. **Hakanen, A., P. Huovinen, P. Kotilainen, A. Siitonen, and H. Jousimies-Somer.** 2002. Quality control strains used in susceptibility testing of *Campylobacter* spp. *J. Clin. Microbiol.* **40:** 2705–2706.
36. **Hakanen, A. J., M. Lehtopolku, A. Siitonen, P. Huovinen, and P. Kotilainen.** 2003. Multidrug resistance in *Campylobacter jejuni* strains collected from Finnish patients during 1995–2000. *J. Antimicrob. Chemother.* **52:**1035–1039.
37. **Harris, N. V., N. S. Weiss, and C. M. Nolan.** 1986. The role of poultry and meats in the etiology of Campylobacter *jejuni/coli* enteritis. *Am. J. Public Health* **76:**407–11.
38. **Heep, M., D. Beck, E. Bayerdorffer, and N. Lehn.** 1999. Rifampin and rifabutin resistance mechanism in *Helicobacter pylori. Antimicrob. Agents Chemother.* **43:**1497–1499.
39. **Heep, M., U. Rieger, D. Beck, and N. Lehn.** 2000. Mutations in the beginning of the *rpoB* gene can induce resistance to rifamycins in both *Helicobacter pylori* and *Mycobacterium tuberculosis. Antimicrob. Agents Chemother.* **44:**1075–1077.
40. **Hunt, R. H., C. A. Fallone, and A. B. Thomson.** 1999. Canadian *Helicobacter pylori* Consensus Conference update: infections in adults. Canadian *Helicobacter* Study Group. *Can. J. Gastroenterol.* **13:**213–217.
41. **Jacob, J., S. Evers, K. Bischoff, C. Carlier, and P. Courvalin.** 1994. Characterization of the *sat4* gene encoding a streptothricin acetyltransferase in *Campylobacter coli* BE/G4. *FEMS Microbiol. Lett.* **120:**13–17.
42. **Jenks, P. J., R. L. Ferrero, and A. Labigne.** 1999. The role of the *rdxA* gene in the evolution of metronidazole resistance in *Helicobacter pylori. J. Antimicrob. Chemother.* **43:**753–758.
43. **Jeong, J. Y., A. K. Mukhopadhyay, D. Dailidiene, Y. Wang, B. Velapatino, R. H. Gilman, A. J. Parkinson, G. B. Nair, B. C. Wong, S. K. Lam, R. Mistry, I. Segal, Y. Yuan, H. Gao, T. Alarcon, M. L. Brea, Y. Ito, D. Kersulyte, H. K. Lee, Y. Gong, A. Goodwin, P. S. Hoffman, and D. E. Berg.** 2000. Sequential inactivation of *rdxA* (HP0954) and frxA (HP0642) nitroreductase genes causes moderate and high-level metronidazole resistance in *Helicobacter pylori. J. Bacteriol.* **182:**5082–5090.
44. **Karmali, M. A., A. E. Simor, M. Roscoe, P. C. Fleming, S. S. Smith, and J. Lane.** 1986. Evaluation of a blood-free, charcoal-based, selective medium for the isolation of *Campylobacter* organisms from feces. *J. Clin. Microbiol.* **23:**456–459.
45. **Kwon, D. H., M. P. Dore, J. J. Kim, M. Kato, M. Lee, J. Y. Wu, and D. Y. Graham.** 2003. High-level beta-lactam resistance associated with acquired multidrug resistance in *Helicobacter pylori. Antimicrob. Agents Chemother.* **47:**2169–2178.
46. **Kwon, D. H., F. A. El Zaatari, M. Kato, M. S. Osato, R. Reddy, Y. Yamaoka, and D. Y. Graham.** 2000. Analysis of *rdxA* and involvement of additional genes encoding NAD(P)H flavin oxidoreductase (FrxA) and ferredoxin-like protein (FdxB) in metronidazole resistance of *Helicobacter pylori. Antimicrob. Agents Chemother.* **44:**2133–2142.

47. **Kwon, S. Y., D. H. Cho, S. Y. Lee, K. Lee, and Y. Chong.** 1994. Antimicrobial susceptibility of *Campylobacter fetus* subsp. *fetus* isolated from blood and synovial fluid. *Yonsei Med. J.* **35:**314–319.
48. **Lachance, N., C. Gaudreau, F. Lamothe, and L. A. Lariviere.** 1991. Role of the beta-lactamase of *Campylobacter jejuni* in resistance to beta-lactam agents. *Antimicrob. Agents Chemother.* **35:**813–818.
49. **Lambert, T., G. Gerbaud, P. Trieu-Cuot, and P. Courvalin.** 1985. Structural relationship between the genes encoding 3′-aminoglycoside phosphotransferases in *Campylobacter* and in gram-positive cocci. *Ann. Inst. Pasteur Microbiol.* **136B:**135–150.
50. **Lee, M. D., S. Sanchez, M. Zimmer, U. Idris, M. E. Berrang, and P. F. McDermott.** 2002. Class 1 integron-associated tobramycin-gentamicin resistance in *Campylobacter jejuni* isolated from the broiler chicken house environment. *Antimicrob. Agents Chemother.* **46:**3660–3664.
51. **Li, C. C., C. H. Chiu, J. L. Wu, Y. C. Huang, and T. Y. Lin.** 1998. Antimicrobial susceptibilities of *Campylobacter jejuni* and *coli* by using E-test in Taiwan. *Scand. J. Infect. Dis.* **30:** 39–42.
52. **Lin, J., L. O. Michel, and Q. Zhang.** 2002. CmeABC functions as a multidrug efflux system in *Campylobacter jejuni*. *Antimicrob. Agents Chemother.* **46:**2124–2131.
53. **Lin, J., O. Sahin, L. O. Michel, and Q. Zhang.** 2003. Critical role of multidrug efflux pump CmeABC in bile resistance and in vivo colonization of *Campylobacter jejuni*. *Infect. Immun.* **71:**4250–4259.
54. **Lucey, B., D. Crowley, P. Moloney, B. Cryan, M. Daly, F. O'Halloran, E. J. Threlfall, and S. Fanning.** 2000. Integronlike structures in *Campylobacter* spp. of human and animal origin. *Emerg. Infect. Dis.* **6:**50–55.
55. **Luo, N., S. Pereira, O. Sahin, J. Lin, S. Huang, L. Michel, and Q. Zhang.** 2005. Enhanced in vivo fitness of fluoroquinolone-resistant *Campylobacter jejuni* in the absence of antibiotic selection pressure. *Proc. Natl. Acad. Sci. USA* **102:**541–546.
56. **Luo, N., O. Sahin, J. Lin, L. O. Michel, and Q. Zhang.** 2003. In vivo selection of *Campylobacter* isolates with high levels of fluoroquinolone resistance associated with *gyrA* mutations and the function of the CmeABC efflux pump. *Antimicrob. Agents Chemother.* **47:**390–394.
57. **Mamelli, L., J. P. Amoros, J. M. Pages, and J. M. Bolla.** 2003. A phenylalanine-arginine beta-naphthylamide sensitive multidrug efflux pump involved in intrinsic and acquired resistance of *Campylobacter* to macrolides. *Int. J. Antimicrob. Agents* **22:**237–241.
58. **Manavathu, E. K., K. Hiratsuka, and D. E. Taylor.** 1988. Nucleotide sequence analysis and expression of a tetracycline-resistance gene from *Campylobacter jejuni*. *Gene* **62:**17–26.
59. **McDermott, P. F., S. M. Bodeis, F. M. Aarestrup, S. Brown, M. Traczewski, P. Fedorka-Cray, M. Wallace, I. A. Critchley, C. Thornsberry, S. Graff, R. Flamm, J. Beyer, D. Shortridge, L. J. Piddock, V. Ricci, M. M. Johnson, R. N. Jones, B. Reller, S. Mirrett, J. Aldrobi, R. Rennie, C. Brosnikoff, L. Turnbull, G. Stein, S. Schooley, R. A. Hanson, and R. D. Walker.** 2004. Development of a standardized susceptibility test for *Campylobacter* with quality-control ranges for ciprofloxacin, doxycycline, erythromycin, gentamicin, and meropenem. *Microb. Drug Resist.* **10:**124–131.
60. **McDermott, P. F., S. M. Bodeis, L. L. English, D. G. White, R. D. Walker, S. Zhao, S. Simjee, and D. D. Wagner.** 2002. Ciprofloxacin resistance in *Campylobacter jejuni* evolves rapidly in chickens treated with fluoroquinolones. *J. Infect. Dis.* **185:** 837–840.
61. **Mendz, G. L., and F. Megraud.** 2002. Is the molecular basis of metronidazole resistance in microaerophilic organisms understood? *Trends Microbiol.* **10:**370–375.
62. **Midolo, P. D., M. G. Korman, J. D. Turnidge, and J. R. Lambert.** 1996. *Helicobacter pylori* resistance to tetracycline. *Lancet* **347:**1194–1195.
63. **Monselise, A., D. Blickstein, I. Ostfeld, R. Segal, and M. Weinberger.** 2004. A case of cellulitis complicating *Campylobacter jejuni* subspecies jejuni bacteremia and review of the literature. *Eur. J. Clin. Microbiol. Infect. Dis.* **23:**718–721.
64. **Moore, R. A., B. Beckthold, S. Wong, A. Kureishi, and L. E. Bryan.** 1995. Nucleotide sequence of the gyrA gene and characterization of ciprofloxacin-resistant mutants of *Helicobacter pylori*. *Antimicrob. Agents Chemother.* **39:**107–111.
65. **Nonaka, L., S. R. Connell, and D. E. Taylor.** 2005. 16S rRNA mutations that confer tetracycline resistance in *Helicobacter pylori* decrease drug binding in *Escherichia coli* ribosomes. *J. Bacteriol.* **187:**3708–3712.
66. **O'Halloran, F., B. Lucey, B. Cryan, T. Buckley, and S. Fanning.** 2004. Molecular characterization of class 1 integrons from Irish thermophilic *Campylobacter* spp. *J. Antimicrob. Chemother.* **53:**952–957.
67. **Occhialini, A., M. Urdaci, F. Doucet-Populaire, C. M. Bebear, H. Lamouliatte, and F. Megraud.** 1997. Macrolide resistance in *Helicobacter pylori:* rapid detection of point mutations and assays of macrolide binding to ribosomes. *Antimicrob. Agents Chemother.* **41:**2724–2728.
68. **Parkhill, J., B. W. Wren, K. Mungall, J. M. Ketley, C. Churcher, D. Basham, T. Chillingworth, R. M. Davies, T. Feltwell, S. Holroyd, K. Jagels, A. V. Karlyshev, S. Moule, M. J. Pallen, C. W. Penn, M. A. Quail, M. A. Rajandream, K. M. Rutherford, A. H. van Vliet, S. Whitehead, and B. G. Barrell.** 2000. The genome sequence of the food-borne pathogen *Campylobacter jejuni* reveals hypervariable sequences. *Nature* **403:**665–668.
69. **Payot, S., L. Avrain, C. Magras, K. Praud, A. Cloeckaert, and E. Chaslus-Dancla.** 2004. Relative contribution of target gene mutation and efflux to fluoroquinolone and erythromycin resistance, in French poultry and pig isolates of *Campylobacter coli*. *Int. J. Antimicrob. Agents* **23:**468–472.
70. **Pinto-Alphandary, H., C. Mabilat, and P. Courvalin.** 1990. Emergence of aminoglycoside resistance genes *aadA* and *aadE* in the genus *Campylobacter*. *Antimicrob. Agents Chemother.* **34:**1294–1296.
71. **Pioletti, M., F. Schlunzen, J. Harms, R. Zarivach, M. Gluhmann, H. Avila, A. Bashan, H. Bartels, T. Auerbach, C. Jacobi, T. Hartsch, A. Yonath, and F. Franceschi.** 2001. Crystal structures of complexes of the small ribosomal subunit with tetracycline, edeine and IF3. *EMBO J.* **20:**1829–1839.
72. **Pratt, A., and V. Korolik.** 2005. Tetracycline resistance of Australian *Campylobacter jejuni* and *Campylobacter coli* isolates. *J. Antimicrob. Chemother.* **55:**452–460.
73. **Pumbwe, L., and L. J. Piddock.** 2002. Identification and molecular characterisation of CmeB, a *Campylobacter jejuni* multidrug efflux pump. *FEMS Microbiol. Lett.* **206:**185–189.
74. **Pumbwe, L., L. P. Randall, M. J. Woodward, and L. J. Piddock.** 2004. Expression of the efflux pump genes *cmeB*, *cmeF* and the porin gene *porA* in multiple-antibiotic-resistant *Campylobacter jejuni*. *J. Antimicrob. Chemother.* **54:**341–347.
75. **Rao, D., J. R. Rao, E. Crothers, R. McMullan, D. McDowell, A. McMahon, P. J. Rooney, B. C. Millar, and J. E. Moore.** 2005. Increased erythromycin resistance in clinical Campylobacter in Northern Ireland—an update. *J. Antimicrob. Chemother.* **55:**395–396.
76. **Ribeiro, M. L., M. M. Gerrits, Y. H. Benvengo, M. Berning, A. P. Godoy, E. J. Kuipers, S. Mendonca, A. H. van Vliet, J. Pe-**

drazzoli, Jr., and J. G. Kusters. 2004. Detection of high-level tetracycline resistance in clinical isolates of *Helicobacter pylori* using PCR-RFLP. *FEMS Immunol. Med. Microbiol.* **40:**57–61.

77. **Ross, J. I., E. A. Eady, J. H. Cove, and W. J. Cunliffe.** 1998. 16S rRNA mutation associated with tetracycline resistance in a gram-positive bacterium. *Antimicrob. Agents Chemother.* **42:**1702–1705.
78. **Schwartz, D., H. Goossens, J. Levy, J. P. Butzler, and J. Goldhar.** 1993. Plasmid profiles and antimicrobial susceptibility of *Campylobacter jejuni* isolated from Israeli children with diarrhea. *Zentralbl. Bakteriol.* **279:**368–376.
79. **Smith, K. E., J. M. Besser, C. W. Hedberg, F. T. Leano, J. B. Bender, J. H. Wicklund, B. P. Johnson, K. A. Moore, and M. T. Osterholm.** 1999. Quinolone-resistant *Campylobacter jejuni* infections in Minnesota, 1992-1998. Investigation Team. *N. Engl. J. Med.* **340:**1525–32.
80. **Stone, G. G., D. Shortridge, R. K. Flamm, J. Versalovic, J. Beyer, K. Idler, L. Zulawinski, and S. K. Tanaka.** 1996. Identification of a 23S rRNA gene mutation in clarithromycin-resistant *Helicobacter pylori. Helicobacter* **1:**227–228.
81. **Tajada, P., J. L. Gomez-Graces, J. I. Alos, D. Balas, and R. Cogollos.** 1996. Antimicrobial susceptibilities of *Campylobacter jejuni* and *Campylobacter coli* to 12 beta-lactam agents and combinations with beta-lactamase inhibitors. *Antimicrob. Agents Chemother.* **40:**1924–1925.
82. **Tankovic, J., D. Lamarque, J. C. Delchier, C. J. Soussy, A. Labigne, and P. J. Jenks.** 2000. Frequent association between alteration of the *rdxA* gene and metronidazole resistance in French and North African isolates of *Helicobacter pylori. Antimicrob. Agents Chemother.* **44:**608–613.
83. **Taylor, D. E., Z. Ge, D. Purych, T. Lo, and K. Hiratsuka.** 1997. Cloning and sequence analysis of two copies of a 23S rRNA gene from *Helicobacter pylori* and association of clarithromycin resistance with 23S rRNA mutations. *Antimicrob. Agents Chemother.* **41:**2621–2628.
84. **Taylor, D. E., Q. Jiang, and R. N. Fedorak.** 1998. Antibiotic susceptibilities of *Helicobacter pylori* strains isolated in the Province of Alberta. *Can. J. Gastroenterol.* **12:**295–298.
85. **Tenover, F. C., S. Williams, K. P. Gordon, C. Nolan, and J. J. Plorde.** 1985. Survey of plasmids and resistance factors in *Campylobacter jejuni* and *Campylobacter coli. Antimicrob. Agents Chemother.* **27:**37–41.
86. **Tomb, J. F., O. White, A. R. Kerlavage, R. A. Clayton, G. G. Sutton, R. D. Fleischmann, K. A. Ketchum, H. P. Klenk, S. Gill, B. A. Dougherty, K. Nelson, J. Quackenbush, L. Zhou, E. F. Kirkness, S. Peterson, B. Loftus, D. Richardson, R. Dodson, H. G. Khalak, A. Glodek, K. McKenney, L. M. Fitzegerald, N. Lee, M. D. Adams, J. C. Venter, et al.** 1997. The complete genome sequence of the gastric pathogen *Helicobacter pylori. Nature* **388:**539–547.
87. **Trieber, C. A., N. Burkhardt, K. H. Nierhaus, and D. E. Taylor.** 1998. Ribosomal protection from tetracycline mediated by Tet(O): Tet(O) interaction with ribosomes is GTP-dependent. *Biol. Chem.* **379:**847–855.
88. **Trieber, C. A., and D. E. Taylor.** 2002. Mutations in the 16S rRNA genes of *Helicobacter pylori* mediate resistance to tetracycline. *J. Bacteriol.* **184:**2131–2140.
89. **van Boven, M., K. T. Veldman, M. C. de Jong, and D. J. Mevius.** 2003. Rapid selection of quinolone resistance in *Campylobacter jejuni* but not in *Escherichia coli* in individually housed broilers. *J. Antimicrob. Chemother.* **52:**719–723.
90. **Versalovic, J., D. Shortridge, K. Kibler, M. V. Griffy, J. Beyer, R. K. Flamm, S. K. Tanaka, D. Y. Graham, and M. F. Go.** 1996. Mutations in 23S rRNA are associated with clarithromycin resistance in *Helicobacter pylori. Antimicrob. Agents Chemother.* **40:**477–480.
91. **Wang, G., M. S. Rahman, M. Z. Humayun, and D. E. Taylor.** 1999. Multiplex sequence analysis demonstrates the competitive growth advantage of the A-to-G mutants of clarithromycin-resistant *Helicobacter pylori. Antimicrob. Agents Chemother.* **43:**683–685.
92. **Wang, G., and D. E. Taylor.** 1998. Site-specific mutations in the 23S rRNA gene of *Helicobacter pylori* confer two types of resistance to macrolide-lincosamide-streptogramin B antibiotics. *Antimicrob. Agents Chemother.* **42:**1952–1958.
93. **Wang, G., T. J. Wilson, Q. Jiang, and D. E. Taylor.** 2001. Spontaneous mutations that confer antibiotic resistance in *Helicobacter pylori. Antimicrob. Agents Chemother.* **45:**727–733.
94. **Wang, Y., W. M. Huang, and D. E. Taylor.** 1993. Cloning and nucleotide sequence of the *Campylobacter jejuni gyrA* gene and characterization of quinolone resistance mutations. *Antimicrob. Agents Chemother.* **37:**457–463.
95. **Wang, Y., and D. E. Taylor.** 1990. Chloramphenicol resistance in *Campylobacter coli:* nucleotide sequence, expression, and cloning vector construction. *Gene* **94:**23–28.
96. **Weisblum, B.** 1995. Erythromycin resistance by ribosome modification. *Antimicrob. Agents Chemother.* **39:**577–585.
97. **Williams, M. D., J. B. Schorling, L. J. Barrett, S. M. Dudley, I. Orgel, W. C. Koch, D. S. Shields, S. M. Thorson, J. A. Lohr, and R. L. Guerrant.** 1989. Early treatment of *Campylobacter jejuni* enteritis. *Antimicrob. Agents Chemother.* **33:**248–250.
98. **Yan, W., and D. E. Taylor.** 1991. Characterization of erythromycin resistance in *Campylobacter jejuni* and *Campylobacter coli. Antimicrob. Agents Chemother.* **35:**1989–1996.

Frontiers in Antimicrobial Resistance: a Tribute to Stuart B. Levy
Edited by D. G. White, M. N. Alekshun, and P. F. McDermott

Chapter 25

Anaerobes

ISABELLE PODGLAJEN, JACQUES BREUIL, AND EKKEHARD COLLATZ

Anaerobes comprise over 500, and perhaps as many as 1,000, species and make up over 99% of the indigenous gastrointestinal or oral microflora (21, 77, 174). Some of them may be pathogenic in humans and responsible for food poisoning or opportunistic infections such as septicemia, abscesses of various localizations or periodontal disease. The major anaerobic genera of clinical interest include *Bacteroides, Prevotella, Porphyromonas, Fusobacterium, Clostridium, Propionibacterium,* and *Peptostreptococcus.* Several reviews have been published in recent years on various aspects concerning the determinants of antimicrobial resistance in these species, their epidemiology and the rates of resistance (62, 114, 122, 129, 162). The most data are available for the *Bacteroides fragilis* group, mainly from the United States, Japan and Europe. Considering the wealth of data, many of which were not obtained under identical experimental conditions, the resistance rates quoted here for the genera mentioned above are certainly selective and do not necessarily compare well with published data not cited here. The comparability of the data should improve in the future with the generalized use of the most recently validated NCCLS reference methods for antibiotic susceptibility testing of anaerobes (61).

With respect to the resistance determinants, data are abundant for the *Bacteroides* group and the clostridia but less so for most of the other genera of interest. Like many other bacteria, these anaerobes harbor a variety of transmissible elements that are involved in the dissemination of tetracycline, erythromycin and clindamycin, metronidazole, and some β-lactam resistance determinants. These elements include mobilizable and conjugative plasmids, composite transposons, mobilizable and conjugative transposons (CTns). In *Bacteroides,* macrolide-lincosamide-streptogramin (MLS_B) and tetracycline resistance determinants are typically found on conjugative transposons (136, 139, 171) and on mobilizable transposons in *Clostridium* (3). The processes involved in the transfer of resistance genes and their dissemination is covered in chapter 33.

RESISTANCE IN BACTEROIDES

β-Lactam Resistance

β-Lactamase production

Nearly all of the clinical isolates of the *B. fragilis* group are resistant to the penicillins. The rates of resistance to this group of antibiotics were high already thirty years ago (141). Resistance to the cefamycin cefoxitin is less common, with yearly rates varying between 10 and 14% in the United States where *B. distasonis,* with rates decreasing from 41 to 25% between 1997 and 2000, was the most frequently resistant (148). In Europe, the current rate reported for the group is 6%, and with 13% highest for *B. thetaiotaomicron* (64). The activity of the β-lactamase inhibitors remains very good, with resistance rates for the most efficient combination (piperacillin-tazobactam) below 2% both in the United States and in Europe (64, 148). Since the isolation of the first imipenem-resistant strains of *B. fragilis* 20 years or more ago (35, 179), resistance to the carbapenems has remained extremely low, with rates of ≤1% (6, 64, 148), despite episodic reports of rates of up to 6% in certain institutions (13).

Among the three β-lactam resistance mechanisms in *Bacteroides*—β-lactamase production, alteration of penicillin-binding-proteins, and altered production of porins—the first is the most widely studied and, as is common in gram-negative bacteria, the most prevalent.

Isabelle Podglajen and Ekkehard Collatz • INSERM U655–Laboratoire de Recherche Moléculaire sur les Antibiotiques, University Paris VI, and University Paris-Descartes, 15, rue de l'Ecole de Médecine, 75270 Paris Cedex 06, France. **Jacques Breuil** • Centre Hospitalier Intercommunal, 94190 Villeneuve Saint Georges, France.

The vast majority, if not all, of the clinical isolates belonging to the species of the *B. fragilis* group are producers of cephalosporinase-type β-lactamases (5, 6), which have been identified and studied to different degrees in some species (Table 1). They confer resistance to most β-lactams, with exception of the cephamycins and carbapenems. Also, these enzymes are generally susceptible to the β-lactamase inhibitors tazobactam, clavulanate and, maybe somewhat less so, sulbactam. The *B. fragilis* strains of the DNA homology group I, accounting for over 95% of this species (70, 107, 135), produce the "endogenous," chromosome-encoded Ambler class A β-lactamase CepA (127,128). A closely related enzyme, with 78% amino acid identity, is encoded on the chromosome of *B. thetaiotaomicron* (173), while *B. uniformis* produces an endogenous cephalosporinase, CblA that shows less than 50% identity with CepA (146).

Carbapenem-resistant strains of *B. fragilis* typically produce a class B, metallo-β-lactamase called CfiA or CcrA (115, 158). An apparently nonmetallo carbapenemase has been described in *B. distasonis* (65) and occasional strains of *B. fragilis* have been reported that are resistant, or of reduced susceptibility, to imipenem but not producing a metallo-β-lactamase (43, 106, 150, 175). Rare high-level imipenem-resistant strains have also been observed in *B. vulgatus* and *B. ovatus*, but the resistance mechanism has not been explored (63). CfiA (CcrA) requires a divalent cation, zinc or cobalt, for activity, confers resistance to all β-lactam antibiotics except the monobactams and is not inhibited by any of the β-lactamase inhibitors in clinical use (35, 177, 179). The crystal structure of the enzyme, with a binuclear zinc center-containing active site, has been elucidated, and its large substrate profile was attributed to several of its structural features, including an active site wider and less restrictive than that of a class A β-lactamase (27, 32). The *cfiA* (*ccrA*) gene, as opposed to *cepA*, is consistently carried on the chromosome of the *B. fragilis* strains that make up the small DNA homology group II (56, 70, 107), but strains harboring both genes have been observed (106). In two thirds or more of the *cfiA*-carrying strains, carbapenem resistance is not expressed or is expressed only at very low levels (45, 102, 107, 150, 175). High-level resistant mutants can be selected in vitro from such strains in one step in which the promoterless carbapenemase gene is activated by an IS-borne promoter inserted immediately upstream (109, 110), an event that has been observed similarly during imipenem therapy (44). In resistant clinical isolates for which MICs of imipenem exceed 32 μg/ml, *cfiA* expression, with few exceptions, is driven by promoters carried on insertion elements. Over ten such elements (including some iso-elements) are known to date (44, 72, 109, 108, 113, 149). The IS-borne promoters conform to the consensus sequence described for the *Bacteroides* promoters (16, 108) and are recognized by the particular primary sigma factor of the *Bacteroidetes* (164). The level of carbapenem resistance is fairly well correlated with the level of *cfiA* transcription (108).

A class A β-lactamase capable of slow cefoxitin but efficient cephalosporin and penicillin degradation has been described in a cefoxitin-resistant strain of

Table 1. β-Lactamases produced by anaerobic bacteria

Genus	Species	β-Lactamase[b]	pI[c]	M_r[c]	Class	Accession no.	Reference(s)
Bacteroides	*B. fragilis* (HG I)[a]	**CepA**	*5.5*	*31.5*	A	L13472, NC_006347	127, 128
	B. thetaiotaomicron	**CepA-like**	*4.8*	*30.5*		NC_006347	173
	B. uniformis	**CblA**	*5.5*	*31.1*	A	L08472	146
	B. fragilis	CfxA-4	*5.8*	*31.7*	A	AY769933	
	B. vulgatus	CfxA	*6.2*	*31.7*		U38243	101
	B. distasonis	CfxA, CfxA-5	*5.8*	*31.7*		AY769933	11
	B. fragilis (HG II)[a]	**CfiA (CcrA)**	*5.2*, 4.8	*25.2*	B	M34831, M63556	115, 158, 177, 179
	B. distasonis	Carbapenemase	6.9	?	Non-B		65
Prevotella	*P. intermedia*	CfxA-2-like	*5.8*, 5.4	*31.7*	A	AF118110	59, 87
						Y753592	
	P. bivia		*6.2*	*31.8*		AF504909	53
	P. denticola		*6.0*	*31.7*		AF504914	
	P. buccae		*5.8*	*31.7*		AF504910	
Fusobacterium	*F. nucleatum*	Penicillinase	4.8/5.2	21–26	A		161
	F. nucleatum	Fus-1	*6.1*	*28.3*	D	AY227054	
Clostridium	*C. butyricum*	Penicillinase	4.4, (4.1)	32	A		60, 73, 88
	C. clostridioforme	Penicillinase	4.2		A		9
	C. ramosum	Penicillinase			A		167

[a] HG I and II: DNA homology group I and II (70).
[b] Species-specific chromosome-encoded β-lactamases are shown in bold.
[c] Values in italics were computed for the proteins after elimination of the putative signal peptides using the ProtParam and SignalP (20) software packages of the ExPASy Proteomics Server (http://www.expasy.org/).

B. vulgatus, where the responsible gene, *cfxA*, was found to be chromosome-encoded and transferable to *B. fragilis* (101) and later found to reside on a mobilizable transposon, MTn*4555* (159, 165). The gene *cfxA*, or homologs thereof, have been observed in *B. fragilis, B. uniformis, B. ovatus,* and *B. distasonis* (Table 1) but there are no comprehensive data on the prevalence of *cfxA* in the *B. fragilis* group.

Cell wall changes

Next to β-lactamase production, penicillin-binding protein (PBP)- and uptake-related changes have been reported as underlying β-lactam resistance in some members of the *B. fragilis* group, but only relatively few strains have been studied in detail and the results are somewhat conflicting. The true contribution of these mechanisms to resistance in clinical isolates is therefore difficult to assess. Seven PBPs are encoded on the *B. fragilis* and *B. thetaiotaomicron* genomes (Table 2) (78, 173). Most of these, and consistently the high-molecular weight proteins, have been observed by several authors who studied PBP-mediated resistance in the *B. fragilis* group (43, 104, 105, 106, 169, 178). In cefoxitin-resistant clinical isolates of *B. fragilis, B. ovatus,* and *B. thetaiotaomicron,* reduced affinity for the drug of one or the other of the two highest molecular weight PBPs was implicated (105), while in *B. fragilis, B. thetaiotaomicron,* and *B. vulgatus* changes in affinity or apparent molecular weight, or both, of similar PBPs were thought to account for cefoxitin resistance (169). In one cefazolin- and cephalotin-, but not cefoperazone- or cefamandole-resistant strain of *B. fragilis,* decreased affinity of a PBP with an apparent molecular mass of 72 kDa for the two former compounds was considered to play an important role (178). Reduced affinity of the high-molecular-weight proteins for imipenem was concluded to contribute to low-level resistance in *B. fragilis* (43).

Moderately decreased activity of β-lactams, particularly when not substantially recovered by β-lactamase inhibitors, has been suspected to be associated with outer membrane changes (18, 48, 62, 105). Pore-forming outer membrane proteins, of the Omp200 complex or OmpA-related, have been identified in *B. fragilis, B. distasonis, Porphyromonas* and *Fusobacterium,* but although the presence or absence of particular Omp proteins in clinical isolates of *Bacteroides* of varying antimicrobial susceptibility was noted, a correlation between the complex Omp changes and β-lactam susceptibility has not been clearly established (168).

Imidazole Resistance

Despite the clinical use of nitroimidazoles since 1960, resistance to these compounds has remained rare. Clinical strains for which the MICs exceeded 16 μg/ml were not found in the United States before 2004, with the exception of one strain presumed to have been imported from Africa (62, 138). However, resistant isolates with MICs ≥32 μg/liter (NCCLS resistance breakpoint) have been reported from Canada (79) Europe (52, 64, 85), the Middle East (134), Asia (29, 156), and Africa (86) where in some cases therapeutic problems were encountered (47, 134). Strains with decreased metronidazole susceptibilities (MICs, 4 to 16 μg/ml) have been encountered in most large-scale susceptibility surveys, including some from the United States (6).

The 5-nitro-imidazoles are inactive prodrugs which enter cells by simple diffusion and are then reduced into a toxic, short-lived radical anion which causes various types of DNA damage that ultimately

Table 2. The penicillin-binding proteins of *B. fragilis* and *B. thetaiotaomicron*

PBP[a]	M_r (thousands)[b]			
	B. fragilis		*B. thetaiotaomicron*	
	Unprocessed	Processed	Unprocessed	Processed
PBP 1A	88.2	84.9	87.7	84.5
PBP 1C	88.2	82.6[c]	86.9	82.6[c]
PBP	77.9	74.6	78.1	75.6
PBP	73.6	71.1	74.3	71.3
PBP2	69.4	65.4[c]	69.7	NSP[d]
PBP[e]	52.5	49.5	51.9	49.6
PBP 2B	30.5	28.6[c]	32.8	NSP

[a] The PBP designations are those given in the annotated genomes of *B. fragilis* YCH46 (NC_006347) and *B. thetaiotaomicron* VPI-5482 (NC_004663).
[b] The molecular weights were calculated for the PBPs before and after cleavage of their predicted signal peptides using the ProtParam and SignalP (20) software packages of the ExPASy Proteomics Server (http://www.expasy.org/).
[c] A signal peptide was not unambiguously predicted.
[d] NSP, no signal peptide was predicted.
[e] D-Ala-D-Ala carboxypeptidase.

lead to cell death. This reaction is mediated by ferredoxin, which receives an electron from the pyruvate-ferredoxin oxidoreductase complex via conversion of pyruvate to acetyl coenzyme A. As a consequence, species with altered, absent, or elevated redox potential pathways are resistant to the imidazoles.

In *Bacteroides* spp., resistance and decreased susceptibility to metronidazole is generally, but not always, associated with the presence of one of the seven primarily plasmid-borne *nim* genes described to date, and most frequently with *nimA* (Table 3) (85, 86). While a strict correlation does not seem to exist between the presence of potentially *nim* gene activating IS elements and metronidazole resistance levels, the presence of this class of genes has been found to be required for what has been called induction of high-level resistance, i.e., the (mostly transient) selection of populations growing in the presence of the drug at 64 to >256 μg/ml, a process for which no link with alterations in *nim* gene sequence or expression has been established (52, 85). It has been thought for some time that the *nim* genes encode 5-nitro-imidazole reductases which use reduced ferredoxin as the electron donor and transform the nitro group of the prodrug into its amine derivative, which is nontoxic (28). This pathway has received recent validation from the work of Leiros et al. (82), who solved the structure of the NimA protein of *Deinococcus radiodurans,* with pyruvate, the electron donor, covalently bound to an active-site histidine residue that is perfectly conserved in all Nim proteins. Several of these proteins, which constitute a distinct family, are also produced by diverse bacterial species not belonging to the *B. fragilis* group, some of them aerobic (82). Imidazole resistance, or reduced susceptibility, apparently not associated with *nim* genes was the first to be described in *B. fragilis* (67) and continues to be observed (52, 85, 102, 155), suggesting that alternative resistance mechanisms exist.

Tetracycline Resistance

Resistance of the *B. fragilis* group isolates to tetracycline and related compounds has been elevated worldwide since the 1970s (ca. 30%) in the clinical setting as well as in the community, and over 90% in the 1990s (136). Current surveys usually provide no more data concerning these drugs. On the other hand, this resistance does not extend to a recent minocycline derivative, tigecycline, and elevated MICs of this drug are observed only exceptionally in clinical isolates (41).

Among the three known tetracycline resistance mechanisms—efflux-mediated resistance, enzymatic tetracycline degradation and ribosomal protection—only the latter is operative in members of the *B. fragilis* group. Of the seven genes conferring this type of resistance in anaerobic bacteria, three, *tetQ, tetM* and *tet36,* have been found in this group (Table 4). The *tetQ* determinant accounts for the vast majority of the resistance observed and is typically located on conjugative transposons, essentially CTnDOT and CTnERL, the former encoding also clindamycin and erythromycin resistance (171). Both carry a regulatory region (*rteA, rteB* and *rteC,* comprising *tetQ*) which mediates a tetracycline induction effect at low concentration that is required for CTnDOT/ERL self-transfer or for the mobilization of coresident *tra*-deficient plasmids or transposons (171). TetM, like others of its homologs, closely resembling elongation factor EF-G, catalyzes the release of tetracycline from the ribosome (26), possibly by triggering a local conformational change in helix 34 of 16S rRNA, which contains the primary tetracycline binding site, as has been suggested for TetO (152).

A tetracycline-inactivating FAD-dependent monooxygenase that requires oxygen for function and therefore does not confer resistance in anaerobes is curiously encoded by the transposon-borne gene *tetX* in some strains of *B. fragilis* (153, 176).

Table 3. Metronidazole resistance genes (*nim*) in anaerobic bacteria

Genus	Species	Gene	Accession no.	Reference
Bacteroides	*B. vulgatus*	*nimA*	X71444	57
	B. fragilis	*nimB*	X71443	57
	B. thetaiotaomicron	*nimC*	X76948	160
	B. fragilis BF239	*nimD*	X76949	160
	B. fragilis ARU6881	*nimE*	AJ244018	155
	B. vulgatus	*nimF*	AJ515145	85
	B. fragilis	*nimG*		52
Prevotella	*P. bivia*	*nimA*		86
Fusobacterium	*F. nucleatum*	*nim*	FN0030/FN1023	71
Clostridium	*C. bifermentans*	*nimA*		86
Propionibacterium	*P. acnes* and sp.	*nimA*		86
Peptostreptococcus	*P. anaerobius, Finegoldia magna, Anaerococcus prevotii, Micromonas micros*	*nimB*		157

Table 4. Tetracycline resistance genes in anaerobic bacteria

Genus	Species	Gene[a]	Accession no.	Reference(s)
Bacteroides	*B. fragilis*	*tetQ*[b]	Z21523	84
	B. fragilis BF-2		Y08615	
	B. thetaiotaomicron		X58717	97
	Bacteroides spp.			83
	B. fragilis	*tetM*[b]		145
	B. ureolyticus			37
	Bacillus sp. 139	*tet36*[b]	AJ514254	172
	B. fragilis	*tetX*[c]	M37699	153
Prevotella	*P. intermedia*	*tetQ*[b]	U73497	
	P. melaninogenica, P. (Treponema) denticola			10
	P. ruminicola		L33696	96
	P. buccae, P. bivia, P. buccalis, P. corporis, P. disiens			83
	Prevotella spp.	*tetM*[b]		163
	Prevotella spp.	*tetW*[b]		163
	P. (Treponema) denticola	*tetB*[d]	S83213	126
Porphyromonas	*P. gingivalis*	*tetQ*[b]		83
Fusobacterium	*F. nucleatum*	*tetM*[b]		83, 90
	F. prausnitzii	*tetW*[b]		140
Clostridium	*C. difficile*	*tetM*[b]	AF333235	94
	C. septicum		AB054984	137
	C. difficile	*tetP*[b, d]		125
	C. perfringens	*tetB(P)*[b]	L20800	143
	C. perfringens	*tetA(P)*[d]	L20800	143
	C. difficile	*tetK*[d]		125
	C. difficile	*tetL*[d]		125
Peptostreptococcus	*Peptostreptococcus* sp.	*tetQ*[b]		83
	Peptostreptococcus spp.	*tetM*[b]		124
	Peptostreptococcus spp.	*tetO*[b]		124
	Peptostreptococcus spp.	*tetK*[d]		124
	Peptostreptococcus spp.	*tetL*[d]		124

[a] Further unreferenced *tet* genes in anaerobes can be found at http://faculty.washington.edu/marilynr/tetweb2.pdf.
[b] *tet* genes mediating ribosomal protection.
[c] *tet* gene mediating drug modification.
[d] *tet* genes mediating drug efflux.

Macrolide Resistance

Despite differences in the treatment of aerobic-anaerobic infections in the United States where clindamycin is the standard and in Europe where it is not, resistance to this drug has increased over the past twenty years. Currently, overall resistance rates of ca. 15% are observed in Europe (64, 76) and of 25 to 30% in the United States (4, 62), but higher rates may be seen in individual species isolated in different institutions and geographic areas (62, 64, 156).

MLS_B-type resistance in the *B. fragilis* group is mediated by the 23S rRNA methylases ErmF, ErmFS, ErmFU, ErmG, ErmB or Erm35 (Table 5) (34, 55, 116) but is not accounted for in all cases by the presence of one of the corresponding genes, which are typically carried by conjugative transposons. To date, the methylase genes *ermQ* and *ermC* identified in anaerobes have not yet been observed in the *B. fragilis* group. In 87% of the cases, *ermF* is linked to the presence of CTnDOT but *ermF*, as well as *ermFS*, has been found also on plasmid-borne composite transposons. The *ermB* and *ermG* determinants, which are thought to have been acquired from gram-positive bacteria recently and which are carried on particular CTn elements, the mobility of which is not tetracycline-dependent, are encountered less frequently (171). MLS_B-type resistance inducible by erythromycin as well as clindamycin has been described in *B. vulgatus* (117). Lincosamide resistance, due to drug modification by an O-nucleotidyltransferase and wide-spread in *Bacteroides* spp., is conferred by the NBU2-borne *lnu(A)* homologue *linA*$_{N2}$, where it is linked to the apparently nonexpressed gene *mefE*$_{N2}$ that codes for an erythromycin efflux pump (165).

Fluoroquinolone Resistance

Interest in fluoroquinolone resistance in anaerobes has arisen since the development of compounds with pronounced activity against this group of bacteria that could be used for the treatment of mixed in-

Table 5. Macrolide resistance genes (*erm*) in anaerobic bacteria

Genus	Species	Gene[a]	Accession no.	Reference(s)
Bacteroides	*B. fragilis*	*ermF*	M14730	31, 116
	B. fragilis	*ermFS*	M17808	144
	B. fragilis	*ermFU*	M62487	58
	B. thetaiotaomicron	*ermG*	L42817	34
	B. ovatus		AJ557257	166
	Bacteroides sp. 139	*erm35*	AF319779	
	Bacteroides spp.	*ermB*	AY345595	55
Prevotella	*P. melaninogenica, P. bivia, P. disiens, P. intermedia, P. nigrescens, P. (Treponema) denticola*	*ermF*		10, 30, 31
Porphyromonas	*P. gingivalis*	*ermF*		30, 31
Fusobacterium	*F. nucleatum*	*ermF*		31
Clostridium	*C. butyricum, C. tertium*	*ermF*		31
	C. difficile	*ermB*	AJ294529	
	C. perfringens		AF109075	50
	C. perfringens, Clostridium spp.	*ermBP*	U18931	22
	C. perfringens	*ermQ*	L22689	23
Propionibacterium	*P. acnes*	*ermX*	AF411029	132
Peptostreptococcus	*Peptostreptococcus* spp.	*ermF*		31
	Peptostreptococcus spp.	*ermTR*		118
	P. anaerobius, Peptoniphilus assacharolyicus, Peptostreptococcus sp.	*ermF*		31, 118
	P. anaerobius	*ermG*		118

[a] Further unreferenced *erm* determinants are listed at http://faculty.washington.edu/marilynr/ermweb4.pdf.

fections. After the withdrawal of several compounds due to their toxic side effects, moxifloxacin is now considered a drug of interest in this context, both in the United States and Europe. Using tentative breakpoints of ≥8 μg/ml, resistance rates vary between 6 and 19% within the *B. fragilis* group in Europe (64). In the United States, resistance rates rose from 30% in 1998 to 43% in 2001 in six *Bacteroides* species (54).

As in most other bacteria, fluoroquinolone resistance in *Bacteroides* may be associated with mutations in the DNA gyrase subunits or alterations of drug accumulation. In many ciprofloxacin-resistant isolates, gyrase mutations were found to occur either in GyrA or in GyrB but so far not in both (Table 6) (98). Among the GyrA mutants, Ser82 substitutions (corresponding to Ser83 in *E. coli*) have been found most frequently, next to substitutions of Asp81 and Tyr86. Recently, GyrB mutants with Glu478Lys or Leu415Val substitutions have been selected in ciprofloxacin-resistant strains under experimental conditions or observed in clinical isolates, respectively (120). These GyrA and GyrB mutations are considered to be second-step mutations, but the nature of the implied first-step mutations remains obscure and resistance-conferring alterations of the topoisomerase IV, a proven fluoroquinolone target in other bacteria, do not seem to have been identified yet in *B. fragilis* (Table 6) (120).

The demonstration that, in strains without gyrase alterations, the addition of efflux pump inhibitors significantly increased the intracellular quinolone concentrations suggested that enhanced efflux was a possible mechanism (99, 120). A NorA/Bmr-type pump that very efficiently mediates efflux of norfloxacin exists in *B. fragilis* (91). A pump of the MATE family, BexA, of which an ortholog exists in *B. fragilis* (78), has been characterized in *B. thetaiotaomicron* (92). However, the extent to which these particular pumps are involved in the acquisition of resistance in clinical isolates requires further investigation.

RESISTANCE IN *PREVOTELLA*, *PORPHYROMONAS*, AND *FUSOBACTERIUM*

Prevotella, *Porphyromonas*, and *Fusobacterium* species are typical anaerobic gram-negative members of the indigenous flora responsible for mouth and periodontal diseases. For a long time, these genera were considered as highly antibiotic susceptible, which is still true with respect to metronidazole, β-lactam–β-lactamase inhibitor combinations and carbapenems (5, 76).

β-Lactam Resistance

Among the three genera, the *Prevotellae* are the most frequent producers of β-lactamases conferring resistance to the penicillins. Penicillin G and amoxicillin resistance rates for *Prevotellae* of 83 and 42% have been reported in the United States and Europe, respectively. For *Porphyromonas* and *Fusobacterium*

Table 6. Amino acid changes reported in gyrase and topoisomerase IV in fluoroquinolone-resistant laboratory mutants and/or clinical or veterinary isolates

Genus	Species	Amino acid substitution				Reference(s)
		GyrA	GyrB	ParC	ParE	
Bacteroides	*B. fragilis*	Ser82Phe Ser82Leu Asp81Asn Asp81Gly Ala118Val	Glu487Lys Leu415Val			12, 98, 120
	B. ovatus	Ser82Phe				
	B. thetaiotaomicron	Tyr86Phe				
	B. uniformis	Ser82Leu				
	B. vulgatus	Ser82Leu				
Porphyromonas	*P. gingivalis*	Ser83Phe				46
Fusobacterium	*F. canifelinum*	Ser79Leu, Gly83Arg				33
Clostridium	*C. perfringens*	Ala119Glu Gly81Cys D87Tyr D82Asn Ser83Leu	D93Tyr Val196Phe Ser89Ile Ala131Ser D502Tyr D88Tyr D11Tyr Val22Phe	Ala431Ser	Glu486Lys	112
	C. difficile	Thr82Ile Thr82Val Asp71Val Thr82Ile Ala118Thr	Asp426Asn Arg447Leu			39

spp., reported resistance rates ranging from 10% or below (5, 51, 76) up to 70% have been reported (156). The association of a β-lactam with a β-lactamase inhibitor is currently active in 100% of isolates, suggesting the absence of other mechanisms of β-lactam resistance.

The β-lactamases described in these genera are listed in Table 1. Slow cefoxitin hydrolysis has been observed in *P. bivia* (*Bacteroides bivius*) by a β-lactamase that was inhibited by clavulanate (89). The β-lactamases described to date in *Prevotellae* as belonging to the CfxA family were not found (as their name would suggest) to confer resistance to cefoxitin or to hydrolyze this compound (87). In *Prevotellae* collected in the United States and Norway, exclusively CfxA-type enzymes were observed (59). Reported production of β-lactamase in *Porphyromonas* and *Fusobacterium* species, based on simple detection with a chromogenic test, appears to be rare. In *F. nucleatum,* a penicillinase has been described, and recently the gene (*fus1*) of a class D beta-lactamase has been reported (Table 1).

Tetracycline Resistance

Current data concerning the rates of tetracycline resistance are scarce. While *Fusobacterium* and *P. gingivalis* were reported to be rather susceptible, resistance rates of up to 50% were observed, particularly in *Prevotella* (74). Tetracycline resistance genes conferring ribosomal protection have been described in the three genera. *tetQ*, *tetM*, and *tetW* have been identified in *Prevotella*, from which *tetQ* transfer to *B. fragilis* is possible under natural conditions and apparently associated with conjugative transposons (96, 163, 171). *tetQ*, and *tetW* have been detected, respectively, in *Porphyromonas* and *F. prausnitzii* (140). *tetM* in *F. nucleatum* may be borne by a Tn*916*-like element (90), and has been shown to be transferable by conjugation to *Peptostreptococcus* as well as *Enterococcus faecalis* (Table 4) (123). The *tetB* gene mediating tetracycline efflux has been described in *Treponema* (*Prevotella*) *denticola* (126).

Macrolide Resistance

The rates of resistance to clindamycin for this group of genera, between 0 and 35%, vary somewhat depending upon the region of isolation and seem to be highest for *Porphyromonas endodontalis* and *F. mortiferum* (5, 8, 76, 156, 170). The only genes conferring macrolide resistance in this group of species are of the *erm* family. The wide-host range element *ermF* has been found in *Prevotella* (10), *Porphyromonas*, and *Fusobacterium* (31), while *ermG* has been reported as being present in *Prevotella* and *Porphyromonas* (Table 5) (122). The chromosome-encoded macrolide efflux protein of *F. nucleatum* could explain the reduced activity of this family of drugs (71). *T. denticola,* which contains two *rrn* operons, is capable of acquiring macrolide resistance through a point mutation, A2058G, in the peptidyl transferase region in domain V of the 23S rRNA (81).

Fluoroquinolone Resistance

Recently developed fluoroquinolones have moderate to good activity against *Prevotella, Porphyromonas* and *Fusobacterium,* with rates of resistance to trovafloxacin of below 5% (5) and 28.5%, 8%, and 2.5%, respectively, to levofloxacin (76). In veterinary isolates of *F. canifelinum,* which apparently do not contain topoisomerase IV, GyrA but no GyrB mutations were identified (Table 6) (33). In a series of in vitro-selected fluoroquinolone-resistant mutants of *P. gingivalis,* only a Ser83Phe substitution in GyrA was observed (46). An AcrB-related multidrug efflux system termed XepCAB has been described in this species, the inactivation of which resulted in only discretely increased susceptibilities to fluoroquinolones, as well as tetracyclines and rifampin (66).

RESISTANCE IN CLOSTRIDIA

Beta-Lactam Resistance

Clostridia are largely susceptible to β-lactam antibiotics. Penicillinase producers have been reported only in *C. ramosum, C. butyricum,* and *C. clostridioforme* (Table 1). *C. perfringens* is normally susceptible to amoxicillin-clavulanate, cefoxitin and imipenem, while *C. difficile* strains are fully susceptible to amoxicillin but intrinsically resistant or of reduced susceptibility to cephalosporins and imipenem but not meropenem (7, 68). The genes of the *Clostridium* penicillinases do not seem to have been identified. The *C. butyricum* enzyme is more efficiently inhibited by clavulanate and sulbactam than the enzyme from *C. clostridioforme* (9, 60).

Imidazole Resistance

With few exceptions, clostridia are susceptible to nitroimidazoles (7, 170). While some recent surveys reported full susceptibility of *C. difficile* to metronidazole (68), a study from one institution in Spain covering 8 years reported an overall resistance rate of 6.3% (103). The only imidazole resistance gene identified so far in clostridia, *nimA,* was reported in a strain of *C. bifermentans* from South Africa (86).

Tetracycline Resistance

Tetracycline resistance rates in *Clostridium difficile* may be decreasing, and values as low as 10% or less have been reported recently (15, 19). A recent survey of veterinary isolates of *C. perfringens* revealed rates varying between 10 and 76% depending upon the country of isolation (69). Among the tetracycline resistance genes identified in *Clostridium* (Table 4), *tetP* is predominant (69). This determinant is made up of the two overlapping genes encoding TetA(P) and TetB(P), an efflux protein and a ribosome protection-type protein, respectively (143). Resistance may also be conferred by *tetM* in *C. perfringens* and *C. difficile,* where the gene resides on different forms of Tn*916*-related structures (3), and by *tetK* and *tetL* in *C. difficile* (125).

Macrolide Resistance

Macrolide resistance is widespread in *C. difficile,* reaching rates of up to 96% (19). Macrolide resistance in *C. perfringens* has been known in the United States and in Europe since the mid 1970s (129), but studies limited to small numbers of strains suggest that resistance rates remain low (14, 75).

As in all other anaerobic bacteria, macrolide resistance in *Clostridium* is conferred by the *erm* family of genes (Table 5). Of the known mechanisms of macrolide resistance—drug modification, drug efflux, 23S rRNA target mutation and 23S rRNA methylase production—only the latter seems to occur in the genus. The corresponding genes, *ermB* in *C. difficile* and *ermB* and *ermQ* in *C. perfringens,* can be transferable and may be chromosome-, plasmid- and/or transposon-borne, (Table 5). The residues essential for methylase function of ErmB from *C. perfringens* have been determined (49). Two copies of *ermB* are carried on transposon Tn*5398* of *C. difficile.* It is not a classical transposon in that it apparently does not encode genes involved in either excision, integration, or conjugation but which seems to be a mobilizable nonconjugative element that has been found to be widespread in isolates from different continents (50). The *erm*-mediated resistance in *Clostridium* does not appear to be linked to tetracycline resistance as is common in the *Bacteroides.* This could partly explain the rather different tetracycline resistance rates in the genus.

Fluoroquinolone Resistance

Among the newer fluoroquinolones of clinical interest, moxifloxacin appears to be the most active against *C. difficile,* although somewhat contradictory data on in vitro activities have been published (98).

Resistance rates between 7 and 26% have been inferred (Table 6) (1, 39). For *C. perfringens* and other *Clostridium* species MIC_{90} values of 0.25 to 0.5 μg/ml have been determined (42, 156), but resistance rates do not seem to have been published.

Moxifloxacin resistance in *C. difficile* is due to either *gyrA* or *gyrB* mutations. The most frequently observed substitution was Thr82Ile in GyrA (the position corresponding to that of Ser83 in *E. coli*), but Asp71Val and exceptionally Ala118Thr substitutions have also been described (2, 39). In GyrB, the substitutions were Asp426Asn and Arg447Leu (39). Apparently, *C. difficile* does not contain topoisomerase IV genes, at least not ones with a sufficient degree of homology to known topoisomerase IV genes to be detected by BLAST or PCR-based procedures (2, 39). On the other hand, *C. perfringens* has the usual sets of gyrase and topoisomerase IV genes, and substitutions in and outside their QRDRs have been found in in vitro generated fluoroquinolone-resistant mutants (Table 6) (112).

As far as resistance by increased drug efflux is concerned, no specific role of a particular system has yet been identified. One *C. difficile* efflux protein conferred some resistance to erythromycin after expression of its gene, *cme*, in *Enterococcus faecalis* (80). The gene *cdeA*, encoding a MATE-type efflux pump, was found to mediate resistance to intercalating agents but not to quinolones (38).

Vancomycin Resistance

The clostridia are considered susceptible to the glycopeptides vancomycin and teicoplanin. An exceptional rate of reduced susceptibility (or intermediate resistance) to vancomycin of 3.1%, with MICs of 8 to 16 μg/ml, has been reported from one institution, but nothing is known about the mechanism responsible for this phenotype (103). Similar levels of intermediate resistance, with concomitant susceptibility to teicoplanin, are characteristic of the intrinsic resistance of *C. innocuum,* which constitutively produces predominantly a pentapeptide precursor with a C-terminal serine and ensuing low affinity for vancomycin (36). In human fecal isolates of high-level vancomycin-resistant *Clostridium* species closely related to *C. innocuum* and *C. symbiosum* and in an unidentified clostridial species, a *vanB* locus similar to that of vancomycin-resistant enterococci was found (154).

RESISTANCE IN PROPIONIBACTERIA

In *P. acnes* isolates from several European countries, collected over a 6-year period, overall rates of resistance to erythromycin, clindamycin and tetracycline were ca. 17%, 15%, and 3%, respectively (100). The corresponding rates were substantially higher in isolates of *P. acnes* and *P. granulosum* from acne patients and their contacts (130). In both sets of isolates there were notable inter-country variations, but all strains were fully susceptible to penicillins and, where tested, also to fusidic acid, vancomycin and linezolid (17, 93, 100). *P. acnes* has otherwise been found to be susceptible to ofloxacin and ciprofloxacin (93, 147).

Propionibacterium species contain one, two or, as in the case of *P. acnes,* three copies of the rRNA operon (25, 131). This is one, if not the, feature which enables them to develop tetracycline and macrolide resistance through target mutations. A single apparent G1058C change in helix 34 was observed in the 16S rRNA of isolates resistant to tetracycline that were more resistant to this compound and to doxycycline than to minocycline (133). As for macrolide resistance, constitutive MLS_B-type resistance, the most frequently observed phenotype (40), was correlated with an A2058G change in the 23S rRNA, low-level erythromycin resistance with a G2057A change and high-level resistance to 14- and 16-membered ring macrolides with an A2059G change. No change in this region was found associated with inducible MLS_B resistance observed in *P. granulosum* (131). High-level resistance to all MLS_B antibiotics, including the ketolide telithromycin, which accounted for almost 10% of erythromycin resistance in propionibacteria from acne patients, was mediated by the constitutively expressed, Tn*5432*-borne *ermX* gene (132).

While the role of the *nimA* genes in the intrinsically metronidazole nonsusceptible *Propionibacterium* spp. is unclear, this genus has been suspected to be a possible reservoir of these genes (86).

RESISTANCE IN PEPTOSTREPTOCOCCI

The peptostreptococci have recently been grouped into the genera *Anaerococcus, Finegoldia, Micromonas, Peptoniphilus,* and *Peptostreptococcus* on the basis of 16S rRNA analysis (151). Resistance rates for the group have been reported to be 6 to 10% for penicillins, 7 to 20% for clindamycin, 5 to 10% for metronidazole (5, 14, 24, 170), and up to 14 and 27% for ciprofloxacin and levofloxacin, respectively, in certain species (5, 76). Noticeably higher rates may be observed in certain regions, with rates of 55% for clindamycin and over 30% for metronidazole resistance (111, 156).

The resistance to β-lactams is likely to depend on PBP alterations, although production of β-lactamases has been suggested, none of which seems to have been characterized satisfactorily, including that of CfxA2 (59, 95). As for metronidazole resistance, the *nimB* gene has been found in a very high proportion of peptostreptococci, although it was concluded to be responsible for high-level resistance in only two strains of *F. magna* and silent in 19 out of 21 susceptible strains (157). On the other hand, several strains of low-level resistant *Peptostreptococcus* and one high-level resistant strain of *Peptoniphilus asaccharolyticus* were devoid of *nim* genes (86). Frequent resistance to tetracycline (24) is correlated with the single or cumulated production of efflux- and ribosomal protection-mediating proteins (Table 4), predominantly *tetM* carried on a transferable Tn*916* transposon (121, 124). Constitutive or inducible MLS_B-type resistance may be conferred by *ermTR*, *ermF* or *ermB* elements. Most predominant is *ermTR*, the *erm* gene most common in *Streptococcus pyogenes*, which can receive it via in vitro transfer from *F. magna* (118, 119).

CONCLUSIONS

Despite the great taxonomic distances between some of the groups of the anaerobic bacteria, the occurrence of similar target mutations in various species is trivial, such as gyrase mutations conferring resistance to fluoroquinolones or ribosomal RNA mutation conferring macrolide resistance in those species that have low copy numbers of the *rrn* operon. More noteworthy is the wide-spread presence of determinants conferring tetracycline and macrolide resistance that are carried by an impressive array of transmissible elements, especially of the conjugative transposons in the *Bacteroides* group (3, 142, 171) (see chapter 33), the dissemination of which is held responsible for the observed increases in resistance rates, notably for clindamycin (4, 5). Dissemination of these elements should, however, not account for the increase in resistance that has been observed also for ampicillin-sulbactam (4, 5). On the other hand, there are observations that may not be as disquieting as many in the realm of antimicrobial resistance. The rates of resistance to carbapenems and imidazoles have remained low overall, as has antimicrobial resistance in *Porphyromonas* spp., *Fusobacteria*, and maybe to a somewhat lesser degree also in *Prevotella* and peptostreptococci. Specifically, the production of the metallo-β-lactamase CfiA has remained confined to a small subgroup of *B. fragilis*, although producers of this enzyme should a priori be easily selectable, not only in the presence of carbapenems but also in the presence of the more commonly prescribed β-lactam–β-lactamase inhibitor combinations. Nevertheless, the absence of increasing rates of carbapenem-resistant isolates should not silence previously expressed concern (114) about the possibility of a less favorable development in the future.

Acknowledgments. We gratefully acknowledge D.W. Hecht and T. Fosse for critical reading of the manuscript and for helpful comments, and M. Lecerf for help with compilation of the tables.

REFERENCES

1. **Ackermann, G., Y. J. Tang, R. Kueper, P. Heisig, A. C. Rodloff, J. Silva Jr., and S. H. Cohen.** 2001. Resistance to moxifloxacin in toxigenic *Clostridium difficile* isolates is associated with mutations in *gyrA*. *Antimicrob. Agents Chemother.* **45:**2348–2353.
2. **Ackermann, G., Y. J. Tang-Feldman, R. Schaumann, J. P. Henderson, A. C. Rodloff, Silva, and S. H. Cohen.** 2003. Antecedent use of fluoroquinolones is associated with resistance to moxifloxacin in *Clostridium difficile. Clin. Microbiol. Infect.* **9:**526–530.
3. **Adams, V., D. Lyras, K. A. Farrow, and J. I. Rood.** 2002. The clostridial mobilisable transposons. *Cell. Mol. Life Sci.* **59:** 2033–2043.
4. **Aldridge, K. E., and M. O'Brien.** 2002. In vitro susceptibilities of the *Bacteroides fragilis* group species: change in isolation rates significantly affects overall susceptibility data. *J. Clin. Microbiol.* **40:**4349–4352.
5. **Aldridge, K. E., D. Ashcraft, K. Cambre, C. L. Pierson, S. G. Jenkins, and J. E. Rosenblatt.** 2001. Multicenter survey of the changing in vitro antimicrobial susceptibilities of clinical isolates of *Bacteroides fragilis* group, *Prevotella, Fusobacterium, Porphyromonas,* and *Peptostreptococcus* species. *Antimicrob. Agents Chemother.* **45:**1238–1243.
6. **Aldridge, K. E., D. Ashcraft, M. O'Brien, and C. V. Sanders.** 2003. Bacteremia due to *Bacteroides fragilis* group: distribution of species, beta-lactamase production, and antimicrobial susceptibility patterns. *Antimicrob. Agents Chemother.* **47:** 148–153.
7. **Alexander, C. J., D. M. Citron, J. S. Brazier, and E. J. Goldstein.** 1995. Identification and antimicrobial resistance patterns of clinical isolates of *Clostridium clostridioforme, Clostridium innocuum,* and *Clostridium ramosum* compared with those of clinical isolates of *Clostridium perfringens. J. Clin. Microbiol.* **33:**3209–3215.
8. **Andres, M. T., W. O. Chung, M. C. Roberts, and J. F. Fierro.** 1998. Antimicrobial susceptibilities of *Porphyromonas gingivalis, Prevotella intermedia,* and *Prevotella nigrescens* spp. isolated in Spain. *Antimicrob. Agents Chemother.* **42:**3022–3023.
9. **Appelbaum, P. C. , S. K. Spangler, G. A. Pankuch, A. Philippon, M. R. Jacobs, R. Shiman, E. J. Goldstein, and D. M. Citron.** 1994. Characterization of a beta-lactamase from *Clostridium clostridioforme. J. Antimicrob. Chemother.* **33:** 33–40.
10. **Arzese, A. R., L. Tomasetig, and G. A. Botta.** 2000. Detection of *tetQ* and *ermF* antibiotic resistance genes in *Prevotella* and *Porphyromonas* isolates from clinical specimens and resident microbiota of humans. *J. Antimicrob. Chemother.* **45:**577–582.
11. **Avelar, K. E., K. Otsuki, A. C. Vicente, J. M. Vieira, G. R. de Paula, R. M. Domingues, and M. C. Ferreira.** 2003. Presence of the *cfxA* gene in *Bacteroides distasonis. Res. Microbiol.* **154:**369–374.
12. **Bachoual, R., L. Dubreuil, C. J. Soussy, and J. Tankovic.** 2000. Roles of *gyrA* mutations in resistance of clinical isolates and in vitro mutants of *Bacteroides fragilis* to the new fluoroquinolone trovafloxacin. *Antimicrob. Agents Chemother.* **44:**1842–1845.
13. **Bandoh, K., K. Ueno, K. Watanabe, and N. Kato.** 1993. Susceptibility patterns and resistance to imipenem in the

Bacteroides fragilis group species in Japan: a 4-year study. *Clin. Infect. Dis.* **16** (Suppl. 4):S382–S386.
14. **Baquero, F., and M. Reig.** 1992. Resistance of anaerobic bacteria to antimicrobial agents in Spain. *Eur. J. Clin. Microbiol. Infect. Dis.* **11**:1016–1020.
15. **Barbut, F., D. Decre, B. Burghoffer, D. Lesage, F. Delisle, V. Lalande, M. Delmee, V. Avesani, N. Sano, C. Coudert, and J. C. Petit.** 1999. Antimicrobial susceptibilities and serogroups of clinical strains of *Clostridium difficile* isolated in France in 1991 and 1997. *Antimicrob. Agents Chemother.* **43**:2607–2611.
16. **Bayley, D. P., E. R. Rocha, and C. J. Smith.** 2000. Analysis of *cepA* and other *Bacteroides fragilis* genes reveals a unique promoter structure. *FEMS Microbiol. Lett.* **193**:149–154.
17. **Behra-Miellet, J., L. Calvet, and L. Dubreuil.** 2003. Activity of linezolid against anaerobic bacteria. *Int. J. Antimicrob. Agents* **22**:28–34.
18. **Behra-Miellet, J., L. Calvet, and L. Dubreuil.** 2004. A Bacteroides thetaiotamicron porin that could take part in resistance to beta-lactams. *Int. J. Antimicrob. Agents* **24**:135–143.
19. **Bendle, J. S. , P. A. James, P. M. Bennett, M. B. Avison, A. P. Macgowan, and K. M. Al-Shafi.** 2004. Resistance determinants in strains of *Clostridium difficile* from two geographically distinct populations. *Int. J. Antimicrob. Agents* **24**:619–621.
20. **Bendtsen, J. D., H. Nielsen, G. Von Heijne, and S. Brunak.** 2004. Improved prediction of signal peptides: SignalP 3.0. *J. Mol. Biol.* **340**:783–795.
21. **Berg, R. D.** 1996. The indigenous gastrointestinal microflora. *Trends Microbiol.* **4**:430–435.
22. **Berryman, D. I., and J. I. Rood.** 1995. The closely related *ermB-ermAM* genes from *Clostridium perfringens, Enterococcus faecalis* (pAM beta 1), and *Streptococcus agalactiae* (pIP501) are flanked by variants of a directly repeated sequence. *Antimicrob. Agents Chemother.* **39**:1830–1834.
23. **Berryman, D. I., M. Lyristis, and J. I. Rood.** 1994. Cloning and sequence analysis of *ermQ*, the predominant macrolide-lincosamide-streptogramin B resistance gene in *Clostridium perfringens. Antimicrob. Agents Chemother.* **38**:1041–1046.
24. **Brazier, J. S., V. Hall, T. E. Morris, M. Gal, and B. I. Duerden.** 2003. Antibiotic susceptibilities of gram-positive anaerobic cocci: results of a sentinel study in England and Wales. *J. Antimicrob. Chemother.* **52**:224–228.
25. **Bruggemann, H., A. Henne, F. Hoster, H. Liesegang, A. Wiezer, A. Strittmatter, S. Hujer, P. Durre, and G. Gottschalk.** 2004. The complete genome sequence of *Propionibacterium acnes*, a commensal of human skin. *Science* **305**:671–673.
26. **Burdett, V.** 1996. Tet(M)-promoted release of tetracycline from ribosomes is GTP dependent. *J. Bacteriol.* **178**:3246–3251.
27. **Carfi, A., E. Duee, R. Paul-Soto, M. Galleni, J. M. Frere, and O. Dideberg.** 1998. X-ray structure of the ZnII beta-lactamase from *Bacteroides fragilis* in an orthorhombic crystal form. *Acta Crystallogr. D. Biol. Crystallogr.* **54**:45–47.
28. **Carlier, J. P., N. Sellier, M. N. Rager, and G. Reysset.** 1997. Metabolism of a 5-nitroimidazole in susceptible and resistant isogenic strains of *Bacteroides fragilis. Antimicrob. Agents Chemother.* **41**:1495–1499.
29. **Chaudhry, R., P. Mathur, B. Dhawan, and L. Kumar.** 2001. Emergence of metronidazole-resistant *Bacteroides fragilis*, India. *Emerg. Infect. Dis.* **7**:485–486.
30. **Chung, W. O., J. Gabany, G. R. Persson, and M. C. Roberts.** 2002. Distribution of *erm(F)* and *tet(Q)* genes in 4 oral bacterial species and genotypic variation between resistant and susceptible isolates. *J. Clin. Periodontol.* **29**:152–158.
31. **Chung, W. O., K. Young, Z. Leng, and M. C. Roberts.** 1999. Mobile elements carrying *ermF* and *tetQ* genes in gram-positive and gram-negative bacteria. *J. Antimicrob. Chemother.* **44**:329–335.
32. **Concha, N. O., B. A. Rasmussen, K. Bush, and O. Herzberg.** 1996. Crystal structure of the wide-spectrum binuclear zinc beta-lactamase from *Bacteroides fragilis. Structure* **4**:823–836.
33. **Conrads, G., D. M. Citron, and E. J. Goldstein.** 2005. Genetic determinant of intrinsic quinolone resistance in *Fusobacterium canifelinum. Antimicrob. Agents Chemother.* **49**:434–437.
34. **Cooper, A. J., N. B. Shoemaker, and A. A. Salyers.** 1996. The erythromycin resistance gene from the Bacteroides conjugal transposon Tcr Emr 7853 is nearly identical to *ermG* from *Bacillus sphaericus. Antimicrob. Agents Chemother.* **40**:506–508.
35. **Cuchural, G. J. Jr., M. H. Malamy, and F. P. Tally.** 1986. Beta-lactamase-mediated imipenem resistance in *Bacteroides fragilis. Antimicrob. Agents Chemother.* **30**:645–648.
36. **David, V., B. Bozdogan, J. L. Mainardi, R. Legrand, L. Gutmann, and R. Leclercq.** 2004. Mechanism of intrinsic resistance to vancomycin in *Clostridium innocuum* NCIB 10674. *J. Bacteriol.* **186**:3415–3422.
37. **de Barbeyrac, B., B. Dutilh, C. Quentin, H. Renaudin, and C. Bebear.** 1991. Susceptibility of *Bacteroides ureolyticus* to antimicrobial agents and identification of a tetracycline resistance determinant related to *tetM. J. Antimicrob. Chemother.* **27**:721–731.
38. **Dridi, L., J. Tankovic, and J. C. Petit.** 2004. CdeA of *Clostridium difficile*, a new multidrug efflux transporter of the MATE family. *Microb. Drug Resist.* **10**:191–196.
39. **Dridi, L., J. Tankovic, B. Burghoffer, F. Barbut, and J. C. Petit.** 2002. *gyrA* and *gyrB* mutations are implicated in cross-resistance to ciprofloxacin and moxifloxacin in *Clostridium difficile. Antimicrob. Agents Chemother.* **46**:3418–3421.
40. **Eady, E. A., J. I. Ross, J. H. Cove, K. T. Holland, and W. J. Cunliffe.** 1989. Macrolide-lincosamide-streptogramin B (MLS) resistance in cutaneous propionibacteria: definition of phenotypes. *J. Antimicrob. Chemother.* **23**:493–502.
41. **Edlund, C., and C. E. Nord.** 2000. In-vitro susceptibility of anaerobic bacteria to GAR-936, a new glycylcycline. *Clin. Microbiol. Infect.* **6**:159–163.
42. **Edlund, C., S. Sabouri, and C. E. Nord.** 1998. Comparative in vitro activity of BAY 12-8039 and five other antimicrobial agents against anaerobic bacteria. *Eur. J. Clin. Microbiol. Infect. Dis.* **17**:193–195.
43. **Edwards, R., and D. Greenwood.** 1996. Mechanisms responsible for reduced susceptibility to imipenem in *Bacteroides fragilis. J. Antimicrob. Chemother.* **38**:941–951.
44. **Edwards, R., and P. N. Read.** 2000. Expression of the carbapenemase gene (*cfiA*) in Bacteroides fragilis. *J. Antimicrob. Chemother.* **46**:1009–1012.
45. **Edwards, R., C. V. Hawkyard, M. T. Garvey, and D. Greenwood.** 1999. Prevalence and degree of expression of the carbapenemase gene (*cfiA*) among clinical isolates of *Bacteroides fragilis* in Nottingham, UK. *J. Antimicrob. Chemother.* **43**:273–276.
46. **Eick, S., A. Schmitt, S. Sachse, K. H. Schmidt, and W. Pfister.** 2004. In vitro antibacterial activity of fluoroquinolones against *Porphyromonas gingivalis* strains. *J. Antimicrob. Chemother.* **54**:553–556.
47. **Elsaghier, A. A., J. S. Brazier, and E. A. James.** 2003. Bacteraemia due to *Bacteroides fragilis* with reduced susceptibility to metronidazole. *J. Antimicrob. Chemother.* **51**:1436–1437.
48. **Fang, H., C. Edlund, M. Hedberg, and C. E. Nord.** 2002. New findings in beta-lactam and metronidazole resistant *Bacteroides fragilis* group. *Int. J. Antimicrob. Agents* **19**:361–370.

49. **Farrow, K. A., D. Lyras, G. Polekhina, K. Koutsis, M. W. Parker, and J. I. Rood.** 2002. Identification of essential residues in the Erm(B) rRNA methyltransferase of *Clostridium perfringens. Antimicrob. Agents Chemother.* **46:**1253–1261.
50. **Farrow, K. A., D. Lyras, and J. I. Rood.** 2001. Genomic analysis of the erythromycin resistance element Tn*5398* from *Clostridium difficile. Microbiology* **147:**2717–2728.
51. **Fosse, T., I. Madinier, C. Hitzig, and Y. Charbit.** 1999. Prevalence of beta-lactamase-producing strains among 149 anaerobic gram-negative rods isolated from periodontal pockets. *Oral Microbiol. Immunol.* **14:**352–357.
52. **Gal, M., and J. S. Brazier.** 2004. Metronidazole resistance in *Bacteroides* spp. carrying *nim* genes and the selection of slow-growing metronidazole-resistant mutants. *J. Antimicrob. Chemother.* **54:**109–116.
53. **Giraud-Morin, C., I. Madinier, and T. Fosse.** 2003. Sequence analysis of *cfxA2*-like beta-lactamases in *Prevotella* species. *J. Antimicrob. Chemother.* **51:**1293–1296.
54. **Golan, Y. , L. A. McDermott, N. V. Jacobus, E. J. Goldstein, S. Finegold, L. J. Harrell, D. W. Hecht, S. G. Jenkins, C. Pierson, R. Venezia, J. Rihs, P. Iannini, S. L. Gorbach, and D. R. Snydman.** 2003. Emergence of fluoroquinolone resistance among *Bacteroides* species. *J. Antimicrob. Chemother.* **52:**208–213.
55. **Gupta, A., H. Vlamakis, N. Shoemaker, and A. A. Salyers.** 2003. A new *Bacteroides* conjugative transposon that carries an *ermB* gene. *Appl. Environ. Microbiol.* **69:**6455–6463.
56. **Gutacker, M., C. Valsangiacomo, and J. C. Piffaretti.** 2000. Identification of two genetic groups in *Bacteroides fragilis* by multilocus enzyme electrophoresis: distribution of antibiotic resistance (*cfiA, cepA*) and enterotoxin (*bft*) encoding genes. *Microbiology* **146:**1241–1254.
57. **Haggoud, A., G. Reysset, H. Azeddoug, and M. Sebald.** 1994. Nucleotide sequence analysis of two 5-nitroimidazole resistance determinants from *Bacteroides* strains and of a new insertion sequence upstream of the two genes. *Antimicrob. Agents Chemother.* **38:**1047–1051.
58. **Halula, M. C., S. Manning, and F. L. Macrina.** 1991. Nucleotide sequence of *ermFU,* a macrolide-lincosamide-streptogramin (MLS) resistance gene encoding an RNA methylase from the conjugal element of *Bacteroides fragilis* V503. *Nucleic Acids Res.* **19:**3453.
59. **Handal, T., I. Olsen, C. B. Walker, and D. A. Caugant.** 2005. Detection and characterization of beta-lactamase genes in subgingival bacteria from patients with refractory periodontitis. *FEMS Microbiol. Lett.* **242:**319–324.
60. **Hart, C. A., K. Barr, T. Makin, P. Brown, and R. W. Cooke.** 1982. Characteristics of a beta-lactamase produced by *Clostridium butyricum. J. Antimicrob. Chemother.* **10:**31–35.
61. **Hecht, D. W.** 2002. Evolution of anaerobe susceptibility testing in the United States. *Clin. Infect. Dis.* **35:**S28–S35.
62. **Hecht, D. W.** 2004. Prevalence of antibiotic resistance in anaerobic bacteria: worrisome developments. *Clin. Infect. Dis.* **39:**92–97.
63. **Hecht, D. W., and J. R. Osmolski.** 2003. Activities of garenoxacin (BMS-284756) and other agents against anaerobic clinical isolates. *Antimicrob. Agents Chemother.* **47:**910–916.
64. **Hedberg, M., and C. E. Nord.** 2003. ESCMID Study Group on Antimicrobial Resistance in Anaerobic Bacteria. Antimicrobial susceptibility of *Bacteroides fragilis* group isolates in Europe. *Clin. Microbiol. Infect.* **9:**475–488.
65. **Hurlbut, S., G. J. Cuchural, and F. P. Tally.** 1990. Imipenem resistance in *Bacteroides distasonis* mediated by a novel beta-lactamase. *Antimicrob. Agents Chemother.* **34:**117–120.
66. **Ikeda, T., and F. Yoshimura.** 2002. A resistance-nodulation-cell division family xenobiotic efflux pump in an obligate anaerobe, *Porphyromonas gingivalis. Antimicrob. Agents Chemother.* **46:**3257–3260.
67. **Ingham, H. R., S. Eaton, C. W. Venables, and P. C. Adams.** 1978. *Bacteroides fragilis* resistant to metronidazole after long-term therapy. *Lancet* **1:**214.
68. **Jamal, W. Y., E. M. Mokaddas, T. L. Verghese, and V. O. Rotimi.** 2002. In vitro activity of 15 antimicrobial agents against clinical isolates of *Clostridium difficile* in Kuwait. *Int. J. Antimicrob. Agents* **20:**270–274.
69. **Johansson, A., C. Greko, B. E. Engstrom, and M. Karlsson.** 2004. Antimicrobial susceptibility of Swedish, Norwegian and Danish isolates of *Clostridium perfringens* from poultry, and distribution of tetracycline resistance genes. *Vet. Microbiol.* **99:**251–257.
70. **Johnson, J. L.** 1978. Taxonomy of the Bacteroides I. Deoxyribonucleic acid homologies among *Bacteroides fragilis* and other *Bacteroides species. Int. J. Syst. Bacteriol.* **28:**245–256.
71. **Kapatral, V, I. Anderson, N. Ivanova, G. Reznik, T. Los, A. Lykidis, A. Bhattacharyya, A. Bartman, W. Gardner, G. Grechkin, L. Zhu, O. Vasieva, L. Chu, Y. Kogan, O. Chaga, E. Goltsman, A. Bernal, N. Larsen, M. D'Souza, T. Walunas, G. Pusch, R. Haselkorn, M. Fonstein, N. Kyrpides, and R. Overbeek.** 2002. Genome sequence and analysis of the oral bacterium *Fusobacterium nucleatum* strain ATCC 25586. *J. Bacteriol.* **184:**2005–2018.
72. **Kato, N., K. Yamazoe, C. G. Han, and E. Ohtsubo.** 2003. New insertion sequence elements in the upstream region of *cfiA* in imipenem-resistant *Bacteroides fragilis* strains. *Antimicrob. Agents Chemother.* **47:**979–985.
73. **Kesado, T., L. Lindqvist, M. Hedberg, K. Tuner, and C. E. Nord.** 1989. Purification and characterization of a new beta-lactamase from *Clostridium butyricum. Antimicrob. Agents Chemother.* **33:**1302–1307.
74. **King, A., J. Downes, C. E. Nord, I. Phillips, and European Study Group.** 1999. Antimicrobial susceptibility of non-*Bacteroides fragilis* group anaerobic gram-negative bacilli in Europe. *Clin. Microbiol. Infect.* **5:**404–416.
75. **Koch, C. L., P. Derby, and V. R. Abratt.** 1998. In-vitro antibiotic susceptibility and molecular analysis of anaerobic bacteria isolated in Cape Town, South Africa. *J. Antimicrob. Chemother.* **42:**245–248.
76. **Koeth, L. M. , C. E. Good, P. C. Appelbaum, E. J. Goldstein, A. C. Rodloff, M. Claros, and L. J. Dubreuil.** 2004. Surveillance of susceptibility patterns in 1297 European and US anaerobic and capnophilic isolates to co-amoxiclav and five other antimicrobial agents. *J. Antimicrob. Chemother.* **53:**1039–1044.
77. **Kroes, I., P. W. Lepp, and D. A. Relman.** 1999. Bacterial diversity within the human subgingival crevice. *Proc. Natl. Acad. Sci. USA* **96:**14547–14552.
78. **Kuwahara, T. , A. Yamashita, H. Hirakawa, H. Nakayama, H. Toh, N. Okada, S. Kuhara, M. Hattori, T. Hayashi, and Y. Ohnishi.** 2004. Genomic analysis of *Bacteroides fragilis* reveals extensive DNA inversions regulating cell surface adaptation. *Proc. Natl. Acad. Sci. USA* **101:**14919–14924.
79. **Lamothe, F., C. Fijalkowski, F. Malouin, A. M. Bourgault, and L. Delorme.** 1986. *Bacteroides fragilis* resistant to both metronidazole and imipenem. *J. Antimicrob. Chemother.* **18:**642–643.
80. **Lebel, S., S. Bouttier, and T. Lambert.** 2004. The *cme* gene of *Clostridium difficile* confers multidrug resistance in *Enterococcus faecalis. FEMS Microbiol. Lett.* **238:**93–100.
81. **Lee, S. Y., Y. Ning, and J. C. Fenno.** 2002. 23S rRNA point mutation associated with erythromycin resistance in *Treponema denticola. FEMS Microbiol. Lett.* **207:**39–42.

82. **Leiros, H. K., S. Kozielski-Stuhrmann, U. Kapp, L. Terradot, G. A. Leonard, and S. M. McSweeney.** 2004. Structural basis of 5-nitroimidazole antibiotic resistance: the crystal structure of NimA from *Deinococcus radiodurans*. *J. Biol. Chem.* **279:** 55840–55849.
83. **Leng, Z., D. E. Riley, R. E. Berger, J. N. Krieger, and M. C. Roberts.** 1997. Distribution and mobility of the tetracycline resistance determinant *tetQ*. *J. Antimicrob. Chemother.* **40:** 551–559.
84. **Lepine, G., J. M. Lacroix, C. B. Walker, and A. Progulske-Fox.** 1993. Sequencing of a *tet*(Q) gene isolated from *Bacteroides fragilis* 1126. *Antimicrob. Agents Chemother.* **37:**2037–2041.
85. **Löfmark, S., H. Fang, M. Hedberg, and C. Edlund.** 2005. Inducible metronidazole resistancee and *nim* genes in clinical *Bacteroides fragilis* isolates. *Antimicrob. Agents Chemother.* **49:**1253–1256.
86. **Lubbe, M. M., K. Stanley, and L. J. Chalkley.** 1999. Prevalence of *nim* genes in anaerobic/facultative anaerobic bacteria isolated in South Africa. *FEMS Microbiol. Lett.* **172:**79–83.
87. **Madinier, I., T. Fosse, J. Giudicelli, and R. Labia.** 2001. Cloning and biochemical characterization of a class A beta-lactamase from *Prevotella intermedia*. *Antimicrob. Agents Chemother.* **45:**2386–2389.
88. **Magot, M.** 1981. Some properties of the *Clostridium butyricum* group beta-lactamase. *J. Gen. Microbiol.* **127:**113–119.
89. **Malouin, F., C. Fijalkowski, F. Lamothe, and J. M. Lacroix.** 1986. Inactivation of cefoxitin and moxalactam by *Bacteroides bivius* beta-lactamase. *Antimicrob. Agents Chemother.* **30:**749–755.
90. **McKay, T. L., J. Ko, Y. Bilalis, and J. M. DiRienzo.** 1995. Mobile genetic elements of *Fusobacterium nucleatum*. *Plasmid* **33:**15–25.
91. **Miyamae, S., H. Nikaido, Y. Tanaka, and F. Yoshimura.** 1998. Active efflux of norfloxacin by *Bacteroides fragilis*. *Antimicrob. Agents Chemother.* **42:**2119–2121.
92. **Miyamae, S., O. Ueda, F. Yoshimura, J. Hwang, Y. Tanaka, and H. Nikaido.** 2001. A MATE family multidrug efflux transporter pumps out fluoroquinolones in *Bacteroides thetaiotaomicron*. *Antimicrob. Agents Chemother.* **45:**3341–3346.
93. **Mory, F., S. Fougnot, C. Rabaud, H. Schuhmacher, and A. Lozniewski.** 2005. In vitro activities of cefotaxime, vancomycin, quinupristin/dalfopristin, linezolid and other antibiotics alone and in combination against *Propionibacterium acnes* isolates from central nervous system infections. *J. Antimicrob. Chemother.* **55:**265–268.
94. **Mullany, P., M. Pallen, M. Wilks, J. R. Stephen, and S. Tabaqchali.** 1996. A group II intron in a conjugative transposon from the gram-positive bacterium, *Clostridium difficile*. *Gene* **174:**145–150.
95. **Murdoch, D. A.** 1998. Gram-positive anaerobic cocci. *Clin. Microbiol. Rev.* **11:**81–120.
96. **Nikolich, M. P., G. Hong, N. B. Shoemaker, and A. A. Salyers.** 1994. Evidence for natural horizontal transfer of *tetQ* between bacteria that normally colonize humans and bacteria that normally colonize livestock. *Appl. Environ. Microbiol.* **60:**3255–3260.
97. **Nikolich, M. P., N. B. Shoemaker, and A. A. Salyers.** 1992. A *Bacteroides* tetracycline resistance gene represents a new class of ribosome protection tetracycline resistance. *Antimicrob. Agents Chemother.* **36:**1005–1012.
98. **Oh, H., and C. Edlund.** 2003. Mechanism of quinolone resistance in anaerobic bacteria. *Clin. Microbiol. Infect.* **9:**512–517.
99. **Oh, H., M. Hedberg, and C. Edlund.** 2002. Efflux-mediated fluoroquinolone resistance in the *Bacteroides fragilis* group. *Anaerobe* **8:**277–282.
100. **Oprica, C., and C. E. Nord** and ESCMID Study Group on Antimicrobial Resistance in Anaerobic Bacteria. 2005. European surveillance study on the antibiotic susceptibility of *Propionibacterium acnes*. *Clin. Microbiol. Infect.* **11:**204–213.
101. **Parker, A. C., and C. J. Smith.** 1993. Genetic and biochemical analysis of a novel Ambler class A beta-lactamase responsible for cefoxitin resistance in *Bacteroides* species. *Antimicrob. Agents Chemother.* **37:**1028–1036.
102. **Paula, G. R. , L. S. Falcao, E. N. Antunes, K. E. Avelar, F. N. Reis, M. A. Maluhy, M. C. Ferreira, and R. M. Domingues.** 2004. Determinants of resistance in *Bacteroides fragilis* strains according to recent Brazilian profiles of antimicrobial susceptibility. *Int. J. Antimicrob. Agents* **24:**53–58.
103. **Pelaez, T., L. Alcala, R. Alonso, M. Rodriguez-Creixems, J. M. Garcia-Lechuz, and E. Bouza.** 2002. Reassessment of *Clostridium difficile* susceptibility to metronidazole and vancomycin. *Antimicrob. Agents Chemother.* **46:**1647–1650.
104. **Piddock, L. J., and R. Wise.** 1986. Properties of the penicillin-binding proteins of four species of the genus *Bacteroides*. *Antimicrob. Agents Chemother.* **29:**825–832.
105. **Piddock, L. J., and R. Wise.** 1987. Cefoxitin resistance in *Bacteroides* species: evidence indicating two mechanisms causing decreased susceptibility. *J. Antimicrob. Chemother.* **19:**161–170.
106. **Piriz, S., S. Vadillo, A. Quesada, J. Criado, R. Cerrato, and J. Ayala.** 2004. Relationship between penicillin-binding protein patterns and beta-lactamases in clinical isolates of *Bacteroides fragilis* with different susceptibility to beta-lactam antibiotics. *J. Med. Microbiol.* **53:**213–221.
107. **Podglajen, I., J. Breuil, I. Casin, and E. Collatz.** 1995. Genotypic identification of two groups within the species *Bacteroides fragilis* by ribotyping and by analysis of PCR-generated fragment patterns and insertion sequence content. *J. Bacteriol.* **177:**5270–5275.
108. **Podglajen, I., J. Breuil, A. Rohaut, C. Monsempes, and E. Collatz.** 2001. Multiple mobile promoter regions for the rare carbapenem resistance gene of *Bacteroides fragilis*. *J. Bacteriol.* **183:**3531–3535.
109. **Podglajen, I., J. Breuil, and E. Collatz.** 1994. Insertion of a novel DNA sequence, 1S1186, upstream of the silent carbapenemase gene *cfiA*, promotes expression of carbapenem resistance in clinical isolates of *Bacteroides fragilis*. *Mol. Microbiol.* **12:**105–114.
110. **Podglajen, I., J. Breuil, F. Bordon, L. Gutmann, and E. Collatz.** 1992. A silent carbapenemase gene in strains of *Bacteroides fragilis* can be expressed after a one-step mutation. *FEMS Microbiol. Lett.* **70:**21–29.
111. **Poulet, P. P., D. Duffaut, and J. P. Lodter.** 1999. Metronidazole susceptibility testing of anaerobic bacteria associated with periodontal disease. *J. Clin. Periodontol.* **26:**261–263.
112. **Rafii, F., M. Park, and J. S. Novak.** 2005. Alterations in DNA gyrase and topoisomerase IV in resistant mutants of *Clostridium perfringens* found after in vitro treatment with fluoroquinolones. *Antimicrob. Agents Chemother.* **49:**488–492.
113. **Rasmussen, B. A., and E. Kovacs.** 1991. Identification and DNA sequence of a new *Bacteroides fragilis* insertion sequence-like element. *Plasmid* **25:**141–144.
114. **Rasmussen, B. A., K. Bush, and F. P. Tally.** 1997. Antimicrobial resistance in anaerobes. *Clin. Infect. Dis.* **24:**S110–S120.
115. **Rasmussen, B. A., Y. Gluzman, and F. P. Tally.** 1990. Cloning and sequencing of the class B beta-lactamase gene (*ccrA*) from *Bacteroides fragilis* TAL3636. *Antimicrob. Agents Chemother.* **34:**1590–1592.
116. **Rasmussen, J. L., D. A. Odelson, and F. L. Macrina.** 1986. Complete nucleotide sequence and transcription of *ermF*, a macrolide-lincosamide-streptogramin B resistance determinant from *Bacteroides fragilis*. *J. Bacteriol.* **168:**523–533.

117. **Reig, M , M. C. Fernandez, J. P. Ballesta, and F. Baquero.** 1992. Inducible expression of ribosomal clindamycin resistance in *Bacteroides vulgatus*. *Antimicrob. Agents Chemother.* **36:**639–642.
118. **Reig, M, J. Galan, F. Baquero, and J. C. Perez-Diaz.** 2001. Macrolide resistance in *Peptostreptococcus* spp. mediated by *ermTR:* possible source of macrolide-lincosamide-streptogramin B resistance in *Streptococcus pyogenes*. *Antimicrob. Agents Chemother.* **45:**630–632.
119. **Reig, M., A. Moreno, and F. Baquero.** 1992. Resistance of *Peptostreptococcus* spp. to macrolides and lincosamides: inducible and constitutive phenotypes. *Antimicrob. Agents Chemother.* **36:**662–664.
120. **Ricci, V., M. L. Peterson, J. C. Rotschafer, H. Wexler, and L. J. Piddock.** 2004. Role of topoisomerase mutations and efflux in fluoroquinolone resistance of *Bacteroides fragilis* clinical isolates and laboratory mutants. *Antimicrob. Agents Chemother.* **48:**1344–1346.
121. **Roberts, M. C.** 1990. Characterization of the Tet M determinants in urogenital and respiratory bacteria. *Antimicrob. Agents Chemother.* **34:**476–478.
122. **Roberts, M. C.** 2003. Acquired tetracycline and/or macrolide-lincosamides-streptogramin resistance in anaerobes. *Anaerobe* **9:**63–65.
123. **Roberts, M. C., and J. Lansciardi.** 1990. Transferable TetM in *Fusobacterium nucleatum*. *Antimicrob. Agents Chemother.* **34:**1836–1838.
124. **Roberts, M. C., and S. L. Hillier.** 1990. Genetic basis of tetracycline resistance in urogenital bacteria. *Antimicrob. Agents Chemother.* **34:**261–264.
125. **Roberts, M. C., L. V. McFarland, P. Mullany, and M. E. Mulligan.** 1994. Characterization of the genetic basis of antibiotic resistance in *Clostridium difficile*. *J. Antimicrob. Chemother.* **33:**419–429.
126. **Roberts, M. C., W. O. Chung, and D. E. Roe.** 1996. Characterization of tetracycline and erythromycin resistance determinants in *Treponema denticola*. *Antimicrob. Agents Chemother.* **40:**1690–1694.
127. **Rogers, M. B., A. C. Parker, and C. J. Smith.** 1993. Cloning and characterization of the endogenous cephalosporinase gene, *cepA*, from *Bacteroides fragilis* reveals a new subgroup of Ambler class A beta-lactamases. *Antimicrob. Agents Chemother.* **37:**2391–2400.
128. **Rogers, M. B., T. K. Bennett, C. M. Payne, and C. J. Smith.** 1994. Insertional activation of *cepA* leads to high-level beta-lactamase expression in *Bacteroides fragilis* clinical isolates. *J. Bacteriol.* **176:**4376–4384.
129. **Rood, J. I.** 1993. Antibiotic resistance determinants of *Clostridium perfringens*, p. 141–155. *In* M. Sebald (ed.), *Genetics and Molecular Biology of Anaerobic Bacteria*. Springer, New York, N.Y.
130. **Ross, J. I. , A. M. Snelling, E. Carnegie, P. Coates, W. J. Cunliffe, V. Bettoli, G. Tosti, A. Katsambas, J. I. Galvan Perez Del Pulgar, O. Rollman, L. Torok, E. A. Eady, and J. H. Cove.** 2003. Antibiotic-resistant acne: lessons from Europe. *Br. J. Dermatol.* **148:**467–478.
131. **Ross, J. I. , E. A. Eady, J. H. Cove, C. E. Jones, A. H. Ratyal, Y. W. Miller, S. Vyakrnam, and W. J. Cunliffe.** 1997. Clinical resistance to erythromycin and clindamycin in cutaneous propionibacteria isolated from acne patients is associated with mutations in 23S rRNA. *Antimicrob. Agents Chemother.* **41:** 1162–1165.
132. **Ross, J. I., E. A. Eady, E. Carnegie, and J. H. Cove.** 2002. Detection of transposon Tn*5432*-mediated macrolide-lincosamide-streptogramin B (MLSB) resistance in cutaneous propionibacteria from six European cities. *J. Antimicrob. Chemother.* **49:**165–168.
133. **Ross, J. I., E. A. Eady, J. H. Cove, and W. J. Cunliffe.** 1998. 16S rRNA mutation associated with tetracycline resistance in a gram-positive bacterium. *Antimicrob. Agents Chemother.* **42:**1702–1705.
134. **Rotimi, V. O., M. Khoursheed, J. S. Brazier, W. Y. Jamal, and F. B. Khodakhast.** 1999. *Bacteroides* species highly resistant to metronidazole: an emerging clinical problem? *Clin. Microbiol. Infect.* **5:**166–169.
135. **Ruimy, R., I. Podglajen, J. Breuil, R. Christen, and E. Collatz.** 1996. A recent fixation of *cfiA* genes in a monophyletic cluster of *Bacteroides fragilis* is correlated with the presence of multiple insertion elements. *J. Bacteriol.* **178:**1914–1918.
136. **Salyers, A. A., A. Gupta, and Y. Wang.** 2004. Human intestinal bacteria as reservoirs for antibiotic resistance genes. *Trends Microbiol.* **12:**412–416.
137. **Sasaki, Y., K. Yamamoto, Y. Tamura, and T. Takahashi.** 2001. Tetracycline-resistance genes of *Clostridium perfringens, Clostridium septicum* and *Clostridium sordellii* isolated from cattle affected with malignant edema. *Vet. Microbiol.* **83:**61–69.
138. **Schapiro, J. M., R. Gupta, E. Stefansson, F. C. Fang, and A. P. Limaye.** 2004. Isolation of metronidazole-resistant *Bacteroides fragilis* carrying the *nimA* nitroreductase gene from a patient in Washington State. *J. Clin. Microbiol.* **42:**4127–4129.
139. **Scott, K. P.** 2002. The role of conjugative transposons in spreading antibiotic resistance between bacteria that inhabit the gastrointestinal tract. *Cell. Mol. Life Sci.* **59:**2071–2082.
140. **Scott, K. P., C. M. Melville, T. M. Barbosa, and H. J. Flint.** 2000. Occurrence of the new tetracycline resistance gene *tet(W)* in bacteria from the human gut. *Antimicrob. Agents Chemother.* **44:**775–777.
141. **Sebald, M.** 1994. Genetic basis for antibiotic resistance in anaerobes. *Clin. Infect. Dis.* **18** (Suppl. 4):S297–S304.
142. **Shoemaker, N. B., H. Vlamakis, K. Hayes, and A. A. Salyers.** 2001. Evidence for extensive resistance gene transfer among *Bacteroides* spp. and other genera in the human colon. *App. Environ. Microb.* **67:**561–568.
143. **Sloan, J., L. M. McMurry, D. Lyras, S. B. Levy, and J. I. Rood.** 1994. The *Clostridium perfringens* Tet P determinant comprises two overlapping genes: *tetA(P)*, which mediates active tetracycline efflux, and *tetB(P)*, which is related to the ribosomal protection family of tetracycline-resistance determinants. *Mol. Microbiol.* **11:**403–415.
144. **Smith, C. J.** 1987. Nucleotide sequence analysis of Tn*4551:* use of *ermFS* operon fusions to detect promoter activity in *Bacteroides fragilis*. *J. Bacteriol.* **169:**4589–4596.
145. **Smith, C. J., M. B. Rogers, and M. L. McKee.** 1992. Heterologous gene expression in *Bacteroides fragilis*. *Plasmid* **27:** 141–154.
146. **Smith, C. J., T. K. Bennett, and A. C. Parker.** 1994. Molecular and genetic analysis of the *Bacteroides uniformis* cephalosporinase gene, *cblA*, encoding the species-specific beta-lactamase. *Antimicrob. Agents Chemother.* **38:**1711–1715.
147. **Smith, M. A., P. Alperstein, K. France, E. M. Vellozzi, and H. D. Isenberg.** 1996. Susceptibility testing of *Propionibacterium acnes* comparing agar dilution with E test. *J. Clin. Microbiol.* **34:**1024–1026.
148. **Snydman, D. R. , N. V. Jacobus, L. A. McDermott, R. Ruthazer, E. J. Goldstein, S. M. Finegold, L. J. Harrell, D. W. Hecht, S. G. Jenkins, C. Pierson, R. Venezia, J. Rihs, and S. L. Gorbach.** 2002. National survey on the susceptibility of *Bacteroides Fragilis* Group: report and analysis of trends for 1997-2000. *Clin. Infect. Dis.* **35:**S126–S134.
149. **Soki, J., E. Fodor, D. W. Hecht, R. Edwards, V. O. Rotimi, I. Kerekes, E. Urban, and E. Nagy.** 2004. Molecular characterization of imipenem-resistant, *cfiA*-positive *Bacteroides*

fragilis isolates from the USA, Hungary and Kuwait. *J. Med. Microbiol.* **53**:413–419.

150. **Soki, J., R. Edwards, E. Urban, E. Fodor, Z. Beer, and E. Nagy.** 2004. Screening of isolates from faeces for carbapenem-resistant *Bacteroides* strains; existence of strains with novel types of resistance mechanisms. *Int. J. Antimicrob. Agents* **24**:450–454.
151. **Song, Y., C. Liu, M. McTeague, A. Vu, J. Y. Liu, and S. M. Finegold.** 2003. Rapid identification of gram-positive anaerobic coccal species originally classified in the genus Peptostreptococcus by multiplex PCR assays using genus- and species-specific primers. *Microbiology* **149**:1719–1727.
152. **Spahn, C. M., G. Blaha, R. K. Agrawal, P. Penczek, R. A. Grassucci, C. A. Trieber, S. R. Connell, D. E. Taylor, K. H. Nierhaus, and J. Frank.** 2001. Localization of the ribosomal protection protein Tet(O) on the ribosome and the mechanism of tetracycline resistance. *Mol. Cell.* **7**:1037–1045.
153. **Speer, B. S., L. Bedzyk, and A. A. Salyers.** 1991. Evidence that a novel tetracycline resistance gene found on two *Bacteroides* transposons encodes an NADP-requiring oxidoreductase. *J. Bacteriol.* **173**:176–183.
154. **Stinear, T. P., D. C. Olden, P. D. Johnson, J. K. Davies, and M. L. Grayson.** 2001. Enterococcal *vanB* resistance locus in anaerobic bacteria in human faeces. *Lancet* **357**:855–856.
155. **Stubbs, S. L., J. S. Brazier, P. R. Talbot, and B. I. Duerden.** 2000. PCR-restriction fragment length polymorphism analysis for identification of *Bacteroides* spp. and characterization of nitroimidazole resistance genes. *J. Clin. Microbiol.* **38**:3209–3213.
156. **Teng, L. J., P. R. Hsueh, J. C. Tsai, S. J. Liaw, S. W. Ho, and K. T. Luh.** 2002. High incidence of cefoxitin and clindamycin resistance among anaerobes in Taiwan. *Antimicrob. Agents Chemother.* **46**:2908–2913.
157. **Theron, M. M., M. N. Janse Van Rensburg, and L. J. Chalkley.** 2004. Nitroimidazole resistance genes (*nimB*) in anaerobic gram-positive cocci (previously *Peptostreptococcus* spp.). *J. Antimicrob. Chemother.* **54**:240–242.
158. **Thompson, J. S., and M. H. Malamy.** 1990. Sequencing the gene for an imipenem-cefoxitin-hydrolyzing enzyme (CfiA) from *Bacteroides fragilis* TAL2480 reveals strong similarity between CfiA and Bacillus cereus beta-lactamase II. *J. Bacteriol.* **172**:2584–2593.
159. **Tribble, G. D., A. C. Parker, and C. J. Smith.** 1999. Genetic structure and transcriptional analysis of a mobilizable, antibiotic resistance transposon from *Bacteroides*. *Plasmid* **42**:1–12.
160. **Trinh, S., A. Haggoud, G. Reysset, and M. Sebald.** 1995. Plasmids pIP419 and pIP421 from *Bacteroides*: 5-nitroimidazole resistance genes and their upstream insertion sequence elements. *Microbiology* **141**:927–935.
161. **Tuner, K., L. Lindqvist, and C. E. Nord.** 1985. Purification and properties of a novel beta-lactamase from *Fusobacterium nucleatum*. *Antimicrob. Agents Chemother.* **27**:943–947.
162. **Vedantam, G., and D. W. Hecht.** 2003. Antibiotics and anaerobes of gut origin. *Curr. Opin. Microbiol.* **6**:457–461.
163. **Villedieu, A., M. L. Diaz-Torres, N. Hunt, R. McNab, D. A. Spratt, M. Wilson, and P. Mullany.** 2003. Prevalence of tetracycline resistance genes in oral bacteria. *Antimicrob. Agents Chemother.* **47**:878–882.
164. **Vingadassalom, D., A. Kolb, C. Mayer, T. Rybkine, E. Collatz, and I. Podglajen.** 2005. An unusual primary sigma factor in the *Bacteroidetes* phylum. *Mol. Microbiol.* **56**:888–902.
165. **Wang, J., N. B. Shoemaker, G. R. Wang, and A. A Salyers.** 2000. Characterization of a Bacteroides mobilizable transposon, NBU2, which carries a functional lincomycin resistance gene. *J. Bacteriol.* **182**:3559–3571.
166. **Wang, Y., G. R. Wang, A. Shelby, N. B. Shoemaker, and A. A. Salyers.** 2003. A newly discovered *Bacteroides* conjugative transposon, CTnGERM1, contains genes also found in gram-positive bacteria. *Appl. Environ. Microbiol.* **69**:4595–4603.
167. **Weinrich, A. E., and V. E. Del Bene.** 1976. Beta-lactamase activity in anaerobic bacteria. *Antimicrob. Agents Chemother.* **10**:106–111.
168. **Wexler, H. M.** 2002. Outer-membrane pore-forming proteins in gram-negative anaerobic bacteria. *Clin. Infect. Dis.* **35** (Suppl. 1):S65–S71.
169. **Wexler, H. M., and S. Halebian.** 1990. Alterations to the penicillin-binding proteins in the *Bacteroides fragilis* group: a mechanism for non-beta-lactamase mediated cefoxitin resistance. *J. Antimicrob. Chemother.* **26**:7–20.
170. **Wexler, H. M., D. Molitoris, S. St John, A. Vu, E. K. Read, and S. M. Finegold.** 2002. In vitro activities of faropenem against 579 strains of anaerobic bacteria. *Antimicrob. Agents Chemother.* **46**:3669–3675.
171. **Whittle, G., N. B. Shoemaker, and A. A. Salyers.** 2002. The role of *Bacteroides* conjugative transposons in the dissemination of antibiotic resistance genes. *Cell. Mol. Life Sci.* **59**:2044–2054.
172. **Whittle, G., T. R. Whitehead, N. Hamburger, N. B. Shoemaker, M. A. Cotta, and A. A. Salyers.** 2003. Identification of a new ribosomal protection type of tetracycline resistance gene, *tet*(36), from swine manure pits. *Appl. Environ. Microbiol.* **69**:4151–4158.
173. **Xu, J. , M. K. Bjursell, J. Himrod, S. Deng, L. K. Carmichael, H. C. Chiang, L. V. Hooper, and J. I. Gordon.** 2003. A genomic view of the human-*Bacteroides thetaiotaomicron* symbiosis. *Science* **299**:2074–2076.
174. **Xu, J., H. C. Chiang, M. K. Bjursell, and J. I. Gordon.** 2004. Message from a human gut symbiont: sensitivity is a prerequisite for sharing. *Trends Microbiol.* **12**:21–28.
175. **Yamazoe, K., N. Kato, H. Kato, K. Tanaka, Y. Katagiri, and K. Watanabe.** 1999. Distribution of the *cfiA* gene among *Bacteroides fragilis* strains in Japan and relatedness of *cfiA* to imipenem resistance. *Antimicrob. Agents Chemother.* **43**:2808–2810.
176. **Yang, W., I. F. Moore, K. P. Koteva, D. C. Bareich, D. W. Hughes, and G. D. Wright.** 2004. TetX is a flavin-dependent monooxygenase conferring resistance to tetracycline antibiotics. *J. Biol. Chem.* **279**:52346–52352.
177. **Yang, Y., B. A. Rasmussen, and K. Bush.** 1992. Biochemical characterization of the metallo-beta-lactamase CcrA from *Bacteroides fragilis* TAL3636. *Antimicrob. Agents Chemother.* **36**:1155–1157.
178. **Yotsuji, A., J. Mitsuyama, R. Hori, T. Yasuda, I. Saikawa, M. Inoue, and S. Mitsuhashi.** 1988. Mechanism of action of cephalosporins and resistance caused by decreased affinity for penicillin-binding proteins in *Bacteroides fragilis*. *Antimicrob. Agents Chemother.* **32**:1848–1853.
179. **Yotsuji, A., S. Minami, M. Inoue, and S. Mitsuhashi.** 1983. Properties of novel beta-lactamase produced by *Bacteroides fragilis*. *Antimicrob. Agents Chemother.* **24**:925–929.

Frontiers in Antimicrobial Resistance: a Tribute to Stuart B. Levy
Edited by D. G. White, M. N. Alekshun, and P. F. McDermott

Chapter 26

Pseudomonas aeruginosa

KEITH POOLE

Pseudomonas aeruginosa is a ubiquitous aerobic gram-negative opportunistic pathogen and one of the most common pathogens causing nosocomial infections in seriously ill patients worldwide. A common cause of hospital-acquired pneumonias and urinary tract and bloodstream infections, *P. aeruginosa* is the major pathogen in cystic fibrosis (CF), where it causes chronic lung infections and is a pathogen in soft tissue, eye, ear and burn infections. Treatment of *P. aeruginosa* infections is complicated by the intrinsic resistance of the organism to many antimicrobial agents and the ready selection of resistant, often highly resistant, isolates during antimicrobial chemotherapy. Long attributed to the presence of a relatively impermeable outer membrane (OM) that restricts the ready entry of antimicrobials into the cell, the intrinsic multidrug resistance of *P. aeruginosa* results, in fact, from the synergistic activity of the outer membrane barrier and the operation of broadly specific multidrug efflux systems that together limit antimicrobial accumulation in this organism (73). And while mutational hyperexpression of these systems and/or OM alterations that further limit drug accumulation contribute to enhanced multidrug resistance that further complicates antipseudomonal chemotherapy (75), it is clear that a multiplicity of mechanisms often contribute to acquired multidrug resistance in *P. aeruginosa* (36). Whatever its cause, however, multidrug resistance is an all-too-common feature of this pathogen, particularly in CF (71) and intensive-care unit (ICU) (29) isolates, where it compromises timely and effective chemotherapy and, thus, increases length of hospital stays, morbidity and mortality (9).

EFFLUX-MEDIATED INTRINSIC AND ACQUIRED MULTIDRUG RESISTANCE

Chromosomally encoded multidrug efflux systems of the resistance-nodulation-division (RND) family appear to be the most significant regarding export of and, thus, resistance to clinically important antimicrobials in *P. aeruginosa* and, indeed, other gram-negative pathogens (78). Typically tripartite and including inner membrane, periplasmic and outer membrane constituents, seven such systems have been described to date in *P. aeruginosa* (reviewed in detail in reference 78) (Table 1), though only MexAB-OprM and MexXY-OprM have been implicated in intrinsic resistance, their loss in otherwise wild type cells enhancing susceptibility to numerous antimicrobials. While MexCD-OprJ and MexEF-OprN are typically quiescent in wild-type cells these are hyperexpressed in lab and clinical multidrug resistant *nfxB* and *nfxC* mutants, respectively, although MexCD-OprJ is inducible by the biocides benzalkonium chloride and chlorhexidine though no antibiotics (60). Recently, too, mutations in a gene, *mexS* (PA2491, *qrh*), very near to the *mexEF-oprN* operon have been identified in laboratory-selected multidrug-resistant *P. aeruginosa* strains overexpressing MexEF-OprN (86a). Mutational hyperexpression of MexAB-OprM and MexXY has also been reported in lab and/or clinical isolates displaying enhanced resistance (78), with hyperexpression of MexAB-OprM seen in *nalB* (i.e., *mexR*), *nalC* and *nalD* (i.e., PA3574) mutants (11, 78, 85a). While hyperexpression of MexXY has been noted in strains carrying mutations in the *mexZ* gene encoding a repressor of *mexXY* expression (49a, 90), such hyperexpression has also been noted in the absence of such mutations (86, 90).

RESISTANCE TO β-LACTAMS

β-lactams with anti-pseudomomal activity include penicillins (e.g., carbenicillin, ticarcillin, piperacillin, azlocillin), cephalosporins (e.g., ceftazidime, cefoperazone, cefotaxime, cefepime, cefpirome), monobactams (aztreonam) and carbapenems (imipenem and meropenem) and while many of these continue to be used in the treatment of *P. aeruginosa* infections,

Keith Poole • Department of Microbiology and Immunology, Queen's University, Kingston, Ontario, Canada K7L 3N6.

Table 1. Efflux determinants of antimicrobial resistance in *P. aeruginosa*[a]

Efflux components[b]					Expression[c]	Substrates[d]
MFP	RND	OMF	MATE	SMR		
MexA	MexB	OprM			wt/+; *nalB*/+++; *nalC*/++; *nalD*/++	BL, FQ, CM, TC, NV, ML, TP, TG, SM, TS
MexC	MexD	OprJ			wt/−; *nfxB*/++; inducible by certain dyes and disinfectants	BL, FQ, CM, TC, NV, TP, TG, ML, TS
MexE	MexF	OprN			wt/−; *nfxC*/++; *qrh*/++	FQ, CM, TP, TS
MexX (AmrA)	MexY (AmrB)	OprM			wt/+; mutant/++; inducible by aminoglycosides, tetracyclines, and macrolides	BL, FQ, AG, TC, ER,TG
MexJ	MexK	OprM			wt/−; mutant/+	TS[e], ER, TC, CIP
MexH	MexI	OpmD			wt/+; mutant/?	NOR
MexV	MexW	OprM			wt/?; mutant/+	FQ, CM, TC, ER
			PmpM		?	FQ, BAC
				EmrE	wt/+	AG
				QacE	plasmid/integron-encoded	BAC
				QacEΔ1	plasmid/integron-encoded	BAC

[a] Modified from Poole (78).
[b] Antimicrobial efflux systems are identified according to the families of bacterial drug efflux systems to which they belong. Resistance-nodulation-division (RND) family pumps are typically tripartite and include as additional components the membrane fusion protein (MFP) and the outer membrane factor (OMF). The multidrug and toxic compound extrusion (MATE) and small multidrug resistance (SMR) family drug pumps described to date in *P. aeruginosa* are single-component pumps.
[c] wt/+, efflux system is known to be expressed in wild-type (wt) cells (under laboratory growth conditions). wt/+ mutant/++, efflux system is expressed in wt cells, but expression is enhanced in resistant strains. wt/− mutant/+, efflux system is not expressed in wt cells but is expressed in resistant strains. In instances where the nature of the mutation leading to enhanced efflux gene expression is known, the gene is indicated along with the relative level of gene expression.
[d] AG, aminoglycosides; BAC, benzalkonium chloride; BL, β-lactams; CM, chloramphenicol; CIP, ciprofloxacin; ER, erythromycin; FQ, fluoroquinolones; ML, macrolides; NOR, norfloxacin; NV, novobiocin; SM, sulfonamides; TC, tetracycline; TG, tigecycline; TS, triclosan; TP, trimethorpim. In instances where only one member of a class of antimicrobial has been tested or is known to be a substrate for a given pump, that member is identified. Where several members of an antimicrobial class are known to be substrates, the class is identified rather than the actual compounds tested.
[e] Efflux of triclosan but not the other antimicrobials is provided by MexJK-OpmH.

resistance is an issue, particularly in CF isolates and isolates from Latin America (Table 2). As with most gram-negative pathogens, the production of β-lactam-hydrolyzing β-lactamases is a major determinant of resistance, although drug efflux or reduced outer membrane permeability can be important contributing factors (see reference 79 for a review on resistance to β-lactams with many references to *P. aeruginosa*).

β-Lactamases

Penicillin-inactivating activity was first reported in *P. aeruginosa* more than 40 years ago, with early reports documenting a cephalosporinase activity now attributed to the chromosomally encoded class C enzyme (AmpC) produced by all *P. aeruginosa* and, indeed, many gram-negative bacteria. In addition to this endogenous β-lactamase, numerous plasmid-encoded β-lactamases have been reported in *P. aeruginosa*, including enzymes of molecular classes A, B and D (Table 3).

AmpC

AmpC enzymes are broad-spectrum β-lactamases that, particularly at high levels, readily hydrolyze and thus provide resistance to most antipseudomonal β-lactams except carbapenems and, generally, the fourth generation cephalosporins (cefepime, cefpirome), though the *P. aeruginosa* AmpC also poorly hydrolyzes carboxypenicillins (i.e., carbenicillin, ticarcillin) (48). Still, the level and range of resistance will vary with AmpC expression (17, 48), with high level production of this enzyme providing resistance to all β-lactams save the carbapenems (48). Like its counterpart in other gram-negative bacteria, the AmpC enzyme of *P. aeruginosa* is inducible by many of its β-lactam substrates, though poorly by many of the β-lactams with antipseudomonal activity (e.g., aztreonam, ceftazidime, cefepime, pipericillin) (17, 48). And while carbapenems are good inducers of this enzyme, their rapid bactericidal activity and stability to AmpC hydrolysis renders them effective against *P. aeruginosa*. Thus, most instances of AmpC-mediated β-lactam resistance in this organism result from mutations that enhance enzyme production (17, 48).

Expression of the *ampC* gene, as in other AmpC-producing organisms, is under the control of the AmpR regulator that apparently represses *ampC* expression in the absence of inducers and activates *ampC* expression when these are present (17). Not surprisingly, then, mutations in *ampR* have been reported in AmpC-producing β-lactam-resistant clinical isolates of *P. aeruginosa* (3). Expression of *ampC* is also negatively

Table 2. Summary of recent studies documenting the incidence of antimicrobial resistance in clinical isolates of *P. aeruginosa*

Location/study[a]	Year(s)	% resistant to:[b]												Reference
		AMI	GEN	TOB	CIP	GAT	LEV	AZT	CPM	CTZ	PIP	IMP	MER	
Asia Pacific (all)														
SENTRY	1997–2000	4.8	15.8	10.4	11.5		12.3	19.1	6.5	14.1	14.4	8.6	5.5	38
Europe (all)														
SENTRY	1997–2000	13.7	28.3	24.2	24.7		24.7	14.9	12.4	15.0	20.4	11.7	9.8	38
L. America (all)														
SENTRY	1997–2000	26.8	38.2	34.5	33.3		33.5	32.9	13.9	27.6	31.8	17.9	15.0	38
N. America (all)														
SENTRY	1997–2000	4.6	15.8	7.8	15.8		17.5	19.9	6.8	13.6	12.9	8.6	4.8	38
Europe (ICU)														
MYSTIC	1997–2000		38.9		36.7					29.4		31.8	23.9	28
Asia-Pacific, Europe, L. America, N. America (all)														
SENTRY	1997–2001	8	19	17	22	24	23		10	18	19	12	10	39
Europe (blood)														
MYSTIC	1997–2001		26.8		30.6					29.4		26.4	19.2	89
L. America (urinary)														
SENTRY	2000	51.5	57.6	54.5	57.6	57.6	57.6	59.7	39.4	27.3	24.2	24.2	18.2	27
N. America (skin/soft tissue)														
SENTRY	2000	2	10.5	6.6	20.4	23	21.1	23	4.6	9.9	14.5	8.6	7.2	80
N. America (respiratory)														
SENTRY	2000	6.3	19.2	9.8	27.6	33	28.5	58	19.5	21.7	17.9	14.4	10.9	35
UK (CF lung)	2000		47	10.1	29.7					39.6	31.9			71
L. America (all)														
SENTRY	2001	34.6	50.4		50.1	53.6	50.4	58.7	45.2	43.7	39.1	37.8	35.6	2
USA (all)														
MYSTIC	2001		17.8	9.1	22.1			21.5	5.4	10.1		9.7	8.4	81
N. America (all)														
MYSTIC	2002		8.4	6.9	22.7			16.5	5.3	9.7		7.5	4.4	62
USA (ICU)														
ICUSS	2002	10		16	32		34	32	25	19	15	23		68a

[a] The geographical regions included in the indicated studies as well as the sites of infection of the isolates examined (in parentheses) are highlighted. L. America, Latin America; N. America, North America; ICU, intensive care unit; MYSTIC, Meropenem Yearly Susceptibility Test Information Collection; SENTRY, SENTRY Antimicrobial Surveillance Program; ICUSS, Intensive Care Unit Surveillance Study database.

[b] Resistance rates were taken directly from the indicated studies, or when studies indicated rates of susceptibility were calculated accordingly. AMI, amikacin; GEN, gentamicin; TOB, tobramycin; CIP, ciprofloxacin; GAT, gatifloxacin; LEV, levofloxacin; AZT, aztreonam; CPM, cefepime; CTZ, ceftazidime; PIP, pipericillin; PIP/TAZ, pipericillin-tazobactam; IMP, imipenem; MER, meropenem.

Table 3. β-Lactamases of P. *aeruginosa*[a]

Representative enzymes reported in *P. aeruginosa*	Type of enzyme	Ambler classification	Significant substrates in *P. aeruginosa*
TEM-1; PSE/CARB series	Restricted-spectrum β-lactamase	A	Pipericillin, carbenicillin, ticarcillin
SHV variants, SHV-2a, -5, -12; TEM variants, TEM-4, -21, -24, -42; VEB-1, -2; PER-1; GES-1, -2 ; IBC-1, -2	Extended-spectrum β-lactamase	A	Pipericillin, carbenicillin, ticarcillin, ceftadizime, cefotaxime, aztreonam, sometimes cefepime
GES-2	Carbapenemase	A	Pipericillin, carbenicillin, ticarcillin, ceftadizime, cefotaxime, aztreonam (poor), cefepime (poor), imipenem (poor)
IMP-1, -2, -4, -7, -9, -10, -11, -13, -16; VIM-1 through -5, -7 through -11; SPM-1; GIM-1	Carbapenemase	B	Pipericillin, carbenicillin, ticarcillin, ceftadizime, cefotaxime, cefepime, imipenem
Chromosomal, inducible AmpC	Expanded-spectrum cephalosporinase	C	Pipericillin, carbenicillin (at high levels of enzyme), ticarcillin (at high levels of enzyme), ceftadizime, cefotaxime, cefepime (at high levels of enzyme), aztreonam
Numerous OXA variants	Narrow-spectrum penicillinase	D	Pipericillin, carbenicillin, ticarcillin
OXA-11, -14, -15, -16, -18, -19, -21, -28, -31, -32, -45	Extended-spectrum β-lactamase	D	Pipericillin, carbenicillin, ticarcillin, ceftazidime, cefotaxime, sometimes aztreonam and/or cefepime

[a]See reference 79 for more details.

affected by the activity of AmpD (44), a cytosolic amidase and, again, mutations in this gene also provide for increased AmpC production and β-lactam-resistance in clinical isolates of *P. aeruginosa* (3). Still, reports of AmpC-hyperproducing isolates lacking mutations in *ampD, ampR, ampC* or their promoter regions (3) indicate that additional genes impact AmpC expression and highlight the complexity of AmpC regulation in *P. aeruginosa.*

AmpC production is a primary determinant of β-lactam resistance in clinical strains of *P. aeruginosa* (79), being the major determinant of resistance to the antipseudomonal cephalosporins (e.g., ceftazidime and cefotaxime) in this organism (6, 15, 20). Despite its comparatively poorer activity against the carboxy-penicillins, it appears to be a significant determinant of ticarcillin resistance as well (12).

Transferable narrow-spectrum β-lactamases

Owing to the general lack of *P. aeruginosa* AmpC activity against carbenicillin, early reports of carbenicillin resistance in this organism were often associated with acquired β-lactamases that were active against this agent and so dubbed carbenicillinases (i.e., enzymes of the PSE/CARB group) (7, 48). Several of these narrow-spectrum enzymes, whose activities are generally limited to the carboxy and ureidopenicillins, have been described in *P. aeruginosa* [CARB-1 (PSE-4), CARB-2 (PSE-1), CARB-3, CARB-4 and CARB-5] (48), and these remain the most common acquired β-lactamases in β-lactam-resistant *P. aeruginosa* (6, 12, 15). Acquired class D enzymes of the OXA (i.e., oxacillinase) group are also commonly found in *P. aeruginosa* (63) and though known for their characteristic activity against oxacillins these, too, are noteworthy for their activity against antipseudomonal carboxy- and ureidopenicillins. Indeed, recent studies of ticarcillin-resistant clinical isolates of *P. aeruginosa* reveal that the PSE/CARB and OXA enzymes predominate in isolates where acquired β-lactamases are implicated in resistance (6, 12). TEM type β-lactamases that are so common-place in β-lactam-resistant *Enterobacteriaceae* are, in contrast, rather uncommon determinants of β-lactam resistance in *P. aeruginosa* though they, too, have been described in, e.g., ticarcillin-resistant clinical isolates of this organism (6). CARB and OXA enzymes have also been reported as determinants of pipericillin resistance in *P. aeruginosa* (7).

Transferable extended-spectrum β-lactamases and carbapenemases

Transferable β-lactamases active against the so-called extended-spectrum β-lactams (e.g., third generation oxyiminocephalosporins like cefotaxime and ceftazidime) are uncommon in *P. aeruginosa* (6) and found, so far, in a limited number of geographic areas (reviewed in 65, 91). These so-called extended-spectrum β-lactamases (ESBLs) include classical ESBLs derived from class A TEM and SHV β-lactamases that are increasingly common in *Enterobacteriaceae* (65, 91), as well as class D OXA ESBLs found mostly in *P. aeruginosa* (63, 65) and the less common VEB, PER,

IBC, and GES enzymes (65) (Table 3). These ESBLs are significant because of their ability to promote resistance to antipseudomonal cephalosporins such as ceftazidime, as well as the antipseudomonal ureido (pipericillin) and carboxy (ticarcillin) penicillins and, sometimes cefepime [e.g., OXA-45 (88)] and/or aztreonam [e.g. OXA-45 (88), VEB and PER (65, 91)]. ESBLs in *P. aeruginosa* may be plasmid- or chromosome-encoded and sometimes associated with integrons carrying additional resistance genes (63, 65, 91).

Enzymes with activity against the carbapenems include class A and D carbapenemases and the class B metallo-β-lactamases, though with one exception (Table 3) only the latter have been described in *P. aeruginosa* (66, 79). Metallo-β-lactamses hydrolyze most β-lactams with the exception of monobactams (aztreonam). Though an infrequent cause of carbapenem resistance in *P. aeruginosa* (84) there are numerous reports of metallo-β-lactamase-mediated resistance in this organism, mostly in Europe and the Far East but also in North and South America (66, 79). Four types of acquired metallo-enzymes have been described to date in *P. aeruginosa,* VIM, IMP (66) and, most recently, SPM-1 (26) and GIM-1 (11a), though the VIM and IMP enzymes are by far the more prevalent and occur most commonly in *P. aeruginosa* (66, 79). These enzymes may be plasmid- or chromosome-encoded and usually associated with integrons carrying addition resistance genes (66, 79). There are currently 16 IMP-type and 11 VIM-type metallo-β-lactamases though not all have been reported in *P. aeruginosa* (Table 3).

Permeability and Efflux

While β-lactamases remain the most prevalent determinants of β-lactam resistance in *P. aeruginosa,* nonenzymatic (12) and 'intrinsic' (15) resistance to these agents has been reported and likely involve impermeability and/or efflux. Indeed, several of the RND type multidrug efflux systems first implicated in fluoroquinolone (FQ) resistance are known to accommodate β-lactams (Table 3), including penicillins (e.g., pipericillin and carbenicillin; MexAB-OprM), third (e.g., cefotaxime; MexAB-OprM) and fourth (e.g., cefepime and cefpirome; MexCD-OprJ) generation cephalosporins, monobactams (e.g., aztreonam; MexAB-OprM) and carbapenems (e.g., several Mex systems of *P. aeruginosa*) (78). While MexAB-OprM, MexCD-OprJ, and MexXY-OprM all demonstrate some ability to promote resistance to carbapenems, none of these impact imipenem resistance (68), and only MexAB-OprM has been implicated in carbapenem (meropenem) resistance in clinical strains (69). An observed increase in OprM in several ticarcillin-resistant isolates of *P. aeruginosa* also supports a contribution by MexAB-OprM and/or MexXY-OprM to resistance in these strains (13). Similarly, several clinical isolates resistant to antipseudomonal β-lactams were shown to overproduce MexAB-OprM (92).

Porin loss is an uncommon cause of β-lactam resistance in *P. aeruginosa,* unlike the *Enterobacteriaceae* were concomitant porin loss and β-lactamase production are frequently reported in β-lactam-resistant clinical isolates, particularly in *Enterobacter* spp. and *Klebsiella pneumoniae* (75). Still, resistance to carbapenems, particularly imipenem, in *P. aeruginosa* is most commonly associated with loss or mutation of outer membrane porin protein OprD, a portal for uptake of basic amino acids that also accommodates carbapenems (49, 69, 75). Such resistance is, however, dependent upon expression of the chromosomal AmpC enzyme (49). The imipenem resistance of *nfxC*-type multidrug resistant strains of *P. aeruginosa,* which overproduce the MexEF-OprN multidrug efflux system, also results from the concomitant shut down of OprD production in such mutants, and not from imipenem efflux (78). The imipenem resistance of multidrug resistant *P. aeruginosa* strains overexpressing MexEF-OprN owing to mutation of the *mexS* (PA2491) gene is similarly due to reduced production of OprD in these mutants (86a).

RESISTANCE TO FLUOROQUINOLONES

Despite the organism's intrinsic resistance to the original quinolone antibiotic, nalidixic acid, the related fluoroquinolones (FQs) have been a potent component of the antipseudomonal armamentarium, though resistance to these agents, particularly ciprofloxacin, threatens to compromise their utility in treating *P. aeruginosa* infections (Table 2). As with most pathogens, resistance to FQs in *P. aeruginosa* is attributable to mutations in the genes encoding the DNA gyrase and/or topoisomerase IV targets of these agents. Although alterations in the so-called quinolone resistance determining region (QRDR) of the GyrA component of gyrase (i.e., residues 83 and 87) and ParC component of topoisomerase IV (i.e., residue 80) predominate in FQ-resistant isolates (1, 34), mutations in other residues also contribute to resistance (1, 34). Much less frequently, FQ resistance due to mutations in the GyrB and ParE components of gyrase and topoisomerase IV, respectively, has also been noted (1, 67). As with other organism, too, highly resistant strains typically carry multiple mutations, often in multiple target genes (1, 34).

Efflux is increasingly appreciated as a significant determinant of FQ resistance in disease-causing microorganisms, with several of the 3-component RND family multidrug efflux systems of *P. aeruginosa* implicated in the efflux of and resistance to FQs [reviewed in (78)] (Table 1). Indeed, FQ selection of these efflux systems is well documented in vitro (72) and explains the well-known cross-resistance of many FQ-resistant lab and clinical isolates of *P. aeruginosa* to multiple antibiotic classes (72). Mutational hyperexpression of the MexAB-OprM [in *nalB* (i.e., *mexR*), *nalC*, and *nalD* mutants], MexCD-OprJ (in *nfxB* mutants), and MexEF-OprN (in nfxC mutants) in particular has been documented in numerous lab and clinical FQ-resistant isolates (72, 73, 85a), in many instances together with target site mutations (34, 67). While the MexXY-OprM, MexVW-OprM, and to a very limited extent, MexHI-OpmD RND type multidrug efflux systems also accommodate FQs (Table 1), there are no reports to date of lab or clinical FQ-resistant isolates hyperexpressing these efflux systems. Recently, a *P. aeruginosa* multidrug efflux system of the MATE family, PmpM, was also shown to accommodate FQs (Table 1), although its contribution to resistance was observed only in strains lacking many of the known RND type efflux systems in this organism (33).

RESISTANCE TO AMINOGLYCOSIDES

Aminoglycosides are used in the treatment of a variety of *P. aeruginosa* infections, particularly pulmonary infections in CF patients (14). Resistance to aminoglycosides with antipseudomonal activity including gentamicin, tobramycin and to a lesser extent, amikacin, is all too common and present in virtually all areas of the world (Table 2). Resistance typically results from drug inactivation by plasmid- or chromosome-encoded enzymes harbored by resistant strains. Enzyme-independent resistance resultant from defects in uptake/accumulation (dubbed impermeability resistance) is also commonplace, however, particularly in CF isolates [see (77) for a recent comprehensive review on aminoglycoside resistance in *P. aeruginosa*].

Modifying Enzymes

Traditionally, aminoglycoside inactivation in resistant strains involves their modification by enzymes that either phosphorylate (aminoglycoside phosphoryltransferase [APH]), acetylate (aminoglycoside acetyltransferase [AAC]) or adenylate (aminoglycoside nucleotydyltransferase [ANT] or aminoglycoside adenyltransferase [AAD]) these antimicrobials and such enzymes are common determinants of aminoglycoside resistance in *P. aeruginosa* (42, 58, 59).

Aminoglycoside acetyltransferases

Acetylation of aminoglycosides can occur at one of several amino groups although enzymes that modify the 3 [3-*N*-aminoglycoside acetyltransferases; AAC(3)] and 6′ [6′-*N*-aminoglycoside acetyltransferases; AAC(6′)] positions are the most prevalent AACs in *P. aeruginosa* (42, 59, 77). The AAC(3)-I enzyme is a common determinant of gentamicin resistance in *P. aeruginosa,* while AAC(3)-II and AAC(3)-III are less commonly described 3-*N*-aminoglycoside acetyltransferases associated with gentamicin as well as tobramycin and netilmicin [AAC(3)-II] or tobramycin and kanamycin [AAC(3)-III] resistance in *P. aeruginosa* (77). The AAC(6′) family of enzymes provides resistance to tobramycin, netilmicin, kanamycin and either amikacin (-I subfamily) or gentamicin (-II) (77). AAC(6′)-II is the predominant AAC(6′) enzyme and, indeed, the predominant AAC in *P. aeruginosa,* where it is a significant determinant of gentamicin and tobramycin resistance (59, 77). The less common AAC(6′)-I is, however, significant for amikacin resistance in this organism (42, 77). Aminoglycoside acetyltransferases in *P. aeruginosa* are often encoded by integrons or transposons that harbor additional resistance determinants (e.g., 82), which may explain the multidrug resistant of many aminoglycoside-resistant isolates.

Aminoglycoside phosphotransferases

Resistance to kanamycin, neomycin and streptomycin in *P. aeruginosa* typically results from their phosphorylation at the 3′-OH by phosphotransferases [APH(3′)] (42, 77). Several APH(3′) enzymes have been described in *P. aeruginosa,* with APH(3′)-I and -II predominating in clinical isolates resistant to kanamycin (and neomycin) (77). Indeed, a chromosomal *aphA*-encoded APH(3′)-II type enzyme, APH(3′)-IIb (30), is likely responsible for the general insensitivity of *P. aeruginosa* to kanamycin. APH(3′) enzymes that provide resistance to other aminoglycosides have also been described in *P. aeruginosa* and include APH(3′)-VI (amikacin and isepamicin) and APH(2″) (gentamicin and tobramycin) (42, 77).

Aminoglycoside nucleotidyltransferases (adenyltransferases)

The most prevalent nucleotidyltransferase found in *P. aeruginosa* is the ANT(2″)-I enzyme, which with AAC(6′) represents the most common determinant of enzyme-dependent aminoglycoside resistance in *P. aeruginosa* (42, 58, 59). The ANT(2″)-I enzyme inactivates gentamicin and tobramycin and is thus found in gentamicin- (10) and tobramycin- (54) resistant clinical isolates. Other adenyltransferases associated with aminoglycoside resistance in *P. aeruginosa* include ANT(3″) (streptomycin resistance) (85) and ANT(4′)-II (amikacin, tobramycin and isepamicin resistance) (83). As with other aminoglycoside-modifying enzymes, genes for ANT enzymes can be integron associated (77).

Impermeability and Efflux

Aminoglycoside resistance independent of inactivating enzymes has been known for some time (8). Characterized by resistance to all aminoglycosides and often associated with reduced aminoglycoside accumulation (8), such resistance was attributed to reduced uptake owing to reduced permeability and, as such, was typically referred to as "impermeability resistance." Numerous studies highlight the importance of impermeability resistance in clinical (42, 59, 77), especially CF (54, 77), isolates resistant to aminoglycosides, which often occurs in conjunction with modifying enzymes to promote broad-spectrum aminoglycoside resistance in *P. aeruginosa* (54). More recent studies of such pan-aminoglycoside-resistant strains have indicated, however, that the reduced aminoglycoside accumulation seen in such strains was likely due to efflux by the RND family multidrug efflux system MexXY (i.e., AmrAB) -OprM (86, 90). MexXY-OprM is the major efflux system exporting aminoglycosides in *P. aeruginosa,* although it actually exports a range of antimicrobials (Table 1) and is, in fact, implicated in intrinsic resistance to tetracyclines, glycylcyclines and macrolides, in addition to aminoglycosides (73). Strikingly, and in contrast to most efflux systems hitherto characterized in *P. aeruginosa,* MexXY is inducible by many of the antimicrobials it exports, including aminoglycosides, macrolides and tetracycline (56). Despite reports of a modest contribution by MexAB-OprM and the SMR family EmrPA efflux systems to aminoglycoside resistance in lab isolates (47), these are unlikely to be significant determinants of aminoglycoside resistance in the clinic.

Adaptive Resistance

The ability of *P. aeruginosa* to adapt to and, thus, grow at elevated levels of aminoglycosides has been known for some time. Characterized by decreased susceptibility to all aminoglycosides and loss of the resistance phenotype in the absence of drug (19) this reversible pan-aminoglycoside resistance is referred to as adaptive resistance. Occurring both in vitro (19, 41) and in vivo (4), resistance typically develops within a few hours of first exposure and disappears several hours after removal of the antibiotic. Intriguingly, resistance appears to result from reduced aminoglycoside accumulation, reminiscent of impermeability resistance (19, 41). Indeed, a recent publication confirms the involvement of drug efflux (by the MexXY system implicated in permeability resistance) in the reduced accumulation that characterizes adaptive aminoglycoside resistance (37), although reduced uptake may also contribute (40).

Outer Membrane

Several early studies documented an apparent role for OM protein OprH in aminoglycoside (and polymyxin) resistance, its expression under low Mg^{2+} growth conditions and in certain mutants being correlated with resistance (64). Still, more recent studies have demonstrated that OprH is actually encoded by the first gene of a 3-gene operon that includes *phoP* (encoding a response regulator) and *phoQ* (encoding a sensor kinase) (53). The low Mg^{++}-dependent and mutational resistance to aminoglycosides previously attributed to OprH are, in fact, related to PhoP-PhoQ expression and activity (52). While the mechanism is as yet unclear, it may involve modification of lipopolysaccharide (LPS) as for polycation resistance (57) (see below).

RESISTANCE TO POLYCATIONS

Although polycation antimicrobials such as polymyxins B and E (i.e., colistin) have been used sparingly in the treatment of *P. aeruginosa,* owing to initial worries about toxicity, the current problems with multidrug resistance in this organism and the apparent effectiveness of colistin in treating multidrug resistant infections, particularly in CF (5), have renewed interest in this family of antimicrobials. Still, resistance to polymyxins in laboratory (61) and clinical (21, 46) isolates of *P. aeruginosa* has been reported and typically involves alterations to LPS (25, 61), including modification of the lipid A with aminoarabinose (61). Interestingly, polymyxin B-resistant lab isolates expressing the aminoarabinose modification carry mutations in the genes encoding the PmrA-PmrB two-component regulatory system (61) previously implicated in adaptive polycation resistance that occurs in response to low Mg^{2+} or the presence of polycations (57). Apparently, PmrA-PmrB (57, 61) and a second two-component system known to regulate *pmrAB* and to mediate the low Mg^{++} and polycation induction of resistance, PhoP-PhoQ (52, 53), both contribute to polycation resistance. This occurs, at least in part, because of a positive influence of these proteins on expression of an operon, PA3552-PA3559, very similar to one responsible for aminoarabinose modification of LPS in polycation-resistant *Salmonella enterica* serovar Typhimurium (57, 61). Significantly, PmrA-PmrB-/PhoP-PhoQ also determine resistance to the so-called cationic antimicrobial peptides of innate immunity that are currently being studied as chemotherapeutic agents (52, 57, 61). Again, this is related in part to aminoarabinose modification of LPS–polymyxin B-resistant *pmrAB* mutants with the aminoarabinose modification showing increased resistance to these peptides (61).

RESISTANCE TO BIOCIDES

P. aeruginosa shows variable susceptibility to biocides (i.e., disinfectants, preservatives and antiseptics) (43a), is particularly resistant to triclosan (16) and chlorhexidine (43), and is able to 'adapt' to high concentrations of quaternary ammonium compounds (QACs) such as benzalkonium chloride (51). The intrinsic insusceptibility of the organism to triclosan is solely attributable to its export by several of the RND family multidrug efflux systems, including MexAB-OprM, MexCD-OprJ, MexXY-OprM and MexJK (16). This property and the fact that triclosan readily selects mutants hyperexpressing these multidrug efflux systems in vitro are of some concern because of the risk of selecting multiple antibiotic resistant strains with this all-too-frequently-used-in-the-community biocide (74). In contrast to triclosan resistance, the nature of the intrinsic or adapted resistance to most biocides, and the identity of the cellular components/factors involved remains somewhat obscure, although changes in membrane composition and ultrastructure have been noted in biocide-resistant *P. aeruginosa* and are suggestive of defects in biocide uptake [reviewed in (74)]. Indeed, one recent study of a mutant adapted for growth on elevated levels of QACs was shown to significantly increase production of a probable OM lipoprotein, OprR, whose expression nicely correlated with QAC resistance, although the exact nature of its contribution to resistance is as yet unknown (87). Resistance to silver sulfadiazone, used in treating burn infections, has also been reported in *P. aeruginosa* (70) and there are very infrequent reports of plasmid-encoded biocide (QAC) efflux mechanisms of the SMR family in this organism (i.e., QacE and QacEΔ1 [Table 1]).

BIOFILMS

Biofilms, surface-attached three-dimensional structures in which bacteria are imbedded in a principally exopolysaccharide matrix, are increasingly recognized as the preferred mode of bacterial growth in nature and infectious disease (31). An important feature of biofilm growth and one that is particularly relevant in a clinical context is its marked resistance to antimicrobial agents (31). In the case of *P. aeruginosa,* it is now recognized that the organisms causing the chronic pulmonary infection that characterizes CF are growing in biofilms, and while biofilm-grown *P. aeruginosa* are notoriously refractory to antimicrobial therapy, the mechanisms behind this resistance are complex and not yet fully elucidated (see reference 22 for a detailed review of antimicrobial resistance of *P. aeruginosa* biofilms).

Several studies of *P. aeruginosa* and other biofilm-forming organisms suggest that restricted penetration of antimicrobials, including biocides, into biofilms is a contributing feature of biofilm resistance, although this seems to be agent-specific (penetration of some but not other agents is restricted) and the data are often contradictory (22). Similarly, reports suggest that *P. aeruginosa* within biofilms are metabolically less active and grow more slowly than cells at the biofilm periphery (owing to limited access to nutrients and oxygen) and this may contribute to increasing biofilm tolerance since antimicrobials often target metabolically active cells (22). Certainly, the demonstration that biofilm-grown *P. aeruginosa* from CF patients are anaerobic is likely to be significant in the context of antimicrobial resistance since many agents are inactive or less active under anaerobiosis (32). Oxygen limitation has also been shown to contribute significantly to the antimicrobial resistance of in vitro-grown *P. aeruginosa* biofilms (7a). An interesting proposal for biofilm resistance is that only a small fraction of biofilm cells (called persisters) actually survive antimicrobial treatment of biofilms and are, in fact, responsible for the high levels of resistance seen in biofilms (45). Still, the specific features of persister organisms that permit this survival remain to be elucidated. In a related study, antibiotic-resistant phenotypic variants of *P. aeruginosa* with enhanced ability to form biofilms have been reported to arise at high frequency in vitro and in the lungs of CF patients, and a gene that controls this conversion, *pvrR,* has been identified (23). Still, while this speaks to a specific link between resistance and conversion to biofilm growth, and again highlights the significance of a subfraction of a bacterial population as responsible for *P. aeruginosa* resistance in biofilms, it still fails to provide insights into the mechanisms behind resistance. Finally, a locus involved in the synthesis of periplasmic glucans, *ndvB,* has recently been implicated in biofilm resistance to several agents, particularly tobramycin (55). These glucans, which are specifically expressed in organisms grown as biofilms, bind to tobramycin in vitro and it has been suggested that similar binding in the periplasm of biofilm cells would restrict this agent's passage into the cytoplasm where its targets lie, thereby promoting resistance.

CONCLUDING REMARKS

Intrinsic and acquired resistance in *P. aeruginosa* is multifactorial, and given the increasing prevalence of multidrug resistance in this organism clearly more needs to be done in terms of both devising strategies for overcoming resistance and undertaking steps to minimize resistance development in the first place. Given the demonstrated role of RND type multidrug efflux systems in intrinsic and acquired resistance to FQs, ongoing efforts to identify efflux inhibitors able to reverse efflux-mediated resistance have shown some success (reviewed in reference 50). Given, too, that the use of selected antipseudomonal antimicrobials is associated with increased resistance development in *P. aeruginosa* (e.g., ceftazidime, ticarcillin and imipenem demonstrate a high resistance potential while cefepime, pipericillin and meropenem demonstrate a low resistance potential [18]) appropriate prescribing practices might well limit resistance development. Indeed, in one retrospective study correlating β-lactam resistance rates with β-lactam use, overall resistance to β-lactams decreased when use of ceftazidime and cefotaxime was curtailed in favor of cefepime (24). Still, the inherent resistance of the organism to so many antimicrobials and its ready selection in patients undergoing antimicrobial chemotherapy suggest that alternative therapies may also need to be considered in controlling infections caused by this major nosocomial pathogen (76).

REFERENCES

1. **Akasaka, T., M. Tanaka, A. Yamaguchi, and K. Sato.** 2001. Type II topoisomerase mutations in fluoroquinolone-resistant clinical strains of *Pseudomonas aeruginosa* isolated in 1998 and 1999, role of target enzyme in mechanism of fluoroquinolone resistance. *Antimicrob. Agents Chemother.* **45:**2263–2268.
2. **Andrade, S. S., R. N. Jones, A. C. Gales, and H. S. Sader.** 2003. Increasing prevalence of antimicrobial resistance among *Pseudomonas aeruginosa* isolates in Latin American medical centres, 5 year report of the SENTRY Antimicrobial Surveillance Program (1997-2001). *J. Antimicrob. Chemother.* **52:**140–141.
3. **Bagge, N., O. Ciofu, M. Hentzer, J. I. Campbell, M. Givskov, and N. Hoiby.** 2002. Constitutive high expression of chromosomal β-lactamase in *Pseudomonas aeruginosa* caused by a new insertion sequence (IS*1669*) located in *ampD*. *Antimicrob. Agents Chemother.* **46:**3406–3411.
4. **Barclay, M. L., E. J. Begg, S. T. Chambers, P. E. Thornley, P. K. Pattemore, and K. Grimwood.** 1996. Adaptive resistance to tobramycin in *Pseudomonas aeruginosa* lung infection in cystic fibrosis. *J. Antimicrob. Chemother.* **37:**1155–1164.
5. **Beringer, P.** 2001. The clinical use of colistin in patients with cystic fibrosis. *Curr. Opin. Pulm. Med.* **7:**434–440.
6. **Bert, F., C. Branger, and N. Lambert-Zechovsky.** 2002. Identification of PSE and OXA β-lactamase genes in *Pseudomonas aeruginosa* using PCR-restriction fragment length polymorphism. *J. Antimicrob. Chemother.* **50:**11–18.
7. **Bert, F., C. Branger, and N. Lambert-Zechovsky.** 2004. Comparative activity of β-lactam agents (carbapenem excepted) against *Pseudomonas aeruginosa* strains with CARB or OXA β-lactamases. *Chemotherapy* (Basel) **50:**31–34.

7a. **Borriello, G., E. Werner, F. Roe, A. M. Kim, G. D. Ehrlich, and P. S. Stewart.** 2004. Oxygen limitation contributes to antibiotic tolerance of *Pseudomonas aeruginosa* in biofilms. *Antimicrob. Agents Chemother.* **48:**2659–2664.

8. **Bryan, L. E., R. Haraphongse, and H. M. Van den Elzen.** 1976. Gentamicin resistance in clinical-isolates of *Pseudomonas aeruginosa* associated with diminished gentamicin accumulation and no detectable enzymatic modification. *J. Antibiot.* (Tokyo) **29:**743–753.
9. **Bukholm, G., T. Tannaes, A. B. Kjelsberg, and N. Smith-Erichsen.** 2002. An outbreak of multidrug-resistant *Pseudomonas aeruginosa* associated with increased risk of patient death in an intensive care unit. Infect. *Control. Hosp. Epidemiol.* **23:**441–446.
10. **Busch-Sorensen, C., M. Sonmezoglu, N. Frimodt-Moller, T. Hojbjerg, G. H. Miller, and F. Espersen.** 1996. Aminoglycoside resistance mechanisms in Enterobacteriaceae and Pseudomonas spp. from two Danish hospitals, correlation with type of aminoglycoside used. *APMIS* **104:**763–768.

10a. **Cao, B., H. Wang, H. Sun, Y. Zhu, and M. Chen.** 2004. Risk factors and clinical outcomes of nosocomial multi-drug resistant *Pseudomonas aeruginosa* infections. *J. Hosp. Infect.* **57:**112–118.

11. **Cao, L., R. Srikumar, and K. Poole.** 2004. MexAB-OprM hyperexpression in NalC type multidrug resistant *Pseudomonas aeruginosa,* identification and characterization of the *nalC* gene encoding a repressor of PA3720-PA3719. *Mol. Microbiol.* **53:**1423–1436.

11a. **Castanheira, M., M. A. Toleman, R. N. Jones, F. J. Schmidt, and T. R. Walsh.** 2004. Molecular characterization of a β-lactamase gene, bla_{GIM-1}, encoding a new subclass of metallo-β-lactamase. *Antimicrob. Agents Chemother.* **48:**4654–4661.

12. **Cavallo, J. D., R. Fabre, F. Leblanc, M. H. Nicolas-Chanoine, and A. Thabaut.** 2000. Antibiotic susceptibility and mechanisms of β-lactam resistance in 1310 strains of *Pseudomonas aeruginosa,* a French multicentre study (1996). *J. Antimicrob. Chemother.* **46:**133–136.
13. **Cavallo, J. D., P. Plesiat, G. Couetdic, F. Leblanc, and R. Fabre.** 2002. Mechanisms of β-lactam resistance in *Pseudomonas aeruginosa,* prevalence of OprM-overproducing strains in a French multicentre study (1997). *J. Antimicrob. Chemother.* **50:**1039–1043.
14. **Cheer, S. M., J. Waugh, and S. Noble.** 2003. Inhaled tobramycin (TOBI), a review of its use in the management of *Pseudomonas aeruginosa* infections in patients with cystic fibrosis. *Drugs* **63:** 2501–2520.
15. **Chen, H. Y., M. Yuan, and D. M. Livermore.** 1995. Mechanisms of resistance to β-lactam antibiotics amongst *Pseudomonas aeruginosa* isolates collected in the UK in 1993. *J. Med. Microbiol.* **43:**300–309.
16. **Chuanchuen, R., R. R. Karkhoff-Schweizer, and H. P. Schweizer.** 2003. High-level triclosan resistance in *Pseudomonas aeruginosa* is solely a result of efflux. *Am. J. Infect. Control.* **31:**124–127.
17. **Ciofu, O.** 2003. *Pseudomonas aeruginosa* chromosomal β-lactamase in patients with cystic fibrosis and chronic lung infection. Mechanism of antibiotic resistance and target of the humoral immune response. *APMIS* **2003**(Suppl.)**:**1–47.

18. **Cunha, B. A.** 2002. *Pseudomonas aeruginosa,* resistance and therapy. *Semin. Respir. Infect.* **17:**231–239.
19. **Daikos, G. L., G. G. Jackson, V. T. Lolans, and D. M. Livermore.** 1990. Adaptive resistance to aminoglycoside antibiotics from first-exposure down-regulation. *J. Infect. Dis.* **162:**414–420.
20. **De Champs, C., L. Poirel, R. Bonnet, D. Sirot, C. Chanal, J. Sirot, and P. Nordmann.** 2002. Prospective survey of β-lactamases produced by ceftazidime-resistant *Pseudomonas aeruginosa* isolated in a French hospital in 2000. *Antimicrob. Agents Chemother.* **46:**3031–3034.
21. **Denton, M., K. Kerr, L. Mooney, V. Keer, A. Rajgopal, K. Brownlee, P. Arundel, and S. Conway.** 2002. Transmission of colistin-resistant *Pseudomonas aeruginosa* between patients attending a pediatric cystic fibrosis center. *Pediatr. Pulmonol.* **34:**257–261.
22. **Drenkard, E.** 2003. Antimicrobial resistance of *Pseudomonas aeruginosa* biofilms. *Microbes. Infect.* **5:**1213–1219.
23. **Drenkard, E., and F. M. Ausubel.** 2002. *Pseudomonas* biofilm formation and antibiotic resistance are linked to phenotypic variation. *Nature* **416:**740–743.
24. **Empey, K. M., R. P. Rapp, and M. E. Evans.** 2002. The effect of an antimicrobial formulary change on hospital resistance patterns. *Pharmacotherapy* **22:**81–87.
25. **Ernst, R. K., E. C. Yi, L. Guo, K. B. Lim, J. L. Burns, M. Hackett, and S. I. Miller.** 1999. Specific lipopolysaccharide found in cystic fibrosis airway *Pseudomonas aeruginosa. Science* **286:**1561–1565.
26. **Gales, A. C., L. C. Menezes, S. Silbert, and H. S. Sader.** 2003. Dissemination in distinct Brazilian regions of an epidemic carbapenem-resistant *Pseudomonas aeruginosa* producing SPM metallo-β-lactamase. *J. Antimicrob. Chemother.* **52:**699–702.
27. **Gales, A. C., H. S. Sader, and R. N. Jones.** 2002. Urinary tract infection trends in Latin American hospitals, report from the SENTRY antimicrobial surveillance program (1997-2000). *Diagn. Microbiol. Infect. Dis.* **44:**289–299.
28. **Garcia-Rodriguez, J. A., and R. N. Jones.** 2002. Antimicrobial resistance in gram-negative isolates from European intensive care units, data from the Meropenem Yearly Susceptibility Test Information Collection (MYSTIC) programme. *J. Chemother.* **14:**25–32.
29. **Goossens, H.** 2003. Susceptibility of multi-drug-resistant *Pseudomonas aeruginosa* in intensive care units, results from the European MYSTIC study group. *Clin. Microbiol. Infect.* **9:**980–983.
30. **Hachler, H., P. Santanam, and F. H. Kayser.** 1996. Sequence and characterization of a novel chromosomal aminoglycoside phosphotransferase gene, *aph (3')-IIb,* in *Pseudomonas aeruginosa. Antimicrob. Agents Chemother.* **40:**1254–1256.
31. **Hall-Stoodley, L., J. W. Costerton, and P. Stoodley.** 2004. Bacterial biofilms, from the natural environment to infectious diseases. *Nat. Rev. Microbiol.* **2:**95–108.
32. **Hassett, D. J., J. Cuppoletti, B. Trapnell, S. V. Lymar, J. J. Rowe, Y. S. Sun, G. M. Hilliard, K. Parvatiyar, M. C. Kamani, D. J. Wozniak, S. H. Hwang, T. R. McDermott, and U. A. Ochsner.** 2002. Anaerobic metabolism and quorum sensing by *Pseudomonas aeruginosa* biofilms in chronically infected cystic fibrosis airways, rethinking antibiotic treatment strategies and drug targets. *Adv. Drug Deliv. Rev.* **54:**1425–1443.
33. **He, G. X., T. Kuroda, T. Mima, Y. Morita, T. Mizushima, and T. Tsuchiya.** 2004. An H^+-coupled multidrug efflux pump, PmpM, a member of the MATE family of transporters, from *Pseudomonas aeruginosa. J. Bacteriol.* **186:**262–265.
34. **Higgins, P. G., A. C. Fluit, D. Milatovic, J. Verhoef, and F. J. Schmitz.** 2003. Mutations in GyrA, ParC, MexR and NfxB in clinical isolates of *Pseudomonas aeruginosa. Int. J. Antimicrob. Agents* **21:**409–413.
35. **Hoban, D. J., D. J. Biedenbach, A. H. Mutnick, and R. N. Jones.** 2003. Pathogen of occurrence and susceptibility patterns associated with pneumonia in hospitalized patients in North America, results of the SENTRY Antimicrobial Surveillance Study (2000). *Diagn. Microbiol. Infect. Dis.* **45:**279–285.
36. **Hocquet, D., X. Bertrand, T. Kohler, D. Talon, and P. Plesiat.** 2003. Genetic and phenotypic variations of a resistant *Pseudomonas aeruginosa* epidemic clone. *Antimicrob. Agents Chemother.* **47:**1887–1894.
37. **Hocquet, D., C. Vogne, F. El Garch, A. Vejux, N. Gotoh, A. Lee, O. Lomovskaya, and P. Plesiat.** 2003. MexXY-OprM efflux pump is necessary for adaptive resistance of *Pseudomonas aeruginosa* to aminoglycosides. *Antimicrob. Agents Chemother.* **47:**1371–1375.

37a.**Islam, S., S. Jalal, and B. Wretlind.** 2004. Expression of the MexXY efflux pump in amikacin-resistant isolates of *Pseudomonas aeruginosa. Clin. Microbiol. Infect.* **10:**877–883.

38. **Jones, R. N., J. T. Kirby, M. L. Beach, D. J. Biedenbach, and M. A. Pfaller.** 2002. Geographic variations in activity of broad-spectrum β-lactams against *Pseudomonas aeruginosa,* summary of the worldwide SENTRY Antimicrobial Surveillance Program (1997-2000). *Diagn. Microbiol. Infect. Dis.* **43:**239–243.
39. **Jones, R. N., H. S. Sader, and M. L. Beach.** 2003. Contemporary in vitro spectrum of activity summary for antimicrobial agents tested against 18569 strains non-fermentative Gram-negative bacilli isolated in the SENTRY Antimicrobial Surveillance Program (1997-2001). *Int. J. Antimicrob. Agents* **22:**551–556.
40. **Karlowsky, J. A., D. J. Hoban, S. A. Zelenitsky, and G. G. Zhanel.** 1997. Altered *denA* and *anr* gene expression in aminoglycoside adaptive resistance in *Pseudomonas aeruginosa. J. Antimicrob. Chemother.* **40:**371–376.
41. **Karlowsky, J. A., M. H. Saunders, G. A. Harding, D. J. Hoban, and G. G. Zhanel.** 1996. In vitro characterization of aminoglycoside adaptive resistance in *Pseudomonas aeruginosa. Antimicrob. Agents Chemother.* **40:**1387–1393.
42. **Kettner, M., P. Milosovic, M. Hletkova, and J. Kallova.** 1995. Incidence and mechanisms of aminoglycoside resistance in *Pseudomonas aeruginosa* serotype O11 isolates. *Infection* **23:**380–383.
43. **Koljalg, S., P. Naaber, and M. Mikelsaar.** 2002. Antibiotic resistance as an indicator of bacterial chlorhexidine susceptibility. *J. Hosp. Infect.* **51:**106–113.

43a.**Lambert, R. J.** 2004. Comparative analysis of antibiotic and antimicrobial biocide susceptibility data in clinical isolates of methicillin-sensitive *Staphylococcus aureus,* methicillin-resistant *Staphylococcus aureus* and *Pseudomonas aeruginosa* between 1989 and 2000. *J. Appl. Microbiol.* **97:**699–711.

44. **Langaee, T. Y., M. Dargis, and A. Huletsky.** 1998. An *ampD* gene in *Pseudomonas aeruginosa* encodes a negative regulator of AmpC β-lactamase expression. *Antimicrob. Agents Chemother.* **42:**3296–3300.
45. **Lewis, K.** 2001. Riddle of biofilm resistance. *Antimicrob. Agents Chemother.* **45:**999–1007.
46. **Li, J., J. Turnidge, R. Milne, R. L. Nation, and K. Coulthard.** 2001. In vitro pharmacodynamic properties of colistin and colistin methanesulfonate against *Pseudomonas aeruginosa* isolates from patients with cystic fibrosis. *Antimicrob. Agents Chemother.* **45:**781–785.
47. **Li, X.-Z., K. Poole, and H. Nikaido.** 2003. Contributions of MexAB-OprM and an EmrE homologue to intrinsic resistance of *Pseudomonas aeruginosa* to aminoglycosides and dyes. *Antimicrob. Agents Chemother.* **47:**27–33.

48. **Livermore, D. M.** 1995. β-lactamases in laboratory and clinical resistance. *Clin. Microbiol. Rev.* **8:**557–584.
49. **Livermore, D. M.** 2001. Of *Pseudomonas,* porins, pumps and carbapenems. *J. Antimicrob. Chemother.* **47:**247–250.
49a.**Llanes, C., D. Hocquet, C. Vogne, D. Benali-Baitich, C. Neuwirth, and P. Plesiat.** 2004. Clinical strains of *Pseudomonas aeruginosa* overproducing MexAB-OprM and MexXY efflux pumps simultaneously. *Antimicrob. Agents Chemother.* **48:** 1797–1802.
50. **Lomovskaya, O., and W. Watkins.** 2001. Inhibition of efflux pumps as a novel approach to combat drug resistance in bacteria. *J. Mol. Microbiol. Biotechnol.* **3:**225–236.
51. **Loughlin, M. F., M. V. Jones, and P. A. Lambert.** 2002. *Pseudomonas aeruginosa* cells adapted to benzalkonium chloride show resistance to other membrane-active agents but not to clinically relevant antibiotics. *J. Antimicrob. Chemother.* **49:** 631–639.
52. **Macfarlane, E. L., A. Kwasnicka, and R. E. Hancock.** 2000. Role of *Pseudomonas aeruginosa* PhoP-phoQ in resistance to antimicrobial cationic peptides and aminoglycosides. *Microbiology* **146:**2543–2554.
53. **Macfarlane, E. L., A. Kwasnicka, M. M. Ochs, and R. E. Hancock.** 1999. PhoP-PhoQ homologues in *Pseudomonas aeruginosa* regulate expression of the outer-membrane protein OprH and polymyxin B resistance. *Mol. Microbiol.* **34:**305–316.
54. **MacLeod, D. L., L. E. Nelson, R. M. Shawar, B. B. Lin, L. G. Lockwood, J. E. Dirk, G. H. Miller, J. L. Burns, and R. L. Garber.** 2000. Aminoglycoside-resistance mechanisms for cystic fibrosis *Pseudomonas aeruginosa* isolates are unchanged by long-term, intermittent, inhaled tobramycin treatment. *J. Infect. Dis.* **181:**1180–1184.
55. **Mah, T. F., B. Pitts, B. Pellock, G. C. Walker, P. S. Stewart, and G. A. O'Toole.** 2003. A genetic basis for *Pseudomonas aeruginosa* biofilm antibiotic resistance. *Nature* **426:**306–310.
56. **Masuda, N., E. Sakagawa, S. Ohya, N. Gotoh, H. Tsujimoto, and T. Nishino.** 2000. Contribution of the MexX-MexY-OprM efflux system to intrinsic resistance in *Pseudomonas aeruginosa. Antimicrob. Agents Chemother.* **44:**2242–2246.
57. **McPhee, J. B., S. Lewenza, and R. E. Hancock.** 2003. Cationic antimicrobial peptides activate a two-component regulatory system, PmrA-PmrB, that regulates resistance to polymyxin B and cationic antimicrobial peptides in *Pseudomonas aeruginosa. Mol. Microbiol.* **50:**205–217.
58. **Miller, G. H., F. J. Sabatelli, R. S. Hare, Y. Glupczynski, P. Mackey, D. Shlaes, K. Shimizu, and K. J. Shaw.** 1997. The most frequent aminoglycoside resistance mechanisms—changes with time and geographic area, a reflection of aminoglycoside usage patterns? Aminoglycoside Resistance Study Groups. *Clin. Infect. Dis.* **24** (Suppl. 1): S46–S62.
59. **Miller, G. H., F. J. Sabatelli, L. Naples, R. S. Hare, and K. J. Shaw.** 1994. Resistance to aminoglycosides in *Pseudomonas.* Aminoglycoside Resistance Study Groups. *Trends Microbiol.* **2:**347–353.
60. **Morita, Y., T. Murata, T. Mima, S. Shiota, T. Kuroda, T. Mizushima, N. Gotoh, T. Nishino, and T. Tsuchiya.** 2003. Induction of *mexCD-oprJ* operon for a multidrug efflux pump by disinfectants in wild-type *Pseudomonas aeruginosa* PAO1. *J. Antimicrob. Chemother.* **5:**991–994.
61. **Moskowitz, S. M., R. K. Ernst, and S. I. Miller.** 2004. PmrAB, a two-component regulatory system of *Pseudomonas aeruginosa* that modulates resistance to cationic antimicrobial peptides and addition of aminoarabinose to lipid A. *J. Bacteriol.* **186:** 575–579.
62. **Mutnick, A. H., P. R. Rhomberg, H. S. Sader, and R. N. Jones.** 2004. Antimicrobial usage and resistance trend relationships from the MYSTIC Programme in North America (1999-2001). *J. Antimicrob. Chemother.* **53:**290–296.
63. **Naas, T., and P. Nordmann.** 1999. OXA-type β-lactamases. *Curr. Pharm. Des* **5:**865–879.
64. **Nicas, T. I., and R. E. W. Hancock.** 1980. Outer membrane protein H1 of *Pseudomonas aeruginosa,* involvement in adaptive and mutational resistance to ethylenediaminetetraacetate, polymyxin B, and gentamicin. *J. Bacteriol.* **143:**872–878.
65. **Nordmann, P., and M. Guibert.** 1998. Extended-spectrum β-lactamases in *Pseudomonas aeruginosa. J. Antimicrob. Chemother.* **42:**128–131.
66. **Nordmann, P., and L. Poirel.** 2002. Emerging carbapenemases in gram-negative aerobes. *Clin. Microbiol. Infect.* **8:**321–331.
67. **Oh, H., J. Stenhoff, S. Jalal, and B. Wretlind.** 2003. Role of efflux pumps and mutations in genes for topoisomerases II and IV in fluoroquinolone-resistant *Pseudomonas aeruginosa* strains. *Microb. Drug Resist.* **9:**323–328.
68. **Okamoto, K., N. Gotoh, and T. Nishino.** 2002. Alterations of susceptibility of Pseudomonas aeruginosa by overproduction of multidrug efflux systems, MexAB-OprM, MexCD-OprJ, and MexXY/OprM to carbapenems, substrate specificities of the efflux systems. *J. Infect. Chemother.* **8:**371–373.
68a.**Obritsch, M. D., D. N. Fish, R. Maclaren, and R. Jung.** 2004. National surveillance of antimicrobial resistance in *Pseudomonas aeruginosa* isolates obtained from intensive care unit patients from 1993 to 2002. *Antimicrob. Agents Chemother.* **48:**4606–4610.
69. **Pai, H., J. Kim, J. Kim, J. H. Lee, K. W. Choe, and N. Gotoh.** 2001. Carbapenem resistance mechanisms in *Pseudomonas aeruginosa* clinical isolates. *Antimicrob. Agents Chemother.* **45:** 480–484.
70. **Pirnay, J. P., D. de Vos, C. Cochez, F. Bilocq, J. Pirson, M. Struelens, L. Duinslaeger, P. Cornelis, M. Zizi, and A. Vanderkelen.** 2003. Molecular epidemiology of *Pseudomonas aeruginosa* colonization in a burn unit, persistence of a multidrug-resistant clone and a silver sulfadiazine-resistant clone. *J. Clin. Microbiol.* **41:**1192–1202.
71. **Pitt, T. L., M. Sparrow, M. Warner, and M. Stefanidou.** 2003. Survey of resistance of *Pseudomonas aeruginosa* from UK patients with cystic fibrosis to six commonly prescribed antimicrobial agents. *Thorax* **58:**794–796.
72. **Poole, K.** 2000. Efflux-mediated resistance to fluoroquinolones in gram-negative bacteria. *Antimicrob. Agents Chemother.* **44:** 2233–2241.
73. **Poole, K.** 2001. Multidrug efflux pumps and antimicrobial resistance in *Pseudomonas aeruginosa* and related organisms. *J. Mol. Microbiol. Biotechnol.* **3:**255–264.
74. **Poole, K.** 2002. Mechanisms of bacterial biocide and antibiotic resistance. *J. Appl. Microbiol.* **92** (Suppl. 1):55S–64S.
75. **Poole, K.** 2002. Outer membranes and efflux, the path to multidrug resistance in gram-negative bacteria. *Curr. Pharm. Biotechnol.* **3:**77–98.
76. **Poole, K.** 2003. Overcoming multidrug resistance in gram-negative bacteria. *Curr. Opin. Investig. Drugs* **4:**139.
77. **Poole, K.** 2004. Aminoglycoside resistance in *Pseudomonas aeruginosa. Antimicrob. Agents Chemother.* **49:**479–487.
78. **Poole, K.** 2004. Efflux-mediated multiresistance in Gram-negative bacteria. *Clin. Microbiol. Infect.* **10:**12–26.
79. **Poole, K.** 2004. Resistance to β-lactam antibiotics. *Cell. Mol. Life Sci.* **61:**2200–2223.
80. **Rennie, R. P., R. N. Jones, and A. H. Mutnick.** 2003. Occurrence and antimicrobial susceptibility patterns of pathogens isolated from skin and soft tissue infections, report from the SENTRY Antimicrobial Surveillance Program (United States and Canada, 2000). *Diagn. Microbiol. Infect. Dis.* **45:**287–293.

81. **Rhomberg, P. R., R. N. Jones, and H. S. Sader.** 2004. Results from the Meropenem Yearly Susceptibility Test Information Collection (MYSTIC) Programme, report of the 2001 data from 15 United States medical centres. *Int. J. Antimicrob. Agents* **23:**52–59.
82. **Riccio, M. L., J. D. Docquier, E. Dell'Amico, F. Luzzaro, G. Amicosante, and G. M. Rossolini.** 2003. Novel 3-N-aminoglycoside acetyltransferase gene, *aac(3)-Ic,* from a *Pseudomonas aeruginosa* integron. *Antimicrob. Agents Chemother.* **47:**1746–1748.
83. **Sabtcheva, S., M. Galimand, G. Gerbaud, P. Courvalin, and T. Lambert.** 2003. Aminoglycoside resistance gene *ant(4′)-IIb* of *Pseudomonas aeruginosa* BM4492, a clinical isolate from Bulgaria. *Antimicrob. Agents Chemother.* **47:**1584–1588.
84. **Sasaki, M., E. Hiyama, Y. Takesue, M. Kodaira, T. Sueda, and T. Yokoyama.** 2004. Clinical surveillance of surgical imipenem-resistant *Pseudomonas aeruginosa* infection in a Japanese hospital. *J. Hosp. Infect.* **56:**111–118.
85. **Shaw, K. J., R. S. Hare, F. J. Sabatelli, M. Rizzo, C. A. Cramer, L. Naples, S. Kocsi, H. Munayyer, P. Mann, G. H. Miller, L. Verbist, H. van Landuyt, Y. Glupczynski, M. Catalano, and M. Woloj.** 1991. Correlation between aminoglycoside resistance profiles and DNA hybridization of clinical isolates. *Antimicrob. Agents Chemother.* **35:**2253–2261.
85a. **Sobel, M. L., D. Hocquet, L. Cao, P. Plesiat, and K. Poole.** 2005. Mutations in PA3574 (*nalD*) lead to increased MexAB-OprM expression and multidrug resistance in lab and clinical isolates of *Pseudomonas aeruginosa. Antimicrob. Agents Chemother.* **49:**1782–1786.
86. **Sobel, M. L., G. A. McKay, and K. Poole.** 2003. Contribution of the MexXY multidrug transporter to aminoglycoside resistance in *Pseudomonas aeruginosa* clinical isolates. *Antimicrob. Agents Chemother.* **47:**3202–3207.
86a. **Sobel, M. L., S. Neshat, and K. Poole.** 2005. Mutations in PA2491 (*mexS*) promote MexT-dependent *mexEF-oprN* expression and multidrug resistance in a clinical strain of *Pseudomonas aeruginosa. J. Bacteriol.* **187:**1246–1253.
87. **Tabata, A., H. Nagamune, T. Maeda, K. Murakami, Y. Miyake, and H. Kourai.** 2003. Correlation between resistance of *Pseudomonas aeruginosa* to quaternary ammonium compounds and expression of outer membrane protein OprR. *Antimicrob. Agents Chemother.* **47:**2093–2099.
88. **Toleman, M. A., K. Rolston, R. N. Jones, and T. R. Walsh.** 2003. Molecular and biochemical characterization of OXA-45, an extended-spectrum class 2d′ β-lactamase in *Pseudomonas aeruginosa. Antimicrob. Agents Chemother.* **47:**2859–2863.
89. **Unal, S., R. Masterton, and H. Goossens.** 2004. Bacteraemia in Europe-antimicrobial susceptibility data from the MYSTIC surveillance programme. *Int. J. Antimicrob. Agents* **23:**155–163.
90. **Vogne, C., J. R. Aires, C. Bailly, D. Hocquet, and P. Plesiat.** 2004. Role of the multidrug efflux system MexXY in the emergence of moderate resistance to aminoglycosides among *Pseudomonas aeruginosa* isolates from patients with cystic fibrosis. *Antimicrob. Agents Chemother.* **48:**1676–1680.
91. **Weldhagen, G. F., L. Poirel, and P. Nordmann.** 2003. Ambler class A extended-spectrum β-lactamases in *Pseudomonas aeruginosa,* novel developments and clinical impact. *Antimicrob. Agents Chemother.* **47:**2385–2392.
92. **Ziha-Zarifi, I., C. Llanes, T. Koehler, J.-C. Pechere, and P. Plesiat.** 1999. In vivo emergence of multidrug-resistant mutants of *Pseudomonas aeruginosa* overexpressing the active efflux system MexA-MexB-OprM. *Antimicrob. Agents Chemother.* **43:**287–291.

Frontiers in Antimicrobial Resistance: a Tribute to Stuart B. Levy
Edited by D. G. White, M. N. Alekshun, and P. F. McDermott

Chapter 27

Antibiotic Resistance in *Salmonella* and *Shigella*

E. J. THRELFALL

Resistance to antimicrobial drugs, and particularly multidrug resistance, is a major problem in *Enterobacteriaceae* in both developing and developed countries throughout the world, affecting a wide range of genera, including *Salmonella enterica*, *Shigella* spp., and *Escherichia coli*. Of particular concern is the increasing occurrence of strains with resistance to key antimicrobials, notably fluoroquinolones and third-generation cephalosporins, often coupled with resistance to conventional antimicrobials. This review concentrates specifically on *Salmonella enterica* and *Shigella* spp.

SALMONELLA ENTERICA

Resistance to frontline antimicrobials is particularly important in the treatment of infections caused by *S. enterica* serovars Typhi and Paratyphi A. The increasing occurrence of multiple resistance in serotypes other than Typhi has also had a profound effect in the treatment of salmonella septicemia in infants and young children in developing countries where multiple resistant strains have been implicated in numerous outbreaks in the community and hospital pediatric units for the past 30 years.

S. enterica Serovar Typhi

Typhoid fever is a systemic illness caused by *S. enterica* serovar Typhi and is a significant cause of morbidity and mortality among children and adults in developing countries. The World Health Organization estimates that 16 million illnesses and 600,000 deaths worldwide are attributable to *S. enterica* serotype Typhi infections annually (19). Infections due to serovar Typhi result in bacteremia characterized by remittent fevers, headache, malaise, abdominal discomfort, constipation or diarrhea, and, in some cases, a distinctive "rose spot" rash (26). *Salmonella* serovar Typhi remains endemic in Africa, South and Central America, and the Indian subcontinent (19). In the United Kingdom or the United States the incidence of serovar Typhi is much lower, with the majority of cases in travellers returning from endemic areas (2, 19, 38).

The choice of the appropriate antibiotic is essential for the treatment of patients with typhoid fever and should commence as soon as clinical diagnosis is made. Until the mid-1970s, chloramphenicol was the undisputed first-line drug. The efficacy of chloramphenicol was undermined in the 1970s following a series of outbreaks in countries as far apart as Mexico and India, caused by strains exhibiting resistance to this antimicrobial (30). A feature of all chloramphenicol-resistant strains from such outbreaks was that, although the strains belonged to different Vi phage types, resistance to chloramphenicol was encoded by a plasmid of the H1 incompatibility (*inc*) group (now termed HI1) and was often found in combination with resistance to streptomycin, sulphonamides, and tetracyclines (R-type CSSuT). Since 1989 outbreaks caused by serovar Typhi strains resistant to chloramphenicol, ampicillin, and trimethoprim, and with additional resistances to streptomycin, sulphonamides, and tetracyclines (R-type ACSSuTTm), have been reported in many developing countries, particularly in the Indian subcontinent (30). The recent emergence of strains with resistance to trimethoprim and ampicillin in addition to chloramphenicol has caused many treatment problems.

Without exception, in all outbreaks involving multiresistant serovar Typhi strains studied to date, the complete spectrum of multiple resistance has been encoded by plasmids of the HI1 incompatibility group (20, 30). Evolutionary diversity within this "Typhi-specific" compatibility group has recently been observed among HI1 plasmids from multiresistant strains from Vietnam over a 10-year time period in the 1990s

E. J. Threlfall • Health Protection Agency, Laboratory of Enteric Pathogens, Centre for Infections, 61, Colindale Avenue, London NW9 5HT, United Kingdom.

(42). Comparison of pHCM1, isolated in 1993 from Typhi in Vietnam with the prototype incH plasmid, R27, isolated in 1961 from *S. enterica* in the United Kingdom, revealed extensive regions of homology (42). However, pHCM1 possessed 18 different predicted antimicrobial and heavy metal resistance genes as compared with R27, which may provide a selective advantage for serovar Typhi isolates in the presence of antimicrobial agents or heavy metals. Ciprofloxacin has become the first-line drug of choice in both developing and developed countries following the emergence of serovar Typhi exhibiting resistance to chloramphenicol, ampicillin, and trimethoprim (21). Regrettably, strains of serovar Typhi with plasmid-encoded resistance to chloramphenicol, ampicillin, and trimethoprim and showing decreased susceptibility to ciprofloxacin (ciprofloxacin MIC, 0.25–1.0 mg/liter) have been increasingly reported. Such strains have caused substantive outbreaks of infection in several developing countries, notably Tajikistan and Vietnam (23, 25, 26), and have also caused treatment problems in developed countries (41). In a recent study of over 400 strains isolated from patients in 10 European countries in the 3-year period 1999-2001, between 20 and 26% exhibited decreased susceptibility to ciprofloxacin, often in combination with resistance to chloramphenicol, ampicillin, and trimethoprim (38). Although the MIC of ciprofloxacin was below the anticipated serum drug level for this antimicrobial following administration at recommended doses, numerous treatment failures have been reported. In strains of serovar Typhi with decreased susceptibility to ciprofloxacin, four different mutations have been identified in *gyrA* (43), indicating that such strains are not clonal in respect of phage type or of ciprofloxacin susceptibility. In infections caused by strains with decreased susceptibility to ciprofloxacin, third-generation cephalosporin antibiotics such as ceftriaxone or cefotaxime have been suggested as possible alternatives (26). Azithromycin, a macrolide antibiotic, has also been evaluated for the treatment of infections caused by multiresistant typhoid, with encouraging results (44). However, the expense of these antimicrobials may preclude their use in developing nations.

S. enterica Serovar Paratyphi A

Infections caused by *Salmonella* serovar Paratyphi A, although in general not as severe as typhoid fever, may also require antimicrobial intervention before the results of susceptibility tests are available. A substantive increase in serovar Paratyphi A isolates with decreased susceptibility to ciprofloxacin has been reported in India since the late 1990s (8). A study conducted in Delhi found the incidence of serovar Paratyphi A in enteric fever cases to be as high as 45 % in 1998 (32). Additionally, multidrug resistant strains have become increasingly common. For example, Chandel et al. found that 45% of serovar Paratyphi A isolates were multidrug resistant (8). They also further noted a decrease in ciprofloxacin susceptibility among these isolates. An increase in strains of serovar Paratyphi A with decreased susceptibility to ciprofloxacin isolated from patients from 10 European countries between 1999 and 2001 has also been observed (37).

Resistance to Fluoroquinolone Antibiotics in *S. enterica* Serovars Typhi and Paratyphi A

The British Society for Antimicrobial Chemotherapy (BSAC) and the National Committee for Clinical Laboratory Standards (NCCLS) zone size equivalents for ciprofloxacin resistance in *Enterobacteriaceae* in disc diffusion tests are 2.0 and 4.0 mg/liter, respectively. Regrettably, testing for ciprofloxacin resistance by disc diffusion at these levels can result in decreased susceptibility (≥0.125 mg/liter) not being detected. Because of the increasing occurrence of treatment failures in patients infected with strains with MICs below these levels, consideration of changing the breakpoint for ciprofloxacin resistance has been suggested (1).

OTHER *SALMONELLA* SEROVARS

Developed Countries

In developed countries such as the majority of countries in Western Europe and the United States, salmonella infections are primarily zoonotic in origin. When resistance is present, it has often been acquired prior to transmission of the organism through the food chain to humans. Such resistance acquisition has been related to the use of antimicrobials in animal husbandry (3, 29, 35). The most important serovars in the United Kingdom and Europe are Enteritidis and Typhimurium and in the United States are Typhimurium, Enteritidis, and more recently, Newport. For all these serovars the main method of spread is through contaminated food. In most cases the clinical presentation is that of mild to moderate enteritis. The disease is usually self-limiting, and antimicrobial therapy is seldom required.

In the United Kingdom for the last four decades, the history of multiple resistance in *Salmonella enterica* has been dominated by three major clones of serovar Typhimurium, namely definitive phage types (DTs) 29, 204/204c/193, and 104 (Fig. 1).

Serovar Typhimurium DT 29 (= DT 29) emerged as the predominant multiresistant (to four or more

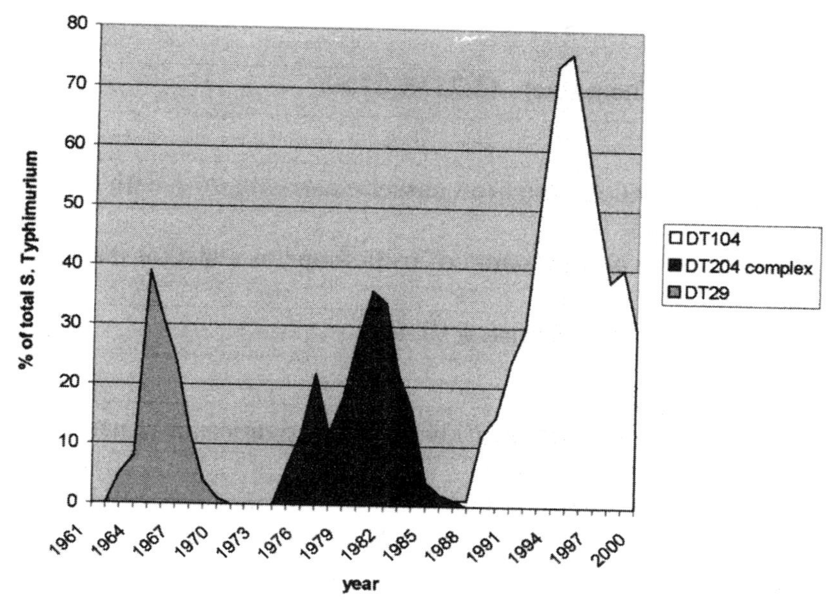

Figure 1. Emergence of multiresistant *S. enterica* serovar Typhimurium in England and Wales.

drugs) strain in humans and bovine animals in the United Kingdom in the mid-1960s (3). The strain was resistant to ampicillin, streptomycin, sulfonamides, tetracyclines, and furazolidone with resistance to ampicillin, streptomycin, sulphonamides, and tetracyclines being plasmid encoded. By 1970, DT 29 had disappeared from bovine animals in Britain. The reasons for this disappearance remain unclear. From 1975 to the mid-1980s, there was a major epidemic in humans and calves of the related phage types DTs 204, 193, and 204c (29). A feature of this epidemic outbreak was the sequential acquisition of plasmids and transposons coding for ampicillin, chloramphenicol, gentamicin, kanamycin, streptomycin, sulphonamides, tetracyclines, and trimethoprim (ACGSSuTTm). All these resistances were plasmid encoded. The acquisition of resistance by strains of these phage types seemed to coincide with the introduction and use of at least some of the antimicrobial agents in attempts to combat infections in calves caused by serovar Typhimurium strains resistant to an increasing range of antibiotics (29). Since 1991 there has been an epidemic in cattle and humans in England and Wales of multiresistant (MR) strains of DT104 of R-type ACSSpSuT (Sp = spectinomycin). In contrast to DTs 29 and 204/193/204c, the ACSSpSuT resistance genes in MR DT104 are chromosomally located (35).

Although some variants have been identified, when studied by PFGE the majority of MR serovar Typhimurium DT104 isolates are characterized by a distinctive *Xba* I-generated macrorestriction fingerprint (35). Over the last 15 years this particular clone has caused outbreaks of infection in food animals and humans in numerous European countries and as far afield as South Africa, the United Arab Emirates, and the Philippines. In 1996 infections with MR DT104 were recognized in cattle and humans in North America, both in Canada and the United States. DT104 has been responsible for many outbreaks in both cattle and humans, particularly in the United States (5). Of particular concern has been the resistance of the organism to a wide range of therapeutic antimicrobials. Furthermore, in some countries there have been reports of an apparent predilection of the organism to cause serious disease (22). This is in contrast to the United Kingdom, where a study of over 70,000 salmonella infections in the 3-year period 1994-1996 demonstrated that MR serovar Typhimurium DT104 appeared no more invasive than other common serotypes and phage types in terms of isolations from blood culture in comparison to fecal isolations (40).

Since 1992 an increasing number of isolates of MR Typhimurium have exhibited decreased susceptibility to ciprofloxacin. This property is chromosomally encoded as a result of mutations in the quinolone resistance determining region (QRDR). Four mutations have been identified in different strains (28), demonstrating that strains of MR DT104 with decreased susceptibility to ciprofloxacin are not clonal. In MR DT104 of R-type ACSSpSuT, resistances are contained in a 43 kb island (salmonella genomic island [SGI]), comprised of integrons containing, respectively, the ASu (bla_{CARB-2} and *sul1*) and SSp (*aadA2*) genes with intervening plasmid-derived genes coding for resistance to chloramphenicol/florfenicol (*flo*) and tetracyclines (*tetG*) (35). All isolates of MR DT104 with the

ACSSpSuT phenotype have contained the same gene cassettes irrespective of source (food animal or human) or country of origin. Of note in recent years has been the identification of SGI1 in several different salmonella serovars, including Agona, Albany, and Paratyphi B variant Java (11–14, 39). Such strains have caused infections in humans and cattle (14, 39), and there is speculation of a connection with ornamental fish originating in the Far East (14).

It should be realized that multiple resistance in salmonellas in developed countries is not confined to serovar Typhimurium DT104 and related strains. In Spain and the United Kingdom, emergent multiresistant strains of *S. enterica* serotype [4,5,12:i:-] have been associated with an increasing number of human infections since the mid-1990s (17). In these strains, resistance has been mediated by an unusual plasmid containing resistance genes located within a class 1 integron and also the *spvA*, *spvB*, and *spvC* serovar Typhimurium plasmid virulence genes (17). In the United States multiple resistance has been reported in serovars Saintpaul, Heidelberg, Newport, and Typhimurium (5, 18, 45). In 2000, the first case of domestically acquired ceftriaxone-resistant *Salmonella* in a case of human salmonellosis in the United States, involving the transmission of a resistant strain from cattle to the child of a veterinarian was reported (15). More recently MDR Newport with plasmid-encoded resistance to ceftriaxone has caused numerous infections in both cattle and humans in North America (27, 45). This organism commonly shows resistance to ACSSuT, with additional resistance to third-generation cephalosporins mediated by the CMY-2 β-lactamase gene (7, 27, 45). Similarly there have been increasing reports of resistance to extended-spectrum β-lactamases in *Salmonella* from humans and food animals in numerous countries worldwide. For example CTX-M-9, -15 and -17 to -18 enzymes have recently been reported in six different serovars isolated from humans in the United Kingdom (4), and CTX-M-like enzymes have been reported in serovar Virchow in Spain (31) and in serovar Anatum in Taiwan (33). In Taiwan a particularly alarming development has been the emergence of a highly virulent strain of serovar Choleraesuis displaying high-level resistance to ciprofloxacin and plasmid-mediated resistance to ceftriaxone (10a). The patient did not respond to treatment and died with septicemia as a result of his salmonella infection. Fortunately, as far as is known the strain did not spread and there have been no further infections with this highly resistant organism.

As antibiotics are not recommended for the treatment of mild to moderate salmonella-induced enteritis in humans, it may be argued that in developed countries drug resistance in nontyphoidal salmonellas is of little consequence to human public health. However, antibiotics are used for the treatment of salmonellosis in immunocompromised patients and sometimes for treating particularly vulnerable patients; in such cases treatment with an appropriate antibiotic is often essential and may be life saving. Whatever the outcome, the increased occurrence of strains of enterica with resistance to fluoroquinolones and/or third-generation cephalosporins in many different countries is an alarming development which has already had fatal consequences.

Developing Countries

In developing countries, particularly in the Indian subcontinent, Southeast Asia, South and Central America, and Africa, serotypes such as Typhimurium, Wien, Johannesburg, and Oranienburg have undergone changes both in their epidemiology and their clinical disease. An additional feature of these strains has been the possession of plasmid-mediated multidrug resistance, often exhibiting resistance to seven or more antimicrobials. With the exception of resistance to furazolidone and, since 1990, to nalidixic acid coupled with decreased susceptibility to ciprofloxacin, resistances in these strains have invariably been plasmid encoded. In contrast to the situation in developed countries, the most common presentation is that of severe gastroenteritis, often accompanied by septicemia (up to 40% in some outbreaks) and with up to 30% mortality. The main method of transmission is by person-to-person spread either in hospitals or in the community. The strains involved are invariably multiresistant, often displaying resistance to up to nine antimicrobial agents (for a review, see reference 36).

A particular property of the majority of these multiresistant strains is the possession of a plasmid of the F_I incompatibility group coding not only for multiple resistance but also for production of the hydroxamate siderophore aerobactin, which is a known virulence factor for some enteric and urinary tract pathogens. This plasmid type, first identified in a strain of serovar Typhimurium DT 208 which caused numerous epidemics in many Middle Eastern countries in the 1970s, was subsequently identified in a strain of Wien responsible for a massive epidemic that began in Algiers in 1969 but spread rapidly thereafter through pediatric and nursery populations in many countries throughout North Africa, Western Europe, the Middle East, and eventually the Indian subcontinent over the next 10 years (36). Similar plasmids have also been identified in several unrelated phage types of serovar Typhimurium that have caused substantial outbreaks in Africa (Kenya, Liberia), India, and Turkey. A 30-year retrospective molecular study of this group of plasmids, from strains of serovar Typhimurium from several countries, has demonstrated that the plasmids have evolved through sequential acquisition of integrons carrying different arrays of antibiotic resistance genes (6). Although not clinically proven, the epidemiological evidence strongly suggests that possession of this class of plasmid has contributed to the virulence and epidemicity of such strains.

SHIGELLA SPP.

Developing Countries

Since 1969 multiresistant strains of *Shigella* spp. and, in particular, *S. dysenteriae* 1 (Shiga's bacillus) have caused extensive outbreaks in many countries in Central America, Africa, and the Indian subcontinent. The first major outbreak was that which occurred in Central America from 1969 to 1972, and the strain was resistant to chloramphenicol, streptomycin, sulphonamides, and tetracyclines (R-type CSSuT). Over 10,000 deaths were reported. In the 1970s a series of outbreaks, not of the same scale as the Central American outbreak but caused by strains with the same resistance pattern, occurred in several countries in the Indian subcontinent (36). In all these outbreaks, resistance to chloramphenicol, streptomycin, sulfonamides, and tetracyclines was invariably plasmid mediated and was encoded by a plasmid of the *inc* B group. The second major international outbreak occurred in Central Africa (Zaire, Rwanda, Burundi) from 1979 to 1982. Over 13,000 cases were reported in Eastern Zaire between 1981 and 1982 with over 1,700 deaths. The strain was of R-type ACSSuT, and although resistances were plasmid encoded, the plasmids were of different incompatibility groups than those identified in the Central American and Indian strains. Following the discontinuation of the use of tetracyclines and the introduction of co-trimoxazole early in 1981, plasmid-mediated resistance to the latter antimicrobial soon emerged. Nalidixic acid was subsequently introduced in Zaire in November 1981 for the treatment of Shiga dysentery, and the use of this antimicrobial resulted in a drop in case fatality rate from 4.6 to 2.0%. Subsequent to these outbreaks in the 1980s, serious epidemics of multiresistant *S. dysenteriae* 1 have been reported in the 1990s in Zimbabwe and Zambia. The causative strains were resistant to a wide range of antimicrobials, including ampicillin, chloramphenicol, tetracyclines, and trimethoprim (36).

Strains of multiresistant *S. dysenteriae* 1 are now becoming commonplace in other developing countries, not only in Central America, India, and Africa but also in the Middle East. For example in the 5-year period 1990-1993 and 1996, *S. dysenteriae* comprised 5% of shigella strains isolated in Kuwait. All isolates were multiresistant. Once more, because of the emergence of multidrug resistance, the therapeutic options for oral therapy were severely restricted.

An increasing problem in many developing countries in recent years has been the appearance and spread of multiresistant strains of *S. flexneri*. Outbreaks with such strains have been reported in many countries in Southeast Asia and the Indian subcontinent, and treatment problems have been recorded in infections caused by multiresistant strains in many other countries. Strains have often been resistant to at least five antimicrobials, including ampicillin, chloramphenicol, and trimethoprim, with an increasing number of strains from the Indian subcontinent showing resistance to nalidixic acid. In strains from Tanzania, resistance to ampicillin has been related to integron-mediated OXA-1 and TEM-1 β-lactamases (24).

Britain

Between 1983 and 1987, the incidence of resistance to ampicillin in *S. dysenteriae, S. flexneri,* and *S. boydii* infections in England and Wales increased from 42 to 65% and the incidence of resistance to trimethoprim from 6 to 64% (16). Furthermore, of 1,524 strains tested in 1995-1996, 46% were resistant to both these antimicrobials. Resistance to nalidixic acid was uncommon, and only a very small number of strains were resistant to ciprofloxacin. On the basis of these observations, it was concluded that if it should be necessary to commence treatment before the results of laboratory-based sensitivity tests were available, the best options would be to use nalidixic acid for children and a fluoroquinolone antibiotic such as ciprofloxacin for adults (10).

In 2002, 10% of shigellas of subgroups A, B, and C, and 13% of isolates of subgroup D (*Shigella sonnei*) from patients in England and Wales were resistant to nalidixic acid (9). These findings suggested that the choice of antimicrobial for the first-line treatment of shigellosis is becoming increasingly limited. Because of the increasing incidence of resistance to nalidixic acid, it has been recommended that that for infections in children under 10 years of age, strains should be tested for resistance to this antimicrobial before commencing treatment. This recommendation is in line with information received from other countries, where in some instances nalidixic acid has been abandoned as a therapeutic option for the treatment of acute shigellosis.

CONCLUSIONS

Multiple drug resistance is now common in pathogenic gut bacteria in both developing and developed countries throughout the world. An increasing number of fatalities caused by multiresistant bacterial pathogens have been reported in developed countries, and there is also concern about the long-term effects caused by infections with such strains. The rapid emergence of resistance to the drugs of choice for diseases such as typhoid fever and bacillary dysentery is of particular concern. Although for the most part resistance is plasmid mediated, a recent development is the emergence of strains with chromosomal resistance, not only to antibiotics such as chloramphenicol, ampicillin, and trimethoprim, but also to nalidixic acid and some of the new fluoroquinolone drugs such as ciprofloxacin, which are now the first-line choice for the treatment of invasive disease. A further alarming development has been the emergence of plasmid-mediated resistance to third-generation cephalosporins in several unrelated genera. In order to preserve the efficacy of such drugs for the treatment of life-threatening infections, it is essential that their usage should be strictly regulated. Whenever possible, antimicrobial drugs such as ciprofloxacin should be reserved for the treatment of severe infections which do not respond to more conventional antimicrobials. In developed countries, where resistance is often associated with the use of antimicrobials in food-producing animals, such usage should be strictly regulated and the unnecessary prophylactic use of antimicrobials avoided wherever possible.

REFERENCES

1. **Aarestrup, F. M., C. Wiuff, K. Molbak, and E. J. Threlfall.** 2003. Is it time to change fluoroquinolone breakpoints for *Salmonella* spp.? *Antimicrob. Agents Chemother.* **47:**827–829.
2. **Ackers, M. L., N. D. Puhr, R. V. Tauxe, and E. D. Mintz.** 2000. Laboratory-based surveillance of *Salmonella* serotype Typhi infections in the United States: antimicrobial resistance on the rise. *JAMA* **283:**2668–2673.
3. **Anderson, E. S.** 1968. Drug resistance in *Salmonella typhimurium* and its implications. *Br. Med. J.* **3:**333–339.
4. **Batchelor, M., K. Hopkins, E. J. Threlfall, F. A. Clifton-Hadley, A. D. Stallwood, R. H. Davies, and E. Liebana.** 2005. bla(CTX-M) genes in clinical Salmonella isolates recovered from humans in England and Wales from 1992 to 2003. *Antimicrob. Agents Chemother.* **49:**1319–1322.
5. **Besser, T. E., M. Goldoft, L. C. Pritchett, R. Khakhria, D. D. Hancock, D. H. Rice, J. M. Gay, W. Johnson, and C. C. Gay.** 2000. Multiresistant *Salmonella typhimurium* DT104 infections of humans and domestic animals in the Pacific Northwest of the United States. *Epidemiol. Infect.* **124:**193–200.
6. **Carattoli, A.** 2003. Plasmid-mediated antimicrobial resistance in Salmonella enterica. *Curr. Issues Mol. Biol.* **5:**113–122.
7. **Carattoli, A., F. Tosini, W. P. Giles, M. E. Rupp, S. H. Hinrichs, F. J. Angulo, T. J. Barrett, and P. D. Fey.** 2002. Characterization of plasmids carrying CMY-2 from expanded-spectrum cephalosporin-resistant *Salmonella* strains isolated in the United States between 1996 and 1998. *Antimicrob. Agents Chemother.* **46:**1269–1272.
8. **Chandel, D. S., R. Chaudhry, B. Dhawan, A. Pandey, and A. B. Dey.** 2000. Drug-resistant *Salmonella enterica* serotype paratyphi A in India. *Emerg. Infect. Dis.* **6:**420–421.
9. **Cheasty, T., M. Day, and E. J. Threlfall.** 2004. Increasing incidence of resistance to nalidixic acid in shigellas from humans in England and Wales: implications for therapy. *Clin. Microbiol. Infect.* **10:**1033–1035.
10. **Cheasty, T., J. A. Skinner, B. Rowe, and E. J. Threlfall.** 1998. Increasing incidence of antibiotic resistance in shigellas from humans in England and Wales: recommendations for therapy. *Microb. Drug Resist.* **4:**57–60.

10a. **Chiu, C., L. H. Su, C. Chu, J. H. Chia, T. L. Wu, T. Y. Lin, Y. S. Lee, and J. T. Ou.** 2004. Isolation of *Salmonella enterica* serotype Choleraesuis resistant to ceftriaxone and ciprofloxacin. *Lancet* **363:**1285–1286.

11. **Doublet, B., D. Boyd, M. R. Mulvey, and A. Cloeckaert.** 2005. The Salmonella genomic island 1 is an integrative mobilizable element. *Mol. Microbiol.* **55:**1911–1924.
12. **Doublet, B., P. Butaye, H. Imberechts, D. Boyd, M. R. Mulvey, E. Chaslus-Dancla, and A. Cloeckaert.** 2004. Salmonella genomic island 1 multidrug resistance gene clusters in *Salmonella enterica* serovar Agona isolated in Belgium in 1992 to 2002. *Antimicrob. Agents Chemother.* **48:**2510–2517.
13. **Doublet, B., R. Lailler, D. Meunier, A. Brisabois, D. Boyd, M. R. Mulvey, E. Chaslus-Dancla, and A. Cloeckaert.** 2003. Variant Salmonella genomic island 1 antibiotic resistance gene cluster in *Salmonella enterica* serovar Albany. *Emerg. Infect. Dis.* **9:**585–591.
14. **Evans, S. J., R. H. Davies, S. H. Binns, T. W. H. Jones, E. Liebana, M. F. Millar, E. J. Threlfall, L. R. Ward, K. L. Hopkins, P. H. S. Mackay, P. J. R. Gayford, and H. Bailie.** 2005. Multiple antimicrobial resistant *Salmonella Paratyphi* B variant Java in cattle: a case report. *Vet. Rec.* **166:**343–346.
15. **Fey, P. D., T. J. Safranek, M. E. Rupp, E. F. Dunne, E. Ribot, P. C. Iwen, P. A. Bradford, F. J. Angulo, and S. H. Hinrichs.** 2000. Ceftriaxone-resistant salmonella infection acquired by a child from cattle. *N. Engl. J. Med.* **342:**1242–1249.
16. **Gross, R. J., E. J. Threlfall, L. R. Ward, and B. Rowe.** 1984. Drug resistance in *Shigella dysenteriae, S flexneri* and *S boydii* in England and Wales: increasing incidence of resistance to trimethoprim. *Br. Med. J.* **288:**784–786.
17. **Guerra, B., I. Laconcha, S. M. Soto, M. A. Gonzalez-Hevia, and M. C. Mendoza.** 2000. Molecular characterisation of emergent multiresistant Salmonella enterica serotype [4,5,12:i:-] organisms causing human salmonellosis. *FEMS Microbiol. Lett.* **190:**341–347.
18. **Holmberg, S. D., J. G. Wells, and M. L. Cohen.** 1984. Animal-to-man transmission of antimicrobial-resistant *Salmonella:* investigations of U.S. outbreaks, 1971-1983. *Science* **225:**833–835.
19. **Ivanoff, B.** 1995. Typhoid fever: global situation and WHO recommendations. *Southeast Asian J. Trop. Med. Publ. Health* **26:**1–6.
20. **Kidgell, C.** 2005. University of London. Genetic variation in *Salmonella enterica* subspecies enterica serotype Typhi.
21. **Mandal, B. K.** 1991. Modern treatment of typhoid fever. *J. Infect.* **22:**1–4.
22. **Molbak, K., D. L. Baggesen, F. M. Aarestrup, J. M. Ebbesen, J. Engberg, K. Frydendahl, P. Gerner-Smidt, A. M. Petersen, and H. C. Wegener.** 1999. An outbreak of multidrug-resistant, quinolone-resistant *Salmonella enterica* serotype typhimurium DT104. *N. Engl. J. Med.* **341:**1420–1425.
23. **Murdoch, D. A., N. Banatvaia, A. Bone, B. I. Shoismatulloev, L. R. Ward, and E. J. Threlfall.** 1998. Epidemic ciprofloxacin-resistant Salmonella typhi in Tajikistan. *Lancet* **351:**339.
24. **Navia, M. M., L. Capitano, J. Ruiz, M. Vargas, H. Urassa, D. Schellemberg, J. Gascon, and J. Vila.** 1999. Typing and characterization of mechanisms of resistance of Shigella spp. isolated

from feces of children under 5 years of age from Ifakara, Tanzania. *J. Clin. Microbiol.* **37:**3113–3117.

25. **Parry, C. M.** 2004. The treatment of multidrug-resistant and nalidixic acid-resistant typhoid fever in Viet Nam. *Trans. R. Soc. Trop. Med. Hyg.* **98:**413–422.
26. **Parry, C. M., T. T. Hien, G. Dougan, N. J. White, and J. J. Farrar.** 2002. Typhoid fever. *N. Engl. J. Med.* **347:**1770–1782.
27. **Rankin, S. C., H. Aceto, J. Cassidy, J. Holt, S. Young, B. Love, D. Tewari, D. S. Munro, and C. E. Benson.** 2002. Molecular characterization of cephalosporin-resistant *Salmonella enterica* serotype Newport isolates from animals in Pennsylvania. *J. Clin. Microbiol.* **40:**4679–4684.
28. **Ridley, A. and E. J. Threlfall.** 1998. Molecular epidemiology of antibiotic resistance genes in multiresistant epidemic *Salmonella typhimurium* DT 104. *Microb. Drug Resist.* **4:**113–118.
29. **Rowe, B., and E. J. Threlfall.** 1984. Drug resistance in gram-negative aerobic bacilli. *Br. Med. Bull.* **40:**68–76.
30. **Rowe, B., L. R. Ward, and E. J. Threlfall.** 1997. Multidrug-resistant *Salmonella typhi:* a worldwide epidemic. *Clin. Infect. Dis.* **24** (Suppl 1):S106–S109.
31. **Simarro, E., F. Navarro, J. Ruiz, E. Miro, J. Gomez, and B. Mirelis.** 2000. *Salmonella enterica* serovar virchow with CTX-M-like beta-lactamase in Spain. *J. Clin. Microbiol.* **38:**4676–4678.
32. **Sood, S., A. Kapil, N. Dash, B. K. Das, V. Goel, and P. Seth.** 1999. Paratyphoid fever in India: an emerging problem. *Emerg. Infect. Dis.* **5:**483–484.
33. **Su, L. H., C. H. Chiu, C. Chu, M. H. Wang, J. H. Chia, and T. L. Wu.** 2003. In vivo acquisition of ceftriaxone resistance in *Salmonella enterica* serotype anatum. *Antimicrob. Agents Chemother.* **47:**563–567.
34. **Teale, C. J., L. Barker, A. P. Foster, E. Liebana, M. Batchelor, D. M. Livermore, and E. J. Threlfall.** 2005. Extended-spectrum beta-lactamase detected in *E. coli* recovered from calves in Wales. *Vet. Rec.* **156:**186–187.
35. **Threlfall, E. J.** 2000. Epidemic salmonella typhimurium DT 104—a truly international multiresistant clone. *J. Antimicrob. Chemother.* **46:**7–10.
36. **Threlfall, E. J.** 2003. Resistant gut bacteria, p. 951–962. *In* G. C. Cook and A. I. Zumla (ed.), *Manson's Tropical Diseases.* W. B. Saunders, London, United Kingdom.
37. **Threlfall, E. J., I. S. Fisher, C. Berghold, P. Gerner-Smidt, H. Tschape, M. Cormican, I. Luzzi, F. Schnieder, W. Wannet, J. Machado, and G. Edwards.** 2003. Antimicrobial drug resistance in isolates of *Salmonella enterica* from cases of salmonellosis in humans in Europe in 2000: results of international multi-centre surveillance. *Euro. Surveill.* **8:**41–45.
38. **Threlfall, E. J., I. S. Fisher, C. Berghold, P. Gerner-Smidt, H. Tschape, M. Cormican, I. Luzzi, F. Schnieder, W. Wannet, J. Machado, and G. Edwards.** 2003. Trends in antimicrobial drug resistance in *Salmonella enterica* serotypes Typhi and Paratyphi A isolated in Europe, 1999-2001. *Int. J. Antimicrob. Agents* **22:**487–491.
39. **Threlfall, E. J., B. Levent, K. L. Hopkins, E. De Pinna, L. R. Ward, and D. J. Brown.** 2005. Multidrug-resistant *Salmonella java. Emerg. Infect. Dis.* **11:**170–171.
40. **Threlfall, E. J., L. R. Ward, and B. Rowe.** 1998. Multiresistant *Salmonella typhimurium* DT 104 and salmonella bacteraemia. *Lancet* **352:**287–288.
41. **Threlfall, E. J., L. R. Ward, J. A. Skinner, H. R. Smith, and S. Lacey.** 1999. Ciprofloxacin-resistant *Salmonella typhi* and treatment failure. *Lancet* **353:**1590–1591.
42. **Wain, J., L. T. Diem Nga, C. Kidgell, K. James, S. Fortune, D. T. Song, T. Ali, O. Gaora, C. Parry, J. Parkhill, J. Farrar, N. J. White, and G. Dougan.** 2003. Molecular analysis of incHI1 antimicrobial resistance plasmids from *Salmonella* serovar Typhi strains associated with typhoid fever. *Antimicrob. Agents Chemother.* **47:**2732–2739.
43. **Walker, R. A., J. A. Skinner, L. R. Ward, and E. J. Threlfall.** 2003. LightCycler gyrA mutation assay (GAMA) identifies heterogeneity in GyrA in Salmonella enterica serotypes Typhi and Paratyphi A with decreased susceptibility to ciprofloxacin. *Int. J. Antimicrob. Agents* **22:**622–625.
44. **Wallace, M. R., A. A. Yousif, N. F. Habib, and D. R. Tribble.** 1994. Azithromycin and typhoid. *Lancet* **343:**1497–1498.
45. **Zhao, S., S. Qaiyumi, S. Friedman, R. Singh, S. L. Foley, D. G. White, P. F. McDermott, T. Donkar, C. Bolin, S. Munro, E. J. Baron, and R. D. Walker.** 2003. Characterization of Salmonella enterica serotype newport isolated from humans and food animals. *J. Clin. Microbiol.* **41:**5366–5371.

Frontiers in Antimicrobial Resistance: a Tribute to Stuart B. Levy
Edited by D. G. White, M. N. Alekshun, and P. F. McDermott

Chapter 28

Antibiotic Resistance in *Escherichia coli*

MARK A. WEBBER AND LAURA J. V. PIDDOCK

INTRODUCTION

Escherichia coli is often considered a model organism due to the enormous historic research effort into the phenotypic and genotypic characteristics of this bacterium (83). As a result more is known about the genes and their function of this species than any other. What is not always appreciated so readily is the role of *E. coli* as a major human pathogen. *E. coli* is a leading cause of urinary tract infection (47, 114), bacteremia (28), and gastroenteritis (51, 86) among others. As a consequence, antibiotic therapy is frequently indicated for the treatment of *E. coli* infections. The ability to acquire and transfer resistance to antimicrobial agents is a characteristic of *E. coli* (122) that has led researchers interested in the mechanisms of antibiotic resistance to again turn to *E. coli* as a model organism; not least among these researchers is Stuart Levy.

The ability of *E. coli* to display concomitant resistance to multiple classes of antibiotic at once has been reported with increasing frequency in recent years (29). Initially this was attributed to the sequential accumulation of both transferable and chromosomal resistance determinants; however, work by Stuart Levy demonstrated the ability of *E. coli* to develop multiple antibiotic resistance (MAR) in one step. The genetic basis behind this MAR phenotype has been a research interest of Professor Levy's since the 1980s, and he was involved in the detection and characterization of the *mar* locus of *E. coli* (3, 32). MarA is a global transcriptional regulator that controls a regulon of over 60 chromosomal *E. coli* genes (7), among these are the efflux system *acrAB*, which is overexpressed, and *micF*, an RNA antisense to *ompF*—a synergistic combination of increased efflux and decreased permeability results in reduced intracellular antibiotic accumulation and consequent decreased susceptibility to a range of antibiotics, dyes, disinfectants, and detergents (64, 88). While the role of the *mar* locus in clinical antibiotic resistance has been the subject of some debate, resistance to certain commonly used antibiotics is clearly of clinical concern; these include the fluoroquinolones and latest β-lactams for which resistance can lead to treatment failure.

Antibiotic resistance is unfortunately an inevitable consequence of antibiotic use. However, to understand the selection and spread of resistance, there is a need to monitor the numbers of resistant isolates in order to anticipate coming problems and to determine the appropriate use of antibiotics in order to minimize the future selection of resistant strains. Several surveillance programs have been initiated since the 1990s to attempt to quantify the current situation regarding antibiotic resistance around the world (9). However, such surveys are often hard to compare against each other and may not include geographical areas of particular concern where resistance may be emerging, e.g., the Asia-Pacific rim. The basic mechanisms that generate and propagate antibiotic resistance can all be found within *E. coli*; understanding the evolving pattern of antibiotic resistance in our favourite model organism will continue to be of importance as we try to understand and prevent the expansion and spread of bacterial antibiotic resistance in the years to come.

Disease Spectrum of *E. coli*

The genetic diversity of different strains of *E. coli* is not always appreciated, as not all *E. coli* are capable of causing the same spectrum of disease (107). Many strains of *E. coli* are part of the natural human commensal flora; these strains may lack virulence determinants required for pathogenesis. Commensal *E. coli* can however play a role in disease in the immunocompromised host or in the presence of indwelling devices (catheters, etc.) or other breakdowns in host de-

Mark. A. Webber and Laura. J. V. Piddock • Antimicrobial Agents Research Group, Division of Immunity and Infection, University of Birmingham, Edgbaston, Birmingham, United Kingdom.

fenses (121). Pathogenic strains of *E. coli* have been classified by various schemes according to the diseases they cause; there are UPEC (uropathogenic *E. coli*), SEPEC (sepsis associated *E. coli*), NEMEC (neonatal meningitis associated *E. coli*) as well as a range of terms to describe distinct groups of intestinal pathogenic *E. coli* (EPEC, EHEC, ETEC, EAEC, EIEC, DAEC) (51). It has been proposed that *E. coli* can be subdivided into two classes—intestinal pathogenic *E. coli* and extraintestinal pathogenic *E. coli* (ExPEC, 107). ExPEC constitute those strains that are usually distinct from those that are responsible for gastroeneritis and have been associated with urinary tract, CNS, bloodstream and respiratory tract infections (99, 107). Indeed, it has been suggested that ExPEC are from the B2 phylogenetic group of *E. coli*, which is the most primitive taxon of *E. coli*. Strains of this branch have acquired numerous virulence determinants post divergence, allowing them to be pathogenic (59). The other phylogenetic groups of *E. coli* (A, B1, D) lack or have fewer of these extraintestinal virulence determinants normally associated with pathogenicity islands.

Antibiotic Therapy for *E. coli* Infections

Antibiotic therapy is often necessary to resolve infections caused by *E. coli*. Life threatening conditions such as meningitis, sepsis and bacteremia all require rapid antibiotic intervention to prevent high levels of mortality. Other conditions such as urinary tract infection, abscess, pneumonia and wound infections also require antibiotic therapy to resolve the infection. In 2003 around 15,000 cases of bacteremia were attributed to *Escherichia* spp in England, Wales, and Northern Ireland (data from the United Kingdom [U.K.] Health Protection Agency). *E. coli* is the single greatest cause of urinary tract infections in the western world, particularly in women (41), with costs to the U.S. economy alone estimated at hundreds of millions of dollars (94, 107). *E. coli* is also a significant causative agent of nosocomially acquired pneumonia in Europe and the United States (U.S.) (29, 127), and the cost of *E. coli* associated pneumonia in the U.S. has been estimated at $75-120 million per year (107). It is also the fourth most commonly isolated organism from surgical wound infections in the U.S. and the cost of these infections has been estimated at $94-252 million per year (107).

E. coli can infect most tissues of the body under appropriate circumstances. Infections caused by *E. coli* range from those that are mild and self-limiting to those that display very high mortality. Subsequently, the use of antimicrobial agents to clear these infections is widespread. The choice of antimicrobial used, however, varies around the world and reflects considerations including cost, availability and local prevalence of resistant strains. A variety of antibiotics are licensed for the treatment of infections caused by *E. coli* in the U.K. and U.S. (Table 1). Around the world there is wide variation in the classes of antibiotics in use and variation in the control (if any) of prescriptions (36).

SURVEILLANCE PROGRAMS

It has been widely accepted that it is desirable and necessary to monitor the prevalence and incidence of antibiotic resistant strains of bacteria (38). Surveillance is required at a local level (within an individual hospital, for example) as well as at national and international levels. Resistance genes and resistant strains do not respect national borders, and it is important to

Table 1. Common conditions where antibiotics are indicated for the treatment of *E. coli* infections

Condition	Antibiotics	Comment[a]
Infection of the urinary tract		
Pyelonephritis	Cephalosporin, quinolone, gentamicin	
Prostatitis	Trimethoprim, quinolone	
Lower UTI infection	Trimethoprim, amoxicillin (or co-amoxiclav), nitrofurantoin, cephalosporin, quinolone, pivmecillinam	High levels of amoxicillin resistance are now seen in the UK, including 56% of *E. coli* isolates from bacteremia in 2002 (BSAC)
Septicemia		
Community acquired	Aminoglycoside and broad-spectrum penicillin, cephalosporin	
Nosocomially acquired	Aminoglycoside and ceftazidime, carbapenem	Multiply resistant strains are more likely to be found within the hospital setting
Gastroenteritis	Ciprofloxacin, trimethoprim	Uncomplicated gastroenteritis is not usually treated with antibiotics except in immunocompromised individuals, elderly patients, or young children.

[a]UK, United Kingdom.

be aware of new and/or novel resistances in other areas before they spread. As a result of the desire to monitor antibiotic resistance a number of surveillance programmes have been initiated (9, 82). These vary in scope; some monitor specific organisms, and others monitor a multitude. The number and classes of antibiotics monitored also varies; some studies are specific to one drug and others survey many. As each study has its own particular aims it is often hard to compare data from one survey with another (9). Additional factors can bias data, including sample isolation, as many studies rely on clinical samples from diagnostic laboratories. This may be misleading; for example, many urinary tract infections may be easily cured by the first antibiotic given, and a urine sample may only be analyzed after failure due to a resistant strain. The criteria used for the interpretation of results can also vary, as can the methodologies for susceptibility testing and breakpoint concentrations used to define susceptibility and resistance (9). A number of large surveillance programs include *E. coli* within their remits; for example, the ECO.SENS project has been collecting data from urinary tract isolates from females in Europe and Canada since 1997 and determines resistance to a range of antibiotics according to the methodologies and breakpoints of the SRGA (Swedish reference group for antibiotics). The SENTRY program has also been active since 1997 in determining the prevalence of resistant strains from differing infections from around the world using Clinical Laboratory Standards Institute (CLSI; formerly NCCLS) methodologies. The Surveillance Network (TSN) is another large program which collates data from hospital laboratories across the US and elsewhere (114), again using CLSI guidelines. As well as these broad surveillance sytstems the MYSTIC program specifically records the prevalence of meropenem resistant isolates (125). The numbers of resistant bacteria from food animals have also been considered; the Spanish VAV network uses CLSI methodology to determine resistance to 18 antimicrobials from a range of food and companion animals, and the DANMAP study also uses CLSI methodology to determine the prevalence of antimicrobial resistance amongst isolates from pigs, cattle and broilers in Denmark (1). Similar initiatives exist in Japan and the U.S. (17, 26).

Resistance to fluoroquinolones and β-lactams is an issue of current concern as these classes are heavily relied upon and rates of resistance are rising, notably in areas of Asia and Latin America where as many as 25% of isolates can be resistant to ciprofloxacin or ceftazidime (Tables 2 and 3).

Advances in molecular biology allow the epidemiology of resistance genes to be studied at a molecular level. This gives further information to that obtained from simpler phenotypic resistance tests, in-

Table 2. Fluoroquinolone resistance rates recorded by surveillance programs around the world[a]

Drug	Source of isolate	Geographical location	Sampling date	Surveillance program	% Strains susceptible, range (mean)	Reference(s)
Ciprofloxacin[b]	UTI (community)	Europe and Canada	1999–2000	ECO.SENS[c]	79.4–100 (94)[d]	47
	UTI (community)	Europe and Canada	1999–2000	ECO.SENS	85.3–100 (97.7)	48
	UTI (hospital)	U.S. and Canada	1997–1998	SENTRY[e]	96.3	70
	UTI (hospital)	U.S.	2000	TRUST/TSN[f]	95.3	52, 53
	UTI (community)	U.S.	1995–2001	TSN	97.5 (2001)–99.3 (1995)	52, 53
	UTI (hospital)	Latin America	2000	SENTRY	82	35
	UTI (hospital)	Asia/Western Pacific	1998–1999	SENTRY	82.3	124
	Blood	U.S.	1997–2000	SENTRY	97	24
	Blood	Latin America	1997–2000	SENTRY	85.4	110, 113
	Skin/tissues	U.S. and Europe	2001	TSN	77.6–94.3[g]	45
	Skin/tissues	U.S. and Canada	2000	SENTRY	87.8	104
	Skin/tissues	U.S. and Canada	1997	SENTRY	97.3	25
	Skin/tissues	Latin America	1997–2000	SENTRY	73.4	112
	Pneumonia	Latin America	1997	SENTRY	93.9	111
	Pneumonia	Latin America	1998	SENTRY	75	63
Levofloxacin	CNS	U.S.	2000–2002	TSN	97.5	45

[a] Care should be taken when comparing data from one study with that from another if different interpretive criteria have been applied; hence, we have indicated which guidelines have been used by each study.
[b] Many studies include data for more than one fluoroquinolone as well as nalidixic acid. Ciprofloxacin is the most commonly used indicator fluoroquinolone.
[c] Data from the ECO.SENS project are interpreted by using recommendations of the Swedish Reference Group for Antimicrobials (SRGA).
[d] The data are for groups of countries put together according to region. The overall rate of 6% includes data from Spain and Portugal, which have much higher levels of resistance than other countries.
[e] Data from the SENTRY program are interpreted using recommendations of the National Committee for Clinical Laboratory Standards (NCCLS).
[f] Data from the TRUST and TSN programs are interpreted using recommendations of the NCCLS.
[g] The highest resistance was in Spain, and the lowest was in Germany.

Table 3. β-Lactam resistance rates reported by surveillance programs around the world

Location	Drug	Source of isolates	% strains susceptible	Comment	Reference(s)
U.S.	Ampicillin	UTI	54–62.1		35,[a] 54,[b] 55,[b] 56,[b] 70[a]
		Blood	39.1		57,[b] 109[a]
		CNS	44.2	Isolates from CSF 2000–2002	46[a]
		Soft tissues	56.6–62.2		25,[a] 45,[a] 104[a]
	Ceftazidime	UTI	98.3–99.3	Data from 1998 and 2000	54,[b] 55,[b] 70[a]
		Blood	95.6–99		24,[a] 57,[b] 110[a]
		CNS	98.1	Isolates from CSF 2000–2002	46[a]
		Soft tissues	95.6–97.3		25,[a] 45,[a] 104[a]
	Imipenem	UTI	100	Data from 1998 and 2000	54,[b] 70[a]
		Blood	100		57,[b] 109[a]
		Soft tissues	100		25,[a] 45,[a] 104[a]
Europe	Ampicillin	UTI	46.1–84.5	Highest resistance was in Spain, lowest was in Sweden; 70.2% was the European average	48,[c] 49,[c] 50[c]
		Soft tissues	34.1–60.1	Isolates from Spain, France, Italy, and Germany	45[a]
	Ceftazdime	Soft tissues	94.9–99.4	Isolates from Spain, France, Italy, and Germany	45[a]
	Imipenem	Soft tissues	100	Isolates from Spain, France, Italy, and Germany	45[a]
Asia-Pacific	Ampicillin	UTI	42.5		124[a]
	Ceftazidime	UTI	96.7		124[a]
	Imipenem	UTI	100		124[a]
Latin America	Ampicillin	RTI	27.3–58.3		63,[a] 111[a]
		Soft tissues	29.3–37	Isolates from 1997–2000	112[a]
	Ceftazidime	RTI	84.8–91.7		63,[a] 111[a]
		Soft tissues	92.8–96.3	Isolates from 1997–2000	112[a]
	Imipenem	RTI	100		63,[a] 111[a]
		Soft tissues	99.4–100	Isolates from 1997–2000	112[a]

[a] Data from the SENTRY program (NCCLS methodology).
[b] Data from the TSN program (NCCLS methodology).
[c] Data from the ECO.SENS program (Swedish Reference Group for Antimicrobials methodology).

cluding the identity of the gene(s) responsible for conferring the resistance and in some cases the specific alleles of these genes involved. The epidemiology of specific clones or strains can be determined as well as the evolution of resistance genes in response to antibiotic pressure by monitoring the spread of specific genes and alleles within and between species. These data provide further information which is extremely useful, and attempts have been made to incorporate molecular tests in certain surveillance programs (96).

EPIDEMIOLOGY OF RESISTANCE MECHANISMS

Mechanisms of Resistance in *E. coli*

Although strains of *E. coli* are usually susceptible to many antibiotics, strains resistant to one or more agents are being isolated with increasing frequency (29). The mechanisms of antibiotic resistance can be classified into four groups: target site mutation, enzymatic inactivation/degradation of the antibiotic, reduced accumulation mediated by decreased cellular permeability or active efflux, and finally metabolic bypass, where a novel antibiotic insensitive allele allows tolerance to the antibiotic. It is worth noting though that *E. coli* can exhibit examples of all these antibiotic resistance mechanisms concomitantly; for example, resistance to quinolones requires mutation in the target site topoisomerase genes, β-lactam resistance is primarily mediated by β-lactamase enzymes, the Tet efflux pumps determine tetracycline resistance, and novel hydrofolate reductase alleles allow tolerance to trimpethoprim. The presence of all these resistance mechanisms in *E. coli* shows the versatility of this pathogen and validates the study of the emergence and dissemination of antibiotic resistance in *E. coli*.

Penicillins, Cephalosporins, and Carbapenems

Resistance to penicillin was first described shortly after its introduction into general use in the 1940s (58). There has been a molecular "arms race" in the last 50 years as more β-lactam classes and derivatives have been developed and introduced. The selective pressure of this array of new and improved β-lactams has led to the consequent selection and spread of resistance genes and strains which has led to a worldwide therapeutic problem (66). The most commonly detected mechanism of β-lactam resistance in *E. coli* is the production of β-lactamase enzymes, which can be chromosomal or plasmid encoded (71). The mobile nature of many β-lactamase genes has facilitated their spread within and between species, often associated with transposable elements, insertion sequences, integrons and plasmids (77).

Resistance to early penicillins emerged after selection of strains with chromosomal β-lactamases and subsequent emergence of plasmid mediated β-lactamases, principally the TEM, SHV, and OXA enzymes (66). Modification of β-lactams to give compounds resistant to enzymatic degradation has led to the selection of strains expressing mutant TEM or SHV genes, encoding the extended spectrum β-lactamases (ESBLs) that are able to attack these newer compounds, including many cephalosporins. There are a number of different ESBL classes. Isolates of *E. coli* producing TEM and SHV ESBLs (see chapter 5) have been recorded from around the world, including Europe (5, 80, 93, 108), Latin America (103), North America (39, 84), and Africa (10, 102) as well as the Asian-Pacific rim, where the proportion of ESBL producing strains is high and has been recorded to be as high as 35% among isolates from China (10). Recently, one class comprising the CTX-M enzymes has dramatically risen to prominence (4).

The CTX-M genes are usually associated with insertion sequences or integrons, and prefer cefotaxime as a substrate. CTX-M genes have been detected in *E. coli* from Latin America (13), CTX-Ms 2, 3, 9, 13, 14, 15, and 17 have been observed in Asia (20, 92, 137, 138, 139). In Europe a variety of CTX-M enzymes have been described from *E. coli*, including CTX-Ms 1, 3, 9, 10, 14, 15, 21, and 27 (6, 8, 14, 91, 108, 115). Moland et al. (81) recently described the first isolation of CTX-M enzymes in the U.S. Although many CTX-M enzymes have been identified from nosocomial outbreaks, recently these genes have been isolated from community isolates (C. J. Munday and P. M. Hawkey, personal communication). The true incidence of CTX-M enzymes may be underestimated when ceftazidime is used as an indicator (12).

Carbapenems appear to be unaffected by ESBLs and are often withheld from use as frontline therapy in order to preserve their excellent activity (65). Recently, resistance to carbapenems mediated by carbapenemase enzymes such as VIM and IMP has been described although there are still few reports of this phenomenon being observed in *E. coli* (79, 118).

The chromosomal AmpC of *E. coli* is normally produced at very low levels; however, plasmid encoded AmpC enzymes and cases of hyperexpression of AmpC in *E. coli* have been reported (89, 119). Such strains are resistant to most cephalosporins.

Although large surveillance programs suggest relatively low rates of resistance (0 to 5%) to cephalosporins and carbapenems (Table 3), these data are often averages for large geographical regions and the impact of localized, transient, outbreaks of resistance may not be reflected. There has been an increase in the numbers of ESBL producing *E. coli* identified around the world, and ESBL producing *Enterobacteriaceae* are commonly encountered in nosocomial outbreaks (5).

Generally the numbers of strains resistant to β-lactams tend to be lowest in the U.S., Australia, and Northern Europe (below 5% of isolates) with higher numbers of resistant strains found in the Mediterranean countries and Latin America (as many as 25% of isolates (10)). There are little data available from Asia, but some publications suggest a major resistance problem in China, neighboring countries, and India (10).

Tetracyclines

Tetracycline and its relatives are broad spectrum antibiotics which have been used widely in the last fifty years. Tetracycline acts as a protein synthesis inhibitor by binding the ribosome (chapter 1). Resistance to tetracycline can be mediated by active efflux of drug by one of the Tet family of efflux pumps found widely in gram-negative bacteria, including *E. coli* (22). The Tet pumps act as antiporters where a proton is changed for a drug molecule, the mechanisms of action of the Tet pumps has been an active research interest of Stuart Levy for over 20 years (32, 73, 62, 87). The most commonly carried resistance determinants detected among resistant *E. coli* isolates are *tetB* (~60%) and *tetA* (~35%) genes, respectively, although numerous other *tet* genes have also been detected in *E. coli* (18). The *mar* operon can also confer decreased susceptibility to tetracyclines, although the contribution of this mechanism to tetracycline resistance in clinical isolates remains unclear (chapters 1 and 14).

The Tet pumps are readily mobilized and often associated with Tn*10* or various plasmids (22); this has allowed the rapid spread of resistance and the consequent loss of efficacy of tetracycline as a broad-spectrum frontline antimicrobial.

Clinical use of tetracycline in humans has been severely compromised due to toxicity concerns and high numbers of resistant strains, particularly in Asia and

Latin America where greater than 50% of *E. coli* strains may be resistant to tetracycline (57.7% in Asia [124] and 45.4 to 62.5% in Latin America [63, 109, 111]). Fewer tetracycline resistant strains are isolated from the U.S. and Europe but percentages can still range between 17.3 to 33.5% (25, 70, 104, 105).

Aminoglycosides

The aminoglycosides are a family of large, broad-spectrum protein synthesis inhibitors that attack the ribosome. Early aminoglycosides were discovered in various members of the *Streptomyces* family; derivation of some semisynthetic aminoglycosides has also been achieved by chemical modification of existing drug backbones (23, 78). Resistance to aminoglycosides can be mediated by ribosomal target site mutations, active efflux, and most commonly by aminoglycoside modifying enzymes which attack aminoglycosides and leave them unable to bind efficiently to the ribosome (67). There are three recognized classes of aminoglycoside modifying enzyme; phosphotransferases (APH), nucleotidyltransferases (ANT) and acetyltransferases (AAC). The most commonly isolated enzymes from resistant strains of *E. coli* belong to the AAC class (128). The AAC enzymes are often plasmid or transposon associated and have disseminated widely across both gram-negative and positive bacterial species (23, 78). Resistance rates to aminoglycosides have remained fairly level in recent years in North America and Europe where typically between 1 and 5% of isolates are resistant (25, 45, 47, 49, 54, 70, 105) although higher rates have been reported from Spain (45). Higher levels of resistance are seen in Asia and Latin America where between 15 and 25% of isolates are resistant to gentamicin (63, 111, 113, 124) although resistance to amikacin is rarer (0 to 5%).

Fluoroquinolone Resistance

Since their introduction, resistance to the quinolones has emerged and spread globally. Quinolone resistance in *E. coli* is mediated by alteration of the target proteins (see chapter 4); DNA gyrase and topoisomerase IV (40, 100, 101) or by impaired antibiotic accumulation mediated by either porin down-regulation or efflux pump over-expression (27, 132). Recently, plasmid mediated quinolone resistance has been described in *Enterobacteriaceae,* but to date this is extremely rare (especially in *E. coli*), and high-level resistance is not conferred by this mechanism (69, 106, 130). Mutations within the genes encoding the target proteins occur within defined quinolone resistance determining regions (QRDRs). DNA gyrase is composed of two GyrA and two GyrB subunits (31); topoisomerase IV is composed of two ParC and two ParE subunits (95). GyrA is the primary target for most fluoroquinolones in *E. coli*, and a number of mutations within the QRDR of *gyrA* conferring resistance have been described, although the most common substitutions occur at serine 83 of GyrA (100). Strains with mutations within *gyrA* may accumulate further mutations within the secondary target ParC (as many as 65% of highly resistant strains may have mutations within GyrA and ParC [27]); mutations within GyrB and ParE are less commonly recorded. Synergistic combinations of reduced accumulation and target site mutations are often seen among highly fluoroquinolone resistant clinical isolates with concomitant increased expression of the *marA* or *soxS* regulons or the AcrAB-TolC efflux system (27, 90, 132).

Although the quinolones remain a vital part of the antibiotic armourarium, rising rates of resistance have become a major concern, as overuse of these compounds can clearly drive resistance (140). Resistance is still relatively rare in the U.S. (0 to 5%), Australia and Northern Europe, but much higher rates (25%) have been observed in Asia, Latin America, Spain and Portugal (Table 2). The potential to develop quinolone resistance is inherent in all *E. coli* due to the chromosomal nature of the resistance determinants. As a result, transfer of resistance genes is not required for resistance; use of quinolones against any *E. coli* strain has the potential to select resistant mutants. Although transfer of resistance genes may not be relevant to quinolone resistance, transfer of resistant strains might be, as establishment of global, highly resistant clones would compromise the therapeutic value of the quinolones.

Sulfonamides/Trimethoprim

The sulfonamides and trimethoprim are synthetic antimicrobials that target dihydropteroate synthase and dihydrofolate reductase, respectively. (42, 120). Trimethoprim is extensively used for the treatment of urinary tract infections due to its broad activity and inexpensive nature. Trimethoprim is usually prescribed in combination with sulfamethoxazole (co-trimoxazole) as the two work synergistically when in combination (42). The numbers of isolates resistant to trimethoprim and sulfonamides has increased in the last 10 years (19). Resistance is mediated by acquisition of novel variants of the target enzymes; two sulfonamide resistance genes (*sulI* and *sulII*) and over 15 trimethoprim resistance genes (dihydrofolate reductase—*dhfr*); the most common of these in *E. coli* are *dhfrI* and *dhfrII* (42). Overexpression of chromosomal enzymes can also lead to reduced susceptibility to trimethoprim. Transfer of *dhfr* genes via plasmids, transposons and integrons has lead to increased numbers of *E. coli* resistant to trimethoprim and has mirrored use of trimethoprim and trimethoprim-sulfamethoxazole (42).

Typically between 10 and 25% of *E. coli* recovered from UTIs in Europe, the U.S., and the Asia Pacific region are resistant to trimethoprim or trimethoprim-sulfamethoxazole and up to 45% of strains from Latin America (35).

Phenicols

Chloramphenicol is a broad spectrum antibiotic that acts as an inhibitor of protein synthesis. Resistance to chloramphenicol can be mediated by several different enzymes that act in different ways. The most common class of enzymes that inactivate chloramphenicol is the acetyltransferases, which have become widespread among bacteria (85). Due to toxicity concerns, chloramphenicol is rarely used in the developed world, and so resistance in *E. coli* is rarely monitored. In the developing world chloramphenicol remains in routine use and resistance rates are typically between 20 and 30% (2, 68, 129). Among Asian isolates resistance to chloramphenicol is often associated with the carriage of integrons (21).

Nitrofurantoin

Nitrofurans (nitrofurantoin and nitrofurazone) are used primarily in the treatment of urinary tract infections (UTIs). Worldwide resistance to nitrofurans has remained rare with few cases reported (35, 48, 124). It has been postulated that this is a result of the multiple mechanisms of action exerted by the reactive intermediates derived from nitrofurans after attack of the compounds by nitroreductases (37, 76). Increasing numbers of resistant strains to alternative treatments for UTIs, including trimethoprim, may lead to increased use of nitrofurantoin for treatment of these infections, while resistance remains low.

Biocide Resistance

Recently concerns about the relationship between biocide use and the emergence of antibiotic resistance have been expressed (30, 33, 34, 60). It has been postulated that certain biocides may share targets with antibiotics and select mutants cross resistant to these antibiotics, may select for efflux mediated multiple antibiotic resistance or may provide selective pressure for integrons or plasmids carrying biocide resistance determinants as well as antibiotic resistance genes. Triclosan, a widely used biocide, has been shown to select for triclosan-resistant *E. coli* readily in the laboratory by mutation of the target site (FabI) or by overexpression of *marA, soxS,* or *acrB* (60, 75). Mutants of *E. coli* O157 selected for resistance to benzylakonium chloride or triclosan have also been found to exhibit decreased susceptibility to a range of antibiotics (16). Currently, it is not clear whether biocide use is a significant factor in the selection of antibiotic resistance and consequently a threat to human health (33, 34, 72). However, continued investigation into the potential dangers of inappropriate biocide use is important in order to formulate informed policies for the use of biocides in clinical settings.

Multiple Antibiotic Resistance in *E. coli*

Increasing numbers of strains are being isolated that are resistant to more than one class of antibiotics; however, the mechanism of this multiple resistance is often unclear. This phenotype has been associated with accumulation of multiple resistance determinants and increased expression of efflux pumps (such as the *acrAB-tolC* system). Efflux pumps can in turn be controlled by global regulators such as *marA* and *soxS* and so are able to mediate reduced susceptibility to a broad range of unrelated compounds (61). Multiple antibiotic resistance conferred by the *mar/sox/acr* loci may lead to a survival advantage to these strains which may then accumulate further resistance determinants. Coresistance has been observed among *E. coli* isolated from urinary tract infection, respiratory tract infection and blood (10, 54, 111, 131, 132). The emergence of isolates with coincident resistance to cephalosporins and quinolones in Asia (10) is a major concern as wide scale spread of these strains would compromise some of the most effective therapeutic options for treatment of life threatening *E. coli* infections.

ANIMALS AND *E. COLI:* SOURCE OF RESISTANCE?

Antimicrobials are used both therapeutically and as growth promoters in food producing animals (although this practice has become controversial and has been discontinued in certain parts of the world, notably within the European union [123]). Antimicrobial drugs are often used in veterinary medicine that belong to the same class as agents used to treat human disease. As mechanisms of resistance to antibiotics typically confer resistance to all or many members of a class of antibiotic, concern has been expressed about the potential for antibiotics used in veterinary medicine to select bacteria resistant to related drugs licensed for use in humans. Many bacteria pathogenic to humans are zoonotic in nature, and contamination of various meats with pathogenic bacteria, including *E. coli*, is common (141). Thus selection of resistant *E. coli* in food animals can possibly lead to transmission to man (126, 136). While it is generally accepted that use of antimicrobials in animals can lead to the selection of resistant *E. coli*, there is still debate as to whether these resistant strains are transferred through the food chain and cause subsequent infections in humans (44, 97, 98, 123). Significant resistance to tetracycline, sulfamethoxazole, cephalothin, and ampicillin has been described among *E. coli* O157 (116). An increasing proportion of resistant strains among clinical *E. coli* O157 isolates suggests transfer via the food chain as antibiotics are rarely used to treat these infections in humans but are used in cattle, the principal reservoir and presumably the transmission vehicle for human infection (134, 135). Resistance has been detected among other clinically significant serotypes of *E. coli* from various animals (11, 117, 133). Cephalosporin resistance has also been reported from *E. coli* isolated from cattle and poultry, which is significant as these compounds are important in human medicine (15, 142).

Even if the transfer of resistance from animals to man is infrequent, new resistance phenotypes and genotypes may appear in animals preferentially to humans as the reservoir of carrier animals is much greater than the human population. Amplification of this resistance may subsequently occur in humans after transfer. Concerns relating to the use and abuse of antimicrobials in humans can be applied equally to veterinary use—selection of resistance strains is a potential risk to human health even if these strains are infrequently transferred to humans. Therefore, even if the strains themselves are not transfered, their resistance determinants certainly could move between animal pathogenic and commensal organisms to human bacterial pathogens.

CONCLUDING REMARKS

Effective antibiotic treatment of *E. coli* infections is vital to prevent mortality and reduce morbidity associated with these infections. Currently, there are a number of antibiotics available that display excellent activity against *E. coli* for a physician to choose from. However, resistance to many antibiotics has been observed among *E. coli* isolates and resistance has emerged and spread as new antimicrobials have become available. As a result the use of certain antibiotics (e.g., ampicillin, tetracyclines) against *E. coli* has been compromised due to the likelihood of resistance. Resistance has emerged less frequently to other agents which are commonly used in the treatment of *E. coli* infections, including quinolones and cephalosporins. This is a concern, as loss of the utility of these compounds would compromise the arsenal of available compounds The emergence of such strains in China and other parts of Asia needs to be monitored closely as importation of highly resistant isolates to other parts of the world is probable.

More information is required to assess the role (if any) of biocides and veterinary antibiotic use as contributors to transfer of resistant strains of *E. coli* to humans. While there are potential risks from biocide use and veterinary antibiotic use, little evidence as to the likelihood of strains being selected and transferred has been presented. However, there is a clear need for further work in this area in order to inform policies regulating use of biocides and antimicrobials in animals.

We can learn much from *E. coli* about the mechanisms of antibiotic resistance and the spread of resistance genes which can be applied to other bacterial species; all known mechanisms of resistance are found in *E. coli*. Research into the evolutionary selection of antibiotic resistance genes and their stability within a population will again provide important information regarding the appropriate use of antibiotics, *E. coli* is an ideal candidate for such research as we know more about this organism than any other.

REFERENCES

1. **Aarestrup, F. M., F. Bager, N. E. Jensen, M. Madsen, A. Meyling, and H. C. Wegener.** 1998. Resistance to antimicrobial agents used for animal therapy in pathogenic-, zoonotic- and indicator bacteria isolated from different food animals in Denmark: a baseline study for the Danish Integrated Antimicrobial Resistance Monitoring Programme (DANMAP). *APMIS* **106:**745–770.
2. **Ahmed, A. A., H. Osman, A. M. Mansour, H. A. Musa, A. B. Ahmed, Z. Karrar, and H. S. Hassan.** 2000. Antimicrobial agent resistance in bacterial isolates from patients with diarrhea and urinary tract infection in the Sudan. *Am. J. Trop. Med. Hyg.* **63:**259–263.
3. **Alekshun, M. N., and S. B. Levy.** 1999. The mar regulon: multiple resistance to antibiotics and other toxic chemicals. *Trends Microbiol.* **7:**410–413.
4. **Alobwede, I., F. H. M'Zali, D. M. Livermore, J. Heritage, N. Todd, and P. M. Hawkey.** 2003. CTX-M extended-spectrum beta-lactamase arrives in the UK. *J. Antimicrob. Chemother.* **51:**470–471.
5. **Arpin, C., V. Dubois, L. Coulange, C. Andre, I. Fischer, P. Noury, F. Grobost, J. P. Brochet, J. Jullin, B. Dutilh, G. Larribet, I. Lagrange, and C. Quentin.** 2003. Extended-spectrum beta-lactamase-producing Enterobacteriaceae in community and private health care centers. *Antimicrob. Agents Chemother.* **47:**3506–3514.
6. **Baraniak, A., J. Fiett, W. Hryniewicz, P. Nordmann, and M. Gniadkowski.** 2002. Ceftazidime-hydrolysing CTX-M-15 extended-spectrum beta-lactamase (ESBL) in Poland. *J. Antimicrob. Chemother.* **50:**393–396.
7. **Barbosa, T. M., and S. B. Levy.** 2000. Differential expression of over 60 chromosomal genes in *Escherichia coli* by constitutive expression of MarA. *J. Bacteriol.* **182:**3467–3474.

8. **Bauernfeind, A., I. Stemplinger, R. Jungwirth, S. Ernst, and J. M. Casellas.** 1996. Sequences of beta-lactamase genes encoding CTX-M-1 (MEN-1) and CTX-M-2 and relationship of their amino acid sequences with those of other beta-lactamases. *Antimicrob. Agents Chemother.* **40:**509–513.
9. **Bax, R., R. Bywater, G. Cornaglia, H. Goossens, P. Hunter, V. Isham, V. Jarlier, R. Jones, I. Phillips, D. Sahm, S. Senn, M. Struelens, D. Taylor, and A. White.** 2001. Surveillance of antimicrobial resistance--what, how and whither? *Clin. Microbiol. Infect.* **7:**316–325.
10. **Bell, J. M., J. D. Turnidge, A. C. Gales, M. A. Pfaller, and R. N. Jones.** 2002. Prevalence of extended spectrum beta-lactamase (ESBL)-producing clinical isolates in the Asia-Pacific region and South Africa: regional results from SENTRY Antimicrobial Surveillance Program (1998-99). *Diagn. Microbiol. Infect. Dis.* **42:**193–198.
11. **Bischoff, K. M., D. G. White, P. F. McDermott, S. Zhao, S. Gaines, J. J. Maurer, and D. J. Nisbet.** 2002. Characterization of chloramphenicol resistance in beta-hemolytic *Escherichia coli* associated with diarrhea in neonatal swine. *J. Clin. Microbiol.* **40:**389–394.
12. **Bonnet, R.** 2004. Growing group of extended-spectrum beta-lactamases: the CTX-M enzymes. *Antimicrob. Agents Chemother.* **48:**1–14.
13. **Bonnet, R., C. Dutour, J. L. Sampaio, C. Chanal, D. Sirot, R. Labia, C. De Champs, and J. Sirot.** 2001. Novel cefotaximase (CTX-M-16) with increased catalytic efficiency due to substitution Asp-240→Gly. *Antimicrob. Agents Chemother.* **45:**2269–2275.
14. **Bonnet, R., C. Recule, R. Baraduc, C. Chanal, D. Sirot, C. De Champs, and J. Sirot.** 2003. Effect of D240G substitution in a novel ESBL CTX-M-27. *J. Antimicrob. Chemother.* **52:**29–35.
15. **Bradford, P. A., P. J. Petersen, I. M. Fingerman, and D. G. White.** 1999. Characterization of expanded-spectrum cephalosporin resistance in *E. coli* isolates associated with bovine calf diarrhoeal disease. *J. Antimicrob. Chemother.* **44:**607–610.
16. **Braoudaki, M., and A. C. Hilton.** 2004. Adaptive resistance to biocides in Salmonella enterica and *Escherichia coli* O157 and cross-resistance to antimicrobial agents. *J. Clin. Microbiol.* **42:**73–78.
17. **Brooks M. B., P. S. Morley, D. A. Dargatz, D. R. Hyatt, M. D. Salman, and B. L. Akey.** 2003. Survey of antimicrobial susceptibility testing practices of veterinary diagnostic laboratories in the United States. *J. Am. Vet. Med. Assoc.* **15:222** (2):168–73.
18. **Bryan A., N. Shapir, and M. J. Sadowsky.** 2004. Frequency and distribution of tetracycline resistance genes in genetically diverse, nonselected, and nonclinical *Escherichia coli* strains isolated from diverse human and animal sources. *App. Env. Micro.* **70:**2503–2507.
19. **Burman, W. J., P. E. Breese, B. E. Murray, K. V. Singh, H. A. Batal, T. D. MacKenzie, J. W. Ogle, M. L. Wilson, R. R. Reves, and P. S. Mehler.** 2003. Conventional and molecular epidemiology of trimethoprim-sulfamethoxazole resistance among urinary *Escherichia coli* isolates. *Am. J. Med.* **115:**358–364.
20. **Chanawong, A., F. H. M'Zali, J. Heritage, J. H. Xiong, and P. M. Hawkey.** 2002. Three cefotaximases, CTX-M-9, CTX-M-13, and CTX-M-14, among Enterobacteriaceae in the People's Republic of China. *Antimicrob. Agents Chemother.* **46:**630–637.
21. **Chang, C. Y., L. L. Chang, Y. H. Chang, T. M. Lee, and S. F. Chang.** 2000. Characterisation of drug resistance gene cassettes associated with class 1 integrons in clinical isolates of *Escherichia coli* from Taiwan, ROC. *J. Med. Microbiol.* **49:**1097–1102.
22. **Chopra, I., and M. Roberts.** 2001. Tetracycline antibiotics: mode of action, applications, molecular biology, and epidemiology of bacterial resistance. *Microbiol. Mol. Biol. Rev.* **65:**232–260.
23. **Davies, J., and G. D. Wright.** 1997. Bacterial resistance to aminoglycoside antibiotics. *Trends Microbiol.* **5:**234–240.
24. **Diekema, D. J., M. A. Pfaller, and R. N. Jones.** 2002. Age-related trends in pathogen frequency and antimicrobial susceptibility of bloodstream isolates in North America: SENTRY Antimicrobial Surveillance Program, 1997-2000. *Int. J. Antimicrob. Agents* **20:**412–418.
25. **Doern, G. V., R. N. Jones, M. A. Pfaller, K. C. Kugler, and M. L. Beach.** 1999. Bacterial pathogens isolated from patients with skin and soft tissue infections: frequency of occurrence and antimicrobial susceptibility patterns from the SENTRY Antimicrobial Surveillance Program (United States and Canada, 1997). SENTRY Study Group (North America). *Diagn. Microbiol. Infect. Dis.* **34:**65–72.
26. **Esaki H., A. Morioka, K. Ishihara, A. Kojima, S. Shiroki, Y. Tamura, and T. Takahashi.** 2004. Antimicrobial susceptibility of Salmonella isolated from cattle, swine and poultry (2001-2002): report from the Japanese Veterinary Antimicrobial Resistance Monitoring Program. *J. Antimicrob. Chemother.* **53:**266–270.
27. **Everett, M. J., Y. F. Jin, V. Ricci, and L. J. Piddock.** 1996. Contributions of individual mechanisms to fluoroquinolone resistance in 36 *Escherichia coli* strains isolated from humans and animals. *Antimicrob. Agents Chemother.* **40:**2380–2386.
28. **Fluit, A. C., F. J. Schmitz, and J. Verhoef.** 2001. Frequency of isolation of pathogens from bloodstream, nosocomial pneumonia, skin and soft tissue, and urinary tract infections occurring in European patients. *Eur. J. Clin. Microbiol. Infect. Dis.* **20:**188–191.
29. **Fluit, A. C., F. J. Schmitz, and J. Verhoef.** 2001. Multiresistance to antimicrobial agents for the ten most frequently isolated bacterial pathogens. *Int. J. Antimicrob. Agents* **18:**147–160.
30. **Fraise, A. P.** 2002. Biocide abuse and antimicrobial resistance—a cause for concern? *J. Antimicrob. Chemother.* **49:**11–12.
31. **Gellert, M., K. Mizuuchi, M. H. O'Dea, and H. A. Nash.** 1976. DNA gyrase: an enzyme that introduces superhelical turns into DNA. *Proc. Natl. Acad. Sci. USA* **73:**3872–3876.
32. **George, A. M., and S. B. Levy.** 1983. Amplifiable resistance to tetracycline, chloramphenicol, and other antibiotics in *Escherichia coli:* involvement of a non-plasmid-determined efflux of tetracycline. *J. Bacteriol.* **155:**531–540.
33. **Gilbert, P., and A. J. McBain.** 2003. Potential impact of increased use of biocides in consumer products on prevalence of antibiotic resistance. *Clin. Microbiol. Rev.* **16:**189–208.
34. **Gilbert, P., A. J. McBain, and S. F. Bloomfield.** 2002. Biocide abuse and antimicrobial resistance: being clear about the issues. *J. Antimicrob. Chemother.* **50:**137–139.
35. **Gordon, K. A., and R. N. Jones.** 2003. Susceptibility patterns of orally administered antimicrobials among urinary tract infection pathogens from hospitalized patients in North America: comparison report to Europe and Latin America. Results from the SENTRY Antimicrobial Surveillance Program (2000). *Diagn. Microbiol. Infect. Dis.* **45:**295–301.
36. **Gould, I. M.** 2002. Antibiotic policies and control of resistance. *Curr. Opin. Infect. Dis.* **15:**395–400.
37. **Guay, D. R.** 2001. An update on the role of nitrofurans in the management of urinary tract infections. *Drugs* **61:**353–364.
38. **Hadley, C.** 2004. Overcoming resistance. *EMBO J.* **5:**550–552.

39. **Heritage, J., F. H. M'Zali, D. Gascoyne-Binzi, and P. M. Hawkey.** 1999. Evolution and spread of SHV extended-spectrum beta-lactamases in gram-negative bacteria. *J. Antimicrob. Chemother.* **44:**309–318.
40. **Hooper, D. C.** 1999. Mechanisms of fluoroquinolone resistance. *Drug Resist. Updates* **2:**38–55.
41. **Hooton, T. M.** 2003. The current management strategies for community-acquired urinary tract infection. *Infect. Dis. Clin. North Am.* **17:**303–332.
42. **Huovinen, P., L. Sundstrom, G. Swedberg, and O. Skold.** 1995. Trimethoprim and sulfonamide resistance. *Antimicrob. Agents Chemother.* **39:**279–289.
43. **Jensen, V. F., J. Neimann, A. M. Hammerum, K. Molbak, and H. C. Wegener.** 2004. Does the use of antibiotics in food animals pose a risk to human health? An unbiased review? *J. Antimicrob. Chemother.* May 12 [Epub ahead of print].
44. **Johnson, J. R., A. Gajewski, A. J. Lesse, and T. A. Russo.** 2003. Extraintestinal pathogenic *Escherichia coli* as a cause of invasive nonurinary infections. *J. Clin. Microbiol.* **41:**5798–5802.
45. **Jones, M. E., D. C. Draghi, J. A. Karlowsky, D. F. Sahm, and J. S. Bradley.** 2003. Prevalence of antimicrobial resistance in bacteria isolated from central nervous system specimens as reported by U.S. hospital laboratories from 2000 to 2002. *Ann. Clin. Microbiol. Antimicrob.* **25:**3.
46. **Jones, R. N., K. C. Kugler, M. A. Pfaller, and P. L. Winokur.** 1999. Characteristics of pathogens causing urinary tract infections in hospitals in North America: results from the SENTRY Antimicrobial Surveillance Program, 1997. *Diagn. Microbiol. Infect. Dis.* **35:**55–63.
47. **Kahlmeter, G.** 2000. The ECO.SENS Project: a prospective, multinational, multicentre epidemiological survey of the prevalence and antimicrobial susceptibility of urinary tract pathogens—interim report. *J. Antimicrob. Chemother.* **46** (Suppl. 1)**:**15–22.
48. **Kahlmeter, G.** 2003. An international survey of the antimicrobial susceptibility of pathogens from uncomplicated urinary tract infections: the ECO.SENS Project. *J. Antimicrob. Chemother.* **51:**69–76.
49. **Kahlmeter, G.** 2003. Prevalence and antimicrobial susceptibility of pathogens in uncomplicated cystitis in Europe. The ECO.SENS study. *Int. J. Antimicrob. Agents* **22** (Suppl 2)**:** 49–52.
50. **Kahlmeter, G., and P. Menday.** 2003. Cross-resistance and associated resistance in 2478 *Escherichia coli* isolates from the Pan-European ECO.SENS Project surveying the antimicrobial susceptibility of pathogens from uncomplicated urinary tract infections. *J. Antimicrob. Chemother.* **52:**128–131.
51. **Kaper, J. B., J. P. Nataro, and H. L. Mobley.** 2004. Pathogenic *Escherichia coli. Nat. Rev. Microbiol.* **2:**123–140.
52. **Karlowsky, J. A., C. Thornsberry, M. E. Jones, and D. F. Sahm.** 2003. Susceptibility of antimicrobial-resistant urinary *Escherichia coli* isolates to fluoroquinolones and nitrofurantoin. *Clin. Infect. Dis.* **36:**183–187.
53. **Karlowsky, J. A., L. J. Kelly, C. Thornsberry, M. E. Jones, A. T. Evangelista, I. A. Critchley, and D. F. Sahm.** 2002. Susceptibility to fluoroquinolones among commonly isolated Gram-negative bacilli in 2000: TRUST and TSN data for the United States. Tracking Resistance in the United States Today. The Surveillance Network. *Int. J. Antimicrob. Agents* **19:** 21–31.
54. **Karlowsky, J. A., L. J. Kelly, C. Thornsberry, M. E. Jones, and D. F. Sahm.** 2002. Trends in antimicrobial resistance among urinary tract infection isolates of *Escherichia coli* from female outpatients in the United States. *Antimicrob. Agents Chemother.* **46:**2540–2545.
55. **Karlowsky, J. A., M. E. Jones, C. Thornsberry, I. Critchley, L. J. Kelly, and D. F. Sahm.** 2001. Prevalence of antimicrobial resistance among urinary tract pathogens isolated from female outpatients across the US in 1999. *Int. J. Antimicrob. Agents* **18:**121–127.
56. **Karlowsky, J. A., M. E. Jones, C. Thornsberry, I. R. Friedland, and D. F. Sahm.** 2003. Trends in antimicrobial susceptibilities among Enterobacteriaceae isolated from hospitalized patients in the United States from 1998 to 2001. *Antimicrob. Agents Chemother.* **47:**1672–1680.
57. **Karlowsky, J. A., M. E. Jones, D. C. Draghi, C. Thornsberry, D. F. Sahm, and G. A. Volturo.** 2004. Prevalence and antimicrobial susceptibilities of bacteria isolated from blood cultures of hospitalized patients in the United States in 2002. *Ann. Clin. Microbiol. Antimicrob.* **3:**7.
58. **Lacey, R. W.** 1984. Antibiotic resistance in *Staphylococcus aureus* and streptococci. *Br. Med. Bull.* **40:**77–83.
59. **Lecointre, G., L. Rachdi, P. Darlu, and E. Denamur.** 1998. *Escherichia coli* molecular phylogeny using the incongruence length difference test. *Mol. Biol. Evol.* **15:**1685–1695.
60. **Levy, S. B.** 2001. Antibacterial household products: cause for concern. *Emerg. Infect. Dis.* **7:**512–515.
61. **Levy, S. B.** 2002. Active efflux, a common mechanism for biocide and antibiotic resistance. *Symp. Ser. Soc. Appl. Microbiol.* 65S–71S.
62. **Levy, S. B., L. M. McMurry, V. Burdett, P. Courvalin, W. Hillen, M. C. Roberts, and D. E. Taylor.** 1989. Nomenclature for tetracycline resistance determinants. *Antimicrob. Agents Chemother.* **33:**1373–1374.
63. **Lewis, M. T., A. C. Gales, H. S. Sader, M. A. Pfaller, and R. N. Jones.** 2000. Frequency of occurrence and antimicrobial susceptibility patterns for pathogens isolated from latin american patients with a diagnosis of pneumonia: results from the SENTRY antimicrobial surveillance program (1998). *Diagn. Microbiol. Infect. Dis.* **37:**63–74.
64. **Li, X. Z., and H. Nikaido.** 2004. Efflux-mediated drug resistance in bacteria. *Drugs* **64:**159–204.
65. **Livermore, D.M., and N. Woodford.** 2000. Carbapenemases: a problem in waiting? *Curr. Opin. Microbiol.* **3:**489–95.
66. **Livermore, D. M.** 1998. Beta-lactamase-mediated resistance and opportunities for its control. *J. Antimicrob. Chemother.* **41** (Suppl D)**:**25–41.
67. **Llano-Sotelo, B., E. F. Azucena, Jr., L. P. Kotra, S. Mobashery, and C. S. Chow.** 2002. Aminoglycosides modified by resistance enzymes display diminished binding to the bacterial ribosomal aminoacyl-tRNA site. *Chem. Biol.* **9:**455–463.
68. **Mansouri, S., and S. Shareifi.** 2002. Antimicrobial resistance pattern of *Escherichia coli* causing urinary tract infections, and that of human fecal flora, in the southeast of Iran. *Microb. Drug Resist.* **8:**123–128.
69. **Martinez-Martinez, L., A. Pascual, and G. A. Jacoby.** 1998. Quinolone resistance from a transferable plasmid. *Lancet* **351:**797–799.
70. **Mathai, D., R. N. Jones, and M. A. Pfaller.** 2001. Epidemiology and frequency of resistance among pathogens causing urinary tract infections in 1,510 hospitalized patients: a report from the SENTRY Antimicrobial Surveillance Program (North America). *Diagn. Microbiol. Infect. Dis.* **40:**129–136.
71. **Matthew, M., R. W. Hedges, and J. T. Smith.** 1979. Types of beta-lactamase determined by plasmids in gram-negative bacteria. *J. Bacteriol.* **138:**657–662.
72. **McBain, A. J., R. G. Ledder, P. Sreenivasan, and P. Gilbert.** 2004. Selection for high-level resistance by chronic triclosan exposure is not universal. *J. Antimicrob. Chemother.* **53:**772–777.

73. **McMurry, L. M., J. C. Cullinane, and S. B. Levy.** 1982. Transport of the lipophilic analog minocycline differs from that of tetracycline in susceptible and resistant *Escherichia coli* strains. *Antimicrob. Agents Chemother.* **22:**791–9.
74. **McMurry, L. M., M. L. Aldema-Ramos, and S. B. Levy.** 2002. $Fe(2^{+})^{-}$ tetracycline-mediated cleavage of the Tn10 tetracycline efflux protein TetA reveals a substrate binding site near glutamine 225 in transmembrane helix 7. *J. Bacteriol.* **184:**5113–5120.
75. **McMurry, L. M., M. Oethinger, and S. B. Levy.** 1998. Overexpression of *marA*, *soxS*, or *acrAB* produces resistance to triclosan in laboratory and clinical strains of *Escherichia coli*. *FEMS Microbiol. Lett.* **166:**305–309.
76. **McOsker, C. C., and P. M. Fitzpatrick.** 1994. Nitrofurantoin: mechanism of action and implications for resistance development in common uropathogens. *J. Antimicrob. Chemother.* **33** (Suppl A):23–30.
77. **Medeiros, A. A.** 1997. Evolution and dissemination of beta-lactamases accelerated by generations of beta-lactam antibiotics. *Clin. Infect. Dis.* **24** (Suppl 1):S19–S45.
78. **Mingeot-Leclercq, M. P., Y. Glupczynski, and P. M. Tulkens.** 1999. Aminoglycosides: activity and resistance. *Antimicrob. Agents Chemother.* **43:**727–737.
79. **Miriagou, V., E. Tzelepi, D. Gianneli, and L. S. Tzouvelekis.** 2003. *Escherichia coli* with a self-transferable, multiresistant plasmid coding for metallo-beta-lactamase VIM-1. *Antimicrob. Agents Chemother.* **47:**395–397.
80. **Miro, E., F. Navarro, B. Mirelis, M. Sabate, A. Rivera, P. Coll, and G. Prats.** 2002. Prevalence of clinical isolates of *Escherichia coli* producing inhibitor-resistant beta-lactamases at a University Hospital in Barcelona, Spain, over a 3-year period. *Antimicrob. Agents Chemother.* **46:**3991–3994.
81. **Moland, E. S., J. A. Black, A. Hossain, N. D. Hanson, K. S. Thomson, and S. Pottumarthy.** 2003. Discovery of CTX-M-like extended-spectrum beta-lactamases in *Escherichia coli* isolates from five US States. *Antimicrob. Agents Chemother.* **47:**2382–2383.
82. **Monnet, D. L.** 2000. Toward multinational antimicrobial resistance surveillance systems in Europe. *Int. J. Antimicrob. Agents* **15:**91–101.
83. **Mori, H.** 2004. From the sequence to cell modelling: comprehensive functional genomics in *Escherichia coli*. *J. Biochem. Mol. Biol.* **37:**83–92.
84. **Mulvey, M. R., E. Bryce, D. Boyd, M. Ofner-Agostini, S. Christianson, A. E. Simor, and S. Paton.** 2004. Ambler class A extended-spectrum beta-lactamase-producing *Escherichia coli* and Klebsiella spp. in Canadian hospitals. *Antimicrob. Agents Chemother.* **48:**1204–1214.
85. **Murray, I. A., and W. V. Shaw.** 1997. O-Acetyltransferases for chloramphenicol and other natural products. *Antimicrob. Agents Chemother.* **41:**1–6.
86. **Nataro, J. P., and J. B. Kaper.** 1998. Diarrheagenic *Escherichia coli*. *Clin. Microbiol. Rev.* **11:**142–201.
87. **Nelson, M. L., and S. B. Levy.** 1999. Reversal of tetracycline resistance mediated by different bacterial tetracycline resistance determinants by an inhibitor of the Tet(B) antiport protein. *Antimicrob. Agents Chemother.* **43:**1719–1724.
88. **Nikaido, H.** 1998. Multiple antibiotic resistance and efflux. *Curr. Opin. Microbiol.* **1:**516–523.
89. **Odeh, R., S. Kelkar, A. M. Hujer, R. A. Bonomo, P. C. Schreckenberger, and J. P. Quinn.** 2002. Broad resistance due to plasmid-mediated AmpC beta-lactamases in clinical isolates of *Escherichia coli*. *Clin. Infect. Dis.* **35:**140–145.
90. **Oethinger, M., I. Podglajen, W. V. Kern, and S. B. Levy.** 1998. Overexpression of the marA or soxS regulatory gene in clinical topoisomerase mutants of *Escherichia coli*. *Antimicrob. Agents Chemother.* **42:**2089–2094.
91. **Oliver, A., J. C. Perez-Diaz, T. M. Coque, F. Baquero, and R. Canton.** 2001. Nucleotide sequence and characterization of a novel cefotaxime-hydrolyzing beta-lactamase (CTX-M-10) isolated in Spain. *Antimicrob. Agents Chemother.* **45:**616–620.
92. **Pai, H., H. J. Lee, E. H. Choi, J. Kim, and G. A. Jacoby.** 2001. Evolution of TEM-related extended-spectrum beta-lactamases in Korea. *Antimicrob. Agents Chemother.* **45:**3651–3653.
93. **Palucha, A., B. Mikiewicz, W. Hryniewicz, and M. Gniadkowski.** 1999. Concurrent outbreaks of extended-spectrum beta-lactamase-producing organisms of the family Enterobacteriaceae in a Warsaw hospital. *J. Antimicrob. Chemother.* **44:**489–499.
94. **Patton, J. P., D. B. Nash, and E. Abrutyn.** 1991. Urinary tract infection: economic considerations. *Med. Clin. North Am.* **75:**495–513.
95. **Peng, H., and K. J. Marians.** 1993. *Escherichia coli* topoisomerase IV. Purification, characterization, subunit structure, and subunit interactions. *J. Biol. Chem.* **268:**24481–24490.
96. **Pfaller, M. A., J. Acar, R. N. Jones, J. Verhoef, J. Turnidge, and H. S. Sader.** 2001. Integration of molecular characterization of microorganisms in a global antimicrobial resistance surveillance program. *Clin. Infect. Dis.* **32**(Suppl 2):S156–S167.
97. **Phillips, I., M. Casewell, T. Cox, B. De Groot, C. Friis, R. Jones, C. Nightingale, R. Preston, and J. Waddell.** 2004. Antibiotic use in animals. *J. Antimicrob. Chemother.* **53:**885.
98. **Phillips, I., M. Casewell, T. Cox, B. De Groot, C. Friis, R. Jones, C. Nightingale, R. Preston, and J. Waddell.** 2004. Does the use of antibiotics in food animals pose a risk to human health? A critical review of published data. *J. Antimicrob. Chemother.* **53:**28–52.
99. **Picard, B., J. S. Garcia, S. Gouriou, P. Duriez, N. Brahimi, E. Bingen, J. Elion, and E. Denamur.** 1999. The link between phylogeny and virulence in *Escherichia coli* extraintestinal infection. *Infect. Immun.* **67:**546–553.
100. **Piddock, L. J.** 1995. Mechanisms of resistance to fluoroquinolones: state-of-the-art 1992-1994. *Drugs* **49**(Suppl 2): 29–35.
101. **Piddock, L. J.** 1999. Mechanisms of fluoroquinolone resistance: an update 1994-1998. *Drugs* **58**(Suppl 2):11–18.
102. **Pitout, J. D., K. S. Thomson, N. D. Hanson, A. F. Ehrhardt, E. S. Moland, and C. C. Sanders.** 1998. beta-Lactamases responsible for resistance to expanded-spectrum cephalosporins in Klebsiella pneumoniae, *Escherichia coli*, and *Proteus mirabilis* isolates recovered in South Africa. *Antimicrob. Agents Chemother.* **42:**1350–1354.
103. **Quinteros, M., M. Radice, N. Gardella, M. M. Rodriguez, N. Costa, D. Korbenfeld, E. Couto, and G. Gutkind.** 2003. Extended-spectrum beta-lactamases in enterobacteriaceae in Buenos Aires, Argentina, public hospitals. *Antimicrob. Agents Chemother.* **47:**2864–2867.
104. **Rennie, R. P., R. N. Jones, and A. H. Mutnick.** 2003. Occurrence and antimicrobial susceptibility patterns of pathogens isolated from skin and soft tissue infections: report from the SENTRY Antimicrobial Surveillance Program (United States and Canada, 2000). *Diagn. Microbiol. Infect. Dis.* **45:**287–293.
105. **Reynolds, R., N. Potz, M. Colman, A. Williams, D. Livermore, and A. MacGowan.** 2004. Antimicrobial susceptibility of the pathogens of bacteraemia in the UK and Ireland 2001-2002: the BSAC Bacteraemia Resistance Surveillance Programme. *J. Antimicrob. Chemother.* **53:**1018–1032.
106. **Rodriguez-Martinez JM, Pascual A, Garcia I, Martinez-Martinez L.** 2003. Detection of the plasmid-mediated quino-

lone resistance determinant *qnr* among clinical isolates of Klebsiella pneumoniae producing AmpC-type beta-lactamase. *J. Antimicrob. Chemother.* **52**:703–706.

107. **Russo, T. A., and J. R. Johnson.** 2003. Medical and economic impact of extraintestinal infections due to *Escherichia coli:* focus on an increasingly important endemic problem. *Microbes. Infect.* **5**:449–456.

108. **Sabate, M., E. Miro, F. Navarro, C. Verges, R. Aliaga, B. Mirelis, and G. Prats.** 2002. Beta-lactamases involved in resistance to broad-spectrum cephalosporins in *Escherichia coli* and Klebsiella spp. clinical isolates collected between 1994 and 1996, in Barcelona (Spain). *J. Antimicrob. Chemother.* **49**: 989–997.

109. **Sader, H. S., D. J. Biedenbach, and R. N. Jones.** 2003. Global patterns of susceptibility for 21 commonly utilized antimicrobial agents tested against 48,440 Enterobacteriaceae in the SENTRY Antimicrobial Surveillance Program (1997-2001). *Diagn. Microbiol. Infect. Dis.* **47**:361–364.

110. **Sader, H. S., M. A. Pfaller, R. N. Jones, G. V. Doern, A. C. Gales, P. L. Winokur, and K. C. Kugler.** 1999. Bacterial Pathogens Isolated from Patients with Bloodstream Infections in Latin America, 1997: Frequency of Occurrence and Antimicrobial Susceptibility Patterns from the SENTRY Antimicrobial Surveillance Program. *Braz. J. Infect. Dis.* **3**:97–110.

111. **Sader, H. S., R. N. Jones, A. C. Gales, P. Winokur, K. C. Kugler, M. A. Pfaller, and G. V. Doern.** 1998. Antimicrobial susceptibility patterns for pathogens isolated from patients in Latin American medical centers with a diagnosis of pneumonia: analysis of results from the SENTRY Antimicrobial Surveillance Program (1997). SENTRY Latin America Study Group. *Diagn. Microbiol. Infect. Dis.* **32**:289–301.

112. **Sader, H. S., R. N. Jones, and J. B. Silva.** 2002. Skin and soft tissue infections in Latin American medical centers: four-year assessment of the pathogen frequency and antimicrobial susceptibility patterns. *Diagn. Microbiol. Infect. Dis.* **44**:281–288.

113. **Sader, H. S., R. N. Jones, S. Andrade-Baiocchi, and D. J. Biedenbach.** 2002. Four-year evaluation of frequency of occurrence and antimicrobial susceptibility patterns of bacteria from bloodstream infections in Latin American medical centers. *Diagn. Microbiol. Infect. Dis.* **44**:273–280.

114. **Sahm, D. F., C. Thornsberry, D. C. Mayfield, M. E. Jones, and J. A. Karlowsky.** 2001. Multidrug-resistant urinary tract isolates of *Escherichia coli:* prevalence and patient demographics in the United States in 2000. *Antimicrob. Agents Chemother.* **45**:1402–1406.

115. **Saladin, M., V. T. Cao, T. Lambert, J. L. Donay, J. L. Herrmann, Z. Ould-Hocine, C. Verdet, F. Delisle, A. Philippon, and G. Arlet.** 2002. Diversity of CTX-M beta-lactamases and their promoter regions from Enterobacteriaceae isolated in three Parisian hospitals. *FEMS Microbiol. Lett.* **209**:161–168.

116. **Schroeder, C. M., C. Zhao, C. DebRoy, J. Torcolini, S. Zhao, D. G. White, D. D. Wagner, P. F. McDermott, R. D. Walker, and J. Meng.** 2002. Antimicrobial resistance of *Escherichia coli* O157 isolated from humans, cattle, swine, and food. *Appl. Environ. Microbiol.* **68**:576–581.

117. **Schroeder, C. M., J. Meng, S. Zhao, C. DebRoy, J. Torcolini, C. Zhao, P. F. McDermott, D. D. Wagner, R. D. Walker, and D. G. White.** 2002. Antimicrobial resistance of *Escherichia coli* O26, O103, O111, O128, and O145 from animals and humans. *Emerg. Infect. Dis.* **8**:1409–1414.

118. **Scoulica, E.V., I. K. Neonakis, A. I. Gikas, and Y. J. Tselentis.** 2004. Spread of bla(VIM-1)-producing *E. coli* in a university hospital in Greece. Genetic analysis of the integron carrying the bla(VIM-1) metallo-beta-lactamase gene. *Diagn. Microbiol. Infect. Dis.* **48**:167–172.

119. **Siu, L. K., P. L. Lu, J. Y. Chen, F. M. Lin, and S. C. Chang.** 2003. High-level expression of *ampC* beta-lactamase due to insertion of nucleotides between -10 and -35 promoter sequences in *Escherichia coli* clinical isolates: cases not responsive to extended-spectrum-cephalosporin treatment. *Antimicrob. Agents Chemother.* **47**:2138–2144.

120. **Skold, O.** 2001. Resistance to trimethoprim and sulfonamides. *Vet. Res.* **32**:261–273.

121. **Sotto, A., C. M. De Boever, P. Fabbro-Peray, A. Gouby, D. Sirot, and J. Jourdan.** 2001. Risk factors for antibiotic-resistant *Escherichia coli* isolated from hospitalized patients with urinary tract infections: a prospective study. *J. Clin. Microbiol.* **39**:438–444.

122. **Teuber, M.** 1999. Spread of antibiotic resistance with foodborne pathogens. *Cell Mol. Life Sci.* **56**:755–763.

123. **Turnidge J.** 2004. Antibiotic use in animals—prejudices, perceptions and realities. *J. Antimicrob. Chemother.* **53**:26–27.

124. **Turnidge, J., J. Bell, D. J. Biedenbach, and R. N. Jones.** 2002. Pathogen occurrence and antimicrobial resistance trends among urinary tract infection isolates in the Asia-Western Pacific Region: report from the SENTRY Antimicrobial Surveillance Program, 1998-1999. *Int. J. Antimicrob. Agents* **20**:10–17.

125. **Unal, S., R. Masterton, and H. Goossens.** 2004. Bacteraemia in Europe—antimicrobial susceptibility data from the MYSTIC surveillance programme. *Int. J. Antimicrob. Agents* **23**: 155–163.

126. **Van den Bogaard, A. E., and E. E. Stobberingh.** 1999. Antibiotic usage in animals: impact on bacterial resistance and public health. *Drugs* **58**:589–607.

127. **Vergis, E. N., C. Brennen, M. Wagener, and R. R. Muder.** 2001. Pneumonia in long-term care: a prospective case-control study of risk factors and impact on survival. *Arch. Intern. Med.* **161**:2378–2381.

128. **Vetting, M., S. L. Roderick, S. Hegde, S. Magnet, and J. S. Blanchard.** 2003. What can structure tell us about in vivo function? The case of aminoglycoside-resistance genes. *Biochem. Soc. Trans.* **31**:520–522.

129. **Vila, J., M. Vargas, J. Ruiz, M. Corachan, M. T. Jimenez De Anta, and J. Gascon.** 2000. Quinolone resistance in enterotoxigenic *Escherichia coli* causing diarrhea in travelers to India in comparison with other geographical areas. *Antimicrob. Agents Chemother.* **44**:1731–1733.

130. **Wang, M., J. H. Tran, G. A. Jacoby, Y. Zhang, F. Wang, and D. C. Hooper.** 2003. Plasmid-mediated quinolone resistance in clinical isolates of *Escherichia coli* from Shanghai, China. *Antimicrob. Agents Chemother.* **47**:2242–2248.

131. **Wang, H., J. L. Dzink-Fox, M. Chen, and S. B. Levy.** 2001. Genetic characterization of highly fluoroquinolone-resistant clinical *Escherichia coli* strains from China: role of *acrR* mutations. *Antimicrob. Agents Chemother.* **45**:1515–1521.

132. **Webber, M. A., and L. J. Piddock.** 2001. Absence of mutations in *marRAB* or *soxRS* in *acrB*-overexpressing fluoroquinolone-resistant clinical and veterinary isolates of *Escherichia coli*. *Antimicrob. Agents Chemother.* **45**:1550–1552.

133. **White, D. G., L. J. Piddock, J. J. Maurer, S. Zhao, V. Ricci, and S. G. Thayer.** 2000. Characterization of fluoroquinolone resistance among veterinary isolates of avian *Escherichia coli*. *Antimicrob. Agents Chemother.* **44**:2897–2899.

134. **White, D. G., S. Zhao, P. F. McDermott, S. Ayers, S. Gaines, S. Friedman, D. D. Wagner, J. Meng, D. Needle, M. Davis, and C. DebRoy.** 2002. Characterization of antimicrobial resistance among *Escherichia coli* O111 isolates of animal and human origin. *Microb. Drug Resist.* **8**:139–146.

135. **White, D. G., S. Zhao, S. Simjee, D. D. Wagner, and P. F. McDermott.** 2002. Antimicrobial resistance of foodborne pathogens. *Microbes. Infect.* **4:**405–412.
136. **Witte, W.** 1997. Impact of antibiotic use in animal feeding on resistance of bacterial pathogens in humans. *Ciba Found. Symp.* **207:**61–71.
137. **Yagi, T., H. Kurokawa, K. Senda, S. Ichiyama, H. Ito, S. Ohsuka, K. Shibayama, K. Shimokata, N. Kato, M. Ohta, and Y. Arakawa.** 1997. Nosocomial spread of cephem-resistant *Escherichia coli* strains carrying multiple Toho-1-like beta-lactamase genes. *Antimicrob. Agents Chemother.* **41:**2606–2611.
138. **Yagi, T., H. Kurokawa, N. Shibata, K. Shibayama, and Y. Arakawa.** 2000. A preliminary survey of extended-spectrum beta-lactamases (ESBLs) in clinical isolates of Klebsiella pneumoniae and *Escherichia coli* in Japan. *FEMS Microbiol. Lett.* **184:**53–56.
139. **Yamasaki, K., M. Komatsu, T. Yamashita, K. Shimakawa, T. Ura, H. Nishio, K. Satoh, R. Washidu, S. Kinoshita, and M. Aihara.** 2003. Production of CTX-M-3 extended-spectrum beta-lactamase and IMP-1 metallo beta-lactamase by five Gram-negative bacilli: survey of clinical isolates from seven laboratories collected in 1998 and 2000, in the Kinki region of Japan. *J. Antimicrob. Chemother.* **51:**631–638.
140. **Zervos, M. J., E. Hershberger, D. P. Nicolau, D. J. Ritchie, L. K. Blackner, E. A. Coyle, A. J. Donnelly, S. F. Eckel, R. H. Eng, A. Hiltz, A. G. Kuyumjian, W. Krebs, A. McDaniel, P. Hogan, and T. J. Lubowski.** 2003. Relationship between fluoroquinolone use and changes in susceptibility to fluoroquinolones of selected pathogens in 10 United States teaching hospitals, 1991-2000. *Clin. Infect. Dis.* **37:**1643–1648.
141. **Zhao, C., B. Ge, J. De Villena, R. Sudler, E. Yeh, S. Zhao, D. G. White, D. Wagner, and J. Meng.** 2001. Prevalence of Campylobacter spp., *Escherichia coli,* and Salmonella serovars in retail chicken, turkey, pork, and beef from the Greater Washington, D.C. area. *Appl. Environ. Microbiol.* **67:**5431–5436.
142. **Zhao, S., D. G. White, P. F. McDermott, S. Friedman, L. English, S. Ayers, J. Meng, J. J. Maurer, R. Holland, and R. D. Walker.** 2001. Identification and expression of cephamycinase bla(CMY) genes in *Escherichia coli* and Salmonella isolates from food animals and ground meat. *Antimicrob. Agents Chemother.* **45:**3647–3650.

Frontiers in Antimicrobial Resistance: a Tribute to Stuart B. Levy
Edited by D. G. White, M. N. Alekshun, and P. F. McDermott

Chapter 29

Epidemiology and Treatment Options for Select Community-Acquired and Nosocomial Antibiotic-Resistant Pathogens

JOHN E. GUSTAFSON AND JOHN D. GOLDMAN

Resistance is a nameless cloud that looms over otherwise controllable infections, but lacks the powerful status of a readily identifiable disease state to spur large scale efforts of control.

—*Levy and Marshall* (74)

Up to 2,000,000 people in the United States (U.S.) suffer from nosocomial infections, which result in 90,000 mortalities (21). How many of these mortalities can be attributed to antimicrobial resistance? Considering that >70% of the bacterial pathogens in U.S. hospitals are resistant to one or more of the antimicrobials commonly used for treatment, it is probable that all antimicrobial resistance genes contribute to overall mortality (23). We believe that Stuart B. Levy deserves the utmost respect for educating both the general public and the medical community on the importance of the judicious use of antimicrobials. In this chapter, we intend to outline the impact of antimicrobial resistance by describing the epidemiology of select antimicrobial-resistant pathogens and the difficulties associated with treatment of infections caused by these organisms. To this end, we will discuss antimicrobial treatment options for infections caused by resistant enterococci, *Staphylococcus aureus, Streptococcus pneumoniae,* and *Mycobacterium tuberculosis.* We choose these pathogens since the former two are the scourge of hospitals around the world, while the latter three pathogens are problems within the community. Furthermore, infections caused by each of the antimicrobial-resistant pathogens that we have selected present their own unique treatment and prevention problems.

STAPHYLOCOCCUS AUREUS

Staphylococcus aureus is the leading cause of nosocomial infections, bacteremias, and surgical wound infections and the second most common agent isolated from prosthetic valve endocarditis cases (56, 78). In the preantimicrobial era, bacteremia caused by this organism reached fatality rates of 80% (123). *S. aureus* endocarditis is 100% fatal in the absence of antimicrobial therapy. Even with effective antimicrobial therapies, mortality due to *S. aureus* bacteremia can still be as high as 32% (92). In 1995 the New York metropolitan area reported 13,550 *S. aureus* infections that led to 1,400 mortalities and cost $435.5 million dollars (112).

Infections with methicillin-resistant *S. aureus* (MRSA) can be associated with higher mortality rates than disease caused by methicillin-susceptible *S. aureus* (112). Nosocomial MRSA infections are associated with prolonged antimicrobial exposure, dialysis, and admission to intensive care units. Approximately 50% of MRSA identified in U.S. hospitals are susceptible only to the glycopeptide antibiotic vancomycin (78).

Methicillin resistance in *S. aureus* is mediated by *mecA*. This gene encodes penicillin-binding protein 2a, which demonstrates reduced affinity for β-lactams and related antimicrobials. The clonal spread of MRSA within hospitals and across international borders is well documented. It appears that a few epidemic MRSA clones are responsible for a large percentage of MRSA disease worldwide (40, 97). Multiple antimicrobial-resistant MRSA (resistant to ≥3 antibiotic-

John E. Gustafson • Department of Biology, New Mexico State University, PO Box 30001, Dept. 3AF, Las Cruces, NM 88003-8001.
John D. Goldman • Department of Internal Medicine, Pinnacle Health Hospitals, 205 S. Front Street, Harrisburg, PA 17104.

classes) are commonly isolated from patients, and vancomycin-intermediate and -resistant *S. aureus* (VISA and VRSA, respectively) have been isolated. The mechanisms leading to intermediate vancomycin resistance in *S. aureus* are complex and possibly involve multiple mutations, while *vanA*-mediated high-level vancomycin resistance in *S. aureus* was acquired from vancomycin-resistant enterococci (for reviews, see references 99 and 113).

Community-Onset Methicillin-Resistant *Staphylococcus aureus*

Recent cases of community-onset MRSA (CO-MRSA) infections have unveiled MRSA strains with unique evolutionary trajectories. Overall, CO-MRSA are not multiply antimicrobial-resistant and are characterized by a unique methicillin resistance determinant (type IV SCC*mec*) and a Panton-Valentine leukocidin gene (PVL), although strains demonstrating exceptions to these rules do occur (96). Strains of CO-MRSA harboring PVL have been reported to cause more severe disease, when compared to infection caused by CO-MRSA without this gene (47, 82). In addition, recent evidence has demonstrated that the presence of *S. aureus* enterotoxin genes (B and C) and an unknown superantigen, may also play an important role in the pathology associated with CO-MRSA infections (114). True CO-MRSA infections, by definition, do not involve patient contact with healthcare workers or facilities (96), and factors associated with CO-MRSA infections include antimicrobial use as well as certain activities such as enjoying group saunas (6).

Treatment of MRSA

Treatment of *S. aureus* infections not only involves antimicrobials, but may also require various surgical debridement procedures, and removal of prostheses and indwelling devices if these items are acting as nuclei for the formation of biofilms and/or infection. In general, MRSA isolated in hospitals should be considered multiply antimicrobial-resistant. Vancomycin is still the antimicrobial of choice to treat serious MRSA infections such as line infections and endocarditis (78). This preference is based on long-term clinical experience, which in turn probably contributes to the lack of clinical data comparing vancomycin to other potential therapeutic agents. For instance, in one study comparing vancomycin to trimethoprim/sulfamethoxazole (TMP-SMX) treatment of *S. aureus* bacteremia, the cure rate for patients treated with TMP-SMX (86%) was slightly less than those treated with vancomycin (98%) (81). TMP-SMX therefore is a cost effective alternative for patients demonstrating vancomycin intolerance. Another trial compared vancomycin and aztreonam to linezolid and aztreonam in the treatment of MRSA pneumonia. The linezolid combination demonstrated clinical cure and survival rates of 59 and 85%, versus 35 and 67% for the vancomycin combination (139). The results of this trial has led some experts to advocate the use of linezolid in MRSA ventilator-associated pneumonia (57). However, because of the increased cost (121) and potential to select for linezolid-resistant MRSA (129), linezolid should probably not be adopted as a first line therapy for MRSA pneumonia. The use of oral linezolid avoids the need for long-term indwelling catheters (89), decreases length of hospital stay, and lowers total medical costs (85). Recent evidence has demonstrated that daptomycin can be used as a secondary agent in the treatment of methicillin-susceptible *S. aureus* infections, as well as MRSA bacteremia/endocarditis and osteomyelitis and joint infections (42, 117). During this small trial, daptomycin was well tolerated even for long treatment durations (up to 42 days). However, daptomycin should not be used as a first line therapy to treat cases of MRSA pneumonia, since this antimicrobial binds to lung surfactants which thereby leads to a reduction in drug availability.

In contrast to health care-associated MRSA, CO-MRSA are often resistant to macrolides and susceptible to the fluoroquinolones, and the inexpensive oral agents clindamycin, doxycycline, minocycline and TMP-SMX, although susceptibility to these agents may vary by geographical area (93). Because of this epidemiology, clindamycin, TMP-SMX or a combination of TMP-SMX and rifampin have been advocated for treatment of CO-MRSA infections. Since inducible clindamycin resistance cannot be detected via standard susceptibility testing protocols, a "D-zone" test should always be performed if clindamycin is considered as a therapeutic option for CO-MRSA infections (75). Vancomycin, which requires IV access, should be reserved for more serious infections (47). Treatment of CO-MRSA infections with oral linezolid might also be considered, but is associated with substantial cost (~$150 per day). When compared to nosocomial infections, the treatment of CO-MRSA infections may be more costly (112).

Therapeutic Options for Infections Caused by Vancomycin-Intermediate and Vancomycin-Resistant *Staphylococcus aureus*

Vancomycin-intermediate *S. aureus* (VISA) are defined as *S. aureus* strains demonstrating a vancomycin MIC of 4 to 16 μg/ml. Approximately, 30 cases of VISA infections have been reported. VISA strains often evolve during extended vancomycin therapy, and patients receiving such treatment should be carefully monitored for VISA emergence. Vancomycin-resistant *S. aureus* (VRSA) are defined as a *S. aureus* strains demonstrating a vancomycin MIC of >32 μg/ml, and to date only three VRSA infections have been reported (for review see references 99 and 113). In addition, clinicians must be wary of treatment failures caused by *S. aureus* strains expressing vancomycin tolerance. Vancomycin-tolerant organisms express vancomycin MICs which make them appear susceptible but express an MBC ≥32 times higher than the MIC (84).

To date VISA infections have been treated with numerous antimicrobials and antimicrobial combinations: ampicillin-sulbactam and arbekacin (18); fusidic acid and rifampin; linezolid; linezolid, chloramphenicol and fusidic acid; linezolid, fusidic acid and rifampin; vancomycin (54); rifampin and TMP-SMX; vancomycin, gentamicin and rifampin (124); and linezolid, doxycycline, and TMP-SMX (50). A VRSA infection was treated successfully with TMP-SMX (22).

Treatment recommendations for VISA and VRSA are difficult to establish, since only limited clinical experience is available. Even though some VISA express low-level vancomycin resistance, vancomycin should not be used for primary treatment, since treatment failures can occur. In vitro susceptibility testing indicates that VISA and VRSA isolates are susceptible to the newer antistaphylococcal antimicrobials (e.g., daptomycin, linezolid and quinupristin-dalfopristin) and well as older antimicrobials. Therapy of VISA and VRSA infections should probably include a newer antistaphylococcal agent in combination with an antimicrobial to which the isolate is also susceptible. If susceptibility data are not readily available, it seems reasonable to empirically treat VISA and VRSA infections with linezolid in combination with an older agent to which local MRSA are usually susceptible. If a clinician chooses this therapeutic option, he or she must also realize that while infrequent today, *S. aureus* strains resistant to both linezolid and quinupristin-dalfopristin have been reported (37, 129).

It has been estimated that 20 to 50% of the population carry *S. aureus* in their nares, and clearance of nasal carriage significantly reduces the risk of developing infection (for a review, see reference 66). Therefore, we advocate active *S. aureus* carriage surveillance and eradication among hospital staff and patients whenever possible. One method to clear carriage is the application of a mupirocin cream to the nasal passages; however, mupirocin-resistant strains of *S aureus* have emerged (28). Lysostaphin is an enzyme that degrades the *S. aureus* cell wall, leading to cell death. A recombinant lysostaphin cream formulated for eradication of *S. aureus* nasal colonization that demonstrated efficacy in an animal model (67), is currently in clinical trials by Biosynexus Inc. (Gaithersburg, Md.).

VANCOMYCIN-RESISTANT ENTEROCOCCI

Enterococci are considered part of the normal flora of the bowel, genital tract and anterior urethra of humans. In general, enterococci cause infections in debilitated patients, then spread clonally within the hospital via direct or indirect contact (83, 133). *Enterococcus faecalis* and *Enterococcus faecium* represent >95% of all clinical enterococcal isolates (107). Enterococci are the third most common cause of infections in intensive care units (ICU), causing 9 and 12% of ICU-associated bacteremias and urinary tract infections, respectively (19).

Enterococci are intrinsically resistant to many antimicrobials, including cephalosporins, penicillins, co-trimoxazole, and clindamycin. The greatest threats to patients are vancomycin-resistant enterococci (VRE), which were first detected in Europe in 1986 (131) ~30 years after the clinical introduction of vancomycin. By 1999, one quarter of all nosocomial enterococcal infections in U.S. hospitals were due to VRE (20). The arrival of VRE is often attributed to the overuse of vancomycin, which has increased about 100-fold in the last 20 years (64), predominantly to treat MRSA, enterococcal, and *Clostridium difficile* infections. The selection and spread of VRE in Europe is also attributed to the extensive use of the glycopeptide avoparcin in animal husbandry. The U.S. is recognized for reporting the greatest percentage of VRE infections in the surveyed world (60). VRE possess acquired glycopeptide resistance determinants and can express high-level resistance to aminoglycosides. In addition, approximately 95% of U.S. VRE isolates are *E. faecium,* almost all of which also express high-level ampicillin resistance (107). Numerous vancomycin resistance determinants have been described in the enterococci (e.g., *vanA, vanB, vanC, vanD, vanE* and *vanG*), and each determinant conveys unique glycopeptide resistance properties. For instance, VRE carrying *vanB* and *vanC* determinants are deemed susceptible to the glycopeptide antimicrobial teicoplanin, while infections caused by VRE expressing *vanA*-mediated vancomycin resistance are refractory to teicoplanin therapy. However, treatment with teicoplanin can select for teicoplanin resistance in VRE harboring the *vanB* determinant by selecting mutants which constitutively express *vanB* (for review see reference 27).

Following VRE gastrointestinal colonization, patients are at increased risk of acquiring VRE infections of surgical wounds, the urinary tract or the intra-abdominal/pelvic region. Enterococci are also known to cause eye or central nervous system shunt infections, and although enterococci can colonize the respiratory tract, they are not considered respiratory pathogens (33). The most important risk factors for

VRE colonization and infection include severe underlying disease, extended hospital stay, and previous antimicrobial exposure. The most consistently recognized antimicrobials that contribute to VRE colonization and infection are vancomycin, cephalosporins, fluoroquinolones and antianaerobic agents. The total antimicrobial volume and treatment or prophylaxis duration are also important VRE acquisition risk factors. Evidence suggests that institutions can decrease VRE infection rates by limiting the use of certain antimicrobials, e.g., cephalosporins (43). Patients with VRE bacteremia generally experience longer hospital stays (~2 weeks) and demonstrate a ~30% mortality rate (for review see 33 and 52). It has been argued that outbreak vancomycin-resistant *E. faecium* (VREF) strains are carried longer than nonepidemic strains; however, this finding may be due to increased antimicrobial pressure in the patients carrying the epidemic strains (83).

Evidence suggests that the gene encoding the enterococcal surface protein (*esp*) carried on a pathogenicity island imparts infectivity on the enterococci, and VREF which carry *esp* are capable of nosocomial dissemination. Interestingly, strains of *E. faecium* resistant to vancomycin and ampicillin carry the virulence genes *esp* and *hyl* with greater frequency than ampicillin-susceptible VREF (83, 133).

Therapeutic Options for Infections Caused by VRE

After identifying VRE infections, the first step in all treatment regimens should be to drain abscesses, debride infected wounds, and remove all foreign items that may serve as a nucleus for infection. In some cases, removal of the offending item can by itself lead to resolution of the infection. While some VRE infections can resolve without active antimicrobial therapy (62), infections in specific patient populations (e.g., liver transplant, hemotologic malignancies) can lead to severe and often fatal disease.

There is no definitive treatment for VRE infections. In many cases, various strategies must be attempted, utilizing both old and new antimicrobials based on in vitro susceptibility data. If the VRE isolate is susceptible to ampicillin, ampicillin is the drug of choice. The use of high dose-ampicillin plus streptomycin, or continuous-ampicillin/sulbactam plus gentamicin infusion, have worked well for infections caused by enterococci expressing relatively low-level ampicillin resistance (34, 88). Chloramphenicol has also been used to treat VRE infections. Due to the interference of the patients underlying illness, however, it is unclear if treatment with this drug reduces patient mortality (106). Following the increased use of chloramphenicol in the treatment of VRE infections, invariably the isolation of chloramphenicol-resistant isolates multiplies. At one institution, chloramphenicol resistance rates in VRE increased from 0 to 11% over a 10-year period (70). Tetracycline has also been used to treat VRE infections (55); however, no major clinical trials evaluating the efficacy of tetracycline for treating VRE infections has been reported. Nitrofurantoin is usually active against VRE, yet its utility is limited to urinary tract infections or prostatitis (62, 128).

Quinupristin-dalfopristin demonstrates activity against vancomycin-resistant *E. faecium* but not *E. faecalis*. In two large studies of quinupristin-dalfopristin therapy, clinical and microbiological response rates of 70.5 to 83%, and 70.5 to 74% were reported (77, 137), while another study demonstrated a 13% decrease in mortality in patients undergoing quinupristin-dalfopristin therapy (76). To complicate matters, superinfection with *E. faecalis*, which is considered intrinsically resistant, does occur during quinupristin-dalfopristin treatment of VREF infections. Linezolid is usually active against both *E. faecalis* and *E. faecium* infections (62). Treatment of VRE infections in solid organ transplant recipients with linezolid resulted in a survival rate of 62.4% (compared to 17% survival in historical controls) (39), while other studies have reported good clinical cure rates (62). Nonetheless, infections caused by linezolid-resistant enterococci have been reported, and resistance to this drug may emerge during therapy (10). A comparative study of linezolid and quinupristin-dalfopristin treatment of VRE-infected cancer patients demonstrated similar cure rates for both drugs (58% for linezolid and 43% for quinupristin-dalfopristin). Thrombocytopenia occurred more often in the linezolid-treated group, and myalgias and athralgias occurred more often in the quinupristin-dalfopristin-treated group (103). Studies have demonstrated that the vast majority of VRE isolates are susceptible to daptomycin (62). In a recent study, treatment of VRE bacteremia and/or endocarditis with daptomycin alone resulted in 1 clinical success out of 5, while overall clinical success, including patients who had received prior vancomycin or linezolid therapy, was apparent in 5 of 11 patients treated with daptomycin (117).

In the absence of literature supporting the use of one agent versus another, linezolid should be used as first line therapy for VRE infections, because it can cover both *E. faecalis* and *E. faecium*. Quinupristin-dalfopristin may be needed for patients who are thrombocytopenic. Furthermore, many clinicians would add a second agent such as doxycycline to decrease the potential for evolution of resistance.

Decolonization is also an important aspect of treating disease caused by VRE, as well as impeding their ability to spread within the hospital environment. While the decolonization potential of numerous antimicrobials (e.g., bacitracin, gentamicin, tetracycline, doxycycline, novobiocin, rifampin, and ramoplanin) have been investigated, agents which are non-absorbable may be preferred for this procedure (for review see reference 62). The non-absorbable lipoglycodepsipeptide antimicrobial ramoplanin, which reaches high fecal concentrations, shows promise as a VRE decolonization agent. Following ramoplanin treatment, however, recolonization with VRE can still occur and decolonization is more likely to succeed in patients who did not receive antianaerobic therapies (90). The problem of VRE colonization is complicated by long-term VRE carriage within individuals where carriage cannot be detected (30), and the inherent difficulties in successfully eradicating any commensal constituent of the intestinal microflora.

STREPTOCOCCUS PNEUMONIAE

S. pneumoniae is a major cause of community-acquired pneumonia (CAP), bacteremia, and meningitis and is responsible for approximately 40% of all otitis media cases (36). Mortality for pneumococcal bacteremia today is ~20% for older adults (136), but before antimicrobial therapy was available, the mortality rate was ~80% in hospitalized patients. In the U.S., ~40,000 deaths per year are due to pneumococcal disease (17). Worldwide, pneumococcal infections cause ~1.2 million deaths per year, and nearly 40% of all pneumonia deaths occur in children <5 years of age (91). There are 2 to 3 million cases of CAP each year (8), requiring 500,000 hospitalizations, at a cost of $8.4-9.7 billion, the majority of which ($7.5 billion) is directed towards in-patient costs (71, 94). *S. pneumoniae* is the most commonly isolated etiologic CAP agent, representing 20-75% of all cases and ~ 66 % of all bacteremic pneumonia cases (7, 53, 79). There is an association between early effective antimicrobial delivery (i.e., 8 h following admission) and decreased mortality for hospitalized pneumonia patients.

Invasive pneumococcal disease (IPD) is defined as the isolation of pneumococcus from a normally sterile site (e.g., blood, cerebrospinal fluid, pleural fluid). Before the introduction of the conjugate vaccine, the incidence of IPD was reported as 23 per 100,000 (110). Overall, ~10% of IPD patients die, and the case fatality rate goes up to as high as 31% among residents of long term care facilities (69). The pneumococci are responsible for approximately one-half of the community-acquired meningitis cases in the U.S. (116), and approximately 50% of patients who survive pneumococcal meningitis suffer neurological sequellae (49).

One problem associated with treatment of pneumococcal disease is the spread of a few successful penicillin-resistant clones that are able to switch their capsular serotypes (65). *S. pneumoniae* are classified as penicillin-susceptible if isolates display a penicillin MIC of <0.1 μg/ml, penicillin-intermediate if an MIC of 0.12 to 1.0 μg/ml is displayed (PISP), and penicillin-resistant if the MIC is ≥2.0 μg/ml (PRSP). While there are 90 serogroups of *S. pneumoniae,* isolates with high-level β-lactam resistance belong to only 5 of these groups (6B, 9V, 14, 19F or 23F), and penicillin resistance is strongly correlated with resistance to other antimicrobial classes (136). Multiple antimicrobial-resistant *S. pneumoniae* emerged in South Africa in 1978 (59), and now 26 multiple antimicrobial- resistant clones have been identified (86). In a 1998 multistate population survey, 24% of invasive pneumococcal isolates were resistant to penicillin and 14% were multiple antimicrobial-resistant isolates (136). In the U.S. it has been estimated that 70% of multiple antimicrobial-resistant *S. pneumoniae* isolates belong to 1 of 9 clonal strains (35).

Treatment of Penicillin-Resistant *Streptococcus pneumoniae*

Treatment recommendations for PRSP and PISP infections are somewhat controversial and depend upon the site of infection. Since serum penicillin levels of 40 to 50 μg/ml are easily achievable, penicillin remains a viable treatment option for cases of pneumonia caused by PISP and low-level PRSP isolates (53). Several studies have reported that infections caused by low-level PRSP have no effect on the mortality rate of patients being treated for bacteremic PRSP pneumonia with penicillin (41, 45). Another report found no effect of streptococcal penicillin resistance on mortality, time to defervescence, or the incidence of suppurative complications (140). However, most of these studies have included a large number of patients with PISP infections. In another study that excluded deaths in the first four days of hospitalization, a higher rate of mortality during infection with PRSP expressing high-level penicillin resistance (MIC >4 μg/ml) was reported (41). This finding is consistent with pharmacodynamic studies that determined treating infections caused by PRSP expressing a penicillin MIC of >4 μg/ml with penicillin would lead to treatment failure (15, 53). It seems that the current NCCLS breakpoints for *S. pneumoniae* β-lactam susceptibility are not predictive of clinical outcome in pneumococcal pneumonia. The isolation of very high-level penicillin-resistant *S. pneumoniae* (MIC ≥8 μg/ml) have also been reported. Factors associated with infections by these multiple antimicrobial-resistant strains include residence in Tennessee and an age of <5 or ≥65 (115).

β-lactams are the drugs of choice for treating PISP and low-level PRSP infections. If a patient is infected with high-level PRSP, treatment should be guided by susceptibility testing. If susceptibility data are unavailable, it is reasonable to empirically treat a patient with CAP using ceftriaxone or cefotaxime until laboratory data are available. The Infectious Diseases Society of America recommends using ceftriaxone, cefotaxime, a broad spectrum fluoroquinolone (gatifloxacin, gemifloxacin, levofloxacin or moxifloxacin) or another agent indicated by the antibiogram (80). The American Thoracic Society emphasizes the use of fluoroquinolones (95). CDC guidelines have recommended reserving fluoroquinolones for situations in which treatment failure with a cephalosporin is evident or when the infection is known to be caused by a high-level PRSP.

Increasing isolation of fluoroquinolone-resistant *S. pneumoniae* has been reported. Recent evidence demonstrates that invasive pneumococcal isolates resistant to levofloxacin demonstrate multiple antimicrobial resistance and clonal spread (101). It is likely that isolation of fluoroquinolone-resistant pneumococci will become more frequent as the use of fluoroquinolones to combat CAP increases. Thus, fluoroquinolones should be reserved for treatment of IPD and CAP caused by high-level PRSP or after treatment failure with β-lactams, in order to preserve the activity of this broad-spectrum antimicrobial class. Anecdotal reports of breakthrough bacteremia due to macrolide-resistant pneumococci among patients receiving macrolide monotherapy illustrates a need for caution on the use of this drug class for treating CAP (111).

Treatment of meningitis with standard doses of penicillin may not produce concentrations in the cerebral spinal fluid (CSF) that exceed the MIC of some PRSP strains. Treatment failures due to infections caused by PRSP have been reported when either penicillin or cephalosporins are used (44, 72, 122, 134). Consequently, a combination of vancomycin and ceftriaxone has been recommended for initial treatment of PRSP meningitis until antibiograms are determined (3, 61). The use of steroids for bacterial meningitis therapy is controversial, yet a recent report suggests that adjunctive steroid therapy is beneficial in adult cases of meningitis (32). Since steroids decrease vancomycin CSF penetration, some experts have recommended the addition of rifampin to ceftriaxone and vancomycin, when steroid therapy is indicated. In the future, fluoroquinolones (68) and daptomycin (29) may also play a role in the treatment of meningitis caused by PRSP.

Treatment of otitis media is complicated by low-level PRSP. A standard dose of amoxicillin (40 to 45 mg/kg of body weight/day) is expected to produce peak levels in the ear that might fail to eradicate PRSP. Consequently, high-dose amoxicillin (80 to 90 mg/kg/day) has been recommended for therapy, when low dose treatment fails (4). Otitis media may also be caused by infections with β-lactamase producing strains of *Haemophilus influenzae* or *Moraxella catarrhalis*. Consequently, an agent used to treat otitis media after amoxicillin treatment failure should cover the possibility of PRSP and β-lactamase production. Recommended agents include amoxicillin-clavulanate, cefuroxime-axetil, or ceftriaxone. TMP-SMX or macrolides, which are traditional second-line agents, are not likely to be effective (34, 36).

MYCOBACTERIUM TUBERCULOSIS

> The poor will remain the breeding ground for the 'white plague' until we realize that it is not only the microbe *Mycobacterium tuberculosis* that is causing the disease, but also the abominable conditions under which hundreds of millions people on this planet are forced to live.
>
> —*Van Helden (132)*

No population or geographical region is immune or isolated from the risk of infection with the primary cause of tuberculosis (TB) *M. tuberculosis*. The pandemic caused by this bacterium represents a global emergency. Every day 5,000 to 8,000 human beings die of TB, leading to ~2 to 3 million deaths each year. It has been estimated that one-third of the world's population is infected with *M. tuberculosis*. Even though tuberculosis may kill more humans than any other infectious disease, on par with malaria and HIV, the U.S. only contributes $8 for TB control compared with $137 for malaria and more than $900 for human immunodeficiency virus (HIV)/AIDS (105, 132). Only about 1 in 10 infected individuals progresses to active TB over their lifetime in the absence of immunosupression. Since malnutrition is the leading cause of immunosuppression worldwide, countries with unfortunate economies and associated circumstances bear the overwhelming brunt of the present TB pandemic. A person that is 10% underweight has a threefold increased risk for developing TB following infection with *M. tuberculosis*. *M. tuberculosis* spreads primarily through respiratory droplets eventually leading to latent infection or pulmonary disease, which then provides a mechanism for the organism to spread and cause disease in any body organ. Another TB agent spread via respiratory droplets, *Mycobacterium bo-*

vis (14), was spread via contaminated milk before the advent of pasteurization.

Due primarily to reactivation of latent TB infections, HIV-positive individuals are at increased risk of developing TB (16). The risk of TB doubles within the first year of HIV infection (125) and likely increases as CD4 cell counts decrease further. TB occurs relatively early in the HIV-related disease spectrum and often before other AIDS-defining conditions (109). Consequently, active TB is considered to be a "herald illness" of AIDS and is the leading cause of death among people with HIV. Today, approximately 14 million individuals live with this coinfection (127). Treatment of rheumatoid arthritis with tumor necrosis factor blocking agents can also lead to active TB in patients harboring latent *M. tuberculosis* infections (25). This latter finding reemphasizes the importance of a functional immune response for keeping this virulent pathogen in check.

In the U.S., a disproportionate number of TB patients are non-Caucasian males living in areas with low income and have publicly funded insurance or no insurance at all. The total costs of TB hospitalizations in the U.S. in 2000 exceeded $385 million (51). During 2003, 14,871 TB cases were reported in the U.S., representing a 1.9% decline in the rate from 2002. Foreign-born persons represented 53.3% of the national case total. Of the U.S.-born population, the largest number of TB cases was reported among the non-Hispanic Black population (3,041 cases) (26).

Population-based genotyping has demonstrated that a small number of *M. tuberculosis* strains seem to cause the majority of TB cases, and it has been suggested that these strains spread more efficiently than others. Compared to other strains, the widely disseminated Beijing strains are thought to be more readily aerosolized, and establish infection with greater speed or progress with greater frequency from infection to disease. The W-Beijing family of strains have been responsible for large outbreaks of MDR-TB cases in the U.S. (9, 12). Simultaneous infections with two or more genetically distinct *M. tuberculosis* strains do occur, even in HIV-negative patients (118). Furthermore, minor alterations in the genetic fingerprints of multiple isolates from one host demonstrate unique subpopulations of a single clone evolving within a single host. The MICs for certain subpopulations can be higher than the cutoff for resistance to anti-TB drugs, and these unique subpopulations may not be identified by routine drug susceptibility testing (118).

The primary goals of anti-TB regimens are to kill the organism rapidly, prevent the evolution of drug resistance, and eliminate persistent bacilli to prevent relapse of disease. With regard to metabolic activity, it has been theorized that there are unique subpopulations of *M. tuberculosis* residing within the TB patient: a large population of rapidly growing extracellular bacilli residing in cavities, small populations of slow growing bacilli within acidic caseous foci, and a dormant subpopulation living within macrophages that demonstrates intermittent spurts of growth (135). The latter two subpopulations are thought to be responsible for TB treatment failures and relapses. The former subpopulation, because of its large size, is thought to give rise to drug- or multidrug-resistant mutants following selection for mutations which resist the activities of the drug, or drugs, being used to treat the patient. Since mutations in the *M. tuberculosis* chromosome leading to single drug resistance occur at mutation frequencies of 10^{-5} to 10^{-8}, anti-TB regimens employ multiple anti-TB drugs. Early on in anti-TB treatment, it was noted that treatment with para-aminosalicylic acid and streptomycin lessened the likelihood of treatment failure and acquisition of drug resistance (87). Theoretically, the reason multiple drugs are used is because mutations leading to resistance to two anti-TB drugs (mutation frequency of 10^{-14}) is unlikely to occur concurrently within *M. tuberculosis* populations (31). Resistance mechanisms to all main anti-TB drugs have been described. Drug resistance in *M. tuberculosis* is primarily due to the accumulation of mutations in drug target genes such as *rpoB*, which encodes the β-subunit of RNA polymerase, the target of rifampin. In addition, mobile genetic elements have been implicated in resistance; IS*6110*-insertional inactivation of the *pncA* gene encoding pyrazinamidase leads to pyrazinamide-resistance. While the molecular mechanisms leading to resistance to many of the anti-TB drugs have been described, the genetic basis for some anti-TB drugs has not been fully elucidated (for review see reference 46).

Treatment of Infections Caused by Drug-Resistant *Mycobacterium tuberculosis*

Today it is common to encounter infections caused by *M. tuberculosis* expressing resistance to both of the first-line anti-TB drugs isoniazid and rifampin, or multidrug-resistant TB (MDR-TB). Patients presenting with drug-resistant TB have either not been treated properly, were non-compliant with regard to therapy, or were infected by an improperly treated or non-compliant patient. It has been estimated that in 1995, 50 million people were infected with drug-resistant strains of *M. tuberculosis* and 273,000 new cases of MDR-TB occurred worldwide in 2000 (38, 138). The advent of drug-resistant *M. tuberculosis* has resulted in changes to the empiric therapy for pulmonary TB that have resulted in more complicated medication schemes of longer duration. Individuals suffering from

HIV are more likely to present with drug-resistant TB (108), which is thought to reflect the spread of MDR-TB in the HIV-seropositive community (1) and/or the lack of an immune response allowing for a larger bacilli population, from which resistant bacilli emerge. However, the most important factor leading to increased MDR-TB in this population is likely to be the complex socio-economic/therapeutic problems that frequently accompany HIV-infection and make TB regimen compliance problematic. Furthermore, it is speculated that HIV patients with concomitant gastrointestinal disease may have poor absorption of anti-TB drugs. In addition, extrapulmonary *M. tuberculosis* infection is more common in HIV patients, perhaps providing compartments that protect the bacilli from antimicrobial activity leading to the selection for drug-resistant or MDR-TB (46).

Before or during TB therapy, it has been recommended that all TB patients have counseling and testing for HIV, since active TB is a marker for the possibility of undiagnosed HIV infection (125). Direct observation of therapy (DOT), involves providing anti-TB drugs directly to a patient and watching the patient swallow the medication. Since anti-TB regimens are of a long duration, DOT not only increases adherence but also allows for aggressive interventions if patients miss doses. In addition, observational studies and meta-analysis in the US indicates that DOT, coupled with individualized case management, leads to the best anti-TB treatment results. The use of DOT for the treatment of latent tuberculosis infections achieved anti-TB regimen completion rates of >80% (102).

Drug regimens for TB have been extensively reviewed and defined (13). During treatment of some cases, these commonly prescribed regimens must often be modified due to drug intolerance/toxicity, or TB caused by drug-resistant *M. tuberculosis*. Regardless, case-appropriate anti-TB regimens are probably among the most complex and protracted therapies presently utilized for the treatment of an infectious disease. Currently, there are 10 drugs approved by the FDA for treating TB, including the first-line drugs—isoniazid, rifampin, rifapentine, ethambutol, and pyrazinamide—and the second-line drugs—cycloserine, ethionamide, *p*-aminosalicylic acid, streptomycin, and capreomycin. The fluoroquinolones moxifloxacin, levofloxacin and gatifloxacin, while not FDA-approved for treating TB, are also considered second-line anti-TB drugs and are commonly used to treat TB caused by drug-resistant *M. tuberculosis,* or are prescribed to patients intolerant to other agents. In small HIV-negative cohort trials testing the efficacy of moxifloxacin for treating smear-positive pulmonary TB, one study reported that moxifloxacin had activity comparable to rifampin (48), while another demonstrated the early bactericidal activities for isoniazid and moxifloxacin were similar (100). It has been suggested that moxifloxacin be assessed for use as part of a short-course regimen for treatment of drug-susceptible TB (48). A study comparing the standard four-drug regimen (see below) against rifampin, isoniazid and ciprofloxacin demonstrated that the latter combination had inferior sterilizing activity (63). Another study demonstrated that TB regimens of 4 to 5 month duration, including the fluoroquinolone ofloxacin for the first three months, resulted in high cure rates (130). Empiric fluoroquinolone monotherapy of smear-negative pulmonary TB cases has been associated with delays in the initiation of appropriate TB therapy and emergence of fluoroquinolone-resistant *M. tuberculosis* (126). Since some *M. tuberculosis* cultures derived from active pulmonary TB cases can be negative, strong clinical and radiographic data must be used to guide the use of fluoroquinolones for the treatment of suspected TB. Other second-line anti-TB drugs not approved for tuberculosis treatment by the FDA include the aminoglycosides amikacin and kanamycin, and rifabutin. The aminoglycosides are used for treating TB caused by drug-resistant *M. tuberculosis,* while rifabutin is used for therapy of patients taking drugs with unacceptable interactions with other rifamycins.

Presently there are four regimens recommended for treating adults with pulmonary TB caused by drug-susceptible organisms (with an initial phase of 2 months) using three to four of the following drugs: isoniazid, rifampin, pyrazinamide, and ethambutol. This initial four-drug cocktail is recommended because of a relatively high proportion of patients presenting with drug-resistant TB. For empirical treatment, streptomycin is not recommended as an initial agent due to increasing global resistance. Ethambutol should be continued until the isolate is known to be susceptible to both isoniazid and rifampin. The initial 2-month regimen is followed by treatment continuation of 4 months or 7 months with isoniazid, rifampin or rifapentine. The 7-month phase is recommended for patients whose initial regimen did not include pyrazinamide, patients with drug-susceptible cavitary pulmonary TB whose sputum is culture-positive after the initial 2-month regimen, and patients being treated once weekly with isoniazid and rifapentine who were culture-positive at the end of their initial 2-month treatment. Currently a 9-month course of isoniazid is recommended for the treatment of latent TB infections (24).

When compared to other TB patients, patients with MDR-TB experience longer treatment with expensive toxic medications, greater productivity losses, and higher hospitalization and mortality rates (104). In a study of patients in Mumbai India, 80% of the isolates were resistant to one agent, and 51% were multidrug resistant (2). The monetary costs of resistance are high. Treatment failure for MDR-TB occurs in 20 to 50% of cases, and TB control programs can spend 30% of their budgets on the <3% of TB patients presenting with MDR-TB (98). In the U.S., overall costs of treating MDR-TB infections averaged $89,594 per survivor and $717,555 for each patient who died (104).

There are no controlled trials for the treatment of TB caused by drug-resistant *M. tuberculosis* or MDR-TB. Consequently, treatment recommendations should be based upon expert advice and consensus opinion. Treatment of TB caused by even single drug-resistant *M. tuberculosis* can be complicated and end with treatment failure. Because the two most potent anti-TB drugs cannot be used to treat MDR-TB cases, regimens with less effective first-line and second-line drugs which are more toxic must be initiated (Table 1). When treating drug-resistant tuberculosis, patients are typically placed on multidrug regimens for periods as along as two years, even though many of these TB patients live under difficult social circumstances. The combination of complex regimens and difficult social situations almost certainly increases the chances for nonadherence, which in turn increases the likelihood for the evolution of MDR-TB. Consequently, DOT is important in all patients presenting with TB caused by drug-resistant *M. tuberculosis*.

SUMMARY

Virtually all bacterial pathogens today express antimicrobial resistance to one, and commonly, multiple antimicrobial classes. In some cases bacterial infections that were once treated with "miracle antimicrobials" are no longer curable. The human and economic costs of these antimicrobial-resistant bacterial infections, while perhaps not yet carefully estimated, are surely staggering (74). Therapy regimens for infections caused by these antimicrobial-resistant pathogens often involve more than one antimicrobial, and more often than not, three antimicrobials. In the case of VRE infections there is no definitive treatment, and appropriate treatment of infections caused by drug-resistant and MDR-TB is often guided by expert advice and consensus opinion. Furthermore, treatment for all infections caused by these organisms should be guided by susceptibility testing, and sometimes on multiple isolates from a single patient, as in the case for drug-resistant *M. tuberculosis* infections.

With the rise of infections caused by multiple antimicrobial-resistant pathogens, new antimicrobials and prevention measures will continually be required. Basic research in conjunction with human ingenuity will always remain at the forefront of our battle against antimicrobial resistance. With regard to infections caused by *S. aureus*, efforts have recently revealed the genes that this organism requires to grow in vivo (11), the products of which may be used to develop novel therapeutic options to treat disease caused by this pathogen. In addition, a *S. aureus* capsular polysaccharide-conjugate vaccine has been reported (119). With further improvements to this vaccine, another tool will be provided to prevent infections caused by this organism. Forty years after the introduction of the last anti-TB drug, a new anti-TB drug is under investigation that potently inhibits both drug-susceptible and -resistant *M. tuberculosis*. This new anti-TB drug (a diarylquinolone) targets the proton pump of adenosine triphosphate synthase, which is mechanistically novel (5). In addition, there is a concerted effort to develop a vaccine which prevents infection by *M. tuberculosis*. However, unlike most organisms for which an effective vaccine exists, *M. tuberculosis* is an intracellular pathogen and resides in macrophages; thus, a cell-mediated immune response rather than a humoral response is needed to control infection. Vaccine candidates that have been developed so far include recombinant *M. bovis* Bacillus Calmette-Guérin; attenuated *M. tuberculosis;* subunit and pooled subunit vaccines; fusion polyproteins; and DNA vaccines (58).

Table 1. Potential regimens for tuberculosis caused by drug-resistant strains of *M. tuberculosis*[a]

Drug resistance pattern	Suggested regimen[b]	Treatment duration (months)
INH (± SM)	RIF, PZA, EMB (a FQN may strengthen the regimen for patients with extensive disease)	6
INH and RIF (± SM)	FQN, PZA, EMB, IA ± alternative agent	18–24
INH, RIF (± SM), and EMB or PZA	FQN (EMB or PZA if active), IA, and two alternative agents	24
RIF	INH, PZA, EMB (a FQN may strengthen the regimen for patients with more extensive disease)	9–12

[a] Adapted from reference 13. Abbreviations: EMB, ethambutol; FQN, fluoroquinolone; IA, injectable agent; INH, isoniazid; PZA, pyrazinamide; RIF, rifampin; and SM, streptomycin.
[b] Injectable agent may include aminoglycosides or the polypeptide capreomycin. Alternative agents: ethionamide, cycloserine, *p*-aminosalicylic acid, clarithromycin, amoxicillin/clavulanate, and linezolid.

Without exception, following the introduction of each new antimicrobial into clinical practice, human pathogens expressing resistance to these antibacterials will appear. Educating health care providers and consumers has been a constant theme in Dr. Levy's career. As noted by Levy and Marshall (74), "The obstacles of few new antimicrobials on the horizon and the increasing frequency of multidrug resistance mean that we must redouble our efforts to preserve the agents at hand, while intensifying the search for new therapeutics." In order to prevent emergence and manage infections caused by antimicrobial-resistant bacteria, Professor Stuart Levy (74) and many of his colleagues (120) advocate appropriate antimicrobial stewardship, tracking resistance frequency; comprehensive applied infection control programs involving patient cohorting; and the introduction of new therapeutic antimicrobials and treatment options. The challenge, while daunting, may not be insurmountable. To again quote Dr. Levy, "We should 'adopt' the action of cows and 'graze', taking on one patch of the problem at a time, as we move together to achieve reversal of the resistance problem" (73).

Acknowledgments. We thank John Segreti (Rush University Medical College) for information on daptomycin trials, Angelo A. Izzo (Colorado State University) for discussions on *M. tuberculosis* vaccine development, and Stephanie A. Cantore for her outstanding job on the administration of references.

REFERENCES

1. **Alland, D., G. E. Kalkut, A. R. Moss, R. A. McAdam, J. A. Hahn, W. Bosworth, E. Drucker, and B. R. Bloom.** 1994. Transmission of tuberculosis in New York City. An analysis of DNA fingerprinting and conventional epidemiologic methods. *N. Engl. J. Med.* **330:**1710–1716.
2. **Almeida, D., C. Rodriques, Z. F. Udwadia, A. Lalvani, G. D. Gothi, P. Mehta, and A. Mehta.** 2003. Incidence of multidrug-resistant tuberculosis in urban and rural India and implications for prevention. *Clin. Infect. Dis.* **36:**152–154.
3. **American Academy of Pediatrics Committee on Infectious Diseases.** 1997. Therapy for children with invasive pneumococcal infections. *Pediatrics* **99:**289–299.
4. **American Academy of Pediatrics Subcommittee on Management of Acute Otitis Media.** 2004. Diagnosis and management of acute otitis media. *Pediatrics* **113:**1451–1465.
5. **Andries, K. P. Verhasselt, J. Guillemont, H. W. H. Göhlman, J. M. Neefs, H. Winkler, J. Van Gestal, P. Timmerman, M. Zhu, E. Lee, P. Williams, D. de Chaffoy, E. Huitric, S. Hoffner, E. Cambau, C. Truffot-Pernot, N. Lounis, and V. Jarlier.** 2005. A diarylquinolone drug active on the ATP synthase of *Mycobacterium tuberculosis. Science* **307:**223–227.
6. **Baggett, H. C., T. W. Hennessy, K. Rudolph, D. Bruden, A. Reasonover, A. Parkinson, R. Sparks, R. M. Donlan, P. Martinez, K. Mongkolrattanothai, and J. C. Butler.** 2004. Community-onset methicillin-resistant *Staphylococcus aureus* associated with antibiotic use and the cytotoxin Panton-Valentine leukocidin during a furunculosis outbreak in rural Alaska. *J. Infect. Dis.* **189:**1565–73.
7. **Bartlett, J. G., and L. M. Mundy.** 1995. Community-acquired pneumonia. *N. Engl. J. Med.* **333:**1618–1624.
8. **Bartlett, J. G., S. F. Dowell, L. A. Mandell, T. M. File, Jr., D. M. Musher, and M. J. Fine.** 2000. Pactice guidelines for the management of community-acquired pneumonia in adults. *Clin. Infect. Dis.* **31:**347–382.
9. **Barnes, P. F., and M. D. Cave.** 2003. Molecular epidemiology of tuberculosis. *New Engl. J. Med.* **349:**1149–1156.
10. **Bassetti, M., P. A. Farrel, D. A. Callan, J. E. Topal, and L. M. Dembry.** 2003. Emergence of linezolid-resistant *Enterococcus faecium* during treatment of enterococcal infections. *Int. J. Antimicrob. Agents.* **21:**593–594.
11. **Benton, B. M., J. P. Zhang, S. Bond, C. Pope, T. Christian, L. Lee, K. M. Winterberg, M. B. Schmid, and J. M. Buysse.** 2004. Large-scale identification of genes required for full virulence of *Staphylococcus aureus. J. Bacteriol.* **186:**8478–8489.
12. **Bifani, P. J., B. Mathema, N. E. Kurepina, and B. N. Kreiswirth.** 2002. Global dissemination of the *Mycobacterium tuberculosis* W-Beijing family strains. *Trends Microbiol.* **10:** 45–52.
13. **Blumberg, H. M., W. J. Burman, R. E. Chaisson, C. L. Daley, S. C. Etkind, L. N. Friedman, P. Fujiwara, M. Grzemska, P. C. Hopewell, M. D. Iseman, R. M. Jasmer, V. Koppaka, R. I. Menzies, R. J. O'Brien, R. R. Reves, L. B. Reichman, P. M. Simone, J. R. Starke, A. A. Vernon; American Thoracic Society, Centers for Disease Control and Prevention and the Infectious Diseases Society.** 2003. American Thoracic Society/Centers for Disease Control and Prevention/Infectious Diseases Society of America: Treatment of tuberculosis. *Am. J. Respir. Crit. Care Med.* **167:**603–662.
14. **Bouvet, E., E. Casalino, G. Mandoza-Sassi, S. Lariven, E. Vallée, M. Pernet, S. Gottot, and F. A. Vachon.** 1993. A nosocomial outbreak of multidrug-resistant *Mycobacterium bovis* among HIV-infected patients, a case-control study. *AIDS* **7:**1453–1460.
15. **Bryan, C. S., R. Talwani, and M. S. Stinson.** 1997. Penicillin dosing for pneumococcal pneumonia. *Chest* **112:**1657–1664.
16. **Bucher, H. C., L. E. Griffith, G. H. Guyatt, P. Sudre, M. Naef, P. Sendi, and M. Battegay** 1999. Isoniazid prophylaxis for tuberculosis in HIV infection: a meta-analysis of randomized controlled trials. *AIDS* **13:**501–507.
17. **Centers for Disease Control and Prevention.** 1997. Prevention of pneumococcal disease: recommendations of the Advisory Committee on Immunization Practices (ACIP). *Morb. Mortal. Wkly. Rep.* **46:**1–24.
18. **Centers for Disease Control and Prevention.** 1997. Reduced susceptibility of *Staphylococcus aureus* to vancomycin—Japan, 1996. *Morb. Mortal. Wkly. Rep.* **46:**624–626.
19. **Centers for Disease Control and Prevention.** 1997. National Nosocomial Infections Surveillance (NNIS) System report: data summary from October 1986-April 1997, issued May 1997. *Am. J. Infect. Control* **25:**477–487.
20. **Centers for Disease Control and Prevention.** 1999. National Nosocomial Infections Surveillance (NNIS) System report: data summary from January 1990-May 1999, issued June 1999. *Am. J. Infect. Control* **27:**520–532.
21. **Centers for Disease Control and Prevention.** 2000. Public health focus: surveillance, prevention and control of nosocomial infections. *Morb. Mortal. Weekly Rep.* **41:**783–787.
22. **Centers for Disease Control and Prevention.** 2002. *Staphylococcus aureus* resistant to vancomycin—United States, 2002. *Morb. Mortal. Wkly. Rep.* **51:**565–576.
23. **Centers for Disease Control and Prevention.** 2002. Campaign to prevent antimicrobial resistance in healthcare settings. Available at: http://www.cdc.gov/drugresistance/healthcare/default.htm.

24. **Centers for Disease Control and Prevention.** 2003. Update: adverse event data and revised American Thoracic Society/CDC recommendations against the use of rifampin and pyrazinamide for treatment of latent tuberculosis infections—United States. *Morb. Mortal. Wkly. Rep.* **52:**735–739.
25. **Centers for Disease Control and Prevention.** 2004. Tuberculosis associated with blocking agents against tumor necrosis factor-alpha—California, 2002-2003. *Morb. Mortal. Wkly. Rep.* **53:**683–686.
26. **Centers for Disease Control and Prevention.** 2004. Trends in tuberculosis—United States, 1998–2003. *Morb. Mortal. Wkly. Rep.* **53:**209–214.
27. **Cetinkaya, Y., Falk, P., and C. G. Mayhall.** 2000. Vancomycin-resistant enterococci. *Clin. Microbiol. Rev.* **13:**686–707.
28. **Chaves F., J. Garcia-Martinez, S. de Miguel, and J. R. Otero.** 2003. Molecular characterization of resistance to mupirocin in methicillin-susceptible and -resistant isolates of *Staphylococcus aureus* from nasal samples. *Antimicrob. Agents Chemother.* **47:**1589–1597.
29. **Cottagnoud, P., M. Pfister, F. Acosta, M. Cottagnoud, L. Flatz, F. Kühn, H-P. Müller, and A. Stucki.** 2004. Daptomycin is highly efficacious against penicillin-resistant and penicillin- and quinolone-resistant pneumococci in experimental meningitis. *Antimicrob. Agents Chemother.* **48:**3928–3933.
30. **D'Agata, E. M. C., S. Gautam, W. K. Green, and Y. W. Tang.** 2002. High rate of false-negative results of rectal swab culture method in detection of gastrointestinal colonization with vancomycin-resistant enterococci. *Clin. Infect. Dis.* **34:**167–172.
31. **David, H. L.** 1970. Probability distribution of drug-resistant mutants in unselected populations of *Mycobacterium tuberculosis. Appl. Microbiol.* **20:**810–814.
32. **de Gans, J., and D. van de Beek.** 2002. Dexamethasone in adults with bacterial meningitis. *N. Engl. J. Med.* **347:**1549–1556.
33. **DeLisle, S., and T. M. Perle.** 2003. Vancomycin-resistant enterococci: A road map on how to prevent the emergence and transmission of antimicrobial resistance. *Chest* **123:**504S–518S.
34. **Dodge, R. A., J. S. Daly, R. Davaro, and R. H. Glew.** 1996. High dose ampicillin plus streptomycin for treatment of a patient with a severe infection due to multi-resistant enterococci. *Clin. Infect. Dis.* **25:**1269–1270.
35. **Doern, G. V., A. B. Brueggemann, M. Blocker, M. Dunne, H. P. Holley, Jr., K. S. Kehl, J. Duval, K. Kugler, S. Putnam, A. Rauch, and M. A. Pfaller.** 1998. Clonal relationships among high-level penicillin-resistant *Streptococcus pneumoniae* in the United States. *Clin. Infect. Dis.* **27:**757–761.
36. **Dowell, S. F., J. C. Butler, G. S. Giebink, M. R. Jacobs, D. Jernigan, D. M. Musher, A. Rakowsky, and B. Schwartz.** 1999. Acute otitis media: management and surveillance in an era of pneumococcal resistance-a report from the Drug-resistant *Streptococcus pneumoniae* Therapeutic Working Group. *Pediatr. Infect. Dis.* **18:**1–9.
37. **Dowzicky, M., G. H. Talbot, C. Feger, P. Prokocimer, J. Etienne, and R. Leclercq.** 2000. Characterization of isolates associated with emerging resistance to quinupristin/dalfopristin (Synercid) during a worldwide clinical program. *Diagn. Microbiol. Infect. Dis.* **37:**57–62.
38. **Dye, C., M. A. Espinal, C. J. Watt, C. Mbiaga, and B. G. Williams.** 2002. Worldwide incidence of multidrug-resistant tuberculosis. *J. Infect. Dis.* **185:**1197–1202.
39. **El-khoury, J., and J. A. Fishman.** 2003. Linezolid in the treatment of vancomycin-resistant *Enterococcus faecium* infections in solid organ transplants: report of a multicenter compassionate-use trial. *Transpl. Infect. Dis.* **5:**121–125.
40. **Enright, M. C., D. A. Robinson, G. Randle, E. J. Feil, H. Grundmann, and B. G. Spratt.** 2002. The evolutionary history of methicillin-resistant *Staphylococcus aureus* (MRSA). *Proc. Natl. Acad. Sci. USA* **99:**7687–7692.
41. **Feikin, D., A. Schuchat, M. Kolczak, N. L. Barrett, L. H. Harrison, L. Lefkowitz, A. McGeer, M. M. Farley, D. J. Vugia, C. Lexau, K. R. Stefonek, J. E. Patterson, and J. H. Jorgensen.** 2000. Mortality from invasive pneumococcal pneumonia in the era of antibiotic resistance, 1995-1997. *Am. J. Public Health.* **90:**223–229.
42. **Finney, M. S., C. W. Crank, and J. Segreti.** 2004. Daptomycin for treatment of drug-resistant Gram-positive bacteremia and infective endocarditis. 42nd Infect. Dis. Soc. Amer. Meeting abstracts, Washington D.C.
43. **Fridkin, S. K., J. R. Edwards, J. M. Courval, H. Hill, F. C. Tenover, R. Lawton, R. P. Gaynes, and J. E. McGowan Jr. Intensive Care Antimicrobial Resistance Epidemiology (ICARE) Project and the National Nosocomial Infections Surveillance (NNIS) System Hospitals.** 2001. The effect of vancomycin and third-generation cephalosporins on prevalence of vancomycin-resistant enterococci in 126 U.S. adult intensive care units. *Ann. Intern. Med.* **135:**175–183.
44. **Friedland, I. R,, S. Shelton, M. Paris, S. Rinderknecht, S. Ehrett, K. Krisher, and G. H. McCracken.** 1993. Dilemmas in diagnosis and management of cephalosporin-resistant *Streptococcus pneumoniae* meningitis. *Pediatr. Infect. Dis. J.* **12:**196–200.
45. **Friedland, I. R.** 1995. Comparison of the response to antimicrobial therapy of penicillin-resistant and penicillin-susceptible pneumococcal disease. *Pediatr. Infect. Dis.* **14:**885–890.
46. **Gillespie, S. H.** 2002. Evolution of resistance in *Mycobacterium tuberculosis:* clinical and molecular perspective. *Antimicrob. Agents Chemother.* **46:**267–274.
47. **Gonzalez, B. E., G. Martinez-Aguilar, K. G. Hulten, W. A. Hammerman, J. Coss-Bu, A. Avalos-Mishaan, E. O. Mason, Jr., and S. L. Kaplan.** 2005. Severe staphylococcal sepsis in adolescents in the era of community-acquired methicillin-resistant *Staphylococcus aureus. Pediatrics* **115:**642–648.
48. **Gosling, R. D., L. O. Uiso, N. E. Sam, E. Bongard, E. G. Kanduma, M. Nyindo, R. W. Morris, and S. H. Gillespie.** 2003. The bactericidal activity of moxifloxacin in patients with pulmonary tuberculosis. *Am. J. Respir. Crit. Care Med.* **168:**1342–1345.
49. **Grimwood, K., P. Anderson, V. Anderson, L. Tan, and T. Nolan T.** 2000. Twelve year outcomes following bacterial meningitis: further evidence for persisting effects. *Arch. Dis. Child.* **83:**111–116.
50. **Hageman, J. C., D. A. Pegues, C. Jepson, R. L. Bell, M. Guinan, K. W. Ward, M. D. Cohen, J. A. Hindler, F. C. Tenover, S. K. McAllister, M. E. Kellum, and S. K. Fridkin.** 2001. Vancomycin-intermediate *Staphylococcus aureus* in a home health-care patient. *Emerg. Infect. Dis.* **7:**1023–1025.
51. **Hansel, N. N., B. Merriman, E. F. Haponik, and G. B. Diette.** 2004. Hospitalizations for tuberculosis in the United States in 2000: predictors of in-hospital mortality. *Chest* **126:**1079–1086.
52. **Harbarth, S., S. Cosgrove, and Y. Carmeli.** 2002. Effects of antibiotics on nosocomial epidemiology of vancomycin-resistant enterococci. *Antimicrob. Agents Chemother.* **46:**1619–1628.
53. **Heffelfinger, J. D., S. F. Dowell, J. H. Jorgensen, K. P. Klugman, L. R. Mabry, D. M. Musher, J. F. Plouffe, A. Rakowsky, A. Schuchat, and C. G. Whitney.** 2000. Management of community-acquired pneumonia in the era of pneumococcal resistance: a report from the Drug-Resistant *Streptococcus pneumoniae* Therapeutic Working Group. *Arch. Intern. Med.* **160:**1399–1408.

54. **Howden, B. P., P. B. Ward, P. G. Charles, T. M. Korman, A. Bak, J. Hurley, P. D. Johnson, A. J. Morris, B. C. Mayall, and M. L. Grayson.** 2004. Treatment outcomes for serious infections caused by methicillin-resistant *Staphylococcus aureus* with reduced vancomycin susceptibility. *Clin. Infect. Dis.* **38:**521–528.
55. **Howe, R. A., M. Robson, A. Oakhill, J. M. Cornish, and M. R. Millar.** 1997. Successful use of tetracycline as therapy of an immunocompromised patient with septicaemia caused by a vancomycin-resistant enterococcus. *J. Antimicrob. Chemother.* **40:**144–145.
56. **Ing, M. B., L. M. Baddour, and A. S. Bayer.** 1997. Bacteremia and infective endocarditis: pathogenesis, diagnosis, and complications, p. 331–354. *In* K. B. Crossley and G. L. Archer (ed.), *The Staphylococci in Human Disease.* Churchill Livingstone, New York, N.Y.
57. **Ioanas, M., and H. Lode.** 2004. Linezolid in VAP by MRSA: a better choice. *Intensive Care Med.* **30:**343–346.
58. **Izzo, A. A., L. Brandt, T. Lasco, A.-P. Kipnis, I. Orme.** NIH pre-clinical screening program: Overview and current status. *Tuberculosis,* in press.
59. **Jacobs, M. R., H. J. Koornhof, R. M. Robins-Browne, C. M. Stevenson, Z. A. Vermaak, I. Freiman, G. B. Miller, M. A. Witcomb, M. Isaacson, J. L. Ward, and R. Austrian.** 1978. Emergence of multiply resistant penumococci. *N. Engl. J. Med.* **299:**735–740.
60. **Jones, M. E., D. C. Draghi, C. Thornsberry, J. A. Karlowsky, D. F. Sahm, and R. P. Wenzel.** 2004. Emerging resistance among bacterial pathogens in the intensive care unit—a European and North American Surveillance study (2000-2002). *Ann. Clin. Microbiol. Antimicrob.* **3:**14–24.
61. **Kaplan, S. L.** 2002. Management of pneumococcal meningitis. *Pediatr. Infect. Dis. J.* **21:**589–591.
62. **Kauffman, C. A.** 2003. Therapeutic and preventative options for the management of vancomycin-resistant enterococcal infections. *J. Antimicrob. Chemother.* **51:**iii23–iii30.
63. **Kennedy, N., R. Fox, G. M. Kisyombe, A. O. Saruni, L. O. Uiso, A. R. Ramsay, F. I. Ngowi, and S. H. Gillespie.** 1993. Early bactericidal and sterilizing activities of ciprofloxacin in pulmonary tuberculosis. *Am. Rev. Respir. Dis.* **148:**1547–1551.
64. **Kirst, H. A., D. G. Thompson, and T. I. Nicas.** 1998. Historical yearly usage of vancomycin. *Antimicrob. Agents Chemother.* **42:**1303–1304.
65. **Klugman, K. P.** 2002. The successful clone: the vector of dissemination of resistance in *Streptococcus penumoniae. J. Antimicrob. Chemother.* **50:**S1–S5.
66. **Kluytmans, J., A. van Belkum, and H. Verbrugh.** 1997. Nasal carriage of *Staphylococcus aureus:* epidemiology, underlying mechanisms, and associated risks. *Clin. Microbiol. Rev.* **10:**505–520.
67. **Kokai-Kun, J. F., S. M. Walsh, T. Chanturiya, and J. J. Mond.** 2003. Lysostaphin cream eradicates *Staphylococcus aureus* nasal colonization in a cotton rat model. *Antimicrob. Agents Chemother.* **47:**1589–1597.
68. **Kuhn, F., M. Cottagnoud, F. Acosta, L. Flatz, J. Entenza, and P. Cottonwood.** 2003. Cefotaxime acts synergistically with levofloxacin in experimental meningitis due to penicillin-resistant pneumococci and prevents selection of levofloxacin-resistant mutants in vivo. *Antimicrob. Agents Chemother.* **47:**2487–2491.
69. **Kupronis, B.A., C. L. Richards, C. G. Whitney; Active Bacterial Core Surveillance Team.** 2003. Invasive pneumococcal disease in older adults residing in long-term care facilities and in the community. *J. Am. Geriatr. Soc.* **51:**1520–1525.
70. **Lautenbach, E., C. V. Gould, L. A. LaRosa, A. M. Marr, I. Nachamkin, W. B. Bilker, and N. O. Fishman.** 2004. Emergence of resistance to chloramphenicol among vancomycin-resistant enterococcal (VRE) blood stream isolates. *Int. J. Antimicrob. Agents.* **23:**200–203.
71. **Lave, J. R., C. J. Lin, M. J. Fine, and P. Hughes-Cromwick.** 1999. Cost of treating patients with community acquired pneumonia. *Sem. Resp. Crit. Care Med.* **20:**189–197.
72. **Leggiadro, R. J.** 1994. Penicillin- and cephalosporin-resistant *Streptococcus pneumoniae:* an emerging microbial threat. *Pediatrics* **93:**500–503.
73. **Levy, S. B.** 2002. The 2000 Garrod Lecture: Factors impacting on the problem of antibiotic resistance. *J. Antimicrob. Chemo.* **49:**25–30.
74. **Levy, S. B., and B. Marshall.** 2004. Antibacterial resistance worldwide: causes challenges and responses. *Nat. Med.* **10:**S122–S129.
75. **Lewis, J. S. II., and J. H. Jorgensen.** 2005. Inducible clindamycin resistance in staphylococci: should clinicians and microbiologists be concerned. *Clin. Infect. Dis.* **40:**280–285.
76. **Linden, P. K., A. W. Pasculle, D. McDevitt, and D. J. Kramer.** 1997. Effect of quinupristin/dalfopristin on the outcome of vancomycin-resistant *Enterococcus faecium* bacteraemia: comparison with a control cohort. *J. Antimicrob. Chemother.* **39:**145–151.
77. **Linden, P. K., R. C. Moellering, R. A. Wood, S. J. Rehm, J. Flaherty, F. Bompart, G. H. Talbot; Synercid Emergency-Use Study Group.** 2001. Treatment of vancomycin-resistant *Enterococcus faecium* with quinupristin-dalfopristin. *Clin. Infect. Dis.* **33:**1816–1823.
78. **Lowy, F. D.** 1998. *Staphylococcus aureus* infections. *N. Engl. J. Med.* **339:**520–532.
79. **Lynch, J. P., 3rd.** 2000. Community-acquired pneumonia: risk factors and specific causes. *J. Resp. Dis.* **21:**457–468.
80. **Mandell, L. A., J. G. Bartlett, S. F. Dowell, T. M. File Jr., D. M. Musher, C. Whitney and Infectious Diseases Society of America.** 2003. Update of practice guidelines for the management of community-acquired pneumonia in immunocompetent adults. *Clin. Infect. Dis.* **37:**1405–1433.
81. **Markowitz, N., E. L. Quinn, and L. D. Saravolatz.** 1992. Trimethoprim-sulfamethoxazole compared with vancomycin for the treatment of *Staphylococcus aureus* infection. *Ann. Int. Med.* **117(5):**390–398.
82. **Martinez-Aguilar G, A. Avalos-Mishaan, K. Hulten, W. Hammerman, E. O. Mason, Jr., and S. L. Kaplan.** 2004. Community-acquired, methicillin-resistant and methicillin-susceptible *Staphylococcus aureus* musculoskeletal infections in children. *Pediatr. Infect. Dis. J.* **23:**701–706.
83. **Mascini, E. M., K. P. Jalink, T. E. M. Kamp-Hopmans, H. E. M. Blok, J. Verhoef, M. J. M. Bonten, and A. Troelstra.** 2003. Acquisition and duration of vancomycin-resistant enterococcal carriage in relation to strain type. *J. Clin. Microbiol.* **41:**5377–5383.
84. **May, J., K. Shannon, A. King, and G. French.** 1998. Glycopeptide tolerance in *Staphylococcus aureus. J. Antimicrob. Chemother.* **42:**189–197.
85. **McCollum, M., D. C. Rhew, and S. Parodi.** 2003. Cost analysis of switching from i.v. to p.o. linezolid for the management of methicillin-resistant *Staphylococcus* species. *Clin. Ther.* **25:**3173–3179.
86. **McGee, L., L. McDougal, J. Zhou, B. G. Spratt, F. C. Tenover, R. George, R. Hakenbeck, W. Hryniewicz, J. C. Lefevre, A. Tomasz, and K. P. Klugman.** 2001. Nomenclature of major antimicrobial-resistant clones of *Streptococcus pneumoniae* defined by the pneumococcal molecular epidemiology network. *J. Clin. Microbiol.* **39:**2565–2571.
87. **Medical Research Council.** 1950. Treatment of pulmonary tuberculosis with streptomycin and para-aminosalicylic acid. *Br. Med. J.* **2:**1073–1085.

88. **Mekonen, E. T., G. A. Noskin, D. M. Hacek, and L. R. Peterson.** 1995. Successful treatment of persistent bacteremia due to vancomycin-resistant, ampicillin-resistant *Enterococcus faecium*. *Microb. Drug Resist.* **1**:249–253.
89. **Moellering, R. C.** 2003. Linezolid: the first oxazolidinone antimicrobial. *Ann. Intern. Med.* **138**:135–142.
90. **Montecalvo, M. A.** 2003. Ramoplanin: a novel antimicrobial agent with the potential to prevent vancomycin-resistant enterococcal infection in high-risk patients. *J. Antimicrob. Chemother.* **51**:iii31–iii35.
91. **Mulholland, E. K.** 1997. A report prepared for the Scientific Advisory Group of Experts, Global Programme for Vaccines and Immunization. Geneva, World Health Organization.
92. **Mylotte, J. M., C. McDermott, and J. A. Spooner.** 1987. Prospective study of 114 consecutive episodes of *Staphylococcus aureus* infections. *Rev. Infect. Dis.* **9**:891–907.
93. **Naimi, T. S., K. H. LeDell, K. Como-Sabetti, S. M. Borchardt, D. J. Boxrud, J. Etienne, S. K. Johnson, F. Vandenesch, S. Fridkin, C. O'Boyle, R. N. Danila, and R. Lynfield.** 2003. Comparison of community- and health care-associated methicillin-resistant *Staphylococcus aureus* infection. *JAMA* **290**: 2976–2984.
94. **Niederman, M. S., J. S. McCombs, A. N. Unger, A. Kumar, R. Popovian.** 1998. The cost of treating community-acquired pneumonia. *Clin. Ther.* **20**:820–837.
95. **Niederman, M. S., L. A. Mandell, A. Anzueto, J. B. Bass, W. A. Broughton, G. D. Campbell, N. Dean, T. File, M. J. Fine, P. A. Gross, F. Martinez, T. J. Marrie, J. F. Plouffe, J. Ramirez, G. A. Sarosi, A. Torres, R. Wilson, V. L. Yu, and American Thoracic Society.** 2001. Guidelines for the management of adults with community-acquired pneumonia. Diagnosis, assessment of severity, antimicrobial therapy, and prevention. *Am. J. Respir. Crit. Care Med.* **163**:1730–1754.
96. **O'Brien, F. G., T. T. Lim, F. N. Chong, G. W. Coombs, M. C. Enright, D. A. Robinson, A. Monk, B. Said-Salim, B. N. Kreiswirth, and W. B. Grubb.** 2004. Diversity among community isolates of methicillin-resistant *Staphylococcus aureus* in Australia. *J. Clin. Microbiol.* **42**:3185–3190.
97. **Oliveira, D. C., A. Tomasz, and H. de Lencastre.** 2002. Secrets of success of a human pathogen: molecular evolution of pandemic clones of methicillin-resistant *Staphylococcus aureus*. *Lancet Infect. Dis.* **2**:180–189.
98. **Pablos-Mendez, A., D. K. Gowda, and T. R. Frieden.** 2002. Controlling multidrug-resistant tuberculosis and access to expensive drugs: a rational frame-work. *Bull. World Health Org.* **80**:489–500.
99. **Pfeltz, R. F., and B. J. Wilkinson.** 2004. The escalating challenge of vancomycin resistance in *Staphylococcus aureus*. *Curr. Drug. Targ. Infect. Disord.* **4**:1–22.
100. **Pletz, M. W., A. De Roux, A. Roth, K. H. Neumann, H. Mauch, and H. Lode.** 2004. Early bactericidal activity of moxifloxacin in treatment of pulmonary tuberculosis: a prospective, randomized study. *Antimicrob. Agents Chemother.* **48**:780–782.
101. **Pletz, M. W., L. McGee, J. Jorgensen, B. Beall, R. R. Facklam, C. G. Whitney, and K. P. Klugman.** 2004. Levofloxacin-resistant invasive *Streptococcus pneumoniae* in the United States: evidence for clonal spread and the impact of conjugate pneumococcal vaccine. *Antimicrob. Agents Chemother.* **48**: 3491–3497.
102. **Priest, D. H., L. F. Vossel Jr., E. A. Sherfy, D. P. Hoy and C. A. Haley.** 2004. Use of intermittent rifampin and pyrazinamide for treatment of latent tuberculosis infection in a targeted testing program. *Clin. Infect. Dis.* **39**:1764–1771.
103. **Raad, I., R. Hachem, H. Hanna, C. Afif, C. Escalante, H. Kantarjian, and K. Rolston.** 2004. Prospective, randomized study comparing quinupristin-dalfopristin with linezolid in the treatment of vancomycin-resistant *Enterococcus faecium* infections. *J. Antimicrob. Chemother.* **53**:646–649.
104. **Rajbhandary, S. S., S. M. Marks, and N. N. Bock.** 2004. Costs of patients hospitalized for multidrug-resistant tuberculosis/int. *J. Tuberc. Lung Dis.* **8**:1012–1016.
105. **Rastogi, N.** 2003. An introduction to mycobacterial taxonomy, structure, drug resistance, and pathogenesis, p. 89–115. *In* D. Dionisio (ed.), *Textbook-Atlas of Intestinal Infections in AIDS*. Springer-Verlag, Milan, Italy.
106. **Ricaurte, J. C., H. W. Boucher, G. S. Turett, R. C. Moellering, V. J. Labombardi, and J. W. Kislak.** 2001. Chloramphenicol treatment for vancomycin-resistant *Enterococcus faecium* bacteremia. *Clin. Microbiol. Infect.* **7**:17–21.
107. **Rice, L. B.** 2001. Emergence of vancomycin-resistant enterococci. *Emerg. Infect. Dis.* **7**:183–187.
108. **Ridzon, R., C. G. Whitney, M. T. McKenna, J. P. Taylor, S. H. Ashkar, A. T. Nitta, S. M. Harvey, S. Valway, C. Woodley, R. Cooksey, and I. M. Onorato.** 1998. Risk factors for rifampin mono-resistant tuberculosis. *Am. J. Respir. Crit. Care Med.* **157**:1881–1884.
109. **Rieder, H. L., G. M. Cauthen, A. B. Bloch, C. H. Cole, D. Holtzman, D. E. Snider Jr., W. J. Bigler, and J. J. Witte.** 1989. Tuberculosis and acquired immunodeficiency syndrome: Florida. *Arch. Intern. Med.* **149**:1268–1273.
110. **Robinson, K. A., W. Baughman, G. Rothrock, N. L. Barrett, M. Pass, C. Lexau, B. Damaske, K. Stefonek, B. Barnes, J. Patterson, E. R. Zell, A. Schuchat, C. G. Whitney; Active Bacterial Core Surveillance (ABCs)/Emerging Infections Program Network.** 2001. Epidemiology of invasive *Streptococcus pneumoniae* infections in the United States, 1995-1998: Opportunities for prevention in the conjugate vaccine era. *J. Am. Med. Assoc.* **285**:1729–1735.
111. **Rothermel, C. D.** 2004. Penicillin and macrolide resistance in pneumococcal pneumonia: does in vitro resistance affect clinical outcomes? *Clin. Infect. Dis.* **38**:S346–S349.
112. **Rubin, R. J., C. A. Harrington, A. Poon, K. Dietrich, J. A. Greene, and A. Moiduddin.** 1999. The economic impact of *Staphylococcus aureus* infection in New York City hospitals. *Emerg. Infect. Dis.* **5**:9–17.
113. **Ruef, C.** 2004. Epidemiology and clinical impact of glycopeptide resistance in *Staphylococcus aureus*. *Infection* **6**:315–327.
114. **Schlievert, P. M.** Personal communication.
115. **Schrag, S. J., L. McGee, C. G. Whitney, B. Beall, A. S. Craig, M. E. Choate, J. H. Jorgensen, R. R. Facklam, and K. P. Klugman, and Active Bacterial Core Surveillance Team.** 2004. Emergence of *Streptococcus pneumoniae* with very-high-level resistance to penicillin. *Antimicrob. Agents Chemother.* **48**: 3016–3023.
116. **Schuchat, A., K. Robinson, J. D. Wenger, L. H. Harrison, M. Farley, A. L. Reingold, L. Lefkowitz, and B. A. Perkins.** 1997. Bacterial Meningitis in the United States in 1995. Active Surveillance Team. *N. Engl. J. Med.* **337**:970–976.
117. **Segreti, J., C. W. Crank, and M. S. Finney.** 2004. Daptomycin for treatment of drug-resistant Gram-positive bacteremia and infective endocarditis. IDSA meeting abstract, Washington D.C.
118. **Shamputa, I. C., L. Rigouts, L. A. Eyongeta, N. A. El Aila, A. van Deun, A. H. Salim, E. Willery, C. Locht, P. Supply, and F. Portaels.** 2004. Genotypic and phenotypic heterogeneity among *Mycobacterium tuberculosis* isolated from pulmonary tuberculosis patients. *J. Clin. Microbiol.* **42**:5528–5536.
119. **Shinefield, H., S. Black, A. Fattom, G. Horwith, S. Rasgon, J. Ordonez, H. Yeoh, D. Law, J. B. Robbins, R. Schneerson, L. Muenz, S. Fuller, J. Johnson, B. Fireman, H. Alcorn, and R. Naso.** 2002. Use of a *Staphylococcus aureus* conjugate vaccine

in patients receiving hemodialysis. *N. Engl. J. Med.* **346**:491–496.

120. **Shlaes, D. M., D. N. Gerding, J. F. John, Jr., W. A. Craig, D. L. Bornstein, R. A. Duncan, M. R. Eckman, W. E. Farrer, W. H. Greene, V. Lorian, S. B. Levy, J. E. McGowan, Jr., S. M. Paul, J. Ruskin, F. C. Tenover, and C. Watanakunakorn.** 1997. Society for Healthcare Epidemiology of America and Infectious Diseases Society of America joint committee on the prevention of antimicrobial resistance: guidelines for the prevention of antimicrobial resistance in hospitals. *Infect. Control Hosp. Epidemiol.* **18**:275–291.
121. **Shorr, A. F., G. M. Sulsa, and M. H. Kollef.** 2004. Linezolid for treatment of ventilator-associated pneumonia: a cost-effective alternative to vancomycin. *Crit. Care Med.* **32**:137–145.
122. **Sloas, M. M., F. F. Barrett, P. J. Chesney, B. K. English, B. C. Hill, F. C. Tenover, and R. J. Leggiadro.** 1992. Cephalosporin treatment failure in penicillin- and cephalosporin-resistant *Streptococcus pneumoniae* meningitis. *Pediatr. Infect. Dis. J.* **11**:662–666.
123. **Smith, I. M., and A. B. Vickers.** 1960. Natural history of 338 treated and untreated patients with staphylococcal septicemia. *Lancet* **1**:1318–1322.
124. **Smith, T. L., M. L. Pearson, K. R. Wilcox, C. Cruz, M. V. Lancaster, B. Robinson-Dunn, F. C. Tenover, M. J. Zervos, J. D. Band, E. White, and W. R. Jarvis.** 1999. Emergence of vancomycin-resistance in *Staphylococcus aureus*. Glycopeptide-intermediate *Staphylococcus aureus* working group. *N. Engl. J. Med.* **340**:493–501.
125. **Sonnenberg, P., J. R. Glynn, K. Fielding, J. Murray, P. Godfrey-Faussett, and S. Shearer.** 2005. How soon after infection with HIV does the risk of tuberculosis start to increase? A retrospective cohort study in South African gold miners. *J. Infect. Dis.* **191**:150–158.
126. **Sterling, T. R.** 2004. The WHO/IUATLD diagnostic algorithm for tuberculosis and emperic fluoroquinolone use: potential pitfalls. *Int. J. Tuberc. Lung Dis.* **8**:1396–1400.
127. **The Stop TB Partnership.** September 14th, 2004. TB/HIV: facts at a glance. Available at: http://www.stoptb.org/events/internationalaidsconference/xv/assets/InfoPack/1GB.pdf.
128. **Taylor, S. E., D. L Paterson, and V. L. Yu.** 1998. Treatment options for chronic prostatitis due to vancomycin-resistant *Enterococcus faecium*. *Eur. J. Clin. Microbiol. Infect. Dis.* **17**:798–800.
129. **Tsiodras, S., H. S. Gold, G. Sakoulas, G. M. Eliopoulos, C. Wennersten, L. Venkataraman, R. C. Moellering, and M. J. Ferraro.** 2001. Linezolid resistance in a clinical isolate of *Staphylococcus aureus*. *Lancet* **358**:207–208.
130. **Tuberculosis Research Centre.** 2002. Shortening short course chemotherapy: a randomised clinical trial for treatment of smear-positive pulmonary tuberculosis with regimens using ofloxacin in the intensive phase. *Ind. J. Tuberc.* **49**:27–38.
131. **Uttley, A. H. C., C. H. Collins, J. Naidoo, and R. C. George.** 1988. Vancomycin-resistant enterococci. *Lancet* **i**:57–58.
132. **van Helden, P. D.** 2003. The economic divide and tuberculosis. Tuberculosis is not just a medical problem, but also a problem of social inequality and poverty. *EMBO Rep.* **4**:S24–S28.
133. **Vankerckhoven, V., T. Van Autgaerden, C. Vael, C. Lammens, S. Chapelle, R. Rossi, D. Jabes, and H. Goossens.** 2004. Development of a multiplex PCR for the detection of *asa1*, *gelE*, *cylA*, *esp*, and *hyl* genes in enterococci and survey for virulence determinants among European hospital isolates of *Enterococcus faecium*. *J. Clin. Microbiol.* **42**:4473–4479.
134. **Vincent, J., V. J Quagliarello, and W. M. Scheld.** 1997. Treatment of bacterial meningitis. *N. Engl. J. Med.* **336**:708–716.
135. **Wayne, L. G.** 1994. Dormancy of *Mycobacterium tuberculosis* and latency of disease. *Eur. J. Clin. Microbiol.* **13**:908–914.
136. **Whitney, C., M. Farley, J. Hadler, L. Harrison, C. Lexau, A. Reingold, L. Lefkowitz, P. Cieslak, M. Cetron, E. Zell, J. Jorgensen, and A. Schuchat.** 2000. Increasing prevalence of multidrug-resistant *Streptococcus penumoniae* in the United States. *N. Engl. J. Med.* **343**:1917–1924.
137. **Winston, D. J., C. Emmanouilides, A. Kroeber, J. Hindler, D. A. Bruckner, M. C. Territo, and R. W. Busuttil.** 2000. Quinupristin-dalfopristin therapy for vancomycin-resistant *Enterococcus faecium*. *Clin. Infect. Dis.* **30**:790–797.
138. **World Health Organization.** 1997. W.H.O. report on the tuberculosis epidemic. World Health Organization, Geneva, Switzerland.
139. **Wunderink, R. G., J. Rello, and S. K. Cammarata.** 2003. Linezolid vs vancomycin: analysis of two double-blind studies of patients with methicillin-resistant *Staphylococcus aureus* nosocomial pneumonia. *Chest* **124**:1789–1797.
140. **Yu, V. L., C. C. Chiou, C. Feldman, A. Ortqvist, J. Rello, A. J. Morris, L. M. Baddour, C. M. Luna, D. R. Snydman, M. Ip, W. C. Ko, M. B. Chedid, A. Andremont, K. P. Klugman, and International Pneumococcal Study Group.** 2003. An international prospective study of pneumococcal bacteremia: correlation with in vitro resistance, antibiotics administered, and clinical outcome. *Clin. Infect. Dis.* **37**:230–237.

Frontiers in Antimicrobial Resistance: a Tribute to Stuart B. Levy
Edited by D. G. White, M. N. Alekshun, and P. F. McDermott

Chapter 30

Drug-Resistant Falciparum Malaria: Mechanisms, Consequences, and Challenges

KAREN HAYTON, RICK M. FAIRHURST, BRONWEN NAUDÉ, XIN-ZHUAN SU, AND THOMAS E. WELLEMS

Malaria is the major parasitic disease in tropical developing countries, causing approximately 300 to 500 million febrile illnesses and at least 1.2 million deaths each year (131). Almost half the world's population runs a significant risk of infection. The disease is caused by species of *Plasmodium* protozoa, with most of the morbidity and mortality attributed to *P. falciparum* in Africa. Current antimalarial drugs and mosquito control programs constitute important measures against malaria, but these have limitations that must be met by research to develop new drugs, vaccines, and insecticides. The spread of drug-resistant malaria is a major obstacle for malaria control (125).

MALARIA MORBIDITY AND MORTALITY IN THE TWENTIETH CENTURY

The incidence of malaria was greatly reduced in the mid-20th century by a worldwide eradication program that employed newly available drugs and insecticides, particularly chloroquine and dichloro-diphenyl-trichloro-ethane (DDT). In Africa, where effective control of malaria transmission by *Anopheles* mosquitoes could not be achieved, widespread use of chloroquine had a major impact on public health, particularly in young children, who suffer most from *P. falciparum*. As chloroquine became increasingly available in the 1970s, death rates from malaria in Africa began to drop and approached half the level of pre-chloroquine years (Fig. 1A) (16). Further progress proved impossible, however, and the incidence of *P. falciparum* malaria in Africa began to increase again. The spread of chloroquine-resistant (CQR) *P. falciparum* strains was a critical factor in this resurgence (16). In the 1980s and 1990s, malaria death rates jumped dramatically as the widespread use of chloroquine led to selection of CQR parasites (Fig. 1B). The impact of chloroquine resistance proved especially devastating in young children, who lack the partially protective immunity that usually develops after repeated parasite infections (113).

THE *PLASMODIUM* LIFE CYCLE

Malaria parasites belong to the Apicomplexa, a phylum consisting entirely of parasitic protozoa characterized by the presence of an apical complex used by invasive stages to enter host cells. Four malaria parasite species cause human malaria: *P. falciparum*, *P. vivax*, *P. malariae*, and *P. ovale*. These parasites are transmitted by female *Anopheles* mosquitoes that are infected while feeding on humans with parasites in their bloodstream (Fig. 2). Malaria parasite sexual stages (gametocytes) emerge from their host red blood cells in the mosquito midgut and become male and female gametes; these then fuse to form a zygote that differentiates into a motile ookinete. The ookinete passes through the midgut wall and encysts on its outer surface, where it develops into an oocyst. After 12 to 15 days, the oocyst ruptures, releasing 1,000 to 10,000 sporozoites that migrate to the salivary glands. Transmission to a subsequent human host occurs when the mosquito takes another blood meal and injects sporozoites along with saliva. Sporozoites injected by mosquitoes travel to the human liver, where they infect hepatocytes and mature into schizonts capable of

Karen Hayton, Rick M. Fairhurst, Bronwen Naudé, Xin-zhuan Su, and Thomas E. Wellems • Laboratory of Malaria and Vector Research, National Institute of Allergy and Infectious Diseases, National Institutes of Health, Bethesda, MD 20892.

A

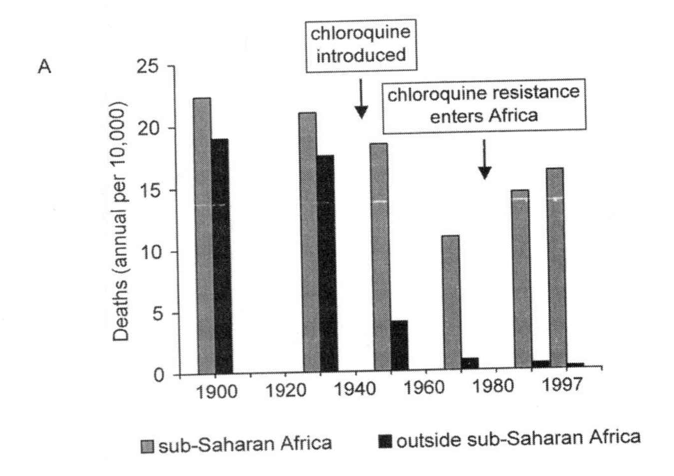

B

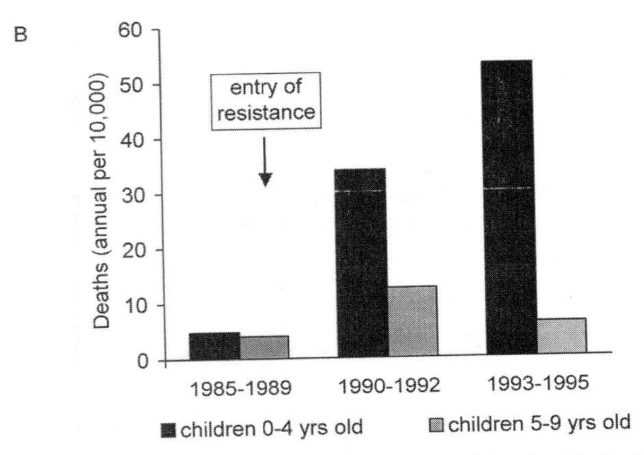

Figure 1. Changes in malaria death rates associated with the introduction of chloroquine and subsequent evolution of chloroquine-resistant malaria strains. (A) Malaria death rates in the 20th century. Dramatic reductions in mortality rates have been achieved outside sub-Saharan Africa. Mortality rates declined after the introduction of chloroquine but have risen again after the spread of chloroquine-resistant parasites across the continent. (Adapted from reference 16). (B) Rise in mortality among children in the village of Mlomp, Senegal. Increased death rates associated with chloroquine resistance occurred chiefly among children <5 years old, the most susceptible age group in highly endemic areas. (Adapted from reference 112.)

releasing 5,000 to 30,000 merozoites into the bloodstream after about 1 to 2 weeks, depending on the species. Upon erythrocyte invasion, merozoites follow a development path of ring stages, trophozoites, and schizonts that divide to produce new merozoites. The asexual erythrocytic stages are propagated by the periodic release (every 48 h for *P. falciparum*) of merozoites that reinvade fresh erythrocytes, a process that establishes a persistent infection, destroys large numbers of erythrocytes, and produces the symptoms of malaria. A subset of merozoites differentiates into gametocytes that infect another mosquito and complete the life cycle.

CHLOROQUINE-RESISTANT *P. FALCIPARUM*

Chloroquine, a synthetic 4-aminoquinoline derivative, is one of the safest, least expensive, and most widely available antimalarial drugs. Shortly after its designation as the drug of choice against malaria at the end of World War II, chloroquine was distributed on a massive scale, to the point that it was even incorporated into the salt distribution programs of some countries. *P. falciparum* infections resistant to chloroquine treatment were reported in the early 1960s in Southeast Asia and South America (48, 72). CQR parasites spread steadily from these locations throughout South America, Asia, and India, arriving in Kenya and Tanzania in the late 1970s (Fig. 3). Chloroquine resistance subsequently swept across Africa in the 1980s and 1990s (8). Independent foci of resistance also occurred in Papua New Guinea and in the Philippines (19, 130). Today, CQR parasites are absent only from some malarious areas of Central America and the Caribbean (128).

Chloroquine acts on erythrocytic-stage malaria parasites that digest host cell hemoglobin. Hemoglobin digestion releases ferriprotoporphyrin IX (heme), a toxic molecule that the parasite normally polymerizes into innocuous hemozoin crystals and sequesters inside its digestive food vacuole (35). Chloroquine interferes with this polymerization process, forming complexes with heme that are toxic to the parasite (22, 134).

CQR *P. falciparum* accumulate less chloroquine than their sensitive counterparts (42). Cell-free extracts prepared from both chloroquine-sensitive and -resistant *P. falciparum* bind similar amounts of drug, however, indicating that resistant parasites possess a mechanism that reduces the interaction of chloroquine with heme (13). The exact mechanism remains unclear but has been variously proposed to involve (i) rapid energy-dependent efflux of chloroquine from resistant but not sensitive parasites (55, 56); (ii) reduced import of chloroquine into resistant relative to sensitive parasites (12, 101); and (iii) increased acidification of the digestive vacuole that may reduce drug accumulation and control toxicity by decreasing the efficiency of the chloroquine-hematin interaction (38).

The Determinant of Chloroquine Resistance, *Pf*CRT

Molecular genetic investigations have definitively linked chloroquine resistance to amino acid substitutions in *Pf*CRT, a putative transporter protein located in the digestive vacuole or membrane (24, 41). Comparisons of the *pfcrt* sequences in chloroquine-sensitive and -resistant *P. falciparum* have shown that resistant isolates obtained from different geographic regions universally carry a key amino acid change, Lys76Thr, in the context of accompanying substitutions else-

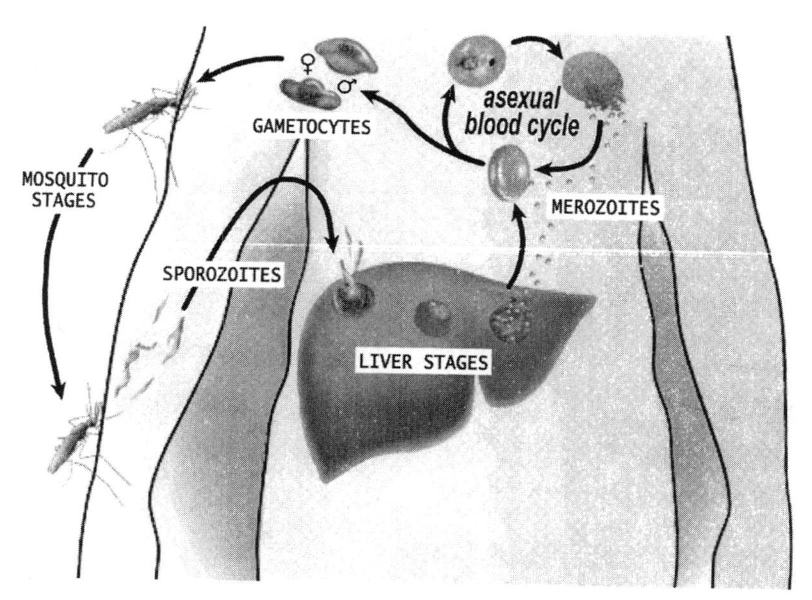

Figure 2. The malaria parasite life cycle (see text for descriptions).

where in *Pf*CRT (19, 30, 41, 123). Allelic replacement studies have demonstrated that chloroquine-sensitive parasites can be converted to the resistant phenotype using DNA transfection methods that introduce codons for these mutations into the expressed sequences of the wild-type *pfcrt* gene (105). The critical role of the Lys76Thr substitution to the chloroquine resistance mechanism is supported by a number of clinical outcome studies that show a clear association of the Lys76Thr marker with chloroquine failures after treatment (123). Loss of the positive charge at *Pf*CRT position 76 appears to be a critical feature of the chloroquine resistance mechanism (24).

The association of accompanying mutations with the Lys76Thr substitution in all foci of chloroquine resistance suggests their involvement in certain accommodative or compensatory changes supporting a critical function of the *Pf*CRT transporter (41). Furthermore, the number and position of these additional mutations and of coselected microsatellite markers in the chromosome segment carrying *pfcrt* have identified at least five chloroquine resistance alleles: two in South America, one in Southeast Asia that subsequently spread through Africa, one or more in Papua New Guinea, and one in the Philippines (Fig. 3) (19, 69, 130). Massive use of chloroquine has driven population sweeps of CQR *P. falciparum* from these locations into most malarious areas, providing a dramatic example of the powerful evolutionary selective force that can be exerted by drug pressure.

Other Genes That May Have a Role in Resistance

Because chloroquine resistance in *P. falciparum* can be partially reversed by the calcium channel blocker verapamil, it was initially proposed that

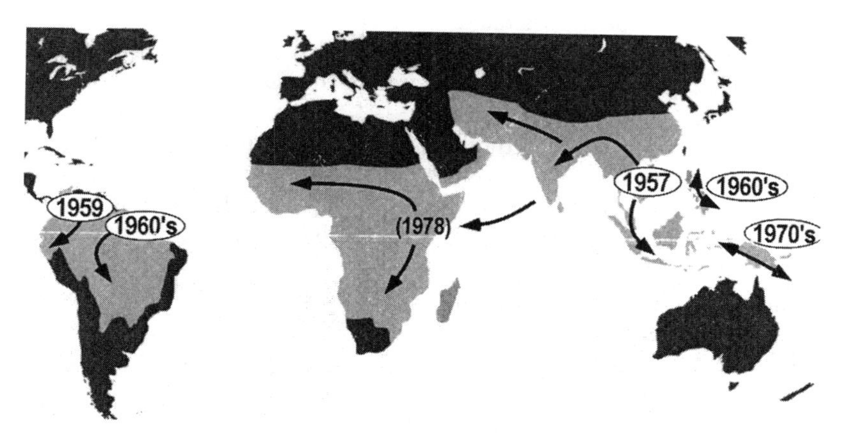

Figure 3. Spread of chloroquine-resistant *P. falciparum* from at least five independent origins under drug pressure.

resistance involved an efflux pump similar to that of ATP-dependent P-glycoproteins found in multidrug-resistant mammalian cancer cells (7, 56, 65, 133). Some studies have provided evidence for an incomplete association between chloroquine resistance and mutations and/or amplification of a gene encoding a P-glycoprotein homologue, *pfmdr1* (26, 37, 45, 126), while others have not (36, 122). In addition to *Pf*CRT, other transporter molecules have been proposed to modulate or contribute to the ability of CQR parasites to decrease chloroquine accumulation (73). Additional field studies will be required to evaluate mutations in these molecules for effects on treatment outcomes and distinguish these effects from other influences on in vitro growth efficiency and drug assay results in the laboratory.

Reemergence of Chloroquine-Sensitive Parasites

Chloroquine-sensitive strains may repopulate regions of CQR malaria when selective pressures from drug administration are removed. Significant drops in treatment failure rates have been reported from China (Hainan Island) and Gabon 5 to 10 years after treatment policy was switched from chloroquine (63, 103). In Malawi, which switched from chloroquine to sulfadoxine-pyrimethamine (SP) in 1993, the prevalence of *Pf*CRT Lys76Thr declined from 85 to 13% by 2000. In the following year, chloroquine cleared 100% of 63 asymptomatic malaria infections, and no infections with the resistant genotype were detected (58). These results suggest an advantage of the native *Pf*CRT over its mutant forms in the absence of drug pressure and point to the possible value of drug rotation strategies in antimalarial policies.

Role of Immunity in Clearance of Drug-Resistant Parasites

The identification of Lys76Thr as a molecular marker for chloroquine resistance made it possible to determine the frequency of individuals carrying CQR parasites prior to receiving drug treatment. In regions of Mali, approximately 40 to 60% of malaria patients harbored CQR parasites, while only ~15% actually failed chloroquine treatment (30, 31). This in vivo clearance of CQR parasites after chloroquine treatment is attributable to host factors such as immunity. Indeed, the ability to clear CQR parasites improves with increasing age, a surrogate for protective immunity (31, 33). Children aged 1 to 5 years were much less likely than older children to clear CQR parasites. These data are consistent with epidemiologic observations on the natural acquisition of antimalaria immunity, as evidenced by increased spontaneous clearances of parasitemia, decreased mortality rates, and increased rates of asymptomatic parasitemia in children aged greater than 5 years (71).

RESISTANCE TO OTHER QUINOLINE DRUGS

Quinine Resistance

Quinine remains an essential drug for the treatment of life-threatening malaria. It does not currently face high levels of resistance and remains effective in all malarious regions, although the spread of increasingly refractory strains has required greater dosing and combination with drugs such as tetracycline or clindamycin to completely clear infections (94). The unpleasant side effects of quinine (particularly cinchonism) and its high cost limit its use to only a small range of cases over which chloroquine was once so successful.

Quinine is an extract of cinchona tree bark, a malaria remedy discovered in South America and brought to Europe in the 1600s. Quinine in its intravenous form is used outside the United States against severe forms of malaria; in the United States, only quinidine is available for intravenous use against severe malaria.

Decreased sensitivity to quinine treatment has been reported in Southeast Asia, South America, and more recently in Africa (28, 47, 95). The molecular mechanisms determining *P. falciparum* quinine response levels in these different regions are complex and probably involve the effects of multiple independent genes (73). Involvement of *pfcrt* as one of these genes is supported by studies showing that substitutions of Lys with Ile or Asn at amino acid position 76 of the transporter change parasite responses to both chloroquine and quinine (24). In some cases, the antimalarial effects of chloroquine and quinine have been antagonistic (107). Other reports have described decreased sensitivities to quinine associated with decreased sensitivities to mefloquine, halofantrine, and chloroquine (4, 10, 11, 120).

Quantitative trait loci (QTL) analysis of a genetic cross between *P. falciparum* clones has distinguished the additive effects of multiple genes in quinine response, including *pfcrt, pfmdr1,* and a third gene that is predicted to encode a sodium-hydrogen exchanger (40). Some of these genes may be subject to pairwise modulatory effects from additional loci (40). The requirement of changes in multiple genes, some of which could be under strong functional constraint, may partly explain why quinine, unlike chloroquine, has still not encountered fully resistant *P. falciparum* and remains effective for malaria treatment despite centuries of use.

Mefloquine Resistance

Mefloquine, a quinoline-methanol derivative introduced in the late 1970s, remains effective for prophylaxis as well as treatment of uncomplicated *P. falciparum* infections in many malarious regions, including Africa and South America. Resistance to mefloquine, however, arose within 6 years of its introduction in Southeast Asia, and widespread resistance is now present in Thailand, Cambodia, and Vietnam (9, 80, 129).

The molecular mechanism of mefloquine resistance is poorly understood. Overexpression of the *pfmdr1* gene (and possibly another related P-glycoprotein, *pfmdr2*) has in some cases been associated with decreased sensitivity to mefloquine (5, 17, 90, 93, 127), and *P. falciparum* clones selected for mefloquine resistance in vitro exhibited amplification and overexpression of *pfmdr1* (25, 86). Mefloquine resistance has also been found to occur in the absence of amplification or increased expression of *pfmdr1*, suggesting the involvement of multiple genes in resistance (62, 99). Artemisinin treatment in vivo selected parasites that have increased resistance to mefloquine yet contain the wild-type *pfmdr1* gene (78). Selection of the wild-type *pfmdr1* in these parasites suggests that this gene can play a role in parasite response to artemisinin and mefloquine. The involvement of *pfmdr1* in mefloquine response is supported by transfection experiments that showed different allelic forms conferred various levels of susceptibility to mefloquine (97). Different balances of *pfmdr1* allele frequencies may therefore be maintained in Southeast Asian parasite populations, depending upon local levels of chloroquine and mefloquine use. Evidence for reciprocal changes in parasite susceptibility to different antimalarial drugs may reflect underlying relationships in response mechanisms that are of potential importance to chemotherapy.

P. FALCIPARUM RESISTANCE TO ANTIFOLATE DRUGS

When pyrimethamine was first introduced as monotherapy against malaria, its use was followed immediately by resistance to treatment in several locations (87); however, joint administration of pyrimethamine plus sulfadoxine later resulted in a much-reduced incidence of resistance and a high level of synergy (98). Nevertheless, resistant strains were eventually selected after the widespread use of the sulfadoxine-pyrimethamine (SP) combination (87). SP resistance spread across Thailand in the 1970s (21, 87, 96), and by the mid-1980s had spread throughout many areas of South America and Southeast Asia. Currently, resistance to SP is consolidating in Africa (100). This creates serious difficulties for malaria control in African countries that have turned to SP as the only affordable alternative to chloroquine (125, 128).

Pyrimethamine and sulfadoxine are structurally similar to dihydrofolate and *p*-aminobenzoic acid (PABA), respectively, and act as competitive inhibitors of the dihydrofolate reductase (DHFR) and dihydropteroate synthase (DHPS) enzymes of *P. falciparum*. As these enzymes function in the folate pathway, their inhibition results in depletion of tetrahydrofolate, an essential cofactor for one-carbon metabolism, thereby inhibiting synthesis of DNA and some amino acids. Other DHFR inhibitors such as cycloguanil or chlorcycloguanil (active metabolites of proguanil and chlorproguanil, respectively) and DHPS inhibitors such as dapsone are also used in combinations against this pathway and have important roles as alternative antimalarial drugs. Some combinations of these drugs (e.g., chloroproguanil-dapsone) may be useful against SP-resistant *P. falciparum* strains (111).

Pyrimethamine Resistance and Dihydrofolate Reductase Mutations

Molecular genetic studies in the 1980s established point mutations in *P. falciparum* DHFR (*Pf*DHFR, encoded by the gene *pfdhfr*) as the basis of pyrimethamine and cycloguanil resistance in natural parasite isolates (27, 44, 88, 89). Certain amino acid substitutions in the enzyme decrease the binding of inhibitors, so that 500- to 1,000-fold increases in drug concentration are required to inhibit activity (18). Evidence pointing to a critical role of the Ser108Asn substitution was subsequently confirmed in DNA transfection experiments that converted pyrimethamine-sensitive to pyrimethamine-resistant parasites by allelic replacement of the endogenous gene (132). In the presence of Ser108Asn, additional mutations can confer increased levels of resistance to pyrimethamine and also high-grade resistance to cycloguanil and chlorocycloguanil (44, 88, 106). In particular, these include the "triple mutant" form Ser108Asn/Asn51Ile/Cys59Arg, which is highly pyrimethamine resistant, and the "quadruple mutant" form Ser108Asn/Asn51Ile/Cys59Arg/Ile164Leu, which is highly resistant to pyrimethamine and both cycloguanils (3, 106). Molecular markers to detect these triple or quadruple mutant parasites are important predictors of resistance to treatment with SP or chloroproguanil-dapsone (59). A relatively infrequent *Pf*DHFR mutant form, Ser108Thr/Ala16Val, has been found in Southeast Asia and is thought to confer high resistance to cycloguanil but not to pyrimethamine (44, 88).

In a recent population survey of *pfdhfr*-resistant alleles from five countries in Southeast Asia, just one was identified at various stages of sequential mutation accumulation (75). Linkage disequilibrium of markers in the chromosome segment containing the *pfdhfr* gene revealed the spread of a single pyrimethamine-resistant allele throughout *P. falciparum* populations, separated by more than 2,000 km. Resistant *pfdhfr* alleles carrying different combinations of amino acid substitutions at positions 51, 59, 108, and 164 were evidently selected from the same ancestral allele, demonstrating the sequential nature of *pfdhfr* mutation accumulation. Similarly, a SP selective sweep was also demonstrated in Southeast Africa, where resistant alleles with a common ancestral origin were found at two sites 4,000 km apart (100).

Sulfadoxine Resistance and Dihydropteroate Synthase Mutations

Correlations between resistance to sulfadoxine and mutations in *P. falciparum* DHPS (*Pf*DHPS, encoded by the gene *pfdhps*) have been more difficult to demonstrate. This has been due to the difficulties associated with in vitro testing for sulfadoxine and SP resistance, in which PABA and folate in the parasite culture media antagonize drug action (121). However, amino acid differences have been identified in *Pf*DHPS from field isolates (14, 114), and their effect on sulfadoxine response has been demonstrated through the expression of functional *Pf*DHPS enzymes containing various combinations of the Ser436Ala/Phe, Ala437Gly, Lys540Glu, Ala581Gly, and Ala613Ser substitutions (115). These altered *Pf*DHPS enzymes decrease sulfadoxine inhibition of enzyme activity by up to three orders of magnitude. Allelic replacements of wild-type *pfdhps* in sensitive parasites with various mutant alleles have confirmed that mutant *pfdhps* confers increased resistance to sulfadoxine (116).

Evolution of Resistance to the Sulfadoxine-Pyrimethamine Combination

A number of in vivo studies have examined *pfdhfr* and *pfdhps* genotypes and their relationship to SP therapeutic failures (6, 59, 82–84, 119). Results from some of these studies suggest that *pfdhfr* mutations arise first under SP selection pressure and that *pfdhps* mutations subsequently increase in prevalence (82, 83, 100). Parasites with the "triple mutant" Ser108Asn/Asn51Ile/Cys59Arg form of *Pf*DHFR and a two-mutant form of *Pf*DHPS (Ala437Gly/Lys540Glu) are associated strongly with SP failure (59). In some studies, mutations in *pfdhps* and *pfdhfr* were a poor predictor of clinical outcome (6, 82, 84), but a lack of correlation between in vivo data and predictions based on genetic information is not uncommon, especially in areas of high transmission, due to the influence of host immunity. A population survey of microsatellite markers flanking the *pfdhfr* and *pfdhps* genes in Southeast African isolates showed that only three *pfdhfr* alleles and just one *pfdhps* allele were selected and driven through *P. falciparum* populations up to 4,000 km apart (100). Furthermore, the dominant triple mutant *pfdhfr* alleles have shared ancestry, indicating that it has arisen only once. These findings provide additional evidence that both *pfdhfr* and *pfdhps* have undergone selection driven by the use of SP combination therapy.

RESISTANCE TO ATOVAQUONE-PROGUANIL

Atovaquone binds to parasite cytochrome *b* and inhibits mitochondrial electron transport, collapsing the mitochondrial membrane potential (108, 109). Atovaquone monotherapy resulted in high treatment failure rates and rapidly selected for resistance (20, 64) caused by a Tyr268Ser substitution in cytochrome *b* (54, 108). Selection of resistance was reduced by partnering atovaquone with proguanil, a combination that was also synergistic (15). This synergy is due to the enhancement of the atovaquone-induced mitochondrial membrane potential collapse by proguanil and is achieved independently from the activity of its metabolite, cycloguanil (110). Although the high cost of atovaquone-proguanil precludes routine use in endemic areas, it is an important prophylactic for nonimmune travelers visiting malaria endemic areas; however, failure of atovaquone-proguanil to clear *P. falciparum* parasites acquired by travelers has recently been reported (43).

TETRACYCLINES

The antimalarial efficacy of the tetracyclines has been exploited for over 30 years (23). Doxycycline is effective for malaria chemoprophylaxis and is often used in combination with other drugs for the treatment of CQR *P. falciparum* infections. Resistance to the tetracyclines in *P. falciparum* has not been reported. However, doxycycline causes gastrointestinal upset, esophageal ulcers, and sun sensitivity and commonly results in vaginal candidiasis. These side effects and the need for daily dosing often result in poor compliance (46). Some studies have suggested that these adverse effects can be minimized using a monohydrate doxycycline salt (85). The use of tetracycline is contraindicated in children <8 years old and in pregnant and breastfeeding women because of effects on bone and tooth development. Clindamycin may be a safe and effective substitute in these patients (60). Novel tetracycline derivatives with reduced side effects and increased efficacy against *P. falciparum* are of keen interest in antimalarial drug discovery (http://www.mmv.org). Such derivatives, especially if they have longer therapeutic half-lives than current tetracyclines, may offer significant advantages for both malaria treatment and prophylaxis.

THE FUTURE OF ANTIMALARIAL CHEMOTHERAPY

Gene flow from relatively few founder events accounts for the distributions of chloroquine- and SP-resistant *P. falciparum* in malarious regions today (75, 100, 130). Strategies of antimalarial use that avoid the low-level, prolonged drug exposure thought to promote resistance, and policies of drug rotation designed to minimize pressures on gene flow may therefore help to extend the useful lives of antimalarial drugs. Implementation of such strategies will require greater ministry-level as well as local attention to the distribution and availability of antimalarial drugs for prophylaxis and treatment and to the ways they are used in programs of mass drug administration.

Mass Drug Administration (MDA)

Malaria control efforts have occasionally employed the administration of antimalarial drugs to whole populations, often in conjunction with insecticides or bed nets (reviewed in reference 117). In one dramatic example, the use of multiple antimalarials and permethrin-treated bed nets eradicated malaria from Aneityum, Vanuatu, a Pacific island of 700 inhabitants (53). While MDA programs have not stopped malaria transmission in general, they have been shown to reduce parasite prevalence and the incidence of clinical malaria episodes when delivered as full therapeutic courses of treatment (34, 50, 71); however, the durable effect of even fully therapeutic MDA programs is typically short lived. In a recent MDA in The Gambia, where SP plus artesunate was administered to 16,400 individuals (85% of the population), a reduced malaria incidence rate in children was achieved that lasted for only two months (118).

MDA programs that deliver inadequate therapeutic doses of drugs—for example, delivery in the form of medicated salt—run the risk of promoting the emergence of drug-resistant parasites by exposing parasites to subtherapeutic drug levels for prolonged periods. The emergence of chloroquine resistance was probably promoted in regions of Southeast Asia and South America, where chloroquinized salt had been widely distributed. In Cambodia, the administration in 1961 of 77 tons of chloroquinized salt reduced parasite prevalence in the area but was quickly followed in 1962 by the appearance of CQR *P. falciparum* (39). Due to the risk for drug resistance, MDA programs are generally restricted to special circumstances, such as immediate use in outbreak settings or seasonal use in areas with extremely short transmission seasons. The efficacy of MDA under these circumstances has not been studied.

Knowledge, Attitude, and Practice Considerations in Malaria Chemotherapy

Proper administration of effective antimalarial drugs is required for parasitologic cure and to minimize the spread of drug-resistant parasites. To this end, the knowledge, attitudes, and practices of the drug consumer are very important. Consumers may seek out ineffective treatments that provide only symptomatic relief (e.g., aspirin or acetaminophen, local herbal remedies) or use effective treatments (e.g., chloroquine) at substandard doses that do not result in parasitologic cure and select drug-resistant parasites (32). Other scenarios that contribute to ineffective malaria treatment include the inability of consumers to recall the correct antimalarial regimen immediately after leaving a physician's office, the common practice of purchasing antimalarial agents from sources that do not provide correct instructions regarding proper dosing, and the tendency of patients not to complete courses of antimalarial drugs because they wish to conserve their supply (29, 32). The difficulty of such problems provides an impetus for the development and use of therapies that can be given in a single dose to cure blood-stage infections.

Ineffective Drug Preparations and Counterfeits

Interventions to ensure the proper use, quality, and availability of antimalarial drugs are as important to the future of malaria chemotherapy as new drug development. Fake or substandard quality drugs sold in national and international markets cause devastating problems. In a recent study, 38% of chloroquine, 74% of quinine, and 12% of SP samples obtained from 132 unofficial vendors in Cameroon were found to have either no active ingredient, an insufficient active ingredient, the wrong active ingredient, or unknown ingredients (2). In 2001, 38% of oral artesunate shop-bought in Vietnam, Cambodia, Laos, and Myanmar did not contain active drug (76). Counterfeiters package their products to closely resemble those made by brand-name manufacturers, even reproducing unique product identifiers (e.g., holograms) meant to confirm the authenticity of the product. Stronger policing is urgently needed to prevent such drug counterfeiters from selling fake antimalarial drugs on international markets (77).

Presumptive Intermittent Treatment

Presumptive intermittent treatment (PIT; also known as intermittent preventative treatment, or IPT) is a form of MDA by which full therapeutic courses of antimalarial drugs are periodically administered to selected subpopulations (typically infants and pregnant women) who are at increased risk of developing severe complications of malaria in high-transmission areas. PIT is administered to all individuals in these subpopulations, without performing diagnostic tests to check whether or not they are parasitemic. Although some individuals are treated unnecessarily, this is more than balanced by the low rates of adverse drug events and the overall public health benefits of malaria control.

PIT may have particular value in infants who suffer multiple episodes of malaria and run a high risk of life-threatening complications. The interval between therapeutic courses in PIT must not, however, be too brief. Infants receiving weekly administration of antimalarial drugs were protected against malarial fevers and anemia but appeared not to develop sufficient natural immunity (61, 70); however, less frequent PIT of infants at the time of routine vaccinations (e.g., amodiaquine every 2 months over a period of 6 months) reduced malaria morbidity by 50 to 65% during the first year of life while still allowing the development of naturally acquired immunity from sufficient exposure to parasites (66, 102). In one study of pregnant women (primigravidae), PIT of one to three SP doses was shown in a randomized, placebo-controlled trial to significantly reduce the incidence of maternal anemia and low birth weight (104).

Artemisinin Drugs and Artemisinin Combination Therapy (ACT)

Treatment of malaria increasingly relies on artemisinin derivatives such as artesunate, artemether, and dihydroartemisinin as previously reliable antimalarial drugs fail against drug-resistant strains (79). In some malarious regions—for example, the borders of Myanmar and Cambodia with Thailand—*P. falciparum* strains progressively acquired resistance to chloroquine, antifolate drugs, low- and high-dose mefloquine monotherapy, and mefloquine-antifolate combinations and also became increasingly refractory to quinine (Fig. 4). On this background of drug resistance, the

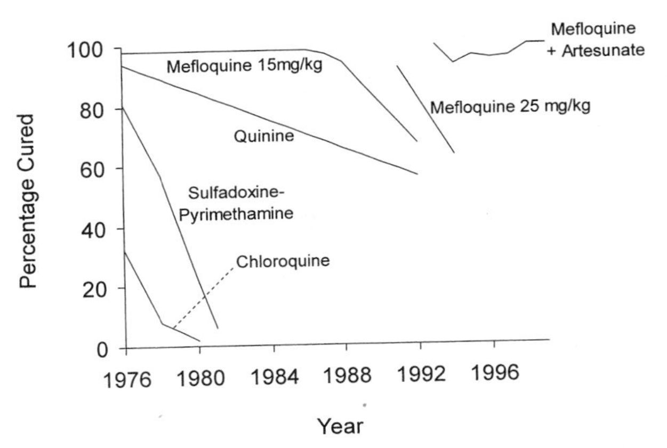

Figure 4. The decline of antimalarial efficacy in Thailand. (Adapted from references 81 and 124.)

wide distribution and use of artemisinin drugs has fortunately not been followed by reports of treatment failures. Artemisinin derivatives have excellent safety records and efficacy (1, 67, 91). A review of 23 trials available in the Cochrane library agreed with conclusions that artemisinin derivatives are at least as effective as quinine for the treatment of severe malaria (68).

Therapeutic levels of artemisinin derivatives are not sustained for significant periods because of their short elimination half-lives in vivo: 1 h for artesunate, 2 h for dihydroartemisinin, and 4 to 5 h for artemether (reviewed in reference 92). Frequent dosing and completion of at least a 7-day course of therapy have been found to be required for parasite eradication (67). Rapid and substantial reduction of parasite burden, along with dramatic clinical improvement, is typically achieved with a 3-day course of therapy, but infections often recrudesce within a few weeks. The likelihood of such recrudescence after a particular course of therapy is thought to depend on initial parasitemia. For example, in one study, recrudescences occurred 9-fold more frequently in patients with parasite densities $>10,000/\mu l$ compared with patients with lower parasite densities (52).

Antimalarial drugs of different classes can ensure complete clearance of infection and no recrudescence when given in combination with 3-day artemisinin treatment (51). Partner drugs with longer half-lives (e.g., piperaquine, mefloquine) may have an advantage in this regard (51). Artemisinin combination therapy (ACT) is also advocated to delay or prevent the development of resistance to either drug in the combination. The validity of this approach has been confirmed in Thailand, where the combination of artesunate plus mefloquine may have halted the progression of mefloquine resistance (81). Measures to prevent the development of artemisinin resistance are particularly important given the value of artemisinin derivatives in the treatment of severe malaria and multidrug-resistant uncomplicated malaria.

Policy decisions to adopt ACT are being increasingly implemented in Africa in the wake of chloroquine and SP resistance (74). Dihydroartemisinin-piperaquine (49) and artemether-lumefantrine (57) are promising ACT combinations co-formulated into tablets that have not encountered drug resistance and are registered for use in several countries. These and other ACT combinations are presently still expensive relative to chloroquine, but measures to offset their expense and eventually reduce their cost will assure their ready availability to all patients in a way that will be of great benefit to malaria control and public health.

CONCLUSIONS

Malaria remains a devastating public health problem, particularly in Africa. The ongoing development of drug-resistant *P. falciparum* will continue to play a major role in the morbidity and mortality of this disease. Broad provision of new artemisinin-based combinations, improved attention to the availability, quality, and use of existing antimalarial drugs, and programs to discover and develop new drugs will lead to badly needed improvements in malaria control.

REFERENCES

1. **Arnold, K., T. H. Tran, T. C. Nguyen, H. P. Nguyen, and P. Pham.** 1990. A randomized comparative study of artemisinine (qinghaosu) suppositories and oral quinine in acute falciparum malaria. *Trans. R. Soc. Trop. Med. Hyg.* **84:**499–502.
2. **Basco, L. K.** 2004. Molecular epidemiology of malaria in Cameroon. XIX. Quality of antimalarial drugs used for self-medication. *Am. J. Trop. Med. Hyg.* **70:**245–250.
3. **Basco, L. K., P. E. de Pecoulas, C. M. Wilson, J. Le Bras, and A. Mazabraud.** 1995. Point mutations in the dihydrofolate reductase-thymidylate synthase gene and pyrimethamine and cycloguanil resistance in *Plasmodium falciparum. Mol. Biochem. Parasitol.* **69:**135–138.
4. **Basco, L. K., and J. Le Bras.** 1992. In vitro activity of halofantrine and its relationship to other standard antimalarial drugs against African isolates and clones of *Plasmodium falciparum. Am. J. Trop. Med. Hyg.* **47:**521–527.
5. **Basco, L. K., J. Le Bras, Z. Rhoades, and C. M. Wilson.** 1995. Analysis of *pfmdr1* and drug susceptibility in fresh isolates of *Plasmodium falciparum* from subsaharan Africa. *Mol. Biochem. Parasitol.* **74:**157–166.
6. **Basco, L. K., R. Tahar, and P. Ringwald.** 1998. Molecular basis of in vivo resistance to sulfadoxine-pyrimethamine in African adult patients infected with *Plasmodium falciparum* malaria parasites. *Antimicrob. Agents Chemother.* **42:**1811–1814.
7. **Bayoumi, R. A., H. A. Babiker, and D. E. Arnot.** 1994. Uptake and efflux of chloroquine by chloroquine-resistant *Plasmodium falciparum* clones recently isolated in Africa. *Acta Trop.* **58:**141–149.
8. **Bloland, P. B., E. M. Lackritz, P. N. Kazembe, J. B. O. Were, R. Steketee, and C. C. Campbell.** 1993. Beyond chloroquine: implications of drug resistance for evaluating malaria therapy efficacy and treatment policy in Africa. *J. Infect. Dis.* **167:** 932–937.
9. **Boudreau, E. F., H. K. Webster, K. Pavanand, and L. Thosingha.** 1982. Type II mefloquine resistance in Thailand. *Lancet* **2:**1335.
10. **Brasseur, P., J. Kouamouo, and P. Druilhe.** 1991. Mefloquine-resistant malaria induced by inappropriate quinine regimens? *J. Infect. Dis.* **164:**625–626.
11. **Brasseur, P., J. Kouamouo, R. Moyou-Somo, and P. Druihe.** 1992. Multi-drug resistant malaria in Cameroon in 1987-1988. II. Mefloquine resistance confirmed in vivo and in vitro and its correlation with quinine resistance. *Am. J. Trop. Med. Hyg.* **46:**8–14.
12. **Bray, P. G., R. E. Howells, G. Y. Ritchie, and S. A. Ward.** 1992. Rapid chloroquine efflux phenotype in both chloroquine-sensitive and chloroquine-resistant *Plasmodium falciparum.* A correlation of chloroquine sensitivity with energy-dependent drug accumulation. *Biochem. Pharmacol.* **44:**1317–1324.
13. **Bray, P. G., M. Mungthin, R. G. Ridley, and S. A. Ward.** 1998. Access to hematin: the basis of chloroquine resistance. *Mol. Pharmacol.* **54:**170–179.
14. **Brooks, D. R., P. Wang, M. Read, W. M. Watkins, P. F. G. Sims, and J. E. Hyde.** 1994. Sequence variation of the hydroxymethyldihydropterin pyrophosphate kinase:dihydropteroate synthase gene in lines of the human malaria parasite, *Plasmodium falciparum,* with differing resistance to sulfadoxine. *Eur. J. Biochem.* **224:**397–405.
15. **Canfield, C. J., M. Pudney, and W. E. Gutteridge.** 1995. Interactions of atovaquone with other antimalarial drugs against *Plasmodium falciparum* in vitro. *Exp. Parasitol.* **80:** 373–381.
16. **Carter, R., and K. N. Mendis.** 2002. Evolutionary and historical aspects of the burden of malaria. *Clin. Microbiol. Rev.* **15:**564–594.
17. **Chaiyaroj, S. C., A. Buranakiti, P. Angkasekwinai, S. Looressuwan, and A. F. Cowman.** 1999. Analysis of mefloquine resistance and amplification of *pfmdr1* in multidrug-resistant *Plasmodium falciparum* isolates from Thailand. *Am. J. Trop. Med. Hyg.* **61:**780–783.
18. **Chen, G. X., C. Mueller, M. Wendlinger, and J. W. Zolg.** 1987. Kinetic and molecular properties of the dihydrofolate reductase from pyrimethamine-sensitive and pyrimethamine-resistant clones of the human malaria parasite *Plasmodium falciparum. Mol. Pharmacol.* **31:**430–437.
19. **Chen, N., D. E. Kyle, C. Pasay, E. V. Fowler, J. Baker, J. M. Peters, and Q. Cheng.** 2003. *pfcrt* allelic types with two novel amino acid mutations in chloroquine-resistant Plasmodium falciparum isolates from the Philippines. *Antimicrob. Agents Chemother.* **47:**3500–3505.
20. **Chiodini, P. L., C. P. Conlon, D. B. Hutchinson, J. A. Farquhar, A. P. Hall, T. E. Peto, H. Birley, and D. A. Warrell.** 1995. Evaluation of atovaquone in the treatment of patients with uncomplicated *Plasmodium falciparum* malaria. *J. Antimicrob. Chemother.* **36:**1073–1078.
21. **Chongsuphajaisiddhi, T., A. Subchareon, S. Puangpartk, and Y. Harinasutra.** 1979. Treatment of falciparum malaria in Thai children. *Southeast Asia J. Trop. Med. Public Health.* **12:** 418–421.
22. **Chou, A. C., R. Chevli, and C. D. Fitch.** 1980. Ferriprotoporphyrin IX fulfills the criteria for identification as the

chloroquine receptor of malaria parasites. *Biochemistry* **19:** 1543–1549.

23. **Clyde, D. F., R. M. Miller, H. L. DuPont, and R. B. Hornick.** 1971. Antimalarial effects of tetracyclines in man. *J. Trop. Med. Hyg.* **74:**238–242.
24. **Cooper, R. A., M. T. Ferdig, X. Z. Su, L. M. Ursos, J. Mu, T. Nomura, H. Fujioka, D. A. Fidock, P. D. Roepe, and T. E. Wellems.** 2002. Alternative mutations at position 76 of the vacuolar transmembrane protein *Pf*CRT are associated with chloroquine resistance and unique stereospecific quinine and quinidine responses in *Plasmodium falciparum*. *Mol. Pharmacol.* **61:**35–42.
25. **Cowman, A. F., D. Galatis, and J. K. Thompson.** 1994. Selection for mefloquine resistance in *Plasmodium falciparum* is linked to amplification of the *pfmdr1* gene and cross-resistance to halofantrine and quinine. *Proc. Natl. Acad. Sci. USA* **91:** 1143–1147.
26. **Cowman, A. F., S. Karcz, D. Galatis, and J. G. Culvenor.** 1991. A P-glycoprotein homologue of *Plasmodium falciparum* is localised on the digestive vacuole. *J. Cell Biol.* **113:**1033–1042.
27. **Cowman, A. F., M. J. Morry, B. A. Biggs, G. A. M. Cross, and S. J. Foote.** 1988. Amino acid changes linked to pyrimethamine resistance in the dihydrofolate reductase-thymidylate synthase gene of *Plasmodium falciparum*. *Proc. Natl Acad. Sci. USA* **85:**9109–9113.
28. **Demar, M., and B. Carme.** 2004. *Plasmodium falciparum* in vivo resistance to quinine: description of two RIII responses in French Guiana. *Am. J. Trop. Med. Hyg.* **70:**125–127.
29. **Depoortere, E., J. P. Guthmann, N. Sipilanyambe, E. Nkandu, F. Fermon, S. Balkan, and D. Legros.** 2004. Adherence to the combination of sulphadoxine-pyrimethamine and artesunate in the Maheba refugee settlement, Zambia. *Trop. Med. Int. Health* **9:**62–67.
30. **Djimde, A., O. K. Doumbo, J. F. Cortese, K. Kayentao, S. Doumbo, Y. Diourte, A. Dicko, X. Z. Su, T. Nomura, D. A. Fidock, T. E. Wellems, C. V. Plowe, and D. Coulibaly.** 2001. A molecular marker for chloroquine-resistant falciparum malaria. *N. Engl. J. Med.* **344:**257–263.
31. **Djimde, A., O. K. Doumbo, R. W. Steketee, and C. V. Plowe.** 2001. Application of a molecular marker for surveillance of chloroquine-resistant falciparum malaria. *Lancet* **358:**890–891.
32. **Djimde, A., C. V. Plowe, S. Diop, A. Dicko, T. E. Wellems, and O. Doumbo.** 1998. Use of antimalarial drugs in Mali: policy versus reality. *Am. J. Trop. Med. Hyg.* **59:**376–379.
33. **Djimde, A. A., O. K. Doumbo, O. Traore, A. B. Guindo, K. Kayentao, Y. Diourte, S. Niare-Doumbo, D. Coulibaly, A. K. Kone, Y. Cissoko, M. Tekete, B. Fofana, A. Dicko, D. A. Diallo, T. E. Wellems, D. Kwiatkowski, and C. V. Plowe.** 2003. Clearance of drug-resistant parasites as a model for protective immunity in *Plasmodium falciparum* malaria. *Am. J. Trop. Med. Hyg.* **69:**558–563.
34. **Doi, H., A. Kaneko, W. Panjaitan, and A. Ishii.** 1989. Chemotherapeutic malaria control operation by single dose of Fansidar plus primaquine in North Sumatra, Indonesia. *Southeast Asian J. Trop. Med. Public Health* **20:**341–349.
35. **Dorn, A., S. R. Vippagunta, H. Matile, A. Bubendorf, J. L. Vennerstrom, and R. G. Ridley.** 1998. A comparison and analysis of several ways to promote haematin (haem) polymerisation and an assessment of its initiation in vitro. *Biochem. Pharmacol.* **55:**737–747.
36. **Dorsey, G., M. R. Kamya, A. Singh, and P. J. Rosenthal.** 2001. Polymorphisms in the *Plasmodium falciparum pfcrt* and *pfmdr-1* genes and clinical response to chloroquine in Kampala, Uganda. *J. Infect. Dis.* **183:**1417–1420.
37. **Duraisingh, M. T., P. Jones, I. Sambou, L. von Seidlein, M. Pinder, and D. C. Warhurst.** 2000. The tyrosine-86 allele of the *pfmdr1* gene of *Plasmodium falciparum* is associated with increased sensitivity to the anti-malarials mefloquine and artemisinin. *Mol. Biochem. Parasitol.* **108:**13–23.
38. **Dzekunov, S. M., L. M. B. Ursos, and P. D. Roepe.** 2000. Digestive vacuolar pH of intact intraerythrocytic *Plasmodium falciparum* either sensitive or resistant to chloroquine. *Mol. Biochem. Parasitol.* **101:**107–124.
39. **Eyles, D. E., C. C. Hoo, M. Warren, and A. A. Sandosham.** 1963. *Plasmodium falciparum* resistant to chloroquine in Cambodia. *Am. J. Trop. Med. Hyg.* **12:**840–843.
40. **Ferdig, M. T., R. A. Cooper, J. Mu, B. Deng, D. A. Joy, X. Z. Su, and T. E. Wellems.** 2004. Dissecting the loci of low-level quinine resistance in malaria parasites. *Mol. Microbiol.* **52:** 985–997.
41. **Fidock, D. A., T. Nomura, A. K. Talley, R. A. Cooper, S. M. Dzekunov, M. T. Ferdig, L. M. Ursos, A. B. Sidhu, B. Naude, K. W. Deitsch, X. Z. Su, J. C. Wootton, P. D. Roepe, and T. E. Wellems.** 2000. Mutations in the *Plasmodium falciparum* digestive vacuole transmembrane protein *Pf*CRT and evidence for their role in chloroquine resistance. *Mol. Cell* **6:**861–871.
42. **Fitch, C. D.** 1970. *Plasmodium falciparum* in owl monkeys: drug resistance and chloroquine binding capacity. *Science* **169:**289–290.
43. **Fivelman, Q. L., G. A. Butcher, I. S. Adagu, D. C. Warhurst, and G. Pasvol.** 2002. Malarone treatment failure and in vitro confirmation of resistance of *Plasmodium falciparum* isolate from Lagos, Nigeria. *Malar. J.* **1:**1–4.
44. **Foote, S. J., D. Galatis, and A. F. Cowman.** 1990. Amino acids in the dihydrofolate reductase-thymidylate synthase gene of *Plasmodium falciparum* involved in cycloguanil resistance differ from those involved in pyrimethamine resistance. *Proc. Natl Acad. Sci. USA* **87:**3014–3017.
45. **Foote, S. J., J. K. Thompson, A. F. Cowman, and D. J. Kemp.** 1989. Amplification of the multidrug resistance gene in some chloroquine-resistant isolates of *Plasmodium falciparum*. *Cell* **57:**921–930.
46. **Fungladda, W., E. R. Honrado, K. Thimasarn, D. Kitayaporn, J. Karbwang, P. Kamolratanakul, and R. Masngammueng.** 1998. Compliance with artesunate and quinine + tetracycline treatment of uncomplicated falciparum malaria in Thailand. *Bull. World Health Organ.* **76**(Suppl. 1):59–66.
47. **Giboda, M., and M. B. Denis.** 1988. Response of Kampuchean strains of *Plasmodium falciparum* to antimalarials: in vivo assessment of quinine and quinine plus tetracycline; multiple drug resistance in vitro. *J. Trop. Med. Hyg.* **91:**205–211.
48. **Harinasuta, T., Migasen, S., and Boonag, D.** 1962. Chloroquine resistance in *Plasmodium falciparum* in Thailand. UNESCO First Regional Symposium on Scientific Knowledge of Tropical Parasites, Singapore University, Singapore.
49. **Hien, T. T., C. Dolecek, P. M. Pham, T. D. Nguyen, T. T. Nguyen, H. T. Le, T. H. Dong, T. T. Tran, K. Stepniewska, N. J. White, and J. Farrar.** 2004. Dihydroartemisinin-piperaquine against multidrug-resistant *Plasmodium falciparum* malaria in Vietnam: randomised clinical trial. *Lancet* **363:**18–22.
50. **Hii, J. L., Y. S. Vun, K. F. Chin, R. Chua, S. Tambakau, E. S. Binisol, E. Fernandez, N. Singh, and M. K. Chan.** 1987. The influence of permethrin-impregnated bednets and mass drug administration on the incidence of *Plasmodium falciparum* malaria in children in Sabah, Malaysia. *Med. Vet. Entomol.* **1:**397–407.
51. **International Artemisinin Study Group.** 2004. Artesunate combinations for treatment of malaria: meta-analysis. *Lancet* **363:**9–17.

52. **Ittarat, W., A. L. Pickard, P. Rattanasinganchan, P. Wilairatana, S. Looareesuwan, K. Emery, J. Low, R. Udomsangpetch, and S. R. Meshnick.** 2003. Recrudescence in artesunate-treated patients with falciparum malaria is dependent on parasite burden not on parasite factors. *Am. J. Trop. Med. Hyg.* **68:**147–152.
53. **Kaneko, A., G. Taleo, M. Kalkoa, S. Yamar, T. Kobayakawa, and A. Bjorkman.** 2000. Malaria eradication on islands. *Lancet* **356:**1560–1564.
54. **Korsinczky, M., N. Chen, B. Kotecka, A. Saul, K. Rieckmann, and Q. Cheng.** 2000. Mutations in *Plasmodium falciparum* cytochrome b that are associated with atovaquone resistance are located at a putative drug-binding site. *Antimicrob. Agents Chemother.* **44:**2100–2108.
55. **Krogstad, D. J., I. Y. Gluzman, B. L. Herwaldt, P. H. Schlesinger, and T. E. Wellems.** 1992. Energy dependence of chloroquine accumulation and chloroquine efflux in *Plasmodium falciparum. Biochem. Pharmacol.* **43:**57–62.
56. **Krogstad, D. J., I. Y. Gluzman, D. E. Kyle, A. M. J. Oduola, S. K. Martin, W. K. Milhous, and P. H. Schlesinger.** 1987. Efflux of chloroquine from *Plasmodium falciparum:* mechanism of chloroquine resistance. *Science* **238:**1283–1285.
57. **Krudsood, S., K. Chalermrut, C. Pengruksa, S. Srivilairit, U. Silachamroon, S. Treeprasertsuk, S. Kano, G. M. Brittenham, and S. Looareesuwan.** 2003. Comparative clinical trial of two-fixed combinations dihydroartemisinin-napthoquine-trimethoprim (DNP) and artemether-lumefantrine (Coartem/Riamet) in the treatment of acute uncomplicated falciparum malaria in Thailand. *Southeast Asian J. Trop. Med. Public Health* **34:**316–321.
58. **Kublin, J. G., J. F. Cortese, E. M. Njunju, R. A. Mukadam, J. J. Wirima, P. N. Kazembe, A. A. Djimde, B. Kouriba, T. E. Taylor, and C. V. Plowe.** 2003. Reemergence of chloroquine-sensitive *Plasmodium falciparum* malaria after cessation of chloroquine use in Malawi. *J. Infect. Dis.* **187:**1870–1875.
59. **Kublin, J. G., F. K. Dzinjalamala, D. D. Kamwendo, E. M. Malkin, J. F. Cortese, L. M. Martino, R. A. Mukadam, S. J. Rogerson, A. G. Lescano, M. E. Molyneux, P. A. Winstanley, P. Chimpeni, T. E. Taylor, and C. V. Plowe.** 2002. Molecular markers for failure of sulfadoxine-pyrimethamine and chlorproguanil-dapsone treatment of *Plasmodium falciparum* malaria. *J. Infect. Dis.* **185:**380–388.
60. **Lell, B., and P. G. Kremsner.** 2002. Clindamycin as an antimalarial drug: review of clinical trials. *Antimicrob. Agents Chemother.* **46:**2315–2320.
61. **Lemnge, M. M., H. A. Msangeni, A. M. Ronn, F. M. Salum, P. H. Jakobsen, J. I. Mhina, J. A. Akida, and I. C. Bygbjerg.** 1997. Maloprim malaria prophylaxis in children living in a holoendemic village in north-eastern Tanzania. *Trans. R. Soc. Trop. Med. Hyg.* **91:**68–73.
62. **Lim, A. S. Y., D. Galatis, and A. F. Cowman.** 1996. *Plasmodium falciparum:* amplification and overexpression of *pfmdr1* is not necessary for increased mefloquine resistance. *Exp. Parasitol.* **83:**295–303.
63. **Liu, D. Q., R. J. Liu, D. X. Ren, D. Q. Gao, C. Y. Zhang, C. P. Qui, X. Z. Cai, C. F. Ling, A. H. Song, and X. Tang.** 1995. Changes in the resistance of *Plasmodium falciparum* to chloroquine in Hainan, China. *Bull. World Health Organ.* **73:**483–486.
64. **Looareesuwan, S., C. Viravan, H. K. Webster, D. E. Kyle, D. B. Hutchinson, and C. J. Canfield.** 1996. Clinical studies of atovaquone, alone or in combination with other antimalarial drugs, for treatment of acute uncomplicated malaria in Thailand. *Am. J. Trop. Med. Hyg.* **54:**62–66.
65. **Martin, S. K., A. M. Oduola, and W. K. Milhous.** 1987. Reversal of chloroquine resistance in *Plasmodium falciparum* by verapamil. *Science* **235:**899–901.
66. **Massaga, J. J., A. Y. Kitua, M. M. Lemnge, J. A. Akida, L. N. Malle, A. M. Ronn, T. G. Theander, and I. C. Bygbjerg.** 2003. Effect of intermittent treatment with amodiaquine on anaemia and malarial fevers in infants in Tanzania: a randomised placebo-controlled trial. *Lancet* **361:**1853–1860.
67. **McGready, R., T. Cho, N. K. Keo, K. L. Thwai, L. Villegas, S. Looareesuwan, N. J. White, and F. Nosten.** 2001. Artemisinin antimalarials in pregnancy: a prospective treatment study of 539 episodes of multidrug-resistant *Plasmodium falciparum. Clin. Infect. Dis.* **33:**2009–2016.
68. **McIntosh, H. M., and P. Olliaro.** 2000. Artemisinin derivatives for treating severe malaria. *Cochrane Database Syst. Rev.* CD000527.
69. **Mehlotra, R. K., H. Fujioka, P. D. Roepe, O. Janneh, L. M. Ursos, V. Jacobs-Lorena, D. T. McNamara, M. J. Bockarie, J. W. Kazura, D. E. Kyle, D. A. Fidock, and P. A. Zimmerman.** 2001. Evolution of a unique *Plasmodium falciparum* chloroquine-resistance phenotype in association with *pfcrt* polymorphism in Papua New Guinea and South America. *Proc. Natl Acad. Sci. USA* **98:**12689–12694.
70. **Menendez, C., E. Kahigwa, R. Hirt, P. Vounatsou, J. J. Aponte, F. Font, C. J. Acosta, D. M. Schellenberg, C. M. Galindo, J. Kimario, H. Urassa, B. Brabin, T. A. Smith, A. Y. Kitua, M. Tanner, and P. L. Alonso.** 1997. Randomised placebo-controlled trial of iron supplementation and malaria chemoprophylaxis for prevention of severe anaemia and malaria in Tanzanian infants. *Lancet* **350:**844–850.
71. **Molineaux, L., and G. Gramiccia.** 1980. The Garki Project. World Health Organization, Geneva.
72. **Moore, D. V., and J. E. Lanier.** 1961. Observations on two *Plasmodium falciparum* infections with abnormal response to chloroquine. *Am. J. Trop. Med. Hyg.* **10:**5–9.
73. **Mu, J., M. T. Ferdig, X. Feng, D. A. Joy, J. Duan, T. Furuya, G. Subramanian, L. Aravind, R. A. Cooper, J. C. Wootton, M. Xiong, and X. Z. Su.** 2003. Multiple transporters associated with malaria parasite responses to chloroquine and quinine. *Mol. Microbiol.* **49:**977–989.
74. **Nafo-Traore, F.** 2004. Response to accusations of medical malpractice by WHO and the Global Fund. *Lancet* **363:**397.
75. **Nair, S., J. T. Williams, A. Brockman, L. Paiphun, M. Mayxay, P. N. Newton, J. P. Guthmann, F. M. Smithuis, T. T. Hien, N. J. White, F. Nosten, and T. J. Anderson.** 2003. A selective sweep driven by pyrimethamine treatment in Southeast Asian malaria parasites. *Mol. Biol. Evol.* **20:**1526–1536.
76. **Newton, P., S. Proux, M. Green, F. Smithuis, J. Rozendaal, S. Prakongpan, K. Chotivanich, M. Mayxay, S. Looareesuwan, J. Farrar, F. Nosten, and N. J. White.** 2001. Fake artesunate in Southeast Asia. *Lancet* **357:**1948–1950.
77. **Newton, P. N., A. Dondorp, M. Green, M. Mayxay, and N. J. White.** 2003. Counterfeit artesunate antimalarials in Southeast Asia. *Lancet* **362:**169.
78. **Ngo, T., M. Duraisingh, M. Reed, D. Hipgrave, B. Biggs, and A. F. Cowman.** 2003. Analysis of *pfcrt, pfmdr1, dhfr,* and *dhps* mutations and drug sensitivities in *Plasmodium falciparum* isolates from patients in Vietnam before and after treatment with artemisinin. *Am. J. Trop. Med. Hyg.* **68:**350–356.
79. **Nosten, F., T. T. Hien, and N. J. White.** 1998. Use of artemisinin derivatives for the control of malaria. *Med. Trop.* **58:**45–49.
80. **Nosten, F., F. ter Kuile, T. Chongsuphajaisiddhi, C. Luxemburger, H. K. Webster, M. Edstein, L. Phaipun, K. L. Thew, and N. J. White.** 1991. Mefloquine-resistant falciparum malaria on the Thai-Burmese border. *Lancet* **337:**1140–1143.
81. **Nosten, F., M. van Vugt, R. Price, C. Luxemburger, K. L. Thway, A. Brockman, R. McGready, F. ter Kuile, S. Looa-**

reesuwan, and N. J. White. 2000. Effects of artesunate-mefloquine combination on incidence of *Plasmodium falciparum* malaria and mefloquine resistance in western Thailand: a prospective study. *Lancet* **356:**297–302.

82. **Nzila, A. M., E. K. Mberu, J. Sulo, H. Dayo, P. A. Winstanley, C. H. Sibley, and W. M. Watkins.** 2000. Towards an understanding of the mechanism of pyrimethamine-sulfadoxine resistance in *Plasmodium falciparum:* genotyping of dihydrofolate reductase and dihydropteroate synthase of Kenyan parasites. *Antimicrob. Agents Chemother.* **44:**991–996.
83. **Nzila, A. M., E. Nduati, E. K. Mberu, C. H. Sibley, S. A. Monks, P. A. Winstanley, and W. M. Watkins.** 2000. Molecular evidence of greater selective pressure for drug resistance exerted by the long-acting antifolate pyrimethamine-sulfadoxine compared with the shorter-acting chlorproguanil/dapsone on Kenyan *Plasmodium falciparum. J. Inf. Dis.* **181:**2023–2028.
84. **Omar, S. A., I. S. Adagu, and D. C. Warhurst.** 2001. Can pretreatment screening for *dhps* and *dhfr* point mutations in *Plasmodium falciparum* infections be used to predict sulfadoxine-pyrimethamine treatment failure? *Trans. R. Soc. Trop. Med. Hyg.* **95:**315–319.
85. **Pages, F., J. P. Boutin, J. B. Meynard, A. Keundjian, S. Ryfer, L. Giurato, and D. Baudon.** 2002. Tolerability of doxycycline monohydrate salt vs. chloroquine-proguanil in malaria chemoprophylaxis. *Trop. Med. Int. Health* **7:**919–924.
86. **Peel, S. A., P. Bright, B. Yount, J. Handy, and R. S. Baric.** 1994. A strong association between mefloquine and halofantrine resistance and amplification, overexpression, and mutation in the p-glycoprotein gene homologue (*pfmdr*) of *Plasmodium falciparum in vitro. Am. J. Trop. Med. Hyg.* **51:**648–658.
87. **Peters, W. (ed.).** 1987. *Chemotherapy and Drug Resistance in Malaria*, 2nd ed., vol. 1 and 2. Academic Press, London, United Kingdom.
88. **Peterson, D. S., W. K. Milhous, and T. E. Wellems.** 1990. Molecular basis to differential resistance to cycloguanil and pyrimethamine in falciparum malaria. *Proc. Natl Acad. Sci. USA* **87:**3018–3022.
89. **Peterson, D. S., D. Walliker, and T. E. Wellems.** 1988. Evidence that a point mutation in dihydrofolate reductase-thymidylate synthase confers resistance to pyrimethamine in falciparum malaria. *Proc. Natl Acad. Sci. USA* **85:**9114–9118.
90. **Price, R., G. Robinson, A. Brockman, A. Cowman, and S. Krishna.** 1997. Assessment of *pfmdr1* gene copy number by tandem competitive polymerase chain reaction. *Mol. Biochem. Parasitol.* **85:**161–169.
91. **Price, R., M. van Vugt, L. Phaipun, C. Luxemburger, J. Simpson, R. McGready, F. ter Kuile, A. Kham, T. Chongsuphajaisiddhi, N. J. White, and F. Nosten.** 1999. Adverse effects in patients with acute falciparum malaria treated with artemisinin derivatives. *Am. J. Trop. Med. Hyg.* **60:**547–555.
92. **Price, R. N.** 2000. Artemisinin drugs: novel antimalarial agents. *Expert. Opin. Investig. Drugs* **9:**1815–1827.
93. **Price, R. N., C. Cassar, A. Brockman, M. Duraisingh, M. van Vugt, N. J. White, F. Nosten, and S. Krishna.** 1999. The *pfmdr1* gene is associated with a multidrug-resistant phenotype in *Plasmodium falciparum* from the western border of Thailand. *Antimicrob. Agents Chemother.* **43:**2943–2949.
94. **Pukrittayakamee, S., A. Chantra, S. Vanijanonta, R. Clemens, S. Looareesuwan, and N. J. White.** 2000. Therapeutic responses to quinine and clindamycin in multidrug-resistant falciparum malaria. *Antimicrob. Agents Chemother.* **44:**2395–2398.
95. **Pukrittayakamee, S., W. Supanaranond, S. Looareesuwan, S. Vanijanonta, and N. J. White.** 1994. Quinine in severe falciparum malaria: evidence of declining efficacy in Thailand. *Trans. R. Soc. Trop. Med. Hyg.* **88:**324–327.
96. **Reacher, M., C. C. Campbell, J. Freeman, E. B. Doberstyn, and A. D. Brandling-Bennett.** 1980. Drug therapy for *Plasmodium falciparum* malaria resistant to pyrimethamine-sulphadoxine (Fansidar). A study of alternate regimes in Eastern Thailand. *Lancet* **2:**1066–1069.
97. **Reed, M. B., K. J. Saliba, S. R. Caruana, K. Kirk, and A. F. Cowman.** 2000. Pgh1 modulates sensitivity and resistance to multiple antimalarials in *Plasmodium falciparum. Nature* **403:**906–909.
98. **Richards, W. H. G.** 1966. Antimalarial activity of sulphonamides and a sulphone, singly and in combination with pyrimethamine, against drug resistant and normal strains of laboratory plasmodia. *Nature* **212:**1494–1495.
99. **Ritchie, G. Y., M. Mungthin, J. E. Green, P. G. Bray, S. R. Hawley, and S. A. Ward.** 1996. In vitro selection of halofantrine resistance in *Plasmodium falciparum* is not associated with increased expression of Pgh1. *Mol. Biochem. Parasitol.* **83:**35–46.
100. **Roper, C., R. Pearce, B. Bredenkamp, J. Gumede, C. Drakeley, F. Mosha, D. Chandramohan, and B. Sharp.** 2003. Antifolate antimalarial resistance in southeast Africa: a population-based analysis. *Lancet* **361:**1174–1181.
101. **Sanchez, C. P., S. Wunsch, and M. Lanzer.** 1997. Identification of a chloroquine importer in *Plasmodium falciparum:* differences in import kinetics are genetically linked with the chloroquine-resistant phenotype. *J. Biol. Chem.* **272:**2652–2658.
102. **Schellenberg, D., C. Menendez, E. Kahigwa, J. Aponte, J. Vidal, M. Tanner, H. Mshinda, and P. Alonso.** 2001. Intermittent treatment for malaria and anaemia control at time of routine vaccinations in Tanzanian infants: a randomised, placebo-controlled trial. *Lancet* **357:**1471–1477.
103. **Schwenke, A., C. Brandts, J. Philipps, S. Winkler, W. H. Wernsdorfer, and P. G. Kremsner.** 2001. Declining chloroquine resistance of *Plasmodium falciparum* in Lambarene, Gabon from 1992 to 1998. *Wien. Klin. Wochenschr.* **113:**63–64.
104. **Shulman, C. E., E. K. Dorman, F. Cutts, K. Kawuondo, J. N. Bulmer, N. Peshu, and K. Marsh.** 1999. Intermittent sulphadoxine-pyrimethamine to prevent severe anaemia secondary to malaria in pregnancy: a randomised placebo-controlled trial. *Lancet* **353:**632–636.
105. **Sidhu, A. B., D. Verdier-Pinard, and D. A. Fidock.** 2002. Chloroquine resistance in *Plasmodium falciparum* malaria parasites conferred by *pfcrt* mutations. *Science* **298:**210–213.
106. **Siriwaraporn, W.** 1998. Dihydrofolate reductase and antifolate resistance in malaria. *Drug Resistance Updates* **1:**397–406.
107. **Skinner-Adams, T. and T. M. Davis.** 1999. Synergistic in vitro antimalarial activity of omeprazole and quinine. *Antimicrob. Agents Chemother.* **43:**1304–1306.
108. **Srivastava, I. K., J. M. Morrisey, E. Darrouzet, F. Daldal, and A. B. Vaidya.** 1999. Resistance mutations reveal the atovaquone-binding domain of cytochrome b in malaria parasites. *Mol. Microbiol.* **33:**704–711.
109. **Srivastava, I. K., H. Rottenberg, and A. B. Vaidya.** 1997. Atovaquone, a broad spectrum antiparasitic drug, collapses mitochondrial membrane potential in a malarial parasite. *J. Biol. Chem.* **272:**3961–3966.
110. **Srivastava, I. K., and A. B. Vaidya.** 1999. A mechanism for the synergistic antimalarial action of atovaquone and proguanil. *Antimicrob. Agents Chemother.* **43:**1334–1339.
111. **Sulo, J., P. Chimpeni, J. Hatcher, J. G. Kublin, C. V. Plowe, M. E. Molyneux, K. Marsh, T. E. Taylor, W. M. Watkins, and P. A. Winstanley.** 2002. Chlorproguanil-dapsone versus sulfadoxine-pyrimethamine for sequential episodes of un-

complicated falciparum malaria in Kenya and Malawi: a randomised clinical trial. *Lancet* **360**:1136–1143.

112. **Trape, J. F., G. Pison, M. P. Preziosi, C. Enel, L. A. Desgrees du, V. Delaunay, B. Samb, E. Lagarde, J. F. Molez, and F. Simondon.** 1998. Impact of chloroquine resistance on malaria mortality. *C. R. Acad. Sci. III* **321**:689–697.
113. **Trape, J. F., C. Rogier, L. Konate, N. Diagne, H. Bouganali, B. Canque, F. Legros, A. Badji, G. Ndiaye, P. Ndiaye, K. Brahimi, O. Faye, P. Druilhe, and L. P. daSilva.** 1994. The Dielmo project: a longitudinal study of natural malaria infection and the mechanisms of protective immunity in a community living in a holoendemic area of Senegal. *Am. J. Trop. Med. Hyg.* **51**:123–137.
114. **Triglia, T., and A. F. Cowman.** 1994. Primary structure and expression of the dihydropteroate synthase gene of *Plasmodium falciparum*. *Proc. Natl Acad. Sci. USA* **91**:7149–7153.
115. **Triglia, T., J. G. T. Menting, C. Wilson, and A. F. Cowman.** 1997. Mutations in dihydropteroate synthase are responsible for sulfone and sulfonamide resistance in *Plasmodium falciparum*. *Proc. Natl Acad. Sci. USA* **94**:13944–13949.
116. **Triglia, T., P. Wang, P. F. G. Sims, J. E. Hyde, and A. F. Cowman.** 1998. Allelic exchange at the endogenous genomic locus in *Plasmodium falciparum* proves the role of dihydropteroate synthase in sulfadoxine-resistant malaria. *EMBO J.* **17**:3807–3815.
117. **von Seidlein, L., and B. M. Greenwood.** 2003. Mass administrations of antimalarial drugs. *Trends Parasitol.* **19**:452–460.
118. **von Seidlein, L., G. Walraven, P. J. Milligan, N. Alexander, F. Manneh, J. L. Deen, R. Coleman, M. Jawara, S. W. Lindsay, C. Drakeley, S. De Martin, P. Olliaro, S. Bennett, M. Schim van der Loeff, K. Okunoye, G. A. Targett, K. P. McAdam, J. F. Doherty, B. M. Greenwood, and M. Pinder.** 2003. The effect of mass administration of sulfadoxine-pyrimethamine combined with artesunate on malaria incidence: a double-blind, community-randomized, placebo-controlled trial in The Gambia. *Trans. R. Soc. Trop. Med. Hyg.* **97**:217–225.
119. **Wang, P., C.-S. Lee, R. Bayoumi, A. Djimde, O. Doumbo, G. Swedberg, L. Duc Dao, H. Mshinda, M. Tanner, W. M. Watkins, P. F. G. Sims, and J. E. Hyde.** 1997. Resistance to antifolates in *Plasmodium falciparum* monitored by sequence analysis of dihydropteroate synthase and dihydrofolate reductase alleles in a large number of field samples of diverse origins. *Mol. Biochem. Parasitol.* **89**:161–177.
120. **Warsame, M., W. H. Wernsdorfer, D. Payne, and A. Bjorkman.** 1991. Susceptibility of *Plasmodium falciparum* in vitro to chloroquine, mefloquine, quinine and sulfadoxine/pyrimethamine in Somalia: relationships between the responses to the different drugs. *Trans. R. Soc. Trop. Med. Hyg.* **85**:565–569.
121. **Watkins, W. M., D. G. Sixsmith, J. D. Chulay, and H. C. Spencer.** 1985. Antagonism of sulfadoxine and pyrimethamine antimalarial activity in vitro by *p*-aminobenzoic acid, *p*-aminobenzoylglutamic acid and folic acid. *Mol. Biochem. Parasitol.* **14**:55–61.
122. **Wellems, T. E., L. J. Panton, I. Y. Gluzman, V. E. do Rosario, R. W. Gwadz, A. Walker-Jonah, and D. J. Krogstad.** 1990. Chloroquine resistance not linked to *mdr*-like genes in a *Plasmodium falciparum* cross. *Nature* **345**:253–255.
123. **Wellems, T. E., and C. V. Plowe.** 2001. Chloroquine-resistant malaria. *J. Infect. Dis.* **184**:770–776.
124. **White, N. J.** 1996. Malaria, p. 1087–1164. *In* G. C. Cook (ed.), *Manson's Tropical Diseases.* W. B. Saunders Company Ltd, London, United Kingdom.
125. **White, N. J., F. Nosten, S. Looareesuwan, W. M. Watkins, K. Marsh, R. W. Snow, G. Kokwaro, J. Ouma, T. T. Hien, M. E. Molyneux, T. E. Taylor, C. I. Newbold, T. K. Ruebush II, M. Danis, B. M. Greenwood, R. M. Anderson, and P. L. Olliaro.** 1999. Averting a malaria disaster. *Lancet* **353**:1965–1967.
126. **Wilson, C. M., A. E. Serrano, A. Wasley, M. P. Bogenschutz, A. H. Shankar, and D. F. Wirth.** 1989. Amplification of a gene related to mammalian *mdr* genes in drug-resistant *Plasmodium falciparum*. *Science* **244**:1184–1186.
127. **Wilson, C. M., S. K. Volkman, S. Thaithong, R. K. Martin, D. E. Kyle, W. K. Milhous, and D. F. Wirth.** 1993. Amplification of *pfmdr-1* associated with mefloquine and halofantrine resistance in *Plasmodium falciparum* from Thailand. *Mol. Biochem. Parasitol.* **57**:151–160.
128. **Wongsrichanalai, C., A. L. Pickard, W. H. Wernsdorfer, and S. R. Meshnick.** 2002. Epidemiology of drug-resistant malaria. *Lancet Infect. Dis.* **2**:209–218.
129. **Wongsrichanalai, C., J. Sirichaisinthop, J. J. Karwacki, K. Congpuong, R. S. Miller, L. Pang, and K. Thimasarn.** 2001. Drug resistant malaria on the Thai-Myanmar and Thai-Cambodian borders. *Southeast Asian J. Trop. Med. Public Health* **32**:41–49.
130. **Wootton, J. C., X. Feng, M. T. Ferdig, R. A. Cooper, J. Mu, D. I. Baruch, A. J. Magill, and X. Z. Su.** 2002. Genetic diversity and chloroquine selective sweeps in *Plasmodium falciparum*. *Nature* **418**:320–323.
131. **World Health Organization.** *The World Health Report* 2003. World Health Organization, Geneva, Switzerland.
132. **Wu, Y. M., L. A. Kirkman, and T. E. Wellems.** 1996. Transformation of *Plasmodium falciparum* malaria parasites by homologous integration of plasmids that confer resistance to pyrimethamine. *Proc. Natl Acad. Sci. USA* **93**:1130–1134.
133. **Yayon, A., Z. I. Cabantchik, and H. Ginsburg.** 1984. Identification of the acidic compartment of *Plasmodium falciparum*-infected human erythrocytes as the target of the antimalarial drug chloroquine. *EMBO J.* **3**:2695–2700.
134. **Zhang, Y., and E. Hempelmann.** 1987. Lysis of malarial parasites and erythrocytes by ferriprotoporphyrin IX-chloroquine and the inhibition of this effect by proteins. *Biochem. Pharmacol.* **36**:1267–1273.

Frontiers in Antimicrobial Resistance: a Tribute to Stuart B. Levy
Edited by D. G. White, M. N. Alekshun, and P. F. McDermott

Chapter 31

Antimicrobial Susceptibility Testing of Bacterial Agents of Bioterrorism: Strategies and Considerations

Fred C. Tenover

Bacillus anthracis, Brucella species, *Burkholderia mallei, Burkholderia pseudomallei, Francisella tularensis,* and *Yersinia pestis* are bacterial species that can produce deadly infections; collectively they have caused epic plagues and human misery for centuries (19, 40). More recently, each of these organisms has been identified as a potential weapon for bioterrorism (BT)(http://www.bt.cdc.gov/agent/agentlist.asp). In response to this impending threat, the Centers for Disease Control and Prevention (CDC), the Working Group on Civilian Biodefense, and others have worked diligently to develop countermeasures should any of these organisms be used as a biological weapon (12, 13, 20, 30, 31). Since vaccination is not a major prevention strategy for the general population for most of these disease agents, it is imperative that the antimicrobial susceptibility patterns of the organisms be known so that effective prophylactic or therapeutic treatment can be administered. Many of these organisms have intrinsic resistance to one or more antimicrobials (2, 8, 12, 14, 17, 18, 25, 41, 47, 61, 68). In addition, reports of engineered resistance in particular strains have appeared in the literature (1, 9, 32, 57); thus, the antimicrobial susceptibility patterns of organisms encountered during a purposeful release of a bacterial BT organism are not necessarily predictable. This makes the availability of standardized methods of antimicrobial susceptibility testing of BT organisms a necessity for guiding antimicrobial use.

Many of the BT organisms are fastidious organisms for which standardized susceptibility testing methods have only recently become available through the National Committee for Clinical Laboratory Standards (NCCLS) (recently renamed as the Clinical and Laboratory Standards Institute or CLSI) (10, 50–52). Interpretive criteria for the results of MIC and disk diffusion data have also been defined within the last 2 years. Prior to issuance of these guidelines, a laboratory could generate an MIC result for a given antimicrobial agent, but the interpretation of that result as susceptible, intermediate, or resistant was not standardized. This chapter will explore the progress made over the last three years to establish both standardized susceptibility testing methods and interpretive criteria (i.e., breakpoints) for *B. anthracis, Y. pestis, B. mallei, B. pseudomallei, F. tularensis,* and *Brucella* species. In establishing these methods, NCCLS/CLSI cautions that susceptibility testing of BT agents may generate aerosols (10). Thus, this testing should take place only under BL-3 safety conditions or BL-2 conditions with additional personal protective equipment (65).

BACILLUS ANTHRACIS

B. anthracis is a pathogen of both animals and humans (14, 31). Disease may be localized or disseminated. Disseminated disease is often fatal if not treated appropriately with antimicrobial agents. Two antimicrobial agents recommended by the CDC for therapeutic use in human anthrax cases are ciprofloxacin and doxycycline (7). *B. anthracis* has already been used as a BT weapon in the United States (7, 34). Fortunately, the antimicrobial susceptibility patterns of *B. anthracis* isolates from a variety of countries had been documented prior to the release of the organism in October 2001 (15, 43, 47). Thus, a review of the antimicrobial susceptibility data from the isolates from Florida, New York, and Connecticut confirmed that they had not been engineered to be resistant to antimicrobial agents (47).

Fred C. Tenover • Division of Healthcare Quality Promotion, Centers for Disease Control and Prevention, 1600 Clifton Rd. NE (MS G-08), Atlanta, GA 30333.

Although the antimicrobial susceptibility patterns of hundreds of isolates had been reported in the medical literature, the susceptibility testing methods used in those studies and the criteria used to interpret the results were often unclear (15, 43). For example, Doganay and Aydin (15) determined the susceptibility patterns of 22 isolates of *B. anthracis* using agar dilution and disk diffusion, but the breakpoints they used to interpret the zone diameters and MICs were not given in the report. Similarly, Lightfoot et al. (43), in a comprehensive study of resistance testing of 70 *B. anthracis* isolates, demonstrated variability in MIC patterns of the strains tested, but the authors did not cite a source for the interpretive criteria used. The breakpoints were similar, but not identical, to those promulgated for other gram-positive organisms by NCCLS and the British Society for Antimicrobial Chemotherapy at that time (51, 67).

In 2002, Mohammed et al. reported a study of the susceptibility patterns of *B. anthracis* in which the CLSI broth microdilution reference method and the Etest method were compared (47). The following year, based in part on the data from the study reported by Mohammed et al., NCCLS addressed the issue of methods and interpretive criteria for *B. anthracis* and approved a standardized MIC testing method using cation-adjusted Mueller-Hinton broth in which the test isolates were incubated for 16 to 18 h in ambient air (50). The exact size of the inoculum for the studies was difficult to calculate because the organism often formed chains that were difficult to disperse (47). Nonetheless, a suspension of the organism that was adjusted to match the turbidity of a 0.5 McFarland standard (as is typically prepared for other MIC tests) was recommended by CLSI. Interpretive criteria, using microbial MIC distribution data and data from the medical literature, were used to establish breakpoints for penicillin, ciprofloxacin, doxycycline, and tetracycline, and these criteria appeared in 2003 in CLSI/NCCLS document M100-S13 (50). Recently, Turnbull and colleagues revisited the effects of several variables of susceptibility test results, including inoculum size, growth temperature, and time of incubation (64). No changes in the testing method were recommended as a result of Turnbull's study.

Although a standardized testing method has been published by CLSI for *B. anthracis,* two issues remain controversial. The first is whether a 0.5 McFarland inoculum contains an adequate number of cells for testing, since colony counts performed from a standardized inoculum are often below 5×10^5 CFU/ml, which is the standard inoculum size for broth microdiution assays. However, according to Turnbull et al. (64), the fast growth rate of the organism compensates for the smaller number of cells in the inoculum. The second issue focuses on the role of the chromosomally encoded β-lactamases produced by *B. anthracis* (8, 46) in mediating penicillin resistance, and whether the beta-lactamases are constitutively produced or inducible in most anthrax strains. Genomic data suggest that most, if not all, strains of *B. anthracis* carry two chromosomal beta-lactamase genes: *bla1,* a penicillinase, and *bla2,* a cephalosporinase (8). While penicillin-resistant strains of *B. anthracis* have been reported in the literature, (41, 43, 54) most organisms are penicillin-susceptible by routine broth microdilution testing methods (11, 43, 47). To address this issue, the CLSI established a relatively low threshold for penicillin resistance in *B. anthracis*, choosing breakpoints of ≤0.25 μg/ml as susceptible and ≥0.5 μg/ml as resistant (50). Based on data from published studies (47), CLSI does not advocate performing a beta-lactamase test with *B. anthracis.*

A naturally occurring macrolide resistance determinant and the in vitro transfer of tetracycline and chloramphenicol resistance determinants via transduction and conjugal transposition have been reported for *B. anthracis* (32, 38, 59). The naturally occurring *ermJ* determinant, which presumably mediated resistance to macrolides, lincosamides, and streptogramin B (MLS_B) drugs, was identified in a strain of *B. anthracis* isolated from soil in Korea (38). Unfortunately, MIC data for the three MLS_B drug classes were not reported. Plasmids encoding either tetracycline or chloramphenicol resistance determinants were introduced into *B. anthracis* Sterne via transduction using the generalized transducing bacteriophage CP-51 (59), suggesting that resistance to these antimicrobial agents may be encountered in naturally occurring strains. Tetracycline resistance was also introduced into *B. anthracis* strain VNR-1 via introduction of Tn*916,* which carries the *tetM* determinant, by filter matings with a *Streptococcus* (*Enterococcus*) *faecalis* donor (32). Thus, the construction of strains of *B. anthracis* resistant to chloramphenicol, erythromycin, and tetracycline using natural exchange methods is well documented. In August 2000, the U.S. Food and Drug Administration (FDA) approved ciprofloxacin hydrochloride for management of postexposure inhalational anthrax. This was the first antimicrobial drug approved by the FDA for treating infection due to a biological agent used intentionally (49). Although *B. anthracis* isolates are usually susceptible to fluoroquinolones (11, 16, 47), several investigators also have been able to select for higher levels of fluoroquinolone, tetracycline, and β-lactam resistance via continued passage of *B. anthracis* on antimicrobial-containing media (1, 9, 57). Athamna and colleagues (1) continuously passaged two strains of *B. anthracis* on antimicrobial-containing media and selected mutants that were resistant to fluoroquinolones, β-lactam drugs, and vancomycin. Although elevated MICs to tetracyclines were noted, the MICs remained in the susceptible range. In similar studies, Price et al. generated ciprofloxacin-resistant variants of the *B. anthracis* Sterne strain through positive selection on ciprofloxacin-containing media (57). The elevated MICs were associated with specific changes in the quinolone resistance-determining regions of *gyrA*, *gyrB*, and *parC*. Rapid methods were also described by Price and colleagues to detect these mutations (57).

Guidelines for treatment and prophylaxis of anthrax have been published by the Centers for Disease Control and Prevention (CDC) and cite the integral role of susceptibility testing in decision making, particularly if penicillin or amoxicillin are consider for use (7). This is particularly critical for the treatment and prophylaxis of children and pregnant women, in whom fluoroquinolones and tetracyclines are contraindicated (7, 24, 31).

YERSINIA PESTIS

Y. pestis is a member of the family *Enterobacteriaceae* and is a cause of severe and often fatal disease (plague) in humans if left untreated (12, 30). There are approximately 1700 cases of plague reported worldwide every year (30). Although streptomycin is the drug of choice for this disease, it is available only in limited quantities. Gentamicin, doxycycline, and ciprofloxacin have been suggested as alternate therapeutic regimens by the Working Group on Civilian Biodefense (30). Penicillins and cephalosporins are considered to be ineffective against plague, although the organisms may appear susceptible to these agents by in vitro tests (12, 60). Boulanger et al. (5) reviewed the medical records of 50 plague patients from New Mexico that were treated with streptomycin, or with gentamicin plus or minus tetracycline. There was no difference in outcome among the patients treated with streptomycin or gentamicin; however, one patient treated only with tetracycline developed complications of his infection. Thus, gentamicin with or without tetracycline should be considered efficacious for treatment of plague infections (5). A small number of patients that received a fluoroquinolone or chloramphenicol also did well, but the numbers were too small for analysis. Patients treated with β-lactams failed to improve until switched to effective therapy. Thus, β-lactam agents should neither be tested nor reported by clinical laboratories for *Y. pestis* infections (30, 60).

Antimicrobial susceptibility testing methods established by CLSI for MIC testing of *Y. pestis* follow those used for other *Enterobacteriaceae* (52); however, in studies conducted by CDC, adequate growth of some isolates was not achieved in Mueller-Hinton broth in 24 hrs (unpublished data). Thus, the latest CLSI/NCCLS standards advocate incubating *Y. pestis* susceptibility tests for up to 48 h if necessary to achieve adequate growth (10). Acquired resistance to ampicillin, chloramphenicol, kanamycin, minocycline, spectinomycin, streptomycin, and tetracycline has been reported in a single strain of *Y. pestis* from Madagascar (21). Other resistance mechanisms have also been described, emphasizing the need for antimicrobial susceptibility testing for guiding antimicrobial therapy (27). Interpretive criteria for chloramphenicol, ciprofloxacin, doxycycline, gentamicin, streptomycin, tetracycline, and trimethoprim-sulfamethoxazole have been established by CLSI (Table 1) (10). The most controversial issue regarding testing of *Y. pestis* was the temperature at which isolates should be tested. An-

Table 1. Testing methods and antimicrobial agents to test and report for bioterrorism agents defined by CLSI/NCCLS[a]

Organism	Broth	Testing conditions	Antimicrobial agents to test
Bacillus anthracis	Cation-adjusted Mueller-Hinton	18–20 h; ambient air	Penicillin, ciprofloxacin, doxycycline, tetracycline
Burkholderia mallei	Cation-adjusted Mueller-Hinton	18–20 h; ambient air	Ceftazidime, doxycycline, imipenem, tetracycline
Burkholderia pseudomallei	Cation-adjusted Mueller-Hinton	18–20 h; ambient air	Amoxicillin-clavulanic acid, ceftazidime, doxycycline, imipenem, tetracycline, trimethoprim-sulfamethoxazole
Brucella species	*Brucella* (anticipated)	48 h; ambient air (anticipated)	Ciprofloxacin, doxycycline, tetracycline (anticipated)
Francisella tularensis	Cation-adjusted Mueller-Hinton + 2% defined supplements	24–48 h; ambient air	Ciprofloxacin, chloramphenicol, doxycycline, gentamicin, levofloxacin, streptomycin, tetracycline
Yersinia pestis	Cation-adjusted Mueller-Hinton	24–48 h; ambient air	Chloramphenicol, ciprofloxacin, doxycycline, gentamicin, streptomycin, tetracycline, trimethoprim-sulfamethoxazole

[a]Reprinted from reference 10 with permission.

timicrobial susceptibility tests of the highly resistant Madagascar strain conducted by the Pasteur Institute were incubated at 28°C, not 35°C (21), the latter of which is the temperature at which most susceptibility tests in the United States are incubated (50). The lower temperature of 28°C is often necessary to maintain the virulence plasmids of *Y. pestis,* which are lost when isolates are incubated at 35°C. It is unclear whether any antimicrobial resistance genes are present on these virulence plasmids, or similar plasmids; if so, the genes may be lost when the isolates are grown at 35°C. Because of the difficulty for clinical laboratories to incubate strains of *Y. pestis* at the unusual temperature of 28°C, and because of the lack of clear evidence of loss of resistance at incubation temperatures of 35°C, CLSI adopted its standard incubation temperature of 35°C for testing *Y. pestis.*

BURKHOLDERIA MALLEI AND *BURKHOLDERIA PSEUDOMALLEI*

B. mallei and *B. pseudomallei* are considered as potential bioterrorism agents and have been included in the category B list of overlap select agents and toxins (http://www.bt.cdc.gov/agent/agentlist.asp). *B. pseudomallei,* the bacterial species responsible for the potentially fatal infection known as melioidosis, is a facultative intracellular pathogen. Treatment of the acute, septicemic form of melioidosis often relies on ceftazidime (66). *B. mallei* and *B. pseudomallei* are unusual clinical isolates in the United States, although they are commonly recovered from patients in Southeast Asia (17, 25, 61). Both organisms are nonfastidious gram-negative bacilli that grow well in Mueller-Hinton broth. Although the resistance profiles of the two organisms are similar, *B. mallei* isolates tend to be susceptible to aminoglycosides, while *B. pseudomallei* are resistant to these agents (25, 33, 36, 61, 68). *B. pseudomallei* is known to harbor a chromosomally encoded β-lactamase and efflux pumps capable of mediating resistance to macrolides and aminoglycosides (48). Interpretive criteria have been established by CLSI for the key antimicrobial agents used for *B. pseudomallei* infections, including amoxicillin-clavulanic acid, ceftazidime, doxycycline, imipenem, tetracycline, and trimethoprim-sulfamethoxazole (Table 1) (10). CLSI interpretive criteria for *B. mallei* are limited to ceftazidime, doxycycline, imipenem, and tetracycline (10).

A recent study from France determined the antimicrobial susceptibilities of 50 isolates of *B. pseudomallei* and 15 isolates of *B. mallei,* from both humans and animals, to 35 antimicrobial agents using the agar dilution method (62). The authors used the testing methods and interpretive criteria, when possible, for *B. cepacia* established by the Comité de l'Antibiogramme de la Société Française de Microbiologie. Among the antimicrobial agents tested, imipenem, ceftazidime, doxycycline, and minocycline showed excellent activity. Piperacillin and piperacillin-tazobactam also showed very good activity, while resistance to amoxicillin, ticarcillin, aztreonam, fluoroquinolones, and aminoglycosides was widespread in *B. pseudomallei. B. mallei* isolates appeared more susceptible to aminoglycosides than *B. pseudomallei* isolates, but these drugs are not considered as therapeutic options for *B. mallei.* One human isolate of *B. pseudomallei* was reported as resistant to ceftazidime, ticarcillin-clavulanic acid, doxycycline, and minocycline (62).

Piliouras et al. (56) noted that there were major discrepancies between the results of MIC methods and disk diffusion testing for co-trimoxazole (trimethoprim-sulfamethoxazole). Among the 80 clinical isolates of *B. pseudomallei* tested, the rate of cotrimoxazole susceptibility by disk diffusion was 41.3%, while by MIC methods the range was 90 to 97.5%. Heine et al. compared the accuracy of Etest results (AB Biodisk, Solna, Sweden) to the results of broth microdilution assays for 11 isolates of *B. mallei* (28). The Etest method gave 10-fold lower MIC results than those of broth microdilution for 16 of 17 drugs evaluated. Etest results for trimethoprim-sulfamethoxazole were 100 to 1,000-fold lower than the broth microdilution results. Based on these results, the authors did not recommend the Etest for susceptibility testing of *B. mallei* isolates.

The disparity between in vitro results and clinical response for several antimicrobial agents used for *B. mallei* and *B. pseudomallei* infections suggests that additional studies correlating clinical outcome with MIC results are needed.

FRANCISELLA TULARENSIS

F. tularensis is another pathogen of both humans and animals, including birds, mammals, amphibians, and fish, that is seen occasionally in North America, Northern Europe, Japan, and the former Soviet Union (18, 29, 55). There are three subspecies of *F. tularensis,* of which the two most important clinically are *F. tularensis* subsp. *tularensis* (also known as type A) and *F. tularensis* subsp. *holarctica* (also known as type B). There are three biovars of *F. tularensis* subsp. *holarctica* of which biovar I is erythromycin-susceptible and biovar II is erythromycin-resistant. The resistance pattern of biovar III is not clear (18). A third subspecies, *F. tularensis* subsp. *mediaasiatica,* is found in central Asia. *F. tularensis* infections are typically treated with aminoglycosides, such as streptomycin or gentamicin (18, 22, 44). Data suggest that fluoroquinolones, such as ciprofloxacin and levofloxacin, are active in vivo against most *F. tularensis* isolates (24, 29, 35, 44, 55); however, chloramphenicol and the tetracyclines show only static activity and failures have been described (55). This organism is considered naturally resistant to azithromycin, extended-spectrum cephalosporins, carbapenems, and sulfa drugs (2, 18, 22, 44).

Antimicrobial susceptibility testing methods used for *F. tularensis* studies vary widely in the literature (2, 22, 29). Studies by Baker et al. suggested that the key component for growing *F. tularensis* was the incorporation of a defined media supplement such as IsoVitalex in addition to supplementing the medium with calcium, magnesium, glucose, and ferric pyrophosphate (2). García del Blanco and colleagues tested 42 isolates of *F. tularensis* subsp. *holartica* using the medium described by Baker et al. (2). Ikäheimo et al. used the Etest method and cysteine heart agar containing 2% hemoglobin incubated for 24 to 48 h in 5% CO_2 (29). More recent studies coordinated by CDC indicated that Mueller-Hinton broth supplemented with 2% Iso-Vitelex or similar defined supplements gave reproducible MIC results consistent with those described in the literature. Thus, Mueller-Hinton broth supplemented with 2% defined supplements was adopted by CLSI as the testing medium of choice for *F. tularensis* (10). Appropriate antimicrobial agents to test against *F. tularensis* isolates include chloramphenicol, ciprofloxacin, doxycycline, gentamicin, streptomycin, and tetracycline (Table 1). One problem encountered by CLSI was the lack of quality control ranges for the antimicrobial agents of choice for *Escherichia coli, Staphylococcus aureus,* and *Pseudomonas aeruginosa* when using the supplemented medium. Thus, a multilaboratory quality control study was undertaken to establish the appropriate quality control ranges (6). Not all *F. tularensis* susceptibility test results could be interpreted after 24 hours of incubation, so CLSI recommended that tests could be held for an additional 24 h if necessary (10). Separate quality control ranges for 24 and 48 h for *E. coli, S. aureus,* and *P. aeruginosa* were developed to accommodate prolonged incubation of isolates (6, 10).

BRUCELLA SPECIES

The most common *Brucella* species that infect humans (i.e., *B. melitensis, B. suis,* and *B. abortus*) are a heterogeneous group of organisms with growth characteristics ranging from slow to highly fastidious (i.e., very slow) (3, 4, 26). Brucellosis is prevalent in many developing countries and occurs in most regions of Southern and Eastern Europe, including certain Mediterranean and Balkan countries, South and Central America, and the Middle East (26, 53). The most common complications of *Brucella* infection include acute granulomatous hepatitis, arthritis, sacroiliitis, and spondylitis, with the latter being the most difficult to treat (26, 52).

Currently, there is no consensus on accepted therapeutic regimens for long-term therapy of these complications; however, various antimicrobial combinations have met with mixed results (26). Standard antimicrobial treatment regimens include the use of tetracyclines, often in combination with streptomycin or other aminoglycosides, rifampin, and trimethoprim-sulfamethoxazole; some macrolides also show activity against *Brucella* species (23, 42, 63). Nonetheless, there is a high percentage of patients who fail treatment and experience residual disability (26, 53).

Sulfonamide resistance is naturally occurring in many isolates (3, 37, 58). Various studies have explored the utility of fluoroquinolones and several other antimicrobial agents for treating *Brucella* infections; however, these studies used differing growth media and incubation conditions so comparing results is difficult (3, 23, 37, 39, 45, 53, 58, 63). Standardized susceptibility testing methods are in development by CLSI for *Brucella* species but will likely utilize *Brucella* broth. The fact that the three major species of *Brucella* that affect humans do not all grow at the same rate complicates the establishment of a single susceptibility testing method. Previous susceptibility testing studies in the literature have used supplemented Mueller-Hinton agar and broth, Iso-Sensitest agar, and *Brucella* broth and agar (3, 4, 45, 58). Nonetheless, it appears the *Brucella* broth provides a sufficiently rich medium to support the growth of all three species when incubation is extended to 48 hours. As with testing of *F. tularensis,* however, it will be necessary to develop quality control ranges for *E. coli, S. aureus,* and *P. aeruginosa* to ensure that the testing of *Brucella* species is accurate.

SUMMARY

Antimicrobial susceptibility testing of bacterial agents of bioterrorism provides critical data for guiding prophylaxis or treatment of patients exposed to or infected with *B. anthracis, Y. pestis, B. mallei, B. pseudomallei, F. tularensis,* and *Brucella* species. The recent standard published by CLSI contains both standardized susceptibility testing methods and interpretive criteria for all of these organisms with the exception of *Brucella* species (10). CLSI has also determined the appropriate antimicrobial agents for inclusion in a routine, primary testing panel, and standard reporting of results for the specific bacterial species (Table 1) (10). Testing methods and interpretive criteria for *Brucella* species are anticipated in January 2006. Because of both naturally occurring resistance to antimicrobial agents and the potential for engineered resistance, standardized susceptibility testing coupled with rapid testing methods for specific resistance genes and mutations associated with antimicrobial resistance will be an integral part of the public health response to a potential release of any of these agents of bioterrorism.

REFERENCES

1. **Athamna, A., M. Athamna, N. Abu-Rashed, B. Madlej, D. J. Bast, and E. Rubenstein.** 2004. Selection of *Bacillus anthracis* isolates resistant to antibiotics *J. Antimicrob. Chemother.* **54:**424–428.
2. **Baker, C. N., D. G. Hollis, and C. Thornsberry.** 1985. Antimicrobial susceptibility testing of *Francisella tularensis* with a modified Mueller-Hinton broth. *J. Clin. Microbiol.* **22:**212–215.
3. **Bosch, J., J. Linares, M. J. Lopez de Goicoechea, J. Ariza, M. C. Cisnal, and R. Martin.** 1986. In vitro activity of ciprofloxacin, ceftriaxone, and five other antimicrobial agents against 95 strains of *Brucella melitensis. J. Antimicrob. Chemother.* **17:** 459–461.
4. **Bodur, H., N. Balaban, S. Aksaray, V. Yetener, E. Akinci, A. Colpan, and A. Erbay.** 2003. Biotypes and antimicrobial susceptibilities of Brucella isolates. *Scand. J. Infect. Dis.* **35:**337–338.
5. **Boulanger, L. L., P. Ettestad, J. D. Fogarty, D. T. Dennis, D. Romig, and G. Mertz.** 2004. Gentamicin and tetracyclines for the treatment of human plague: review of 75 cases in New Mexico, 1985-1999. *Clin. Infect. Dis.* **38:**663–669.
6. **Brown, S. D., K. Krisher, and M. M. Traczewski.** 2004. Broth microdilution susceptibility testing of *Francisella tularensis:* quality control limits for nine antimicrobial agents and three standard quality control strains. *J. Clin. Microbiol.* **42:**5877–5880.
7. **Centers for Disease Control and Prevention.** 2001. Update: Investigation of bioterrorism-related anthrax and interim guidelines for exposure management and antimicrobial therapy, October 2001. *Morb. Mortal. Wkly. Rep.* **50:**909–918.
8. **Chen, Y., J. Succi, F. C. Tenover, and T. Koehler.** 2003. β-lactamase genes of the penicillin-susceptible *Bacillus anthracis* Sterne strain. *J. Bacteriol.* **185:**823–830.
9. **Choe, C. H., S. S. Bouhaouala, I. Brook, T. B. Elliot, and J. Knudson.** 2000. In vitro development of resistance to ofloxacin and doxycycline in *Bacillus anthracis* Sterne. *Antimicrob. Agents Chemother.* **44:**1766.
10. **Clinical and Laboratory Standards Institute.** 2005. *Performance Standards for Antimicrobial Susceptibility Testing. Fifteenth Informational Supplement M100-S15.* Clinical and Laboratory Standards Institute, Wayne, Pa.
11. **Coker, P. R., K. L. Smith, and M. E. Hugh-Jones.** 2002. Antimicrobial susceptibilities of diverse *Bacillus anthracis* isolates. *Antimicrob. Agents Chemother.* **46:**3843–3845.
12. **Dennis, D. T., and J. M. Hughes.** 1997. Multidrug resistance in plague. *N. Engl. J. Med.* **337:**702–704.
13. **Dennis, D. T., T. V. Inglesby, D. A. Henderson, J. G. Bartlett, M. A. Ascher, E. Eitzen, A. D. Fine, A. M. Friedlander, J. Hauer, M. Layton, S. R. Lillibridge, J. E. McDade, M. T. Osterholm, T. O'Toole, G. Parker, T. M. Perl, P. K. Russell, K. Tonat, for the Working Group on Civilian Biodefense.** 2001. Tularemia as a biological weapon. *JAMA.* **285:**2763–2773.
14. **Dixon, T. C., M. Meselson, J. Guillemin, and P. C. Hanna.** 1999. Anthrax. *N. Engl. J. Med.* **341:**815–826.
15. **Doganay, M., and N. Aydin.** 1991. Antimicrobial susceptibility of *Bacillus anthracis. Scand. J. Infect. Dis.* **23:**333–335.
16. **Drago, L., E. De Vecchi, A. Lombardi, L. Nicola, M. Valli, and M. R. Gismondo.** 2002. Bactericidal activity of levofloxacin, gatifloxacin, penicillin, meropenem, and rokitamycin against *Bacillus anthracis* clinical isolates. *J. Antimicrob. Chemother.* **50:** 1059–1063.
17. **Eickhoff, T. C., J. V. Bennett, P. S. Hayes, and J. Feeley.** 1970. *Pseudomonas pseudomallei* susceptibility to chemotherapeutic agents. *J. Infect. Dis.* **121:**95–102.
18. **Ellis, J., P. C. F. Oyston, M. Green, and R. W. Titball.** 2002. Tularemia. *Clin. Microbiol. Rev.* **15:**631–646.
19. **Evans, A. S., and P. S. Brachman (ed.).** 1998. *Bacterial Infections of Humans: Epidemiology and Control.* Plenum Publishing Corporation, New York, N.Y.
20. **Friedlander, A. M., S. L. Welkos, M. L. M. Pitt, J. W. Ezzell, P. L. Worsham, K. J. Rose, B. E. Ivins, J. R. Lowe, G. B. Howe, P. Mikesell, and W. B. Lawrence.** 1993. Postexposure prophylaxis against experimental inhalation anthrax. *J. Infect. Dis.* **167:** 1239–1242.
21. **Galimand, M., A. Guiyoule, G. Gerbaud, B. Rasoamanana, S. Chanteau, E. Carniel, and P. Courvalin.** 1997. Multidrug resistance in *Yersinia pestis* mediated by a transmissible plasmid. *N. Engl. J. Med.* **337:**677–680.
22. **Garcia del Blanco, N., C. B. Gutiérrez Martin, V. A. de la Puenta Redondo, and E. F. Rodríguez Ferri.** 2004. In vitro susceptibility of field isolates of *Francisella tularensis* subsp. *holarctica* recovered in Spain to several antimicrobial agents. *Res. Vet. Sci.* **76:**195–198.
23. **Garcia-Rodriguez, J. A., J. L. Monoz Bellido, M. J. Fresnadillo, and I. Trujillano.** 1993. In vitro activities of new macrolides and rifapentine against *Brucella* spp. *Antimicrob. Agents Chemother.* **37:**911–913.
24. **Gilbert, D. N., R. C. Moellering, Jr., G. M. Eliopoulos, and M. A Sande.** 2004. *The Sandford Guide to Antimicrobial Therapy—2004.* Antimicrobial Therapy, Inc., Hyde Park, Vt.
25. **Godfrey, A. J., S. Wong, D. A. B. Dance, W. Chaowagul, and L. E. Bryan.** 1991. *Pseudomonas pseudomallei* resistance to β-lactam antibiotics due to alterations in the chromosomally encoded β-lactamase. *Antimicrob. Agents Chemother.* **35:**1635–1640.
26. **Gotuzzo, E., and C. Carrillo.** 2004. Brucella, p. 1717–1724. *In* S. L. Gorbach, J. G. Bartlett, and N. R. Blacklow (ed.), *Infectious Diseases,* 3rd ed. Lippincott Williams, & Wilkins, Philadelphia, Pa.
27. **Guiyoule, A., B. Rasoamanana, C. Buchrieser, P. Michel, S. Chanteau, and E. Carniel.** 1997. Recent emergence of new variants of *Yersinia pestis* in Madagascar. *J. Clin. Microbiol.* **35:** 2826–2833.
28. **Heine, H. S., M. J. Englan, D. M. Wang, and W. R. Byrne.** 2001. In vitro antibiotic susceptibilities of *Burkholderia mallei* (causative agent of glanders) determined by broth microdilution and Etest. *Antimicrob. Agents Chemother.* **45:**2119–2121.
29. **Ikäheimo, I., H. Syrjälä, J. Karhukorpi, R. Schildt, and M. Koskela.** 2000. In vitro antibiotic susceptibility of *Francisella tularensis* isolated from humans and animals. *J. Antimicrob. Chemother.* **46:**287–290.
30. **Inglesby, T. V., D. T. Dennis, D. A. Henderson, J. G. Bartlett, M. S. Ascher, E. Eitzen, A. D. Fine, A. M. Friedlander, J. Hauer, J. F. Koerner, M. Layton, J. McDade, M. T. Osterholm, T. O'Toole, G. Parker, T. M. Perl, P. K. Russell, M. Schoch-Spana,**

and **K. Tonat.** 2000. Plague as a biological weapon: medical and public health management. *JAMA.* **283:**2281–2290.

31. **Inglesby, T. V., D. A. Henderson, J. G. Bartlett, M. S. Ascher, E. Eitzen, A. M. Friedlander, J. Hauer, J. McDade, M. T. Osterholm, T. O'Toole, G. Parker, T. M. Perl, P. K. Russell, and K. Tonat.** 1999. Anthrax as a biological weapon: medical and public health management. *JAMA.* **281:**1735–1745.
32. **Ivans, B. E., S. I. Welkos, G. B. Knudson, and D. J. LeBlanc.** 1988. Transposon Tn*916* mutagenesis in *Bacillus anthracis. Infect. Immun.* **56:**176–181.
33. **Jenny, A. W., G. Lum, D. A. Fisher, and B. J. Currie.** 2001. Antibiotic susceptibility of *Burkholderia pseudomallei* from tropical northern Australia and implications for therapy of melioidosis. *Int. J. Antimicrob. Agents* **17:**109–113.
34. **Jernigan, J. A., D. S. Stephens, D. A. Ashford, C. Omenaca, M. S. Topiel, M. Galbraith, M. Tapper, T. L. Fisk, S. Zaki, T. Popovic, R. F. Meyer, C. P. Quinn, S. A. Harper, S. K. Fridkin, J. J. Sejvar, C. W. Shepard, M. McConnell, J. Guarner, W. J. Shieh, J. M. Malecki, J. L. Gerberding, J. M. Hughes, B. A. Perkins, and the Anthrax Bioterrorism Investigation Team.** 2001. Bioterrorism-related inhalational anthrax: the first 10 cases reported in the United States. *Emerg. Infect. Dis.* **7:**933–944.
35. **Johansson A., L. Beglund, L. Gothefors, A. Sjostedt, and A. Tarnvik.** 2000. Ciprofloxacin for treatment of tularemia in children. *Pediatr. Infect. Dis. J.* **19:**449–453.
36. **Kenny, D. J., P. Russell, D. Rogers, S. M. Eley, and R. W. Titball.** 1999. In vitro susceptibilities of *Burkholderia mallei* in comparison to those of other pathogenic *Burkholderia* spp. *Antimicrob. Agents Chemother.* **43:**2773–2775.
37. **Khan, M.Y., M. Dizon, and F. W. Kiel.** 1989. Comparative in vitro activities of ofloxacin, difloxacin, ciprofloxacin, and other selected antimicrobial agents against *Brucella melitensis. Antimicrob. Agents Chemother.* **33:**1409–1410.
38. **Kim, H-S., E-C. Choi, and B-K. Kim.** 1993. A macrolide-lincosamide-streptogramin B resistance determinant from *Bacillus anthracis* 590: cloning and expression of *ermJ. J. Gen. Microbiol.* **139:**601–607.
39. **Kocagoz, S., M. Akova, B. Altun, D. Gur, and G. Hascelik.** 2001. In vitro activities of new quinolones against *Brucella melitensis* isolated in a tertiary care hospital in Turkey. *Clin. Microbiol. Infect.* **8:**240–242.
40. **Klietman, W. F., and K. L. Ruoff.** 2001. Bioterrorism: implications for the clinical microbiologist. *Clin. Microbiol. Rev.* **14:** 364–381.
41. **Lalitha, M. K., and M. K. Thomas.** 1997. Penicillin resistance in *Bacillus anthracis. Lancet* **349:**1522.
42. **Landinez, R., J. Linares, E. Loza, J. Martinez-Beltran, and F. Baquero.** 1992. In vitro activity of azithromycin and tetracycline against 358 clinical isolates of *Brucella melitensis. Eur. J. Clin. Microbiol.* **11:**265–267.
43. **Lightfoot, N. F., R. J. D. Scott, and P. C. B. Turnbull.** 1990. Antimicrobial susceptibility of *Bacillus anthracis. Salisbury Med. Bull.* **68**(Suppl.):95–98.
44. **Limaye, A. P., and C. J. Hooper.** 1999. Treatment of tularemia with fluoroquinolones: two cases and a review. *Clin. Infect. Dis.* **29:**922–924.
45. **Lopez-Merino, A., A. Contreras-Rodriguez, R. Migranas-Ortiz, R. Orrantia-Gradin, G. M. Hernandez-Oliva, A. T. Gutierrez-Rubio, and O. Cardenosa.** 2004. Susceptibility of Mexican brucella isolates to moxifloxacin, ciprofloxacin and other antimicrobials used in the treatment of human brucellosis. *Scand. J. Infect. Dis.* **36:**636–638.
46. **Materon, I. C., A. M. Queenan, T. M. Koehler, K. Bush, and T. Palzkill.** 2003. Biochemical characterization of β-lactamase Bla1 and Bla2 from *Bacillus anthracis. Antimicrob. Agents Chemother.* **47:**2040–2042.
47. **Mohammed, M. J., C. K. Marston, T. Popovic, R. S. Weyant, and F. C. Tenover.** 2002. Antimicrobial susceptibility testing of *Bacillus anthracis*: comparison of results obtained by using the National Committee for Clinical Laboratory Standards broth microdilution reference and Etest agar gradient diffusion methods. *J. Clin. Microbiol.* **40:**1902–1907.
48. **Moore, R. A., D. DeShazer, S. Reckseidler, A. Wissman, and D. E. Woods.** 1999. Efflux-mediated aminoglycoside and macrolide resistance in *Burkholderia pseudomallei. Antimicrob. Agents Chemother.* **43:**465–470.
49. **Meyerhoff, A., R. Albrecht, J. M. Meyer, P. Dionne, K. Higgins, and D. Murphy.** 2004. US Food and Drug Administration approval of ciprofloxacin hydrochloride for management of postexposure inhalational anthrax. *Clin. Infect. Dis.* **39:**303–308.
50. **National Committee for Clinical Laboratory Standards.** 2003. *Methods for Dilution Antimicrobial Susceptibility Tests for Bacteria That Grow Aerobically.* 6th ed., vol. 20, no. 2. *Approved Standard M7-A6.* National Committee for Clinical Laboratory Standards, Wayne, Pa.
51. **National Committee for Clinical Laboratory Standards.** 2001. *Performance Standards for Antimicrobial Susceptibility Testing. Eleventh Informational Supplement M100-S11.* National Committee for Clinical Laboratory Standards, Wayne, Pa.
52. **National Committee for Clinical Laboratory Standards.** 2004. *Performance Standards for Antimicrobial Susceptibility Testing. Fourteenth Informational Supplement M100-S14.* National Committee for Clinical Laboratory Standards, Wayne, Pa.
53. **Pappas, G., S. Seitaridis, N. Akritidis, and E. Tsianos.** 2004. Treatment of brucella spondylitis: lessons from an impossible meta-analysis and initial report of efficacy of a fluoroquinolone-containing regimen. *Int. J. Antimicrob. Agents.* **24:**502–507.
54. **Patra G., J. Vaissaire, M. Weber-Levy, C. Le Doujet, and M. Mock.** 1998. Molecular characterization of *Bacillus* strains involved in outbreaks of anthrax in France in 1997. *J. Clin. Microbiol.* **36:**3412–3414.
55. **Perez-Castrillon, J. L., P. Bachiller-Luque, M. Martin-Luquero, F. J. Mena-Martin, and V. Herreros.** 2001. Tularemia epidemic in Northwestern Spain; clinical description and therapeutic response. *Clin. Infect. Dis.* **33:**573–576.
56. **Piliouras, P., G. C. Ulett, C. Ashhurst-Smith, R. G. Hirst, and R. E. Norton.** 2002. A comparison of antibiotic susceptibility testing methods for cotrimoxazole with *Burkholderia pseudomallei. Int. J. Antimicrob. Agents.* **19:**427–429.
57. **Price, L. B., A. Volger, T. Pearson, and P. Keim.** 2003. In vitro selection and characterization of *Bacillus anthracis* mutants with high-level resistance to ciprofloxacin. *Antimicrob. Agents Chemother.* **47:**2362–2365.
58. **Rubinstein, E., R. Lang, B. Shasha, B. Hagar, L. Diamanstein, G. Joseph, M. Anderson, and K. Harrison.** 1991. In vitro susceptibility testing of *Brucella melitensis* to antibiotics. *Antimicrob. Agents Chemother.* **35:**1925–1927.
59. **Ruhfel, R. E., N. J. Robillard, and C. B. Thorne.** 1984. Interspecies transduction of plasmids among *Bacillus anthracis, B. cereus,* and *B. thuringiensis. J. Bacteriol.* **157:**708–711.
60. **Smith, M. D., D. X. Vinh, T. T. Nguyen, J. Wain, D. Thung, and N. J. White.** 1995. In vitro antimicrobial susceptibilities of strains of *Yersinia pestis. Antimicrob. Agents Chemother.* **39:**2153–2154.
61. **Sookpranee, T., M. Sookpranee, M. A. Mellencamp, and L. C. Preheim.** 1991. *Pseudomonas pseudomallei,* a common pathogen in Thailand that is resistant to the bactericidal effects of

many antibiotics. *Antimicrob. Agents Chemother.* **35:**484–489.

62. **Thibault, F. M., E. Hernandez, D. R. Vidal, M. Girardet, and J. D. Cavallo.** 2004. Antibiotic susceptibility of 65 isolates of *Burkholderia pseudomallei* and *Burkholderia mallei* to 35 antimicrobial agents. *J. Antimicrob. Chemother.* **54:**1134–1138.
63. **Trujillano-Martin, I., E. Garcia-Sanchez, M. J. Fresnadillo, J. E. Garcia-Sanchez, and J. A. Garcia-Rodriguez.** 1999. In vitro activities of five new antimicrobial agents against *Brucella melitensis. Int. J. Antimicrob. Agents* **12:**185–186.
64. **Turnbull, P. C. B., N. M. Sirianni, C. I. LeBron, M. N. Samaan, F. N. Sutton, A. E. Reyes, and L. F. Peruski.** 2004. MICs of selected antibiotics for *Bacillus anthracis, Bacillus cereus, Bacillus thuringiensis,* and *Bacillus mycoides* from a range of clinical and environmental sources as determined by Etest. *J. Clin. Microbiol.* **42:**3626–3634.
65. **U. S. Department of Health and Human Services.** 1999. Biosafety in microbiology and biomedical laboratories. J. Y. Richmond and R. W. McKinney (ed.), U.S. Government Printing Office, Washington, D.C.
66. **White, N.** 2003. Meliodosis. *Lancet.* **361:**1715–1722.
67. **Working Party of the British Society for Antimicrobial Chemotherapy.** 1988. Breakpoints in in-vitro antibiotic sensitivity testing. *J. Antimicrob. Chemother.* **21:**701–710.
68. **Yamamoto, T., P. Naigowit, S. Dejsirilert, D. Chiewsilp, E. Kondo, T. Yokota, and K. Kanai.** 1990. In vitro susceptibilities of *Pseudomonas pseudomallei* to 27 antimicrobial agents. *Antimicrob. Agents Chemother.* **34:**2027–2029.

PART V. ECOLOGY AND FITNESS OF DRUG RESISTANCE

SECTION EDITOR: David G. White

EFFORTS TO MAINTAIN THE EFFECTIVENESS OF ANTIMICROBIALS in both human and veterinary medicine must address the ecological consequences of their use. Although much scientific information is available on this subject, many ecological aspects of the development and dissemination of antimicrobial resistance remain to be described.

The selective pressures exerted by antibiotics in plants, animals, and humans have led to reservoirs of resistance in our immediate environments. The following chapters within this section, though varied, all share a common theme, namely, that there are insights to be gained from viewing antibiotic resistance and gene spread from a broad ecological perspective. The interface of different microbial ecosystems, the shared gene pool of commensal and pathogenic bacteria, and the dissemination of resistant organisms in the food chain are each highlighted.

Frontiers in Antimicrobial Resistance: a Tribute to Stuart B. Levy
Edited by D. G. White, M. N. Alekshun, and P. F. McDermott

Chapter 32

Fitness Traits in Soil Bacteria

FABRICE N. GRAVELAT, STEVEN R. STRAIN, AND MARK W. SILBY

INTRODUCTION

Soil, a Heterogenous Interface with a Dynamic Structure

The term "soil" refers to the interface between the lithosphere, which provides minerals, and the biosphere, which provides organic compounds. In addition to supporting the life of plants and animals, soil also supports an abundant microflora: one gram of soil is estimated to hold more than 10^9 bacteria, comprising 4,000 to 7,000 different genomes. The bacterial biodiversity is increasingly regarded as the greatest guarantee for the stability and richness of soil. Since environmental conditions constantly evolve, no single species of microorganism is redundant, even if several fulfill the same function at a given time (3).

One function of soil bacteria is the structuring of the soil under a principle of reciprocity (51): while bacteria alter the parent rock and stabilize the aggregates (i.e., the basic units of the structure of soils), soil colonization by bacteria is a function of soil structure. For instance, poorly ventilated (argillaceous) soils or soils with a low water holding capacity (sandy) are not favorable for bacterial colonization and survival, and therefore they lack the bacterial activities necessary for the formation and stability of aggregates. Indeed, soil microaggregates (under 250 μm in diameter) are the consequence of the cementation of the sand and silt particles (chemically inert), with the clay particles (layers of phyllosilicates negatively charged, and more or less separated by hydrated interfoliaceous spaces), the humic acids (negatively charged and source of nutrients for some bacteria), and the bacteria (biofilm formation and polysaccharide excretion) (43). Therefore the composition of the soil combined with bacterial growth determines the friability, the water infiltration and retention rates, and the erodibility of the soil (21).

Dr. Levy's Early Involvement

Pseudomonas spp. have been of interest for their production of a broad range of antifungals and metabolic enzymes, and for their near-ubiquitous presence, suggesting that they were well-adapted to soils and thus represent good candidate biocontrol and bioremediation agents.

In the late 1980s, Stuart Levy's lab began studies of soil pseudomonads, as part of a National Science Foundation move to support basic science directed at understanding microbial ecology in soils. A general problem when bacteria are applied to soil is the failure to establish a population in the soil. Using antibiotic resistance markers presented a means by which to monitor the population of introduced bacteria. Building on vast experience in the study of antibiotic resistance under laboratory and field conditions, Stuart Levy initiated experiments designed to examine the utility of rifampin resistance (Rif^r) as a marker for monitoring *Pseudomonas putida* and *P. fluorescens* in soils. These studies found that in *P. fluorescens,* Rif^r isolates could be grouped in two general categories—one in which resistance had no impact on fitness and another in which there was a general reduction in competitive fitness associated with rifampin resistance (11). This study was important for a number of reasons. First, it indicated that rifampin was an appropriate monitoring tool, provided experiments were done to confirm the fitness of the marked strain. In addition, the generation of less fit mutants gave support to the possibility of using genetics to understand soil ecology. Finally, the results of colonization experiments in which fit strains were inoculated into soil with a previously established population showed limited colonizing success, underscoring the idea that colonization is an active process rather than a passive result of inoculation, and that there are only a limited number of sites available for coloniza-

Fabrice N. Gravelat and Mark W. Silby • Center for Adaptation Genetics and Drug Resistance, Department of Molecular Biology and Microbiology, Tufts University School of Medicine, Boston, MA 02111. **Steven R. Strain** • Department of Biology, Slippery Rock University of Pennsylvania, Slippery Rock, PA 16057.

tion. Soon after this early study on soil fitness, Levy's interests in soil ecology focused on using genetics to examine the role of adhesion in soil survival (17, 37), and more recently, on the use of genetic screens to identify genes that may be important in soil survival and persistence (53, 55).

BACTERIAL COLONIZATION OF SOIL: COMPETITION IN A HARSH ENVIRONMENT

Inoculation of Soils with Bacteria: Interests and First Difficulties

The abundance of biological functions among soil bacteria led scientists to consider the inoculation of selected bacteria into soils to benefit human welfare and activities. It could allow, for instance, (i) improvement in supply of nutrients to plants, especially of nitrogenous nutrients supplied by atmospheric nitrogen fixing bacteria (e.g., *Rhizobium, Azotobacter,*); (ii) production of plant growth-stimulating hormones by plant-growth-promoting-rhizobacteria (PGPR); (iii) control of soil-borne plant diseases (e.g., *Pseudomonas fluorescens* versus *Pythium* fungi); (iv) formation of soil aggregates responsible for the stability of the soil structure, and for the soil hydration and ventilation properties—thus a soil rich in aggregates favorable for agriculture and silviculture; (v) degradation of some xenobiotic pollutants (64).

One major obstacle for inoculation of bacteria in soil is the rapid loss of cells in the first days (64). If the bacteria have a fitness advantage, a light inoculation is enough to ensure significant colonization (e.g., nodulating bacteria, xenobiotic degraders). But generally, large inocula are needed to overcome (i) competition with indigenous bacteria, (ii) predation by protists, and (iii) heterogeneity of soil: spatial (texture, pH, temperature, moisture, substrate availability) and temporal (e.g., *Pythium* attacks germinative seeds, thus bacteria used for biocontrol are needed at a particular place at a specific time).

Distribution of Bacteria in Soil

Pores and ecological niches

The distribution of bacteria is neither homogenous nor random. More than 80% of the bacteria are located in soil pores (even if only 4 to 10% of the pore space of an aggregate are colonized) (43, 51), and biodiversity decreases around plant roots and increases in bulk soil, even though microbial activity is lower in bulk soil than in the rhizosphere (3). The first reason for the heterogeneity of the bacterial distribution in soil is its structure. Indeed, the cations and organic matter are mainly concentrated in the micropores of the soil (<10 μm in diameter) while the macropores (>10 μm) are more oxygenated but subject to strong drying (51). Micropores also reduce the grazing pressure on bacteria by protozoans and nematodes (which may favor a particular bacterial population and thus may influence the balance of bacterial activities in the soil) (30), and the dispersal of bacteria in soil by larger, mobile animals (mainly the lumbricids) (20). The second reason for the heterogeneity of the distribution of bacteria in soil is that they modify the surrounding environment by their development and their activity. As bacteria exploit an available source of nutrients around which they established, they induce soil microaggregate formation, decrease in oxygen rate, pH variations, etc. This creates an ecological niche, in which the chemistry is different from the rest of the soil (43). The capacity of an inoculated bacterium to colonize the niche depends on its ability to tolerate the local chemistry and to cooperate or compete with the indigenous bacteria.

The case of rhizosphere: a nutrient rich niche

The zone immediately surrounding the plant root is called the rhizosphere. Roots naturally exude organic matter (mainly amino acids, monosaccharides, and organic acids) and release dead root hairs and epidermial cells into the rhizosphere, thus creating a niche in which the chemistry is different than that of bulk soil and the concentration of bacteria and bacterial activity levels are significantly higher (16, 35).

Despite the relative nutrient-richness of the rhizosphere, the competition between microorganisms remains, since high metabolic activities of plants and microbes can cause certain nutrients to become limiting. For instance, the oxygen concentration dramatically drops in the rhizosphere in comparison with bulk soil, up to conditions of complete anoxia, thus favoring bacteria able to use alternate mechanisms of respiration (2). Another example is the acquisition of ferric iron for which plants and microflora compete, thus making the Fe^{3+} concentration lower in the rhizosphere than in the bulk soil. To overcome this limitation, many microorganisms use an active acquisition system, based on siderophores with high affinity for Fe^{3+} and corresponding receptors to uptake ferri-siderophore complexes (39). Bacteria can also compete for the uptake of siderophores produced by other strains (16, 39).

Taxis and motility

Bacteria utilize information processing systems to constantly monitor their surroundings for important changes; these processes are called taxis and are used to guide bacteria toward more favorable environments and away from unfavorable ones (2, 8, 52), either by motility or, for non-flagellated bacteria, by gliding, swarming, or twitching over solid surfaces. In soil, bacteria face seasonal fluctuations due to the plant biology (dead leaves, growth of roots), which creates a situation of "feast and famine" for the microorganisms (2). Moreover, the rhizosphere is an environment rich in organic matter but poor in oxygen, thus creating strong gradients of carbon sources, oxygen, and redox potential between it and the bulk soil. Finally, in the bulk soil, the free diffusion of water, soluble nutrients and oxygen in the macropores creates other gradients between macropores and micropores. In such an heterogeneous environment, taxis can be seen as a major trait for colonization (as suggested by the fact that, in soil, the majority of bacteria are motile [22]) and as an immediate survival strategy (less than one second is necessary between the binding of a chemical stimulus to a receptor on the cell surface and a change in the swimming direction [2]).

Chemical gradients are the basis of the most studied system, namely chemotaxis. Chemotactic systems are built around one sensor kinase that possesses environmentally regulated autophosphorylation activity (CheA) and two response regulators (CheY and CheB) with an activity that is modulated by phosphoryl groups transferred from the kinase (8). In flagellated bacteria, phosphorylated CheY binds to the flagella proteins and induces flagella movements. Studies in phylogenically distant bacteria showed that pathways in unrelated bacteria involve homologous proteins, but their interactions may be totally different (52), i.e., the regulatory strategy remains the same among studied Eubacteria and Archaebacteria (8), suggesting the importance of chemotaxis as a survival strategy.

Another kind of taxis, energy taxis, involves FAD cofactor associated sensors that detect gradients of terminal electron acceptors, light, redox chemicals and metabolizable substrates that act as donors of reducing equivalents to the electron transport system (2). This signal is transduced in a cascade manner by interacting with CheA and CheW. Although energy generation is one of the most important parameters for the survival of microorganisms in their environment, and energy taxis has been known for a long time, the contribution of energy-taxis to the ecology of microorganisms remains poorly understood. Only recently energy-tactic mutants of *Azospirillum brasilense* were found to have reduced fitness in soil (1).

Studying the fitness traits of *Pseudomonas fluorescens* in soil, Stuart Levy and colleagues demonstrated the importance of the AdnA transcriptional factor in spreading of the bacterium in soil (17, 18, 37, 53). AdnA has 83% identity with FleQ, a transcriptional factor of *P. aeruginosa* which belongs to the NtrC subfamily of transcriptional activators and regulates adhesion and flagellar gene expression in interaction with the alternative sigma factor RpoN (4, 34). The insertional inactivation of AdnA results in mutants that are non-motile, and biofilm, chemotaxis and adhesion deficient (Fig. 1) (53). AdnA activates the transcription of at least seven genes, named *aba* genes, the products of which are homologs of methyl-accepting chemotaxis proteins, flagella structural proteins, or perosamine synthetase (involved in the synthesis of

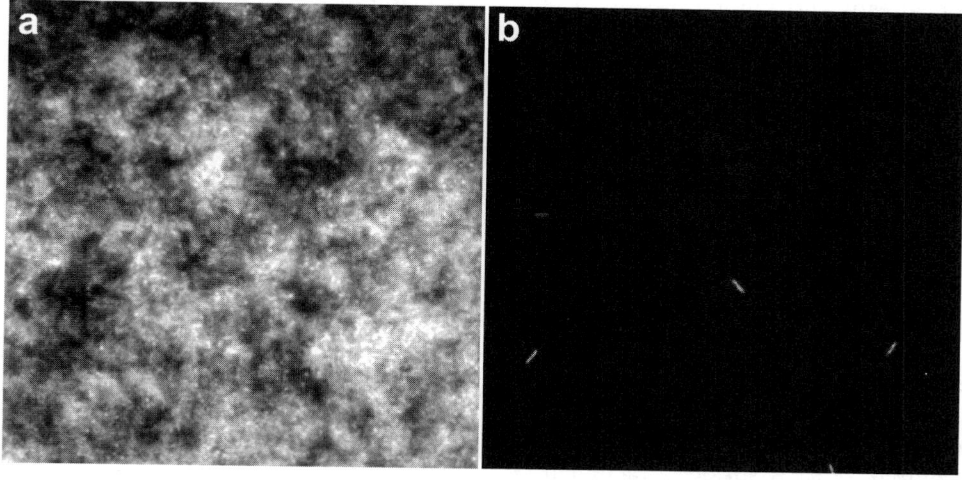

Figure 1. Biofilm formation by the wild-type (a) and an *adnA* mutant (b) *Pseudomonas fluorescens* Pf0-1. Strains possessing a GFP-expressing plasmid were incubated for 24 h in the presence of a glass cover-slip. The cover-slip was gently washed with distilled water, and adherent cells were visualized by fluorescence microscopy at 600x magnification. After 24 h wild-type Pf0-1 has formed a thick, structured biofilm, while disruption of *adnA* profoundly reduced the ability to initiate growth as a biofilm. The importance of biofilms for soil fitness is suggested by the correlation between the biofilm and soil-persistence defects associated with *adnA* mutants.

the LPS, which plays a role in the assembly of the flagella). These genes are not contiguous and rather belong to a regulon. Field studies showed that the loss of *adnA* reduces the ability of *P. fluorescens* to spread and persist in soil, suggesting the transcriptional control by AdnA over a number of genes involved in bacterial fitness in soil, and confirming chemotaxis and motility as crucial traits for the competitiveness of a bacterium in soil (37).

Adhesion and Biofilm Formation

Given that soil has a myriad of surfaces, both biotic and abiotic, it is not surprising that adhesion and then biofilm formation play such an important role in the bacterial colonization of soil. A general model to explain the progression from free-living to a mature biofilm (44) proposes a three stage process: (i) initial attachment to a surface (establishment of a monolayer); (ii) growth and recruitment to form microcolonies; and (iii) maturation into a spatially heterogenous, exopolysaccharide (EPS)-surrounded biofilm.

How is initial colonization achieved?

The first stage of biofilm formation is the development of a monolayer of cells, which was recently shown to be a genetically distinct stage of the biofilm pathway in *Vibrio cholerae* (41). Environmental cues (such as nutritional or physical changes) are thought to trigger the initial transition from free-living to attached growth, while the type of surface available may influence the attachment mechanism used, rather than whether or not to attach per se.

Flagellation, already known for its involvement in motility, also plays a role in attachment by bringing the bacteria into close proximity with surfaces and facilitating movement across the surface. Flagella also play a direct role in temporary adhesion to surfaces. The presence of flagella modifies the electrical charge of bacteria, allowing closer interaction of flagellar distal adhesive proteins (35). However, ongoing research has demonstrated cases in which flagella can become dispensable. For example, in *P. fluorescens*, flagella-independent biofilm formation has been seen in various mutants (46, 53). In fact, enhanced biofilm formation has been noted in non-motile *P. fluorescens* Pf0-1 (*adnA*$^-$ mutant) upon introduction of an additional mutation (53).

Factors required for early surface attachment include LPS (66), the ClpP protease (46), and pili. In particular, the role of pili has been well studied. Pili appear to be essential, but with different roles: *E. coli* absolutely requires type I pili for attachment, whereas the role of pili in *P. aeruginosa* seems to be more for allowing movement across the surface (using "twitching motility").

A recent report implicated an ABC transporter and a large cell-surface protein as important in the progress of *P. fluorescens* WCS365 from transient surface attachment to 'irreversible' attachment, which is a prerequisite for further development of the surface attached community (32). Indeed, protease treatment prevents *P. fluorescens* biofilm development and lends support to the importance of cell surface and possibly secreted proteins for biofilm formation (46).

Formation of microcolonies

In *P. aeruginosa*, genetic experiments and microscopic observations clearly indicate that pili-mediated twitching motility across the surface is an important process for microcolony formation. Rather than resting and dividing once they are in contact with a surface, cells move toward each other and aggregate as microcolonies. Bacteria incapable of twitching motility, like *P. fluorescens*, must solve the problem of how to aggregate into microcolonies using a different mechanism (44). Interestingly, it appears that oxygen sensing is important for transition from a monolayer to the later stages of biofilm growth in *Agrobacterium tumefaciens* (50). The speed at which microcolonies form provides additional evidence that they are not simply a result of multiplication, but rather are the product of a distinct, active process.

The mature biofilm

The switch to attachment and growth on a surface is made in response to cues from the environment, and presumably provides significant benefits to the organism(s) participating in the development of the structure. These benefits will be fully realized upon maturation of the biofilm structure.

The final step in the development of a mature biofilm, as revealed by microscopic and genetic studies, is the formation of a complex, EPS-surrounded, three-dimensional structure. The appearance of mature biofilms is often described as resembling mushrooms protruding from a surface. Whether this is really a genetically distinct stage of development or a continuation of the microcolony stage is not totally resolved. Two key elements seem to be of importance in progression to the fully mature biofilm of *P. aeruginosa*, namely, quorum sensing and EPS. Quorum sensing (*lasI*) mutants fail to develop the characteristic 3-D architecture; these mutants resemble pili deficient mutants, which fail to form microcolonies. The importance of EPS seems evident since it surrounds mature biofilms. For instance, an *E. coli* strain unable to form colanic acid (its major EPS) can attach to surfaces but cannot form the complex biofilm structures (13). Interestingly, upregulated expression of genes for the synthesis of alginate, a significant component of "mucoid" biofilms formed by *P. aeruginosa* in cystic fibrotic lungs but not in soil (67), is concomitant with reduced flagella gene expression, indicating changes in expression of processes that are important at different developmental stages (26). The common association of EPS with biofilms points to its importance in the structures, and possible functions include protection from antibacterial compounds (a physical barrier), environmental stresses (e.g., osmotic shock, dessication), and predation.

Development of a mature biofilm is a complex process, requiring a significant investment on the part of the bacteria. This is supported by the identification of a large number of genes that are induced in biofilms, as shown by microarray experiments (65), and by using an "in biofilm expression technology" promoter trap (24). Given the apparent cost of making a biofilm and the common occurrence of biofilms, it seems intuitive that this mode of life presents considerable benefits for bacteria, and thus the ability to grow as a biofilm is a significant fitness trait. It should be noted that to respond rapidly to conditions, it is also likely that well-adapted bacteria have devised efficient and well-regulated pathways to dissolve the biofilm community if need be. These dissolution pathways are only just beginning to be explored using genetic tools.

The broad outline above has been delineated by studying single species biofilms, but in the wild it is likely that multispecies biofilms are prevalent. These would be structurally similar, and at least as complex in terms of the mechanism underlying their development and maintenance.

Environmental (nutritional) cues influence biofilm formation

During screens for mutants that could not form biofilms, certain mutants did not fit easily into the development of a model for biofilm formation. For example, Monds et al isolated mutants in the phosphate specific transport (Pst) system in the biocontrol strain *P. aureofaciens* PA147-2, which could not form biofilms under normal culture conditions (40). Further analysis demonstrated that *P. aureofaciens* was able to sense the phosphate concentration, and when inorganic phosphate was very low, biofilm formation was negatively regulated by the PhoB/R two-component regulatory pair. This supports the contention that planktonic growth or biofilm formation, as alternative growth cycles, result from cellular responses to environmental signals. Intriguingly, the biofilm defect in *pst* mutants of *P. aureofaciens* was accompanied by a loss of antifungal activity (40). Inhibition of fungal growth is an important fitness trait enabling exclusion of competitors from an environment, and coregulation of these two traits further illustrates the complexity of fitness in soil environments. Based upon the phenotype of a Crc (catabolite repression control protein) mutant of *P. aeruginosa,* it has been suggested that Crc integrates nutritional cues as part of a pathway that regulates biofilm formation (45).

Additional evidence that bacteria make lifestyle choices based upon the nutritional environment comes from studies on biofilm formation by *P. fluorescens* Pf0-1. This strain was able to form strong biofilms when using glucose as a carbon source; however, Pf0-1 grown with citrate or glutamate as the sole source of carbon was unable to form biofilms. Further adjusting the medium by adding $MgSO_4$ and $CaCl_2$ rescued the biofilm defect. Finally, adding $FeSO_4$ had no effect on biofilms formed in glutamate (plus $MgSO_4$ and $CaCl_2$), but abolished biofilms if the amended medium had citrate rather than glutamate (Casaz and Levy, unpublished). These findings in the wildtype Pf0-1 are supported by the rescue of biofilm growth in various *P. fluorescens* WCS365 mutants by altered nutrient conditions (46).

"Real World"—the Use of IVET Technology

In addition to detailed, laboratory-based studies of surface colonization, several laboratories have identified important traits for rhizosphere colonization by using experimental systems involving bacteria-plant interactions, and work in the Levy laboratory is beginning to apply similar tools to ecological success in bulk soil.

Screening for transposon mutants with defects in root colonization or soil survival requires considerable effort. In addition, genes with subtle effects on fitness could easily be missed in a transposon hunt. One can hypothesize that at least a subset of genes whose expression is upregulated in situ would be important for growth in that environment. Such reasoning was behind the development of in vivo expression technology (IVET) by the Mekalanos group in 1993 (36). Recently, IVET has been adapted to allow the identification of rhizosphere- and soil-induced genes. This genetic approach has yielded in excess of 80 sequences that are induced in the rhizosphere or soil, and therefore represent genes that may encode fitness traits specific to those environments. Among these, many of the genes identified have no known function, indicating that IVET is indeed uncovering novel aspects of microbial biology relating to fitness in soil-based environments. In addition, recent work in Stuart Levy's laboratory found that 10 of 22 soil-induced genes are actually encoded on the opposite strand of DNA to known genes and therefore overlap those known genes (55). Previously, only a limited number of sequences arranged in this way had been reported. The soil-induced sequences that overlap known genes are novel, and represent a new feature of genome organization. Clearly, fitness in soil is a complex process, and laboratory studies have only just begun to discover the underlying mechanisms.

Although IVET studies identify large numbers of sequences, assessing the relative importance of each gene requires more traditional genetic studies. In an effort to add value to IVET-derived results, Paul Rainey's group examined the competitive fitness of a mutant with a transposon insertion in the *wss* operon. This locus consists of 10 genes that encode an acetylated cellulose polymer (57), and was isolated in their IVET screen (25). This mutant showed reduced competitive fitness (relative to wild type) in rhizosphere colonization experiments, and an even more dramatic reduction in fitness in phyllosphere colonization (25). No defect was seen in bulk soil, confirming the niche-specific nature of the selected sequences. In our soil studies, mutations in each of three soil-induced sequences resulted in a defect in soil growth in the first day after inoculation (56). The results with mutants support the hypothesis that sequences induced in a given environment specify potential fitness traits and suggest that IVET studies will yield a rich collection of traits important within various soil-based niches.

Evolution and Adaptation—Bacterial Facelifts

Genetic diversity is a critical factor for bacterial adaptation to environmental changes. During the stationary phase of the bacterial life cycle, the expression of the "growth advantage in stationary phase" phenotype depends on the appearance of new mutations in the population that confer a competitive advantage to mutated cells, allowing them to take over the population (69). Processes increasing the mutability of bacteria may thus lead to increased adaptability and survival in soil and other environments.

Mutator alleles

It has been argued that because most newly arising mutations are neutral or deleterious, organisms have evolved mechanisms to keep the mutation frequency as low as possible (about 5×10^{-10} in *E. coli*). Indeed, a plethora of enzymes ensure the maintenance of genetic information. These are enzymes for protection or repair of DNA, hydrolysis of modified nucleotides (e.g., MutT that degrades 8-oxo-dGTP), the 3′ to 5′ exonuclease activity of DNA polymerases that have a proofreading activity, and the post-replicative methyl-directed mismatch repair system (MRS). But in natural bacterial populations, up to one per cent of the isolates are "mutator" clones, i.e., clones that have constitutively high mutation rates (increased by up to 10,000-fold) due to the inactivation of any of the enzymatic systems listed above (62).

Despite the cost due to a heavy deleterious mutation load, mutator alleles can spread in asexual populations after an adaptation process if they have generated mutations with which they remain associated and which compensate for the cost of deleterious mutations (61). Conversely, the fate of the fitness-improving allele thereby generated is not necessarily dependent upon maintenance of the mutator allele, and can persist in the bacterial population even if the mutator allele happens to revert. The fact that high mutation rates can lead to the loss of vital functions when the mutator allele does not revert supports a model in which the evolution of bacterial populations occurs through alternating periods of high (new environment, introduction of a stress factor) and low (bacterial stasis) mutation rates (62). The factors favoring the fixation of a mutator allele are (i) a large population, (ii) a strong fitness advantage, (iii) a succession of favorable mutations, and (iv) a limited rate of genetic exchanges (mutator alleles constitute a second-order selection which is dramatically shut down after the introduction into a population of bacterial conjugation or transformation, despite the rarity of these genetic exchanges [61]). An evolutionary strategy could lay in inducible mutators, with a mutation rate increase induced by environmental changes, but the existence of such an inducible mutator still remains to be clearly demonstrated (62).

Site-specific recombination and colony phase variation

Inactivation of genes homologous to *xerC* (*E. coli*) or *sss* (*Pseudomonas aeruginosa*), which specify enzymes belonging to the λ-integrase family of site-specific recombinases, leads to competitive defects in root colonization in *P. fluorescens* WCS365 (19, 35). In pathogenic bacteria these enzymes have previously been implicated in DNA rearrangements involved in phenotype switching for some surface antigens, like pili, flagella, LPS, or lipoprotein, switches that help the bacterium to colonize new host tissues or evade the immune system.

The demonstration that an *sss* like mutant of *P. fluorescens* is impaired in competitive root-tip colonization of gnotobiotically grown potato, radish, wheat, and tomato, as well as in colonization of potting soil, suggests that phase variation is important for root colonization in an analogous manner to the suspected importance in pathogenesis (19). The role in rhizosphere colonization is further highlighted by a recent study on phase variation in *P. fluorescens* F113. A homogenous population of *P. fluorescens* F113 was shown to diversify into three distinct phenotypic variants after colonization of the alfalfa rhizosphere. One variant was indistinguishable from the wild type, while the other types (F and S variants) preferentially colonized distal parts of roots and had a range of altered traits, including increased flagellin synthesis and increased flagella, increased siderophore production, loss of exoprotease and cyanide production, and altered LPS profile (54). These observations indicate that subpopulations with particular niche preferences are generated by phase variation, and it is possible that such sub-populations provide an advantage during periods of environmental change, in that they may be more competitive in the new environment than their parent.

Adaptive (or stationary-phase) mutation, in which mutations occur more often when selected than when not, is a still controversial process that would be promoted by recombination (9). But whether this kind of process is generally true, or specific to a few strains of *E. coli*, is unclear, and any occurrence in soil environments has yet to be shown.

SOS polymerase

The intrinsic instability of DNA and the reactive intermediates of cellular metabolism and environmental agents cause damaged DNA bases. During replication, the replication fork stalls when encountering a damaged base on the template DNA, thus threatening the continuity of the replication and the survival of the cell. To overcome this threat, bacteria benefit from the coordinated expression of the SOS system, regulated by RecA and LexA proteins, which ensures the resumption of replication even if mutations occur during this process at a higher rate than "normal." Indeed, SOS system DNA polymerases IV and V either misincorporate a nucleotide (Pol V) or induce a frameshift (Pol IV) in front of the damaged DNA template base while resuming the replication (28). Another SOS system DNA polymerase, Pol II, also allows the cell to overcome a stalling of the replication fork, but this polymerase harbors an active 3′ exonuclease and synthesizes DNA accurately. The action of Pol II thus counterbalances Pol IV and V, creating in the cell with damaged DNA a dynamic balance between conserving the genetic information (Pol II) and generating genetic diversity (Pol IV and V). Interestingly, under certain conditions, an error-prone polymerase can catalyze an error-free reaction, while an error-free polymerase will catalyze an error-prone reaction (28).

It is estimated that when the SOS polymerases are derepressed, the mutation rate increases by 100-fold, thus generating an increased mutational load. In a favorable environment, the cost of high mutation frequency outweighs the benefit, but in inhospitable environments the risks may be more than offset by an increased relative fitness, paving the road toward adaptation and evolution, especially during the stationary phase of the bacterial life cycle (28). For instance, 95% of the mutations that arise in aging colonies are RecA dependant and LexA independent (7), and a mutant deficient for one or several SOS DNA polymerases exhibits reduced fitness compared to the wild-type strain (68). The importance of the genes coding for Pol II, IV, and V for the survival of the cell and as a fitness factor is also supported by the fact that these loci appear to be as highly conserved among bacteria as housekeeping genes (7).

Horizontal gene transfer

Horizontal gene transfer (HGT), or the flow of genetic information between organisms, is a rich source of new genetic information in recipient cells. Genomic sequence data suggest that HGT is one of the major forces in shaping prokaryotic genomes as we know them today (15, 27). Genes can be transferred between prokaryotes by phage-mediated transduction, natural transformation, or by conjugation. The latter is regarded as the most potent means for genes to spread. The benefits for a bacterium obtaining genes in this way are obvious—a biochemical function can be obtained far more rapidly than by de novo evolution. There seems to be little if any barrier preventing the HGT between microbial species—indeed, *E. coli* can even act as a gene donor in conjugation with the yeast *Saccharomyces cerevisiae* (31).

Adaptation by the acquisition of new genes can be particularly useful for a newcomer in a niche that presents particular challenges, in which there is a well-established indigenous population. An important example of adaptation by HGT is the spread of antibiotic resistance genes, and subsequent selection of bacteria harboring those resistance genes. Soil is likely to be a reservoir of a range of antibiotic resistance genes. The evolution of antibiotic resistance by HGT is discussed elsewhere in this volume. An excellent example of the acquisition of an advantageous trait by HGT involves the spread among rhizobia of nitrogen fixation and symbiosis genes on plasmids and "symbiosis islands" (see below). Other well-documented examples of the

HGT of useful traits, both in bulk soil and in the rhizosphere, involve antibiotic and heavy metal (especially mercury) resistance, as well as xenobiotic degradation. Research indicates that both conjugation and transformation occur at relatively high levels within biofilms, partly because the number of cells in those communities means that cells are in close proximity (reviewed in reference 14).

HGT case study—acquisition of "symbiosis island"

The "symbiosis island" has been described for the bacteria belonging to the rhizobia, a group of bacteria split between the α- and β-subdivision of the *Proteobacteria,* and defined by their ability to enter into a symbiotic relationship with leguminous plants (38, 42). As part of the symbiosis, the bacterial partner takes up residence in plant tissue-derived nodules where it fixes atmospheric nitrogen into forms usable by the plant, while the plant provides the bacteria with metabolizable photosynthetic carbon as well as shelter from the soil environment. The nitrogen fixation has an extended agricultural and environmental importance. Symbiotic nitrogen fixation is responsible for the majority of biological nitrogen fixation in terrestrial ecosystems and is crucial for replenishing nitrogen lost by denitrification processes (5).

The initial establishment and nitrogen-fixing effectiveness of the symbiosis requires a large number of genes carried by the bacterial partner (principally *nod, nif,* and *fix*) (48), which are generally carried on one or more symbiotic plasmids (38). However, in species of *Bradyrhizobium* and in *Mesorhizobium loti* strain ICMP3153, the genes have been found to be located on the bacterial chromosome (29, 58).

Although the frequency with which horizontal gene transfers occur under natural conditions remains controversial, there is considerable evidence demonstrating that such transfers play a significant role in shaping the evolution and population genetic structure of the rhizobia, even if efforts to directly detect plasmid transfer under field conditions have generally met with little success (10, 33). Interestingly, horizontal gene flow among the rhizobia is not restricted to the conjugative transfer of plasmids, but also may involve chromosomal genes (23, 58, 59, 63). In a recent phylogenetic study, van Berkum et al. (63) found that using sequences of the 16S rRNA, the 23S rRNA, and the internally transcribed spacer (ITS) region located between the genes coding for the rRNAs each produced distinct phylogenetic trees, and obtained evidence to support the occurrence of chromosomal recombination events between small portions of the 16S rRNA genes of *Bradyrhizobium elkanii* and *Mesorhizobium* sp., and between *Sinorhizobium* and *Mesorhizobium.*

Sullivan et al. (58) confirmed such chromosomal gene transfers after inoculating a stand of *Lotus corniculatus* with a single strain (ICMP3153) of *Mesorhizobium loti,* in a field which contained no indigenous rhizobia capable of nodulating this plant host. After several years a collection of genetically diverse mesorhizobial strains, recovered from *Lotus* nodules, possessed a chromosomally located region of DNA that was identical to a region in strain ICMP3153. It was demonstrated that the genetically diverse mesorhizobia became capable of symbiosis due to the transfer and chromosomal integration of a 502-kb DNA element from strain ICMP3153 into a phe-tRNA gene of nonsymbiotic mesorhizobia present in the soil (59). Located at the left end of the element is a P4-type integrase gene (59) which is required for chromosomal integration and excision of the element (60). Laboratory matings have confirmed that the symbiosis gene cluster could be transmitted to nonsymbiotic mesorhizobia resulting in transconjugants that were able to nodulate *L. corniculatus* (59). Due to its similarity to pathogenicity islands, clusters of virulence-related genes found in many gram-negative bacterial pathogens but not found in avirulent strains, this element has come to be called a "symbiosis island" (59).

By comparative sequence analysis with strain R7A, a derivative of strain ICMP3153, a putative symbiosis island has also been identified in *M. loti* strain MAFF303099, but it is not known whether it is transmissible and functions as a conjugative transposon as does the R7A island (60). Similarly, a 410-kb cluster of potentially symbiosis-related genes has been identified on the chromosome of *Bradyrhizobium japonicum* USDA 110 (29).

To what extent does horizontal symbiosis-related gene transfer between rhizobia affect their ability to survive in the soil environment? In addressing this question, it is necessary to keep in mind that the rhizobia exist either living free in soil or as symbionts within the sheltering tissues of the host plant. Symbiotic rhizobia need the ability to persist in the environment as free-living saprophytes, both before the infection of the host plant's roots and after plant senescence at the end of the growing season (5). As the persistence of relatively large numbers of nonsymbiotic rhizobia in soil is well recognized (58), the possession of symbiotic genes by a rhizobial strain is unlikely to be the sole, or even a primary, determinant of saprophytic survival. In a study comparing *Rhizobium leguminosarum* bv. *viciae* CT0370 (lacking a symbiotic plasmid) and a derivative strain carrying pRL1JI (a symbiotic plasmid) it was shown that the strain possessing the symbiotic plasmid outnumbered the plasmid-free strain in the rhizosphere and rhizoplane, while the inverse occurred in bulk soil (10).

Therefore, the advantages of possessing symbiosis-related DNA, apart from enabling the bacterium to initiate the symbiosis per se, may be manifest primarily in the rhizosphere soil. It has been proposed that perhaps one of the soil survival functions of symbiotic DNA is to extend the metabolic capabilities of the rhizobia in the rhizosphere and increase microbial access to diverse carbon and nitrogen sources, thus increasing saprophytic competence with a concomitant increase in relative abundance. For instance, evidence for the existence of genes coding transport systems and diverse catabolic pathways has been obtained for the symbiosis island of *Mesorhizobium loti* (60) and the symbiotic DNA region in *Bradyrhizobium japonicum* (29). As additional sequencing data for other rhizobia becomes available, the role of symbiotic DNA in enhancing the survival of the bacteria in the soil environment may be clarified.

SUMMARY AND SPECULATION

To cope with an ever-changing environment, soil-dwelling bacteria use a range of strategies. Motility and taxis contribute to survival by allowing them to sense and move to more hospitable environments and to avoid less favorable conditions. Colonization of a given environment is likely to involve growth of a biofilm, which is thought to afford protection from a number of hostile environmental factors. Environmental sensing is probably used by bacteria in making the decision whether to attach to a surface and grow as a biofilm, and whether to remain attached or to dissolve the biofilm and seek a new niche. Soil microbes can also adapt to the conditions around them. Mutation generates genetic diversity, leading to improved survival in difficult environments. Phase variation can allow bacteria to undergo a "phenotypic switch," so that appropriate genes are expressed to allow the best survival under the prevailing conditions. Finally, the acquisition of genes by horizontal transfer is a rapid means by which bacteria can gain new abilities, some of which might contribute to improved survival or competitive fitness.

The environment is a reservoir of pathogenic bacteria which are capable of living free from a host (6, 12, 47, 49) and is likely to harbor many opportunistic pathogens. A good example is *Burkholderia cepacia*, a species formerly described as a rhizosphere inhabitant and a pathogen of onion, but which has more recently been isolated from patients suffering from cystic fibrosis (6). To colonize their host, opportunistic and bona fide pathogens may use some of the following traits: chemotactism and motility, adhesion and formation of biofilms, secretion of siderophores and of hydrolytic enzymes, and interspecies communication. It should not escape our attention that these traits are important in environmental persistence, and further study of bacterial ecology outside of hosts may give insight into the mechanisms opportunists use to gain a foothold when infecting a patient.

REFERENCES

1. **Alexandre, G., S. E. Greer, and I. B. Zhulin.** 2000. Energy taxis is the dominant behavior in *Azospirillum brasilense*. *J. Bacteriol.* **182:**6042–6048.
2. **Alexandre, G., S. Greer-Phillips, and I. B. Zhulin.** 2004. Ecological role of energy taxis in microorganisms. *FEMS Microbiol. Rev.* **28:**113–126.
3. **Andrén, O., and J. Balandreau.** 1999. Biodiversity and soil functioning - from black box to can of worms? *Appl. Soil Ecol.* **13:**105–108.
4. **Arora, S. K., B. W. Ritchings, E. C. Almira, S. Lory, and R. Ramphal.** 1997. A transcriptional activator, FleQ, regulates mucin adhesion and flagellar gene expression in *Pseudomonas aeruginosa* in a cascade manner. *J. Bacteriol.* **179:**5574–5581.
5. **Atlas, R. M., and R. Bartha.** 1998. Microbial ecology: fundamentals and applications. Benjamin Cummings Publishing Company, Inc., Menlo Park, Calif.
6. **Bevivino, A., S. Tabacchioni, L. Chiarini, M. V. Carusi, M. Del Gallo, and P. Visca.** 1994. Phenotypic comparison between rhizosphere and clinical isolates of *Burkholderia cepacia*. *Microbiology* **140:**1069–1077.
7. **Bjedov, I., O. Tenaillon, B. Gerard, V. Souza, E. Denamur, M. Radman, F. Taddei, and I. Matic.** 2003. Stress-induced mutagenesis in bacteria. *Science* **300:**1404–1409.
8. **Bourret, R. B., and A. M. Stock.** 2002. Molecular information processing: lessons from bacterial chemotaxis. *J. Biol. Chem.* **277:**9625–9628.
9. **Bull, H. J., G. J. McKenzie, P. J. Hastings, and S. M. Rosenberg.** 2000. Evidence that stationary-phase hypermutation in the *Escherichia coli* chromosome is promoted by recombination. *Genetics* **154:**1427–1437.
10. **Clark, I. M., T. A. Mendum, and P. R. Hirsch.** 2002. The influence of the symbiotic plasmid pRL1JI on the distribution of GM rhizobia in soil and crop rhizospheres, and implications for gene flow. *Antonie van Leeuwenhoek* **81:**607–616.
11. **Compeau, G., B. J. Al-Achi, E. Platsouka, and S. B. Levy.** 1988. Survival of rifampin-resistant mutants of *Pseudomonas fluorescens* and *Pseudomonas putida* in soil systems. *Appl. Environ. Microbiol.* **54:**2432–2438.
12. **Dance, D. A.** 2000. Ecology of *Burkholderia pseudomallei* and the interactions between environmental *Burkholderia* spp. and human-animal hosts. *Acta Trop.* **74:**159–168.
13. **Danese, P. N., L. A. Pratt, and R. Kolter.** 2000. Exopolysaccharide production is required for development of *Escherichia coli* K-12 biofilm architecture. *J. Bacteriol.* **182:**3593–3596.
14. **Davison, J.** 1999. Genetic exchange between bacteria in the environment. *Plasmid* **42:**73–91.
15. **de la Cruz, F., and J. Davies.** 2000. Horizontal gene transfer and the origin of species: lessons from bacteria. *Trends Microbiol.* **8:**128–133.
16. **de Weger, L. A., A. J. van der Bij, L. C. Dekkers, M. Simons, C. A. Wijffelman, and B. J. J. Lugtenberg.** 1995. Colonization of the rhizosphere of crop plants by plant-beneficial pseudomonads. *FEMS Microbiol. Ecol.* **17:**221–227.
17. **DeFlaun, M., B. Marshall, E. Kulle, and S. Levy.** 1994. Tn5 insertion mutants of Pseudomonas fluorescens defective in

adhesion to soil and seeds. *Appl. Environ. Microbiol.* **60:**2637–2642.

18. **DeFlaun, M., A. Tanzer, A. McAteer, B. Marshall, and S. Levy.** 1990. Development of an adhesion assay and characterization of an adhesion-deficient mutant of *Pseudomonas fluorescens*. *Appl. Environ. Microbiol.* **56:**112–119.
19. **Dekkers, L. C., C. C. Phoelich, L. van der Fits, and B. J. Lugtenberg.** 1998. A site-specific recombinase is required for competitive root colonization by *Pseudomonas fluorescens* WCS365. *Proc. Natl. Acad. Sci. USA* **95:**7051–7056.
20. **Dighton, J., H. E. Jones, C. H. Robinson, and J. Beckett.** 1997. The role of abiotic factors, cultivation practices and soil fauna in the dispersal of genetically modified microorganisms in soils. *Appl. Soil Ecol.* **5:**109–131.
21. **Dixon, J. B.** 1991. Roles of clays in soils. *Applied Clay Science* **5:**489–503.
22. **Fenchel, T.** 2002. Microbial behavior in a heterogeneous world. *Science* **296:**1068–1071.
23. **Finan, T. M.** 2002. Evolving insights: symbiosis islands and horizontal gene transfer. *J. Bacteriol.* **184:**2855–2856.
24. **Finelli, A., C. V. Gallant, K. Jarvi, and L. L. Burrows.** 2003. Use of in-biofilm expression technology to identify genes involved in *Pseudomonas aeruginosa* biofilm development. *J. Bacteriol.* **185:**2700–2710.
25. **Gal, M., G. M. Preston, R. C. Massey, A. J. Spiers, and P. B. Rainey.** 2003. Genes encoding a cellulosic polymer contribute toward the ecological success of *Pseudomonas fluorescens* SBW25 on plant surfaces. *Mol. Ecol.* **12:**3109–3121.
26. **Garrett, E. S., D. Perlegas, and D. J. Wozniak.** 1999. Negative control of flagellum synthesis in *Pseudomonas aeruginosa* is modulated by the alternative sigma factor AlgT (AlgU). *J. Bacteriol.* **181:**7401–7404.
27. **Gogarten, J. P., W. F. Doolittle, and J. G. Lawrence.** 2002. Prokaryotic evolution in light of gene transfer. *Mol. Biol. Evol.* **19:**2226–2238.
28. **Goodman, M. F.** 2002. Error-prone repair DNA polymerases in Prokaryotes and Eukaryotes. *Annu. Rev. Biochem.* **71:**17–50.
29. **Göttfert, M., S. Röthlisberger, C. Kündig, C. Beck, R. Marty, and H. Hennecke.** 2001. Potential symbiosis-specific genes uncovered by sequencing a 410-kilobase DNA region of the *Bradyrhizobium japonicum* chromosome. *J. Bacteriol.* **183:**1405–1412.
30. **Hassink, J., L. A. Bouwman, K. B. Zwart, and L. Brussaard.** 1993. Relationships between habitable pore space, soil biota and mineralization rates in grassland soils. *Soil Biol. Biochem.* **25:**47–55.
31. **Heinemann, J. A., and G. F. Sprague, Jr.** 1989. Bacterial conjugative plasmids mobilize DNA transfer between bacteria and yeast. *Nature* **340:**205–209.
32. **Hinsa, S. M., M. Espinosa-Urgel, J. L. Ramos, and G. A. O'Toole.** 2003. Transition from reversible to irreversible attachment during biofilm formation by *Pseudomonas fluorescens* WCS365 requires an ABC transporter and a large secreted protein. *Mol. Microbiol.* **49:**905–918.
33. **Hirsch, P. R., and J. D. Spokes.** 1994. Survival and dispersion of genetically modified rhizobia in the field and genetic interactions with native strains. *FEMS Microbiol. Ecol.* **15:**147–160.
34. **Jyot, J., N. Dasgupta, and R. Ramphal.** 2002. FleQ, the major flagellar gene regulator in *Pseudomonas aeruginosa,* binds to enhancer sites located either upstream or atypically downstream of the RpoN binding site. *J. Bacteriol.* **184:**5251–5260.
35. **Lugtenberg, B. J., L. Dekkers, and G. V. Bloemberg.** 2001. Molecular determinants of rhizosphere colonization by *Pseudomonas. Annu. Rev. Phytopathol.* **39:**461–490.
36. **Mahan, M. J., J. M. Slauch, and J. J. Mekalanos.** 1993. Selection of bacterial virulence genes that are specifically induced in host tissues. *Science* **259:**686–688.
37. **Marshall, B., E. A. Robleto, R. Wetzler, P. Kulle, P. Casaz, and S. B. Levy.** 2001. The adnA transcriptional factor affects persistence and spread of *Pseudomonas fluorescens* under natural field conditions. *Appl. Environ. Microbiol.* **67:**852–857.
38. **Martínez-Romero, E., and J. Caballero-Mellado.** 1996. *Rhizobium* phylogenies and bacterial genetic diversity. *Crit. Rev. Plant Sci.* **15:**113–140.
39. **Mirleau, P., S. Delorme, L. Philippot, J. Meyer, S. Mazurier, and P. Lemanceau.** 2000. Fitness in soil and rhizosphere of *Pseudomonas fluorescens* C7R12 compared with a C7R12 mutant affected in pyoverdine synthesis and uptake. *FEMS Microbiol. Ecol.* **34:**35–44.
40. **Monds, R. D., M. W. Silby, and H. K. Mahanty.** 2001. Expression of the Pho regulon negatively regulates biofilm formation by *Pseudomonas aureofaciens* PA147-2. *Mol. Microbiol.* **42:**415–426.
41. **Moorthy, S., and P. I. Watnick.** 2004. Genetic evidence that the Vibrio cholerae monolayer is a distinct stage in biofilm development. *Mol. Microbiol.* **52:**573–587.
42. **Moulin, L., A. Munive, B. Dreyfus, and C. Boivin-Masson.** 2001. Nodulation of legumes by members of the beta-subclass of Proteobacteria. *Nature* **411:**948–950.
43. **Nunan, N., K. Wu, I. M. Young, J. W. Crawford, and K. Ritz.** 2003. Spatial distribution of bacterial communities and their relationships with the micro-architecture of soil. *FEMS Microbiol. Ecol.* **44:**203–215.
44. **O'Toole, G., H. B. Kaplan, and R. Kolter.** 2000. Biofilm formation as microbial development. *Annu. Rev. Microbiol.* **54:**49–79.
45. **O'Toole, G. A., K. A. Gibbs, P. W. Hager, P. V. Phibbs, Jr., and R. Kolter.** 2000. The global carbon metabolism regulator Crc is a component of a signal transduction pathway required for biofilm development by *Pseudomonas aeruginosa. J. Bacteriol.* **182:**425–431.
46. **O'Toole, G. A., and R. Kolter.** 1998. Initiation of biofilm formation in *Pseudomonas fluorescens* WCS365 proceeds via multiple, convergent signalling pathways: a genetic analysis. *Mol. Microbiol.* **28:**449–461.
47. **Parke, J. L., and D. Gurian-Sherman.** 2001. Diversity of the *Burkholderia cepacia* complex and implications for risk assessment of biological control strains. *Annu. Rev. Phytopathol.* **39:**225–258.
48. **Perret, X., C. Staehelin, and W. J. Broughton.** 2000. Molecular basis of symbiotic promiscuity. *Microbiol. Mol. Biol. Rev.* **64:**180–201.
49. **Rahme, L. G., F. M. Ausubel, H. Cao, E. Drenkard, B. C. Goumnerov, G. W. Lau, S. Mahajan-Miklos, J. Plotnikova, M. W. Tan, J. Tsongalis, C. L. Walendziewicz, and R. G. Tompkins.** 2000. Plants and animals share functionally common bacterial virulence factors. *Proc. Natl. Acad. Sci. USA* **97:**8815–8821.
50. **Ramey, B. E., A. G. Matthysse, and C. Fuqua.** 2004. The FNR-type transcriptional regulator SinR controls maturation of *Agrobacterium tumefaciens* biofilms. *Mol. Microbiol.* **52:**1495–1511.
51. **Ranjard, L., and A. Richaume.** 2001. Quantitative and qualitative microscale distribution of bacteria in soil. *Research Microbiol.* **152:**707–716.
52. **Rao, C. V., J. R. Kirby, and A. P. Arkin.** 2004. Design and diversity in bacterial chemotaxis: a comparative study in *Escherichia coli* and *Bacillus subtilis. PLoS Biol.* **2:**E49. Epub 2004 Feb 17.

53. **Robleto, E. A., I. Lopez-Hernandez, M. W. Silby, and S. B. Levy.** 2003. Genetic analysis of the AdnA regulon in *Pseudomonas fluorescens:* nonessential role of flagella in adhesion to sand and biofilm formation. *J. Bacteriol.* **185:**453–460.
54. **Sanchez-Contreras, M., M. Martin, M. Villacieros, F. O'Gara, I. Bonilla, and R. Rivilla.** 2002. Phenotypic selection and phase variation occur during alfalfa root colonization by *Pseudomonas fluorescens* F113. *J. Bacteriol.* **184:**1587–1596.
55. **Silby, M. W., P. B. Rainey, and S. B. Levy.** 2004. IVET experiments in *Pseudomonas fluorescens* reveal cryptic promoters at loci associated with recognizable overlapping genes. *Microbiology* **150:**518–520.
56. **Silby, M. W., and S. B. Levy.** 2004. Use of in vivo expression technology to identify genes important in growth and survival of *Pseudomonas fluorescens* Pf0-1 in soil: discovery of expressed sequences with novel genetic organization. *J. Bacteriol.* **186:**7411–7419.
57. **Spiers, A. J., J. Bohannon, S. M. Gehrig, and P. B. Rainey.** 2003. Biofilm formation at the air-liquid interface by the *Pseudomonas fluorescens* SBW25 wrinkly spreader requires an acetylated form of cellulose. *Mol. Microbiol.* **50:**15–27.
58. **Sullivan, J. T., H. N. Patrick, W. L. Lowther, D. B. Scott, and C. W. Ronson.** 1995. Nodulating strains of *Rhizobium loti* arise through chromosomal symbiotic gene transfer in the environment. *Proc. Natl. Acad. Sci. USA* **92:**8985–8989.
59. **Sullivan, J. T., and C. W. Ronson.** 1998. Evolution of rhizobia by acquisition of a 500-kb symbiosis island that integrates into a phe-tRNA gene. *Proc. Natl. Acad. Sci. USA* **95:**5145–5149.
60. **Sullivan, J. T., J. R. Trzebiatowski, R. W. Cruickshank, J. Gouzy, S. D. Brown, R. M. Elliot, D. J. Fleetwood, N. G. McCallum, U. Rossbach, G. S. Stuart, J. E. Weaver, R. J. Webby, F. J. de Bruijn, and C. W. Ronson.** 2002. Comparative sequence analysis of the symbiosis island of *Mesorhizobium loti* strain R7A. *J. Bacteriol.* **184:**3086–3095.
61. **Tenaillon, O., H. Le Nagard, B. Godelle, and F. Taddei.** 2000. Mutators and sex in bacteria: conflict between adaptive strategies. *Proc. Natl. Acad. Sci. USA* **97:**10465–10470.
62. **Tenaillon, O., F. Taddei, M. Radman, and I. Matic.** 2001. Second-order selection in bacterial evolution: selection acting on mutation and recombination rates in the course of adaptation. *Res. Microbiol.* **152:**11–16.
63. **van Berkum, P., Z. Terefework, L. Paulin, S. Suomalainen, K. Lindström, and B. D. Eardly.** 2003. Discordant phylogenies within the rrn loci of rhizobia. *J. Bacteriol.* **185:**2988–2998.
64. **van Veen, J. A., L. S. van Overbeek, and J. D. van Elsas.** 1997. Fate and activity of microorganisms introduced into soil. *Microbiol. Mol. Biol. Rev.* **61:**121–135.
65. **Whiteley, M., M. G. Bangera, R. E. Bumgarner, M. R. Parsek, G. M. Teitzel, S. Lory, and E. P. Greenberg.** 2001. Gene expression in *Pseudomonas aeruginosa* biofilms. *Nature* **413:**860–864.
66. **Williams, V., and M. Fletcher.** 1996. *Pseudomonas fluorescens* adhesion and transport through porous media are affected by lipopolysaccharide composition. *Appl. Environ. Microbiol.* **62:** 100–104.
67. **Wozniak, D. J., T. J. Wyckoff, M. Starkey, R. Keyser, P. Azadi, G. A. O'Toole, and M. R. Parsek.** 2003. Alginate is not a significant component of the extracellular polysaccharide matrix of PA14 and PAO1 *Pseudomonas aeruginosa* biofilms. *Proc. Natl. Acad. Sci. USA* **100:**7907–7912.
68. **Yeiser, B., E. D. Pepper, M. F. Goodman, and S. E. Finkel.** 2002. SOS-induced DNA polymerases enhance long-term survival and evolutionary fitness. *Proc. Natl. Acad. Sci. USA* **99:**8737–8741. Epub 2002 Jun 11.
69. **Zambrano, M. M., D. A. Siegele, M. Almiron, A. Tormo, and R. Kolter.** 1993. Microbial competition: *Escherichia coli* mutants that take over stationary phase cultures. *Science* **259:**1757–1760.

Frontiers in Antimicrobial Resistance: a Tribute to Stuart B. Levy
Edited by D. G. White, M. N. Alekshun, and P. F. McDermott

Chapter 33

Ecology of Antibiotic Resistance Genes

ABIGAIL A. SALYERS, HERA VLAMAKIS, AND NADJA B. SHOEMAKER

Microbiologists have long been interested in antibiotic resistance genes and their transmission, but until the 1990s, the main focus of research on antibiotic resistant bacteria was on mechanisms of resistance and mechanisms of horizontal gene transfer in a few bacterial systems. Little attention was paid to the ways in which and the extent to which antibiotic resistance genes were being spread in nature. It would be gratifying to be able to say that scientists suddenly realized that the ecology of resistance genes, i.e., their distribution in real world settings and the factors that gave rise to this distribution, was a fascinating scientific question that merited in-depth research. However, it took the public debates over the safety of genetically modified (GM) plants and the possible consequences of widespread agricultural use of antibiotics to stimulate at long last an interest in the ecology of antibiotic resistance genes. In the process, scientists discovered that the ecology of antibiotic resistance genes was a handy indicator of the extent of bacterial gene transfer in general and was, after all, a fascinating new problem worthy of further study. This article will start with a brief history of the growing public interest in the ecology of resistance genes and will then move on to a survey of some of the conceptual problems that have emerged. No attempt will be made to cover completely the growing literature in this area; rather, the aim is to highlight the issues that are emerging in this exciting and important area. In particular, this review will focus on a few groups of bacteria that are major players in the oral and intestinal ecosystems of humans and animals, the obligate anaerobes. Work on enterococci has been recently reviewed elsewhere (47, 53). A caveat is that although the bacteria that are the focus of this chapter are the numerically predominant bacteria in the ecosystems surveyed, they are not necessarily the only important players. Lower abundance species such as *Escherichia coli* and *Enterococcus* species may well play a leading role in some settings, and the ecology of these groups is certainly worthy of attention.

NEW INTEREST IN THE ECOLOGY OF RESISTANCE GENES

Perhaps the first modern example of the sudden importance of understanding the ecology of antibiotic resistance genes arose in connection with the debate over the safety of GM plants. In this case, the burning question was the ultimate fate of the bacterial antibiotic resistance genes that were introduced into the plant along with the gene of interest, the *Bacillus thuringiensis* insecticidal toxin gene (5, 15, 28, 35). Was it possible that DNA released from the plant cells could be taken up by human intestinal bacteria, which might later cause disease in humans?

Eventually, virtually all of the scientists involved in the seemingly interminable debate about this issue accepted the conclusion that such an event would be extremely rare and would probably be clinically irrelevant if it occurred. Entry of resistance genes released from plant cells into human intestinal bacteria would be rare because few intestinal bacteria have been shown to be capable of natural transformation, the process responsible for DNA uptake in nature.

Moreover, if such transfers occurred they would be irrelevant from the medical point of view since the specific resistance genes used in cloning are "old" resistance genes that were cloned in the 1970s. Since then, extensively mutated forms of these genes have arisen naturally among many bacterial species in the community and in hospitals. These "new" resistance genes are the ones that are causing problems in the treatment of human infections. Moreover, in the case of the ampicillin resistance gene found on cloning vec-

Abigail A. Salyers, Hera Vlamakis, and Nadja B. Shoemaker • Department of Microbiology, University of Illinois, Urbana, IL 61801.

tors, modern ampicillin formulations such as Augmentin (amoxicillin-clavulanate) contain not only the β-lactam antibiotic, but also a compound that inhibits the activity of the type of β-lactamase that is encoded by the ampicillin resistance gene.

During the long period of analysis and debate that led to this conclusion, it became obvious that although some natural transformation systems in pathogenic bacteria had been studied intensively, little was known about the extent to which natural transformation occurs in intestinal or soil bacteria. Moreover, studies of natural transformation in a few model organisms such as *Streptococcus pneumoniae, Haemophilus influenzae,* and *Neisseria gonorrhoeae* had led to the conclusion that the spread of DNA by natural transformation would likely be limited to transfers between members of the same species. This conclusion was suggested by the fact that homologous recombination was the most efficient way for a recipient bacterium to incorporate the incoming DNA into its genome.

Subsequent studies have shown that many soil bacteria are naturally transformable, although it is still unclear what significance this fact has in the overall ecology of antibiotic resistance genes. It is not clear to what extent bacteria that can take up DNA by natural transformation are able to retain it, if they lack DNA sequences of sufficient homology to allow recombination. Similarly, the efficiency of illegitimate recombination is poorly understood in most organisms. This process, in which end joining of nonhomologous pieces of DNA occurs, is efficient in some bacteria, such as *Mycobacterium tuberculosis* (14), and may be equally efficient in some other species of bacteria. Nonetheless, there is as yet no coherent body of knowledge about the distribution of bacteria capable of efficient illegitimate recombination and about the importance of this type of DNA incorporation in natural ecosystems.

The subsequent debate over agricultural use of antibiotics underscored even more dramatically the paucity of information about DNA transfer in nature. In this case, scientists had established that antibiotic resistant bacteria from animals on farms where antibiotic use was extensive were entering the food supply and thus, presumably, the human intestine (37, 38). It is true that the meats carrying these bacteria are usually cooked before eating, but the widespread occurrence of foodborne diseases transmitted by these same sources illustrates all too graphically how many holes were in that particular safety net. Also, the smoking of meat does not kill all the bacteria on the meat, and smoked meats are frequently eaten without further treatment.

The concern was not so much that the foodborne bacteria themselves would cause disease, but rather that they might transfer antibiotic resistance genes, probably by conjugation, to human intestinal or other swallowed oral bacteria, during their passage through the human intestinal tract (Fig. 1). Conjugal transfer of DNA is well-known to occur across genus lines, and this process can take place in a matter of hours. Even bacteria that do not colonize the colon remain within that organ for 24 to 48 h before they are excreted (dotted lines in Fig. 1). The colon contains abundant nutrients and many surfaces, such as plant cell wall fragments, where biofilms could form. These conditions should be conducive to conjugative transfer of DNA.

Why would the transfer of antibiotic resistance genes to innocuous intestinal bacteria pose a health

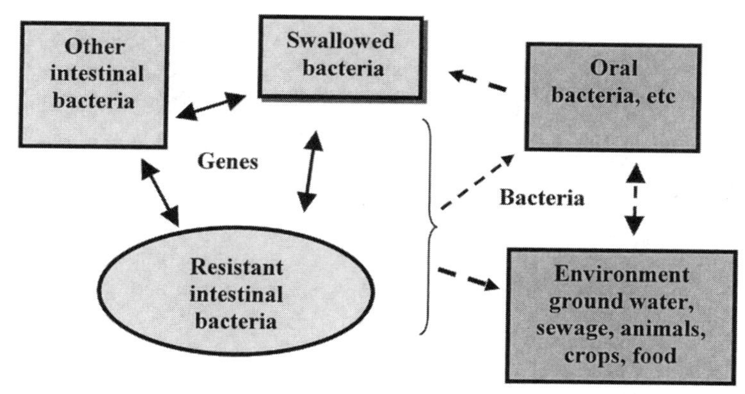

Figure 1. The intestinal bacteria as a resistance gene reservoir and the possible cycle of the bacteria and genes in the environment. Bacteria that normally reside in the human colon (on the left) are normally benign and can transfer resistance genes among themselves. Bacteria that pass through the colon will be in transit long enough to transfer or acquire genes by conjugation and/or transformation or even transduction (shown by solid arrows). The cycle on the right, indicated by dashed lines, follows bacteria that are excreted from the colon in high numbers. Once excreted the bacteria can enter the environment and/or can return to or contaminate other human sites such as skin or the mouth and interact with the microbial communities in these niches before they once again pass through the intestinal tract.

problem? Many intestinal bacteria, although they are normally innocuous as long as they remain in the colon, are capable of causing life-threatening infections after surgery or other trauma to the abdominal area (16). Thus, the higher the incidence of antibiotic resistance in a person's intestinal bacteria, the greater the risk of a difficult to treat postsurgical infection. The question arose as to how likely it was that conjugal transfer of resistance genes occurs between intestinal bacteria and bacteria that are just passing through the colon? If the answer to this question is that such gene transfers occur fairly frequently, the concern expands beyond the issue of agricultural use of antibiotics. Many serious human pathogens such as *Streptococcus pneumoniae* and *Staphylococcus aureus* regularly pass through the colon when they are swallowed from the sites in the throat or upper respiratory tract where they normally reside (Fig. 1). If such bacteria can acquire resistance genes from human colonic bacteria, the colon might well be an important site for dissemination of antibiotic resistance genes in nature (again the dotted lines in Fig. 1). The most numerous bacterial species in the colon would be expected to be the major players in such gene transfer events. Studies have been done of gene transfer agents in *Bacteroides* species, but it is still the case that virtually nothing is known about gene transfer systems and resistance gene distribution in such numerically major, but little-studied, members of the human colonic microbiota such as the *Clostridium coccoides* and the *Clostridium leptum* groups (52).

These examples merely illustrate the potential importance of antibiotic resistance genes in the microbiota of the human and animal body. This normal microbiota also protects against disease and, in the intestinal tract, contributes to human and animal nutrition. Disruptions of the microbiota of the intestine lead to diarrhea and worse. Disruptions of the microbiota of the vaginal tract may even play a role in premature birth (20, 21). The microbiota of the mouth has been implicated in diseases ranging from peridontal disease to heart disease and premature birth (4, 39, 40). In all parts of the body, the microbiota helps to shape the immune responses and physiology of the host animal (7, 22, 23). The changing distribution of resistance genes in this bacterial population will surely have multiple impacts on the functioning of the microbiota.

The Organization That Roared

Focusing scientific interest on an area has proven, unfortunately, to involve more than making a well-reasoned case for its importance. Publicizing the area and making available information that helps to attract new investigators is at least as important. This was particularly critical in the case of the normal microbiota, because the literature in this area is widely scattered in many different journals and government reports. The first systematic attempt made to collect and highlight information about antibiotic resistance genes in commensal bacteria was a project initiated by Stuart Levy's organization, APUA (Alliance for Prudent Use of Antibiotics), and given the name ROAR (Reservoirs of Antibiotic Resistance). This project was the brain child of Stuart Levy and Abigail Salyers. APUA now runs the ROAR website and network, which was supported from its inception by a grant from the U.S. National Institutes of Health. Stuart Levy was one of the first scientists to argue that more attention should be paid to resistance in the bacteria that normally inhabit the human body. As has already been mentioned, these bacteria could serve as reservoirs of antibiotic resistance genes, a concept illustrated in Fig. 1 for human intestinal bacteria. Also, many members of the normal human microbiota are capable of causing serious disease if they are introduced into normally sterile areas of the body by surgery, catheters or wounds. In fact, the majority of hospital-acquired infections probably arise from members of the patient's own microbiota or that of caretakers (6).

APUA oversaw the development of the ROAR web site and network (http://www.tufts.edu/med/apua/ROAR/roarhome.htm), whose purpose is to collect and make available literature and observations on antibiotic resistance genes in bacteria from the mammalian body and the external environment. APUA has now gone on to award grants to support further work in this area.

An interesting development that accompanied the establishment of ROAR highlights the extent to which political pressures have been driving the renaissance of interest in antibiotic resistance gene ecology. Originally, the aim of ROAR was to connect the basic scientists who were working on antibiotic resistance genes and resistance gene transfer in nature. As word of ROAR began to spread, it became clear that there was great interest in this subject on the part of scientists in industry and in regulatory agencies, scientists who were embroiled in the debates about safety of GM plants and agricultural use of antibiotics. In the end, ROAR got to spend very little time in the ivory tower of academia and was soon immersed in the real world.

TOWARD A MOLECULAR ECOLOGY OF RESISTANCE GENES

The first attempts to understand the ecology of antibiotic resistant strains focused on the strains rather than the genes that conferred resistance. Stuart Levy and his colleagues conducted some of the earliest studies in which they monitored the effect of feeding antibiotics to chickens (29, 30). They found that when chickens were fed tetracycline or other antibiotics, the concentration of antibiotic resistant *E. coli* rose significantly. It subsequently decreased when administration of the antibiotic stopped, if the chickens were placed in a setting in which susceptible strains were available to compete with the resistant strains.

Important as these early experiments were in highlighting the practical consequences of antibiotic administration, we now know that they had some limitations. First, administration of the antibiotic was short-term. More recently, Lenski and colleagues have shown, in a classic set of laboratory experiments involving longer antibiotic exposures, that while a strain that has recently become resistant to streptomycin will be rapidly out-competed by susceptible strains, antibiotic exposure of resistant strains over many generations will result in compensatory mutations that make the resistance trait more stable and no longer out-competed by the susceptible strains (27). This observation may explain why cessation of use of an antibiotic in some areas of the world leads at first to a reduction in the incidence of resistance strains, but then the incidence levels off to expose a "hard core" group of resistant strains that do not go away (36).

A second limitation of the early Levy studies, which was not really a flaw in the sense that the technology was not yet available to remedy it, was that only phenotypic resistance, not the resistance genes, was followed. The problem with following phenotype rather than genotype is that genes move freely from one organism to another; and changes in phenotypic resistance may reflect a complex interplay of acquisition and loss of different resistance genes. Today, Levy and others have been quick to take advantage of the molecular revolution and the increasing amount of information about resistance gene transfer mechanisms to develop a more sophisticated way of viewing the ecology of resistance genes.

A third problem with the early Levy experiments was the focus on *E. coli*. In the time during which these early experiments were done, *E. coli* was still believed to be the predominant intestinal bacterial species. And even if one admitted that there were other genera present, *E. coli* was considered to be a quite satisfactory model for their physiology and genetics. Today, we realize that the microbial world is a lot bigger than *E. coli,* but only in recent years has it become feasible to follow the resistance ecology of the many other genera that play an important role in the larger antibiotic resistance picture. Levy has been in the forefront of scientists who have embraced the new understanding of microbial diversity and who are leading the way to the development of a model of the ecology of resistance genes and resistant bacteria that takes account of the vast diversity of the microbial world.

Given that this chapter focuses on and advocates the new molecular ecology of resistance genes, rather than simply monitoring the presence of resistant bacterial strains, it is worth asking a question that is still asked by many older scientists: why is molecular analysis preferable simply to following the resistance phenotype of different strains? The main reason is that phenotypic changes in resistance usually reveal no information about the identity of the genes involved or the source of the change, in which genetic exchange can play a significant role. Also, genotype contains a much more detailed "fingerprint" of resistant strains and should thus be of greater assistance in understanding the roles of various selective pressures and other factors that lead to increases in the incidence of strains that are resistant to antibiotics. Genotype, for example, may help to explain why one strain's resistance to erythromycin may disappear rather quickly when erythromycin use is discontinued whereas another strain's resistance to erythromycin may persist even in the absence of the use of the antibiotic.

A second consideration is that a focus on phenotypic change tends to support the assumption that all resistance changes result from mutation and not from acquisition of foreign DNA. Although mutation to resistance clearly plays some role in the emergence of resistant strains of bacteria, acquisition of DNA from other bacteria undoubtedly plays a much greater role in most species. The reason for this is evident if one looks at the problem from the bacterium's point of view. Mutation to resistance means that the bacteria undergo repeated rounds of general mutational changes in their genomes. Not only can this be a time consuming process, but it is a process in which many bacteria die. Contrast this with acquisition of a preformed resistance gene from another bacterium. This is as close as bacteria come to a no-risk change in their genome. The acquisition of a resistant phenotype is rapid, at most requiring only a few mutations in a promoter region of the gene or insertion of an IS element to activate expression if the gene is not expressed already. Little wonder that acquisition of resistance genes by horizontal gene transfer proves to be the rule rather than the exception.

TRANSFER OF RESISTANCE GENES AMONG HUMAN INTESTINAL BACTERIA

If one is convinced of the importance of understanding the ecology of resistance genes, including the mode of distribution and the factors that stimulate distribution of these genes, the question naturally arises as to how one can study resistance gene ecology, especially in humans. An obvious way to do this is to feed bacteria containing resistance genes and antibiotics to laboratory animals and monitor the fate of the resistance genes over time. Such experiments can be quite expensive to perform because no one knows what the time scale is likely to be. Certainly the process would have to be monitored for months, even years. This is the main reason why relatively few prospective studies of resistance gene transfer in laboratory rodents have been done. Perhaps more important, it is easy to discount laboratory rodents as a model for humans since their microbiota is different from that of humans (23, 43). No ethics oversight committee is going to approve such a study in human volunteers. Thus, studies of the movement of resistance genes among bacteria in the human microbiota have to be retrospective studies, i.e., studies that take a "snapshot" of the current distribution of resistance genes and attempt to deduce from this information how the current distribution of resistance genes arose. The question that heads this section will be used as an example of questions that can be and have been raised, but there are many others that relate to specific conditions, such as the use of particular antibiotics as therapeutic, prophylactic or growth promoting agents. However, this question provides a good example of the problems faced by research scientists who attempt to answer such questions.

The human colon provides an environment that should be quite conducive for horizontal transfer of genes. The environment is warm, wet, and rich in nutrients. Perhaps more importantly, the concentrations of bacteria are among the highest found anywhere in nature, 10^{11} to 10^{12} per gram (11, 32). Put in a more direct way, bacteria account for at least a third of the volume of colon contents. Moreover, there are numerous surfaces in colon contents, including mucin aggregates and plant particles. Gene transfer systems, especially those that mediate conjugative transfer of DNA, tend to work most efficiently when bacteria are in close proximity to each other on surfaces.

Since the sorts of exchanges depicted in Fig. 1 are most interesting if they occur across species or genus lines, our interest immediately focused on gene transfers mediated by conjugation, the type of gene transfer associated with broad host range transfers. Initially, however, the search for evidence that such gene transfer events had or had not occurred used an approach that was independent of whether transfer occurred by transformation, phage transduction or conjugation. Simply put, the initial approach was to look for genes with identical or virtually identical DNA sequences in intestinal strains of bacteria from different species or genera.

The reason for demanding that the genes found in different isolates have virtually identical DNA sequences (arbitrarily defined for the purposes of this chapter as greater than 95% DNA sequence identity) is that convergent evolution, the evolution of genes in independent settings under similar selection pressures, occurs at the level of the gene product. Two proteins can have identical amino sequences but still have DNA sequences than differ by as much as 20% because of third base wobble.

This simple approach of looking for sequence identity of resistance genes in diverse bacteria has won favor because of its simplicity. One merely needs to amplify a particular gene from the DNA of a number of different organisms and obtain the DNA sequence of that gene. Yet simplicity is not a guarantee of perfection. A problem with this approach is that it is retrospective and qualitative. It is retrospective in the sense that one is using strains already isolated. Thus, in a sense, it is the equivalent of performing an archeological analysis of events that happened in the past. The data is qualitative in the sense that it only establishes that an event that may have been a horizontal gene transfer event occurred at some point in the past. and does not yield rates. If the information on the time and place of strain isolation is properly documented, one can establish a rate of sorts but usually in decades rather than in months. The lack of information about rates of transfer is of particular concern to those who are trying to assess the human health risks associated with antibiotic use and subsequent resistance emergence and dissemination.

The next stage in such a retrospective analysis is to try to ascertain how a resistance gene has spread among bacteria. The most direct way of doing this is to demonstrate that the resistance gene is on a transmissible element, whether plasmid, phage, or conjugative transposon. Showing that the transmissible element, or at least parts of it, is found in two diverse bacteria adds confidence to the assertion that this element in fact mediated the gene transfer. If the element is a type that is transferred by conjugation, then there is strong evidence that conjugation was the route of spread. If it is transferred by phage, as in lysogenic conversion, phage clearly played an important role. The problem that arises with this approach is that it may be difficult to ascertain which organism was the donor and which the recipient. If, however, the form of the element in organism A is still active for transfer and if the form in B is no longer active, organism A can be identified tentatively as the origin of the transferred resistance gene. Of course, the rate of transfer is still unknown. Some recent studies conducted in our laboratory will serve as examples of our attempts to carry out this type of analysis and what we have learned from the experience.

In our initial retrospective study of the distribution of resistance genes in human colonic bacteria, we focused on the genus *Bacteroides* (42). *Bacteroides* species account for an estimated 20% of human intestinal bacteria, and are thus present in concentrations that would make them a group in which gene transfer events would most likely occur. The results of this study are summarized in Table 1.

We chose five antibiotic resistance genes as our focus in this survey: *tetQ, ermB, ermF, ermG,* and *cfxA*. The *tetQ* gene was chosen because, given the high percentage of *Bacteroides* clinical isolates that were resistant to tetracycline, it was likely that a gene like *tetQ* would be widespread. In fact, we found that already in the pre-1970 period, 20 to 30% of human colonic isolates, both community and clinical isolates, carried *tetQ*. By the 1990s, over 80% of *Bacteroides* isolates carried *tetQ*. Internal sequences from a subset of these genes showed that the DNA sequences were over 95% identical. *tetQ* has been found in at least 10 different *Bacteroides* species, some of which are quite distantly related to each other. This implies that *tetQ* had been spread widely among *Bacteroides* species by horizontal transfer, and not through clonal expansion. Our findings suggested two further conclusions. First, *tetQ* in every *Bacteroides* strain characterized so far has proved to be carried on a conjugative transposon (50). In many cases, *tetQ* is carried on a conjugative transposon (CTn) of the CTnDOT family. In fact, we were able to show by Southern blot analysis that *tetQ* in selected strains from our survey was actually linked with a CTnDOT type element. The transfer of this type of CTn is stimulated 100 to 1,000-fold in the presence of the antibiotic tetracycline. Thus, we can hypothesize that the increase in carriage of *tetQ* by *Bacteroides* strains, from the pre-1970s to the 1990s, may well have been stimulated by tetracycline use. Tetracycline was widely used in the 1960-1980 period to treat a number of infections. Today, due to widespread resistance to tetracyclines, such use has become less common, but dermatologists still administer tetracycline routinely to treat skin conditions such as acne or rosaceia. This treatment can continue for months or years. Also, tetracycline is still commonly used in animal husbandry and agriculture. The finding that most *tetQ* genes that were spreading in *Bacteroides* species were on CTnDOT type elements (41) thus gave a clue as to one factor that might have propelled their spread—widespread tetracycline use in medicine which not only selected for maintenance of *tetQ* but also stimulated the conjugative transfer of this gene.

Not all CTns found in *Bacteroides* species exhibit inducible transfer. Several have been found that appear to transfer constitutively, i.e., transfer is not stimulated by the antibiotics that select for resistance genes carried on the CTn. Examples of these elements are CTn12256, CTn7853, CTnBST, and CTnGERM1 (19, 34, 44, 45). These CTns are much less commonly seen in *Bacteroides* strains than CTnDOT type CTns and appear to be relatively recent entries into *Bacteroides* species. The challenge in future studies will be to find out their origin.

A surprising conclusion from our survey of *Bacteroides* isolates was that both *tetQ* and the CTnDOT elements that carried it were remarkably stable. They were found in many people who had no recent history of antibiotic use. This is yet another illustration of the principle that antibiotic resistant strains are not necessarily lost rapidly in the absence of antibiotic selection and in some cases appear to be extremely stably maintained. A summary of evidence for the stability of many antibiotic resistance genes even in the absence of selection has been published in an earlier review (36).

Widespread transmission of antibiotic resistance genes among human commensals and pathogens has not been restricted to *Bacteroides* species. Some of the many other examples in which identical genes have been found in very different bacterial genera are provided in Table 2. It is interesting to note that in many cases these genes, especially *tetM, tetQ,* and *ermG,* have been found mainly on CTns. Of course some of these, including *ermG* and *ermB,* have also been found on plasmids. In fact, we found *tetQ* in a strain of *Prevotella ruminocola* that was carried on a plasmid rather than a CTn (33, 42). In all cases, conjugation rather than transformation or phage transduction has been a common theme.

Table 1. Percentage of *Bacteroides* strains that contained antibiotic resistance genes based on a survey by Shoemaker et al. (41)

Source of isolates	No. of isolates	Percentage of isolates carrying:				
		tetQ	*ermB*	*ermF*	*ermG*	*cfxA*
Community (pre-1970)	69	32	0	0	0	0
Clinical (pre-1970)	23	22	0	9	0	4
Community (1996–1997)	102	81	3	15	1	3
Clinical (1980–1995)	87	86	3	23	7	14

Table 2. Examples of genes that have been found in both gram-positive and gram-negative commensal bacteria

Gene	Some of the genera carrying the gene
tetM	*Enterococcus, Staphylococcus, Streptococcus, Neisseria,* but rarely in other gram-negative organisms (9)
tetQ	*Bacteroides, Porphyromonas, Prevotella,* and rarely in gram-positive organisms, which are usually anaerobes, such as *Clostridium* (9)
ermB	*Streptococcus, Staphylococcus, Enterococcus,* and most gram-positive organisms; recently in *Bacteroides* (9, 41)
ermG	*Bacillus, Bacteroides,* and recently in *Streptococcus* (12, 31, 46)

BEYOND THE HUMAN COLON—ANTIBIOTIC-RESISTANT BACTERIA IN THE ORAL CAVITY

Although the human colon is clearly a hotbed of resistance gene transfer, other bacteria in and on the human body have also participated in the distribution of resistance genes. An interesting case in point is the population of bacteria that makes its home in the periodontal pocket, the region between the gums and the roots of the teeth. This population, in contrast to the bacteria that colonize the surface of teeth and oral mucosal cells, are obligate anaerobes, many of which are relatives of the colonic *Bacteroides* species.

Prominent among these are the *Porphyromonas* and *Prevotella* species. These bacteria have been of particular interest in dentistry because they are thought to be instrumental in the development of periodontal disease, the main cause of tooth loss in adults. Currently, the treatment for periodontal disease is surgery that cuts into the gums, exposing the buried surface of the teeth to allow scraping of the plaque that has accumulated there and is causing inflammation. The realization that bacteria were a primary factor in the development of gum disease has led to the hope that periodontal surgery might be replaced by antibiotic treatment. Although this concept is still controversial, especially among periodontal surgeons, it has focused attention on the antibiotic resistance profiles of the oral anaerobes. Not surprisingly, *tetQ*, the main form of tetracycline resistance gene found in human colonic *Bacteroides* species, is found in *Porphyromonas* and *Prevotella* species, although it is not yet as widespread as in *Bacteroides* species (3, 10, 26).

Does the presence of *tetQ* in oral anaerobes mean that there is genetic exchange between these bacteria and colonic anaerobes? Certainly, this type of transfer is possible in theory and has been shown to occur in the laboratory (17, 18). Oral bacteria are constantly being swallowed and would come into regular contact with colonic *Bacteroides*. Fecal-oral transmission could return them to the mouth (dotted lines in Fig. 1). Moreover, some of the genetic vectors developed for use in manipulating *Bacteroides* species have proven to be useful in the genetic manipulation of the oral anaerobes. In fact, recent genome sequences of *Porphyromonas* strains have revealed the existence of genes associated with the conjugative transposons found originally in *Bacteroides* species. Since genes like *tetQ* are more widely distributed in the colonic *Bacteroides* species than in the oral anaerobes, it is tempting to speculate that the resistance genes found in the oral anaerobes came originally from the colonic *Bacteroides* species, but this direction of transfer has not been proven. Possibly, the transfer, if it occurred, was in the other direction, but for some reason the genes and the elements that carried them proved to be more stably maintained in bacteria growing in the colonic environment.

A CRUISE ON THE PIG LAGOON

Earlier in this chapter, the possibility was considered that resistance genes in bacteria from the intestinal tracts of animals, which have entered the human food chain, might be transferred to human intestinal bacteria. This is an interesting speculation, but an important missing link in this argument has been the lack of information about what types of resistance genes are found in animal intestinal isolates.

A recent study at the USDA laboratories in Peoria, Ill., examined genes of intestinal bacteria found in the feces of pigs and in an under barn manure lagoon (46). The pigs on this farm had been fed tylosin, a macrolide antibiotic, so the focus of the study was on bacteria that were resistant to the macrolides erythromycin and tylosin. Almost all of the resistant bacteria isolated in this study were gram-positive bacteria. This is not surprising, since even in the human intestine, the majority of the bacteria present are gram positive. The gram-negative anaerobes such as *Bacteroides, Porphyromonas,* and *Prevotella* species have been given what might be considered an unfair amount of attention because of their known propensity for causing opportunistic infections. Thus, the manure lagoon study offers an unusual look at the resistance gene profile of the gram-positive bacteria. A similar study, long overdue, of the resistance gene profile of the human colonic gram-positive anaerobes is currently underway.

Not surprisingly, many of the genes found in the porcine isolates were old familiar friends, such as *ermB* and *tetM*. Several of the strains also contained the *Bacteroides tetQ* and the relatively rare *ermG*. There was even the intriguing finding of DNA that cross-hybridized with DNA adjacent to the *ermG* on CTnGERM1 (45) found in our earlier surveys of human colonic bacteria. However, it was clear that this hybridization was to a small region and not to the entire element. In other words, there was no "smoking gun" horizontal gene transfer element that links the animal and human isolates that carry the same genes. Another interesting, but daunting, finding was that indications of some previously unknown resistance *erm* genes such as *ermT* (49) and *erm35*, (T. R. Whitehead, unpublished) and ribosomal protection *tet* genes, e.g., *tet36* (51) were found. For those scientists who are willing to try to account for all of the resistance found in bacteria in a site, the reward—if that is the right word for it—may well turn out to be a bewildering new variety of resistance genes. When one considers that few such comprehensive studies of human intestinal or oral bacteria have been attempted, the potential for expanding diversity in human bacteria may be just as large.

We think that the swine resistance gene study was a useful first step toward further ecological studies of resistance genes in animal intestinal bacteria, but it is important to point out a key flaw in this study. Ideally, bacteria would have been isolated from different farms at different times and from many different locations on the same farm. This study covered only one part of one farm at one time, and the number of isolates was limited. The study was initiated based on the need for analysis of the community of organisms found at the site, and the identification of some of the antibiotic resistance genes carried by these organisms based on PCR (13, 48). However, it would have been much better if samples could have been taken from surrounding non-farm areas, and perhaps even from workers who handled the animals, but this was beyond the resources of this study. Thus, it is important not to draw too many conclusions from this study. Its importance, however, is that it shows how easy such a study can be, and it gives an idea of the types of genes that might be found. There have been more extensive studies, using PCR typing methods, on DNA isolated from various sites of swine facilities and from feed sources. These studies focused on tetracycline resistance genes, although in most cases the presence of the genes was determined but which genes were carried by which organisms was not determined (1, 2, 8). So, just as the much earlier "flawed" studies of Levy and his coworkers stimulated further and improved efforts to comprehend the dynamic distribution of antibiotic resistant bacteria, we hope that our "flawed" pig lagoon study will stimulate other groups to refine and apply molecular techniques to understand the ecology of antibiotic resistance genes.

FUTURE GOALS

The push to understand the molecular ecology of antibiotic resistance genes is still in its infancy. A beginning has been made in an attempt to determine not only that a particular gene is being spread, but to learn how that spread is mediated. Is the gene spread by conjugation or by other means? What is the element that is responsible for its transfer? Recently, we noted that *ermG* and *ermB* seem to have entered *Bacteroides* species only recently. That is, they have been found only in modern *Bacteroides* isolates (41). We are currently examining the DNA sequence of the CTn that carries *ermB* in *Bacteroides* strains in an attempt to determine whether this CTn might have come into *Bacteroides* species from gram-positive bacteria, where *ermB* is normally found.

A particularly interesting question is whether there are any limits to the spread of resistance genes. This question can be formulated in two ways. First, are there barriers to the spread of resistance genes, that is, are there phylogenetic barriers? It is intriguing that *tetQ* seems to have stayed primarily within the *Bacteroides* phylum, which includes *Porphyromonas* and *Prevotella*. However, *tetQ* has been reported in a few gram-positive isolates, which indicates transfer from the *Bacteroides* group to select gram-gram positive. By contrast, although *tetM* has been found in a variety of gram-positive and gram-negative bacteria, it appears not yet to have penetrated the *Bacteroides* group (24, 25). Of course, these apparent limitations in spread could be an artifact of the relatively small number of strains surveyed to date. Future surveys should reveal what, if any, phylogenetic barriers exist and, if they exist, what factors prevent spread of resistance genes across them.

A second way to formulate the question of whether there are limitations in the spread of resistance genes is to ask whether there are geographical limits to this spread. Evidence is emerging that resistance genes can move through the food supply from farm animals to humans. We found examples of *tetQ* in human oral and intestinal bacteria, as well as in farm animals. The world is very connected from a microbiological perspective. In fact, it could be argued that there is a fourth mode of gene transfer, in addition to conjugation, transformation and phage transduction: the interstate and transnational transportation systems.

A final question needs to be answered if a full accounting of the ecology of resistance genes is to be gained: do resistance genes move through the bacterial world independently of other genes or do they sometimes act as fellow travelers? It is well established in many cases that a transmissible element such as a plasmid or CTn will carry more than one type of resistance gene. In the case of the *Bacteroides* CTns, *tetQ* is frequently found linked to *erm* genes. Similar linkages have been seen as well with the gram-positive CTns and with conjugal plasmids from a variety of bacterial genera.

Yet, reports of resistance genes moving together with virulence genes or other genes that might contribute to survival of the bacteria in different settings where antibiotics are virtually nonexistent. Do resistance genes really move independently of genes that confer other traits or is this apparent lack of linkage due to the fact that scientists who are interested in antibiotic resistance genes do not look for linked virulence genes and vice versa? There is a widespread misconception that antibiotic-resistant bacteria are more virulent than susceptible bacteria. So far, the evidence tends to support the hypothesis that becoming resistance to antibiotics does not make a bacterial strain more able to cause disease, except in the sense that it is harder to eliminate from the human body with antibiotic therapy. Is it possible, however, that this idea might be incorrect in some cases? Are there cases in which transmissible elements that carry resistance genes also confer added virulence factors on the strains that receive the transmissible element? If so, how common are such occurrences? And, if they seldom occur, what mechanism(s) are responsible for this restriction? Clearly a sophisticated model for analyzing resistance gene ecology—or any model for that matter—is badly needed.

Acknowledgments. Work described in this article was supported by a grant from the U.S. National Institutes of Health (AI/GM 22383) and a grant from the Ellison Foundation. Above all, we want to acknowledge the continued inspiration gained from our contacts, whether personal or through the literature, with Stuart Levy, an amazing and charismatic scientist.

REFERENCES

1. **Aminov, R. I., J. C. Chee-Sanford, N. Garrigues, B. Teferedegne, I. J. Krapac, B. A. White, and R. I. Mackie.** 2002. Development, validation, and application of PCR primers for detection of tetracycline efflux genes of gram-negative bacteria. *Appl. Environ. Microbiol.* **68:**1786–1793.
2. **Aminov, R. I., N. Garrigues-Jeanjean, and R. I. Mackie.** 2001. Molecular ecology of tetracycline resistance: Development and validation of primers for detection of tetracycline resistance genes encoding ribosomal protection proteins. *Appl. Environ. Microbiol.* **67:**22–32.
3. **Arzese, A. R., L. Tomasetig, and G. A. Botta.** 2000. Detection of *tetQ* and *ermF* antibiotic resistance genes in *Prevotella* and *Porphyromonas* isolates from clinical specimens and resident microbiota of humans. *J. Antimicrob. Chemother.* **45:**577–582.
4. **Berbari, E. F., F. R. Cockerill, 3rd, and J. M. Steckelberg.** 1997. Infective endocarditis due to unusual or fastidious microorganisms. *Mayo Clin. Proc.* **72:**532–542.
5. **Beringer, J.** 1999. Keeping watch over genetically modified crops and foods. *Lancet* **353:**605–606.
6. **Brennan, M. T., F. Bahrani-Mougeot, P. C. Fox, T. P. Kennedy, S. Hopkins, R. C. Boucher, and P. B. Lockhart.** 2004. The role of oral microbial colonization in ventilator-associated pneumonia. *Oral Surg. Oral Med. Oral Pathol. Oral Radiol. Endod.* **98:**665–672.
7. **Cash, H. L., and L. V. Hooper.** 2005. Commensal bacteria shape intestinal immune system development. *ASM News* **71:** 77–83.
8. **Chee-Sanford, J. C., R. I. Aminov, I. J. Krapac, N. Garrigues-Jeanjean, and R. I. Mackie.** 2001. Occurrence and diversity of tetracycline resistance genes in lagoons and groundwater underlying two swine production facilities. *Appl. Environ. Microbiol.* **67:**1494–1502.
9. **Chopra, I., and M. Roberts.** 2001. Tetracycline Antibiotics: Mode of Action, Applications, Molecular Biology, and Epidemiology of Bacterial Resistance. *Microbiol. Mol. Biol. Rev.* **65:**232–260.
10. **Chung, W. O., J. Gabany, G. R. Persson, and M. C. Roberts.** 2002. Distribution of erm(F) and tet(Q) genes in 4 oral bacterial species and genotypic variation between resistant and susceptible isolates. *J. Clin. Periodontol.* **29:**152–158.
11. **Conway, P.** 1996. Development of intestinal microbiota, p. 3–38. *In* R. I. Mackie, B. A. White, and R. E. Isaacson (ed.), *Gastrointestinal Microbiology,* vol. 2. Chapman and Hall, London, United Kingdom.
12. **Cooper, A. J., N. B. Shoemaker, and A. A. Salyers.** 1996. The erythromycin resistance gene from the *Bacteroides* conjugal transposon TcrEmr 7853 is nearly identical to *ermG* from *Bacillus sphaericus. Antimicrob. Agents Chemother.* **40:**506–508.
13. **Cotta, M. A., T. R. Whitehead, and R. L. Zeltwanger.** 2003. Isolation, characterization, and comparison of bacteria from swine feces and manure storage pits. *Environ. Microbiol.* **5:**737–745.
14. **Della, M., P. L. Palmbos, H. M. Tseng, L. M. Tonkin, J. M. Daley, L. M. Topper, R. S. Pitcher, A. E. Tomkinson, T. E. Wilson, and A. J. Doherty.** 2004. Mycobacterial Ku and ligase proteins constitute a two-component NHEJ repair machine. *Science* **306:**683–685.
15. **editorial, N.** 1996. Distrust in genetically altered foods, p. 559, *Nature,* vol. 383.
16. **Finegold, S. M.** 1995. Overview of clinically important anaerobes. *Clin. Infect. Dis.* **20**(Suppl. 2):S205–S207.
17. **Guiney, D. G., and K. Bouic.** 1990. Detection of conjugal transfer systems in oral, black-pigmented *Bacteroides* spp. *J. Bacteriol.* **172:**495–497.
18. **Guiney, D. G., and P. Hasegawa.** 1992. Transfer of conjugal elements in oral black-pigmented *Bacteroides* (*Prevotella*) spp. involves DNA rearrangements. *J. Bacteriol.* **174:**4853–4855.
19. **Gupta, A., H. Vlamakis, N. Shoemaker, and A. A. Salyers.** 2003. A new *Bacteroides* conjugative transposon that carries an *ermB* gene. *Appl. Environ. Microbiol.* **69:**6455–6463.
20. **Hauth, J. C., R. L. Goldenberg, W. W. Andrews, M. B. DuBard, and R. L. Copper.** 1995. Reduced Incidence of preterm delivery with metronidazole and erythromycin in women with bacterial vaginosis. *N. Engl. J. Med.* **333:**1732–1736.
21. **Hillier, S. L., R. P. Nugent, D. A. Eschenbach, M. A. Krohn, R. S. Gibbs, D. H. Martin, M. F. Cotch, R. Edelman, J. G.**

Pastorek, A. V. Rao, D. McNellis, J. A. Regan, J. C. Carey, M. A. Klebanoff, and The Vaginal Infections and Prematurity Study Group. 1995. Association between bacterial vaginosis and preterm delivery of a low-birth-weight infant. *N. Engl. J. Med.* **333:**1737–1742.

22. **Hooper, L. V.** 2004. Bacterial contributions to mammalian gut development. *Trends Microbiol.* **12:**129–134.
23. **Hooper, L. V., T. Midtvedt, and J. I. Gordon.** 2002. How host-microbial interactions shape the nutrient environment of the mammalian intestine. *Annu. Rev. Nutr.* **22:**283–307.
24. **Lacroix, J. M., and C. B. Walker.** 1995. Detection and incidence of the tetracycline resistance determinant tet(M) in the microflora associated with adult periodontitis. *J. Periodontol.* **66:** 102–108.
25. **Lacroix, J. M., and C. B. Walker.** 1996. Detection and prevalence of the tetracycline resistance determinant Tet Q in the microbiota associated with adult periodontitis. *Oral Microbiol. Immunol.* **11:**282–288.
26. **Leng, Z., D. E. Riley, R. E. Berger, J. N. Krieger, and M. C. Roberts.** 1997. Distribution and mobility of the tetracycline resistance determinant *tetQ*. *J. Antimicrob. Chemother.* **40:**551–559.
27. **Lenski, R. E., S. C. Simpson, and T. T. Nguyen.** 1994. Genetic analysis of a plasmid-encoded, host genotype-specific enhancement of bacterial fitness. *J. Bacteriol.* **176:**3140–3147.
28. **Letourneau, D. K., G. S. Robinson, and J. A. Hagen.** 2003. Bt crops: predicting effects of escaped transgenes on the fitness of wild plants and their herbivores. *Environ. Biosafety Res.* **2:**219–246.
29. **Levy, S. B., G. B. FitzGerald, and A. B. Macone.** 1976. Changes in intestinal flora of farm personnel after introduction of a tetracycline-supplemented feed on a farm. *N. Engl. J. Med.* **295:**583–588.
30. **Levy, S. B., G. B. FitzGerald, and A. B. Macone.** 1976. Spread of antibiotic-resistant plasmids from chicken to chicken and from chicken to man. *Nature* **260:**40–42.
31. **Monod, M., S. Mohan, and D. Dubnau.** 1987. Cloning and analysis of *ermG,* a new macrolide-lincosamide-streptogramin B resistance element from *Bacillus sphaericus*. *J. Bacteriol.* **169:** 340–350.
32. **Moore, W. E., and L. V. Holdeman.** 1974. Human fecal flora: the normal flora of 20 Japanese-Hawaiians. *Appl. Microbiol.* **27:**961–979.
33. **Nikolich, M. P., G. Hong, N. B. Shoemaker, and A. A. Salyers.** 1994. Evidence for natural horizontal transfer of *tetQ* between bacteria that normally colonize humans and bacteria that normally colonize livestock. *Appl. Environ. Microbiol.* **60:**3255–3260.
34. **Nikolich, M. P., N. B. Shoemaker, G. R. Wang, and A. A. Salyers.** 1994. Characterization of a new type of *Bacteroides* conjugative transposon, Tcr Emr 7853. *J. Bacteriol.* **176:**6606–6612.
35. **Salyers, A. A.** 1996. The real threal from antibiotics. *Nature* **384:**304.
36. **Salyers, A. A., and C. F. Amabile-Cuevas.** 1997. Why are antibiotic resistance genes so resistant to elimination? *Antimicrob. Agents Chemother.* **41:**2321–2325.
37. **Salyers, A. A., A. Gupta, and Y. Wang.** 2004. Human intestinal bacteria as reservoirs for antibiotic resistance genes. *Trends Microbiol.* **12:**412–416.
38. **Salyers, A. A., and P. McManus.** 2001. Possible impact on antibiotic resistance in human pathogens due to agricultural use of antibiotics. *In* D. Hughes and D. I. Andersson (ed.), *Antibiotic Development and Resistance.* Taylor and Francis Inc, New York, N.Y.
39. **Salyers, A. A., and J. A. Shipman.** 2002. 11: Getting in touch with your prokaryotic self: mammal-microbe interactions, p. 315–341. *In* J. T. Staley and A.-L. Reysenbach (ed.), *Biodiversity of Microbial Life.* Wiley-Liss, Inc.
40. **Scannapieco, F. A., and J. M. Mylotte.** 1996. Relationships between periodontal disease and bacterial pneumonia. *J. Periodontol.* **67:**1114–1122.
41. **Shoemaker, N. B., H. Vlamakis, K. Hayes, and A. A. Salyers.** 2001. Evidence for extensive resistance gene transfer among *Bacteroides spp* and between *Bacteroides* and other genera in the human colon. *Appl. Environ. Microbiol.* **67:**561–568.
42. **Shoemaker, N. B., G. R. Wang, and A. A. Salyers.** 1992. Evidence for natural transfer of a tetracycline resistance gene between bacteria from the human colon and bacteria from the bovine rumen. *Appl. Environ. Microbiol.* **58:**1313–1320.
43. **Tannock, G. W.** 1997. Normal microbiota of the gastrointestinal tract of rodents, p. 187–215. *In* R. I. Mackie, B. A. White, and R. E. Isaacson (ed.), *Gastrointestinal Microbiology: Gastrointestinal Microbes and Host Interactions,* vol. 2. Chapman and Hall, New York, N.Y.
44. **Valentine, P. J., N. B. Shoemaker, and A. A. Salyers.** 1988. Mobilization of *Bacteroides* plasmids by *Bacteroides* conjugal elements. *J. Bacteriol.* **170:**1319–1324.
45. **Wang, Y., G.-R. Wang, A. Shelby, N. B. Shoemaker, and A. A. Salyers.** 2003. A newly discovered *Bacteroides* conjugative transposon, CTnGERM1, contains genes also found in Gram-positive bacteria. *Appl. Environ. Microbiol.* **69:**4595–4603.
46. **Wang, Y., G.-R. Wang, N. B. Shoemaker, T. R. Whitehead, and A. A. Salyers.** Distribution of the *ermG* gene in bacterial isolates from porcine intestinal contents. *Appl. Environ. Microbiol.,* in press.
47. **Wegener, H. C., F. M. Aarestrup, L. B. Jensen, A. M. Hammerum, and F. Bager.** 1999. Use of antimicrobial growth promoters in food animals and *Enterococcus faecium* resistance to therapeutic antimicrobial drugs in Europe. *Emerg. Infect. Dis.* **5:**329–335.
48. **Whitehead, T. R., and M. A. Cotta.** 2001. Characterisation and comparison of microbial populations in swine faeces and manure storage pits by 16S rDNA gene sequence analyses. *Anaerobe* **7:**181–187.
49. **Whitehead, T. R., and M. A. Cotta.** 2001. Sequence analyses of a broad host-range plasmid containing *ermT* from a tylosin-resistant *Lactobacillus* sp. isolated from swine feces. *Curr. Microbiol.* **43:**17–20.
50. **Whittle, G., N. B. Shoemaker, and A. A. Salyers.** 2002. The role of *Bacteroides* conjugative transposons in the dissemination of antibiotic resistance genes. *Cell. Mol. Life Sci.* **59:**2044–2054.
51. **Whittle, G., T. R. Whitehead, N. Hamburger, N. B. Shoemaker, M. A. Cotta, and A. A. Salyers.** 2003. Identification of a new ribosomal protection type of tetracycline resistance gene, *tet*(36), from swine manure pits. *Appl. Environ. Microbiol.* **69:**4151–4158.
52. **Wilson, K. H., J. S. Ikeda, and R. B. Blitchington.** 1997. Phylogenic placement of community members of human colonic biota. *Clin. Infect. Dis.* **25**(Suppl 2)**:**S114–S116.
53. **Witte, W.** 1998. Medical consequences of antibiotic use in agriculture. *Science* **279:**996–997.

Frontiers in Antimicrobial Resistance: a Tribute to Stuart B. Levy
Edited by D. G. White, M. N. Alekshun, and P. F. McDermott

Chapter 34

Resistance in the Food Chain and in Bacteria from Animals: Relevance to Human Infections

VINCENT PERRETEN

Since their introduction in the 1930s, antibiotics have been widely used to treat infectious diseases in humans and animals. Subtherapeutic doses of antibiotics have also been used as feed additives in animal husbandry to improve growth rate. In 1968, it was reported that the massive use of antibiotics in animal farming had selected for a specific multidrug resistant *Salmonella* strain, which represented a potential risk for human health (8). One year later in an effort to limit the transfer of antibiotic resistant bacteria from animals to humans, the Swann Committee of the United Kingdom, made up of scientists and physicians, recommended that antibiotics used for therapeutic purposes or for generating cross-resistances to therapeutically used antibiotics should not be used as growth promoters (175). In 1976, Stuart Levy was among the first to confirm that antibiotic resistant bacteria could be transferred from animals to humans (113). Transfer of antibiotic resistant *Escherichia coli* was shown to have occurred from chicken to chicken and from chicken to humans. In addition, antibiotic resistant *E. coli* were recovered from people that were not taking antibiotic medication, but that were in close contact with animals receiving antibiotics (112, 113). These studies emphasized the public health hazards that can arise from the wide use of antibiotics in animal husbandry. Since then, many studies have been conducted to analyze the situation of antibiotic resistance in bacteria of animal origin and their modes of transmission to humans (181, 195), leading to WHO recommendations (213, 214, 216). It is now known that bacteria can be transmitted by direct contact between humans and animals or indirectly via different ecosystems such as sewage, water, soil, and plants (222) and, in a more direct way, via the food chain (187, 195).

This chapter gives a general overview of the antibiotic resistance situation in bacteria of both animal and food origin and analyzes whether these bacteria represent a greater threat to human health when they harbor antibiotic resistance genes. Important zoonotic bacteria, food-borne pathogens and commensal bacteria from animal and food sources will be considered in order to evaluate the impact that antibiotic resistant bacteria may have on human health.

FOOD AS A VEHICLE FOR TRANSFER OF BACTERIA FROM ANIMALS TO HUMANS

Food has been suspected to be one of the main vehicles for the transfer of antibiotic resistant bacteria and resistance genes from animals to humans (139, 184, 186, 210, 221, 222). In slaughterhouses or during milking processes, bacteria from animal origin can contaminate meat or milk. Eggs and egg shells can also be a source of bacterial contamination (87, 218). Thus, bacteria can be found in the retail raw products such as milk, meat and eggs or in food products made with unpasteurized raw material, including fermented meat, dairy products, and products containing raw eggs. Some of these microorganisms may bring about desirable changes as a result of growth in the foods with which they are associated as a starter culture, while others may cause food spoilage or food-borne disease.

Bacterial Food-Borne Disease and Antibiotic Treatment

2.1 million people worldwide died from diarrheal diseases that were due mainly to contaminated food or drinking water in 2000. In industrialized countries, food borne illnesses can affect up to 30% of the population each year (215). It is estimated that each year bacterial food-borne diseases cause approximately 4,176,000 illnesses, 36,500 hospitalizations and 1,300

Vincent Perreten • Institute of Veterinary Bacteriology, Faculty of Veterinary Medicine, University of Berne, CH-3001 Berne, Switzerland.

deaths in the United States (U.S.) (128). In England (Engl.) and Wales, 607,950 cases, 20,129 hospitalizations and 395 deaths were reported in 2000 (4, 155). The most important pathogens capable of causing lethal infections are *Salmonella* (30%), *Campylobacter* (21% Engl., 5.5% U.S.), *Clostridium perfringens* (22% Engl., 0.4% U.S.), *Listeria monocytogenes* (17% Engl., 28% U.S.) and the verocytotoxic *E. coli* O157 (5.4% Engl., 2.9% U.S.). In the U.S. the Food-Borne Diseases Active Surveillance Network (FoodNet) (http://www.cdc.gov/foodnet) was put in place to reduce the incidence of food-borne infections (30) while in the European Union (EU), the EuroSurveillance is dedicated to the surveillance, prevention and control of infectious and communicable diseases (http://www.eurosurveillance.org).

Bacterial food-borne illnesses may be caused by the consumption of food contaminated with either bacteria or their toxins. Uncooked or undercooked food like meat, eggs, milk, or food prepared with such raw items that are contaminated with zoonotic pathogens like *Salmonella, Campylobacter, Listeria,* and *E. coli* are mainly incriminated (180). The bacterial food-borne infections are usually self-limiting diseases causing diarrhea, abdominal cramps, nausea, vomiting, headache and fever. Antibiotic treatment in these cases is usually not necessary unless the infection persists or spreads from the intestines. Severe infections can often occur in the elderly, newborns and young children, pregnant women, and immunocompromized persons, and antibiotic therapy would then be necessary. For these patients, such infections can have fatal consequences if not treated in a timely manner. Nevertheless, it has been advised that the use of antibiotics to treat young children suffering from bloody diarrhea caused by *E. coli* O157 should be avoided as such antibiotics were suggested to increase hemolytic uremic syndrome, which is specific to O157 infections (189, 223).

Processed food may be contaminated with bacteria such as *Staphylococcus aureus, Clostridium perfringens,* and *Bacillus cereus* that can grow and produce toxins if the food product is not refrigerated appropriately or if refrigeration is interrupted. During subsequent reheating of the contaminated food, the bacteria will be killed, but their toxins, which are heat-stable, will survive the heat treatment and therefore reach the consumers. Antibiotic treatment in these instances would be inappropriate. Toxin-producing bacteria, however, can also reach the consumers if the food products are consumed uncooked. These bacteria can then colonize their new hosts but without directly affecting their health. Thus, the consumers become healthy carriers of antibiotic resistant bacteria that may develop into nosocomial pathogens in the case of hospitalization. Indeed, bacteria such as *S. aureus, C. perfringens,* and *B. cereus* are commonly involved in nosocomial infections, and a prior acquisition of an antibiotic resistance gene(s) would make them difficult to treat (11, 80, 199).

Besides the pathogenic bacteria, bacteria such as the *Lactobacillus,* the *Lactococcus,* and *Enterococcus,* which are part of the natural flora of animals, can also be present in food products, and they play a major role in the acquisition and distribution of antibiotic resistance determinants to other bacteria including human pathogens.

ANTIBIOTIC USE IN ANIMALS: A SELECTIVE PRESSURE FOR RESISTANCE DEVELOPMENT

The emergence and spread of antibiotic resistance depends on a variety of factors. In what he called "the drug resistance equation," Stuart Levy set out two main factors that contribute to the emergence of resistance, antibiotics and resistance determinants. In his equation, he indicated that the combination of the antibiotics themselves and of the presence of resistance determinants affected the proportion of the antibiotic resistance problem (109, 110). He also considered the spread of antibiotic resistant bacteria from one host to another and the transfer of antibiotic resistance determinants between bacteria as additional factors (111). Thus, the amount and the way that antibiotics are used in the animal husbandry have an impact on the selection of an antibiotic resistant flora in animals (221) and humans (170). Low levels of antibiotic doses, such as those used in growth promoting feed preparations, favor the development and the transfer of antibiotic resistance determinants. The amount of antibiotics used worldwide is difficult to quantify as the exact amount of antibiotic effectively used is not always available and, when published, the data diverge from one source to the other. For example, in the U.S., a survey of the members of the Animals Health Institute estimated that in 1998, 14.7 million pounds of antibiotics were used for the prevention and treatment of infectious diseases in animals and a further 3.1 million pounds were used for growth promotion (127). However, the Union of Concerned Scientists estimated an amount of 27.5 million of pounds of antibiotics for the prevention and growth promotion and 2 million of pounds for the treatment of diseases in animals, whereas ca. 4.5 million pounds are used annually in humans. The European Federation of Animal Health (FEDESA) (http://www.fedesa.be) estimated that human consumption represented 65% (8,528 tons [18.8 million pounds]) of the total volume of antibiotics prescribed in EU in 1999, while animals consumed 35% (4,688 tons [10.3 million pounds]).

In the late 1990s, the EU and Switzerland decided to ban the use of antibiotics for nontherapeutic purposes (136) (see below). As a result of the ban, the consumption of antibiotics used as growth promoters dropped by almost 50% from 1997 (1,599 tons [3.5 million pounds]) to 1999 (786 tons [1.7 million pounds]). The amount of therapeutic antibiotics used for animals in the EU, however, increased 11% during the same period; 3,494 tons [7.7 million pounds] in 1997 to 3,902 tons [6.8 million pounds] in 1999.

Monitoring systems that record the amount of antimicrobial agents used for humans and animals and carry out surveillance of antimicrobial resistance in bacteria from food-producing animals have been in place for some years in and are published annually in Denmark (46), Sweden (174), Norway (132), and The Netherlands (122). These programs have kept the level of resistances in these countries low. For example, Sweden has introduced a pharmacy system for the sale of antibiotics in both veterinary and human medicine, thereby recording the amount of antibiotics sold during the year. The experience acquired by these countries should enable optimal systems to be found and applied to other countries where the incidence of antibiotic resistant bacteria is very high.

BAN OF ANTIMICROBIAL GROWTH PROMOTERS IN EUROPE AND ITS CONSEQUENCES

In Europe, authorization for the use of antibiotics promoting growth in animals and that belong to classes that are used therapeutically or have the propensity of generating cross-resistance to therapeutically used antibiotics was withdrawn in the late 1990s (136). Avoparcin, a glycopeptide antibiotic like vancomycin, had been used as a feed additive for more than 20 years, while at the same time vancomycin was reserved in human medicine as the last weapon to treat infections caused by methicillin resistant *S. aureus* (MRSA). The authorization for avoparcin was then withdrawn in all EU countries as a preventive measure to avoid the spread of vancomycin resistance to important human pathogens. It was also decided in January 1998 not to renew the authorization of ardacin, another glycopeptide antibiotic used in animal feed. Similarly, the authorization to use tylosin and spiramycin (macrolides), virginiamycin (a streptogramin), and bacitracin as growth promoting feed additives was withdrawn on 1 July 1999. In view of their possible adverse effect on human health (genotoxicity), the authorizations for the growth promotants olaquindox and carbadox were withdrawn 31 August 1999. By the end of 1999, four antibiotics were still authorized: avilamycin, flavophospholipol (bambermycin), monensin and salinomycin. In March 2002, the European Commission presented new proposals to prohibit any antibiotic use for the purposes of growth-promotion. Under these proposals, the four remaining authorized antibiotics are to be phased out by 1 January 2006.

Since the ban on the use of antibiotics for nontherapeutic purposes was put in place in Europe, several studies have shown that the number of resistant isolates from animals (2, 24, 102, 192) and from meat (50, 134) have decreased but have not disappeared completely. In the absence of selective pressure, antibiotic resistant bacterial populations can be challenged with new populations of antibiotic susceptible bacteria. A subpopulation of antibiotic resistant bacteria, however, still persists in the animal host even in the absence of an antimicrobial pressure (97), and any use of antibiotic has been shown to rapidly select a resistant population. Indeed, a recent study showed that the relative number of macrolide resistant enterococci increased from 18 to 100% within a few days after calves were fed milk from cows treated with spiramycin. All these strains carried an *erm*(B) gene conferring resistance to macrolides, lincosamides and streptogramins B antibiotics (the MLS_B phenotype). On the other hand, the percentage of macrolide resistant enterococci remained the same in calves fed milk from untreated cows (226). These findings suggest that sucklings rapidly acquire an intestinal flora that already contains antibiotic resistant bacteria and any specific antimicrobial selective pressure rapidly selects resistant bacteria from this population. The persistence of antibiotic resistance in *Enterococcus* spp. may be explained by the presence on enterococcal plasmids (166a) of an antitoxin/toxin system which promotes stabilization of plasmids in the absence of antimicrobial selective pressure and kills cells that have lost these plasmids (128a).

The decreased use of antibiotics in animal husbandry has curbed the amount of antibiotic resistant bacteria in food producing animals, thus limiting the number of resistant bacteria that could reach the human population via the food chain. Unfortunately, bacterial populations resistant to several antibiotics have already been selected due to long-term intensive use and misuse of antibiotics, and these organisms persist in animals and in the environment. Only the prudent and well directed use of antibiotics will help limit the reemergence of antibiotic resistant bacteria and keep the number of resistant organisms low. With such actions, the selection and spread of antibiotic resistance genes to pathogenic bacteria could be suppressed.

SPREAD OF RESISTANCE GENES WITH FOOD-BORNE COMMENSALS AND OPPORTUNISTIC PATHOGENS

Bacterial fermentation has been used to alter and preserve food products for about 5,000 years. Microorganisms that are part of the normal flora of animals and plants were found to be useful in food fermentation. Indeed, spontaneous fermentation occurs in the presence of such bacteria in the raw material, and these bacteria are used today as starter cultures (183). It has been proposed that bacteria, used as starter cultures for the food industry or as probiotics in food, should be free of antibiotic resistance genes (http://europa.eu.int/comm/food/fs/sc/scf/out178_en.pdf). The bacteria used in food preparation are mainly gram-positive and include *Lactococcus, Lactobacillus, Pediococcus, Leuconostoc, Carnobacterium, Enterococcus, Micrococcus, Streptococcus, Staphylococcus,* and *Propionibacterium.* Fermented products made with raw material such as raw milk cheese and raw meat products, however, still contain starter cultures originating directly from animals. During antibiotic treatment of their animal hosts, they are also subject to the antimicrobial selective pressure even if they are not the primary targets, forcing them to acquire specific antibiotic resistance genes to survive in such a hostile environment. These bacteria then act as a large reservoir of transmissible antibiotic resistance genes that can be shared with pathogenic bacteria. Among them are *Lactococcus* and *Lactobacillus,* which are also the two most widely used genera of lactic acid bacteria in the dairy industry (183), and *Enterococcus,* which can be an opportunistic pathogen and has the ability to easily acquire and redistribute antibiotic resistance genes to other bacteria.

Enterococcus

Enterococci are commensal bacteria found in the intestines of humans and animals and have been used as indicators for fecal contamination in food. Enterococci occur in uncooked food and can persist and multiply in fermented food products made with raw material such as sausages and raw milk cheese. They have been used as starter cultures for the production of certain cheeses (66) and as probiotics, and in the 1990s they turned out to be important pathogens (117, 92). Today, they are among the leading causes of nosocomial infections (182). The two most frequent species causing infections in humans, *E. faecalis* and *E. faecium,* are intrinsically resistant to many antibiotics and have the ability to easily acquire and spread antibiotic resistance genes to other bacteria. They possess mobile DNA molecules such as plasmids and transposons that can carry one or more antibiotic resistance genes. Thus, enterococci from food-producing animals were able to acquire a variety of genes conferring resistance to the drugs used in animal husbandry. Some of these belonged to the same classes of important drugs used in human medicine and selected for cross-resistance. Cross-resistance occurs between the glycopeptides avoparcin and vancomycin, the streptogramins virginiamycin and quinupristin-dalfopristin, the macrolides tylosin/spiramycin and erythromycin, and the oligosaccharide antibiotics avylamycin and evernimicin.

Antibiotic resistant enterococci in animals and the food chain

The enterococci possess a large variety of antibiotic resistance mechanisms, making some of them resistant to most antibiotics currently available including vancomycin (168). The emergence of vancomycin resistance in the enterococci is of special concern because it increases the difficulty in treating enterococcal infections and because of the potential for the plasmid and transposon-mediated vancomycin resistance to be transferred to other gram-positive bacteria, especially to MRSA. Vancomycin alone or in combination with gentamicin is the last weapon to treat MRSA infections, and the acquisition of vancomycin resistance genes would transform this pathogen into an invulnerable strain exhibiting resistance to all the currently approved antibiotics. Today, MRSA with decreased susceptibility to vancomycin have already emerged in some countries in America, Asia, and Europe (31, 116). The risk of an in vivo transfer of a vancomycin resistant gene between an *Enterococcus* species and *S. aureus* is highly probable, as this event can occur in vitro since both vancomycin-resistant enterococci (VRE) and *S. aureus* have been found to coexist in the intestinal tract of hospitalized patients (150). Furthermore, although the intestinal colonization of VRE does not directly affect the health of the carriers, it may serve as a source of contamination and of transmission to patients at risk of infection (135).

In 1993, the first glycopeptide resistant enterococci were found in animals and in the environment in Europe due to the use of avoparcin, a glycopeptide antibiotic similar to vancomycin, in animal husbandry (21, 101). Since then, several studies have confirmed that vancomycin resistant *E. faecium* are widespread in animals, food and humans (173). In the U.S., glycopeptide antibiotics were never licensed for animal use, yet the prevalence of vancomycin resistant clinical isolates is high. This is due to a relatively high oral and intravenous use of vancomycin in U.S. hospitals since the late 1970s (99, 206). Vancomycin-resistant *E. faecium,* however, have also been recovered from non-hospitalized patients (43). Compared to the European situation, vancomycin-resistant enterococci are not as widespread in healthy human and in the environment in North America (40). Indeed, the prevalence of VRE in the intestinal flora of healthy persons reached up to 12% in one region of Germany (100) and in The Netherlands (192), before the authorization for avoparcin was withdrawn in 1996. After the ban, the incidence of VRE in healthy people decreased in the same region of Germany to 3% (100) and to 6% in The Netherlands in 1997 and was 5 % in Switzerland in 1999 (17). Vancomycin-resistant enterococci, however, tend to persist to a certain level in animal environments despite the suspension of glycopeptide use in animal husbandry. This phenomenon results from genetic linkages that exist between glycopeptide resistance genes and other antibiotic resistance genes such as the aminoglycoside resistance determinant *aac*(6′)-*I*-*aph*(2″) (225) or the MLS_B resistance gene *erm*(B). Both the *vanA* and the *erm*(B) genes were found to be located on the same plasmid in *E. hirae* isolates from poultry (25) and in *E. faecium* from broilers and pigs (1). The use of the macrolide tylosin in medicated preparations simultaneously selects for both macrolide and glycopeptide resistances if such a tandem element is widespread in the enterococcal population (2). Additionally, the *erm*(B) resistance gene was also found to be linked, on a plasmid, with the streptogramin A resistance gene *vat*(E). The identical gene cluster was found among isolates from poultry manure, poultry meat, human stool samples, and hospitals, suggesting the spread of such a gene cluster via the food chain (208). The streptogramin A and B combination dalfopristin-quinupristin (Synercid) and linezolid are drugs effective against vancomycin-resistant enterococci infections today, with the exception of *E. faecalis*, which is intrinsically resistant to the streptogramins (3). The emergence of *vat*(D) and *vat*(E), conferring resistance to quinupristin-dalfopristin in the enterococci, was linked to the use of virginiamycin, a mixture of the group A streptogramin virginiamycin M and the group B streptogramin virginiamycin S, which was authorized as a therapeutic and as a growth promoter in animal husbandry in both the U.S. and Europe. Indeed, streptogramin resistant *E. faecium* isolates were found in food products and could be recovered from human feces before this drug combination was used in human medicine (81, 90, 126). Its use as a growth promoter has been banned in Europe since 1999. Yet, even in the absence of a specific selection for streptogramin A antibiotics, the genes conferring resistance to these drugs can be maintained in the bacterial population, like the glycopeptides resistance genes, by the use of macrolide antibiotics in food-producing animals because of the existence of gene linkages with *erm*(B). Furthermore, *erm*(B) is particularly widespread among enterococci of animal origin, such as pigs and pig carcasses (125), and of food origin (185). Its presence in animal isolates has been effectively maintained by the use of tylosin, which was used for both therapeutic and growth promoting purposes in animal husbandry. The choice of an antibiotic for the treatment of animals can be a determining factor in the selection of resistance genes. Antibiotics belonging to a class of drugs that is of major importance for the treatment of infectious diseases in humans should not be used in animals because animal bacteria with acquired resistance can be transmitted to humans via the food chain.

Enterococci harboring resistance to vancomycin, quinupristin-dalfopristin, macrolides, and tetracycline are all present in the food chain (78, 169, 172, 207) and may reach consumers (77). Traditional food products made with raw material and intended to be eaten raw represent a large reservoir of live enterococci. Among these products, cheeses and fermented dry sausage have been found to harbor enterococci displaying resistance to gentamicin, tetracycline, erythromycin, chloramphenicol penicillin, fusidic acid, and rifampin (141, 142, 185). Additionally, vancomycin resistant *E. faecium* isolates carrying the *vanA* gene were found in cheeses and meat products (38, 50, 67). Vancomycin and tetracycline resistant enterococci can persist and grow in fermented sausages (84) where they can also pass their resistance genes to other enterococci even without antimicrobial selective pressure (38). The same transfer phenomenon was observed during cheese fermentation (38), with plasmids carrying either the vancomycin resistance gene *vanA* or the tetracycline resistance gene *tet*(M). The *tet*(M) gene was found to be predominant in tetracycline resistant *Enterococcus* from European cheese where it occurred mainly on mobile elements of the Tn*916*-Tn*1545* family (89). A Tn*916*-like transposon (Tn*FO1*) carrying the tetracycline resistance determinant Tet(M) was found in *E. faecalis* FO1 isolated from a raw milk cheese in Switzerland. This transposon was very similar to the well-characterized transposon Tn*916* originating from *E. faecalis* of clinical origin. Tn*FO1* was shown to be transferred to *E. faecalis, S. aureus, Listeria innocua,* and to *Lactococcus lactis* by conjugation (138). Fermented dry sausages may also harbor enterococci carrying multidrug resistance conjugative plasmids. For example, the broad-host range plasmid pRE25 from *E. faecalis* RE25, isolated from a raw meat dry sausage, contains five antibiotic resistance genes that protect the strain against 12 different antimicrobial substances. The genes consisted of a chloramphenicol acetyltransferase (cat_{pIP501}), *erm*(B), and aminoglycodide 6-adenylyltransferase (*ant(6)-Ia*), a streptothricin acetyltransferase (*sat4*), and an aminoglycodide phosphotransferase (*aph(3')-III*). Plasmid pRE25 may be transferred, by conjugation, to *E. faecalis* and *L. innocua* and integrated into the chromosome of *L. lactis* conferring multidrug resistance in these new hosts (166a).

The enterococci of animal and food origin represent a large reservoir of transferable antibiotic resistance genes, including genes conferring resistance to drugs essential to human medicine such as vancomycin, erythromycin, and quinupristin-dalfopristin.

Lactococcus

Lactococci, which are naturally found on plants, are spread throughout grass and are also present in the mouths and on the udders of cows. They may transferred to milk during milking and occur in cheeses made with unpasteurized milk. Intramammary dry cow therapy is a common practice whereby an antibiotic is injected into the udder of cows to avoid the development of mastitis during the nonlactating period. This practice affects the entire flora surrounding the udders and in time, commensal bacteria have developed resistances to the antibiotics that have been used for these purposes. *Lactococcus lactis* and *Lactococcus garviae* displaying resistance to antibiotics such as tetracycline, clindamycin, chloramphenicol, and quinupristin-dalfopristin are regularly isolated from milk samples (unpublished data from the Institute of Veterinary Bacteriology, Berne, Switzerland). One strain, *Lactococcus lactis* K214, is a good example of a strain originating from animals and now present in the food chain. This strain, isolated from a raw milk cheese, was found to harbor a mosaic plasmid (pK214). This plasmid confers resistance to tetracycline, streptomycin, and chloramphenicol and could be transferred to *E. faecalis* JH2-2 by electrotransformation (139). In addition, plasmid pK214 was shown to harbor a gene encoding a new membrane-spanning efflux protein [Mdt(A)] that confered decreased susceptibility to macrolides, lincosamides, streptogramins, and tetracyclines in *L. lactis* and in *E. coli* (140). The plasmid had a mosaic structure, as it contained fragments derived from other gram-positive bacteria, and the fragments are clearly separated by IS elements that were involved in the recombination of the plasmid. It is composed of a staphylococcal-derived fragment containing a streptomycin adenylase, a chloramphenicol acetyltransferase, and a tetracycline resistance gene (*tet*(S)) from *Listeria monocytogenes* (34). The presence of *tet*(S) has also reported in the enterococci (35). The mosaic plasmid pK214 is a good demonstration of the genetic exchange that can occur between pathogenic and nonpathogenic bacteria in a food environment.

Lactobacillus

The *Lactobacillus* family contains a large variety of different species. They are a part of the oral, intestinal, and vaginal flora of animals and humans, and are also present in plant material. Lactobacilli are a common component of meat and meat products, milk, cheeses, fermented dairy products, and fermented vegetables.

Most of the lactobacilli used in the food industry have shown decreased susceptibility to bacitracin, cefoxitin, fusidic acid, metronidazole, nitrofurantoin, sulfadiazine, trimethoprim-sulfamethoxazole, ciprofloxacin, kanamycin, streptomycin, teicoplanin, and vancomycin. Since variations in antibiotic susceptibility has been observed between the different species (45), the origin of these resistances remains to be elucidated in order to determine whether they are naturally occurring in the lactobacilli. Besides these "intrinsic" resistances, lactobacilli have the ability to acquire antibiotic resistance genes encoded on extrachromosomal genetic elements such as plasmids. Plasmids conferring resistance to either MLS_B antibiotics or chloramphenicol were characterized in different *Lactobacillus* strains from animal origin. The chloramphenicol resistance plasmid pTC82 has been found in a *Lactobacillus reuteri* isolate from chicken intestines (115). Plasmids carrying genes sharing identity to *erm*(B), *erm*(C), and *erm*(T) and conferring resistance to MLS_B antibiotics were found in *L. reuteri* of poultry origin and in *Lactobacillus* sp. and *Lactobacillus fermentum* from pig feces (179) (114) (212) (59). The *erm*(B), *erm*(C) genes are widespread among the enterococci, streptococci and staphylococci suggesting that gene transfer occurred between these species in vivo. Experiments in vivo have demonstrated that *erm*(B)-carrying plasmids, pAMβ1 and pIP501, could be transferred between *Lactobacillus, Lactococcus,* and *Enterococcus* species (104, 178).

Lactobacillus strains isolates from food products are also able to spread their resistance genes to other gram-positive bacteria. *Lactobacillus* strains from fermented dry sausage were shown to transfer the tetracycline resistance gene *tet*(M) to *E. faecalis, S. aureus,* and *L. lactis* by conjugation (62). These bacteria can survive the different processes carried out during the production of fermented raw sausages (63). Dairy products such as raw milk cheeses are also a source of tetracycline resistant *L. plantarum* (185). In raw milk products, a *L. fermentum* strain was found to harbor the plasmid pLME300, which carried both an *erm*(T)-like and a *vat*(E) gene conferring resistance to MLS_B antibiotics and dalfopristin, respectively (64).

In some cases, lactobacilli can be opportunistic pathogens and cause human infections, yet they can still be treated and are in themselves rarely fatal (12, 88). However, their intrinsic resistance to several antibiotics including vancomycin and their ability to acquire other resistance genes could complicate the antibiotic therapy. Vigilance is therefore recommended to avoid the emergence of multidrug resistant *Lactobacillus* in hospitals (131).

ANTIBIOTIC RESISTANCE IN ZOONOTIC BACTERIA

Enterobacteriaceae

Members of the family *Enterobacteriaceae* are ubiquitous, and many of them are pathogenic for animals and humans. They can cause diarrheal disease as well as extraintestinal infections, including bacteremia, meningitis, and urinary, respiratory and wound infections. The nosocomial infections are mainly caused by species such as *E. coli, Klebsiella, Enterobacter, Proteus, Providencia,* and *Serratia marcescens.*

Two important zoonotic members of the *Enterobacteriaceae,* namely *E. coli* and *Salmonella,* have emerged in animal husbandry with only a few antibiotic resistances but were able to acquire, within two decades, a set of different genes making them resistant to the most important classes of drugs.

Salmonella

Salmonella enterica serovar Enteritidis and serovar Typhimurium are the two most prevalent salmonella serotypes, and both are zoonotic organisms. They are found primarily in the intestines of humans and animals and have been isolated from many food products, including meat (poultry, beef, and pork), dairy, egg, vegetables, and fruit products. Salmonellosis in humans occurs after the ingestion of contaminated food and causes nausea, vomiting, and diarrhea. Generally, about 10^8 cells of *S. enterica* must be ingested to cause disease, although in some cases as few as 100 cells are able to cause gastroenteritis. About 5% of the human population is estimated to be healthy carriers of *Salmonella* spp. and are likely to be a source of contamination when handling food. Animals, however, remain the main reservoir of zoonotic salmonella, and the intensive use of antibiotics in animal husbandry has selected for antibiotic resistant strains. In 1968, Anderson reported the outbreaks, occurring in the U.K. from 1964 to 1966, of a multidrug-resistant serovar Typhimurium phage type 29 originating from cattle (8). At that time, some type 29 isolates were resistant to ampicillin, chloramphenicol, kanamycin, neomycin, streptomycin, sulfonamides, tetracycline, and furazolidone. Later, other antibiotic resistant serovar Typhimurium strains belonging to phage types 204, 193, and 204c spread throughout Europe (188). In the late 1980s, the multidrug-resistant serovar Typhimurium DT104 emerged in cattle in the U.K. and rapidly disseminated worldwide (48, 68, 86, 147, 188). Since then, it has frequently occurred in poultry, pigs, and sheep and is one of the most common strains causing salmonellosis in human besides serovar Enteritidis phage type 4, which originated in poultry. Most of the serovar Typhimurium DT104 strains display a pattern of resistance to five different antibiotics (ampicillin, chloramphenicol, streptomycin, sulfonamides, and tetracycline), but epidemic strains harboring resistance to nine antibiotics have been found (15, 29). Multidrug-resistant serovar Typhimurium DT104 was involved in several outbreaks linked to the consumption of different foods of animal origin. Raw milk and raw milk cheeses have been implicated in infections caused by multidrug-resistant serovar Typhimurium DT104, indicating, that the strains originated from local dairy cows (39, 202). Meat products such as pork and retail ground meat can also harbor multidrug-resistant serovar Typhimurium DT104 (166, 211, 229), and some meat products have been responsible for salmonellosis outbreaks (47, 129). Serovar Typhimurium DT104 with decreased susceptibility to fluoroquinolones, the drugs of choice for the treatment of salmonellosis in adults, has already been involved in community outbreaks following the consumption of raw milk or meat products (129, 203). Serovar Typhimurium DT104 has also been found within intact eggs (218). Recently, a multidrug-resistant *S. enterica* serotype Newport has emerged and has been involved in severe food-borne outbreaks associated with exposure to raw or undercooked ground beef and raw milk cheese in the U.S. (74) and with ground horse meat consumed in France (55). Like serovar Typhimurium DT104, the serovar Newport strains showed resistance to ampicillin, streptomycin, sulfonamide, tetracycline, and chloramphenicol and were also resistant to cephalosporins. In the U.S., the resistance to extended-spectrum cephalosporins has been attributed to a plasmid-mediated β-lactamase gene (*bla_{CMY-2}*) (65), and the serovar Newport strains carrying this gene (serovar Newport-MDR AmpC) are today the most prevalent in cattle and in humans. The emergence of serovar Newport-MDR AmpC in human correlated with the emergence of serovar Newport-MDRAmpC in cattle (74). Extended-spectrum-cephalosporin resistance has also emerged in *S. enterica* from animal origin (70). The emergence of multidrug-resistant *Salmonella* spp. that also display resistance to expanded-spectrum cephalosporins is of major concern, since cephalosporins such as ceftriaxone and cefotaxime are the drugs of choice to treat invasive *Salmonella* infections in children. Indeed, ceftriaxone resistant serovar Newport-MDR AmpC from cattle has already been shown to be responsible for salmonellosis in a child (58).

The emergence and the spread of a variety of multidrug-resistant *Salmonella* strains originating from animals and causing food-borne outbreaks becomes more than an alarming threat for public health, and adequate measures should be taken rapidly to avoid the dissemination of such strains and outbreaks of incurable salmonellosis.

E. coli

E. coli is part of the normal flora of humans and animals. Some *E. coli* strains harbor specific virulence factors that are responsible for enteritis in animals and humans and for a variety of extraintestinal infectious diseases in humans, such as urinary tract infections and septicemia. Infected humans and animals are the main reservoir of enteropathogenic *E. coli*, and transmission occurs by direct contact or via contaminated food products or water. The most important zoonotic *E. coli* causing the most severe outbreaks are the Shiga-toxin-producing (STEC) strains. Only a few cells of STEC, from 10 to 100, are sufficient to cause infection. Before the 1990s, the degree of antibiotic resistance in clinical isolates of pathogenic *E. coli* was not considered alarming, and one study at that time showed that in the U.S., less than 3% of the STEC O157 clinical isolates were resistant to an antibiotic (149). Since then antibiotic resistances have emerged in STEC O157 of human, animal and food origins, but the overall prevalence of resistant organisms generally remain low. The most prevalent resistances found are those to sulfamethoxazole (10 to 100%), tetracycline (4 to 8%) chloramphenicol (up to 3%) and streptomycin (2%) (165, 176, 191). Some outbreaks and sporadic cases where antibiotic resistant enteropathogenic *E. coli* isolates, including O157, were involved have been reported (41, 44, 121, 165, 217). Although the use of antibiotics for the treatment of O157 infections is still controversial, in general it is not recommended (189) as the use of subinhibitory concentrations of antibiotics in vitro such as co-trimoxazole, trimethoprim, azithromycin, and gentamicin induced the release of verotoxin. Furthermore, with other antibiotics such as β-lactams, ciprofloxacin, streptomycin, fosfomycin, and sulfomethoxazole, the release of verotoxin was dependent on the strains involved (71). Recently, cell wall inhibitors such as ampicillin and fosfomycin have been shown to increase the release of the shiga toxin Stx 2e in *E. coli* O139, which causes edema disease in pigs (190). In non-O157 STEC causing animal diseases, a higher percentage of antibiotic resistance has been detected than in O157. Resistance to sulfamethoxazole, tetracycline and streptomycin are the most common (105, 163, 209), but resistance to ampicillin, trimethoprim, gentamicin, kanamycin, enrofloxacin are also emerging in other *E. coli* serotypes which cause infectious diseases in animals (105, 227). The use of antibiotics in animal husbandry may select for specific resistances and invasive strains that can rapidly spread among animals and be transferred to humans. The presence of such specific multidrug-resistant STEC isolates has already been reported. Indeed, STEC O118:[H16] strains from cattle and humans in different countries of Europe were found to belong to the same genetic clone, and some of them displayed resistance to up to 8 different antibiotics (120). Even if some specific *E. coli* serotypes are pathogenic for animals only and do not affect human health, they can act as a reservoir of antibiotic resistance genes that may be transferred to human pathogens. Indeed, the acquired antibiotic resistance genes found in human isolates are generally closely related to the genes found in animal isolates. Recently, a new sulfonamide resistance gene (*sul3*) was detected in *E. coli* strains causing infectious diseases in pigs in Switzerland (137). Since its discovery, the *sul3* gene was also found in pathogenic *E. coli* from pigs, cattle, and poultry (73); in *S. enterica* from poultry, pigs, and meat products in Germany (72); and in *E. coli* from poultry and pigs in Spain (158). In humans, this gene has emerged in *E. coli* isolates from urinary tract infections in Sweden (69) and from feces in Spain (158). In these two countries, the *sul3* gene was also present among animal isolates (158; M. Grape and C. Greko, personal communication). The relationship between the human and animal isolates has yet to be elucidated, but the presence of *sul3* genes in retail meat products strongly suggests that the food chain acts as link between food-producing animals and humans.

Commensal nonpathogenic *E. coli* strains of human, animal and food origin also display resistance to many different antibiotics (49, 158, 164). In Spain, the isolates were shown to harbor specific genes conferring resistance to drugs such as the quinolones (nalidixic acid and ciprofloxacin), ampicillin, tetracycline, chloramphenicol, kanamycin, streptomycin, sulfonamides and trimethoprim. Here again, the resistance genes are similar to those found in pathogenic *E. coli*. Of note, the majority of the analyzed strains also harbored mutations in the multiple antibiotic resistance (*mar*) locus (158).

Pathogenic and non-pathogenic *E. coli* strains represent a large reservoir of antibiotic resistance genes that can be exchanged within the *Enterobacteriaceae* (23) and spread from animals to humans and vice versa (193, 220).

Yersinia enterocolitica

Y. enterocolitica is widely distributed in aquatic and animal environments, but pigs represent one of the main reservoirs for human pathogenic *Y. enterocolitica*. Human infections with *Y. enterocolitica* result mainly from the ingestion of contaminated food or water, causing gastroenterititis, especially in young children, and from blood transfusions (26, 27). A large amount of cells (10^9) is usually necessary to cause gastroenteritis in humans, but *Y. enterocolitica*, like *Listeria monocytogenes*, can survive and multiply in refrigerated food. Food of animal origin such as milk and meat products have been incriminated in several outbreaks (27). Food-borne yersiniosis is mainly associated with specific *Y. enterocolitica* serogroups O:8, O:13a, O13:b, and O3. The antibiotic resistance situation in *Y. enterocolitica* from animal origin is poorly documented. *Y. enterocolitica* serogroups O:3 and O:9 produce two chromosomally mediated β-lactamases (types A and B) conferring resistance to ampicillin, cephalotin, carbenicillin, and penicillin, whereas the serogroup O:5,27 produces only the type B β-lactamase having higher activity against cephalosporins. Serogroup O:8 is susceptible to ampicillin, and its susceptibility to carbenicillin and cephalothin is variable. Besides these intrinsic β-lactam resistances, *Y. enterocolitica* isolates from pig and pork meat products in Canada (103) and *Y. enterocolitica* strains of serogroup O:5 from raw milk in France were susceptible to other antibiotics (201). Some studies have determined the antibiotic resistance profile of strains belonging to these serogroups and originating from human intestines samples. In Spain, *Y. enterocolitica* belonging to serogroup O:3, isolated between 1995 and 1998, were resistant to streptomycin (72%), sulfonamides (45%), the combination trimethoprim-sulfamethoxazole (70%), chloramphenicol (60%), and nalidixic acid (5%) (146). In Poland, strains tested were primarily susceptible to the antibiotics mentioned above (148). A correlation between the antibiotic resistant strains from humans and those of animal origin is difficult to establish as not enough epidemiological data are available for this zoonotic pathogen.

Campylobacter

Campylobacter spp. are widely distributed in warm blooded animals. *Campylobacter jejuni* is frequently found as a commensal of the intestinal tract of cattle and poultry, and *Campylobacter coli* is a frequent commensal of the intestines of poultry and pigs. In the last decade, the degree of antibiotic resistant *Campylobacter* has increased dramatically in both animals and humans, especially in the case of resistance to the fluoroquinolones (159). The emergence of fluoroquinolone resistant *Campylobacter* strains of human origin correlated with the introduction of fluoroquinolone antibiotics for the treatment of food-producing animals (54). In The Netherlands, the percentage of quinolone resistant *Campylobacter* isolates from poultry and humans increased from 0 to 14% and from 0 to 11%, respectively, between 1982 and 1989 (53); enrofloxacin was authorized in this country for use in poultry therapy in 1987. In Spain, the prevalence of quinolone resistant *Campylobacter* isolated from humans increased from 3 up to 50% between 1989 and 1991, correlating with the introduction of enrofloxacin for veterinary purposes in 1990 (151, 160, 200). In Germany, 45% of human and chicken isolates were resistant to ciprofloxacin in 2001, whereas in 1991 the rates of resistance were 27% for chicken isolates and only 5% for human isolates (119). In the U.S., the same phenomenon was observed after the introduction, in poultry production, of the fluoroquinolones sarafloxacin in 1995 and enrofloxacin in 1996. The amount of fluoroquinolone resistant *Campylobacter* dramatically increased among patients in the mid-1990s (83, 130, 171). The increased resistance was attributed to the transmission of resistant strains from poultry and to acquired campylobacteriosis during foreign travel (96, 171). Campylobacter not only has the ability to rapidly acquire fluoroquinolone resistance, but the resistant strains have been suggested to have the fitness potential to persist in animal environments even in the absence of fluoroquinolone use (228). Besides fluoroquinolones resistance, some *Campylobacter* strains of animal origin are today resistant to many different antibiotics and are also present in food products (196).

Because antibiotic resistant *Campylobacter* is widespread in food-producing animals, it has the potential to reach consumers via the food chain. A low number of cells (<500) are necessary to cause campylobacteriosis in humans, provoking diarrhea and possibly bloody stool, abdominal cramps, fever, and nausea. In case of severe infection, antibiotic treatment may be necessary. The antibiotic resistance situation has therefore also been assessed for *Campylobacter* present in retail meat products.

In the U.S., 82% of the strains isolated from raw poultry meat displayed resistance to tetracycline antibiotics, 54% to erythromycin, 41% to nalidixic acid, 35% to ciprofloxacin, and none to gentamicin. Coresistance to ciprofloxacin and erythromycin was found in 26% of the isolates (61). Decreased susceptibility to antibiotics was also observed in European countries. In Switzerland, resistance to ciprofloxacin (29%) was the most frequent resistance observed in *Campylobacter* from retail raw chicken meat, followed by resistances to tetracycline (13%), sulfonamides (12%), ampicillin

(10%), and to erythromycin (1%) (106). In Northern Ireland, 35% of the campylobacters isolated from whole chicken were resistant to tetracycline, 19% to nalidixic acid, 11% to ciprofloxacin, 6% to erythromycin, and 2% to gentamicin (219). In Germany, 46% of campylobacters isolated from chicken meat were resistant to ciprofloxacin, 21% to ampicillin, 24% to tetracycline, and 1.5% to erythromycin (119). While the rate of ciprofloxacin resistant campylobacter approached 40% in both the U.S. and Europe, the prevalence of erythromycin resistant campylobacter is significantly higher in the former than in the latter.

These studies have demonstrated that the campylobacter found in the food chain harbors a considerable number of antibiotic resistances, including resistances to the fluoroquinolones and macrolides, which are the drugs used to treat severe campylobacter infections in humans (6). The number of ciprofloxacin resistant campylobacter, however, has reached such high levels that, today, erythromycin or newer macrolides are usually prescribed to treat campylobacter infections. Yet some macrolide antibiotics are widely used in animal production, which increases the potential risk of selecting for macrolide resistant *Campylobacter.* The "enrofloxacin experience" should convince public health authorities to take adequate measures in order to avoid the sudden emergence and a rapid increase of macrolide resistant campylobacter in animals. Thus, macrolides should not be used in animals such as poultry, where the prevalence of campylobacter is high. Each year, an alarming number of antibiotic resistant *Campylobacter* strains reach consumers' kitchens via retail raw meat. Infections with domestically acquired quinolone resistant *C. jejuni,* attributed to poultry consumption, have been reported (9). Good hygiene and adequate cooking of meat are necessary to limit the risks of human infections with campylobacter.

Emergence of Antibiotic Resistance in *Arcobacter*

Arcobacter are aerotolerant *Campylobacter*-like organisms that grow in aerobic environments and at lower temperatures than campylobacter, which can grow at 42°C. *Arcobacter* are frequently isolated from poultry, pork, beef, and lamb products (153). *Arcobacter butzleri* and *Arcobacter cryaerophilus* are the most common *Arcobacter* that have been associated with enteritis and bacteremia in humans (98, 108, 133, 198). In addition, *A. butzleri* isolates have been involved in food-borne outbreaks causing recurrent abdominal cramps (197). As with *Campylobacter,* diarrhea caused by *Arcobacter* is usually self-limiting, but antibiotic treatment, generally with erythromycin or ciprofloxacin, may be necessary in the most severe and persistent cases (85, 198). Recently, an alarming number of *Arcobacter* strains have been reported in meat products in Germany, where 37% of poultry meat products surveyed tested positive for *Arcobacter* (156). The presence of *A. butzleri* has also been reported in meat products in Australia (153) and in Japan (95). In these countries, the strains showed resistance to nalidixic acid, cephalothin, sulfamethoxazole-trimethoprim, and chloramphenicol and were susceptible to tetracycline, erythromycin, ampicillin, streptomycin, and kanamycin (95). In a Belgian study, resistance to ciprofloxacin and to erythromycin was found among poultry isolates (85). Closer attention now needs to be paid to avoid the rapid emergence of fluoroquinolone and macrolide resistant *Arcobacter* in meat products and to limit the spread of this emerging pathogen to humans. If not, future outbreaks caused by multidrug resistant *Arcobacter* may become a new threat to human health.

Listeria monocytogenes

Listeria spp. are ubiquitous bacteria commonly found in soil, plants, water, and the feces of healthy animals. Listeriosis, however, can occur in animals and humans after the ingestion of feed or food contaminated with *Listeria monocytogenes. L. monocytogenes* has been shown to cause encephalitis in animals that were previously fed with contaminated silage (118). This bacterium can also cause mastitis in sheep and cows (28, 91, 162). Thus, the affected udders represent a direct source of contaminated milk. Dairy products made with unpasteurized milk, such as soft cheeses, are a potential source of *L. monocytogenes,* as are raw meat products and ready-to-eat foods. In these products, listeria can survive and multiply because they are able to tolerate food preservatives such as salt and nitrite and can grow at 4°C, the standard temperature of a refrigerator. During the last two decades, *L. monocytogenes* was involved in several large food-borne outbreaks of human listeriosis. The main food products incriminated were meat products, dairy products, and ready-to-eat foods, including hot dogs, deli meats, coleslaw, rice, corn salad, and shrimp (161). Food preparations made with raw or under cooked products from animal or plant origin represent a potential risk to human health if contaminated with *L. monocytogenes.* Invasive listeriosis may be fatal and needs to be treated with antibiotics such as the combination ampicillin or penicillin plus gentamicin or cotrimoxazole as an alternative for patients allergic to penicillin. Vancomycin can be used for the treatment of primary bacteremia, and erythromycin can be used for the treatment of listeriosis in pregnant women (42, 82, 94). Of note, vancomycin therapy failure in

L. monocytogenes meningitis and peritonitis has already been reported, even when the *Listeria* strain involved was susceptible to the antibiotic in vitro (16, 18, 52, 152).

Antibiotic resistance was first reported in a *L. monocytogenes* strain of clinical origin in 1988 (144). The strain was resistant to tetracycline, chloramphenicol, erythromycin, and streptomycin. The genes conferring resistance to chloramphenicol, erythromycin, and streptomycin were found on a large plasmid, pIP811, that could be transferred to *L. monocytogenes, Enterococcus,* and *S. aureus* by conjugation. *L. monocytogenes* from environmental, food, and clinical samples collected worldwide (33) were then tested for antibiotic susceptibility. Resistances to tetracycline and trimethoprim were observed. Several resistance genes were characterized and were shown to be related to genes commonly found in the enterococci and staphylococci, two organisms that are inhabitants of animal environments such as the intestines, udders, and skin. The *tet*(M) gene is the most common tetracycline resistance gene found among *Listeria*, followed by *tet*(S), which was originally detected in a clinical *L. monocytogenes* isolate (34), and *tet*(L) (145). The trimethoprim resistance gene *dfrD,* which was first detected in a *Staphylococcus haemolyticus* isolate, has been found in an environmental strain of *L. monocytogenes* (32). The gene was located on a small plasmid, pIP823, that can be disseminated among gram-positive and gram-negative bacteria by conjugative mobilization (36). The emergence of trimethoprim resistance in *L. monocytogenes* of environmental origin is alarming as the combination trimethoprim-sulfamethoxazole is a second drug of choice for the treatment of listeriosis. Another *L. monocytogenes* strain from a patient in Switzerland was found to harbor a large multidrug resistance plasmid. This plasmid, pWDB100, which carried a chloramphenicol acetyltransferase (*cat*), *erm*(B), and *tet*(M) could also be transferred to other gram-positive bacteria by conjugation (75). Additional studies have been performed in vitro to demonstrate the ability of *L. monocytogenes* to acquire and transfer genes conferring resistance to clinically important antibiotics such as vancomycin. A plasmid mediated vancomycin resistant gene *vanA* was transferred from *E. faecium* to *L. monocytogenes* by conjugation (22).

Antibiotic-resistant *Listeria* spp. from food

L. monocytogenes isolates from food are primarily susceptible to antibiotics used in human and veterinary medicine (14, 60). Some isolates have been found to show decreased susceptibility to different antibiotics. The prevalence of antibiotic resistant strains varies from one country to another and depends on the food products analyzed. In Ireland, 0.6% of *L. monocytogenes* and 19.5% of nonpathogenic *Listeria innocua* from retail products including meat products, salads, ice cream, and haddock were found to exhibit resistance to one or more antibiotics. Besides the two *L. monocytogenes* isolates that displayed resistance to tetracycline, the other resistant isolates were identified as *L. innocua.* Among these *L. innocua* strains, resistance to tetracycline, found in 64 isolates (6.7%), and to penicillin (3.7%) were the most frequently observed, followed by isolated cases of streptomycin and erythromycin resistances. Two strains exhibited resistance to vancomycin and one strain was resistant to gentamicin (204). In Taiwan, a large variety of meat products, vegetables, seafood, and frozen semiready foods were screened for antibiotic resistant *L. monocytogenes.* With the exception of 14.5% of the isolates found to be resistant to tetracycline, the isolates were susceptible to the major antibiotic classes (224). In Portugal, the number of antibiotic resistant isolates from poultry carcasses obtained in butcher shops and canteens was significantly higher with 73% of *L. monocytogenes* strains found to be resistant to one or more antibiotics (10). In Italy, *Listeria* harboring resistance to antibiotics such as tetracycline, erythromycin, cotrimoxazole, sulfamethoxazole, kanamycin, streptomycin and clindamycin were recovered in meat products and soft cheeses (19, 56). In general, resistance to tetracycline was the most commonly found resistance in *Listeria* strains from food. Tetracycline resistance in food borne isolates is mainly due to the presence of either *tet*(M) or *tet*(S) (37), and in some cases *tet*(K) (57). *L. innocua* and *L. monocytogenes* from food origin are able to transfer *tet*(M) to *L. monocytogenes* and *Listeria ivanoii* and to *E. faecalis* (57, 143). In addition, an *erm*(C) gene conferring resistance to MLS_B antibiotics was found on the chromosome of *L. monocytogenes* and *L. innocua* isolated from food. This gene was transferable to *L. monocytogenes* and *L. ivanoii* and to *E. faecalis* by conjugation (154).

Clostridium perfringens

Clostridium perfringens is widespread in nature occurring primarily in soil and in the intestines of animals. *C. perfringens* type A is responsible for gastroenteritis occurring after the consumption of contaminated meat. *C. perfringens* vegetative cells produce enterotoxins during sporulation in the intestinal tract causing abdominal cramps and diarrhea. A relatively high number of cells (10^8) must be ingested to cause gastroenteritis in humans, and antibiotic treatment is not usually necessary. Cases of necrotizing enteritis due to type C and type A organisms, however, require antibiotic treatment. *C. perfringens* type A is also associated with wound infections (gas gangrene) in both humans and animals. Antibiotics such as penicillin, vancomycin, clindamycin, rifampin, and tetracycline are usually indicated for the treatment of wound infections in humans. Without intensive therapy, the mortality rate is very high and the emergence of resistant strains would have dramatic consequences. Very few studies have been conducted to determine the antibiotic resistance situation of *C. perfringens* from animal and environmental origins and its relationship to human infections. However, *C. perfringens* is known to harbor antibiotic resistance genes such as *tetA*(P) and *tetB*(P) for tetracycline resistance, *catP* and *catQ* for chloramphenicol resistance, and *erm*(B) for MLS_B resistance, and all are located on mobile genetic elements (5). Resistance to these drugs has been found in pig isolates (157), and tetracycline and/or lincomycin resistance has been reported in poultry isolates (93, 124, 205). To date, no incidence of metronidazole or vancomycin resistant *C. perfringens* has been reported in food-producing animals. Metronidazole resistance has appeared in *C. perfringens* isolated from dogs (123). Further studies are now necessary to determine whether *C. perfringens* exhibiting resistance to antibiotics relevant to human medicine is emerging in animal environments. Such surveillance studies would help prevent a sudden emergence of multidrug resistant *C. perfringens* isolates and monitor *C. perfringens* diseases in both humans and animals.

IMPACT ON HUMAN HEALTH: DIRECT AND INDIRECT CONSEQUENCES

The transfer of antibiotic resistant bacteria from animals to humans occurs by direct contact or via the food chain. Farmers and other people working in close contact with food-producing animals are predominantly exposed to bacteria present in animal populations and can acquire antibiotic resistant flora when antibiotics are used in animal feed (13, 193). For the rest of the population, bacteria of animal sources are acquired through the consumption of contaminated food from animal origin (51, 107). Appropriate cooking of the food products and good hygiene in the kitchen help to prevent the ingestion of live bacteria. Food products made with raw material of animal origin, however, may still contain a large number of live bacteria that reach consumers. The production of these products requires strict hygiene, and microbiological quality control is necessary to avoid food-borne infections and the spread of antibiotic-resistant bacteria. Large numbers of antibiotic-resistant bacteria in the food chain increases the risk of therapy failures in the case of infectious diseases requiring antibiotic treatment.

Antibiotic-resistant food-borne bacteria of animal origin can affect human health in different ways: directly if an infection is caused by a multidrug resistant food-borne pathogen and antibiotic treatment is required and indirectly if food contains opportunistic bacteria that may be involved in nosocomial infections or if food harbors bacteria carrying resistance genes that can be transferred to pathogenic bacteria, making them unresponsive to antibiotics. The direct consequences can be illustrated by the fact that multidrug resistant *Salmonella* strains acquired from animals have already caused human fatalities (20, 129, 177). Furthermore, the rate and duration of hospitalization and mortality has increased with antibiotic-resistant bacteria (20, 79). The indirect consequences are difficult to quantify, and the real impact of multidrug resistance in opportunistic pathogens and commensal bacteria may still come and may lead to an increase in infections with multidrug-resistant bacteria in hospitals. In summary, the widespread and inappropriate use of antibiotic in agriculture has created a large reservoir of antibiotic-resistant bacteria that can affect public health (7, 76, 167, 194).

To stop the cascade of multidrug resistance and its consequences for human health, adequate measures should be taken in all fields where antibiotics are used, including agriculture and veterinary and human medicine. Considering the ability of bacteria to develop antibiotic resistance, one should not forget that antibiotics are drugs with a limited effective lifespan. Antibiotics of major importance in human health and newer drugs should be reserved for use in human medicine. Surveillance of antibiotic resistance should be correlated with surveillance of antimicrobial usage in order to develop a prescription system that allows adequate treatment and avoids random and overuse of antibiotics. Furthermore, the implementation of hazard analysis critical control point (HACCP) procedures throughout the husbandry and the food industries (30), impeccable hygiene in commercial and household kitchens, and appropriate cooking methods will help reduce the frequency of bacterial food-borne infections and limit the spread of antibiotic resistant bacteria from animals to humans.

Acknowledgments. This work is dedicated to my mentors, Stuart Levy and Michael Teuber.

REFERENCES

1. **Aarestrup, F. M.** 2000. Characterization of glycopeptide-resistant *Enterococcus faecium* (GRE) from broilers and pigs in Denmark: genetic evidence that persistence of GRE in pig herds is associated with coselection by resistance to macrolides. *J. Clin. Microbiol.* 38:2774–2777.

2. **Aarestrup, F. M., A. M. Seyfarth, H. D. Emborg, K. Pedersen, R. S. Hendriksen, and F. Bager.** 2001. Effect of abolishment of the use of antimicrobial agents for growth promotion on occurrence of antimicrobial resistance in fecal enterococci from food animals in Denmark. *Antimicrob. Agents Chemother.* **45:**2054–2059.
3. **Acar, J., M. Casewell, J. Freeman, C. Friis, and H. Goossens.** 2000. Avoparcin and virginiamycin as animal growth promoters: a plea for science in decision-making. *Clin. Microbiol. Infect.* **6:**477–482.
4. **Adak, G. K., S. M. Long, and S. J. O'Brien.** 2002. Trends in indigenous foodborne disease and deaths, England and Wales: 1992 to 2000. *Gut* **51:**832–841.
5. **Adams, V., D. Lyras, K. A. Farrow, and J. I. Rood.** 2002. The clostridial mobilisable transposons. *Cell. Mol. Life Sci.* **59:**2033–2043.
6. **Allos, B. M.** 2001. *Campylobacter jejuni* infections: update on emerging issues and trends. *Clin. Infect. Dis.* **32:**1201–1206.
7. **Anderson, A. D., J. M. Nelson, S. Rossiter, and F. J. Angulo.** 2003. Public health consequences of use of antimicrobial agents in food animals in the United States. *Microb. Drug Resist.* **9:**373–379.
8. **Anderson, E. S.** 1968. Drug resistance in *Salmonella typhimurium* and its implications. *BMJ* **3:**333–339.
9. **Anonymous.** 1999. Food-borne antibiotic-resistant *Campylobacter* infections. *Nutr. Rev.* **57:**224–227.
10. **Antunes, P., C. Reu, J. C. Sousa, N. Pestana, and L. Peixe.** 2002. Incidence and susceptibility to antimicrobial agents of *Listeria* spp. and *Listeria monocytogenes* isolated from poultry carcasses in Porto, Portugal. *J. Food Prot.* **65:**1888–1893.
11. **Appelbaum, P. C., and B. Bozdogan.** 2004. Vancomycin resistance in *Staphylococcus aureus*. *Clin. Lab. Med.* **24:**381–402.
12. **Arpi, M., M. Vancanneyt, J. Swings, and J. J. Leisner.** 2003. Six cases of *Lactobacillus* bacteraemia: identification of organisms and antibiotic susceptibility and therapy. *Scand. J. Infect. Dis.* **35:**404–408.
13. **Aubry-Damon, H., K. Grenet, P. Sall-Ndiaye, D. Che, E. Cordeiro, M. E. Bougnoux, E. Rigaud, Y. Le Strat, V. Lemanissier, L. Armand-Lefevre, D. Delzescaux, J. C. Desenclos, M. Lienard, and A. Andremont.** 2004. Antimicrobial resistance in commensal flora of pig farmers. *Emerg. Infect. Dis.* **10:**873–879.
14. **Aureli, P., A. M. Ferrini, V. Mannoni, S. Hodzic, C. Wedell-Weergaard, and B. Oliva.** 2003. Susceptibility of *Listeria monocytogenes* isolated from food in Italy to antibiotics. *Int. J. Food Microbiol.* **83:**325–330.
15. **Baggesen, D. L., D. Sandvang, and F. M. Aarestrup.** 2000. Characterization of *Salmonella enterica* serovar Typhimurium DT104 isolated from Denmark and comparison with isolates from Europe and the United States. *J. Clin. Microbiol.* **38:**1581–1586.
16. **Baldassarre, J. S., M. J. Ingerman, J. Nansteel, and J. Santoro.** 1991. Development of *Listeria* meningitis during vancomycin therapy: a case report. *J. Infect. Dis.* **164:**221–222.
17. **Balzereit-Scheuerlein, F., and R. Stephan.** 2001. Prevalence of colonisation and resistance patterns of vancomycin-resistant enterococci in healthy, non-hospitalised persons in Switzerland. *Swiss Med. Wkly* **131:**280–282.
18. **Banerji, C., D. C. Wheeler, and J. R. Morgan.** 1994. *Listeria monocytogenes* CAPD peritonitis: failure of vancomycin therapy. *J. Antimicrob. Chemother.* **33:**374–375.
19. **Barbuti, A., A. Maggi, and C. Casoli.** 1992. Antibiotic resistance in strain of *Listeria* spp. from meat and meat products. *Lett. Appl. Microbiol.* **15:**56–58.
20. **Barza, M., and K. Travers.** 2002. Excess infections due to antimicrobial resistance: the "Attributable Fraction". *Clin. Infect. Dis.* **34**(Suppl 3)**:**S126–S130.
21. **Bates, J., Z. Jordens, and J. B. Selkon.** 1993. Evidence for an animal origin of vancomycin-resistant enterococci. *Lancet* **342:**490–491.
22. **Biavasco, F., E. Giovanetti, A. Miele, C. Vignaroli, B. Facinelli, and P. E. Varaldo.** 1996. In vitro conjugative transfer of VanA vancomycin resistance between *Enterococci* and *Listeriae* of different species. *Eur. J. Clin. Microbiol. Infect. Dis.* **15:**50–59.
23. **Blake, D. P., K. Hillman, D. R. Fenlon, and J. C. Low.** 2003. Transfer of antibiotic resistance between commensal and pathogenic members of the Enterobacteriaceae under ileal conditions. *J. Appl. Microbiol.* **95:**428–436.
24. **Boerlin, P., A. Wissing, F. M. Aarestrup, J. Frey, and J. Nicolet.** 2001. Antimicrobial growth promoter ban and resistance to macrolides and vancomycin in enterococci from pigs. *J. Clin. Microbiol.* **39:**4193–4195.
25. **Borgen, K., M. Sørum, Y. Wasteson, H. Kruse, and H. Oppegaard.** 2002. Genetic linkage between *erm*(B) and *vanA* in *Enterococcus hirae* of poultry origin. *Microb. Drug Resist.* **8:**363–368.
26. **Bottone, E. J.** 1997. *Yersinia enterocolitica:* the charisma continues. *Clin. Microbiol. Rev.* **10:**257–276.
27. **Bottone, E. J.** 1999. *Yersinia enterocolitica:* overview and epidemiologic correlates. *Microb. Infect.* **1:**323–333.
28. **Bourry, A., B. Poutrel, and J. Rocourt.** 1995. Bovine mastitis caused by *Listeria monocytogenes:* characteristics of natural and experimental infections. *J. Med. Microbiol.* **43:**125–132.
29. **Breuil, J., A. Brisabois, I. Casin, L. Armand-Lefevre, S. Fremy, and E. Collatz.** 2000. Antibiotic resistance in salmonellae isolated from humans and animals in France: comparative data from 1994 and 1997. *J. Antimicrob. Chemother.* **46:**965–971.
30. **Centers for Disease Control and Prevention.** 2004. Preliminary FoodNet data on the incidence of infection with pathogens transmitted commonly through food—selected sites, United States, 2003. *Morb. Mortal. Wkly. Rep.* **53:**338–343.
31. **Centers for Disease Control and Prevention.** 2004. Vancomycin-resistant *Staphylococcus aureus*—New York, 2004. *Morb. Mortal. Wkly. Rep.* **53:**322–323.
32. **Charpentier, E., and P. Courvalin.** 1997. Emergence of the trimethoprim resistance gene *dfrD* in *Listeria monocytogenes* BM4293. *Antimicrob. Agents Chemother.* **41:**1134–1136.
33. **Charpentier, E., and P. Courvalin.** 1999. Antibiotic resistance in *Listeria* spp. *Antimicrob. Agents Chemother.* **43:**2103–2108.
34. **Charpentier, E., G. Gerbaud, and P. Courvalin.** 1993. Characterization of a new class of tetracycline-resistance gene tet(S) in *Listeria monocytogenes* BM4210. *Gene* **131:**27–34.
35. **Charpentier, E., G. Gerbaud, and P. Courvalin.** 1994. Presence of the *Listeria* tetracycline resistance gene tet(S) in *Enterococcus faecalis*. *Antimicrob. Agents Chemother.* **38:**2330–2335.
36. **Charpentier, E., G. Gerbaud, and P. Courvalin.** 1999. Conjugative mobilization of the rolling-circle plasmid pIP823 from *Listeria monocytogenes* BM4293 among gram-positive and gram-negative bacteria. *J. Bacteriol.* **181:**3368–3374.
37. **Charpentier, E., G. Gerbaud, C. Jacquet, J. Rocourt, and P. Courvalin.** 1995. Incidence of antibiotic resistance in *Listeria* species. *J. Infect. Dis.* **172:**277–281.
38. **Cocconcelli, P. S., D. Cattivelli, and S. Gazzola.** 2003. Gene transfer of vancomycin and tetracycline resistances among *Enterococcus faecalis* during cheese and sausage fermentations. *Int. J. Food Microbiol.* **88:**315–323.

39. **Cody, S. H., S. L. Abbott, A. A. Marfin, B. Schulz, P. Wagner, K. Robbins, J. C. Mohle-Boetani, and D. J. Vugia.** 1999. Two outbreaks of multidrug-resistant *Salmonella* serotype typhimurium DT104 infections linked to raw-milk cheese in Northern California. *JAMA* **281:**1805–1810.
40. **Coque, T. M., J. F. Tomayko, S. C. Ricke, P. C. Okhyusen, and B. E. Murray.** 1996. Vancomycin-resistant enterococci from nosocomial, community, and animal sources in the United States. *Antimicrob. Agents Chemother.* **40:**2605–2609.
41. **Crampin, M., G. Willshaw, R. Hancock, T. Djuretic, C. Elstob, A. Rouse, T. Cheasty, and J. Stuart.** 1999. Outbreak of *Escherichia coli* O157 infection associated with a music festival. *Eur. J. Clin. Microbiol. Infect. Dis.* **18:**286–288.
42. **Crum, N. F.** 2002. Update on *Listeria monocytogenes* infection. *Curr. Gastroenterol. Rep.* **4:**287–296.
43. **D'Agata, E. M., J. Jirjis, C. Gouldin, and Y. W. Tang.** 2001. Community dissemination of vancomycin-resistant *Enterococcus faecium. Am. J. Infect. Control* **29:**316–320.
44. **Dalton, C. B., E. D. Mintz, J. G. Wells, C. A. Bopp, and R. V. Tauxe.** 1999. Outbreaks of enterotoxigenic *Escherichia coli* infection in American adults: a clinical and epidemiologic profile. *Epidemiol. Infect.* **123:**9–16.
45. **Danielsen, M., and A. Wind.** 2003. Susceptibility of *Lactobacillus* spp. to antimicrobial agents. *Int. J. Food Microbiol.* **82:**1–11.
46. **DANMAP.** 2003. Use of antimicrobial agents and occurrence of antimicrobial resistance in bacteria from food animals, foods and humans in Denmark. Danish Integrated Antimicrobial Resistance Monitoring and Research Programme, Copenhagen, Denmark.
47. **Davies, A., P. O'Neill, L. Towers, and M. Cooke.** 1996. An outbreak of *Salmonella typhimurium* DT104 food poisoning associated with eating beef. *Commun. Dis. Rep. CDR Rev.* **6:**R159–R162.
48. **Davis, M. A., D. D. Hancock, and T. E. Besser.** 2002. Multiresistant clones of *Salmonella enterica:* The importance of dissemination. *J. Lab. Clin. Med.* **140:**135–141.
49. **DeFrancesco, K. A., R. N. Cobbold, D. H. Rice, T. E. Besser, and D. D. Hancock.** 2004. Antimicrobial resistance of commensal *Escherichia coli* from dairy cattle associated with recent multi-resistant salmonellosis outbreaks. *Vet. Microbiol.* **98:**55–61.
50. **Del Grosso, M., A. Caprioli, P. Chinzari, M. C. Fontana, G. Pezzotti, A. Manfrin, E. D. Giannatale, E. Goffredo, and A. Pantosti.** 2000. Detection and characterization of vancomycin-resistant enterococci in farm animals and raw meat products in Italy. *Microb. Drug Resist.* **6:**313–318.
51. **Donabedian, S. M., L. A. Thal, E. Hershberger, M. B. Perri, J. W. Chow, P. Bartlett, R. Jones, K. Joyce, S. Rossiter, K. Gay, J. Johnson, C. Mackinson, E. DeBess, J. Madden, F. Angulo, and M. J. Zervos.** 2003. Molecular characterization of gentamicin-resistant *Enterococci* in the United States: evidence of spread from animals to humans through food. *J. Clin. Microbiol.* **41:**1109–1113.
52. **Dryden, M. S., N. F. Jones, and I. Phillips.** 1991. Vancomycin therapy failure in *Listeria monocytogenes* peritonitis in a patient on continuous ambulatory peritoneal dialysis. *J. Infect. Dis.* **164:**1239.
53. **Endtz, H. P., G. J. Ruijs, B. van Klingeren, W. H. Jansen, T. van der Reyden, and R. P. Mouton.** 1991. Quinolone resistance in campylobacter isolated from man and poultry following the introduction of fluoroquinolones in veterinary medicine. *J. Antimicrob. Chemother.* **27:**199–208.
54. **Engberg, J., F. M. Aarestrup, D. E. Taylor, P. Gerner-Smidt, and I. Nachamkin.** 2001. Quinolone and macrolide resistance in *Campylobacter jejuni* and *C. coli:* resistance mechanisms and trends in human isolates. *Emerg. Infect. Dis.* 7:24–34.
55. **Espié, E., H. De Valk, V. Vaillant, N. Quelquejeu, F. Le Querrec, and F. X. Weill.** 2005. An outbreak of multidrug-resistant *Salmonella enterica* serotype Newport infections linked to the consumption of imported horse meat in France. *Epidemiol. Infect.* **133:**373–376.
56. **Facinelli, B., E. Giovanetti, P. E. Varaldo, P. Casolari, and U. Fabio.** 1991. Antibiotic resistance in foodborne *Listeria. Lancet* **338:**1272.
57. **Facinelli, B., M. C. Roberts, E. Giovanetti, C. Casolari, U. Fabio, and P. E. Varaldo.** 1993. Genetic basis of tetracycline resistance in food-borne isolates of *Listeria innocua. Appl. Environ. Microbiol.* **59:**614–616.
58. **Fey, P. D., T. J. Safranek, M. E. Rupp, E. F. Dunne, E. Ribot, P. C. Iwen, P. A. Bradford, F. J. Angulo, and S. H. Hinrichs.** 2000. Ceftriaxone-resistant salmonella infection acquired by a child from cattle. *N. Engl. J. Med.* **342:**1242–1249.
59. **Fons, M., T. Hege, M. Ladire, P. Raibaud, R. Ducluzeau, and E. Maguin.** 1997. Isolation and characterization of a plasmid from *Lactobacillus fermentum* conferring erythromycin resistance. *Plasmid* **37:**199–203.
60. **Franco Abuin, C. M., E. J. Quinto Fernandez, S. C. Fente, J. L. Rodriguez Otero, R. L. Dominguez, and S. A. Cepeda.** 1994. Susceptibilities of *Listeria* species isolated from food to nine antimicrobial agents. *Antimicrob. Agents Chemother.* **38:**1655–1657.
61. **Ge, B., D. G. White, P. F. McDermott, W. Girard, S. Zhao, S. Hubert, and J. Meng.** 2003. Antimicrobial-resistant *Campylobacter* species from retail raw meats. *Appl. Environ. Microbiol.* **69:**3005–3007.
62. **Gevers, D., G. Huys, and J. Swings.** 2003. In vitro conjugal transfer of tetracycline resistance from *Lactobacillus* isolates to other gram-positive bacteria. *FEMS Microbiol. Lett.* **225:**125–130.
63. **Gevers, D., L. Masco, L. Baert, G. Huys, J. Debevere, and J. Swings.** 2003. Prevalence and diversity of tetracycline resistant lactic acid bacteria and their tet genes along the process line of fermented dry sausages. *Syst. Appl. Microbiol.* **26:**277–283.
64. **Gfeller, K. Y., M. Roth, L. Meile, and M. Teuber.** 2003. Sequence and genetic organization of the 19.3-kb erythromycin- and dalfopristin-resistance plasmid pLME300 from *Lactobacillus fermentum* ROT1. *Plasmid* **50:**190–201.
65. **Giles, W. P., A. K. Benson, M. E. Olson, R. W. Hutkins, J. M. Whichard, P. L. Winokur, and P. D. Fey.** 2004. DNA sequence analysis of regions surrounding bla_{CMY-2} from multiple *Salmonella* plasmid backbones. *Antimicrob. Agents Chemother.* **48:**2845–2852.
66. **Giraffa, G.** 2003. Functionality of enterococci in dairy products. *Int. J. Food Microbiol.* **88:**215–222.
67. **Giraffa, G., A. M. Olivari, and E. Neviani.** 2000. Isolation of vancomycin-resistant *Enterococcus faecium* from Italian cheeses. *Food Microbiol.* **17:**671–677.
68. **Glynn, M. K., C. Bopp, W. Dewitt, P. Dabney, M. Mokhtar, and F. J. Angulo.** 1998. Emergence of multidrug-resistant *Salmonella enterica* serotype typhimurium DT104 infections in the United States. *N. Engl. J. Med.* **338:**1333–1338.
69. **Grape, M., L. Sundstrom, and G. Kronvall.** 2003. Sulphonamide resistance gene *sul3* found in *Escherichia coli* isolates from human sources. *J. Antimicrob. Chemother.* **52:**1022–1024.
70. **Gray, J. T., L. L. Hungerford, P. J. Fedorka-Cray, and M. L. Headrick.** 2004. Extended-spectrum-cephalosporin resistance in *Salmonella enterica* isolates of animal origin. *Antimicrob. Agents Chemother.* **48:**3179–3181.

71. **Grif, K., M. P. Dierich, H. Karch, and F. Allerberger.** 1998. Strain-specific differences in the amount of shiga toxin released from enterohemorrhagic *Escherichia coli* O157 following exposure to subinhibitory concentrations of antimicrobial agents. *Eur. J. Clin. Microbiol. Infect. Dis.* **17:**761–766.
72. **Guerra, B., E. Junker, and R. Helmuth.** 2004. Incidence of the recently described sulfonamide resistance gene *sul3* among German *Salmonella enterica* strains isolated from livestock and food. *Antimicrob. Agents Chemother.* **48:**2712–2715.
73. **Guerra, B., E. Junker, A. Schroeter, B. Malorny, S. Lehmann, and R. Helmuth.** 2003. Phenotypic and genotypic characterization of antimicrobial resistance in German *Escherichia coli* isolates from cattle, swine and poultry. *J. Antimicrob. Chemother.* **52:**489–492.
74. **Gupta, A., J. Fontana, C. Crowe, B. Bolstorff, A. Stout, S. Van Duyne, M. P. Hoekstra, J. M. Whichard, T. J. Barrett, and F. J. Angulo.** 2003. Emergence of multidrug-resistant *Salmonella enterica* serotype Newport infections resistant to expanded-spectrum cephalosporins in the United States. *J. Infect. Dis.* **188:**1707–1716.
75. **Hadorn, K., H. Hächler, A. Schaffner, and F. H. Kayser.** 1993. Genetic characterization of plasmid-encoded multiple antibiotic resistance in a strain of *Listeria monocytogenes* causing endocarditis. *Eur. J. Clin. Microbiol. Infect. Dis.* **12:**928–937.
76. **Hamer, D. H., and C. J. Gill.** 2002. From the farm to the kitchen table: the negative impact of antimicrobial use in animals on humans. *Nutr. Rev.* **60:**261–264.
77. **Hammerum, A. M., C. H. Lester, J. Neimann, L. J. Porsbo, K. E. Olsen, L. B. Jensen, H. D. Emborg, H. C. Wegener, and N. Frimodt-Moller.** 2004. A vancomycin-resistant *Enterococcus faecium* isolate from a Danish healthy volunteer, detected 7 years after the ban of avoparcin, is possibly related to pig isolates. *J. Antimicrob. Chemother.* **53:**547–549.
78. **Hayes, J. R., L. L. English, P. J. Carter, T. Proescholdt, K. Y. Lee, D. D. Wagner, and D. G. White.** 2003. Prevalence and antimicrobial resistance of *Enterococcus* species isolated from retail meats. *Appl. Environ. Microbiol.* **69:**7153–7160.
79. **Helms, M., P. Vastrup, P. Gerner-Smidt, and K. Molbak.** 2002. Excess mortality associated with antimicrobial drug-resistant *Salmonella typhimurium*. *Emerg. Infect. Dis.* **8:**490–495.
80. **Hernaiz, C., A. Picardo, J. I. Alos, and J. L. Gomez-Garces.** 2003. Nosocomial bacteremia and catheter infection by *Bacillus cereus* in an immunocompetent patient. *Clin. Microbiol. Infect.* **9:**973–975.
81. **Hershberger, E., S. Donabedian, K. Konstantinou, and M. J. Zervos.** 2004. Quinupristin-dalfopristin resistance in gram-positive bacteria: mechanism of resistance and epidemiology. *Clin. Infect. Dis.* **38:**92–98.
82. **Hof, H.** 2003. Listeriosis: therapeutic options. FEMS Immunol. *Med. Microbiol.* **35:**203–205.
83. **Hooper, D. C.** 2001. Emerging mechanisms of fluoroquinolone resistance. *Emerg. Infect. Dis.* **7:**337–341.
84. **Houben, J. H.** 2003. The potential of vancomycin-resistant enterococci to persist in fermented and pasteurised meat products. *Int. J. Food Microbiol.* **88:**11–18.
85. **Houf, K., L. A. Devriese, F. Haesebrouck, O. Vandenberg, J. P. Butzler, J. Van Hoof, and P. Vandamme.** 2004. Antimicrobial susceptibility patterns of *Arcobacter butzleri* and *Arcobacter cryaerophilus* strains isolated from humans and broilers. *Microb. Drug Resist.* **10:**243–247.
86. **Humphrey, T.** 2001. *Salmonella* Typhimurium definitive type 104. A multi-resistant *Salmonella*. *Int. J. Food Microbiol.* **67:**173–186.
87. **Humphrey, T. J.** 1994. Contamination of egg shell and contents with *Salmonella enteritidis*—A Review. *Int. J. Food Microbiol.* **21:**31–40.
88. **Husni, R. N., S. M. Gordon, J. A. Washington, and D. L. Longworth.** 1997. *Lactobacillus* bacteremia and endocarditis: review of 45 cases. *Clin. Infect. Dis.* **25:**1048–1055.
89. **Huys, G., K. D'Haene, J. M. Collard, and J. Swings.** 2004. Prevalence and molecular characterization of tetracycline resistance in *Enterococcus* isolates from food. *Appl. Environ. Microbiol.* **70:**1555–1562.
90. **Jensen, L. B., A. M. Hammerum, F. M. Aerestrup, A. E. van den Bogaard, and E. E. Stobberingh.** 1998. Occurrence of *satA* and *vgb* genes in streptogramin-resistant *Enterococcus faecium* isolates of animal and human origins in the Netherlands. *Antimicrob. Agents Chemother.* **42:**3330–3331.
91. **Jensen, N. E., F. M. Aarestrup, J. Jensen, and H. C. Wegener.** 1996. *Listeria monocytogenes* in bovine mastitis. Possible implication for human health. *Int. J. Food Microbiol.* **32:**209–216.
92. **Jett, B. D., M. M. Huycke, and M. S. Gilmore.** 1994. Virulence of enterococci. *Clin. Microbiol. Rev.* **7:**462–478.
93. **Johansson, A., C. Greko, B. E. Engstrom, and M. Karlsson.** 2004. Antimicrobial susceptibility of Swedish, Norwegian and Danish isolates of *Clostridium perfringens* from poultry, and distribution of tetracycline resistance genes. *Vet. Microbiol.* **99:**251–257.
94. **Jones, E. M., and A. P. MacGowan.** 1995. Antimicrobial chemotherapy of human infection due to *Listeria monocytogenes*. *Eur. J. Clin. Microbiol. Infect. Dis.* **14:**165–175.
95. **Kabeya, H., S. Maruyama, Y. Morita, T. Ohsuga, S. Ozawa, Y. Kobayashi, M. Abe, Y. Katsube, and T. Mikami.** 2004. Prevalence of *Arcobacter* species in retail meats and antimicrobial susceptibility of the isolates in Japan. *Int. J. Food Microbiol.* **90:**303–308.
96. **Kassenborg, H. D., K. E. Smith, D. J. Vugia, T. Rabatsky-Ehr, M. R. Bates, M. A. Carter, N. B. Dumas, M. P. Cassidy, N. Marano, R. V. Tauxe, and F. J. Angulo.** 2004. Fluoroquinolone-resistant *Campylobacter* infections: eating poultry outside of the home and foreign travel are risk factors. *Clin. Infect. Dis.* **38:**S279–S284.
97. **Khachatryan, A. R., D. D. Hancock, T. E. Besser, and D. R. Call.** 2004. Role of calf-adapted *Escherichia coli* in maintenance of antimicrobial drug resistance in dairy calves. *Appl. Environ. Microbiol.* **70:**752–757.
98. **Kiehlbauch, J. A., D. J. Brenner, M. A. Nicholson, C. N. Baker, C. M. Patton, A. G. Steigerwalt, and I. K. Wachsmuth.** 1991. *Campylobacter butzleri* sp. nov. isolated from humans and animals with diarrheal illness. *J. Clin. Microbiol.* **29:**376–385.
99. **Kirst, H. A., D. G. Thompson, and T. I. Nicas.** 1998. Historical yearly usage of vancomycin. *Antimicrob. Agents Chemother.* **42:**1303–1304.
100. **Klare, I., D. Badstubner, C. Konstabel, G. Bohme, H. Claus, and W. Witte.** 1999. Decreased incidence of VanA-type vancomycin-resistant enterococci isolated from poultry meat and from fecal samples of humans in the community after discontinuation of avoparcin usage in animal husbandry. *Microb. Drug Resist.* **5:**45–52.
101. **Klare, I., H. Heier, H. Claus, and W. Witte.** 1993. Environmental strains of *Enterococcus faecium* with inducible high-level resistance to glycopeptides. *FEMS Microbiol. Lett.* **106:**23–29.
102. **Kruse, H., B. K. Johansen, L. M. Rorvik, and G. Schaller.** 1999. The use of avoparcin as a growth promoter and the occurrence of vancomycin-resistant *Enterococcus* species in Norwegian poultry and swine production. *Microb. Drug Resist.* **5:**135–139.
103. **Kwaga, J., and J. O. Iversen.** 1990. In vitro antimicrobial susceptibilities of *Yersinia enterocolitica* and related species iso-

lated from slaughtered pigs and pork products. *Antimicrob. Agents Chemother.* **34:**2423–2425.

104. **Langella, P., and A. Chopin.** 1989. Conjugal transfer of plasmid pIP501 from *Lactococcus lactis* to *Lactobacillus delbruckii* subsp. *bulgaricus* and *Lactobacillus helveticus. FEMS Microbiol. Lett.* **51:**149–152.
105. **Lanz, R., P. Kuhnert, and P. Boerlin.** 2003. Antimicrobial resistance and resistance gene determinants in clinical *Escherichia coli* from different animal species in Switzerland. *Vet. Microbiol.* **91:**73–84.
106. **Ledergerber, U., G. Regula, R. Stephan, J. Danuser, B. Bissig, and K. D. Stärk.** 2003. Risk factors for antibiotic resistance in *Campylobacter* spp. isolated from raw poultry meat in Switzerland. *BMC Public Health* **3:**1–9.
107. **Lee, J. H.** 2003. Methicillin (oxacillin)-resistant *Staphylococcus aureus* strains isolated from major food animals and their potential transmission to humans. *Appl. Environ. Microbiol.* **69:**6489–6494.
108. **Lerner, J., V. Brumberger, and V. Preacmursic.** 1994. Severe diarrhea associated with *Arcobacter butzleri. Eur. J. Clin. Microbiol. Infect. Dis.* **13:**660–662.
109. **Levy, S. B.** 1994. Balancing the drug-resistance equation. *Trends Microbiol.* **2:**341–342.
110. **Levy, S. B.** 1997. Antibiotic resistance: an ecological imbalance. *Ciba Found. Symp.* **207:**1–9.
111. **Levy, S. B.** 2002. The 2000 Garrod lecture. Factors impacting on the problem of antibiotic resistance. *J. Antimicrob. Chemother.* **49:**25–30.
112. **Levy, S. B., G. B. FitzGerald, and A. B. Macone.** 1976. Changes in intestinal flora of farm personnel after introduction of a tetracycline-supplemented feed on a farm. *N. Engl. J. Med.* **295:**583–588.
113. **Levy, S. B., G. B. FitzGerald, and A. B. Macone.** 1976. Spread of antibiotic-resistant plasmids from chicken to chicken and from chicken to man. *Nature* 260:40–42.
114. **Lin, C. F., and T. C. Chung.** 1999. Cloning of erythromycin-resistance determinants and replication origins from indigenous plasmids of *Lactobacillus reuteri* for potential use in construction of cloning vectors. *Plasmid* **42:**31–41.
115. **Lin, C. F., Z. F. Fung, C. L. Wu, and T. C. Chung.** 1996. Molecular characterization of a plasmid-borne (pTC82) chloramphenicol resistance determinant (*cat*-TC) from *Lactobacillus reuteri* G4. *Plasmid* **36:**116–124.
116. **Linares, J.** 2001. The VISA/GISA problem: therapeutic implications. *Clin. Microbiol. Infect.* **7:**8–15.
117. **Low, D. E., B. M. Willey, S. Betschel, and B. Kreiswirth.** 1994. Enterococcis: pathogens of the 90s. *Eur. J. Surg. Suppl.* **154:** 19–24.
118. **Low, J. C., and C. P. Renton.** 1985. Septicaemia, encephalitis and abortions in a housed flock of sheep caused by *Listeria monocytogenes* type 1/2. *Vet. Rec.* **116:**147–150.
119. **Luber, P., J. Wagner, H. Hahn, and E. Bartelt.** 2003. Antimicrobial resistance in *Campylobacter jejuni* and *Campylobacter coli* strains isolated in 1991 and 2001-2002 from poultry and humans in Berlin, Germany. *Antimicrob. Agents Chemother.* **47:**3825–3830.
120. **Maidhof, H., B. Guerra, S. Abbas, H. M. Elsheikha, T. S. Whittam, and L. Beutin.** 2002. A multiresistant clone of Shiga toxin-producing *Escherichia coli* O118:[H16] is spread in cattle and humans over different European countries. *Appl. Environ. Microbiol.* **68:**5834–5842.
121. **Makino, S., H. Asakura, T. Obayashi, T. Shirahata, T. Ikeda, and K. Takeshi.** 1999. Molecular epidemiological study on tetracycline resistance R plasmids in enterohaemorrhagic *Escherichia coli* O157:H7. *Epidemiol. Infect.* **123:**25–30.
122. **MARAN.** 2003. Monitoring of antimicrobial resistance and antibiotic usage in animals in the Netherlands in 2003. Lelystad, The Netherlands.
123. **Marks, S. L., and E. J. Kather.** 2003. Antimicrobial susceptibilities of canine *Clostridium difficile* and *Clostridium perfringens* isolates to commonly utilized antimicrobial drugs. *Vet. Microbiol.* **94:**39–45.
124. **Martel, A., L. A. Devriese, K. Cauwerts, K. De Gussem, A. Decostere, and F. Haesebrouck.** 2004. Susceptibility of *Clostridium perfringens* strains from broiler chickens to antibiotics and anticoccidials. *Avian Pathol.* **33:**3–7.
125. **Martel, A., L. A. Devriese, A. Decostere, and F. Haesebrouck.** 2003. Presence of macrolide resistance genes in streptococci and enterococci isolated from pigs and pork carcasses. *Int. J. Food Microbiol.* **84:**27–32.
126. **McDonald, L. C., S. Rossiter, C. Mackinson, Y. Y. Wang, S. Johnson, M. Sullivan, R. Sokolow, E. DeBess, L. Gilbert, J. A. Benson, B. Hill, and F. J. Angulo.** 2001. Quinupristin-dalfopristin-resistant *Enterococcus faecium* on chicken and in human stool specimens. *N. Engl. J. Med.* **345:**1155–1160.
127. **McEwen, S. A., and P. J. Fedorka-Cray.** 2002. Antimicrobial use and resistance in animals. *Clin. Infect. Dis.* **34:**S93–S106.
128. **Mead, P. S., L. Slutsker, V. Dietz, L. F. McCaig, J. S. Bresee, C. Shapiro, P. M. Griffin, and R. V. Tauxe.** 1999. Food-related illness and death in the United States. *Emerg. Infect. Dis.* **5:** 607–625.

128a. **Meinhart, A., J. C. Alonso, N. Sträter, and W. Saenger.** 2003 Crystal structure of the plasmid maintenance system ε/ζ: functional mechanism of toxin ζ and inactivation by $\varepsilon_2\,\zeta_2$ complex formation. *Proc. Natl. Acad. Sci. USA* **100:**1661–1666.

129. **Molbak, K., D. L. Baggesen, F. M. Aarestrup, J. M. Ebbesen, J. Engberg, K. Frydendahl, P. Gerner-Smidt, A. M. Petersen, and H. C. Wegener.** 1999. An outbreak of multidrug-resistant, quinolone-resistant *Salmonella enterica* serotype typhimurium DT104. *N. Engl. J. Med.* **341:**1420–1425.
130. **Nachamkin, I., H. Ung, and M. Li.** 2002. Increasing fluoroquinolone resistance in *Campylobacter jejuni*, Pennsylvania, USA, 1982-2001. *Emerg. Infect. Dis.* **8:**1501–1503.
131. **Nelson, R. R.** 1999. Intrinsically vancomycin-resistant gram-positive organisms: clinical relevance and implications for infection control. *J. Hosp. Infect.* **42:**275–282.
132. **NORM_NORM-VET.** 2003. Usage of antimicrobial agents and occurrence of antimicrobial resitance in Norway. Norwegian Monitoring Program for Resistance in Microbes, Tromso / Oslo, Norway.
133. **On, S. L., A. Stacey, and J. Smyth.** 1995. Isolation of *Arcobacter butzleri* from a neonate with bacteraemia. *J. Infect.* **31:**225–227.
134. **Pantosti, A., M. Del Grosso, S. Tagliabue, A. Macri, and A. Caprioli.** 1999. Decrease of vancomycin-resistant enterococci in poultry meat after avoparcin ban. *Lancet* **354:**741–742.
135. **Patel, R.** 2003. Clinical impact of vancomycin-resistant enterococci. *J. Antimicrob. Chemother.* **51**(Suppl. S3):iii13–iii21.
136. **Perreten, V.** 2003. Use of antimicrobials in food-producing animals in Switzerland and the European Union (EU). *Mitt. Lebensm. Hyg.* **94:**155–163.
137. **Perreten, V., and P. Boerlin.** 2003. A new sulfonamide resistance gene (*sul3*) in *Escherichia coli* is widespread in the pig population of Switzerland. *Antimicrob. Agents Chemother.* **47:**1169–1172.
138. **Perreten, V., B. Kollöffel, and M. Teuber.** 1997. Conjugal transfer of the Tn*916*-like transposon Tn*FO1* from *Enterococcus faecalis* isolated from cheese to other Gram-positive bacteria. *System. Appl. Microbiol.* **20:**27–38.

139. **Perreten, V., F. Schwarz, L. Cresta, M. Boeglin, G. Dasen, and M. Teuber.** 1997. Antibiotic resistance spread in food. *Nature* **389:**801–802.
140. **Perreten, V., F. V. Schwarz, M. Teuber, and S. B. Levy.** 2001. Mdt(A), a new efflux protein conferring multiple antibiotic resistance in *Lactococcus lactis* and *Escherichia coli. Antimicrob. Agents Chemother.* **45:**1109–1114.
141. **Perreten, V., and M. Teuber.** 1995. Antibiotic resistant bacteria in fermented dairy products—a new challenge for raw milk cheeses? p. 144–148. *In Residues of Antimicrobial Drugs and Other Inhibitors in Milk.* International Dairy Federation, Brussels, Belgium.
142. **Peters, J., K. Mac, H. Wichmann-Schauer, G. Klein, and L. Ellerbroek.** 2003. Species distribution and antibiotic resistance patterns of enterococci isolated from food of animal origin in Germany. *Int. J. Food Microbiol.* **88:**311–314.
143. **Pourshaban, M., A. M. Ferrini, V. Mannoni, B. Oliva, and P. Aureli.** 2002. Transferable tetracycline resistance in *Listeria monocytogenes* from food in Italy. *J. Med. Microbiol.* **51:**564–566.
144. **Poyart-Salmeron, C., C. Carlier, P. Trieu-Cuot, A. L. Courtieu, and P. Courvalin.** 1990. Transferable plasmid-mediated antibiotic resistance in *Listeria monocytogenes. Lancet* **335:**1422–1426.
145. **Poyart-Salmeron, C., P. Trieu-Cuot, C. Carlier, A. MacGowan, J. McLauchlin, and P. Courvalin.** 1992. Genetic basis of tetracycline resistance in clinical isolates of *Listeria monocytogenes. Antimicrob. Agents Chemother.* **36:**463–466.
146. **Prats, G., B. Mirelis, T. Llovet, C. Munoz, E. Miro, and F. Navarro.** 2000. Antibiotic resistance trends in enteropathogenic bacteria isolated in 1985-1987 and 1995-1998 in Barcelona. *Antimicrob. Agents Chemother.* **44:**1140–1145.
147. **Rabatsky-Ehr, T., J. Whichard, S. Rossiter, B. Holland, K. Stamey, M. L. Headrick, T. J. Barrett, F. J. Angulo, and the NARMS Working Group.** 2004. Multidrug-resistant strains of *Salmonella enterica* Typhimurium, United States, 1997-1998. *Emerg. Infect. Dis.* **10:**795–801.
148. **Rastawicki, W., R. Gierczynski, M. Jagielski, S. Kaluzewski, and J. Jeljaszewicz.** 2000. Susceptibility of Polish clinical strains of *Yersinia enterocolitica* serotype O3 to antibiotics. *Int. J. Antimicrob. Agents* **13:**297–300.
149. **Ratnam, S., S. B. March, R. Ahmed, G. S. Bezanson, and S. Kasatiya.** 1988. Characterization of *Escherichia coli* serotype O157:H7. *J. Clin. Microbiol.* **26:**2006–2012.
150. **Ray, A. J., N. J. Pultz, A. Bhalla, D. C. Aron, and C. J. Donskey.** 2003. Coexistence of vancomycin-resistant enterococci and *Staphylococcus aureus* in the intestinal tracts of hospitalized patients. *Clin. Infect. Dis.* **37:**875–881.
151. **Reina, J., N. Borrell, and A. Serra.** 1992. Emergence of resistance to erythromycin and fluoroquinolones in thermotolerant *Campylobacter* strains isolated from feces 1987-1991. *Eur. J. Clin. Microbiol. Infect. Dis.* **11:**1163–1166.
152. **Richards, S. J., C. M. Lambert, and A. C. Scott.** 1992. Recurrent *Listeria monocytogenes* meningitis treated with intraventricular vancomycin. *J. Antimicrob. Chemother.* **29:**351–353.
153. **Rivas, L., N. Fegan, and P. Vanderlinde.** 2004. Isolation and characterisation of *Arcobacter butzleri* from meat. *Int. J. Food Microbiol.* **91:**31–41.
154. **Roberts, M. C., B. Facinelli, E. Giovanetti, and P. E. Varaldo.** 1996. Transferable erythromycin resistance in *Listeria spp.* isolated from food. *Appl. Environ. Microbiol.* **62:**269–270.
155. **Rocourt, J., G. Moy, K. Vierk, and J. Schlundt.** 2003. The present state of foodborne disease in OECD countries, p. 1–39. World Health Organization, Geneva, Switzerland.
156. **Rohder, A., J. Kleer, and G. Hildebrandt.** 2003. Einsatz von modifiziertem kulturellen Nachweis nach JOHNSON & MURANO und Multiplex PCR nach HARMON & WESLEY zur Ermittlung des Vorkommens von *Arcobacter spp.* bei Frischgeflügel und Rindergehacktem aus dem Berliner Einzelhandel. Proceedings, p. 169–174. 44. Arbeitstagung des Arbeitsgebietes Lebensmittelhygiene der DVG, Garmisch-Partenkirchen.
157. **Rood, J. I., J. R. Buddle, A. J. Wales, and R. Sidhu.** 1985. The occurrence of antibiotic resistance in *Clostridium perfringens* from pigs. *Aust. Vet. J.* **62:**276–279.
158. **Saenz, Y., L. Brinas, E. Dominguez, J. Ruiz, M. Zarazaga, J. Vila, and C. Torres.** 2004. Mechanisms of resistance in multiple-antibiotic-resistant *Escherichia coli* strains of human, animal, and food origins. *Antimicrob. Agents Chemother.* **48:**3996–4001.
159. **Sam, W. I., M. M. Lyons, and D. J. Waghorn.** 1999. Increasing rates of ciprofloxacin resistant campylobacter. *J. Clin. Pathol.* **52:**709.
160. **Sanchez, R., V. Fernandezbaca, M. D. Diaz, P. Munoz, M. Rodriguezcreixems, and E. Bouza.** 1994. Evolution of susceptibilities of *Campylobacter* spp. to quinolones and macrolides. *Antimicrob. Agents Chemother.* **38:**1879–1882.
161. **Schlech, W. F., III.** 2000. Foodborne listeriosis. *Clin. Infect. Dis.* **31:**770–775.
162. **Schoder, D., P. Winter, A. Kareem, W. Baumgartner, and M. Wagner.** 2003. A case of sporadic ovine mastitis caused by *Listeria monocytogenes* and its effect on contamination of raw milk and raw-milk cheeses produced in the on-farm dairy. *J. Dairy Res.* **70:**395–401.
163. **Schroeder, C. M., J. Meng, S. Zhao, C. DebRoy, J. Torcolini, C. Zhao, P. F. McDermott, D. D. Wagner, R. D. Walker, and D. G. White.** 2002. Antimicrobial resistance of *Escherichia coli* O26, O103, O111, O128, and O145 from animals and humans. *Emerg. Infect. Dis.* **8:**1409–1414.
164. **Schroeder, C. M., D. G. White, B. Ge, Y. Zhang, P. F. McDermott, S. Ayers, S. Zhao, and J. Meng.** 2003. Isolation of antimicrobial-resistant *Escherichia coli* from retail meats purchased in Greater Washington, D.C., USA. *Int. J. Food Microbiol.* **85:**197–202.
165. **Schroeder, C. M., C. Zhao, C. DebRoy, J. Torcolini, S. Zhao, D. G. White, D. D. Wagner, P. F. McDermott, R. D. Walker, and J. Meng.** 2002. Antimicrobial resistance of *Escherichia coli* O157 isolated from humans, cattle, swine, and food. *Appl. Environ. Microbiol.* **68:**576–581.
166. **Schroeter, A., B. Hoog, and R. Helmuth.** 2004. Resistance of salmonella isolates in Germany. *J. Vet. Med. B Infect. Dis. Vet. Public Health* **51:**389–392.
166a. **Schwarz, F. V., V. Perreten, and M. Teuber.** 2001. Sequence of the 50-kb conjugative multiresistance plasmid pRE25 from *Enterococcus faecalis* RE25. *Plasmid* **46:**170–187.
167. **Shea, K. M.** 2003. Antibiotic resistance: what is the impact of agricultural uses of antibiotics on children's health? *Pediatrics* **112:**253–258.
168. **Shepard, B. D., and M. S. Gilmore.** 2002. Antibiotic-resistant enterococci: the mechanisms and dynamics of drug introduction and resistance. *Microb. Infect.* **4:**215–224.
169. **Simjee, S., D. G. White, J. Meng, D. D. Wagner, S. Qaiyumi, S. Zhao, J. R. Hayes, and P. F. McDermott.** 2002. Prevalence of streptogramin resistance genes among *Enterococcus* isolates recovered from retail meats in the Greater Washington D.C. area. *J. Antimicrob. Chemother.* **50:**877–882.
170. **Smith, D. L., A. D. Harris, J. A. Johnson, E. K. Silbergeld, and J. G. Morris, Jr.** 2002. Animal antibiotic use has an early but important impact on the emergence of antibiotic resistance in

human commensal bacteria. *Proc. Natl. Acad. Sci. USA* **99**:6434–6439.

171. **Smith, K. E., J. M. Besser, C. W. Hedberg, F. T. Leano, J. B. Bender, J. H. Wicklund, B. P. Johnson, K. A. Moore, M. T. Osterholm, and Investigation Team.** 1999. Quinolone-resistant *Campylobacter jejuni* infections in Minnesota, 1992-1998. *N. Engl. J. Med.* **340**:1525–1532.
172. **Soltani, M., D. Beighton, J. Philpott-Howard, and N. Woodford.** 2000. Mechanisms of resistance to quinupristin-dalfopristin among isolates of *Enterococcus faecium* from animals, raw meat, and hospital patients in Western Europe. *Antimicrob. Agents Chemother.* **44**:433–436.
173. **Sundsfjord, A., G. S. Simonsen, and P. Courvalin.** 2001. Human infections caused by glycopeptide-resistant *Enterococcus* spp: are they a zoonosis? *Clin. Microbiol. Infect.* **7**:16–33.
174. **SVARM.** 2003. Swedish veterinary antimicrobial resistance monitoring. The National Veterinary Institute, Uppsala, Sweden.
175. **Swann Committee.** 1969. Report of joint committee on the use of antibiotics in animal husbandry and veterinary medicine. Her Majesty's Stationary Office, London, United Kingdom.
176. **Swartz, M. N.** 2002. Human diseases caused by foodborne pathogens of animal origin. *Clin. Infect. Dis.* **34**:S111–S122.
177. **Tacket, C. O., L. B. Dominguez, H. J. Fisher, and M. L. Cohen.** 1985. An outbreak of multiple-drug-resistant *Salmonella enteritis* from raw milk. *JAMA* **253**:2058–2060.
178. **Tannock, G. W.** 1987. Conjugal transfer of plasmid pAM beta 1 in *Lactobacillus reuteri* and between lactobacilli and *Enterococcus faecalis. Appl. Environ. Microbiol.* **53**:2693–2695.
179. **Tannock, G. W., J. B. Luchansky, L. Miller, H. Connell, S. Thode-Andersen, A. A. Mercer, and T. R. Klaenhammer.** 1994. Molecular characterization of a plasmid-borne (pGT633) erythromycin resistance determinant (*ermGT*) from *Lactobacillus reuteri* 100-63. *Plasmid* **31**:60–71.
180. **Tauxe, R. V.** 2002. Emerging foodborne pathogens. *Int. J. Food Microbiol.* **78**:31–41.
181. **Teale, C. J.** 2002. Antimicrobial resistance and the food chain. *J. Appl. Microbiol. Symp. Suppl.* **92**:85S–89S.
182. **Tendolkar, P. M., A. S. Baghdayan, and N. Shankar.** 2003. Pathogenic enterococci: new developments in the 21st century. *Cell. Mol. Life Sci.* **60**:2622–2636.
183. **Teuber, M.** 1995. The genus *Lactococcus,* p. 173–234. *In* B. J. B. Wood and W. H. Holzapfel (ed.), *The Genera of Lactic Acid Bacteria.* Blackie Academic & Professional, London, United Kingdom.
184. **Teuber, M.** 1999. Spread of antibiotic resistance with foodborne pathogens. *Cell. Mol. Life Sci.* **56**:755–763.
185. **Teuber, M., L. Meile, and F. Schwarz.** 1999. Acquired antibiotic resistance in lactic acid bacteria from food. *Antonie Van Leeuwenhoek* **76**:115–137.
186. **Teuber, M., and V. Perreten.** 2000. Role of milk and meat products as vehicles for antibiotic-resistant bacteria. *Acta Vet. Scand. Suppl.* **93**:75–87.
187. **Thorns, C. J.** 2000. Bacterial food-borne zoonoses. *Rev. Sci. Tech.* **19**:226–239.
188. **Threlfall, E. J.** 2002. Antimicrobial drug resistance in *Salmonella:* problems and perspectives in food- and water-borne infections. *FEMS Microbiol. Rev.* **26**:141–148.
189. **Todd, W. T., and S. Dundas.** 2001. The management of VTEC O157 infection. *Int. J. Food Microbiol.* **66**:103–110.
190. **Uemura, R., M. Sueyoshi, Y. Taura, and H. Nagatomo.** 2004. Effect of antimicrobial agents on the production and release of shiga toxin by enterotoxaemic *Escherichia coli* isolates from pigs. *J. Vet. Med. Sci.* **66**:899–903.
191. **Vali, L., K. A. Wisely, M. C. Pearce, E. J. Turner, H. I. Knight, A. W. Smith, and S. G. Amyes.** 2004. High-level genotypic variation and antibiotic sensitivity among *Escherichia coli* O157 strains isolated from two Scottish beef cattle farms. *Appl. Environ. Microbiol.* **70**:5947–5954.
192. **van den Bogaard, A. E., N. Bruinsma, and E. E. Stobberingh.** 2000. The effect of banning avoparcin on VRE carriage in The Netherlands. *J. Antimicrob. Chemother.* **46**:146–148.
193. **van den Bogaard, A. E., N. London, C. Driessen, and E. E. Stobberingh.** 2001. Antibiotic resistance of faecal *Escherichia coli* in poultry, poultry farmers and poultry slaughterers. *J. Antimicrob. Chemother.* **47**:763–771.
194. **van den Bogaard, A. E., and E. E. Stobberingh.** 1999. Antibiotic usage in animals: impact on bacterial resistance and public health. *Drugs* **58**:589–607.
195. **van den Bogaard, A. E., and E. E. Stobberingh.** 2000. Epidemiology of resistance to antibiotics. Links between animals and humans. *Int. J. Antimicrob. Agents* **14**:327–335.
196. **Van Looveren, M., G. Daube, L. De Zutter, J. M. Dumont, C. Lammens, M. Wijdooghe, P. Vandamme, M. Jouret, M. Cornelis, and H. Goossens.** 2001. Antimicrobial susceptibilities of *Campylobacter* strains isolated from food animals in Belgium. *J. Antimicrob. Chemother.* **48**:235–240.
197. **Vandamme, P., P. Pugina, G. Benzi, R. Vanetterijck, L. Vlaes, K. Kersters, J. P. Butzler, H. Lior, and S. Lauwers.** 1992. Outbreak of recurrent abdominal cramps associated with *Arcobacter butzleri* in an Italian school. *J. Clin. Microbiol.* **30**:2335–2337.
198. **Vandenberg, O., A. Dediste, K. Houf, S. Ibekwem, H. Souayah, S. Cadranel, N. Douat, G. Zissis, J. P. Butzler, and P. Vandamme.** 2004. *Arcobacter* species in humans. *Emerg. Infect. Dis.* **10**:1863–1867.
199. **Vedantam, G., and D. W. Hecht.** 2003. Antibiotics and anaerobes of gut origin. *Curr. Opin. Microbiol.* **6**:457–461.
200. **Velazquez, J. B., A. Jimenez, B. Chomon, and T. G. Villa.** 1995. Incidence and transmission of antibiotic resistance in *Campylobacter jejuni* and *Campylobacter coli. J. Antimicrob. Chemother.* **35**:173–178.
201. **Vidon, D. J., and C. L. Delmas.** 1981. Incidence of *Yersinia enterocolitica* in raw milk in eastern France. *Appl. Environ. Microbiol.* **41**:355–359.
202. **Villar, R. G., M. D. Macek, S. Simons, P. S. Hayes, M. J. Goldoft, J. H. Lewis, L. L. Rowan, D. Hursh, M. Patnode, and P. S. Mead.** 1999. Investigation of multidrug-resistant *Salmonella* serotype typhimurium DT104 infections linked to raw-milk cheese in Washington State. *JAMA* **281**:1811–1816.
203. **Walker, R. A., A. J. Lawson, E. A. Lindsay, L. R. Ward, P. A. Wright, F. J. Bolton, D. R. Wareing, J. D. Corkish, R. H. Davies, and E. J. Threlfall.** 2000. Decreased susceptibility to ciprofloxacin in outbreak-associated multiresistant *Salmonella typhimurium* DT104. *Vet. Rec.* **147**:395–396.
204. **Walsh, D., G. Duffy, J. J. Sheridan, I. S. Blair, and D. A. Mcdowell.** 2001. Antibiotic resistance among *Listeria,* including *Listeria monocytogenes,* in retail foods. *J. Appl. Microbiol.* **90**:517–522.
205. **Watkins, K. L., T. R. Shryock, R. N. Dearth, and Y. M. Saif.** 1997. In-vitro antimicrobial susceptibility of *Clostridium perfringens* from commercial turkey and broiler chicken origin. *Vet. Microbiol.* **54**:195–200.
206. **Wegener, H. C.** 1998. Historical yearly usage of glycopeptides for animals and humans: the American-European paradox revisited. *Antimicrob. Agents Chemother.* **42**:3049.
207. **Wegener, H. C., M. Madsen, N. Nielsen, and F. M. Aarestrup.** 1997. Isolation of vancomycin resistant *Enterococcus faecium* from food. *Int. J. Food Microbiol.* **35**:57–66.

208. **Werner, G., B. Hildebrandt, I. Klare, and W. Witte.** 2000. Linkage of determinants for streptogramin A, macrolide-lincosamide-streptogramin B, and chloramphenicol resistance on a conjugative plasmid in *Enterococcus faecium* and dissemination of this cluster among streptogramin-resistant enterococci. Int. *J. Med. Microbiol.* **290:**543–548.
209. **White, D. G., S. Zhao, P. F. McDermott, S. Ayers, S. Gaines, S. Friedman, D. D. Wagner, J. Meng, D. Needle, M. Davis, and C. DebRoy.** 2002. Characterization of antimicrobial resistance among *Escherichia coli* O111 isolates of animal and human origin. *Microb. Drug Resist.* **8:**139–146.
210. **White, D. G., S. Zhao, S. Simjee, D. D. Wagner, and P. F. McDermott.** 2002. Antimicrobial resistance of foodborne pathogens. *Microbes Infect.* **4:**405–412.
211. **White, D. G., S. Zhao, R. Sudler, S. Ayers, S. Friedman, S. Chen, P. F. McDermott, S. McDermott, D. D. Wagner, and J. Meng.** 2001. The isolation of antibiotic-resistant salmonella from retail ground meats. *N. Engl. J. Med.* **345:**1147–1154.
212. **Whitehead, T. R., and M. A. Cotta.** 2001. Sequence analyses of a broad host-range plasmid containing *erm*T from a tylosin-resistant *Lactobacillus* sp. Isolated from swine feces. *Curr. Microbiol.* **43:**17–20.
213. **WHO.** 1997. The medical impact of the use of antimicrobials in food animals. Report of a WHO Meeting. Berlin, Germany. 13-17 October, 1997.
214. **WHO.** 1998. Use of quinolones in food animals and potential impact on human health. Report of a WHO Meeting. Geneva, Switzerland. 2-5 June, 1998.
215. **WHO.** 2002. Food safety and foodborne illness. Fact sheet 237.
216. **WHO-APUA.** 2001. Antibiotic resistance: synthesis of recommendations by expert policy groups. World Health Organization.
217. **Wilkerson, C., M. Samadpour, N. van Kirk, and M. C. Roberts.** 2004. Antibiotic resistance and distribution of tetracycline resistance genes in *Escherichia coli* O157:H7 isolates from humans and bovines. *Antimicrob. Agents Chemother.* **48:**1066–1067.
218. **Williams, A., A. C. Davies, J. Wilson, P. D. Marsh, S. Leach, and T. J. Humphrey.** 1998. Contamination of the contents of intact eggs by *Salmonella typhimurium* DT104. *Vet. Rec.* **143:** 562–563.
219. **Wilson, I. G.** 2003. Antibiotic resistance of *Campylobacter* in raw retail chickens and imported chicken portions. *Epidemiol. Infect.* **131:**1181–1186.
220. **Winokur, P. L., D. L. Vonstein, L. J. Hoffman, E. K. Uhlenhopp, and G. V. Doern.** 2001. Evidence for transfer of CMY-2 AmpC beta-lactamase plasmids between *Escherichia coli* and *Salmonella* isolates from food animals and humans. *Antimicrob. Agents Chemother.* **45:**2716–2722.
221. **Witte, W.** 1998. Medical consequences of antibiotic use in agriculture. *Science* **279:**996–997.
222. **Witte, W.** 2000. Ecological impact of antibiotic use in animals on different complex microflora: environment. *Int. J. Antimicrob. Agents* **14:**321–325.
223. **Wong, C. S., S. Jelacic, R. L. Habeeb, S. L. Watkins, and P. I. Tarr.** 2000. The risk of the hemolytic-uremic syndrome after antibiotic treatment of *Escherichia coli* O157:H7 infections. *N. Engl. J. Med.* **342:**1930–1936.
224. **Wong, H. C., W. L. Chao, and S. J. Lee.** 1990. Incidence and characterization of *Listeria monocytogenes* in foods available in Taiwan. *Appl. Environ. Microbiol.* **56:**3101–3104.
225. **Woodford, N., B. L. Jones, Z. Baccus, H. A. Ludlam, and D. F. Brown.** 1995. Linkage of vancomycin and high-level gentamicin resistance genes on the same plasmid in a clinical isolate of *Enterococcus faecalis*. *J. Antimicrob. Chemother.* **35:** 179–184.
226. **Würgler-Aebi, I.** 2004. Entwicklung von Resistenzen gegen Makrolid-Antibiotika bei Enterokokken im Kot von Kälbern, gefüttert mit antibiotikahaltiger Milch. Inaugural-Dissertation. Veterinär-Medizinische Fakultät der Universität Bern, Schweiz.
227. **Yang, H., S. Chen, D. G. White, S. Zhao, P. McDermott, R. Walker, and J. Meng.** 2004. Characterization of multiple-antimicrobial-resistant *Escherichia coli* isolates from diseased chickens and swine in China. *J. Clin. Microbiol.* **42:**3483–3489.
228. **Zhang, Q., J. Lin, and S. Pereira.** 2003. Fluoroquinolone-resistant *Campylobacter* in animal reservoirs: dynamics of development, resistance mechanisms and ecological fitness. *Anim. Health Res. Rev.* **4:**63–71.
229. **Zhao, T., M. P. Doyle, P. J. Fedorka-Cray, P. Zhao, and S. Ladely.** 2002. Occurrence of *Salmonella enterica* serotype typhimurium DT104A in retail ground beef. *J. Food Prot.* **65:**403–407.

Frontiers in Antimicrobial Resistance: a Tribute to Stuart B. Levy
Edited by D. G. White, M. N. Alekshun, and P. F. McDermott

Chapter 35

Antimicrobial Use in Plant Agriculture

ANNE K. VIDAVER

Stuart Levy's sense of social responsibility caused him to form the Alliance for the Prudent Use of Antibiotics (APUA). Soon after he formed APUA, I urged Stuart to include the agricultural use of antibiotics in conceptualizing the risk of antibiotic resistance in the global environment. His comprehensive understanding of the issue allowed him to look past the general bias of focusing only on human and animal antibiotic usage, and he invited me to join the APUA Scientific Advisory Board. I have served on this Board with pleasure for many years as the sole plant pathologist, and I am honored by this request to summarize the current state of both antibiotic and fungicide use in agriculture.

Plants are beset with a wide diversity of pathogens. Fungi and viruses cause the most common and devastating diseases, while bacteria, nematodes, phytoplasmas, and viroids are generally more localized and less damaging, with some notable exceptions. Antifungal agents or fungicides are used heavily, with worldwide use approaching 5 billion dollars (13). Antibiotics have been used since the 1960s (12) to control several economically important bacterial diseases of plants. Antibiotic, as used here, is defined by the Centers for Disease Control and Prevention as a type of antimicrobial agent derived from molds or bacteria that kills or inhibits the growth of other microbes, specifically bacteria (http://www.cdc.gov/drugresistance/glossary.htm). Viruses are dealt with principally through management of the vector, plant resistance breeding, timing of planting, and, to a limited extent, use of transgenic plants. This review will focus on the antibiotics and fungicides in agricultural use, problems of resistance, and effects on other organisms, particularly humans.

ANTIBIOTIC USE IN AGRICULTURE

Antibiotics have been used in the United States for over 40 years to prevent infection and spread of bacteria that cause diseases in plants, especially high value fruit and vegetable crops. The most common antibiotics used in plant agriculture are streptomycin and oxytetracycline, either in its hydrochloride or calcium form (12, 23). These antibiotics are generally sprayed on apple, pear, and peach trees during blossom time as preventive treatments against infection by *Erwinia amylovora,* the causal agent of fire blight, and occasionally for control of *Pseudomonas syringae* pv. *papulans,* the causal agent of apple blister spot. In 2003, data from the U.S. Department of Agriculture (http://www.usda.gov/nass/pubs/pubs.htm) showed that 7,500 kg (16,500 lb) of streptomycin was used on about 16% of the apple and 32% of the pear acreage, typically occurring in two applications per season. Total amounts of oxytetracycline use were the same as for streptomycin but applied to 6% of the apple acreage, 22% of the peaches, and 32% of the pears. Its use also averaged two applications per season. Streptomycin quantities decreased and oxytetracycline increased over previous years, apparently due to streptomycin resistant bacteria found in apple and pear orchards. These quantities are estimated to be between 0.1 and 0.5% of the total antibiotics used in the United States for all purposes and have been fairly constant over the last decade (13).

Reliable data on antibiotics used in plant agriculture are often years behind in being made available to the public, if at all. In the United States, the USDA's National Agricultural Statistics Survey (NASS; http://www.usda.gov/nass/pubs/pubs.htm) normally

Anne K. Vidaver • Department of Plant Pathology, University of Nebraska, Lincoln, NE 68583-0722.

issues reports biennially on actual amounts and use patterns of antibiotics and fungicides used on field crops in even years and fruit crops in alternate years. This appears to be the source for the most complete and recent data available in the United States. The U.S. Geological Survey determines water pollution by pesticides, including antibiotics and fungicides, by estimating actual use; however, their last report (1998) was based on data from 1992. The Environmental Protection Agency's (EPA) 2002 report on Pesticide Industry Sales and Usage employs data from 1998 and 1999. The EPA has legal jurisdiction of antibiotics and fungicides used on plants. The state of California has more detailed and timely information than any other state (http://www.pesticideinfo.org/Search_Use.jsp). In addition to antimicrobial usage information, the California report also provides information mandated by the state on phytotoxicity of fungicides and bactericides, but human toxicity, especially to those performing the pesticide applications, is not reported separately from other data on pesticide toxicity.

However limited the data available in the United States, they are considered the most accurate. Data from other countries are often either incomplete or inaccurate, as evidenced by the United Nation's Food and Agriculture Organization (FAO) disclaimer (http://apps.fao.org/notes/datasources-e.htm) on the accuracy of its data.

Some countries use antibiotic treatments that are different from those used in the United States and for other crops. For example, Israel uses oxolinic acid, a synthetic quinolone antibiotic, Latin America uses gentamicin, and in Asia, kasugamycin is used as both a bactericide and fungicide (12, 23). Analyses of global antibiotic usages are also complicated by the fact that these data are grouped with fungicides in reporting documents. This is presumably because both the EPA and FAO group antibiotics and fungicides together in their definitions of pesticides.

In addition to canonical antibiotics, antimicrobial agents also include bacteriocins, specialized, restricted host range antibiotics produced by bacteria. Bacteriocins produced by commercially available live cultures of a biological control agent help protect transplants of roses and fruit trees against subsequent infection by the crown gall bacterium *Agrobacterium tumefaciens*. Two strains of *Agrobacterium radiobacter*, K84 and its derivative, K1026, carry plasmids encoding the same bacteriocin (22). Strain K1026 is the first commercial genetically engineered bacterium used in agriculture: the *tra+* gene is deleted from the K84 plasmid, preventing plasmid transfer by conjugation to other bacteria (22). Although occasional resistance of *A. tumefaciens* to the bacteriocin has been reported, it is not considered a serious concern.

RESISTANCE TO ANTIBIOTICS

Resistance to antibiotics may be either inherent or acquired. Some microorganisms are not sensitive to the agent tested without any known prior exposure, whereas others become resistant in a population that was previously sensitive. In the latter case, acquired resistance may occur quickly or in multiple selection events. Both types of resistance are found in plant pathogenic bacteria.

The emergence of widespread resistance to streptomycin has begun to limit its use in plant agriculture (12). Two mechanisms of resistance have been described. The first mechanism is chromosomal point mutations in the *rpsL* gene, which alters the ribosomal binding protein S12. These mutations, which differ among strains, alter the target site of the protein. The second mechanism is through the acquisition of streptomycin-modifying enzymes encoded by a *strA-strB* two gene complex. These genes may arrive in the bacteria on a transposable element, transposon Tn*5393* (13) or on an integron of p5TR1, a large (75kb) plasmid from *Shigella flexneri* (19). These mechanisms of resistance are similar to those reported in strains isolated in clinical medicine and some environmental isolates. Low conjugal transfer in vitro from nonpathogenic streptomycin-resistant strains to pathogens and vice versa may suggest that little transfer likely occurs in nature (18). However, in orchards at least, once resistance occurs, it may linger for decades even in the absence of antibiotic use (12). Reasons for such retention of resistance are speculative.

Where streptomycin resistance of bacteria has become a problem, oxytetracycline has sometimes been used instead. Because current EPA regulatory policy requires that resistance to streptomycin in a target pathogen be documented before oxytetracycline can be used by special exemption, some scientists believe that the initial streptomycin resistance might have been delayed or avoided if rotation of the antibiotics were approved (2). Oxytetracycline resistance among plant pathogens has not yet been reported in nature. However, some environmental bacteria found in orchards, including those with no history of antibiotic use, may display antibiotic resistance (12), raising the possibility that some plant pathogens may be able to acquire tetracycline resistance in the future.

There are few options for growers when bacterial plant pathogens become resistant to antibiotics. Other methods of disease management may include the use of chemical or biocontrol agents, use of disease resistant plants, or crop rotation. However, these options are not often available. For example, copper sprays have been used for control of several bacterial plant pathogens, but development of resistance to copper has also occurred. Determinants for copper resistance can be on the same plasmids that confer antibiotic resistance (12).

IMPLICATION OF ANTIBIOTIC USE ON HUMAN HEALTH

Antibiotic resistance is of universal concern to human health. The antibiotics used in plant agriculture, except for kasugamycin, are also used in human clinical medicine. However, there are no documented reports of resistant bacterial infections in humans arising from agricultural uses of antibiotics. Reasons for this finding may include the fact that streptomycin does not persist in the environment for an appreciable time (11), and relatively low quantities of about 50 to 100 μg/ml per 600 gallons of water are applied per acre. Furthermore, people who perform antibiotic applications do so infrequently, typically 1 to 4 times per season, and are required to wear protective clothing and masks (for streptomycin) during application. Also, all personnel are restricted by the EPA from re-entering sprayed areas to prevent exposure to aerosols (12).

Strains of some infectious antibiotic-resistant bacteria of plants can also cross kingdoms into humans. For example, strains of *Pseudomonas aeruginosa* are minor plant pathogens (18, 21, 24) but are a major cause of nosocomial infections (18, 21). *Burkholderia* (*Pseudomonas*) *cepacia*, originally isolated from rotting onions (6), is commonly found in cystic fibrosis patients (14). Bacterial strains isolated from humans have been shown to be infectious in plants (18, 21), although the opposite has not been found. The basis of such human-plant crossover by the same strains is not yet known. Reassuringly, however, there are as yet no data that demonstrate transfer of these antibiotic resistance determinants from plant pathogens to bacteria causing human disease or vice versa under natural conditions (12, 19).

FUNGICIDE USE IN AGRICULTURE

The fungicides currently used in plant pathogen control are primarily synthesized through chemistry, although some are produced by microbes. There are approximately 135 fungicides sold worldwide in 40 chemical classes, according to the International Fungicide Resistance Action Committee (FRAC; http://www.frac.info/publications/frac_list01.html) of Crop Life International, a federation of 87 countries representing the plant science industry. The major chemical classes of fungicides include azoles, benzimidazoles, dicarboximides, phenylamides, aminopyrimidines, and morpholines, which have all been introduced in the last 20 to 35 years (17). New classes of fungicides are in development (9). Azoles, used in human clinical medicine, are particularly popular in plant agriculture, because they are relatively cheap, have broad spectrum systemic action in plants for both preventive and curative effects, and are relatively stable (5). The azoles differ in their imidazole or triazole ring or in the side chain. The target site in all cases is the same active site in a fungal enzyme, lanosterol 14 α-demethylase. The azoles inhibit the enzyme's transformation of lanosterol to ergosterol, which is an essential constituent of most fungal cytoplasmic membranes (2, 25). Azoles are considered fungistatic, inhibiting reproduction and multiplication, rather than fungicidal (12), resulting in de facto cell death.

Plant fungicides are typically produced as dispensable powders with inert ingredients, including spreaders and stickers for adherence to plant surfaces. They fall into two broad classes: contact fungicides that are effective externally, and systemic fungicides, which have various degrees of movement within tissues and act internally (systemically). Both types are principally used as preventive rather than curative treatments, although systemics can provide some curative effects. Effective concentrations generally range from 50 to 300 ppm (50 to 300 μg/ml). Common application methods include seed coating before planting, aerosol spraying by ground level and aerial specialized equipment, or by soil drenches for systemic fungicides, in which they are taken up by the roots and may be dispersed throughout the plant.

The quantities of fungicides used worldwide are enormous. Tons of fungicides are used, with approximately 24,000 metric tons (26,000 tons) used in the United States per year (see FAO website). This usage may be overestimated, however, since the EPA assumes 100% of a crop's acreage is treated if data are not available. However, fungicide treatment data show from 2 to 90% of a crop's acreage may be treated, depending upon the crop, crop prices, pesticide costs, and costs for alternative pest management, if available. Western Europe is the world's largest consumer of fungicides, using nearly 40% of the world's total (13).

In recent years, the amounts of active ingredients in fungicide formulations have decreased and efficacy has increased, but the total amount of fungicides used to treat plant disease has not decreased (3). The decrease in the amounts of active ingredients is due to regulatory and social policy changes; increased efficacy, at lower concentrations, of the active compounds; the phase-out of fungicides considered harmful to the environment by regulatory agencies; and the withdrawal of some fungicides from use due to pathogen resistance.

RESISTANCE TO FUNGICIDAL AGENTS

Resistance to fungicidal or antifungal agents is relatively common. By 1988, resistance was recorded for 12 fungicidal groups affecting 60 genera, including those affecting plants and humans (9). More recent surveys of antifungal drug resistance are lacking (8). Fungicide resistance in agriculture is monitored globally by the Fungicide Resistance Action Committee (http://www.frac.info).

Currently, at least four mechanisms of resistance are known for azoles (12, 15), a class of fungicides also used in clinical medicine. They are (i) upregulation of proteins controlling drug efflux, (ii) alterations in sterol synthesis, (iii) decreased affinity of azoles for the cellular target, and (iv) increased levels of the cellular target. Understanding the basis of resistance is necessary for achieving countermeasures. Thus, additional research and development will be required as resistance to current agents becomes more prevalent. Fungicides for plants will continue to play a major, although controversial, role in plant disease control.

Management of fungal resistance in agriculture is relatively easier than with bacterial resistance. Growers are advised to alternate use of fungicides from different chemical classes and to limit the number of applications in a growing season. Prudence in pesticide use is advised by international agreement, with the promulgation by the United Nations FAO Code of Conduct on the Distribution and Use of Pesticides (http://www.croplife.org/librarypublications.aspx). Whenever possible, crop rotation is advised, as is planting of crops inherently resistant to damaging pathogens. The creation of transgenic commercial crops and ornamental plants is expected to increase the inherent resistance of plants to fungal pathogens, although the public may not broadly accept their use. Ornamental plants, such as roses resistant to black spot, may be more accepted than edible plants modified to be resistant to fungi.

IMPLICATIONS OF FUNGICIDE USE ON HUMAN HEALTH

The incidence of invasive fungal infections in humans is growing rapidly (8, 15). Although the absolute numbers of infections remain low (8), the number of fungal species responsible is over 300 (20). In the last 20 years or so, these fungal infections in humans (mycoses) have increased sharply due to more organ transplants and their accompanying need for immunosuppressive drugs, as well as the immune system deterioration accompanying AIDS. Over 10,000 species of fungi are estimated to cause diseases in plants worldwide (1), and some of these species are within the same taxonomic groups that affect humans (1, 20). However, there are no studies that have examined whether a fungal strain isolated from a plant can infect an animal or human or vice versa, as has been demonstrated for some bacterial pathogens (see above). The question thus remains: are strains from different cross-infective hosts the same or not? The number of fungal strains for which genomic sequences have been determined are so few that no informative comparisons between plant and animal pathogens can be made. If certain strains are shown to infect both plants and animals, then humans, especially immunocompromised individuals, may need to be more wary of fungal exposure in different environments, such as gardening, horticulture, agriculture, golfing, and diagnostic laboratory work.

As is the case in plants, azole resistance is beginning to be a problem in the human population. Azoles in particular have dual function as medically important antifungal or antimycotic drugs and as fungicides in agriculture (http://europa.eu.int/comm/health/ph_risk/committees/sccp/documents/out207_en.pdf). However, there are conflicting opinions as to the severity of the resistance problem: as of 2002, the European Union (http://europa.eu.int/comm/food/fs/sc/ssc/out278_en.pdf) was not particularly concerned about resistance in mycoses affecting humans, due to the limited known means of horizontal gene transfer to related fungi and essentially no transfer to unrelated fungi. Others believe fungal resistance in clinical medicine warrants investigation (2, 5, 8). *Aspergillus fumigatus* and *Fusarium* species, the most prevalent mycotic agents seen in clinical medicine, appear to be intrinsically resistant to some azoles (see E.U. site above). Thus, treatment of mycotic infections remains a challenge. Currently, there is no cross-over use between specific antifungal agents in agriculture and those in clinical medicine. Most plant antifungal agents used in agriculture are considered toxic to humans and hence have not been used to treat human mycoses.

TRENDS AND RECOMMENDATIONS

The availability and use of antimicrobials in agriculture is affected by disease prevalence and severity, economic need, public perception, scientific findings, and political climate. Concerns about pesticide residues in food led to the passage of the Food Quality Protection Act (FQPA) in 1996, a modification of the Federal Insecticide, Fungicide and Rodenticide Act (FIFRA) of 1947 (16). Little consideration was given to the benefits of pesticide use in the passage of this act, such as food availability and affordability or improved quality contributing to healthy plants and human health, nor to improved crop yields (16). Ramifications of this act include the possibility that streptomycin and oxytetracycline will not be reregistered as agricultural pesticides in the United States due to the increased data collection required by this act and relatively small market for these antibiotics. This may present an insurmountable challenge to some farmers, as no new agricultural antibiotics are expected to be marketed in the foreseeable future. Bacterial diseases, with few exceptions, are considered of erratic appearance and insufficiently severe to warrant research and development of new antibiotics. The increased stringency of regulatory requirements for plant protection and the environment will further hinder antibiotic development.

A potential approach to obtain new alternative antibiotics for plant agriculture might be a federal program analogous to the Orphan Drug Act. This act was passed to recognize that many human disease conditions affect small numbers of people. It provides incentives for drug companies to conduct research and market limited products (http://www.fda.gov/orphan/oda.htm). An international agreement on such policies and procedures for agriculture is also possible, but unlikely, given the amount of distrust of pesticide development in many other countries.

There is a mechanism in the United States for registering previously approved products for minor use, which is based on acreage of particular crops, principally fruits, vegetables, and ornamental plants. This program is called the IR-4 Interregional Project and is sponsored by the USDA in partnership with the State Agricultural Experiment Stations at Land Grant Universities (http://ir4.rutgers.edu/). Working with grower groups, the program plays a critical role in developing data to support registrations for minor uses when there is no economic incentive for the chemical industry to do so (16).

Whereas one could argue that antibiotics are of limited use in agriculture, fungicides are considered vital, necessary, and highly profitable (13) for worldwide plant pathogen control. The FQPA, in addition to its effects on the use of antibiotics, is expected to lower fungicide use and development as well. Disease management will become much more challenging if fungicides in use today become unavailable and if additional fungicides are not developed. Other ramifications of this act are that new antimicrobial agents for fruits and vegetable crops considered to be minor crops based on acreage are unlikely to be marketed unless assisted by the IR-4 Project. Costs of fruits and vegetables are therefore likely to increase and availability to decrease (16).

Transgenic varieties of major crops, such as corn, rice, soybeans, and wheat will become available, based on knowledge of disease resistance determined through genomic and proteomic studies. This will decrease profitability of fungicides for field crops. The investments required for achieving resistance through transgenic transfer of appropriate genes, regulatory compliance, and uncertain consumer acceptance currently are major deterrents to businesses and many governments.

Understanding the risks of pesticide use to human health is essential. There is a paucity of data on comparative activities of azole fungicides against target enzymes of fungi and humans (15). Azoles are among the 10 antimycotic drugs currently approved by the U.S. Food and Drug Administration for therapy of human systemic fungal infections (2, 5). Such comparative information is needed to determine human health risks of these popular antimycotic agents. Although no current data implicate fungicide use in agriculture as a source of resistance in human mycotic infections, many research questions remain unanswered as to whether we are creating an environment that will increase the likelihood of this scenario. The medical community needs to undertake a large-scale epidemiological survey of the extent of antifungal drug resistance (8) to determine applicability of lessons learned with the use of antibiotics for bacterial pathogens.

Data acquisition, usage, and interpretation for managing antimicrobial resistance in plant agriculture and human medicine should be coordinated. APUA is a model for integrating information and communication across disciplines. Federal and international organizations should do so as well. Such cooperation is essential to minimize and cope with microbial resistance in both plant and human pathogens. Both antibiotics and antifungal agents must be used appropriately in agriculture and in clinical medicine to ensure the health of our nation and the world.

Acknowledgment. I thank Patricia Lambrecht for her superb assistance with the preparation of the manuscript.

REFERENCES

1. **Agrios, G. N.** 1997. *Plant Pathology,* 4th ed. Academic Press, San Diego, Calif.
2. **Dismukes, W. E.** 2000. Introduction to antifungal drugs. *Clinical Infect. Dis.* **30:**653–657.
3. **Epstein, L., and S. Bassein.** 2003. Patterns of pesticide use in California and the implications for strategies for reduction of pesticides. *Ann. Rev. Phytopathol.* **41:**351–375.
4. **He, J., R. L. Baldini, E. Deziel, M. Saucier, Q. Zhang, N. T. Liberati, D. Lee, J. Urbach, H. M. Goodman, and L. G. Rahme.** 2004. The broad host range pathogen Pseudomonas aeruginosa strain PA14 carries two pathogenicity islands harboring plant and animal virulence genes. *Proc. Natl. Acad. Sci. USA* **101:**2530–2535.

5. **Hof, H.** 2001. Critical annotations to the use of azole antifungals for plant protection. *Antimicrob. Agents Chemother.* **45:**2987–2990.
6. **Holmes, A., J. Govan, and R. Goldstein.** 1998. Agricultural use of Burkholderia (Pseudomonas) cepacia: a threat to human health? *Emerg. Infect. Dis.* **4:**221–227.
7. **Huang, T.-C., and T. J. Burr.** 1999. Characterization of plasmids that encode streptomycin-resistance in bacterial epiphytes of apple. *J. Appl. Microbiol.* **86:**741–751.
8. **Hudson, M. M.** 2001. Antifungal resistance and over-the-counter availability in the UK: a current perspective. *J. Antimicrob. Chemother.* **48:**345–350.
9. **Knight, S. C., V. M. Anthony, A. M. Brady, A. J. Greenland, S. P. Heaney, D. C. Murray, K. A. Powell, M. A. Schulz, C. A. Spinks, P. A. Worthington, and D. Youle.** 1997. Rationale and perspectives on the development of fungicides. *Annu. Rev. Phytopathol.* 35:349–372.
10. **Lindow, S. E.** 1995. Control of epiphytic ice-nucleation-active bacteria for management of plant frost injury, p. 239–256. *In* R. E. Lee, Jr., G. J. Warren, and L. V. Gusta (ed.), *Biological Ice Nucleation and Its Applications.* APS Press, St., Paul, Minn.
11. **Lupetti, A., R. Danesi, M. Campa, M. Del Tacca, and S. Kelly.** 2002. Molecular basis of resistance to azole antifungals. *Trends Mol. Med.* **8:**76–81.
12. **McManus, P. S., V. O. Stockwell, G. W. Sundin, and A. L. Jones.** Antibiotic use in plant agriculture. *Ann. Rev. Phytopathol.* 40:443–465.
13. **Meister Publishing Company.** 2004. Global outlook, p. A3. *Meister Pro Crop Protection Handbook.* Meister Publishing Co., Willoughby, Ohio.
14. **Nzula, S., P. Vandamme, and J. R. Govan.** 2002. Influence of taxonomic status on the in vitro antimicrobial susceptibility of the *Burkholderia cepacia* complex. *J. Antimicrob. Chemother.* 50:265–269.
15. **Perea, S., and T. F. Patterson.** 2002. Antifungal resistance in pathogenic fungi. *Clinical Infect. Dis.* **35:**1073–1080.
16. **Ragsdale, N. N.** 2000. The impact of the Food Quality Protection Act on the future of plant disease management. *Annu. Rev. Phytopathol.* **38:**577–596.
17. **Russell, P.** 1999. Fungicide resistance management: into the next millennium. *Pesticide Outlook* **10:**213–215.
18. **Silo-Suh, L., S. J. Suh, P. A. Sokol, and D. E. Ohman.** 2002. A simple alfalfa seedling infection model for *Pseudomonas aeruginosa* strains associated with cystic fibrosis shows AlgT (sigma-22) and RhlR contribute to pathogenesis. *Proc. Natl. Acad. Sci. USA* **99:**15699–15704.
19. **Sundin, G. W.** 2002. Distinct recent lineages of the StrA-StrB streptomycin resistance genes in clinical and environmental bacteria. *Curr. Microbiol.* **45:**63–69.
20. **Taylor, L. H., S. M. Latham, and M. E. Woolhouse.** 2001. Risk factors for human disease emergence. *Phil. Trans. R. Soc. Lond. B.* **356:**983–989.
21. **Van Eldere, J.** 2003. Multicentre surveillance of *Pseudomonas aeruginosa* susceptibility patterns in nosocomial infection. *J. Antimicrob. Chemother.* **51:**347–352.
22. **Vicedo, B., R. Penalver, M. J. Asins, and M. M. Lopez.** 1993. Biological control of *Agrobacterium tumefaciens,* colonization, and pAGK84 transfer with *Agrobacterium radiobacter* K84 and the Tra(sup$^-$) mutant strain K1026. *Appl. Environ. Microbiol.* **59:**309–315.
23. **Vidaver, A. K.** 2002. Uses of antimicrobials in plant agriculture. *Clinical Infect. Dis.* **34**(Suppl. 3):S107–S110.
24. **Walker, T. S., H. P. Bais, E. Deziel, H. P. Schweizer, L. G. Rahme, R. Fall, and J. M. Vivanco.** 2004. *Pseudomonas aeruginosa*-plant root interactions. Pathogenicity, biofilm formation, and root exudation. *Plant Physiol.* **134:**320–331.
25. **Zarn, J. A., B. J. Bruschweiler, and J. R. Schlatter.** 2003. Azole fungicides affect mammalian steroidogenesis by inhibiting sterol 14α-demethylase and aromatase. *Environ. Health Perspect.* **111:**255–261.

PART VI. DRUG RESISTANCE IN CANCER CELLS

SECTION EDITOR: Michael N. Alekshun

WITH RESPECT TO DRUG EFFLUX, parallels between the development of antibiotic resistance in bacteria and resistance to chemotherapy in cancerous cells are now evident. With his training as a physician in the field of hematology and oncology, it seemed natural for Stuart Levy to apply the recently discovered principles of efflux in bacteria to the field of cancer chemotherapy. This segue was immediately rewarding, as the Levy group became one of the first laboratories to discover a multidrug resistance-associated protein (MRP) in murine erythroleukemia cells using vincristine as a selective agent. The chapters within this section provide the reader with an overview of the mechanisms involved in resistance to anticancer agents as well as the paths used to get there.

Frontiers in Antimicrobial Resistance: a Tribute to Stuart B. Levy
Edited by D. G. White, M. N. Alekshun, and P. F. McDermott

Chapter 36

Mechanisms of Resistance to Anticancer Agents

MICHAEL P. DRAPER, GRAHAM K. JONES, CHRISTOPHER J. GOULD, AND DAVID E. MODRAK

Drug resistance, either intrinsic or acquired, remains one of the primary causes of failure for cancer chemotherapy. We now know that genetic instability and microenviromental effects represent powerful forces fostering the appearance of resistant cancer cells. While it may be possible to develop therapies or therapeutic regimens to prevent the development of resistance, it is also possible that an understanding of the end result of the selection process will elucidate points of attack in the war against cancer drug resistance. Reversing anticancer drug resistance, however, requires a clear understanding of the cellular mechanisms that are relevant to clinical drug resistance. In the past few decades, great strides have been made in identifying some of these resistance mechanisms, mainly using laboratory based models. Most of the known mechanisms for small molecule associated drug resistance fall into the categories shown in Fig. 1. While additional mechanisms are likely to be discovered in the future, one of the greatest challenges is determining the clinical relevance of the currently known resistance mechanisms. Only with a clear understanding of the mechanisms operating at the clinical level can therapies targeted at circumventing this problem be developed.

DRUG CARRIER-MEDIATED MECHANISMS OF RESISTANCE

Early in the study of resistance, it was realized that transport of drugs across the cell membrane by carrier-mediated processes was important for effective therapy (128, 345). The intracellular nature of the target of most chemotherapeutic agents requires that the drug cross the cell membrane. As transport must occur prior to interaction with the drug target, this transport step can regulate both the rate and extent of the drugs' pharmacological action. Two of the most important classes of chemotherapeutic agents are the antifolate and nucleoside analog compounds. Drugs belonging to the antifolate or nucleoside classes diffuse through the lipid bilayer of the cell membrane slowly because of their hydrophilic nature. Therefore, these and other hydrophilic drugs can only enter cells by utilizing transporters normally used for physiological substrates, such as amino acids, fatty acids, sugars, peptides, organic cations, nucleotides or nucleosides, metals and vitamins. Unlike the ATP binding cassette (ABC) transporters, these facilitative carriers achieve the transport of drug both into and out of the cell by a process that is not directly linked to ATP hydrolysis. These carriers can achieve drug equilibrium between the outside and inside of the cell by facilitating the uptake of drugs, without the generation of a transmembrane gradient. They can also concentrate drug inside the cell by linking transport to the flow of another substrate, such as sodium ions or protons (13).

Antifolates and Carrier-Mediated Resistance

The antifolates were first used clinically to treat cancer over 50 years ago (350). The biological target for the antifolate drugs is the enzyme dihydrofolate reductase (DHFR). DHFR plays a crucial role in the synthesis of thymidylate, which is necessary for the synthesis of DNA. As a consequence of inhibition of DHFR, essential cofactors for the biosynthesis of thymidylate and purines are depleted. The result is inhibition of DNA synthesis and cell death. One of the most clinically used antifolates is methotrexate (MTX), and much of what is known about the class is the result of studies performed using this drug (Fig. 2).

The folates are a class of physiologically important substrates that are essential for cell survival (16). The

Michael P. Draper and Graham K. Jones • Paratek Pharmaceuticals, 75 Kneeland Street, Boston, MA 02111. **Christopher J. Gould** • Brandeis University Department of Biochemistry, MS009, Waltham, MA 02454-9110. **David E. Modrak** • Garden State Cancer Center, 520 Belleville Avenue, Belleville, NJ 07109.

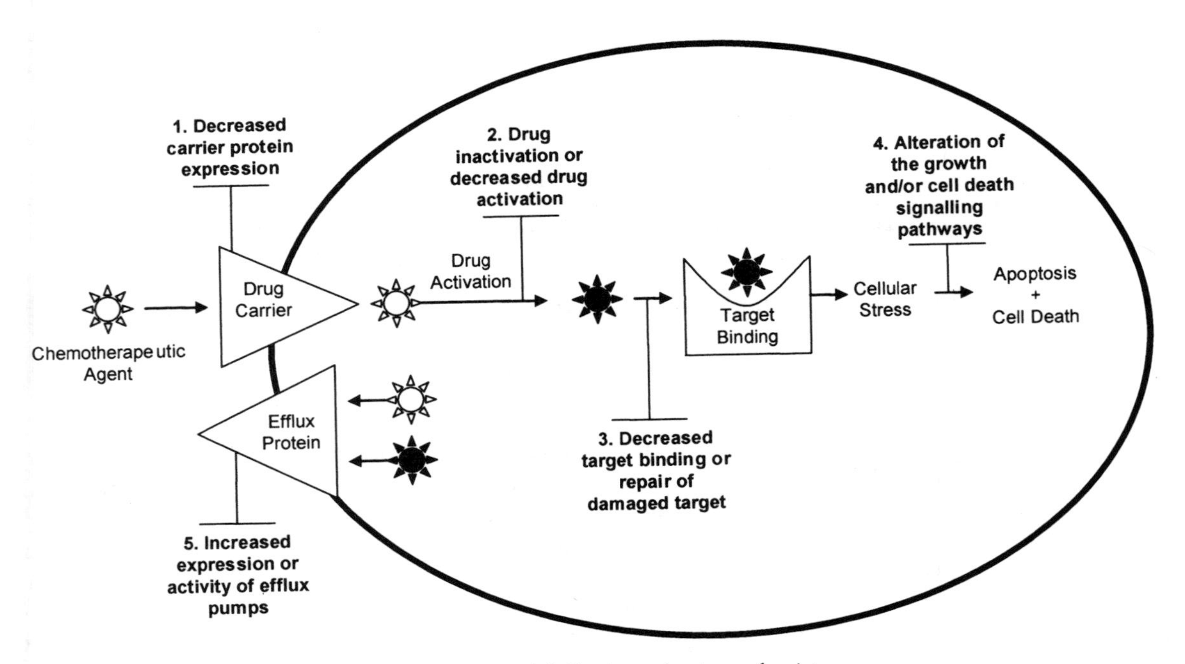

Figure 1. Overview of the basic mechanisms of resistance.

most predominant plasma folate is 5-methyl tetrahydrofolate (5-CH_3-THF), although other cellular folates exist and are important for proper cellular function (139). These THF-cofactors play a key role in the provision of one-carbon moieties for a variety of biosynthetic pathways. As folates are naturally present within the body and participate in many biosynthetic pathways, cells have evolved ways of acquiring these molecules through transport. Folates in the blood are usually present as monoglutamates, but once transported into cells they are converted to the polyglutamyl form by the enzyme folylpolyglutamate synthetase (FPGS) (15, 17). This biochemical transformation results in the retention of these polyglutamyl derivatives within the cells because these forms are effluxed from the cell less efficiently than the non-glutamyl and monoglutamyl forms (17, 227). MTX and many of the other 4-amino antifolate chemotherapeutic drugs are also substrates for transport and subject to the addition of polyglutamyl groups.

Studies have demonstrated that there are multiple mechanisms responsible for intrinsic and acquired resistance to the antifolate drugs (197, 361). The important determinants of resistance to 4-amino antifolate inhibitors of DHFR include increased DHFR expression, decreased affinity of DHFR for drug, altered membrane transport regulating drug influx or efflux, and impaired polyglutamylation of drug (197, 361). While this section will focus on alterations in the carrier mediated transport systems that are responsible for intrinsic and acquired resistance, these other mechanisms are of equal importance with regard to anticancer resistance. Many of these additional mechanisms are reviewed elsewhere in this chapter.

Several studies have demonstrated that cellular accumulation of the antifolate drug is the most critical parameter correlating with successful treatment (229, 316, 347). At physiological pH, most antifolate drugs carry two negative charges in the glutamate portion of the molecule. As a result, they are hydrophilic and only enter the cell slowly by passive diffusion. Efficient internalization of these drugs therefore requires the use of specific membrane transporters, some of which are bidirectional and others that are unidirec-

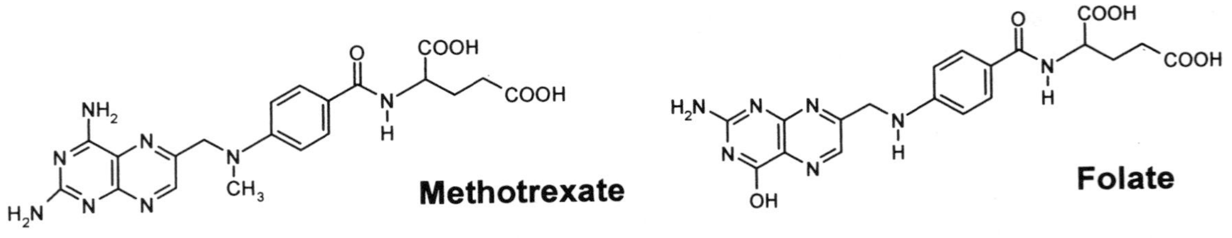

Methotrexate

Folate

Figure 2. Chemical structure of folic acid (folate), a naturally occurring vitamin necessary for cells, presented opposite the anticancer drug methotrexate.

tional importers or exporters. The final intracellular level of active drug is therefore determined by the activity of all these different transport systems.

The reduced folate carrier (RFC) is the most well characterized antifolate transport system and a member of the major facilitator superfamily. Uphill folate transport is generated through an anion exchange mechanism linked to a transmembrane organic phosphate gradient (355). Several lines of evidence suggest a role for this transporter in acquired or intrinsic resistance. In the laboratory, mutations in the transmembrane domain (TMD) of RFC have been noted in numerous resistant cell lines (34, 70, 71, 128, 214). In addition, in human breast cancer and leukemia cell lines, alterations in the regulation of RFC gene expression have been demonstrated (265, 266, 352). While the mechanism responsible for this altered level of RFC expression is unclear, changes in the methylation status of the promoter and altered expression of transcription factors that regulate its expression have been noted.

While there is significant evidence from the laboratory using cell lines suggesting a role of the RFC in resistance, only a limited number of studies have attempted to correlate changes in RFC with clinical resistance. Osteosarcoma is a disease requiring treatment with high doses of MTX because of intrinsic resistance concerns. In a study involving 42 patients with osteosarcomas, on initial biopsy 65% of the patients' tumors had low levels of RFC expression. In all, 36% of the patients who responded well to therapy had tumors with low levels of RFC and 65% of the patients with a poor response had low levels of RFC message (122). In addition to this study, several other studies have correlated low RFC expression with resistance. These include studies in patients with acute lymphocyctic leukemia (ALL) and additional studies in patients with osteosarcoma (122, 141, 189, 296). In another study examining leukemia specimens from patients, inactivating mutations in the RFC were not found to play any significant role in intrinsic or acquired resistance (157). Thus, it appears that changes in RFC expression are more likely to be of clinical importance than inactivating mutations. In addition to changes in expression, sequence alterations in the RFC have been observed in samples from osteosarcoma patients (356). The significance of these polymorphisms on methotrexate transport and resistance remains to be established. If future studies demonstrate a correlation between a particular polymorphism and clinical outcome, it may be possible to identify patients whose tumors will respond favorably to methotrexate therapy prior to, or shortly after, initiating therapy.

While the clinical relevance of RFC mediated resistance is still emerging, drug discovery efforts to circumvent RFC mediated resistance have forged ahead. Lipophilic analogs that do not require transport systems have been developed. Raltitrexed (ZD1694, Tomudex, [AstraZeneca, Sweden]) is a hydrophilic thymidylate synthase (TS) inhibitor, active against transport-deficient and DHFR overproducing cells, and has been approved for the treatment of colon cancer in Europe (57, 143, 216). Raltitrexed represents a step forward for antifolates, since MTX is well known to be ineffective in colon cancer; thus, raltitrexed extends the tumor range of antifolates. Other new antifolate drugs have been described and are in various stages of preclinical and clinical development (361).

Nucleoside Analogs and Carrier-Mediated Resistance

Nucleotide building blocks are critical for the synthesis of DNA and RNA, and they serve as key chemical intermediates for many biosynthetic pathways. The chemotherapeutic nucleoside analogs, which serve as antimetabolites, exploit these critical cellular pathways. Incorporation of nucleoside analog drugs, or their metabolic products, into DNA results in chain termination of DNA synthesis and ultimately cell death. In addition, some of the nucleoside analogs have been shown to inhibit DNA polymerase alpha or ribonucleotide reductase, the latter of which is responsible for the production of deoxynucleotides required for DNA synthesis (222). There are currently a number of different nucleoside agents in clinical use, including the pyrimidine analogs cytarabine, gemcitabine, capecitabine and the purine analogs fludarabine and cladrabine (Fig. 3). Commonalities between the compounds include the need for transport into the cell and activation by kinases. Each of the compounds, however, possesses some drug specific properties with regard to pharmacokinetics and target interaction (100, 148, 213).

As with the antifolate drugs, a single, or a combination of different, mechanism(s) can mediate resistance to nucleoside anticancer agents. In addition to changes to the target (DNA polymerase or ribonucleotide reductase) or defective induction of apoptosis (100), the failure of drug to accumulate to sufficiently high concentrations remains a significant barrier to effective treatment. This can result from reduced cellular uptake, reduced levels of activating enzymes, increased drug degradation, or expansion of the deoxynucleotide (dNTP) pools. The hydrophilic nature of both nucleosides and nucleoside anticancer agents necessitates some form of transport into the cell. This transport is accomplished by membrane-associated nucleoside transporter proteins. There are two major classes of nucleoside transporters, equilibrative nucleoside transporters (ENTs) and concentrative nucleoside transporters (CNTs). The ENTs are facilitated carrier

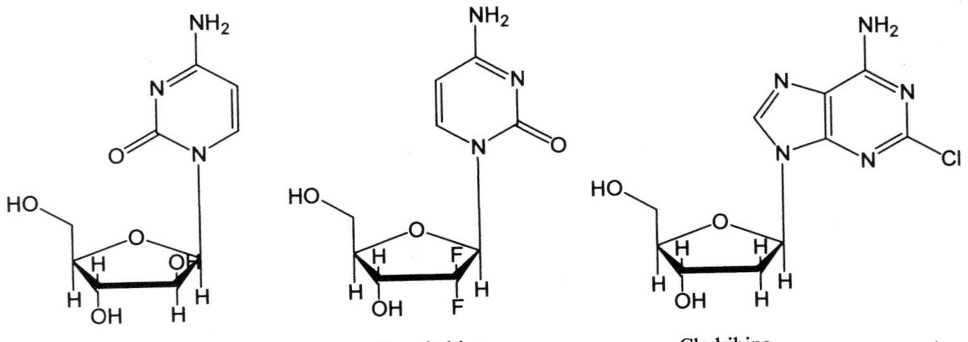

Figure 3. Chemical structures for clinically used nucleoside drugs.

proteins and the CNTs are Na^+-dependent secondary active transporters. In humans, there are a number of known ENTs (designated hENT1-4). It has been established that hENT1 and hENT2 mediate the transport of purine and pyrimidine nucleosides down a concentration gradient (171). The role of hENT3 and hENT4 in nucleoside transport has yet to be established. The CNTs are divided into two classes, those that are sensitive to inhibition by nanomolar concentrations of nitrobenzylmercaptopurine ribonucleoside (NBMPR), and those that are insensitive. The genes for those transporters sensitive to NBMPR are yet to be identified (121). Insensitive transporters are the Na^+ nucleoside cotransporters known as hCNT1, hCNT2, and hCNT3. The transporter hCNT1 accepts as a substrate pyrimidine nucleosides, while hCNT2 accepts purine nucleosides, and hCNT3 accepts both pyrimidine and purine substrates (20, 336).

In vitro studies have demonstrated that defects in nucleoside transport can result in drug resistance. A number of groups have demonstrated that hENT1 deficient cells are resistant to nucleoside analogs (102, 107, 206). In a study using the 50 cell lines included in the National Cancer Institute's (NCI) Anticancer Drug Screen Panel, an attempt was made to correlate the pattern of mRNA levels for nucleoside transporters with the pattern of cytotoxicity of anticancer drugs in the NCI drug screen database. The closest correlations between drug cytotoxicity patterns and transporter gene expression patterns (where increased expression corresponded to increased sensitivity) included those between CNT1 and O6-methylguanine and between hENT2 and hydroxyurea (202).

Most of the clinical studies performed have attempted to establish a link between the nucleoside transporter hENT1 and resistance. In patients with acute myeloid leukemia (AML), low transport rates correlated with poor clinical responsiveness to cytarabine therapy (349). In a separate study, reduced expression of hENT1 in leukemic blasts was correlated with clinical outcome (101). Recently, in a study examining primary breast cancer, low expression of hENT1 was seen in 25 to 30% of the samples, compared to normal breast epithelial cells (205). While not conclusive, these data suggest that resistance, caused by low expression of hENT1, can limit drug activity.

ENERGY-DEPENDENT EFFLUX PUMPS

Reduced cellular accumulation of anticancer agents is one of the most studied forms of multidrug resistance (MDR). This effect is often termed "classical" MDR, where enhanced expression of an adenosine triphosphate (ATP) dependant drug efflux pump causes resistance. These ATP binding cassette (ABC) type efflux pumps act on a variety of cancer drugs at the cell surface. So far, 48 human ABC genes have been identified, which fall into seven distinct subfamilies (ABCA-ABCG), determined through domain organization and sequence homology (63). These genes code for pumps that are members of the largest superfamily of membrane proteins and are also known by the names P-glycoproteins and traffic ATPases (287). The minimal ABC transporter structure contains two nucleotide binding domains (NBDs) and two transmembrane domains (TMDs) (7). In humans, many of these transporters have been implicated in diseases, reinforcing the fact that these proteins are ordinarily expressed in cells, and act on endogenous substrates (19, 21, 27, 45, 109, 146, 172, 180, 181, 256, 284, 314, 348). In terms of cellular resistance to anticancer agents, at least 10 family members have been implicated in drug resistance (MRP1-7, MDR1, BCRP, and ABCA2), and four of these have been shown to generate a resistance phenotype in cultured cells (6). Of all the family members, only two have been examined in a clinical setting in an effort to directly correlate laboratory and clinical data (120).

These are the MDR1 (ABCB1) protein, and the MRP1 (multidrug resistance protein, ABCC1) protein, which are the main focus of this section. More research will determine whether other members of this family of proteins play a role in clinical forms of drug resistance.

P-Glycoprotein

P-glycoprotein (P-gp) was first discovered nearly 30 years ago as a result of employing selective pressure on cancer cells in culture with natural product anticancer drugs (120, 153). Classes of drug that serve as substrates for P-gp include the vinca alkaloids, anthracyclines, and epipodophyllotoxins (Table 1). Many researchers have attempted to define the critical substrate attributes necessary for transport by P-gp, but nothing highly conserved has been identified (5, 10, 288, 290). It is generally accepted that P-gp substrates are hydrophobic in character and fall in the range of 300 to 2,000 Da (92). Anionic compounds are not substrates, and models also suggest that hydrogen bonding and/or spatial separation of electron donor groups determine protein substrate interaction (75, 288, 289). Besides cancer, P-gp is also normally found expressed in the cells of the intestinal mucosa, kidneys, testes, liver, and blood-brain barrier, where it has been implicated in altering the biodistribution and pharmacokinetics of pharmaceutical compounds (52, 280, 354). Also, it is important to note that P-gp and the cytochrome P450 enzymes seem to cooperatively interact and possess many of the same substrates (283).

P-gp is a 170-kDa membrane protein, glycosylated at three sites (Asn 91, Asn 94, and Asn 99) which are present in the first extracellular loop (279). These sites are important for trafficking purposes but are not needed for protein function. The protein also contains phosphorylation sites on serine residues (Ser 661, Ser 667, Ser 671, and Ser 683). However, a phosphorylation deficient P-gp retained drug efflux activity, suggesting that phosphorylation events are unnecessary for function (111). The *mdr1* gene has been localized to chromosome 7q21, and an extensive review of genetic polymorphisms and haplotypes of the gene has been recently compiled (6). The gene for *mdr1* contains 28 exons, and codes for a 4.5 kb cDNA. The putative protein structure of P-gp includes 12 TMDs and two NBDs (44). No crystallographic data are available. However, an electron micrograph study has been

Table 1. General characteristics, common substrates, and normal tissue expression patterns for ABC transporters implicated in drug transport

Common name(s)	Name	Size (tms)[a]	Chemotherapy substrates	Nonchemotherapy substrates	Tissue locations
P-gp, MDR1, PGY1, GP170	ABCB1	1,279 (12)	Anthracyclines, epipodophyllotoxins, *Vinca* alkaloids, taxols, Gleevec (STI-571), and other xenobiotics	Neutral and cationic compounds, many drugs, glucosylceramide, platelet activating factor	Intestine, liver, kidney, placenta, blood-brain barrier
MDR2, PGY3	ABCB4	1,279 (12)	Paclitaxel, vinblastine	Phosphatidylcholine	Liver hepatocytes
MRP1	ABCC1	1,531 (17)	Doxorubicin, epirubicin, etoposide, vincristine, methotrexate	Glutathione conjugates, organic cations, LTC_4	Ubiquitous
MRP2, cMOAT, cMRP	ABCC2	1,545 (17)	Methotrexate, etoposide, doxorubicin, cisplatin, vincristine, mitoxantrone	Bilirubin-glucuronides, glutathione, acidic bile salts	Liver, intestine, kidney
MRP3, MLP2, MOATD, cMOAT2	ABCC3	1,527 (17[b])	Etoposide, teniposide, methotrexate, cisplatin, vincristine, doxorubicin	Glucuronate and glutathione conjugates, bile acids	Pancreas, kidney, intestine, liver, adrenal glands
MRP4, MOATB	ABCC4	1,325 (12[b])	Methotrexate, thiopurines, nucleotide analogs	Organic anions	Prostate, testis, ovary, intestine, pancreas, lung
MRP5, MOATC, pABC11	ABCC5	1,437 (11[b])	6-Mercaptopurine, 6-thioguanine, nucleotide analogs	Cyclic nucleotides, organic anions	Ubiquitous
MXR, BCRP, ABCP	ABCG2	655 (6)	Doxorubicin, daunorubicin, mitoxantrone, topotecan, SN-38	Prazosin	Placenta, liver, intestine, breast
BSEP, SPGP, PGY4	ABCB11	1,321 (12[b])	Paclitaxel (low resistance)	Bile salts	Liver
ABCA2	ABCA2	2,436 (12[b])	Estramustine	Steroid derivatives, lipids	Brain, kidney, lung, heart

[a] Size indicates amino acid number. The numbers in parentheses indicate the number of transmembrane segments of the protein.
[b] The actual number of transmembrane segments has not been experimentally measured.

completed (261). A bacterial member of the ABC superfamily has been recently crystallized, which has led to the generation of a theoretical model of ABC pump function (41, 42).

Current research has focused closely on the catalytic cycle, activation energies, and coupling between ATP hydrolysis and drug transport (276, 294). Functionally, P-gp transports one substrate utilizing two nonsimultaneous ATP-hydrolysis events (295). It has been postulated that binding to the transmembrane regions stimulates ATPase activity, and this induces a conformational change, resulting in the expulsion of the substrate to the outer membrane leaflet or extracellular space, and that the second ATP hydrolysis event returns the transporter to its initial conformation (250, 275). The substrates are likely presented directly from the inner lipid bilayer to the transmembrane regions of the transporter (5). This transport activity has been referred to as a lipid flippase, or "hydrophobic vacuum cleaner," function because of its ability to interact with substrate molecules intercalated into the lipid inner membrane leaflet and extrude them into either the outer leaflet or extracellular milieu (190, 251). Since the NBDs are not necessary for binding, it has become clear that binding to the TMDs is an independent phenomenon from transport (199, 200). The location, size, and number of binding sites on the protein are controversial (110, 331). However, there appear to be multiple sites that overlap one another. A study, using cysteine scanning mutagenesis to localize the drug binding site, has recently been summarized (198). It is important to note that this result is tentative, due to the great deal of rearrangement in the placement of the TMDs suggested to occur during the catalytic cycle (27). In general, it appears that mutations affecting substrate specificity commonly are in TMDs 4, 5, 6, 10, 11, and 12, which supports the role of these TMDs in substrate interaction (5, 137, 198).

Clinically, P-gp is highly expressed in colonic, kidney, adrenocortical, and hepatocellular cancers (90, 118). The most concrete association between clinical resistance and P-gp expression is with AML, where both a higher incidence of refractory disease and reduced complete remission rate correlate with P-gp expression (182, 184, 220, 332). A meta-analysis of P-gp expression in breast cancer concluded that, despite a pretherapy incidence range of 0 to 80%, in the sum of the studies, expression increased after therapy (241). Of note, P-gp expression seems to be a later-stage development in cancer cell resistance, correlating with exposure to relatively high drug concentrations. (302). Initially, the MDR phenotype was thought to be a singular phenomenon, caused by only one membrane pump, and the possibility of reversing the MDR effect raised hopes of circumventing all manifestations of cancer drug resistance. This optimism resulted in many compounds being investigated in the clinic for an ability to reverse P-gp induced resistance in human cancer. These include the first generation compounds verapamil and cyclosporin A and the second generation compounds PSC-833(Novartis Pharma, Basel, Switzerland), GF-902128, and VX-710 (Vertex Pharmaceuticals, Cambridge, Mass.). All of these function as competitive substrates, and all have failed clinically to improve survival (304). The failures were largely the result of excessive and unpredictable toxicities caused by the inhibition of P-gp found in normal tissues and altered excretion and reduced detoxification by cytochrome P450 of other chemotherapeutic agents being administered. In addition, the studies were further complicated by the heterogeneous expression of P-gp in tumors, expression of P-gp in normal tissue, and a lack of surrogate markers for successful effects (32, 87-89, 163). Third generation compounds have recently been identified. These third generation compounds have a high specificity for P-gp while lacking cross-reactivity with cytochrome P450 but may still have some of the limitations of earlier P-gp inhibitors. For example, the third generation compound Tariquidar (XR9576 [Xenova, Cambridge, U.K.]) increases ^{99m}Tc-sestamibi accumulation in drug-resistant tumors, but also in normal liver (4). Zosuquidar (LY335979 [Lily Research Laboratories, Indianapolis, Ind.]), at doses which inhibited P-gp function in circulating natural killer cells, resulted in a modest decrease in doxorubicin clearance with a concomitant increase in area under the curve (272). This was associated with enhanced leukopenia and thrombocytopenia but was without demonstrable clinical significance.

Multidrug Resistance Proteins

The second ATP efflux pump directly implicated in "classical" anticancer drug resistance is MRP1 (ABCC1), a 190 kDa N-glycosylated phosphoprotein. Functionally, it recognizes neutral and anionic compounds, transporting glutathione conjugated drugs and other drug conjugates. In addition, a cotransport phenomenon may occur linked to the transport of free glutathione (GSH) (147, 193, 194, 230). This cotransport is referred to as being a glutathione-X conjugate pump. Originally, it seemed clear that free cellular GSH was not needed for transport, but recently, this idea has changed as several glutathione conjugate substrates have shown stimulation by free GSH (187, 249, 270). Similar to P-gp, both MRP1 and MRP2 (commonly referred to as cMOAT) are capable of changing their substrate specificity through single amino acid substitutions in the protein (142, 187, 359, 360). Basic residues in TMDs 6, 9, 16, and 17 are thought to be involved in substrate binding, and a residue in TMD 11 is involved in stable expression of MRP2 (269).

MRP1 was first discovered in 1992 using a cell line selected for resistance in a laboratory setting (50). The discovery of MRP1 led to the search for other homologous genes, uncovering another nine members of this ABCC subfamily of membrane proteins (29, 133). Seven of these family members (MRP1-7) have been shown to confer some type of drug resistance in cell culture (27, 28, 130). The substrate profiles of these newer members of the MRP family are still being defined, and relevancy to cancer resistance has not been determined.

Structurally, the family falls into two categories: the first is similar to P-gp and contains 12 transmembrane segments, while the second is made up of 17 membrane spanning segments (two domains of six transmembrane segments plus an additional N-terminal domain of five transmembrane segments). This latter class of MRPs contains MRP1, 2, 3, and 6 (29). No known function for this additional transmembrane domain (termed TMD_0) has been determined, as it is not needed for either function or intracellular routing (12). Another segment (termed L_0) shared by all MRPs, found in the N terminus of the shorter class, and connecting TMD_0 with the rest of the molecule, however, is essential for catalytic function and seems to associate with the lipid bilayer (262). In terms of mechanism, a relatively simple model of transport by MRP1 has been developed that details two separate binding sites, termed the G and D sites. The two have opposing properties, the G-site having high affinity for GSH, and a low affinity for drug, and the D-site, vice versa (83). While this model has limitations, it represents a first step to understanding how the MRP protein seems to function in a cooperative manner, as opposed to the independent, but allosterically interacting sites of P-gp.

MRP1 is found expressed in leukemias, esophageal carcinoma, and non-small-cell lung cancers (NSCLC). Expression in noncancerous tissues is much more ubiquitous than MDR1, making many of the published results of clinical cancer manifestations difficult to interpret (234). Only a few studies have examined MRP expression pre- and posttherapy (186). In AML, two of three studies found a reduced incidence of MRP expression after therapy, while the other study found an increase in expression. In cancers of the bladder and CNS, a higher number of cancers expressed MRP after therapy.

The chemotherapeutic substrate profile for the MRP1 pump is overlapping with that of P-gp and includes etoposide, doxorubicin, vincristine, and methotrexate (28, 49, 147, 194, 230). Verapamil is weakly inhibitory and variably effective, and other, stronger P-gp inhibitors are either nonfunctional or only functional at very high concentrations (26, 84, 173, 195). Many of the MRP pump inhibitors are nonspecific or toxic, and therefore not useful clinically; they include sulfinpyrazone, probenecid, several flavinoids, and benzbromarone (135, 188). Nonsteroidal anti-inflammatory drugs (NSAIDs) also function to inhibit this pump, independent of their effect on cyclooxygenase (69, 74). Current research has focused on acidicly shielded prodrugs such as MK571 (a leukotriene receptor antagonist) and certain tricyclic isoxazoles as potential modulators of in vivo function (27).

Other Efflux Pumps

The last ATP dependent efflux pump that has been implicated in cases of cancer resistance is the breast cancer resistance protein, BCRP (also known as MXR, ABCG2, and ABC-P). Structurally, it appears to function as a homodimer, made up of two segments, each with six TMDs and an amino terminal NBD (77). This transporter is potentially as important in anticancer drug resistance as P-gp, as it has a broad substrate specificity profile (33, 191). It has also been shown that a single amino acid substitution (Arg 482 to Gly or Thr) can alter BCRP to confer much higher resistance to doxorubicin, and further widen the substrate specificity profile of the pump (134, 170). Several potent inhibitors of this pump have been found among the fumitremorgin-type indole alkaloids. Three other ABC pumps worth mentioning in reference to cancer resistance are ABCB11, the bile salt exporter pump (BSEP), which can confer paclitaxel resistance after transfection (47) and can transport paclitaxel, and vinblastine (30, 303, 363), MDR3 (sometimes known as MDR2), and ABCA2 an ABC-type pump that is up-regulated in estramustine-resistant cells (177, 337). The clinical relevance of these pumps to drug resistance is currently unknown.

FAILURE OF DRUG TO ACTIVATE AND METABOLIC DRUG INACTIVATION

Anticancer chemotherapy drugs can become inactivated by enzymatic modification prior to reaching their targets within the cell, resulting in decreased drug efficacy. This can occur as part of the normal processing of biological intermediates in the pathway that the compound is designed to inhibit (as in the case for nucleoside analog inactivation) or it may be the result of the body's detoxification systems attempting to clear the foreign compound from circulation (using both phase I and phase II metabolic excretion systems). Examples of nucleoside analog modifications and of detoxification mechanisms resulting in anticancer drug inactivation are discussed below.

Failure of Drug To Activate—Modifications of Nucleoside Analogs

Nucleoside analogs function as anti-cancer therapeutics by direct competition with their physiological counterparts. Rapidly dividing cells incorporate modified nucleoside analogs into the growing DNA chain resulting in disruption of DNA and RNA polymerases (209, 258). Several purine (e.g., fludarabine, cladribine) and pyrimidine (e.g., cytarabine, gemcitabine, capecitabine) nucleoside analogs are used clinically to treat a variety of malignancies. These compounds require modification by cellular enzymes to become active therapeutics. The activities of these modifying enzymes, or of enzymes acting to reverse the modifications, can affect cellular resistance to the compounds.

Following uptake into the cytoplasm of the target cell, cellular kinases activate nucleoside analogs to their active triphosphate forms. Cytarabine and gemcitabine, which are both cytosine analogues with 2′ sugar modifications, become 5′ phosphorylated to the monophosphate form by the enzyme deoxycytidine kinase (dCK). Studies have shown that deoxycytidine kinase (dCK) is the rate-limiting enzyme in the process of nucleoside activation and that a reduction in dCK activity can result in a reduction in drug efficacy (1, 226). Mutational inactivation of dCK has been observed in transformed cell lines (239, 268), and more recently the inactivation of dCK by the formation of alternatively spliced dCK transcripts has been postulated as a mechanism of AraC resistance in patients with resistant AML.

In contrast to dCK, 5′-nucleotidase (5′-NU) dephosphorylates nucleoside monophosphates, thereby opposing the activity of dCK. The net accumulation of a nucleoside analog likely depends on both the levels and activity of both dCK and 5′-NU. Drug resistance has been associated both with increased levels of 5′-NU mRNA (99) and with enzyme activity (160). AML patients with blasts expressing 5′-NU mRNA have a poorer prognosis when compared with patients whose blasts lack the transcript (99).

Cytarabine and gemcitabine can also become inactivated prior to the initial phosphorylation modification by cytidine deaminase (178). Several studies have correlated increases in deaminase activity with drug resistance in AML patients (51, 145, 282). Fludarabine and cladribine were designed as adenosine deaminase resistant arabinosides (105, 201). These agents have found extensive use in the treatment of hematological diseases. In addition, they are resistant to repair after incorporation into DNA and, thus, use of these drugs in combination with DNA damaging agents is currently being explored (8, 232).

Ribonucleotide reductase (RR) functions in the S-phase of the cell cycle to modulate dNTP levels (53). The heterodimeric RR protein catalyzes the reduction of NTPs to dNTPs, which are then incorporated into the nascent DNA chain. The diphosphate form of gemcitabine inhibits RR activity, thereby decreasing the concentration of physiological dNTPs or increasing the relative concentration of cytotoxic gemcitabine triphosphate. At least one study reports a gemcitabine-resistant phenotype associated with increased expression and activity of RR (114). The increased concentration of RR overcomes the inhibitory effects of gemcitabine diphosphate and facilitates the production of physiological dNTPs to compete with the toxic nucleotide.

Failure of Drug to Activate—Defective Polyglutamation of Antifolate Drugs

The biological activity of many antifolate drugs is dependant upon polyglutamylation, which determines the cellular concentration (59, 85, 96, 150, 151, 264, 281). The polyglutamylated form of an antifolate, such as methotrexate, ultimately becomes the predominant form of the drug bound to, and inhibiting the activity of, DHFR (212, 263, 277, 346). Since polyglutamation is a key determinant of antifolate activity, defects in the process can result in drug resistance. Using cultured cells it has been demonstrated that changes in the expression of FPGS or mutations that affect the activity of the enzyme can result in resistance (144, 305, 341, 362). Only a limited number of clinical studies have been conducted examining the role of FPGS in clinical resistance. One study measuring FPGS activity in liver metastases found significantly higher levels of FPGS activity in fluorouracil responsive versus nonresponsive metastases (43). Preliminary in vitro and clinical data have shown that the FPGS is a promising tool for identifying responsive tumors (80). Antifolates, which are not dependent on polyglutamylation for activity, have been developed. For example, ZD9331 (AstraZeneca, Sweden) is a nonglutamylated TS inhibitor, has a manageable toxicity profile, and shows some evidence of activity in patients with relapsed or refractory NSCLC, ovarian and breast cancer (123). These non-polyglutamated antifolates, however, are rapidly cleared from cells and require more frequent administration.

Cytochrome P450-Mediated Drug Inactivation

Cytochrome P450 monooxygenases comprise a family of approximately 50 proteins in humans, which modify exogenous substances through oxidation. Their activity is commonly referred to as phase I metabolism, preceding the phase II conjugative modifications discussed below. P450 enzymes are typically found in liver cells, where they encounter and modify intestinally absorbed exogenous substances at their port of entry into the body. Cytochrome P450-mediated reactions often lead to the production of more cytotoxic metabolites as is the case with the drugs mitomycin and doxorubicin (115). While drug modification by P450s can increase the potency of certain drugs, it is also a possible drug resistance mechanism whereby cytotoxic compounds become oxidized to less effective analogues at the level of the target cell (112, 231).

Using an in vitro cell viability system, the detoxifying potential of P450 for Vinca alkaloids has been shown (357). CHO cells co-expressing CYP3A4 and P450 reductase (the P450 electron donor) show significant resistance to vincristine or vinblastine (357). Additionally, a selective effect of vinblastine for CYP3A43-expressing cells, after 3 days of exposure of mixed cultures to high concentrations of the drug has been demonstrated (357). Similarly treatment of the human colon adenocarcinoma cell line LS174T with vinblastine resulted in the selection of a cell population with higher levels of endogenous CYP3A4 (357). Other cytotoxic compounds for which P450 has been implicated in the deactivation process include paclitaxel, docetaxel, mitoxantrone (55), etoposide (161), and tamoxifen (260). For a more comprehensive discussions on P450-dependent metabolism and inactivation of anticancer drugs, see Crommentuyn et al. (55) and McFadyen et al. (215).

Phase II Metabolism-Mediated Drug Inactivation

Phase II metabolic enzymes catalyze detoxifying conjugations which increase solubility and excretion of a range of toxins. Examples of phase II enzymes are glutathione-S-transferase, UDP-glucuronosyltransferase, N-acetyltransferase, and sulfotransferases. Conjugative modifications can have implications for drug resistance, as exemplified below.

The glutathione-S-transferase (GST) family of proteins function to conjugate glutathione to both endogenous and exogenous toxins (323, 327). GST catalyzes the conjugation of glutathione (GSH) to a diverse spectrum of electrophilic toxins, usually increasing their solubility and facilitating their excretion. Resistance to several different anticancer agents that are substrates for GST is associated with increases in GST levels (e.g., chlorambucil, melphalan). Also, depletion of the conjugant glutathione from drug-resistant ovarian cancer cells with buthionine sulfoximine increases the cytotoxicity of melphalan and cisplatin (124). Interestingly, increased GST levels have been shown to be associated with resistance to non-GST-substrate anticancer agents. The reason for this apparent inconsistency probably lies in the more recently identified role for GST proteins as MAP kinase pathway inhibitors. Monomeric GSTπ complexes with Jun N-terminal kinase (JNK) in unstressed cells, sequestering the kinase from participation in apoptotic signaling and preventing the cell from entering into an apoptotic response as a result of exposure to drug (2).

The circumvention of drug resistance associated with GST has been explored clinically. Buthionine sulfoximine and ethacryinic acid have been used to reduce glutathione levels, thereby reducing metabolic inactivation (106, 240). Ethacrynic acid, an inhibitor that binds to the GST substrate-binding site, has demonstrated chemosensitizing activity in cancer patients (236, 242). The novel peptidomimetic glutathione analog TLK199 (Telik, Palo Alto, Calif.) has shown promise as a chemosensitizer and is currently undergoing phase I clinical trials in patients with myelodysplastic syndromes (82, 204).

UDP-glucuronosyltransferases (UGTs) are a family of proteins that facilitate the conjugation of hydrophobic substrates with glucuronic acid prior to excretion. Approximately 50% of absorbed substances are eliminated from the body by the action of UGTs (82). UGT activity has been implicated in the resistance phenotype of both selected and unselected cell lines. Irinotecan (CPT-11), a new drug used to treat colorectal cancer, is modified by cellular esterases to release the active metabolite topoisomerase inhibitor SN-38 (113, 267, 271). The irinotecan resistant lung cancer cell line PC-7/CPT expresses considerable increases in UGT protein, message, and SN-38 glucuronide-forming activity as compared to the wild type PC-7 cell line (317). Glucuronidation-based resistance to irinotecan and to NU/ICRF 505 in both HT29 cells and clinical colorectal samples has been shown (56). Inactivation by glucuronidation of the guanine biosynthesis-inhibiting compound mycophenolic acid (MPA) has also been demonstrated in the non-selected human adenocarcinoma cell line HT29 (93).

ALTERED TARGETS

Qualitative or quantitative changes to the specific cellular target of an anticancer drug can result in resistance. Qualitative changes often imply direct mutation of a particular cellular protein, while quantitative differences indicate either an up- or down-regulation of target protein levels. Known examples of resistance mediated by target alteration can be found in the context of the topoisomerase targeting drugs (epipodophyllotoxins, anthracyclines and camptothecins), the dihydrofolate reductase (DHFR) targeting drugs (antifolates and fluoropyrimidines), the thymidylate synthase (TS) targeting drugs, the tubulin targeting drugs (taxanes), and, most recently, the BCR-ABL targeted drug imatinib mesylate (Gleevec, [Novartis Pharmaceuticals, East Hanover, N.J.]).

Topoisomerase Alterations

Human topoisomerases are essential cellular proteins that are important for the maintenance of proper chromatin structure (338). Topoisomerase types I and II change the linking number of a DNA strand by either one (single strand break) or two (double strand break) units, respectively (339). In terms of cancer cell resistance, we shall only be concerned with three specific human DNA topoisomerase types: topoisomerase I (TopoI), and the two isoforms of topoisomerase II (TopoIIα and TopoIIβ. These enzymes are the target of many anticancer drugs because of their major role in the cellular replication processes (108, 192).

Camptothecins are cytotoxic to cells, because of their interaction with the TopoI "cleavable complex," wherein TopoI covalently binds to DNA by a phosphodiester linkage between the 3′ phosphate group of the scissile DNA strand and Tyr-723 of the enzyme. During the catalytic cycle, this strand is then passed through the other strand, re-ligated and released (40). Topoisomerase poisons, such as the camptothecins, function through stabilizing the cleavable complex and preventing re-ligation. This drug stabilized complex will eventually collide with a DNA replication fork, resulting in double strand DNA breaks followed by apoptosis (162). Commonly used camptothecins include topotecan and irinotecan. Cell culture based experiments have elucidated several point mutations in the TopoI enzyme, which are responsible for resistance (Fig. 4). It should be noted that mutations in TopoI produce resistance to a class of molecules but do not always grant cross-resistance to other TopoI inhibitors (179). A recent crystal structure of TopoI has been resolved, and mapping of the resistance causing mutations has identified three important areas of the protein (48, 245, 252, 312). One of these is located close to the Tyr-723 active site, while two others are found at the "lip" regions of the enzyme that likely function in DNA binding and undergo conformational movements during catalysis (86, 164, 310). Clinically, no definitive correlations have been proven to exist between mutations in the TopoI enzyme and resistance (237, 318). One study attempting to detect TopoI mutations in surgically excised tumor samples from eight patients with NSCLC, who had received preoperative chemotherapy consisting of irinotecan and cisplatin, found nu-

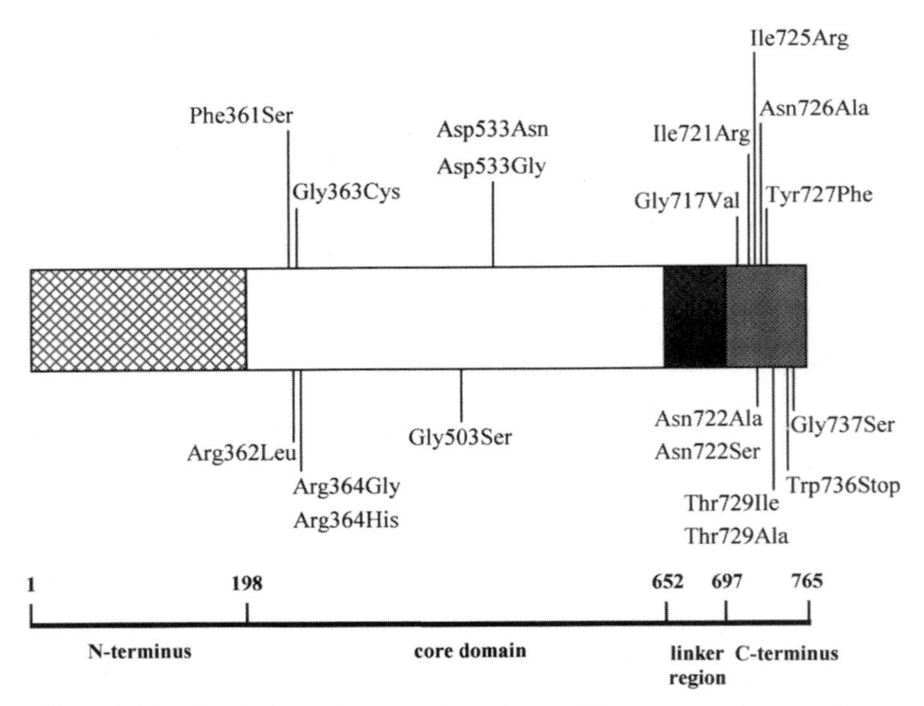

Figure 4. Mutations in the topoisomerase I gene that result in increased resistance to drug.

cleotide substitutions resulting in a Trp736stop and Gly737Ser in one tumor specimen (329). These mutations were located near a site in TopoI that has previously been reported to give resistance in cultured PC7 lung cancer cells when mutated (329). The results strongly suggest that mutations in TopoI can occur and are selected for during chemotherapy. The true significance of this clinical finding, however, requires additional study.

DNA topoisomerases can also be structurally covalently modified by phosphorylation or attachment of poly(ADP)-ribose tails. Phosphorylation and poly-ADP ribosylation have been implicated in the susceptibility of TopoI to inhibition (155, 244), and lowered phosphorylation of TopoI is a direct reason for reduced sensitivity of L5178Y-S cells to camptothecin (311). The clinical significance of these findings has yet to be explored.

In addition to qualitative changes to the TopoI enzyme, quantitative changes may be of importance in TopoI related resistance. A correlation between the sensitivity of cultured cells to camptothecins and the levels of TopoI has been demonstrated (79, 179, 315). These decreases in protein levels can be the result of either changes in gene copy number, gene rearrangement or gene specific methylation (97, 140, 218, 306, 321). In addition to changes in TopoI protein levels, alterations in the cellular localization of TopoI may play a role in resistance (22, 35, 61, 76, 224). The clinical relevance of these types of changes is currently unknown.

The TopoIIα and TopoIIβ enzymes are targeted by the epipodophyllotoxins and the anthracyclines. The mechanisms mediating resistance to topoisomerase II poisons are similar to those for TopoI poisons (18, 132, 162, 246, 333). Several different mutations in the TopoII gene cause resistance in cultured cell lines (Fig. 5). These mutations usually cluster around the catalytic Tyr-804 residue and the ATP-binding domains (48, 95, 129, 138, 333). Hypophosphorylation of TopoII, resulting in an attenuated ability of TopoII to form drug-stabilized DNA cleavable complexes, has also been implicated in several cases of resistance (104, 257). Unlike TopoI, TopoII mutations usually exhibit cross resistance to most, if not all, TopoII poisons, including epipodophyllotoxins and anthracyclines. This form of cross resistance is termed atypical multidrug resistance (at-MDR) (125). Decreased TopoII protein levels are the most often found, and investigated, form of at-MDR both in vitro and in vivo (259). Endogenous TopoIIα expression is cell cycle dependant, while TopoIIβ is not. This has led to investigations into the cell cycle dependent expression of either isoform, and findings seem to be somewhat relevant to levels in resistant tumor cells (330). TopoIIα is necessary for cell survival, unlike TopoIIβ, which has been found in some cases to be strongly downregulated or even absent in resistant cells (81, 165, 247). Clinically speaking, no clear correlation exists between resistance and in vitro results (62, 158). Mutations identical to those found in cancer cell lines, however, have been noted in the tumors of patients refractory to therapy (175, 259).

Dihydrofolate Reductase Alterations

The thymidylate synthase and DHFR proteins are integral components of the cellular system of purine biosynthesis and are critical for the survival of mammalian cells (361). Drugs targeting these enzymes, commonly termed antifolates (e.g., methotrexate) or fluoropyrimidines (e.g., 5-fluorouracil), are some of the oldest forms of antimetabolites used in cancer chemotherapy, dating back to the 1940s. DHFR, the

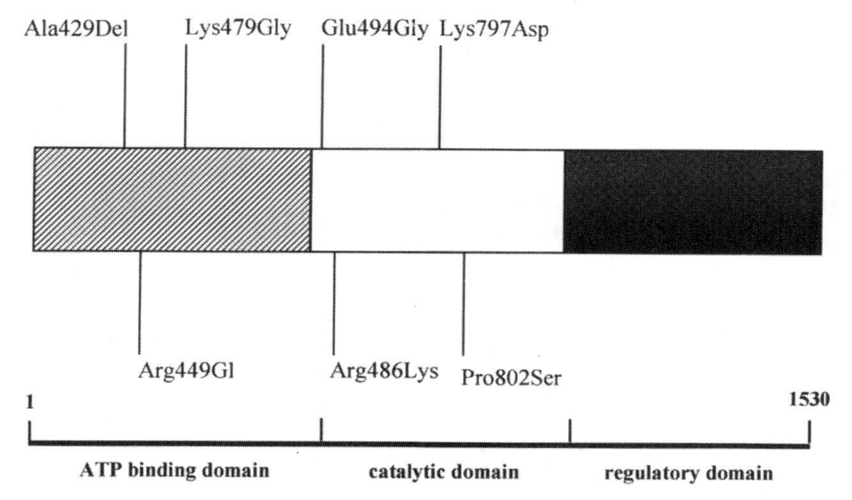

Figure 5. Topoisomerase II mutations that result in increased drug resistance.

target of the antifolates, is an attractive anticancer target mainly because of its necessity for cellular activity and replication (243). DHFR functions to catalyze the reaction of either folate or 7,8-dihydrofolate to the reduced form 5,6,7,8-tetrahydrofolate (24). In cultured cells, inhibition of DHFR activity results in termination of the tetrahydrofolate cofactor-dependent reactions in minutes (292, 344). The levels of expression and activity of DHFR can be altered through increased transcription, gene amplification, and mutation. The higher the DHFR level, the smaller the fraction of enzyme needed for proper cellular function, and, thus, a higher degree of saturation is needed to suppress the reaction. Due to the limitations of concentrative methotrexate transport, extremely high extracellular drug levels are needed to saturate the amplified DHFR target responsible for conferring resistance. This type of resistance also exists with other, newer members of the antifolate class of drugs (361). Gene amplification has been shown clinically to be causative of resistance, as it has been found in patients with acute lymphoblastic leukemia (ALL) and ovarian carcinoma (39, 59, 116, 119, 136, 328). It is produced by an increase in gene copy number, localized to chromosome 21, or in microsatellites causing up-regulation of both mRNA and enzyme (58, 67, 156, 221).

The affinity of DHFR for an antifolate is also an important determinant for activity of the cancer drug. Laboratory studies using cultured cells have identified DHFR mutations that correlate with the emergence of drug resistance (65, 217, 219, 300, 309). The methotrexate binding site contains most of the resistance producing mutations of the enzyme, but there are some exceptions (66, 358). While these mutations yield resistance in the laboratory setting, however, the considerable gene amplification to offset lowered catalytic efficiency, and the need for two-base changes for most substitutions, makes the clinical relevance of these mutations doubtful (308).

Thymidylate Synthase Alterations

Production of deoxythymidylate by the enzyme thymidylate synthase (TS) is the cell's only source for deoxythymidylate used in DNA replication and has been an important target for chemotherapy since the introduction of 5-fluorouracil in the 1950s (126). Inhibition of the TS enzyme occurs through the formation of a ternary complex between the active metabolite of 5-fluorouracil, a reduced folate and the 36-kDa dimeric TS protein (273). Many of the newer antimetabolites are drugs that potently inhibit and specifically target TS (361). Similar to DHFR, TS is often overexpressed in resistant laboratory cultured cells (68, 94, 144, 166, 235, 324). Intrinsic resistance and lack of clinical response have been clearly linked to high TS levels in different cancers, possibly as the result of polymorphisms in the TS promoter (9, 149, 183, 185, 248). However, the relationship between TS protein levels and the degree to which the cells are resistant is not linear (196, 285). Posttranscriptional changes of TS have been implicated in increased stability of the protein, potentially leading to increased levels (166, 167). Mutations of the enzyme have resulted in resistance, and these changes usually map to regions in, or around, the substrate binding site of the protein (14, 207, 254, 325, 342). An especially interesting application of the discovery and phenotypic analysis of mutations in the DHFR or TS enzymes is the transduction of these mutant forms of DHFR and TS into bone marrow cells to protect them from the deadly effects of antimetabolite chemotherapy drugs (37, 98, 274, 307, 319, 320). This engineering of noncancerous host cells will potentially allow for more aggressive treatment with DHFR or TS inhibitory drugs while minimizing side effects.

Microtubule Alterations

Microtubule targeting agents such as the taxanes, *Vinca* alkaloids, and colchicines, including such drugs as paclitaxel, docetaxel, vincristine, vinblastine, and colchicine, are especially important in the chemotherapy of cancer (152). A common form of resistance that is encountered with the taxanes is an alteration to their target, the tubulins (54). Tubulins are the basic subunits making up microtubules. The microtubules are structural components of cells originating near the nucleus and involved in the intracellular movement of organelles, cell shape, cell movement and cell division (64). Microtubules are important chemotherapeutic targets, in particular because of their necessity in the movement of chromosomes during metaphase (64, 299). Anti-tubulin drugs act by either stabilizing (i.e., taxanes) or destabilizing (i.e., *Vinca* alkaloids) tubulin polymerization. In the laboratory, cell lines resistant to microtubule targeting agents have been created. Changes in these cell lines include altered tubulin isotype expression, tubulin mutations that affect assembly, mutations that affect the binding of regulatory proteins, alterations of tubulin through posttranslational modifications and altered expression or posttranslational modification of tubulin or assembled microtubule regulatory proteins (203, 211, 238, 291, 334, 335). The majority of clinical studies have focused on establishing the relevance of tubulin mutations to clinical cancer chemotherapy. These studies have been complicated by the existence of pseudogenes for tubulin. A detailed understanding of the clinically relevant mechanisms of resistance awaits further studies.

BCR-ABL Alterations

The tightly regulated *c-ABL* gene encodes for a nuclear kinase. When fused with the BCR gene through a reciprocal translocation between chromosomes 9 and 22 (the Philadelphia chromosome) a constitutively active tyrosine kinase is created. The oncogenic BCR-ABL tyrosine kinase targets and phosphorylates proteins inducing various signal pathways which, in turn, increase cellular proliferation, prevent apoptosis, and decrease cellular adhesion (60, 127). Imatinib mesylate (Gleevec) is a recent addition to the chemotherapeutic arsenal and has shown incredible promise as a therapy for leukemia, with a hematologic response rate of 95% (154). Imatinib mesylate was developed as a specific inhibitor of the BCR-ABL oncoprotein and, as such, it represents a potential change in how antineoplastic agents are identified (117, 233, 351). The majority of conventional anticancer agents are cytotoxic agents that target cancer cells as a result of their somewhat selective effect on cells undergoing rapid division. In contrast, these new drugs represent a more targeted approach to anticancer therapy, specifically exploiting the changes associated with the cancerous cell.

Through the development of specific cell lines, resistance mechanisms have been identified, including amplification and mutation of BCR-ABL (208, 301, 322, 343). Gene amplification has been identified in cell lines and in several patient samples (36, 103, 119, 131, 228). Mutational analysis has revealed the existence of three sites on the protein that when mutated can give rise to resistance. These three sites are located at the ATP binding domain, the catalytic loop, and the activation loop (72, 298, 322). A crystal structure of the ABL protein complexed to imatinib mesylate has been resolved and has allowed for some useful interpretations of these mutations. It is currently thought that resistance yielding mutations act to destabilize an inactive, autoinhibited conformation of BCR-ABL that imatinib mesylate preferentially binds to (278). Imatinib mesylate thus exemplifies the concept of a selectively targeted anticancer strategy designed to target specific pathways involved in antiapoptotic, proliferative, or angiogenic processes. The fact that the mechanisms of resistance seen with this drug mimic many of the others discussed above is prophetic and a sign of the persistence of resistant forms of cancer.

APOPTOSIS AND DRUG RESISTANCE

Down-Regulation of Apoptosis as a Prerequisite for Tumor Growth

In normal tissues, the rate of cell proliferation is equal to the rate of death. In tumors, however, this equilibrium is unbalanced, with the rate of proliferation exceeding the rate of death. This condition may result from excessive proliferation, reduced cell death or a combination of both. The latter possibility seems more likely, since in solid tumors there is both increased mitotic figures and apoptotic cells are common (353). Several apoptosis-related genes have been found to be altered in cancer, including, Bcl-2, Bcl-xL, Bax, Apaf-1, XIAP, ML-IAP, Fas receptor, TRAIL receptors, c-FLIP, and caspases 8 and 10 (159). Most of these are weakly oncogenic alone but can enhance tumor formation in combination with other oncogenes. For instance, Bcl-2 mutant transgenic mice develop tumors only after an extended period of time (12 to 18 months), whereas c-myc transgenic mice develop tumors after 10 weeks. However, in c-myc^{+}/bcl-2^{-} mice, tumors develop when the mice are only days old, indicating that inhibition of apoptosis can greatly enhance tumorigenesis (313).

The Role of Apoptotic Factors in Drug Resistance

The presence or absence of both positive and negative regulators of apoptosis has the potential to modify the cellular response to anticancer therapy. Increasing the concentration of pro-apoptotic molecules leads to enhanced sensitivity, while increasing the concentration of anti-apoptotic and/or mitogenic molecules has the opposite effect. Thus, the intrinsic resistance of cells to chemotherapy often resides in the balance between the activities of pro- and anti-apoptotic molecules.

p53 and Resistance

Central to tumor formation are several oncogenes and tumor suppressor proteins affecting the cell cycle, and one of the most important of these is p53 (78). p53 is mutated in >50% of cancers at diagnosis and in an even higher percentage at the time of emergence of drug resistance. Under normal circumstances the function of p53 is to halt cell division after DNA damage or cytotoxic stress, in order to allow repair to occur, or to induce apoptosis in heavily damaged cells. Thus, cells without functional p53 do not enter apoptosis normally and will continue to divide without pausing to repair damage. Uncontrolled growth, besides leading to tumor formation, will also lead to genetic instability, with the possibility of beneficial mutations arising in the tumor cell population. The importance of p53 in tumorigenesis is highlighted by the frequency at which it is found to be mutated in cancer. p53 has long been known to be central in suppressing tumor growth. DNA damage, oncogenic signaling, hypoxia and nucleotide depletion can cause accumulation and activation of p53. Several proteins are known to activate p53, including ATM, Chk1, Chk2, ATR, ARF, and HIF1, while MDM2, MDMX, and Pirh2 are negative regulators of p53 function. The activation of p53 in response to chemotherapeutics results in cell cycle arrest and repair of cellular damage, or, if the damage is more extensive, in apoptosis. Loss of p53 function, either through direct mutation of the gene or through activation of repressors, results in the loss of this cell cycle regulator and in the inability to initiate apoptosis via this pathway. In culture, loss of p53 functionality leads to 2- to 10-fold resistance to chemotherapeutic drugs (23, 31).

The Bcl-2 Family of Proteins and Resistance

In mammals, there are approximately 20 different Bcl-2 family members that can be divided into three groups (253). The first group, containing Bcl-2 and Bcl-x_L, is anti-apoptotic. The proteins of this group contain multiple Bcl-2 homology domains (BH domains) involved in homo- and heterotypic protein-protein interactions. The second group, containing Bax and Bak, are pro-apoptotic and also contain multiple BH domains. The last group, containing Bim, Bik, Bid, and several others, is also pro-apoptotic, but possesses a single BH domain and is known as "BH3-only proteins." Our current understanding of the role Bcl-2 family members play in apoptosis is that pro-apoptotic members (e.g., Bax and Bak) oligomerize to form pores in the outer mitochondrial membrane through which pro-apoptotic Apaf-1, cytochrome c, and AIF are released (46). Apaf-1 and cytochrome c go on to oligomerize with cytosolic caspase 9 to form the "apoptosome." Bcl-2 prevents apoptosis by directly preventing the formation of pores and, thus, preventing the release of pro-apoptotic components. Most chemotherapeutics and ionizing radiation induce apoptosis through the Bcl-2-inhibitable, or mitochondrial, pathway. Bcl-2 was originally identified as an oncogene in B cell leukemia and transfection of Bcl-2 can lead to multidrug resistance (223). Alterations in the ratios of pro- and anti-apoptotic Bcl-2 family members have been correlated with clinical drug resistance (253). High Bcl-2 levels are associated with poorer prognosis in AML, ALL, non-Hodgkin's lymphoma, and prostate cancer while high bcl-x_L levels are associated with poorer prognosis in lymphomas, sarcomas and squamous cell carcinoma (3, 11, 168).

Role of Caspases and IAPs in Resistance

Caspases are cysteine proteases involved in the dismantling of cells. They cleave important cellular proteins, such as DNA repair proteins and cytoskeletal/structural proteins. Inhibitors of apoptosis proteins (IAPs) bind to, and inhibit, caspase activity. In mammalian cells, this family includes XIAP, cIAP1, cIAP2, ML-IAP, and others. The evidence for a contribution to drug resistance by the IAPs is not strong; however, overexpression of XIAP is able to block apoptosis and is found to be overexpressed in some cancers (73). Clinical overexpression of XIAP is associated with a poor prognosis and resistance to drug and radiation therapies (176).

Death Receptors and Resistance

Activation of cytokine receptors of the tumor necrosis factor (TNF) family can lead to the initiation of apoptosis. The receptors for TNF (TNFR1), FasL (Fas), and TRAIL (DR4, DR5) oligomerize in the presence of their ligands and recruit several factors, including caspase 8, to the plasma membrane to form a death inducible signaling complex (DISC). This form of apoptosis (i.e., the extrinsic pathway) is rarely inhibitable by overexpression of Bcl-2. Several proteins related to the death receptor-mediated apoptotic pathway are altered in tumors, including Fas and TRAIL receptors. One role for alteration of these receptors is to allow the cancer cells to escape immune surveillance (210, 340). Reduced sensitivity of cancer cells to either Fas or TRAIL stimulation by T cells results in reduced cell death and, thus, tumor growth.

Sphingolipids and Resistance

The sphingolipids, ceramide, sphingosine and sphingosine-1-phosphate have emerged as potential lipid regulators of cellular homeostasis (169, 255). The idea of a "sphingolipid rheostat" has been suggested in which the balance between pro-apoptotic ceramide and sphingosine and mitogenic sphingosine-1-phosphate (S1P) determines the fate of the cell. Ceramide is typically produced through the activation of sphingomyelinases and subsequent hydrolysis of sphingomyelin, although in some cases, activation of the de novo biosynthetic pathway is noted. Ceramide formation occurs upon exposure of cells to most chemotherapeutics, ionizing radiation, and some cytokines. Direct targets of ceramide include phosphatases, kinases, and proteases. Recently, mitochondria have been shown to play a central role in ceramide-mediated apoptosis as both a site of generation and a target of ceramide. Incubation of cells with synthetic and natural ceramides results in the induction of apoptosis. Compounds, which alter the rate of formation or catabolism of ceramide, also affect the induction of apoptosis and drug resistance. For instance, blocking formation of ceramide through loss of sphingomyelinases or increased metabolism through increased glycosylceramide synthase activity results in drug resistance. In these cases, resistance to multiple drugs is affected; that is, modulation of ceramide signaling results in the MDR phenotype. Up regulation of sphingosine kinase also results in MDR through the increased production of S1P.

Reversing Drug Resistance through Modulation of Cell Cycle and Apoptotic Pathways

Resistance to apoptosis can be viewed as one of the early critical first steps in carcinogenesis and can be expected to affect sensitivity to most antimetabolite and natural product drugs. As such, pharmacological manipulation of the pathway should result in greater efficacy. Current research is aimed at inhibiting pro-survival proteins, inhibiting anti-apoptosis proteins, modulating lipid second messenger levels, and altering cell signaling pathways.

Bcl-2 family member proteins are overexpressed in many types of cancers. Oblimersen sodium (Genasense) is a phosphorothioate bcl-2 antisense oligonucleotide in phase II/III trials for a number of different cancers (Genta, Berkeley Heights, N.J.). In patients with advanced melanoma treated with Oblimersen sodium plus dacarbazine, progression-free survival was greater compared with patients treated with dacarbazine alone (74 days versus 49 days, respectively; $P = 0.0003$). However, overall survival did not achieve statistical significance ($P = 0.17$). Early data from patients with leukemia and non-Hodgkin's lymphoma indicate that Oblimersen sodium may have efficacy. Another antisense oligonucleotide, which has just entered clinical trials, is AEG35156 (Aegera Therapeutics, Ottawa, Ontario). AEG35156 is an IAP antisense oligonucleotide and has shown preclinical efficacy, both as a single agent and in combination with front-line chemotherapeutic agents (docetaxel, carboplatin, cisplatin), producing a dose-dependent reduction in tumor burden in models of human ovarian, prostate, and colon carcinomas.

Safingol (L-threo-dihyrosphingosine) is a protein kinase C inhibitor that enhances apoptosis. In a phase I trial, safingol administration did not affect doxorubicin pharmacokinetics and resulted in minor responses in patients with pancreatic cancer (286).

Cellular ceramide has also been targeted in preclinical and early clinical studies (255, 297). Ceramide is a pro-apoptotic lipid second messenger generated in response to most chemotherapeutics. Fenretinide (4-hydroxyphenylretinamide) is a synthetic retinoid which reduces cell proliferation and enhances apoptosis through the increased production of ceramide (255). It is currently in phase I trials for the treatment of solid tumors and phase III trials for the prevention of breast cancer, where it shows protection in premenopausal women (326). Other preclinical data point to inhibition of glucosylceramide synthase (GCS) and augmentation of the cellular sphingomyelin pools as potential targets for altering the apoptotic response of cancer cells. GCS reduces ceramide levels by shunting ceramide into the ganglioside biosynthetic pathway. Preclinical studies suggest that inhibition of GCS by tamoxifen or antisense oligonucleotides can reverse multidrug resistance by increasing ceramide levels (25). Apoptosis levels can also be increased in animal models of colon cancer by supplementing chemotherapy with exogenous sphingomyelin (225).

Cyclin dependent kinases (Cdks) participate in cell cycle progression. Inhibition of Cdk1 and Cdk2 by UNC-01 or flavopiridol induces G_1 arrest and enhances chemosensitivity. UNC-01 has been observed to induce partial responses in melanoma and lymphoma (293). Flavopiridol has shown limited clinical activity in phase I trials. Two new Cdk inhibitors, BMS 387032 (Bristol-Myers Squibb, Princeton, N.J.) and R-Roscovitine (Cyclacel Limited, United Kingdom), have just entered phase I trials.

Although not directly activating apoptotic pathways, suppression of pro-survival signal transduction via the *ras* oncogene has been targeted using inhibitors of farnesyl transferase, an enzyme necessary for activation of *ras*. Three inhibitors (BMS-214662 (Bristol-Myers Squibb, Princeton, N.J.), SCH66336 (Schering-Plough, Kenilworth, N.J.) and R115777 (Johnson & Johnson Pharmaceutical Research and Development, Titusville, N.J.) have been introduced into clinical trials (38). Both BMS-214662 and SCH66336 were associated with severe gastrointestinal toxicity and a limited number of partial responses. R115777 is better tolerated; however, a recently completed, large phase III trial, comparing gemcitabine plus R115777 versus gemcitabine plus placebo in advanced pancreatic cancer, failed to demonstrate any survival benefit in the R115777 arm.

CONCLUDING REMARKS

The arsenal that we have to treat cancer has increased considerably over the last 50 years. Despite advances in drug development, however, resistance continues to be the main problem resulting in our inability to make major strides in increasing overall survival rates for cancer patients. While many patients will have a very favorable initial response to chemotherapy, the likelihood of relapse for cancer patients is high. Laboratory studies have made it clear that there are many mechanisms through which a cancer cell can acquire resistance to a chemotherapeutic agent. However, despite our success at the bench we have yet to understand completely the clinical relevance of these findings. Clinically, it can be very difficult to discern the mechanism(s) responsible for drug resistance, even in the case of single agent resistance. This is due to the relatively low level of resistance needed to overcome therapy, since as little as two- to threefold increased resistance may result in a poor clinical outcome (91, 174). A better understanding of the mechanisms operating at the clinical level is crucial for the development of drugs aimed at circumventing resistance or therapeutic regimens for preventing the emergence of resistance. Currently our therapeutic approach to beat resistance is limited. However, just as the development of more targeted cancer chemotherapeutics has arisen because of a better understanding of the clinical mechanisms responsible for tumorigenesis, our ability to beat resistance awaits a clearer understanding of the clinical mechanisms. With this understanding will follow new drugs that circumvent specific resistance mechanisms and significant advances in long-term survival rates for cancer patients. This is the challenge for the future.

REFERENCES

1. **Abbruzzese, J. L., R. Grunewald, E. A. Weeks, D. Gravel, T. Adams, B. Nowak, S. Mineishi, P. Tarassoff, W. Satterlee, and M. N. Raber.** 1991. A phase I clinical, plasma, and cellular pharmacology study of gemcitabine. *J. Clin. Oncol.* **9:** 491–498.
2. **Adler, V., Z. Yin, S. Y. Fuchs, M. Benezra, L. Rosario, K. D. Tew, M. R. Pincus, M. Sardana, C. J. Henderson, C. R. Wolf, R. J. Davis, and Z. Ronai.** 1999. Regulation of JNK signaling by GSTp. *EMBO J.* **18:**1321–1334.
3. **Aebersold, D. M., A. Kollar, K. T. Beer, J. Laissue, R. H. Greiner, and V. Djonov.** 2001. Involvement of the hepatocyte growth factor/scatter factor receptor c-met and of Bcl-xL in the resistance of oropharyngeal cancer to ionizing radiation. *Int. J. Cancer* **96:**41–54.
4. **Agrawal, M., J. Abraham, F. M. Balis, M. Edgerly, W. D. Stein, S. Bates, T. Fojo, and C. C. Chen.** 2003. Increased 99mTc-sestamibi accumulation in normal liver and drug-resistant tumors after the administration of the glycoprotein inhibitor, XR9576. *Clin. Cancer Res.* **9:**650–656.
5. **Ambudkar, S. V., S. Dey, C. A. Hrycyna, M. Ramachandra, I. Pastan, and M. M. Gottesman.** 1999. Biochemical, cellular, and pharmacological aspects of the multidrug transporter. *Annu. Rev. Pharmacol. Toxicol.* **39:**361–398.
6. **Ambudkar, S. V., C. Kimchi-Sarfaty, Z. E. Sauna, and M. M. Gottesman.** 2003. P-glycoprotein: from genomics to mechanism. *Oncogene* **22:**7468–7485.
7. **Ambudkar, S. V., I. H. Lelong, J. Zhang, C. O. Cardarelli, M. M. Gottesman, and I. Pastan.** 1992. Partial purification and reconstitution of the human multidrug-resistance pump: characterization of the drug-stimulatable ATP hydrolysis. *Proc. Natl. Acad. Sci. USA* **89:**8472–8476.
8. **Armitage, J. O., K. Tobinai, D. Hoelzer, and M. J. Rummel.** 2004. Treatment of indolent non-Hodgkin's lymphoma with cladribine as single-agent therapy and in combination with mitoxantrone. *Int. J. Hematol.* **79:**311–321.
9. **Aschele, C., D. Debernardis, S. Casazza, G. Antonelli, G. Tunesi, C. Baldo, R. Lionetto, F. Maley, and A. Sobrero.** 1999. Immunohistochemical quantitation of thymidylate synthase expression in colorectal cancer metastases predicts for clinical outcome to fluorouracil-based chemotherapy. *J. Clin. Oncol.* **17:**1760–1770.
10. **Bain, L. J., J. B. McLachlan, and G. A. LeBlanc.** 1997. Structure-activity relationships for xenobiotic transport substrates and inhibitory ligands of P-glycoprotein. *Environ. Health Perspect.* **105:**812–818.
11. **Bairey, O., Y. Zimra, M. Shaklai, E. Okon, and E. Rabizadeh.** 1999. Bcl-2, Bcl-X, Bax, and Bak expression in short- and long-lived patients with diffuse large B-cell lymphomas. *Clin. Cancer Res.* **5:**2860–2866.
12. **Bakos, E., R. Evers, G. Szakacs, G. E. Tusnady, E. Welker, K. Szabo, M. de Haas, L. van Deemter, P. Borst, A. Varadi, and B. Sarkadi.** 1998. Functional multidrug resistance protein (MRP1) lacking the N-terminal transmembrane domain. *J. Biol. Chem.* **273:**32167–32175.
13. **Baldwin, S. A., J. R. Mackey, C. E. Cass, and J. D. Young.** 1999. Nucleoside transporters: molecular biology and implications for therapeutic development. *Mol. Med. Today* **5:**216–224.
14. **Barbour, K. W., D. K. Hoganson, S. H. Berger, and F. G. Berger.** 1992. A naturally occurring tyrosine to histidine replacement at residue 33 of human thymidylate synthase confers resistance to 5-fluoro-2′-deoxyuridine in mammalian and bacterial cells. *Mol. Pharmacol.* **42:**242–248.
15. **Barrueco, J. R., D. F. O'Leary, and F. M. Sirotnak.** 1992. Facilitated transport of methotrexate polyglutamates into lysosomes derived from S180 cells. Further characterization and evidence for a simple mobile carrier system with broad specificity for homo- or heteropeptides bearing a C-terminal glutamyl moiety. *J. Biol. Chem.* **267:**19986–19991.
16. **Baugh, C. M., and C. L. Krumdieck.** 1971. Naturally occurring folates. *Ann. N. Y. Acad. Sci.* **186:**7–28.
17. **Baugh, C. M., C. L. Krumdieck, and M. G. Nair.** 1973. Polygammaglutamyl metabolites of methotrexate. *Biochem. Biophys. Res. Commun.* **52:**27–34.
18. **Beck, W. T., M. K. Danks, J. S. Wolverton, M. Chen, B. Granzen, R. Kim, and D. P. Suttle.** 1994. Resistance of mammalian tumor cells to inhibitors of DNA topoisomerase II. *Adv. Pharmacol.* **29B:**145–169.
19. **Belinsky, M. G., and G. D. Kruh.** 1999. MOAT-E (ARA) is a full-length MRP/cMOAT subfamily transporter expressed in kidney and liver. *Br. J. Cancer* **80:**1342–1349.
20. **Belt, J. A., N. M. Marina, D. A. Phelps, and C. R. Crawford.** 1993. Nucleoside transport in normal and neoplastic cells. *Adv. Enzyme Regul.* **33:**235–252.
21. **Bergen, A. A., A. S. Plomp, E. J. Schuurman, S. Terry, M. Breuning, H. Dauwerse, J. Swart, M. Kool, S. van Soest, F. Baas, J. B. ten Brink, and P. T. de Jong.** 2000. Mutations in ABCC6 cause pseudoxanthoma elasticum. *Nat. Genet.* **25:** 228–231.
22. **Bharti, A. K., M. O. Olson, D. W. Kufe, and E. H. Rubin.** 1996. Identification of a nucleolin binding site in human topoisomerase I. *J. Biol. Chem.* **271:**1993–1997.
23. **Blagosklonny, M. V., P. Giannakakou, M. Wojtowicz, L. Y. Romanova, K. B. Ain, S. E. Bates, and T. Fojo.** 1998. Effects of

p53-expressing adenovirus on the chemosensitivity and differentiation of anaplastic thyroid cancer cells. *J. Clin. Endocrinol. Metab.* **83**:2516–2522.

24. **Blakley, R. L., and S. J. Benkovic.** 1984. Folates and pterins, vol. v.1. Wiley, New York, N.Y.
25. **Bleicher, R. J., and M. C. Cabot.** 2002. Glucosylceramide synthase and apoptosis. *Biochim. Biophys. Acta* **1585**:172–178.
26. **Bohme, M., G. Jedlitschky, I. Leier, M. Buchler, and D. Keppler.** 1994. ATP-dependent export pumps and their inhibition by cyclosporins. *Adv. Enzyme Regul.* **34**:371–380.
27. **Borst, P., and R. O. Elferink.** 2002. Mammalian ABC transporters in health and disease. *Annu. Rev. Biochem.* **71**:537–592.
28. **Borst, P., R. Evers, M. Kool, and J. Wijnholds.** 2000. A family of drug transporters: the multidrug resistance-associated proteins. *J. Natl. Cancer Inst.* **92**:1295–1302.
29. **Borst, P., R. Evers, M. Kool, and J. Wijnholds.** 1999. The multidrug resistance protein family. *Biochim. Biophys. Acta* **1461**:347–357.
30. **Borst, P., N. Zelcer, and A. van Helvoort.** 2000. ABC transporters in lipid transport. *Biochim. Biophys. Acta* **1486**:128–144.
31. **Boyer, J., E. G. McLean, S. Aroori, P. Wilson, A. McCulla, P. D. Carey, D. B. Longley, and P. G. Johnston.** 2004. Characterization of p53 wild-type and null isogenic colorectal cancer cell lines resistant to 5-fluorouracil, oxaliplatin, and irinotecan. *Clin. Cancer Res.* **10**:2158–2167.
32. **Bradshaw, D. M., and R. J. Arceci.** 1998. Clinical relevance of transmembrane drug efflux as a mechanism of multidrug resistance. *J. Clin. Oncol.* **16**:3674–3690.
33. **Brangi, M., T. Litman, M. Ciotti, K. Nishiyama, G. Kohlhagen, C. Takimoto, R. Robey, Y. Pommier, T. Fojo, and S. E. Bates.** 1999. Camptothecin resistance: role of the ATP-binding cassette (ABC), mitoxantrone-resistance half-transporter (MXR), and potential for glucuronidation in MXR-expressing cells. *Cancer Res.* **59**:5938–5946.
34. **Brigle, K. E., M. J. Spinella, E. E. Sierra, and I. D. Goldman.** 1995. Characterization of a mutation in the reduced folate carrier in a transport defective L1210 murine leukemia cell line. *J. Biol. Chem.* **270**:22974–22979.
35. **Buckwalter, C. A., A. H. Lin, A. Tanizawa, Y. G. Pommier, Y. C. Cheng, and S. H. Kaufmann.** 1996. RNA synthesis inhibitors alter the subnuclear distribution of DNA topoisomerase I. *Cancer Res.* **56**:1674–1681.
36. **Campbell, L. J., C. Patsouris, K. C. Rayeroux, K. Somana, E. H. Januszewicz, and J. Szer.** 2002. BCR/ABL amplification in chronic myelocytic leukemia blast crisis following imatinib mesylate administration. *Cancer Genet. Cytogenet.* **139**:30–33.
37. **Capiaux, G. M., T. Budak-Alpdogan, N. Takebe, P. Mayer-Kuckuk, D. Banerjee, F. Maley, and J. R. Bertino.** 2003. Retroviral transduction of a mutant dihydrofolate reductase-thymidylate synthase fusion gene into murine marrow cells confers resistance to both methotrexate and 5-fluorouracil. *Hum. Gene Ther.* **14**:435–446.
38. **Caponigro, F., M. Casale, and J. Bryce.** 2003. Farnesyl transferase inhibitors in clinical development. *Expert Opin. Investig. Drugs* **12**:943–954.
39. **Carman, M. D., J. H. Schornagel, R. S. Rivest, S. Srimatkandada, C. S. Portlock, T. Duffy, and J. R. Bertino.** 1984. Resistance to methotrexate due to gene amplification in a patient with acute leukemia. *J. Clin. Oncol.* **2**:16–20.
40. **Champoux, J. J.** 2001. DNA topoisomerases: structure, function, and mechanism. *Annu. Rev. Biochem.* **70**:369–413.
41. **Chang, G.** 2003. Multidrug resistance ABC transporters. *FEBS Lett.* **555**:102–105.
42. **Chang, G., and C. B. Roth.** 2001. Structure of MsbA from E. coli: a homolog of the multidrug resistance ATP binding cassette (ABC) transporters. *Science* **293**:1793–1800.
43. **Chazal, M., S. Cheradame, J. L. Formento, M. Francoual, P. Formento, M. C. Etienne, E. Francois, H. Richelme, M. Mousseau, C. Letoublon, D. Pezet, H. Cure, J. F. Seitz, and G. Milano.** 1997. Decreased folylpolyglutamate synthetase activity in tumors resistant to fluorouracil-folinic acid treatment: clinical data. *Clin. Cancer Res.* **3**:553–557.
44. **Chen, C. J., J. E. Chin, K. Ueda, D. P. Clark, I. Pastan, M. M. Gottesman, and I. B. Roninson.** 1986. Internal duplication and homology with bacterial transport proteins in the mdr1 (P-glycoprotein) gene from multidrug-resistant human cells. *Cell* **47**:381–389.
45. **Chen, Z. S., K. Lee, and G. D. Kruh.** 2001. Transport of cyclic nucleotides and estradiol 17-beta-D-glucuronide by multidrug resistance protein 4. Resistance to 6-mercaptopurine and 6-thioguanine. *J. Biol. Chem.* **276**:33747–33754.
46. **Cheng, E. H., M. C. Wei, S. Weiler, R. A. Flavell, T. W. Mak, T. Lindsten, and S. J. Korsmeyer.** 2001. BCL-2, BCL-X(L) sequester BH3 domain-only molecules preventing BAX- and BAK-mediated mitochondrial apoptosis. *Mol. Cell.* **8**:705–711.
47. **Childs, S., R. L. Yeh, D. Hui, and V. Ling.** 1998. Taxol resistance mediated by transfection of the liver-specific sister gene of P-glycoprotein. *Cancer Res.* **58**:4160–4167.
48. **Chrencik, J. E., B. L. Staker, A. B. Burgin, P. Pourquier, Y. Pommier, L. Stewart, and M. R. Redinbo.** 2004. Mechanisms of camptothecin resistance by human topoisomerase I mutations. *J. Mol. Biol.* **339**:773–784.
49. **Cole, S. P., G. Bhardwaj, J. H. Gerlach, J. E. Mackie, C. E. Grant, K. C. Almquist, A. J. Stewart, E. U. Kurz, A. M. Duncan, and R. G. Deeley.** 1992. Overexpression of a transporter gene in a multidrug-resistant human lung cancer cell line. *Science* **258**:1650–1654.
50. **Cole, S. P., and R. G. Deeley.** 1993. Multidrug resistance-associated protein: sequence correction. *Science* **260**:879.
51. **Colly, L. P., W. G. Peters, D. Richel, M. W. Arentsen-Honders, C. W. Starrenburg, and R. Willemze.** 1987. Deoxycytidine kinase and deoxycytidine deaminase values correspond closely to clinical response to cytosine arabinoside remission induction therapy in patients with acute myelogenous leukemia. *Semin. Oncol.* **14**:257–261.
52. **Cordon-Cardo, C., J. P. O'Brien, J. Boccia, D. Casals, J. R. Bertino, and M. R. Melamed.** 1990. Expression of the multidrug resistance gene product (P-glycoprotein) in human normal and tumor tissues. *J. Histochem. Cytochem.* **38**:1277–1287.
53. **Cory, J. G., and A. Sato.** 1983. Regulation of ribonucleotide reductase activity in mammalian cells. *Mol. Cell Biochem.* **53-54**:257–266.
54. **Cragg, G. M., and D. J. Newman.** 2004. A tale of two tumor targets: topoisomerase I and tubulin. The Wall and Wani contribution to cancer chemotherapy. *J. Nat. Prod.* **67**:232–244.
55. **Crommentuyn, K. M., J. H. Schellens, J. D. van den Berg, and J. H. Beijnen.** 1998. In-vitro metabolism of anti-cancer drugs, methods and applications: paclitaxel, docetaxel, tamoxifen and ifosfamide. *Cancer Treat. Rev.* **24**:345–366.
56. **Cummings, J., B. T. Ethell, L. Jardine, G. Boyd, J. S. Macpherson, B. Burchell, J. F. Smyth, and D. I. Jodrell.** 2003. Glucuronidation as a mechanism of intrinsic drug resistance in human colon cancer: reversal of resistance by food additives. *Cancer Res.* **63**:8443–8450.
57. **Cunningham, D.** 1998. Mature results from three large controlled studies with raltitrexed ('Tomudex'). *Br. J. Cancer* **77** (Suppl 2):15–21.

58. **Curt, G. A., D. N. Carney, K. H. Cowan, J. Jolivet, B. D. Bailey, J. C. Drake, K. S. Chien Song, J. D. Minna, and B. A. Chabner.** 1983. Unstable methotrexate resistance in human small-cell carcinoma associated with double minute chromosomes. *N. Engl. J. Med.* **308**:199–202.
59. **Curt, G. A., J. Jolivet, D. N. Carney, B. D. Bailey, J. C. Drake, N. J. Clendeninn, and B. A. Chabner.** 1985. Determinants of the sensitivity of human small-cell lung cancer cell lines to methotrexate. *J. Clin. Invest.* **76**:1323–1329.
60. **Daley, G. Q., R. A. Van Etten, and D. Baltimore.** 1990. Induction of chronic myelogenous leukemia in mice by the P210bcr/abl gene of the Philadelphia chromosome. *Science* **247**:824–830.
61. **Danks, M. K., K. E. Garrett, R. C. Marion, and D. O. Whipple.** 1996. Subcellular redistribution of DNA topoisomerase I in anaplastic astrocytoma cells treated with topotecan. *Cancer Res.* **56**:1664–1673.
62. **Danks, M. K., M. R. Warmoth, E. Friche, B. Granzen, B. Y. Bugg, W. G. Harker, L. A. Zwelling, B. W. Futscher, D. P. Suttle, and W. T. Beck.** 1993. Single-strand conformational polymorphism analysis of the M(r) 170,000 isozyme of DNA topoisomerase II in human tumor cells. *Cancer Res.* **53**:1373–1379.
63. **Dean, M., A. Rzhetsky, and R. Allikmets.** 2001. The human ATP-binding cassette (ABC) transporter superfamily. *Genome Res.* **11**:1156–1166.
64. **Desai, A., and T. J. Mitchison.** 1997. Microtubule polymerization dynamics. *Annu. Rev. Cell Dev. Biol.* **13**:83–117.
65. **Dicker, A. P., M. Volkenandt, B. I. Schweitzer, D. Banerjee, and J. R. Bertino.** 1990. Identification and characterization of a mutation in the dihydrofolate reductase gene from the methotrexate-resistant Chinese hamster ovary cell line Pro-3 MtxRIII. *J. Biol. Chem.* **265**:8317–8321.
66. **Dicker, A. P., M. C. Waltham, M. Volkenandt, B. I. Schweitzer, G. M. Otter, F. A. Schmid, F. M. Sirotnak, and J. R. Bertino.** 1993. Methotrexate resistance in an in vivo mouse tumor due to a non-active-site dihydrofolate reductase mutation. *Proc. Natl. Acad. Sci. USA* **90**:11797–11801.
67. **Dolnick, B. J., R. J. Berenson, J. R. Bertino, R. J. Kaufman, J. H. Nunberg, and R. T. Schimke.** 1979. Correlation of dihydrofolate reductase elevation with gene amplification in a homogeneously staining chromosomal region in L5178Y cells. *J. Cell Biol.* **83**:394–402.
68. **Drake, J. C., C. J. Allegra, R. G. Moran, and P. G. Johnston.** 1996. Resistance to tomudex (ZD1694): multifactorial in human breast and colon carcinoma cell lines. *Biochem. Pharmacol.* **51**:1349–1355.
69. **Draper, M. P., R. L. Martell, and S. B. Levy.** 1997. Indomethacin-mediated reversal of multidrug resistance and drug efflux in human and murine cell lines overexpressing MRP, but not P-glycoprotein. *Br. J. Cancer* **75**:810–815.
70. **Drori, S., G. Jansen, R. Mauritz, G. J. Peters, and Y. G. Assaraf.** 2000. Clustering of mutations in the first transmembrane domain of the human reduced folate carrier in GW1843U89-resistant leukemia cells with impaired antifolate transport and augmented folate uptake. *J. Biol. Chem.* **275**:30855–30863.
71. **Drori, S., H. Sprecher, G. Shemer, G. Jansen, I. D. Goldman, and Y. G. Assaraf.** 2000. Characterization of a human alternatively spliced truncated reduced folate carrier increasing folate accumulation in parental leukemia cells. *Eur. J. Biochem.* **267**:690–702.
72. **Druker, B. J.** 2003. Imatinib alone and in combination for chronic myeloid leukemia. *Semin. Hematol.* **40**:50–58.
73. **Duckett, C. S., V. E. Nava, R. W. Gedrich, R. J. Clem, J. L. Van Dongen, M. C. Gilfillan, H. Shiels, J. M. Hardwick, and C. B. Thompson.** 1996. A conserved family of cellular genes related to the baculovirus iap gene and encoding apoptosis inhibitors. *EMBO J.* **15**:2685–2694.
74. **Duffy, C. P., C. J. Elliott, R. A. O'Connor, M. M. Heenan, S. Coyle, I. M. Cleary, K. Kavanagh, S. Verhaegen, C. M. O'Loughlin, R. NicAmhlaoibh, and M. Clynes.** 1998. Enhancement of chemotherapeutic drug toxicity to human tumour cells in vitro by a subset of non-steroidal anti-inflammatory drugs (NSAIDs). *Eur. J. Cancer* **34**:1250–1259.
75. **Ecker, G., M. Huber, D. Schmid, and P. Chiba.** 1999. The importance of a nitrogen atom in modulators of multidrug resistance. *Mol. Pharmacol.* **56**:791–796.
76. **Edwards, T. K., A. Saleem, J. A. Shaman, T. Dennis, C. Gerigk, E. Oliveros, M. R. Gartenberg, and E. H. Rubin.** 2000. Role for nucleolin/Nsr1 in the cellular localization of topoisomerase I. *J. Biol. Chem.* **275**:36181–36188.
77. **Ejendal, K. F., and C. A. Hrycyna.** 2002. Multidrug resistance and cancer: the role of the human ABC transporter ABCG2. *Curr. Protein Pept. Sci.* **3**:503–511.
78. **El-Deiry, W. S.** 2003. The role of p53 in chemosensitivity and radiosensitivity. *Oncogene* **22**:7486–7495.
79. **Eng, W. K., F. L. McCabe, K. B. Tan, M. R. Mattern, G. A. Hofmann, R. D. Woessner, R. P. Hertzberg, and R. K. Johnson.** 1990. Development of a stable camptothecin-resistant subline of P388 leukemia with reduced topoisomerase I content. *Mol. Pharmacol.* **38**:471–480.
80. **Etienne, M. C., T. Guillot, and G. Milano.** 1996. Critical factors for optimizing the 5-fluorouracil-folinic acid association in cancer chemotherapy. *Ann. Oncol.* **7**:283–289.
81. **Evans, C. D., S. E. Mirski, M. K. Danks, and S. P. Cole.** 1994. Reduced levels of topoisomerase II alpha and II beta in a multidrug-resistant lung-cancer cell line. *Cancer Chemother. Pharmacol.* **34**:242–248.
82. **Evans, W. E., and M. V. Relling.** 1999. Pharmacogenomics: translating functional genomics into rational therapeutics. *Science* **286**:487–491.
83. **Evers, R., M. de Haas, R. Sparidans, J. Beijnen, P. R. Wielinga, J. Lankelma, and P. Borst.** 2000. Vinblastine and sulfinpyrazone export by the multidrug resistance protein MRP2 is associated with glutathione export. *Br. J. Cancer* **83**:375–383.
84. **Evers, R., M. Kool, A. J. Smith, L. van Deemter, M. de Haas, and P. Borst.** 2000. Inhibitory effect of the reversal agents V-104, GF120918 and Pluronic L61 on MDR1 Pgp-, MRP1- and MRP2-mediated transport. *Br. J. Cancer* **83**:366–374.
85. **Fabre, I., G. Fabre, and I. D. Goldman.** 1984. Polyglutamylation, an important element in methotrexate cytotoxicity and selectivity in tumor versus murine granulocytic progenitor cells in vitro. *Cancer Res.* **44**:3190–3195.
86. **Fan, Y., J. N. Weinstein, K. W. Kohn, L. M. Shi, and Y. Pommier.** 1998. Molecular modeling studies of the DNA-topoisomerase I ternary cleavable complex with camptothecin. *J. Med. Chem.* **41**:2216–2226.
87. **Ferry, D. R., H. Traunecker, and D. J. Kerr.** 1996. Clinical trials of P-glycoprotein reversal in solid tumours. *Eur. J. Cancer* **32A**:1070–1081.
88. **Fisher, G. A., B. L. Lum, J. Hausdorff, and B. I. Sikic.** 1996. Pharmacological considerations in the modulation of multidrug resistance. *Eur. J. Cancer* **32A**:1082–1088.
89. **Fisher, G. A., and B. I. Sikic.** 1995. Clinical studies with modulators of multidrug resistance. *Hematol. Oncol. Clin. North Am.* **9**:363–382.
90. **Fojo, A. T., K. Ueda, D. J. Slamon, D. G. Poplack, M. M. Gottesman, and I. Pastan.** 1987. Expression of a multidrug-resistance gene in human tumors and tissues. *Proc. Natl. Acad. Sci. USA* **84**:265–269.
91. **Fojo, T., and S. Bates.** 2003. Strategies for reversing drug resistance. *Oncogene* **22**:7512–7523.

92. **Ford, J. M., and W. N. Hait.** 1990. Pharmacology of drugs that alter multidrug resistance in cancer. *Pharmacol. Rev.* **42:**155–199.
93. **Franklin, T. J., V. Jacobs, P. Bruneau, and P. Ple.** 1995. Glucuronidation by human colorectal adenocarcinoma cells as a mechanism of resistance to mycophenolic acid. *Adv. Enzyme Regul.* **35:**91–100.
94. **Freemantle, S. J., A. L. Jackman, L. R. Kelland, A. H. Calvert, and J. Lunec.** 1995. Molecular characterisation of two cell lines selected for resistance to the folate-based thymidylate synthase inhibitor, ZD1694. *Br. J. Cancer* **71:**925–930.
95. **Freire, R., F. d'Adda Di Fagagna, L. Wu, G. Pedrazzi, I. Stagljar, I. D. Hickson, and S. P. Jackson.** 2001. Cleavage of the Bloom's syndrome gene product during apoptosis by caspase-3 results in an impaired interaction with topoisomerase IIIalpha. *Nucleic Acids Res.* **29:**3172–3180.
96. **Fry, D. W., J. C. Yalowich, and I. D. Goldman.** 1982. Rapid formation of poly-gamma-glutamyl derivatives of methotrexate and their association with dihydrofolate reductase as assessed by high pressure liquid chromatography in the Ehrlich ascites tumor cell in vitro. *J. Biol. Chem.* **257:**1890–1896.
97. **Fujimori, A., Y. Hoki, N. C. Popescu, and Y. Pommier.** 1996. Silencing and selective methylation of the normal topoisomerase I gene in camptothecin-resistant CEM/C2 human leukemia cells. *Oncol. Res.* **8:**295–301.
98. **Galipeau, J., E. Benaim, H. T. Spencer, R. L. Blakley, and B. P. Sorrentino.** 1997. A bicistronic retroviral vector for protecting hematopoietic cells against antifolates and P-glycoprotein effluxed drugs. *Hum. Gene Ther.* **8:**1773–1783.
99. **Galmarini, C. M., E. Cros, K. Graham, X. Thomas, J. R. Mackey, and C. Dumontet.** 2004. 5 -(3)-nucleotidase mRNA levels in blast cells are a prognostic factor in acute myeloid leukemia patients treated with cytarabine. *Haematologica* **89:**617–619.
100. **Galmarini, C. M., J. R. Mackey, and C. Dumontet.** 2001. Nucleoside analogues: mechanisms of drug resistance and reversal strategies. *Leukemia* **15:**875–890.
101. **Galmarini, C. M., X. Thomas, F. Calvo, P. Rousselot, A. E. Jafaari, E. Cros, and C. Dumontet.** 2002. Potential mechanisms of resistance to cytarabine in AML patients. *Leuk. Res.* **26:**621–629.
102. **Galmarini, C. M., X. Thomas, F. Calvo, P. Rousselot, M. Rabilloud, A. El Jaffari, E. Cros, and C. Dumontet.** 2002. In vivo mechanisms of resistance to cytarabine in acute myeloid leukaemia. *Br. J. Haematol.* **117:**860–868.
103. **Gambacorti-Passerini, C. B., R. H. Gunby, R. Piazza, A. Galietta, R. Rostagno, and L. Scapozza.** 2003. Molecular mechanisms of resistance to imatinib in Philadelphia-chromosome-positive leukaemias. *Lancet Oncol.* **4:**75–85.
104. **Ganapathi, R., A. Constantinou, N. Kamath, G. Dubyak, D. Grabowski, and K. Krivacic.** 1996. Resistance to etoposide in human leukemia HL-60 cells: reduction in drug-induced DNA cleavage associated with hypophosphorylation of topoisomerase II phosphopeptides. *Mol. Pharmacol.* **50:**243–248.
105. **Gandhi, V., and W. Plunkett.** 2002. Cellular and clinical pharmacology of fludarabine. *Clin. Pharmacokinet.* **41:**93–103.
106. **Gate, L., and K. D. Tew.** 2001. Glutathione S-transferases as emerging therapeutic targets. *Expert Opin. Ther. Targets* **5:**477–489.
107. **Gati, W. P., A. R. Paterson, A. R. Belch, V. Chlumecky, L. M. Larratt, M. J. Mant, and A. R. Turner.** 1998. Es nucleoside transporter content of acute leukemia cells: role in cell sensitivity to cytarabine (araC). *Leuk. Lymphoma* **32:**45–54.
108. **Gellert, M.** 1981. DNA topoisomerases. *Annu. Rev. Biochem.* **50:**879–910.
109. **Gerloff, T., B. Stieger, B. Hagenbuch, J. Madon, L. Landmann, J. Roth, A. F. Hofmann, and P. J. Meier.** 1998. The sister of P-glycoprotein represents the canalicular bile salt export pump of mammalian liver. *J. Biol. Chem.* **273:**10046–10050.
110. **Germann, U. A.** 1996. P-glycoprotein—a mediator of multidrug resistance in tumour cells. *Eur. J. Cancer* **32A:**927–944.
111. **Germann, U. A., T. C. Chambers, S. V. Ambudkar, T. Licht, C. O. Cardarelli, I. Pastan, and M. M. Gottesman.** 1996. Characterization of phosphorylation-defective mutants of human P-glycoprotein expressed in mammalian cells. *J. Biol. Chem.* **271:**1708–1716.
112. **Gibson, P., J. H. Gill, P. A. Khan, J. M. Seargent, S. W. Martin, P. A. Batman, J. Griffith, C. Bradley, J. A. Double, M. C. Bibby, and P. M. Loadman.** 2003. Cytochrome P450 1B1 (CYP1B1) is overexpressed in human colon adenocarcinomas relative to normal colon: implications for drug development. *Mol. Cancer Ther.* **2:**527–534.
113. **Glimelius, B., R. Ristamaki, M. Kjaer, P. Pfeiffer, T. Skovsgaard, K. M. Tveit, T. Linne, J. E. Frodin, B. Boussard, D. Oulid-Aissa, and S. Pyrhonen.** 2002. Irinotecan combined with bolus 5-fluorouracil and folinic acid Nordic schedule as first-line therapy in advanced colorectal cancer. *Ann. Oncol.* **13:**1868–1873.
114. **Goan, Y. G., B. Zhou, E. Hu, S. Mi, and Y. Yen.** 1999. Overexpression of ribonucleotide reductase as a mechanism of resistance to 2,2-difluorodeoxycytidine in the human KB cancer cell line. *Cancer Res.* **59:**4204–4207.
115. **Goeptar, A. R., E. J. Groot, H. Scheerens, J. N. Commandeur, and N. P. Vermeulen.** 1994. Cytotoxicity of mitomycin C and adriamycin in freshly isolated rat hepatocytes: the role of cytochrome P450. *Cancer Res.* **54:**2411–2418.
116. **Goker, E., M. Waltham, A. Kheradpour, T. Trippett, M. Mazumdar, Y. Elisseyeff, B. Schnieders, P. Steinherz, C. Tan, and E. Berman.** 1995. Amplification of the dihydrofolate reductase gene is a mechanism of acquired resistance to methotrexate in patients with acute lymphoblastic leukemia and is correlated with p53 gene mutations. *Blood* **86:**677–684.
117. **Goldman, J. M., and J. V. Melo.** 2001. Targeting the BCR-ABL tyrosine kinase in chronic myeloid leukemia. *N. Engl. J. Med.* **344:**1084–1086.
118. **Goldstein, L. J., H. Galski, A. Fojo, M. Willingham, S. L. Lai, A. Gazdar, R. Pirker, A. Green, W. Crist, G. M. Brodeur, M. Lieber, J. Cossman, M. M. Gottesman, and I. Pastan.** 1989. Expression of a multidrug resistance gene in human cancers. *J. Natl. Cancer Inst.* **81:**116–124.
119. **Gorre, M. E., M. Mohammed, K. Ellwood, N. Hsu, R. Paquette, P. N. Rao, and C. L. Sawyers.** 2001. Clinical resistance to STI-571 cancer therapy caused by BCR-ABL gene mutation or amplification. *Science* **293:**876–880.
120. **Gottesman, M. M., T. Fojo, and S. E. Bates.** 2002. Multidrug resistance in cancer: role of ATP-dependent transporters. *Nat. Rev. Cancer* **2:**48–58.
121. **Gray, J. H., R. P. Owen, and K. M. Giacomini.** 2004. The concentrative nucleoside transporter family, SLC28. *Pflugers Arch.* **447:**728–734.
122. **Guo, W., J. H. Healey, P. A. Meyers, M. Ladanyi, A. G. Huvos, J. R. Bertino, and R. Gorlick.** 1999. Mechanisms of methotrexate resistance in osteosarcoma. *Clin. Cancer Res.* **5:**621–627.
123. **Hainsworth, J., I. Vergote, and J. Janssens.** 2003. A review of phase II studies of ZD9331 treatment for relapsed or refractory solid tumours. *Anticancer Drugs* **14**(Suppl 1):S13–S19.
124. **Hamilton, T. C., M. A. Winker, K. G. Louie, G. Batist, B. C. Behrens, T. Tsuruo, K. R. Grotzinger, W. M. McKoy,**

R. C. Young, and R. F. Ozols. 1985. Augmentation of adriamycin, melphalan, and cisplatin cytotoxicity in drug-resistant and -sensitive human ovarian carcinoma cell lines by buthionine sulfoximine mediated glutathione depletion. *Biochem. Pharmacol.* **34:**2583–2586.

125. **Hande, K. R.** 1998. Clinical applications of anticancer drugs targeted to topoisomerase II. *Biochim. Biophys. Acta* **1400:**173–184.

126. **Heidelberger, C., N. K. Chaudhuri, P. Danneberg, D. Mooren, L. Griesbach, R. Duschinsky, R. J. Schnitzer, E. Pleven, and J. Scheiner.** 1957. Fluorinated pyrimidines, a new class of tumour-inhibitory compounds. *Nature* **179:**663–666.

127. **Heisterkamp, N., G. Jenster, J. ten Hoeve, D. Zovich, P. K. Pattengale, and J. Groffen.** 1990. Acute leukaemia in bcr/abl transgenic mice. *Nature* **344:**251–253.

128. **Hill, B. T., B. D. Bailey, J. C. White, and I. D. Goldman.** 1979. Characteristics of transport of 4-amino antifolates and folate compounds by two lines of L5178Y lymphoblasts, one with impaired transport of methotrexate. *Cancer Res.* **39:**2440–2446.

129. **Hinds, M., K. Deisseroth, J. Mayes, E. Altschuler, R. Jansen, F. D. Ledley, and L. A. Zwelling.** 1991. Identification of a point mutation in the topoisomerase II gene from a human leukemia cell line containing an amsacrine-resistant form of topoisomerase II. *Cancer Res.* **51:**4729–4731.

130. **Hipfner, D. R., R. G. Deeley, and S. P. Cole.** 1999. Structural, mechanistic and clinical aspects of MRP1. *Biochim. Biophys. Acta* **1461:**359–376.

131. **Hochhaus, A., S. Kreil, A. S. Corbin, P. La Rosee, M. C. Muller, T. Lahaye, B. Hanfstein, C. Schoch, N. C. Cross, U. Berger, H. Gschaidmeier, B. J. Druker, and R. Hehlmann.** 2002. Molecular and chromosomal mechanisms of resistance to imatinib (STI571) therapy. *Leukemia* **16:**2190–2196.

132. **Hochhauser, D., and A. L. Harris.** 1993. The role of topoisomerase II alpha and beta in drug resistance. *Cancer Treat. Rev.* **19:**181–194.

133. **Holland, I. B.** 2003. *ABC Proteins: from Bacteria to Man.* Academic Press, London.

134. **Honjo, Y., C. A. Hrycyna, Q. W. Yan, W. Y. Medina-Perez, R. W. Robey, A. van de Laar, T. Litman, M. Dean, and S. E. Bates.** 2001. Acquired mutations in the MXR/BCRP/ABCP gene alter substrate specificity in MXR/BCRP/ABCP-overexpressing cells. *Cancer Res.* **61:**6635–6639.

135. **Hooijberg, J. H., H. J. Broxterman, M. Kool, Y. G. Assaraf, G. J. Peters, P. Noordhuis, R. J. Scheper, P. Borst, H. M. Pinedo, and G. Jansen.** 1999. Antifolate resistance mediated by the multidrug resistance proteins MRP1 and MRP2. *Cancer Res.* **59:**2532–2535.

136. **Horns, R. C., Jr., W. J. Dower, and R. T. Schimke.** 1984. Gene amplification in a leukemic patient treated with methotrexate. *J. Clin. Oncol.* **2:**2–7.

137. **Hrycyna, C. A.** 2001. Molecular genetic analysis and biochemical characterization of mammalian P-glycoproteins involved in multidrug resistance. *Semin. Cell Dev. Biol.* **12:**247–256.

138. **Hsiung, Y., M. Jannatipour, A. Rose, J. McMahon, D. Duncan, and J. L. Nitiss.** 1996. Functional expression of human topoisomerase II alpha in yeast: mutations at amino acids 450 or 803 of topoisomerase II alpha result in enzymes that can confer resistance to anti-topoisomerase II agents. *Cancer Res.* **56:**91–99.

139. **Huennekens, F. M., T. H. Duffy, and K. S. Vitols.** 1987. Folic acid metabolism and its disruption by pharmacologic agents. *NCI Monogr.* **5:**1–8.

140. **Husain, I., J. L. Mohler, H. F. Seigler, and J. M. Besterman.** 1994. Elevation of topoisomerase I messenger RNA, protein, and catalytic activity in human tumors: demonstration of tumor-type specificity and implications for cancer chemotherapy. *Cancer Res.* **54:**539–546.

141. **Ifergan, I., I. Meller, J. Issakov, and Y. G. Assaraf.** 2003. Reduced folate carrier protein expression in osteosarcoma: implications for the prediction of tumor chemosensitivity. *Cancer* **98:**1958–1966.

142. **Ito, K., S. L. Olsen, W. Qiu, R. G. Deeley, and S. P. Cole.** 2001. Mutation of a single conserved tryptophan in multidrug resistance protein 1 (MRP1/ABCC1) results in loss of drug resistance and selective loss of organic anion transport. *J. Biol. Chem.* **276:**15616–15624.

143. **Jackman, A. L., and A. H. Calvert.** 1995. Folate-based thymidylate synthase inhibitors as anticancer drugs. *Ann. Oncol.* **6:**871–881.

144. **Jackman, A. L., L. R. Kelland, R. Kimbell, M. Brown, W. Gibson, G. W. Aherne, A. Hardcastle, and F. T. Boyle.** 1995. Mechanisms of acquired resistance to the quinazoline thymidylate synthase inhibitor ZD1694 (Tomudex) in one mouse and three human cell lines. *Br. J. Cancer* **71:**914–924.

145. **Jahns-Streubel, G., C. Reuter, U. Auf der Landwehr, M. Unterhalt, E. Schleyer, B. Wormann, T. Buchner, and W. Hiddemann.** 1997. Activity of thymidine kinase and of polymerase alpha as well as activity and gene expression of deoxycytidine deaminase in leukemic blasts are correlated with clinical response in the setting of granulocyte-macrophage colony-stimulating factor-based priming before and during TAD-9 induction therapy in acute myeloid leukemia. *Blood* **90:**1968–1976.

146. **Jedlitschky, G., B. Burchell, and D. Keppler.** 2000. The multidrug resistance protein 5 functions as an ATP-dependent export pump for cyclic nucleotides. *J. Biol. Chem.* **275:**30069–30074.

147. **Jedlitschky, G., I. Leier, U. Buchholz, K. Barnouin, G. Kurz, and D. Keppler.** 1996. Transport of glutathione, glucuronate, and sulfate conjugates by the MRP gene-encoded conjugate export pump. *Cancer Res.* **56:**988–994.

148. **Johnson, S. A.** 2000. Clinical pharmacokinetics of nucleoside analogues: focus on haematological malignancies. *Clin. Pharmacokinet.* **39:**5–26.

149. **Johnston, P. G., H. J. Lenz, C. G. Leichman, K. D. Danenberg, C. J. Allegra, P. V. Danenberg, and L. Leichman.** 1995. Thymidylate synthase gene and protein expression correlate and are associated with response to 5-fluorouracil in human colorectal and gastric tumors. *Cancer Res.* **55:**1407–1412.

150. **Jolivet, J., and B. A. Chabner.** 1983. Intracellular pharmacokinetics of methotrexate polyglutamates in human breast cancer cells. Selective retention and less dissociable binding of 4-NH2-10-CH3-pteroylglutamate4 and 4-NH2-10-CH3-pteroylglutamate5 to dihydrofolate reductase. *J. Clin. Invest.* **72:**773–778.

151. **Jolivet, J., R. L. Schilsky, B. D. Bailey, J. C. Drake, and B. A. Chabner.** 1982. Synthesis, retention, and biological activity of methotrexate polyglutamates in cultured human breast cancer cells. *J. Clin. Invest.* **70:**351–360.

152. **Jordan, M. A., and L. Wilson.** 2004. Microtubules as a target for anticancer drugs. *Nat. Rev. Cancer* **4:**253–265.

153. **Juliano, R. L., and V. Ling.** 1976. A surface glycoprotein modulating drug permeability in Chinese hamster ovary cell mutants. *Biochim. Biophys. Acta* **455:**152–162.

154. **Kantarjian, H., C. Sawyers, A. Hochhaus, F. Guilhot, C. Schiffer, C. Gambacorti-Passerini, D. Niederwieser, D. Resta, R. Capdeville, U. Zoellner, M. Talpaz, B. Druker, J. Goldman, S. G. O'Brien, N. Russell, T. Fischer, O. Ottmann, P. Cony-Makhoul, T. Facon, R. Stone, C. Miller, M. Tallman, R. Brown, M. Schuster, T. Loughran, A. Gratwohl, F. Mandelli,

G. Saglio, M. Lazzarino, D. Russo, M. Baccarani, and E. Morra. 2002. Hematologic and cytogenetic responses to imatinib mesylate in chronic myelogenous leukemia. *N. Engl. J. Med.* 346:645–652.

155. **Kasid, U. N., B. Halligan, L. F. Liu, A. Dritschilo, and M. Smulson.** 1989. Poly(ADP-ribose)-mediated post-translational modification of chromatin-associated human topoisomerase I. Inhibitory effects on catalytic activity. *J. Biol. Chem.* **264:** 18687–18692.
156. **Kaufman, R. J., P. C. Brown, and R. T. Schimke.** 1979. Amplified dihydrofolate reductase genes in unstably methotrexate-resistant cells are associated with double minute chromosomes. *Proc. Natl. Acad. Sci. USA* **76:**5669–5673.
157. **Kaufman, Y., S. Drori, P. D. Cole, B. A. Kamen, J. Sirota, I. Ifergan, M. W. Arush, R. Elhasid, D. Sahar, G. J. Kaspers, G. Jansen, L. H. Matherly, G. Rechavi, A. Toren, and Y. G. Assaraf.** 2004. Reduced folate carrier mutations are not the mechanism underlying methotrexate resistance in childhood acute lymphoblastic leukemia. *Cancer* **100:**773–782.
158. **Kaufmann, S. H., J. E. Karp, R. J. Jones, C. B. Miller, E. Schneider, L. A. Zwelling, K. Cowan, K. Wendel, and P. J. Burke. 1994.** Topoisomerase II levels and drug sensitivity in adult acute myelogenous leukemia. *Blood* **83:**517–530.
159. **Kaufmann, S. H., and D. L. Vaux.** 2003. Alterations in the apoptotic machinery and their potential role in anticancer drug resistance. *Oncogene* **22:**7414–7430.
160. **Kawasaki, H., C. J. Carrera, L. D. Piro, A. Saven, T. J. Kipps, and D. A. Carson.** 1993. Relationship of deoxycytidine kinase and cytoplasmic 5′-nucleotidase to the chemotherapeutic efficacy of 2-chlorodeoxyadenosine. *Blood* **81:**597–601.
161. **Kawashiro, T., K. Yamashita, X. J. Zhao, E. Koyama, M. Tani, K. Chiba, and T. Ishizaki.** 1998. A study on the metabolism of etoposide and possible interactions with antitumor or supporting agents by human liver microsomes. *J. Pharmacol. Exp. Ther.* **286:**1294–1300.
162. **Kellner, U., M. Sehested, P. B. Jensen, F. Gieseler, and P. Rudolph.** 2002. Culprit and victim —DNA topoisomerase II. *Lancet Oncol.* **3:**235–243.
163. **Kerr, D. J., J. Graham, J. Cummings, J. G. Morrison, G. G. Thompson, M. J. Brodie, and S. B. Kaye.** 1986. The effect of verapamil on the pharmacokinetics of adriamycin. *Cancer Chemother. Pharmacol.* **18:**239–242.
164. **Kerrigan, J. E., and D. S. Pilch.** 2001. A structural model for the ternary cleavable complex formed between human topoisomerase I, DNA, and camptothecin. *Biochemistry* **40:**9792–9798.
165. **Khelifa, T., M. R. Casabianca-Pignede, B. Rene, and A. Jacquemin-Sablon.** 1994. Expression of topoisomerases II alpha and beta in Chinese hamster lung cells resistant to topoisomerase II inhibitors. *Mol. Pharmacol.* **46:**323–328.
166. **Kitchens, M. E., A. M. Forsthoefel, K. W. Barbour, H. T. Spencer, and F. G. Berger.** 1999. Mechanisms of acquired resistance to thymidylate synthase inhibitors: the role of enzyme stability. *Mol. Pharmacol.* **56:**1063–1070.
167. **Kitchens, M. E., A. M. Forsthoefel, Z. Rafique, H. T. Spencer, and F. G. Berger.** 1999. Ligand-mediated induction of thymidylate synthase occurs by enzyme stabilization. Implications for autoregulation of translation. *J. Biol. Chem.* **274:**12544–12547.
168. **Kohler, T., C. Schill, M. W. Deininger, R. Krahl, S. Borchert, D. Hasenclever, S. Leiblein, O. Wagner, and D. Niederwieser.** 2002. High Bad and Bax mRNA expression correlate with negative outcome in acute myeloid leukemia (AML). *Leukemia* **16:**22–29.
169. **Kok, J. W., and H. Sietsma.** 2004. Sphingolipid metabolism enzymes as targets for anticancer therapy. *Curr. Drug Targets* **5:**375–382.
170. **Komatani, H., H. Kotani, Y. Hara, R. Nakagawa, M. Matsumoto, H. Arakawa, and S. Nishimura.** 2001. Identification of breast cancer resistant protein/mitoxantrone resistance/placenta-specific, ATP-binding cassette transporter as a transporter of NB-506 and J-107088, topoisomerase I inhibitors with an indolocarbazole structure. *Cancer Res.* **61:**2827–2832.
171. **Kong, W., K. Engel, and J. Wang.** 2004. Mammalian nucleoside transporters. *Curr. Drug Metab.* **5:**63–84.
172. **Kool, M., M. van der Linden, M. de Haas, G. L. Scheffer, J. M. de Vree, A. J. Smith, G. Jansen, G. J. Peters, N. Ponne, R. J. Scheper, R. P. Elferink, F. Baas, and P. Borst.** 1999. MRP3, an organic anion transporter able to transport anticancer drugs. *Proc. Natl. Acad. Sci. USA* **96:**6914–6919.
173. **Kornberg, A., and T. A. Baker.** 1992. *DNA Replication*, 2nd ed. W.H. Freeman, New York, N.Y.
174. **Kruh, G. D.** 2003. Introduction to resistance to anticancer agents. *Oncogene* **22:**7262–7264.
175. **Kubo, A., A. Yoshikawa, T. Hirashima, N. Masuda, M. Takada, J. Takahara, M. Fukuoka, and K. Nakagawa.** 1996. Point mutations of the topoisomerase IIalpha gene in patients with small cell lung cancer treated with etoposide. *Cancer Res.* **56:**1232–1236.
176. **LaCasse, E. C., S. Baird, R. G. Korneluk, and A. E. MacKenzie.** 1998. The inhibitors of apoptosis (IAPs) and their emerging role in cancer. *Oncogene* **17:**3247–3259.
177. **Laing, N. M., M. G. Belinsky, G. D. Kruh, D. W. Bell, J. T. Boyd, L. Barone, J. R. Testa, and K. D. Tew.** 1998. Amplification of the ATP-binding cassette 2 transporter gene is functionally linked with enhanced efflux of estramustine in ovarian carcinoma cells. *Cancer Res.* **58:**1332–1337.
178. **Laliberte, J., and R. L. Momparler.** 1994. Human cytidine deaminase: purification of enzyme, cloning, and expression of its complementary DNA. *Cancer Res.* **54:**5401–5407.
179. **Larsen, A. K., and A. Skladanowski.** 1998. Cellular resistance to topoisomerase-targeted drugs: from drug uptake to cell death. *Biochim. Biophys. Acta* **1400:**257–274.
180. **Le Saux, O., Z. Urban, C. Tschuch, K. Csiszar, B. Bacchelli, D. Quaglino, I. Pasquali-Ronchetti, F. M. Pope, A. Richards, S. Terry, L. Bercovitch, A. de Paepe, and C. D. Boyd.** 2000. Mutations in a gene encoding an ABC transporter cause pseudoxanthoma elasticum. *Nat. Genet.* **25:**223–227.
181. **Lecureur, V., D. Sun, P. Hargrove, E. G. Schuetz, R. B. Kim, L. B. Lan, and J. D. Schuetz.** 2000. Cloning and expression of murine sister of P-glycoprotein reveals a more discriminating transporter than MDR1/P-glycoprotein. *Mol. Pharmacol.* **57:**24–35.
182. **Legrand, O., G. Simonin, A. Beauchamp-Nicoud, R. Zittoun, and J. P. Marie.** 1999. Simultaneous activity of MRP1 and Pgp is correlated with in vitro resistance to daunorubicin and with in vivo resistance in adult acute myeloid leukemia. *Blood* **94:**1046–1056.
183. **Leichman, C. G., H. J. Lenz, L. Leichman, K. Danenberg, J. Baranda, S. Groshen, W. Boswell, R. Metzger, M. Tan, and P. V. Danenberg.** 1997. Quantitation of intratumoral thymidylate synthase expression predicts for disseminated colorectal cancer response and resistance to protracted-infusion fluorouracil and weekly leucovorin. *J. Clin. Oncol.* **15:**3223–3229.
184. **Leith, C. P., K. J. Kopecky, I. M. Chen, L. Eijdems, M. L. Slovak, T. S. McConnell, D. R. Head, J. Weick, M. R. Grever, F. R. Appelbaum, and C. L. Willman.** 1999. Frequency and clinical significance of the expression of the multidrug resistance proteins MDR1/P-glycoprotein, MRP1, and LRP in acute myeloid leukemia: a Southwest Oncology Group Study. *Blood* **94:**1086–1099.

185. **Lenz, H. J., C. G. Leichman, K. D. Danenberg, P. V. Danenberg, S. Groshen, H. Cohen, L. Laine, P. Crookes, H. Silberman, J. Baranda, Y. Garcia, J. Li, and L. Leichman.** 1996. Thymidylate synthase mRNA level in adenocarcinoma of the stomach: a predictor for primary tumor response and overall survival. *J. Clin. Oncol.* **14:**176–182.
186. **Leonard, G. D., T. Fojo, and S. E. Bates.** 2003. The role of ABC transporters in clinical practice. *Oncologist* **8:**411–424.
187. **Leslie, E. M., K. Ito, P. Upadhyaya, S. S. Hecht, R. G. Deeley, and S. P. Cole.** 2001. Transport of the beta-O-glucuronide conjugate of the tobacco-specific carcinogen 4-(methylnitrosamino)-1-(3-pyridyl)-1-butanol (NNAL) by the multidrug resistance protein 1 (MRP1). Requirement for glutathione or a non-sulfur-containing analog. *J. Biol. Chem.* **276:**27846–27854.
188. **Leslie, E. M., Q. Mao, C. J. Oleschuk, R. G. Deeley, and S. P. Cole.** 2001. Modulation of multidrug resistance protein 1 (MRP1/ABCC1) transport and atpase activities by interaction with dietary flavonoids. *Mol. Pharmacol.* **59:**1171–1180.
189. **Levy, A. S., H. N. Sather, P. G. Steinherz, R. Sowers, M. La, J. A. Moscow, P. S. Gaynon, F. M. Uckun, J. R. Bertino, and R. Gorlick.** 2003. Reduced folate carrier and dihydrofolate reductase expression in acute lymphocytic leukemia may predict outcome: a Children's Cancer Group Study. *J. Pediatr. Hematol. Oncol.* **25:**688–695.
190. **Ling, V.** 1992. Charles F. Kettering Prize. P-glycoprotein and resistance to anticancer drugs. *Cancer* **69:**2603–2609.
191. **Litman, T., T. E. Druley, W. D. Stein, and S. E. Bates.** 2001. From MDR to MXR: new understanding of multidrug resistance systems, their properties and clinical significance. *Cell. Mol. Life Sci.* **58:**931–959.
192. **Liu, L. F.** 1989. DNA topoisomerase poisons as antitumor drugs. *Annu. Rev. Biochem.* **58:**351–375.
193. **Loe, D. W., R. G. Deeley, and S. P. Cole.** 1996. Biology of the multidrug resistance-associated protein, MRP. *Eur. J. Cancer* **32A:**945–957.
194. **Loe, D. W., R. G. Deeley, and S. P. Cole.** 1998. Characterization of vincristine transport by the M(r) 190,000 multidrug resistance protein (MRP): evidence for cotransport with reduced glutathione. *Cancer Res.* **58:**5130–5136.
195. **Loe, D. W., R. G. Deeley, and S. P. Cole.** 2000. Verapamil stimulates glutathione transport by the 190-kDa multidrug resistance protein 1 (MRP1). *J. Pharmacol. Exp. Ther.* **293:**530–538.
196. **Longley, D. B., P. R. Ferguson, J. Boyer, T. Latif, M. Lynch, P. Maxwell, D. P. Harkin, and P. G. Johnston.** 2001. Characterization of a thymidylate synthase (TS)-inducible cell line: a model system for studying sensitivity to TS- and non-TS-targeted chemotherapies. *Clin. Cancer Res.* **7:**3533–3539.
197. **Longo-Sorbello, G. S., and J. R. Bertino.** 2001. Current understanding of methotrexate pharmacology and efficacy in acute leukemias. Use of newer antifolates in clinical trials. *Haematologica* **86:**121–127.
198. **Loo, T. W., and D. M. Clarke.** 2002. Location of the rhodamine-binding site in the human multidrug resistance P-glycoprotein. *J. Biol. Chem.* **277:**44332–44338.
199. **Loo, T. W., and D. M. Clarke.** 1999. Merck Frosst Award Lecture 1998. Molecular dissection of the human multidrug resistance P-glycoprotein. *Biochem. Cell Biol.* **77:**11–23.
200. **Loo, T. W., and D. M. Clarke.** 1999. The transmembrane domains of the human multidrug resistance P-glycoprotein are sufficient to mediate drug binding and trafficking to the cell surface. *J. Biol. Chem.* **274:**24759–24765.
201. **Lotfi, K., G. Juliusson, and F. Albertioni.** 2003. Pharmacological basis for cladribine resistance. *Leuk. Lymphoma* **44:**1705–1712.
202. **Lu, X., S. Gong, A. Monks, D. Zaharevitz, and J. A. Moscow.** 2002. Correlation of nucleoside and nucleobase transporter gene expression with antimetabolite drug cytotoxicity. *J. Exp. Ther. Oncol.* **2:**200–212.
203. **Luduena, R. F.** 1998. Multiple forms of tubulin: different gene products and covalent modifications. *Int. Rev. Cytol.* **178:**207–275.
204. **Lyttle, M. H., A. Satyam, M. D. Hocker, K. E. Bauer, C. G. Caldwell, H. C. Hui, A. S. Morgan, A. Mergia, and L. M. Kauvar.** 1994. Glutathione-S-transferase activates novel alkylating agents. *J. Med. Chem.* **37:**1501–1507.
205. **Mackey, J. R., L. L. Jennings, M. L. Clarke, C. L. Santos, L. Dabbagh, M. Vsianska, S. L. Koski, R. W. Coupland, S. A. Baldwin, J. D. Young, and C. E. Cass.** 2002. Immunohistochemical variation of human equilibrative nucleoside transporter 1 protein in primary breast cancers. *Clin. Cancer Res.* **8:**110–116.
206. **Mackey, J. R., R. S. Mani, M. Selner, D. Mowles, J. D. Young, J. A. Belt, C. R. Crawford, and C. E. Cass.** 1998. Functional nucleoside transporters are required for gemcitabine influx and manifestation of toxicity in cancer cell lines. *Cancer Res.* **58:**4349–4357.
207. **Mahdavian, E., H. T. Spencer, and R. B. Dunlap.** 1999. Kinetic studies on drug-resistant variants of Escherichia coli thymidylate synthase: functional effects of amino acid substitutions at residue 4. *Arch. Biochem. Biophys.* **368:**257–264.
208. **Mahon, F. X., M. W. Deininger, B. Schultheis, J. Chabrol, J. Reiffers, J. M. Goldman, and J. V. Melo.** 2000. Selection and characterization of BCR-ABL positive cell lines with differential sensitivity to the tyrosine kinase inhibitor STI571: diverse mechanisms of resistance. *Blood* **96:**1070–1079.
209. **Major, P. P., E. M. Egan, G. P. Beardsley, M. D. Minden, and D. W. Kufe.** 1981. Lethality of human myeloblasts correlates with the incorporation of arabinofuranosylcytosine into DNA. *Proc. Natl. Acad. Sci. USA* **78:**3235–3239.
210. **Mapara, M. Y., and M. Sykes.** 2004. Tolerance and cancer: mechanisms of tumor evasion and strategies for breaking tolerance. *J. Clin. Oncol.* **22:**1136–1151.
211. **Martello, L. A., H. M. McDaid, D. L. Regl, C. P. Yang, D. Meng, T. R. Pettus, M. D. Kaufman, H. Arimoto, S. J. Danishefsky, A. B. Smith, 3rd, and S. B. Horwitz.** 2000. Taxol and discodermolide represent a synergistic drug combination in human carcinoma cell lines. *Clin. Cancer Res.* **6:**1978–1987.
212. **Matherly, L. H., D. W. Fry, and I. D. Goldman.** 1983. Role of methotrexate polyglutamylation and cellular energy metabolism in inhibition of methotrexate binding to dihydrofolate reductase by 5-formyltetrahydrofolate in Ehrlich ascites tumor cells in vitro. *Cancer Res.* **43:**2694–2699.
213. **Matsuda, A., and T. Sasaki.** 2004. Antitumor activity of sugar-modified cytosine nucleosides. *Cancer Sci.* **95:**105–111.
214. **Mauritz, R., G. J. Peters, D. G. Priest, Y. G. Assaraf, S. Drori, I. Kathmann, P. Noordhuis, M. A. Bunni, A. Rosowsky, J. H. Schornagel, H. M. Pinedo, and G. Jansen.** 2002. Multiple mechanisms of resistance to methotrexate and novel antifolates in human CCRF-CEM leukemia cells and their implications for folate homeostasis. *Biochem. Pharmacol.* **63:**105–115.
215. **McFadyen, M. C., W. T. Melvin, and G. I. Murray.** 2004. Cytochrome P450 CYP1B1 activity in renal cell carcinoma. *Br. J. Cancer* **91:**966–971.
216. **McGuire, J. J.** 2003. Anticancer antifolates: current status and future directions. *Curr. Pharm. Des.* **9:**2593–2613.
217. **McIvor, R. S., and C. C. Simonsen.** 1990. Isolation and characterization of a variant dihydrofolate reductase cDNA from methotrexate-resistant murine L5178Y cells. *Nucleic Acids Res.* **18:**7025–7032.

218. **McLeod, H. L., and W. N. Keith.** 1996. Variation in topoisomerase I gene copy number as a mechanism for intrinsic drug sensitivity. *Br. J. Cancer* **74:**508–512.
219. **Melera, P. W., J. P. Davide, C. A. Hession, and K. W. Scotto.** 1984. Phenotypic expression in Escherichia coli and nucleotide sequence of two Chinese hamster lung cell cDNAs encoding different dihydrofolate reductases. *Mol. Cell. Biol.* **4:**38–48.
220. **Michieli, M., D. Damiani, A. Ermacora, P. Masolini, D. Raspadori, G. Visani, R. J. Scheper, and M. Baccarani.** 1999. P-glycoprotein, lung resistance-related protein and multidrug resistance associated protein in de novo acute non-lymphocytic leukaemias: biological and clinical implications. *Br. J. Haematol.* **104:**328–335.
221. **Mini, E., S. Srimatkandada, W. D. Medina, B. A. Moroson, M. D. Carman, and J. R. Bertino.** 1985. Molecular and karyological analysis of methotrexate-resistant and -sensitive human leukemic CCRF-CEM cells. *Cancer Res.* **45:**317–324.
222. **Miura, S., and S. Izuta.** 2004. DNA polymerases as targets of anticancer nucleosides. *Curr. Drug Targets* **5:**191–195.
223. **Miyashita, T., and J. C. Reed.** 1992. bcl-2 gene transfer increases relative resistance of S49.1 and WEHI7.2 lymphoid cells to cell death and DNA fragmentation induced by glucocorticoids and multiple chemotherapeutic drugs. *Cancer Res.* **52:**5407–5411.
224. **Mo, Y. Y., Y. Yu, Z. Shen, and W. T. Beck.** 2002. Nucleolar delocalization of human topoisomerase I in response to topotecan correlates with sumoylation of the protein. *J. Biol. Chem.* **277:**2958–2964.
225. **Modrak, D. E., M. D. Rodriguez, D. M. Goldenberg, W. Lew, and R. D. Blumenthal.** 2002. Sphingomyelin enhances chemotherapy efficacy and increases apoptosis in human colonic tumor xenografts. *Int. J. Oncol.* **20:**379–384.
226. **Momparler, R. L., and G. A. Fischer.** 1968. Mammalian deoxynucleoside kinase. I. Deoxycytidine kinase: purification, properties, and kinetic studies with cytosine arabinoside. *J. Biol. Chem.* **243:**4298–4304.
227. **Moran, R. G., W. C. Werkheiser, and S. F. Zakrzewski.** 1976. Folate metabolism in mammalian cells in culture. I Partial characterization of the folate derivatives present in L1210 mouse leukemia cells. *J. Biol. Chem.* **251:**3569–3575.
228. **Morel, F., M. J. Bris, A. Herry, G. L. Calvez, V. Marion, J. F. Abgrall, C. Berthou, and M. D. Braekeleer.** 2003. Double minutes containing amplified bcr-abl fusion gene in a case of chronic myeloid leukemia treated by imatinib. *Eur. J. Haematol.* **70:**235–239.
229. **Moscow, J. A.** 1998. Methotrexate transport and resistance. *Leuk. Lymphoma* **30:**215–224.
230. **Muller, M., C. Meijer, G. J. Zaman, P. Borst, R. J. Scheper, N. H. Mulder, E. G. de Vries, and P. L. Jansen.** 1994. Overexpression of the gene encoding the multidrug resistance-associated protein results in increased ATP-dependent glutathione S-conjugate transport. *Proc. Natl. Acad. Sci. USA* **91:**13033–13037.
231. **Murray, G. I., M. C. McFadyen, R. T. Mitchell, Y. L. Cheung, A. C. Kerr, and W. T. Melvin.** 1999. Cytochrome P450 CYP3A in human renal cell cancer. *Br. J. Cancer* **79:**1836–1842.
232. **Nabhan, C., R. B. Gartenhaus, and M. S. Tallman.** 2004. Purine nucleoside analogues and combination therapies in B-cell chronic lymphocytic leukemia: dawn of a new era. *Leuk. Res.* **28:**429–442.
233. **Nagar, B., W. G. Bornmann, P. Pellicena, T. Schindler, D. R. Veach, W. T. Miller, B. Clarkson, and J. Kuriyan.** 2002. Crystal structures of the kinase domain of c-Abl in complex with the small molecule inhibitors PD173955 and imatinib (STI-571). *Cancer Res.* **62:**4236–4243.
234. **Nooter, K., A. M. Westerman, M. J. Flens, G. J. Zaman, R. J. Scheper, K. E. van Wingerden, H. Burger, R. Oostrum, T. Boersma, and P. Sonneveld.** 1995. Expression of the multidrug resistance-associated protein (MRP) gene in human cancers. *Clin. Cancer Res.* **1:**1301–1310.
235. **O'Connor, B. M., A. L. Jackman, P. H. Crossley, S. E. Freemantle, J. Lunec, and A. H. Calvert.** 1992. Human lymphoblastoid cells with acquired resistance to C2-desamino-C2-methyl-N10-propargyl-5,8-dideazafolic acid: a novel folate-based thymidylate synthase inhibitor. *Cancer Res.* **52:**1137–1143.
236. **O'Dwyer, P. J., F. LaCreta, S. Nash, P. W. Tinsley, R. Schilder, M. L. Clapper, K. D. Tew, L. Panting, S. Litwin, and R. L. Comis.** 1991. Phase I study of thiotepa in combination with the glutathione transferase inhibitor ethacrynic acid. *Cancer Res.* **51:**6059–6065.
237. **Ohashi, N., Y. Fujiwara, N. Yamaoka, O. Katoh, Y. Satow, and M. Yamakido.** 1996. No alteration in DNA topoisomerase I gene related to CPT-11 resistance in human lung cancer. *Jpn. J. Cancer Res.* **87:**1280–1287.
238. **Orr, G. A., P. Verdier-Pinard, H. McDaid, and S. B. Horwitz.** 2003. Mechanisms of Taxol resistance related to microtubules. *Oncogene* **22:**7280–7295.
239. **Owens, J. K., D. S. Shewach, B. Ullman, and B. S. Mitchell.** 1992. Resistance to 1-beta-D-arabinofuranosylcytosine in human T-lymphoblasts mediated by mutations within the deoxycytidine kinase gene. *Cancer Res.* **52:**2389–2393.
240. **Ozols, R. F., P. J. O'Dwyer, and T. C. Hamilton.** 1993. Clinical reversal of drug resistance in ovarian cancer. *Gynecol. Oncol.* **51:**90–96.
241. **Pakos, E. E., and J. P. Ioannidis.** 2003. The association of P-glycoprotein with response to chemotherapy and clinical outcome in patients with osteosarcoma. A meta-analysis. *Cancer* **98:**581–589.
242. **Petrini, M., A. Conte, F. Caracciolo, A. Sabbatini, B. Grassi, and G. Ronca.** 1993. Reversing of chlorambucil resistance by ethacrynic acid in a B-CLL patient. *Br. J. Haematol.* **85:**409–410.
243. **Picciano, M. F., E. L. R. Stokstad, and J. F. Gregory.** 1990. *Folic Acid Metabolism in Health and Disease*, vol. 13. Wiley-Liss, New York, N.Y.
244. **Pommier, Y., D. Kerrigan, K. D. Hartman, and R. I. Glazer.** 1990. Phosphorylation of mammalian DNA topoisomerase I and activation by protein kinase C. *J. Biol. Chem.* **265:**9418–9422.
245. **Pond, C. D., X. G. Li, E. H. Rubin, and L. R. Barrows.** 1999. Effects of mutations in the F361 to R364 region of topoisomerase I (Topo I), in the presence and absence of 9-aminocamptothecin, on the Topo I-DNA interaction. *Anticancer Drugs* **10:**647–653.
246. **Prost, S.** 1995. Mechanisms of resistance to topoisomerases poisons. *Gen. Pharmacol.* **26:**1673–1684.
247. **Prost, S., and G. Riou.** 1994. A human small cell lung carcinoma cell line, resistant to 4′-(9-acridinylamino)-methanesulfon-m-anisidide and cross-resistant to camptothecin with a high level of topoisomerase I. *Biochem. Pharmacol.* **48:**975–984.
248. **Pullarkat, S. T., J. Stoehlmacher, V. Ghaderi, Y. P. Xiong, S. A. Ingles, A. Sherrod, R. Warren, D. Tsao-Wei, S. Groshen, and H. J. Lenz.** 2001. Thymidylate synthase gene polymorphism determines response and toxicity of 5-FU chemotherapy. *Pharmacogenomics J.* **1:**65–70.
249. **Qian, Y. M., W. C. Song, H. Cui, S. P. Cole, and R. G. Deeley.** 2001. Glutathione stimulates sulfated estrogen transport by multidrug resistance protein 1. *J. Biol. Chem.* **276:**6404–6411.
250. **Ramachandra, M., S. V. Ambudkar, D. Chen, C. A. Hrycyna, S. Dey, M. M. Gottesman, and I. Pastan.** 1998. Human

P-glycoprotein exhibits reduced affinity for substrates during a catalytic transition state. *Biochemistry* **37**:5010–5019.

251. **Raviv, Y., H. B. Pollard, E. P. Bruggemann, I. Pastan, and M. M. Gottesman.** 1990. Photosensitized labeling of a functional multidrug transporter in living drug-resistant tumor cells. *J. Biol. Chem.* **265**:3975–3980.
252. **Redinbo, M. R., L. Stewart, P. Kuhn, J. J. Champoux, and W. G. Hol.** 1998. Crystal structures of human topoisomerase I in covalent and noncovalent complexes with DNA. *Science* **279**:1504–1513.
253. **Reed, J. C.** 1995. Regulation of apoptosis by bcl-2 family proteins and its role in cancer and chemoresistance. *Curr. Opin. Oncol.* **7**:541–546.
254. **Reilly, R. T., A. M. Forsthoefel, and F. G. Berger.** 1997. Functional effects of amino acid substitutions at residue 33 of human thymidylate synthase. *Arch. Biochem. Biophys.* **342**:338–343.
255. **Reynolds, C. P., B. J. Maurer, and R. N. Kolesnick.** 2004. Ceramide synthesis and metabolism as a target for cancer therapy. *Cancer Lett.* **206**:169–180.
256. **Ringpfeil, F., M. G. Lebwohl, A. M. Christiano, and J. Uitto.** 2000. Pseudoxanthoma elasticum: mutations in the MRP6 gene encoding a transmembrane ATP-binding cassette (ABC) transporter. *Proc. Natl. Acad. Sci. USA* **97**:6001–6006.
257. **Ritke, M. K., N. R. Murray, W. P. Allan, A. P. Fields, and J. C. Yalowich.** 1995. Hypophosphorylation of topoisomerase II in etoposide (VP-16)-resistant human leukemia K562 cells associated with reduced levels of beta II protein kinase C. *Mol. Pharmacol.* **48**:798–805.
258. **Riva, C. M., and Y. M. Rustum.** 1985. 1-beta-D-arabinofuranosylcytosine metabolism and incorporation into DNA as determinants of in vivo murine tumor cell response. *Cancer Res.* **45**:6244–6249.
259. **Robert, J., and A. K. Larsen.** 1998. Drug resistance to topoisomerase II inhibitors. *Biochimie* **80**:247–254.
260. **Rochat, B., J. M. Morsman, G. I. Murray, W. D. Figg, and H. L. McLeod.** 2001. Human CYP1B1 and anticancer agent metabolism: mechanism for tumor-specific drug inactivation? *J. Pharmacol. Exp. Ther.* **296**:537–541.
261. **Rosenberg, M. F., R. Callaghan, R. C. Ford, and C. F. Higgins.** 1997. Structure of the multidrug resistance P-glycoprotein to 2.5 nm resolution determined by electron microscopy and image analysis. *J. Biol. Chem.* **272**:10685–10694.
262. **Rosenberg, M. F., Q. Mao, A. Holzenburg, R. C. Ford, R. G. Deeley, and S. P. Cole.** 2001. The structure of the multidrug resistance protein 1 (MRP1/ABCC1). crystallization and single-particle analysis. *J. Biol. Chem.* **276**:16076–16082.
263. **Rosenblatt, D. S., V. M. Whitehead, M. M. Dupont, M. J. Vuchich, and N. Vera.** 1978. Synthesis of methotrexate polyglutamates in cultured human cells. *Mol. Pharmacol.* **14**:210–214.
264. **Rosenblatt, D. S., V. M. Whitehead, N. Vera, A. Pottier, M. Dupont, and M. J. Vuchich.** 1978. Prolonged inhibition of DNA synthesis associated with the accumulation of methotrexate polyglutamates by cultured human cells. *Mol. Pharmacol.* **14**:1143–1147.
265. **Rothem, L., A. Aronheim, and Y. G. Assaraf.** 2003. Alterations in the expression of transcription factors and the reduced folate carrier as a novel mechanism of antifolate resistance in human leukemia cells. *J. Biol. Chem.* **278**:8935–8941.
266. **Rothem, L., M. Stark, Y. Kaufman, L. Mayo, and Y. G. Assaraf.** 2004. Reduced folate carrier gene silencing in multiple antifolate-resistant tumor cell lines is due to a simultaneous loss of function of multiple transcription factors but not promoter methylation. *J. Biol. Chem.* **279**:374–384.
267. **Rothenberg, M. L.** 2001. Irinotecan (CPT-11): recent developments and future directions--colorectal cancer and beyond. *Oncologist* **6**:66–80.
268. **Ruiz van Haperen, V. W., G. Veerman, S. Eriksson, E. Boven, A. P. Stegmann, M. Hermsen, J. B. Vermorken, H. M. Pinedo, and G. J. Peters.** 1994. Development and molecular characterization of a 2′,2′-difluorodeoxycytidine-resistant variant of the human ovarian carcinoma cell line A2780. *Cancer Res.* **54**:4138–4143.
269. **Ryu, S., T. Kawabe, S. Nada, and A. Yamaguchi.** 2000. Identification of basic residues involved in drug export function of human multidrug resistance-associated protein 2. *J. Biol. Chem.* **275**:39617–39624.
270. **Sakamoto, H., H. Hara, K. Hirano, and T. Adachi.** 1999. Enhancement of glucuronosyl etoposide transport by glutathione in multidrug resistance-associated protein-overexpressing cells. *Cancer Lett.* **135**:113–119.
271. **Saltz, L. B., J. V. Cox, C. Blanke, L. S. Rosen, L. Fehrenbacher, M. J. Moore, J. A. Maroun, S. P. Ackland, P. K. Locker, N. Pirotta, G. L. Elfring, and L. L. Miller.** 2000. Irinotecan plus fluorouracil and leucovorin for metastatic colorectal cancer. Irinotecan Study Group. *N. Engl. J. Med.* **343**:905–914.
272. **Sandler, A., M. Gordon, D. P. De Alwis, I. Pouliquen, L. Green, P. Marder, A. Chaudhary, K. Fife, L. Battiato, C. Sweeney, C. Jordan, M. Burgess, and C. A. Slapak.** 2004. A Phase I trial of a potent P-glycoprotein inhibitor, zosuquidar trihydrochloride (LY335979), administered intravenously in combination with doxorubicin in patients with advanced malignancy. *Clin. Cancer Res.* **10**:3265–3272.
273. **Santi, D. V., C. S. McHenry, and H. Sommer.** 1974. Mechanism of interaction of thymidylate synthetase with 5-fluorodeoxyuridylate. *Biochemistry* **13**:471–481.
274. **Sauerbrey, A., J. P. McPherson, S. C. Zhao, D. Banerjee, and J. R. Bertino.** 1999. Expression of a novel double-mutant dihydrofolate reductase-cytidine deaminase fusion gene confers resistance to both methotrexate and cytosine arabinoside. *Hum. Gene Ther.* **10**:2495–2504.
275. **Sauna, Z. E., and S. V. Ambudkar.** 2000. Evidence for a requirement for ATP hydrolysis at two distinct steps during a single turnover of the catalytic cycle of human P-glycoprotein. *Proc. Natl. Acad. Sci. USA* **97**:2515–2520.
276. **Sauna, Z. E., M. M. Smith, M. Muller, K. M. Kerr, and S. V. Ambudkar.** 2001. The mechanism of action of multidrug-resistance-linked P-glycoprotein. *J. Bioenerg. Biomembr.* **33**:481–491.
277. **Schilsky, R. L., B. D. Bailey, and B. A. Chabner.** 1980. Methotrexate polyglutamate synthesis by cultured human breast cancer cells. *Proc. Natl. Acad. Sci. USA* **77**:2919–2922.
278. **Schindler, T., W. Bornmann, P. Pellicena, W. T. Miller, B. Clarkson, and J. Kuriyan.** 2000. Structural mechanism for STI-571 inhibition of abelson tyrosine kinase. *Science* **289**:1938–1942.
279. **Schinkel, A. H., R. J. Arceci, J. J. Smit, E. Wagenaar, F. Baas, M. Dolle, T. Tsuruo, E. B. Mechetner, I. B. Roninson, and P. Borst.** 1993. Binding properties of monoclonal antibodies recognizing external epitopes of the human MDR1 P-glycoprotein. *Int. J. Cancer* **55**:478–484.
280. **Schinkel, A. H., E. Wagenaar, C. A. Mol, and L. van Deemter.** 1996. P-glycoprotein in the blood-brain barrier of mice influences the brain penetration and pharmacological activity of many drugs. *J. Clin. Invest.* **97**:2517–2524.
281. **Schirch, V., and W. B. Strong.** 1989. Interaction of folylpolyglutamates with enzymes in one-carbon metabolism. *Arch. Biochem. Biophys.* **269**:371–380.

282. **Schroder, J. K., C. Kirch, S. Seeber, and J. Schutte.** 1998. Structural and functional analysis of the cytidine deaminase gene in patients with acute myeloid leukaemia. *Br. J. Haematol.* **103:**1096–1103.
283. **Schuetz, E. G., W. T. Beck, and J. D. Schuetz.** 1996. Modulators and substrates of P-glycoprotein and cytochrome P4503A coordinately up-regulate these proteins in human colon carcinoma cells. *Mol. Pharmacol.* **49:**311–318.
284. **Schuetz, J. D., M. C. Connelly, D. Sun, S. G. Paibir, P. M. Flynn, R. V. Srinivas, A. Kumar, and A. Fridland.** 1999. MRP4: A previously unidentified factor in resistance to nucleoside-based antiviral drugs. *Nat. Med.* **5:**1048–1051.
285. **Schultz, R. M., V. J. Chen, J. R. Bewley, E. F. Roberts, C. Shih, and J. A. Dempsey.** 1999. Biological activity of the multitargeted antifolate, MTA (LY231514), in human cell lines with different resistance mechanisms to antifolate drugs. *Semin. Oncol.* **26:**68–73.
286. **Schwartz, G. K., D. Ward, L. Saltz, E. S. Casper, T. Spiess, E. Mullen, J. Woodworth, R. Venuti, P. Zervos, A. M. Storniolo, and D. P. Kelsen.** 1997. A pilot clinical/pharmacological study of the protein kinase C-specific inhibitor safingol alone and in combination with doxorubicin. *Clin. Cancer Res.* **3:**537–543.
287. **Scotto, K. W.** 2003. Transcriptional regulation of ABC drug transporters. *Oncogene* **22:**7496–7511.
288. **Seelig, A.** 1998. A general pattern for substrate recognition by P-glycoprotein. *Eur. J. Biochem.* **251:**252–261.
289. **Seelig, A.** 1998. How does P-glycoprotein recognize its substrates? *Int. J. Clin. Pharmacol. Ther.* **36:**50–54.
290. **Seelig, A., and E. Landwojtowicz.** 2000. Structure-activity relationship of P-glycoprotein substrates and modifiers. *Eur. J. Pharm. Sci.* **12:**31–40.
291. **Seidman, A. D., H. I. Scher, D. Petrylak, D. D. Dershaw, and T. Curley.** 1992. Estramustine and vinblastine: use of prostate specific antigen as a clinical trial end point for hormone refractory prostatic cancer. *J. Urol.* **147:**931–934.
292. **Seither, R. L., D. F. Trent, D. C. Mikulecky, T. J. Rape, and I. D. Goldman.** 1989. Folate-pool interconversions and inhibition of biosynthetic processes after exposure of L1210 leukemia cells to antifolates. Experimental and network thermodynamic analyses of the role of dihydrofolate polyglutamylates in antifolate action in cells. *J. Biol. Chem.* **264:**17016–17023.
293. **Senderowicz, A. M.** 2003. Small-molecule cyclin-dependent kinase modulators. *Oncogene* **22:**6609–6620.
294. **Senior, A. E., M. K. al-Shawi, and I. L. Urbatsch.** 1995. The catalytic cycle of P-glycoprotein. *FEBS Lett.* **377:**285–289.
295. **Senior, A. E., and S. Bhagat.** 1998. P-glycoprotein shows strong catalytic cooperativity between the two nucleotide sites. *Biochemistry* **37:**831–836.
296. **Serra, M., G. Reverter-Branchat, D. Maurici, S. Benini, J. N. Shen, T. Chano, C. M. Hattinger, M. C. Manara, M. Pasello, K. Scotlandi, and P. Picci.** 2004. Analysis of dihydrofolate reductase and reduced folate carrier gene status in relation to methotrexate resistance in osteosarcoma cells. *Ann. Oncol.* **15:**151–160.
297. **Shabbits, J. A., Y. Hu, and L. D. Mayer.** 2003. Tumor chemosensitization strategies based on apoptosis manipulations. *Mol. Cancer Ther.* **2:**805–813.
298. **Shah, N. P., J. M. Nicoll, B. Nagar, M. E. Gorre, R. L. Paquette, J. Kuriyan, and C. L. Sawyers.** 2002. Multiple BCR-ABL kinase domain mutations confer polyclonal resistance to the tyrosine kinase inhibitor imatinib (STI571) in chronic phase and blast crisis chronic myeloid leukemia. *Cancer Cell* **2:**117–125.
299. **Sharp, D. J., G. C. Rogers, and J. M. Scholey.** 2000. Microtubule motors in mitosis. *Nature* **407:**41–47.
300. **Simonsen, C. C., and A. D. Levinson.** 1983. Isolation and expression of an altered mouse dihydrofolate reductase cDNA. *Proc. Natl. Acad. Sci. USA* **80:**2495–2499.
301. **Sirulink, A., R. T. Silver, and V. Najfeld.** 2001. Marked ploidy and BCR-ABL gene amplification in vivo in a patient treated with STI571. *Leukemia* **15:**1795–1797.
302. **Slapak, C. A., J. C. Daniel, and S. B. Levy.** 1990. Sequential emergence of distinct resistance phenotypes in murine erythroleukemia cells under adriamycin selection: decreased anthracycline uptake precedes increased P-glycoprotein expression. *Cancer Res.* **50:**7895–7901.
303. **Smit, J. J., A. H. Schinkel, R. P. Oude Elferink, A. K. Groen, E. Wagenaar, L. van Deemter, C. A. Mol, R. Ottenhoff, N. M. van der Lugt, and M. A. van Roon.** 1993. Homozygous disruption of the murine mdr2 P-glycoprotein gene leads to a complete absence of phospholipid from bile and to liver disease. *Cell* **75:**451–462.
304. **Smith, A. J., U. Mayer, A. H. Schinkel, and P. Borst.** 1998. Availability of PSC833, a substrate and inhibitor of P-glycoproteins, in various concentrations of serum. *J. Natl. Cancer Inst.* **90:**1161–1166.
305. **Sohn, K. J., F. Smirnakis, D. N. Moskovitz, P. Novakovic, Z. Yates, M. Lucock, R. Croxford, and Y. I. Kim.** 2004. Effects of folylpolyglutamate synthetase modulation on chemosensitivity of colon cancer cells to 5-fluorouracil and methotrexate. *Gut* **53:**1825–1831.
306. **Sorensen, M., M. Sehested, and P. B. Jensen.** 1995. Characterisation of a human small-cell lung cancer cell line resistant to the DNA topoisomerase I-directed drug topotecan. *Br. J. Cancer* **72:**399–404.
307. **Sorrentino, B. P., J. A. Allay, and R. L. Blakley.** 1999. In vivo selection of hematopoietic stem cells transduced with DHFR-expressing retroviral vectors. *Prog. Exp. Tumor Res.* **36:**143–161.
308. **Spencer, H. T., B. P. Sorrentino, C. H. Pui, S. K. Chunduru, S. E. Sleep, and R. L. Blakley.** 1996. Mutations in the gene for human dihydrofolate reductase: an unlikely cause of clinical relapse in pediatric leukemia after therapy with methotrexate. *Leukemia* **10:**439–446.
309. **Srimatkandada, S., B. I. Schweitzer, B. A. Moroson, S. Dube, and J. R. Bertino.** 1989. Amplification of a polymorphic dihydrofolate reductase gene expressing an enzyme with decreased binding to methotrexate in a human colon carcinoma cell line, HCT-8R4, resistant to this drug. *J. Biol. Chem.* **264:**3524–3528.
310. **Staker, B. L., K. Hjerrild, M. D. Feese, C. A. Behnke, A. B. Burgin, Jr., and L. Stewart.** 2002. The mechanism of topoisomerase I poisoning by a camptothecin analog. *Proc. Natl. Acad. Sci. USA* **99:**15387–15392.
311. **Staron, K., B. Kowalska-Loth, and I. Szumiel.** 1996. Lowered phosphorylation of topoisomerase I is a direct reason for reduced sensitivity of L5178Y-S cells to camptothecin. *Ann. N. Y. Acad. Sci.* **803:**321–323.
312. **Stewart, L., M. R. Redinbo, X. Qiu, W. G. Hol, and J. J. Champoux.** 1998. A model for the mechanism of human topoisomerase I. *Science* **279:**1534–1541.
313. **Strasser, A., A. W. Harris, and S. Cory.** 1993. E mu-bcl-2 transgene facilitates spontaneous transformation of early pre-B and immunoglobulin-secreting cells but not T cells. *Oncogene* **8:**1–9.
314. **Struk, B., L. Cai, S. Zach, W. Ji, J. Chung, A. Lumsden, M. Stumm, M. Huber, L. Schaen, C. A. Kim, L. A. Goldsmith, D. Viljoen, L. E. Figuera, W. Fuchs, F. Munier, R. Ramesar,

D. Hohl, R. Richards, K. H. Neldner, and K. Lindpaintner. 2000. Mutations of the gene encoding the transmembrane transporter protein ABC-C6 cause pseudoxanthoma elasticum. *J. Mol. Med.* **78:**282–286.

315. **Sugimoto, Y., S. Tsukahara, T. Oh-hara, T. Isoe, and T. Tsuruo.** 1990. Decreased expression of DNA topoisomerase I in camptothecin-resistant tumor cell lines as determined by a monoclonal antibody. *Cancer Res.* **50:**6925–6930.
316. **Synold, T. W., M. V. Relling, J. M. Boyett, G. K. Rivera, J. T. Sandlund, H. Mahmoud, W. M. Crist, C. H. Pui, and W. E. Evans.** 1994. Blast cell methotrexate-polyglutamate accumulation in vivo differs by lineage, ploidy, and methotrexate dose in acute lymphoblastic leukemia. *J. Clin. Invest.* **94:**1996–2001.
317. **Takahashi, T., Y. Fujiwara, M. Yamakido, O. Katoh, H. Watanabe, and P. I. Mackenzie.** 1997. The role of glucuronidation in 7-ethyl-10-hydroxycamptothecin resistance in vitro. *Jpn. J. Cancer Res.* **88:**1211–1217.
318. **Takatani, H., M. Oka, M. Fukuda, F. Narasaki, R. Nakano, K. Ikeda, K. Terashi, A. Kinoshita, H. Soda, T. Kanda, E. Schneider, and S. Kohno.** 1997. Gene mutation analysis and quantitation of DNA topoisomerase I in previously untreated non-small cell lung carcinomas. *Jpn. J. Cancer Res.* **88:**160–165.
319. **Takebe, N., S. Nakahara, S. C. Zhao, D. Adhikari, A. U. Ural, M. Iwamoto, D. Banerjee, and J. R. Bertino.** 2000. Comparison of methotrexate resistance conferred by a mutated dihydrofolate reductase (DHFR) cDNA in two different retroviral vectors. *Cancer Gene Ther.* **7:**910–919.
320. **Takebe, N., S. C. Zhao, D. Adhikari, S. Mineishi, M. Sadelain, J. Hilton, M. Colvin, D. Banerjee, and J. R. Bertino.** 2001. Generation of dual resistance to 4-hydroperoxycyclophosphamide and methotrexate by retroviral transfer of the human aldehyde dehydrogenase class 1 gene and a mutated dihydrofolate reductase gene. *Mol. Ther.* **3:**88–96.
321. **Tan, K. B., M. R. Mattern, W. K. Eng, F. L. McCabe, and R. K. Johnson.** 1989. Nonproductive rearrangement of DNA topoisomerase I and II genes: correlation with resistance to topoisomerase inhibitors. *J. Natl. Cancer Inst.* **81:**1732–1735.
322. **Tauchi, T., and K. Ohyashiki.** 2004. Molecular mechanisms of resistance of leukemia to imatinib mesylate. *Leuk. Res.* **28** (Suppl 1)**:**39–45.
323. **Tew, K. D.** 1994. Glutathione-associated enzymes in anticancer drug resistance. *Cancer Res.* **54:**4313–4320.
324. **Tong, Y., X. Liu-Chen, E. A. Ercikan-Abali, G. M. Capiaux, S. C. Zhao, D. Banerjee, and J. R. Bertino.** 1998. Isolation and characterization of thymitaq (AG337) and 5-fluoro-2-deoxyuridylate-resistant mutants of human thymidylate synthase from ethyl methanesulfonate-exposed human sarcoma HT1080 cells. *J. Biol. Chem.* **273:**11611–11618.
325. **Tong, Y., X. Liu-Chen, E. A. Ercikan-Abali, S. C. Zhao, D. Banerjee, F. Maley, and J. R. Bertino.** 1998. Probing the folate-binding site of human thymidylate synthase by site-directed mutagenesis. Generation of mutants that confer resistance to raltitrexed, Thymitaq, and BW1843U89. *J. Biol. Chem.* **273:**31209–31214.
326. **Torrisi, R., A. Decensi, F. Formelli, T. Camerini, and G. De Palo.** 2001. Chemoprevention of breast cancer with fenretinide. *Drugs* **61:**909–918.
327. **Townsend, D. M., and K. D. Tew.** 2003. The role of glutathione-S-transferase in anti-cancer drug resistance. *Oncogene* **22:**7369–7375.
328. **Trent, J. M., R. N. Buick, S. Olson, R. C. Horns, Jr., and R. T. Schimke.** 1984. Cytologic evidence for gene amplification in methotrexate-resistant cells obtained from a patient with ovarian adenocarcinoma. *J. Clin. Oncol.* **2:**8–15.
329. **Tsurutani, J., T. Nitta, T. Hirashima, T. Komiya, H. Uejima, H. Tada, N. Syunichi, A. Tohda, M. Fukuoka, and K. Nakagawa.** 2002. Point mutations in the topoisomerase I gene in patients with non-small cell lung cancer treated with irinotecan. *Lung Cancer* **35:**299–304.
330. **Turley, H., M. Comley, S. Houlbrook, N. Nozaki, A. Kikuchi, I. D. Hickson, K. Gatter, and A. L. Harris.** 1997. The distribution and expression of the two isoforms of DNA topoisomerase II in normal and neoplastic human tissues. *Br. J. Cancer* **75:**1340–1346.
331. **Ueda, K., Y. Taguchi, and M. Morishima.** 1997. How does P-glycoprotein recognize its substrates? *Semin. Cancer Biol.* **8:**151–159.
332. **van der Kolk, D. M., E. G. de Vries, W. J. van Putten, L. F. Verdonck, G. J. Ossenkoppele, G. E. Verhoef, and E. Vellenga.** 2000. P-glycoprotein and multidrug resistance protein activities in relation to treatment outcome in acute myeloid leukemia. *Clin. Cancer Res.* **6:**3205–3214.
333. **Vassetzky, Y. S., G. C. Alghisi, and S. M. Gasser.** 1995. DNA topoisomerase II mutations and resistance to anti-tumor drugs. *Bioessays* **17:**767–774.
334. **Verdier-Pinard, P., F. Wang, B. Burd, R. H. Angeletti, S. B. Horwitz, and G. A. Orr.** 2003. Direct analysis of tubulin expression in cancer cell lines by electrospray ionization mass spectrometry. *Biochemistry* **42:**12019–12027.
335. **Verdier-Pinard, P., F. Wang, L. Martello, B. Burd, G. A. Orr, and S. B. Horwitz.** 2003. Analysis of tubulin isotypes and mutations from taxol-resistant cells by combined isoelectrofocusing and mass spectrometry. *Biochemistry* **42:**5349–5357.
336. **Vijayalakshmi, D., and J. A. Belt.** 1988. Sodium-dependent nucleoside transport in mouse intestinal epithelial cells. Two transport systems with differing substrate specificities. *J. Biol. Chem.* **263:**19419–19423.
337. **Vulevic, B., Z. Chen, J. T. Boyd, W. Davis, Jr., E. S. Walsh, M. G. Belinsky, and K. D. Tew.** 2001. Cloning and characterization of human adenosine 5′-triphosphate-binding cassette, sub-family A, transporter 2 (ABCA2). *Cancer Res.* **61:**3339–3347.
338. **Wang, J. C.** 2002. Cellular roles of DNA topoisomerases: a molecular perspective. *Nat. Rev. Mol. Cell. Biol.* **3:**430–440.
339. **Wang, J. C.** 1996. DNA topoisomerases. *Annu. Rev. Biochem.* **65:**635–692.
340. **Wang, S., and W. S. El-Deiry.** 2003. TRAIL and apoptosis induction by TNF-family death receptors. *Oncogene* **22:**8628–8633.
341. **Wang, Y., R. Zhao, and I. D. Goldman.** 2003. Decreased expression of the reduced folate carrier and folypolyglutamate synthetase is the basis for acquired resistance to the pemetrexed antifolate (LY231514) in an L1210 murine leukemia cell line. *Biochem. Pharmacol.* **65:**1163–1170.
342. **Webber, S., C. A. Bartlett, T. J. Boritzki, J. A. Hillard, E. F. Howland, A. L. Johnston, M. Kosa, S. A. Margosiak, C. A. Morse, and B. V. Shetty.** 1996. AG337, a novel lipophilic thymidylate synthase inhibitor: in vitro and in vivo preclinical studies. *Cancer Chemother. Pharmacol.* **37:**509–517.
343. **Weisberg, E., and J. D. Griffin.** 2000. Mechanism of resistance to the ABL tyrosine kinase inhibitor STI571 in BCR/ABL-transformed hematopoietic cell lines. *Blood* **95:**3498–3505.
344. **White, J. C., and I. D. Goldman.** 1976. Mechanism of action of methotrexate. IV. Free intracellular methotrexate required to suppress dihydrofolate reduction to tetrahydrofolate by Ehrlich ascites tumor cells in vitro. *Mol. Pharmacol.* **12:**711–719.
345. **White, J. C., J. P. Rathmell, and R. L. Capizzi.** 1987. Membrane transport influences the rate of accumulation of cytosine

arabinoside in human leukemia cells. *J. Clin. Invest.* **79:**380–387.

346. **Whitehead, V. M.** 1977. Synthesis of methotrexate polyglutamates in L1210 murine leukemia cells. *Cancer Res.* **37:**408–412.
347. **Whitehead, V. M., D. S. Rosenblatt, M. J. Vuchich, J. J. Shuster, A. Witte, and D. Beaulieu.** 1990. Accumulation of methotrexate and methotrexate polyglutamates in lymphoblasts at diagnosis of childhood acute lymphoblastic leukemia: a pilot prognostic factor analysis. *Blood* **76:**44–49.
348. **Wijnholds, J., C. A. Mol, L. van Deemter, M. de Haas, G. L. Scheffer, F. Baas, J. H. Beijnen, R. J. Scheper, S. Hatse, E. De Clercq, J. Balzarini, and P. Borst.** 2000. Multidrug-resistance protein 5 is a multispecific organic anion transporter able to transport nucleotide analogs. *Proc. Natl. Acad. Sci. USA* **97:**7476–7481.
349. **Wiley, J. S., M. B. Snook, and G. P. Jamieson.** 1989. Nucleoside transport in acute leukaemia and lymphoma: close relation to proliferative rate. *Br. J. Haematol.* **71:**203–207.
350. **Wolff, J. A.** 1991. History of pediatric oncology. *Pediatr. Hematol. Oncol.* **8:**89–91.
351. **Woolfrey, J. R., and G. S. Weston.** 2002. The use of computational methods in the discovery and design of kinase inhibitors. *Curr. Pharm. Des.* **8:**1527–1545.
352. **Worm, J., A. F. Kirkin, K. N. Dzhandzhugazyan, and P. Guldberg.** 2001. Methylation-dependent silencing of the reduced folate carrier gene in inherently methotrexate-resistant human breast cancer cells. *J. Biol. Chem.* **276:**39990–40000.
353. **Wyllie, A. H.** 1985. The biology of cell death in tumours. *Anticancer Res.* **5:**131–136.
354. **Xie, R., M. Hammarlund-Udenaes, A. G. de Boer, and E. C. de Lange.** 1999. The role of P-glycoprotein in blood-brain barrier transport of morphine: transcortical microdialysis studies in mdr1a (−/−) and mdr1a (+/+) mice. *Br. J. Pharmacol.* **128:**563–568.
355. **Yang, C. H., F. M. Sirotnak, and M. Dembo.** 1984. Interaction between anions and the reduced folate/methotrexate transport system in L1210 cell plasma membrane vesicles: directional symmetry and anion specificity for differential mobility of loaded and unloaded carrier. *J. Membr. Biol.* **79:**285–292.
356. **Yang, R., R. Sowers, B. Mazza, J. H. Healey, A. Huvos, H. Grier, M. Bernstein, G. P. Beardsley, M. D. Krailo, M. Devidas, J. R. Bertino, P. A. Meyers, and R. Gorlick.** 2003. Sequence alterations in the reduced folate carrier are observed in osteosarcoma tumor samples. *Clin. Cancer Res.* **9:**837–844.
357. **Yao, D., S. Ding, B. Burchell, C. R. Wolf, and T. Friedberg.** 2000. Detoxication of vinca alkaloids by human P450 CYP3A4-mediated metabolism: implications for the development of drug resistance. *J. Pharmacol. Exp. Ther.* **294:**387–395.
358. **Yu, M., and P. W. Melera.** 1993. Allelic variation in the dihydrofolate reductase gene at amino acid position 95 contributes to antifolate resistance in Chinese hamster cells. *Cancer Res.* **53:**6031–6035.
359. **Zhang, D. W., S. P. Cole, and R. G. Deeley.** 2001. Identification of a nonconserved amino acid residue in multidrug resistance protein 1 important for determining substrate specificity: evidence for functional interaction between transmembrane helices 14 and 17. *J. Biol. Chem.* **276:**34966–34974.
360. **Zhang, D. W., S. P. Cole, and R. G. Deeley.** 2001. Identification of an amino acid residue in multidrug resistance protein 1 critical for conferring resistance to anthracyclines. *J. Biol. Chem.* **276:**13231–13239.
361. **Zhao, R., and I. D. Goldman.** 2003. Resistance to antifolates. *Oncogene* **22:**7431–7457.
362. **Zhao, R., S. Titus, F. Gao, R. G. Moran, and I. D. Goldman.** 2000. Molecular analysis of murine leukemia cell lines resistant to 5, 10-dideazatetrahydrofolate identifies several amino acids critical to the function of folylpolyglutamate synthetase. *J. Biol. Chem.* **275:**26599–26606.
363. **Zhou, Y., M. M. Gottesman, and I. Pastan.** 1999. Studies of human MDR1-MDR2 chimeras demonstrate the functional exchangeability of a major transmembrane segment of the multidrug transporter and phosphatidylcholine flippase. *Mol. Cell. Biol.* **19:**1450–1459.

Frontiers in Antimicrobial Resistance: a Tribute to Stuart B. Levy
Edited by D. G. White, M. N. Alekshun, and P. F. McDermott

Chapter 37

Development of Resistance to Anticancer Agents

DAVID E. MODRAK, GRAHAM K. JONES, AND MICHAEL P. DRAPER

Cancer is the result of aberrant activity of the cell division and apoptotic machinery. While development in multicellular animals results from the coordination of cellular division and apoptosis, most animals are at risk of having these essential processes malfunction in a manner which results in cells that are unrestrained with regard to growth. Over the last few decades numerous chemotherapy drugs and targeted therapies have been developed, each aimed at limiting cancer cell proliferation or destroying the cancerous cells. In some instances great strides have been made, such as in testicular cancer. With most cancer types, however, survival figures have largely remained unchanged in recent decades. We are faced with a spectrum of diseases ranging from the highly drug sensitive to the intrinsically resistant, such as pancreatic cancer. Within this spectrum, most cancers are initially susceptible to chemotherapy but become resistant in a relatively short period. This phenomenon is known as acquired resistance. Ultimately, either intrinsic or acquired resistance is responsible for the failures of our current therapies. Elucidation of the mechanisms responsible for resistance holds the promise of improved survival outcomes through the development of resistance-modulating therapies. In recent years there has been great progress made in identifying new mechanisms of resistance. From these studies, it has become clear that genomic instability and the tumor microenvironment strongly influence the development of resistance and play a major role in the evolution of resistance.

RESISTANCE AND THE ROLE OF GENOMIC INSTABILITY

Maintaining the integrity and stability of the genome is a critical function of the cell, and loss of genome stability due to DNA damage is likely to result in cell death. From the perspective of the entire organism, however, it is the lower probability, but self-selecting event of tumorigenesis that underlines the importance of genome stability. The increased frequency of nuclear anomalies in a cell with compromised genome-stabilizing mechanisms can result in its transformation into an unregulated and rapidly dividing cancer cell. Further, while it is anticipated that even highly chemosensitive cancers, such as leukemias and lymphomas, will become resistant to therapy through the accumulation of further mutations, normal tissues will not become resistant. Thus, somatic mutations and genomic plasticity are the foundations upon which the acquisition of drug resistance occurs (58).

Genome stability is maintained by addressing DNA damage which might otherwise lead to the incorporation of mutations, regulating the process of cell division (DNA replication and segregation), and maintaining the integrity of telomeres. In addition, stability may be influenced by the expression of viral proteins, the presence of fragile sites and, possibly, by the development of instability in the genome of parents. In a worst case scenario, a cell which has incurred excessive DNA damage will die rather than allow that damaged genome to persist, and, thus, apoptosis itself also may be considered a mechanism by which cells maintain genomic stability.

DNA Replication Controls

DNA replication is controlled by a complex network of signaling molecules that alternatively promote or inhibit movement through the cell cycle. At specific points in the cell cycle, the G1/S and G2/M "checkpoints," DNA integrity is evaluated before moving on to replication or division, respectively. It is at the checkpoints that the cyclin family of proteins exert their influence by binding to specific cyclin dependent kinases (CDKs). At the G1/S restriction point, CDK4/6 and CDK2 bind cyclins D and E, respectively, and

David E. Modrak • Garden State Cancer Center, 520 Belleville Avenue, Belleville, NJ 07109. **Graham Jones and Michael P. Draper** • Paratek Pharmaceuticals, 75 Kneeland Street, Boston, MA 02111.

then phosphorylate the retinablastoma protein (Rb), releasing Rb from its blockade of the E2F transcription factor. E2F is responsible for the transcription of S-phase genes. The activities of the cyclin/CDK complexes are regluated via several pathways responsive to DNA damage, growth factor withdrawal, replicative senescence and contact inhibition. Key components of these pathways include p53, ATM, ATR, Smad3/Smad4, p16^{INK4}, p15^{INK4}, p27^{Kip1}, p21^{Cip1}, and cdc25A (29, 91). At the G2/M transition, cdc25B/C phosphatase activates cdc2/cyclin B kinase, which had been kept previously in an inactive state by the Myt1 and Wee1 kinases. The activity of cdc2/cyclin B can be inhibited by DNA damage sensing pathways: the ATM and ATR kinases activate Chk1 and Chk2 to inactivate cdc25B/C, leading to a rapid inhibition of cdc2/cyclin B, and the ATM and ATR kinases can activate directly and indirectly through Chk2, p53, which also leads to inhibition of cdc2/cyclin B as well as the induction of DNA repair genes (55, 88).

The p53 protein is critically important to the chemotherapeutic response (30). It is activated in response to many different stimuli, including, radiation, chemotherapy, oncogene expression and hypoxia. Once activated, it stimulates or represses the transcription of genes involved in the cell cycle, stress response, DNA repair, angiogenesis and apoptosis. It has been found that cells with a wild type p53 are generally more sensitive to chemotherapy than cells with mutated p53, taxol being a notable exception (125). The p53 protein mutations can allow the cell to avoid apoptosis and, thus, live with mutation causing damage. Inactivating p53 mutations are frequently found in cancer and thus have a role in genome stability and the selective process for resistance development.

DNA Repair Mechanisms

DNA exhibits many different types of damage upon exposure to both environmental and physiological insults. For example, UV light causes cyclobutane pyrimidine dimers and other photoproducts (6), and the oxidative stress from free radicals produced by normal cellular metabolism can cause more than one hundred different kinds of DNA damage (7). The diversity of genetic damage has resulted in the evolution of several mechanisms for DNA repair. These systems are highly conserved in both prokaryotes and eukaryotes, almost certainly because DNA integrity is vital to survival at a fundamental level. Double-stranded breaks caused, for example, by ionizing radiation or by the normal cellular process of V(D)J recombination are repaired mainly by the homologous recombination or the non-homologous end joining (NHEJ) mechanisms (68). A key component common to both mechanisms is the ATM (ataxia telangiectasia mutated) protein, which coordinates DNA-breakage-sensing, cell cycle arrest, and repair functions (reviewed in references 46, 100, and 127). Homologous recombination is the most effective and preferred of the two mechanisms as it uses the second intact allele to recover the correct sequence. In contrast, NHEJ can result in nucleotide loss and therefore frame-shift or deletion errors. Other types of damage such as single stranded breaks, uracil incorporation, or guanidine oxidation are repaired using the base-excision repair mechanism. Individual bases are removed by DNA glycosylases and replaced with the correct base using a combination of polymerases and ligases (70). Two different nucleotide-excision repair mechanisms function to repair UV damaged DNA or to remove bulky adducts from environmental insults (46, 70). Nucleotide-excision repair is most often associated with environmental damage, whereas base-excision repair is associated with damage resulting from endogenous causes. The mismatch repair (MMR) mechanism is employed for the repair of A-G and T-C mismatch, insertion, and deletion errors. Six human MMR genes (hMSH2, hMSH3, hMSH6, hMLH1, hPMS1, and hPMS2) have been identified, encoding the monomeric constituents of various heterodimeric proteins which can recognize and initiate repair of the lesions (reviewed in references 46, 80, 84, and 126).

Several human conditions are known to be associated with defective DNA repair mechanisms (46). For example, inborn genetic defects in the nucleotide-excision repair gene XPD, a 5′-3′ helicase (103), result in the UV-hypersensitivity conditions xeroderma pigmentosum, Cockayne syndrome, and trichothiodystrophy (46, 61). In contrast to the latter two conditions, xeroderma pigmentosum causes severe skin changes and a greatly elevated incidence of sun-induced skin cancer. Preliminary data also suggest that somatic mutations in the XPD gene might lead to basal cell carcinoma in normal populations (28, 122), supporting the notion that mutations disrupting repair mechanism genes can lead to cancer directly. MMR defects factor heavily in hereditary nonpolyposis colorectal cancer. Mutations in hMLH1, hMSH2 and, to a lesser extent, hMSH6 can be found in 50% of these patients, with some of the other redundant MMR genes contributing to a lesser degree (51, 75, 76, 80, 93). Ataxia telangiectasia, an autosomal recessive disease marked in part by a predisposition to cancer, involves mutation of the ATM gene, which is involved in double stranded base-repair (46, 100, 127). Other diseases involving double stranded base-repair defects predispose patients to cancer include Nijmegan breakage syndrome, Werner syndrome, Bloom syndrome, and Rothmund Thomas syndrome (46).

Telomeres

Telomeres are the nucleoprotein complexes found at the ends of chromosomes where they function to protect the ends from being recognized as broken or damaged DNA (41). In humans, the telomeres are 4 to 15 kbp of the repeated TTAGGG hexanucleotide sequence in association with specific proteins, including hTRF1, hTRF2, and hRAP1, followed by a capping structure, the T-loop. The T-loop is an additional 100 to 150 bases of single stranded DNA composed of TTAGGG repeats that is folded back on itself and is associated with additional proteins (POT1, Ku, Mre11, Rad50, and Nbs1, the latter three being involved also with NHEJ). TRF2 is essential for telomere integrity, as loss of DNA binding ability leads to uncapping of telomeres, G-strand overhang degradation, telomere fusions and anaphase bridges. In tumor cells, uncapping induces a p53 and ATM dependent DNA damage response with cell cycle arrest and apoptosis, while in human fibroblasts, uncapping leads to p53 and p16/Rb dependent senescence.

As normal cells divide, their telomeres become shorter with each division since telomerase is not active in most normal cells. At some point, the telomeres become critically short and, unless telomerase expression is increased, a DNA damage response via p53 and ATM is illicited, which is similar to the uncapped telomere response in that the ends of telomeres fuse leading to anaphase bridges. Patients with dyskeratosis congenita have reduced levels of telomerase, leading to shortened telomeres, and suffer from bone marrow failure and increased incidence of myeloid cancers.

Fragile Sites

Chromosome fragile sites are nonrandom gaps or breaks that can be induced under specific culture conditions (98). Their relevance to disease in vivo continues to be a matter of debate. Based on the observation that common fragile sites are located at or near structural defects known in various cancers, it is postulated that fragile sites are the boundaries of chromosomal rearrangements, amplifications and deletions. For instance, in familial clear cell renal carcinoma, the FRA3B fragile site has been found to be the breakpoint in chromosome 3 of the t(3;8) translocation (89). In addition, fragile sites may be preferred sites of viral integration (118).

Mutation Phenotype

DNA mutations that escape repair are very low probability events. The relatively low mutation rate of a somatic cell (1.4×10^{-10} nucleotides/cell/division) is considered by some to be unable to explain the thousands of mutations present in a typical tumor cell (18, 71). Accounting for this discrepancy is a current area of scientific debate. The mutator phenotype hypothesis postulates that mutations in key genome-stabilizing genes occur as an early event in tumor progression and result in a significantly increased mutation rate (71, 72). The alternative to the mutator phenotype hypothesis argues that the increased number of mutations in the cancer cell is due to selective pressures and clonal expansion and is sufficient to explain tumorigenesis (77). The explanation for the increased mutation levels in tumor cells will likely include elements of both enhanced mutation rates and clonal selection.

Microsatellite and Chromosomal Instability

Genome instability is divided into two classes: microsatellite instability (MIN) and chromosomal instability (CIN) (64). MIN, recognizable as altered lengths of small tandem repetitive DNA sequences called microsatellites, was originally identified by PCR analysis of human colon tumor DNA (92). MIN is usually associated with mutations in MMR genes, but the exceptions to this generalization imply other as yet unknown causes of the phenotype (32, 33, 60).

CIN, the more prevalent of the two classes, involves errors in chromatin maintenance and segregation, resulting in aneupoidy, translocations, deletions, and amplifications. Unlike MIN mutations, CIN mutations are dominant, requiring the disruption of just one allele to express the phenotype, e.g., those that occur in the mitotic checkpoint protein gene, hBub1 (8, 63, 64).

Radiation-Induced Transgenerational Genomic Instability

One fascinating, if still controversial, aspect of genomic instability is the evidence linking instability in offspring to that in their parents (23). In studies of irradiated male mice, it was found that their offspring were more sensitive to developing carcinogen-induced cancers than nonirradiated controls and that different types of cancers were seen between the two groups. Futhermore, maternal alleles from nonirradiated females could become destabilized as measured in the somatic cells of the first generation offspring, and mutation rates were equivalent in F_1 and F_2 populations. These data suggest that genomic instability may be a heritable trait, possibly through an epigenetic mechanism.

Consequences of Genomic Instability on Anticancer Therapy

Tumor cells exhibit a high degree of heterogeneity in chromosome number, arrangement and polymorphisms within a given tumor. The genetic instability that has generated this enormous amount of diversity also presents a significant problem for therapy, the phenomenon of drug resistance. The cancer cell genome is undisputedly much less stable than that of a normal cell, influencing not only the progression of the disease state but also the tumor's responsiveness to therapy. Furthermore, the genomic plasticity observed is overlaid onto the genetic individualism of the patient, resulting in a great diversity of cells within the tumor. This diversity allows for selection of aggressively dividing, metastasizing and/or drug resistant subpopulations, in an almost Darwinian "survival of the fittest" manner.

Chemotherapy often fails because of the emergence of the multidrug-resistance phenotype, in which tumors acquire resistant simultaneously to a range of drugs. The same destabilized genome that causes tumorigenesis also activates chemotherapy resistance mechanisms. Whereas tumorigenesis is a self-selecting event which may simultaneously result in intrinsic resistance, acquired drug-resistance requires the selective pressure of the drug itself to expand the resistant tumor cell population into the major cell type of the tumor. Thus, cancer should be viewed not as a single disease, nor even a single disease within an individual; each tumor should be seen as a collection of diseases potentially with different characteristics and chemosensitivities. What kills 99% of the cancer cells may not kill the 1% which will regrow several months later.

RESISTANCE AND THE ROLE OF THE TUMOR MICROENVIRONMENT

A large number of scientific studies have focused on the identification and characterization of mechanisms of drug resistance using tissue culture based models. These culture based models are largely unicellular in nature and for the most part fail to consider the fact that the environment, in which a cancerous cell is multiplying, is extremely heterogeneous. In recent years, the term microenvironment has been used to describe the fact that each cancer cell is within its own unique environment. Within the host, each cancer cell is subjected to varying levels of growth factors, cytokines, and components of the extracellular matrix (ECM) (10, 50). Additionally, there is the added consideration that even within a given solid tumor there are environmental compartments that vary with regard to additional factors such as hypoxia, pH, vascularization and necrosis (123). These microenvironment effects are now thought to be important in the response to drug and in the emergence of acquired resistance. In general, the mechanisms of microenvironment-related drug resistance can be subdivided into those that lead to reduced drug effect (e.g., damage or inhibition of the primary target) and those mechanisms that provide increased tolerance to damage.

Microenvironmental Effects That Minimize Damage or Inhibition of the Primary Target

A prerequisite for effective chemotherapeutic activity is the need for the drug to reach the targeted cancer cell. Before a chemotherapeutic drug reaches the intended target it needs to make its way through the tumor vasculature, extravasate across the vessel wall, and then diffuse through the ECM (50). Therefore, tumor blood supply is thought to be a key determinant in effective antitumor therapy as it controls the delivery of the drug. The vascular network in a tumor consists of the original host network of vessels and additional vessels that result from angiogenic processes initiated as a result of the cancer. In contrast to normal vessels, the vasculature of a tumor is highly disorganized and vessels are dilated and tortuous, with uneven diameters and excessive branching (66, 67, 112). This disorganized structure is thought to be the result of an uneven distribution of angionenic factors such as vascular endothelial growth factor (VEGF) (9). Consequently, tumor blood flow is chaotic and variable, leading to hypoxic and acidic regions within the tumor. In addition to these gross morphological differences, the tumor vasculature differs greatly in terms of its ultrastructure. The walls of the vessels have numerous openings, widened interendothelial junctions and a discontinuous or absent basement membrane. These defects ultimately make the vessels more leaky, although this leakiness is not distributed evenly throughout the tumor vasculature (42, 49).

However, while the vessels themselves may be leaky, diffusive and convective forces primarily govern the movement of molecules out of the vessels and into the interstitium of tumors. An examination of pressure gradients in experimental tumors has suggested that the movement of drugs and particulates out of tumor blood vessels into the extra-vascular compartment is remarkably limited. This has been attributed to a higher-than-expected interstitial fluid pressure, in part due to a lack of functional lymphatic drainage, coupled with lower intravascular pressure (4).

Once a drug has found its way out of the tumor vasculature, there is the additional problem of movement across the tumor interstitium. Movement of the drug is thought to be dependent mainly on factors affecting passive diffusion. Factors that affect passive diffusion include the molecular weight of the drug, protein binding of the drug, and the composition of the ECM (35). In addition, interstitial pressure tends to be higher at the center of solid tumors, diminishing toward the periphery, creating a mass flow movement of fluid and drug away from the central region of the tumor (35). Ultimately, all of the factors mentioned above can influence the concentration of drug at any particular site within the tumor. The result is that some cells within the tumor are exposed to more or less drug. These levels may be too low to effectively destroy the cancer cells and may contribute to selection of cells with low to moderate levels of drug resistance.

Microenvironmental Effects That Provide Increased Tolerance to Damage

It has been known for many years that there are specific mechanisms associated with tumors and multicellular spheroids that act to protect cancer cells from cytotoxic drugs and radiation. This was first noted in vitro by Durand and Sutherland, who observed that multicellular spheroids of tumor cells were more resistant to anticancer agents than the corresponding monolayer cultures (25). Subsequent work demonstrated that tumors made resistant to alkylating agents in vivo were sensitized once removed from the animal (116). It has now become clear that a variety of microenvironmental interactions, including the interaction of a cancer cell with other cells, growth factors, and the ECM, can dramatically alter the efficacy of a chemotherapeutic agent.

The term multicellular resistance (MCR) has been used to describe resistance mediated as a result of the interaction or association of a cancer cell with other cells. It is likely that MCR occurs in part because of the physical conditions that exist within a tumor. As mentioned, the altered vascular network within a tumor can limit the availability of oxygen and nutrients and, therefore, cells in this environment are less likely to be proliferating. Since most current cancer treatments interfere with the basic machinery of DNA synthesis or cell division, these quiescent cells are intrinsically more resistant to drug. There is also evidence that the altered growth conditions within a tumor can stimulate the expression of known drug resistance factors (123). The precise mechanism responsible for this decreased drug sensitivity is unknown, but there is evidence that suggests that the transcription factor HIF1 (hypoxia inducible factor 1) may play a role as it can promote the expression of stress-related genes (109). In addition, some of the resistance to anticancer drugs and ionizing radiation could also be due to a phenomenon termed regrowth resistance. After therapy, a great proportion of well oxygenated and rapidly dividing cells die. Subsequently, cells, which were once quiescent due to their physical location within a portion of the tumor where growth was limited as a result of hypoxia or nutrient deprivation, find themselves in a more a favorable environment. These cells resume rapid growth and quickly repopulate the tumor (26). Ironically, the harsh conditions within the interior of the tumor may have resulted in a selection of cells resistant to apoptotic signals giving these newly dividing cells protection from further drug treatments.

In contrast to resistance resulting from the physical conditions within a tumor, there is ample evidence that in small multicellular spheroids, which are avascular in nature and do not contain the different environmental compartments present within a large tumor, that cell-cell interactions can mediate resistance (19, 25, 115, 116). While the precise mechanisms responsible for this type of resistance have yet to be determined, it is not thought to be the result of an altered ability of drug to penetrate the spheroids. Instead there is evidence that E-cadherin, a homophilic cell-cell adhesion protein, can meditate G1 cell cycle arrest by an increase of the cyclin dependent kinase inhibitor $p27^{KIP1}$ (15, 36). In a panel of mouse and human carcinoma cells, a consistent upregulation of $p27^{KIP1}$ was noted when the cells were transferred from monlayer culture to three-dimensional culture. In EMT-6 mammary tumor cells, antisense mediated downregulation of $p27^{KIP1}$ reduced cellular adhesion, increased proliferation and sensitized cells to drugs and radiation (15). These data suggest that cell cycle related mechanisms might be important in MCR and represent good drug targets for circumventing resistance.

The tumor microenvironment consists of a diverse milieu of growth factors, cytokines, and hormones. In myeloma cells, the cytokine IL-6 has been shown to increase resistance to chemotherapeutic agents by preventing the apoptotic response (12). Numerous studies have shown that activation of the IL-6 receptor can affect cell survial and apoptosis through effects on the JAK/STAT, MAPK and PI3-K/AKT pathways (10, 12, 31, 44, 57). Other studies have shown that growth factors such as IGF-1, EGF and bFGF can influence the sensitivity of cancer cells to chemotherapeutic drugs (11, 14, 39, 81, 104).

The ECM is a complex assembly of collagen, proteoglycans and other molecules. This matrix is an essential component of normal tissues and provides important cues for cell development, migration, adhesion, proliferation and survival (52, 53, 94). Tumor cells are surrounded by ECM, which is produced either by neighboring stromal cells or directly by the cancer cells. Adhesion of tumor cells to the ECM protein fibronectin via β_1 integrins can cause increased resistance of cancer cells to chemotherapeutic drugs (10, 47, 57, 79, 107). This form of resistance is referred to as cell adhesion mediated drug resistance (CAM-DR). A large number of studies in the laboratory have provided evidence that CAM-DR may be important in clinical resistance. For example, human 8226 myeloma cells pre-adhered to fibronectin were found to have decreased sensitivity to doxorubicin and melphalan compared with the same cells grown in suspension (16). In another study, the leukemic cell line K562, when bound to fibronectin, demonstrated increased resistance to the apoptosis inducing agent and BCR/ABL kinase inhibitor AG957, the DNA damaging agents mitoxantrone and melphalan, the microtubule assembly inhibitor vincristine, the nucleotide analog Ara-C, and radiation (17). The precise mechanism for CAM-DR is unknown at this time though studies suggest that it is a multi-factorial process. One study with the human promonocytic cell line U937 demonstated that CAM-DR correlated with decreased drug-induced DNA double-strand breaks caused by the topoisomerase II inhibitory drugs mitoxantrone and etoposide. This decrease in DNA damage was associated with changes in the nuclear pool of topoisomerase II (17). Changes in topoisomerase II levels and activity are clearly not responsible for all cases of CAM-DR, as groups have demonstrated cellular protection via β_1 integrin binding without changes in topoisomerase levels or activity (17, 107). Recently, the perturbation of Fas ligand-induced apoptotic signaling has been implicated as a mediator of CAM-DR in myeloma cells (108). The Fas receptor is a transmembrane protein in the TNFα family that is involved in the apoptotic signaling process during the maturation and selection of hematologic cell types (85). These data suggest that CAM-DR is linked to the apoptotic process. In a study with small cell lung cancer (SCLC) cells, ECM proteins were found to protect the cells from chemotherapy induced apoptosis. Importantly, incubation with a function blocking β_1 integrin antibody or protein tyrosine kinase inhibitors could inhibit these effects, demonstrating that apoptosis resistance was mediated through binding of the integrins to the ECM and implicating tyrosine phosphorylation downstream of integrin activation (99). Lastly, studies have shown that increases in the $p27^{KIP1}$ cyclin dependent kinase inhibitor can mediate G1 cell cycle arrest and apoptosis resistance in myeloma cells adhered to fibronectin (43).

Role of the Microenvironment and Evolution of Resistance

In order to circumvent resistance it is critical that we understand not only the dominant mechanism(s) involved, but also the process by which it arose. While much work has been done over the past two decades to understand resistance, the effort has yielded little in the way of therapeutically useful strategies. One explanation for this failure is that the experimental models used to elucidate the mechanisms of resistance were flawed. Many studies were performed on cells whose resistance was induced in vitro by long term exposure to increasingly higher concentrations of drug. The resistance level of these laboratory-selected cells was often 10 to 1,000 times higher than that of the original parent cell line. By contrast, changes of as little as twofold in the sensitivity of a cell are likely to result in chemotherapeutic failure in vivo. While the focus on these highly resistant cell lines facilitated the elucidation of the mechanisms behind high level drug resistance, it is possible that the initial lines in which resistance was two- to fourfold, relative to the original parent cell line, contained the secrets to understanding clinically relevant cancer drug resistance. Additionally, it now seems likely that the use of cells largely removed from normal tumor environment obscured the finding of many relevant mechanisms of resistance. Microenvironment-mediated resistance represents a novel mechanism that is likely clinically important for successful chemotherapy. The initial cytoprotection afforded by MCR or CAM-DR may allow for the survival and selection of cells possessing higher levels of resistance. In the laboratory, it is well established that, in the selection of high level resistant cell lines, a process of stepping up resistance by increasingly exposing the cells to ever higher levels of drug over the course of many months is necessary. This evolutionary process is likely operating in vivo, where the selection of resistance is proceeding from low to high and the role of the tumor microenvironment in this process is only now just beginning to be understood. Ultimately, the identification of the cellular factors that mediate this form of resistance may provide novel strategies for circumventing both intrinsic and acquired drug resistance.

EVOLUTION OF DRUG RESISTANCE IN CANCER

It was clear almost from the beginning that multidrug resistance in vitro and in vivo involved multiple mechanisms acting in concert to modulate chemosensitivity, although often this was not readily appreciated. In several laboratories, stepwise selection for resistant cell lines resulted in a series of cell lines, each more resistant than the previous. It could be demonstrated in these cell lines that levels of resistance did not correlate with expression of known drug resistance mechanisms, specifically the cellular efflux transporter P-glycoprotein (Pgp). In fact, in some series of cell lines, Pgp was never expressed, even though resistance was increased and drug accumulation was decreased. Many researchers were keenly aware of this discrepancy and as a result the multidrug resistance associated protein (MRP) was identified (13). This act broke the seeming intransigence of some in the research community to the

Table 1. PC4 cell lines resulting from adriamycin selection[a]

Cell line	Fold resistance[b]			Percent accumulation[c]	Pgp expression[d]
	Adriamycin	Vincristine	Etoposide		
PC4-wt	1.0	1.0	1.0	100	None
PC4-A5	4.9	1.0	8.3	ND	None
PC4-A10	9.8	1.1	25	ND	None
PC4-A20	15	1.5	62	99	None
PC4-A40	44	6.3	72	76	Low
PC4-A80	83	7.6	86	50	Moderate
PC4-A160	180	21	133	ND	High

[a] Reprinted from *Biochemical Pharmacology* (83) with permission from the publisher.
[b] Fold resistance was defined as the IC_{50} for the cell line divided by the IC_{50} of the parental (wt) cell line.
[c] Percent accumulation was defined as the amount of tritiated daunorubicin present in resistant cells relative to the amount present in the parental cells after a 30 min incubation at 37°C. For some cell lines, adriamycin accumulation was not determined (ND).
[d] *mdr3* expression was assayed by reverse transcriptase-polymerase chain reaction (RT-PCR).

idea that multidrug resistance could arise by overexpression of an efflux transporter other than Pgp.

This also redirected attention to the fact that while cells lines will develop resistance to the primary drug of exposure in predictable and incremental steps, cross-resistance is gained stochastically. For example, in the PC4-A series of PC4 cells selected for increasing resistance to adriamycin developed in Stuart Levy's laboratory (113), the expected increases in adriamycin resistance were seen throughout the series, but vincristine and etoposide cross-resistances were acquired at different rates and to different levels during selection (Table 1). In this series of PC4 cells, Pgp was expressed at low levels in PC4-A40 cells with higher levels in the PC4-A80 and -A160 cells, and this expression seemed to be responsible for reduced adriamycin accumulation in these cells (83). MRP was expressed in the parental cells and up to the PC4-A40 cells at very low but consistent levels. However, MRP was not overexpressed in the more drug resistant cell lines in this series (114). While vincristine resistance was proportional to Pgp expression, adriamycin and etoposide resistance were not. In fact, the early cell lines, even though selected for adriamycin resistance, demonstrated greater increases in resistance to etoposide. The later Pgp-expressing cell lines showed only moderate increases in etoposide resistance. Additional studies found that in the PC4-A5, and all subsequent cell lines, there was a down regulation in topoisomerase II activity (83) and increase in c-fos expression (3). It is likely that other mutations are present in the cell lines, especially PC4-A10 and -A20, which await identification.

Different results were obtained in another series of PC4 cells developed by exposure to increasing concentrations of vincristine (114). Like the cell lines derived with adriamycin selection, vincristine resistance increased incrementally, but resistance to other drugs, such as adriamycin and etoposide, increased stochastically (Table 2). In contrast, low MRP expression was seen in the PC4-V10 cell line and expression was high in the subsequent cell lines. Pgp expression was not seen until very high vincristine resistance was attained.

Table 2. PC4 cell lines resulting from vincristine selection[a]

Cell line	Fold resistance[b]			Percent accumulation[c]	MRP expression[d]	Pgp expression[d]
	Adriamycin	Vincristine	Etoposide			
PC4-wt	1.0	1.0	1.0	100%	None	None
PC4-V5	1.0	5.5	2.8	ND	ND	ND
PC4-V10	1.1	11	3.8	67%	Low	ND
PC4-V20	1.8	17	18	ND	High	ND
PC4-V40	6.0	42	28	56%	High	None
PC4-V80	5.8	96	38	ND	High	Very low
PC4-V160	31	215	25	35%	High	Moderate

[a] Reprinted from *Cancer Research* (114) with permission from the publisher.
[b] Fold resistance was defined as the IC_{50} for the cell line divided by the IC_{50} of the parental (wt) cell line.
[c] Percent accumulation was defined as the amount of tritiated vincristine present in resistant cells relative to the amount present in the parental cells after a 30 min incubation at 37°C. For some cell lines, vincristine accumulation and/or gene expression was not determined (ND).
[d] *mrp* and *mdr3* expression was assayed by Northern blot.

Cell lines expressing either MRP and/or Pgp were MDR but, interestingly, MRP ceased to be a significant resistance mechanism in the PC4-V160 cell line, instead reliquishing that role to Pgp (21). What additional changes in cellular protein expression occurred that contributed to the increase in resistance remain to be elucidated.

Multiple Mechanisms Interact in the Multidrug Resistance Phenotype

As numerous investigators have shown through their derived series of cell lines, multidrug resistance (MDR) is complex and results from contributions of several resistance mechanisms simultaneously. In some ways, then, MDR might be more accurately refered to as multifactorial resistance. In the PC4 series of cell lines, the first detectable mutation in the adriamycin selected lines was a down reguation of topoisomerase II, the presumed target of adriamycin, while in the vincristine selected lines, upregulation of c-fos was found. In each series, these changes were maintained while other changes were added in order to increase resistance. Changes in expression of, or mutations within, specific genes are likely to have a component founded on genomic instability that allows for increased likelihood of non-faithful DNA replication.

Primary drug targets effecting cross-resistance

One of the most frequent responses in a cancer cell's biology when faced with the stress of growth in the presence of a chemotherapeutic agent is to alter and/or modify the protein or pathway the drug targets. One of the best studied systems, in this regard, is the response of cells to anti-folate therapy (128). The antifolate methotrexate (MTX) binds dihydrofolate reductase (DHFR) and inhibits the formation of tetrahydrofolate. Inhibition of DHFR results in a rapid depletion of reduced folates and the cessation of tetrahydrofolate dependent reactions (i.e., purine and dTMP synthesis and the regeneration of methionine from homocysteine). MTX, like natural folic acid, is transported into cells and glutamylated. It can then be polyglutamylated, which increases its affinity for DHFR, or effluxed. The initial response of a cancer cell to MTX is to increase DHFR expression, a response which does not require the selection of a mutation. If MTX concentrations are sufficiently high as to start killing cells, selection will occur. Several possible mutations within the folate-dependent metabolic reactions may be acquired, including mutations which decrease the affinity of DHFR for MTX or decrease the glutamylation of MTX. In addition, DHFR gene amplification may occur. Many of these mutations have the added benefit that they render other anti-folate therapies less effective as well.

For other antimetabolites, changes in their targets are common. Cell lines with mutated deoxycytidine kinase are resistant to the deoxyadenosine analogs, fludarabine (20) and cladribine (74), and the deoxycytosine analog, gemcitabine (87). Targets of antimetabolites are not the only drug targets in which mutations are found to decrease the effectiveness of the drug. Mutations in and/or lower expression of topoisomerase I are found to confer resistance to camptothecins (96), while changes in topoisomerase II are found to confer resistance to anthracyclines (86), epipodophyllotoxins (121), and other topoisomerase II poisons. Inhibition of microtubule function by the *Vinca* alkaloids and taxanes can be abrogated by mutations in the α and β tubulin genes (2, 5, 22, 90).

It has long been known that alterations in cellular biochemistry and protein expression can impact resistance to antineoplastic agents. The possible mechanisms available to tumor cells to modulate sensitivity run the gamut of every conceivable cellular process. Resistance generating mechanisms include alterations and/or modifications of the protein or pathway the drug targets reduced drug uptake, increased drug efflux, increased repair of damage, enhanced detoxification, and global gene transcription realignments. No tumor cell, it seems, is incapable of mounting some survival response that will result in the regrowth of drug resistant tumors. This is likely why chemotherapy is very capable of extending the life of patients without effecting a cure and why new drugs are likely to meet the same obstacles and impediments as current therapies (34).

Mutation, or other target alterations, while common, are not the only changes that occur in tumor cells in response to chemotherapeutic insult. Secondary mutations arise that can further reduce sensitvity to therapy. Often, these secondary mutations can confer resistance to other, previously unused, drugs. This type of resistance is known as MDR and is particularly difficult to deal with clinically, as it can severely limit alternative therapy options.

Factors affecting drug accumulation

Mutations in nucleoside transporters can reduce the cellular uptake of antifolates (128) and nucleoside analogs (74, 87) while changes in passive transport, or possibly a poorly-defined active transport mechanism, can modulate alkylating agent uptake (111). Much more attention and interest have been given to active efflux mechanisms.

Active transporting, or effluxing, proteins belong to the class of molecules known as ATP-binding cassette (ABC) transporters, of which 49 have been identified in humans to date (http://nutrigene.4t.com/humanabc.htm). Approximately 14 of these are known to, or presumed to, efflux therapeutically beneficial drugs from cancer cells (1, 59). Expression of these proteins confers MDR against various drugs, with each protein having a specific portfolio of drugs upon which they act. Pgp (ABCB1) was the first such protein identified and is one of the best characterized mechanisms (1, 69). Pgp is composed of two halfs, each with six transmembrane domains and an ATP-binding domain. Although glycosylated and phosphorylated, these modifications do not appear to impact the function of the protein. Pgp hydrolyzes two ATPs per drug molecule effluxed and is capable of effluxing basic and neutral drugs. These drugs include many of the most commonly used in the clinic, including anthracyclines, *Vinca* alkaloids, epipodophyllotoxins, taxanes, and Gleevec (STI-571). Generally, the mechanism by which drug efflux occurs is refered to as the lipophilic vaccuum cleaner model. In this model, Pgp reduces drug accumulation by catching drugs as they pass through the membrane by passive diffusion and effluxing them into the surrounding media in an energy-dependent manner. Pgp is normally expressed in the intestine, liver, and kidney and at the blood-brain barrier. Cancers arising from these tissues typically express Pgp, although Pgp expression is not restricted to these cancer types.

Pgp, however, is not the only energy-dependent, drug effluxing pump in cells. MRP1 (ABCC1) also effluxes drugs in an energy-dependent mechanism (40, 59). MRP1 effluxes glutathione and glucuronate conjugates of anthracyclines, *Vinca* alkaloids, epipodophyllotoxins, camptothecins, and methotrexate, although often these drugs are not expelled from the cell as conjugates, but are cotransported with glutathione. MRP1 belongs to a family of ABC transporters of which there are at least nine members (http://nutrigene.4t.com/humanabc.htm). These members each have their own profile of drugs, which are effluxed from the cell. MRP2 (cMOAT/ABCC2) effluxes the same drugs as MRP1 but also mitoxantrone and cis-platin. Other MRP family members are not as well characterized but can confer resistance to many of these drugs and/or nucleoside analogs. MRPs are found in most normal tissues.

Apoptosis and the cellular response to DNA damage

Another common response to DNA damaging agents is to increase intrinsic DNA repair mechanisms. Resistance to *cis*-platin (111), camptothecin (96), mitomycin C (24), and other alkylating agents is associated with increased DNA repair. Interestingly, resistance via increased DNA repair appears to be specific to the alkylating agent, in that cross-resistance to other alkylating agents is uncommon (102).

Related to this is the role of p53 in regulating the cell cycle in response to DNA damage (30). Numerous drugs affect the integrity of DNA, including the alkylating agents, camptothecins, anthracyclines, and epipodophyllotoxins. These agents induce the activation of p53 to halt the cell cycle and allow for time to repair the damage. Alternatively, if the damage is too extensive, p53 will invoke apoptosis to eliminate the cell. Cancer cells often have a deleted, mutated or otherwise inactivated p53. Inactivation of p53 by mutation or increased expression of a negative regulator, such as MDM2, allows the cell to bypass the cell cycle checkpoints and become tolerate to DNA damage. This also leads to genome instability as previously discussed.

A reduced potential for apoptosis can be viewed as a prerequisite for cancer (54). Several apoptosis-related genes have been found to be altered in various cancers, including Bcl-2, Bcl-x_L, Bax, Apaf-1, XIAP, ML-IAP, Fas receptor, TRAIL receptors, c-FLIP, and caspases 8 and 10 (54). Alterations in the expression of these and other proteins involved in mitogenic and apoptotic signaling can lead to intrinsic resistance. Most chemotherapeutics, ionizing radiation, and p53-activating drugs induce apoptosis through the Bcl-2-inhibitable, or mitochondrial, pathway. Alterations in the ratios of pro- and anti-apoptotic Bcl-2 family members have been correlated with clinical drug resistance (97).

Detoxification and drug inactivation

All drugs are metabolically processed and eliminated to some degree. The activity of these detoxification processes often lead to variability in pharmacokenetics between patients. Some drugs are directly inactivated, for instance, camptothecin by irinotecan carboxylesterase-converting enzyme (101), bleomycin by bleomycin hydrolase (82), 6-mercaptopurine and other purine antimetabolites by thiopurine methyltransferase (124), and gemcitabine and other pyrimidine antimetabolites by 5′-nucleotidase (87). The alkylating agents, and other drugs for which the active form is electrophilic, can be inactivated by reaction with glutathione (119). Other drugs are glucuronidated (camptothecins [48]) or conjugated to glutathione (anthracyclines [27], mitoxantrone [27], bleomycin [120]) before excretion. Importantly, MRP can facilitate the excretion of conjugated natural products (40).

Transcriptional control of resistance

Controlling transcription may also govern drug resistance. Down regulation of topoisomerase II leads to resistance against anthracyclines and epipodophyllotoxins (45, 73), and reduced expression of folate receptors leads to resistance to anti-folates (128). Exposure to chemotherapeutic agents can also lead to the transient expression of resistance mechanism(s) (105, 128).

The action of p53, HIF1, NF-κB, AP-1 (fos/jun), Sp1/Sp3, and a host of other transcription factors can lead to increased cell survival and drug resistance. The genes specifying the drug-resistance conferring ABC proteins of humans are TATA-less, and significant work has led to the proposal of the MDR1 enhancesome (105). The enhancesome is composed of NF-Y and Sp1/Sp3, covers a 5′ region containing a CCAAT box and a GC element, and recruits P/CAF (histone acetyltransferase) to the region. It is presumed that P/CAF remodels the chromatin locally to transiently activate transcription. Continuous exposure to drug may stimulate further chromatin remodeling resulting in constitutive expression.

This brief review only touched on the variety of ways cancer cells can become drug resistant. Other mechanisms are known, for instance, reducing necessary metabolic activation, as with mitomycin C (24), or altering the intracellular localization, as with the camptothecins by sumoylation (96). What is certain is that the methods employed by cancer cells to escape, avoid and confound our attempts to eradicate them will not cease to amaze researchers for at least the forseeable future.

Multiple Mechanisms Interact in the Multidrug Resistance Phenotype

Tumor cells are not wedded to a single mechanism for overcoming resistance in vitro (62, 78). Instead, cancer cells will utilize multiple mechanisms to increase resistance. As discussed previously, in the PC4-A series of cells, topoisomerase II expression was reduced initially, and other mechanisms, including Pgp overexpression, were added onto this (83), while the PC4-V series accumulated changes in the expression of c-fos, Pgp, and MRP1, as well as other mechanisms (21, 114). Additional alterations have been found in other established MDR cell lines. The KB-3-1 cell lines selected for resistance to either vincristine or adriamycin were initially found to express Pgp (110). Recent proposals that sphingolipid metabolism may contribute to multidrug resistance (56, 95, 106) prompted a reexamination of these cells and the identification of a possible contribution of altered sphingolipid metabolism to the resistance found in these cell lines (38).

Evolution of Clinical Resistance

Determining which resistance mechanisms are at play clinically has proved challenging. Numerous and diverse mechanisms are observed to increase drug resistance up to several hundred fold higher than the parental cell line, in vitro. This may be the result of the experimental protocols used to select resistant cell lines, where the cell cultures are exposed for long periods of time to constant levels of drug. In vivo, the situation may be completely different. As opposed to continuous drug exposure, drug concentrations rise and fall during therapy. In addition, the tumor microenvironement makes it likely that cells are exposed to very heterogeneous levels of drug during therapy. In this situation, an increase of only two- to threefold may confer enough resistance to allow the tumor to escape and regrow. This can be achieved, potentially, through the transcriptional activation or enhancement of drug resistance genes. Increased transcription may or may not be retained when the tumor cells are isolated and examined in vitro. For example, the expression of DHFR in the case of antifolate therapy, DNA repair enzymes in the case of DNA damaging therapies, or ABC transporters in the case of therapies using natural products, can be increased to provide temporary relief to the tumor cell, but these changes may not be retained once the tumor cells are cultured. Furthermore, the significance of drug resistance in clinical samples can be obscured by the lack of meaningful controls (i.e., pre- and posttherapy tumor samples), sample contamination with normal tissue and technical difficulties (i.e., antibody cross-reactivity, for instance, in immunohistochemical analysis).

Experimental evidence from clinical samples suggests that relapsed tumors are capable of expressing a variety of mechanisms for achieving resistance. For patients with ALL, 13% of untreated compared with 70% of methotrexate treated, relapsed cancers had impaired folate transport systems (37). In the majority of these patients, the folate receptor mRNA was decreased, while in the other cases, decreased function suggested that mutations in the receptor were responsible. This and other observations have led to the development of newer antifolate therapies with altered pharmacological properties and which are currently in clinical trials (128).

There is a great deal of interest in the ABC transporters as targets for modulation to increase chemotherapy efficacy, and several trials have been undertaken to assess the utility of this approach. Generally, Pgp expression is found to be higher in relapsed cancer when compared to samples taken at presentation; however, not all studies find this to be so (65). For MRP1, there is no evidence that resistance in clinical samples correlates with increased MRP1 levels (1). Attempts to inhibit Pgp in order to increase efficacy have met with very limited success and only a few, of the many, studies have shown an improved outcome (65). This was attributed to the expression of Pgp in normal tissues, which necessitated the reduction in drug dose, and to the substantial non-Pgp related toxicities of the modulators themselves. Newer, nontoxic Pgp modulators have been developed and await clinical evaluation (117).

SUMMARY

Genomic instability underlies the induction of cancer and in doing so, underlies the acquisition of resistance. Some events within a cell leading to cancer are themselves capable of imparting some measure of resistance, while other, subsequent, events related to instability may further tumor resistance. The environment in which the tumor grows can also influence resistance through the heterogeneous exposure of cancer cells to drug, autocrine release of growth factors, cytokines, and hormones or stimulation of cell growth through contact with the ECM or cell adhesion proteins of neighboring cells. Lastly, the exposure of tumor cells to drugs may, in conjuction with genomic instability, lead to highly resistant cancer cells through selective pressures. It is also clear that tumor cells are capable of adding multiple layers of resistance upon themselves. The challenge in the future will be to develop therapuetics which can selectively inhibit tumor resistance without increasing toxicity.

As highlighted above, in the laboratory setting the creation of highly resistant cell lines requires a step-wise selection process. At each step in the process, additional beneficial changes occur, increasing the level of resistance. Genomic instability is a powerful force that can increase the likelihood of a resistance yielding mutation occurring. The tumor microenvironment provides a heterogeneous environment, with regards to drug exposure levels and cellular phenotype, where cells with favorable first step mutations bestowing low levels of drug resistance can be exposed to nonlethal selective pressures. Ultimately genomic instability and the microenviroment act in concert to increase the likelihood of resistant cancer cells emerging. In the development of therapies or therapeutic regimens to combat the emergence of resistance, it is clear that careful attention should be paid to these elements.

REFERENCES

1. **Ambudkar, S. V., C. Kimchi-Sarfaty, Z. E. Sauna, and M. M. Gottesman.** 2003. P-glycoprotein: from genomics to mechanism. *Oncogene* **22:**7468–7485.
2. **Bhalla, K. N.** 2003. Microtubule-targeted anticancer agents and apoptosis. *Oncogene* **22:**9075–9086.
3. **Bhushan, A., C. A. Slapak, S. B. Levy, and T. R. Tritton.** 1996. Expression of c-fos precedes MDR3 in vincristine and adriamycin selected multidrug resistant murine erythroleukemia cells. *Biochem. Biophys. Res. Commun.* **226:**819–821.
4. **Boucher, Y., L. T. Baxter, and R. K. Jain.** 1990. Interstitial pressure gradients in tissue-isolated and subcutaneous tumors: implications for therapy. *Cancer Res.* **50:**4478–4484.
5. **Cabral, F.** 2001. Factors determining cellular mechanisms of resistance to antimitotic drugs. *Drug Resist. Update* **4:**3–8.
6. **Cadet, J., O. Anselmino, T. Douki, and L. Voituriez.** 1992. Photochemistry of nucleic acids in cells. *J. Photochem. Photobiol B.* **15:**277–298.
7. **Cadet, J., M. Berger, T. Douki, and J. L. Ravanat.** 1997. Oxidative damage to DNA: formation, measurement, and biological significance. *Rev. Physiol. Biochem. Pharmacol.* **131:** 1–87.
8. **Cahill, D. P., C. Lengauer, J. Yu, G. J. Riggins, J. K. Willson, S. D. Markowitz, K. W. Kinzler, and B. Vogelstein.** 1998. Mutations of mitotic checkpoint genes in human cancers. *Nature* **392:**300–303.
9. **Carmeliet, P., and R. K. Jain.** 2000. Angiogenesis in cancer and other diseases. *Nature* **407:**249–257.
10. **Catlett-Falcone, R., T. H. Landowski, M. M. Oshiro, J. Turkson, A. Levitzki, R. Savino, G. Ciliberto, L. Moscinski, J. L. Fernandez-Luna, G. Nunez, W. S. Dalton, and R. Jove.** 1999. Constitutive activation of Stat3 signaling confers resistance to apoptosis in human U266 myeloma cells. *Immunity* **10:**105–115.
11. **Chang, F., L. S. Steelman, J. T. Lee, J. G. Shelton, P. M. Navolanic, W. L. Blalock, R. A. Franklin, and J. A. McCubrey.** 2003. Signal transduction mediated by the Ras/Raf/MEK/ERK pathway from cytokine receptors to transcription factors: potential targeting for therapeutic intervention. *Leukemia* **17:** 1263–1293.
12. **Chauhan, D., S. Kharbanda, A. Ogata, M. Urashima, G. Teoh, M. Robertson, D. W. Kufe, and K. C. Anderson.** 1997. Interleukin-6 inhibits Fas-induced apoptosis and stress-activated protein kinase activation in multiple myeloma cells. *Blood* **89:**227–234.
13. **Cole, S. P., G. Bhardwaj, J. H. Gerlach, J. E. Mackie, C. E. Grant, K. C. Almquist, A. J. Stewart, E. U. Kurz, A. M. Duncan, and R. G. Deeley.** 1992. Overexpression of a transporter gene in a multidrug-resistant human lung cancer cell line. *Science* **258:**1650–1654.
14. **Coleman, A. B.** 2003. Positive and negative regulation of cellular sensitivity to anti-cancer drugs by FGF-2. *Drug Resist. Update* **6:**85–94.
15. **Croix, B. S., J. W. Rak, S. Kapitain, C. Sheehan, C. H. Graham, and R. S. Kerbel.** 1996. Reversal by hyaluronidase of adhesion-dependent multicellular drug resistance in mammary carcinoma cells. *J. Natl. Cancer Inst.* **88:**1285–1296.
16. **Damiano, J. S., A. E. Cress, L. A. Hazlehurst, A. A. Shtil, and W. S. Dalton.** 1999. Cell adhesion mediated drug resistance (CAM-DR): role of integrins and resistance to apoptosis in human myeloma cell lines. *Blood* **93:**1658–1667.
17. **Damiano, J. S., L. A. Hazlehurst, and W. S. Dalton.** 2001. Cell adhesion-mediated drug resistance (CAM-DR) protects the K562 chronic myelogenous leukemia cell line from apoptosis induced by BCR/ABL inhibition, cytotoxic drugs, and gamma-irradiation. *Leukemia* **15:**1232–1239.
18. **DeMars, R., and K. R. Held.** 1972. The spontaneous azaguanine-resistant mutants of diploid human fibroblasts. *Humangenetik* **16:**87–110.

19. **dit Faute, M. A., L. Laurent, D. Ploton, M. F. Poupon, J. C. Jardillier, and H. Bobichon.** 2002. Distinctive alterations of invasiveness, drug resistance and cell-cell organization in 3D-cultures of MCF-7, a human breast cancer cell line, and its multidrug resistant variant. *Clin. Exp. Metastasis.* **19**:161–168.
20. **Dow, L. W., D. E. Bell, L. Poulakos, and A. Fridland.** 1980. Differences in metabolism and cytotoxicity between 9-beta-D arabinofuranosyladenine and 9-beta-D-arabinofuranosyl-2-fluoroadenine in human leukemic lymphoblasts. *Cancer Res.* **40**:1405–1410.
21. **Draper, M. P., R. L. Martell, and S. B. Levy.** 1997. Indomethacin-mediated reversal of multidrug resistance and drug efflux in human and murine cell lines overexpressing MRP, but not P-glycoprotein. *Br. J. Cancer.* **75**:810–815.
22. **Drukman, S., and M. Kavallaris.** 2002. Microtubule alterations and resistance to tubulin-binding agents (review). *Int. J. Oncol.* **21**:621–628.
23. **Dubrova, Y. E.** 2003. Radiation-induced transgenerational instability. *Oncogene* **22**:7087–7093.
24. **Dulhanty, A. M., M. Li, and G. F. Whitmore.** 1989. Isolation of Chinese hamster ovary cell mutants deficient in excision repair and mitomycin C bioactivation. *Cancer Res.* **49**:117–122.
25. **Durand, R. E., and R. M. Sutherland.** 1972. Effects of intercellular contact on repair of radiation damage. *Exp. Cell. Res.* **71**:75–80.
26. **Durand, R. E., and C. Aquino-Parsons.** 2001. Clinical relevance of intermittent tumour blood flow. *Acta. Oncol.* **40**:929–936.
27. **Dusre, L., E. G. Mimnaugh, C. E. Myers, and B. K. Sinha.** 1989. Potentiation of doxorubicin cytotoxicity by buthionine sulfoximine in multidrug-resistant human breast tumor cells. *Cancer Res.* **49**:511–515.
28. **Dybdahl, M., U. Vogel, G. Frentz, H. Wallin, and B. A. Nexo.** 1999. Polymorphisms in the DNA repair gene XPD: correlations with risk and age at onset of basal cell carcinoma. *Cancer Epidemiol. Biomarkers Prev.* **8**:77–81.
29. **Dyson, N.** 1998. The regulation of E2F by pRB-family proteins. *Genes Dev.* **12**:2245–2262.
30. **El-Deiry, W. S.** 2003. The role of p53 in chemosensitivity and radiosensitivity. *Oncogene* **22**:7486–7495.
31. **Fan, M., and T. C. Chambers.** 2001. Role of mitogen-activated protein kinases in the response of tumor cells to chemotherapy. *Drug Resist. Update* **4**:253–267.
32. **Fishel, R., M. K. Lescoe, M. R. Rao, N. G. Copeland, N. A. Jenkins, J. Garber, M. Kane, and R. Kolodner.** 1994. The human mutator gene homolog MSH2 and its association with hereditary nonpolyposis colon cancer. *Cell* **77**:167.
33. **Fleisher, A. S., M. Esteller, G. Tamura, A. Rashid, O. C. Stine, J. Yin, T. T. Zou, J. M. Abraham, D. Kong, S. Nishizuka, S. P. James, K. T. Wilson, J. G. Herman, and S. J. Meltzer.** 2001. Hypermethylation of the hMLH1 gene promoter is associated with microsatellite instability in early human gastric neoplasia. *Oncogene.* **20**:329–335.
34. **Fojo, T., and S. Bates.** 2003. Strategies for reversing drug resistance. *Oncogene* **22**:7512–7523.
35. **Galmarini, C. M., and F. C. Galmarini.** 2003. Multidrug resistance in cancer therapy: role of the microenvironment. *Curr. Opin. Investig. Drugs.* **4**:1416–1421.
36. **Gil-Gomez, G.** 2004. Measurement of changes in apoptosis and cell cycle regulatory kinase cdk2. *Methods Mol. Biol.* **282**:131–144.
37. **Gorlick, R., E. Goker, T. Trippett, P. Steinherz, Y. Elisseyeff, M. Mazumdar, W. F. Flintoff, and J. R. Bertino.** 1997. Defective transport is a common mechanism of acquired methotrexate resistance in acute lymphocytic leukemia and is associated with decreased reduced folate carrier expression. *Blood* **89**:1013–1018.
38. **Gouaze, V., J. Y. Yu, R. J. Bleicher, T. Y. Han, Y. Y. Liu, H. Wang, M. M. Gottesman, A. Bitterman, A. E. Giuliano, and M. C. Cabot.** 2004. Overexpression of glucosylceramide synthase and P-glycoprotein in cancer cells selected for resistance to natural product chemotherapy. *Mol. Cancer Ther.* **3**:633–639.
39. **Guo, Y. S., G. F. Jin, C. W. Houston, J. C. Thompson, and C. M. Townsend, Jr.** 1998. Insulin-like growth factor-I promotes multidrug resistance in MCLM colon cancer cells. *J. Cell Physiol.* **175**:141–148.
40. **Haimeur, A., G. Conseil, R. G. Deeley, and S. P. Cole.** 2004. The MRP-related and BCRP/ABCG2 multidrug resistance proteins: biology, substrate specificity and regulation. *Curr. Drug Metab.* **5**:21–53.
41. **Harrington, L.** 2004. Those dam-aged telomeres! *Curr. Opin. Genet Dev.* **14**:22–28.
42. **Hashizume, H., P. Baluk, S. Morikawa, J. W. McLean, G. Thurston, S. Roberge, R. K. Jain, and D. M. McDonald.** 2000. Openings between defective endothelial cells explain tumor vessel leakiness. *Am. J. Pathol.* **156**:1363–1380.
43. **Hazlehurst, L. A., J. S. Damiano, I. Buyuksal, W. J. Pledger, and W. S. Dalton.** 2000. Adhesion to fibronectin via beta1 integrins regulates p27kip1 levels and contributes to cell adhesion mediated drug resistance (CAM-DR). *Oncogene* **19**:4319–4327.
44. **Hideshima, T., N. Nakamura, D. Chauhan, and K. C. Anderson.** 2001. Biologic sequelae of interleukin-6 induced PI3-K/Akt signaling in multiple myeloma. *Oncogene* **20**:5991–6000.
45. **Hochhauser, D., C. A. Stanway, A. L. Harris, and I. D. Hickson.** 1992. Cloning and characterization of the 5′-flanking region of the human topoisomerase II alpha gene. *J. Biol. Chem.* **267**:18961–18965.
46. **Hoeijmakers, J. H.** 2001. Genome maintenance mechanisms for preventing cancer. *Nature* **411**:366–374.
47. **Hoyt, D. G., R. J. Mannix, M. E. Gerritsen, S. C. Watkins, J. S. Lazo, and B. R. Pitt.** 1996. Integrins inhibit LPS-induced DNA strand breakage in cultured lung endothelial cells. *Am. J. Physiol.* **270**:L689–694.
48. **Iyer, L., C. D. King, P. F. Whitington, M. D. Green, S. K. Roy, T. R. Tephly, B. L. Coffman, and M. J. Ratain.** 1998. Genetic predisposition to the metabolism of irinotecan (CPT-11). Role of uridine diphosphate glucuronosyltransferase isoform 1A1 in the glucuronidation of its active metabolite (SN-38) in human liver microsomes. *J. Clin. Invest.* **101**:847–854.
49. **Jain, R. K.** 1987. Transport of molecules across tumor vasculature. *Cancer Metastasis Rev.* **6**:559–593.
50. **Jain, R. K.** 1998. The next frontier of molecular medicine: delivery of therapeutics. *Nat. Med.* **4**:655–657.
51. **Jiricny, J., and M. Nystrom-Lahti.** 2000. Mismatch repair defects in cancer. *Curr. Opin. Genet. Dev.* **10**:157–161.
52. **Kadar, A., A. M. Tokes, J. Kulka, and L. Robert.** 2002. Extracellular matrix components in breast carcinomas. *Semin Cancer Biol.* **12**:243–257.
53. **Kalluri, R.** 2002. Discovery of type IV collagen non-collagenous domains as novel integrin ligands and endogenous inhibitors of angiogenesis. *Cold Spring Harb. Symp. Quant. Biol.* **67**:255–266.
54. **Kaufmann, S. H., and D. L. Vaux.** 2003. Alterations in the apoptotic machinery and their potential role in anticancer drug resistance. *Oncogene* **22**:7414–7430.
55. **Kohn, K. W.** 1999. Molecular interaction map of the mammalian cell cycle control and DNA repair systems. *Mol. Biol. Cell.* **10**:2703–2734.

56. **Kolesnick, R.** 2002. The therapeutic potential of modulating the ceramide/sphingomyelin pathway. *J. Clin. Invest.* **110:** 3–8.
57. **Kraus, A. C., I. Ferber, S. O. Bachmann, H. Specht, A. Wimmel, M. W. Gross, J. Schlegel, G. Suske, and M. Schuermann.** 2002. In vitro chemo- and radio-resistance in small cell lung cancer correlates with cell adhesion and constitutive activation of AKT and MAP kinase pathways. *Oncogene* **21:**8683–8695.
58. **Kruh, G. D.** 2003. Introduction to resistance to anticancer agents. *Oncogene* **22:**7262–7264.
59. **Kruh, G. D., and M. G. Belinsky.** 2003. The MRP family of drug efflux pumps. *Oncogene* **22:**7537–7552.
60. **Kuismanen, S. A., M. T. Holmberg, R. Salovaara, P. Schweizer, L. A. Aaltonen, A. de La Chapelle, M. Nystrom-Lahti, and P. Peltomaki.** 1999. Epigenetic phenotypes distinguish microsatellite-stable and -unstable colorectal cancers. *Proc. Natl. Acad. Sci. USA* **96:**12661–12666.
61. **Lehmann, A. R.** 2001. The xeroderma pigmentosum group D (XPD) gene: one gene, two functions, three diseases. *Genes Dev.* **15:**15–23.
62. **Lehnert, M.** 1996. Clinical multidrug resistance in cancer: a multifactorial problem. *Eur. J. Cancer* **32A:**912–920.
63. **Lengauer, C., K. W. Kinzler, and B. Vogelstein.** 1997. Genetic instability in colorectal cancers. *Nature* **386:**623–627.
64. **Lengauer, C., K. W. Kinzler, and B. Vogelstein.** 1998. Genetic instabilities in human cancers. *Nature* **396:**643–649.
65. **Leonard, G. D., T. Fojo, and S. E. Bates.** 2003. The role of ABC transporters in clinical practice. *Oncologist* **8:**411–424.
66. **Less, J. R., T. C. Skalak, E. M. Sevick, and R. K. Jain.** 1991. Microvascular architecture in a mammary carcinoma: branching patterns and vessel dimensions. *Cancer Res.* **51:**265–273.
67. **Less, J. R., M. C. Posner, T. C. Skalak, N. Wolmark, and R. K. Jain.** 1997. Geometric resistance and microvascular network architecture of human colorectal carcinoma. *Microcirculation* **4:**25–33.
68. **Liang, F., M. Han, P. J. Romanienko, and M. Jasin.** 1998. Homology-directed repair is a major double-strand break repair pathway in mammalian cells. *Proc. Natl. Acad. Sci. USA* **95:**5172–5177.
69. **Lin, J. H., and M. Yamazaki.** 2003. Clinical relevance of P-glycoprotein in drug therapy. *Drug Metab. Rev.* **35:**417–454.
70. **Lindahl, T., and R. D. Wood.** 1999. Quality control by DNA repair. *Science* **286:**1897–1905.
71. **Loeb, L. A.** 1991. Mutator phenotype may be required for multistage carcinogenesis. *Cancer Res.* **51:**3075–3079.
72. **Loeb, L. A.** 2001. A mutator phenotype in cancer. *Cancer Res.* **61:**3230–3239.
73. **Lok, C. N., A. J. Lang, S. E. Mirski, and S. P. Cole.** 2002. Characterization of the human topoisomerase IIbeta (TOP2B) promoter activity: essential roles of the nuclear factor-Y (NF-Y)- and specificity protein-1 (Sp1)-binding sites. *Biochem. J.* **368:** 741–751.
74. **Lotfi, K., G. Juliusson, and F. Albertioni.** 2003. Pharmacological basis for cladribine resistance. *Leuk. Lymphoma* **44:**1705–1712.
75. **Lynch, H. T., and T. Smyrk.** 1998. An update on Lynch syndrome. *Curr. Opin. Oncol.* **10:**349–356.
76. **Lynch, H. T., T. Smyrk, and J. F. Lynch.** 1998. Molecular genetics and clinical-pathology features of hereditary nonpolyposis colorectal carcinoma (Lynch syndrome): historical journey from pedigree anecdote to molecular genetic confirmation. *Oncology* **55:**103–108.
77. **Marx, J.** 2002. Debate surges over the origins of genomic defects in cancer. *Science* **297:**544–546.
78. **Mattern, J.** 2003. Drug resistance in cancer: a multifactorial problem. *Anticancer Res.* **23:**1769–1772.
79. **Maubant, S., S. Cruet-Hennequart, L. Poulain, F. Carreiras, F. Sichel, J. Luis, C. Staedel, and P. Gauduchon.** 2002. Altered adhesion properties and alphav integrin expression in a cisplatin-resistant human ovarian carcinoma cell line. *Int. J. Cancer* **97:**186–194.
80. **Miturski, R., M. Bogusiewicz, C. Ciotta, M. Bignami, M. Gogacz, and D. Burnouf.** 2002. Mismatch repair genes and microsatellite instability as molecular markers for gynecological cancer detection. *Exp. Biol. Med. (Maywood)* **227:** 579–586.
81. **Miyake, H., I. Hara, K. Gohji, K. Yoshimura, S. Arakawa, and S. Kamidono.** 1998. Expression of basic fibroblast growth factor is associated with resistance to cisplatin in a human bladder cancer cell line. *Cancer Lett.* **123:**121–126.
82. **Miyaki, M., T. Ono, S. Hori, and H. Umezawa.** 1975. Binding of bleomycin to DNA in bleomycin-sensitive and -resistant rat ascites hepatoma cells. *Cancer Res.* **35:**2015–2019.
83. **Modrak, D. E., M. P. Draper, and S. B. Levy.** 1997. Emergence of different mechanisms of resistance in the evolution of multidrug resistance in murine erythroleukemia cell lines. *Biochem. Pharmacol.* **54:**1297–1306.
84. **Modrich, P.** 1997. Strand-specific mismatch repair in mammalian cells. *J. Biol. Chem.* **272:**24727–24730.
85. **Nagata, S., and P. Golstein.** 1995. The Fas death factor. *Science* **267:**1449–1456.
86. **Nielsen, D., C. Maare, and T. Skovsgaard.** 1996. Cellular resistance to anthracyclines. *Gen. Pharmacol.* **27:**251–255.
87. **Obata, T., Y. Endo, D. Murata, K. Sakamoto, and T. Sasaki.** 2003. The molecular targets of antitumor 2′-deoxycytidine analogues. *Curr. Drug Targets* **4:**305–313.
88. **Ohi, R., and K. L. Gould.** 1999. Regulating the onset of mitosis. *Curr. Opin. Cell Biol.* **11:**267–273.
89. **Ohta, M., H. Inoue, M. G. Cotticelli, K. Kastury, R. Baffa, J. Palazzo, Z. Siprashvili, M. Mori, P. McCue, T. Druck, and et al.** 1996. The FHIT gene, spanning the chromosome 3p14.2 fragile site and renal carcinoma-associated t(3;8) breakpoint, is abnormal in digestive tract cancers. *Cell* **84:**587–597.
90. **Orr, G. A., P. Verdier-Pinard, H. McDaid, and S. B. Horwitz.** 2003. Mechanisms of Taxol resistance related to microtubules. *Oncogene* **22:**7280–7295.
91. **Pavletich, N. P.** 1999. Mechanisms of cyclin-dependent kinase regulation: structures of Cdks, their cyclin activators, and Cip and INK4 inhibitors. *J. Mol. Biol.* **287:**821–828.
92. **Peinado, M. A., S. Malkhosyan, A. Velazquez, and M. Perucho.** 1992. Isolation and characterization of allelic losses and gains in colorectal tumors by arbitrarily primed polymerase chain reaction. *Proc. Natl. Acad. Sci. USA* **89:**10065–10069.
93. **Peltomaki, P., and H. F. Vasen.** 1997. Mutations predisposing to hereditary nonpolyposis colorectal cancer: database and results of a collaborative study. The International Collaborative Group on Hereditary Nonpolyposis Colorectal Cancer. *Gastroenterology* **113:**1146–1158.
94. **Pupa, S. M., S. Menard, S. Forti, and E. Tagliabue.** 2002. New insights into the role of extracellular matrix during tumor onset and progression. *J. Cell. Physiol.* **192:**259–267.
95. **Radin, N. S.** 2001. Killing cancer cells by poly-drug elevation of ceramide levels: a hypothesis whose time has come? *Eur. J. Biochem.* **268:**193–204.
96. **Rasheed, Z. A., and E. H. Rubin.** 2003. Mechanisms of resistance to topoisomerase I-targeting drugs. *Oncogene* **22:** 7296–7304.
97. **Reed, J. C.** 1995. Regulation of apoptosis by bcl-2 family proteins and its role in cancer and chemoresistance. *Curr. Opin. Oncol.* **7:**541–546.
98. **Richards, R. I.** 2001. Fragile and unstable chromosomes in cancer: causes and consequences. *Trends Genet.* **17:**339–345.

99. **Rintoul, R. C., and T. Sethi.** 2002. Extracellular matrix regulation of drug resistance in small-cell lung cancer. *Clin. Sci.* (London). **102:**417–424.
100. **Rotman, G., and Y. Shiloh.** 1998. ATM: from gene to function. *Hum. Mol. Genet.* **7:**1555–1563.
101. **Saijo, N., K. Nishio, N. Kubota, F. Kanzawa, T. Shinkai, A. Karato, Y. Sasaki, K. Eguchi, T. Tamura, Y. Ohe, and et al.** 1994. 7-Ethyl-10-[4-(1-piperidino)-1-piperidino] carbonyloxy camptothecin: mechanism of resistance and clinical trials. *Cancer Chemother. Pharmacol.* **34**(Suppl):S112–S117.
102. **Schabel, F. M., Jr., M. W. Trader, W. R. Laster Jr., G. P. Wheeler, and M. H. Witt.** 1978. Patterns of resistance and therapeutic synergism among alkylating agents. *Antibiot. Chemother.* **23:**200–215.
103. **Schaeffer, L., V. Moncollin, R. Roy, A. Staub, M. Mezzina, A. Sarasin, G. Weeda, J. H. Hoeijmakers, and J. M. Egly.** 1994. The ERCC2/DNA repair protein is associated with the class II BTF2/TFIIH transcription factor. *EMBO J.* **13:**2388–2392.
104. **Schmidt, M., and R. B. Lichtner.** 2002. EGF receptor targeting in therapy-resistant human tumors. *Drug Resist. Update* **5:**11–18.
105. **Scotto, K. W.** 2003. Transcriptional regulation of ABC drug transporters. *Oncogene* **22:**7496–7511.
106. **Senchenkov, A., D. A. Litvak, and M. C. Cabot.** 2001. Targeting ceramide metabolism—a strategy for overcoming drug resistance. *J. Natl. Cancer Inst.* **93:**347–357.
107. **Sethi, T., R. C. Rintoul, S. M. Moore, A. C. MacKinnon, D. Salter, C. Choo, E. R. Chilvers, I. Dransfield, S. C. Donnelly, R. Strieter, and C. Haslett.** 1999. Extracellular matrix proteins protect small cell lung cancer cells against apoptosis: a mechanism for small cell lung cancer growth and drug resistance in vivo. *Nat. Med.* **5:**662–668.
108. **Shain, K. H., T. H. Landowski, and W. S. Dalton.** 2002. Adhesion-mediated intracellular redistribution of c-Fas-associated death domain-like IL-1-converting enzyme-like inhibitory protein-long confers resistance to CD95-induced apoptosis in hematopoietic cancer cell lines. *J. Immunol.* **168:**2544–2553.
109. **Shannon, A. M., D. J. Bouchier-Hayes, C. M. Condron, and D. Toomey.** 2003. Tumour hypoxia, chemotherapeutic resistance and hypoxia-related therapies. *Cancer Treat. Rev.* **29:**297–307.
110. **Shen, D. W., C. Cardarelli, J. Hwang, M. Cornwell, N. Richert, S. Ishii, I. Pastan, and M. M. Gottesman.** 1986. Multiple drug-resistant human KB carcinoma cells independently selected for high-level resistance to colchicine, adriamycin, or vinblastine show changes in expression of specific proteins. *J. Biol. Chem.* **261:**7762–7770.
111. **Siddik, Z. H.** 2003. Cisplatin: mode of cytotoxic action and molecular basis of resistance. *Oncogene* **22:**7265–7279.
112. **Sivridis, E., A. Giatromanolaki, and M. I. Koukourakis.** 2003. The vascular network of tumours—what is it not for? *J. Pathol.* **201:**173–180.
113. **Slapak, C. A., J. C. Daniel, and S. B. Levy.** 1990. Sequential emergence of distinct resistance phenotypes in murine erythroleukemia cells under adriamycin selection: decreased anthracycline uptake precedes increased P-glycoprotein expression. *Cancer Res.* **50:**7895–7901.
114. **Slapak, C. A., P. M. Fracasso, R. L. Martell, D. L. Toppmeyer, J. M. Lecerf, and S. B. Levy.** 1994. Overexpression of the multidrug resistance-associated protein (MRP) gene in vincristine but not doxorubicin-selected multidrug-resistant murine erythroleukemia cells. *Cancer Res.* **54:**5607–5613.
115. **St Croix, B., and R. S. Kerbel.** 1997. Cell adhesion and drug resistance in cancer. *Curr. Opin. Oncol.* **9:**549–556.
116. **Teicher, B. A., T. S. Herman, S. A. Holden, Y. Y. Wang, M. R. Pfeffer, J. W. Crawford, and E. Frei, 3rd.** 1990. Tumor resistance to alkylating agents conferred by mechanisms operative only in vivo. *Science* **247:**1457–1461.
117. **Thomas, H., and H. M. Coley.** 2003. Overcoming multidrug resistance in cancer: an update on the clinical strategy of inhibiting p-glycoprotein. *Cancer Control.* **10:**159–165.
118. **Thorland, E. C., S. L. Myers, D. H. Persing, G. Sarkar, R. M. McGovern, B. S. Gostout, and D. I. Smith.** 2000. Human papillomavirus type 16 integrations in cervical tumors frequently occur in common fragile sites. *Cancer Res.* **60:**5916–5921.
119. **Townsend, D. M., and K. D. Tew.** 2003. The role of glutathione-S-transferase in anti-cancer drug resistance. *Oncogene* **22:**7369–7375.
120. **Tsuruo, T., T. C. Hamilton, K. G. Louie, B. C. Behrens, R. C. Young, and R. F. Ozols.** 1986. Collateral susceptibility of adriamycin-, melphalan- and cisplatin-resistant human ovarian tumor cells to bleomycin. *Jpn. J. Cancer Res.* **77:**941–945.
121. **Vassetzky, Y. S., G. C. Alghisi, and S. M. Gasser.** 1995. DNA topoisomerase II mutations and resistance to anti-tumor drugs. *Bioessays* **17:**767–774.
122. **Vogel, U., M. Hedayati, M. Dybdahl, L. Grossman, and B. A. Nexo.** 2001. Polymorphisms of the DNA repair gene XPD: correlations with risk of basal cell carcinoma revisited. *Carcinogenesis.* **22:**899–904.
123. **Wartenberg, M., F. Donmez, F. C. Ling, H. Acker, J. Hescheler, and H. Sauer.** 2001. Tumor-induced angiogenesis studied in confrontation cultures of multicellular tumor spheroids and embryoid bodies grown from pluripotent embryonic stem cells. *FASEB J.* **15:**995–1005.
124. **Weinshilboum, R.** 2001. Thiopurine pharmacogenetics: clinical and molecular studies of thiopurine methyltransferase. *Drug Metab. Dispos.* **29:**601–605.
125. **Weinstein, J. N., T. G. Myers, P. M. O'Connor, S. H. Friend, A. J. Fornace Jr., K. W. Kohn, T. Fojo, S. E. Bates, L. V. Rubinstein, N. L. Anderson, J. K. Buolamwini, W. W. van Osdol, A. P. Monks, D. A. Scudiero, E. A. Sausville, D. W. Zaharevitz, B. Bunow, V. N. Viswanadhan, G. S. Johnson, R. E. Wittes, and K. D. Paull.** 1997. An information-intensive approach to the molecular pharmacology of cancer. *Science* **275:** 343–349.
126. **Wood, R. D., M. Mitchell, J. Sgouros, and T. Lindahl.** 2001. Human DNA repair genes. *Science* **291:**1284–1289.
127. **Yang, J., Z. P. Xu, Y. Huang, H. E. Hamrick, P. J. Duerksen-Hughes, and Y. N. Yu.** 2004. ATM and ATR: sensing DNA damage. *World J. Gastroenterol.* **10:**155–160.
128. **Zhao, R., and I. D. Goldman.** 2003. Resistance to antifolates. *Oncogene* **22:**7431–7457.

PART VII. POLICY, EDUCATION, AND EXPLORATION

SECTION EDITOR: Michael N. Alekshun

INVESTIGATING THE MECHANISMS OF DRUG RESISTANCE at the fundamental level in the laboratory is a central component toward understanding how future resistance will arise and how it might be curtailed. It is, however, equally important to appreciate that the injudicious use of antibiotics in society (e.g., in our hospitals, communities, and farms) has a profound impact on the development and dissemination of antibiotic-resistant organisms. These scenarios threaten to undermine antibiotic efficacy. While many within the pharmaceutical industry have tried to overcome antibiotic-resistant bacteria by producing more potent antibiotics, most efforts have been lackluster in specifically addressing the resistance mechanisms themselves.

From the beginning, Stuart Levy has been a passionate supporter of appropriate and rational antibiotic use. The chapters that follow depict his contributions to policy and education as well as the development of new anti-infective therapies. The Alliance for the Prudent Use of Antibiotics is a worldwide nonprofit organization that was founded by Stuart Levy in the early 1980s with a mission to educate the public on appropriate antibiotic use in hopes of maximizing their future utility. The chapter on antimicrobial use in animals succinctly describes our current understanding of the issues at hand as well as the regulatory issues facing new and approved drugs. The main section of the book ends with a description of Stuart Levy's foray into the pharmaceutical industry at Paratek Pharmaceuticals, Inc., a company he co-founded with Nobel laureate Walter Gilbert, whose mission is to develop novel therapies to prevent, combat, and cure serious diseases.

Frontiers in Antimicrobial Resistance: a Tribute to Stuart B. Levy
Edited by D. G. White, M. N. Alekshun, and P. F. McDermott

COMMENTARY

The Birth of the Alliance for the Prudent Use of Antibiotics (APUA)

ELLEN KOENIG

The scene was typical—a group of microbiologists at the yearly American Society for Microbiology meeting sitting around talking about their projects and trying to decide where to have their next plasmid meeting. Previously it had been held at a castle in Europe, on Mediterranean beaches, and at other exciting places. Intrepidly I suggested the Dominican Republic. What better place, I suggested, since antibiotics were sold freely in the pharmacies and were included in animal feed and the weather was warm when the United States was deep in winter. The year was 1980.

To my surprise and delight, the decision was made to further explore this possibility; I would head up the local arrangements, and my brother Stuart Levy and Royston Clowes would handle the scientific part of the meeting. And so we began. I contacted one of the better hotels at that time and was given an acceptable rate. I called on local companies to offer welcoming gifts to the participants and to provide food and drinks during the afternoon poster sessions. Finally, as we were on a tight budget, we secured the buses from the armed forces to transport the participants from the airport to the hotel, complete with military personnel as baggage handlers.

Medical students from my microbiology class served as the welcoming committee, working endless shifts to accommodate the different flights that came in and resolve problems such as those people arriving without the necessary visa. In particular, I remember the Grecian delegate, Dr. Polyxeni Kontomichalou, who was assured that with her medal showing membership in the Greek Academy of Science she would have no problem entering the country, even though the visa we sent to her in England did not arrive. What a miscalculation. At midnight I was informed that they were holding her in a room at the airport and she would be put on the 6 a.m. plane out of the country. My husband quickly moved into action, finding the name and phone number of the head of the Dominican government organization that had the power to grant her entrance into the Dominican Republic. Because it was the night of January 3, he was in good spirits, although disturbed to be awakened by such a call during the holidays; he gladly gave blanket permission to all coming to the conference.

Our plan was to engage the participants in organized meetings in the morning and early evening with the poster sessions beginning in the late afternoon. In this way, those who came could maximize their attendance at the sessions but could spend the afternoons enjoying the Caribbean sun with their families or exploring Santo Domingo, one of the oldest cities, having numerous historic areas, in the New World. Nearby beaches could also be reached in an hour, and scuba diving was available at a fraction of the U.S. cost. Most importantly, we hoped that the relaxed area around the pool would encourage informal interchanges between participants as well.

Had delegates gone to the pharmacies, they would have seen the ever-present evidence of the need to control antibiotics, especially in developing countries. Anyone could and still can walk into a pharmacy and buy antibiotics, and if he or she is not sure which one to get, the pharmacist will gladly suggest a drug. In addition, combinations were particularly prevalent. Antibiotics with vitamins, aspirin, and cough suppressants could be found, as well as in combinations with other antibiotics in preparations not registered in the more-industrialized world. Antibiotics were also present in the animal food supply.

The meetings were full of important information from the then high-tech information-gathering tools, as well as the reconfirmation of what was appearing at that time: penicillin-resistant gonorrhea. This organism was not just an Asian phenomenon but relevant to the rest of the world. By this time it had started its spread throughout the planet.

Ellen Koenig • Dominican Institute of Virological Studies (IDEV), Dr. Pineyro 211, Zona Universitario, Santo Domingo, Dominican Republic.

Plenum Press published all the presentations from the meeting as well as extended abstracts in their volume *Molecular Biology, Pathogenicity, and Ecology of Bacterial Plasmids*. At the end is included the "statement regarding worldwide antibiotic misuse" signed by all the participants representing over 25 different countries which formed the basis for the Alliance for the Prudent Use of Antibiotics (APUA). Also, as though cooperating with the meeting's purpose, the local environment gave the close to 200 participants something to study among themselves, gastroenteritis of the tourist, commonly known as "turista." Of the 114 who responded to the questionnaire, 67 had suffered some form of "turista," the first case appearing on the second day of the conference and the peak being seen at days 4 and 5. Two cases of multiple resistant *Shigella sonnei* were isolated. The eating of cold salad was the factor most highly associated with illness. Investigators who had traveled before were less likely to eat salad and therefore were less likely to become infected. The findings led to the conclusion that "plasmid investigators should travel more and eliminate cold salads from their diets." The presence of plasmids was not determined at that time, but I am sure all of the participants in this volume will agree with these conclusions from 24 years ago.

From this beginning and Stuart's ever present fear of the bacterial victory in the antibiotic war was born APUA. This organization is really Stuart's first child. He has nurtured it, worried about it, pampered it, and watched it grow into the force it is today as a reference for antibiotic use and misuse. Single-handedly, I would venture, he kept it going when the odds against it looked indomitable.

Stuart's passion for this area of investigations has been passed on to many, including me. In the Dominican Republic, with, as I have mentioned, its lax laws concerning antibiotic use and no watchdog organization, antibiotic misuse is rampant. For example, with Stuart's tutelage we examined prescriptions at rural clinics near Santo Domingo and found that almost all prescriptions for antibiotics were given without cultures and antibiograms. Furthermore, antibiotics were being prescribed for scabies or even Steven-Johnson's Disease, not to mention the errors in dosage, incorrect routes of administration, or presentation. Furthermore, we found a specific medicine prescribed on various occasions when it was obviously having no effect on the infection.

To correct such occurrences Stuart, through APUA, has tried to take antibiotic education to countries of the world through its local chapters. Pamphlets for physicians' offices have been disseminated, and local laws now control the dispensing of the medicines. Unfortunately, in the Dominican Republic, as in much of the world, physicians continue to prescribe antibiotics for the common cold, diarrhea, and other ills just to satisfy their patients. Ignorance leads to mothers saving half the prescribed antibiotic from one child and starting it on the next and then running to the doctor as the illness in the second child gets worse. The ignorance is not only on the part of the parents but the physicians as well. They continue to believe that antibiotics are a cure-all, even in the case of nonbacterial infections. Furthermore, the use of broad-spectrum antibiotics, antibiotics without antibiograms, and medicines with severe side effects although less toxic ones are now on the market continues today. The work of APUA is just scratching the surface in countries such as the Dominican Republic and must continue.

Time can only tell whether these efforts will succeed, but we would be a lot further behind and less likely to succeed had Stuart not been so active in advocating his belief that a group of interested individuals can and will make a difference in this area, in helping to create APUA, and in maintaining his vision for the past 25 years.

Frontiers in Antimicrobial Resistance: a Tribute to Stuart B. Levy
Edited by D. G. White, M. N. Alekshun, and P. F. McDermott

Chapter 38

Alliance for the Prudent Use of Antibiotics: Scientific Vision and Public Health Mission

KATHLEEN T. YOUNG AND THOMAS F. O'BRIEN

September 11, 2001 was to be a milestone for Stuart Levy and the scientific and public health organization he had founded 20 years earlier, the Alliance for the Prudent Use of Antibiotics (APUA). That morning, at a national press conference in Washington, D.C., Levy and APUA executive director Kathleen Young were scheduled to join leaders from the World Health Organization (WHO), the World Bank, and the U.S. Centers for Disease Control and Prevention (CDC) to focus the world's attention on the global increase of antimicrobial resistance. Levy had been asked to lead the 10 a.m. conference to announce the 2001 release of the WHO's *Global Strategy for Containment of Antimicrobial Resistance*—a landmark document in which APUA had played a major contributing role (18).

But APUA's moment in the spotlight would have to wait. That morning, as military personnel ran toward the burning Pentagon, Levy and the APUA staff hired a car for an exhausting, circuitous, 20-hour drive back to Boston. The WHO, which was about to alert the world to antimicrobial resistance as one of the three major public health threats of the 21st century, released its monumental report without the scheduled fanfare.

Three weeks later, anthrax mysteriously broke out in the United States, spread by highly refined spores sent through the mail. Americans were urged to stock up on ciprofloxacin, a touted treatment for the infection. As events unfolded, Stuart Levy and APUA were voices of reason and calm, explaining why stockpiling antibiotics and prophylactic use were unwise. Meanwhile, the APUA website (http://www.apua.org) monitored the epidemic, offering insight and reassurance to thousands of citizens. Both in the rich scientific analysis provided to the WHO in their global strategy and in the nimble response to a public health crisis, APUA did what it does best: marshal timely scientific evidence and convey it thoroughly, thoughtfully, and speedily to those who need it.

APUA was launched as a global nonprofit organization in 1981 with the goals of improving antimicrobial use and containing drug resistance worldwide.

APUA's stewardship role is based on the premise that each individual antimicrobial use has broader ecological and community impact. Dovetailing with WHO's longstanding Essential Medicines program, APUA emphasizes the need for dissemination of guidelines and resistance information, independent of commercial interests. The organization aims to improve current infectious disease treatment today while prolonging the effective life span of antibiotics for the children of tomorrow.

THE COSTS OF ANTIMICROBIAL MISUSE

Antimicrobial resistance complicates treatment choices and places affected patients at increased risk of adverse outcomes. Resistance also adds to ever-rising health care costs. Multidrug-resistant tuberculosis (TB), for instance, is 100 times more costly to treat than drug-susceptible TB. In the United States, more than half of all hospital *Staphylococcus aureus* infections are methicillin resistant, a spiraling problem that has led to increased morbidity and mortality, higher pharmacy and drug monitoring costs, and even ward closures. Overall in the United States, drug-resistant microbes are estimated to cost $30 billion per year (5) (Table 1).

According to a recent WHO report of 10 studies undertaken at teaching hospitals worldwide, between 40 and 91% of antibiotics were inappropriately prescribed (18). In numerous analyses from rich and poor

Kathleen T. Young • Alliance for the Prudent Use of Antibiotics, 75 Kneeland Street, Boston, MA 02111. **Thomas F. O'Brien** • Medical Director, Microbiology Laboratory, Brigham & Women's Hospital, 75 Francis Street, Boston, MA 02115-6110.

Table 1. Annual costs of common infectious diseases to the United States[a]

Disease	Estimated annual cost[b]
HIV/AIDS	$3 billion in Public Health Service Funds
Tuberculosis	$343 million in Public Health Service funds, $700 million in direct treatment costs
Nosocomial (hospital-acquired) infections	$10 billion in direct treatment costs
Sexually transmitted diseases (excluding AIDS)	$5 billion in direct treatment costs
Intestinal infections	$23 billion in direct medical and lost productivity costs
Drug-resistant infections	$4 billion (and increasing) in treatment costs
Influenza	$5 billion in direct treatment costs, 12 billion in lost productivity costs

[a] Adapted from CISET (7a).
[b] Costs of monitoring or containing bioterrorist events are not included.

nations, APUA and its chapters have documented massive misuse of these powerful life saving agents and demonstrated that improving antibiotic use and curtailing antimicrobial resistance saves money without discernable adverse health outcomes (see http://www.apua.org). The APUA-Chile chapter made the case for government enforcement of a moribund prescription law and then documented the resulting 30 to 53% decrease in consumption of various antibiotics nationwide within the following 3 years. The APUA-Russia chapter demonstrated that an antibiotic education and formulary program cut $2.2 million in hospital spending on antibiotics and led to decreased resistance. In Sweden, the institution of a national education and awareness program by the local APUA chapter resulted in a reduction of 22% in inappropriate antibiotic use from 1993 to 1997.

As the "go to" place for drug resistance information for nearly a quarter of a century, APUA deals with the threats of resistant infectious diseases with the dedication and scarce resources that characterize many public health efforts.

AN ANSWER TO ESCALATING EPIDEMICS

APUA's organizational reach and longstanding dedication, however, could never have coalesced without the zeal and commitment of its creator. In 1981, when Dr. Levy launched APUA, new and emerging infectious diseases were suddenly looming on the horizon. Starting in the mid-1970s, a new public health vocabulary had taken hold: Legionnaires' Disease, Lyme disease, toxic shock syndrome, *Escherichia coli* O157:H7 infections, sexually transmitted diseases (STDs), Ebola virus, etc. In the realm of bacterial diseases, where drug-resistance was a well-known phenomenon, the danger seemed to be accelerating (Table 2). While investigators had isolated the first penicillin-resistant pneumococcus in 1967, only 10 years later drug-resistant strains of the organism spread to South Africa. In the 1970s, other common infections also began to change. Within a year after the introduction of methicillin, researchers found staphylococci that defied this new synthetic agent. *Haemophilus influenzae* turned penicillin resistant. Gonorrhea caused by penicillin-resistant strains of *Neisseria gonorrhoeae* spread from brothels in Vietnam. Penicillin-resistant pneumococci had begun to be reported in a few countries. Gram-negative bacilli resistant to gentamicin and many other antimicrobials had emerged over the previous 5 years and were causing hospital outbreaks. Bacterial resistance to trimethoprim was increasing. Multidrug-resistant *Shigella dysenteriae* would claim hundreds of lives in Zaire, while the United States faced an epidemic of multidrug-resistant *Salmonella* (7a).

And then, a June 1981 CDC report found a strange cluster of fatal symptoms among five gay men in Los Angeles, Calif. AIDS, as we now know, would go on to become the most deadly infection in human history. Indeed, at this pivotal moment in the history of infectious disease, just as Stuart Levy was launching in Boston a frontline public health organization devoted to antibiotic resistance, his twin brother, Dr. Jay Levy, was making groundbreaking discoveries about AIDS in San Francisco, Calif.

COUNTRIES COALESCE AND SHED THE SHAME

Realizing that there was no organized channel for communicating drug-resistance data to guide infectious disease policy and practice globally, Stuart Levy envisioned a worldwide alliance able to act at the grassroots level and engage researchers, prescribers, and governments as allies in the battle against resistance. In January 1981, more than 200 clinicians and scientists convened in Santo Domingo, Dominican Republic, to discuss concerns about antibiotic treatment problems and the ecology of resistance. The messages

Table 2. Top seven antibiotic resistance problems in the U.S. in 2004[a]

Resistant bacterium[b]	Impact	Estimated trends
MRSA	Major cause of hospital-acquired infections, including surgical infections and septicemia.	50–60% of all hospital *S. aureus* are MRSA
CA-MRSA	Strains of MRSA traditionally found in hospitals have unexpectedly emerged in the community, exposing the broad population to resistant infections. Increased morbidity and mortality in hospitals; higher pharmacy and drug monitoring costs for last-line antibiotics; ward closures. CA-MRSA threatens the broad nonhospitalized population.	CA-MRSA now represents 10% of all MRSA infections in the U.S. It is epidemic in certain community populations.
VISA	Strains of MRSA with reduced susceptibility to vancomycin were first described in Japan in 1996.	Signs that resistance to this "antibiotic of last resort" are now emerging in the U.S.
VRSA	Recently, two strains of MRSA fully resistant to vancomycin (the antibiotic of last resort) have appeared in U.S. Failure of 'last-line' antibiotics. Few new agents on the horizon; need to use potentially less safe, more expensive new antibiotics.	
VRE	VRE cause serious infections in transplant, cancer, and dialysis patients. Like MRSA, causes hospital "epidemics," which require very large resources to control.	Now present in most U.S. hospitals
DRSP	Pneumococci are responsible for most middle ear and sinus infections, acute and chronic bronchitis, pneumonia, and bacterial meningitis. Require more-expensive first-line treatments; may fail, resulting in hospitalization; 50% increase in costs of hospitalization for resistant pneumonia.	28–35% prevalence; rose 25% over the last decade
MDR *Mycobacterium tuberculosis*	MDR TB is rising, especially in inner-city, foreign-born, and HIV patients. Further spread of these strains can have an impact for up to 100 years, as acquisition in childhood may not clinically manifest itself until old age. Need for prolonged treatment (12 months) with very expensive combined alternative therapy. A failure of treatment leads to wider spread of disease.	Multidrug resistance among foreign born individuals rose from 31% in 1993 to 72% in 2002
Resistant *Haemophilus influenzae* and *Moraxella catarrhalis*	These bacteria account for most other middle ear and sinus infections and acute and chronic bronchitis. More expensive first-line treatments required; failed treatment results in hospitalization.	25% resistance in *H. influenzae,* 90% resistance in *M catarrhalis*
MDR *E. coli*	MDR *E. coli* is now common in hospitals and in the community. These organisms cause resistant urinary tract infections, abdominal surgery infections, and septicemia. Animals and food are suspected as potential sources for some strains. More expensive and broader spectrum antibiotics needed for treatment; resistance to our strongest agents, such as fluoroquinolones, is rising.	*E. coli* resistance to first -line agents now approaches 25% in some areas, and this level rose 10-fold in past 5 years. Ciprofloxacin resistance is currently at 7% and rising.

[a] Compiled by the staff of the Alliance for the Prudent Use of Antibiotics from various sources (1, 5, 6, 7, 7a, 12).
[b] Abbreviations: MRSA, methicillin-resistant *S. aureus;* CA-MRSA, community-acquired methicillin-resistant *S. aureus;* VISA, vancomycin-intermediate (resistant) *S. aureus;* VRSA, vancomycin-resistant *S. aureus;* VRE, vancomycin-resistant *E. faecalis* and *E. faecium;* DRSP, drug-resistant *S. pneumoniae;* MDR, multidrug resistant.

were clear: antibiotic resistance was universal (although manifested in different ways in different countries), and antibiotics were being misused everywhere.

Out of this meeting came an "Antibiotic Misuse Statement" endorsed by hundreds of clinicians, researchers, and scientific societies. "We are faced with a worldwide public health problem," it declared (12). On 4 August 1981, the statement drafted by Levy and translated into several languages was formally presented at a press conference that took place simultaneously in Boston, Mass; Santo Domingo, Dominican Republic; Mexico City, Mexico; and Sao Paulo, Brazil. The public responded with a flurry of questions, and the issue was immediately picked up by *Newsweek, Time, The New York Times, The Washington Post,* and other media outlets, as well as by scientific journals. Suddenly, antibiotic resistance was on the public's radar screen.

TENACITY AND COLLABORATION: THE APUA STYLE

While APUA focuses on a single therapeutic modality, its work affects myriads of medical therapies and millions of people of all ages, economic strata, and nations. Drug resistance is a concept that is more difficult to explain and understand than the dangers of any single disease. Dr. Levy and APUA have turned out to be masters at communicating this complexity (12).

As a far-flung extension of the meeting in Santo Domingo, APUA set up headquarters in Boston at a small desk outside of Dr. Levy's lab in 1981 and became registered in Massachusetts as a nonprofit 501(c)(3) organization. For the first 15 years, the organization was more or less run out of Stuart Levy's back pocket with meager financial support and an annual operating budget rarely exceeding the low four-figure mark. With little public recognition of the problem, it is unlikely that APUA could have survived, let alone accomplished as much as it did in those days, without the commitment and tenacity of Stuart Levy. He relished any David-and-Goliath challenge before him and offered the staff and board the optimism that spurs on the organization.

The complex causes of antimicrobial resistance can easily engender entrenched antagonists. With humor and patience, Stuart Levy is often able to gain well-deserved trust, defuse antagonisms, and pave the way for progress. He has a unique leadership style that enables him to bridge deep divides between stakeholders in this highly contentious arena of government and industry and trade association interests. Dr. Richard Besser, medical director of the CDC's "Get Smart" campaign, has described Levy's early work with APUA as a lone voice cautioning more careful use of antibiotics. Now most professional and public health organizations include drug resistance as one of their top priorities (10).

APUA EVOLVES AND DEVELOPS KEY PARTNERSHIPS

From 1981 to 1996, Stuart Levy ran APUA with the help of a part-time secretary. He wrote occasional press releases and was widely cited in the media. Joining APUA in those early days as vice president was Dr. Thomas O'Brien, an infectious disease specialist, microbiologist, and leading authority on antimicrobial resistance surveillance. Dr. O'Brien and Dr. John Stelling codeveloped the WHONET software and are now codirectors of the WHO Collaborating Center on Antimicrobial Resistance Surveillance at Brigham and Women's Hospital in Boston, Mass., an organization helping to track resistance trends worldwide.

By 1996, the board of directors realized that APUA's media outreach, public advocacy, and informal networking were not enough. To broaden its reach, APUA needed to expand its programs and links to government and professional societies. Soon thereafter, a dedicated space was found in an adjacent building on the Tufts University Health Sciences Campus, and Kathy Young, a veteran in healthcare policy and organizational development, was hired as APUA's first's full-time staff person and executive director. Young developed an organizational structure and established professional divisions dedicated to research, education, and international field work. Her other priorities were development of key multiyear partnerships with the CDC, WHO, and the U.S. Agency for International Development (USAID) and positioning the organization to tackle controversial policy issues to broaden APUA's reach in the developing world (Fig. 1 and 2).

Today, the world is beset by a widening array of antimicrobial threats—with emerging resistance to treatments for HIV/AIDS drugs, tuberculosis, and malaria, as well as the emergence of multidrug-resistant strains of *S. aureus, Streptococcus pneumoniae,* and *E. coli.* More than two decades after its humble start, APUA now stands at the forefront of antimicrobial research and interventions in over 50 countries and as the "go-to" group for timely data, rigorous analysis, and effective collaborative action.

With deep roots in bench science and extensive field experience, the APUA staff, advisory board, and chapters serve on the front lines of the global battle against infectious disease and provide grassroots leadership for implementation of the United States and WHO strategic drug resistance plans (17, 18). In 2002, APUA expanded its scope in response to pressing global epidemics of HIV, malaria, and tuberculosis. Now serving on the APUA Scientific Advisory Board are three Nobel Prize winners as well as Dr. Paul Farmer, medical director, Clinique Bon Saveur and the Institute for Health and Social Justice in Haiti and professor in the Program in Infectious Disease and Social Change at Harvard Medical School; Dr. Thomas Wellems, cochief, Laboratory of Malaria and Vector Research at NIH in Bethesda, Md.; Dr. Jay Levy, director of the Laboratory for Tumor and AIDS Virus Research at the University of California, San Francisco School of Medicine; and other international experts.

ROAR: RESERVOIRS OF ANTIBIOTIC RESISTANCE

APUA was established with strong roots in the American Society for Microbiology (ASM), the world's largest scientific organization. A number of past ASM presidents, including Abigail Salyers of the

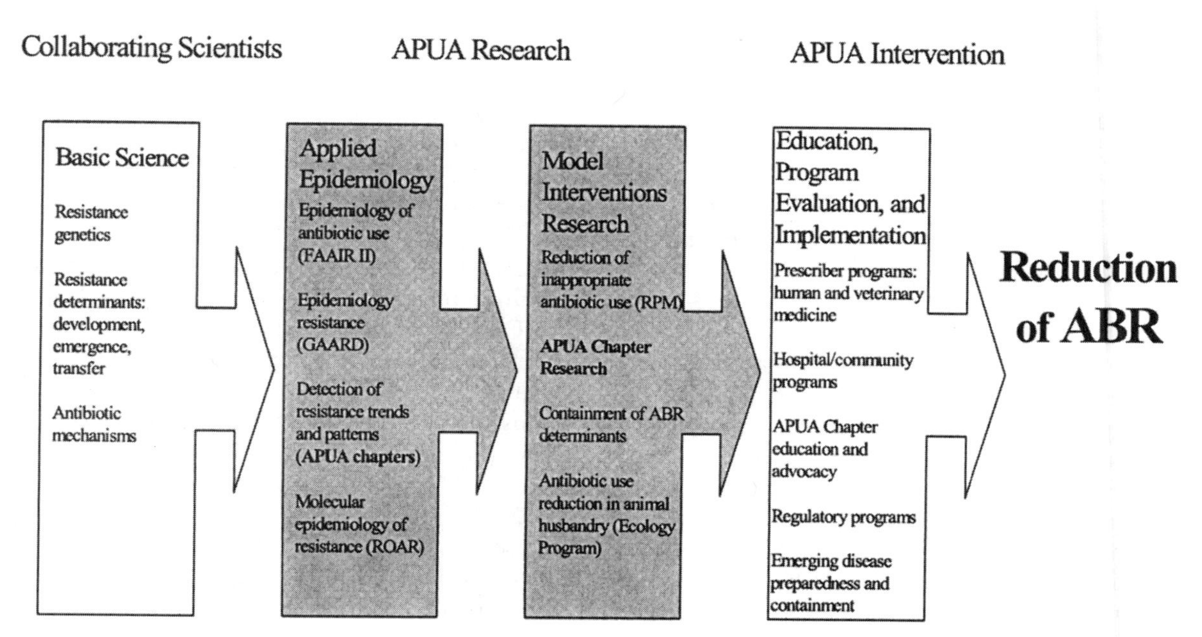

Figure 1. APUA's antibiotic effectiveness program.

University of Illinois; Julian Davies of the University of British Columbia in Vancouver, Canada; James Tiedje of Michigan State University; and Stuart Levy himself have all contributed to APUA's work. In 1998, Abigail Salyers joined with APUA to apply for a modest grant from the National Institute for Allergies and Infectious Diseases of Health (NIAID). The resulting project, known as ROAR (Reservoirs of Antibiotic Resistance), called for formation of a committee of scientists as well as some internet communications vehicles to explore antibiotic resistance in commensal organisms (search http://www.apua.org).

In its typically ambitious fashion, APUA charged forward with minimal funding to establish the leading

Figure 2. APUA country chapters. Chapters are located in Argentina, Australia, Bangladesh, Belarus, Bolivia, Brazil, Bulgaria, Chile, China, Columbia, Costa Rica, Croatia, Cuba, Dominican Republic, Ecuador, El Salvador, Fiji Islands, Georgia, Greece, Guatemala, India, Italy, Kazakhstan, Kenya, Lebanon, Mexico, Moldova, Nepal, Nicaragua, Nigeria, Pakistan, Panama, Paraguay, Peru, Philippines, Poland, Russia, Senegal, Serbia and Montenegro, South Korea, Spain, Sweden, Taiwan, Turkey, Ukraine, United Kingdom, Uganda, Uruguay, Venezuela, Vietnam, and Zambia.

international scientific network organization investigating resistance in commensals. Not long after the initial grant award, Senators Edward Kennedy and Bill Frist convened a hearing on antibiotic resistance before their Senate Committee on Health, Education, Labor and Pensions. Testifying for APUA, Dr. Levy noted the paucity of national attention to this increasing problem. Dr. Anthony Fauci, director of the National Institute for Allergy and Infectious Diseases, noted that the NIH was aware of the problem and active in the antibiotic resistance field. He displayed a chart with a list of activities highlighting ROAR, APUA's tiny, self-named project, as one of the government's flagship programs on antibiotic resistance.

The initial grant for the ROAR project did indeed signal an increase in NIH commitment to antimicrobial resistance research that heretofore could be characterized as an "orphan" research area. Through the leadership of the NIAID director, spurred on by four talented public servants, antimicrobial resistance was elevated on the national agenda. In a unique U.S. government interagency effort, Marissa Miller of the NIAID (now senior advisor for Public Health and Emergency Preparedness for Department of Health and Human Services) and her counterparts, Murray Lumpkin of the U.S. Food and Drug Administration (FDA), David Bell at the CDC, and Paula Cray at USDA, succeeded in bringing the resistance issue into the government limelight (17). ROAR has since grown into a 5-year, $3 million cooperative agreement with NIAID, involving many of the world's top investigators studying resistance transfer from nonpathogenic bacteria to disease causing ones, producing an international database for analysis of phenotypic and genotypic data, testing risk assessment tools, and serving as a resource for investigators worldwide.

GAARD: GLOBAL ADVISORY ON ANTIBIOTIC RESISTANCE DATA NETWORK

While disease surveillance is considered the foundation of any public health effort, the world still lacks a coordinated system of antibiotic resistance surveillance. To fill this void, APUA founded the GAARD program to help coordinate efforts of commercially and publicly sponsored resistance surveillance to help inform public health programming worldwide (search http://www.apua.org). The commitment of millions of dollars to terrorist surveillance throughout the world is a model for what it may take to prevent the emergence and spread of "terrorist" pathogens. In a 1998 article in the *APUA Newsletter,* APUA vice president Thomas F. O'Brien laid out the problem, stating that "huge sample size is needed to discriminate epidemiological detail whether for infection control or for tracing the spread of distinctive resistance phenotypes . . . The world's bacteria collaborate efficiently to speed dissemination of resistance. The world's humans should do the same to slow it." (14).

Today, APUA's GAARD project is chaired by Stuart Levy, staffed by APUA scientists, and operated as a unique public-private partnership which includes WHONET, the WHO, and the CDC as advisors and the major commercial companies tracking antibiotic resistance worldwide. The company surveillance programs include Bayer HealthCare AG (TARGETed), Glaxo-SmithKline (Alexander Network), Bristol-Myers Squibb (SENTRY), AstraZeneca (MYSTIC), and Ortho-McNeil Pharmaceuticals (TRUST). By combining data sets for special studies and facilitating joint analyses by its steering committee, GAARD allows cross-validation of individual system findings and detection of resistance trends that may otherwise be missed or misinterpreted. In the fall of 2004, APUA convened a wide range of infectious disease surveillance experts to produce the world's first comprehensive report on global antimicrobial resistance in HIV, malaria, TB, and other priority bacterial pathogens, titled, "Shadow Epidemic: The Growing Menace of Drug Resistance 2005 GAARD Report." A summary of the report was presented at ICAAC 2004, and the full report will be published in the September 2005 CID journal supplement (3).

THE APUA CHAPTERS—LOCAL CHAMPIONS AROUND THE WORLD

APUA began in 1981 with a few of Dr. Levy's colleagues in nine nations. Now APUA has members in over 100 nations and 50 country-based affiliated organizations, including 25 in the developing world. These chapters tailor APUA's message to the cultural and economic conditions of their country and report on local antimicrobial resistance patterns. Through advocacy, treatment guidelines, and education, these chapters play a critical role and provide a ready resource to support programs such as the United Nations Global Fund in the rational distribution and use of antimicrobials in developing countries.

Over the past 5 years, APUA's international chapter program director, Dr. Anibal Sosa, has worked tirelessly to project APUA's reach to every continent. A Venezuelan-born physician and microbiologist, he has spearheaded APUA's action plan in Latin America and has been a catalyst for much of the ambitious work currently taking place in that region. In Africa, Dr. Sosa works with APUA scientific advisory board member, Dr. Iruka Okeke of Nigeria and Haverford College, to meet the uniquely challenging conditions of African countries.

Worldwide, it is often the APUA chapter that first highlights antimicrobial resistance for a health facility or government official by providing statistical evidence and advocating for policy changes. When Dr. Adebayo Lamikanra convened the APUA-Nigeria chapter-founding meeting in 2001, health care workers traveled for days from remote regions to share their concerns about antimicrobial resistance. Doctors, nurses, and community health workers heard out each others' challenges and agreed to contact the Nigerian government to gain more attention to the increase in untreatable infections in their country. Today, resistance surveillance and an active APUA chapter are centerpieces of public health efforts in the Lusaka, Nigeria region—a model of what can be done even under dire circumstances. APUA also coordinates North-South partnerships to complement scarce resources in the developing world. For example, APUA United Kingdom, which operates as part of the British Society of Antimicrobial Chemotherapy (BSAC), is working with APUA headquarters in Boston to address infectious disease epidemics in Africa. The APUA website highlights many other examples of the influence of strong, voluntary local APUA leaders who are spurred on through their affiliation and support from the APUA network.

APUA-Zambia and APUA-Nepal conduct education of the local drug and therapeutic committees and are engaged in development of antibiotic prescription and policy guidelines. The work is conducted under a 5-year subcontract with the U.S. Agency for International Development (USAID) in the Rational Pharmaceutical Management Plus (RPM+) Program (2).

IGNORANCE—THE MAIN ENEMY

In many developing countries, neither specialized medical journals nor Internet access are available to the health providers who struggle with the worst problems of resistance. Dr. Thomas O'Brien, APUA's vice president, who often travels to the developing world, tells of the worn copies of the *APUA Newsletter* he has seen passed from reader to reader and the eagerness of those not in such a chain to find out how they can get a subscription. To address this vacuum, the APUA quarterly newsletter has been published continuously since 1981, with the goal of disseminating timely scientific and clinical information to over 7,000 subscribers in 100 nations. Translated into five languages and disseminated free of charge in the developing world, the *APUA Newsletter* synthesizes data and reports often too detailed and voluminous for caregivers to digest. Edited by Madeline Drexler, a scientific journalist, the newsletter attracts articles from the world's top experts and develops feature stories. Articles in recent editions included "Quality Control of Antibiotic Use: . . . in Brooklyn, New York", "DOTS (directly observed treatment, short course): The Jewel of Global Public Health", "Chloroquine-Resistant Malaria: Evolution and Future Prospects," and "How is Multidrug-Resistant Tuberculosis Managed in Resource-Poor Settings?" (search archives at http://www.apua.org).

Acclaimed in the ASM magazine and several biotechnology journals, the APUA website together with the *APUA Newsletter* stand as unique and noncommercial sources of authoritative information for scientists, clinicians, policy makers, and the public and press.

THE FAAIR PROJECT: FACTS ABOUT ANTIBIOTICS IN AGRICULTURE

In 2000, APUA was approached by a consortium of environmental funders requesting that APUA join a campaign to demand an end to the use of antibiotic growth promoters in food animals. The APUA board deliberated and decided that APUA should stick to what it does best: developing unbiased, scientific evidence to inform the policy debates. APUA staff members Dr. Stephen DeVincent and Christina Viola convened an expert panel which met over 2 years, resulting in the production of a scientific report, with over 500 references. "The Need to Improve Antimicrobial Use in Agriculture: Ecological and Human Health Consequences" was published in *Clinical Infectious Diseases* in 2002 (4). Project cochairs, Dr. Sherwood Gorbach of Tufts University School of Medicine and Dr. Michael Barza, medical director of Carney Hospital worked tirelessly and, in the end, pressed the scientific panel to state the implications of their research. After great debate and consternation, the committee reached consensus and concluded that "(1.) All uses of antimicrobials . . . contribute to the global problem of antimicrobial resistant genes in the environment; (2.) Antimicrobial use in food animal production selects for resistance strains and amplifies their resistance and dissemination; and (3.) Antimicrobial agents should not be used in agriculture in the absence of disease." Comprised of the world's most knowledgeable scientific experts and former Institute of Medicine participants, the panel presented its bottom line, concluding that elimination of nontherapeutic use of antimicrobials in food animals and agriculture would lower the burden of antimicrobial resistance in the environment with consequent benefits to human and animal health.

The key findings were presented by APUA at the National Press Club in June 2000 and have caused a ripple effect in professional societies, government, and the animal health industry to this day. The research has been cited by the Institute of Medicine (7a) and in the 2004 court decision to withdraw government approval for the use of the fluoroquinolone Baytril in poultry (16). It was also influential in McDonald's Corporation's decision in 2003 to require its meat suppliers to stop using antibiotics important in human medicine to promote animal growth. As a follow-up project, APUA convened another advisory group, cochaired by Dr. Paula Cray of the USDA and by Dr. Scott McEwen of Toronto, Canada, and involving advocacy groups, industry, and government to determine likely sources of data to better inform U.S. agricultural antimicrobial use policy. The resulting APUA FAAIR II Report is expected to be published as a supplement in *Preventive Veterinary Medicine* in late 2005.

LOOKING TO THE FUTURE

As Stuart Levy so boldly envisioned in 1981, APUA has kept a laser-like focus on a topic that often gets subjugated to shifting priorities. As new disease threats abound, public health vigilance is even more necessary to forestall a drug resistant bioterrorism event and resistant infectious disease epidemics. Dr. Philip G. Jenkins, of the World Health Organization (WHO), has characterized APUA's network as representing the broadest grassroots field network for global efforts to contain antimicrobial resistance. Now in its third decade of groundbreaking work, with the required expertise, partnerships, and infrastructure in place, APUA is well suited to help lead the way.

At its 20th anniversary in 2003, APUA received scores of professional accolades and letters of commendation. Dr. Linda Tollefson, deputy director of the Center for Veterinary Medicine noted "the critical role that APUA plays in coordinating diverse interest groups from industry, academia and advocacy organizations and for providing scientific evidence to inform . . . policy." Dr. Joseph F. John, Jr., chief of the Medical Specialty Service at the Department of Veterans Affairs observed, "Since responses to antibiotic resistance are bigger than the scope of any one entity, it is important to have a central coordinating body such as APUA."

Increasing attention from government and professional groups such as the Infectious Diseases Society of American (IDSA) now provides a broader platform for APUA's work (10). The recent creation of an NIH Drug Discovery and Mechanisms of Antimicrobial Resistance (DDR) Study Section signifies a new government commitment to investigation of new antimicrobial agents that could be deployed in bioterrorism and to study of evolution mechanisms, resistance transmission, and strategies for prevention of antimicrobial resistance.

Recognizing the essential need for collaboration in the battle against global resistance, the APUA Annual Leadership Award was established in 2001 to honor extraordinary leadership in control of antimicrobial resistance worldwide. Honorees for the past 3 years have included Dr. Frank M. Aarestrup and Dr. Henrik C. Wegener, representing the Danish Veterinary Institute for scientific leadership in improving antibiotic use in agriculture; Robert L. Langer, representing McDonald's Corporation for that company's policy of encouraging its food animal suppliers not to use antibiotics in growth promotion; Dr. David Bell (CDC), Dr. Marissa Miller (NIAID-NIH), and Dr. Murray Lumpkin (FDA) for the U.S. Interagency Group which developed the "U.S. Interagency Antimicrobial Resistance Plan"; and Dr. Rosamund Williams of the WHO for spearheading the "WHO Global Strategy to Contain Antimicrobial Resistance."

For Stuart Levy—scientific pioneer and passionate public health advocate—APUA stands as a testament to the commitment, courage, and compassion it takes to realize such an improbable dream. As a global organization now sustainable beyond his tenure, APUA has achieved Dr. Stuart Levy's ultimate vision. With Dr. Levy's continuing devotion and APUA's staying power and expanding collaborations, the world can share his optimism in the achievement of APUA's goal—preserving the power of antibiotics for generations to come.

REFERENCES

1. **American Lung Association.** Trends in tuberculosis morbidity and mortality. November 2003. American Lung Association.
2. **APUA, The Alliance for the Prudent Use of Antibiotics.** April 2003. *Framework for Use of Antimicrobial Resistance Surveillance in the Development of Standard Treatment Guidelines.* Under subcontract with the Rational Pharmaceutical Management Plus Program at the Management Sciences for Health, Arlington, Va.
3. **APUA, The Alliance for the Prudent Use of Antibiotics.** September 2005. The Alliance for the Prudent Use of Antibiotics GAARD Report: Global Antimicrobial Resistance. *Clin. Infect. Dis.* **41**(Suppl. 3), in press.
4. **Barza, Michael, S. L. Gorbach (ed.).** 1 June 2002. *A Report of the Alliance for the Prudent Use of Antibiotics. The Need to Improve Antimicrobial Use in Agriculture: Ecological and Human Health Consequences. Clin. Infect. Dis.* **34** (Suppl. 3).
5. **Brower, J., and P. Chalk.** 2003. *The Global Threat of New and Reemerging Infectious Diseases: Reconciling U.S. National Se-*

curity and Public Health Policy, p. 69–70. RAND, Santa Monica, Calif.

6. **Centers for Disease Control and Prevention.** 1997. Reduced susceptibility of Staphylococcus aureas to vancomycin—Japan, 1996. *Morb. Mortal. Wkly. Rep.* **46:**624–626.
7. **Centers for Disease Control and Prevention.** 2000 *Staphlylococcus aureus* with reduced susceptibility to vancomycin—Illinois, 1999. *Morb. Mortal. Wkly. Rep.* **48:**1165–1167.

7a. **Committee on International Science, Engineering, and Technology (CISET).** 25 July 1995. *Global Microbial in the 1990s. Report of the NSTC Committee on International Science, Engineering, and Technology (CISET) Working Group on Emerging and Re-emerging Infectious Diseases.* National Science and Technology Council.

8. **Drexler, M.** 2002. *Secret Agents: The Menace of Emerging Infections,* p. 119–157. Joseph Henry Press, Washington, D.C.
9. **Hamburg, M. A., J. Lederberg, and M. S. Smolinski (ed.).** 2003. *Microbial Threats to Health.* The National Academies Press, Washington, D.C.
10. **Infectious Diseases Society of America (IDSA).** 2004. Bad bugs, no drugs: as antibiotic discovery stagnates . . . A public health crisis brews. IDSA White Paper. [Online.] http://www.idsociety.org. Accessed 5 April 2005.
11. Reference deleted.
12. **Levy, S. B.** 2002. From tragedy the antibiotic age is born, p. 1–14, 305. *In The Antibiotic Paradox: How the Misuse of Antibiotics Destroys Their Curative Powers.* Perseus Publishing, Cambridge, Mass.
13. **Murray, B. E.** 2000. Vancomycin-resistant enterococcal infections. *N. Engl. J. Med.* **342:**710–721.
14. **O'Brien, T. F.** 1998. Towards coordinated surveillance of antimicrobial resistance: The need to merge databases, *APUA Newsl.* **16:**2.
15. **Rotun, S. S., V. McMath, D. J. Schoonmaker, P. S. Maupin, F. C. Tenover, B. C. Hill, and D. A. Ackman.** *Staphylococcus* with reduced susceptibility to vancomycin isolated from a patient with fatal bacteremia. *Emerg. Infect. Dis.* 1999;**5:**147–149.
16. U.S. Food and Drug Administration, Center for Veterinary Medicine. FDA Administrative Law Judge Daniel J. Davidson's decision issued March 18, 2004. See www.fda.gov/cvm/index/updates/baytrilup.htm.
17. **U.S. Interagency Task Force on Antimicrobial Resistance.** 2001. Public health action plan to combat antimicrobial resistance. [Online.] http://www.cdc.gov/drugresist/actionplan/aractionplan/pdf. Accessed 28 June 2004.
18. **World Health Organization.** 2001. Global strategy for containment of antimicrobial resistance, Geneva, Switzerland (WHO/CDC/CSR/DRS/2001.2) [Online.] http://www.who.int/csr/resources/publications/drugresist/en/EGlobal_Strat.pdf. Accessed 28 June 2004.
19. **World Health Organization.** 2000. Overcoming antimicrobial resistance, Geneva, Switzerland. [Online.] http://www.who.int/infectious-disease-report/2000/. Accessed 28 June 2004.

Frontiers in Antimicrobial Resistance: a Tribute to Stuart B. Levy
Edited by D. G. White, M. N. Alekshun, and P. F. McDermott

Chapter 39

Antimicrobial Use in Animals in the United States: Developments in Policy and Practice

STEPHEN J. DEVINCENT AND CHRISTINA VIOLA

Since the early 1950s, antimicrobials have been used in companion, sport, and food animals to treat bacterial infections and to control or prevent their spread throughout populations. In some sectors of food animal production, antimicrobials have also been administered to animals in feed or water to promote growth or enhance feed efficiency. Such antimicrobial use is often referred to as "subtherapeutic" because the dose and duration of use are typically lower and longer relative to the quantity and period of use that would be administered to treat bacterial disease in the same species under similar circumstances. Technically, subtherapeutic describes relative dosage rather than indication or purpose of use, and the term therefore can be misleading. As an alternate term, some authors have employed the term "nontherapeutic" to describe antimicrobial use for the purpose of animal growth promotion. However, use of the term is also not standardized; some stakeholders consider nontherapeutic use to also include disease prevention (22). In any case, the exact biological mechanism(s) by which long-term, low-dose antimicrobial use promotes animal growth and enhances feed efficiency remains incompletely understood (24). Among microbiologists, however, it is widely agreed that such use is more likely to promote the emergence and spread of antimicrobial resistance because the selective pressure is exerted over a longer period of time and the lower dosage may increase the likelihood that bacteria with intermediate resistance could survive.

While it seems clear from even the most conservative industry estimates that the total amount of antimicrobials administered to animals in the United States is not insignificant relative to use in human medicine (6, 7, 8), the exact quantity administered annually remains unknown. Nonetheless, a number of concerns have been raised about antimicrobial use in animals, and subtherapeutic use in food animals in particular, with respect to public health. The first expert report to address this issue was published in 1969, and the debate continues to the present day (Table 1).

Food-borne transfer of resistant pathogens from animals to humans is perhaps the most widely recognized threat to human health posed by long-term, low-dose antimicrobial use in animals. Common food-borne pathogens include strains of *Campylobacter, Salmonella, Escherichia coli, Listeria, Vibrio,* and *Yersinia.* Most strains of these bacteria are part of the commensal flora in one or more food animal species, only rarely causing disease in the animal. The concern is that antimicrobial use in animals selects for an increased prevalence of resistance genes in these bacterial populations, resulting in an increased prevalence of resistant food-borne infections in humans. So-called "trace back" studies of food-borne illness outbreaks have provided direct evidence for the transfer of resistant pathogens via the food chain (25), leaving little room for doubt that such transfer can occur. The remaining questions concern the frequency with which it does occur and the significance of food-borne bacterial infections from a public health standpoint.

The transfer of resistant pathogens through the food chain, however, is not the only concern associated with antimicrobial use in food animals. Potentially more alarming is ecosystem-mediated transfer of resistant pathogens and/or resistance genes and an increase in the total pool or "reservoir" of transferable genetic elements that confer resistance. Dr. Stuart Levy was one of the first researchers to emphasize the ecological dimensions of antimicrobial resistance. In a prospective study, Levy and co-workers (19) observed that tetracycline resistance increased among

Stephen J. DeVincent • (Corresponding address) 26 Montgomery Street, Boston, MA 02116.

Table 1. Reports on antimicrobial use in animals and the impact on human health[a, b]

Title	Source	Date	Conclusions/Recommendations
The Report to Parliament by The Joint Committee on Antibiotic Uses in Animal Husbandry and Veterinary Medicine ("The Swann Report")	English Parliament	1969	Recommendation for reclassification and use restriction of antibiotics in therapeutic (prescription) and nonprescription feed additives. Committee also recommended a ban on "nutritional" use of antibiotics.
The Effects on Human Health of Subtherapeutic Use of Antimicrobials in Animal Feed	National Research Council (NRC)	1980	Literature is currently insufficient to assess relationship between subtherapeutic use in animals and human health.
Antibiotics in Animal Feeds, Report 88	Council for Agricultural Science & Technology (CAST)	1981	In humans evidence of prevalence of resistant infections is dose related; therefore, it is irrational to ban subtherapeutic use without also banning therapeutic use in food animals.
Human Health Risks with the Subtherapeutic Use of Penicillin or Tetracyclines in Animal Feed	Institute of Medicine (IOM)	1981	Current evidence indicates that there is selection pressure for antimicrobial resistance with subtherapeutic use, but it does not establish definite human health risk.
Human Health Risks with the Subtherapeutic Use of Penicillin or Tetracyclines in Animal Feed	Institute of Medicine (IOM)	1989	Evidence not sufficient to directly link antimicrobial use in animal feeds to deaths from salmonellosis but further study is recommended.
The Medical Impact of the Use of Antimicrobials in Food Animals	World Health Organization (WHO)	1997	Antimicrobials should not be used for animal growth promotion if they are used in human medicine or if known to select for cross-resistance to an antimicrobial used in human medicine.
A Review of Antimicrobial Resistance in the Food Chain	UK Ministry of Agriculture, Fisheries, and Food (MAFF)	1998	Selection for resistance in animal pathogens due to antimicrobial use and transfer of antimicrobial resistant infections from animals to humans are well documented.
The Use of Drugs in Food Animals: Benefits and Risks	National Research Council (NRC), and Institute of Medicine (IOM)	1999	Use of antimicrobials in food animal production is "not without some problems and concerns, but it does not appear to constitute an immediate public health concern; additional data might alter this conclusion."
The Agricultural Use of Antibiotics and Its Implications for Human Health	General Accounting Office (GAO)	1999	"While research has linked the use of antibiotics in agriculture to the emergence of antibiotic-resistant foodborne pathogens, agricultural use is only one of several factors that contributes to antibiotic resistance in humans for pathogens that are not foodborne."
Microbial Antibiotic Resistance in Relation to Food Safety	UK Advisory Committee on the Microbiological Safety of Food (ACMSF)	1999	There is conclusive evidence that administering antibiotics to food animals results in the emergence of some resistant bacteria that infect humans.
The Use and Misuse of Antibiotics in UK Agriculture. Part 2: Antibiotic Resistance and Human Health	UK Soil Association	1999	"Routine prophylaxis with therapeutic antibiotics poses as great a threat as the use of growth-promoting antibiotics and a much greater threat than full therapeutic treatment for short periods."
The Use of Antibiotics in Food-Producing Animals: Antibiotic-Resistant Bacteria in Animals and Humans	Joint Expert Advisory Committee on Antibiotic Resistance (JETACAR)	1999	Evidence for (i) emergence of resistant bacteria in humans and animals following antibiotic use; (ii) spread of resistant animal bacteria in humans; (iii) transfer of antibiotic resistance genes from food animal bacteria to human pathogens; and (iv) resistant strains of animal bacteria causing human disease.
Opinion of the Scientific Steering Committee on Antimicrobial Resistance, 28 May 1999	European Commission (EC)	1999	"Veterinary antimicrobial use may constitute a threat to human health but the impact of resistance among zoonotic bacteria and the risk of transfer of resistance determinants between animal and human pathogens is still an unquantifiable hazard."

Continued on following page

Table 1. *Continued*

Title	Source	Date	Conclusions/Recommendations
Effect of the Use of Antimicrobials in Food-Producing Animals on Pathogen Load: Systematic Review of the Published Literature	Exponent (prepared for FDA Center for Veterinary Medicine)	2000	"Although the majority of studies indicate antibiotics do not increase the pathogen load, the concern remains that the use of antibiotics may contribute to the prevalence of antimicrobial resistance."
The Need to Improve Antimicrobial Use in Agriculture: Ecological and Human Health Consequences ("FAAIR Report")	Alliance for the Prudent Use of Antibiotics (APUA)	2002	Elimination of nontherapeutic use in food animals will lower the burden of antimicrobial resistance in the environment, with consequent benefits to human and animal health.
Uses of Antimicrobials in Food Animals in Canada: Impact on Resistance and Human Health	Veterinary Drugs Directorate, Health Canada	2002	"While the precise magnitude of the public health impact [of antimicrobial use in animals] is unknown, it is known that resistance is a serious problem in bacterial infections of humans that originate in animals."
Joint FAO/OIE/WHO Expert Workshop on Non-Human Antimicrobial Usage and Antimicrobial Resistance: Scientific Assessment (Geneva, 1–5 December 2003)	World Organization for Animal Health (OIE), World Health Organization (WHO), and Food and Agriculture Organization of the United Nations (FAO)	2004	"There is clear evidence of adverse human health consequences due to resistant organisms resulting from non-human usage of antimicrobials."
Federal Agencies Need To Better Focus Efforts To Address Risk to Humans from Antibiotic Use in Animals	General Accounting Office (GAO)	2004	Foodborne transfer of resistance does occur, but researchers disagree on the extent and importance to human health.

[a] Inclusion criteria: (i) report content exclusively or primarily focused on antimicrobial use in animals and impact on human health; (ii) substantial portion of report dedicated to review of peer-reviewed scientific evidence; (iii) scope of evidence considered in report is relatively broad (i.e., risk assessments for a single drug-outcome pair excluded); (iv) English language publication accessible to the public; and (v) authors of report independent of regulatory agency.
[b] Table format based on NRC 1999.

fecal *E. coli* in chickens within 24 to 36 h following the introduction of tetracycline-supplemented feed to the flock. After three months, the treated chickens were excreting bacteria resistant to more than one antibiotic. The prevalence of tetracycline resistance and multidrug resistance in fecal coliforms from the farm family was elevated significantly relative to neighbors on untreated farms within 4 to 6 months (19).

A second study by Levy et al. (19) examined the spread of *E. coli* with a marked tetracycline resistance plasmid through chicken populations. The marked plasmid spread quickly among *E. coli* from caged chickens fed subtherapeutic levels of tetracycline and were stably maintained for 9 weeks despite the presence of other unmarked tetracycline resistance genes. Although the marked bacteria were also introduced to a control group not fed tetracycline, the plasmid was never detected in those chickens. Perhaps most surprisingly of all, however, was that it was detected in a group of tetracycline-fed birds housed at least 50 feet from any population to which it had been introduced (19).

Collectively, these studies demonstrated that subtherapeutic antibiotic use could lead to the spread of antibiotic resistance plasmids among environmental bacteria. As Dr. Levy later argued, this and other evidence strongly suggests that there are interacting reservoirs of resistance determinants in the environment, including reservoirs at the level of the host (in this case, chickens and humans) and the environmental bacteria (in this case, *E. coli,* which may serve as a source for resistance genes in other members of the *Enterobacteriaceae*) (20). The evidence that subtherapeutic antimicrobial use increases the total pool of transferable resistance elements in the environment has troubling implications for antibiotic effectiveness over the long term.

Possible implications for environmental health from antimicrobial resistance are also troubling. For example, in a 1999-2000 survey of 139 streams across 30 states, the U.S. Geological Survey detected at least one of 22 different antibiotic compounds in over half of the samples collected (18). The presence of drug residues in waterways is likely to exert a selective pressure for resistance in aquatic bacteria, further amplifying the pool of resistance elements in the environment, with unknown consequences on microbial genetics and biodiversity.

REGULATORY FRAMEWORK FOR ANTIMICROBIAL USE IN ANIMALS IN THE UNITED STATES

The Federal Food, Drug and Cosmetic Act (FFDCA) of 1938 forms the basis for antimicrobial regulation in the United States. When first enacted, FFDCA required manufacturers to provide proof of safety for both human and animal drugs, and subsequent amendments added quality control provisions to ensure consumer safety. Most therapeutic uses of antimicrobials in human medicine were eventually designated as prescription only, but antimicrobial use in animal husbandry was included under the Food Additives Amendment of 1958, which applied to additives in animal feeds as well as foods intended for direct consumption by humans. The Kefauver-Harris Drug Amendments of 1962 added more stringent pre-market clearance and licensing requirements. The Animal Drug Amendments of 1968 consolidated these regulations and established a unique procedural mechanism for review and clearance of new animal drugs that remains in effect. Under this mechanism, potential drug sponsors are required to file a detailed New Animal Drug Application (NADA) to establish the safety and efficacy of each drug product and indication(s) for which they seek approval. For drugs intended for use in food animals, the NADA requires additional data on safety for the target animal as well as environmental safety, food safety, and occupational user safety. An approved NADA is required for interstate commerce involving the drug product.

The Food and Drug Administration (FDA) is responsible for determining whether a submitted NADA demonstrates that the proposed use of the drug product meets its effectiveness and safety requirements. Upon approval, FDA issues a separate regulation for each NADA that specifies the sponsor, active ingredients, intended species, dose(s) and route(s) of administration, indications, required preslaughter withdrawal times, and any additional restrictions or limitations it imposes on use of the drug. Another regulation may specify the residue tolerance or safe concentrations permitted in edible tissues. If the drug is approved for mixing with feed, maximum allowable concentrations are also defined.

The NADA regulation also specifies the marketing status for the drug. As a default position, animal drugs may be sold without a prescription ("over-the-counter") unless FDA determines that adequate directions cannot be written in a manner that will ensure safe use by laypersons. Prior to 1996, animal drugs intended for use in feed were always assigned to this marketing class, in part because certain state laws requiring prescription drugs to be dispensed by a licensed practitioner were considered burdensome to producers (17). To address this problem, the Animal Drug Availability Act (ADAA) of 1996 created a new marketing class, the Veterinary Feed Directive (VFD), which is analogous to prescription-only status for animal drugs administered in feed. ADAA also introduced a variety of changes designed to increase the flexibility of the drug review process (17).

The Abbreviated New Animal Drug Application (ANADA) process, introduced in 1988 with the Generic Animal Drug and Patent Term Restoration Act, allows FDA to approve generic copies of previously approved veterinary drugs after the patent or exclusivity period has expired. ANADA sponsors must demonstrate that a generic product is chemically and biologically equivalent to a previously approved animal drug product.

Finally, the 1994 Animal Medicinal Drug Use Clarification Act (AMDUCA) gave veterinarians the flexibility to prescribe approved human and animal drugs in species and/or for indications other than those specified in the NADA approval (commonly referred to as "extra-label" prescribing). This use must occur within the context of a valid veterinarian-client-patient relationship, defined generally as one in which the veterinarian has assumed responsibility for the health of the animal, has sufficient knowledge of the animal to initiate diagnosis, is available for follow-up, and the client has agreed to follow the instructions of the veterinarian (15). Furthermore, extra-label use must not result in detectable residues (defined as any compound detected in edible tissues as a result of drug use), and it must comply with other stringent requirements (17). Significantly, however, certain drugs are specifically excluded from extra-label use in food animals; these include drug products containing chloramphenicol as well as all fluoroquinolones (e.g., ciprofloxacin and enrofloxacin) and glycopeptides (e.g., vancomycin). Extra-label use of approved human or animal drug products in feed is illegal.

MONITORING AND SURVEILLANCE OF ANTIMICROBIAL USE AND RESISTANCE

Surveillance of antimicrobial-resistant bacteria in the food supply and elsewhere is a crucial element of antimicrobial regulation. Surveillance data allow regulators to keep track of trends in the prevalence of resistance in different bacterial species, helping them to identify and assess emerging threats to human health. The data also serve as a baseline against which the results of any change in antimicrobial use policy and practice can be evaluated.

In the United States, antimicrobial resistance surveillance is conducted through the National Antimicrobial Resistance Monitoring System—Enteric Bacteria (NARMS). The program was established in 1996 as a collaborative effort between the Centers for Disease Control and Prevention (CDC), Food and Drug Administration Center for Veterinary Medicine (CVM), and the U.S. Department of Agriculture (USDA). NARMS monitors changes in antimicrobial drug susceptibilities of selected bacteria in food animals, humans, and retail meats in relation to antimicrobial agents of importance in human and food animal medicine. Indicator species monitored by NARMS include strains of *Campylobacter, E. coli, Enterococcus, Listeria, Salmonella,* and *Vibrio*. Bacterial isolates are collected from clinical specimens (human and veterinary), healthy livestock, postslaughter food products, and retail meats. A pilot study of animal feed ingredients collected at slaughterhouses is also being conducted. Stated objectives of the NARMS program include (i) providing descriptive data on the extent and trends over time of antimicrobial susceptibility in *Salmonella* spp. and other enteric bacterial organisms from food animal and human populations and retail meats; (ii) facilitating the identification of antimicrobial resistance as it occurs in humans, livestock, and retail meats; and (iii) providing timely and useful information to the human and animal health communities on antimicrobial resistance drug patterns (10). In addition to NARMS, the Food Borne Diseases Active Surveillance Network (FoodNet) of the CDC conducts population-based studies to estimate origins and impact of specific diseases in select states (8).

While antimicrobial resistance surveillance data are intrinsically valuable, their use in risk assessment would be greatly facilitated by the availability of accurate antimicrobial use surveillance data. However, no comprehensive system currently exists in the United States to provide quantitative data on antimicrobial use in food animals. The only official data on antimicrobial use in animals available to the public are annual surveys conducted by the National Animal Health Monitoring System (NAHMS) of USDA. NAHMS reports typically focus on a single commodity sector and NAHMS survey data may be generally useful in estimating the extent of antimicrobial use (expressed as a percentage of a specific population of animals) as well as average duration of use in a specific year. However, total use during the lifetime of an animal (or group of animals) cannot be determined from NAHMS survey data, limiting their use in risk assessment (34).

A 2001 report, *A Public Health Action Plan to Combat Antimicrobial Resistance*, identified establishing an animal antimicrobial use surveillance system as one of its top-priority action items (1). Numerous nongovernmental organizations in the United States have also stressed the need to address this lack of data, as have distinguished committees convened by the Congressional Office of Technology Assessment (29) and Institute of Medicine (16). Internationally, the European Union (EU), as well as the World Health Organization (WHO), Food and Agriculture Organization of the United Nations, and World Organization for Animal Health, has recognized the need and called for the collection of such data.

In 1999, a panel of experts was convened by the nonprofit organization, the Alliance for the Prudent Use of Antibiotics (APUA), to consider antimicrobial use in food animals and its impact on human health. In its 2001 report, the committee highlighted the absence of comprehensive animal antimicrobial use data in the United States as a major data gap (6). APUA established an Advisory Committee on Animal Antimicrobial Use Data Collection in the United States in the spring of 2002 in order to address methodological issues surrounding domestic food animal antimicrobial use surveillance. The committee identified (i) methodological options for the collection of antimicrobial use data, (ii) potential limitations in these methods, and (iii) priorities and a general "strategy" for data collection (34).

RECENT DEVELOPMENTS IN LEGISLATIVE, REGULATORY, ADMINISTRATIVE, AND PRIVATE-SECTOR POLICY

In the last few years, bills have been introduced in both houses of the U.S. Congress that would severely restrict antimicrobial use in animals. As of this writing, the most recent version of the proposed legislation is entitled the "Preservation of Antibiotics for Medical Treatment Act of 2003" (S. 1460 and H.R. 2932). The focus of this bill is on withdrawal of approval for all "nontherapeutic" use of antimicrobials in food animals deemed important in human medicine. The legislation defines "nontherapeutic" use as use of any "critical antimicrobial animal drug" (defined to include penicillins, tetracyclines, bacitracin, macrolides, lincomycins, streptogramins, aminoglycosides, sulfonamides, quinolones and any other drug or derivative intended for use in humans) in feed or water in the absence of clinical signs of disease in the animal. Significantly, the bill does not allow for any form of antimicrobial prophylaxis when disease is present in the herd or flock, nor does it distinguish companion and sport animals from animals intended for food. Versions of the bill failed to make it to a floor vote in both 2002 and 2003, but proponents consider its introduction by a bipartisan group of Senators in 2003 to be a symbolic victory.

At the request of Congress, the General Accounting Office (GAO) recently conducted a review of (i) federal efforts to address the human health risks of antimicrobial use in animal agriculture; (ii) current U.S. use of antimicrobials in animal agriculture compared with their use by key agricultural trading partners/international competitors; and (iii) the potential impact, if any, of current U.S. policy regarding the use of antimicrobials in animal agriculture on U.S. farmers' ability to export animal products. The GAO released a report entitled *Antibiotic Resistance: Federal Agencies Need to Better Focus Efforts to Address Risk to Humans from Antibiotic Use in Animals* in May 2004 (33). Recommendations included expedited risk assessment of antimicrobials used in food animals important to human health and for the development and implementation of a plan to collect animal antimicrobial use data for use in risk assessment.

As the regulatory agency with primary responsibility for animal antimicrobial drug approval, FDA has also attempted to improve its policies without a legislative mandate. Towards this end, the agency has adopted a new framework for assessing potential adverse effects to human health from antimicrobial resistance that develops from use of new animal antimicrobial drugs. The purpose of the *Guidance for Industry #152: Evaluating the Safety of Antimicrobial New Animal Drugs with Regard to Their Microbiological Effects on Bacteria of Human Health Concern* ("Guidance") (32) is to outline a suggested approach to risk assessment for use by New Animal Drug Application (NADA) sponsors in demonstrating the microbial food safety of antimicrobial new animal drugs. Under the guidance, each new drug is evaluated qualitatively with respect to three risk parameters that are combined to produce an overall risk estimate. The Guidance also outlines a range of risk management options available to FDA based upon the risk assessment results. However, the document clearly states that this option of risk estimate for drug sponsors in the NADA approval process does not automatically determine approval status; FDA retains the option to exercise discretion in deciding whether or not there is a "reasonable certainty of no harm" on a case-by-case basis.

The semi-qualitative approach to risk assessment described in the Guidance is both more flexible and less burdensome to potential drug sponsors than would be a purely quantitative approach. Given the considerable uncertainty surrounding the transfer of antimicrobial-resistant bacteria from food animals to humans, this approach seems appropriate at present. However, in light of the evidence for multiple, ecological effects of antimicrobial use in animals, the exclusive focus on a single pathway (direct food-borne transfer of resistant pathogens) is troubling (33). It also remains to be seen whether the framework will be applied to re-evaluate previously approved drugs in a timely fashion.

FDA has also taken other actions to limit the potential human health risk from antimicrobial use in food animals. Most notably, in 2000, FDA requested that manufacturers voluntarily withdraw approval of the fluoroquinolone drugs sarafloxacin and enrofloxacin for use in poultry (13). The decision by FDA was based upon evidence that poultry is the principal source of domestically acquired fluoroquinolone-resistant *Campylobacter* infections in the United States (1). Although fluoroquinolones had been used in human medicine since 1986, a national retrospective study of *Campylobacter* infections between 1989 and 1990 found no *Campylobacter jejuni* to be resistant to fluoroquinolones. However, in 2001, following approval for use in poultry of sarafloxacin in 1995 and enrofloxacin in 1996, a second study found that 18% of *C. jejuni* isolates were flouroquinolone-resistant (9). The FDA determined that the use of these drugs in poultry causes the development of fluoroquinolone-resistant *Campylobacter* in poultry, which, when transferred to humans through the food supply, may cause fluoroquinolone resistant *Campylobacter* infections in humans.

At the time of the initial request by FDA for voluntary withdrawal, two animal pharmaceutical manufacturers were marketing fluoroquinolone drugs for use in poultry. Abbott Laboratories complied with the request, but the Bayer Corporation opted not to withdraw their product, Baytril, from the market. In February 2002, FDA announced that administrative hearings would take place on their recommendation to withdraw approval of enrofloxacin in poultry (14). Representatives of FDA and Bayer and other scientists entered both oral and written testimony into evidence. In March 2004, an administrative law judge issued a decision in support of FDA that resulted in the withdrawal of the NADA approval of enrofloxacin for use in poultry. However, the matter remains undecided as Bayer can appeal the decision.

Citing concerns for food safety, some multinational food-industry corporations have also become involved in issues relating to antimicrobial use in food animals in recent years. Most notably, in 2003, the McDonald's Corporation (McDonald's) implemented its *Global Policy on Antibiotic Use in Food Animals* that required meat suppliers to phase-out use of antimicrobials for animal growth promotion by the end of 2004 if they were also approved for use in human medicine (21). The McDonald's policy also encourages suppliers to comply with its detailed "Guiding

Principles for Sustainable Use." McDonald's developed its *Global Policy* in collaboration with a coalition of organizations and corporations, including Environmental Defense, Elanco Animal Health, Tyson Foods, and Cargill. Although the announcement itself does not describe the procedures by which McDonald's intends to enforce the policy, its worldwide scope implies that the potential impact could be significant. This action by McDonald's and similar actions by other global companies demonstrate the increasing importance of consumer concern and influence with respect to food safety.

ALTERNATE PATHS: THE EUROPEAN APPROACH

At present, at least 17 classes of antimicrobials are approved for animal growth promotion in the United States (1). In contrast, in 1988 the EU banned the use of four antimicrobial feed additives for growth promotion that were in the same classes as human antimicrobials (bacitracin zinc, spiramycin, virginiamycin, and tylosin phosphate) (11). By 2006, the four remaining antimicrobial feed additives used for growth promotion (monensin sodium, salinomycin sodium, avilamycin, and flavophospholipol) are scheduled to be phased out (12). What accounts for this vast difference in policy outcomes? In all likelihood, the discrepancy reflects differing views with respect to levels of acceptable risk in matters concerning food safety. Specifically, the United States and EU differ over the extent to which the so-called "precautionary principle" should be applied to approval and use of antimicrobial drugs that are deemed of importance in human medicine. U.S. regulatory policy has long maintained that drugs should be evaluated for safety and effectiveness on an individual basis. In contrast, the EU invoked the precautionary principle in the termination of use of the four remaining antimicrobial growth promoters as feed additives based upon a level of protection it determined was proportional to the risk to its citizens. The EU chose such action although the four drugs do not belong to antimicrobial classes currently used in humans. The failed "Preservation of Antibiotics for Medical Treatment Act of 2003" in the United States could be interpreted as reflecting a level of precaution comparable to the EU when it defined bacitracin, a drug of minimal importance to human medicine, as a "critical antimicrobial drug."

The EU, in 1997, banned the antimicrobial growth promoter, avoparcin, a member of the glycopeptide class of antimicrobials, due to concerns about glycopeptide-resistant enterococci (including vancomycin-resistant enterococci or VRE). The decision regionalized a 1995 decision by Denmark to ban the use of avoparcin as an additive to animal feed. Due to the fact that virtually no antimicrobial growth promoters have been used in Denmark since 1999, the impacts of antimicrobial growth promoter termination in Denmark are of international interest and speculation. According to data from the respected Danish Integrated Antimicrobial Resistance Monitoring and Research Programme (DANMAP), the ban on antimicrobial growth promoters appears to have resulted in substantial reduction of overall antimicrobial use in food animals in Denmark (5, 26) and reduction of the food animal reservoir of enterococci resistant to these growth promoters (27).

With respect to animal health and welfare, the removal of antimicrobial growth promoters in Denmark was associated with a reduction in growth rate and increase in morbidity and mortality in young pigs (27). In poultry, necrotic enteritis was a minor problem for broiler health after the termination of the growth promoters, due largely to the approved continued use of ionophores for prophylaxis against necrotic enteritis and coccidiosis (28). Ionophores (e.g., monensin) are a class of antimicrobials that are not used in human medicine and for which current evidence indicates no direct or environmentally mediated risk to human health. In regard to Danish food animal production, the effects of termination of the drug on poultry production appeared to be small and limited to decreased feed efficiency, which was partially offset by savings in the cost of the antimicrobial growth promoters. In swine, there was loss of productivity in weanlings, but there was no major effect on productivity or feed efficiency in finishing pigs.

A panel of experts convened by the WHO in November 2002 reviewed the overall impact on human health, animal health and welfare, animal production, and national economy resulting from Denmark's ban on the use of antimicrobial growth promoters in food animal production. The panel concluded that "under situations similar to those found in Denmark, the use of antimicrobials for the sole purpose of growth promotion can be discontinued" (35). In particular, the authors noted that there was a small but negative effect on the Danish economy from the termination of use of growth promoters in swine and poultry production. However, the panel concluded that costs associated with modification of production systems could be substantial for some producers. The authors also noted that the Danish experience may not be fully applicable to other countries, which would include the United States, because of factors such as varied scales of production, infrastructure, and drug availability. In Denmark, for example, approximately 22.5 million pigs and 130 million broilers are raised annually (7, 28), whereas in the U.S., population estimates are of 61.4 million pigs and 8.6 billion broilers (30, 31).

CONCLUDING REMARKS

On a global scale, the use of antimicrobials in food animals has caused increasing concern regarding the potential adverse consequences to human health. As in human medicine, the use of antimicrobials in food animals results in the emergence and dissemination of antimicrobial-resistant bacteria. Scientific evidence has demonstrated that transfer of antimicrobial-resistant bacteria or their genetic determinants from food animals to humans through the food supply has clinical implications. Controversy has focused on the use of antimicrobial growth promoters and its greater likelihood to increase the prevalence of antimicrobial resistance through long-term use at low dosage levels. There is significant disagreement, however, among researchers of the extent of harm to human health from this transference. While most studies have determined that the use of antimicrobials in food animals potentially poses significant risks to human health, a small number of studies put forth data that show that the public health risks are minor or negligible. The discordance demonstrates the need for more comprehensive and conclusive data to assess the actual extent of antimicrobial resistance in humans and animals and potential ecological reservoirs of resistance.

In addition to antimicrobial resistance surveillance data, knowledge of the quantity of antimicrobial use in the United States is necessary for a more complete understanding of human health consequences from the use of antimicrobials in food animals. While other countries have developed national systems for monitoring animal antimicrobial use, no such surveillance system exists in the United States. The recommendation by GAO and others to develop and implement a plan to collect food animal antimicrobial usage data on an ongoing basis in the United States should be heeded.

However, in the collective effort to decrease antimicrobial resistance in food-borne bacteria, animal producers should not bear disproportionate burden. Antimicrobial resistance is a multifactorial problem and thus requires a multidisciplinary solution. It should not be forgotten that the use of antimicrobials in humans likely contributes more to the crisis of antimicrobial resistance in the United States than use in food animals. Access to an adequate supply of antimicrobials in the treatment of disease in food animals, and prophylaxis where necessary, are critical to safeguarding the health and well being of food animals, the economic livelihood of producers, and the food supply of the nation.

REFERENCES

1. **Anderson, A. D., J. M. Nelson, S. Rossiter, and F. J. Angulo.** 2003. Public health consequences of use of antimicrobial agents in food animals in the United States. *Microb. Drug Resist.* **9:**373–379.
2. **Animal Health Institute.** 2001. Press Release: New data on antibiotic use in animals available.
3. **Animal Health Institute.** 2000. Press Release: Survey indicates most antibiotics used in animals are for treating and preventing disease.
4. **Animal Health Institute.** 2002. News Release. Survey Shows Decline in Antibiotic Use in Animals. September 29, 2002. [Online] (http://www.ahi.org/mediaCenter/pressReleases/surveyShowsDecline.asp)
5. **Bager, F., V. F. Jensen, and E. Jacobsen.** 2002. Presented at the International Invitational Symposium: Beyond Antimicrobial Growth Promoters in Food Animal Production, Foulum, Denmark, 6–7 November 2002.
6. **Barza, M., and S. Gorbach (ed.).** 2002. The need to improve antimicrobial use in agriculture: ecological and human health consequences. *Clin. Infect. Dis.* **34**(Suppl. 3):S1–S144.
7. **Callesen, J.** 2002. Presented at the International Invitational Symposium; Beyond Antimicrobial Growth promoters in Food Animal Production, Foulum, Denmark, 6-7 November 2002.
8. **Centers for Disease Control and Prevention.** 2004. Foodborne Diseases Active Surveillance Network (FoodNet). [Online] http://www.cdc.gov/foodnet/.
9. **Centers for Disease Control and Prevention.** 2000. *National Antimicrobial Resistance Monitoring System (NARMS) 2001 Annual Report.* [Online.] http://www.cdc.gov/narms/.
10. **Centers for Disease Control and Prevention.** 2004. *National Antimicrobial Resistance Monitoring System: Enteric Bacteria.* http://www.cdc.gov/narms/.
11. **European Commission.** 2001.*Council Regulation no. 2821/98, As Regards Withdrawal of the Authorisation of Certain Antibiotics, Directive 70/524/EEC Concerning Additives in Feedstuffs.* 17 December 1998, Official Journal L 351, 29/12/1998 P. 0004–0008.
12. **European Commission.** 2003. *Council and Parliament Prohibit Antibiotics as Growth Promoters: Commissioner Byrne Welcomes Adoption of Regulation on Feed Additives.* Press release IP/03/1058, 22 July 2003, Brussels, DN:IP/03/1058. [Online.] http://europa.eu.int/rapid/pressReleasesAction.do?reference=IP/03/1058&format=HTML&aged=0&language=EN&guiLanguage=en
13. **Federal Register.** 2000. Enrofloxacin for poultry; opportunity for hearing. **65:**64954–64965.
14. **Federal Register.** 2002. Enrofloxacin for poultry; notice of hearing. **67:**7700–7701.
15. **Federal Register.** 1994. Animal medicinal drug use clarification act of 1994. (vol. 1) (108 Stat. 4153; 3 pages) S. 455/P.L. 103–397.
16. **Institute of Medicine.** 1989. *Human Health Risks with Subtherapeutic Use of Penicillin or Tetracyclines in Animal Feed.* National Academy Press, Washington, D.C.
17. **Kaplan, A. H.** 1995. Fifty years of drug amendments revisited: in easy-to-swallow capsule form. *Food Drug Law J.* **50**(Spec.): 179–196.
18. **Kolpin, D. W., E. T. Furlong, M. T. Meyer, E. M. Thurman, S. D. Zaugg, L. B. Barber, and H. T. Buxton.** 2002. Pharmaceuticals, hormones, and other organic wastewater contaminants in U.S. streams, 1999-2000: a national reconnaissance. *Environ. Sci. Technol.* **36:**1202–1211.

19. **Levy, S. B., G. B. FitzGerald, and A. B. Macone.** 1976. Spread of antibiotic-resistant plasmids from chicken to chicken and from chicken to man. *Nature* **260:**40–42.
20. **Levy, S. B.** 1987. Antibiotic use for growth promotion in animals: ecologic and public health consequences. *J. Food Prot.* **50:** 616–620.
21. **McDonald's Corporation.** 2003. Global policy on antibiotic use in food animals. http://www.mcdonalds.com/corp/values/socialrespons/market/antibiotics/global_policy.html.
22. **Mellon, M., C. Benbrook, and K. L. Benbrook.** 2001. *Hogging It!: Estimates of Antimicrobial Use in Livestock.* UCS Publications, Cambridge, Mass.
23. **Miller, M. A., and W. T. Flynn.** 2000. Regulation of antibiotic use in animals, p. 760–761. *In* J. F. Prescott, J. D. Baggot, and R. D. Walker (ed.), *Antimicrobial Therapy in Veterinary Medicine,* 3rd ed. Iowa State University Press, Ames, Iowa.
24. **Shryock, T. J.** 2000. Growth promotion and feed antibiotics, p. 735–743. *In* J. F. Prescott, J. D. Baggot, and R. D. Walker (ed.), *Antimicrobial Therapy in Veterinary Medicine,* 3rd ed. Iowa State University Press, Ames, Iowa.
25. **Spika, J. S., S. H. Waterman, G. W. Hoo, M. E. St Louis, R. E. Pacer, S. M. James, M. L. Bissett, W. Mayer, J. Y. Chiu, and B. Hall.** 1987. Chloramphenicol-resistant *Salmonella newport* traced through hamburger to dairy farms. A major persisting source of human salmonellosis in California. *N. Engl. J. Med.* **316:**565–570.
26. **Statens Serum Institute.** 2001. DANMAP 2000–Consumption of antimicrobial agents and occurrence of antimicrobial resistance in bacteria from food animals, foods and humans in Denmark. http://www.svs.dk.
27. **Statens Serum Institute.** 2002. DANMAP 2001. Use of antimicrobial agents and occurrence of antimicrobial resistance in bacteria from food animals, foods and humans in Denmark. [Online.] http://www.vetinsk.dk.
28. **Tornoe, N.** 2002. Presented at the International Invitational Symposium; Beyond Antimicrobial Growth Promoters in Food Animal Production, Foulum, Denmark, 6–7 November 2002.
29. **U.S. Congress, Office of Technology Assessment.** 1995. *Impacts of Antibiotic-Resistant Bacteria, OTA-H-629.* Office of Technology Assessment, Washington, D.C.
30. **United States Department of Agriculture.** 2004. *Briefing Room, Poultry and Eggs: Background.* Economic Research Service, U.S. Department of Agriculture, Washington, D.C. [Online.] http://www.ers.usda.gov/Briefing/Poultry/background.htm.
31. **United States Department of Agriculture.** 2004. *Quarterly Hogs and Pigs.* National Agricultural Statistics Service, Agricultural Statistics Board, U.S. Department of Agriculture, Washington, D.C. [Online.] http://usda.mannlib.cornell.edu/reports/nassr/livestock/php-bb/2004/.
32. **U.S. Food and Drug Administration.** 2003. *Guidance for Industry #152: Evaluating the Safety of Antimicrobial New Drugs with Regard to Their Microbiological Effects on Bacteria of Human Health Concern.* U.S. Food and Drug Administration, Washington, D.C. October 23, 2003. [Online.] http://www.fda.gov/cvm/antimicrobial.html.
33. **United States General Accounting Office.** 2004. *Antibiotic Resistance: Federal Agencies Need To Better Focus Efforts to Address Risk to Humans from Antibiotic Use in Animals, GAD-04-490.* United States General Accounting Office, Washington, D.C.
34. **Viola, C., and S. DeVincent.** Overview of Issues Pertaining to the Manufacture, Distribution, and Use of Antimicrobials in Animals and other Information Relevant to Animal Antimicrobial Use Data Collection in the United States., in press. Preventive Veterinary Medicine, The Netherlands.
35. **World Health Organization.** 2003. *Impacts of Antimicrobial Growth Promoter Termination in Denmark. WHO/CDS/CPE/ZFK/2003.1.* World Health Organization, Geneva, Switzerland. [Online.] http://www.who.int/salmsurv/links/gssamrgrowthreportstory/en/.

Frontiers in Antimicrobial Resistance: a Tribute to Stuart B. Levy
Edited by D. G. White, M. N. Alekshun, and P. F. McDermott

Chapter 40

From Perplexing Proteins to Paratek Pharmaceuticals: One Scientific Front Forged by Stuart B. Levy

Mark L. Nelson and S. Ken Tanaka

It was an intersection of three different disciplines of research in Stuart Levy's laboratory—the study of R-factors (21), the classification of different resistance determinants causing tetracycline resistance (33), and the studies by Laura McMurry and her expertise in transport kinetics and the biochemistry of tetracycline resistance (22)—that led to a cornerstone paper in molecular microbiology that for the first time implicated a novel mechanism of antibiotic resistance, the energy-dependent removal of antibiotics from within bacterial cells. With the publication in 1980 of "Active efflux of tetracycline encoded by four genetically different resistance determinants in *Escherichia coli*" in the *Proceedings of the National Academy of Sciences* (30), they reported incontrovertible evidence that a third mechanism of antibiotic resistance existed, the active or energy-dependent removal of antibiotics, in their case, the active efflux of tetracycline. The protein they reported on removed tetracyclines from the interior of cells via the action of membrane-bound efflux "pumps" to the extracellular media, decreased intracellular concentrations of the drug, and allowed the cell to survive. The experiments were conducted by following the kinetics of ^{3}H-Tc uptake in tetracycline resistant and susceptible cells and, in a model of membrane transporter function developed by Rosen (41), the everted membrane vesicle assay, which relied on studying the isolated phenotype away from interfering cellular machinery. The everted vesicle was also an ideal way to study transport proteins in bacteria and bacterial membranes.

In a bacterium, membrane-bound Tet proteins efflux tetracyclines outward, away from target ribosomes, decreasing the intracellular concentration of tetracycline. However, the everted membrane vesicle preparation relies on French-press lysis of *E. coli* cells expressing active Tet proteins, resulting in the membrane fractions reorienting inside out during their recoalescing, forming liposome-like vesicles so that the proteins now pump tetracyclines into the vesicle interior or lumen. The Levy group found that vesicles from *E. coli* bearing the resistance determinant R222 and expressing the functioning Tet protein accumulated much more ^{3}H-Tc, while Tc susceptible cells and their membrane preparations did not accumulate ^{3}H-Tc (Fig. 1). By using both resistant and sensitive vesicles the transport kinetics of ^{3}H-Tc and the biochemistry of tetracycline transport was readily followed and first described. Tetracycline transport in resistant vesicles was found to require an energy source such as Li-lactate, NADH or ATP for active uptake, while energy inhibitors such as CCCP or DNP caused uptake to cease (30, 50).

Since the Levy laboratory reported their findings on tetracycline efflux, it has been found that all bacteria can and will express antibiotic efflux proteins. Among the tetracycline resistant bacteria, the number of distinct efflux determinants has grown to over 30 (12), where each determinant is given a letter or number designation to describe the class of transferable genetic determinant responsible for encoding the antibiotic efflux proteins (23, 24). The tetracycline family of efflux proteins are transmembrane antiporters and are proton-motive-force dependent, exchanging a proton for a tetracycline-cation complex removing tetracyclines via the inner periplasmic membrane in gram-negative bacteria and the cell membrane of gram-positive bacteria such as *Staphylococcus aureus*. Tet proteins have also been found to be highly conserved and homologous in primary sequence with slight amino-acid composition differences and are structurally composed of 12-membrane-spanning α-helices in a two-domain complex separated by a large cytoplasmic loop (12, 31). Phylogenetically, Tet proteins emerged from the major-facilitator superfamily of transport proteins and have evolved as a multimer in membranes to specifically remove tetracyclines from within bacterial cells (32, 42).

Mark L. Nelson and S. Ken Tanaka • Paratek Pharmaceuticals, 75 Kneeland Street, Boston, MA 02111.

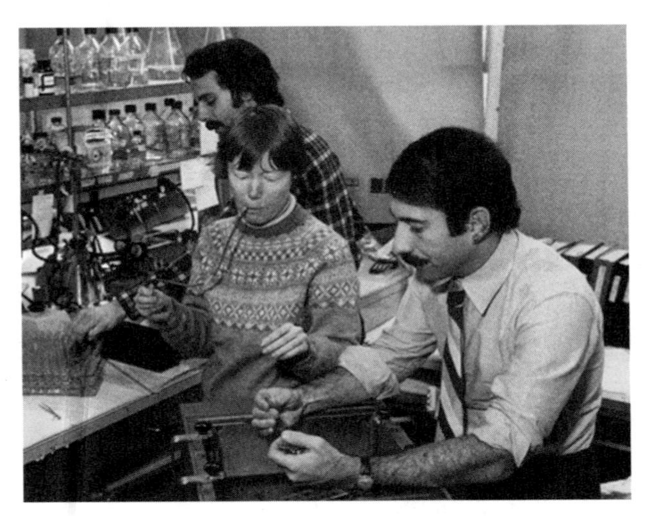

Figure 1. (back to front) Richard Petrucci, Laura McMurry and Stuart Levy in the laboratory performing the everted membrane vesicle assay sometime in the 1970s.

EFFLUX INHIBITORS AND THE CHEMICAL EVOLUTION OF THE TETRACYCLINES

During the 1980s, the study of antibacterial efflux proteins became a scientific front that continues to this day, where bacterial efflux pumps are responsible, at least in part, for antibiotic resistance across most, if not all, of the clinically relevant antibiotics. As a graduate student (M. Nelson) in medicinal chemistry at Temple University in Philadelphia in 1987, I was writing my thesis one late night and I caught the televised image of an energetic scientist with a bow tie describing how "we (society) have been caught with our pants down" and that "the wonder drugs just do not work as well as they used to, antibiotic resistance will become a huge problem." He appeared on a NOVA documentary entitled "When Wonder Drugs Don't Work," and I watched the story unfold. The documentary showed that because of the use and, particularly, the misuse of antibiotics, deadly bacteria were thriving in their presence, unaffected by their activity or potency. This was not simply a laboratory phenomenon being described; many people were affected, chemotherapy costs were rising, and, even more frightening, mortality rates were climbing. The premise of the show postulated a return to the preantibiotic era, when wonder drugs did not exist, and, even more controversial, was that research in the field of new antibiotics was lagging, and few new antibiotics were being discovered. I asked my colleagues, some with experience and jobs in pharmaceutical houses surrounding the Philadelphia area, about antibiotic resistance and about how antibiotics were losing their activity via new drug resistance mechanisms, and their reply was unenthusiastic. According to their managers, we had all the antibiotics we needed, and drug discovery programs were focused in "blockbuster" areas, in cardiology, neurology, and psychiatry—mostly chronic and more profitable diseases. Those in academia saw it differently, however. My mentors at Temple University School of Medicine spoke of drug resistance mechanisms and new mechanisms and proteins that were just beginning to be identified, much less understood in function. To them it was a black box, waiting to be opened and studied, especially in the discipline of medicinal chemistry where the structure activity relationships of resistance and antibiotics were far less known. In eukaryotic cells and cancer cells, there was considerable research into the function of P-glycoproteins that were identified and found to act as nonspecific efflux pumps, removing cellular metabolites, toxins and anticancer drugs alike (6, 8). But bacterial resistance mechanisms, especially efflux, were further behind in their study and offered a golden opportunity for a medicinal chemist to enter and begin a research career. Upon answering an advertisement in the employment section of the journal *Science,* placed in November of 1987 for a postdoctoral medicinal chemist to study the inhibition of efflux proteins in bacteria, I met the scientist with the bow tie who was the featured activist in the NOVA show, Stuart B. Levy, from Tufts University School of Medicine, Boston, Mass.

Inhibition of Tetracycline Efflux Proteins

Dr. Levy had hired me (M. Nelson) in the winter of 1988 to continue studying the structure-versus-activity relationships (SAR) of the tetracyclines against the Tet(B) efflux protein expressed in *E. coli* cells. By then their laboratory, via the expertise of Laura McMurry, had begun using the everted membrane vesicle assay as a means for assessing the activity of other tetracycline analogs as substrates for the efflux pump system and determining whether or not analogs acted as inhibitors. Compounds that inhibited intravesicular uptake of ^{3}H-Tc (without nonspecifically damaging vesicle pH gradient formation) were deemed efflux protein inhibitors and could be used to design more potent inhibitors, while observing the SAR of the tetracycline-Tet protein interaction. It was hypothesized that when a competitive inhibitor or a substrate of the efflux protein was encountered by the Tet protein, the net result would be a decrease in accumulated ^{3}H-Tc in functioning vesicles. Subsequent experiments found that the vesicular uptake of ^{3}H-Tc followed Michealis-Menton kinetics within certain limitations, and was useful for finding molecules capable of blocking Tet protein-mediated efflux and thus could reverse

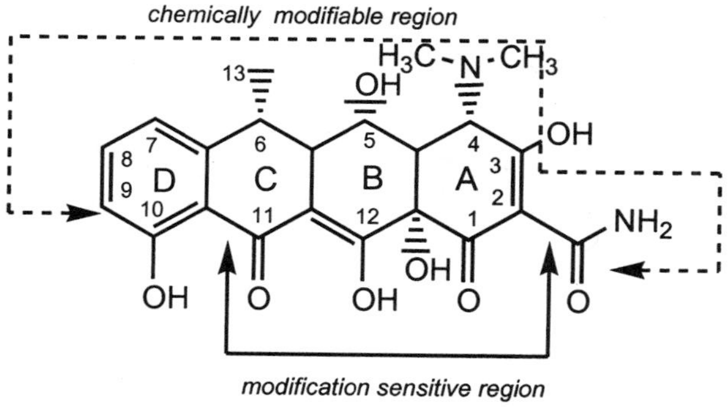

Figure 2. The structural locants of the tetracycline naphthacene ring system and structural features needed for optimal efflux inhibition in vesicles and *E. coli* possessing Tet(B) proteins. Region outlined in solid lines indicates features needed for optimal efflux inhibition. Dashed line shows region that can be chemically modified for optimal efflux inhibition.

efflux-mediated resistance. The goal of the project, which at the time was sponsored by the Upjohn Co. of Kalamazoo, Mich., was to discover a tetracycline analog or compound to inhibit the resistance efflux protein that would allow improved intracellular accumulation and increased activity of antibiotics normally subject to efflux. Increased accumulation of the drug would inhibit bacterial growth and constitute an effective approach to reversing tetracycline resistance, changing a resistant cell back into a susceptible one.

Using derivatives of tetracycline that were commercially available, as well as compounds that represented specific chemical changes along the positions of the tetracycline naphthacene ring, we sought to develop a potent inhibitor of tetracycline efflux while determining the molecular requirements for efflux inhibition of the tetracycline efflux pump. We examined numerous derivatives of tetracyclines, along with tetracycline substructures representing segments of the ABCD naphthacene ring system for their ability to inhibit accumulation of ^{3}H-Tc into everted membrane vesicles prepared from Tc-resistant *Escherichia coli* D1-209 possessing the plasmid R222, bearing the class B determinant on transposon Tn*10*, and their IC_{50}s were calculated (36). Class B-derived vesicle preparations were chosen, as class B proteins are found in many bacterial genera and have the most pronounced and stable transport activity of all the Tet proteins examined.

After examining over 47 tetracycline positional variants, tetracycline substructures, and partial ring systems, we found that an intact naphthacene ring was needed for optimal efflux inhibition (Fig. 2). The positional changes around the scaffold indicated that the presence of an intact lower peripheral region consisting of phenolic-keto enol substructures spanning carbons C10, C11, C12, and C1 were required for inhibitory activity. The lower peripheral region is required for metal complexation with calcium and magnesium, where the C11-C12 carbonyl enolate region is the primary ligand-metal attachment point and necessary for both efflux inhibition and antibacterial activity. Chemical modifications of the tetracycline scaffold at positions 2, 4, 5, 6, 7, 8, 9, and 13 produced efflux pump inhibitors of variable activity. In particular, 13-thiol derivatives of methacycline possessed the most potent activity in inhibiting efflux function (Fig. 3).

We found that tetracycline derivatives were scarce. In order to further a detailed SAR for inhibition of Tc efflux, especially for upper periphery derivatives, we began a semisynthesis program in 1988 that continues to the present day at Paratek Pharmaceuticals, Inc., in Boston, Mass. In order to perform chemistry, we began work at Tufts University, in the Department of Chemistry, Medford, Mass., in collaboration with Professor Emeritus Vlasios Georgian (who was R. B. Woodward's first graduate student and a synthetic and

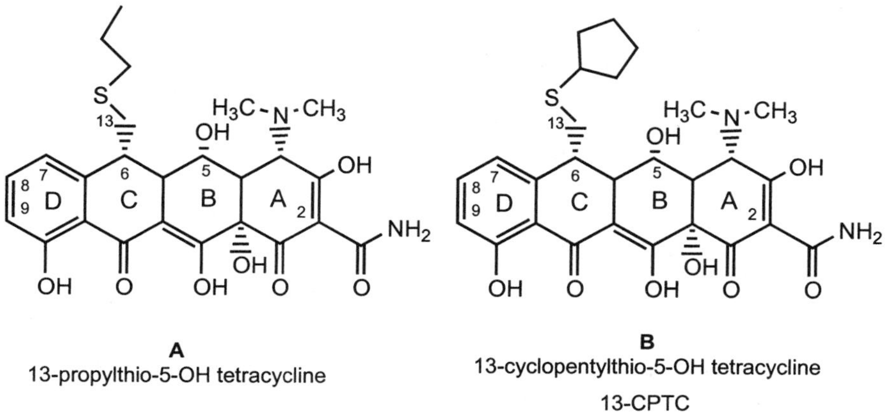

Figure 3. 13-Alkylthio derivatives of methacycline. A) 13-propylthio-5-OH tetracycline and B) 13-cyclopentylthio-5-OH tetracycline (13-CPTC).

natural products chemist). His laboratory was reopened at Tufts, and an organic synthesis laboratory was built from the ground up. Glassware, rotavap apparatuses, chemical reagents, and tetracyclines were obtained, and a crucial chromatography apparatus for tetracycline purification was used, at first using traditional flash chromatography techniques and then moving to high-pressure liquid and reverse-phase C18 chromatography. Given the chemical nature of the tetracyclines and some of the difficulties encountered during their chemical modification and purification, traditional chromatography and analytical and preparative HPLC soon became an integral part in the discovery phase of tetracycline research.

In the Levy laboratory we had different assays set up to study the antibacterial properties of the new derivatives as well as their effects in the everted membrane vesicle system. We had obtained numerous bacterial strains possessing different Tet efflux pumps in both gram-positive and gram-negative bacteria, as well as clinical strains and deep-rough mutants of *E. coli,* which were used to assess the scope of activity and other biological aspects of antimicrobial action. We further developed assays to measure efflux inhibition in whole cells following the uptake of ^{3}H-Tc, while other assays were set up to study the macromolecular synthesis pathways affected by the analogs. We had in place a laboratory dedicated to the study of antibiotic resistant bacteria via efflux-mediated and ribosomal protection, and in a short time we had opened a chemistry laboratory to synthesize new tetracycline analogs.

The initial SAR of Tet(B) protein efflux inhibition had pointed to the exocyclic C13 thiol derivatives of methacycline as the most potent inhibitors, compounds that were synthesized by the free-radical catalyzed reaction of mercaptans with methacycline (7) in the presence of a radical initiator, and we began the synthesis of an array of derivatives at this position. After the synthesis of over 40 derivatives, the SAR versus Tet efflux protein inhibition began to emerge (35). In particular, 13-alkylthio derivatives, exemplified by 13-propylthio-5-OH tetracycline (Fig. 2), were the most potent inhibitors found, with the 13-propyl derivative exhibiting an IC_{50} of 0.7 μM, in contrast to the K_m of tetracycline (15 to 20 μM) and doxycycline (3.2 μM). Within the C13-alkylthio series, polar derivatives were far less active and that derivatives exhibited maximum activity that was dependent on the steric attributes, lipophilicity, and polarity of the C13 substituent. The correlation of the IC_{50}s from the bioassay results with Verloop-Hoogenstraten STERIMOL values (49), which measure the steric attributes of a substituent, suggested that the more potent efflux protein inhibitors had steric length values between 4.2 and 6.2 Å, while optimal width values ranged between 3.0 and 4.2 Å at the C13 position. Polar substituents that had similar steric values were inactive, as were compounds not falling within a discreet range of lipophilicity π values of 1.0 to 2.5. We concluded that the inhibition of efflux in everted membrane vesicles was optimal when a C13 substituent was clustered in size and lipophilicity, pointing to an area within the Tet(B) protein that contained a complementary hydrophobic pocket able to accommodate C13 derivatives. Within the similar size and lipophilicity clusters was one potent analog, 13-cyclopentylthio-5-OH tetracycline (13-CPTC, Fig. 2), which had an IC_{50} of 0.4 μM. It was chosen for further study for its effect on Tet(B) efflux protein kinetics in vesicles and in bacteria possessing tetracycline efflux proteins.

The effect of 13-CPTC on everted vesicles showed that it acted by competitive inhibition, where it had a greater affinity and was competing with tetracycline for the binding site on the Tet(B) protein (37). In whole cells of susceptible *E. coli,* ^{3}H-Tc accumulation was not impeded by 13-CPTC, resistant *E. coli* with Tet(B) showed a rapid accumulation of ^{3}H-Tc upon preincubation with or addition of 13-CPTC, and in both susceptible and resistant *E. coli,* 13-CPTC alone had little antibacterial activity (Fig. 4). When combined with doxycycline, synergy was exhibited, indicating a reversal of efflux-mediated resistance in tetracycline-resistant strains. 13-CPTC alone had activity against gram-positive bacteria, including *S. aureus* possessing TetK and TetL determinants and ribosomal protection mediated by the TetM resistance determinant. By screening compounds against the Tet(B) efflux protein we found we could modulate antibiotic transport and resistance phenotypes in both gram-negative and gram-positive bacteria, changing a resistant cell to a susceptible one, reversing resistance.

SYNTHESIS OF NOVEL TETRACYCLINE DERIVATIVES

While we were exploring the general chemistry of the tetracyclines, semisynthetic reactions were applied to favorable antibacterial positions of the naphthacene ring, in particular the aromatic D-ring and the upper peripheral region spanning carbons C5 to C9. By 1992, we began synthesizing C5 tetracycline derivatives in collaboration with the Chas. Pfizer Co. and then C5-C13 substituted derivatives, using methacycline as a starting scaffold (Fig. 3). We found that some of the compounds were potent as efflux protein inhibitors and as synergists with doxycycline, while other compounds with upper periphery substitutions were extremely potent alone against gram-positive bacteria, including *S. aureus* strains that were resistant to methicillin (MRSA) and enterococci that were resistant

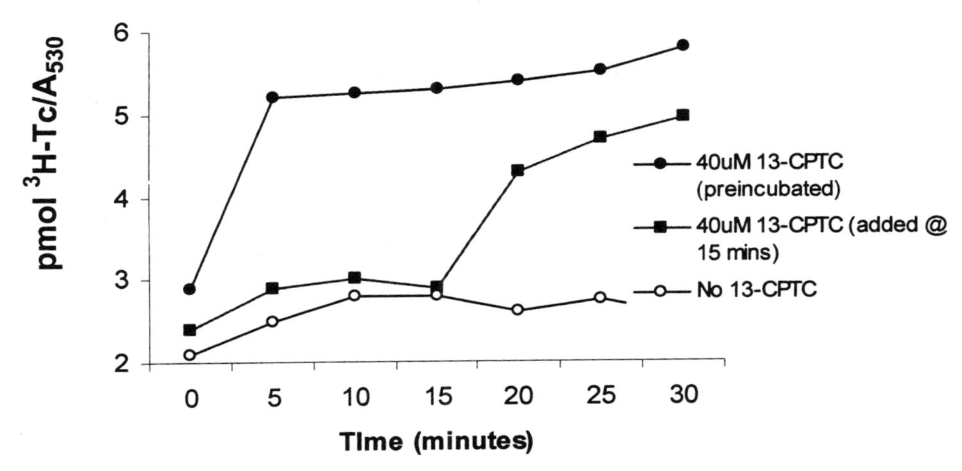

Figure 4. Reversing ^{3}H-Tc efflux in E. coli D1-209 cells possessing the Tet(B) protein pre-incubated with 13-CPTC (40 μM, ●), or without 13-CPTC (○). A rapid increase of intracellular ^{3}H-Tc occurs after addition of 13-CPTC (40 μM, ■) in glucose activated cells.

to vancomycin (VRE). We were now synthesizing compounds that had excellent activity against resistant bacteria, regardless of whether they possessed tetracycline efflux-mediated resistance.

We realized that there had been few new tetracyclines reported in over 30 years except for Wyeth's research on the glycylcyclines (44). We felt that the tetracyclines had been neglected in the area of semisynthesis and were confident that even more potent tetracyclines could be synthesized, particularly against antibiotic-resistant bacteria. It was the application of more modern reactions to the tetracyclines from our medicinal chemistry laboratory and biology advances derived from the Levy laboratory and the study of the multiple antibiotic resistance (MAR) locus that led to the start of Paratek Pharmaceuticals in Boston, Mass., in the summer of 1996.

THE BIRTH OF PARATEK PHARMACEUTICALS, INC.

Stuart and Walter "Wally" Gilbert had been friends for the last 20 or so years, and one night over dinner Stuart told him of the progress we had made in synthesizing new tetracyclines, particularly against MRSA and VRE strains, the "superbugs" that were gaining notoriety in the public press. We were also regularly meeting with other pharmaceutical companies, venture capitalists, and experts in the field, pitching our scientific and business plans for a new and improved tetracycline antibiotic and were seeking further funding to continue. Wally was so intrigued by the prospect of reviving the tetracycline family of antibiotics that he proposed forming a company to expand the chemical and biological research necessary to produce a clinical candidate and a new commercial antibiotic. This area of business was not uncharted territory for the Nobel laureate and cofounder of Biogen (now Biogen-Idec), one of the largest and most successful biotechnology companies in the United States.

In the summer of 1996 and within days of the start of Paratek Pharmaceuticals, Inc., the small synthesis laboratory in the Tufts University School of Veterinary Medicine was moved into laboratory space once utilized by Genzyme in downtown Boston, and we began actively recruiting chemists and biologists to begin work.

Synthesis of the Tetracycline Library at Paratek

With the purchase of a high-field NMR and other analytical instruments we began to explore the types of reactions that could chemically modify the tetracycline core scaffold without modification of the many labile functional groups and substructures within the tetracycline nucleus. An initial semisynthesis breakthrough occurred in the summer of 1997. One pathway to novel substituted tetracyclines that was examined early on was the semisynthesis of C7 and C9 aminotetracyclines, used at Wyeth to produce the glycylcyclines and Tigecycline (44, 45). The Wyeth scientists used the amine functional group as a linker subunit, attaching reactive carbonyl compounds and forming a α halo-carbonyl substituent whose reactive halogen atom could be further derivatized by nucleophilic amine compounds. In the past, Cyanamid chemists had reacted the intermediate amine with reagents forming a diazo compound followed with nucleophilic reagents producing tetracycline derivatives of little or no antibacterial activity (19, 40). We chose a different synthetic strategy using the inherent reactivity of the diazo group and palladium metal catalysis to couple alkene compounds directly to the C9 carbon, forming a carbon-carbon bond at the C9 position and the terminal carbon of the alkenyl substituent and giving rise to C9 alkenyl derivatives of the tetracyclines (24). Using this new synthetic pathway, 9-amino doxycycline was quickly and quantitatively diazotized, and the reactive intermediate 9-diazo doxycycline was coupled with acrylic acid, ethyl ester forming 9-acrylic acid-ethyl ester doxycycline via palladium metal catalysis. The application of transition and palladium metal-based reactions to the tetracycline family of antibiotics has never been reported, and the alkenyl tetracycline derivative prepared that day was the first of hundreds synthesized using palladium-based chemical reactions (Fig. 5) (24, 43).

Other reactions and novel tetracycline products soon followed. C7 diazo tetracyclines reacted similarly, yielding C7 alkenyl derivatives, while substituting phenylboronic acids led to other novel classes of tetracyclines, the C7 and C9 phenyl tetracyclines. Soon we began to synthesize a diverse array of tetracycline derivatives based upon their palladium-catalyzed reactions with alkenes, alkynes, phenylboronic acids, carbon monoxide, and other reagents typically used in transition metal chemistry and coupling procedures (Fig. 5).

Other reactive intermediates besides C7 and C9 diazotetracyclines could be used for coupling reactions, and some were found to facilitate the workup and increase the yields of tetracycline derivatives. Iodinated C7 and C9 tetracycline reactive intermediates were found reactive for coupling chemistries, but they formed regioisomeric mixtures at C7 and C9 that had to be laboriously separated by preparative C18 reverse-phase preparative HPLC. By changing reagents

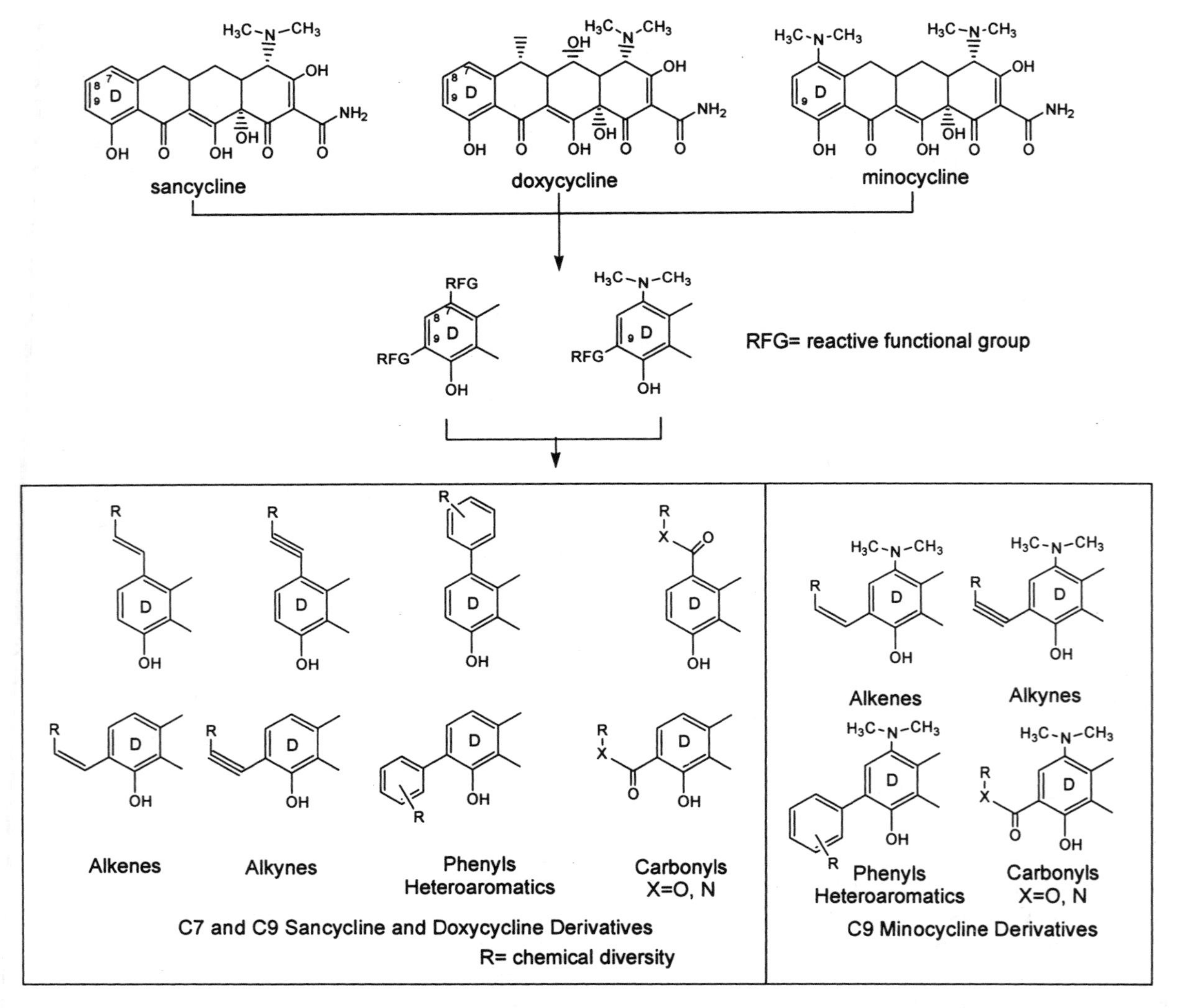

Figure 5. Novel tetracyclines synthesized by transition metal catalyzed reactions producing C7 and C9 derivatives possessing structural diversity. Intermediate tetracyclines can form alkenes, alkynes, phenyl and heteroaromatic and carbonyl derivatives, among others.

and reaction conditions, we produced regiospecific C7 or C9 iodotetracyclines selectively with few or no competing by-products. The success of this reaction led to the semisynthesis of C7 derivatives of sancycline possessing alkenyl, alkynyl, and aromatic and other coupled substituents. Similar derivatization schemes also produced C9 derivatives of minocycline, producing for the first time a novel array of C9 minocycline derivatives based on alkene, alkyne (and their reduction products, alkanes), phenyl, and heteroaromatic couplings as well as carbonylated products and analogs (38). The same chemistries were applied to the C8 position, whereby 8-halo tetracyclines were efficiently coupled with alkenes, alkynes, phenylboronic, and heteroaromatic boronic acids to produce C8 derivatives. In another semisynthesis pathway, methacycline, with an exocyclic double bond, was found to efficiently couple to phenylboronic acids via transition metal coupling, forming C13-phenyl derivatives of methacycline in high yield, one of the few reactions that can be successfully applied at this position of the methacycline scaffold (38).

Since the initial discovery and application of transition metal based chemistry applied to the tetracyclines, chemists at Paratek have synthesized over 1,500 novel tetracyclines at most of the positions along the core tetracycline scaffold. This collection represents a new wave of third-generation tetracyclines for study, while our biologists are currently evaluating them for activity in vitro and efficacy in vivo against bacteria, parasites, and nonantibiotic uses related to inflammation and neurodegeneration. Currently, Stuart is Chief Scientific Officer at Paratek.

Nontetracycline Projects at Paratek Pharmaceuticals, Inc.: Inhibiting the MAR Locus

Paratek is also focusing part of its drug discovery program on novel small molecules that prevent a bacterium from causing infection by inhibiting a major regulator of virulence, the multiple adaptation response (MAR) protein (1, 2). MAR regulatory proteins, first described by Stuart and colleagues at Tufts University (13, 17, 18), has homologs in most bacteria examined so far, where it controls the ability of bacteria to initiate infections, resist antibiotics, and adapt to hostile environments. When MAR is expressed and activated, bacterial defenses go up. Paratek scientists have recently identified small molecules that inhibit MAR in a variety of human pathogens, disarming the bacterium and thus preventing the expression of virulence factors and infection in animal models (1, 2). As a target for chemotherapeutic intervention, MAR appears distinct in bacteria only, and no homologs exist in humans.

Paratek's MAR program represents a new paradigm in chemotherapy, using small molecules to prevent bacteria from causing infection. Past approaches have focused on blocking individual virulence factors, such as pili or toxins (4), but have suffered from narrow-spectrum applications and the presence of complementary, redundant, or compensatory pathways (3). The MAR inhibitor strategy differs from these by blocking the central controlling pathway for bacterial defense, preventing the expression of the virulence cascade and multiple virulence mechanisms. MAR proteins control the pathogenic functions in many different medically important bacteria and are, therefore, broad-spectrum targets not confined to a specific species or genera.

As therapeutic agents, MAR inhibitors can be envisioned as broad-spectrum "acute vaccines" when given in a number of clinically relevant settings where prevention of infection is important to patient care, e.g., recurrent urinary tract and pulmonary infections, and in high-risk settings such as ventilator, catheterized, and surgical patients. By lowering bacterial defenses, MAR inhibitors will improve the effectiveness of current standards of care in difficult to treat infections and as adjunctive therapy with known antibacterial agents.

Discovery of the Aminomethylcyclines (AMCs): Antibacterial Tetracyclines for the Next Millennium

The primary goal of Paratek was always and foremost to develop a novel tetracycline for use against infections, with improved potency against tetracycline- and antibiotic-resistant bacteria. While developing new semisynthesis reactions for chemically modifying the tetracycline scaffold based on transition metal-catalyzed coupling, we also tried other semisynthetic reactions and techniques. Tetracyclines can react with certain electrophilic agents at positions C7 and C9 (19, 26), and this synthetic strategy is the origin of Wyeth's new tetracycline, tigecycline, in which an intermediate tetracycline with a 9-amino functional group attached to minocycline affords the core scaffold (44, 45). In the past, Cyanamid chemists had also reacted sancycline with the reagent N-hydroxymethylphthalimide, generating C7 and C9 methylphthalimide derivatives (26). The phthalimide groups added no increased antibacterial activity, but could serve as a reactive synthon via its nitrogen functional group if cleaved to produce a reactive primary amine functional group. A tetracycline primary amine at C7 or C9 would serve as a scaffold for the production of a diverse array of derivatives produced through the reaction of the primary amine with amine-modifying reagents. This was understood by Cyanamid, but they were never able to produce the reactive

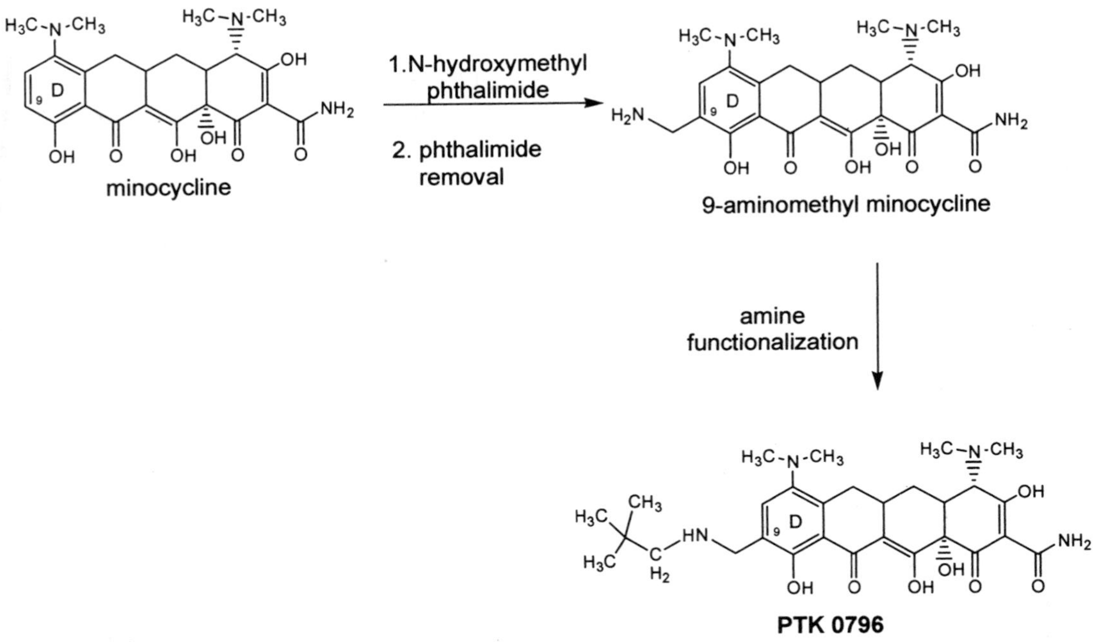

Figure 6. Synthesis pathway for the 9-aminomethyl minocyclines and the synthesis of derivatives, notably 7-dimethylamino-2,2-(dimethylpropyl)aminomethyl minocycline, designated PTK 0796.

primary amine; instead, they produced intractable products, and there was no further research reported on this reaction pathway. We had attempted to synthesize the primary amine numerous times either in the early days at Tufts University, as well as at Paratek, and this reaction and its possibilities were always in the back of our minds. During a period of expansion of the Paratek tetracycline library by transition metal chemistry, one chemistry meeting stood out in 1999, where Mohamed Ismail again thought it was possible to remove the phthalimido group using other reaction conditions and reagents (Fig. 6). This way the primary aminomethyl group attached to a tetracycline could be isolated and further functionalized, affording derivatives not synthetically accessible by transition metal catalysis coupling. Soon, Mohamed, along with Todd Bowser, worked out different reaction conditions to produce the C9-aminomethyl derivative of minocycline and the C7 and C9 aminomethyl derivatives of sancycline in quantities sufficient to allow the further exploration of its reactivity and chemistry (Fig. 6). In the hands of chemist Laura Honeyman, the primary amine maintained reactivity with a diverse array of aldehydes under reductive alkylation conditions, producing 9-alkylaminomethyl derivatives of minocycline and C7 and C9 alkylaminomethyl derivatives of sancycline. The aminomethylation reaction was applicable to all of the tetracycline scaffolds tried, and other derivatization reactions with acid chlorides, isocyanates, chloroformates, and other amine reactive reagents confirmed the synthetic versatility of the aminomethyl group. Soon hundreds of new derivatives were synthesized by Paratek chemists. Biologically, numerous derivatives had pronounced activity against Tc-susceptible and -resistant bacteria from both gram-positive and gram-negative bacteria, including MRSA and VRE, and an SAR began to emerge. One compound in particular, 7-dimethylamino-2,2-(dimethylpropyl)-aminomethyl minocycline (Fig. 6), an amimomethyl derivative of minocycline, was extremely active both in vitro and in vivo and became a preclinical designated as PTK 0796.

PTK 0796

The discovery of the aminomethylcyclines and the understanding of the SAR of this novel class (5, 25), led Paratek scientists to identify the company's first preclinical candidate, PTK 0796 (7-dimethylamino-2,2-(dimethylpropyl)aminomethyl minocycline) (Fig. 6). PTK 0796 has undergone extensive evaluation in vitro and in vivo, the results of which indicate the potential clinical utility of this compound in the treatment of serious bacterial infections.

Activity of PTK 0796 in vitro

PTK 0796 has excellent activity against gram-positive, gram-negative, anaerobic, and atypical bacteria. By virtue of its ability to overcome the most common forms of tetracycline resistance, PTK 0796 exhibits excellent activity against many of the most prob-

Table 1. In vitro activity of PTK 0796 (5, 25)

Organism (*n*)	Compounds	MIC (μg/ml)[a]		
		Range	MIC_{50}	MIC_{90}
S. aureus MSSA (16)	PTK 0796	≤0.06–0.25	0.125	0.125
	Vancomycin	0.25–0.5	0.5	0.5
	Linezolid	1.0–2.0	1.0	2.0
	Cefotaxime	1.0–2.0	2.0	2.0
	Levofloxacin	≤0.06–4.0	0.125	0.125
S. aureus MRSA MultiR (10)	PTK 0796	0.25–0.5	0.5	0.5
	Vancomycin	0.5–1.0	1.0	1.0
	Linezolid	0.5–2.0	1.0	2.0
	Cefotaxime	NA	NA	NA
	Levofloxacin	NA	NA	NA
E. faecium VRE (19)	PTK 0796	0.125–0.5	0.25	0.5
	Vancomycin	NA	NA	NA
	Linezolid	0.5–4.0	2.0	2.0
	Cefotaxime	NA	NA	NA
	Levofloxacin	1.0–>64	64	>64
E. faecalis (31)	PTK 0796	0.125–0.5	0.25	0.5
	Vancomycin	0.5–8.0	1.0	2.0
	Linezolid	1.0–4.0	1.0	2.0
	Cefotaxime	NA	NA	NA
	Levofloxacin	0.5–64	1.0	32
S. pneumoniae PRSP MultiR (18)	PTK 0796	≤0.06	≤0.06	≤0.06
	Vancomycin	0.125–0.25	0.125	0.25
	Linezolid	0.5–1.0	1.0	1.0
	Cefotaxime	0.5–8.0	1.0	8.0
	Levofloxacin	0.5–1.0	0.5	1.0
E. coli (23)	PTK 0796	0.5–2.0	1.0	2.0
	Vancomycin	NA	NA	NA
	Linezolid	NA	NA	NA
	Cefotaxime	≤0.06–0.5	≤0.06	0.125
	Levofloxacin	≤0.06–16	≤0.06	4.0
K. pneumoniae (23)	PTK 0796	1.0–8.0	2.0	4.0
	Vancomycin	NA	NA	NA
	Linezolid	NA	NA	NA
	Cefotaxime	≤0.06–>64	≤0.06	32
	Levofloxacin	≤0.06–64	≤0.06	32
H. influenzae (53)	PTK 0796	0.5–8.0	1.0	2.0
	Vancomycin	NA	NA	NA
	Linezolid	NA	NA	NA
	Cefotaxime	≤0.06–1.0	≤0.06	≤0.06
	Levofloxacin	≤0.06	≤0.06	≤0.06

[a] NA, not applicable. Not clinically useful.

lematic bacteria found in serious infections, including multidrug-resistant *S. aureus, Streptococcus pneumoniae* (and other streptococci), *Enterococcus* species (both *E. faecalis* and *E. faecium*), *E. coli, Klebsiella pneumoniae,* and *Haemophilus influenzae* (48). The activity of PTK 0796 is summarized in Table 1.

Important highlights of the activity of PTK 0796 include its superior activity against MRSA, VRE, penicillin-resistant *S. pneumoniae, E. coli,* and *K. pneumoniae* compared to many of the currently available antibiotics, particularly vancomycin, linezolid, cephalosporins, and tetracyclines. In addition, PTK 0796 is active against intracellular pathogens such as *Chlamydia* and *Legionella* and anaerobes such as *Bacteroides fragilis,* consistent with its evolution and clearly a differentiating characteristic from vancomycin, linezolid, and the β-lactams (data not shown). This activity reflects both the antibacterial potency based on the ability of PTK 0796 to inhibit protein synthesis (Table 2) and its ability to do so even in the presence of tetracycline resistance, regardless of whether the resistance is mediated by efflux pumps or ribosome protection (Table 3) (47).

Pharmacology

The pharmacokinetics of PTK 0796 were evaluated in mice, rats, and cynomolgous monkeys (Table 4) (10). Following single intravenous administration,

Table 2. Mechanism of action of PTK 0796 versus *S. aureus* RN450[a] (47, 48)

Compound	IC_{50} (μ/ml)			
	Protein	RNA	DNA	Peptidoglycan
PTK 0796	≤0.03	>32	>32	11.6
Tetracycline	0.04	31.4	25.7	8.8
Rifampin	0.01	0.01	>32	>32
Ciprofloxacin	14.0	>32	0.4	>32
Fosfomycin	>32	>32	>32	7.8

[a] IC_{50}, concentration of compound that inhibits by 50% the incorporation of radiolabelled precursors into TCA precipitable material following incubation of mid-log cultures of *S. aureus* RN450.

Table 3. Inhibition of protein synthesis in *S. aureus* resistant to tetracycline

Compound	Strain	Tetracycline resistance determinant	MIC (μg/ml)	IC_{50} protein synthesis (μg/ml)
PTK 0796	RN450	None	0.125	≤0.03
	RN4250	*tet*K	0.25	0.19
	MRSA5	*tet*M	0.125	0.11
Tetracycline	RN450	None	≤0.06	0.04
	RN4250	*tet*K	32	13.8
	MRSA5	*tet*M	>64	1.8

Table 4. Pharmacokinetics of a single dose (10 mg/kg) of PTK 0796 (11)

Animal	Dose (mg/kg)	λz (h)[a]	AUC (h•μg/ml)	C_{max} (μg/ml)	V_d[a] (liters/kg)
Mouse (CD-1)	10	5.6	6.23	2.4	12.1
Rat (S-D)	10	3.8	5.6	1.7	7.4
Monkey (Cynomolgus)	25	11.3	72.8	79.7	4.4

[a] λz, elimination half-life; V_d, volume of distribution.

Table 5. Tissue pharmacokinetics (rat) of PTK 0796 (11)

Dose (mg/kg)	Tissue	λz (h)	AUC (h•μg/g)	C_{max} (μg/ml)	Time (h) >1.0 μg/g
5.0	Kidney	3.2	31.9	18.0	7.7
	Lung	4.1	18.9	7.4	6.2
10.0	Kidney	3.6	59.0	26.9	12.8
	Lung	3.9	33.7	20.7	8.5

Table 6. Protein binding of PTK 0796[a] (10, 11)

Drug	Concn	% Unbound drug		
		Mouse plasma	Rat plasma	Monkey plasma
PTK 0796	1.0 μg/ml	98.0	100	100
Minocycline		17.9	15.9	19.5
Doxycycline		24.0	13.9	17.8
PTK 0796	0.1 μg/ml	80.0	82.0	86.0
Minocycline		26.5	21.7	25.0
Doxycycline		21.8	18.0	28.0

[a] Plasma samples spiked at indicated concentrations and extent of binding in microequilibrium dialysis determined by LC-MS/MS.

PTK 0796 exhibited a long half-life, consistent with projections of once-daily dosing in humans and high concentrations in tissue with long residence time, most notably in the lungs (Table 5) (11). Protein binding was less than 1%, indicating that most of the drug will be available to exert antibacterial activity (Table 6), and no metabolism by liver microsomes was observed (Table 7).

Table 7. Stability of PTK 0796 in liver microsomes (11)[a]

Compound	% Unchanged compound at t = 60 min		
	Mouse liver microsomes	Rat liver microsomes	Monkey liver microsomes
PTK 0796	100	100	99.0
Minocycline	100	100	100
Doxycycline	100	92.5	91.0
7-Ethoxy coumarin	30.0	57.0	18.0

[a] Liver microsome samples with 25 μM compound, incubated at 37°C for 60 min and unchanged compound remaining in reactions determined by LC-MS/MS.

Table 8. Efficacy of PTK 0796 in murine models of systemic infections[a] (9, 27-29)

Model	Compound	MIC (μg/ml)	PD_{50} (mg/kg)
S. aureus MSSA	PTK 0796	0.5	0.4
	Linezolid	2.0	3.5
	Vancomycin	0.5	0.4
S. pneumoniae PSSP	PTK 0796	0.125	0.09
	Linezolid	1.0	3.5
	Vancomycin	0.25	1.4
S. pneumoniae PRSP	PTK 0796	0.25	0.14
	Linezolid	0.5	7.1
	Vancomycin	0.25	0.14
S. pneumoniae PSSP lung (immunocompromised)	PTK 0796	≤0.06	11.0
	Linezolid	1.0	>40
	Vancomycin	0.5	7.2
S. pneumoniae PRSP lung (immunocompromised)	PTK 0796	0.25	8.5
	Linezolid	1.0	>40
	Vancomycin	0.5	5.4

[a] Infections were lethal in untreated controls. Efficacy (dose protective of 50% of animals) was determined on the basis of survival 7 days after infection.

Table 9. Efficacy of PTK 0796 in murine tissue-based models of infection[a] (9, 15, 16)

Model	Compound	MIC (μg/ml)	ED_{50} (mg/kg)
S. pneumoniae PSSP thigh	PTK 0796	0.125	0.75
	Linezolid	1.0	>40
	Vancomycin	0.25	10.2
S. pneumoniae PRSP thigh	PTK 0796	0.25	0.14
	Linezolid	0.5	>40
	Vancomycin	0.25	10.2
S. pneumoniae PSSP lung	PTK 0796	≤0.06	7.4
	Linezolid	1.0	>20
	Vancomycin	0.5	>40
H. influenzae lung	PTK 0796	1.0	4.7
	Ciprofloxacin	≤0.06	1.0
	Azithromycin	1.0	31.6
E. faecalis kidney	PTK 0796	0.5	4.5
	Linezolid	1.0	14.3
	Vancomycin	2.0	70.3

[a] Efficacy (ED_{50}) based on mean 100-fold reduction in viable bacteria per gram of tissue 22 h after treatment.

Consistent with this pharmacokinetic profile, PTK 0796 exhibited excellent efficacy in both systemic (Table 8) and tissue-based (Table 9) murine infection models (27–29). PTK 0796 was effective at equal or lower doses compared to standards of therapy in systemic infections and was markedly superior to the standards in tissue-based infections, consistent with its antibacterial activity and tissue distribution. PTK 0796 was consistently efficacious at lower doses than linezolid in normal and immunocompromised animals.

As a result of these evaluations by Paratek and others (9, 10, 15, 16, 27–29), Paratek is now committed to develop and commercialize PTK 0796 on a worldwide basis as an intravenous agent for use in the treatment of infections in patients requiring hospitalization.

THE PAST AND THE FUTURE OF A SCIENTIFIC FRONT

The impact of antibacterial resistance on our supply of antibiotics initiated the search for the mechanisms of antibiotic resistance, and advances in our understanding of these mechanisms have led to the development of more potent drugs across most, if not all, antibiotic families. The efforts of Stuart and his colleagues have, over the years, made significant advances in the understanding of resistance and in the identification and characterization of efflux of antibiotics by resistant bacteria. By first understanding the function of Tet efflux proteins, the Levy laboratory paved the way for the future of the tetracycline antibiotics, which in the past had been neglected by researchers studying their chemistry because "they were too old" or because we "had all the antibiotics we needed." Stuart not only had the foresight to realize that antibiotics were endangered, he also had the vision to realize the need for newer antibiotics that could thwart or bypass resistance by efflux. Most important was his determination to begin the long process of blending the disciplines of molecular microbiology, medicinal chemistry, and organic chemistry to the point where they are today, with the establishment of Paratek and its commitment to the study of the chemistry and biology of antiinfectives. This has led to chemical revisions and a renaissance of the tetracycline family of molecules for future generations, beginning with our first clinical candidate, PTK 0796, poised to enter human clinical trials for the treatment of serious infections. What began as the study of perplexing membrane proteins has resulted in a scientific front that has changed the chemical structure and biological activity profile of the tetracycline family of antibiotics, a drive that continues to this day with the efforts of dedicated scientists from around the world, and with our colleagues at Paratek.

REFERENCES

1. **Alekshun, M. N., V. J. Bartlett, T. Bowser, A. K. Verma, M. Grier, T. Warchol, K. Ohemeng, S. B. Levy, and S. K. Tanaka.** 2004. Novel anti-infection agents: small molecule transcription factor modulators. Presented at the 44th Interscience Conference on Antimicrobial Agents and Chemotherapy. American Society for Microbiology, Washington, D.C.
2. **Alekshun, M. N., and S. B. Levy.** 2004. The *Escherichia coli* mar locus: antibiotic resistance and more. *ASM News* **70:**451–456.
3. **Alekshun, M. N., and S. B. Levy.** Targeting virulence to prevent infection: to kill or not to kill? *Drug Discov. Today,* in press.
4. **Alksne, L. E., and S. J. Projan.** 2000. Bacterial virulence as a target for antimicrobial chemotherapy. *Curr. Opin. Biotechnol.* **11:** 625–636.
5. **Bhatia, B., T. Bowser, J. Chen, M. Ismail, L. McIntyre, R. Mechiche, M. Nelson, K. Ohemeng, and A. Verma.** 2003. PTK0796 and other novel tetracycline derivatives exhibiting potent in vitro and in vivo activities against antibiotic resistant gram-positive bacteria, abstr. 2420. *In Programs and Abstracts of the 43rd Interscience Conference on Antimicrobial Agents and Chemotherapy.* American Society for Microbiology, Washington, D.C.
6. **Beck, W.T.** 1987. The cell biology of multidrug resistance. *Biochem. Pharmacol.* **36:**2879–2887.
7. **Blackwood, R. K., and C. R. Stephens.** 1962. 6-Methylenetetracyclines. II. Mercaptan adducts. *J. Am. Chem. Soc.* **84:** 4157–4159.
8. **Bradley, G., Juranka, P. F., and V. Ling.** 1988. Mechanism of multidrug resistance. *Biochim. Biophys. Acta.* **948:**87–128.
9. **Broetz-Oesterhelt, H., R. Endermann, C.H. Ladel, and H. Labischinski.** 2004. Superior Efficacy of BAY 73–7388, A novel aminomethylcycline, compared with linezolid and vancomycin in murine sepsis caused by susceptible or multiresistant staphylococci. 14th European Congress of Clinical Microbiology and Infectious Diseases. Prague, Czech Republic. P930.
10. **Cannon, E. P., N. M. White, P. Chaturvedi, C. Esposito, J. Koroma, and S. K. Tanaka.** 2003. Pharmacokinetics of PTK 0796 in mouse, rat, and cynomolgous monkey. 41st Interscience Conference on Antimicrobial Agents and Chemotherapy, Chicago, IL. Abst. 2655.
11. **Chaturvedi, P., C. Esposito, J. Koroma, E. P. Cannon, and S. K. Tanaka.** In vitro assessment of plasma protein binding and metabolic stability of PTK 0796. 2003. 41st Interscience Conference on Antimicrobial Agents and Chemotherapy, Chicago, IL. Abst. 2675.
12. **Chopra, I., and M. Roberts.** 2001. Tetracycline antibiotics: mode of action, applications, molecular biology and epidemiology of bacterial activities. *Micro. and Mol. Biol. Rev.* **65:**232–260.
13. **Cohen, S. P., H. Hächler, and S. B. Levy.** 1993. Genetic and functional analysis of the multiple antibiotic resistance (*mar*) locus in *Escherichia coli. J. Bacteriol.* **175:**1484–1492.
14. **Dieck, H. A., and R. F. Heck.** 1974. Organophosphinepalladium complexes as catalysts for vinylic hydrogen substitution reactions. *J. Am. Chem. Soc.* **96:**1133.
15. **Endermann, R., C. H. Ladel, H. Broetz-Oesterhelt, and H. Labischinski.** 2004. BAY 73-7388 is highly efficacious in animal models of intra-abdominal infections caused by a range of aerobic and anaerobic organisms, including VRE. 14th European Congress of Clinical Microbiology and Infectious Diseases. Prague, Czech Republic. P928.
16. **Endermann, R. C. H. Ladel, H. Broetz-Oesterhelt, and H. Labischinski.** 2004. BAY 73-7388, a novel aminomethylcycline, is highly active in vivo in a murine model of pneumococcal pneumonia. 14th European Congress of Clinical Microbiology and Infectious Diseases. Prague, Czech Republic. P931.
17. **George, A. M., and S. B. Levy.** 1983. Amplifiable resistance to tetracycline, chloramphenicol, and other antibiotics in *Escherichia coli:* Involvement of a non-plasmid-determined efflux of tetracycline. *J. Bacteriol.* **155:**531–540.
18. **George, A. M., and S. B. Levy.** 1983. Gene in the major cotransduction gap of the *Escherichia coli* K-12 linkage map required for the expression of chromosomal resistance to tetracycline and other antibiotics. *J. Bacteriol.* **155:**541–548.

19. **Hlavka, J. J., Schneller, A., Krazinski, H. M., and J. H. Boothe.** 1962. The 6-deoxytetracyclines. III. Electrophilic and nucleophilic substitution. *J. Am. Chem. Soc.* **84:**1426–1430.
20. **Ladel, C.H., R. Endermann, H. Broetz-Oesterhelt, and H. Labischinski.** 2004. BAY 73-7388 demonstrates greater activity than linezolid in a range of murine models of skin and soft tissue infection. 14th European Congress of Clinical Microbiology and Infectious Diseases. Prague, Czech Republic. P929.
21. **Levy, S. B., and L. McMurry.** 1974. Detection of an inducible membrane protein associated with R-factor-mediated tetracycline resistance. *Biochem. Biophys. Res. Comm.* **56:**1060–1068.
22. **Levy, S. B., and L. McMurry.** 1978. Plasmid-mediated resistance involves alternative transport systems for tetracycline. *Nature.* **276:**90–92.
23. **Levy, S. B., L. M. McMurry, V. Burdett, P. Courvalin, W. Hillen, M. C. Roberts, and D. E. Taylor.** 1989. Nomenclature for tetracycline resistance determinants. *Antimicrob. Agents Chemother.* **33:**1373–1374.
24. **Levy, S. B., L. M. McMurry, T. M. Barbosa, V. Burdett, P. Courvalin, W. Hillen, M. C. Roberts, J. I. Rood, and D. E. Taylor.** 1999. Nomenclature for new tetracycline resistance determinants. *Antimicrob. Agents Chemother.* **43:**1523–1524.
25. **Macone, A., J. Donatelli, T. Dumont, S. B. Levy, and S. K. Tanaka.** 2003. In vitro activity of PTK 0796 against Gram-positive and Gram-negative organisms. 41st Interscience Conference on Antimicrobial Agents and Chemotherapy, Chicago, IL. Abst. 2439.
26. **Martell, M. J., Ross, A. S., and J. H. Boothe.** 1967. The 6-Deoxytetracyclines. IX. Imidomethylation. *J. Med. Chem.* **10:** 485–486.
27. **McKenney, D., J. M. Quinn, C. L. Jackson, J. L. Guilmet, J. A. Landry, S. K. Tanaka, and E. P. Cannon.** 2004. BAY 73-7388, a novel aminomethylcycline, exhibits potent efficacy in pulmonary murine models of infection. 14th European Congress of Clinical Microbiology and Infectious Diseases. Prague, Czech Republic. P927.
28. **McKenney, D., J. M. Quinn, C. L. Jackson, J. L. Guilmet, J. A. Landry, S. K. Tanaka, and E. P. Cannon.** 2003. Evaluation of PTK 0796 in experimental models of infections caused by Gram-positive and Gram-negative pathogens. 41st Interscience Conference on Antimicrobial Agents and Chemotherapy, Chicago, IL. Abst. 2627.
29. **McKenney, D., J. M. Quinn, C. L. Jackson, J. L. Guilmet, J. A. Landry, S. K. Tanaka, and E. P. Cannon.** 2003. The efficacy of PTK 0796 in murine models of *Streptococcus pneumoniae* infections. 41st Interscience Conference on Antimicrobial Agents and Chemotherapy, Chicago, IL. Abst. 2637.
30. **McMurry, L., Petrucci, R., and S. B. Levy.** 1980. Active efflux of tetracycline encoded by four genetically different tetracycline resistance determinants in *E. coli. Proc. Natl. Acad. Sci. USA* 77:3974–3977.
31. **McMurry, L. M., and S. B. Levy.** 2000. Tetracycline resistance in gram-positive bacteria. p. 660–677. *In* V. A. Fischetti, R. P. Novick, J. J. Ferretti, D. A. Portnoy, and J. I. Rood (ed.), *Gram-Positive Pathogens.* American Society for Microbiology, Washington, D.C.
32. **McNicholas, P., M. McGlynn, G. G. Guay, and D. M. Rothstein.** 1995. Genetic analysis suggests functional interactions between the N- and C-terminal domains of the TetA(C) efflux pump encoded by pBR322. *J. Bacteriol.* **177:**5355–5357.
33. **Mendez, B., Tachibana, C., and S. B. Levy.** 1980. Heterogeneity of tetracycline resistance determinants. *Plasmid* **3:**99–108.
34. **Nelson, M. L.** The chemistry and biology of the tetracyclines. *Annu. Rev. Med. Chem.* **37:**105–115.
35. **Nelson, M. L., B. H. Park, J. S. Andrews, V. A. Georgian, R. C. Thomas, and S. B. Levy.** 1993. Inhibition of the tetracycline efflux antiport protein by 13-thiosubstituted -5-hydroxy-6-deoxytetracyclines. *J. Med. Chem.* **36:**370–377.
36. **Nelson, M. L., B. H. Park, and S. B. Levy.** 1994. Molecular requirements for the inhibition of the tetracycline antiport protein and the effect of potent inhibitors on growth of tetracycline resistant bacteria. *J. Med. Chem.* **37:**1355–1361.
37. **Nelson, M. L., and S. B. Levy.** 1999. Reversal of tetracycline resistance mediated by different bacterial resistance determinants by an inhibitor of the Tet(B) antiport protein. *Antimicrob. Agents Chemother.* **43:**1523–1524.
38. **Nelson, M. L., M. Y. Ismail, L. McIntyre, B. Bhatia, P. Viski, G. Rennie, D. Andorsky, D. Messersmith, K. Stapleton, J. Dumornay, P. Sheahan, A. K. Verma, T. Warchol, and S. B. Levy.** 2003. Versatile and facile synthesis of diverse semi synthetic tetracycline derivatives via Pd-catalyzed reactions. *J. Org. Chem.* **68:**5838–5851.
39. **Paulsen, I. T., M. H. Brown, and R. A. Skurray.** 1996. Proton-dependent multidrug efflux systems. *Microbiol. Rev.* **60:**575–608.
40. **Petisi, J., J. L. Spencer, J. J. Hlavka, and J. H. Boothe.** 1962. 6-Deoxytetracyclines.III Nitrations and subsequent reactions. *J. Med. Pharm. Chem.* **5:**538.
41. **Rosen, B. P., and J. S. McClees.** 1974. Active transport of calcium in everted membrane vesicles in *Eschericia coli. Proc. Natl. Acad. Sci. USA* **71:**5942–5946.
42. **Rubin, R. A., and S. B. Levy.** 1991. Interdomain hybrid Tet proteins confer tetracycline resistance only when they are derived from closely related members of the *tet* gene family. *J. Bacteriol.* **173:**4503–4509.
43. **Sengupta, S., and S. Bhattacharyya.** 1997. Palladium-catalyzed cross-coupling of arenediazonium salts with arylboronic acids. *J. Org. Chem.* **62:**3405.
44. **Sum, P.-E., V. J. Lee, R. T. Testa, J. J. Hlavka, G. A. Ellestad, J. D. Bloom, Y. Gluzman, and F. P. Tally.** 1994. Glycylcyclines. 1. A new generation of potent antibacterial agents through modification of 9-aminotetracyclines. *J. Med. Chem.* **37:**184–188.
45. **Sum, P. E., and P. Petersen.** 1999. Synthesis and structure-activity relationship of novel glycylcycline derivatives leading to the discovery of GAR-936. *Bioorg. Med. Chem. Lett.* **9:**1459–1462.
46. **Traczewski, M. M., and S. D. Brown.** 2003. PTK 0796: In vitro potency and spectrum of activity compared to ten other antimicrobial compounds. 41st Interscience Conference on Antimicrobial Agents and Chemotherapy, Chicago, IL. Abst. 2458.
47. **Weir, S., A. Macone, J. Donatelli, C. Trieber, D. E. Taylor, S. K. Tanaka, and S.B. Levy.** 2003. The mechanism of action of PTK 0796. 41st Interscience Conference on Antimicrobial Agents and Chemotherapy, Chicago, IL. Abst. 2473.
48. **Weir, S., A. Macone, J. Donatelli, C. Trieber, D. E. Taylor, S. K. Tanaka, and S. B. Levy.** 2003. The activity of PTK 0796 against tetracycline resistance. 41st Interscience Conference on Antimicrobial Agents and Chemotherapy, Chicago, IL. Abst. 2611.
49. **Verloop, A.** 1987. The STERIMOL approach to drug design. Marcel Dekker, Inc., New York, N.Y.
50. **Yamaguchi, A., T. Udagawa, and T. Sawai.** 1990. Transport of divalent cations with tetracycline as mediated by the transposon Tn*10*-encoded tetracycline resistance protein. *J. Biol. Chem.* **268:**4809–4813.

Frontiers in Antimicrobial Resistance: a Tribute to Stuart B. Levy
Edited by D. G. White, M. N. Alekshun, and P. F. McDermott

AFTERWORD

Learning and Teaching: a Personal Reflection

JULIAN DAVIES

In spite of the earlier discovery of penicillin, it was perhaps the discovery of streptomycin for the treatment of tuberculosis that was the most important and most dramatic event in the history of infectious diseases. The introduction of an effective treatment for the "white plague" was of paramount and worldwide importance. Antibiotic treatments for other diseases in history such as cholera, plague, and syphilis soon followed, and the golden age of antibiotics began. The problems of antibiotic resistance were minimalized at first, because laboratory studies showed that mutations associated with antibiotic resistance, while possible, were rare (streptomycin was the model antibiotic at the time) and that resistant mutants appeared at such low frequencies that they would not be expected to be an impediment to therapeutic antibiotic use. The bacterial geneticists could not have realized how wrong they were! Streptomycin-resistant *Mycobacterium tuberculosis,* sulfonamide-resistant *Streptococcus pneumoniae,* and other antibiotic-resistant pathogens appeared in the clinic and were associated with treatment failure and increased mortality. In fact, the famed writer and socialist George Orwell died when his *M. tuberculosis* infection no longer responded to streptomycin.

Tetracycline, chloramphenicol, and other antibiotics (including methicillin, the first semi-synthetic compound to be designed specifically for the treatment of resistant infections) were soon discovered, developed, and approved (with limited interference from government regulation) for human use. The use of such drugs had a remarkable effect on a worldwide scale, and diseases such as cholera, *Shigella* dysentery, and gonorrhea appeared to be controllable with just a few shots. The appearance of antibiotic resistance was inexorable, though, as illustrated by the detailed studies of the late Maxwell Finland and others.

In the late 1950s came the first reports of infectious multidrug resistance from Japan, followed by Germany and the United Kingdom. This was an unacceptable concept to old-fashioned bacteriologists who poured scorn on the reports. The initial manuscript submissions of Watanabe and others were delayed in publication, finally appearing in print (in English) in 1960. In 1963, Watanabe published the first detailed review of the phenomenon, and R-factor determined resistance to several antibiotics was subsequently identified in U.S. hospitals (1966).

Thus the field of plasmid biology was born, and since that time, this innovative characteristic of bacteria has been widely studied. After half a century of studies of the molecular epidemiology and biochemistry of plasmid-encoded antibiotic resistance, virulence, biocatalysis, symbiosis, and antibiotic biosynthesis, this form of horizontal gene transfer (HGT) is well recognized and better understood. For many years, work on horizontal gene transfer was largely the work of microbiologists (especially bacteriologists), but two important events changed this. First was the demonstration that plasmids could be used to bring about the functional transfer of specific genes from any living organism into bacterial hosts. This became the conceptual foundation of modern biotechnology. Later, as more and more partial microbial genome sequences were obtained, it became apparent that the range of horizontal gene transfer in nature was much greater than previously suspected. Horizontal gene transfer was recognized as a key factor in microbial evolution and (probably) also important in the evolution of more complex genomes. In 20 years or so, the topic of HGT went from ridicule to gospel!

Plasmids are frequently found as components of many genomes (some plasmids are larger than the smaller bacterial genomes), and more is known of their functions than older "plasmidologists" would ever have expected. Nonetheless, there remain significant gaps in the understanding of how HGT takes place in vivo, particularly with respect to the acquisition of a resistance gene and its expression. That is,

Julian Davies • Department of Microbiology and Immunology, University of British Columbia, Vancouver, British Columbia, Canada.

how does an open reading frame become a functional antibiotic resistance gene? Can we assume that DNA fragments are fluid in microbial populations? A DNA transfer event may only rarely be of selective advantage to a recipient organism and only rarely fixed in the recipient genome. DNA being transferred may or may not be of value to the recipient, depending on the nature of the selection pressure (which may be weak). Additionally, cryptic (putative) resistance genes can be found in bacterial genomes, but their true function in their hosts may not be associated with resistance. For example, genes encoding enzymes related to aminoglycoside-modifying enzymes are found in many species, yet they could have other functionalities. Recent soil metagenomic studies show that potential antibiotic-resistance determinants are widely spread, but it is not known if they are "professional."

Heterologous gene expression is the key to effective HGT. The events involved in antibiotic resistance development may be complex; following the introduction of a novel antimicrobial agent into clinical practice, there is usually a period of some 2 to 3 years before strains displaying decreased susceptibility are identified in patient specimens. What happens during this time that allows a resistance genotype to emerge? In modern biotechnology, techniques have been developed to express almost any open reading frame in any host. It is simply a matter of having the right transcription and translation signals. The heterologous gene product may not be produced in large amounts, but there are ways to tinker with genes and the translation system of the host such that a protein is overexpressed. It is thought that differences in G+C content of a sequence would be a constraint, but in fact a small number of base changes may suffice. Nature is the ultimate tinkerer, and in terms of functional gene expression, anything is possible. The word "cannot" is not in the dictionary of life.

What is required to generate a resistance gene? That is, what are the functions to be added to a protein-coding DNA fragment to make a functional protein? Admittedly this is somewhat simplistic because the production of a defined string of amino acids does not mean that a functional enzyme or structural protein is the result. At this stage of the discussion, two important points must be made. First, the identification of a resistance gene in a clinical isolate is the final step of the process. What went before is a black box (and there may be many black boxes!). The events necessary to make a functional resistance determinant and the hosts (how many?) in which the changes (adaptations) took place are unknown. Second, the organisms involved in the multiple transfers are always under some type of stress. Many (if not all) stress responses induce mutagenesis or a hypermutable state. Billions of kilos of toxic organic compounds, antibiotics, and biocides have been disseminated into the biosphere in the past 200 years. Even the process of HGT induces a stress response. As Darwin stated, "It is not the strongest of species that survive, nor the most intelligent, but the ones most responsive to change." In microbes, responsiveness is mutation, recombination, and HGT.

At least a dozen biochemical mechanisms of resistance are recognized at this time. In principal, the genetic determinants for all of these are capable of dissemination by HGT. Many of these mechanisms are plasmid determined or carried by other mobile elements, and in some cases may be present as components of genomic islands. These resistance determinants (and many orthologs) can be found in gram-positive and gram-negative pathogens, human and animal commensals, and environmental bacteria. Many mechanisms of HGT have been characterized and, in all probability, there are others waiting to be discovered. These, also, are likely to be variations on the themes of type IV secretion or bacteriophage infection, but who knows?

As mentioned earlier, the R-plasmid encoded resistance genes isolated in clinical situations are the end of a process of evolution of which very little can be reconstructed. According to Watanabe, there was no evidence for multidrug resistance in *Shigella* sp. before the discovery of R-factors. Therefore, the nascent R-plasmids must have been present in other bacterial species before that time (but in what form?) and may have evolved through a variety of bacterial genera. Since no antibiotics were being employed in animal feeds in Japan at this time, such use could not have been a factor. In retrospect, one cannot eliminate the possibility that R factor strains were brought into Japan from another country. The use of antibiotics as growth promotants in animal feeds was tested in North America as early as 1949 and was employed generally by the mid-1950s. It should be noted that the clinical identification of R-factors in Europe was in 1962. Animal use was then prevalent. (Was DNA contamination of these preparations perhaps a causal factor?) R factors were reportedly found in strains collected by Joshua Lederberg in 1949.

Getting DNA from here to there and elsewhere is not the limiting factor but "making" a functional gene is likely to be a constraint. Studies of antibiotic resistance in the laboratory can identify most of the molecular and biochemical aspects of the process, but we are ill equipped to predict what happens in a complex microbial population that possesses mechanisms to evolve rapidly in response to even minor environmental changes. It cannot be stated with any confidence what might be the origin of antibiotic resistance determinants, although in certain cases the possibilities are limited. Clearly, we can state with some confidence that a particular gene from one pathogen has been transferred to another (for example ampicillin resistance of *H. influenzae* and *N. gonorrhoeae,* or the *vanR* resistance cluster found in VRE and GRSA); these examples illustrate HGT only! The study of the evolution of antibiotic resistance genes (or pathogenicity or biotransformation genes) in toto is a fascinating problem, and it needs more attention if we are to comprehend resistance development in a productive manner; that is, as an aid to overcoming the problem.

The extensive use of antibiotics has created a situation wherein effective antibiotic treatment of infectious diseases in the industrialized or developing world, in hospitals, or in the community can no longer be taken for granted. Older antibiotics are declining in efficacy, and vanishingly few new compounds are being discovered to take their place. In the 1950-1960 period, antibiotics were discovered and rapidly introduced into human clinical practice. In the 2000s, discovery is more difficult and approval for use is expensive and tortuous to obtain. Penicillin, tetracycline, erythromycin, and similar antibiotics would probably not be approved for human use had they been discovered in the current regulatory and legal climate. In developing nations, antibiotics are being used that are not acceptable in Europe and North America, for example, cycloserine and capreomycin. Valuable antibiotics most likely still exist in nature to be discovered, but the search is not committed. There is increasing concern in the infectious disease community over the increase in resistance combined with the lack of interest of the pharmaceutical industry in finding new drugs. Perhaps we need to reexamine the problem in different ways; looking at resistance in an evolutionary context is worthy of consideration.

Throughout the half century of research since the discovery of transmissible antibiotic resistance, certain scientists stand out for their contributions, not only in scientific leadership but also in their recognition of the great value of antibiotic therapy and the way in which imprudent use is "killing the goose that laid the golden egg." Stuart Levy is one such person and it has been a pleasure and a privilege to have known him, worked with him, and played with him. I first met Stuart in the early 1970s, at a NATO plasmid course in Greece. He made an immediate impression, or at least his bow tie did! I learned that he did take it off, especially when after a day of exciting science and much wine at dinner, we all went for a nude dip in the Aegean with the entire flight crew of a Quantas jumbo jet. Rumor has it that we just missed getting arrested by the local police, who frowned on this sort of thing. I have had many interesting experiences with Stuart since that time; he was always up for a bit of fun and I learned that his bow tie was not a permanent fixture. At a plasmid meeting in (communist) Slovakia we were taken to visit a local health clinic for the afternoon off. There was a swimming pool at the resort, but we were discouraged from using it because it was too cold. However, Stuart and I and a few others insisted on changing clothes in someone's office and dashed out to the pool only to find that they were draining the water out! We splashed around like idiots for a while and gave up when there was no water left. It has always been a pleasure to attend meetings with Stuart, and we have co-organized several over the years. He often took charge of the entertainment at the closing party, and he is etched in our memories for writing and directing the first performance of the "Plasmid Song" at a meeting in Berlin in 1978 and at numerous plasmid meetings subsequently. (See below for the catchy lyric). Both Stuart and I liked to play tennis, and we sneaked off for a game on several occasions when there was free time or when the session was not to our liking. We played against each other in several different countries and the result was always the same: I won and Stuart complained that he had a bad back!

We are all fully aware of the fact that Stuart Levy's contributions go beyond his good science; he has taken the lead in pushing for the proper use of antibiotics, recognizing all the time their worldwide impact on infectious disease control. He, above all, has been our conscience. When APUA was formed, it seemed like a small effort. But under Stuart's leadership and the assistance of a group of committed helpers, APUA has become the leading voice for rational antibiotic use. Long may it flourish!

GENETIC CAPERS

Written and organized by Stuart Levy,
Karen Ippen-Ihler & others (1980-1983)
(Tune: Lemon Tree by Will Holt)

When plasmids came into this world
Who'd have known that they would be
A problem for the doctors
And a research grant for me
We see them keep on growing
World-widely they abound
Collecting genes like garbage
And passing them around.

When I was just a little gene
Upon the chromosome,
A plasmid sat down next to me
Said–don't be a "stay at home"
Excise with me, I have the stuff
My TRA genes are intact
The time has come for you to learn
About the mating act.

I am a lively pilus
On the surface of a cell
I have the reputation
As a sexy organelle
It's my good luck, I'm not FIN^+
I'm always out to play
When I attach, oh what a match
We have a holiday.

A horny streptococcus
Encultured with a male
Exudes her sexy pheromone
And leaves a perfumed trail
He likes her smell
It turns him on
He sprouts a fuzzy coat
Once they begin, their friends join in
A gang-coccal love boat

I'm losing my stability
My vector's in arrears
My cop is low, my rep is gone
I fear the end is near
Don't hold me back I've got to go
It's clear I have two P
My neat repeat, provides a seat
It's in or out for me

I am a special entity
Oh happy, happy me
My life is full of fun and games
And promiscuity
Who else can flit so easily
From one mate to the next
No care about the species
And even less the sex.

There are new circles on the block
That we would like to meet
They hang around eukaryotes
In sites that are discrete
Let's have a blast with a chloroplast
Let's teach them how we play
They're more complex, but do love sex
At least that's what they say

I am a mitochondrion
My job is ATP
Which I can make when I am grande
But now when I'm petit(e)
In days of yore, I used to score
But now I must depend
On my gross host, who can but boast
Of fusion in the end.

I. Conjugation, penetration, consummation
Oh that's great
Derepress your inhibition
Find a mate and aggregate
Now penetrate the membrane
Right on to the chromosome
If you miss, you're just a plasmid
Try to be an episome.

II. Stimulation, titulation,
CIA will turn you on
You'll become a hairy fellow
What a joy to gaze upon
It's a positive experience
For an *S. faecalis* male
Forget it, you gram-negatives
You're only doomed to fail

III. When you're in a bad position
Transposition is the thing
Look around for a new target
Grab ahold and move right in
Though you're needy don't be greedy
Just one copy is enough
If you amplify your genome
You will just much up the stuff

IV. Illegitimate recombination, transformation
Are just two
Of the acts that we do daily
In and out of each of you
We are nature's way of giving
Special functions which you'll use
To survive the contamination
In the stuff they call their foods.

V. Give me water, give me sunlight
Give me you and CO_2
It may not be conjugation
But I'll give my all for you
I'll expand and fill my belly
With lamelli, oh so sweet
When I'm green you know I'm ready
And mature enough to eat.

VI. A, T, ...

Repeat VI. A, T, ...
Now I've learned to read a plastid
In the eukaryotic way
A, T, ...A, T.
Will I ever find my way
To a more exciting sequence
Fore I get to TAA

Frontiers in Antimicrobial Resistance: a Tribute to Stuart B. Levy
Edited by D. G. White, M. N. Alekshun, and P. F. McDermott

Concluding Remarks

Jay A. Levy

Congratulations to David White, Michael Alekshun, and Patrick McDermott for assembling such an outstanding volume of comprehensive reviews in the field of drug resistance. My brother, Stuart's, efforts to increase scientific knowledge and public awareness of this public health threat began, as noted by several authors, over 30 years ago. He became intrigued by transferable genes through the early work of Joshua Lederberg and later that of Tsutomu Watanabe. Stuart surmised that these transferable plasmids would become increasingly important obstacles to infectious disease treatment.

Stuart's first focus was on bacterial resistance to tetracyclines, which led to his discovery of the efflux mechanism of resistance, in this case, mediated by the Tet protein. Therefore, this volume appropriately begins with a very fine description of efflux mechanisms by Laura McMurry and her colleagues. Laura has been Stuart's long-term associate and has helped to bring major contributions to this field. Other chapters by Marilyn Roberts, Mark Nelson, and Stephen Projan dealing with this feature of tetracycline resistance complete the picture.

Section II on single antimicrobial resistance traits illustrates the tremendous advancements made in this field, with reference to a large number of microbes and many common antibiotics. Readers are treated to very cogent and helpful discussions on the mechanisms involved and can compare the mechanisms of resistance to tetracyclines with those associated with resistance to other antibiotics.

Multiple antimicrobial resistance is a major challenge to infectious diseases, and that topic is discussed authoratively by Anthony George in his overview, followed by other contributions in the third section of this book. These authors discuss subjects such as resistance to heavy metals (Anne Summers), whose determinants may be co-associated with antibiotic resistance. There is also consideration of resistance to surface disinfectants—the target of triclosan being a notable discovery by Stuart's laboratory. We learn that such products may encourage the emergence of organisms that are resistant to them and to antibiotics as well. It is a warning to approaches directed at "sterilizing" environments. One of the biggest players in this multidrug resistance phenotype is the *mar* locus, originally found in *Escherichia coli* but also found in other bacteria. It differs from other genes for resistance in being a regulatory locus. The traits affected by this locus provide valuable insights into drug resistances and the infectious process itself.

Certainly efflux pumps, first described for tetracycline, are now recognized as a common means by which bacteria avoid the effects of antibiotics. Many of these are able to pump out antibiotics with different structures. Thus, the chapters in this section by Hiroshi Nikaido and Glenn Kaatz on gram-negative and gram-positive bacteria provide further important information on the role of these membrane structures in mediating resistance to various antibiotics among all types of bacteria. It is also noteworthy that a similar type of mechanism has been recognized for tumor cell resistance to chemotherapy. The chapters by Mike Draper and colleagues in section VI are appropriate and valuable for understanding the role of efflux processes in the broader context of drug resistance.

The editors have wisely focused section IV on the problem of drug resistance in selective pathogens. This extensive section, dealing with a wide variety of gram-positive and gram-negative bacteria including anaerobes, provides the reader with a comprehensive appreciation of the level at which resistance can now be identified in the microbial world. The view is expanded in the chapter by Tom Wellems, who cochaired a Keystone Conference on drug resistance mechanisms with Stuart in 1996. This chapter describes the current

Jay A. Levy • Department of Medicine, School of Medicine, University of California, San Francisco, San Francisco, CA 94143.

status of knowledge on resistance in the malaria parasite, *Plasmodium*.

The ecology and fitness of organisms with drug resistance are well handled in section V. The abilities of bacteria to exist within the environment (e.g., soil) and evolve with environmental pressures are important considerations. Moreover, the role of commensals in warding off infections as well as in serving as reservoirs of resistance genes is a topic that still raises challenging questions to researchers in the field. In this latter category comes the knowledge of determining how food, animals, and plants can be sources of resistance gene transmission because of the use of antimicrobials in agriculture and horticulture. This point was the subject of Stuart's prospective studies in the 1970s which examined the consequences of low-dose tetracycline use on farm animals and workers.

The final section of this volume, covering policy, education, and exploration, highlights the establishment of APUA as a nonprofit organization that champions the value of antibiotics and ways to curb bacterial resistance. Improper use of "miracle" drugs encourages resistance and the consequent failure of these valuable drugs. To avoid this, public policy is most important. The chapter by Stephen DeVincent and Kathleen Young addresses these issues appropriately. Yet another approach to the resistance problem is the development of new antibiotics. Stuart and associates used the knowledge and technology he developed at Tufts regarding tetracycline resistance and the *mar* locus to create Paratek Pharmaceuticals. The company's goal is to develop drugs that circumvent the various types of drug resistances and to find ways to treat as well as prevent infections. With the exit of large pharmaceutical companies from antibiotic discovery, the smaller biotech/biopharm companies are valuable commodities. Efforts must be made to accept the struggle of macrobe versus microbe and find further approaches to dispel the harm that microbial resistance brings to human populations.

I would be remiss without commenting on the relationship of this volume to therapy of other organisms, such as viruses. We now face similar problems of drug resistance in human immunodeficiency virus (HIV) as a result of improper use of antiretroviral therapy. This AIDS-causing retrovirus mutates through normal replicative cycles, and selection of progeny occurs within drug-containing environments (much like resistance occurs with bacteria). Unless the anti-HIV drugs are given appropriately and under the correct conditions (with adherence), development of resistance takes place. In addition, similar scenarios are occurring with drugs for herpesviruses.

In this regard, we need an HIV database to plotting antiviral resistances around the world, as is being established by APUA for other microbial resistances. Fortunately, HIV loses some of its fitness when it becomes drug resistant, but eventually, as is now being seen in certain resistant bacteria, the drug-resistant strains could adapt with restored virulence, leading to failed treatment.

We should also find directions for blocking viral resistance similar to what is being considered by researchers and companies reporting on other microbes in this volume. Toward this objective I would like to emphasize that the immune response of the host can be a means by which resistant organisms can be controlled without the need for antibiotics. After all, in most parts of the world, it is this host defense that determines the ultimate fate of microbial infections. For a virus like HIV, immune responses via immunization strategies can be enhanced without a danger to other organisms that might be present and are of value in the host. With bacteria, however, it is extremely important to be able to target the pathogenic bacteria without affecting the commensal organisms. This challenge depends on the creativity of the antibacterial vaccine scientist. This aim has been successfully accomplished by the conjugated vaccines for pneumococci and *Haemophilus influenzae*. Importantly, vaccines or immune-enhancing drugs offer a different direction from antimicrobial therapy and should be pursued with greater emphasis in the years ahead along with the discovery of new effective antibiotics.

I would like to end with a personal note. Being an identical twin is a unique and enjoyable experience. Stuart and I have had lots of fun exchanging places, fooling friends, and essentially taking advantage of a birth situation that we obviously had not planned. During our early years, it was evident that nature and science would be part of our life. We lived in the suburbs of Wilmington, Del., where we were surrounded by a variety of animals and natural habitats. We encountered situations that challenged us to pursue explanations in books or from our father, a practicing physician. While we both considered other careers, such as law for Stuart and international affairs for myself, we could not disregard our continual passion for the study and understanding of microbes. It is interesting that Stuart turned towards bacteria and I towards viruses. Moreover, our approaches to these infectious organisms have been different. Stuart has emphasized molecular directions and I have focused on biologic features.

Our close interactions continued through our education at different colleges and medical schools and travels through many parts of the world. (We even secretly exchanged colleges, Wesleyan/Williams, for a week.) It was convenient as well to have our younger sister, Ellen, nearby at Mt. Holyoke College and fortuitous that she also chose a career in infectious diseases. Our research on different microbes certainly provided us all with many opportunities to meet at scientific and medical conferences, to compare notes on our research studies, and to derive information that was beneficial to our respective fields.

It is thus timely that a book dedicated to Stuart's efforts to resolve the threat of antibiotic resistance should overlap with my own concerns regarding the eventual emergence of HIV strains highly resistant to current antiviral therapies. It is our respect for this challenge between microbes and humans that brings us both to emphasize the appropriate use of antimicrobial drugs.

Stuart and I remain very close, communicating by phone but also by email. It has been gratifying to see our careers develop in parallel and overlap sufficiently that we get to visit often despite our locations on opposite coastlines. He is my best critic and my closest friend.

INDEX

Abbreviated New Animal Drug Application (ANADA), 531
ABC transporters, 152, 154, 158, 160, 281
Acinetobacter baumannii, fluoroquinolone resistance in, 47
AcrB, 267–270
Actinomycetes spp., aminoglycoside resistance in, 90
ADAA (Animal Drug Availability Act), 531
Agriculture, plant, 465–469
  antibiotics
    implication of use on human health, 467
    resistance to, 466–467
    use of, 465–466
  fungicides
    implication of use on human health, 468
    resistance to, 468
    use of, 467
  trends and recommendations, 468–469
Agrobacterium tumefaciens
  bacteriocins, 466
  biofilm formation, 428
  chloramphenicol resistance, 128, 130
Alliance for the Prudent Use of Antibiotics (APUA)
  Advisory Committee on Animal Antimicrobial Use Data Collection, 522
  antibiotic effectiveness program, 523
  Antibiotic Misuse Statement, 521
  APUA Newsletter, 525
  birth of, 517–518
  chapters, 523, 524–525
  FAAIR (Facts About Antibiotics in Agriculture) project, 525–526
  future directions, 526
  GAARD (Global Advisory on Antibiotic Resistance Data) program, 524
  partnerships, development of, 522
  ROAR (Reservoirs of Antibiotic Resistance), 438, 523–524
  scientific vision and public health mission, 519–526
  style, tenacity and collaboration, 522
AMDUCA (Animal Medicinal Drug Use Clarification Act), 531
Aminoglycoside resistance
  Campylobacter spp., 87, 335
  Enterococcus spp., 87, 91, 92, 95, 299–300, 301, 302–304
  Escherichia coli, 90, 93, 95, 379
  inhibition of resistance, 93–95
  mechanisms
    aminoglycoside acetyltransferases (AAC), 90–92, 304, 360
    aminoglycoside kinases, 87–89
    aminoglycoside nucleotidyltransferases (ANT), 89–90, 304, 360
    aminoglycoside phosphotransferases (APH), 87–89, 303–304, 314, 316, 360
    efflux, 93, 361
    ribosomal methylation, 92–93
  Pseudomonas aeruginosa, 90, 92–93, 360–361
  Streptococcus pneumoniae, 315–316
Aminoglycosides
  discovery, 85
  entry into cells, 87
  mode of action, 85, 87
  structure, 86
Aminomethylcyclines (AMCs), 543–548
Amoxicillin resistance, in *Helicobacter pylori,* 332
ANADA (Abbreviated New Animal Drug Application), 531
Anaerobes, 340–349. *See also specific organisms*
  Bacteroides, 340–345
  Clostridium, 347–348
  peptostreptococci, 348–349
  Prevotella, Porphyromonas, and Fusobacterium, 345–347
  Propionibacterium, 348
Animal Drug Availability Act (ADAA), 531
Animal Medicinal Drug Use Clarification Act (AMDUCA), 531
Animals, antibiotic use in, 528–535
  bacterial food-borne disease, 446–447
  ban on growth promoters in Europe, 448
  biocide resistance, 181–182
  European approach, 448, 534
  food as transfer vehicle of bacteria, 446–447
  impact on human health, 457
  monitoring and surveillance of, 531–532
  overview, 528, 530
  policy, recent developments in, 532–533
  regulatory framework in the United States, 531
  reports of use and impact, table of, 529–530
  resistance gene spread in food-borne commensals and opportunistic pathogens, 449–452
    Enterococcus, 449–451

Lactobacillus, 451–452
Lactococcus, 451
selective pressure for resistance development, 447–448
swine, antibiotic-resistant bacteria in, 442–443
Animals Health Institute, 447
Antibiotic resistance problems, table of, 521
Anticancer agents, resistance to
evolution of, 505–510
apoptosis and, 508
clinical resistance, 509–510
detoxification, 508
drug accumulation, factors affecting, 507–508
multidrug resistance (MDR), 507, 509
primary targets effecting cross-resistance, 507
transcriptional control of resistance, 509
genomic instability, role of, 500–503
consequences on anticancer therapy, 502–503
DNA repair mechanisms, 501
DNA replication controls, 500–501
fragile sites, 502
microsatellite and chromosomal instability, 502
mutation phenotypes, 502
radiation-induced, 502
telomeres, 502
mechanisms of resistance
apoptosis and, 485–487
drug carrier-mediated, 473–476
drug inactivation, 481
energy-dependent efflux pumps, 476–479
failure of drug to activate, 479–480
multidrug resistance proteins, 478–479
target alteration, 481–485
tumor microenvironmental effects, 503–505
evolution of resistance, 505
minimizing damage, 503–504
tolerance to damage, 504–505
Antifolate resistance
in cancer
carrier-mediated resistance, 473–475
defective polyglutamation, 480–481
dihydrofolate reductase alterations, 483–484
in *Plasmodium falciparum,* 405–406
Antimicrobial susceptibility testing
of bioterrorism agents, 414–419
Bacillus anthracis, 414–416
Brucella spp., 418
Burkholderia spp., 417
Francisella tularensis, 417–418
Yersinia pestis, 416–417
Campylobacter spp., 330–331
Helicobacter pylori, 331
Antimony resistance, 169–170
Apoptosis and drug resistance, 485–487, 508
Bcl-2, 486
caspases, 486
death receptors, 486
p53 and resistance, 485–486, 508
reversing through modulation of cell cycle, 487
sphingolipids, 486–487
APUA. *See* Alliance for the Prudent Use of Antibiotics (APUA)
AraC/XylS family, 202, 205, 265
Arcanobacterium pyogenes, tetracycline resistance in, 24–25
Arcobacter, resistance in, 455
Arsenic resistance, 169–170
Artemisinin, 408–409
Atovaquone-proguanil resistance, in *Plasmodium falciparum,* 406
Aureomycin (chlortetracycline), 30, 31
Avilamycin resistance, in *Enterococcus* spp., 308
Avoparcin, 448
Azithromycin, 66, 67

Bacillus anthracis, antimicrobial susceptibility testing in, 414–416
Bacillus cereus, β-lactamase in, 54, 57
Bacillus licheniformis, β-lactam resistance in, 53, 249
Bacillus pumilus, chloramphenicol resistance in, 128
Bacillus subtilis
Bmr and Blt, 278–279
multiple drug resistance (MDR) efflux, 154, 156, 158, 278–279, 281
OhrR, 252–253
ScoC, 200
tetracycline resistance, 11
YusO, 256–257
Bacillus thuringiensis, 436
Bacitracin resistance, in *Enterococcus* spp., 308
Bacteriocins, 466
Bacteroides spp., 340–345
β-lactamase, 54, 340–342
chloramphenicol resistance, 128, 130
distribution of resistance genes, 441
fluoroquinolone resistance, 344–345
imidazole resistance, 342–343
macrolide resistance, 344
tetracycline resistance, 13, 19, 22–25, 343–344, 441
Bcl-2, 486
BCR-ABL alterations, 485
β-lactamase, 53–60
Bacteroides, 54, 340–342
classification schemes, 53–54
clinical detection of, 59–60
factors modifying expression, 58–59
functional groups, 54
genes, table of, 55
introduction of new, 55–57
multiplicity of, 57
origins, 54–55
prevalence
bacterial species, 57
geographic distribution, 57–58
Pseudomonas aeruginosa, 356–359
β-lactam resistance
Campylobacter spp., 336
Clostridium spp., 347
Enterococcus spp., 54, 299–300
Escherichia coli, 54, 59, 377

Helicobacter pylori, 57
Pseudomonas aeruginosa, 57, 59–60, 355–359
Staphylococcus aureus, 53–55, 249
Streptococcus pneumoniae, 53, 57, 318–319
Biocides, resistance to
activity of biocides, 176
cross-resistance to antibiotics, 177
elevated tolerance, 183
environmental residues, 183–184
home use, 184
laboratory studies of resistance, 178–180
adaptation to tolerance, 178
in *Escherichia coli,* 179
in mycobacteria, 179
mechanism of action of biocides, 176
mechanism of resistance, 176–178
Pseudomonas aeruginosa, 178, 180, 362
quaternary ammonium compounds (QACS), 175–176
resistance in the environment, 180–183
in clinical setting, 180–181
failure to demonstrate resistance, 183
in food and agriculture industries, 181–182
household studies, 182
in oral cavity, 182–183
prospective assays, 182–183
retrospective studies, 180–182
triclocarban, 174–175
triclosan, 174–175, 178–180
Biofilm
fluoroquinolone resistance development and, 49
Pseudomonas aeruginosa, 362–363
soil bacteria, 428–429
environmental (nutritional) cues influencing biofilm formation, 429
initial colonization, 428
mature biofilm, 428–429
microcolony formation, 428
Bioterrorism agents, antimicrobial susceptibility testing of, 414–419
Blt, 278–279
Bmr, 278–279
Bradyrhizobium, 432–433
Brucella spp., antimicrobial susceptibility testing in, 418
Burkholderia spp.
aminoglycoside resistance, 93
antimicrobial susceptibility testing, 417
chloramphenicol resistance, 135, 136
Butyrivibrio fibrisolvens
MarR family members, 252
tetracycline resistance, 24–25

Cadmium resistance, 169
Campylobacter spp.
aminoglycoside resistance, 87, 335
antimicrobial susceptibility testing, 330–331
β-lactam resistance, 336
chloramphenicol resistance, 130
ciprofloxacin resistance, 334–335
fluoroquinolone resistance, 45, 46
macrolide resistance, 335
tetracycline resistance, 22, 335
zoonotic bacteria, resistance in, 454–455
Caspases, 486
Cell adhesion mediated drug resistance, 504–505
Cell cycle, modulation of, 487
Chloramphenicol, 124–139
mode of action, 125–127
prevalence of resistance, 137–138
resistance mechanisms, 127–137
chloramphenicol drug exporters, 131–136
enzyme inactivation, 128–131
multidrug exporters, 134–136
Type A chloramphenicol acetyltransferases (CATs), 128–130
Type B chloramphenicol acetyltransferases (CATs), 130–131
spectrum of activity, 125–127
structure, 125
toxicities, 124, 126
in vitro susceptibility testing parameters, 127
Chloramphenicol acetyltransferases (CATs), 127–131
Type A, 128–131
Type B, 130–131
Chloramphenicol resistance
Bacteroides spp., 128, 130
Campylobacter spp., 130
Enterococcus spp., 307
Escherichia coli, 128, 133, 135–138, 380
Pseudomonas aeruginosa, 132, 135–136
Streptococcus pneumoniae, 322–323
Chloroquine resistance, in *Plasmodium falciparum,* 402–404
Chlortetracycline (Aureomycin), 30, 31
Ciprofloxacin resistance, in *Campylobacter* spp., 334–335
Clarithromycin resistance, in *Helicobacter pylori,* 331–332
Clindamycin, 67, 68, 316
Clostridium perfringens, resistance in, 456–457
Clostridium spp., 347–348
β-lactam resistance, 347
chloramphenicol resistance, 130
fluoroquinolone resistance, 347–348
imidazole resistance, 347
macrolide resistance, 347
tetracycline resistance, 12, 19, 23, 347
vancomycin resistance, 348
Cobalt resistance, 169
Colony phase variation, in soil bacteria, 431
Comamonas testosteroni, 252
Commensals, resistance in, 449–452
Enterococcus, 449–451
Lactobacillus, 451–452
Lactococcus, 451
Community acquired infections
Mycobacterium tuberculosis, 392–395
Streptococcus pneumoniae, 391–392
Conjugal transfer of DNA, 437–438
Copper resistance, 168–169
Corynebacterium spp.
chloramphenicol resistance, 134

fluoroquinolone resistance, 46
tetracycline resistance, 12
Costs of antimicrobial misuse, 519–520
Co-trimoxazole. *See* Trimethoprim-sulfamethoxazole
Cytochrome P450-mediated drug inactivation, 481

DHFR (dihydrofolate reductase), 306–307, 321, 405–406, 473–474, 483–484, 507
DHPS (dihydropteroate synthase), 321–322, 405–406
DNA gyrase, 41–44, 46–49, 305–306, 319
DNA repair mechanisms, 501
DNA replication controls, 500–501
Doxycycline, 32–33

Ecology of antibiotic resistance genes, 436–444
GM (genetically modified) plants and, 436
intestinal bacteria, transfer of resistance genes among, 437, 439–441
molecular, 438–439
in oral cavity, 442
ROAR (Reservoirs of Antibiotic Resistance), 438, 523–524
Efflux
aminoglycoside resistance, 93, 361
anticancer agent resistance, 476–479
multidrug resistance proteins, 478–479
P-glycoprotein, 477–478
arsenic and antimony resistance, 169
ATP dependent drug efflux pumps, 476, 478
biocide resistance, 179, 180
cadmium, cobalt, nickel, zinc, and lead resistance, 169
chloramphenicol resistance, 131–136
copper resistance, 168
fluoroquinolone resistance, 46
Mar phenotype and, 218
MLSKO (macrolides, lincosamides, streptogramins, ketolides, and oxazolidinones) resistance, 73, 75, 77
multiple drug resistance (MDR)
ABC transporters, 152, 154, 158, 160, 281
binding site in efflux pumps, 159–161
Escherichia coli, 154, 156, 159–161
inhibitors, 282
MATE (multidrug and toxic compound extrusion) transporters, 152, 154, 281
MFS (major facilitator superfamily) transporters, 152, 153, 157, 160, 277–280
Bmr and Blt, 278–279
EmeA, 280
LmrP, 280
MdeA, 280
NorA, 277–278
PmrA, 280
QacA, 279
multidrug efflux pumps in gram-negative bacteria, 261–270
multidrug efflux pumps in gram-positive bacteria, 275–282
Neisseria gonorrhoeae, 157
overview, 275–276
Pseudomonas aeruginosa, 155–156, 160
regulators *versus* transporters, 157–159
RND (resistance-nodulation-cell division) transporters, 152, 153, 157, 158, 160, 263–266, 267, 281
Escherichia coli, 263–264
inhibitors, 266
intrinsic resistance, 263
overexpression, 265–266
overview, 261–263
Pseudomonas aeruginosa, 264–265
structural basis, 267–270
SMR (small multidrug resistance) transporters, 152–154, 156, 159, 160, 266–267, 281
Staphylococcus aureus, 156, 157, 159
silver resistance, 168
tetracycline resistance, 3–12, 537–541
EmeA, 280
Enterobacteriaceae
β-lactamase, 55, 57, 59–60
chloramphenicol resistance, 130, 135
fluoroquinolone resistance, 46–47, 49
MarB and MarC, 204
regulation of multiple antibiotic resistance, 247–249
zoonotic bacteria, resistance in, 452
Enterobacter spp.
β-lactam resistance, 54, 59
chloramphenicol resistance, 133, 136
mar mutant identification in clinical isolates, 231–232
Enterococcus spp.
acquired resistance, 301–308
aminoglycoside resistance, 87, 91, 92, 95, 299–300, 301, 302–304
β-lactam resistance, 54, 299–300
chloramphenicol resistance, 307
commensal, resistance in, 449–451
fluoroquinolone resistance, 305–306
glycopeptide resistance, 101–118, 301
intrinsic resistance, 299–301
lincosamide resistance, 301, 304–305
macrolide resistance, 304–305
MDR pump, 280
oxazolidinone resistance, 307
penicillin resistance, 301–302
SlyA-like protein, 256
streptogramin resistance, 301, 304–305
tetracycline resistance, 13, 21, 22, 306
trimethoprim resistance, 306–307
trimethoprim-sulfamethoxazole resistance, 300
VRE (vanocomycin-resistant enterococci), 290–291, 389–391
Environmental health, implications of antimicrobial use in animals, 530
Environmental Protection Agency (EPA), 175
Erwinia chrysanthemi, PecS and, 200, 250
Erythromycin, 66, 67
Escherichia coli, 374–381
aminoglycoside resistance, 90, 93, 95, 379
antibiotic therapy for infections, 375
arsenic resistance, 169
biocide resistance, 178, 179

β-lactam resistance, 54, 59, 377
chloramphenicol resistance, 128, 133, 135–138, 380
copper resistance, 168
disease spectrum, 374–375
epidemiology of resistance mechanisms
aminoglycosides, 379
animals, antimicrobial use in, 380–381
biocides, 380
fluoroquinolones, 379
multiple antibiotic resistance, 380
nitrofurantoin, 380
penicillins, cephalosporins, and carbapenems, 377–378
phenicols, 380
tetracyclines, 378–379
trimethoprim-sulfamethoxazole, 379–380
fluoroquinolone resistance, 44–47, 49, 376, 379
MAR and, 198–204, 247–249, 252, 253–254
mar mutant identification in clinical isolates, 224–231
multiple drug resistance (MDR) efflux, 154, 156, 159–161
oxidative stress response, 191–195
RND transporters, 263–264
surveillance program, 375–377
zoonotic bacteria, resistance in, 453
European Federation of Animal Health (FEDESA), 447
European Union
antimicrobial use in animals, 447–448, 534
Evernimicin resistance, in *Enterococcus* spp., 308
Evolution
molecular evolution of MarA, 205
of resistance to anticancer agents, 505–510
soil bacteria, 430–433
colony phase variation, 431
horizontal gene transfer, 431–433
mutator alleles, 430
site-specific recombination, 430–431
SOS polymerase, 431
symbiosis island, 432–433

FAAIR (Facts About Antibiotics in Agriculture) project, 525–526
FDA (Food and Drug Administration), 531, 533
Federal Food, Drug and Cosmetic Act (FFDCA), 531
Federal Insecticide, Fungicide and Rodenticide Act (FIFRA), 468
FEDESA (European Federation of Animal Health), 447
FFDCA (Federal Food, Drug and Cosmetic Act), 531
FIFRA (Federal Insecticide, Fungicide and Rodenticide Act), 468
Fis protein, 202, 203
Florfenicol, 124–127, 132–133, 138
Fluoroquinolone resistance, 41–49
Bacteroides spp., 344–345
Campylobacter spp., 45, 46
in clinical setting, 46–48
nonfermentative gram-negative bacilli, 47
Staphylococcus aureus, 48
Streptococcus pneumoniae, 47–48
traveler's diarrhea, 46–47
Clostridium spp., 347–348
Enterococcus spp., 305–306
Escherichia coli, 44–47, 49, 376, 379
factors favoring emergence of, 48–49
Helicobacter pylori, 334
mechanisms
efflux, 46
membrane permeability, 44, 46
plasmid-mediated, 46
topoisomerase mutations, 44
Pseudomonas aeruginosa, 44, 45, 47–49, 359–360
Streptococcus pneumoniae, 45, 47–48, 319–321, 392
Fluoroquinolones
classification of commercialized quinolones, 42
mechanism of action, 41–43
structure-activity relationships, 41–43
use in poultry, 533
Food and Drug Administration, 531, 533
Food Borne Diseases Active Surveillance Network (FoodNet), 532
Food chain, resistance of bacteria in
antibiotic use in animals
ban on growth promoters in Europe, 448
selective pressure for resistance development, 447–448
bacterial food-borne disease, 446–447
biocide resistance, 181–182
food as transfer vehicle of bacteria, 446–447
impact on human health, 457
resistance gene spread in food-borne commensals and opportunistic pathogens, 449–452
Enterococcus, 449–451
Lactobacillus, 451–452
Lactococcus, 451
resistance in zoonotic bacteria
Arcobacter, 455
Campylobacter, 454–455
Clostridium perfringens, 456–457
Enterobacteriaceae, 452
Escherichia coli, 453
Listeria monocytogenes, 455–456
Salmonella, 452–453
Yersinia enterocolitica, 454
Food Quality Protection Act (FQPA), 468–469
FQPA (Food Quality Protection Act), 468–469
Fragile DNA sites, 502
Francisella tularensis, antimicrobial susceptibility testing in, 417–418
Fungicides in plant agriculture
implication of use on human health, 468
resistance to, 468
use of, 467
Fusobacterium, antibiotic resistance in, 345–347

GAARD (Global Advisory on Antibiotic Resistance Data) program, 524
Genomic instability, role in resistance to anticancer agents, 500–503
Gentamicin, 85–86. *See also* Aminoglycosides
use in plant agriculture, 466

Glycopeptide resistance in *Enterococcus* spp., 101–106, 101–118, 301
genetic background of *van* operons, 114–117
*van*A gene cluster, 114–116
*van*B gene cluster, 116–117
*van*G gene cluster, 117
glycopeptide-dependent strains, 111
mechanisms, 101–106, 102–106
removal of susceptible target, 105–106
target modification, 105
origin of *van* genes, 117–118
regulation of resistance, 112–113
target, 101–102
teicoplanin resistance acquisition by VanB-type enterococci, 113–114
types of resistance, 106–111
VanA, 106–108
VanB, 108
VanC, 110
VanD, 108–110
VanE, 110–111
VanG, 111
Glycylcyclines, 36
GM (genetically modified) plants, 436
Greenberg, Jean T., 191
Growth promoters, ban by European Union, 448
gyr genes/proteins, 41, 44–48, 305–306, 319–320

Haemophilus influenzae
β-lactamase, 54
chloramphenicol resistance, 128, 136
Heavy metals. *See* Metal resistance
Helicobacter pylori
amoxicillin resistance, 332
antimicrobial susceptibility testing, 331
β-lactam resistance, 57
clarithromycin resistance, 331–332
fluoroquinolone resistance, 334
metronidazole resistance, 333
rifampin resistance, 333–334
tetracycline resistance, 13, 332–333
Horizontal gene transfer (HGT)
history, 550–552
soil bacteria, 431–433

Imidazole resistance
Bacteroides spp., 342–343
Clostridium spp., 347
Intestinal bacteria, transfer of resistance genes among, 437, 439–441
In vitro susceptibility testing parameters, for chloramphenicol, 127
Iron-associated genes, SoxS and, 219
IVET (in vivo expression technology) studies of soil bacteria, 429–430

Kanamycin, 85–87. *See also* Aminoglycosides
Kasugamycin, use in plant agriculture, 466
Ketolides, 66. *See also* MLSKO (macrolides, lincosamides, streptogramins, ketolides, and oxazolidinones)

Klebsiella oxytoca, β-lactamase of, 54, 57, 59
Klebsiella pneumoniae
aminoglycoside resistance, 92
β-lactam resistance, 54–55, 57, 59–60
chloramphenicol resistance, 133
fluoroquinolone resistance, 46
mar mutant identification in clinical isolates, 231

Lactobacillus, resistance in, 451–452
Lactococcus
commensal, resistance in, 451
multiple drug resistance (MDR) efflux, 159, 160, 280, 281
Lead resistance, 169
Legionella pneumophila, aminoglycoside resistance in, 89
Levy, Stuart
Antibiotic Misuse Statement, 521
antibiotic resistance transfer studies, 438–439, 446, 448, 528, 530
APUA and, 438, 465, 517–518, 519–524, 526
biocide resistance studies, 178–179, 180
Julian Davies and, 552
Mark Nelson and, 538
multiple antibiotic resistance and, 151
NOVA documentary, 538
Paratek Pharmaceuticals, Inc., founding of, 541
Plasmid Song, 552
soil pseudomonads, study of, 425–426
tetracycline research, 537–541
Lincomycin, 67, 68
Lincosamides, 67. *See also* MLSKO (macrolides, lincosamides, streptogramins, ketolides, and oxazolidinones)
resistance in *Enterococcus* spp., 301, 304–305
Linezolid, 67, 71
Listeria monocytogenes
from food, 456
zoonotic bacteria, resistance in, 455–456
Listeria spp.
MarR family members, 252
tetracycline resistance, 22–24
LmrA and LmrP, 280, 281
Lotus corniculatus, 432

Macrolide-lincosamide-streptogramin B (MLSB) resistance
Streptococcus pneumoniae, 316–317
Macrolide resistance
Bacteroides spp., 344
Campylobacter spp., 335
Clostridium spp., 347
Enterococcus spp., 304–305
Macrolides, 66. *See also* MLSKO (macrolides, lincosamides, streptogramins, ketolides, and oxazolidinones)
Malaria. *See Plasmodium falciparum*
mar
identification of mutants in clinical isolates, 224–232
Enterobacter, 231–232
Escherichia coli, 224–231
Klebsiella, 231

Mycobacterium, 232
Salmonella, 231
oxidative stress response and, 192–195
MAR (multiple antibiotic resistance)
biocide resistance, 178–179, 185–186
MarA, 202–204
binding sites, position and orientation of, 236–237
DNA-MarA-RNAP interactions, 216–217
MarA-RNAP scanning complexes, 238–240
mechanisms for transcriptional activation by MarA, 214–216
molecular evolution of, 205
orthologs, 242
paralogs, 241–242
promoters activated by, 240–241
regulation, 209, 210–211, 242
as repressor, 241
RNAP, interaction with, 237–240
specificity of binding, 236
structure, 235–236
transcriptional repression by MarA, 216
MarB, 204
MarC, 204
mar locus, 198–206
characterization of mutants, 198–199
clinical implications, 206
conservation of locus in other bacteria, 205
gene arrangement and regulation of *marRAB* locus, 199–200
genetic elements
MarA, 202–204
MarB, 204
MarC, 204
MarR, 200–202
molecular evolution of MarA, 205
MarR, 200–202, 209–211
DNA binding sites, 248
mutations in, 225–227, 230–231
mar regulon, 209–220
defining, 212–214
DNA-MarA-RNAP interactions, 216–217
marbox, 214–216
Mar phenotype, contribution of individual genes in, 217–219
cell permeability, reduction of, 217–218
detoxification, 218–219
efflux of toxic compounds, 218
iron-associated genes, 219
isoenzyme switch, 218–219
reduction of redox-active prosthetic groups, 219
sugar transport and catabolism, 219
mechanisms for transcriptional activation by MarA, 214–216
physiology of, 242
regulatory circuits, 209–212
MarA, 210–211
Rob, 212
SoxS, 211
transcriptional repression by MarA, 216
MarR family members, functions of, 247–253
interaction with small organic molecules, 252–253
Bacillus subtilis, 252–253
Butyrivibrio fibrisolvens, 252
Comamonas testoseroni, 252
Escherichia coli, 252
Xanthomonas campestris pv. *phaseoli,* 252–253
regulation of multiple antibiotic resistance, 247–249
Enterobacteriaceae, 247–249
Neisseria gonorrhoeae, 249
Pseudomonas aeruginosa, 249
Staphylococcus aureus, 249
regulation of single antibiotic resistances, 249–250
β-lactam resistance in *Staphylococcus aureus* and *Bacillus licheniformis,* 249
teicoplanin resistance in *Staphylococcus aureus,* 249
roles in pathogenesis and global gene regulation, 250–252
Erwinia chrysanthemi, 250
Listeria monocytogenes, 252
Mycobacterium, 252
NorR, 251
Rot, 251
Salmonella, 250
SarA, 250–251
Staphylococcus aureus, 250–252
TcaR, 251–252
Yersinia enterocolitica, 250
MarR family members, structure of, 253–257
Bacillus subtilis, 256
Enterococcus faecalis, 255, 256
Escherichia coli, 253–254
Methanococcus jannaschii, 256–257
Pseudomonas aeruginosa, 253–254
Staphylococcus aureus, 255, 256
MATE (multidrug and toxic compound extrusion) transporter, 152, 154, 281
McDonald's Corporation's *Global Policy on Antibiotic Use in Food Animals,* 533–534
MdeA, 280
MDR. *See* Multiple drug resistance (MDR)
Mefloquine resistance, in *Plasmodium falciparum,* 405
Megasphaera elsdenii, tetracycline resistance in, 19
Membrane permeability, fluoroquinolone resistance and, 44, 46
Mercury resistance, 166–168
Mesorhizobium, 432–433
Metal resistance, 165–172
antibiotic resistance, connection to, 170–171
arsenic and antimony, 169–170
biological uses of metals, 165–166
cadmium, cobalt, nickel, zinc, and lead, 169
copper, 168–169
Enterococcus spp., 307
genomic considerations, 171
mercury, 166–168
silver, 168
table of plasmid-determined systems, 167
tellurite, 170
Methacycline, 32–33
Methanococcus jannaschii, 257

Methicillin. *See* MRSA (methicillin resistance in *Staphylococcus aureus*)
Methotrexate, 473–475, 507
Metronidazole resistance, in *Helicobacter pylori,* 333
MexR, in *Pseudomonas aeruginosa,* 200, 249, 253–254, 261–266
MFS (major facilitator superfamily) transporter, 152, 153, 157, 160, 261, 277–280
  Bmr and Blt, 278–279
  EmeA, 280
  LmrP, 280
  MdeA, 280
  NorA, 277–278
  PmrA, 280
  QacA, 279
MIC (minimum inhibitory concentration), biocide, 176, 180–181
Microsatellite and chromosomal instability, 502
Microtubule alterations, 484–485
MLSKO (macrolides, lincosamides, streptogramins, ketolides, and oxazolidinones)
  mechanisms of action, 66–78
  resistance, 66–78
    bacterial species with resistance genes, table of, 69–70
    efflux, 73, 75, 77
    inactivating genes, 73–74, 77–78
    mobile genetic elements, role of, 75–78
    mutational, 74
    rRNA methylation, 72–73, 75–77
  structures, 67–68
Molecular ecology of antibiotic resistance genes, 438–439
Molecular evolution of MarA, 205
Monocycline, 35–36
MRSA (methicillin resistance in *Staphylococcus aureus*)
  biocide resistance, 180–181
  community onset, 388
  nosocomial infection, 387–388
  synthesis of new tetracyclines agents and, 540–541
  treatment, 388
  vancomycin resistance in, 289–296
Multidrug efflux pumps in gram-negative bacteria
  overview, 261–263
  RND transporters, 263–266, 267
    Escherichia coli, 263–264
    inhibitors, 266
    intrinsic resistance, 263
    overexpression, 265–266
    overview, 261–263
    Pseudomonas aeruginosa, 264–265
    structural basis, 267–270
  SMR transporters, 266–267
Multidrug efflux pumps in gram-positive bacteria, 275–282
  ABC transporters, 281
  inhibitors, 282
  MATE transporters, 281
  MFS transporters, 277–280
    Bmr and Blt, 278–279
    EmeA, 280
    LmrP, 280
    MdeA, 280
    NorA, 277–278
    PmrA, 280
    QacA, 279
  overview, 275–276
  RND transporters, 281
  SMR transporters, 281
Multidrug resistance (MDR) of anticancer agents, 478–479, 507, 509
Multiple adaptation response (MAR), Paratek Pharmaceuticals research on, 543
Multiple drug resistance (MDR), 151–161
  definition, 151–152
  efflux systems
    ABC transporters, 152, 154, 158, 160, 281
    binding site in efflux pumps, 159–161
    Escherichia coli, 154, 156, 159–161
    in gram-negative bacteria, 261–270
    in gram-positive bacteria, 275–282
    inhibitors, 282
    MATE (multidrug and toxic compound extrusion), 152, 154, 281
    MFS (major facilitator superfamily), 152, 153, 157, 160, 277–280
    Neisseria gonorrhoeae, 157
    Pseudomonas aeruginosa, 155–156, 160
    regulators *versus* transporters, 157–159
    RND (resistance-nodulation-cell division), 152, 153, 157, 158, 160, 263–266, 267, 281
    SMR (small multidrug resistance), 152–154, 156, 159, 160, 266–267, 281
    Staphylococcus aureus, 156, 157, 159
Mutator alleles, in soil bacteria, 430
Mycobacterium spp.
  aminoglycoside resistance, 91
  biocide resistance, 179
  β-lactamase, 54
  mar mutant identification in clinical isolates, 232
  MarR family members, 252
  tetracycline resistance, 12, 21, 23
Mycobacterium tuberculosis, community acquired infections of, 392–395

NADA (New Animal Drug Application), 531, 533
NAHMS (National Animal Health Monitoring System), 532
NARMS (National Antimicrobial Resistance Monitoring System), 532
National Animal Health Monitoring System (NAHMS), 532
National Antimicrobial Resistance Monitoring System (NARMS), 532
Neisseria gonorrhoeae
  β-lactamase, 54
  chloramphenicol resistance, 130
  multiple drug resistance (MDR) efflux, 157, 249
Neomycin, 85–87. *See also* Aminoglycosides
New Animal Drug Application (NADA), 531, 533

Nickel resistance, 169
NorA, in *Staphylococcus aureus,* 156, 157, 249, 276, 277–278
Nosocomial infections, 387–396
  Staphylococcus aureus, 387–389
  vanocomycin-resistant enterococci, 389–391
Nucleoside analogs resistance
  carrier-mediated, 475–476
  modification and failure to activate, 480

Oral cavity
  antibiotic-resistant bacteria in, 442
  biocide-resistant bacteria in, 182–183
Oxazolidinones, 71. *See also* MLSKO (macrolides, lincosamides, streptogramins, ketolides, and oxazolidinones)
  resistance in *Enterococcus* spp., 307
Oxidative stress response, 191–195
Oxolinic acid, use in plant agriculture, 466
Oxytetracycline (Terramycin), 30–33
  resistance in plant agriculture, 466
  use in plant agriculture, 465–466, 469

p53, 485, 508
Paratek Pharmaceuticals, Inc., 541–548
  aminomethylcyclines (AMCs), 543–548
  founding of, 541
  MAR locus, inhibition of, 543
  nontetracycline projects, 543
  PTK 0796, 543–548
  tetracycline library, synthesis of, 541–543
par genes/proteins, 41, 44–49, 305–306, 319–320
Pasteurella multocida, chloramphenicol resistance in, 130, 137–138
PBP (penicillin-binding protein), 342
Penicillin resistance
  Enterococcus spp., 301–302
  Streptococcus pneumoniae, 391–392
Peptidoglycan
  β-lactams and, 318–319
  vancomycin and, 101–106, 290–292
Peptostreptococci, antibiotic resistance in, 348–349
P-glycoprotein (P-gp), 477–478, 505–507, 508
Phase II metabolism-mediated drug inactivation, 481
Phenicols, 124–139
  mode of action, 125–127
  prevalence of resistance, 137–138
  resistance mechanisms, 127–137
    chloramphenicol drug exporters, 131–136
    enzyme inactivation, 128
    multidrug exporters, 134–136
    Type A chloramphenicol acetyltransferases (CATs), 128–130
    Type B chloramphenicol acetyltransferases (CATs), 130–131
  spectrum of activity, 125–127
  structure, 125
  in vitro susceptibility testing parameters, 127
Photobacterium damselae subsp. *piscicida,* chloramphenicol resistance in, 128, 133

Plant agriculture, 465–469
  antibiotics
    implication of use on human health, 467
    resistance to, 466–467
    use of, 465–466
  fungicides
    implication of use on human health, 468
    resistance to, 468
    use of, 467
  trends and recommendations, 468–469
Plasmid-mediated resistance
  animal use of antibiotics and, 530
  biocide, 176
  β-lactams, 54, 57–58
  chloramphenicol resistance, 130, 132–133, 137–139
  fluoroquinolone, 46
  glycopeptide resistance, 106, 112, 116, 118
  history of, 550–552
  metals, 165–172
  MLSKO (macrolides, lincosamides, streptogramins, ketolides, and oxazolidinones), 75–78
  Streptococcus pneumoniae and, 315
Plasmid Song, 552–553
Plasmodium falciparum
  antifolate resistance, 405–406
  atovaquone-proguanil resistance, 406
  chloroquine resistance, 402–404
  future chemotherapy, 407–409
    artemisinin combination therapy (ACT), 408–409
    ineffective drug preparations and counterfeits, 407–408
    mass drug administration (MDA), 407
    presumptive intermittent treatment (PIT), 408
  life cycle, 401–402
  mefloquine resistance, 405
  morbidity and mortality from malaria, 401
  quinine resistance, 404
  tetracycline use, 407
PmrA, 280
Polycations, resistance to, in *Pseudomonas aeruginosa,* 361–362
Porphyromonas, antibiotic resistance in, 345–347, 442
Preservation of Antibiotics for Medical Treatment Act of 2003, 532, 534
Prevotella, antibiotic resistance in, 345–347, 442
Propionibacterium, antibiotic resistance in, 13, 348
Pseudomonas aeruginosa
  aminoglycoside resistance, 90, 92–93, 360–361
  biocide resistance, 178, 180, 362
  biofilm production, 362–363, 428–429
  β-lactam resistance, 57, 59–60, 355–359
  chloramphenicol resistance, 132, 135–136
  fluoroquinolone resistance, 44, 45, 47–49, 359–360
  MexR, 200, 249, 253–254, 261—266
  multiple drug resistance (MDR) efflux, 155–156, 160
  polycations, resistance to, 361–362
  quorum sensing, 49
  RND transporters, 264–265
  tetracycline resistance, 13
Pseudomonas aureofaciens, 429

Pseudomonas cepacia, multiple antibiotic resistance in, 205
Pseudomonas fluorescens
biofilm, 427–429
in soils, 425–431
Pseudomonas putida
chloramphenicol resistance, 135
in soils, 425
Pseudomonas stutzeri, chloramphenicol resistance in, 135
Pseudomonas syringae, copper resistance in, 168
PTK 0796
activity in vitro, 543–545
pharmacology, 545–548
Pyrimethamine, 405–406

QacA, 279
Quaternary ammonium compounds (QACS), 175–176, 181–182
Quinine resistance, in *Plasmodium falciparum,* 404
Quinolones. *See* Fluoroquinolones
Quinupristin-dalfopristin, 390
Quorum sensing
fluoroquinolone resistance development and, 49
in *Pseudomonas aeruginosa,* 49

Radiation-induced transgenerational genomic instability, 502
Ralstonia metallidurens, metal resistance in, 169
Rhizosphere, 426
Rhodococcus spp., chloramphenicol resistance in, 134
Ribosomes
aminoglycosides and, 85, 87, 92–93
chloramphenicol effect on, 125–126, 322
MLSKO drugs and, 71–72, 316–317
tetracyclines and, 3, 13, 19–25, 318
Rifampin resistance, in *Helicobacter pylori,* 333–334
RNA polymerase (RNAP), 202, 203, 216–217
RND (resistance-nodulation-cell division) transporter, 152, 153, 157, 158, 160, 263–266, 267
Escherichia coli, 263–264
inhibitors, 266
intrinsic resistance, 263
multidrug efflux pumps in gram-positive bacteria, 281
overexpression, 265–266
overview, 261–263
Pseudomonas aeruginosa, 264–265
structural basis, 267–270
ROAR (Reservoirs of Antibiotic Resistance), 438, 523–524
Rob protein
promoters activated by, 240–241
regulation, 212
structure, 192–194, 209
Rofloxacin, 533
RovA, in *Yersinia enterocolitica,* 200, 250
rRNA methylation, 72–73, 75–77

Salmonella spp.
biocide resistance, 178
chloramphenicol resistance, 126, 130, 133, 138
*mar*C, 204
mar mutant identification in clinical isolates, 231
oxidative stress response, 195
resistance in developed countries, 368–370
resistance in developing countries, 370–371
S. enterica, 178, 195, 367, 452
S. enterica serovar Paratyphi A, 368
S. enterica serovar Typhi, 367–368
silver resistance, 168
SlyA, 250
zoonotic bacteria, resistance in, 452–453
Sancycline, 35–36
Sarafloxacin, 533
Serratia marcescens, aminoglycoside resistance in, 92
Shigella spp.
resistance in Britain, 371
resistance in developing countries, 371
Silver resistance, 168
Site-specific recombination, in soil bacteria, 430–431
SMR (small multidrug resistance) transporter, 152–154, 156, 159, 160, 266–267, 281
Soil bacteria, 425–433
adhesion and biofilm formation, 428–429
environmental (nutritional) cues influencing biofilm formation, 429
initial colonization, 428
mature biofilm, 428–429
microcolony formation, 428
colonization of soil, 426–433
distribution of bacteria in soils, 426–428
pores and ecological niches, 426
rhizosphere niche, 426
taxis and mobility, 426–428
evolution and adaptation, 430–433
colony phase variation, 431
horizontal gene transfer, 431–433
mutator alleles, 430
site-specific recombination, 430–431
SOS polymerase, 431
symbiosis island, 432–433
heterogeneous interface, 425
inoculation of soils with bacteria, 426
IVET (in vivo expression technology) studies, 429–430
Stuart Levy's studies on, 425–426
SOS polymerase, in soil bacteria, 431
soxRS genes, 191–195
SoxS
efflux of toxic compounds, 218
iron-associated genes, 219
isoenzyme switch, 218–219
promoters activated by, 240–241
reduction in cell permeability and, 217–218
regulation, 211
RND transporter, 266
structure, 209
sugar transport and catabolism, 219
Sphingolipids, 486–487
Staphylococcus aureus
aminoglycoside resistance, 87, 89, 92
β-lactam resistance, 53–55, 249

chloramphenicol resistance, 128, 136
fluoroquinolone resistance, 45, 48
MAR (multiple antibiotic resistance) family members, 205, 249–252
MdeA, 280251
metal resistance in, 169
multiple drug resistance (MDR) efflux, 156, 157, 159, 249, 276–280
NorA, 156, 157, 249, 277–278
NorR, 251
nosocomial infections, 387–389
QacA, 279
Rot, 251
SarA, 250–251, 255, 256
TcaR, 251–252
teicoplanin resistance, 250
vancomycin resistance, 112, 289–296
Staphylococcus intermedius, tetracycline resistance in, 21
Staphylococcus spp.
aminoglycoside resistance, 87, 89, 92
biocide resistance, 180–181
chloramphenicol resistance, 133
Stemotrophomonas maltophilia, fluoroquinolone resistance in, 47
Streptococcus equi, β-lactam resistance in, 57
Streptococcus pneumoniae, 314–323
aminoglycoside resistance, 315–316
β-lactam resistance, 53, 57, 318–319
chloramphenicol resistance, 322–323
community acquired infections, 391–392
fluoroquinolone resistance, 45, 47–48, 319–321
macrolide-lincosamide-streptogramin B (MLSB) resistance, 316–317
multiple drug resistance (MDR) efflux, 156, 280
tetracycline resistance, 22, 24, 318
trimethoprim-sulfamethoxazole resistance, 321–322
Streptococcus pyogenes, tetracycline resistance in, 21, 23, 24
Streptococcus spp., aminoglycoside resistance in, 87
Streptogramins, 67–68, 70. *See also* MLSKO (macrolides, lincosamides, streptogramins, ketolides, and oxazolidinones)
resistance in *Enterococcus* spp., 301, 304–305
Streptomyces aureofaciens, Aureomycin (chlortetracycline) production by, 30, 31
Streptomyces griseus, streptomycin production in, 85
Streptomyces rimosus
Terramycin (oxytetracycline) production, 30, 31, 33
tetracycline resistance, 12
Streptomyces spp.
tetracycline resistance, 19, 21–22
tetracyclines, production of, 34–35
Streptomyces venezuelae, chloramphenicol production by, 124, 134, 136
Streptomycin, 85–86. *See also* Aminoglycosides
resistance in plant agriculture, 466
for tuberculosis treatment, 550
use in plant agriculture, 465–466, 469
Sulfadoxine, 405, 406
Swine, antibiotic-resistant bacteria in, 442–443
Symbiosis island, 432–433

Taxis and mobility in soil bacteria, 426–428
Teicoplanin, 101–102, 105–106, 108, 110–111, 113–114, 250
Telithromycin, 66, 67, 316
Tellurite resistance, 170
Telomeres, 502
Terramycin (oxytetracycline), 30–33
TetA protein
crystal structure, 7–8
cytoplasmic interdomain loop, role in resistance, 8
efflux mechanism, 6–11
historical perspective, 3–6
interaction between N-terminal and C-terminal domains, 8
multimerization, 8–9
multiple sequence alignment of group 1 proteins, 4–5
mutational analysis, 9
purification, 7
regulation of *tet* transcription, 11
residues involved in conformational changes, 109
residues involved in proton transport, 11
residues involved in substrate binding, 10
topology and functional relationships, 7
Tet(B) efflux protein, structure-*versus* activity relationships of tetracyclines and, 538–540
tet(W) gene, 19–22, 24–25
Tet(K) protein, 11–12
Tet(L) protein, 11–12
Tet(M) protein, 19–25
Tet(O) protein, 19–24
Tetracycline(s), 1–38
action, 3
activity of glycyclines and tigecycline, 36
biosynthesis, 31–32
discovery, 29–31
for *Plasmodium falciparum* infection, 407
semisynthesis
doxycycline, 32–33
methacycline, 32–33
monocycline, 35–36
sancycline, 35–36
synthesis of novel
aminomethylcyclines (AMCs), 543–548
in Levy laboratory, 540–541
at Paratek Pharmaceuticals, Inc., 541–548
Tetracycline resistance
by active efflux, 3–12
historical perspective, 3–6
mechanism (group 1), 6–11
mechanism (group 2), 11–12
mechanism (other groups), 12
overview, 3
research in Levy laboratory, 537–541
transmembrane α-helices, 6
animal use of antibiotics and, 528, 530
Bacteroides spp., 13, 19, 22–25, 343–344, 441
Campylobacter spp., 22, 335
Clostridium spp., 12, 19, 23, 347

by degradative inactivation, 12–13
Enterococcus spp., 13, 21, 22, 306
Helicobacter pylori, 13, 332–333
ribosomal 16S RNA mutations, 13
by ribosomal protection proteins, 19–25
distribution of protection genes, 21–22
G+C content, 19–20, 22–24
mechanism of resistance, 19–21
mobile elements and gene linkages, 22–25
Streptococcus pneumoniae, 22, 24, 318
TetX protein, 12–13
Thiamphenicol, 124–126, 128
Thymidylate synthase alterations, 484
Tigecycline, 36
TolC, 268–270
Topoisomerase alterations, 482–483
Topoisomerase IV, 41–43, 48, 49, 305–306, 319
Transcriptional control of resistance, 509
Transporters. *See also* Efflux
ABC transporters, 152, 154, 158, 160, 281
MATE (multidrug and toxic compound extrusion), 152, 154, 281
MFS (major facilitator superfamily), 152, 153, 157, 160, 277–280
RND (resistance-nodulation-cell division), 152, 153, 157, 158, 160, 263–267, 281
SMR (small multidrug resistance), 152–154, 156, 159, 160, 266–267, 281
Transposon
β-lactamase, 54–55
chloramphenicol resistance, 128, 130, 133, 134
conjugal, 441, 443–444
glycopeptide resistance, 101, 104, 106–109, 112, 114, 116, 118
MLSKO (macrolides, lincosamides, streptogramins, ketolides, and oxazolidinones) resistance, 75–78
Streptococcus pneumoniae and, 315
tetracycline resistance, 21–25, 441, 444
Triclocarban, 174–175
Triclosan
inhibition of FabI, 180
mechanism of action, 178–179
studies in *Escherichia coli,* 179
studies in mycobacteria, 179
uses, 174–175
Trimethoprim resistance, in *Enterococcus* spp., 306–307
Trimethoprim-sulfamethoxazole resistance
Enterococcus spp., 300
Streptococcus pneumoniae, 321–322
Tumor microenvironment, role in resistance to anticancer agents, 503–505

Union of Concerned Scientists, 447

Ureaplasma urealyticum, tetracycline resistance in, 19

Vancomycin
dependent strains, 111
resistance in *Clostridium,* 348
resistance in *Enterococcus* spp., 101–118, 290–291, 389–391, 542
Vancomycin resistance in *Staphylococcus aureus,* 289–296
emergence, 289–290, 291
hetero-VISA, 294–296
clinical significance, 295–296
detection, 296
heterogeneity of population, 294–295
incidence of mutation, 294
mechanism of action, 290–294
genetic basis, 292–294
horizontal transfer and mutation, 290–291
VISA, 291–292
therapeutic options for infections, 388–389
Veillonella sp., tetracycline resistance in, 22
Vibrio spp.
chloramphenicol resistance, 128, 130, 131, 133
tetracycline resistance, 12, 13
ToxT, 202
VISA (vancomycin-intermediate *Staphylococcus aureus*), 289–296, 388–389
VRE (vancomycin-resistant enterococci), 101–118, 290–291
nosocomial infections, 389–391
synthesis of new tetracyclines agents and, 541

World Health Organization (WHO)
Essential Medicines program, 519

Xanthomonas campestris pv. *phaseoli,* 252–253
Xylaria digitata, tetracycline resistance in, 12

Yersinia enterocolitica
RovA, 200, 250
zoonotic bacteria, resistance in, 454
Yersinia pestis, antimicrobial susceptibility testing in, 416–417

Zinc resistance, 169
Zoonotic bacteria, resistance in
Arcobacter, 455
Campylobacter, 454–455
Clostridium perfringens, 456–457
Enterobacteriaceae, 452
Escherichia coli, 453
Listeria monocytogenes, 455–456
Salmonella, 452–453
Yersinia enterocolitica, 454